3rd International Conference on Water Resource and Environment (WRE 2017)

IOP Conference Series: Earth and Environmental Science
Volume 82

Qingdao, China
26 – 29 June 2017

ISBN: 978-1-5108-4772-9
ISSN: 1755-1307

Printed from e-media with permission by:

Curran Associates, Inc.
57 Morehouse Lane
Red Hook, NY 12571

Printed by Curran Associates, Inc. (2017)

Institute of Physics
Dirac House, Temple Back
Bristol BS1 6BE UK

Phone: 44 1 17 929 7481
Fax: 44 1 17 920 0979

techtracking@iop.org

Additional copies of this publication are available from:

Curran Associates, Inc.
57 Morehouse Lane
Red Hook, NY 12571 USA
Phone: 845-758-0400
Fax: 845-758-2633
Email: curran@proceedings.com
Web: www.proceedings.com

Table of contents

Volume 82

3rd International Conference on Water Resource and Environment (WRE 2017)
26–29 June 2017, Qingdao, China

Accepted papers received: 18 August 2017
Published online: 31 August 2017

Preface

Papers

IOP Conf. Series: Earth and Environmental Science **82** (2017) 011001 doi:10.1088/1755-1315/82/1/011001

Preface

The 3rd International Conference on Water Resource and Environment (WRE 2017) was held from June 26th to 29th, 2017 in Qingdao, one of the most famous tourist cities in China. The technical program comprised four keynote speakers, fifty invited speakers, five oral presentation sessions and an interesting poster session.

Water resource and environment are key elements guaranteeing the sustainability of societies. WRE 2017 aimed to provide an academic platform for researchers and scholars to present and discuss their latest findings about water resource and environment. It gathered over 140 participants from 30 different countries. The main subjects of the conference were "Water Resources and Hydrology", "Water Quality and Environment Science", "Water, Climate and Ecological Environment", "Water Pollution and Wastewater Treatment" and "Soil, Groundwater and Hydrology".

This volume records the proceedings of WRE 2017 and contains 93 rigorously selected manuscripts submitted to the WRE 2017 conference. All the manuscripts were selected depending on their quality and relevance to the conference.

The Organizing Committee would like to thank all the authors who contributed to WRE 2017 and also the reviewers who provided their valuable comments and suggestions.

The 4th International Conference on Water Resource and Environment (WRE 2018) will be held in Taiwan. Everybody is welcomed to submit high quality papers to WRE 2018! More information will be available at the conference website: http://www.wreconf.org/.

Editor

Prof. Peiyue Li

Chang'an University, China

IOP Conf. Series: Earth and Environmental Science **82** (2017) 011002 doi:10.1088/1755-1315/82/1/011002

Peer review statement

All papers published in this volume of *IOP Conference Series: Earth and Environmental Science* have been peer reviewed through processes administered by the proceedings Editors. Reviews were conducted by expert referees to the professional and scientific standards expected of a proceedings journal published by IOP Publishing.

IOP Conf. Series: Earth and Environmental Science **82** (2017) 012001 doi:10.1088/1755-1315/82/1/012001

Assessing water quality of the Chesapeake Bay by the impact of sea level rise and warming

P Wang[1,8], L Linker[2], H Wang[3], G Bhatt[4], G Yactayo[5], K Hinson[6] and R Tian[7]

[1]Senior Research Scientist, Virginia Institute Marine Science, Gloucester P, VA, USA
[2]Modeling Coord, USEPA Chesapeake Bay Program Office, Annapolis, MD, USA
[3]Professor, Virginia Institute of Marine Science, Gloucester Point, VA, USA
[4]Research Scientist, Penn State University, University Park, PA, USA
[5]Environmental Engineer, Maryland Dept. of Environment, Baltimore, MD, USA
[6]Staffer, Chesapeake Research Consortium, Edgewater, MD, USA
[7]Research Sceintist, Univ of Maryland Center Envir. Sc, Cambridge, MD, USA

E-mail: pwang@chesapeakebay.net

Abstract. The influence of sea level rise and warming on circulation and water quality of the Chesapeake Bay under projected climate conditions in 2050 were estimated by computer simulation. Four estuarine circulation scenarios in the estuary were run using the same watershed load in 1991-2000 period. They are, 1) the Base Scenario, which represents the current climate condition, 2) a Sea Level Rise Scenario, 3) a Warming Scenario, and 4) a combined Sea Level Rise and Warming Scenario. With a 1.6-1.9°C increase in monthly air temperatures in the Warming Scenario, water temperature in the Bay is estimated to increase by 0.8-1°C. Summer average anoxic volume is estimated to increase 1.4 percent compared to the Base Scenario, because of an increase in algal blooms in the spring and summer, promotion of oxygen consumptive processes, and an increase of stratification. However, a 0.5-meter Sea Level Rise Scenario results in a 12 percent reduction of anoxic volume. This is mainly due to increased estuarine circulation that promotes oxygen-rich sea water intrusion in lower layers. The combined Sea Level Rise and Warming Scenario results in a 10.8 percent reduction of anoxic volume. Global warming increases precipitation and consequently increases nutrient loads from the watershed by approximately 5-7 percent. A scenario that used a 10 percent increase in watershed loads and current estuarine circulation patterns yielded a 19 percent increase in summer anoxic volume, while a scenario that used a 10 percent increase in watershed loads and modified estuarine circulation patterns by the aforementioned sea level rise and warming yielded a 6 percent increase in summer anoxic volume. Impacts on phytoplankton, sediments, and water clarity were also analysed.

1. Introduction

To protect the water quality and ecosystems of the Chesapeake Bay a TMDL (total maximum daily load) policy was enacted by the USEPA in December of 2010 [1]. A suite of computer models including the Watershed Model (WSM), estuarine Water Quality and Sediment Transport Model (WQSTM) and ancillary living resource models were used in the development of the 2010 TMDL to establish nutrient and sediment reduction targets to achieve the Chesapeake water quality standards of dissolved oxygen (DO), chlorophyll-*a*, and water clarity [2]. The 2010 TMDL documentation [1] and the 2014 Chesapeake Watershed Agreement [3] also called for an assessment of the impacts of

IOP Conf. Series: Earth and Environmental Science **82** (2017) 012001 doi:10.1088/1755-1315/82/1/012001

changing climate on Chesapeake Bay water quality and living resources. The assessment will include projected changes in land use, precipitation, air temperature, water temperature, sea level rise, wetland loss, and water quality in the tidal Bay for 2025 and 2050. The climate change analysis will use the same methodology as that employed in the Chesapeake 2010 TMDL assessment, which applies the WQSTM estuarine model to estimate the influence exerted upon Chesapeake Bay circulation and water quality due to changed climate conditions. This study only involves the impact on water quality due to sea level rise and temperature change, as well as changes in watershed loads due to altered precipitation and temperature in a projected 2050 climate condition.

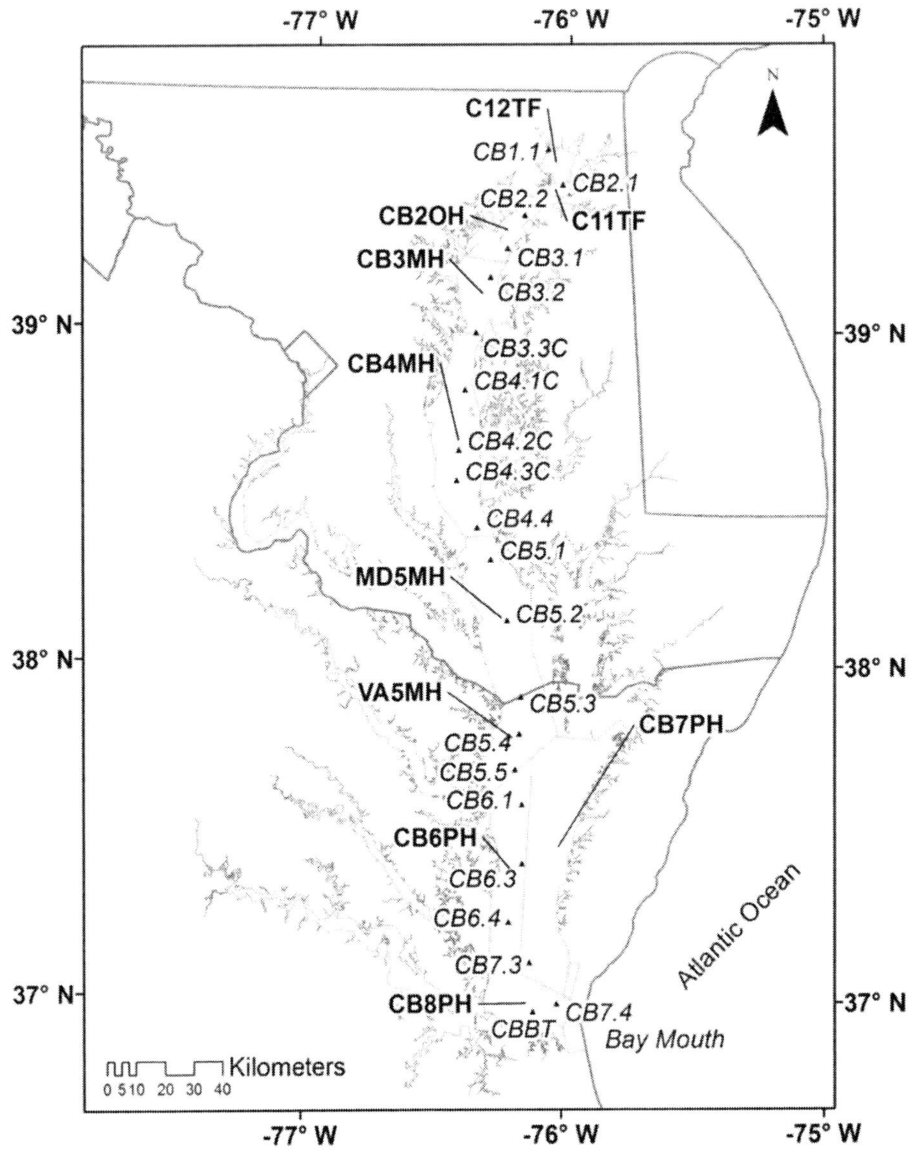

Figure 1. The Chesapeake Bay estuary, its mainstem CB segments (in bold), and key monitoring stations (in italic) along the main channel.

Climate conditions in 2050 have been estimated by various researchers [4-7]. Average air temperatures in 2050 will be higher than the current condition due to global warming (GW), and are estimated to increase by 1.5-2°C [7, 8]. According to a linear regression of observed water elevations at the Chesapeake Bay Bridge Tunnel station (CBBT) near the Bay mouth (figure 1), sea level rise has occurred at a rate of 0.63 cm/year over the past 30 years [9]. Under a continuation of this trend, local sea level is expected to rise approximately 32 cm between 2000 and 2050. However, considering the acceleration of sea level rise in later years [5], this estimate may increase to 40-50 cm by 2050. This value is consistent with an estimate produced by Parris et al. [5] and Boesch et al. [10]. Associated with historical sea level observations in the Chesapeake Bay, Boesch et al. [10] estimated sea level rise in 2050 (from 2000) to reach about 0.3-0.7 m, with a most likely estimate of 0.4 m. Considering 1995 as the midpoint of the 1991-2000 hydrology used to establish the Chesapeake TMDL, this work uses an estimated 0.5 m sea level rise for 2050 (from 1995) that is consistent with recommendations put forth by the Chesapeake Scientific and Technology Advisory Committee [11].

Sea level rise (SLR) increases salt intrusion into the Bay. Associated with the sea level rise in the past 30 years, the observed salinity in both surface water and bottom water at the Bay mouth increased by about 4 ppt. This is consistent with model simulations completed by Hilton et al [12]. Hong and Shen [13] simulated salinity in the Chesapeake Bay estuary under a 0.5 meter sea level rise for years 2001 and 2003, and produced similar results [13]. The proper setting of the ocean boundary salinity is an important consideration in this type of model simulation. The assumption of a 0.5 meter sea level rise in this work enables the utilization of Hong and Shen's [13] estimated salinity boundary condition, and the ability to compare simulations [13].

Summer hypoxia/anoxia and its resultant cascade of biogeochemical and ecological effects is a key threat to the Bay's living resources. Hypoxic and anoxic waters are located at regions of stratified deep water. The existence of stratification prevents exchange of DO between the saltier, denser lower layer and the fresher, less dense upper layer and acts to preserve hypoxic conditions in deep waters. The pycnocline, with abrupt changes in density, separates waters in the upper and lower layers. In the 2010 Chesapeake Bay TMDL for DO compliance, DO standards were set for three depth-categories of designated-use, i.e., *Open Water* (OW), *Deep Water* (DW) and *Deep Channel* (DC), individually. The *Open Water* and *Deep Water* designated-uses are separated by the upper boundary of the pycnocline, and the Deep Channel designated-use is below the lower boundary of the pycnocline. Of particular concern is DO attainment in *Deep Water* and *Deep Channel* designated-uses. Nutrient and sediment loads from the watershed and subsequent Bay water quality responses were assessed for the TMDL evaluation using model results in a 10-year (1991-2000) average hydrology period, in which the current climate condition was used. This study will analyze water quality of the estuary in a modified 1991-2000 hydrology due to sea level rise and warming. Three types of climate change scenarios will be conducted: 1) a 0.5-m Sea Level Rise (SLR) scenario, 2) a warming (GW, global warming) scenario (e.g., about 1.6-1.9°C increase in monthly air temperature), and 3) a combined Sea Level Rise and Warming (SLR+GW) scenario. In these three scenarios, the watershed load will be the same as that in the 1991-2000 calibration of the WQSTM under the current climate condition. The differences in results are therefore related to altered estuarine hydrodynamics. Besides the impact on estuarine circulation and water quality due to changes in sea level and air temperature, the watershed nutrient loading would also be changed in the 2050 climate condition. Therefore, additional model scenarios will also be analyzed that include changes in watershed loads due to an altered climate in 2050 and use the estuarine hydrodynamics with or without sea level rise and warming.

2. Method

This work uses an estuarine model to estimate water quality responses to sea level rise and warming. A proper setting of the model is critical. The following describes the related methods.

2.1. *The estuarine hydrodynamic and water quality model*

The WQSTM, a coupled 3-dimensional CH3D hydrodynamic model and CE-QUAL-ICM water quality and sediment transport model [14] was used. The CH3D model simulates hydrodynamic circulation, which will be impacted by changes in either sea level or water temperature. The WQSTM model grid utilizes a z-type depth structure. The model grid has up to 19 layers depending upon the depth of water, and the thickness of each layer is fixed. The thickness of the surface layer is 2.13 meters (i.e., 7 feet), and the thickness of other layers is 1.53 meters (i.e., 5 feet). The model grid consists of 11,064 surface cells, and a total of 56,920 cells. The model boundary with the ocean is adjacent to the Bay mouth, and extends 10 km out to the ocean. The model calibration run is called the Base Scenario as a reference to analyze changes in estuarine conditions due to climate scenarios.

2.2. *Model setting of water elevation at ocean boundary*

The water elevation at the ocean boundary for model calibration in 1991-2000 was based on the observed hourly water level at the Chesapeake Bay Bridge Tunnel (CBBT) station (figure 1). For the sea level rise scenario, a necessary simplifying assumption presumed that tidally influenced fluctuations of hourly water level in the 10-year hydrology period would be approximated by the observed hourly water level fluctuations in 1991-2000. Under this assumption, the water level rise in the estimated 0.5-meter sea level rise condition can be accomplished in the simulation by increasing the vertical depth of the model grid by 0.5 meters, while still using observed hourly water elevations for the boundary.

There are three potential methods by which the model grid depth can be increased by 0.5 m. One method involves adding an additional layer of 0.5 m below the surface layer. The second method involves adding an additional layer of 0.5 m below a layer other than the surface layer, such as the bottom layer. The third method available would increase the surface layer by 0.5 m.

The second method involves significant modification of the algorithm for horizontal transport, because there would be different thicknesses of water in the same model layer throughout the model domain. The first method still involves a reconstruction of linkages among faces of the model grid, modification of source code, and an associated slowdown in computation, but is easier to implement than the second method. The difference in depth resolution in the first layer between the first and third methods is only a quarter of a cell thickness. The two methods of grid modification produced insignificant differences [9]. Because the third method is easy to implement and yields results comparable to the first method, the third method was adopted.

2.3. *Salinity simulation and model settings of salinity boundary conditions*

The model computes 3-D salinity fields in the following way [15]:

$$\frac{\partial(S)}{\partial t} = -R_O\left(\frac{\partial(uS)}{\partial x} + \frac{\partial(vS)}{\partial y} + \frac{\partial(wS)}{\partial z}\right) + \frac{Ek_H}{\Pr_H}\left(\frac{\partial(K_H\partial(S)/\partial x)}{\partial x} + \frac{\partial(K_H\partial(S)/\partial y)}{\partial y}\right) + \frac{Ek_V}{\Pr_V}\frac{\partial(K_V\partial S/\partial z)}{\partial z} \quad (1)$$

Where,
S = salinity;
z = thickness;
u, v, w = x, y, z component of velocity
Ro = Rossby number
K = turbulent eddy coefficients (horizontal or vertical: H, v)
Ek = Ekman number (horizontal or vertical: H, v)
Pr = Prandtl number (horizontal or vertical: H, v)

The salinity at the ocean boundary for model calibration in 1991-2000 was based on bi-weekly observations at three monitoring stations, CB7.4, CB7.4N, and CB8.1E, at the Bay mouth (figure 1). Due to the influence of freshwater, the salinity at the Bay mouth is lower than that in the pelagic ocean. Sea level rise increases salt intrusion to the Bay from the ocean, increasing the salinity at the Bay mouth [13]. An adjustment of salinity at the ocean boundary is needed in the simulation of sea level rise. The model simulation of 0.5 meter sea level rise by Hong and Shen [13] used a model grid with a boundary in the pelagic ocean, 100 km away from the Bay mouth. The salinity changes in the Chesapeake estuary, including at the mouth, in the current condition and the 0.5 meter sea level rise condition were estimated for hydrology years 2001 and 2003 by Hong and Shen [13], as well as 2000 (Shen, personal communication). In courtesy of Shen's detailed information, we adjusted salinities at the WQSTM boundary near the Bay mouth for a 0.5 m sea level rise scenario [9]. The adjustment was conducted by trial testing the boundary salinity modifications for the sea level rise simulation using the WQSTM, and simulated salt intrusion results were obtained, similar to Hong and Shen [13] for hydrology years 2000, 2001, and 2003 with a 0.5 m increase in sea level rise. This work uses the same method employed by Wang et al [9] to set salinity at the boundary for 1991-2000 hydrology in the Sea Level Rise Scenario. The modifications to the ocean boundary involved three steps: 1) projecting bi-weekly observed salinity values recorded in the centroid of cells in the normal Base Scenario grid to the new Sea Level Rise Scenario grid with increased total depth, 2) calculating salinity values at the centroid of cells of the new grid, and 3) increasing salinity by 0.4 ppt. The salinity adjustment is also consistent with the salinity increase at the Bay mouth in sea level rise scenarios completed by Hilton et al [12].

2.4. Key parameters used in water temperature simulation

Circulation in the estuary also redistributes water temperature in model cells, which is simulated in a manner similar to the salinity computation (equation (1)). Other than advection in tidal waters, thermal diffusion and heat exchange between air and water is the main process used to increase water temperature in the WQSTM simulation under warmer conditions. Changes in temperature due to the air-water exchange is considered proportional to the temperature difference between the water surface and a theoretical equilibrium temperature [16]:

$$\partial Tw / \partial t = K\ (Te - T) / (\rho CpdZ) \tag{2}$$

Where:

Tw=water temperature;
Z= depth
Te: Equilibrium temperature
K: heat exchange coefficient (Watt m^{-2} $C^{o\ -1}$)
Cp: specific heat of water (4200 Watt sec kg^{-1} $C^{o\ -1}$)
ρ: water density (1000 kg m^{-3})

For a water temperature simulation in the 2050 Warming Scenario, it is essential to estimate air temperature, as well as the theoretical equilibrium temperature (Te) and heat exchange coefficient (K) in year 2050.

2.5. Estimate of 2050 air temperature

Air temperature will increase by 2050 due to global warming [7, 10]. The air temperature in 2050 was estimated based on averaged values from six global circulation model (GCM) scenarios from phase 3 of the Coupled Model Intercomparison Project (CMIP3). The six GCM scenarios were: 1) BCM2.0 of BCCR (Bjerknes Centre for Climate Research, Norway), 2) Mk3.0 of CSIRO (Commonwealth Scientific and Industrial Research Organization, Australia), 3) Mk3.5 of CSIRO, 4) CM3.0 of INM (Institute for Numerical Mathematics, Russia), 5) MIROC3.2 of NIES (National Institute for Environmental Studies, Japan), and 6) ncar_ccsm3_0, i.e., CCSM3 of NCAR (National Center for Atmospheric Research, USA). Monthly temperature changes between 1995 (the simulation midpoint year used to establish the Chesapeake TMDL) and 2050 conditions were estimated [17, 18]. After noting abrupt changes between December and January a running-three-month average method was used to smooth the data (blue diamonds in figure 2(b)) and then adjust daily air temperatures for scenarios involving a warming climate. The increase of monthly air temperatures for 2050 in this work ranges from 1.6 to 1.9°C as shown in figure 2(b) (blue dots).

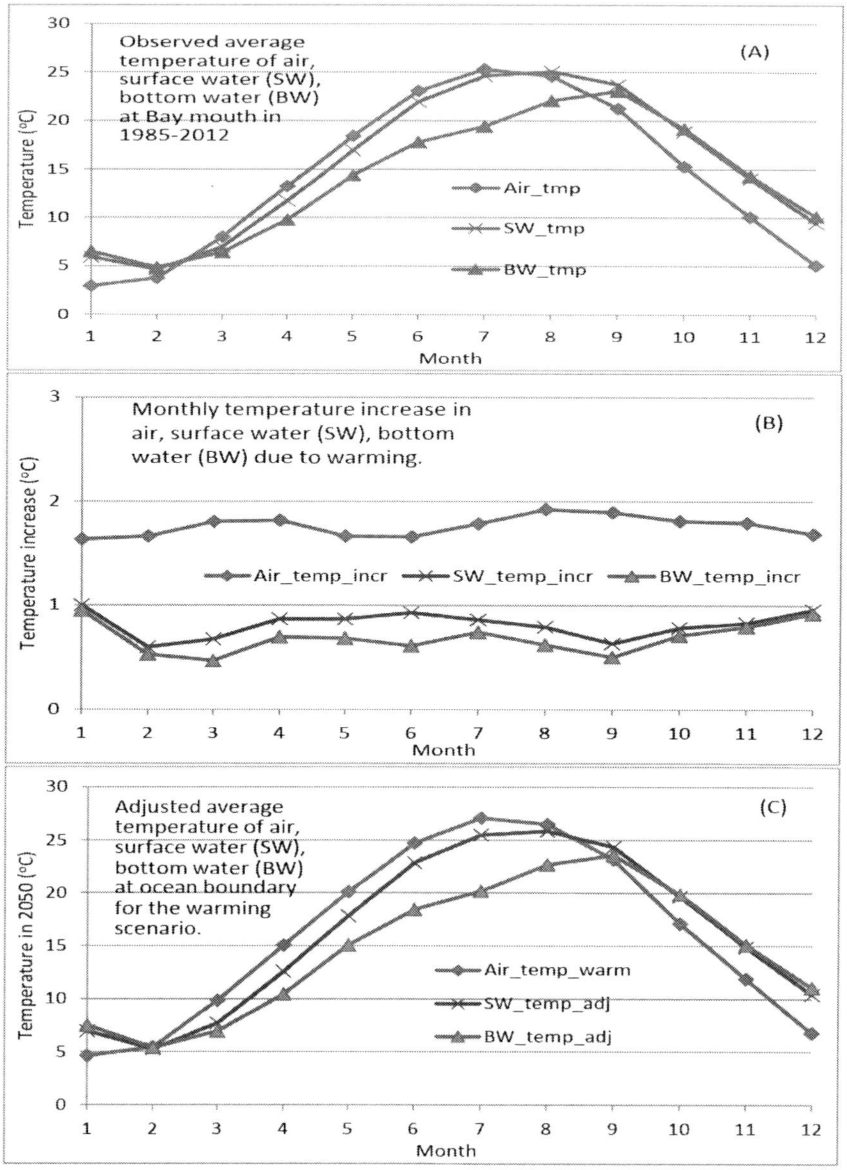

Figure 2. Illustration of adjustments of monthly water temperature at the boundary due to changes in air temperature in 2050. a) Observed average monthly temperatures of air, surface water, and bottom water at the Bay mouth from 1985-2012, b) increases of monthly temperature in the air, surface water, and bottom water due to warming in 2050, based on slopes of temperature changes in panel (a); c) averaged monthly temperatures in the air, surface water, and bottom water at the ocean boundary in the projected 2050 condition.

2.6. Simulation of water temperature in 2050

2.6.1. Te and K for simulation of water temperature in warming scenarios. The theoretical equilibrium temperature (Te) and heat exchange coefficient (K) are required to simulate water temperature (equation (2)). Te and K can be estimated from daily air temperature, dew-point temperature (or wet-bulb temperature), wind speed, and cloud cover or solar radiation [16, 19, 20]. By assuming no change in humidity for the same day in the Warming Scenario and the Base Scenario conditions and using the estimated air temperature in the warming condition, a dew-point temperature for the Warming Scenario can be obtained. By further assuming no change in wind and cloud cover conditions in the warming and the base conditions, Te and K were calculated for the Warming Scenario. Using the estimated Te, K, and air temperature due to warming, water temperature in the Warming Scenario can be simulated.

The temperature of water changes more slowly than air temperature because water has a higher heat capacity. About one unit of heat energy is needed to warm the air one degree Celsius, while about four times more heat energy is needed to warm the water one degree Celsius. Morrill et al [21] showed that water temperature in the upland streams of their study area increased by approximately 0.6- 0.8°C for every 1°C increase in air temperature.

The Chesapeake estuary has a larger ratio of water volume to surface area than that of shallower rivers. For a 1°C increase in air temperature, there is expected to be a lesser temperature increase in the Bay's water than the stream estimates produced by Morrill et al [21]. A simplified scoping warming scenario for the Chesapeake estuary was run with an increase in monthly air temperature of 1.6-1.9°C (figure 2(b)), which resulted in about 0.8-1°C increase in the estuarine water, i.e., an approximate 0.5°C increase in water temperature for every 1°C increase in air temperature [9]. This result indicates that the CH3D simulation of water temperature and the estimated Te and K for the warming condition is acceptable. Note: in the simplified scoping warming scenario, the water temperature of the ocean boundary and freshwater inputs used the current climate condition. For more accurate calculations in warming scenarios, water temperature at the boundary and the input freshwater needed to be adjusted accordingly, and this point is addressed in the next two subsections.

2.6.2. Adjustment of water temperature at the ocean boundary. Figure 2(a) shows observed averaged monthly water temperatures at both the surface and the deepest bottom layer (about 18 meters deep) at the Bay mouth from 1985-2000, along with the observed averaged air temperatures from the Patuxent Naval Air Station. In high winter, the deep water had a higher temperature than the surface water, which in turn had a higher temperature than the air. In the summer, the air had a higher temperature than the surface water, which in turn had a higher temperature than the bottom water. The changes of air and water temperatures among months were nonlinear, showing hysteresis in the rising and falling limbs, consistent with observations made by Lee and Cho [22]. Based on the monthly change in air temperature (the blue diamonds in figure 2(b)) and considering slower changes in water temperature in comparison with changes in air temperature from the scoping warming scenario, relative changes in surface water and bottom water temperature in each month were calculated (figure 2(b)). These values were used to adjust the bi-weekly observed water temperature at the Bay mouth boundary for use in the warming scenarios (figure 2(c)).

2.6.3. Adjustment of water temperature in daily freshwater discharge. The model uses daily inputs of freshwater discharges. The increase of water temperature in rivers in the Warming Scenario can refer to the increases of temperature in the estuarine surface water (figure 2(b), red crosses) with adjustment by a factor. Considering greater temperature changes in stream water compared to that in the estuary resultant from air temperature changes and referring to Morrill et al [21], a factor of 1.2 was applied to produce slightly greater increases of monthly temperature from stream water. These values were interpolated for daily temperature increments for freshwater discharged from the watershed in the warming scenarios.

2.7. Simulations of primary production, dissolved oxygen, and other water quality parameters

The CE-QUAL-ICM water quality model simulates 36 state variables. They include various species of N-P nutrients and carbon as well as 3 grain sizes of fixed solids. Nutrient, light, and temperature dependent algal growth are simulated for three generalized groups, i.e., green algae, blue-green algae, and spring diatoms. A full carbon based simulation of DO that considers aeration, diffusion, mixing, temperature, algal activities, and processes involving materials of biological and chemical oxygen

IOP Conf. Series: Earth and Environmental Science **82** (2017) 012001 doi:10.1088/1755-1315/82/1/012001

demand is also applied [14]. Transport forcing is provided by the CH3D hydrodynamic model. The daily constituent loads from the watershed are based on the simulation by the Chesapeake Bay Watershed Model (CBWSM) Phase 6.0 [2] in the 1991-2000 calibration run that used actual climate conditions, land uses, and management practices in 1991-2000. Table 1 lists the estuarine model (WQSTM) scenarios applied in this study.

Table 1. List of scenarios and the modification of model settings.

Scenario name	Depth of surface layer	Boundary salinity	T of ocean bound,	T of input freshwater	Heat exchange parameters	Watershed load
Base Case Scenario	As initially designed	bi-weekly observations	bi-weekly observations	Daily, observations/calibration	Based on observed	Daily load from watershed model,
Sea Level Rise Scenario (SLR)	Surface layer + 50 cm	readjust + increase	Same as the Base	Same as the Base	Same as the Base	calibrated for the years 1991-2000
Warming Scenario (GW)	Same as the Base	Same as the Base	Increase, vary monthly	Increase, vary daily	Based on increased air T	
Sea Level Rise & Warming Scenario (SLR+GW)	Same as SLR	Same as SLR	Same as GW	Same as GW	Same as GW	
5% load increase (inf. 2050 load) Scenario	Same as the Base	Same as the Base	Same as the Base	Same as the Base	Same as the Base	5% increase in watershed loads compared to the Base,
5% load increase and SLR+GW Scenario	Same as SLR	Same as SLR	Same as GW	Same as GW	Same as GW	approximate the load in 2050.
10% load increase Scenario	Same as the Base	Same as the Base	Same as the Base	Same as the Base	Same as the Base	10% increase in watershed
10% load increase and SLR+GW Scenario	Same as SLR	Same as SLR	Same as GW	Same as GW	Same as GW	loads compared to the Base, resembling a worst possiblecondition in 2050.

A Base Scenario is based on the observed conditions in 1991-2000. It is the same as the calibration run and is used as a reference to analyze relative differences or changes in the estuarine water quality due to warming and/or sea level rise. The same watershed load used in the Base Scenario was also employed in three climate change scenarios: a Sea Level Rise Scenario (SLR), a Warming Scenario (GW, i.e., global warming), and a Sea Level Rise and Warming Scenario (SLR+GW).

Under the projected 2050 climate condition, due to global warming and modified precipitation and evapotranspiration, watershed loads would increase [17]. Compared to the Base Scenario, the loads of total nitrogen (TN), total phosphorus (TP) and total suspended sediment (TSS) from the watershed to the Bay were estimated to increase by 5 to 7 percent. Two additional scenarios for the estuary were run for 1991-2000, in which the watershed load was increased 5 percent to roughly represent estimates of 2050 watershed loads, but with different estuarine circulation patterns; one is identical to the Base scenario, and the other is the same as the SLR+GW scenario. Considering the potential influence of model errors motivated us to include two additional counterpart scenarios, in which the watershed loads were increased by 10 percent, in order to represent a dramatic change in conditions in 2050 for both aforementioned circulation scenarios.

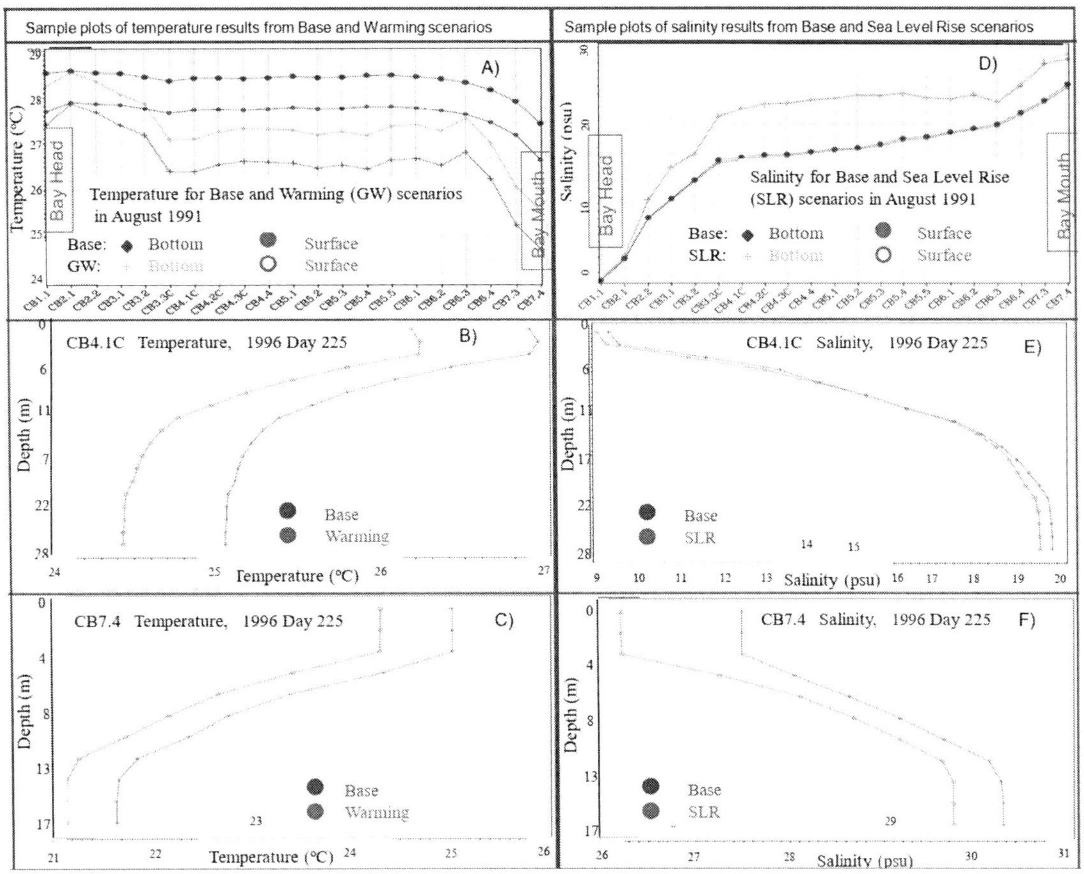

Figure 3. Sample results of temperature and salinity simulation in the Warming (GW) and Sea Level Rise (SLR) Scenarios. A) Temperatures of surface and bottom water in channel stations from the Bay head to mouth in Warming Scenario; B and C) Temperature profiles in stations CB4.1C and CB7.4; D). Salinity of surface and bottom water in channel stations from the Bay head to mouth in Sea Level Rise Scenario; E and F) Salinity profiles in stations CB4.1C and CB7.4.

3. Result and discussion

3.1. Response of circulation due to sea level rise and warming

This section uses the responses of salinity, temperature, and strength of stratification to analyse the influence on circulation by sea level rise or warming.

3.1.1. Response of water temperature and stratification strength by warming. The most significant physical impact produced by warming is upon water temperature and then, secondly on stratification. Without sea level change, the Warming Scenario does not cause significant changes in salt flux and salinity in the Bay. Compared to the Base Scenario, the Warming Scenario (GW) caused both the bottom and surface water temperatures to increase by about 0.6 -1°C in the summer in all stations along the channel, from the Bay head to the Bay mouth (figure 3(a)). The increase in temperature is further evident from figures 3(b) and 3(c), which display temperatures at vertical profiles in 2 channel stations on an arbitrarily chosen summer day (August 12, 1996).

The stratification strength was assessed based on the square of the Brunt-Vaisala frequency, i.e., the N^2 value [23], at a maximum density gradient of a depth profile: $N^2=-(g/\rho)(d\rho/dz)$, where ρ is density, g is the acceleration of gravity, and z is depth. Heat exchanges with the air is the primary method by which water temperatures are increased in the Warning Scenario, and results in greater increases in water temperature for shallower depths in summer months. The density of the surface water was reduced to a greater extent under higher temperatures. This led to a greater density gradient and a greater stratification strength in the Warming Scenario compared to the Base Scenario, as shown in figure 4(b) (grey dots) at station CB4.1C. The increase of stratification strength was also detected at other stations, e.g. CB2.2, CB3.1, CB5.1, CB6.1, and CB7.1. Accompanied with the warming, the depth of the maximum density gradient moved upwards in the water column to a slightly shallower position.

3.1.2. Response in salt flux, salinity fields, and stratification by sea level rise. Among the most significant physical impacts on the estuary produced by sea level rise are changes in salinity and stratification. Sea level rise slightly increases salt flux into the Bay. The change in monthly salt flux ranges from -1 to +3 kg/s. The increase of flux into the Bay by sea level rise occurs mainly through the boundary at deeper levels, which has positive influxes in every month (figure 5(d)), with greater influxes from late fall to early spring and lesser influxes from late spring to early fall. This pattern is reversed for the net flux through the upper boundary layers (figure 5(c)). In the winter and spring months the boundary salinity response is more sensitive to salt intrusion by sea level rise than the comparatively dryer seasons of summer and fall. Although elevated sea level retards the outflow of freshwater in tidal fresh and oligohaline regions, the increased salt intrusion from lower layers generates an upwelling at the head of the salt wedge where it occurs, and prompts a greater returned down-bay-ward force along the surface. This force exerted along the surface generates more outflow through the upper boundary overall. Generally, sea level rise promote the two-layer estuarine circulation, which can be seen more clearly from a cross section of the salt flux through the mid-Bay (figures 5(a) and 5(b), near station CB5.1 (figure 1). In the Base Scenario both the influx in the lower layer and the outflux in the upper layer were greater in the wetter winter-spring months than in the dryer summer months. Under a regimen of sea level rise, the monthly influxes in the lower layer increased (i.e., more positive, figure 5(b) and, concordantly, the monthly outflux in the upper layer also increased (i.e., more negative, figure 5(a)). The changes in salt influx or outflux were greater in the winter-spring months than in the summer months.

Along with elevated salt intrusion due to sea level rise, the average bottom and surface salinity increases, as shown in figure 3(d) at monitoring stations along the channel, from the Bay head to the Bay mouth in an arbitrarily chosen summer. The salinity has a greater increase at the Bay mouth, e.g., CB7.4, figure 3(f), than at the inner Bay (e.g., CB4.1C, figure 3(e)). Generally, salinity increases throughout the entire depth profile, but in some upper Bay stations there are occasions wherein (e.g. CB2.2 on August 8, 1994) the surface salinity may show a decrement (figure not shown). These decrements can be attributed to conditions a few days prior to specific events. In this instance (August

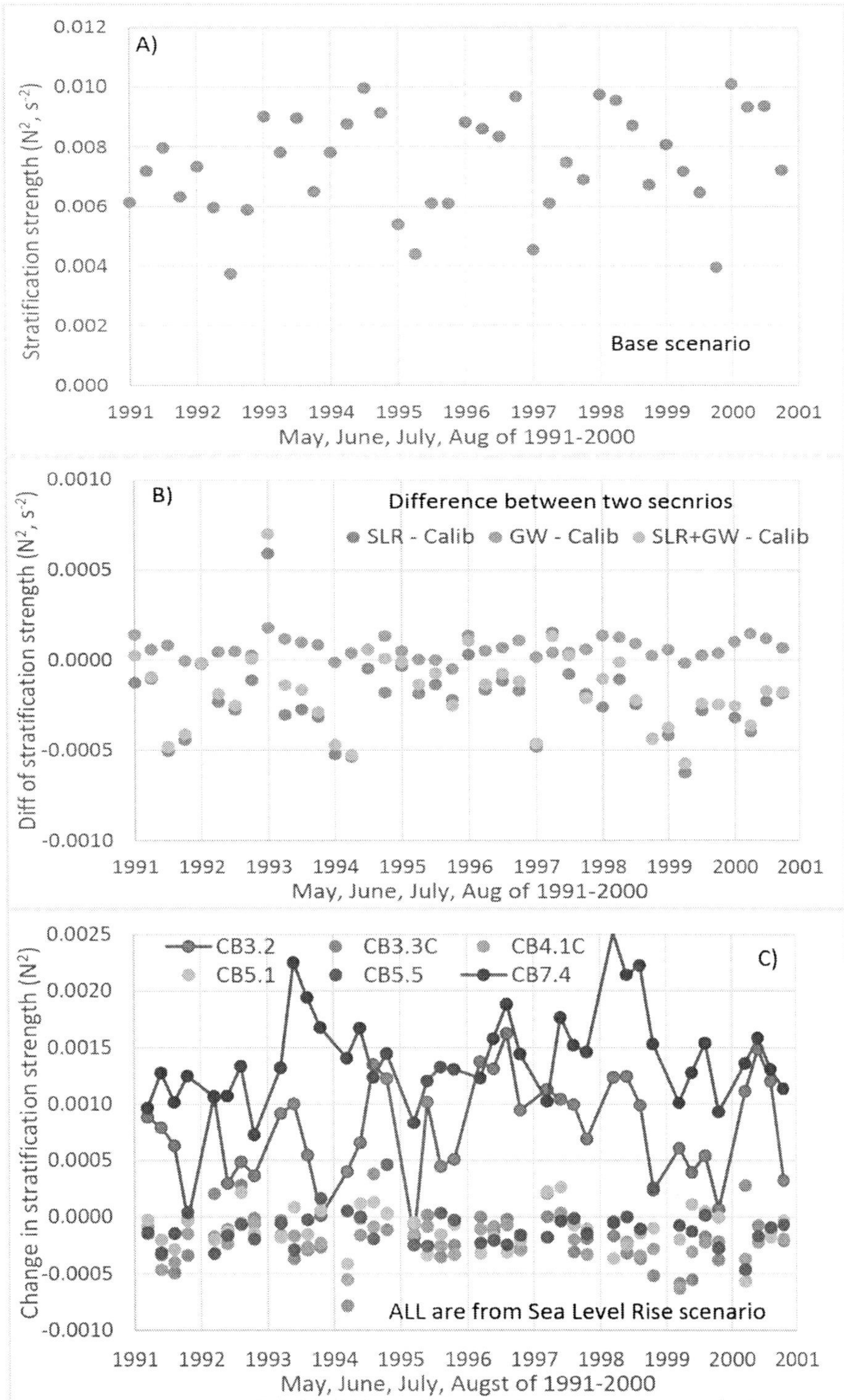

Figure 4. Stratification strengths in May, June, July and August of 1991-2000. A) stratification strength in CB4.1C in the Base Case Scenario; B) Difference of stratification strengths in CB4.1C between three climate scenarios and the Base Case Scenario; C) Difference of stratification strengths in selected channel stations between the Sea Level Rise scenario and the Base Case Scenario.

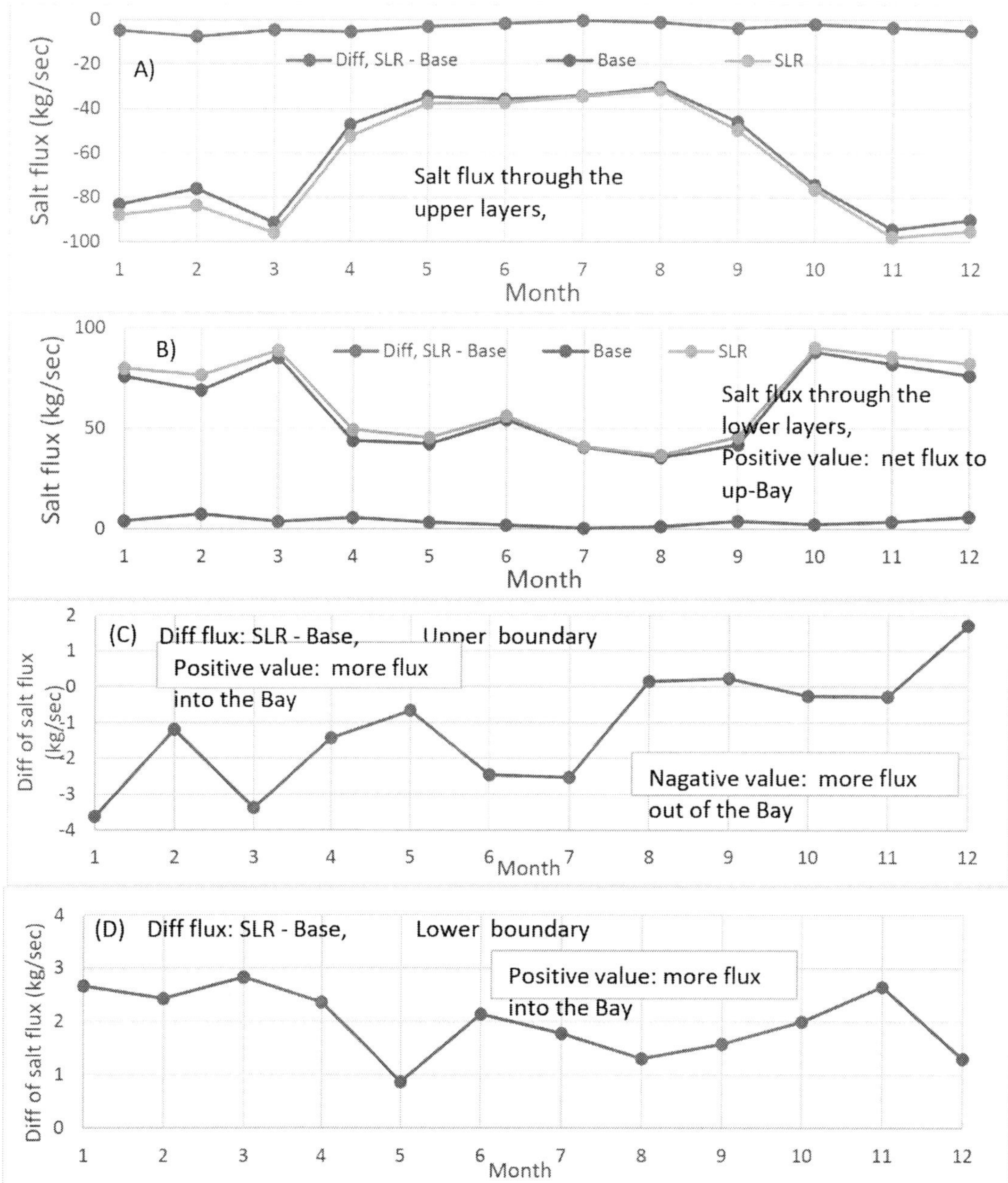

Figure 5. Monthly salt flux in the Base Case Scenario (orange) and the Sea Level Rise (grey) scenarios, and difference of salt flux between the two scenarios (blue). a) through the upper layer at a mid-Bay cross section near CB5.1 (figure 1); b) through the lower layer at the mid-Bay cross section; c) through the upper layer at the Bay mouth; d) through the lower layer at the Bay mouth. Note: a positive sign denotes net flux is up-bay-ward to the north or into the Bay, and negative sign denotes net flux is down-bay-ward to the south or out of the Bay.

8, 1994) dry conditions and a lack of freshwater discharge contributed to strong salt intrusion in the Base Scenario, but the Sea Level Rise Scenario did not generate enough salt intrusion on that day to greatly affect the surface water salinity near the Bay head.

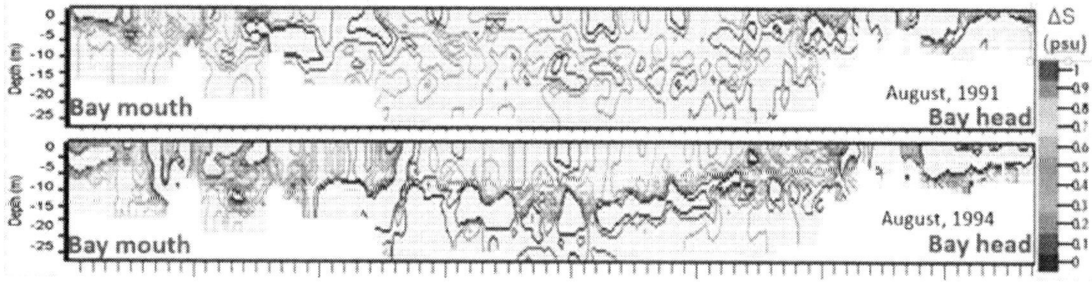

Figure 6. Change of salinity (ΔS, psu) along the channel of the mainstem Bay in August of 1991 and August of 1994 by 50-cm sea level rise versus the Base Case Scenario.

Figures 6(a) and 6(b) are contour maps of salinity difference between the sea level rise and the Base Scenario along the mid channel from Bay mouth to Bay head in August of 1991 and 1994, respectively. Under the same amount of salt intrusion with sea level rise, the relative increase, i.e. percent increase, compared to the Base Scenario, of salinity in a wet time period, e.g. August 1994, are greater than that in a normal or dryer time period, e.g. August 1991. Thus, the changes in salinity are greater in August 1994 (figure 6(b)) than those in August 1991 (figure 6(a)), which is in agreement with model experiments conducted by Hong and Shen [13]. In both cases, salinity increases more in the mid-depth (at about 10 m where pycnoclines usually occur) than in the surface or bottom layers. There, the maximum density gradient in the Base Scenario condition was mostly situated around the upper boundary of the pycnoclines, at somewhat less than 10 m. In the Sea Level Rise Scenario, the depth of greater salinity increases is below the maximum density gradient, thereby increasing stratification strength in most places.

However, in deep regions that have a deep channel designated-use such as CB4.1C, the maximum density gradients in the Base Scenario for such a location lies at approximately 12 m in depth at most times of the simulated summer periods, near the lower boundary of the pycnocline. The simulated increase of salinity at the aforementioned mid-depth (about 10 meters) above the usual location of the maximum density gradient reduces the overall density gradient, thus the stratification strength is reduced. Accordingly, in most of the simulated summer months in the Sea Level Rise Scenario, stratification strength in CB4.1 decreases (figure 4(b)), while in a few months, such as August 1994, the stratification increases. The response of stratification strength to sea level rise in the deep stations near CB4.1, like CB5.1, is similar (figure 4(c)). However, in most other channel stations (such as CB3.2 and its northern stations, CB7.4 and CB6.1 and its southern stations) the sea level rise caused stratification strengths in most months to increase (figure 4(c)). The aforementioned phenomena in these stations have the potential to change when freshwater discharge is significantly altered.

3.1.3. Response of stratification strength by combined sea level rise and warming. In comparison with the Base Scenario, stratification at station CB4.1 in the Sea Level Rise and Warming (SLR+GW) Scenario decreases in most cases, but is slightly greater than that in the Sea Level Rise Scenario (figure 4(b)). This indicates that the impact on stratification is mainly controlled by sea level rise, and less so by temperature change.

3.2. Impact on phytoplankton, dissolved oxygen and water clarity by sea level rise and warming

Chlorophyll-*a*, DO and water clarity are three key water quality components in the Chesapeake Bay TMDL (USEPA, 2010). This section analyzes their responses to warming and sea level rise.

3.2.1. Impact of water quality by warming. Increase of temperatures in the spring increases metabolism of algae. In the Warming Scenario spring chlorophyll-a concentration increases broadly in the estuary. However, the increase is only slight, ranging from 0 - 0.11 μg/L (figure 7(b)), and is approximately a 0.01 μg/L increase in most areas (note: not scaled in the map). Only in areas closest to freshwater inputs (which are more influenced by warmed terrestrial sources of water) is the increase of chlorophyll-a greater than 0.11 μg/L. The response of summer chlorophyll-a is similar to that in the spring, with slightly greater increases in summer chlorophyll-a (figure 7(e)), around 0.02 μg/L increase in most areas. These can also be seen from figures 8(b) and 8(c) (orange dots). The average chlorophyll-a concentration in the upper- and middle-Bay increases slightly, i.e., 0.03 percent and 0.12 percent in the spring and summer, respectively (table 2).

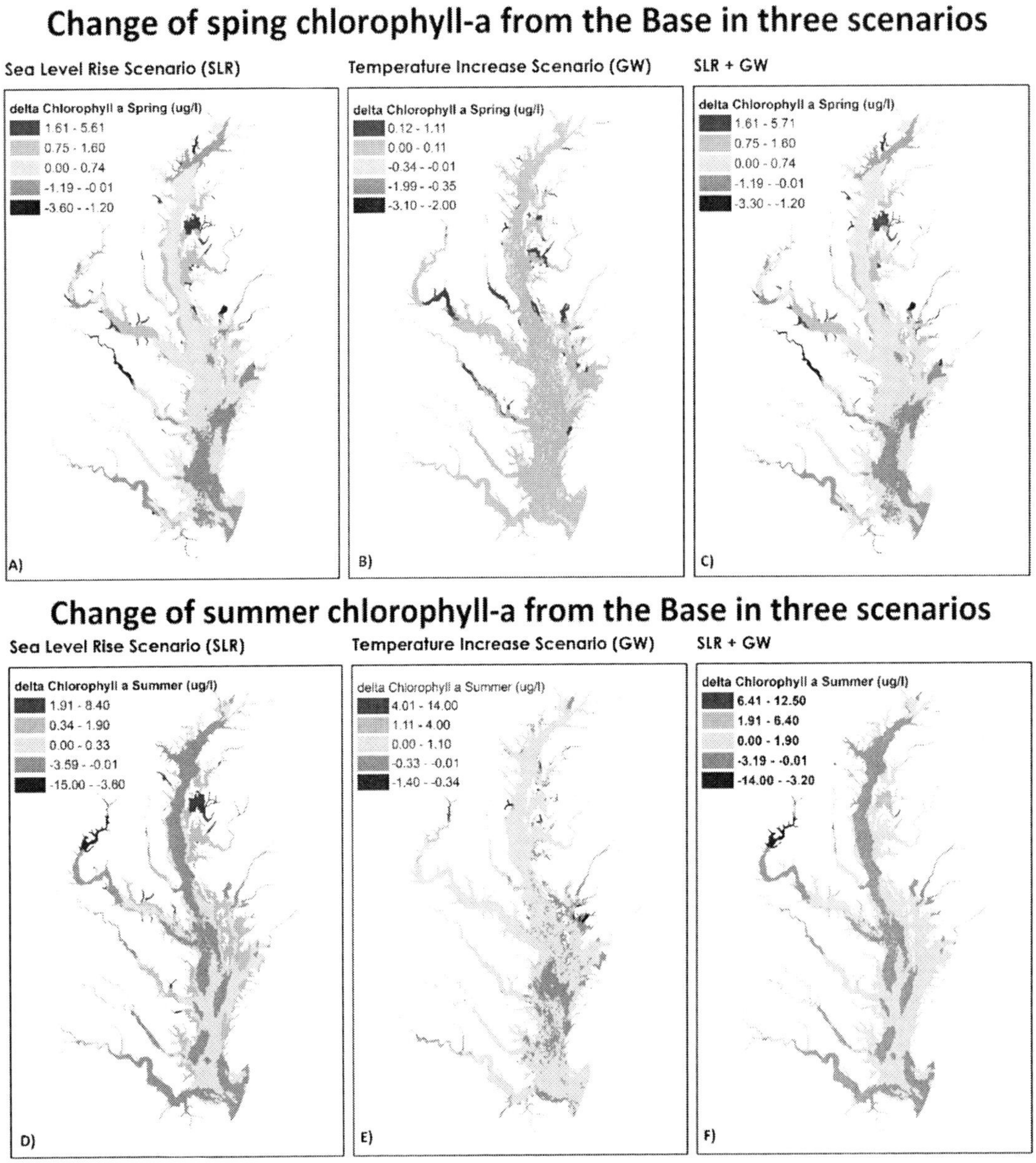

Figure 7. Change of spring and summer chlorophyll-*a* concentrations in three climate scenarios from the Base Scenario. A) spring ΔChl (SLR Scenario - Base Scenario); B) spring ΔChl (Warming Scenario - Base Scenario); C) spring ΔChl (Sea Level Rise and Warming Scenario - Base Case Scenario); D) summer ΔChl (Sea Level Rise Scenario - Base Case Scenario); E) summer ΔChl (Warming Scenario - Base Case Scenario); F) summer ΔChl (Sea Level Rise and Warming Scenario - Base Scenario).

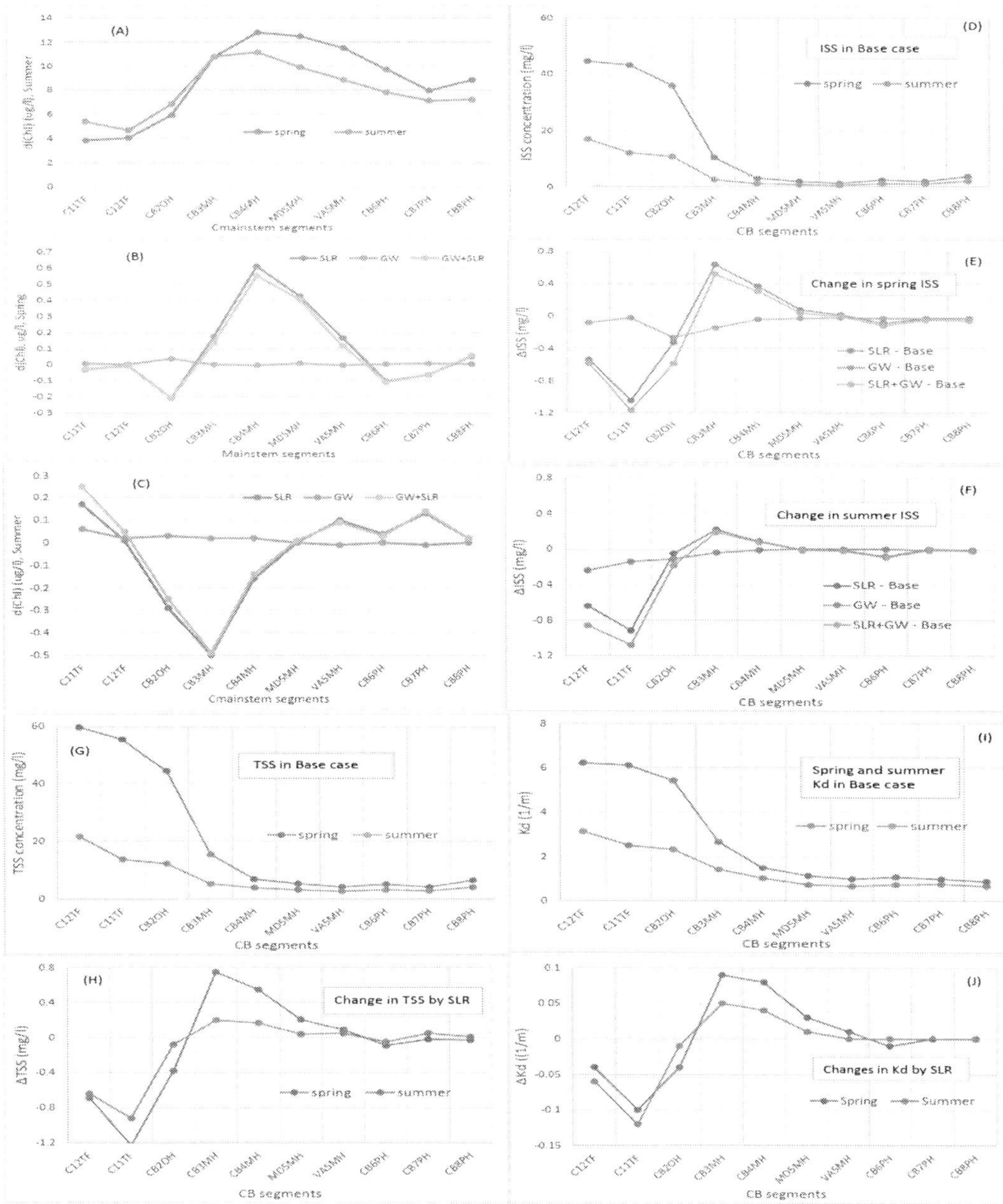

Figure 8. Response of surface chlorophyll-*a* (a-c), inorganic suspended solids (ISS) (d-f), total suspended solids (TSS) (g-h) and diffused light attenuation coefficient (Kd) (i-j) in mainstem segments. a) Spring and summer chlorophyll-*a* in the Base Scenario, b) changes of spring chlorophyll-*a* by three climate scenarios compared to the Base Scenario, and c) changes of summer chlorophyll-*a* by climate scenarios compared to the Base Scenario; d) Spring and summer ISS in the Base Case Scenario, e) changes of spring ISS by climate change scenarios compared to the Base Scenario, f) changes of summer ISS by climate change scenarios compared to the Base Scenario. Surface water total suspended solids (TSS) concentration and diffused light attenuation coefficient (Kd) in mainstem segments. g) Spring and summer TSS in the Base Scenario, h) changes of TSS by sea level rise; i) spring and summer Kd in the Base Scenario, j) changes of Kd by sea level rise.

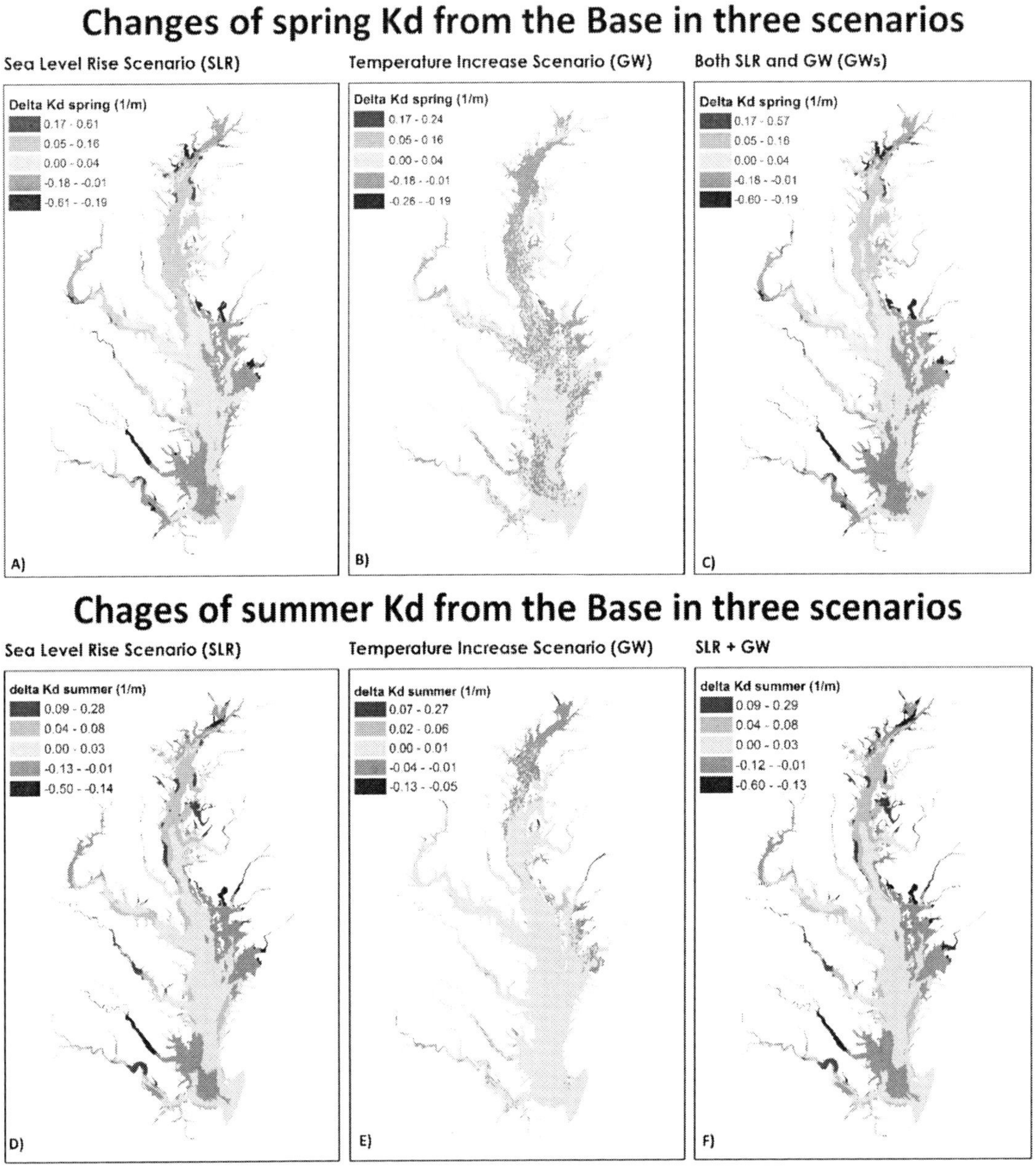

Figure 9. Change of spring and summer Kd in three climate scenarios compared to the Base Scenario. A) spring ΔKd (Sea Level Rise Scenario - Base Case Scenario); B) spring ΔKd (Warming Scenario - Base Case Scenario); C) spring ΔKd (Sea Level Rise and Warming Scenario - Base Scenario); D) summer ΔKd (Sea Level Rise Scenario - Base Scenario; E) summer ΔKd (Warming Scenario - Base Scenario); F) summer ΔKd (Sea Level Rise and Warming Scenario - Base Scenario).

Light attenuation by water (denoted by Kd as the diffused light attenuation coefficient) is a function of concentrations of chlorophyll-*a* and other organic particulates, inorganic suspended solid (ISS) and color of dissolved organic matter [24]. The warming climate increased chlorophyll-*a*, leading to increased Kd. However, the increased temperature of surface water decreased its density and increased the settlement of fine ISS, e.g., clays and silts, thereby decreasing ISS concentration in the surface layer (figures 8(e) and 8(f), orange dots). Kd increased slightly in the spring and summer in mid and lower Bay in the Warming Scenario (figure 9(b) and 9(e)), where the changes in chlorophyll-*a* were the main causes of changes in Kd. However, in the upper Bay closer to the mouth of the Susquehanna River, which was influenced significantly by terrestrial water inputs (segments CB1TF and CB2OH), the influences warming water had on increasing sinking rates of ISS became more important in controlling changes in Kd, and drove decreases of Kd. The average Kd in the mainstem Bay decreased by about 0.2 percent in the spring and summer in the Warming Scenario (table 2).

Table 2. Summary of relative changes in anoxic volume, average surface chlorophyll, and Kd in the upper and middle Bay by climate change scenarios versus the Base Case Scenario.

Scenario name	Watershed load	Estuarine circulation	Summer anoxic volume (% change)	Spring surface chlorophyll (% change)	Summer surface chlorophyll (% change)	Spring surface Kd (% change)	Summer surface Kd (% change)
Base Case	Present	Present	---	---	---	---	---
SLR (Sea level rise)	Present	SLR	-14.4%	1.93%	-1.23%	1.23%	1.15%
GW (Global warming)	Present	GW	2.26%	0.03%	0.12%	-0.245%	-0.21%
SLR+GW	Present	SLR+GW	-11.6%	1.84%	-1.07%	0.85%	0.81%
5% increase of load	1.05 times the Base	Present	9.65%	2.51%	2.27%	0.72%	0.52%
5% increase of load, and SLR+GW	1.05 times the Base	SLR+GW	-2.12%	4.80%	1.34%	1.60%	1.38%
10% increase of load	1.1 times the Base	Present	18.7%	4.750%	4.26%	4.81%	3.35%
10% increase of load, and SLR+GW	1.1 times the Base	SLR+GW	6.35%	6.95%	3.21%	5.67%	4.14%

There are three main physical and chemical processes related to warming that influence the Bay's anoxia: 1) the increased water temperature increases respiration and oxygen consumption, 2) the increased stratification (figure 4(b), grey dots) helps to retain the anoxic condition, and 3) increased temperature reduces oxygen solubility and DO concentrations. Biologically, warming helps algal blooms in the spring and summer (figures 7(b) and 7(e)), and results in higher biological oxygen demands. Compared to the anoxic volume in the Base Scenario, the influence on anoxia by the stated physiochemical and biological processes is small, thus there is only an increase in anoxic volume of 1.34 percent produced by the Warming Scenario (table 2 and figure 10). Figure 11 is a GIS plot of difference in DO (ΔDO), between a climate change scenario and the Base Scenario in water greater than 12 meters in depth. In the Warming Scenario, ΔDO decreases in most areas (figure 11(b)), consistent with the increases of spring and summer chlorophyll-*a* (figures 7(b) and 7(e)) and 8(b) and 8(c)). The algae produced in the surface water of a segment may drift to other segments before they settle and decay in bottom layers. Therefore, changes in bottom DO are not always consistent with changes in surface chlorophyll-*a* [25]. The ΔDO has certain degrees of variability over these areas. The factors driving an increase of DO in the area near the Bay mouth is unclear, but may be due to warmer sea water on the bottom that reduces stratification.

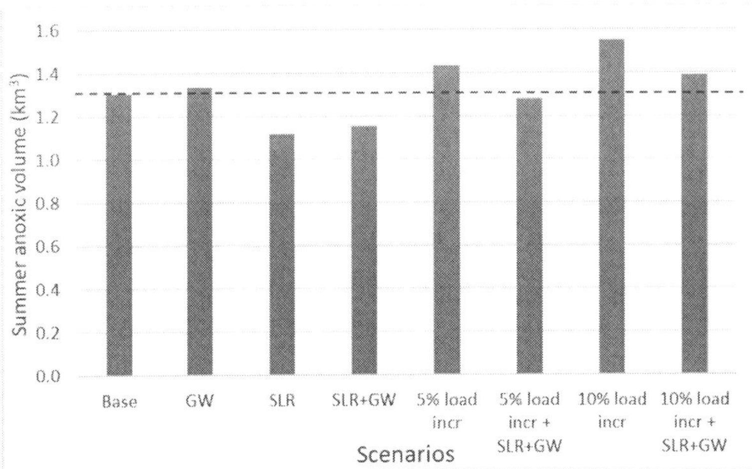

Figure 10. Summer average anoxic volume (km^3) from the Base Scenario and climate change scenarios.

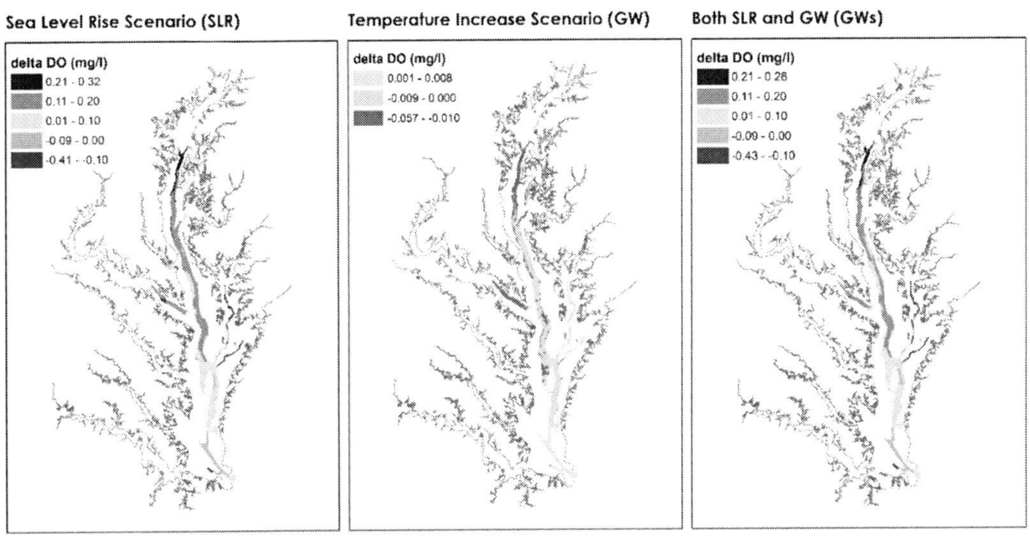

Figure 11. Change of summer DO concentration (ΔDO, mg/L) for water depth>=12 m in three climate scenarios from the Base Case Scenario. A) ΔDO (Sea Level Rise Scenario - Base Scenario); B) ΔDO (Warming Scenario - Base Scenario); C) ΔDO (Sea Level Rise and Warming Scenario - Base Scenario).

3.2.2. Impact on water quality by sea level rise. Nutrients and light are two main limiting factors of algal growth in the Chesapeake, while light attenuation is dependent upon concentrations of ISS, chlorophyll-a, and other matter that determine water clarity. Sea level rise influences transport and redistributes the concerned components among CB segments. Comparing these components together can help to better understand the influences of sea level rise on water quality.

- Response in transport of suspended solids and water clarity in sea level rise

The influence on circulation by sea level rise affects transport of inorganic suspended solids. In the Base Scenario, the surface ISS decreases rapidly from the Bay head to the Bay mouth (figure 8(d)) as settling continues during down-Bay ward transport. Concentrations of ISS at the Bay mouth (e.g., CB7.4) are slightly greater than those in inner stations, e.g., CB7.3, due to the oceanic source.

In the shallow fresher areas (i.e., the tidal fresh and oligohaline environments) of the upper Bay, the slower downstream transport of fresh water in the Sea Level Rise Scenario documented by Hong and Shen (2012) generates a greater volume of water than that observed in the Base Scenario and reduces ISS concentration in segments C12TF, C11TF, and CB2OH (figures 8(e) and 8(f), blue dots). Similarly, in the lower Bay near the Bay mouth (e.g., segments CB6PH, CB7PH and CB8PH) the sea water intrusion also reduces ISS concentrations, though the influence is weak and the reduction of ISS concentrations is slight. However, in both the lower and upper Bay the increase in salinity from sea level rise may also benefit the suspension of fine ISS in surface waters and thereby increase ISS concentrations. Still, this may not be the key controlling factor in most of the lower and upper Bay areas; rather the increased surface water volume or changes in water movement may be more critical in determining ISS concentrations in the Sea Level Rise Scenario.

The situation is different in the deep water areas of the mid-Bay, such as that observed in segments CB3MH and CB4MH. There the increase in surface water volume becomes a less important factor, while the retention of materials in the Bay influenced by sea level rise becomes more important in governing ISS levels in surface water. Overall, surface ISS concentrations increase in the Sea Level Rise Scenario. In addition to a greater amount of material remaining within the Bay, the increased salinity increases water density and retards sinking of ISS from the surface.

TSS consists of ISS and particulate organic matter. The distribution and patterns in changes of TSS among the CB segments in the Sea Level Rise Scenario (Figure 8h, blue and orange dots) are similar to changes in ISS (Figure 8, e-f, blue dots). This suggests that ISS is the main component of TSS, while the contributions made by changes in algae due to sea level rise are relatively small compared to the changes in ISS concentrations. The changes in Kd among the CB segments due to sea level rise (figure 8(j), blue and orange dots) have similar patterns as those found for TSS (figure 8(h)). The average spring and summer Kd in the upper- and middle-Bay increases by about 1.2 percent (table 2).

- The response of chlorophyll-*a* to sea level rise

Changes in spring chlorophyll-*a* in the Sea Level Rise Scenario have greater ranges than those produced in the Warming Scenario (Figure7a vs. 7b, or 8b blue vs. orange dots). In the mid-Bay area (segments CB3-CB5), the range of increases in the Sea Level Rise Scenario vary between 0.2-0.6 μg/L (figure 8(b), blue dots). Spring chlorophyll-*a* decreases 0.1-0.3 μg/L in the upper and lower Bay, seen in segments C12TF, C11TF, CB2OH, CB6PH and CB7PH. Accompanied with the increased salt flux into the Bay, sea level rise facilitated the retention of nutrients in the estuary for a longer period of time, causing larger areas in the Bay, especially the mid-Bay, to grow more algae in the spring. Simultaneously, the increased influx from the ocean may dilute nutrient concentrations. This action is more prominent in areas nearer to the Bay mouth, where nutrient concentrations are relatively low and are more affected by ocean water dilution. Thus, spring chlorophyll concentrations in the lower Bay are decreased. The cause of decrease in chlorophyll concentration in the upper Bay is similar to the cause of decreased ISS concentrations, namely dilution produced by increased water volume in the upper Bay as a result of sea level rise [13].

In the summer, the response of chlorophyll-*a* to sea level rise in the CB segments (Figure 8c, blue dots) is the reverse of that observed in the spring (Figure 8b, blue dots), apparently to compensate for the nutrient uptake occurring in the spring. The average chlorophyll-*a* concentration in the upper- and middle-Bay increases by about 2 percent in the spring, while decreases of 1.2 percent occur in the summer (table 2). In the mid Bay (segments CB3MH and CB4MH) the excessive algal growth in the spring in the Sea Level Rise Scenario lessened the amount of nutrients remaining in the summer and reduced summer chlorophyll. In the upper Bay (segments C11TF and C12TF) and lower Bay (segments CB6PH and CB7PH), the weaker algal growth in the spring in the Sea Level Rise Scenario left more nutrients available in the summer, increasing summer chlorophyll. The segments between the two areas of increasing and decreasing chlorophyll-*a*, (e.g. CB2OH, VA5MH, and MD5M) do not show a reversed direction of change for chlorophyll-*a* between spring and summer. Because they are in transition between the areas of different trends in chlorophyll-*a* changes, their nutrient and chlorophyll-*a* concentrations are influenced by adjacent segments.

- Response of anoxic volume to sea level rise

Sea level rise increases residence times of nutrients and organic carbon in the estuary. This can promote algal blooms, increase oxygen demand, and increase anoxic volume. However, the anoxic volume in the simulated 10 summers of the Sea Level Rise Scenario were about 0.2-0.4 km^3 less (14.5 percent, table 2) than those in the Base Scenario (figure 10). This suggests that the change in summer hypoxic conditions in the Sea Level Rise Scenario may be more closely related to alterations of physical processes rather than algal processes.

There are two physical processes in sea level rise that may reduce anoxic volume: A) The promoted salt intrusion in the lower layer (figure 5(b)) brings oxygen-rich sea water into the Bay's anoxic area, as is represented by the cross section near station CB5.1 at the mid-Bay. At this cross section, the variation of monthly oxygen influxes through the lower layer or the outward fluxes through the upper layer (figures 12(b) and 12(a)) are consistent with the variation of salt fluxes (figures 5(b) and 5(a)). Despite the increase of oxygen influx through the lower layer in the summer being less than that in the winter, such an increase can significantly reduce the Bay's anoxic condition. B) The reduced stratification in the anoxic center promotes mixing and increases bottom DO. It should be noted, as presented in Section 3.1, that sea level rise promotes estuarine circulation and increases the overall stratification in the Bay. However, in the Bay's anoxic center, i.e., the deep water region of segments CB3MH, CB4MH, and CB5MH, the increase in salinity at its maximum density gradients in most summer months was less than that produced at shallower depths, causing a reduction of stratification strength (figure 4(c)) and reducing anoxic conditions.

There are cases showing that the reduction of stratification in the anoxic center by sea level rise may not be the main driver in the reduction of summer anoxia. For example, 1996 had a relatively wet summer in the time period 1991-2000. The decrease of stratification in the anoxic center by sea level rise in 1996 was weaker than that in other years (figure 4). However, the reduction of anoxic volume in 1996 was not the lowest among the simulated 10 years (figure 13). In another instance, the reduction of anoxia by sea level rise is greatest in 1993 (figure 13), yet the decrease of stratification in the anoxic center was also not the greatest (figure 4). The years 1991, 1992, and 1997 all had relatively dry summers, and the Bay was relatively saltier. In these years, the salt influx in the Sea Level Rise Scenario was less than that seen in comparatively wetter years. The associated intrusion of oxygen-rich sea water to the Bay's anoxic center was relatively weaker, thus the anoxia reduction was less in these years than in other years (figure 13). In this case, salt water intrusion with relatively higher DO may have played an important role in reducing anoxia.

Summer hypoxic/anoxic volume reflects DO conditions in the bulk Bay. Still, changes in average DO concentrations in individual model cells can provide information on its spatial variation and provide some clues to analyze the related mechanisms. Figure 10(a) shows changes of average DO concentrations in water deeper than 12 m in the Sea Level Rise Scenario. Segments CB4MH, CB5MH, and CB3MH cover the Bay's anoxic center. There, DO is greater in the Sea Level Rise Scenario than in the Base Scenario. However, in segment CB7PH near the ocean boundary, DO is lower in the Sea Level Rise Scenario. Under the promoted bottom salt intrusion during the up-Bay-ward movement the particulate organic matter moved more slowly than the dissolved oxygen. Much of the particulates may remain in the lower Bay and continue to consume oxygen. Segment CB7PH is closer to the Bay mouth and has much greater DO than the anoxic zone. The relative change in DO by salt intrusion is not as large as that in the Bay's anoxic center. The oxygen consumption of the retained particulate organic matters plays a more important role in regulating DO in segment CB7PH, producing lower DO concentrations in the Sea Level Rise Scenario than the Base Scenario. The changes in DO concentrations during sea level rise are controlled by multiple factors, detailed analyses of which are beyond the scope of this paper.

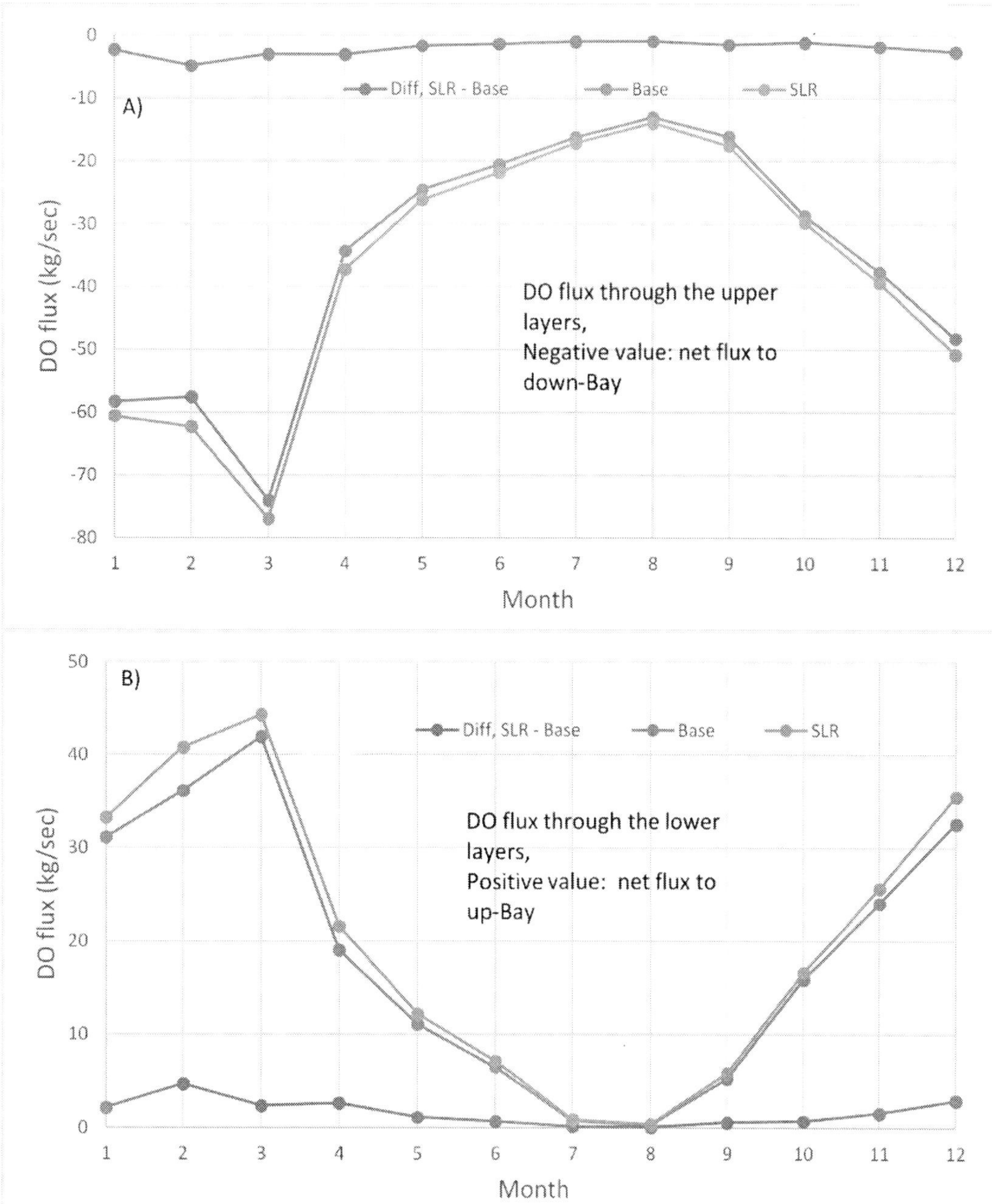

Figure 12. DO flux (kg/s) through the mid-Bay cross-section in the Base Case Scenario and Sea Level Rise Scenario. A) through the upper layers, B) through the lower layers.

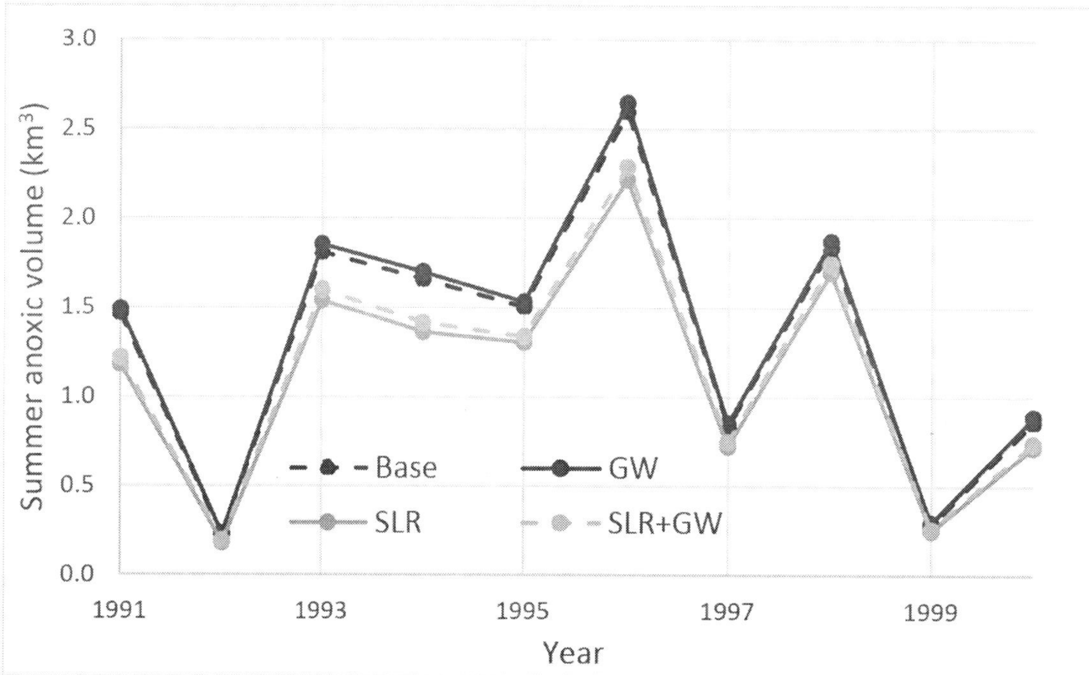

Figure 13. Summer anoxic volume of the mainstem Bay in the Base Case Scenario and three climate scenarios.

- Influence on water quality by combined sea level rise and warming

The summer anoxic volume in the combined Sea Level Rise and Warming (SLR+GW) Scenario is 11.7 percent less than the Base Scenario (table 2), but is slightly greater than the Sea Level Rise Scenario (figure 10). These findings suggest that the DO change in the Sea Level Rise and Warming Scenario compared to the Base Scenario is mainly influenced by sea level rise, while the influence of warming is small. The influence on chlorophyll-*a* (figure 8(b) and 8(c)) and water clarity (referring to ISS, figures 8(e) and 8(f)) of combined sea level rise and warming is also closer to that produced by sea level than that produced by warming.

3.3. Response of water quality due to increased watershed loads in a projected 2050 climate condition

Scenarios of a 5 percent increase in watershed loads (compared to the Base scenario) approximate the loading condition in 2050. The increase of nutrient loads causes an increase of spring and summer chlorophyll-a concentrations (about 2.4 percent) and light attenuation (about 0.6 percent) (table 2). The increase in summer anoxia is mainly due to increased spring (January – May) nutrient loads, which has a high correlation with summer hypoxia and anoxia [25, 26]. Under the same estuarine circulation as the Base Scenario, summer anoxic volume will increase by 9.65 percent, but is reduced slightly (-2.12 percent) when the estuarine circulation considers sea level rise (figure 10). The promoted oxygen-rich sea water intrusion by sea level rise has a greater influence on summer bottom DO than increasing phytoplankton concentrations in 2050 due to higher watershed loads or increasing temperatures.

Scenarios of a 10 percent increase in watershed loads are used to assess potential worst conditions in 2050. This scenario yielded about 4.5 percent increase of spring and summer chlorophyll concentrations and light attenuation (table 2). Consistent with the scenarios using watershed loads of the Base condition, sea level rise and warming increases chlorophyll-a concentrations in spring and promotes greater light attenuation in the spring and summer and the retaining of suspended sediment in the Bay due to sea level rise, while less summer chlorophyll-a is produced due to the consumption

IOP Conf. Series: Earth and Environmental Science 82 (2017) 012001 doi:10.1088/1755-1315/82/1/012001

of nutrients by algae in the spring. In scenarios of a 10 percent increase in watershed loads, summer anoxic volume increases by 18.7 percent when the estuarine circulation uses the Base condition, but only increases by 6.35 percent when sea level rise is considered for estuarine circulation. Thus, for management preparation in 2050, more nutrient controls than the current effort would be needed to reach water quality attainment goals.

4. Conclusions

In a projected warming condition of 2050 which employs an ensemble of GCMs estimating a 1.5-1.9°C increase in monthly air temperatures, the water temperature in the Chesapeake Bay is estimated to increase by about 0.8-1°C. The increased temperature promotes the metabolism of phytoplankton, leading to increases in spring and summer chlorophyll-*a* concentrations as well as increased oxidation rates, causing increased summer oxygen consumption and hypoxia. Our model experiment of a Warming Scenario results in lower summer DO in deep waters and greater anoxic volume. Besides the influence of algal growth in the warming condition, the increase of anoxia is also related to increases in stratification. The temperature of surface water increases more than temperatures at greater depths, resulting in lighter density in the surface layers which slightly increases stratification strengths.

In the 0.5 meter Sea Level Rise Scenario, due to the interactions of the increased sea water intrusion and fresh water outflow, the changes in nutrient retention varied among different Bay areas and CB segments. In some areas, spring or summer chlorophyll-*a* increases, while it simultaneously decreases in others. The changes in summer chlorophyll-*a* concentrations among the CB segments are reversed compared to the pattern of changes in spring chlorophyll, suggesting nutrient availability in the summer is related to algal consumption in the spring. In the Sea Level Rise Scenario, summer dissolved oxygen in deep water increases and anoxic volume is reduced. Sea level rise promotes estuarine circulation and increases oxygen-rich seawater intrusion, mainly through the lower layers, increasing bottom DO concentrations and reducing hypoxia. The stratification strengths in the majority of the Bay are increased under a scenario of sea level rise. However, in the Bay's anoxic center which has a Deep Channel designated-use, the maximum density gradient is primarily located near the lower boundary of the pycnocline. The larger increase of salinity in the pycnocline may reduce the salinity gradient and weaken stratification. This promotes mixing of deep water and surface water, thereby reducing hypoxia in the affected waters. While, the main cause of reducing anoxia in sea level rise is mainly due to the promoted oxygen-rich sea water intrusion into the Bay's anoxic area.

The combined Sea Level Rise and Warming Scenario also results in reduced anoxia, because the reduction of anoxia by sea level rise is much more significant than the increase of anoxia by warming in this model scenario.

A warming climate promotes algal metabolism and increases chlorophyll-*a*, leading to increased Kd. However, in the upper Bay closer to the mouth of the Susquehanna River, which is influenced significantly by water inputs from terrestrial sources, the influences of warming water on sinking rates of ISS may be more important than the increase of chlorophyll-*a* to control changes in decreasing Kd.

Sea level rise's influence on circulation also affects the transport of inorganic suspended solids. In the simulated conditions, ISS is the main component of TSS and affects Kd, while the contributions by the changes in algal production are relatively small compared to the changes in ISS concentrations. In the upper Bay, accompanied with a slower downstream transport of fresh water in the Sea Level Rise Scenario, a greater volume of water than that observed in the Base Case Scenario reduces ISS concentrations in segments C12TF, C11TF and CB2OH, reducing Kd. In the deep water areas in the mid-Bay, the importance of increasing surface water volume becomes less important, while the retention of materials in the Bay due to sea level rise becomes more critical in governing ISS in surface waters. Therefore, we see ISS and Kd increases in the Sea Level Rise Scenario.

Overall, the estimated impact on the Bay's water quality is slight in the simulated warming condition if there is no nutrient load change; the impact on water quality is greater in our simulated 0.5 meter sea level rise. Under the condition of both sea level rise and warming, the water quality response to their combined effects and the changes in chlorophyll-*a* result in Kd and summer anoxia values that are closer to those produced by a condition of sea level rise. This indicates that in our simulated climate change condition, the influence on water quality is controlled to a greater extent by sea level rise than by warming.

The above impacts on water quality by the three climate scenarios were compared with the Base Scenario, and all scenarios used the same watershed loads as those in the current condition but utilized different estuarine hydrodynamics. The scenario of a 5 percent increase in nutrient loads due to climate change with no change in estuarine circulation (the Base Scenario condition) yielded a summer anoxic volume that was 10 percent higher than the Base Scenario condition. The scenario of a 5 percent increase in nutrient loads that also used a modified estuarine circulation due to sea level rise and warming yielded an extent of summer anoxic volume that was 2 percent less than the Base Scenario condition. This indicates that sea level rise can potentially benefit the reduction of summer anoxia and hypoxia. However, water quality relating to chlorophyll-*a* and water clarity will still be degraded. If watershed loads increase by 10 percent from their current condition, (a potential worst case for 2050) summer anoxic volumes would increase, from 6 to 19 percent, when using estuarine circulation in a sea level rise or current condition, respectively. Greater efforts are needed in controlling nutrient loads to prepare for the challenges of an altered climate in 2050. The estuarine water quality model used in this work does not simulate the function of wetlands in nutrient retention. Sea level rise may also contribute to an increased loss of wetlands that may negatively impact levels of DO, an important issue for management consideration that is not simulated in this work.

References

[1] USEPA (U.S. Environmental Protection Agency) 2010 *Chesapeake Bay Total Maximum Daily Load for Nitrogen, Phosphorus and Sediment* (US: Environmental Protection Agency Chesapeake Bay Program Office)

[2] Linker C L, Batiuk R A, Shenk G W and Cerco C F 2013 Development of the Chesapeake Bay watershed total maximum daily load allocation *JAWRA* **49** 986-1006

[3] CEC (Chesapeake Executive Council) 2014 *Chesapeake Watershed Agreement* (Annapolis, MD, USA)

[4] USACOE (U.S. Army Corps of Engineers) 2011 *Sea-level Change Considerations for Civil Works Programs* (Circular 1165-2-212. U.S. Army Corps of Engineers, Washington, DC)

[5] Parris A, *et al* 2012 *Global Sea Level Rise Scenarios for the United States National Climate Assessment* (NOAA Technical Report OAR CPO-1. National Oceanic and Atmospheric Administration, Silver Spring, Maryland)

[6] Sallenger A H, Doran K S and Howd P A 2012 Hotspot of accelerated sea-level rise on the Atlantic coast of North America *Nat. Climate Change* **2** 884-8

[7] Najjar R G, *et al* 2010 Potential climate-change impacts on the Chesapeake Bay *Estuar. Coast. Shelf Sci.* **86** 1-20

[8] Walsh J, *et al* 2014. Chapter 2: Our changing climate *Climate Change Impacts in the United States: The Third National Climate Assessment* ed J M Melillo, T C Richmond and G W Yohe (U.S. Global Change Research Program) pp 19-67

[9] Wang P, Linker L, Wang H, Cerco C, Bhatt G and Tian R 2014 Simulation of Chesapeake Bay hydrodynamics for 2050 climate condition Chesapeake Community Model Program Symp. (May 28-29, 2014, Annapolis, MD)

[10] Boesch D F, *et al* 2013 Updating Maryland's sea-level rise projections *Special Report of the Scientific and Technical Working Group to the Maryland Climate Change Commission* (University of Maryland Center for Environmental Science, Cambridge, MD)

[11] Johnson Z, Bennett M, Linker L, Julius S, Najjar R, Mitchell M, Montali D and Dixon R 2016 *The Development of Climate Projections for Use in Chesapeake Bay Program Assessments* (STAC Publication Number16-006, Edgewater, MD)

[12] Hilton T W, Najjar R G, Zhong L and Li M 2008 Is there a signal of sea-level rise in Chesapeake Bay salinity? *J. Geophys. Res.* **113** C09002

[13] Hong B and Shen J 2012 Response of estuarine salinity and transport processes to potential future sea-level rise in the Chesapeake Bay *Estuar. Coast. Shelf Sci.* **104-105** 33-45
[14] Cerco C F, Kim S C and Noel M 2010 The 2010 *Chesapeake Bay Eutrophication Model* (A report to USEPA Chesapeake Bay Program, US Army ERDC, Vicksburg, MS)
[15] Johnson B, Heath R, Hsieh B, Kim K and Butler L 1991 *User's Guide for a Three-dimensional Numerical Hydrodynamic, Salinity, and Temperature Model of Chesapeake Bay* (Technical Report HL-91-20, Department of the Army Waterways Experiment Station, Vicksburg, MS)
[16] Edinger J and Geyer J 1965 *Heat Exchange in the Environment* (Dept of Sanitary Engr and water resources, Research Project No. 49, The Johns Hopkins University, Baltimore, MD)
[17] Bahatt G, Yactayo G, Hinson K, Claggett P, Shenk G and Linker L 2015 Towards an integrated climate change analysis of the Chesapeake Bay watershed (Abstract) *Proc. Coastal and Estuarine Research Federation Biannual Conference* (Seatle, WA)
[18] Wang P, Linker L, Bhatt G, Yachato G and Tian R 2015 Accessing impacts of 2050 sea level rise and temperature increases on Chesapeake water quality (Abstract) *Proc. Coastal and Estuarine Research Federation Biannual Conference* (Seatle, WA)
[19] Edinger J, Brady D and Geyer J 1974 *The Heat Exchange and Transport in the Environment* (Report 14, Dept of Geography and Environmental Engineering, Johns Hopkins University, Baltimore, MD)
[20] Dake J M K and Harlemen D R F 1969 *An Analytical and Experimental Investigation of Thermal Stratification in Lakes and Ponds* (Hydrodynamics Laboratory Rept No. 99, Mass Inst of Technology, Cambridge, MA)
[21] Morrill J C, Bales R C and Conklin M H 2014 Estimating stream temperature from air temperature: implications for future water quality (Merced, CA: University of California, School of Engineering) https://eng.ucmerced.edu/people/rbales/CV/PubsM/98)
[22] Lee K H and Cho H Y 2015 Projection of climate-induced future water temperature for the aquatic environment *J. Environ. Engr.* doi: 10.1061/(ASCE)EE.1943-7870.0000974
[23] Knauss J A 1997 *Introduction to Physical Oceanography* 2nd or later editions (upper Saddle River, NJ: Prentice-Hall)
[24] Gallegos L C 2001 Calculating optical water quality targets to restore and protect submersed aquatic vegetation: Overcoming problems in partitioning the diffused attenuation coefficient for photosynthetically active radiation *Estuaries* **24(3)** 381-97
[25] Wang P, Linker L and Shenk G 2015 Using geographically isolated loading scenarios to analyze nitrogen and phosphorus exchanges and explore tailored nutrient control strategies for efficient management *Environ. Model. Assess.* Doi 10.1007/s10666-015-9487-x, 1-18
[26] Hagy J D, Boynton W R, Keefe C W and Wood K V 2004 Hypoxia in Chesapeake Bay, 1950–2001: long-term changes in relation to nutrient loading and river flow *Estuaries* **27** 634-58

IOP Conf. Series: Earth and Environmental Science 82 (2017) 012002 doi:10.1088/1755-1315/82/1/012002

Development of ion-exchange properties of bamboo charcoal modified with concentrated nitric acid

S Khandaker[1,2], T Kuba[1], Y Toyohara[1], S Kamida[1] and Y Uchikawa[1]

[1]Department of Urban and Environmental Engineering, Graduate School of Engineering, Kyushu University, 744 Motooka Nishi-ku, Fukuoka 819-0395, Japan

E-mail: s.khandaker@duet.ac.bd

Abstract. The surface chemistry and the structural properties of activated carbon can be altered by the acidic modification. The objective of this study is to investigate the changes occurring in bamboo charcoal (BC) during activation with concentrated nitric acid. Low temperature (500°C) carbonized BC has been prepared and oxidized with 70% concentrated boiling nitric acid (BC-AC). The porous properties of the BC are analyzed with nitrogen adsorption isotherm at 77 K. The surface structure is observed by Field emission scanning electronic microscope (FESEM) and the surface functional groups are examined by Fourier transform infrared (FTIR) spectroscopy, X-ray photoelectron spectroscopy (XPS) and the pH of the point of zero charge (pH_{PZC}). The results reveal that severe oxidation with HNO_3 considerably decreases the surface area of BC with enhanced pore widening and FESEM observation demonstrates the erosive effect of oxidation. The FTIR analysis detects that some absorption bands are assigned for carboxyl, aldehyde and ketone groups on BC-AC. The XPS analysis also clearly shows that the ratio of oxygen and acidic functional groups has been enriched significantly on the BC-AC. The low pH_{PZC} value of BC-AC confirms that the surface is highly acidic for the fixation of acidic functional groups on surface. In general, the existence of the abundant amount of acidic functional groups on adsorbents enhances the sorption of heavy metals ions in aqueous solution. Therefore, it is strongly expected that the modified BC, activated under the proposed conditions would be a promising ion exchanger in aqueous solution and can be applied for the adsorption of different heavy metal ions and radioactive materials from effluent.

1. Introduction

Activated carbon had been used in the field of water and contaminated water purification over the years for the uptake of a wide range of contaminants dissolved in wastewater due to the reason of its highly porous structure. Large external area, pore volume, and pore dimensions scattering are strongly interconnected to the adsorption capacity of activated carbon [1]. In general, adsorbents are physically modified by heating under a determined atmosphere and under high temperature to enhance the creations of more micropores and mesopores which are important for physisorption [2]. Moreover, acidic functional groups are spoiled and increased basicity on the carbon surface at elevated temperature. On the other hand in chemical modification process, adsorbents are soaked into certain chemical solutions to generate surface functional groups for chemisorption [3]. Chemical modification is usually more effective and economically viable for the adsorption of metallic ions compared to the activation through heating at higher temperature.

The principal objective of the chemical oxidation of carbonaceous materials is to develop the acidic functional groups on the external surface that possess ion-exchange properties in the adsorption process. A number of chemicals have been used as oxidizing agents such as nitric acid, sulfuric acid, sodium hypo chlorite, hydrogen peroxide, phosphoric acid, potassium permanganate etc. [4-6]. Different functional groups such as acidic, basic and neutral are generated due to the activation of these chemicals. Among the different chemical treatments, HNO_3 oxidation is commonly used in order to upsurge the total surface acidity and surface acidic functionalities [7]. Acidic functional groups are produced when surface of the activated carbon is oxidized. Adsorption of metal ions or polar substances such as agricultural chemicals from aqueous solution may be associated with the ion exchange mechanism with adsorbents. Acidic functional groups on the modified carbon may perform as ion exchangers and enhance the adsorption though modifying its selectivity to metal cation species in aqueous solution [8]. Radovic et al [9] also proposed that the acidic modification makes the donor-acceptor complexes that are positioned on the larger micropore. Their intensity on the surface of modified carbon favors the adsorption process. The specific acidic functional groups are basically important for the uptake of heavy metals and alkali earth metals cations for the reason of their chelating capability. This phenomenon can be demonstrated from the following equation [10].

$$M^{n+} + n\,(\text{-COOH}) \leftrightarrow (\text{-COO})\,nM + nH^{+}$$

The reaction takes places for the reason of cation exchange mechanism where the metal cations exchange with the hydrogen of carboxylic groups. The generation of more number of acidic groups can be promoted to exchange the higher number of cations during adsorption process.

However, activated carbon is generally prepared from the non-natural sources of coal which can inhibit its uses in future. Hence, further research is deserved to introduce renewable adsorbents for sustainable use in the field of wastewater treatment. Bamboo is echo-friendly, naturally abundant, renewable and an inexpensive bioresource; mostly found in south-east Asian countries. In recent years, bamboo charcoal has attracted a great deal of attention as an adsorbent substance for the adsorption of different contaminants from polluted and drinking water due to its exclusive micro porous structure and 4 to 10 times greater surface area than those in wood charcoal [11, 12]. Therefore, the use of bamboo charcoal has increased remarkably in the adsorption of a large variety of heavy metals [13, 14]. Proper oxidation of bamboo charcoal can make it one of the most selective ion exchangers by developing of the acidic functional groups on its surface and this property can be applied successfully in wide range of purposes as an adsorbent. This paper presents the partial work of our research in this area. The objective of this study was the preparation and characterization of concentrated boiling nitric acid modified-bamboo charcoal. The changes of physiochemical properties of the bamboo charcoal upon the treatment were investigated by nitrogen adsorption, Field emission scanning electronic microscope (FESEM), Fourier transform infrared (FTIR) spectroscopy, X-ray photoelectron spectroscopy (XPS). The acidic surface behavior was determined through the pH of the point of zero charge (pHzpc).

2. Materials and methods

The preparation and surface modification of the bamboo charcoal are briefly presented below.

2.1. The procedure of bamboo charcoal preparation

Properly dried mosso bamboo (*Phyllostachys pubescens*) was cut into small pieces. Bamboo chips were carbonized at 500°C for 3 h with a heating rate of 5°C per min under nitrogen (N_2) gas flow of 150 cm^3/min in a muffle furnace (Yamato FP32, Japan). Carbonized bamboo charcoal was crushed by a ball mills machine (Mixer Mill MM 400, RETSCH, Japan) and sieved at about 75 μm. The obtained raw bamboo charcoal was identified as BC.

2.2. Surface modification

Raw BC was chemically modified by using 70% concentrated nitric acid. 10 g of dried BC was allowed to boil at 120°C in 100 ml HNO_3 (solid liquid ratio=1:10) for 6 h in a hot plate and washed carefully several times with deionized ultrapure water (Elix Millipore, Japan) using a vacuum filter for several times until the pH was unchanged. Modified BC was oven dried at 110°C for 12 h and this was identified as BC-AC.

2.3. Characterization methods
The different physical properties analyzing methods that were used in this study are briefly presented below.

2.3.1. Surface area and porosity analysis. Physical characterization such as specific surface area, pore dimension and average pore size of BC and BC-AC were estimated based on the nitrogen gas (N_2) adsorption-desorption isotherms at -196°C (77 K). The analysis was carried out by using BELSORP-mini II, Japan according to the BET (Brunauer, Emmett and Teller) methods [15]. The total pore volume was analyzed at a relative pressure of 0.99. Before to the analysis, BC and modified BC samples were outgassed at 120°C for 3 h in the presence of nitrogen gas (N_2) flow to remove impurities adhered onto the surface.

2.3.2. Field emission scanning electron microscope (FESEM). The surface physical properties and porous structure were observed by using an Ultra-high resolution field emission scanning electronic microscope (UHR FE-SEM, model: SU8000, HITACHI, Japan). All images were taken at accelerating the voltage of 1.0 kV.

2.3.3. Fourier transform infra-red (FTIR) spectroscopy analysis. In order to understand the characteristics of surface functional groups, BC and BC-AC samples were investigated by using a FTIR spectrometer (Model: FT/IR-620, JASCO, Japan) in potassium bromide (KBr) pellet method. Properly dried samples were amalgamated with KBr powder maintaining the ratio of 1:100 (weight ratio of samples to KBr). After grinding the samples into homogenous powder, thin pellets were prepared for spectrum analysis. The absorbance of IR spectra was recorded from 4000 to 400 cm^{-1}.

2.3.4. X-ray photoelectron spectroscopy (XPS) analysis. The XPS analysis was performed to identify the elemental composition (carbon, oxygen and nitrogen), types of acidic functional groups and also their relative percentage on the surface of BC and BC-AC. The XPS spectra were obtained using an AXIS-ULTRA^{DLD} instrument (Shimadzu, Japan). A monochromatic X-ray source (Al Kα, hv = 1486.6 eV) and a discharge source (He I, hv = 21.2 eV) were used during the analysis. The base pressure of the analysis chamber was about 5×10^{-9} mbar. Wide scan spectra in the 1000-0 eV binding energy (BE) range were recorded with pass energy of 50 eV for the samples analysis.

2.3.5. pH at the point of zero charge (pH_{PZC}). The pH at the point of zero charge (pH_{PZC}) of the BC and BC-AC was measured following the procedure depicted by Yang et al [16]. Briefly, a solution of 0.005 M $CaCl_2$ was boiled to remove CO_2 and then cooled to room temperature. The different initial pH values of the six solutions (2.0, 4.0, 6.0, 8.0, 10.0, and 12.0) were adjusted by adding 0.1 M NaOH or HCl. 60 mg of samples were put into 20 ml of the pH-adjusted solution into 100 mL Erlenmeyer flasks and allowed to equilibrate for 24h at 20°C (± 1) in a reciprocating shaker (EYELA Multi shaker MMS–3010, Japan) at 140 rpm. The final pH was recorded and a diagram was plotted between the final pH against the initial pH. The crossing point of the plotted curve and the $pH_{initial} = pH_{final}$ line is considered as pH_{PZC}.

3. Results and discussion
The analytical results obtained through the different physicochemical study of BC and BC-AC and their relevant explanations are given below.

3.1. Porosity and surface area analysis
Textural properties including surface area, pore size distribution, and pore volume are essential

characteristics of adsorbents. The surface physical parameters of BC and BC-AC attained from the nitrogen adsorption isotherms are presented in table 1. The results demonstrate that the BET surface area, total pore volume and pore dimension reformed remarkably for the reason of the boiling nitric acid treatment. In comparison with BC, the BET surface area and pore volume of BC-AC declined from 312.50 m^2/g to 2.27 m^2/g and 0.1518 cm^3/g to 0.0079 cm^3/g, respectively. However, the calculated average pore size increased significantly from 1.94 nm for BC to 13.96 nm for BC-AC. The results suggest that the boiling nitric acid could shrink the surface area and pore volume. During nitric acid treatment some oxygen containing functional groups were introduced on the BC and this may have mainly blocked the pores and decreased the micropore volume of BC-AC [17]. Moreover, Molina-Sabio et al [18] reported that the deposition of acidic groups on the entrance of the pores in the treated activated carbon could increase the weight up to 6 to 13 % depending upon the materials and the treatment conditions. However, the widening of the pore size of BC-AC indicates that the effective action of HNO_3 causes significant damage to micropores of their structure and changes the surface area and also the pore volume [19]. Therefore, it can be concluded that liquid phase boiling nitric acid oxidation leads to fixation of acidic functional groups on modified BC surface and in the same time the partial destruction of porous structure has been occurred. It is to be expected that the modified BC enriched with oxygen–carbon groups on surface might have an enhanced affinity for chemisorption.

Table 1. The BET results of BC and BC-AC.

Samples name	**Surface area**	**Total pore volume (cm^3/g)**	**Avg. pore size(nm)**
BC	312.50	0.1518	1.94
BC-AC	2.27	0.0079	13.96

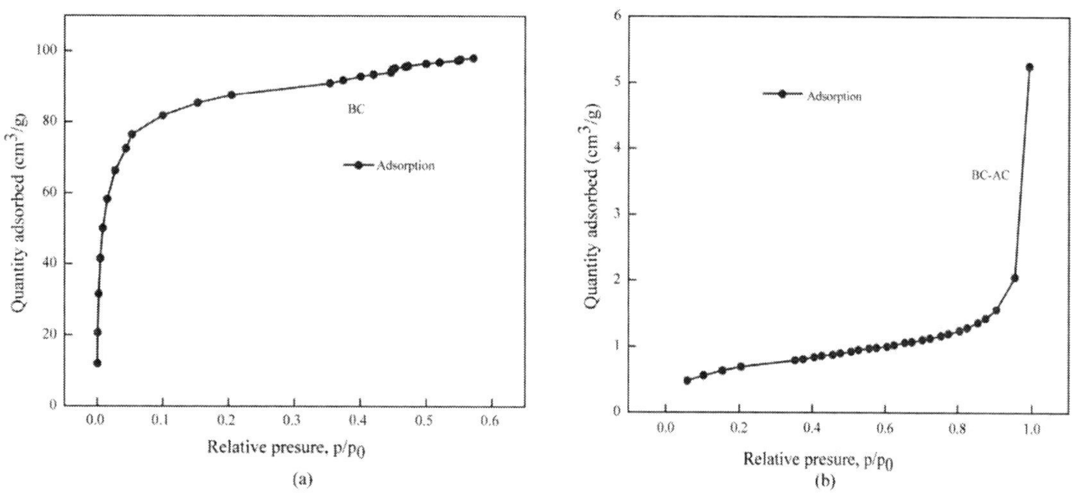

Figure 1. N_2 adsorption isotherm of (a) BC and (b) BC-AC.

The N_2 adsorption isotherms at -196°C (77 K) of BC and BC-AC are presented in figure 1. The nature of the isotherms curve was analyzed by using the IUPAC classification [20]. The character of the adsorption isotherm generally represents qualitative information on the adsorption mechanism and porous structure of the adsorbents. It is observed that the isotherm shape of BC increases at low relative pressure and tends to plateau even at high relative pressure. In this case, after filling the micropores of the BC with N_2 gas, there was no further increase in the isotherm curve. This shape clearly indicates the Type I isotherm on the basis of IUPAC classification. This type of isotherm actually demonstrates materials characterized with microporous arrangement. On the other hand, the shape of the isotherm of BC-AC slightly increases with relative pressure and that indicates the loss of micropores on the surface. Therefore, N_2 adsorption isotherm of BC-AC can be considered Type II isotherm which generally focuses on the non-porous or macro-porous material.

3.2. FESEM observations
Figure 2 demonstrates the morphological image of raw and modified bamboo charcoal. It can be obviously perceived that the surface of BC is highly porous in structure and pores are well developed. However, the surface morphology of the BC was modified due to the nitric acid oxidation. In the BC-AC sample, it can be observed that the porous structure of the surface almost disappeared due to clogging. This phenomenon may have led to the reduction of the surface porous structure which is already proved by the surface area analysis.

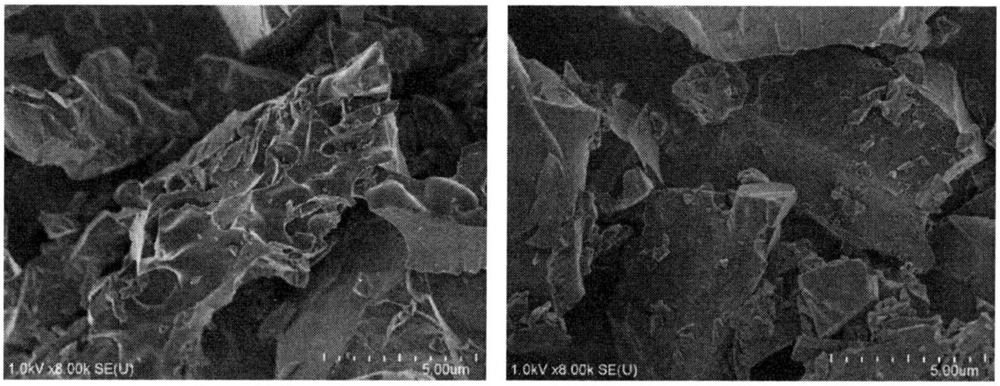

(a) (b)

Figure 2. FESEM images of (a) BC and (b) BC-AC.

3.3. FTIR analysis
In the chemisorption process, adsorption of cations is generally enhanced by the existence of the higher amount of acidic functional groups on the adsorbent surface. FTIR analysis is important for qualitative postulation of the nature of functional groups [21, 22]. FTIR spectra for raw and modified BC were presented in figure 3 and the changes and formation of new spectra were discussed. There are some IR spectra that were shifted; extricated and new peaks were seemed on the modified BC compared with raw one.

The weak spectra in the broad region of 3413.39 cm^{-1} and 3667.94 cm^{-1} on BC are related to hydroxyl groups (–OH) that were shifted in 3444.24 cm^{-1} and 3744.12 cm^{-1} on BC-AC, respectively. Moreover, a small bending vibration in 3856.93 cm^{-1} and 2481.94 cm^{-1} on BC-AC were attributed due to the appearance of the same type of functional groups [23].

The most important peaks were formed at 1711.61 cm^{-1}, and 1611.23 cm^{-1} region on BC-AC for the presence of carboxylic acid (–COOH), aldehyde, and ketone (C=O) functional groups, respectively. Haniyeh et al [24] reported that they observed the carboxylic and ketone groups in 1711 cm^{-1} and 1638 cm^{-1} region on nitric acid modified-carbon nanotubes. The same clarifications were also noted by Sellitti et al [25] and, Shim et al [17] for HNO_3 oxidized activated carbon. Furthermore, the medium peak or stretching vibration on BC in the region of 1593.88 cm^{-1}, 1376.93 cm^{-1}, and 1249.65 cm^{-1} are the characteristic of carbonyl groups, single C-H bond and alcohol or ester groups that were shifted after nitric acid modification in 1535.06 cm^{-1}, 1339.32 cm^{-1}, and 1240.97 cm^{-1}, respectively.

However, in finger print region (below 1000 cm^{-1}), some minor peaks for alkene (C-H) bending were moved where adsorption process could not be clearly allocated for specific stretching vibration because of their complex relating vibration structures [26]. The possible elucidation of the alteration in a spectra, reduction, and vanishing of peaks might be due to the fact that the strong oxidation changes their surface chemistry.

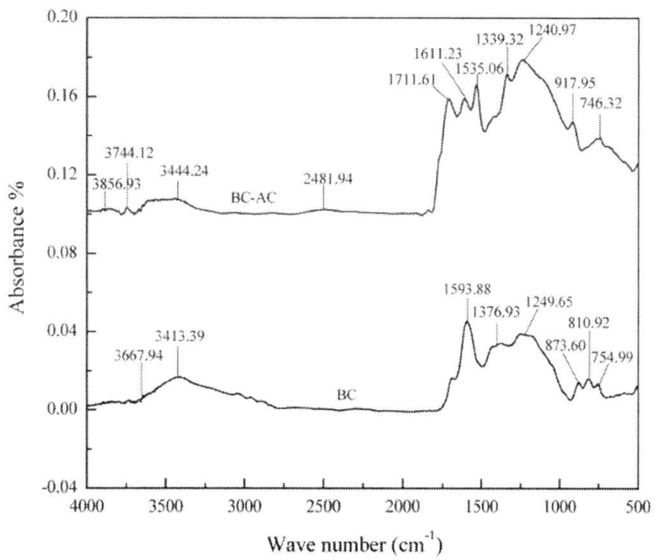

Figure 3. FTIR Spectra of BC and BC-AC.

3.4. XPS analysis

XPS analysis has been used to estimate the surface chemical features of BC and HNO_3 modified BC. The surface chemical composition and functional groups which have attained from XPS analysis is shown in table 2. According to the results, the HNO_3 treatment reduced the ratio of carbon composition significantly from 90.31% to 68.48%. However, it was surprising to see that about 228 % oxygen was introduced onto the surface after concentrated acidic treatment. In addition, the percentage of nitrogen was also increased significantly from 0.39% to 1.0% on the surface of modified BC. The change of C, O, and N after modification of BC is also clearly observed in the spectra of the samples (figure 4). The decrease of carbon and rise of the surface oxygen may be due to the damage of carbon surface from severe oxidation with nitric acid. Importantly, development of surface oxygen on BC may lead to increase of the oxygen-containing surface complexes that can enhance the ion exchange properties during adsorption. It should be noted that XPS analysis can only estimate the sample structure over a depth of few nanometers.

Table 2. The percentage of elements and acidic functional groups on the surface from XPS results.

Samples	**C**	**O**	**N**	**Graphite**	**C-C**	**C-O**	**C=O**	**COO$^-$**	**π-π***
Peak (eV)	-	-	-	284.4	285.0	286.5	288.0	288.9	290.2
BC (%)	90.31	9.30	0.39	71.12	4.98	15.09	0.52	2.70	5.59
BC-AC (%)	68.48	30.52	1.00	40.20	32.76	10.58	6.79	6.86	2.80

The high-regulation spectra of C1s were presented for further analysis using a curve fitting procedure. Deconvolution of C1s spectra represents six peaks of surface functional groups at different binding energy (BE) as reported [25, 26]. These are graphitic carbon (284.4 eV), C-C (285.0 eV), C-O (286.5eV), C=O (288.0eV), COO$^-$, (288.9 eV), and π-π* (290.2 eV). These groups are illustrated in figures 5(a) and 4(b) and the percentage of each group is listed in table 2.

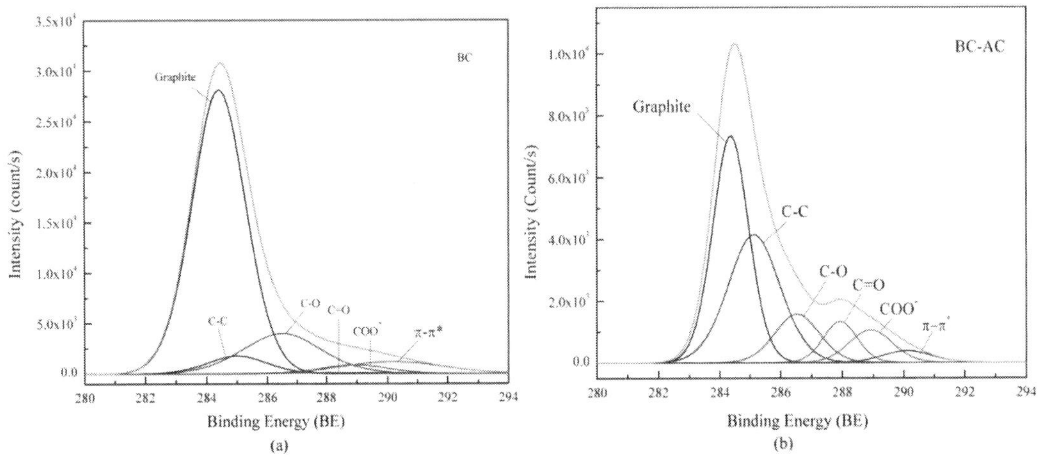

Figure 4. Acidic functional groups from XPS analysis for (a) BC and (b) BC-AC.

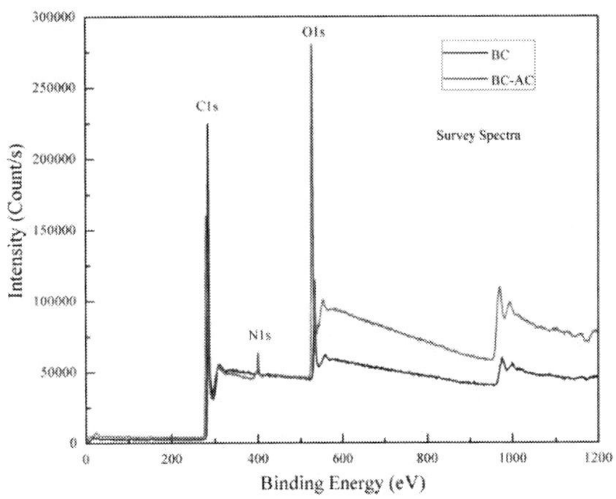

Figure 5. XPS spectrum of BC and BC-AC with their C1s, O1s and N1s % atomic concentration.

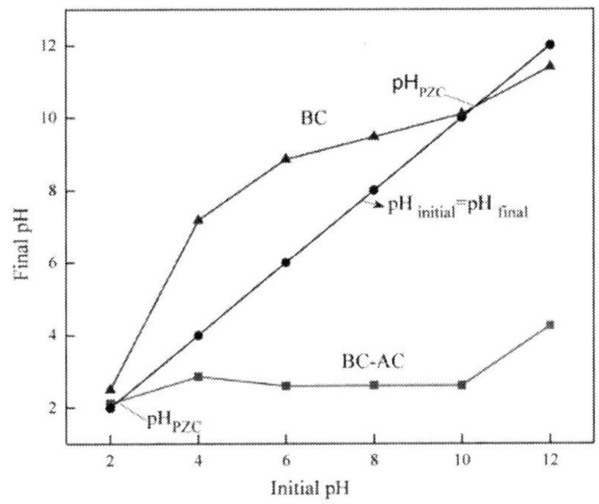

Figure 6. pH_{ZPC} of BC and BC-AC.

The results revealed that a great quantity of the carbon atoms on the surface is graphite configuration. C-O is the predominant essential of oxygen holding functional groups. The surface modification leads to increase in the substantial portion of aldehyde/ketone(C=O) and carboxyl (COO^-) groups at 0.52 % and 2.70 % in BC to 6.79 % and 6.86% in BC-AC, respectively. Similar results were also experienced for HNO_3 modified activated carbon by Chen and Shunnian [29]. It can be estimated that the significant growth of oxygen holding functional groups (C=O and COO^-) on the BC surface can be introduced it as a auspicious adsorbents to uptake heavy metals and radioactive materials from aqueous solution by ion exchange process.

3.5. pH at the point of zero charge (pH_{PZC})

The pH at the point of zero charge of adsorbents indicates the chemical and electrical potential of the functional groups on solid surface [8]. Generally, the surface charge of the adsorbents is positive when the pH of the adsorbate solution is lower than the pH_{PZC}. Positive charged surface always inhibited the attraction of metal cations from the adsorbate solution because of the repulsive force between protons and metal cations. Figure 6 shows that the pH at the point of zero charges of BC and BC-AC. The pH_{PZC} of raw BC and modified BC were 10.20 and 2.10, respectively. Therefore, BC and BC-AC might have enhanced adsorption capability of cations at above their pH_{PZC} value. Furthermore, the extremely low pH_{PZC} on BC-AC might be the reason of the deposition of sufficient acidic functional groups on the surface during oxidation.

4. Conclusions

Activation of BC with concentrated nitric acid is one of the promising approaches to produce oxygen containing functional groups on adsorbents although it can damage their textural properties. In our study, low temperature carbonized BC was activated with concentrated boiling nitric acid. The results demonstrated that a huge number of oxygen containing surface functional groups deposited on the modified BC surface. In FTIR analysis, results showed that new adsorption peak for carboxylic and aldehyde/ketone groups were generated on the surface. Moreover, XPS results also clearly exhibit that a higher percentage of carboxylic and aldehyde/ketone functional groups developed on BC-AC. The pH at the point of zero charge (pH_{PZC}) of BC-AC indicated that its surface was strongly acidic due to the deposition of acidic functional groups on surface. These acidic functional groups could generally improve the ion exchange capacity of modified BC. Therefore, it can be strongly expected that the modified BC under the proposed conditions would be a promising ion exchanger in aqueous solution. It is also estimated that the modified BC could be applied to the adsorption of different heavy metals and radioactive materials from wastewater and drinking water. The use of this adsorbent for the removal of such cations would be really interesting in practice. Recently our research group used this adsorbent for the adsorption of cesium from aqueous solution. The obtained results reveal that about 100% of cesium is possible to remove from a wide range of cesium contaminated (20-400 mg/L) wastewater [30].

Acknowledgements

The authors are thankful to Dr. Tsuyohiko Fujigaya, Department of Applied Chemistry, Graduate School of Engineering, Kyushu University, Japan for his support for surface area measurement and XPS analysis. The authors are also indebted to Dr. Midori Watanabe, Center of Advanced Instrumental Analysis, Ito campus, Kyushu University, Japan for her cordial support to FTIR analysis and FESEM image observation.

References

[1] Biniak S, Szymański G, Siedlewski J and Świątkowski A 1997 The characterization of activated carbons with oxygen and nitrogen surface groups *Carbon* **35** 1799-810
[2] Yin C Y, Aroua M K and Daud W M A W 2007 Review of modifications of activated carbon for enhancing contaminant uptakes from aqueous solutions *Sep. Puri. Tech.* **52** 403-15
[3] Pan B and Xing B 2008 Adsorption mechanisms of organic chemicals on carbon nanotubes *Environ. Sci. Tech.* **42** 9005-13
[4] El-Hendawy A N A 2003 Influence of HNO_3 oxidation on the structure and adsorptive properties of corncob-based activated carbon *Carbon* **41** 713-22
[5] Kutics K and Suzuki M 1990 Adsorption of organics on surface modified activated carbon fibers *The 2nd Korea–Japan Symposium on Separation Technology* (Seoul Korea) pp 395-8
[6] Ermolenko I N, Lyyubliner I P and Gulko N V 1990 *Chemically Modified Carbon Fibres and Their Application* (New York) Chap. 6 p 155
[7] Horváth G and Kawazoe K 1983 Method for the calculation of effective pore size distribution in molecular sieve carbon *J. Chemi. Eng. Japan* **16** 470-5
[8] Song X, Liu H, Cheng L and Qu Y 2010 Surface modification of coconut-based activated carbon by liquid-phase oxidation and its effects on lead ion adsorption *Desalination* **255** 78-83.
[9] Radovic L R, Moreno-Castilla C and Rivera-Utrilla J 2001 Carbon materials as adsorbents in aqueous solutions *Chem. Phy. Carbon* **27** 227-406
[10] Monser L and Adhoum N 2002 Modified activated carbon for the removal of copper, zinc, chromium and cyanide from wastewater *Sep. Puri. Techn.* **26** 137-46

[11] Wang S Y, Tsai M H, Lo S F and Tsai M J 2008 Effects of manufacturing conditions on the adsorption capacity of heavy metal ions by Makino bamboo charcoal *Bioresour. Technol.* **99** 7027-33
[12] Zhao D, Zhang J, Duan E and Wang J 2008 Adsorption equilibrium and kinetics of dibenzothiophene from n-octane on bamboo charcoal *Appl. Surf. Sci.* **254** 3242-7
[13] Tan Z, Niu G and Chen X 2015 Removal of elemental mercury by modified bamboo carbon *Chinese J. Chem. Eng.* **2** 1875-80
[14] Wang Y, Wang X, Wang X, Liu M, Yang L, Wu Z and Zhao J 2012 Adsorption of Pb (II) in aqueous solutions by bamboo charcoal modified with $KMnO_4$ via microwave irradiation *Coll. & Surf. A: Physicochemical and Engineering Aspects* **414** 1-8
[15] Brunauer S, Emmett P H and Teller E 1938 Adsorption of gases in multimolecular layers *J. Ame. Chem. Soci.* **60** 309-19
[16] Yang Y, Chun Y, Sheng G and Huang M 2004 pH-dependence of pesticide adsorption by wheat-residue-derived black carbon *Langmuir* **20** 6736-41
[17] Shim J W, Park S J and Ryu S K 2001 Effect of modification with HNO_3 and NaOH on metal adsorption by pitch-based activated carbon fibers *Carbon* **39** 1635-42
[18] Molina-Sabio M, Muñecas-Vidal M A and Rodriguez-Reinoso F 1991 Modification in porous texture and oxygen surface groups of activated carbons by oxidation *Surf. Sci. Cataly.* **62** 329-39
[19] Choma J, Burakiewicz-Mortka W, Jaroniec M, Li Z and Klinik J 1999 Monitoring changes in surface and structural properties of porous carbons modified by different oxidizing agents *J. Coll. Interface Sci.* **214** 438-46
[20] Sing K S 1985 Reporting physisorption data for gas/solid systems with special reference to the determination of surface area and porosity (Recommendations 1984) *Pure Appl. Chem.* **57** 603-19
[21] Ramachandra T V 2005 Biosorption of chromium (VI) from aqueous solutions by the husk of Bengal gram (*Cicer arientinum*) *Electronic J. Biotech.* **8** 258-64
[22] Sheng P X, Ting Y P, Chen J P and Hong L 2004 Sorption of lead, copper, cadmium, zinc, and nickel by marine algal biomass: characterization of biosorptive capacity and investigation of mechanisms *J. Coll. Interf. Sci.* **275** 131-41
[23] Foo K Y and Hammed B H 2012 Textural porosity, surface chemistry and adsorptive properties of durian shell derived activated carbon prepared by microwave assisted NaOH activation *Chem. Eng. J.* **187** 53-62
[24] Haniyeh F, Abdollah A and Abolghasem D 2013 Controlled growth of well-aligned carbon nanotubes, electrochemical modification and electro deposition of multiple shapes of gold nanostructures *Mat. Sci. Appl.* **4** 667-78
[25] Sellitti C, Koenig J L and Ishida H 1990 Surface characterization of graphitized carbon fibers by attenuated total reflection Fourier transform infrared spectroscopy *Carbon* **28** 221-8
[26] Yazıc H, Kılıç M and Solak M 2008 Biosorption of copper (II) by *Marrubium globosum* subsp. Globosum leaves powder: Effect of chemical pretreatment *J. Hazardous Mat.* **151** 669-75
[27] Zhou J H, Sui Z J, Zhu J, Li P, Chen D, Dai Y C and Yuan W K 2007 Characterization of surface oxygen complexes on carbon nanofibers by TPD, XPS and FT-IR *Carbon* **45** 785-96
[28] Barroso-Bujans F, Fierro J L G, Rojas S, Sánchez-Cortes S, Arroyo M and López-Manchado M A 2007 Degree of functionalization of carbon nanofibers with benzenesulfonic groups in an acid medium *Carbon* **45** 1669-78
[29] Chen J P and Wu S 2004 Acid/base-treated activated carbons: Characterization of functional groups and metal adsorptive properties *Langmuir* **20** 2233-42
[30] Khandaker S, Kuba T, Kamida S and Uchikawa Y 2017 Adsorption of cesium from aqueous solution by raw and concentrated nitric acid– modified bamboo charcoal *J. Environ. Chem. Eng.* **5** 1456-64

IOP Conf. Series: Earth and Environmental Science **82** (2017) 012003 doi:10.1088/1755-1315/82/1/012003

Modelling the salinization of a coastal lagoon-aquifer system

N Colombani[1,3] and M Mastrocicco[2]

[1]Department of Life Sciences and Biotechnology, University of Ferrara, Via L. Borsari 46, 44121, Ferrara, Italy
[2]DISTABIF, University of Campania "Luigi Vanvitelli", Via Vivaldi 43, 81100 Caserta, Italy

E-mail: nicolo.colombani@unife.it

Abstract. In this study, a coastal area constituted by alternations of saline-brackish lagoons and freshwater bodies was studied and modelled to understand the hydrological processes occurring between the lagoons, the groundwater system of the Po River Delta (Italy) and the Adriatic Sea. The contribution of both evaporation and anthropogenic factors on groundwater salinization was assessed by means of soil, groundwater and surface water monitoring. High-resolution multi-level samplers were used to capture salinity gradients within the aquifer and surface water bodies. Data were employed to calibrate a density-dependent numerical transport model implemented with SEAWAT code along a transect perpendicular to the coast line. The results show that the lagoon is hydraulically well connected with the aquifer, which provides the major source of salinity because of the upcoming of paleo-seawater from the aquitard laying at the base of the unconfined aquifer. On the contrary, the seawater (diluted by the freshwater river outflow) creates only a limited saltwater wedge. The increase in groundwater salinity could be of serious concern, especially for the pinewood located in the dune near the coast, sensitive to salinity increases. This case study represents an interesting paradigm for other similar environmental setting, where the assumption of classical aquifer salinization from a saltwater wedge intruding from the sea is often not representative of the actual aquifer's salinization mechanisms.

1. Introduction

Modelling the salinization of coastal water resources is one of the key issues to properly understand the mechanisms at the base of water quality deterioration induced by climate changes [1, 2]. Nevertheless, the mutual influence of groundwater and surface waters in coastal environments is influenced by many factors [3] and site specific models need to be employed to gain a comprehensive conceptual model of the studied area [4, 5]. So far, most of the attentions of the scientific literature have been focused on the ecological status of surface waters [6, 7]; however, the ecological status of a certain environment is often dependent on groundwater–surface water interactions [8].

In coastal environments, the most important mechanisms that control water salinization are [9]: the composition and amount of the atmospheric precipitation, the evapotranspiration processes, entrapped paleo seawaters, anthropogenic salt sources and biogeochemical water interactions with soils and aquifers. Numerical flow and transport models are a powerful tool to quantify and distinguish among the above-mentioned processes, which could potentially insist in the same environment [10]. In particular, density dependent groundwater flow models are quite promising in evaluating reactive mixing processes due to tidal induced fluctuations [11, 12] or to constant and variable pumping conditions [13-15]. While coupled saturated-unsaturated flow and transport models can give insights into groundwater recharge and exchanges with surface water bodies [16, 17].

In this study, a coastal area pertaining to the Po River Delta (Italy) was studied and modelled by means of a density driven groundwater flow and transport model to identify the sources, and the processes of salinization, occurring between a saline lagoon, the shallow aquifer of the Po River Delta (Italy) and the Adriatic Sea.

2. Site description

The investigated site is positioned within the coastal floodplain belonging to the Po di Volano River, a terminal branch of the Po River located in Northern Italy (figure 1).

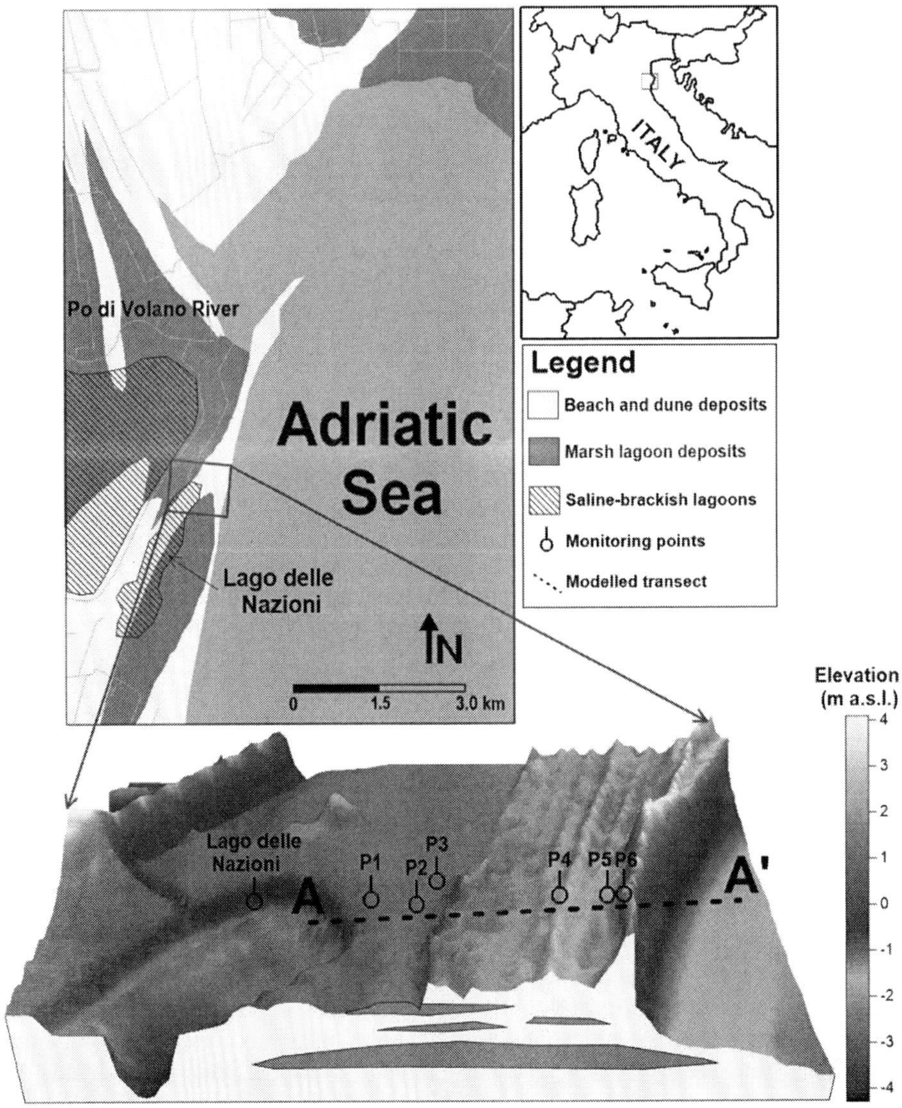

Figure 1. Geological sketch of the investigated site and distribution of the sampling and monitoring points. The density dependent groundwater flow model was built up along transect A-A' and the DTM was obtained interpolating bathymetric data with a LIDAR survey.

The peculiar landscape of the Po Delta is characterized by important environmental and ecological values and is often subject to land use changes [18]. Due to the flat ground elevation respect to the sea, these lands underwent intensive reclamation works [19]. The surface hydrographic network has a double use, first as drainage system serving a set of pumping stations, and for irrigation during the growing season. The shallow coastal aquifer is principally formed by sandy dune sediments (10-12 m thick) interbedded with silty-clay lenses (1-2 m thick), which locally create semi-confined conditions [20]. At the base of the aquifer, the prodelta wedge (4-15 m thick) formed by clay and silty-clay lenses acts as a saline aquitard. The groundwater flow is mainly oriented in the vertical direction and it is generally converging towards the drainage network [21]. Lago delle Nazioni is a brackish-saline lagoon of artificial origin, it is a stretched basin oriented parallel to the coast line of approximately 3 km length and 500 m wide (about 97 ha). The lagoon was in direct contact with the sea until 1960', then in the mid-sixties the Lago delle Nazioni basin was created for recreational purposes. Thus the pre-existing Valle di Volano, a brackish back-barrier lagoon formed since the Middle Ages, was completely reshaped and excavated. The sediments that form the bottom of Lago delle Nazioni range from silty-loam to sandy while along the shore there are prevalently sandy sediments, suggesting a good degree of connection between the lagoon and the shallow aquifer [20]. The average water depth is about 4.0 m, with maximum depth of 5.5 m and large areas of 1.5-2.0 m [22]. Water exchange and recirculation are guaranteed by an artificial channel connected with the Adriatic Sea by means of a siphon and a pumping station. Within this complex hydrogeological framework, the shallow aquifer, a brackish-saline lagoon (Lago delle Nazioni) and the Adriatic Sea, have been modelled to understand and quantify their mutual relationships.

3. Materials and methods

3.1. Field sampling

One multi-level monitoring well belonging to the monitoring network of the Emilia-Romagna Region (P2 in figure 1), and five shallow monitoring wells (P1, P3, P4, P5 and P6 in figure 1) were used to monitor groundwater quality and depth. The shallow monitoring wells (from 1.5 to 2 m depth) were drilled using a hands driven auger along a flow line and were positioned approximately perpendicular to the coast line (figure 1). Soil samples were collected every 20 cm from the ground surface during the shallow monitoring wells installation. Porewater salinity of each sample was measured in laboratory after extraction with double deionized water with a sediment to water ratio of 1/5 and correcting the value for the sediment volumetric water content. The monitoring wells were located in actual dunes characterized by medium to fine sandy deposits and brackish marsh lagoon deposits, typical of back barrier environments. The Lago delle Nazioni, P2, P3, P5 and P6 points were monitored for water level, temperature and electrical conductivity every 30 days during 2015 employing a multi-parameter sonde (Hydrolab MS-5).

Groundwater was sampled at various depths in May and November 2015 from P1 using a low pressure pneumatic straddle-packers system (Solinst Canada Ltd.). The packers' system comprised: a two inflatable packers, a 20 cm screen placed between the packers and a 12 V electrical pump employed for well purging and sampling.

3.2. Numerical modelling

Through the designated transect A-A', which is normal to the coast line and to the Lago delle Nazioni (figure 1), a conservative flow and transport model was employed to explore the salinization processes taking place within the unconfined aquifer. The simulations were performed by means of SEAWAT 4.0 [23], to mimic the interconnections between the lagoon, groundwater and the sea. SEAWAT 4.0 was selected since it is able to deal with variable-density processes. Chemical reactions were not considered here because the aim of this study was to explore the physical control on salinity transport, although in some cases neglecting this part can lead to errors in estimating model parameters [24].

The stratigraphic architecture of this area was reconstructed using the most recent literature data available for this zone [25].

The bi-dimensional numerical model was formed by 250 columns with a width of 2 m for a whole length of 500 m. The vertical direction was simulated using 26 layers of variable thickness to permit a fine delineation of the topography and the geometry of local lenses (figure 2). The first layer was thicker than the others, with the purpose of contain the groundwater table variations. In this way, grid cells wetting and rewetting was avoided, thus eliminating the numerical oscillations and solver non-convergence connected with this process.

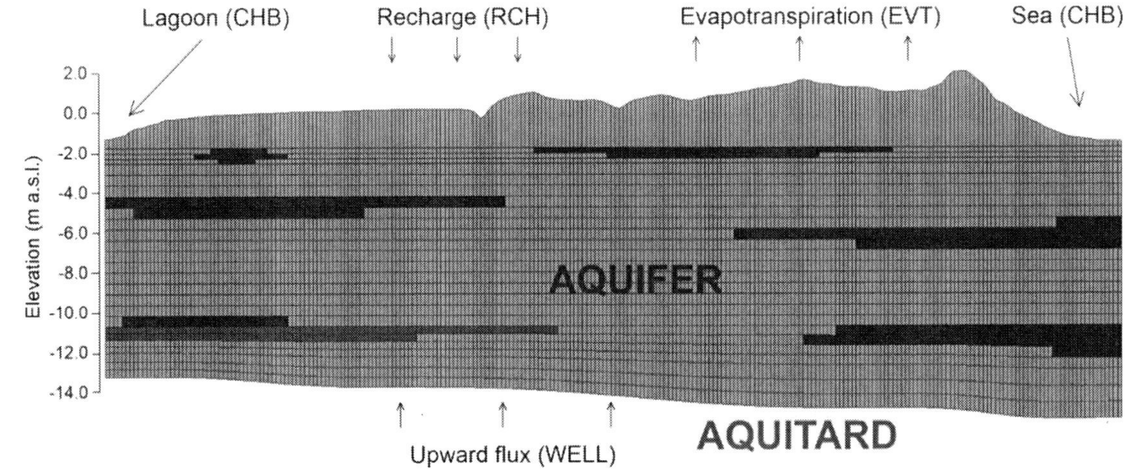

Figure 2. SEAWAT model boundaries and grid dimensions of the A-A' transect.
The colours refer to the properties of the different units described in table 1. Vertical exaggeration is 1:10.

Table 1. Model parameters used in the flow and transport model transect represented in figure 2.

Parameter	Value
Green sand k (m/d)	2.0
Green sand Ss (1/m)	$1e^{-4}$
Green sand Sy (-)	0.2
Blue silty-loam k (m/d)	0.1
Blue silty-loam Ss (1/m)	$1e^{-5}$
Blue silty-loam Sy (-)	0.1
Red sand k (m/d)	5.0
Red sand Ss (1/m)	$1e^{-4}$
Red sand Sy (-)	0.2
Longitudinal dispersivity (m)	0.2
Vertical transverse dispersivity (m)	0.01
Evapotranspiration (mm/y)	360
Recharge (mm/y)	80

The vertical hydraulic conductivity (k_v) was assumed to be 5 times lower than the horizontal one (k_h) for all layers [16], while the effective porosity was taken as 0.25 for the entire model [21]. The specific yield (S_y) and specific storage (S_s) were taken from literature values [26]. The diffusion coefficient was supposed to be equivalent to $1x10^{-4}$ m^2/d.

A fully transient simulation was performed reproducing the changes occurred during the last 160 years. Here the time was segmented into three stress periods to mimic the change of the salinity in the back barrier environment. In the first stress period (50 years long), a large freshwater recharge from the Po di Volano River was set in the flat topographic zone beneath the coastal dunes, to attain a characteristic distribution of heads and salinity inside the aquifer until 1910. In the second stress period (50 years long), the Lagoon' Constant Head Boundary (CHD) in layer 1 was extended in all the area topographically below the actual sea level to achieve a representative distribution of heads and salinity within the aquifer until 1960. In the third stress period (60 years long), the actual configuration was considered, restricting the Lago delle Nazioni to the first 10 cells at the left side of the model. To account for seawater input in the aquifer from the Adriatic Sea, a CHD was set in the cells of layer 1 located on the right corner of the modelled transect (figure 2). The WELL Package was employed to mimic the ascending flux provoked by the reclamation system; the specific discharge rate was considered employing a published method [27]. The Recharge Package (RCH) was employed to mimic the effective infiltration averaging the yearly recharge value of 155 mm/y for this area as initial estimate [21]. The average salinity concentrations in the porewater of the upper soil samples were employed as salinity concentration of the recharging waters. The evapotranspiration was simulated with the Evapotranspiration Package (EVT), the maximum evapotranspiration rate was initially set to 3 mm/d with an extinction depth of -1 below ground level. The maximum transpirable salinity was set to 5 g/l due to the presence of a pinewood [28].

Dispersivity, recharge, evapotranspiration and hydraulic conductivity parameters were tuned via trial and error calibration to fit the observed heads and salinities collected in May and November 2015, considered representative of the mean water table conditions.

4. Results and discussion

Figure 3 shows the calibration results of measured versus computed hydraulic heads and salinities in all the monitoring wells. Relatively good calibration was achieved for both heads and salinities throughout the aquifer. The plots show the presence of brackish to saline groundwater in the model domain and small heads variations imputable to the flat topographic surface. The trial and error calibration procedure lead to an absolute mean error between simulated and observed heads and salinities of 5.1 cm and 2.5 g/l, respectively. These errors resemble to the 6.9% of the observed piezometric range and to the 7.8% of the observed salinity range throughout the examined time frame. Table 1 reports the calibrated parameters; in general, they are in agreement with previous model parameters applied in this coastal aquifer [2, 5, 21].

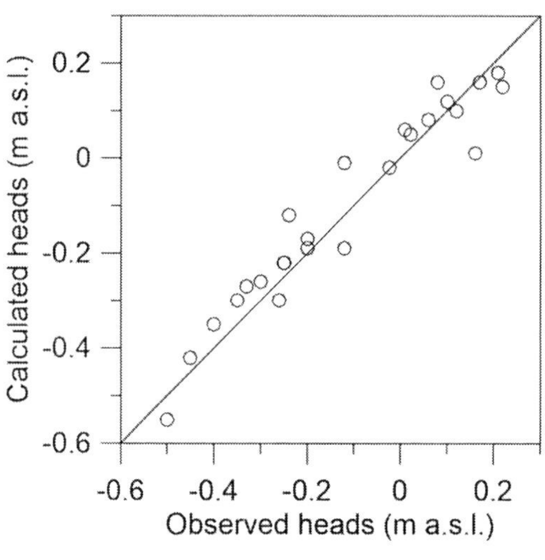

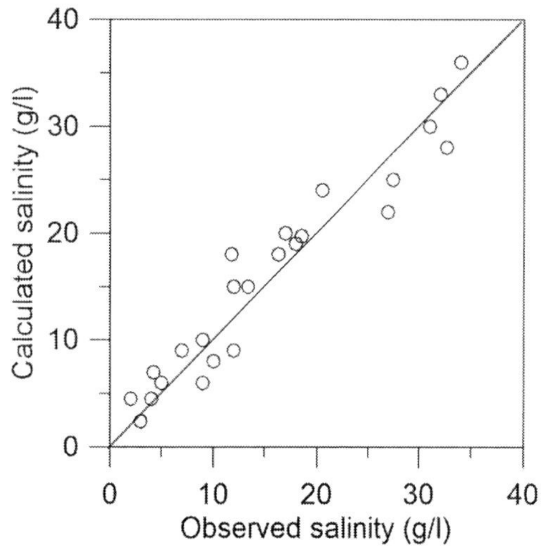

Figure 3. Scatter diagrams of observed heads versus calculated ones (left plot) and Scatter diagrams of observed salinities versus calculated ones (right plot).

The unevenness of the recorded groundwater heads and salinities during the monitored period was elevated, due to recharge events, storm events and summer droughts. Thus, to precisely replicate all the variations due to small fluctuations in groundwater heads and salinities a much higher time-based discretization would be need. In any case, the aim of this model was to reproduce the general piezometric head trend and salinity distribution within the aquifer. In particular, figure 4 shows the piezometric heads decrease induced by the Lago delle Nazioni that constantly drains groundwater from the coastal aquifer. The sand dune belt acts as recharge area, constituting a groundwater divide between the Adriatic Sea and the back barrier environment. The groundwater heads map shows a complex flow pattern, mainly driven by differences in groundwater densities, lenses of fine sediments and the upward flux from the aquitard.

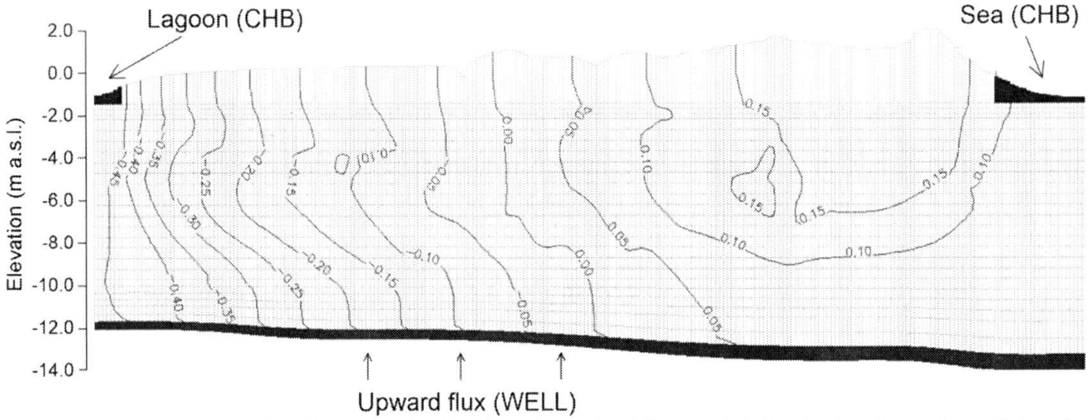

Figure 4. Groundwater heads map calculated at the end of the model simulation (actual conditions). The model grid and model boundaries are also depicted. Vertical exaggeration is 1:10.

To get a more robust idea of the different flow component, table 2 shows the flow budget between the various boundaries. The storage term within the aquifer is quite limited, while approximately 70 l/d flow from the aquifer towards the Lago delle Nazioni. From the Adriatic Sea the actual seawater intrusion is limited to approximately 2.4 l/d, while submarine groundwater discharge is 18 l/d. The upward flux from the aquitard is in general quite limited, although it slowly contributes to salinize the aquifer. The recharge is the prevailing term of this model with 110 l/d infiltrating through the simulated transect, while evapotranspiration accounts for 27 l/d, becoming the second term of the flow budget per magnitude. The model budget is perfectly balanced as denoted by the difference between the inflows and outflows, guaranteeing that the estimated magnitude of the flow components is reliable.

Table 2. Model flow budget at the end of the simulation (actual conditions).

Flow term	IN (m^3/d)	OUT (m^3/d)	IN-OUT (m^3/d)
Storage	$8.52e^{-03}$	$4.58e^{-04}$	$-3.73e^{-04}$
Lagoon	0.00	$6.96e^{-02}$	$-6.96e^{-02}$
Sea	$2.39e^{-03}$	$1.82e^{-02}$	$-1.58e^{-02}$
Upward flux	$2.50e^{-03}$	0.00	$2.5e^{-03}$
Recharge	$1.1e^{-01}$	0.00	$1.1e^{-01}$
Evapotranspiration	0.00	$2.68e^{-02}$	$2.68e^{-02}$
Sum	$1.15e^{-01}$	$1.15e^{-01}$	$9.53e^{-08}$

Figure 5 shows the model results for the simulated salinity distribution within the aquifer cross section at the end of each stress periods. In the upper panel of figure 5 the reconstructed salinity distribution at the beginning of the 1900 is shown, with a thick freshwater lens near to the actual Lago delle Nazioni. This was due to the Po di Volano River that was flowing in a slight different position respect to the actual river mouth [29]. The results of the second stress period are shown in the middle panel of figure 5, where the salinization of the shallow part of the coastal aquifer is evident, due to the connection between the Adriatic Sea and the lagoon. The freshwater lens still persists, but it is pushed down by the saline waters of the embayment. Finally, the present conditions, reported in the bottom panel of figure 5, are a complex mixture of the different conditions that have acted in the last century within this portion of coastal aquifer. Brackish groundwater fingers develop in the model domain, producing a very complex pattern that highly limits the capability to distinguish between modern seawater intrusion processes, evapotranspiration processes and paleo-seawater upconing.

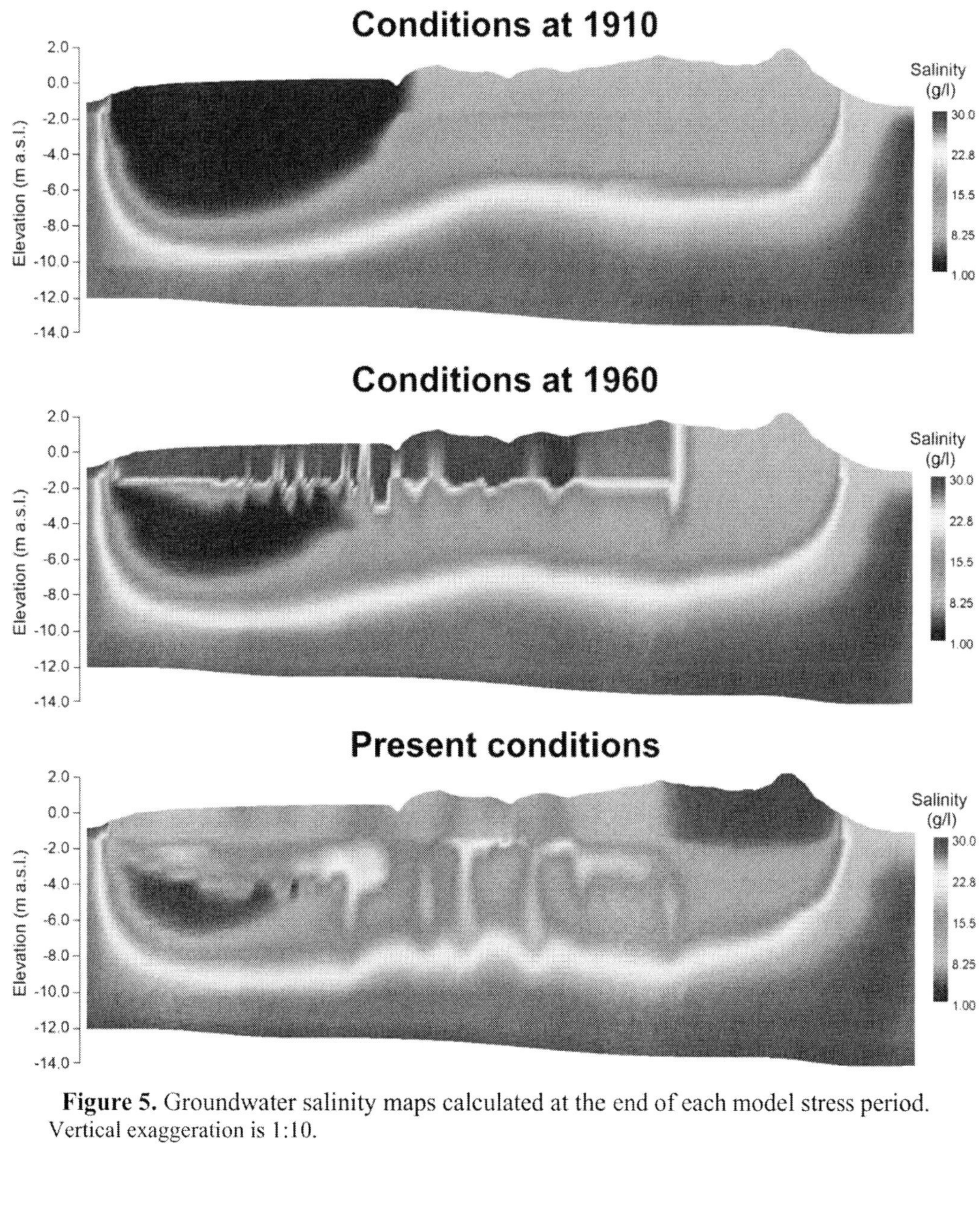

Figure 5. Groundwater salinity maps calculated at the end of each model stress period. Vertical exaggeration is 1:10.

One of the key points in the difficulty of precisely locating the actual seawater intrusion wedge is the anomalous presence of diluted seawater in the near shore portion of the Adriatic Sea in this area. This is due to freshwater plumes originating from the Po Delta river branches, which constantly lower the Adriatic Sea salinity at concentrations that hardly reach 20 g/l in this shallow sea. Another major issue is the evapotranspiration capacity of vegetation in the dune environment; in fact, since the water table is quite shallow, vegetation makes use of groundwater and consequently increases the aquifer salinity. Moreover, vegetation may also increase the unsaturated soil salinity and thus increasing the salinity of the infiltrating water from precipitations. For instance, the measured porewater salinity was always greater than 1 g/l in all the sediment profiles and the highest shallow porewater salinities were recovered within the pinewood, with concentrations ranging from 3.5 to 11.3 g/l. The increase in salinity of groundwater could be of serious concern, especially for the pinewood located in the dune near the coast, sensitive to salinity increases. In fact, even little changes in the hydrological balance due to climate variability or sea level rise could seriously affect this fragile ecotone.

5. Conclusions

To understand and quantify the hydrological processes occurring between the lagoons, the groundwater system pertaining to the Po River Delta (Italy) and the Adriatic Sea, a density driven numerical model (SEAWAT 4.0) was set up and calibrated through a cross section perpendicular to the coast line. The contribution of anthropogenic and natural factors on groundwater salinization were assessed by past reclamation activities and actual groundwater and surface water monitoring. The results of the actual flow budget show that the lagoon (Lago delle Nazioni) is hydraulically well connected with the aquifer, and its salinity is kept constant by saline groundwater. The major sources of actual salinity are: (i) the upconing of paleo-seawater from the aquitard situated immediately below the unconfined aquifer; (ii) the evapotranspiration processes that increase soil and shallow groundwater salinity. In contrast, the Adriatic Sea (diluted by the freshwater outflowing from the Po River branches) creates only a limited brackish water wedge. This case study represents an interesting paradigm for other similar environmental setting, where the assumption of classical aquifer salinization from a saltwater wedge intruding from the sea is often not representative of the actual aquifer's salinization mechanisms. Thus more accurate hydrogeological analyses must be performed to accurately distinguish between actual and paleo seawater intrusion processes.

References

[1] Oude Essink G H P, Van Baaren E S and Louw P G 2010 Effects of climate change on coastal groundwater systems: A modeling study in the Netherlands *Water Resour. Res.* **46(10)**

[2] Colombani N, Osti A, Volta G and Mastrocicco M 2016 Impact of climate change on salinization of coastal water resources *Water Resour. Manage.* **30(7)** 2483-96

[3] Bratton J F, Böhlke J K, Krantz D E and Tobias C R 2009 Flow and geochemistry of groundwater beneath a back-barrier lagoon: The subterranean estuary at Chincoteague Bay, Maryland, USA *Mar. Chem.* **113(1)** 78-92

[4] Beekman H E, Eggenkamp H G M and Appelo C A J 2011 An integrated modelling approach to reconstruct complex solute transport mechanisms–Cl and δ 37 Cl in pore water of sediments from a former brackish lagoon in the Netherlands *Appl. Geochem.* **6(3)** 257-68

[5] Caschetto M, Colombani N, Mastrocicco M, Petitta M and Aravena R 2016 Estimating groundwater residence time and recharge patterns in a saline coastal aquifer *Hydrol Proces* **30(22)** 4202-13

[6] Newton A, Icely J, Cristina S, Brito A, Cardoso A C, Colijn F and Ivanova K 2014 An overview of ecological status, vulnerability and future perspectives of European large shallow, semi-enclosed coastal systems, lagoons and transitional waters *Estuar. Coast. Shelf Sci.* **140** 95-122

[7] Basset A, Elliott M, West R J and Wilson J G 2013 Estuarine and lagoon biodiversity and their natural goods and services *Estuar. Coast. Shelf Sci.* **132** 1-4

[8] Miller D C and Ullman W J 2004 Ecological consequences of ground water discharge to Delaware Bay, United States *Ground Water* **42(7)** 959-70

[9] Hart B T, Bailey P, Edwards R, Hortle K, James K, McMahon A, Meredith C and Swadling K 1990 Effects of salinity on river, stream and wetland ecosystems in Victoria, Australia *Water Res.* **24(9)** 1103-17

[10] Anderson M P, Woessner W W and Hunt R J 2015 *Applied Groundwater Modeling: Simulation of Flow and Advective Transport* (London UK: Academic Press)

[11] Mao X, Enot P, Barry D A, Li L, Binley A and Jeng D S 2006 Tidal influence on behaviour of a coastal aquifer adjacent to a low-relief estuary *J. Hydrol.* **327(1)** 110-27

[12] Campana C and Fidelibus M D 2015 Reactive-transport modelling of gypsum dissolution in a coastal karst aquifer in Puglia, southern Italy *Hydrogeol. J.* **23(7)** 1381-98

[13] Qahman K and Larabi A 2006 Evaluation and numerical modeling of seawater intrusion in the Gaza aquifer (Palestine) *Hydrogeol. J.* **14(5)** 713-28

[14] Mastrocicco M, Colombani N and Petitta M 2011 Modelling the density contrast effect on a chlorinated hydrocarbon plume reaching the shore line *Water Air Soil Poll.* **220(1-4)** 387-98

[15] Bobba A G 2002 Numerical modelling of salt-water intrusion due to human activities and sea-level change in the Godavari Delta, India *Hydrol Sci J* **47(S1)** S67-80

[16] Colombani N, Di Giuseppe D, Faccini B, Ferretti G, Mastrocicco M and Coltorti M 2016 Inferring the interconnections between surface water bodies, tile-drains and an unconfined aquifer–aquitard system: A case study *J. Hydrol.* **537** 86-95

[17] Wilson A M and Gardner L R 2006 Tidally driven groundwater flow and solute exchange in a marsh: Numerical simulations *Water Resour. Res.* **42(1)**

[18] Gaglio M, Aschonitis V G, Gissi E, Castaldelli G and Fano E A 2017 Land use change effects on ecosystem services of river deltas and coastal wetlands: case study in Volano–Mesola–Goro in Po river delta (Italy) *Wetlands Ecol. Manage.* **25(1)** 67-86

[19] Correggiari A, Cattaneo A and Trincardi F 2005 The modern Po Delta system: Lobe switching and asymmetric prodelta growth *Mar. Geol.* **222** 49-74

[20] Colombani N, Giambastiani B M S and Mastrocicco M 2017 Impact of climate variability on the salinization of the coastal wetland-aquifer system of the Po Delta, Italy *J. Water Suppl. Res. T.* (in press)

[21] Colombani N, Giambastiani B M S a nd Mastrocicco M 2015 Predicting salinization trends in a lowland coastal aquifer: Comacchio (Italy) *Water Resour. Manage.* **29(2)** 603-18

[22] Mistri M and Rossi R 1999 Structure and production of the benthic macroinvertebrate community in a shallow lagoon of the Po River Delta *It. J. Zool.* **66(4)** 361-8

[23] Langevin C D, Thorne D T Jr, Dausman A M, Sukop M C and Guo W 2007 *SEAWAT Version 4: A Computer Program for Simulation of Multi-Species Solute and Heat Transport: U.S. Geological Survey Techniques and Methods* book 6 Chap. A22 pp 1-39

[24] Mastrocicco M, Prommer H, Pasti L, Palpacelli S and Colombani N 2011 Evaluation of saline tracer performance during electrical conductivity groundwater monitoring *J. Cont. Hydrol.* **123(3)** 157-66

[25] Amorosi A, Centineo M C, Colalongo M L and Fiorini F 2005 Millennial-scale depositional cycles from the Holocene of the Po Plain, Italy *Mar. Geol.* **222** 7-18

[26] Fetter C W 2001 *Applied Hydrogeology (4th); Upper Saddle River* (NJ: Prentice Hall, 598)

[27] De Louw P G, Eeman S, Oude Essink G H P, Vermue E and Post V E A 2013 Rainwater lens dynamics and mixing between infiltrating rainwater and upward saline groundwater seepage beneath a tile-drained agricultural field *J. Hydrol.* **501** 133-45

[28] Antonellini M and Mollema P N 2010 Impact of groundwater salinity on vegetation species richness in the coastal pine forests and wetlands of Ravenna, Italy *Ecol. Eng* **36(9)** 1201-11

[29] Simeoni U and Corbau C 2009 A review of the Delta Po evolution (Italy) related to climatic changes and human impacts *Geomorphol.* **107(1)** 64-71

A comparison of anaerobic 2, 4-dichlorophenoxy acetic acid degradation in single-fed and sequencing batch reactor systems

P Elefsiniotis[1], D G Wareham[2,3] and P Fongsatitukul[1]

[1]Department of Sanitary Engineering, Mahidol University, 420/1Ratchavithee Road, Ratchathevee, Bangkok 10400, Thailand
[2]Department of Civil and Natural Resources Engineering, University of Canterbury, Private Bag 4800, Christchurch, New Zealand

E-mail: david.wareham@canterbury.ac.nz

Abstract. This paper compares the practical limits of 2, 4-dichlorophenoxy acetic acid (2,4-D) degradation that can be obtained in two laboratory-scale anaerobic digestion systems; namely, a sequencing batch reactor (SBR) and a single-fed batch reactor (SFBR) system. The comparison involved synthesizing a decade of research conducted by the lead author and drawing summative conclusions about the ability of each system to accommodate industrial-strength concentrations of 2,4-D. In the main, 2 L liquid volume anaerobic SBRs were used with glucose as a supplemental carbon source for both acid-phase and two-phase conditions. Volatile fatty acids however were used as a supplemental carbon source for the methanogenic SBRs. The anaerobic SBRs were operated at an hydraulic retention time of 48 hours, while being subjected to increasing concentrations of 2,4-D. The SBRs were able to degrade between 130 and 180 mg/L of 2,4-D depending upon whether they were operated in the acid-phase or two-phase regime. The methanogenic-only phase did not achieve 2,4-D degradation however this was primarily attributed to difficulties with obtaining a sufficiently long SRT. For the two-phase SFBR system, 3.5 L liquid-volume digesters were used and no difficulty was experienced with degrading 100 % of the 2,4-D concentration applied (300 mg/L).

1. Introduction

With an estimated annual worldwide use of over 2.2 million tonnes [1]; it is not surprising that engineers and scientists are increasingly interested in the fate of pesticides, particularly since they have been found in the earth's water and soil environments. Over and above their known potential to specifically pollute the earth's water ways, there is an increasing concern about the pesticide impact on public health [2, 3]. In particular, one of the most ubiquitous pesticides is the organic herbicide 2, 4-dichlorophenoxy acetic acid (2,4-D). This pesticide is used to enhance agricultural productivity (mainly for crops of the cereal variety); however, it is also the active constituent in several herbicides used to manage broadleaf weeds.

Within the New Zealand context, agriculture is a major revenue earner with a third of all commodity exports arising from the agricultural-based sector [4]. Such industries rely heavily on pesticides with 2,4-D and/or others of the phenoxy herbicide category making up 68 % of all New Zealand herbicides [5]. It is hardly surprising therefore that 2,4-D contamination of water has been reported not only in New Zealand but also around the world [6]. Pollution of aqueous environments however not only arises from surface run-off and aerial drift (i.e. crop spraying); it also arises from pesticide discharges originating from chemical manufacturing plants. In particular, it has been reported that concentrations of chlorinated organic compounds in industrial streams can range from ten to several hundreds of mg/L, depending upon the industry [7].

In recent years, biological processes have received increasing attention as both cost-effective and technically-efficient methods for the treatment of xenobiotic substances [8]. Notwithstanding, limited success has been achieved with conventional activated sludge systems, particularly regarding their ability to degrade certain pesticides [9]. Because of this, there exists a continual need to conduct further research into alternative biological options for the treatment of recalcitrant compounds [10, 11]. More specifically, no study to date has reported on the 2,4-D degradation efficiency of a sequencing batch reactor (SBR) system as compared to a single-fed batch reactor (SFBR) system under anaerobic conditions. Of major interest is their ability to accommodate industrial strength concentrations.

The prime function of this paper is to synthesize together (and comment on) several published studies that have recently examined the biodegradation potential of 2,4-D under anaerobic conditions. These disparate studies (conducted mainly over the last decade in the lead author's previous laboratory (the University of Auckland, New Zealand)) have built on each other to explore the practical upper limits of both SBR and SFBR systems anaerobically degrading 2,4-D. The anaerobic SBR work was mainly done using glucose as an extra carbon source; while, as part of this report, the SBR also explored the impact of phase separation (acid-phase vs. two-phase digestion) on 2,4-D removal.

2. Materials and methods

2.1. Reactor operation and configuration

This research was conducted at the laboratory-scale, with the SBRs having total and liquid volumes of 3 L and 2 L, while the SFBRs had total and liquid volumes of 4 L and 3 L respectively. Ports were incorporated into each SBR for feeding, decanting, sample collection/wastage, nitrogen gas addition and biogas collection. To maintain anaerobic conditions, nitrogen gas was injected into the head space of the SBR, with bolted O-ring stoppers sealing the reactors. In contrast, sealing of the SFBRs was by a rubber cork, while a pipe throughout the cork connected to an anaerobic respirator system in order to monitor gas production.

Both systems were mixed by magnetic stirrers, while feeding and decanting (of the SBR) was by peristaltic pumps. When methane or two-phase digestion was desired in the SBR, heated coils controlled by a voltmeter held the temperature generally in the lower mesophilic range ($32 \pm 2^{o}C$). When acid-phase conditions were desired, the pH was maintained between 4.5 and 5.0 by an aqueous solution of 0.05 N H_2SO_4. When methane or two-phase conditions were desired, 0.05 N NaOH was added to maintain the pH around 7.0.

The experimental program consisted of 10 runs which were broken down into two stages; namely, 7 SBR runs and 3 SFBR runs. Table 1 shows the experimental matrix as well as the operating conditions. Note that the mixed liquor volatile suspended solids (MLVSS) values are average values for the SBR while they are initial values for the SFBR. In selected runs, samples were sequentially extracted and measured for residual 2,4-D, in order to track the degradation behavior. Previous laboratory research had established that biological degradation was the main mechanism of 2,4-D removal, rather than biosorption [12]. In addition, a control reactor had definitively proven that there were no abiotic losses.

Runs 1 to 3 (table 1) explored the impact of increasing 2,4-D concentration in the acid-phase SBR while runs 4 and 5 examined one 2,4-D concentration only, in the methane-phase SBR. Runs 6 and 7 investigated two different 2,4-D concentrations in the SBR under two-phase anaerobic conditions, while runs 8 to 10 considered increasing 2,4-D concentrations in the two-phase conditions of the SFBR. Given that this research investigated industrial-strength concentrations of 2,4-D (i.e. 20 – 300 mg/L); a relatively long hydraulic retention time (HRT) of 48 hrs was chosen for the SBRs, while nominal solids retention times (SRTs) were targeted at either 10 d (acid-phase SBRs) or in the 12-20 d range (methane and two-phase SBRs).

Table 1. Operating conditions of experimental design.

Run	2,4-D Feed [ ]	Phase	Temp (°C)	Length of Run (day)	MLVSS
1 – SBR	20	Acid	22 ± 2	129	2450 ± 440
2 – SBR	100	Acid	22 ± 2	33	2630 ± 110
3 – SBR	200	Acid	23 ± 2	26	2350 ± 60
4 – SBR	20	Methane	33 ± 2	116	3150 ± 766
5 – SBR	20	Methane	38 ± 2	42	2318 ± 262
6 – SBR	100	Two-phase	30 ± 2	62	4640 ± 309
7 – SBR	300	Two-phase	30 ± 2	31	4910 ± 273
8 – SFBR	100	Two-phase	32 ± 2	73	10,780 ± 330
9 – SFBR	200	Two-phase	32 ± 2	73	10,680 ± 170
10 – SFBR	300	Two-phase	32 ± 2	73	10,730 ± 270

2.2. *Feed and biomass characteristics*

The SBR reactors were seeded with a mixture (3:1 ratio by weight) of anaerobically digested sludge from a nearby domestic Auckland sewage treatment plant. Biomass was obtained from an on-going University of Auckland research program studying biomass exposure to 2,4-D. For the acid and two-phase SBRs (Runs 1-3, 6-7), the feed consisted of various concentrations of 2,4-D (table 1) and a target value of supplemental substrate; namely, 1000 mg/L of COD in the form of glucose. For the SBR methane phase (Runs 4,5), a target value of 1000 mg/L supplemental COD in the form of a VFA mixture (60 % acetate, 30 % propionate and 10 % butyrate) was used. The SFBRs were also seeded with digested sludge, however the feed was basically primary sludge with no supplemental carbon substrate added. In all cases, 2,4-D was added using small pellets (98 % purity) which had previously been dissolved in water. The concentrations of the 2,4-D stock solution, was about 500 mg/L (just to keep it well below the solubility limit of about 900 mg/L). In general, 2,4-D pellets dissolve very poorly even at concentrations well below the solubility limit.

Acclimation allows the bacteria time to express the enzyme-inducing genes to induce the pesticide's biodegradation; and, as such, was defined as the time from the first application of 2,4-D (or the next increment in 2,4-D concentration where applicable) until the beginning of consistent 2,4-D removal. This was defined as removal over 10 %; that is, 2 or 3 % removal occasionally occurred, however this was generally ignored until the removal reached $\geq$ 10 %.

2.3. *Analytical and sampling methods*

The reactor contents, influent and effluent from each reactor were collected and measured at least three times per week. During in-depth track studies, sampling was done primarily on an hourly basis. Chemical parameters included glucose, 2,4-D, MLVSS, pH, temperature, and VFAs. All samples, except solids were centrifuged at 4,000 rpm for 8 min and the supernatant was filtered through a 0.2 μm nylon filter membrane and acidified to a pH of 2 with 9.8 % H_2SO_4, before further analysis. Solids were filtered using 55-mm-diameter glass microfiber filters (GF/C 1822 055). Glucose was analyzed by the American Chemical Society (ACS) method using anthrone as a reagent [13]. 2,4-D was measured by a Dionex® high performance liquid chromatograph (HPLC) equipped with a Microsorb® MV 100-5 C18 column (24 cm) and an UV/FIS detector (wavelength 240 nm, temperature 40°C, elution with acetonitrile/water (60/40) at 1.2 mL/min). VFAs (i.e. acetic, propionic, butyric, and valeric acids) were determined by a Hewlett-Packard® HP-6890 gas chromatograph (GC) using helium as carrier gas (inlet split at 5:1, temperature 240°C, split flow 9.0 mL/min at 22.16 psi, total flow 13.2 mL/min). The system was equipped with a EC-1000 column (30 cm) and an flame ionization detector (FID) (temperature 300°C, hydrogen flow 45.0 mL/min, air flow 400 mL/min, constant makeup flow of nitrogen 32.0 mL/min). An 1 mmol VFA standard solution was used for calibration purposes.

All other parameters were measured according to Standard Methods for the Examination of Water and Wastewater [14].

3. Results and discussions

3.1. Degradation of 2,4-D: SBR reactor configuration

Table 2 indicates the percentage of degradation achieved during each run along with the amount of acclimation time needed to begin degradation. What is clear from the data is that a reasonably long time frame (i.e. 100 days) was needed before the acid phase anaerobic SBR showed any 2,4-D degradation (Run 1). However once acclimated, subsequent runs (2 and 3) quickly adjusted to the new, increased concentration; thereby shortening the acclimation period (11 and 7 days respectively). It can be seen however that Run 3 never achieved full degradation of the 200 mg/L of 2,4-D. It is reasonable therefore to suggest that at least for the conditions studied in this research, acid-phase digestion has an upper removal bound of around 130 mg/L of 2,4-D.

Table 2. Length of time needed for acclimation and % 2,4-D degradation.

Run	2,4-D Feed [ ]	Phase	Length of Acclimation	% 2,4-D Degraded
1 - SBR	20	Acid	100 days	100
2 - SBR	100	Acid	11 days	100
3 - SBR	200	Acid	7 days	65
4 - SBR	20	Methane	Unachieved	0
5 - SBR	20	Methane	Unachieved	0
6 - SBR	100	Two-phase	46 days	100
7 - SBR	300	Two-phase	12 days	59
8 – SFBR	100	Two-phase	25 days	100
9 – SFBR	200	Two-phase	25 days	100
10 - SFBR	300	Two-phase	30 days	100

When the SBR was operated in the methane phase (Run 4), it was observed that even a small concentration of 2,4-D (20 mg/L), experienced no degradation. From a reporting perspective, since the SBR never acclimated to the pesticide, the percent 2,4-D degradation achieved was 0 %. Similarly, Run 5 did not achieve 2,4-D acclimation (again 0 % degradation), even though to encourage 2,4-D degradation, it was operated at a higher temperature (table 1). Due to the apparent lack of success, Run 5 was terminated after 42 days (table 1).

Run 6 operated as a two-phase digester in that the acid- and methane-phases were allowed to find their natural equilibrium in terms of co-existence. As can be seen, it took 46 days to acclimate (i.e. longer than the acid-phase only SBR (at an equivalent concentration of 100 mg/L)) but this was not surprising, since its initial starting concentration was so much higher. However, Run 6 performed better than the methane-only phase digesters (Run 4 and 5) which did not acclimate. This does suggest that an additional reason for the long acclimation period of Run 6 (i.e. as compared to Run 2) was the presence of a significant number of methanogenic bacteria. When the 2,4-D concentration was increased to 300 mg/L (Run 7), again a radically shortened acclimation period was experienced, even though the step-change in concentration was quite significant (i.e. the 200 mg/L concentration level was eliminated). As table 2 indicates, 100 % of the 2,4-D was removed at a concentration of 100 mg/L (Run 6); however, only 59 % was removed at the new concentration of 300 mg/L (Run 7). Figure 1 indicates that 59 % removal was a mean value for Run 7.

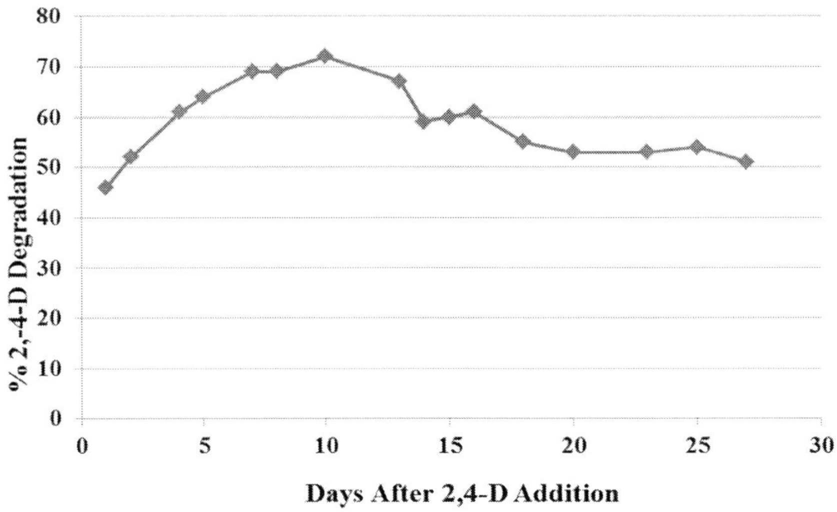

Figure 1. Example of Stability of 2,4-D Degradation: Run7.

3.2. Degradation of 2,4-D: SFBR reactor configuration

Runs 8, 9 and 10 (table 2) in the SFBR systems achieved 100 % degradation for each increment in concentration, taking approximately the same amount of time to acclimate in each case. A comparison of the SFBR to the SBRs is not quite valid here however; since certainly any initial decline in 2,4-D can be accounted for by dilution. That is, during each sampling day, a certain volume of liquid was taken from the 4 L reactors in order to measure VFAs, 2,4-D, and solids etc. Since it is impossible to do this for 73 days without running out of liquid, an equivalent amount of tap water had to be added to each reactor on sampling days. This effect however has been taken into account by applying appropriate "dilution factors" to normalize all data obtained.

The SFBR appeared to be a better system than the SBRs since it removed 100 % of the 300 mg/L 2,4-D concentration, whereas, the SBR only removed 59 % (table 2). However, this comparison may not be strictly valid for two reasons. Firstly, the SFBRs experienced a lower total amount of 2,4-D added, since they were fed 300 mg/L of 2,4-D only at the start of the experiment. In contrast, the SBRs were fed that amount on a daily basis as part of the SBR feeding regime. Undoubtedly, the SBR's continual exposure to incoming 2,4-D may have caused cumulative stress to the bacteria , despite the fact that they were actively-growing in response to the daily influx of glucose. Secondly, it has to be acknowledged that the SFBRs carried a much higher solids loading (table 1). As such, what might appear to be inhibition in the SBR might have been overcome by just straight adaption in the SFBR, simply because of the larger concentration of bacteria present.

3.3. Specific carbon consumption rates: Effect of 2,4-D concentration

Specific rates are not strictly applicable to an SFBR system primarily because it is considered as a "single-cycle" operation. However, rates can be readily calculated for the SBR system and; as such, Table 3 shows the specific carbon consumption rates for each of the SBR runs. As can be seen, the specific glucose consumption rate for Runs 1 to 3 seem to be independent of the 2,4-D concentration. However, the specific 2,4-D consumption rate clearly increased substantially between the 20 mg/L and 100 mg/L 2,4-D concentration. Increasing the 2,4-D concentration further to 200 mg/L however did not seem to increase the specific 2,4-D consumption rate. As such, the existence of a "rate" plateau suggests a possible limitation in the biomass' ability to degrade 2,4-D.

As mentioned, Runs 4 and 5 (operating in the methanogenic phase) did not remove 2,4-D, thus no specific 2,4-D rates are calculable. The lack of removal may be attributed to the potential toxicity of

2,4-D on methanogenic bacteria; however, in addition, at times, the SRT dropped perilously low (5 to 8 days) which meant there was difficulty in keeping the biomass in the system. The observed increase in the specific VFA utilisation rate between Runs 4 and 5, is likely due to the fact that Run 5 was operated at the higher temperature. Finally, as can be seen, the specific glucose consumption rate dropped substantially from Run 6 to Run 7 which indicates a negative effect on the glucose degradation pattern (caused likely by inhibition associated with the increased concentration of 2,4-D). However once again, the 2,4-D consumption rate reached a plateau which explains the upper bound on the 2,4-D removal (180 mg/L).

Table 3. Specific rates for carbon consumption.

Run	2,4-D Feed [ ]	Phase	Supplementary Substrate (mg/mg VSS·d)	2,4-D Rate (mg/mg VSS·d)
1 – SBR	20	Acid	2.45 ± 0.29	0.007 ± 0.001
2 – SBR	100	Acid	2.73 ± 0.22	0.047 ± 0.006
3 – SBR	200	Acid	2.74 ± 0.33	0.048 ± 0.005
4 – SBR	20	Methane	0.04 ± 0.003	0.000
5 - SBR	20	Methane	0.07± 0.002	0.000
6 - SBR	100	Two-phase	0.32 ± 0.03	0.015 ± 0.003
7 - SBR	300	Two-phase	0.19 ± 0.03	0.018 ± 0.004

3.4. Pattern of sequential carbon consumption: SBRs and SFSBRs

Figure 2 plots the % carbon removed for Run 2 of the SBRs as an example of the sequential pattern obtained in terms of carbon removal. That is, in an acid-phase anaerobic digester fed with both glucose and 2,4-D, complete degradation of both carbon substrates occurred in a sequential manner when the 2,4-D concentration was 100 mg/L or less. This pattern was consistently observed throughout this study, and implies sequential utilization known as "diauxic growth", a hallmark of systems treating toxic substances in the presence of readily biodegradable substrates [15, 16].

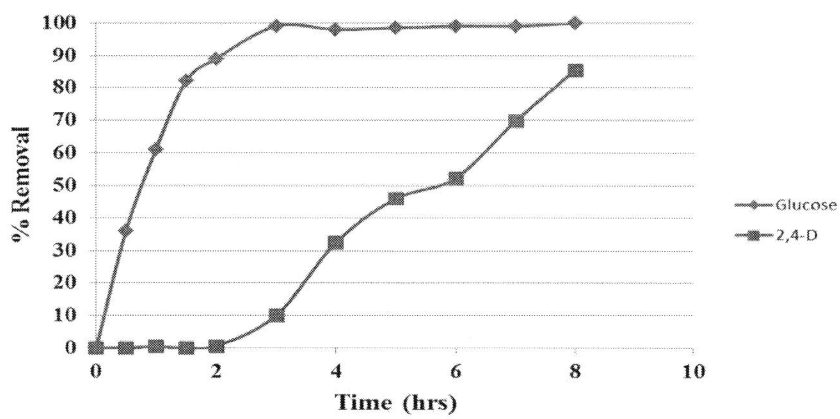

Figure 2. Example of sequential carbon degradation pattern in SBR: Run2.

Further to this, in Runs 6 and 7, VFAs were generated, with acetic and propionic acids being the major species (making up at least 90 % of the total VFAs). This pattern of speciation is characteristic of wastewaters low in protein and high in carbohydrate [17]. In Run 6, the VFAs were totally degraded within the first 8 hrs of the cycle, representing a stable relationship between acidogens and methanogens in the reactor. However, when the 2,4-D concentration was increased to 300 mg/L, a modest rise in total VFAs was observed followed by a 4 hr prolongation for total VFA consumption.

This confirms the deterioration in the performance of the anaerobic SBR at the largest 2,4-D concentration.

Figure 3 indicates the biodegradation pattern of the 2,4-D inside the SFBRs. It can be seen that after an acclimation period of between 25 and 35 days (a function of the amount of 2,4-D added), complete biodegradation of 2,4-D was accomplished.

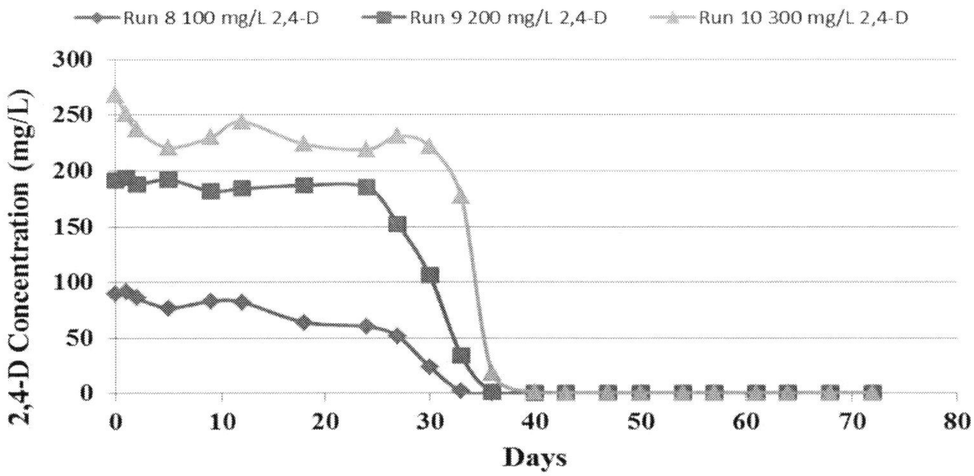

Figure 3. 2,4-D Biodegradation pattern: Runs 8-10.

4. Conclusions

This study indicated that after biomass acclimation, anaerobic biodegradation of industrial-strength concentrations of 2,4-D was accomplished in both the SBRs (except under methanogenic-only conditions) and the SFBRs. Under the experimental conditions of this research, the SBRs degraded between 130 and 180 mg/L of 2,4-D depending upon whether the reactors were operating in acid-phase or two-phase conditions. In contrast, the SFBRs degraded 100 % of the 2,4-D up to a concentration of 300 mg/L.

References

[1] U.S. Environmental Protection Agency *Pesticide Market Estimates 2000-2001*: *Usage* http://www.epa.gov/oppbead1/pestsales/01pestsales/usage2001/ (accessed March 2008)

[2] Dalvie M A., White N, Raine R, Myers J E, London L, Thompson M and Christiani D C 1999 Long term respiratory health effects of the herbicide, paraquat, among workers in the Western Cape *Occup. Environ. Med.* **56** 391-6

[3] Kleter G A, Harris C, Stephenson G and Unsworth J 2008 Comparison of herbicide regimes and the associated potential environmental effects of glyphosate-resistant crops and what they replace in Europe *Pest Manag. Sci.* **64** 479-88

[4] Elefsiniotis P and Li W 2008 Biodegradation behavior of agricultural herbicides in anaerobic batch reactors *J. Environ. Sci. Health Pt B* **43** 172-8

[5] Holland P and Anis R 1999 *Pesticide Trends in New Zealand* (Wellington, New Zealand: Ministry of Agriculture Technical) pp 1-53

[6] Gaw S, Close M E and Flintoft M J 2008 Fifth national survey of pesticides in groundwater in New Zealand *New Zealand J. Marine Freshwater Res.* **42** 397-407

[7] de Blanc-Knowles J and Motaval J 2003 Clean and green: New Zealand is a study in environmental contrasts *The Environ. Magazine* **14** 21

[8] Mitchell C, Brodie J and White I 2005 Sediments, nutrients and pesticide residues in event flow conditions in streams of the Mackay Whitsunday Region, Australia *Mar. Pollut. Bull.* **51** 23-

IOP Conf. Series: Earth and Environmental Science **82** (2017) 012004 doi:10.1088/1755-1315/82/1/012004

36
[9] Otto S, Vianello M, Infantino A, Zanin G and Di Guardo A 2008 Effect of a full-grown vegetative filter strip on herbicide runoff: Maintaining of filter capacity over time *Chemosphere* **71** 74-82
[10] Lee C Y, Shin H S, Chae S R, Nam S Y and Paik B C 2002 Nutrient removal using anaerobically fermented leachate of food wastes in the BNR process *Water Sci. Technol.* **47** 159-65
[11] Hernández-Leal L, Vieno N, Temmink H, Zeeman G and Buisman C J N 2010 Occurrence of xenobiotics in gray water and removal in three biological treatment systems *Environ. Sci. Technol.* **44** 6835-42
[12] Elefsiniotis P and Wareham D G 2013 Biodegradation of industrial-strength 2,4-D wastewaters in the presence of glucose in aerobic and anaerobic sequencing batch reactors *Environ. Technol.* **34** 1167-74
[13] ACS 2000 *Reagent Chemicals: Anthrone, ACS Specifications* 9th ed. (Columbus, OH: American Chemical Society)
[14] APHA, AWWA and WEF 2005 *Standard Methods for the Examination of Water and Wastewater.* 21st ed, (Washington, DC: American Public Health Association)
[15] Adour L, Couriol C and Amrane A 2005 Diauxic growth of Penicillium camembertii on glucose and arginine *Enzyme Microb. Technol.* **36** 198-202
[16] Chang H L and Alvarez-Cohen L 1995 Model for the cometabolic biodegradation of chlorinated organics *Environ. Sci. Technol.* **29** 2357-67
[17] Maharaj I and Elefsiniotis P 2001 The role of HRT and low temperature on the acid-phase anaerobic digestion of municipal and industrial wastewaters *Bioresource Technol.* **76** 191-7

IOP Conf. Series: Earth and Environmental Science **82** (2017) 012005 doi:10.1088/1755-1315/82/1/012005

The effect of drinking water quality on the health and longevity of people-A case study in Mayang, Hunan Province, China

J Lu[1,2] and F Yuan[1]

[1]Department of Engineering and Safety, UiT The Arctic University of Norway, N-9037 Tromsø, Norway

E-mail: Jinmei.lu@uit.no

Abstract. Drinking water is an important source for trace elements intake into human body. Thus, the drinking water quality has a great impact on people's health and longevity. This study aims to study the relationship between drinking water quality and human health and longevity. A longevity county Mayang in Hunan province, China was chosen as the study area. The drinking water and hair of local centenarians were collected and analyzed the chemical composition. The drinking water is weak alkaline and rich in the essential trace elements. The daily intakes of Ca, Cu, Fe, Se, Sr from drinking water for residents in Mayang were much higher than the national average daily intake from beverage and water. There was a positive correlation between Ni and Pb in drinking water and Ni and Pb in hair. There were significant correlations between Cu, K in drinking water and Ba, Ca, Mg, Sr in the hair at the 0.01 level. The concentrations of Mg, Sr, Se in drinking water showed extremely significant positive relation with two centenarian index 100/80% and 100/90% correlation. Essential trace elements in drinking water can be an important factor for local health and longevity.

1. Introduction

Trace elements can not be manufactured by human body itself, and they must be taken from the natural environment. Water is a major source of trace elements necessary for the growth of biological organisms. The composition of trace elements in water has a significant impact on human health. Changes in drinking water and groundwater sources can lead to significant changes in health risk related with trace elements [1].

Insufficient or excessive trace elements in water can lead to the occurrence of certain diseases. Liu XJ et al. found that the concentrations of Cu, Fe, Sr, Ti and V in the water samples from area with high incidence of gastric cancer were significantly higher than those in the area with low incidence of gastric cancer [2]. Another research on the relationship between the concentration of trace elements in drinking water and gastric cancer showed that Se and Zn can significantly prevent the development of gastric cancer [3]. Kikuchi H. et al. studied the relationship between the levels of trace elements in water and age-adjusted incidence of colon and rectal cancer, and the results showed that the incidence of colon cancer and rectal cancer was significantly related with the concentration of Au, Mg, Na, P, Sn,

Sr in water [4]. Studies on the relationship between Se and I in soil, water, and rice in an endemic thyroid area in Sri Lanka and the incidence of thyroid disease have shown that low incidence of thyroid disease in northern arid areas is associated with higher I concentrations in local drinking water [5]. In arid and semi-arid areas, the incidence of thyroid disease increases due to drinking deep groundwater with high iodine concentration [6]. P C Ferreira et al have shown that too high concentration of Al in water can increase the risk of neurological disorders in the older persons [7]. Research on the Ca concentration in the drinking water from a high incidence area of esophageal cancer showed that low Ca concentration in drinking water may be one reason for the high incidence of esophageal cancer, and Ca has a certain inhibitory effect on the occurrence of esophageal cancer [8]. Another research on the relationship between drinking water quality and the incidence of Kashin-Beck disease showed significantly higher concentrations of Cu and K and significantly lower concentrations of Al, Fe, Se in a Kashin-Beck disease endemic area than those in the non-disease areas [9]. Studies on the Cu concentration in the water and human health have shown that the acute exposure to Cu in drinking water can cause nausea and vomiting, and the chronic exposure of Cu in drinking water can affect the liver in the sensitive the population [10].

China has the phenomenon of regional health and longevity. From 2000 onwards, China started to evaluate and nominate the longevity county throughout the country. So far, 13 longevity counties have been nominated. To become a longevity county, an area should meet certain standards. The three basic assessment criteria for longevity county are: representativity of longevity, i.e. the number of centenarians in 100000 shall reach the international standard of 7/1000000; the integrity of longevity, i.e. the average life expectancy in the region should be at least 3 years higher than the national average; the sustainability of longevity, i.e. people over 80 years old should account for 1.4% of the total population. Mayang is one of the longevity counties that are nominated by the Chinese government.

High-quality water environment is an extremely important factor for the formation of a longevity region. Currently only random research have been conducted on the relationship between drinking water quality and the regional longevity phenomenon in China. The drinking water in the longevity area of Dujiangyan City in Sichuan province is from the densely vegetated Aba forest, and the concentration of harmful elements in the water is especially low. The spring water from Aba Dagu glacier is the world's oldest water, and has small water molecular group, low deuterium, high dissolved oxygen, a variety of natural mineral elements, weak alkaline, no pollution, no degradation, the vitality of life, and meet the needs of human nutrition and health standards for drinking water [11]. Research on the drinking water from another longevity region Yunlong county in Yunnan province showed that the spring water contained six macro elements Ca, K, Mg, Na, P, S and 8 trace elements Cu, Cr, Fe, Zn, Mn, Mo, Se, Sr [12]. Drinking water from Changxing Island, another longevity region, has slightly higher concentration of Mn and Fe, which is conducive to longevity [13].

In order to interpret the effect of drinking water quality on local health and longevity, a longevity county Mayang was selected as the study area in this study. The drinking water samples and hair samples of local centenarians was collected and tested the concentrations of macro and microelements. The drinking water quality was evaluated. The correlation coefficients between the elements in drinking water quality and hair and the correlation coefficients between the drinking water quality and different longevity and centenarian indices were analyzed and discussed in this manuscript.

2. Study area

Mayang County is located in the northwest of Huaihua City, western part of Hunan Province (figure 1). It is located between the Wuling Mountain and Xuefeng Mountain, with a total area of 1568 square kilometers and a total population of 3.73 million. Mayang belongs to subtropical monsoon climate zone. It has a mild climate, four distinct seasons, adequate light and heat, and abundant rainfall. It has an average yearly temperature of 17.1 ℃, an average relative humidity of 78%, annual average sunshine of 1294.1 hours, and an average frost-free period of 297 days. The highest elevation is 1405 meters at the south Xihuang Mountain, and the lowest elevation is 130 meters at Maojiatan. Mayang has high life expectancy. According to the latest statistics, the average life expectancy in Mayang is

75.6. According to the fifth census, the county has 48320 people who are more than 60 years old, accounting for 13% of the total population; 5430 people at the age of 80-99 years old, accounting for 1.46% of the total population; 41 people who are hundred or more than hundred years old, accounting for 11/100,000 of the total population. All the indicators are well above the longevity standard specified by the United Nations Population Organization and China. It was formally named as a longevity region in China in December 2007 by the Chinese Society of Gerontology.

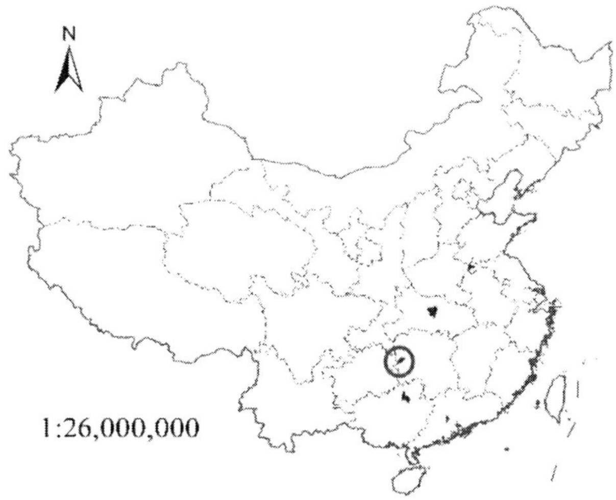

Figure 1. Study area.

3. Materials and methods

3.1. Water sampling and analysis

The drinking water samples were collected according to the distribution of centenarians in the county. As well water is the main drinking water sources for the local people. On average, there is a well in each centenarian's home. The drinking water samples were collected with a mineral water bottle in the centenarian's home. The pH of the sample was measured directly after the collection of the water samples by a hand-held pH meter. Afterwards 10 drops of 1: 1 HNO_3 were added to each sample to prevent the adsorption of soluble metal ions to the inner wall of the bottle and to reduce the microbial activity after sampling. The samples were immediately sent to the laboratory and stored in the refrigerator at 4°C until analysis. A total of 23 water samples were collected.

The concentrations of elements Al, Ba, Ca, Cd, Co, Cr, Cu, Fe, K, Li, Mg, Mn, Mo, Na, Ni, P, Pb, Se, Si, Sr, V, Zn in the water samples were analyzed with ICP-MS and ICP-AES according to the standard method GB/T 2050.6-2006. The concentration of F in the water samples was determined with a national standard F ion-selective electrode method GB/T 2050.5-2006.

3.2. Hair samples sampling and analysis

The occipital hair samples from the selected centenarians were collected with a clean stainless steel scissor. The collected samples were stored in a clean paper bag. A total of 17 hair samples were collected from the centenarians.

The hair samples were pretreated and digested according to the methods developed by K S Rao et al [14]. The hair samples were soaked in 1% detergent (volume percentage) for 24 hours. The oil and grease were removed by stirring several times during the soaking. The samples were washed with pure water afterwards until no foam, then rinsed with ultrapure water for 2 to 3 times. The washed samples were placed on clean filter papers and air dried for two days. Then the hair was cut it to 2mm with Teflon scissors, put into a clean paper bag and stored in a dry place until analysis.

0.2 g of hair sample was weighed into a 100ml glass beaker and added 1.0 mL H_2O_2 and 5.0ml superpure HNO_3. The glass beaker was placed on the hot plate for digestion. The temperature of the digestion was set from low to high. The temperature was controlled between 140-180℃. The digestion will continue until the liquid became light yellow and the residue was less than 0.5 mL. Then the beaker was taken away from the hot plate and cooled to room temperature. The residue was transferred from the beaker into a tube and diluted with 1% HNO_3 to 15 ml. The concentrations of elements Al, Ba, Ca, Fe, K, Mg, Mn, Mo, Na, P, Sr, V and Zn in the mixture were analyzed by ICP-OES [14], and the concentrations of Cd, Co, Cr, Cu, Li, Ni, Pb, Se were determined by ICP-MS [15].

3.3. Quality control

To control the analysis quality, 3 national standard samples were tested simultaneously together with each batch of samples. In addition, several random parallel samples were selected to test for each batch of samples. The standard sample for water samples is GBW08607 and the standard sample for hair sample is GBW09101b. To reduce system error, 2-3 blank samples were set for each batch of samples.

3.4. Statistical analysis

All the statistical analyses in this study are conducted by SPSS 13.0 for windows.

4. Results

4.1. The water quality in Mayang

Table 1. The water analysis results and national and WHO guideline values.

	Mayang	National standard		WHO Guideline values	
		Value	CR (%)	Value	CR (%)
pH	7.21	6.5-8.5	100	6.5-9.5	100
F (mg/L)	0.072		–	≤1.5	100
Ba (mg/L)	0.196	≤0.7	100		–
Cd (mg/L)	ND	≤0.005	100	≤0.003	100
Co (mg/L)	ND	≤1	100		–
Cr (mg/L)	ND	≤0.05	100	≤0.05	100
Cu (mg/L)	0.018	≤1	100	≤2	100
Fe (mg/L)	0.074	≤0.3	93.33	≤2	100
Mn (mg/L)	ND	≤0.1	100	≤0.4	100
Mo(mg/L)	ND	≤0.07	100	≤0.07	100
Na (mg/L)	15.70	≤200	100	≤200	100
Ni (mg/L)	0.004	≤0.02	100	≤0.07	100
Pb(mg/L)	ND	≤0.01	100	≤0.01	100
Se (μg/L)	0.872	≤10	100	≤10	100
Zn(mg/L)	0.009	≤1	100	≤3	100

Note: ND denotes "not detected"; CR denotes "compliance rate".

The average value of the tested parameters in the drinking water samples from Mayang was compared with China's national drinking water standard GB5479-1985 and WHO guideline values [16]. The results are shown in table 1. All the tested items are in line with China's national drinking water standards and WHO guidance values. The drinking water has an average pH of 7.21, which is between 7 and 8 and weak alkaline. Research showed that weak alkaline mineral water is conducive to human health, which can enhance blood circulation and cell metabolism, reduce the production of acidic substances, neutralize gastric acid, eliminate acidosis, and increase the pH of the stomach acid to meet the normal body needs [17]. Weak alkaline water has an almost same pH as human blood, which can maintain the blood vessels of the elderly soft and not hardened, the blood pressure low, and the pulse normal [18]. Alkaline ionized water improves the hyperacidity and abnormal intestinal fermentation [19]. The appropriate mineral elements and trace element content is also an important factor in promoting human health.

Since the existence of the extreme values (either extremely high or low values) may obscure the true range of the element, the probability that the measured sample falls within the range of China's national drinking water standard and WHO guideline value was further calculated. The results are also shown in table 1. Except Fe, which has a very small percentage out of the standard, all the other tested parameters are 100% in line with the national standard and the WHO standard. The drinking water quality is good in Mayang. This may be an important factor for local longevity.

4.2. The effect of element concentrations in drinking water on local health and longevity

The elements in drinking water can get into the human body through the daily dietary intake. Thus, the element concentration in drinking water has an important impact on the corresponding element content in the human body.

4.2.1. Effect of drinking water quality on daily element intake of residents. The elderly in Mayang drink local water in the long-term. Drinking water is an important source of daily elements intake for local people. The daily element intake from drinking water for the residents of Mayang was calculated using the USEPA reference value, and the daily drinking water consumption was 2.00 L [20]. The daily element intake from drinking water can be calculated with the following formula:

$$I_i = C_i * D$$

Where I_i is the daily intake of element i from drinking water, unit mg; C_i is the concentration of element i in drinking water, unit mg/L; D is the amount of water that people consume every day and it is a constant 2.00 L.

The daily element intake from drinking water for the residents in Mayang was calculated and compared with the national average daily intake from beverage and water [21]. The results were shown in table 2. The daily intakes of Ca, Cu, Fe, Se, Sr from drinking water for residents in Mayang were higher than the national average daily intake from beverage and water. The drinking water in Mayang was rich in these elements. Local residents can get enough elements for normal human body needs just by drinking 2 L of water every day.

Table 2. Daily element intake from drinking water for the residents in Mayang.

Element	Daily intake amount	National average daily intake from beverage and water
Ba (mg)	0.37	0.064
Ca (mg)	134.21	30.8
Cr (ug)	-	3.2
Cu (mg)	0.04	0.01
Fe (mg)	0.11	0.1
K (mg)	8.78	10.2

Mg (mg)	31.96	16.3
Mn (mg)	-	0.18
Mo (ug)	-	3
P (mg)	0.08	3
Sr (mg)	1.13	0.18
Zn (mg)	0.007	0.18
Se (ug)	1.77	0.5

4.2.2. Correlation of elements in drinking water and hair. The economic conditions are relatively backward in Mayang and the centenarians normally live in the place where they were born throughout their whole life. As water is an important source for various macro and microelements in human body, the concentrations of elements in drinking water have an important impact on the concentration of corresponding elements in human hair.

There are other sources for human exposure to these macro and micro elements, such as different hair care products shampoos, hair conditioners, hair dryers, etc. As all the hair samples are from centenarians living in the villages. Their living style is quite old fashioned. They use shampoos to wash hair sometimes, but not quite often. They normally don't use hair conditioners, hair dryers etc. Therefore, the effect of these hair care products on element concentration in hair is not taken into account in this study.

The correlation coefficients between the concentrations of elements in drinking water and in human hair were calculated using town as a statistical unit. There are totally 12 towns taken into statistics. The results were shown in table 3. The corresponding descriptive statistics were listed in table 4. There was a positive correlation between Ni and Pb in drinking water and Ni and Pb in hair, with a correlation coefficient of 0.392 and 0.567 respectively. There were significant correlations between Cu, K in the drinking water and Ba, Ca, Mg, Sr in the hair at the 0.01 level. The K, Ca and Mg exhibit the same characteristics in the hair [22]. Sr rarely exists as independent minerals. Sr normally exists in the Ca containing minerals in isomorphic forms due to the similar geochemical properties of Sr and Ca [23]. The results indicated that the element concentrations in drinking water will affect the concentrations of corresponding elements in the hair to a certain extent. Drinking water is an important source for the trace elements in hair.

Table 3. The correlation coefficients between elements in drinking water and elements in hair.

Water \ Hair	Ba	Ca	Cu	Fe	K	Mg	Ni	Pb	Sr	Se
Ba	-	-0.351	**0.788****	0.167	-	-	0.311	0.222	-	-
Ca	-	-0.486	**0.866****	0.152	-	-	0.384	0.273	-	-
Cu	0.090	0.375	0.076	0.134	-0.142	0.494	-	-	0.400	0.400
Fe	-	0.044	-0.184	-	-0.069	-	-	-	-	-
K	0.051	-0.080	-0.002	-	-0.279	-	0.137	-	-	-
Mg	-	-	**0.958****	0.404	-	-	0.472	0.272	-	-
Ni	0.296	0.403	-0.204	0.316	0.219	0.485	**0.392**	-	0.571	0.393
Pb	-	-0.043	0.146	0.189	-0.264	0.332	-	**0.567**	0.397	0.208
Sr	-	-	**0.899****	0.200	-	-	0.510	0.329	-	-
Se	-	-0.308	0.411	-	-0.350	-	-	0.359	-	-

Note: ** Indicates a significant correlation at $p = 0.01$ level;
* Indicates a significant correlation at $p = 0.05$ level

Table 4. The statistical results of element concentrations in drinking water and human hair (Take town as the statistical unit).

	Min	Max	Mean	SD
Hair Ba	0.95	6.85	3.93	1.74
Hair Ca	884.9	2847.8	1929	605.1
Hair Cu	3.52	7.99	5.26	1.28
Hair Fe	7.91	41.54	16.75	9.97
Hair K	0.85	64.77	25.09	17.20
Hair Mg	83.71	296.33	194.4	60.74
Hair Mn	0.141	1.484	0.382	0.379
Hair Ni	0.940	17.211	5.298	4.783
Hair Pb	2.069	7.198	5.301	1.550
Hair Sr	0.106	0.563	0.378	0.111
Hair Se	0	0.52	0.19	0.13
Water Ba	2.29	95.54	58.88	30.13
Water Ca	0.017	0.020	0.019	0.001
Water Cu	0	0.19	0.04	0.06
Water Fe	0.46	14.71	3.31	4.34
Water K	0.47	29.25	12.52	9.71
Water Mg	0.003	0.005	0.004	0.001
Water Mn	0	0.004	0.002	0.001
Water Ni	0.018	1.622	0.425	0.549
Water Pb	0.038	1.117	0.571	0.340
Water Sr	0.95	6.85	3.93	1.74
Water Se	884.9	2847.8	1929	605.1

Note: SD is the standard deviation; The concentrations of elements in hair is in the unit of mg/kg; For drinking water, all the elements are in the unit of mg/L, except; Se which is in the unit of μg/L.

4.2.3. The correlation between element concentrations in drinking water and longevity index. In order to further analyze the relationship between concentrations of elements in drinking water and local longevity, the correlation coefficients between concentrations of macro and microelements in drinking water and different longevity population index in Mayang were calculated with town as the statistical unit. The longevity population index calculated in this study include: the proportion of people over 80 years old in people over 60 years old (80/60%); the proportion of people over 90 years old in people over 60 years old (90/60%); the proportion of people over 100 years old in people over 80 years old (100/80%); the proportion of people over 100 years old in people over 90 years old (100/80%) (100/90%).

The distribution of the aged population and longevity level in a region are mainly affected by social factors such as the birth rate, mortality rate, economic level and medical level. Centenarians, as a group of very special people, may be affected by different factors other than that of people in other age range. Therefore the population index are subdivided into two categories: longevity index, which mainly reflects the longevity level of a region, including 80/60% and 90/60%; and the centenarian index, which mainly reflects the level of super-longevity in a region, including 100/80% and 100/90%. The correlation coefficients between concentrations of elements in drinking water and different population indices were shown in table 5.

The concentrations of Mg, Sr, Se in drinking water showed extremely significant positive relation with two centenarian index 100/80% and 100/90% and very weak negative correlation with the two longevity index 80/60% and 90/60% (table 5). The concentrations of K and Ca in drinking water also showed a high positive relationship with the two centenarian index 100/80% and 100/90%. Centenarians, as a special group of people, were significantly related with the concentrations of trace elements in drinking water. This illustrates that high-quality water environment is an extremely important factor for the emergence of a large number of centenarians in this region.

Table 5. The correlation coefficient between concentrations of elements in drinking water and different population indices.

Concentrations in water	80/60%	90/60%	100/80%	100/90%
Ba	0.334	0.296	0.436	0.329
Ca	0.079	-0.050	0.471	0.494
Cu	0.290	0.468	-0.314	-0.420
Fe	0.329	0.271	0.292	0.273
K	0.105	-0.292	0.361	0.491
Mg	-0.204	-0.166	**0.710****	**0.717****
Ni	0.322	-0.153	-0.168	0.062
Pb	-0.013	-0.052	-0.227	-0.253
Sr	-0.344	-0.411	**0.643***	**0.749****
Se	-0.006	-0.169	**0.717****	**0.731****

Note: ** Indicates a significant correlation at p = 0.01 level; * Indicates a significant correlation at p = 0.01 level.

5. Discussion

The chemical analysis results on the drinking water samples from local region showed that the water is weak alkaline and high quality. The water is rich in elements Ca, Cu, Fe, Se and Sr. The daily intake of these elements from drinking water is higher than the national average intake from beverage and water. The concentrations of Mg, Se and Sr had an extremely significant positive relationship with the two centenarian index 100/80% and 100/90%.

The concentration of Ca in drinking water has a direct impact on mineral contents in bone, and drinking water with high Ca concentration in the long-term is conducive to human health [24]. Research have shown that the concentration of Ca in drinking water have significant preventive effects on rectal cancer, gastric cancer, breast cancer and acute myocardial injury [25-29].

Iron Fe is an important trace element in the body. The deficiency of Fe can cause severe chronic diarrhea [30]. In addition, because of diarrhea, gastrointestinal loss of trace element will increase.

Selenium has antioxidant effects and can prevent excessive peroxide damage to the body. In the meantime, selenium can also protect the vision and the blood vessels. Selenium is the main anti-cancer element in the immune system and can directly kill tumor cells. Selenium and Se compounds are considered the most potent anticancer agent for prostate cancer [31]. Selenium can also prevent and treat cardiovascular disease. Drinking water that is rich in Se in the long term can enhance body's immunity to different diseases and thus is conducive to prolong lifespan.

Strontium is a component of human skeletal and teeth. It can promote bone growth and development. It can compete with Na in the intestine [32]. It can reduce Na absorption in the intestinal tract to a certain extent and reduce the negative effect of Na on health and longevity [33]. Strontium can prevent high blood pressure, prevention and treatment of diabetes, high blood cholesterol gallstone and other diseases.

6. Conclusion

Element content in natural environmental media has an important impact on the element intake into human body. Of the different environmental medium, drinking water is an important source for element intake. The drinking water in longevity county Mayang is weak alkaline and rich in the essential trace elements, and has low concentrations of heavy metals. The drinking water is of high quality according to WHO and China national drinking water standard. The daily intakes of Ca, Cu, Fe, Se, Sr from drinking water for residents in Mayang were much higher than the national average daily intake from beverage and water. There are significant positive correlations between element concentrations in drinking water and in hair of local centenarians. The concentrations of Mg, Sr, Se in drinking water showed extremely significant positive relation with two centenarian index 100/80% and 100/90% correlation. High drinking water quality is an important factor for the formation of local longevity phenomenon.

References

[1] Mills C F 1996 Geochemical aspects of the aetiology of trace element related diseases *Geol. Soc. Sp.* **113** 1-5
[2] Liu X J, Liu Q Q, Cheng J, Zhang L P, Mu B F, Xue Z N and Cheng G P 2000 Comparison of trace elements in drinking water between high and low incidence districts of gastric carcinoma *Nucl. Instrum. Meth. B.* **160** 167-71
[3] Nakaji S, Fukuda S, Sakamoto J, Sugawara K, Shimoyama T, Umeda T and Baxter D 2001 Relationship between mineral and trace element concentrations in drinking water and gastric cancer mortality in *Japan Nutr. Cancer* **40** 99-102
[4] Kikuchi H, Iwane S, Munakata A, Tamura K, Nakaji S and Sugawara K 1999 Trace element levels in drinking water and the incidence of colorectal cancer *Tohoku J. Exp. Med.* **188** 217-25
[5] Fordyce F M, Johnson C C, Navaratna U R B, Appleton J D and Dissanayake C B 2000 Selenium and iodine in soil, rice and drinking water in relation to endemic goitre in Sri Lanka *Sci. Total Environ.* **263** 127-41
[6] He Y, *et al* 2008 The enlightenment from the relationship between trace elements and human health (in Chinese) *Geophysical & Geochemical Exploration* **32** 70-4
[7] Ferreira P C, Tonani K A D, Juliao F C, Cupo P, Domingo J L and Segura-Munoz S I 2009 Aluminum concentrations in water of elderly people's houses and retirement homes and its relation with elderly health *B. Environ. Contam. Tox.* **83** 565-9
[8] Li W 2008 Study on impact factors of esophageal cancer among residents in Ningyang county of the main Trunk-stream areas of Dawen river, Shangdong province, China (in Chinese) (Jinan: Shangdong University)
[9] Sha R, *et al* 1992 Correlative analysis of 21 elements in the external environment of Kashin-Beck disease and activity index in the endemic area (in Chinese) *Bull. Endemic Dis.* **7** 80-4
[10] Damgaard M 2000 *Copper in Drinking Water* (Washington D.C.: National Academy Press)
[11] Qin J 2008 Study on China's senior residents aged at and above 100 years old (in Chinese) *Guangdong Trace Elements Science* **15** 15-31
[12] Chen Y, *et al* 1999 Elemental analysis of rocy soil and spring water in a Baizu Longevity area *Trace Elements and Health Research* **16** 61-3
[13] Hua X, *et al* 1991 The investigation on the natural environment and longevity in Changxing island (in Chinese) *Shanghai Environmental Science* **10** 46-8

[14] Rao K S, Balaji T, Rao T P, Babu Y and Naidu G R K 2002 Determination of iron, cobalt, nickel, manganese, zinc, copper, cadmium and lead in human hair by inductively coupled plasma-atomic emission spectrometry *Spectrochim Acta B* **57** 1333-8
[15] Rodushkin I and Axelsson M D 2000 Application of double focusing sector field ICP-MS for multielemental characterization of human hair and nails. Part I. Analytical methodology *Sci Total Environ* **250** 83-100
[16] WHO 2008 *Guidelines for Drinking Water Quality* (Geneva: WHO Press, World Health Organization)
[17] Marchione V 2015 6 Health Benefits of Drinking Alkaline Water
[18] Wang Q, *et al* 2007 Characteristics of dietary consumption of longevity population and its enlightment to dietary adjustment in China (in Chinese) *Chinese Food and Nutrition* **7** 52-4
[19] Lu H X 2008 The study on the macrobiotic geographic distribution and environmental chemical element of Bama (Nanning: Guangxi Medical University)
[20] Agency U E E P 1997 *Exposure Factors Handbook* (Washington)
[21] Zhu H D, *et al* 2000 Element concentrations in food and dietary intakes of elements for Chinese *Chin J Radiol Med Prot* **20** 378-84
[22] Shu P and Zheng X Q 1993 Relationship between the elemental content and Ph of human hair *Sci Total Environ* **128** 151-6
[23] Lu J and Yuan F 2016 The assessment of heavy metal distribution in the sediment of Eastern Chongming tidal flat, China *WIT Transactions on Ecology and the Environment* **203** 91-100
[24] Mazess R B and Barden H S 1991 Bone-density in premenopausal women - effects of age, dietary-intake, physical-activity, smoking, and birth-control pills *Am J Clin Nutr* **53** 132-42
[25] Yang C Y, Chiu H F, Chiu J F, Tsai S S and Cheng M F 1997 Calcium and magnesium in drinking water and risk of death from colon cancer *Jpn J Cancer Res* **88** 928-33
[26] Yang C Y, Cheng M F, Tsai S S and Hsieh Y L 1998 Calcium, magnesium, and nitrate in drinking water and gastric cancer mortality *Jpn J Cancer Res* **89** 124-30
[27] Yang C Y and Chiu H F 1998 Calcium and magnesium in drinking water and risk of death from rectal cancer *Int J Cancer* **77** 528-532.
[28] Yang C Y, Chiu H F, Cheng B H, Hsu T Y, Cheng M F and Wu T N 2000 Calcium and magnesium in drinking water and the risk of death from breast cancer *J Toxicol Env Heal A* **60** 231-41
[29] Yang C-Y, Chang C-C, Tsai S-S and Chiu H-F 2006 Calcium and magnesium in drinking water and risk of death from acute myocardial infarction in Taiwan *Environmental Research* **101** 407-11
[30] Solomons N W, Rosenberg I H and Sandstead H H 1976 Zinc nutrition in celiac sprue *Am J Clin Nutr* **29** 371-5
[31] Nadiminty N and Gao A C 2008 Mechanisms of selenium chemoprevention and therapy in prostate cancer *Mol Nutr Food Res* **52** 1247-60
[32] Chen S Y, *et al* 2005 Longevity county-trace element investigation in Rugao, Jiangsu province (in Chinese) *Guangdong Trace Elements Science* **12** 13-8
[33] Zou X Y 2010 Investigation on ecological and chemical characteristics in a longevity county in Xiayi, Henan province, China (in Chinese) (Beijing: Institute of Geographic Sciences and Natural Resources Research)

IOP Conf. Series: Earth and Environmental Science 82 (2017) 012006 doi:10.1088/1755-1315/82/1/012006

Iterative circle-inserting algorithm CST3D-OC of truly orthogonal curvilinear grid for coastal or river modelling

H Kim[1], S Lee[1,3], J Lee[1] and H-S Lim[2]

[1]Kookmin University, Seoul, Korea
[2]Korea Institute of Ocean Science and Technology, Ansan, Korea

E-mail: stork@kookmin.ac.kr

Abstract. A geometric method to generate orthogonal curvilinear grid is proposed here. Elliptic partial differential equations have frequently been solved to find orthogonal grid positions, but questions on orthogonality have remained so far. Algebraic methods have also been developed to improve orthogonality, but their applications have been limited to special situations. When two confronting boundary lines of the quadrilateral boundaries are straight, and their positions are known, and we assume that some degree of freedom exists on the other two confronting boundary curves under the condition that the each curve passes through a point, we can assign a set of latitudinal curves in the domain using polynomials. The curves are expected not to fold on their own. The grid positions along longitudinal curves are found by inserting circles between two neighbouring latitudinal curves one by one. If the two curves are straight, the new grid point above the grid point of interest can be found geometrically. This algorithm involves iterations because the curves are not straight lines. The present new algorithm is applied to a domain, and produced almost perfect orthogonality, and similar aspect ratio compared to an existing partial differential equation approach. The present algorithm also can express almost quadrant domain. The present algorithm seems useful for generation of orthogonal curvilinear grids along coasts or rivers. Some example grids are demonstrated.

1. Introduction

Some governing equations for fluid flow use other coordinates from Cartesian coordinates. When boundary shape is not straight, and still flow behaviour near boundary plays an unimportant role, boundary fitted grid becomes more useful. Flows at rivers or coasts are often horizontally two-dimensional, and requires curvilinear grid. Furthermore, if terms for non-orthogonal grid are to be excluded, orthogonality of a curvilinear grid becomes very important, and affects the accuracy of the whole computation.

Methods to generate orthogonal curvilinear grid may be evaluated on three quantities, i.e. orthogonality, aspect ratio, and computational speed of generation. We don't pay attention to the computational speed here. The relative importance of the orthogonality and the aspect ratio depends on the situation or modellers preference, if one value erodes the other value. The uneven aspect ratio has been treated by numerical schemes, while the poor orthogonality cannot be recovered without adopting other governing equations which include terms for non-orthogonality. Here we assume a situation that the orthogonality is the ultimate target, the aspect ratio is the second priority.

Orthogonal grid generation has often been conducted by solving an elliptic partial differential equation (PDE) like Laplace equation or Poisson equation [1-3]. However, there have been arguments

IOP Conf. Series: Earth and Environmental Science 82 (2017) 012006 doi:10.1088/1755-1315/82/1/012006

that the PDE methods don't guarantee orthogonality of the generated grids. Interim results towards asymptotic solution from the PDE method show very poor orthogonality.

On the other hand algebraic methods may have advantage to generate orthogonal grids due to its inherent mathematical attribute. An algebraic grid generation method by Chibisov et al [4] uses circle-rolling method with invariant radiuses to produce almost orthogonal grid, but their method is limited to situations when orthogonal grid is needed around an obstacle or a boundary, and cannot choose the other outside or inside boundary limit. In that case, their method is not ideal for resolving coastal areas which are clearly defined by surrounding land and offshore boundaries. Conti et al [5] proposed a modified PDE method by including an interpolation, but their results still contain poor orthogonality.

We assume that the distribution of coordinate lines has some degree of freedom as Duraiswami and Prosperetti [6] argued. We divide the lines into longitudinal lines and latitudinal lines for convenience. If we can arrange non-intersecting latitudinal lines at first, then the following longitudinal lines will not intersect at all. We attempt to generate grid points for a quadrilateral domain, two confronting boundaries of which are straight, and each of the other two boundary lines passes through a known position.

2. Circle-inserting grid generation algorithm

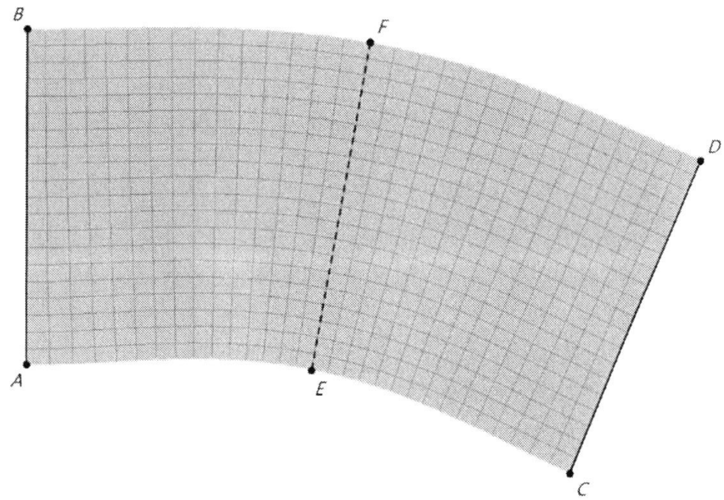

Figure 1. Definition of one domain.

(a) Coast (b) River

Figure 2. Use of orthogonal curvilinear grids.

If a domain is surrounded by two opposite straight lines like figure 1, it is convenient to attach another domain with the same straight line at one end, see figure 2. This is ideal for describing coastal zones or rivers. Block border lines are recommended to sit at relatively straight areas. Separation of a large domain into several small domains is useful if the whole domain is distorted with a wide angle. Even a circular domain like coast surrounding an island can be described by four quadrants.

We start from choosing two straight confronting boundary lines AB and CD, then, we choose another straight line EF between the two boundary straight lines, which provides mid points of the latitudinal lines. We call AB, CD and interim quasi-parallel lines "longitudinal lines", and all other orthogonal lines from AB to CD lines "latitudinal lines". We can choose arbitrary ratio for selection of points along the three straight line segments, AB, CD, and EF. Each latitudinal line has 5 constraints: point positions and gradients at two side boundaries, and a point position between them. Therefore, a fourth order polynomial function for each latitudinal line is adequate to contain these five constraints.

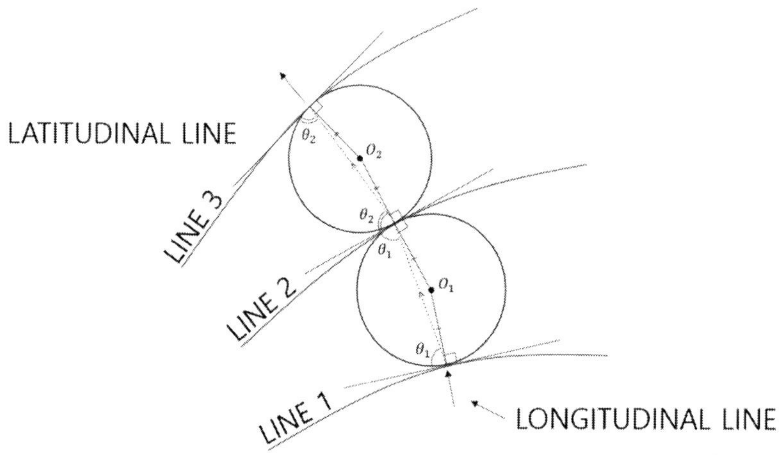

Figure 3. Circle-inserting between curves.

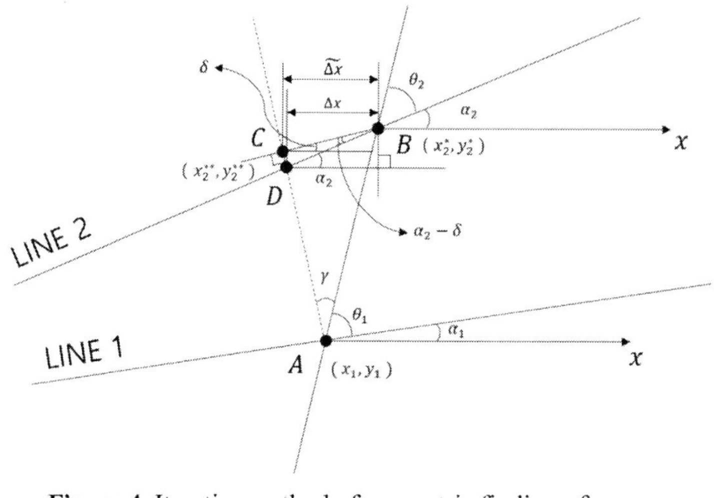

Figure 4. Iterative method of geometric finding of new.

When the latitudinal lines are chosen, longitudinal lines should be found to decide the final grid points. However, we won't try to find any full algebraic equation for a longitudinal line. Instead we choose the grid point positions along the bottom latitudinal line. Now we start to find the first longitudinal line. We insert a circle between the bottom latitudinal line and the next latitudinal line, so that the two latitudinal lines are tangent to the inner circle, and make the two intersection angles equal, e. g. θ_1 or θ_2, see figure 3. If the two latitudinal lines are straight, the new point on the upper latitudinal line can be found geometrically, see figure 4. The interaction point B chosen as the first prediction should be replaced by the correct point D as:

$$\Delta x = \frac{d \sin \gamma \cos \alpha_2}{\cos \delta} \tag{1}$$

$$\gamma = \frac{\theta_1 + \theta_2 - \pi}{2} \tag{2}$$

$$\delta = \frac{\alpha_2 - \alpha_1}{2} \tag{3}$$

where α_1 is the gradient of the bottom line, α_2 is the gradient of the upper line, θ_1 is the angle between the bottom latitudinal line and the longitudinal line, and θ_2 is the angle between the upper latitudinal line and the longitudinal line.

Because the lower and upper latitudinal lines are not strictly straight, the above correction is not complete at the first trial, and requires repetition until a given allowance is met. This geometric iteration could also be replaced by other algebraic iteration method like Newton-Rhapson method.

3. Applications

The present algorithm is applied to two simple domains, a square-like one, and a partial quadrant (donut). Domain A is surrounded by three straight lines, and a polynomial curve, see figure 5.

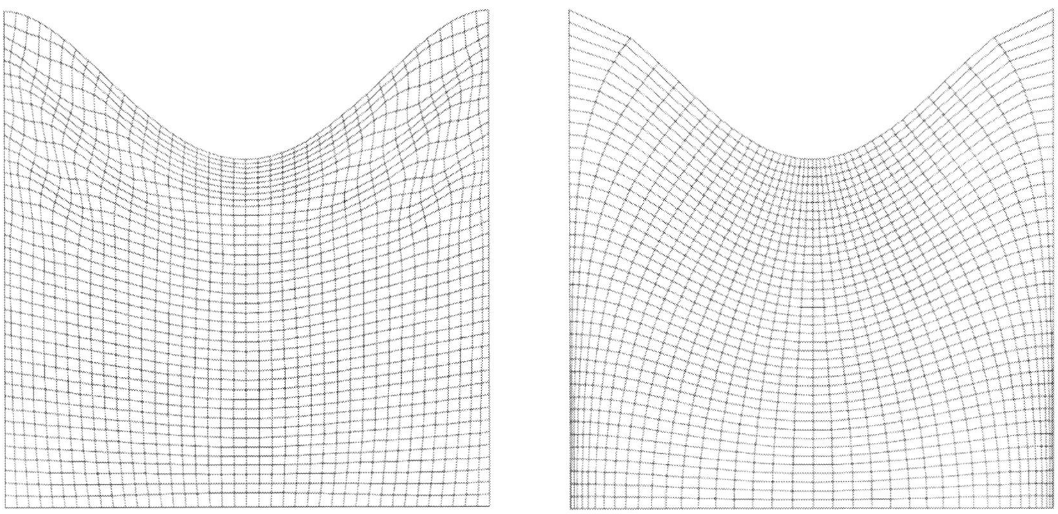

(a) PDE method after 200 iterations; (M=41, N=41) (b) present algorithm; (M=41, N=41)

Figure 5. Domain A (boundary passes (0,0), (1,0), (1,1), (0.5, 0.7), (0, 1)).

An existing PDE method [3] is compared with the present method for Domain A. Four properties are compared for evaluation of the two methods, i.e. the maximum deviation from orthogonality (O_{max}), the mean deviation from orthogonality (O_{mean}), the maximum grid aspect ratio (R_{max}), and the mean grid aspect ratio (R_{mean}).

The maximum deviation from orthogonality is computed by:

$$O_{max} = \max\left(\left|90^{\circ} - \theta_{i,j}\right|\right) \tag{4}$$

The mean deviation from orthogonality is computed by:

$$O_{mean} = \frac{1}{(M-2)(N-2)} \sum_{i=2}^{M-1} \sum_{i=2}^{N-1} \left| 90^{\circ} - \theta_{i,j} \right| \tag{5}$$

The maximum grid aspect ratio is computed by:

$$R_{max} = \max\left(\max\left(\frac{h_{\xi_{i,j}}}{h_{\eta_{i,j}}}, \frac{h_{\eta_{i,j}}}{h_{\xi_{i,j}}}\right)\right) \tag{6}$$

And mean grid aspect ratio is computed by:

$$R_{mean} = \frac{1}{(M-2)(N-2)} \sum_{i=2}^{M-1} \sum_{i=2}^{N-1} \max\left(\frac{h_{\xi_{i,j}}}{h_{\eta_{i,j}}}, \frac{h_{\eta_{i,j}}}{h_{\xi_{i,j}}}\right) \tag{7}$$

$$\theta_{i,j} = \arccos\left(\frac{g_{12}}{\sqrt{g_{11}}\sqrt{g_{22}}}\right)_{i,j} \tag{8}$$

Computed four properties of the two methods are shown in table 1.

Table 1. Comparison of a PDE method and the present method.

	O_{max}	O_{mean}	R_{max}	R_{mean}
PDE	0.474	0.203	2.88	1.43
Present	0.121	0.000252	9.97	1.76

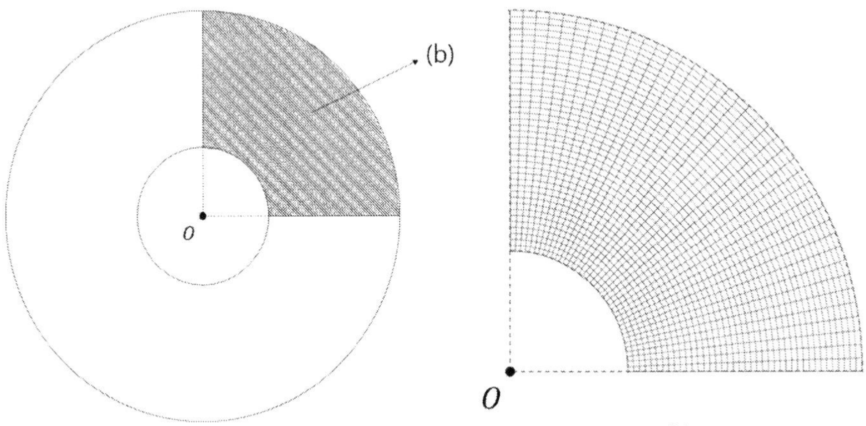

(a) (b)

Figure 6. Domain B.

Results demonstrate that the present circle-inserting method produces almost orthogonal grid. Conceptually the present method should produce perfectly orthogonal grid, but small deviation from perfect orthogonality appears due to the finite sizes of the grid cells. On the other hand both methods show similar order of aspect ratios, although the present method produces slightly larger aspect ratios for this specific case. Next the present method was applied to a circular domain, Domain B, see figure 6. The present method reproduces a quadrant, and the maximum error in radius is 0.354%, which is thought sufficiently small. This test result demonstrates a full circular domain can be expressed by four partial quadrants, each of which can be represented by the grid produced by the present method.

The present method is relatively free from choosing the internals latitudinal lines, see figure 7(a). If we want a zoomed zone in the vertical direction, an option of vertical zooming can be used, see figure 7(b). Similarly a horizontal zooming zone can be assigned, see figure 7(c). Zooming on both directions can also be used, see figure 7(d). The unzoomed and zoomed zones are separated, and the size increment experience discontinuity, and this problem can be overcome by using the transitional zooming, see figure 7(e). If a zoomed zone is attached to a border line like a coastline, the zoomed zone can situated at a side, see figure 7(f). If a domain has complicated geometry, it can be manually split and joined, see figure 7(g). The discretization between two adjacent block domains can be smoothly transitioned, see figure 7(h), and a blog site [7].

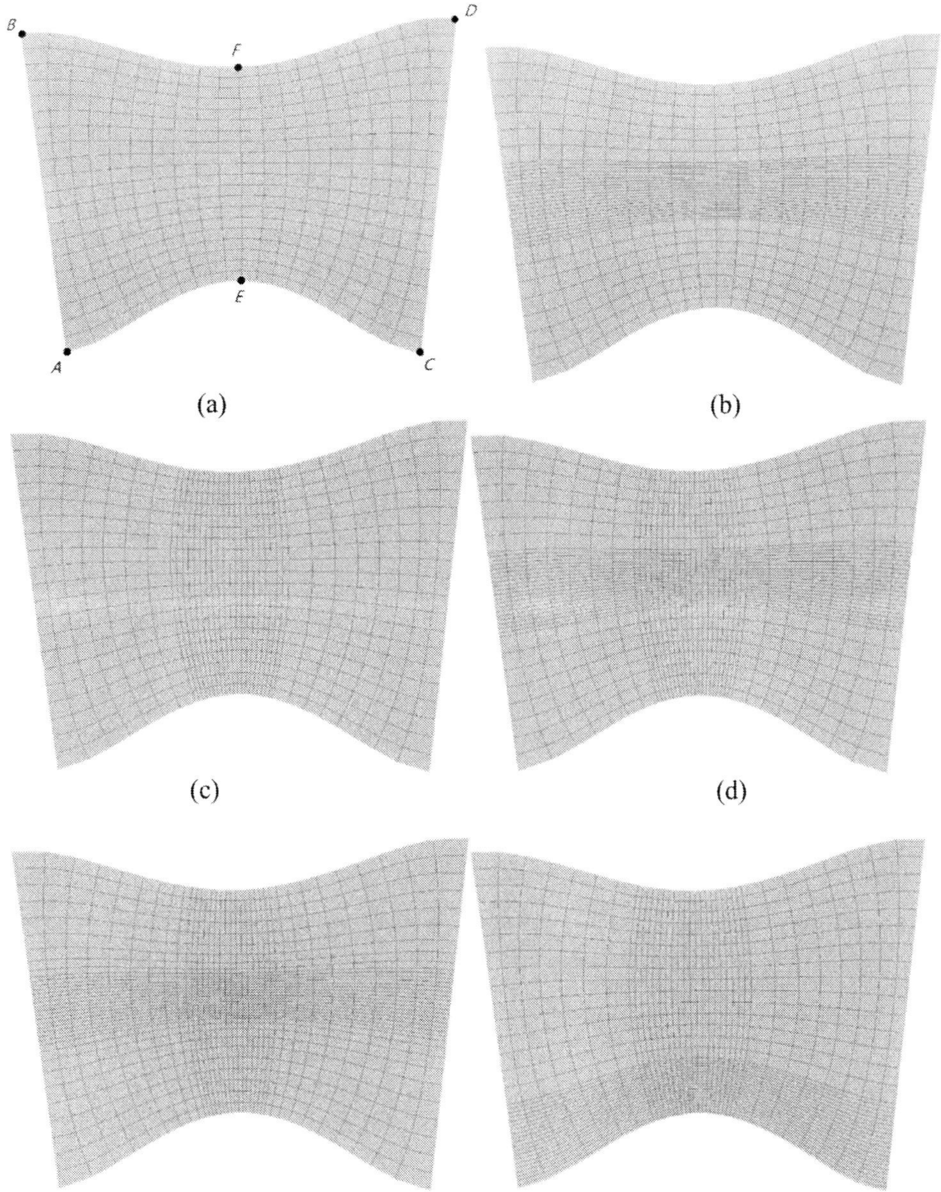

(a) (b)

(c) (d)

(e) (f)

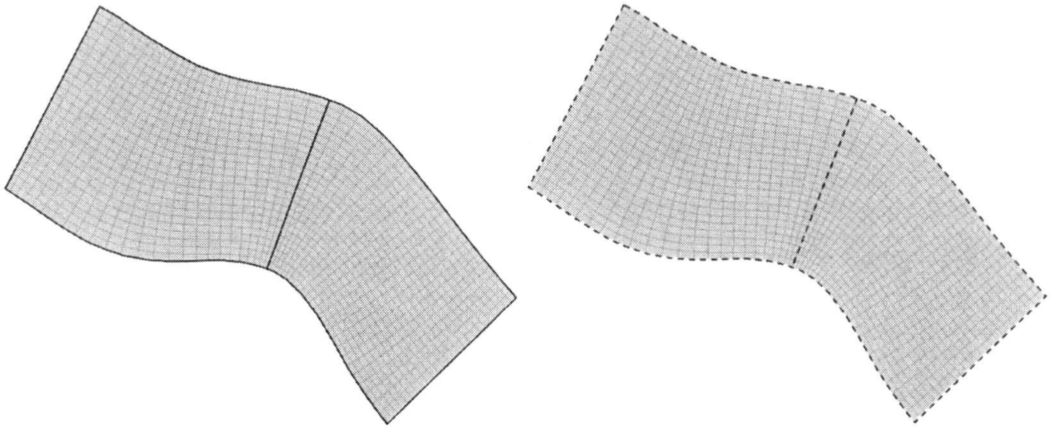

(g) (h)

Figure 7. Various applications.

4. Conclusions

A geometric iterative method to generate orthogonal curvilinear grid was proposed. The present algorithm guarantees orthogonality of the grid due to circle-inserting concept, i.e. the grid generated is almost orthogonal as inner circles are tangential to neighbouring latitudinal lines. The grid points are found successively from one end to the other end of the latitudinal lines. The method can also provide reasonable aspect ratios compared to an existing PDE method. The method is useful for resolving coastal or river zones which could be expressed as combination of Domain A or B.

Acknowledgments

This is a result of the research project titled "Development of Coastal Erosion Control Technology" funded by the Ministry of Oceans and Fisheries, Korea.

References

[1] Ryskin G and Leal L G 1983 Orthogonal mapping *J. Comput. Phys.* **50** 71-100

[2] Thompson J F, Soni B K and Weatherill N P 1999 *Handbook of Grid Generation* (Boca Raton: CRC Press)

[3] Akcelik V, Jaramaz B and Chattas O 2001 Nearly orthogonal two-dimensional grid generation with aspect ratio control *J. Comput. Phys.* **171** 805-21

[4] Chibisov D, Ganzha V, Mayr E W and Vorozhtsov E V 2005 Generation of orthogonal grids on curvilinear trimmed regions in constant time computer algebra in scientific computing 2005 *Lecture Notes in Computer Science* **3718** 105-14

[5] Conti C, Morandi R and Spitaleri R M 2005 An algebraic-elliptic algorithm for boundary orthogonal grid generation *Appl. Math. Comput.* **162** 15-27

[6] Duraiswami R and Prosperetti A 1992 Orthogonal mapping in two dimensions *J. Comput. Phys.* **98** 254-68

[7] Him K 2017 *Coastal Sediment Transport in 3 Dimensions* Retrieved from http://blog.naver.com/ cst3d

IOP Conf. Series: Earth and Environmental Science **82** (2017) 012007 doi:10.1088/1755-1315/82/1/012007

Similarity index between irrigation water and soil saturation extract in the experimental field of Yachay University, Ecuador

D V Carrera-Villacrés[1,2,3], V P Sánchez-Gómez[1], O A Portilla-Bravo[1] and D R Bolaños-Guerrón[1]

[1]Universidad de las Fuerzas Armadas ESPE, Departamento de Ciencias de la Tierra y Construcción, Carrera de Ingeniería Geográfica y del Medio Ambiente, Grupo de Investigación en Contaminación Ambiental (GICA), Sangolquí, Pichincha, Ecuador
[2]Universidad Central del Ecuador - Facultad de Ingeniería Geología Minas Petróleo y Ambiental (FIGEMPA). Carrera de Ingeniería Ambiental. Quito, Pichincha. Ecuador

E-mail: dvcarrera@espe.edu.ec

Abstract. Soil monitoring is a job that demands a lot of time and money. therefore, measuring the same parameters in the water becomes simple because it can be done in situ. The objective of this work was to find a similarity index for the validation of mathematical correlation models based on physicochemical parameters to verify if there is a balance between irrigation water and soil saturation extract in the experimental field Yachay that is known as the city of knowledge that is located in Imbabura province, Ecuador, for which, the sampling of water was carried out in two representative periods (dry and rainy). Sampling of 10 soil profiles was also performed, covering the total area; these samples were obtained results of Electrical Conductivity (EC), pH and total dissolved salts (TDS). With correlation models between soils and water, it is possible to predict concentrations of elements in the irrigation water. It was concluded that there is a balance between soil and water, so that the salts present in the soil are highly soluble, in addition, there is a high probability that the elements in the irrigation water are in the soil. In sample water, the same concentrations were found in the soil, at their saturation point, and very close to the field capacity.

1. Introduction

The experimental field of Yachay is located in San Miguel de Urcuquí in Imbabura, Ecuador. It is a research and higher education center and also a developing industrial park, where one of the main topics is agricultural research. About 1 000 hectares will be destined to the development of experimental agriculture and agribusiness, which will support the change of the productive matrix in the Ecuador [1].

The relationship between soil and irrigation water represents one of the major limitations for agricultural production in many parts of the world [2]. In response to this limitation, models have been generated to predict the dynamics and chemical, physical and biological interactions that exist in an agricultural system.

The characteristics of irrigation water, in terms of salinity level, have a direct impact on the composition of salts present in the soil due to retention capacity of this. Contrariwise, inversely there is also an important relationship, since the soil presents highly soluble compounds, which will be present mainly in the waters coming from agricultural drains [2].

A case study carried out in Northeast of China stablishes the problem in the farmland due to the accumulation of salts, and performs a salinity analysis in the irrigation water to identify soil composition, based on the Close soil-water relationship exists; specifically uses data from water conditions to perform a model of soil salinity dispersion [3].

Mathematical correlation models when evaluating the association or independence degree between tow quantitative variables [4] become a tool used as a quality control method, this is due to the fact that they facilitate the monitoring in field of measurements of parameters that are of interest for the agricultural activity, obtaining in a fast and efficient way the quality of soil and water [5].

The chemical composition and physical-chemical properties of the soil solution are important for the investigation of the processes that occur in it, as well as for crop nutrition [6], in turn, is important as well to analyse the irrigation water, so that it can be identified if there is a balance in the water-soil relationship.

Content of salts in soil or water can be estimated by the electrical conductivity (EC), both in a sample of water and soil extract. Currently one of the classic problems of land degradation has been to control, prevent or improve soils affected by salinity [7].

The problem with salinity can be intensified with other sources as the use of fertilizers and the quality of the water of irrigation. The relative significance of the contribution of each salt supply source depends on soil conditions, drainage effectiveness, irrigation water quality, land/soil overexploitation and agronomic management practices [6].

There are two chemical indicators of salinity in the soil, EC and pH [8, 9]; Whose parameters allow to separate the soils affected by salinity in three groups (table 1)

Table 1. Classification of soils by salinity.

Clasification	pH	EC	Process
Saline	<8.5	> 4	Salinization
Sodic	>8.5	<4	Sodification
Sodium-Saline	<8.5	>4	Salinization – Sodification

Source: [8, 9]

The accumulation of soluble salts occurs when annual precipitation is lower than evaporation of soil including evapotranspiration of plants, which is associated to a low drainage soil. Concentration of salts causes a soil management problem since its high concentrations of salts generate problems for development of susceptible crops [10].

Percentage of these total dissolved salts changes according to concentration of extract soil/water, depending on the correlation obtained it is possible to predict which percentage of salts pass from solid phase of soil to liquid phase of soil by increasing soil-water ratio. Salinity of irrigation water determines salinity of soils affecting to crops. TDS is a parameter of soils and irrigation water to determine theirs quality and theirs effect over crops [11].

Relationship physic and chemical between soil and water are very useful in formulation of river basin management and management plans, a process to which current territorial ordering is oriented, taking into account basin as a basic unit of environmental development. In addition, this knowledge will provide criteria with scientific and technical justification for development of vegetation exploitation activities and their appropriate use in development projects in which it is involved, either for protection or exploitation purposes [12].

By determining that same soil characteristics can be identified in water, analysis costs required in agricultural soil studies can be minimized. Water is an indicator of lower cost and easier to characterize than soil, which is an advantage when carrying out studies that require the identification of physic-chemical characteristics, hence the importance of this study.

The aim of this work was to determine similarity index for validation of mathematical correlation models of pH, EC and TDS, in order to verify if there is a balance between the irrigation water and saturation extract of the soil in experimental field Yachay.

2. Materials and methods

2.1. Sampling: water

Sampling of water was carried out in basin along Ambi River, in tributaries, ditches and ravines close to YACHAY City of Knowledge. Samplings were performed at two representative times, determined from precipitation data provided by National Institute of Hydrology and Meteorology of Ecuador, INHAMI.

First sampling was conducted in October 2014, 25 samples were taken (dry season) and in the second sampling, April 11, 11 samples (rainy season) were taken (figure 1). Samples were taken according to NTE INEN 2169 (1998) [13, 14].

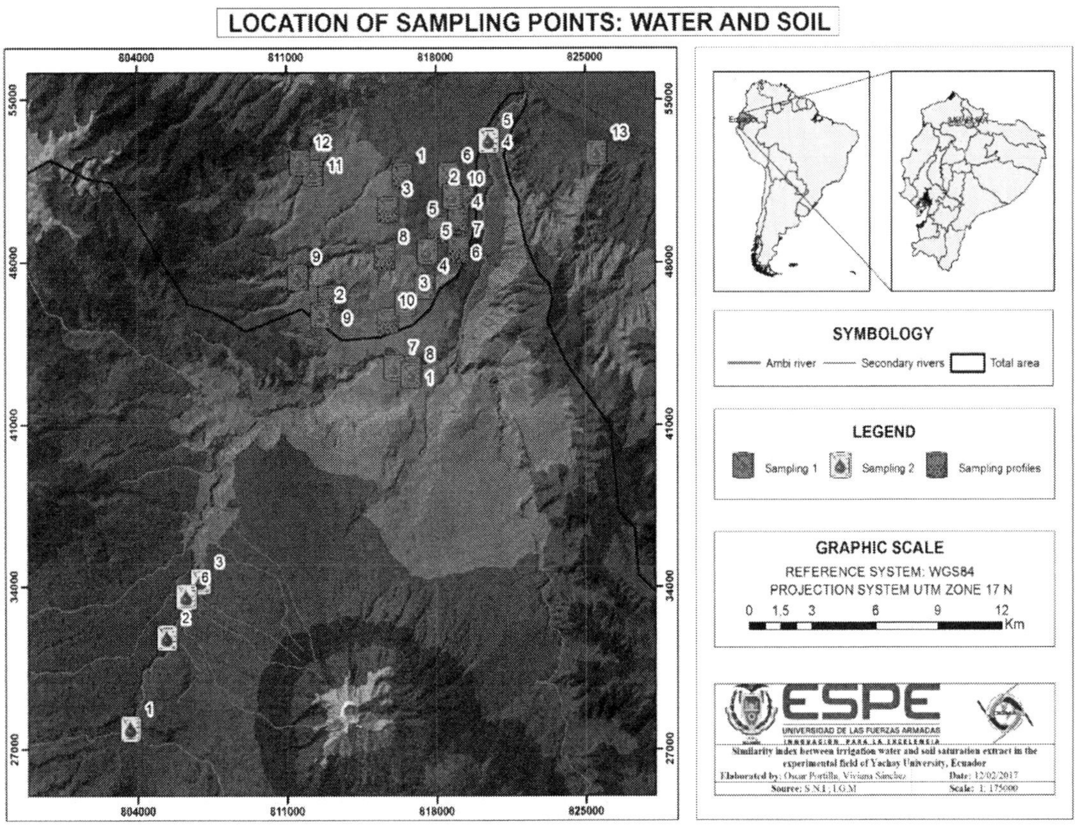

Figure 1. Location map of the sampling points made on the experimental field area Yachay.

2.2. Sampling: soil

Ten soil profiles were taken covering most of Yachay (figure 1). Locations of points were done using zig-zag method according to NOM-021-SEMARNAT-2000 [15]. Soils were monitored covering the entire Northern extension corresponding to agricultural fields to be irrigated, with a simple sample type that represents specific conditions of population in the time that was collected [16]. Samples were taken at depths of 20 cm soil profiles, up to 40 cm depending on soil uniformity in order to determine the concentrations of the different parameters in ratios 1:5, 1:10 and the saturation extract (1:0.2 ≈ 1:0.6).

2.3. Analysis method
The analysis of the water and soil samples were done based on the Standard Methods 20th edition [17], following the procedure described therein, for each water sample were analysed 13 parameters and 16 parameters for soils, from which 3 parameters (pH, EC, TDS) were chosen for correlation analysis. Water samples from first sampling were sent to the HAVOC Analytical Laboratory in October 2014, while samples from the second sampling were analysed at the Environmental Laboratory of the University of the Armed Forces (ESPE) in 2015.

2.4. Correlation models
To generate the correlation models, a spreadsheet was used, where both the mathematical dependence equations between the values obtained in the different parameters of the soil and water samples and the linear regressions between the first and third degree equations were determined.

Clouds of points or dispersion diagram for each variable were obtained in soil and water. Thus, it can interpret intensity of relationship between both variables. Additionally, was obtained a correlation coefficient; when dispersion diagram shows a cloud of highly clustered points around a line, that is mean a strong linear relationship. Correlation coefficient is measure of that relationship [18].

In addition, we proceeded to validate the Model taking as a criterion of goodness of fit the Similarity Index (SI).

2.5. Elaboration of map of location
Thematic maps of location of study area were made by an elaboration of the geodatabase that were entered into GIS software, moved to the geospatial information. Finally the map was generated; Part of the methodology is proposed by Soriano [20].

For analysis of the results, only saturation extract was used because it is considered as medium that gives the best representation of current conditions of soil with respect to natural conditions [21] and this was demonstrated in Yachay.

3. Results
Water sampling was carried out along Ambi river, 2 samples were obtained; at the same time, 10 soil profiles were sampled throughout Yachay's area (figure 1).

Data obtained from the analysis of soil saturation extracts and the first sampling of irrigation water is indicated in table 2.

Table 2. pH values, electrical conductivity and total dissolved salts in water samples and soil profiles.

SAMPLING 1 – WATERS							
N°	**pH**	**EC ($dS\ m^{-1}$)**	**TDS ($mg\ L^{-1}$)**	**P-PO_4 ($mg\ L^{-1}$)**	**N-NO_3 ($mg\ L^{-1}$)**	**East**	**North**
1	8,37	1,4	1048	0,50	0,30	816808	43382
2	7,68	0,5	360	0,50	0,30	812633	45710
3	7,44	0,6	486	0,50	0,30	816561	46224
4	8,45	0,9	700	0,50	0,30	817487	46958
5	8,46	0,9	690	0,50	0,30	817600	48469
6	8,06	2,0	1581	0,50	0,30	818602	51719
7	8,12	1,3	970	0,50	0,30	816044	43447
8	8,18	1,2	950	0,50	0,30	816846	43134
9	8,26	1,2	795	0,50	0,30	811535	47344
10	8,41	1,8	1143	0,60	0,30	818802	50836
11	8,26	3,2	1741	0,50	0,30	812269	51845

12	8,49	4,8	2337	0,50	0,30	811641	52281
13	9,37	0,3	190	0,50	0,30	825545	52697
SOIL SAMPLING PROFILES							
N°	**pH**	**EC ($dS\ m^{-1}$)**	**TDS ($mg\ L^{-1}$)**	**$P-PO_4$ ($mg\ L^{-1}$)**	**$N-NO_3$ ($mg\ L^{-1}$)**	**East**	**North**
1	7,11	1,06	621,23	9,66	6,66	816439	51708
2	7,28	0,70	470,48	10,33	6,62	819297	50775
3	6,71	0,78	488,15	8,09	6,74	815782	50247
4	7,57	1,10	655,75	11,47	6,57	819064	49787
5	7,55	0,82	501,33	11,39	6,57	818109	49402
6	7,82	0,64	446,78	13,00	6,52	819001	48600
7	8,03	0,64	422,70	11,97	6,47	819082	48533
8	8,16	0,92	616,73	14,12	6,45	815610	48209
9	8,2	0,83	531,87	14,37	6,44	812971	46466
10	7,06	0,77	503,32	9,47	6,67	815744	45453

Mathematical correlation models are used for determinate water quality through systematic measurements of EC, data for obtaining correlation models were processed using a second and third degree polynomial equation that passes through the origin of Coordinates y=b1 X+b2 X2, y=b1 X+b2 X2+b3 X3 since these showed the best values of correlation coefficient, and according to [16] when there are values of "r" between 0.8 is a correlation "Very High" (table 3).

Table 3. Correlation coefficients.

r=1	**Perfect Correlation**
$0,8 < r < 1$	Very High Correlation
$0,6 < r < 0,8$	High Correlation
$0,4 < r < 0,6$	Moderate Correlation
$0,2 < r < 0,4$	Low Correlation
$0 < r < 0,2$	Very Low Correlation
r=0	Null Correlation

Source: [19]

Two different samples were taken for irrigation water, and the physical-chemical parameters were identified for both cases, however, the best correlation was obtained with values of the first, in table 4 is presented the correlation equations as well as the R^2, calculated for both samples.

Table 4. Equations and correlation coefficients. Extracts of saturation vs. Irrigation water.

Sampling	**Parameter**	**Order**	**Equation**	**R^2**	**Correlation**
1	pH	1 order	$y = 0,8553x + 0,4729$	R^2=0,7815	HIGH
		2 order	$y = -0,2777x^2 + 5,5113x - 18,987$	R^2=0,8306	VERY HIGH
		3 order	$y = -0,8725x^3 + 21,706x^2 - 178,43x + 492,13$	R^2=0,8857	VERY HIGH
	EC	1 order	$y = 0,1054x + 0,664$	R^2=0,895	VERY HIGH
		2 order	$y = -0,0215x^2 + 0,2118x + 0,5819$	R^2=0,9657	VERY HIGH
		3 order	$y = -0,0002x^3 - 0,0202x^2 + 0,2095x + 0,5829$	R^2=0,9657	VERY HIGH

	STD	1 order	$y = 0{,}1137x + 412{,}18$	R^2=0,9704	VERY HIGH
		2 order	$y = -2E\text{-}05x^2 + 0{,}1565x + 392{,}41$	R^2=0,9808	VERY HIGH
		3 order	$y = -2E\text{-}08x^3 + 5E\text{-}05x^2 + 0{,}0836x + 410{,}98$	R^2=0,9852	VERY HIGH
2	**Parameter**	**Order**	**Equation**	**R^2**	**Correlation**
	pH	1 order	$y = 0{,}7266x + 2{,}5664$	R^2=0,7886	HIGH
		2 order	$y = 0{,}2994x^2 - 3{,}7715x + 19{,}389$	R^2=0,8133	VERY HIGH
		3 order	$y = 0{,}6993x^3 - 15{,}345x^2 + 112{,}6x - 268{,}4$	R^2=0,8341	VERY HIGH
	EC	1 order	$y = 1{,}6209x - 0{,}4348$	R^2=0,6451	HIGH
		2 order	$y = -0{,}507x^2 + 2{,}4996x - 0{,}8031$	R^2=0,6463	HIGH
		3 order	$y = 56{,}318x^3 - 144{,}73x^2 + 123{,}06x - 33{,}725$	R^2=0,7882	HIGH
	STD	1 order	$y = 2{,}6705x - 674{,}25$	R^2=0,6898	HIGH
		2 order	$y = -0{,}0071x^2 + 10{,}354x - 2718{,}8$	R^2=0,7076	HIGH
		3 order	$y = 0{,}0005x^3 - 0{,}7294x^2 + 389{,}9x – 68446$	R^2=0,9886	VERY HIGH

3.1. Electrical conductivity of irrigation water and soil saturation extract from the experimental field Yachay.

EC is measured to know if the amount of soluble salts is enough to affect the development of plants and level of water absorption. In saturation extracts of Yachay profiles an average EC of 0,83 dS.m-1 (figure 2) was identified. The regulation (NOM021-RECNAT, 2000) [15] proposes a series of effects related to EC. Considering that measurements obtained from soil extracts are lower than 1.0, which means that this soil is non-saline or with negligible salinity. According to regulations TULAS (2014) Book VI Annex 2 [14] and according to the soil quality criteria for agricultural use, the maximum allowable value of 2 dS.m-1 for this parameter is considered.

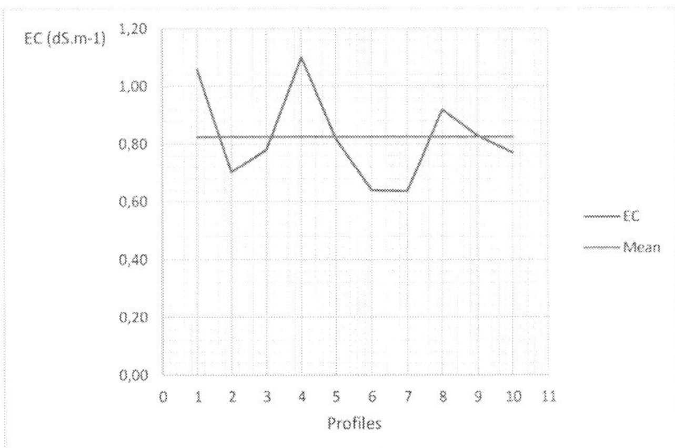

Figure 2. Values of EC in soil profiles.

EC in water samples averaged 1.5 dS.m-1 (figure 3).

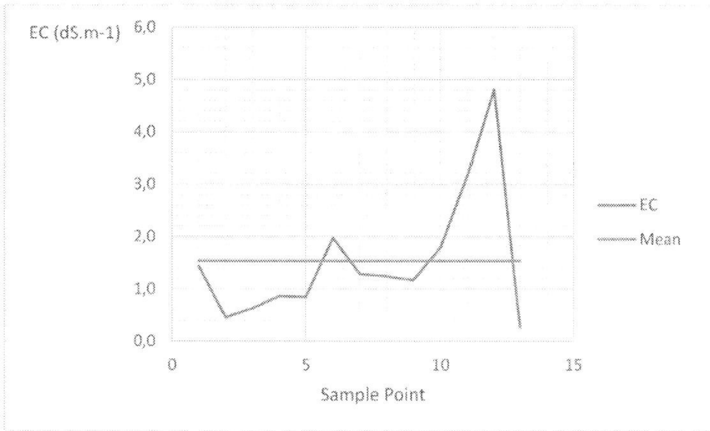

Figure 3. Values of EC in water.

After this correlation the equations were obtained (figure 4). This analysis resulted in a high correlation.

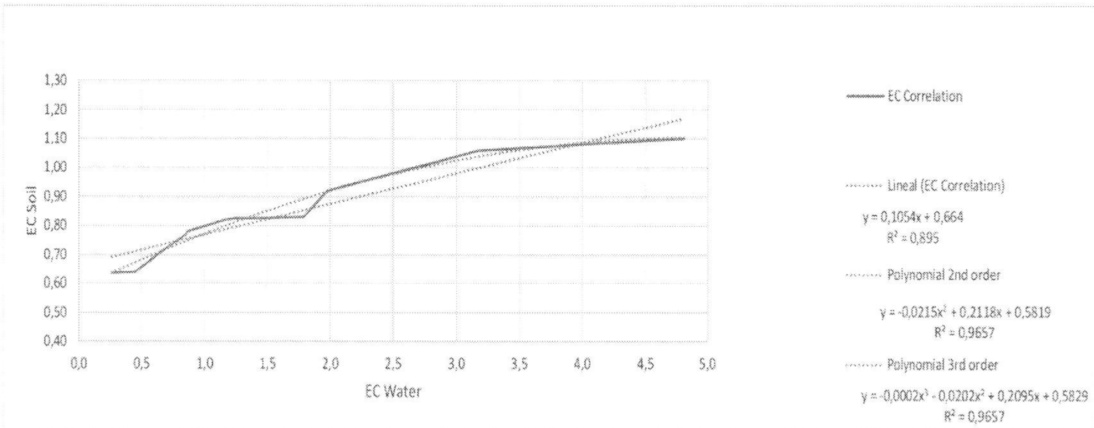

Figure 4. Correlation of electrical conductivity in water samples and soil profiles.

3.2. pH of irrigation water and soil saturation extract of the experimental field Yachay.

In 1909 a Danish chemist Serenasen defined the hydrogen potential (pH) as the negative logarithm of the molar concentration (more exactly of the molar activity) of the hydrogens, ie: $pH = -\log [H^+]$

pH influences to soil in different aspects; the main relevance is availability of nutrients, such as (Phosphorus, Potassium, Iron, Copper, Boron, among others). At pH levels higher than 6.5 in soils, phosphorus with calcium precipitates are present and when acid soils with pH below 6.0 are present, reactions with iron, aluminium and manganese are present, while phosphorus is not available [22].

Results obtained in both soils and waters are shown in figures 5 and 6, resulting in an average pH for soils of 7.54 and for water of 8.27, in summary, medium alkaline soils [15]. According to regulations TULAS (2014) Book VI Annex 2 and the soil quality criteria for agricultural use, the maximum permissible values between 6-8 for pH are considered [14].

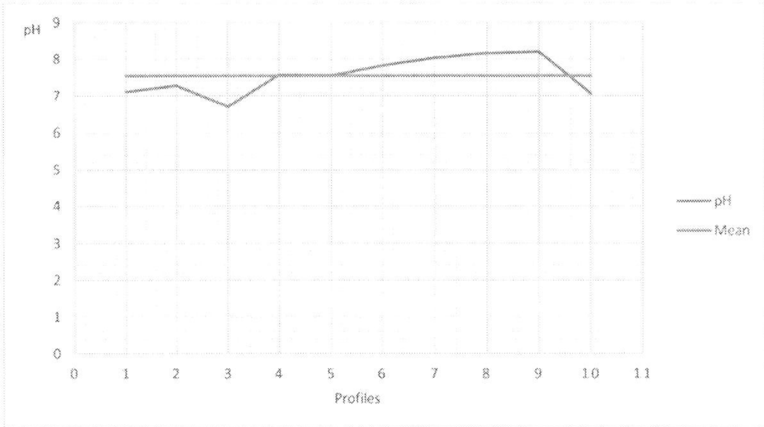

Figure 5. pH in soil profiles.

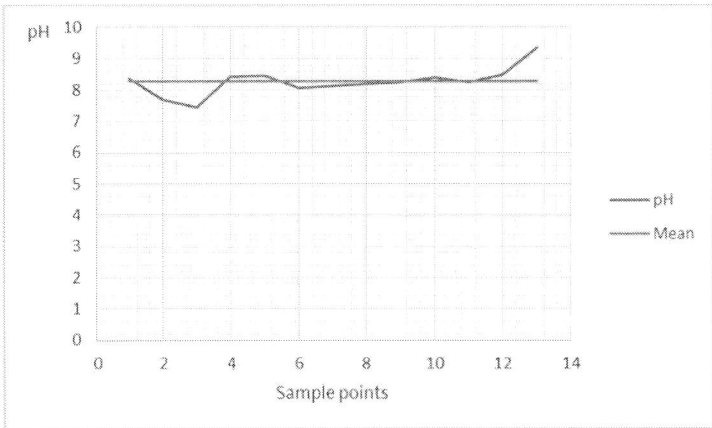

Figure 6. pH in water samples.

Linear, second and third order correlation equations were calculated. Analysis of pH in waters and soils resulted in a high correlation (figure 7).

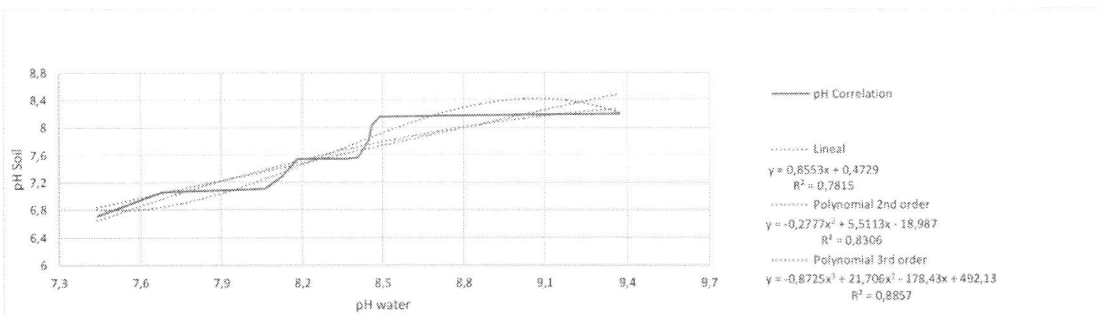

Figure 7. Correlation of pH in water samples and soil profiles.

3.3. TDS of irrigation water and extract of soil saturation of experimental field Yachay.

The analysis of soluble salts in soils and waters tries to establish whether there are sufficient quantities of salts in them to interfere with normal germination of seeds, with growth of plants or with intake of water through themselves [23].

Total Dissolved Salts were 525 mg L^{-1} for soils and 999 mg L^{-1} for water (figures 8 and 9). Both normative [14] and [15] propose that TDS is a parameter that is closely related to EC and pH properties. Like other parameters, mathematical modeling was performed with first, second and third degree equations, resulting in a very high correlation between variables (figure 10).

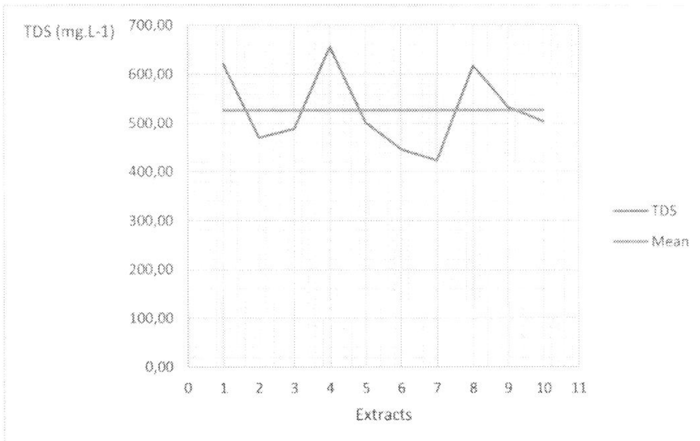

Figure 8. TDS in soil profiles.

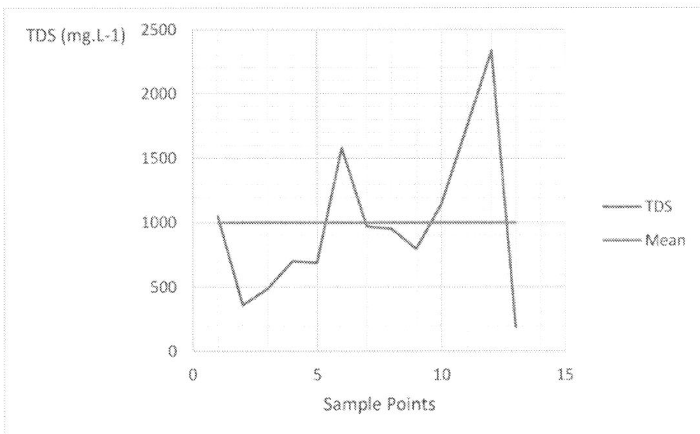

Figure 9. TDS salts in water.

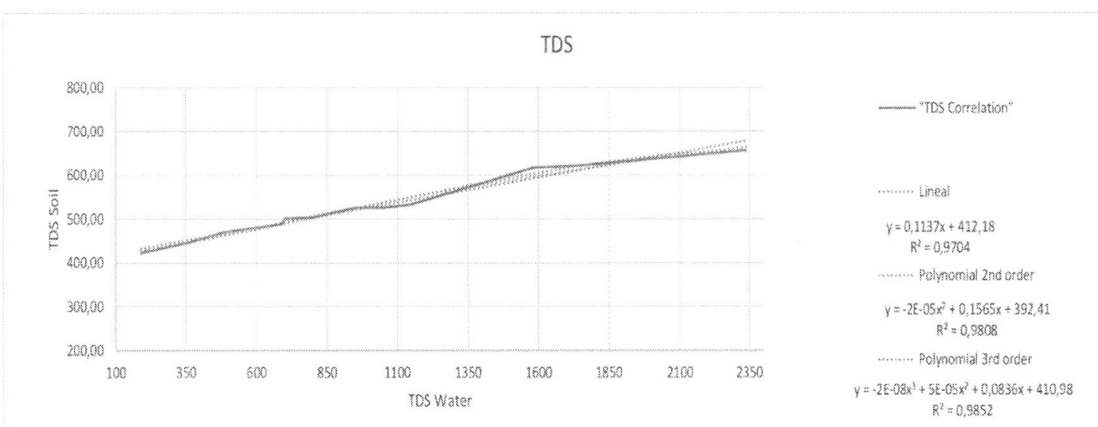

Figure 10. Correlation of dissolved total salts in water samples and soil profiles.

Correlation obtained has R^2 from 0.78 to 0.985, which means that there is a balance between water and soil. This is an indicator that the type of salts present in the soil is very soluble, for this reason correlation is high.

4. Conclusions

Due to high correlation between the analysed parameters in soils and waters, it is determined that there is a balance between the two and it is possible to measure in water irrigation and to predict in soils, and thus can reduce costs and optimize time, so as to facilitate analysis of physicochemical parameters in soils and waters.

Mainly parameter in water is the pH that it is measured in aqueous solutions, since a high correlation was obtained, this can be translated into a minimization of costs in the processes of water and soil characterization, since it is only necessary analyse one of these.

Salts present in soil are very soluble, so that the same ions can be found in matrices, soil and water. With this analysis it can be predicted that soil studied in the future will be poor, because the irrigation water leaks all nutrients. According to the slope that has Yachay, these nutrients will be taken to the lower areas of basin that pass through this sector as is the case of Ambi River.

In experimental field of Yachay University it is possible to measure elements in irrigation water with a high probability that these elements are also in similar concentrations in soil, at its point of saturation, and very close to field capacity.

Acknowledgments

Universidad de las Fuerzas Armadas – ESPE, Departamento de Ciencias de la Tierra y Construcción, Carrera de Ingeniería Geográfica y del Medio Ambiente.

References

[1] Yachay 2016 Yachay, la ciudad del conocimiento. Qué es Yachay?. http://www.yachay.gob.ec/yachay-la-ciudad-del-conocimiento/
[2] Letey J and Otros Y 2011 Evaluation of soil salinity leaching requirement guidelines *Agric. Water Manage.t* **98** 502-6
[3] Chen W and Otros Y 2010 Evaluating salinity distribution in soil irrigated with saline water in arid regions *Agric. Water Manage.* **97** 2001-8
[4] Canela A and Monge L 2007 Cómo realizar “paso a paso” un contraste de hipótesis con SPSS para Windows: (III) Relación o asociación y análisis de la dependencia (o no) entre dos variables cuantitativas Correlación y regresión lineal simple. Fabis, p. 15
[5] Rodríguez F and Otros Y 1992 Caracterización hidroquímica y control de la calidad de las aguas del Carso de la cuenca Zapata. Taller Internacional sobre cuencas experimentales en el Karst, p. 9
[6] Otero L, Francisco A, Gálvez V, Morales R, Sánchez I, Labaut M and Rivero L 2007 Ministerio de Agricultura, Instituto de Suelos. Caracterización y evaluación de la salinidad. Capdevila, Boyeros, Cuba: CENDA
[7] Kovda A V 1964 *The Origine of Saline and Their Regime* vol 1 (USSR Academy of Sciences Publishing House)
[8] Kochba M, Ritvo G and Avnimelech Y 2004 The effect of municipal solid waste compost (msw) on the replacement of sodium in sodic soil models *Soil Sci.* **169(8)** 567-72
[9] Mueller T G, Mijatovic B, Sears B G, Pusuluri N and Stombaugh T S 2004 Soil electrical conductivity map quality *Soil Sci.* **169(12)** 841-51
[10] Solís J N 2000 Fundamentos de Edafología (3ra Edición). Costa Rica: EUNED.
[11] Torres A and Acevedo E 2008 El problema de la salinidad en los recursos suelo y agua que afectan el riego y cultivos en los valles de Lluta y Azapa en el norte de Chile IDESIA 31-44
[12] Villegas J C 2004 Análisis del conocimiento en la relación agua-suelo-vegetación para el departamento de Antioquia. EIA, n/d

IOP Conf. Series: Earth and Environmental Science **82** (2017) 012007 doi:10.1088/1755-1315/82/1/012007

[13] NTE INEN 2169 (1998): Agua Calidad del agua. Muestreo. Manejo y Conservación de Muestras
[14] TULAS Texto Unificado de Legislación Ambiental Secundaria 2014 Obtained online: http://www.industrias.ec/archivos/CIG/file/CARTELERA/Reforma%20Anexo%2028%20feb%202014%20FINAL.pdf
[15] NOM021-RECNAT-2000 Norma oficial mexicana
[16] Sepúlveda T V, Velasco J A and De la Rosa D A 2005 *Suelos Contaminados por metales y metaloides: Muestreo y Alternativas para su remediación (1era edición)* (México: INE-SEMARNAT)
[17] APHA 1995 *Standard M. D. Methods for the Examination of Water and Waste Water* 19th ed (Baltimore: American Public Health Association-Port City Press)
[18] Katz R D 2013 Correlación y Regresión, Santa Fe Argentina: UNR
[19] Bisquerra R 1989 Introducción conceptual al Análisis Multivariable Un enfoque informático con los paquetes SPSS-X, BMDP, LISREL y SPAD (Vol I y Vol II) Barcelona: PPU.
[20] Soriano M A 1995 Methodology for elaboration of a hazard map of alluvial doline. Dpto. de Geología. Facultad de Ciencias. Universidad de Zaragoza. Enseñanza de las Ciencias de la Tierra **3.3** 184-90
[21] SIAN Sistema de Información Agrícola Nacional 1985 FONAIAP Divulga. Revista. Venezuela. Obtained online: http://www.sian.inia.gob.ve/repositorio/manualessuelos/metodos_analiticos_suelos/VIII.pdf
[22] Hassan 1998 f. Microirrigation management and maintenance. Agro Industrial Management Fresno CA 233
[23] Richards 1959 Suelos Salinos y Sódicos Personal de laboratorio de salinidad de los Estados Unidos de América *Manual de Agricultura* **60** 172

Capability of microalgae for local saline sewage treatment towards biodiesel production

K-C Wu[1], Y-H Yau[1,2] and K-C Ho[1,2,3]

[1]School of Science & Technology, The Open University of Hong Kong, Hong Kong, China
[2]Centre for Research in Environmental Science, The Open University of Hong Kong, Hong Kong SAR, China

E-mail: kcwu@ouhk.edu.hk

Abstract. Seawater flushing was introduced in Hong Kong since 1950's. High salinity has an inhibitory effect on nitrification and biological phosphorus uptake of microorganisms. Therefore, saline sewage has impact on traditional biological wastewater treatment. Saline conditions of domestic wastewater then pose opportunity to use algal technology in wastewater treatment. During the treatment (phycoremediation), biodiesel can be produced. This study aims to give an in-depth investigation and development on application of local microalgal strains on biodiesel production. *Dunaliella tertiolecta* was selected the appropriate algal species with high potential for phycoremediation then biodiesel production. *D.tertiolecta* was further investigated by optimizing its growth in different process condition in preliminary effluent as based medium. The optimized process condition were acclimated culture with medium initial cell number (5.0×10^5 cells mL^{-1}), under 5% CO_2 aerations in preliminary effluent adjust to 15 psu (denoting practical salinity unit). Results showed that lipid content increased from 30.2% to 42%, and biomass productivity reached 463.3 mg $L^{-1}day^{-1}$ by Fatty acid Methyl Ester (FAME) profile was found for biodiesel production in optimized stage. The treatment period of preliminary effluent was shortened from 15d in original design (un-acclimated culture, low initial cell number (5.0×10^5 cells mL^{-1}), without CO_2 aeration) to 4d.

1. Introduction

In the twenty-first century, the rapid growth of human development has increased large input of nitrogen and phosphorus into receiving water bodies and caused eutrophication, which resulting in red tide occurrence in coastal regions [1-7].

By means of different types of treatment processes and advanced technologies, pollutants, toxic materials and bacteria inside the sewage was removed to a level meeting the environmental standards. According to Hong Kong Drainage Services Department (DSD) sustainability report (2014-15), DSD is operating 297 sewage treatment facilities, including 70 sewage treatment works (STWs) and 277 sewage pumping stations. Among the treatment with these processes of treatment, the most capacity were Stonecutters Island, Sha tin and Tai Po in terms of treatment flow. Their treatment processes often up to secondary/biological treatment except for Stonecutter Island.

Among the STWs, most of the discharge standard of remaining nutrient level in terms of Total nitrogen (TN) and Biological Oxygen Demand-5 days (BOD_5) was 20 mg L^{-1} in the effluent. Part of them even not required for nutrient removal in the standard. Therefore, the effluents from these STWs contain a certain amount of nutrient.

On the other hand, most of the STWs treat saline sewage except Shek Wu Hui STWs. As seawater flushing was introduced into the catchment areas of large-scale municipal treatment works in local area, for example, Tai Po and Sha Tin STWs in 1995 and 1996, respectively. Nevertheless, it was believed that the kinetic rates of nitrification and denitrification in secondary treatment were further affected by the variation (daily and seasonal) in salinity levels [8-10]. Some studies showing that good nutrient removal reached when salinity of influent were low (0.3 psu to 2 psu NaCl) but showed effect at high salinity levels of 5psu [11, 12]. Thus, the effect of interaction between salinity and pollutants in sewage increased the difficulty of saline wastewater treatment. Saline sewage has salt effect, lowering overall effective microbial metabolic rate and selectivity on nutrient removal. Therefore, microbial treatment capacity on per unit volume of saline wastewater is not high [8, 10-12].

Phycoremediation by using marine microalgae may be one of the efficiency and energy effective way to reduce these challenges [4, 13, 14]. Phycoremediation is a type of bioremediation by using microalgae metabolism [15, 16]. To improve treatment efficiency of STWs, wastewater can be acted as a nutrient source for microalgal biomass production. The biomass produced from the process can further produce bioenergy such as biodiesel. Microalgae convert nutrients or pollutants to organic materials (proteins, carbohydrates, fats) for growth through photosynthesis. Among these materials, triacylglycerols (TAG) and free fatty acid within microalgal cells are main raw materials for biodiesel production [17-22]. On the other hand, in microalgal biodiesel production, cultivation costs accounted for more than 70 percent of total costs, which is the major constrains of its industry development [22, 23]. Therefore, to complement both of saline wastewater treatment and microalgal biodiesel production, wastewater can used as nutrient sources for microalgae cultivation [14, 20, 24-28]. For example, in our previous study, *D.tertiolecta* was cultivated in the preliminary effluent successfully by acclimated cultivation and further biodiesel production [14, 28] .

In this study, preliminary effluent from Tai Po Sewage Treatment Works (TPSTW) and a local isolated microalgal species (*D.tertiolecta*) was used as a model to evaluate potential of phycoremediation associated with biodiesel production in local area. The selected species was adopted to cultivate in the wastewater based medium by acclimated cultivation in previous study[14, 28]. The following investigation then focuses on minimizing the cultivating days and maxizing the biomass production. Cultivating days indicate the day of the microalgal strain reach the maximum capacity and nutrient removal in specific conditions. The cultivation days is highly related to hydraulic retention time of microalgal wastewater treatment when the biomass productivity was important for biodiesel production.

Concentrated cultures require smaller cultivation areas, or reduced cultivation period, both have potential advantages for microalgal biomass production. Some studies stated that higher of algal density, better of growth and nutrient removal efficiency [29-33]. If the seeding cell density of selected species is not high enough, the microalgal growth will significantly affected by indigenous microorganisms and was not able to survive eventually from non-sterile wastewater [33]. However, in a relatively turbid medium, algal cells in wastewater may limit by light, carbon or nutrients rather than by plant, which called self-shading [31, 34]. Self-shading limits the operational maximum cell capacity. For example, *Chlorella* spp., in treating wastewater [35] have been investigated in high algal density for auto inhibitory effects [36]. A trade-off point between algal density, microalgal growth and treatment efficiency should consider in an algal system [37, 38].

2. Material & methods

2.1. Wastewater collection

The preliminary effluent (PE) in this study was collected directly from the Tai Po Sewage Treatment Works (TPSTW) in Hong Kong which studied in previous study [14]. PE was the discharge after screening treatment, solids which larger than 6 mm (diameter) were removed from raw wastewater.

The wastewater samples were taken in summer (June-September). Wastewater samples were stored in 10 liter buckets at 4°C and used within 2 hours after sampling before each experiment. The physicochemical characteristics of PE were shown in table 1. The BOD5-to-PO4 ratio is found relatively high. The PO4 to TP, BOD5: TP ratio and TN: TP ratios are found at lower range for domestic wastewater. The PE used in this study is then considered to be favorable for available nutrient removal with enhanced growth of microalgae.

Table 1. Physicochemical characteristics of preliminary effluent.

Characteristics	Range	Average	Unit
pH	5.72-5.8	5.76	/
Dissolved oxygen	3.35-3.97	3.66	mg L^{-1}
Turbidity	679.5-707	693.3	NTU
Total dissolved solid	6.33-6.83	6.58	mg L^{-1}
Total suspended solid	330-460	395	
Salinity	5.8-8	6.9	psu
Total nitrogen (TN)	55-60	57.5	mg L^{-1}
Ammonium-Nitrogen (NH_4^+)	19-28	23.5	
Nitrate- Nitrogen (NO_3^-)	5.5-13	9.25	
Total phosphorus (TP)	13.6-20.1	16.85	
Ortho-Phosphate (PO_4^{3-})	11.8-18.7	15.25	
Biological Oxygen	240-302	271	
Demand(BOD_5)	255-350		
Chemical Oxygen	6.61-8.95	302.5	
Demand(COD)	0.87-0.93		
N:P ratio*	15.0-17.6	7.78	/
PO_4:TP ratio	16.1-20.3	0.9	
BOD_5:TP ratio		16.3	
BOD_5: PO_4 ratio		18.2	
Metallic ions			
Cadmium	<0.01	<0.01	
Chromium	<0.01	<0.01	
Copper	0.06-0.12	0.09	mg L^{-1}
Iron	0.05-0.17	0.11	
Lead	<0.01	<0.01	
Zinc	<0.01	<0.01	

*Atomic ratio calculated by TN and TP

2.2. Algal culture and growth conditions

The microalgal strain was isolated from local algal blooming samples from coastal water in this study. *D. tertiolecta* is motile and unicellular (8-12 μm) which is a genus of *Dunaliellaceae*. It can commonly found in Hong Kong costal area. *D. tertiolecta* cells are cultivated in L1 medium prepared using autoclaved seawater. The cultures were transferred to new medium weekly at 1:10 v/v in order to keeping in the exponential growth phase. Microalgal cells in the exponential growth phase were inoculated into artificial medium (L1) or preliminary effluent. The microalgae were cultivated in 800ml preliminary effluent in conical flasks with 12:12 light dark cycle. 125 μmol photons $m^{-2}day^{-1}$ was provided by cool white fluorescent tubes in a convison growth chamber. Constant temperature at 24°C was provided for the cultivation.

2.3. Experimental set-up

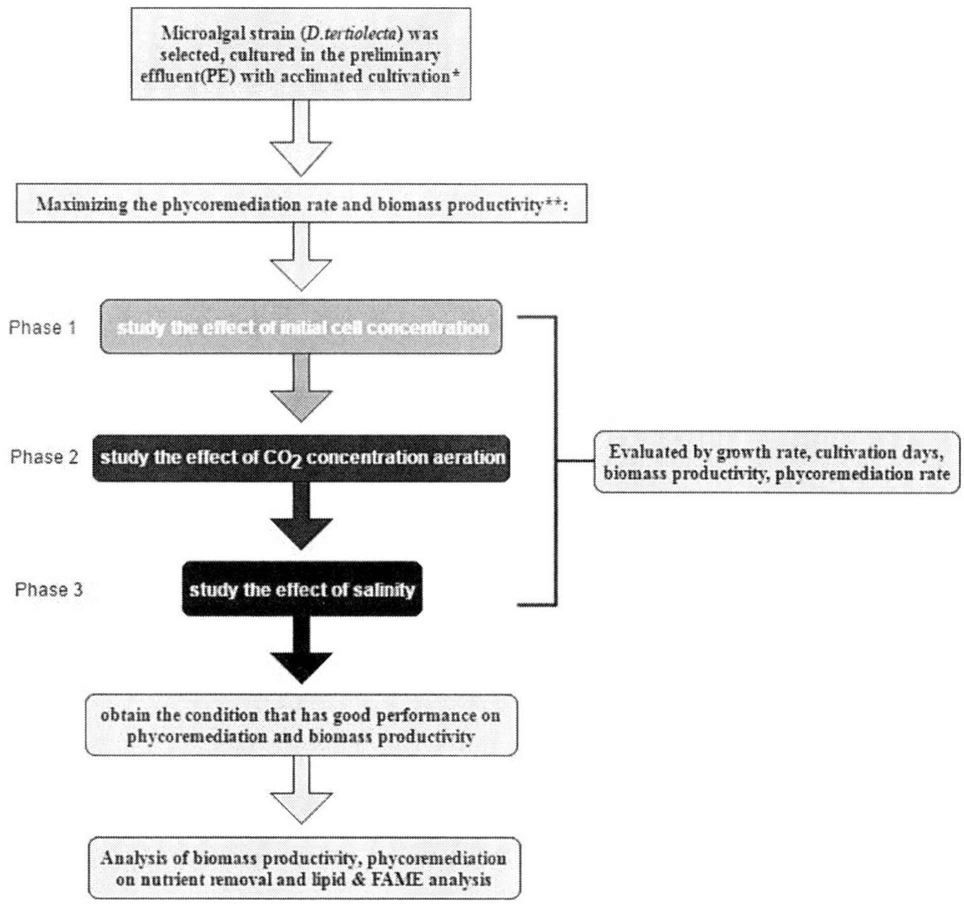

Figure 1. Flow diagram of experimental processes involved in the present study of microalgae cultivation in domestic wastewater [14, 28] present study.

Table 2. Related cell density and biomass concentration with different cell density of inoculums.

Different initial cell density					
Related concentration	Low	Median	Concentrated	Hyper-concentrated	Ultra- concentrated
Cell density (cells mL^{-1})	1×10^{5}	5×10^{5}	1×10^{6}	5×10^{6}	1×10^{7}
Biomass concentration (g dry weight L^{-1})	0.15	0.75	1.5	7.5	14.5

The experiments were carried out in three phase (figure 1). Each experiment was done for triplicate. In the first phase, different cell density of inoculums ranged from 1×10^{5} to 1×10^{7} cells mL^{-1} presenting dry weight of biomass range from 0.15 to 14.5 g dry weight L^{-1} (table 2). Each set was namely as 'low', 'median', 'concentrated', 'hyper-concentrated' and 'ultrahigh concentrated' cultures, respectively. An algal biomass larger than 1.5 dry weight g L^{-1} was regarded as hyper-concentrated cultures. Ultrahigh celldensity culture describes cultures with cell density above 10 g dry weight L^{-1}. The control was settled as without any inoculums. The different cell densities of inoculum in the culture were prepared by concentrated by centrifuge (Eppendorf® 5804) at 3000rpm for 2 mins, and sub-cultured to related medium. In the second phase, gas provided as different concentrations of CO_2 mixed with ambient air was prepared with a volumetric percentage of CO_2 (provided by Hong Kong oxygen®) to give CO_2 concentrations of air (0.04%), 1%, 5%, 10%, or 100%. Microalgal cultures were aerated continuously at the bottom atrate of 100 mL min^{-1} (i.e. 0.125 vvm, volume gas /volume/min). In this phase, the culture without any aeration was acted as control. In the third phase, the saline sewage was adjust salinity by adding sodium chloride from Sigma (USA) in (8 psu, 10 psu, 12 psu, 15 psu and 30 psu) equivalent to 0.1 M to 0.25 M NaCl while the original salinity of PE was 6 psu.

2.4. Cell Counting and calculation of specific growth rates

A benchtop instrument was developed by The Countess™. It uses the standard trypan blue technique for cell number determination. Digital image was captured and a sophisticated image analysis program determines the cell count within the medium [39]. Cell density was determined every two days at the same time. Each point on graph shows average of triplicate results. Growth rate (μ) in exponential growth phase was calculated by following equation:

$$\mu = \frac{\ln N_0 - \ln N_1}{t_0 - t_1}$$

Where N_0 and N_1 are the cell concentration at time t_0 and t_1.

2.5. Biomass concentration determination

The dry weight of microalgae was measured though filtering 10 mL culture suspension through pre-weighed GF/C glassfiber filters (Whatman GF/F, pore size 0.7 μm). Vacuum pressure was kept at 35 to 55 mm Hg during filtration [40]. After rinsing with natural seawater which filtered with 0.45 μm nylon membrane, the filters were dried at 75°C for 24 hrs until there was no further lot of weight.

2.6. Wastewater analysis

10 ml samples were collected from each trial in the experiments. The physicochemical characteristics of the wastewater including, salinity, pH ammonium (NH_4^+), nitrate (NO_3^-), phosphate (PO_4^{3-}), and chemical oxygen demand (COD), were analyzed. Salinity was checked with Atago Master-S/Mill refractometer. Residual free NO_3^-, NH_4^+ and PO_4^{3-} concentrations in culture were determined by using commercial spectrometry-based assay systems according to manufacturer's instruction menu. Assays were done by using DRB 200 reactor and DR 2800 spectrophotometer (Hach Co., Loveland, CO, USA).Nutrient removal efficiencies were obtained according to the following equation:

$$R_{\%} = \frac{S_0 - S_t}{t(d)}$$

where: $R_{\%}$ represents removal efficiency of substrate i (NO_3^-, NH_4^+, or PO_4^{3-}); S_0 and S_t are defined as the mean values of substrate; i concentration at initial time t_0 and time t_i.

2.7. Lipid content determination

Modified method from Bligh and Dyer, 1959 [40] were used for total lipids extraction. Microalgal lipids were extracted by chloroform–methanol (1:1, v/v) Chloroform and aqueous methanol layers then separated by addition of 5% NaCl solution to give a final solvent ratio of chloroform: methanol: 5% NaCl solution of 2:2:1. The chloroform layer was collected and dryness by nitrogen flow. The total lipids were determined gravimetrically.

2.8. Fatty acid methyl ester determination

Fatty acids profile was examined (in triplicate) by the modified transesterification method described by Lepage and Roy [41]. Fifty-milligram samples were mixed with the transesterification reagent (i.e. 100.0 ml methanol containing 2.2 ml of sulphuric acid with 98% purity) and then stirred for 2 hours at 90°C under reflux. Afterwards the reaction product is extracted with hexane/water mixture (1/1v/v %). Fatty acids profile was determined by gas chromatography using Gas Chromatograph (Perkin Elmer

GC 600) equipped with flame ionization detector (FID) and a fused silica capillary column OmegawaxTM 250 (30 m×0.25 mm×0.25 μm film thickness). The injections volume was 1 μl of triplicate in each extraction. FAME was identified by comparing retention times with standards (Sigma-Aldrich R, USA). Analysis programs were: injector temperature and detector temperature were 250 and 260°C respectively. For thermal program, 150°C was held for 16 min firstly, 2°C min^{-1} was increased to 180°C for 25 mins. 210°C was reached up of 5°C min^{-1} and held for more than 25 min. 1.3 mL min^{-1} helium was used as carrier gas. Make up gas was nitrogen gas as 30 ml min^{-1}, flow of hydrogen gas was provided at 30 ml min^{-1}. Synthetic air was provided at 300 ml min^{-1}.

2.9. Statistical analysis

The measurements of nutrient concentration were averaged in three samples of influent and effluent. Every data point on graph represented average of triplicate results. Standard deviations (+SD) were plotted as error bars in the curve.

3. Results and discussion

3.1. Effect of seeding cell density on phycoremediation and biomass productivity

Variation in seeding concentration influenced microalgal growth and further nutrient removal efficiencies. Figures 2 and 3 show the change in cell density, biomass productivity and growth rates at various seeding cell density. From the results, biomass productivity and growth rates of low, median, concentrated and hyper-concentrated cultures were positive value while ultra-concentrated was negatively value (death rate). The highest of biomass productivity was achieved by median (74.7 mg L^{-1} day^{-1}) when the second is low (64 mg L^{-1} day^{-1}) in PE after cultivate for 14 d. The control, without inoculum, showed no biomass productivity in terms of microalgae.

Table 3. Nutrient removal rate and removal percentage in different seeding cell density.

Seeding cell density	Ortho-phosphate		Nitrate		Ammonium	
	Removal rate (mgL^{-1} day^{-1})	Removal percentage (%)	Removal rate (mgL^{-1} day^{-1})	Removal percentage (%)	Removal rate (mgL^{-1} day^{-1})	Removal percentage (%)
Low	1.56 ±0.2	86.4 ±7.2	3.81 ±0.9	92.6 ±0.7	4.46±1.22	95.9 ±5.1
Median	2.52 ±0.14	87.1 ±8.3	4.04 ±1.5	87.6 ±7	5.40±0.85	96.4 ±3.2
Concentrated	2.74 ±0.11	87.8±2.04	4.69 ±2.6	85.2 ±8.6	4.54±1.33	97.6 ±1.8
Hyper concentrated	2.55 ±0	80.3 ±8.3	4.48 ±1.8	98.1 ±0.6	3.31±0.94	94.6 ±5.6
Ultra concentrated	3.13 ± 0.1	63.9 ±9.6	2.50 ±0.5	85.2 ±0	4.95±1.41	99.4 ±0.4
Control	/	2.7 ±0	/	-1.25 ±0	/	N.D.

*The removal rate and removal percentage were calculated from the nutrient removed at the end of experiment (14 d).

At the end of experiment (14 d), almost complete removal was showed in each initial density (table 3). For orthophosphate, ultra-concentrated can remove higher than 80% in first four days and thus the phycoremediation rate was highest while the median, concentrated and hype concentrated show similar phycoremediation rate and reach 2.5-2.7 mg $L^{-1}day^{-1}$. Both algal and bacterial cells were able to assimilate phosphorus from preliminary effluent. It is suggested that phosphorus precipitation may also be contributed for phosphate removal [4]. Removal rate of nitrate in each seeding cell density was remarkable and reached 90% and reach nearly 4.5 mg $L^{-1}day^{-1}$ except for hyper-concentrated. Nitrate

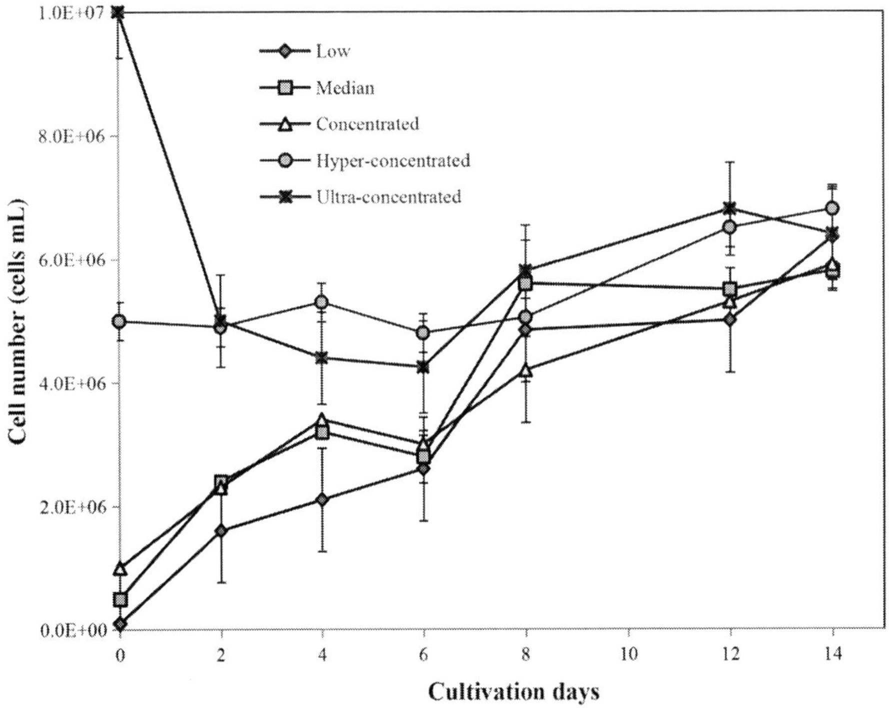

Figure 2. Variation of cell concentration with different seeding cell density.

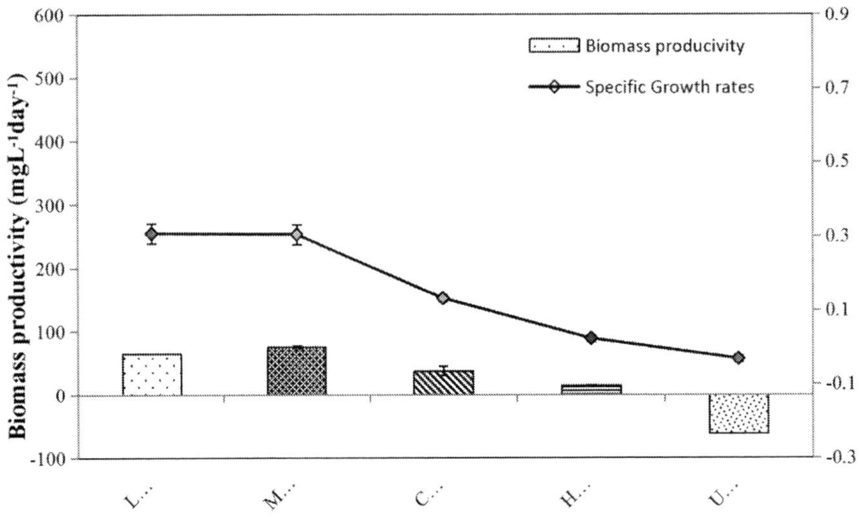

Figure 3. Biomass productivity and specific growth rate depending upon seeding cell concentration.

* Negative value indicate the decay rate of microalga cell concentration

** The specific growth rates and biomass productivity were calculated from the biomass produced and cell number at the end of experiment (14d) is essential nitrogen sources rather than ammonium for *D.tertiolecta* growth [14]. Therefore, it shows relatively good removal performance of nitrate in PE. Concentration shows the best removal and higher than ultrahigh concentrated in ammonium). About 4.48 and 4.69 mg L-1 ammonium were removed by hyper- and concentrated culture each day. The degree of ammonia stripping was showed in control (without any algae) that indicated a similar alkalization. About 5% ammonium was lost in control in first 4 cultivation days (data not shown). The removal percentage decrease with an average of about 1% ammonium was loss throughout the experiment. It is then suggested that ammonia stripping has been considered as one of the mechanism to remove ammonium from PE.

In other similar studies, for example, Lau et al 1995 [31] studied relationship between inoculum concentration and nutrient removal efficiency in primary settled effluent. They stated that super-concentrated inoculum of C.vulgaris (1.2×10^7 cells mL^{-1}) did not promote any self-shading that will limit removal of nutrient. Super-concentrated culture was also considered to optimal level for nutrient removal.

Previous studies stated that higher of inoculum concentration, better of growth and nutrient removal efficiency [31, 42]. Impacts of cultivation such as accumulation of auto-inhibitors and self-shading may not present in initial phase of low to concentrated inoculums. Some study also claimed that higher of inoculum concentrations reduce toxic effect of ammonium to microalgae. Competition between microalgae with other microorganisms was also reduced when cultivating in the wastewater [43]. *S. obliquus* showed high nitrogen removal for 1.9 g L^{-1} inoculum compared to lower density inoculum of 0.5 g L^{-1} [29]. They also stated that removal rate of ammonium and phosphorous were related to microalgal concentration and obvious light limitation is independent of due to self-shading. However, growth of *D.tertiolecta* shows better in median and low initial inoculums in first 5 d than others in this study. Limitation of gas exchange or availability of inorganic carbon such as hyper-concentrated, ultra-concentrated could be one of the key limiting factors [44, 45]. Accumulation of auto-inhibitory substances (e.g. chlorellin) [46] in very dense algal cultures may also affect microalgal growth. Different concentrated culture showed similar nutrient removal performance. The cultures showed a pattern of an exponential removal while those of high, hyper-concentrated and ultrahigh concentrated ones were more hyperbolic.

In the view of microalgal biomass productivity, median concentrated cultures show considerable phycoremediation. Therefore, median concentration was chosen for the following experiment.

3.2. Effect of CO_2 concentration on phycoremediation and biomass productivity

Table 4. Nutrient removal rate and removal percentage in different CO_2 concentration.

CO_2 concentration	Ortho-phosphate		Ammonium		Nitrate		pH
	Removal rate (mgL^{-1} day^{-1})	Removal percentage (%)	Removal rate (mgL^{-1} day^{-1})	Removal percentage (%)	Removal rate (mgL^{-1} day^{-1})	Removal percentage (%)	
Air	3.21 +0.77	86.6 ±6.6	1.87 ±0.16	94 ±0.5	3.2 ±0.67	61.5 ±2.7	7.6-8.6
CO_2 (1%)	3.01 +0.72	66.7 ±0.1	1.77 ±0.15	86.6 ±0.1	1.88 ±0.4	50.7 ±7.1	6.4-7.1
CO_2 (5%)	3.54 +0.74	96.3 ±0.53	1.92 ±0.17	94 ±0.5	3.29 ±0.69	91.7 ±2.2	5.9-7.8
CO_2 (10%)	3.2 +0.67	62.7 ±1	1.03 ±0.08	71.5 ±0.1	2.41 ±0.51	46.4 ±0.5	5.1-6.2
CO_2 (100%)	2.17 + 0.45	33.3 ±8.1	0.9 ±0.07	44.1 ±2.7	0.33 ±0.07	-15.2 ±3.9	5.1-4.8
Control	/	92.3 ±4.1	/	72.4 ±3.4	/	61.5 ±4.9	6.9-7.8

*The removal rate and removal percentage were calculated from the nutrient removed at the end of experiment (6 d).

From the results (table 4), amount of CO_2 dissolved in water varies with pH, (i.e. pH decrease resulting from addition of CO_2). The optimum growth and metabolism of various microalgae was also affected by pH value. In air aeration, the pH had not shown any drop. For the groups aerated with CO_2

enriched air, initial pH value rapidly dropped in day 1. With the assimilation of CO_2 by microalgae, pH value increased due to CO_2 acclimation rate was higher than the supply rate. Nevertheless, when CO_2 supplied rate was higher than the acclimation rate, pH value would be dropped. The low pH also leads the cell death or lower growth rate and further decrease nutrient acclimation. This maybe one of the reason, biomass productivity and growth rate were low in high concentration of CO_2 aeration [47, 48].

As microalgae fix the carbon dioxide, hydroxyl ions will then accumulate which often increasing the pH value. During pH values larger than 9, inorganic carbon form carbonate (CO_3^{2-}) which microalgae cannot assimilate. Lower the availability of CO_2 then limit algal growth [49, 50]. Microalgal species favour neutral pH while some species such as *S.platensis* are tolerant to high pH(e.g. pH 9) [51] or lower pH 4 such as *C.littorale* [52]. There is relationship between pH and carbon dioxide concentration in cultivation, underlying balance among such the form as CO_2, H_2CO_3, HCO_3^- and CO_3^{2-}.

For the results of microalgal growth (figures 4 and 5), it was expected that the microalgae could not be survival under 100% CO_2 concentration [53]. At the aeration of 5% CO_2, *D.tertiolecta* increased most rapidly at growth rate of 0.593 d^{-1} and markedly fell to be 0.293 d^{-1} and 0.148 d^{-1} when cultures were provided with 10% and 100% CO_2. These results also showed in Tang's study [47]. In their experiments, it showed that highest growth rate can be reached in the range of 2-6% CO_2 in air while 1 and 10% showed lower growth rates. These results are consistent with studies that optimal condition for *D.tertiolecta* owards highest growth rate was 6% CO_2 [54, 55].

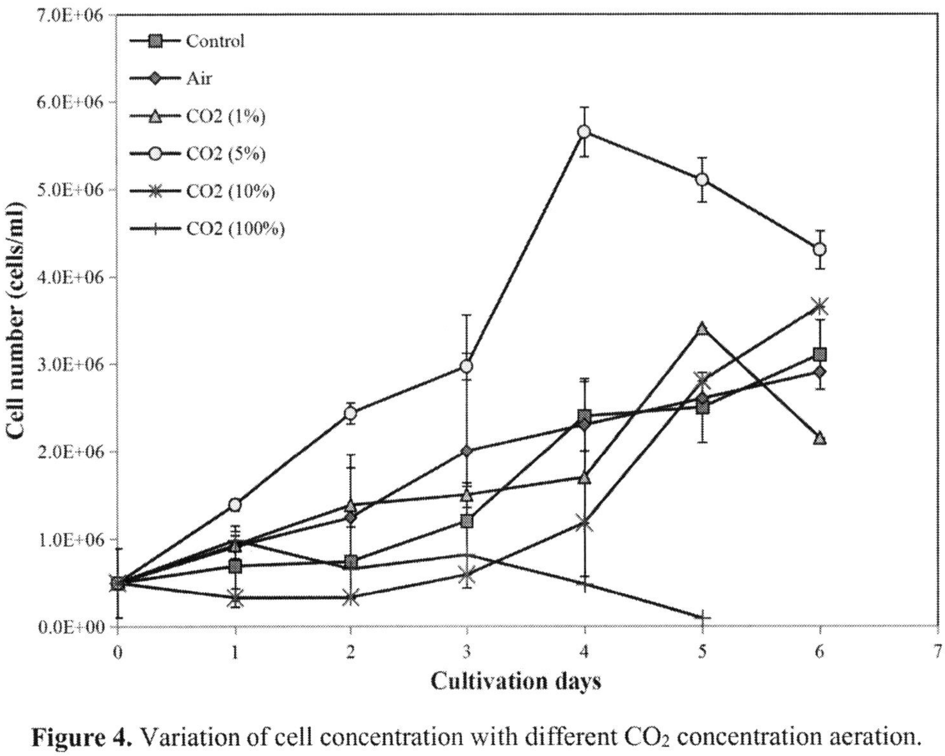

Figure 4. Variation of cell concentration with different CO_2 concentration aeration.

In nitrate removal, the best was CO_2 (5%) and reached 3.29 mg $L^{-1}day^{-1}$ and second high was air and reached 3.20 mg $L^{-1}day^{-1}$. Moreover, removal of ammonium was better in the 5% CO_2 and air compared to others treatments which reached 1.92 mg $L^{-1}day^{-1}$ and 1.87 mg $L^{-1}day^{-1}$ respectively. However, the ammonium removal of 5% was also relatively low when compare to previous experiment even it show the best growth in this study. In contrast, the nitrate removals of 5% CO_2

show much higher value (3.29 mg L^{-1} day^{-1}) than the ammonium one (1.92 mg L^{-1} day^{-1}). This is because the strain could use ammonium than nitrate as the nitrogen sources as previous discussion. Ammonium removal efficiency in culture using air aeration was higher than carbon dioxide enriched air aeration except 5% aeration in our study. This can be explained by pH differences triggered ammonium stripping. Some studies reported that the volatile NH_3 of total NH_4-N is a function between pH and temperature [56, 57]. Therefore, it is suggested that depletion of NH_4^+-N was not attributed to nitrogen uptake by algae only, but also volatilization of ammonia. Moreover, as BOD within PE was relatively high, the heterotrophic microorganisms such as bacteria may also contribute to the ammonium removal. For phosphorus removal, it were stated that sedimentation and direct cellular absorption by microorganisms under aerobic conditions. More than 90% of phosphorus removal was showed in culture with CO_2 aeration and air aeration control even the pH value was not high in this experiment. As pH value of the culture increased to 8.7, which was high enough to cause phosphorus to precipitate [58, 59]. Therefore, the orthophosphate within the CO_2-enriched air aeration was assimilation mainly though phycoremediation but not sedimentation. Increasing CO_2 levels in air provided the higher growth rates which were demonstrated in many studies. However, a balance should be established between acclimation of CO_2 and pH value during the phycoremediation. 5% CO_2 was found to be the optimum percentage for selected strains on phycoremediation in this study either the growth and nutrient removal also show a relatively good performance.

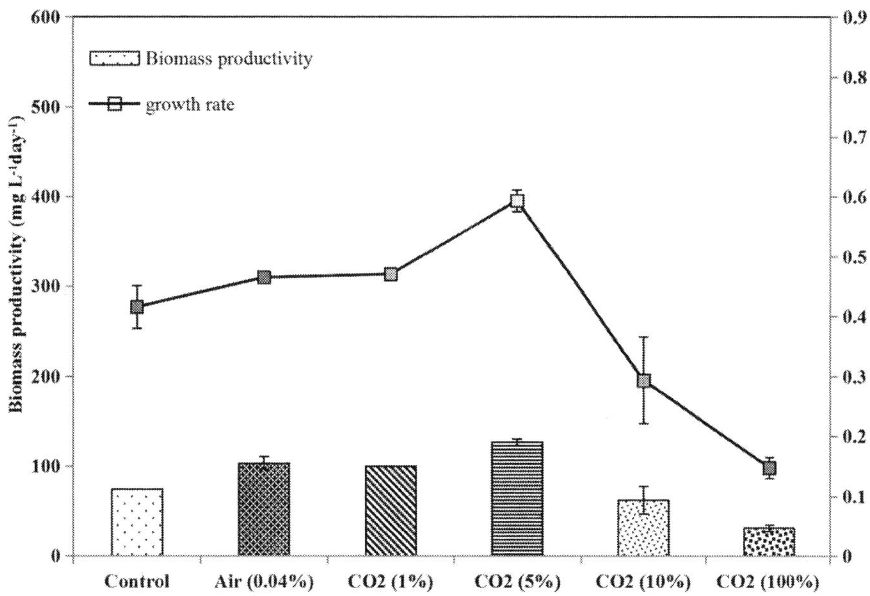

Figure 5. Specific growth rate and biomass productivity depending upon CO_2 concentration aeration.

*The specific growth rates and biomass productivity were calculated from the biomass produced and cell number at the end of experiment (6d)

3.3. Effect of salinity on phycoremediation and biomass productivity

In this phase, algal inoculums were cultivated with different salinity of preliminary effluent to real applicate the salinity adjustment as one of the major optimized process condition (figures 6 and 7). The additions of salt provide niche opportunity for the selected microalgal strains and show more superiority in 12-15 psu for *D.tertiolecta* growth. The maximum capacity in biomass concentration was also reached in 15 psu which was 1.09 g dry weight L^{-1} in day 3, while the lowest maximum capacity was 6 psu in day 4 (0.58 g dry weight L^{-1}). Optimum salinity for nutrient removal in PE was

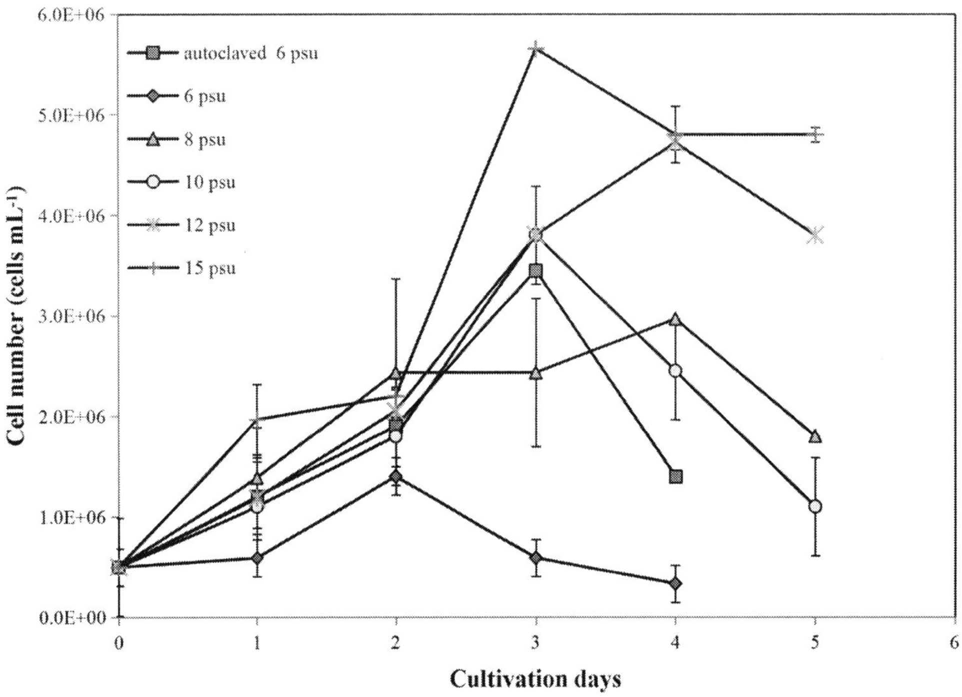

Figure 6. Variation of cell concentration with different salinity.

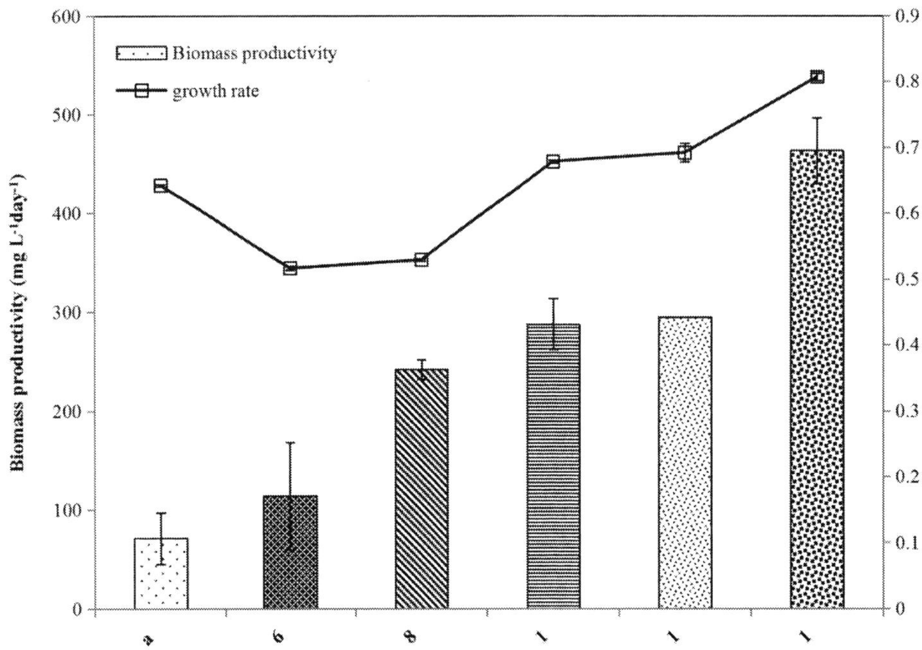

Figure 7. Biomass productivity and specific growth rate depending upon salinity.
*The specific growth rates and biomass productivity were calculated from cell number and the biomass produced at the end of experiment.

Table 5. Nutrient removal rate and removal percentage in different salinity.

Salinity	Total Phosphorus		Ortho-phosphate	
	Removal rate (mg L^{-1} day^{-1})	Removal percentage (%)	Removal rate (mg L^{-1} day^{-1})	Removal percentage (%)
Autoclaved 6 psu	1.64 ±0.03	72.9 ±9.1	2.42 ±0.19	97.0 ±3.54
6 psu	1.08 ±0.08	59.7 ±1.5	2.2 ±0.11	83.3 ±5.6
8 psu	1.8 ±0.14	90.4 ±0.1	1.28 ±0.46	96.3 ±0.07
10 psu	4.98 ±0.27	90.0 ±2.2	3.15 ±0.21	96.5 ±0.37
12 psu	4.52 ±1.32	65.7 ±3.5	2.35 ±0.1	94.7 ±0.52
15 psu	7.19 ±2.5	86.2 ±0.1	3.02 ±0.31	97.1 ±0.28

Total nitrogen		Ammonium		Nitrate	
Removal rate (mg L^{-1} day^{-1})	Removal percentage (%)	Removal rate (mg $L^{-1}day^{-1}$)	Removal percentage (%)	Removal rate (mgL^{-1} day^{-1})	Removal percentage (%)
6.25 ±1.12	69.4 ±0.3	6.75 ±0.66	54.5 ±1.45	2.63 ±0.53	71.9 ±4.4
6.01 ± 0.1	66.6 ±0.1	3.75 ±0.1	37.7 ±3.04	2.55 ±0.19	91.4 ±3.6
5.01 ±0.82	26.5 ±0	2.63±0.53	15.52 ±0.1	3.53 ±0.11	87.5 ±0.24
6.25 ±0.35	71.4 ±2.8	8.25 ±1.32	70.96 ±1.39	2.36 ±0.16	93.9 ±4.1
6.75±1.9	82.1 ±0.5	7.01±1	56.95 ±4.69	3.75 ±0.35	76.8 ±2.5
6.38 ±0.53	83.7 ±0.47%	7.88 ±0.13	79.07 ±3.95	3.02 ±0.31	98.4 ±2.15

*The removal rate and removal percentage were calculated from the nutrient removed at the end of experiment (5 d).

found in 10 psu & 15 psu when the P-related nutrients were almost 80% in first three days and reached 3.17 mg L^{-1} day^{-1} and 3.58 mg L^{-1} day^{-1} (table 5). Rao et al concluded that phosphate removal was because used by freshwater microalga *B. braunii*, which salinity concentration did not influence the removal [60]. This conform to this study that most of phosphorus in preliminary effluent was used for the synthesis biomass of *D.tertiolecta*. For N-related nutrient, the nitrate removal reached 80% in first three days when ammonium also showed considerable amount (6-8.5 mg $L^{-1}day^{-1}$).

Compositions are very variable in wastewater. Even though acclimation, initial cell number and CO_2 enriched air aeration improve growth of microalgae and efficiency of phycoremediation, interaction between microalgae and other microorganisms during wastewater treatment may affect the stability [20, 55, 61]. Although sterilization such as UV radiation, autoclaved provide a solution for this problem, high energy and cost usage are required and further affect the treatment process [20, 35]. Therefore, phycoremediation in real application may need to be consistently enriched using low cost and energy consumption method to reduce microbial activity that impacts the microalgal growth. Microorganisms have their own ecological niche [62, 63]. Salinity, for instance, is a condition that limits the growths and reproductions of many aquatic microorganisms, different organisms tolerate different ranges of salinity [64]. This range is one dimension of an organism's ecological niche. To further improve the growth of selected microalgal species for phycoremediation, optimal niche opportunity can be provided. For example, adjust the salinity to an optimal value for the selected microalgal species.

3.4. Effects of salinity on total Lipid content and FAME composition

Figures 8 and 9 show the comparison of different salinity on lipid production. Under unfavorable environments, algal species produce a large amount of storage lipid as triglyceride [64-66]. However, it may not the case in current study. The lipid content of *D. tertiolecta* increased from 6 psu to 15 psu ranging from 32.2% to 43.5%. As the growth rate and biomass productivity were also better in higher salinity, therefore, increased lipid content following the salinity was not due to salt stress. The reasons behind may base on optimum niche was reached and further to attain a relativity good yield of lipid [67].

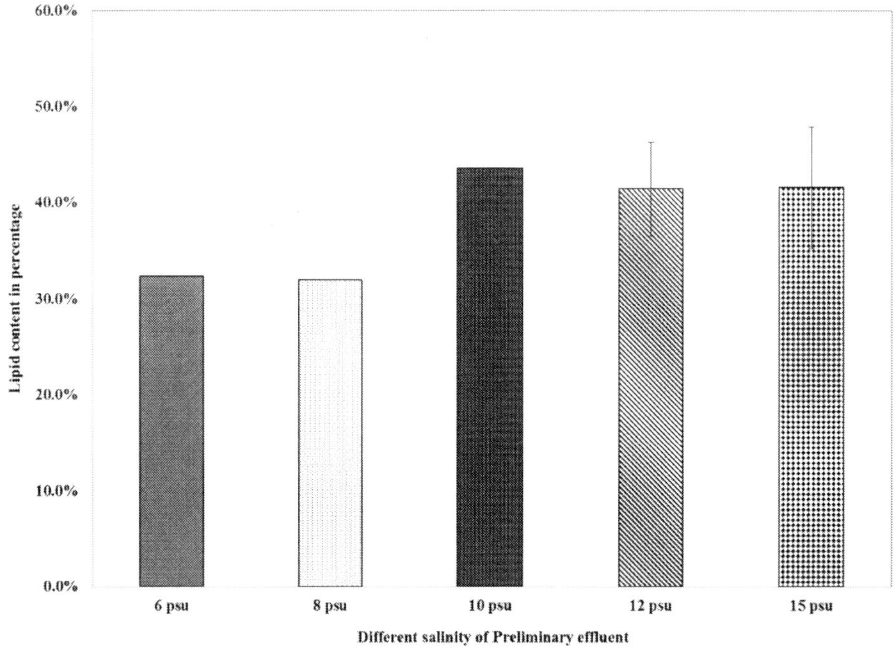

Figure 8. Lipid content of D.tertiolecta on correlated cultivation day under different salinity.

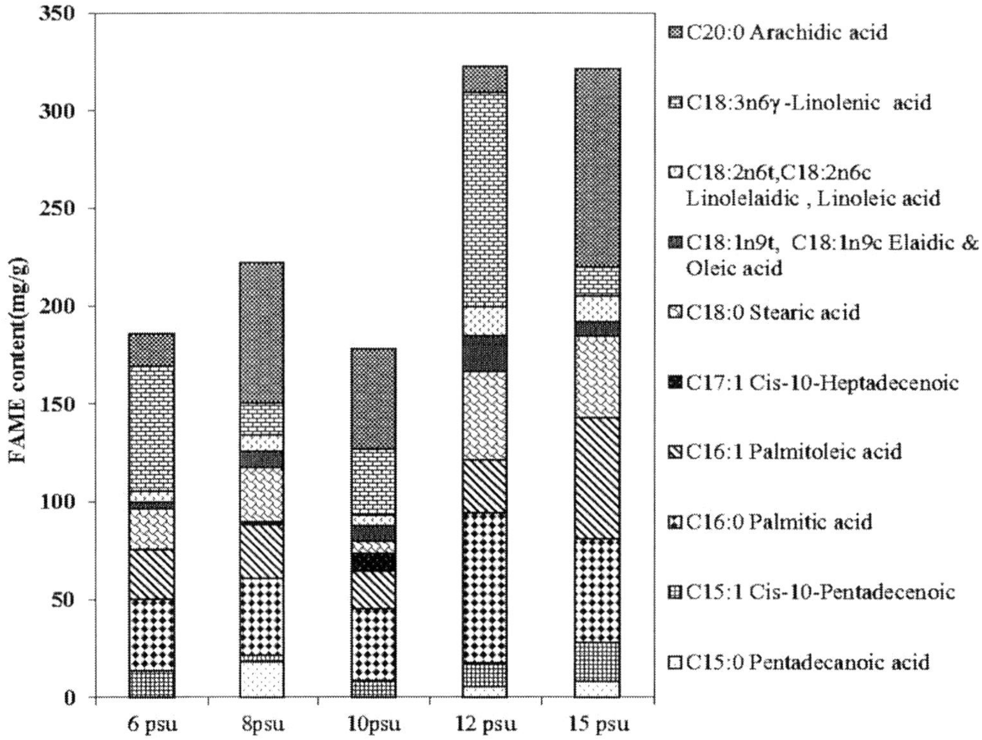

Figure 9. Fatty acid profiles (mg/g) of D.tertiolecta grown under different salinity.

A study state that salinity affect the growth of *Chlamydomonas mexicana* cultivating in municipal wastewater and lipid content in microalgal biomass due to the accumulation of glycerol in responding osmotic pressure [68]. Their results showed that inorganic carbon transfer to glycerol during photosynthesis in *C. mexicana* improving lipid production with increasing NaCl concentrations in wastewater.

Salinity is an intricate biotic factor affecting physiological and biochemical metabolism associated with microalgal growth. Salinity can also lead to improvement in the FAME content of microalgae in changing in fatty acid metabolism [69, 70].

FAME produced from *D.tertiolecta* in different psu contains fatty acid profiles comprising mainly C_{16} and C_{18} fatty acids (figure 3). Fatty acid composition was studied at maximum capacity (end of exponential phase) to reduce the variation in the distribution of fatty acids. The 15 psu conditions also show highest FAME productivity while 6 psu show the least. This was also similar to the growth pattern of each salinity condition in terms of biomass production. Salinity does not seem to affect significantly overall structure but amount of production. The current results show that key fatty acid compositions of *D.tertiolecta* cultivated in different salinity preliminary effluent were Palmitic acid (C16:0), Palmitoleic acid (C16:1), Stearic acid (C18:0), Linolelaidic acid (C18:2n6t), Linoleic acid (C18:2n6) and γ -Linolenic acid (C18:3n6) alongside several minor fatty acid components with carbon number which is lower than C_{16} and higher than C_{18}. Although the major profiles among different salinity were not different, amounts of each fatty acid were variable. In briefly summary, the C_{16} and C_{18} groups still contain for about 97% of the total FAME. Therefore, microalgal lipid produced from *D.tertiolecta* cultivating in enhanced salinity of PE based medium has high potential for producing biodiesel.

4. Conclusion

It was found that in acclimates culture with 5.0×10^5 cells mL^{-1} initial cell densities under 5% CO_2 aeration when culturing with 15 psu salinity of preliminary effluent show the best improvement in phycoremediation of *D.tertiolecta* of preliminary effluent. It show 0.463 g L^{-1} day^{-1} biomass production with 41.6 % total lipid content, and remove most of the nutrient in three cultivation days by the combination of optimized cultivation method. Phycoremediation show the potential insert into the current sewage treatment plant to act as one of treatment unit in local area.

Acknowledgments

The author would like to thank his supervisor, Prof. Kin-chung Ho for his guidance, support, and encouragement during his research project for Doctor of philosophy in Energy and Environment at the Open University of Hong Kong. The author would also like to thank Dr Y H Yau for his support and assistance with this project.

References

[1] Hodgkiss I J and Ho K C 1997 *Are changes in N: P ratios in coastal waters the key to increased red tide blooms? Asia-Pacific Conf. on Science and Management of Coastal Environment* (Springer)
[2] Ho K C and Hodgkiss I J 1991 *Asian Marine Biol.* **8** 5-23
[3] Ho K C and Hodgkiss I J *Asian Marine Biol.* 1993 **10** 77-94
[4] Lau P S, Tam N F Y and Wong Y S 1997 *Environ. Technol.* **18** 945-51
[5] Xu J, Yin K, Liu H, Lee J H, Anderson D M, Ho A Y and Harrison P J 2010 *J. Marine Syst.* **83** 276-86
[6] Lee C M, Jiao J J, Luo X and Moore W S 2012 *Sci. Total Environ* .**433** 427-33
[7] Luo X, Jiao J J, Moore W S and Lee C M 2014 *Mar. Pollut. Bull.* **82** 144-54
[8] Yu S M, Leung W Y, Ho K M, Greenfield P F and Eckenfelder W W 2002 *Water Sci. Technol.* **46** 209-16
[9] Moussa M S, Sumanasekera D U, Ibrahim S H, Lubberding H J, Hooijmans C M, Gijzen H J and Van Loosdrecht M C M 2006 *Water Res.* **40** 1377-88
[10] Lay W C, Liu Y and Fane A G 2010 *Water Res.* **44** 21-40

[11] Uygur A and Kargı F 2004 *Enzyme Microb. Tech.* **34** 313-8
[12] Intrasungkha N, Keller J and Blackall L L 1999 *Water Sci. Technol.* **39** 183-90
[13] Tam N F Y and Wong Y S 2000 *Environ. Pollut.* **107** 145-51
[14] Wu K C, Ho K C and Yau Y C 2015 *Mod. Environ. Sci. Eng.* **1** 225-34
[15] Golueke C G and Oswald W J 1965 *J. Water Pollut. Con. F.* **37** 471-98
[16] Nurdogan Y and Oswald W J 1995 *Water Sci. Technol.* **31** 33-43
[17] Chisti Y 2007 *Biotechnol. Adv.* **25** 294-306
[18] Chisti Y 2008 *Trends Biotechnol.* **26** 126-31
[19] Grima E M, Belarbi E H, Fernández F A, Medina A R and Chisti Y 2003 *Biotechnol. Adv.* **20** 491-515
[20] Rawat I, Kumar R R, Mutanda T and Bux F 2011 *Appl. Energ.* **88** 3411-24
[21] Singh V, Tiwari A and Das M 2016 *Fuel* **173** 90-7
[22] Prajapati S K, Kaushik P, Malik A and Vijay V K 2013 *Biotechnol. Adv.* **31** 1408-25
[23] Ventura J R S, Yang B, Lee Y W, Lee K and Jahng D 2013 *Bioresource Technol.* **137** 302-10
[24] Rbib Z, Ruiz J, Álvarez-Díaz P, Garrido-Perez C and Perales J A 2014 *Water Res.* **49** 465-74
[25] Chen C Y, Yeh K L, Aisyah R, Lee D J and Chang J S 2011 *Bioresource Technol.* **102** 71-81
[26] Chisti Y 2013 *J. Biotechnol.* **167** 201-14
[27] Chen G, Zhao L and Qi Y 2015 *Appl. Energ.* **137** 282-91
[28] Wu K C, Ho K C and Yau Y C 2013 *Benefical Uses of Algal Biomass* (Hong Kong: Environmental Publication House Hong Kong)
[29] Lavoie A and Noüe J De la 1985 *Water Res.* **19** 1437-42
[30] Hashimoto S and Furukawa K 1989 *J. Ferment Bioeng.* **67** 62-9
[31] Tam N F Y and Wong Y S 1995 *Environ. Pollut.* **89** 59-66
[32] Pittman J K, Dean A P and Osundeko O 2011 *Bioresource Technol.* **102** 17-25
[33] Pienkos P T and Darzins A 2009 Biofuels, Bioprod *Biorefin* **3** 431-40
[34] Ruiz-Marin A, Mendoza-Espinosa L G and Stephenson T 2010 *Bioresource Technol.* **101** 58-64
[35] Li Y, Chen Y F, Chen P, Min M, Zhou W, Martinez B and Ruan R 2011 *Bioresource Technol.* **102** 5138-44
[36] Javanmardian M and Palsson B O 1992 *Biotechnol. Bioeng.* **39** 487-97
[37] Guieysse B, Borde X, Muñoz R, Hatti-Kaul R, Nugier-Chauvin C, Patin H and Mattiasson B 2002 *Biotechnol. Lett.* **24** 531-8
[38] Chevalier P, Proulx D, Lessard P, Vincent W F and De la Noüe J 2000 *J. Appl. Phycol.* **12** 105-12
[39] Sarrafzadeh M H, La H J, Seo S H, Asgharnejad H and Oh H M 2015 *J. Biotechnol.* **216** 90-7
[40] Bligh E G and Dyer W J 1959 *Can. J. Biochem. Physiol.* **37** 911-7
[41] Lepage G and Roy C C 1986 *J. Lipid Res.* **27** 114-20
[42] Park J, Jin H F, Lim B R, Park K Y and Lee K 2010 *Bioresource Technol.* **101** 8649-57
[43] Mezzomo N, Saggiorato A G, Siebert R, Tatsch P O, Lago M C, Hemkemeier M and Colla L M 2010 *Food Sci. Technol. (Campinas)* **30** 173-8
[44] Talbot P and J De la Noüe 1993 *Water Res.* **27** 153-9
[45] Talbot P, Thébault J M, Dauta A and De la Noüe J 1991 *Water Res.* **25** 465-72
[46] Safonova E and Reisser W 2005 *Phycolog. Res.* **53** 189-93
[47] Tang H, Chen M, Garcia M E D, Abunasser N, Ng K Y and Salley S O 2011 *Biotechnol. Bioeng.* **108** 2280-7
[48] Lee J, Gillis J M and Hwang J Y B 2003 *Kor. Chem. Soc.* **24** 1763-6
[49] Moss B 1973 *J. Ecol* .157-77
[50] Trimborn S, Lundholm N, Thoms S, Richter K U, Krock B, Hansen P J and Rost B 2008 *Physiol. Plantarum* **133** 92-105
[51] Qiang H, Zarmi Y and Richmond A 1998 *Eur. J. phycol.* **33** 165-71
[52] Pesheva I, Kodama M, Dionisio-Sese M L and Miyachi S 1994 *Plant Cell Physiol.* **35** 379-87
[53] Goldman J C and Graham S J 1981 *Appl. Environ. Microb.* **41** 60-70

[54] Suzuki T, Matsuo T, Ohtaguchi K and Koide K 1995 *J. Chem. Technol. Biot.* **62** 351-8
[55] Tang H, Abunasser N, Garcia M E D, Chen M, Ng K S and Salley S O 2011 *Appl. Energ.* **88** 3324-30
[56] Wong C, Cheung R and Wong M 1999 *Environ. Pollut.* **105** 175-83
[57] Cheung K C, L M Chu and M H Wong 1997 *Water Air Soil Poll.* **94** 209-21
[58] Larsdotter K, Jansen J L C and Dalhammar G 2007 *Environ. Technol.* **28** 953-60
[59] Delgadillo-Mirquez L, Lopes F, Taidi B and Pareau D 2016 *Biotech. Rep.* **11** 18-26
[60] Rao A R, Dayananda C, Sarada R, Shamala T R and Ravishankar G A 2007 *Bioresource Technol.* **98** 560-4
[61] Garcia J, Mujeriego R and Hernandez-Marine M 2000 *J. Appl. Phycol.* **12** 331-9
[62] Atlas R and Bartha R 1981 *Ecological Aspects of Biodeterioration Control; Soil, Water and Waste Management* (Reading, MA: Addison-Wesley) pp 382-414
[63] Prosser J I, Bohannan B J, Curtis T P, Ellis R J, Firestone M K, Freckleton R and Osborn A M 2007 *Nature R. Microbiol.* **5** 384-92
[64] Bartley M L, Boeing W J, Corcoran A A, Holguin F O and Schaub T 2013 *Biomass Bioenerg.* **54** 83-8
[65] Rodolfi L, Chini Zittelli G, Bassi N, Padovani G, Biondi N, Bonini G and Tredici M R 2009 *Biotechnol. Bioeng.* **102** 100-12
[66] Guschina I A and Harwood J L 2006 *Prog. Lipid Res.* **45** 160-86
[67] Yeesang C and Cheirsilp B 2011 *Bioresource Technol.* **102** 3034-40
[68] Hu Q, Sommerfeld M, Jarvis E, Ghirardi M, Posewitz M, Seibert M and Darzins A 2008 *The Plant J.* **54** 621-39
[69] Goswami R C D and Kalita M 2011 *J. Algal. Biomass Utln.* **2** 42-9
[70] Schwenk D, Seppälä J, Spilling K, Virkki A, Tamminen T, Oksman-Caldentey K M and Rischer 2013 *Aquat. Ecol.* **47** 415-24

The utilization of leftover as acid catalyst to catalyse the transesterification and esterification reactions

K K Leung[1] and Y H Yau[1,2,3]

[1]School of Science and Technology, The Open University of Hong Kong, 30 Good Shepherd Street, Ho Man Tin, Kowloon, Hong Kong SAR, China
[2]Centre for Research in Environmental Science, The Open University of Hong Kong, Hong Kong SAR, China

E-mail: yhyau@ouhk.edu.hk

Abstract. Biodiesel (Fatty Acid Methyl Ester, FAME) is a green and renewable energy. It is carbon neutral and produces less air pollutants in combustion. In my project, the selected feedstock of biodiesel production is grease trap oil (GTO). It is extracted from restaurants, and needs pre-treatment. The triglycerides and free fatty acid (FFA) are the main components of GTO. Both triglycerides and free fatty acid can be converted to biodiesel (Fatty Acid Methyl Ester) by transesterification and esterification, through reaction with alcohol (methanol) and catalyst. In the processes, acidic catalyst is chosen to speed up the reactions. The catalyst used In the study, a heterogeneous solid acid is applied. It is waste cooked rice (WCR) collected from leftover. The WCR powder is pyrolysed in 400°C furnace 15 hours and blown with nitrogen gas (incomplete carbonization). The WCR black powder is then mixed with concentrated sulphuric acid and heat in 160°C furnace 15 hours and continuous blown with nitrogen gas (sulphonation). This heterogeneous solid acid is used in the both transesterification and esterification to produce FAME. Moreover, in the optimal reaction conditions, this catalyst offers a stable catalytic effect. After 20 times usage in optimal reaction condition, the catalytic activity remains unchanged.

1. Introduction

Hong Kong is one of the most modern cities in the earth. It suffers from a lot of problems as same as other modern cities. Waste, air pollution, energy exhaustion and global warming are some of the important problems.

In Hong Kong, the year 2014, domestic solid waste is generated 6418 tonnes per day and disposed to landfill site finally [1]. There, the food waste is the major part and wasted cooked rice is one of the components.

Air Pollution in Hong Kong is worse year by year and the major source of the pollutants was derived from motor vehicles, marine vessels and power plants. They induce a problem which is the street–level pollution and region smog [2]. The first source of the pollutants is mainly caused by diesel vehicles, truck, buses and light buses. The other source is power plant. Subsequently, these pollutants are transformed to smog. It is happened under sun light. The pollutants include hydrocarbons, carbon monoxide, sulphur oxides, nitrogen compounds and Ozone. Especially in street level these pollutants are emitted from vehicle engines mainly [3].

IOP Conf. Series: Earth and Environmental Science **82** (2017) 012009 doi:10.1088/1755-1315/82/1/012009

The last and the most important problem is the global warming. It is proved that the carbon dioxide concentration level increased is one of the important factors to induce the global warming [4]. The byproduct of fuel combustion is carbon dioxide which emits into atmosphere and increases the carbon dioxide level. The usage of biofuel (biodiesel) is one of the ways to reduce the carbon dioxide emission because the carbon dioxide emitted from combustion of biodiesel is carbon neutral.

Biodiesel is derived from renewable biological materials such as plant. They pass through chemical and biological process to transform to biofuel. It is another form of indirectly utilized the solar energy. Moreover, the biofuel is as same as solar battery to store the solar energy and uses any time. It is because plant absorbs sunlight, carbon dioxide and water to process the photosynthesis and the result is sunlight energy is captured. Nowadays, the sources of biofuel mainly are derived from oil extracted from plant and non-edible oil especially the wasted cooked oil due to not induce a competition in food, land and other human resources.

Biodiesel is Fatty Acid Methyl Ester (FAME) which is monoalkyl ester of long chain fatty acid. It can be produced by transesterification reaction and esterification reaction. The transesterification reaction is the reaction of triglyceride TG and alcohol (methanol is the most common) to produce FAME and glycerol. TG is the combination of glycerol and three long chain fatty acids FAs. The esterification reaction is that carboxylic acid (Fatty Acids FAs) reacts with alcohol to produce FAME and water.

The chains of these fatty acids in triglyceride include saturated and unsaturated. The fatty acid chain without carbon-carbon double bonds is called saturated and then the present of carbon-carbon double is called unsaturated. The fat is saturated TG with higher viscosity and unsaturated TG is oil with lower melting point and is in liquid form in room temperature.

- Transesterification reaction

$$\underset{\text{TG}}{\text{RCOOR'}} + \underset{\text{alcohol}}{\text{R''OH}} \xrightleftharpoons[]{\text{Catalyst } H^+/OH^-} \underset{\text{FAME}}{\text{RCOOR''}} + \underset{\text{Glycerol}}{\text{R'OH}}$$

- Esterification reaction

$$\underset{\text{FA}}{\text{R'COOH}} + \underset{\text{alcohol}}{CH_3OH} \xrightleftharpoons[]{H^+} \underset{\text{FAME}}{\text{R'COOCH}_3} + \underset{\text{Water}}{H_2O}$$

In order to shift the equilibrium to the right hand side to increase the yield, an excess amount of alcohol is added. For transesterification, methanol, and ethanol are suggested choices of alcohol [5]. The lowest priced methanol is the best choice for production.

The waste cooking oil is selected as feedstock of the FAME production according to statistic of Environmental Protection Department [1]. 2027 metric tonnes of waste oil are collected and needed to be treated and grease trap oil is included. Thus grease trap oil is selected as feedstock to produce biodiesel in my project. The grease trap oil must be under a pretreatment to be separated from the grease trap waste water. The collected grease trap oil includes a lot of mixture. There are mainly triglyceride, diglyceride, monoglyceride, fatty acid and a little amount of soapy water.

2. Literature review

In the production of biodiesel, a lot of processes can be applied. As early as year 1900 in Paris Exhibition, a peanut oil was directly used in engine for several hours by renowned scientist Sir Rudolf Diesel. It shows that the vegetable oil is an old diesel fuel used in diesel engine. Moreover, it was used as emergency fuel as in World War II. The vegetable oils used in the compression ignition engines, its exhaust emission are almost free from sulphur oxide [6]. But the usage of vegetable oil in engine induces several problems. At first, the physical properties of the vegetable oil are related high viscosity and with low volatility. They cause the damages of engine system. The vegetable oil will block the fuel pipe and injectors, carbon deposit on the piston rings, and fouling on the piston head [7-10].

2.1. Catalyst

A catalyst has an important role to speed up the chemical reaction in biodiesel [11]. Nowadays, catalysts can be considered in transesterification and esterification to produce biodiesel is a lot. Generally, they include biotechnology methods (enzyme), base and acid catalyst to speed up the reaction.

2.1.1. Biocatalyst. Biotechnology is a technology to produce biodiesel by using enzyme as catalyst and lipases are generally used in biodiesel production passing through the process of esterification and transesterification [12, 13]. In biodiesel production of usage of enzyme as catalyst can be divided into several types. They are mobilized lipases; immobilized lipases and whole-cell biocatalyst. Although the development of immobilized lipases, whole-cell biocatalyst, new lipase, new fermentation optimization to increase the yield and reduce the cost [14, 15]. But their cost and reaction time is still relative higher and longer [16]. Moreover, If a short chain alcohols (e.g. methanol and ethanol applied in the biodiesel production, these short chain alcohol will cause denature in enzyme. In addition, glycerol is by-product of transesterification reaction; it can also induce a serious negative effect on the enzyme [17-19].

2.1.2. Chemical catalysts. Chemical catalysts can be divided into two main categories, which are base catalysts and acid catalysts. As early as 1984, Freeman et al studied the catalysts both acid and alkaline in the biodiesel production and concluded that the reaction time of alkaline catalyst is more rapid than the acid catalyst. The reaction conditions are NaOH in 1%, methanol to soybean oil molar ratio 6:1, temperature 60°C and in one hour. The FAME yield is 95%. In usage of 1 % sulphuric acid catalyst, the yield of FAME is above 90%. But it needed 60 hours at 65°C and molar ratio of methanol to oil is 30:1 [20, 21].

2.1.3. Alkaline Catalyst. In base catalysts, they consist of homogenous and heterogeneous catalysts. The common homogenous catalysts contain potassium hydroxide, potassium methoxide, sodium hydroxide and sodium methoxide. They are the widely to be selected in transesterification reaction.

In the homogenous alkaline catalysts such as potassium hydroxide (KOH) and is one of the common catalysts to be investigated. In the reaction of transesterification, KOH It is used to convert the karanja oil into biodiesel (FAME) and the optimum reaction condition includes the usage of KOH in 1%, methanol to oil molar ratio 6:1, temperature 65°C and in two hours. The FAME yield is 97% approximately [11]. Moreover, another scientist Karmee develops a better method, which is a usage of a co-solvent tetrahydrofuran (THF). It can increase the yield of FAME from 92 to 95% (feedstock is crude pongamis pinnata) and the reaction requirements are KOH 1%, methanol to oil ratio 10:1, temperature 60°C and 1.5 hours [22]. In the usage of KOH as catalyst, when lower amount of KOH and extra amount of methanol were chosen, it caused the soap formation and yield reducing. In there, the feedstock is carinata oil and the reaction conditions are in KOH 1.4%, methanol to oil ratio 4.6:1, temperature 20-45°C and 0.5 hours [23].

Potassium methoxide (CH_3OK) is an alkaline catalyst. It can be applied in organic chemistry such as process of transesterification. But it is lesser applied in biodiesel production. In the investigation, the performance of CH_3OK and KOH were compared. The reaction used CH_3OK as catalyst the yield is 99.2%. The reaction conditions are 0.455% of CH_3OK, methanol to vegetable oil 8.5:1 and the reaction time is one hour. In the catalyst of KOH, the yield is 98.1%. The reaction conditions are 0.382% KOH, methanol to vegetable oil ratio 11.3 :1 and in three hours [24].

Sodium Hydroxide (NaOH) is the more recommended than KOH and CH_3OK. It is because the NaOH is cheaper in price, easer in separation of product and with lesser emulsification happened within the transesterification reaction. The yield of FAME is as high as 85.5% that the catalyst is 3%

NaOH, temperature 55°C, methanol 35% and reaction time is 1.5 hours [25, 26]. In other research, the yield of biodiesel can be more than 97% when the NaOH in 1%, methanol to sunflower oil 6:1, temperature 65°C and reaction time is 60 mins [27]. In the feedstocks of crude palm oil and coconut oil, the investigation shows that although the oils containing Free Fatty Acid (FFA) in little amount, the yield can be dropped to 80% of palm oil and 55% of coconut oil. The reaction conditions are 1% NaOH, methanol 28%. It shows that the FFA is an important role to affect transesterification reaction yield. In the research of Crude Jatropha Curcas Seed Oil (JCJO) Berchmans found that a higher FFA content (15%) in CJCO sample, the yield is only 55% when directly use alkaline catalyst. But used two step reaction to treat the FFA, the yield of 90% is obtained. The 1.4% of NaOH is applied [28].

Sodium methoxide (CH_3ONa) is less recommended to apply in the commercial biodiesel production. It is because the price of the CH_3ONa is high, although the performance is higher than the catalyst of sodium hydroxide. The advantages of the CH_3ONa consist of does not form water in FAME production compare with the application of NaOH and KOH. Moreover, it is more effective than the usage NaOH in the production and it just only requires the 50% of amount of NaOH [29].

In the comparison of these four alkaline catalysts (potassium hydroxide, sodium hydroxide, potassium methoxide, and sodium methoxide), Singh finds the performance of methoxide catalysts are better than the hydroxide and with the better yield. In the reaction, it is not only to form lesser water but also elevated the conversion equilibrium. In the investigation, the most optimal combination is potassium methoxide as catalyst at 0.2 mol/mol (1.59%wt), reaction temperature 50°C, and methanol-to-oil molar ratio 4.5:1, which can obtain the max. yield as high as 95.8% and with min. soap formation 0.75% [30].

Heterogeneous base catalysts, this kind of alkaline catalyst is very effective. The heterogeneous catalyst means the catalyst composes unrelated or different part of body it also means with different phase. Such as usage the homogeneous base catalyst, a separation of the liquid state of FAME and the liquid catalyst can be needed. In the other hand, the Heterogeneous base catalysts are in solid state, they can be more simply to be separated from product. Moreover, these catalysts can be reused many time any just a simple filtration to be applied. The most common solid base catalysts include alkaline earth metal oxide, Zeolite, KNO_3 loaded on Al_2O_3 KNO_3/Al_2O_3, BaO, SrO, CaO, MgO etc. [31, 32].

In other investigation shows the usage of CaO (8%) as catalyst can obtain FAME above 95% [33]. The reaction conditions are molar ratio of methanol to oil 12:1, temperature 65°C and 2.03% water content in methanol. Moreover, the catalyst activities can maintain after repeat usage in 20 cycles productions.

At the same period, another investigation found that the catalysts of $CaTiO_3$, $CaMnO_3$, $Ca_2Fe_2O_5$, $CaZrO_3$, and $CaO–CeO_2$ shown high activities to give high yield of about 90% and the reaction conditions are at 60°C, molar ratio of methanol to oil 6:1, and reaction time is 10 hours. In the investigation, the catalyst of $CaZrO_3$ and $CaO–CeO_2$ show high durability and suitable to be used in biodiesel production processes as heterogeneous base catalysts [34].

2.1.4. Acid Catalyst. The homogenous acid catalysts can also be used in the FAME production. They have their advantages in the production of FAME but they also include disadvantages when used in the FAME production.

In the above explanation, the limitation of basic catalyst is that they have high catalytic power in biodiesel production and show high yield, but the feedstock content must be controlled in low concentration of water and free fatty acid FFA. Otherwise, the basic catalysts are consumed and also the saponification will be induced. Moreover, the best feed stock of biodiesel is grease trap oil, which always contains water and FFA and in related high concentration.

According to the characteristic of grease trap oil with high water content and high concentration of FFA, acid catalysts are the best selection in biodiesel production. Because the esterification process it is not only to consume the FFA but also produce FAME from FFA. Thus a two steps biodiesel production is induced. At the first an esterification is initial to produce FAME and consume FFA then a basic catalyst is applied in this feedstock which without FFA or FFA in low concentration to process transesterification [28, 35-37].

The most common homogeneous acid catalyst is sulphuric acid [38-40]. In the usage of sulphuric acid catalyst, this can get the yield as high as 99%. The reaction conditions are 1 mol % of sulphuric acid, molar ration 30:1, temperature 65°C and 50 hours. If the butanol (in 117°C) and ethanol (in 78°C) are instead the reaction time can be reduced to 3 hours and 18 hours respectively [38]. In the other research of usage crude palm oil to produce FAME, the reaction conditions are varies in 70, 80, 95°C but the sulphuric acid concentration, molar ratio of oil to alcohol are kept constant. Then the experiment can conclude the reaction rate be increased by temperature increased. The 99.7% yield is controlled by 5% sulphuric acid and oil to butanol molar ratio 40:1 when the temperature is kept in 95°C and the react time can be reduced to 9 hours [41].

Finally, the usage of the homogeneous acid catalysts will induce some problems [42, 43]. They include:

- higher amount of alcohol be used;
- consume more energy by needed higher reaction temperature;
- lower reaction rate; corrosive problem in reactor; and
- environmental problem in waste acid treatment before disposal.

In heterogeneous acid catalysts, they not only have the advantages of homogeneous acids type catalysts, but also have less disadvantages of homogeneous acid catalyst in biodiesel productions.

In some researches, the heterogeneous acid catalysts are investigated e.g. sulphated zirconis (SO_4^{2-}/ZrO_2), sulphated tin oxide (SO_4^{2-}/SnO_2) and sulphated titanium oxide (SO_4^{2-}/TiO_2). All of them have good catalytic activities performances in the processes of esterification and transesterification simultaneously. In addition, the stability of them is also high. Unfortunately, they are not chosen in the production of biodiesel in industries. The reasons are the high catalyst costs and hard to be recycle by their small size. Thus they are difficultly to be separated by simple filtration [44-46].

In other research, various heterogeneous solid acids are investigated in production of FAME converted from canola oil which with 10% content of FFA. In there, a technology of immobilization of Keggin-type heterpolyacid (HPA) on different support e.g. hydrous zirconia, silica, alumina, and activated carbon is investigated. The hydrous Zirconia support induces to the best catalytic performance [47].

Finally, the interested catalysts are introduced. These novel strong solid acids are developed by Takagaki. They select a carbon material (D-glucose) as catalyst resource. At first, the glucose is under incomplete carbonization to break down some their bonds and then deals with sulphonation to bind with SO_3H group then a novel strong solid acid is formed. In the incomplete carbonation the glucose is pyrolyzed at 400°C in 15 hours and blown with nitrogen continuously. The polycyclic aromatic carbon is obtained. In the sulphonation, the polcyclic aromatic carbon is heat concentrated sulphuric acid in 150°C, 15 hours and under nitrogen. The new solid acid is formed [48].

The D-glucose derived catalyst is compared with other catalysts, which including sulphate zirconia, amberlyst-15 and niobic acid. The D-glucose derived catalyst shows the best result. They also concluded that the glucose derived catalyst is very suitable to replace sulphuric acid and other acid catalysts as a green catalyst to produce biodiesel from waste oil with high acid value [49].

A further study of that type of acid catalyst is set up to focus on the carbohydrate compounds. The D-glucose, sucrose, cellulose and starch are converted to heterogeneous solid acid catalyst by passing through incomplete carbonization and sulphonation. In the investigation, all of them can act as acid catalysts to produce biodiesel and the starch is the most high efficiency, high recyclable and environmental friendly catalyst to produce biodiesel from waste oil and with High acid value [50].

From the above references, the biodiesel production should select the waste cooking oil (grease trap oil) as feedstock. Moreover, the catalyst should consider the carbohydrates derived substances. Waste cooked rice (WCR) is a good direction to be investigated to act as a heterogeneous solid acid in the production of biodiesel.

3. Experimental

3.1. Materials
The sulphuric acid was purchased from Riedel-DeHaen, (Extra Pure). The Oleic acid and Methanol were purchased from Sigma-Aidrich (analytical standard). The peanut oil was purchased from Knife Pure Peanut oil. The grease trap oil was collected from Chinese restaurant and was heated to 60°C and the top brown oil was separated easily. After the pretreatment, the grease trap oil contained mainly triglyceride, 38.7% free fatty acid and 0.08% of water. The waste cooked rice was collected from leftover of high school after lunch and treated with washing and drying in oven.

3.2. Catalyst preparation
At first, the treated dry waste cooked rice (WCR) was grinded in powder form and put 20 g of rice powder into the reactor to pyrolyze in temperature 400°C for 15 hours and well mixing. Nitrogen was blown into the reactor continually within the reaction period. In the end of the reaction some rice compound bonding were broken and the rice powder was changed to black color and in lump form. After cooling, the incomplete carbonization of product was grinded in powder by mortar and pestle. The black grinded powder was put into the reactor again and mixed with concentrated sulphuric acid heating to 160°C for 15 hours. Simultaneously, nitrogen was continually blown into the reactor until the end of the reaction. At the end, the product of mixture was cooled to 30°C and diluted with 800 ml DI water. In there, the black precipitate was settled into the bottom and the upper liquid was removed. These black precipitate was washed with > 80°C DI water continually until the pH of the washed DI water changed to neutral. Finally, the wet black precipitate was dry in oven at 65°C oven and the heterogeneous solid acid catalyst was formed.

3.3. Catalyst activity performance
At the first, the WCR derived catalyst activity was investigated in esterification reaction. The WCR derived catalyst was used to catalyze the reaction of oleic acid and methanol. Before the reaction the WCR derived catalyst was dried in oven 100°C for one hour to remove the moisture on the catalyst surface. All of them were transferred to reactor. The 10% weight of catalyst, methanol, oleic acid (molar ratio 20:1) were transferred in to reactor form 8 hours and the temperature was 80°C. In each hour, 400 μL sample was withdrawn and put into centrifuge (Microcentrifuge) for 10 min at 10000 rpm. At the same time, the vacuum was applied and kept in 70°C. After removed the catalyst, each sample FAME yield was determined from GC.

A similar investigation was applied to investigate the esterification and transesterification. The different was the oleic acid changed with neat peanut oil and mixed with 38.7% oleic acid.

3.4. Optimum parameters of biodiesel production reaction
Attempt was made to optimize parameters such as the reaction temperature, catalyst concentration and molar ratio of methanol to oil in the biodiesel productions. Each parameter was tested in turns, with the other variables controlled at constant values. For example, in investigation of the effect of reaction temperature, different reaction temperatures were tested respectively. While the other parameters such as methanol to oil molar ratio, catalyst concentration etc. were kept constant. Thus the relationship between the reaction temperature and the production yield becomes clear.

3.5. Production yield of FAME converted by grease trap oil
After the optimum reaction parameters of the WCR derived catalyst were obtained, the yield of biodiesel was investigated according the optimum parameters. In the investigation the feedstock was the grease trap oil collected from Chinese restaurant. In there the esterification and transesterification were conducted simultaneously. The samples were withdrawn from each hour to investigate the FAME yield percentage. It was used to consider the reaction rate.

3.6. Stability of WCR derived catalyst

Finally, the stability of the RCW derived catalyst was investigated. The experiment was based on the section 3.5. But the FAME yield percentage was only considered at the end of reaction. Each reaction time was set to five hours. The catalyst was reused 20 cycles under same reaction condition. In there, the catalyst will be recovered by dry at 100°C oven two hours between each experiment. It is because a little amount of catalyst will be lost at each cycle of recovering, thus a suitable amount of catalyst were added to fulfill the catalyst in 10wt % weight in each cycle.

4. Results and discussion

4.1. Catalytic activity of waste cooked rice (WCR) derived catalyst

A number of solid acid catalysts derived from carbohydrate had been investigated by other researchers [50]. They included starch, cellulose, sucrose and D-glucose, and etc. They were converted to solid acid catalysts, and their usage in biodiesel production was compared. They were processed through incomplete carbonization, and pyrolysis in 400°C, 15 hours and blown with nitrogen, in order to achieve dehydration and dissociation of bonds in –C–O–C–. During the reaction, polycyclic aromatic carbon ring was form. This polycyclic aromatic carbon ring (semi-product) takes part in the subsequent sulphonation, in which the semi-product was mixed with concentrated sulphuric acid, heated to 150°C for 15 hours, and blown with nitrogen simultaneously. The SO_3H group, derived from the concentrated sulphuric acid, can bind into the polycyclic aromatic carbon ring to form heterogeneous solid acid catalyst. It had been reported that starch (polysaccharide), among the different substances tested, shows an excellent catalyst performance. Inspired by this information, waste cooked rice (WCR) is selected as the feed stock, and it is to be treated with similar reactions. The waste cooked rice indeed also shows high catalyst activity performance. The high catalyst activity performance can be attributed to similarity between rice and starch; rice contains approximately 89-94% endosperm, which comprises a lot of polygonal starch granules and some protein body.

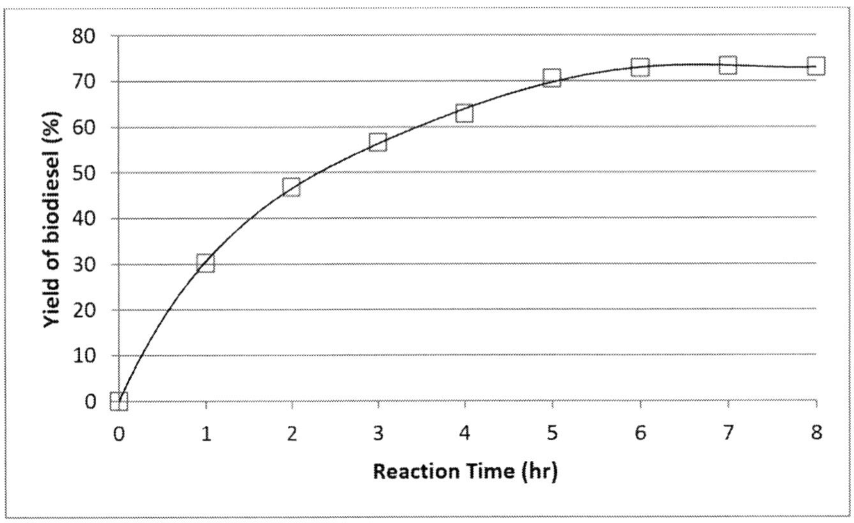

Figure 1. Yield of Biodiesel (FAME) produced form Oleic acid (FFA) and catalyzed by waste cooked rice (WCR) derived catalyst. Reaction condition: 80°C; molar ratio of methanol to oil 20: 1; catalyst 10wt%.

In figure 1 the yield of biodiesel (FAME) is 70.5% in 5 hours. The waste cooked rice (WCR) derived catalyst shows a high reaction rate. WCR derived catalyst is demonstrated capable of catalyzing the esterification between oleic acid and methanol. The role of the oleic acid is to substitute the FFA in grease trap oil (GTO).

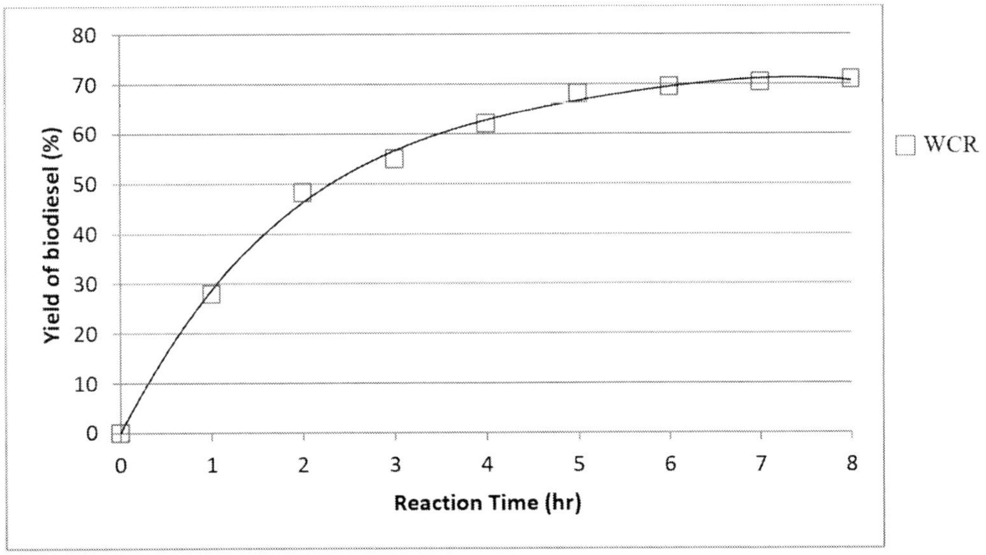

Figure 2. The yield of Biodiesel (FAME) from neat peanut oil and with 38.7wt% of oleic acid catalyzed by Waste cooked rice (WCR) derived catalyst. Reaction condition: 80°C; Molar ratio of methanol to oil with oleic acid 20:1; catalyst 10wt%.

In figure 2 the neat peanut oil is mixed with 38.7% oleic acid, to simulate the composition of grease trap oil. WCR derived catalyst also shows good catalyst activity in transesterification. The FAME production yield can be as high as 68.1% in 80°C; molar ratio of methanol to oil is 20 : 1; catalyst 10 wt%. Comparing figure 1 and figure 2 the yield in figure is higher than the yield of figure 2. In figure 2, the WCR derived catalyst needs to catalyse two different type reactions (transesterification & esterification reaction) and thus the performance might be different. In addition, the transesterification reaction is believed to be more complicated than the esterification.

4.2. Optimum parameters of WCR derived catalyst in biodiesel production

In order to find out the optimum parameters to induce the best WCR derived catalyst activities. Several investigations about the parameters are carried out. The parameters include reaction temperature; molar ratio of methanol to neat peanut oil with 38.7% oleic acid; and catalyst concentration.

4.2.1. Reaction Temperature. Reaction temperature plays an important role in chemical reaction. In the investigation higher temperature increases both the reaction rate and product yield. In figure 3 the temperatures are set to 65°C, 80°C, and 95°C. The reaction conditions: Molar ratio of methanol to oil mixed with 38.7% oleic acid is 20:1, and catalyst concentration is 10wt%. The best result is obtained when reaction temperature is set at 95°C. The reaction rate is the fastest and the production yield is the highest (71.8%). When the reaction temperature is kept at 65°C, the reaction is the slowest and the product yield is the lowest (62.2%). At 80°C, the performance is intermediate. The reaction rate and yield are just marginally below those at 95°C reaction temperature. Thus the optimum reaction temperature is chosen as 80°C. Because it can save the energy to reduce the production cost.

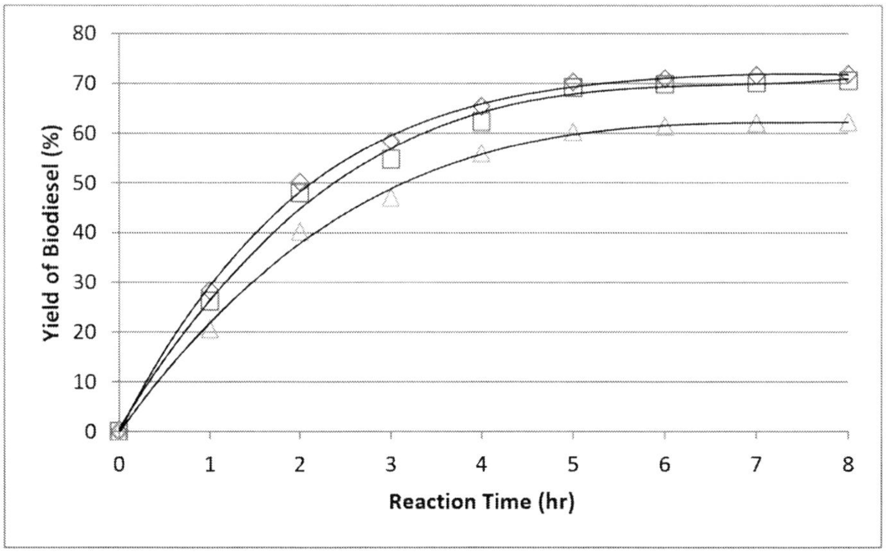

Figure 3. The yield of Biodiesel (FAME) from neat peanut oil and with 38.7wt% of oleic acid catalyzed by Waste cooked rice (WCR) derived catalyst in different temperature. Reaction condition: Molar ratio of methanol to oil with oleic acid 20:1; catalyst 10wt%.

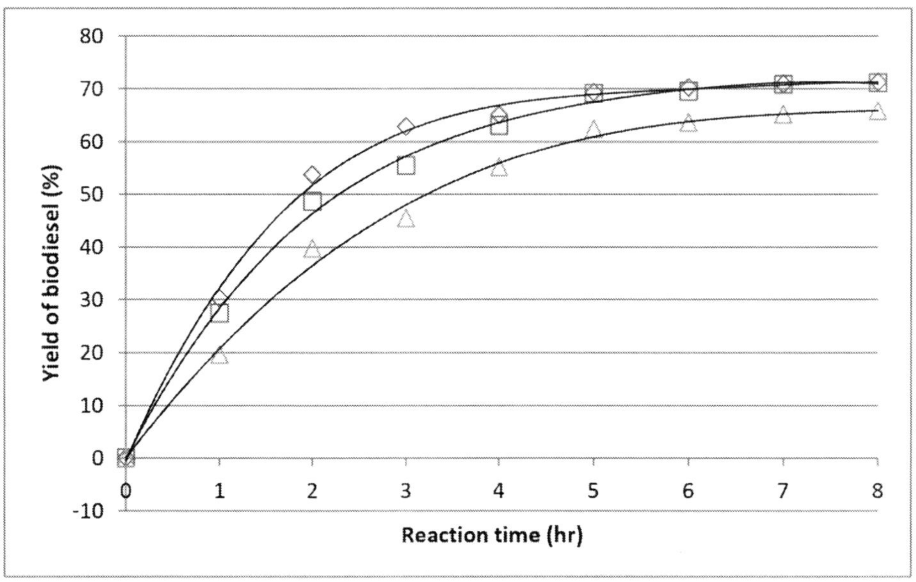

Figure 4. The yield of Biodiesel (FAME) from neat peanut oil and with 38.7wt% of oleic acid catalyzed in different weight % of Waste cooked rice (WCR) derived catalyst. Reaction condition: 80°C; Molar ratio of methanol to oil with oleic acid 20:1.

4.2.2. Catalyst Concentration. The catalyst loading is another important parameter. When the catalyst loading increases, the reaction rate and production yield are increased. Different catalyst loadings were tested in the investigation; these were 5 wt%, 10 wt% and 15 wt% respectively. The other parameters in the investigation consist of reaction temperature (80°C), the molar ratio of methanol to oil with 38.7 wt% oleic acid (20:1). Figure 4 shows, when catalyst concentration increases from 5 wt% to 10 wt%, the production yield is increased sharply from 65.8 to 71.1%. However, when the catalyst concentration further increases to 15 wt%, the production yield becomes stationary, or increases slightly at best (Use data for 5 hours reaction time). From this test, the suitable catalyst concentration for conversion of biodiesel is chosen to be 10 wt%, as increasing the catalyst concentration above 10 wt% does not increase the yield much.

4.2.3. Molar ratio of methanol to peanut oil with 38.7wt% oleic acid. Theoretically, in the esterification reaction, one mole of oleic acid should react with one mole of methanol. In transesterification, one mole of triglyceride will react with three moles of methanol to produce biodiesel. In reality, the methanol should be increased to shift the equilibrium toward the right hand side to favour the product. Furthermore, the transesterification reaction is reversible. Thus in order to keep a high product yield, the methanol should largely increase. However, to keep the production cost low, usage of the methanol should be as less as possible. In the investigation of figure, different mole ratios of methanol to neat peanut oil with 38.7% oleic acid are set, specifically 5:1; 10:1; 15:1; 20:1; 25:1; and 30:1. The other parameters: reaction temperature 80°C; catalyst concentration10wt% and the reaction time 5 hours. In figure 5 the yield increases together with the increase of methanol. But the optimum molar ratio of the methanol to oil is the 20:1. It is because the further increase the molar ration of methanol to oil 30:1, the yield is just slightly increased from 69.3% (20:1) to 71.9%. Thus the best molar ratio of methanol to oil in the conversion of FAME is 20:1.

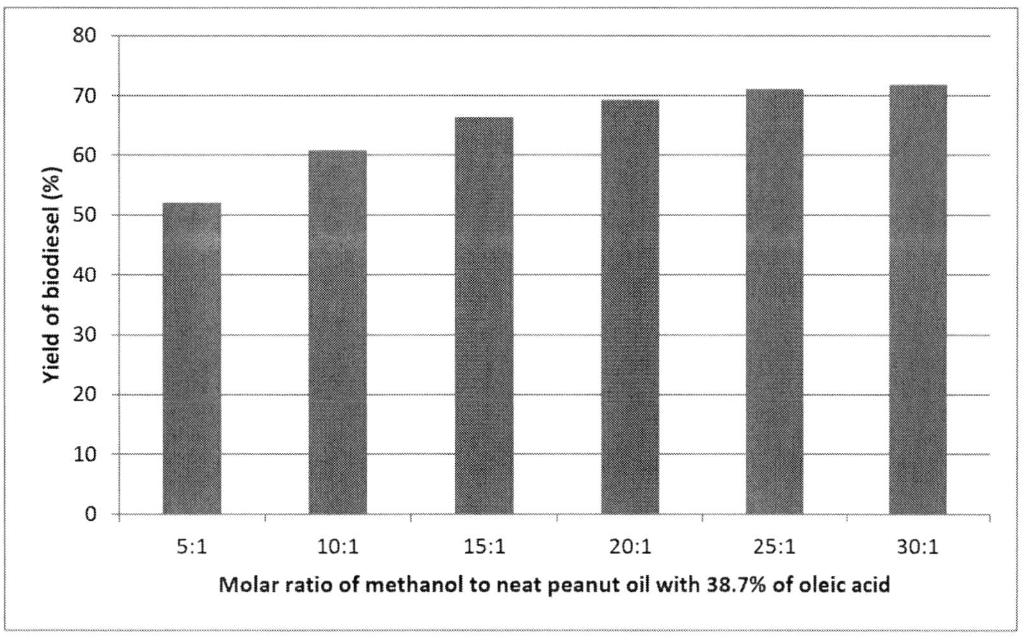

Figure 5. The yield of FAME from neat peanut oil and with 38.7wt% of oleic acid in different molar ratio of methanol to neat peanut oil with oleic acid. Reaction condition 80°C; catalyst 10wt%; and 5 hours.

4.3. Biodiesel converted from GTO and catalyzed by WCR derived catalyst
In the above investigations, the optimum parameters are determined. When the feedstock is changed to cheap grease trap oil (GTO), these optimum parameters are still used. The first step is a pretreatment, in which the collected GTO is heated and impurities separated. In figure 6, the conversion of GTO to biodiesel, the reaction conditions are: temperature 80°C; WCR derived catalyst 10wt%; molar ratio methanol to FTO 20:1 and the reaction is 5 hours. The yield is 63.9% and it is similar to experimental yield 68.1%. The difference is 4.2%. The difference may be caused by the feedstock. In the experiment, the oil is the neat peanut oil and the oleic acid is analytical standard, thus the yield can be

as high as 68.1% in 5 hours. In practical investigation, the feedstock is changed to cheap GTO. Although the collected GTO is pretreated, it may contain vegetable oil, seed oil, or animal fat. The most important difference is that animal fat is always in solid form in room temperature due to their long saturated chains and the other oils such as vegetable oil and seed oil are in liquid form in room temperature, as their chains are relatively short and are unsaturated. The long saturated chain is more compacted and non- polar while the unsaturated chains are packed more loosely. These structures of animal fats are hard to be mixed to with polar methanol. Thus the production yield converted from GTO is lower than the neat peanut oil, may be the GTO contains animal fats.

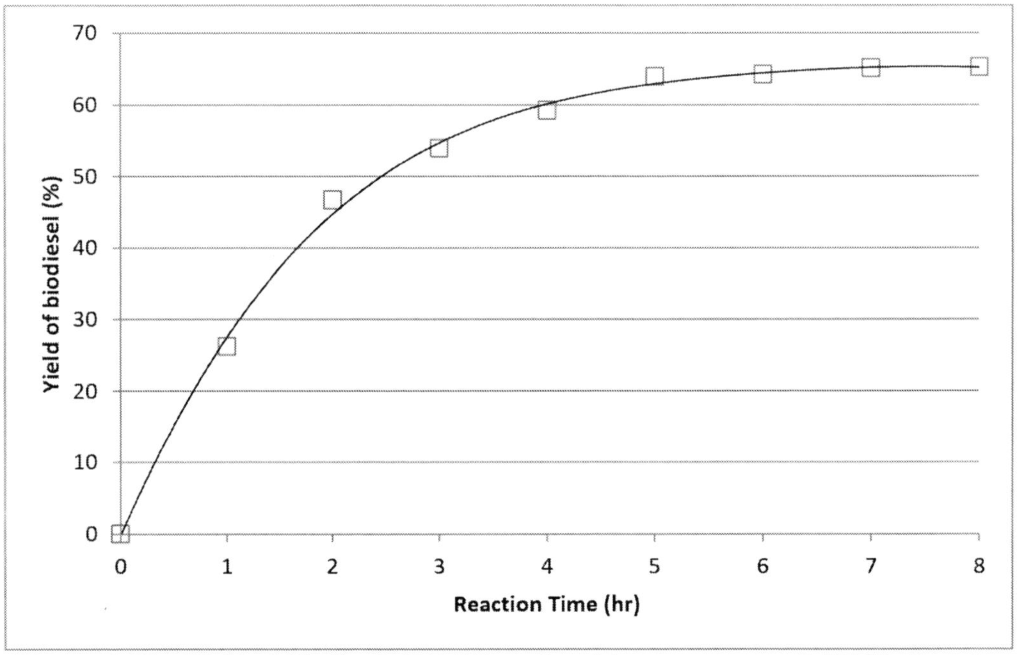

Figure 6. The Yield of Biodiesel (FAME) produced by grease trap oil (GTO). Reaction condition: 80°C; catalyst 10wt%; molar ratio methanol to GTO 20:1.

4.4. Stability of WCR derived catalyst in biodiesel production

The other important factor in the production cost of biodiesel is the reusability and stability of the catalyst. The higher number of catalyst recycle and better catalyst activities stability will reduce the production cost and conserve our resources. Thus the same catalyst is repeatedly used in the biodiesel production to investigate the WCR derived catalyst stability. In figure 7, the yields of biodiesel in twenty cycles are similar and the reaction conditions are reaction temperature 80°C; catalyst concentration 10wt%; molar ratio of methanol to GTO 20:1 and reaction time is five hours. It means the stability of WCR derived catalyst is high and can maintain the catalyst activities within 20 cycles in the optimum reaction condition.

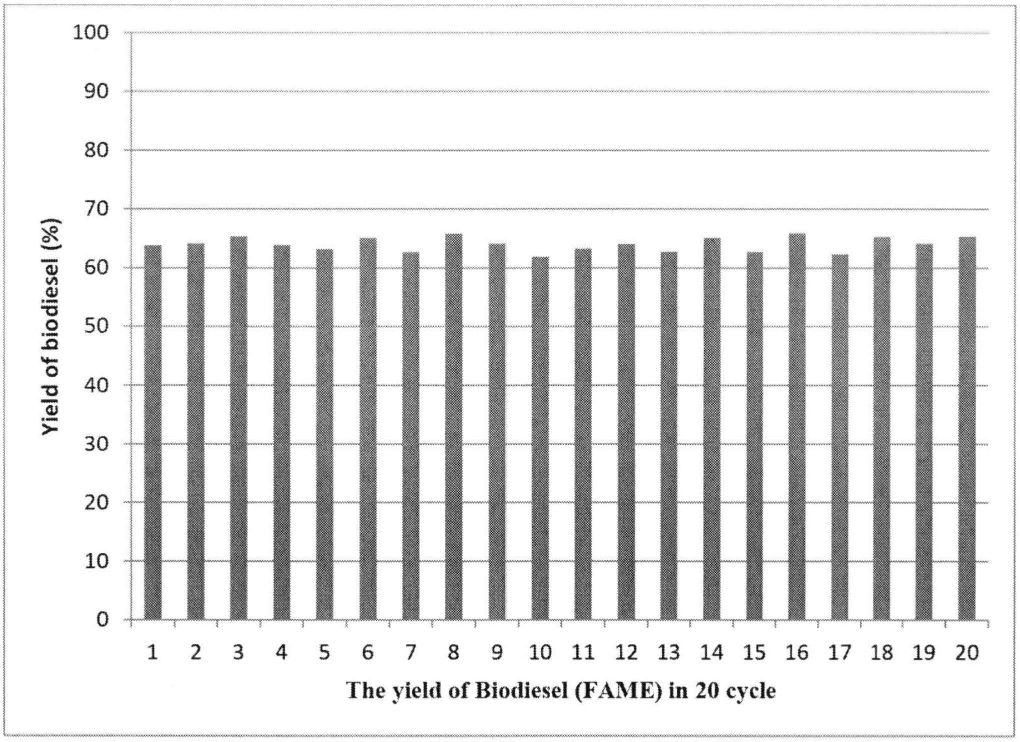

Figure 7. The stability test of the WCR derived catalyst in 20 cycles of biodiesel (FAME) productions. Reaction condition: 80°C; catalyst 10wt%, molar ratio 20:1; and 5 hrs.

5. Conclusion

A new catalyst is developed for biodiesel production from waste cooked rice (WCR). It is a heterogeneous solid catalyst, which can catalyse esterification and transesterification reactions at the same time. Moreover, the catalyst is a solid. It can be easily separated from the product and can be recycled. The WCR derived catalyst shows a good catalyst activity in the production of biodiesel (FAME) from cheap feedstock of grease trap oil (GTO) with 38.7% free fatty acid (FFA). It also shows a stable performance. The yield maintains similar, even when the catalyst has been reused 20 times. The WCR derived catalyst has great potential to be applied in industrial biodiesel production in the future.

Acknowledgments

The first author would like to acknowledge his supervisor, Dr Y H Yau and others who provide advices and technical support to his research during the DEng studies.

References

[1] http://www.epd.gov.hk/epd/english/environmentinhk/waste/data/stat_treat.html (Retrieved on 2016-08-03)
[2] http://www.epd.gov.hk/epd/msw/htm_en/ch02/main.htm (Retrieved on 2016-08-03)
[3] Lin C Y and Lin H A 2006 Diesel engine performance and emission characteristics of biodiesel produced by the peroxidation process *Fuel* **85(3)** 298-305
[4] Shakun J D et al 2012 Global warming preceded by increasing carbon dioxide concentrations during the last deglaciation *Nature* **484(7392)** 49-54
[5] Ma F and Hanna M A 1999 Biodiesel production: A review *Bioresour. Technol.* **70(1)** 1-15
[6] Yahya A and Marley S J 1994 Performance and exhaust emissions of a compression ignition engine operating on ester fuels at increased injection pressure and advanced timing *Biomass*

Bioenerg. **6(4)** 297-319
[7] Nwafor O M I and Rice G 1996 Performance of rapeseed oil blends in a diesel engine *Appl. Energy* **54(4)** 345-54
[8] Korus R A, Jo J and Peterson C L 1985 A rapid engine test to measure injector fouling in diesel engines using vegetable oil fuels *J. Am. Oil Chem. Soc.* **62(11)** 1563-4
[9] Engler C R, Johnson L A, Lepori W A and Yarbrough C M 1983 Effects of processing and chemical characteristics of plant oils on performance of an indirect-injection diesel engine *J. Am. Oil Chem. Soc.*.**60(8)** 1592-6
[10] Bruwer J J *et al* 1981 Utilization of sunflower seed oil as a renewable fuel for diesel engines *ASAE Publ* (United States) vol 2
[11] Meher L C, Sagar D V and Naik S N 2006 Technical aspects of biodiesel production by transesterification—a review *Renew. Sustainable Energy Rev.* **10(3)** 248-68
[12] Noureddini H, Gao X and Philkana R S 2005 Immobilized Pseudomonas cepacia lipase for biodiesel fuel production from soybean oil *Bioresour. Technol.* **96(7)** 769-77
[13] Orçaire O, Buisson P and Pierre A C 2006 Application of silica aerogel encapsulated lipases in the synthesis of biodiesel by transesterification reactions *J. Mol. Catal. B Enzym.* **42(3)** 106-13
[14] Du W, Li W, Sun T, Chen X and Liu D 2008 Perspectives for biotechnological production of biodiesel and impacts *Appl. Microbiol. Biotechnol.* **79(3)** 331-7
[15] Röttig A, Wenning L, Bröker D and Steinbüchel A 2010 Fatty acid alkyl esters: perspectives for production of alternative biofuels *Appl. Microbiol. Biotechnol.* **85(6)** 1713-33
[16] Leung D Y, Wu X and Leung M K H 2010 A review on biodiesel production using catalyzed transesterification *Appl. Energy* **87(4)** 1083-95
[17] Dossat V, Combes D and Marty A 1999 Continuous enzymatic transesterification of high oleic sunflower oil in a packed bed reactor: Influence of the glycerol production *Enzyme Microb. Technol.* **25(3)** 194-200
[18] Watanabe Y, Shimada Y, Sugihara A, Noda H, Fukuda H and Tominaga Y 2000 Continuous production of biodiesel fuel from vegetable oil using immobilized Candida antarctica lipase *J. Am. Oil Chem. Soc.* **77(4)** 355-60
[19] Köse Ö, Tüter M and Aksoy H A 2002 Immobilized Candida antarctica lipase-catalyzed alcoholysis of cotton seed oil in a solvent-free medium *Bioresource Technology* **83(2)** 125-9
[20] Freedman B E H P, Pryde E H and Mounts T L 1984 Variables affecting the yields of fatty esters from transesterified vegetable oils *J. Am. Oil Chem. Soc.* **61(10)** 1638-43
[21] Freedman B, Butterfield R O and Pryde E H 1986 Transesterification kinetics of soybean oil 1 *J. Am. Oil Chem. Soc.* **63(10)** 1375-80
[22] Karmee S K and Chadha A 2005 Preparation of biodiesel from crude oil of Pongamia pinnata *Bioresour. Technol.* **96(13)** 1425-9
[23] Dorado M P, Ballesteros E, López F J and Mittelbach M 2004 Optimization of alkali-catalyzed transesterification of Brassica C arinata oil for biodiesel production *Energy Fuels* **18(1)** 77-83
[24] Casas A, Fernández C M, Ramos M J, Pérez Á and Rodríguez J F 2010 Optimization of the reaction parameters for fast pseudo single-phase transesterification of sunflower oil *Fuel* **89(3)** 650-8
[25] Math M C and Irfan G 2007 Optimization of restaurant waste oil methyl ester yield *J. Sci. Ind. Res.* **66(9)** 772
[26] Saifuddin N and Chua K H 2004 Production of ethyl ester (biodiesel) from used frying oil: Optimization of transesterification process using microwave irradiation *Malaysian Journal of Chemistry* **6(1)** 77-82
[27] Rashid U, Anwar F, Moser B R and Ashraf S 2008 Production of sunflower oil methyl esters by optimized alkali-catalyzed methanolysis *Biomass Bioenergy* **32(12)** 1202-5
[28] Berchmans H J and Hirata S 2008 Biodiesel production from crude Jatropha curcas L. seed oil

with a high content of free fatty acids *Bioresour. Technol.* **99(6)** 1716-21
[29] Shahid E M and Jamal Y 2011 Production of biodiesel: A technical review *Renew. Sustainable Energy Rev.* **15(9)** 4732-45
[30] Singh A, He B, Thompson J and Van Gerpen J 2006 Process optimization of biodiesel production using alkaline catalysts *Appl. Eng. Agric.* **22(4)** 597-600
[31] Xie W, Peng H and Chen L 2006 Transesterification of soybean oil catalyzed by potassium loaded on alumina as a solid-base catalyst *Appl. Catal. A.* **300(1)** 67-74
[32] Suppes G J, Dasari M A, Doskocil E J, Mankidy P J and Goff M J 2004 Transesterification of soybean oil with zeolite and metal catalysts *Appl. Catal. A.* **257(2)** 213-23
[33] Liu X, He H, Wang Y, Zhu S and Piao X 2008 Transesterification of soybean oil to biodiesel using CaO as a solid base catalyst *Fuel* **87(2)** 216-21
[34] Kawashima A, Matsubara K and Honda K 2008 Development of heterogeneous base catalysts for biodiesel production *Bioresour. Technol.* **99(9)** 3439-43
[35] Hancsók J, Kovács F and Krár M 2004 Production of vegetable oil fatty acid methyl esters from used frying oil by combined acidic/alkali transesterification *Pet. Coal* **46(3)** 36-44
[36] Canakci M and Van Gerpen J 2001 Biodiesel production from oils and fats with high free fatty acids *Trans. ASAE* **44(6)** 1429
[37] Talens L, Villalba G and Gabarrell X 2007 Exergy analysis applied to biodiesel production *Resour. Conserv. Recy.* **51(2)** 397-407
[38] Freedman B E H P, Pryde E H and Mounts T L 1984 Variables affecting the yields of fatty esters from transesterified vegetable oils *J. Am. Oil Chem. Soc.* **61(10)** 1638-43
[39] Freedman B, Butterfield R O and Pryde E H 1986 Transesterification kinetics of soybean oil *J. Am. Oil Chem. Soc.* **63(10)** 1375-80
[40] Upadhyay A K 2012 Biodiesel production and fuel quality-review *Res. J. Eng. Technol.* **3(3)** IV
[41] Crabbe E, Nolasco-Hipolito C, Kobayashi G, Sonomoto K and Ishizaki A 2001 Biodiesel production from crude palm oil and evaluation of butanol extraction and fuel properties *Process Biochem.* **37(1)** 65-71
[42] Sinha S, Agarwal A K and Garg S 2008 Biodiesel development from rice bran oil: Transesterification process optimization and fuel characterization *Energy Convers. Manage.* **49(5)** 1248-57
[43] Peng B X, Shu Q, Wang J F, Wang G R, Wang D Z and Han M H 2008 Biodiesel production from waste oil feedstocks by solid acid catalysis *Process Saf. Environ. Prot.* **86(6)** 441-7
[44] Furuta S, Matsuhashi H and Arata K 2004 Biodiesel fuel production with solid superacid catalysis in fixed bed reactor under atmospheric pressure *Catal. Commun.* **5(12)** 721-3
[45] Jitputti J, Kitiyanan B, Rangsunvigit P, Bunyakiat K, Attanatho L and Jenvanitpanjakul P 2006 Transesterification of crude palm kernel oil and crude coconut oil by different solid catalysts *Chem. Eng. J.* **116(1)** 61-6
[46] Kiss A A, Dimian A C and Rothenberg G 2006 Solid acid catalysts for biodiesel production-towards sustainable energy *Adv. Synth. Catal.* **348(1-2)** 75-81
[47] Kulkarni M G, Gopinath R, Meher L C and Dalai A K 2006 Solid acid catalyzed biodiesel production by simultaneous esterification and transesterification *Green Chem.* **8(12)** 1056-62
[48] Takagaki A, Toda M, Okamura M, Kondo J N, Hayashi S, Domen K and Hara M 2006 Esterification of higher fatty acids by a novel strong solid acid *Catal. Today* **116(2)** 157-61
[49] Zong M H, Duan Z Q, Lou W Y, Smith T J and Wu H 2007 Preparation of a sugar catalyst and its use for highly efficient production of biodiesel *Green Chem.* **9(5)** 434-7
[50] Lou W Y, Zong M H and Duan Z Q 2008 Efficient production of biodiesel from high free fatty acid-containing waste oils using various carbohydrate-derived solid acid catalysts *Bioresour. Technol.* **99(18)** 8752-8

Modelling the transport of oil after a proposed oil spill accident in Barents Sea and its environmental impact on Alke species

J Lu[1,2], F Yuan[1], J D Mikkelsen[1], C Ohm[1], E Stange[1] and M Holand[1]

[1]Department of Engineering and Safety, UiT The Arctic University of Norway, N-9037 Tromsø, Norway

E-mail: Jinmei.lu@uit.no

Abstract. Accidental oil spills can have significant effect on the coastal and marine environment. As the oil extraction and exploration activities increase in the Barents Sea, it is of increasingly importance to investigate the potential oil spill incidents associated with these activities. In this study, the transport and fate of oil after a proposed oil spill incident in Barents Sea was modelled by oil spill contingency and response model OSCAR. The possibility that the spilled oil reach the open sea and the strand area was calculated respectively. The influence area of the incident was calculated by combining the results from 200 simulations. The possibility that the spilled oil reach Alke species, a vulnerable species and on the National Red List of birds in Barents Sea, was analyzed by combining oil spill modelling results and the Alke species distribution data. The results showed that oil is dominated with a probability of 70-100% in the open sea to reach an area in a radius of 20km from the release location after 14 days of release. The probability reduces with the increasing distances from the release location. It is higher possibility that the spilled oil will reach the Alke species in the strand area than in the open sea in the summer. The total influence area of the release is 11 429 km^2 for the surface water and 1528 km^2 for the coastal area.

1. Introduction

Among the different types of marine pollution, oil is a major threat to the sea ecosystems [1]. Accidental oil spills to the marine environment has received great concern due to their potential impacts on the marine ecosystems and marine activities [1, 2]. There are serious consequences associated with an acute discharge, when large amounts of oil suddenly released into the natural environment and reach the vulnerable ecosystem. Examples of accidental oil spills involve vessels that come in distress or collide, oil well blowouts, pipeline ruptures, and explosions at storage facilities etc. [3]. This has happened several times in history. Historically, the largest accidental oil spill worldwide was a blowout at the Ixtoc-1 well that released 480000 tons crude oil into the Gulf of Mexico over a 10-month period from June 1979 to February 1980 [3]. The Prestige sank in northwestern Spain on 19 November 2002 and contaminated the Galician coastline with 60000 tons of oil [4]. The incident caused serious ecological damage along 900 km of coastline, from northern Portugal to southern France, and some areas of the coast are still not yet fully clean after two years of the incident happened [4]. A total of 12,700 cubic meters of oil suddenly released into the environment in an uncontrolled blowout accident happened in Ekofisk field in 1977 [5], which got high international attention, and became the basis for the strict environmental requirements on the Norwegian continental shelf [6].

Another acute oil spill incident from the grounded oil tanker "Exxon Valdez" outside Alaska in 1989, resulted in a sudden spill of 42 000 cubic meters of oil to the environment and 23 species and habitats were affected. After 25 years there were only 13 species that could be considered fully recovered [7]. The Deepwater Horizon drill rig explosion and subsequent sinking in 2010 resulted in the largest offshore oil spill in US history. This spill lasted for 3 months, and resulted in serious consequences to the Gulf of Mexico, its coastal zone and living marine resources [8-10].

As oil extraction and exploration activities increase in the Barents Sea, the probability for an acute oil spill incident increases thereby. Since accidental oil spills can cause serious consequences to the recipient environment and the ecosystems living in the recipient environment, an environmental risk assessment is of great importance for oil companies in the planning and implementation of various activities. Modelling the transport and fate of oil after the oil spill incident is an important step in evaluating the potential risk to the exposed ecosystems.

In this study, the transport and fate of oil after a proposed oil spill incident in Barents Sea was modelled with OSCAR model. The possibility that oil will reach the open sea and shoreline and the influence area following the oil spill accident was investigated. The possibility that the spilled oil will reach Alke, an important species in Barents Sea, was also investigated.

2. Methodology

2.1. Oil spill transport model-OSCAR

The OSCAR (Oil Spill Contingency and Response) model is developed by SINTEF in Norway and is a state of the art, 3-dimensional dynamic simulation tool for planning and response for oil spill. OSCAR model can simulate the transport, fate and effects of oil from an oil spill incident. OSCAR system has several key components: OWM (SINTEF's data-based oil weathering model), FATES (a three-dimensional oil trajectory and chemical fates model) and an oil spill COMBAT model, and exposure model for fish and ichthyic plankton, birds and marine mammals (figure 1) [11]. FATES model calculates both the total and the dissolved concentration of oil in the water column and in the sediment, and the total area of the sea and strand area that is contaminated by oil. Therefore this model needs the sea data and wind data as input parameters. The exposure model is able to analyze which sensitive areas will be affected by oil spill and how the species in the water will be affected [12].

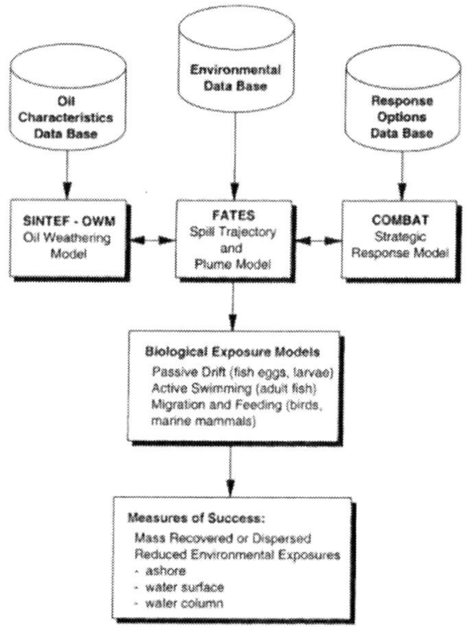

Figure 1. Overview of the OSCAR model [11].

IOP Conf. Series: Earth and Environmental Science **82** (2017) 012010 doi:10.1088/1755-1315/82/1/012010

OSCAR model can support both 2D and 3D modelling. The 2D modeling is for surface release. 3D modelling can be used to investigate the oil concentrations, amount of oil in the sediment and in the strand. OSCAR employs surface spreading, advection, entrainment, emulsification, and volatilization algorithms to determine transport and fate at the surface.

OSCAR model can also be used to perform stochastic simulations. The scenario can be simulated hundreds or thousands of times. Afterwards the software can export several factors linked to oil release, for example average oil amount, shortest arrival time and possibility to reach the area, to the user defined grid. The software support GIS. Different data set with geographical information can be imported into the software, such as the bird and fish species data in the context of an oil spill. The analysis results can also export as GIS supported format.

2.2. Python

Python is an interpreted, interactive, object-oriented programming language [13]. The programming was designed by Guido van Rossum, and first published in February 1991. Python is an open software and can be used to treat large amounts of data [13].

In this study, Python version 3.5.0 is used to link the stochastic data from OSCAR and bird population data. The exported data from OSCAR contains 74218 rows, which is time consuming to connect each cell of this data with the corresponding cell for bird population manually. The stochastic data from OSCAR was linked to the data from bird population by their common parameter IDCell. By this way, each cell with different oil spill amount in the scenario is linked together with the corresponding number of Alke species (figure 2).

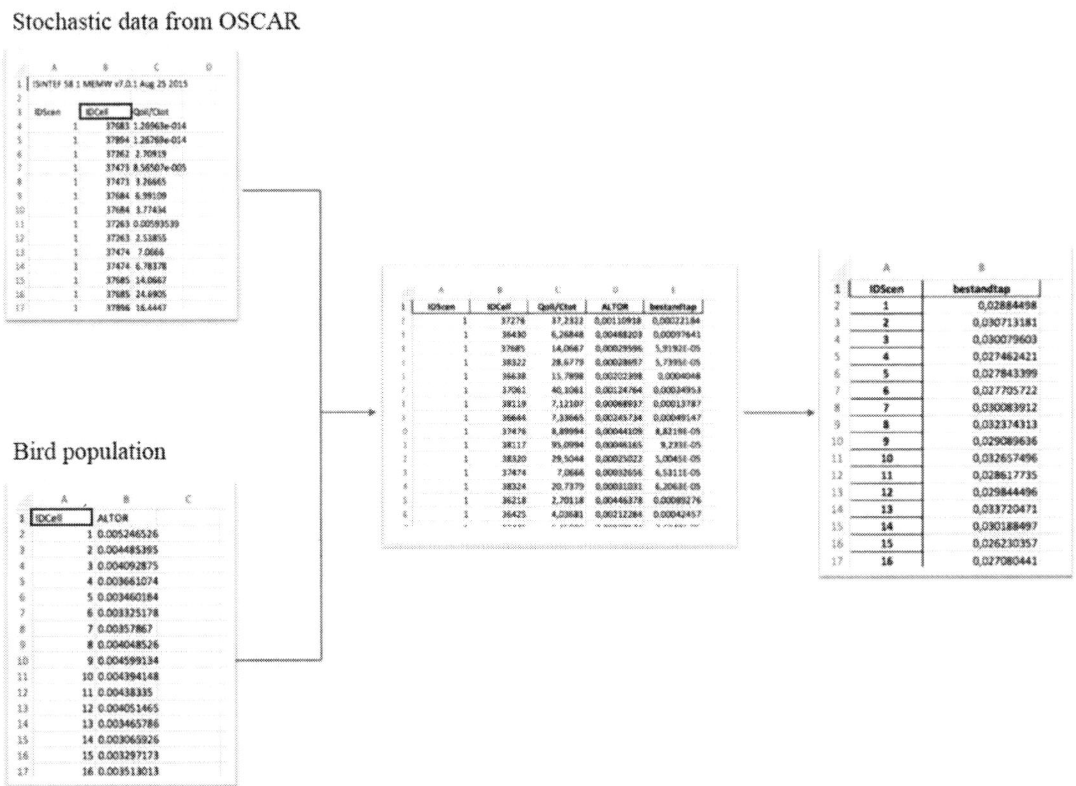

Figure 2. Simplified interpretation on link the data from OSCAR and data from species population.

2.3. ArcGIS
ArcGIS is a geographic information system (GIS) software that deals with maps and other geographic information. It is developed by ESRI (Environmental Systems Research Institute). ArcGIS can be used to make and present map, to put together different geographic data, to analyze and maintain geographic information in the databases.

In this study, ArcGIS version 10.0 is used to present the results from oil disperse modelling and combine the Alke species data with oil disperse modelling results.

3. Methodology

3.1. Location for oil disperse modelling
Area PL 764 is selected as the study area. It is relatively close to the shore. Therefore, it is higher possibility for the oil to reach the strand area in case of an accidental oil spill. The assumed location of oil spill in the current study is 70.80°N, 19.67°E as shown in figure 3. The shortest distance to the shore from the location is 57 km.

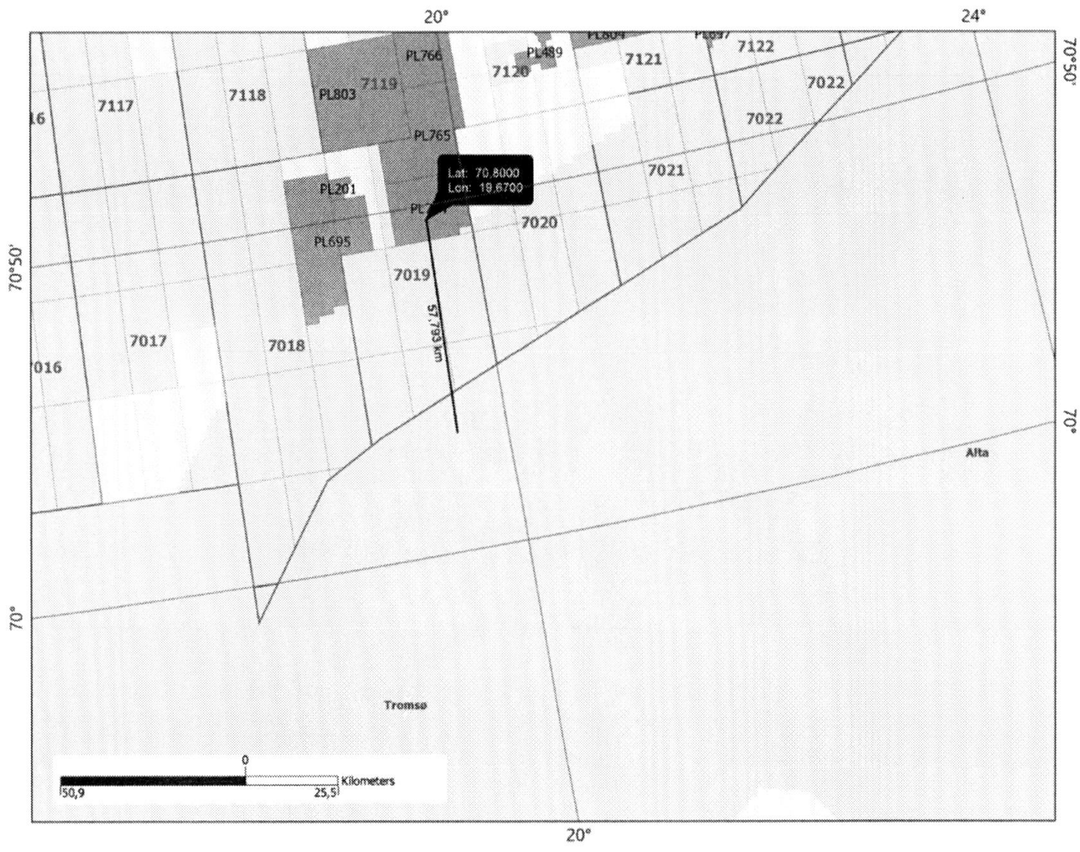

Figure 3. Location of oil spill incident.

3.2. Spilled oil types used in the model
PL 764 is an area that hasn't started the test drilling. There is no previous data about what types of oil exist there. Due to its relatively short distance to Goliat-field, the oil type Goliat Blend was selected first. However, Goliat Blend oil doesn't exist in the OSCAR system. Therefore another oil type Alpine, which has similar properties as Goliat Blend oil, was selected in this study to do the simulation.

Alpine is a paraffinic crude oil. It contains large amounts of light molecular compounds. In case of an oil spill, about 25% of the total volume of spilled oil will be evaporated within 24 hours, either during the summer season or during the winter season. The oil has a pour point of -18°C. At a wind speed of 10 m/s, the pour point increases significantly to 10°C in winter and 15°C in summer with weathering [14].

3.3. Data collection

3.3.1. Ocean current. Ocean current data is collected from Institute of Marine Research and modelled with NorKyst-800 model. NorKyst-800 (Norwegian Coast 800 m) is a numerical ocean modelling system suitable for reproduction of physical variables as sea level, temperature, salinity, and currents for all coastal areas in Norway and adjacent seas [15]. The full NorKyst-800 bathymetric grid consists of 2600×900 grid cells with each cell having an area of 800×800 m [15].

The modelled data from 01.05.2010 to 31.07.2010 was collected. This will be sufficient to cover an oil dispersion duration of 70 days. Figure 4 shows the area modelled in NorKyst-800.

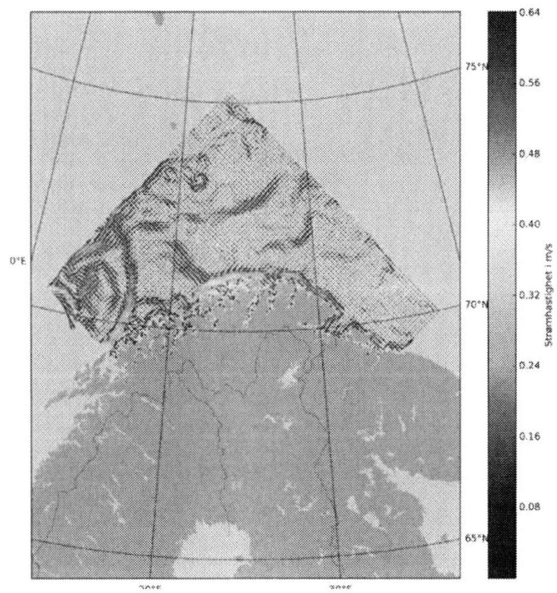

Figure 4. Average value for ocean current in the modelling period.

In oil dispersion modelling, the data used have a time resolution of one hour. Therefore the ocean data is modelled for each hour in the period. Sakshaug et al mentioned that the currents in the Barents Sea is low compared with the upstream speed of the Norwegian Sea, which is between 0.5-1 m/s [16]. The Norwegian coastal current has a near-coastal speed in excess of 0.25 m/s [16]. Our current data (figure 4) is in consistent with their study.

3.3.2. Wind. The wind data is collected from the Meteorological Institute. The data is Hindcast over that area. Hindcast data is generated by running the numerical model with the historical data. In contrast to weather observation stations, the Hindcast model can cover a much bigger area than individual observations. In such a way, the wind data for the area without monitoring stations can also be generated, for example the Barents Sea [17].

The wind data is from HIRLAM10 model, the High Resolution Limited Area Model. The model has a resolution of 10-11 km. HIRLAM10 atmospheric downscaling and wave hindcast is performed based on the ERA40 reanalysis covering the Norwegian Sea, the North Sea and the Barents Sea [17].

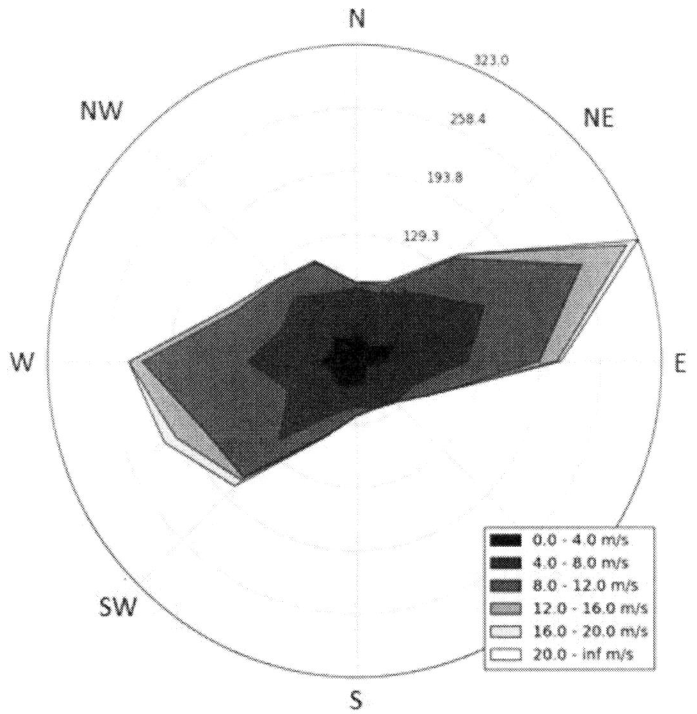

Figure 5. Wind rose from 70.81°N, 19.75°E in the period 2010.05.01 to 2010.07.31 with HIRLAM10-wind data.

Wind data for the corresponding period and the same time resolution as the ocean currents is used (figure 5). Figure 5 shows that the dominating wind direction in the study area is from southwest, west and between east and northeast. The numbers in figure 5 are the number of measurements made on the wind speed and wind direction. The size of the areas with different colors indicates the number of measurements made at the corresponding wind speed. The bigger the area is, the more measurements that has been made at that wind speed category. The data set consists of 2208 measurements at different wind directions. The average wind speed is 7.07 m/s.

3.3.3. Bird Species. Data for bird species is collected from SEAPOP. SEAPOP is organized and carried out by personnel from the Norwegian Institute for nature research (NINA), Norwegian Polar Institute (NP) and Tromsø Museum, University Museum (TMU).

Alke is a pelagic diving seabird. It can dive down to 100 meters. According to the estimates in 2013, the population of Alke species in Norway was just under 55,000 [18]. This represents 5% of the total population in Europe. Of these, it is estimated that just under 45,000 couples living in the Barents Sea. Alke is on the National Red List of birds that are highly endangered (EN). Since the species has had a 50-80% reduction in the last three generations, it is now classified as highly endangered [19]. Since Alke is a very vulnerable and sensitive species in the study area, Alke is selected in this study to investigate the impact on it from the proposed oil spill incident.

The Alke species data in the summer is collected from SEAPOP. The definition of summer period in SEAPOP is between 1st April and31st July. This is consistent with the time period for the ocean current data and wind data. The number of each species is modelled and presented in a 10×10 km network. The distribution of the Alke species in the open sea and along the coast is analyzed.

- The distribution of Alke at open sea

The distribution of Alke at open sea is shown in figure 6. From figure 6, the number of alke species at open sea is quite low.

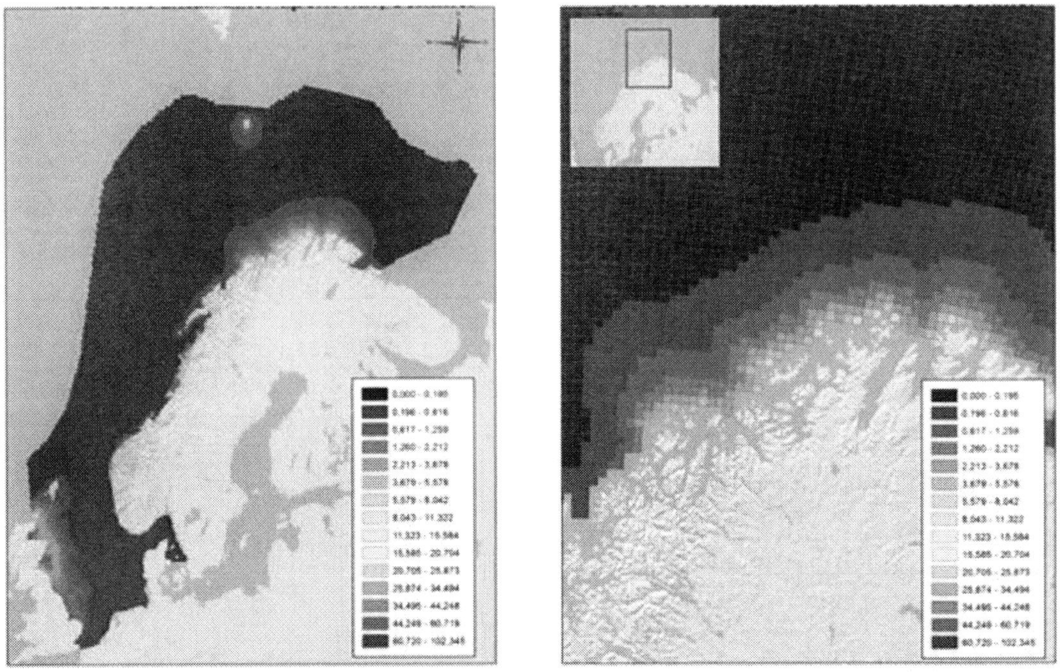

Figure 6. Distribution of Alke species at open sea.

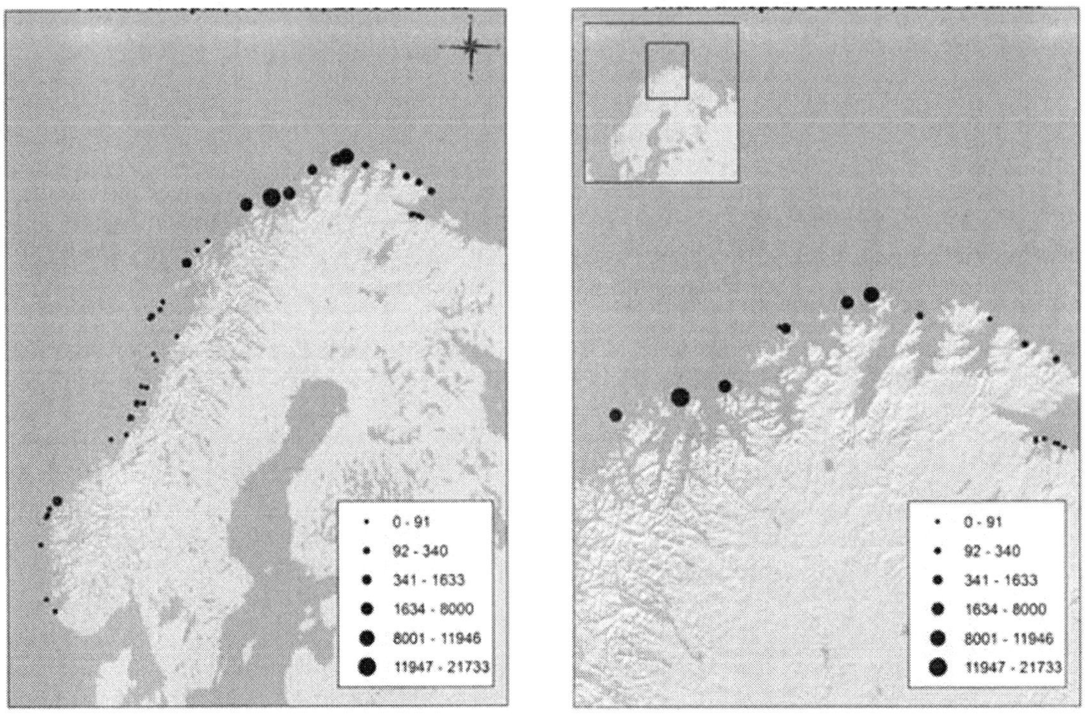

Figure 7. Estimated Alke species in summer 2013 along the coast.

- The distribution of Alke along the coast

The population size of Alke species along the coast in the breeding period is mapped. The number is calculated by counting the number of breeding couples from flight, boat and from land [20]. The localities that SEAPOP count on are distributed at different locations in Norway as shown in figure 7. In our study for the coastal area, we used the estimated number of Alke couples in the summer period in 2013, as shown in figure 7.

4. Results

4.1. Probability that oil reach open sea and the strand area at an accidental oil spill in Barents Sea

The probability that oil will reach the strand area and the open sea 14 days after the accidental spill is simulated with the OSCAR model. The results are combined with estimated Alke species number in summer 2013 and shown in figures 8 and 9 with the help of ArcGIS.

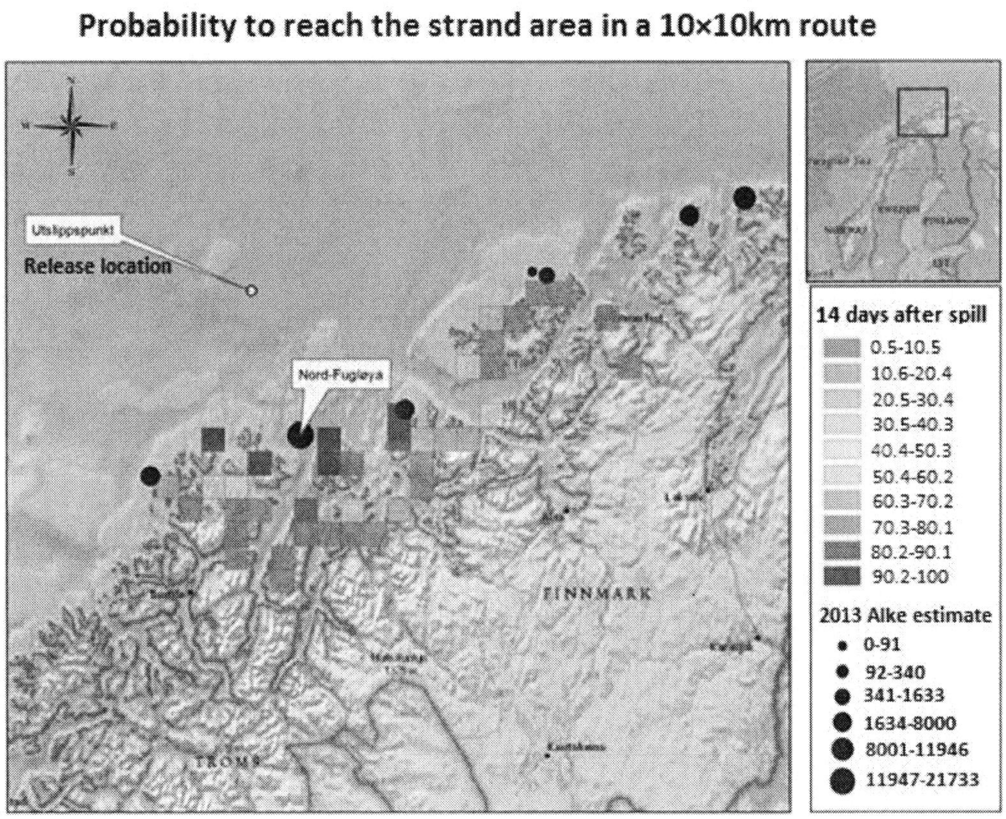

Figure 8. The probability that oil reach the strand area.

As it is shown in the figures, it is much higher number of Alke species in the strand than in the open sea in our analysis period. The breeding of Alke species is ongoing in the summer season. Therefore the population distribution of Alke species corresponding to this period dominates in the strand area since Alke reside in the strand during their breeding. The probability that the spilled oil will reach the Alke species is higher in the strand area than in the open sea.

Oil is dominated with a probability of 70-100% in the open sea to reach an area in a radius of 20km from the release location after 14 days of release. However the probability to reach the strand varies between 1-20%, with the exception of some points with 90-100% probability. The probability reduces with the increasing distances from the release location.

Nord-Fugløya is estimated to be the largest breeding place for Alke. It is zero probability that the spilled oil will reach this place. Reason to this maybe the resolution of the model is too low. However, it is very high probability that oil will reach the surrounding area of this place (figure 8). More modelling with high resolution is needed therefore.

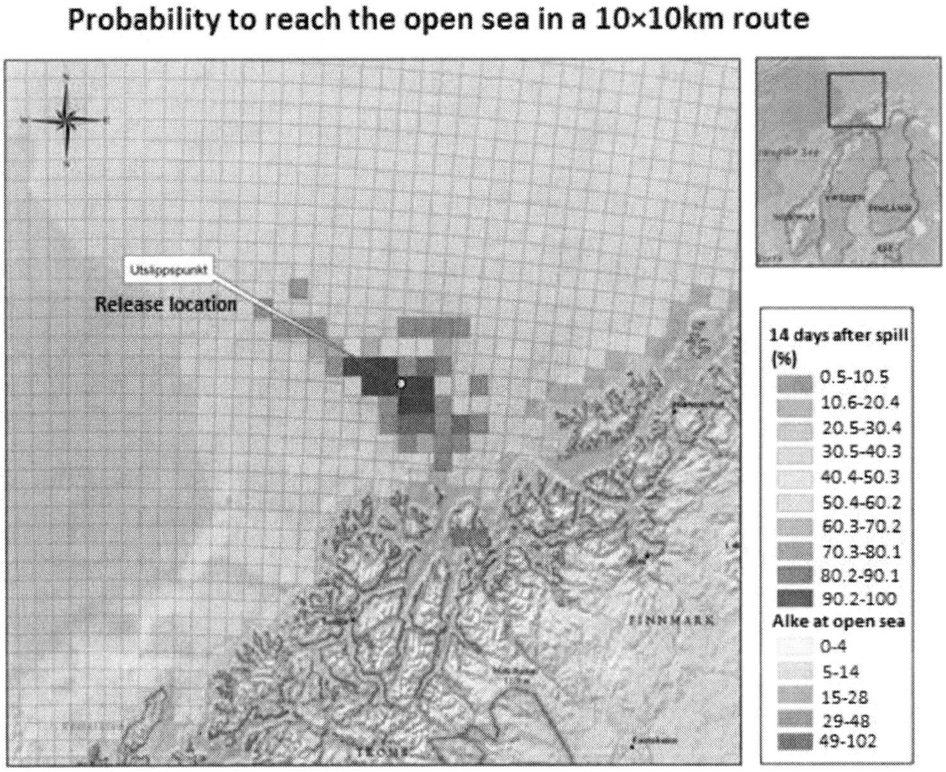

Figure 9. Probability that oil reach open sea.

4.2. The influence areas following the accidental oil spill

To illustrate how the influence area is calculated, single simulation during our modelling period is shown. By looking at three simulations at different times with identical parameters, the influence of the wind and current data on the scope of oil dispersion is analyzed.

Simulated oil release on 1st May:

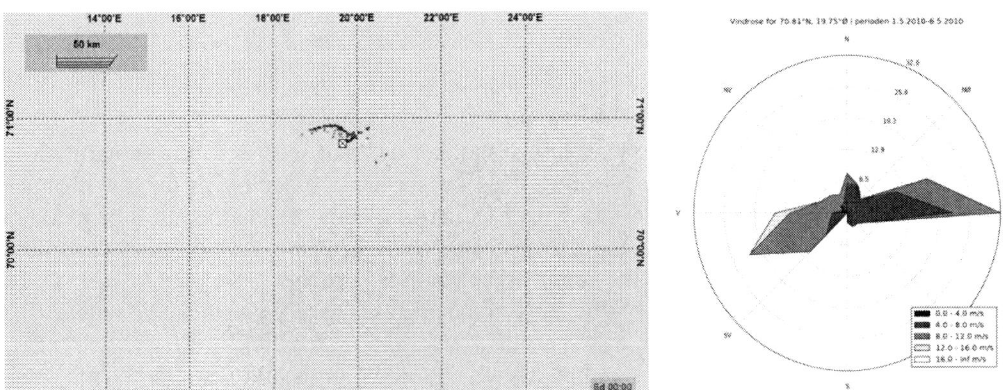

Figure 10. Simulated oil disperse after six days of release, and the corresponding wind rose.

IOP Conf. Series: Earth and Environmental Science **82** (2017) 012010 doi:10.1088/1755-1315/82/1/012010

Simulated oil release on 1st June:

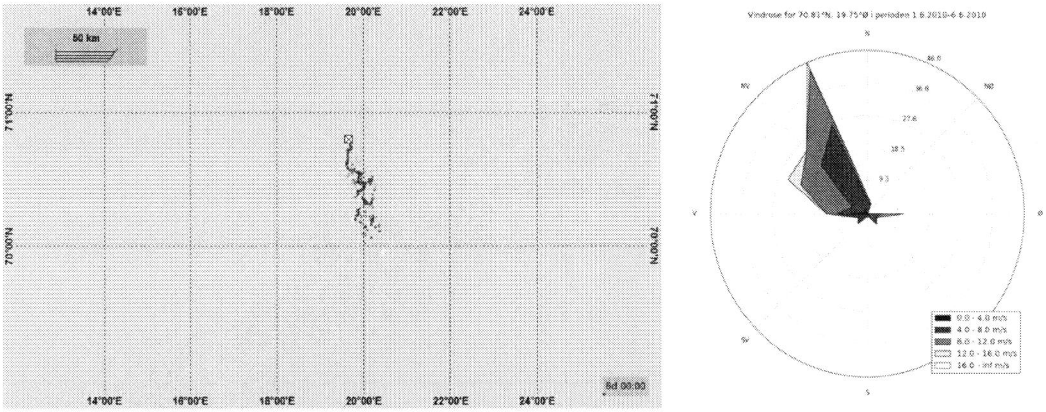

Figure 11. Simulated oil disperse after six days of release, and the corresponding wind rose.

Simulated oil release on 1st July:

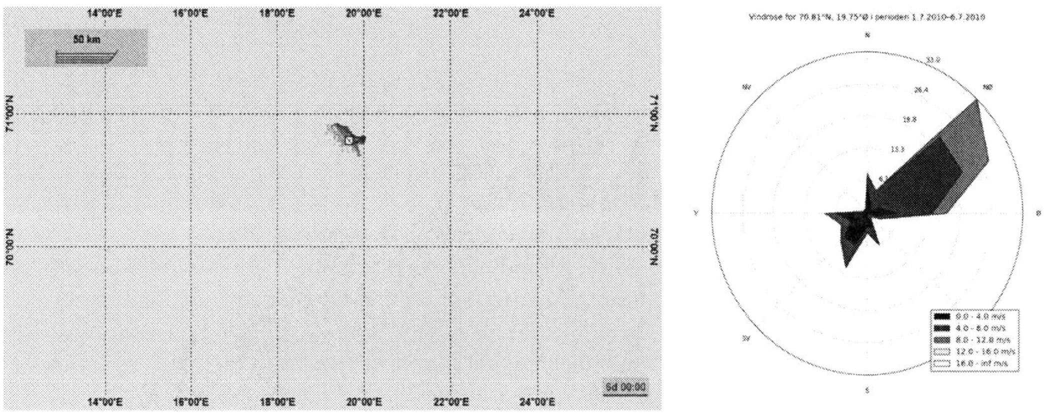

Figure 12. Simulated oil disperse after six days of release, and the corresponding wind rose.

Results from the single simulations are compared with the corresponding wind rose for the same period. Figures 10-12 illustrates how the oil on the sea surface will be affected by the dominating wind direction. The figures showed how the current and wind direction will affect the transportation of the spilled oil. In addition, the influence of currents counteracts the spread of oil from the prevailing wind direction is shown in figure 12.

The potential influence area of the release in this study is calculated by combining the results of all the 200 simulations to one total risk picture (figure 13). Oil dispersion calculations by OSCAR formed the basis for the influence area.

The total influence area of the release is 11 429 km^2 for the surface water and 1528 km^2 for the coastal area.

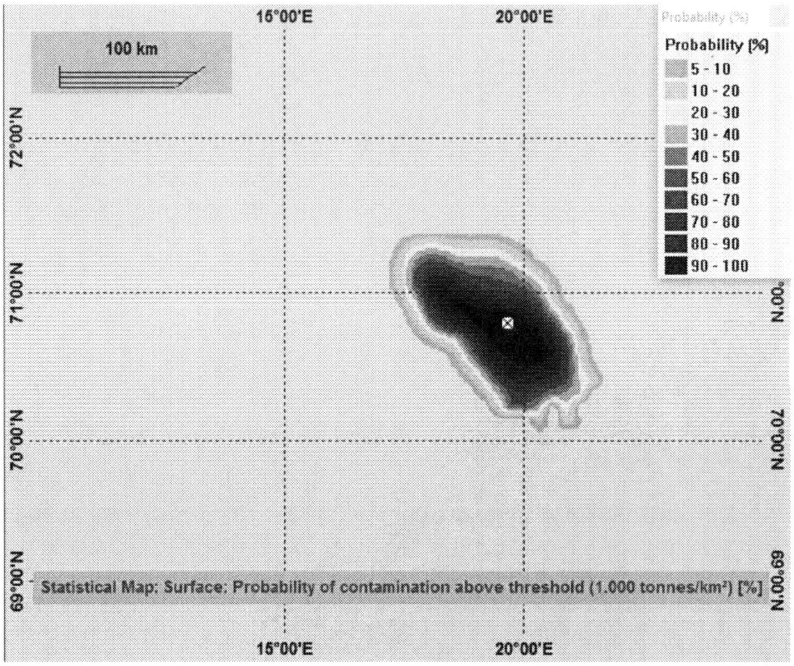

Figure 13. Influence area for our release.

5. Discussions

After an oil spill incident, the spilled oil will float on the surface of the water. Therefore species living at the surface of the water is most vulnerable to oil spill. The bird species must regularly pass through the air-water interface to breathe, are particularly vulnerable to oil exposure [21].

There are more Alke species reside in the strand area in the summer. So it is high possibility that the spilled oil will reach the species in the strand area than in the sea. The species are more affected in the places with high oil coverage on the shoreline. The shoreline surveillance on the environmental impact of the Exxon Valdez oil spill also indicated that high local mortality of species occurred at the place where oil coverage exceeded 90% [22].

The dominating wind direction has a significant effect on the transportation of fate of spilled oil. The effects of wind on the advection of a parcel of oil floating on the sea surface have been studied extensively [23]. Al-rabeh A.H. estimated the wind-induced surface oil spill speed and computed the associated deflection angle suitable for the Arabian Gulf [23].

Several data bases used in our study are based on the operations in Nordsjøen and Norwegian sea [24, 25]. Therefore it will be certain degree of uncertainty linked to them when we use these data to analyze an oil spill incident at Barents Sea.

6. Conclusions

The transportation and fate of oil in case of a proposed oil spill incident in Barents Sea is modelled with oil spill contingency and response model OSCAR. The results showed that it is higher probability that oil will reach the open sea than the strand. Oil is dominated with a probability of 70-100% in the open sea and between 1-20% probabilities to reach the strand. There are more Alke species distributed in the strand than that in the open sea in the summer. Therefore it is higher probability that the spilled oil will reach the Alke in strand than in the open sea. More Alke species will be affected in the strand than in the open sea. The total influence area of the release is 11 429 km^2 for the surface water and 1528 km^2 for the coastal area. More emergency response and management strategies should be developed for the strand area to counteract the potential oil spill accidents in the Barents Sea.

Acknowledgments
We would like to thank SINTEF to give the license for OSCAR and Meteorological Institute and the Institute of Marine research to assist with the essential data.

References

[1] Topouzelis K N 2008 Oil spill detection by SAR images: Dark formation detection, feature extraction and classification algorithms *Sensors* **8** 6642-59
[2] Gin K Y H, Huda K, Lim W K and Tkalich P 2001 An oil spill-food chain interaction model for coastal waters *Mar. Pollut. Bull.* **42** 590-7
[3] Burgherr P 2007 In-depth analysis of accidental oil spills from tankers in the context of global spill trends from all sources *J. Hazard Mater.* **140** 245-56
[4] Vieites D R, Nieto-Roman S, Palanca A, Ferrer X and Vences M 2004 European Atlantic: The hottest oil spill hotspot worldwide *Naturwissenschaften* **91** 535-8
[5] Oljeindustriens Landsforening. 2003 *Oljevernberedskap* available: https://www.norskoljeogga ss.no/PageFiles/8795/oljevernberedskap.pdf
[6] Norsk olje og gass 2010 *Olje- og gasshistorien* available: https://www.norskoljeogga ss.no/no/Faktasider/Oljehistorie/
[7] Grøsvik B E, Olsen E, Vikebø F B and Meier S 2014 *25 år etter Exxon Valdez* available: http://forskning.no/meninger/kronikk/2014/03/25-ar-etter-exxon-valdez
[8] Jernelov A 2010 The threats from oil spills: Now, then, and in the future *Ambio* **39** 353-66
[9] Hu C M *et al* 2011 Did the northeastern Gulf of Mexico become greener after the Deepwater Horizon oil spill? *Geophys. Res. Lett.* **38** 1633-44
[10] Mearns A J, Reish D J, Oshida P S and Ginn T 2010 Effects of Pollution on Marine Organisms *Water Environ. Res.* **82** 2001-46
[11] Reed M, Ekrol N, Rye H and Turner L 1999 Oil Spill Contingency and Response (OSCAR) analysis in support of environmental impact assessment offshore Namibia *Spill Sci. & Technol. Bull.* **5** 29-38
[12] Reed M, Aamo O M and Daling P S 1995 Quantitative-analysis of alternate oil-spill response strategies using oscar *Spill Sci. & Technol. Bull.* **2** 67-74
[13] Foundation P S *General Python FAQ* available: https://docs.python.org/2/faq/general.html
[14] Leirvik F 2002 *Weathering Properties of Endicott, Milne Point Unit, High Island Composite, the Alpine Composite, the Neptune Field Composite, and North Star Oil Samples* ed SINTEF Applied Chemistry, Marine Environmental Technology
[15] Albretsen J, Sperrevik A K, Staalstrøm A, Sandvik A D, Vikebø F and Asplin L 2011 NorKyst-800 report no. 1 User manual and technical descriptions (Institute of Marine Research)
[16] Sakshaug E, Johnsen G H and Kovacs K M 2009 *Ecosystem Barents Sea* (Tapir Academic Press)
[17] Reistad M, Breivik Ø, Haakenstad H, Aarnes O J and Furevik B R 2009 A high-resolution hindcast of wind and waves for the North Sea, the Norwegian Sea and the Barents Sea (Norwegian Meteorological Institute)
[18] SEAPOP 2016 *Population size (in Norwegian)* available: http://www.seapop.no/no/aktiviteter/ bestandsstorrelser/bestandsstorrelser.html
[19] Henriksen S and Hilmo O 2015 Veileder til rødlistevurdering for: Norsk rødliste for arter 2015 (Norge: Artsdatabanken)
[20] SEAPOP 2016 *Kartleggingsmetodikk på kysten* available: http://www.seapop.no/no/metoder/ kartlegging-kyst/index.html
[21] Chang S E, Stone J, Demes K and Piscitelli M 2014 Consequences of oil spills: A review and framework for informing planning *Ecol. Soc.* **19** 26
[22] Maki A W 1991 The Exxon Valdez oil spill: initial environmental impact assessment. Part 2 *Environ. Sci. Technol.* **25** 24-9
[23] Alrabeh A H 1994 Estimating surface oil-spill transport due to wind in the Arabian gulf *Ocean Engineering* **21** 461-5

[24] Frigstad S, Hoffman P, Andersen L A and Aspholm O Ø 2006 Frekvenser for uhellsutslipp av olje i Barentshavet DNV2006 available: https://www.regjeringen.no/globalassets/upload/kilde/md/rap/2005/0008/ddd/pdfv/275907-frekvens_for_utslipp_i_barentshavet_endelig_vers jon.pdf.

[25] Årstad I, Kristensen V, Wiencke H S, Roed W, Haver K and Eidesen K 2010 Frekvenser for akutte utslipp fra petroleumsvirksomheten available: http://evalueringsportalen.no/evaluering/frekvenser-for-akutte-utslipp-fra-petroleumsvirksomheten/Frekvenser%20for%20akutte%20utslipp%20fra%20petroleumsvirksomheten%20-%20Memorandum%20memonr%20PS%200357%20ME%2002.pdf/@@inline.

Prediction of soil water erosion risk within GIS- case study of Beni Amrane Dam catchment (North of Algeria)

S Touahir[1,4], K Khenter[2], B Remini[3] and H Saad[1]

[1]Hydraulic Department, Civil Engineering and Hydraulics Faculty, Hassiba Benbouali University of Chlef, Chlef, 02000, Algeria
[2]National High School for Hydraulics of Blida, Blida, 09000, Algeria
[3]Department of Water and Environmental Sciences, Faculty of Technology, University of Blida 1, Blida, 09000, Algeria

E-mail: s.berbache@ensh.dz

Abstract. Isser River is one of North Algeria's major resources. It is vulnerable to water soil erosion because of favourable conjunctions of different geomorphological, hydro-climatic and lithologic factors. This case study has been carried out on the Beni Amrane dam Catchment, which is located in the bottom of Isser River, in North Algeria. The study involves a mapping of main factors of water erosion: rainfall erosivity, soil erodibility, slope and land use. Essentially a data mapping specification analysis shows, on each factor, how to identify the areas that are prone to water erosion. 04 classes of multifactorial vulnerability to water erosion have been identified: areas with low vulnerability (10 per cent); area with middle vulnerability (49 per cent); areas with high and very high vulnerability (38 per cent and 3 per cent). This could be a first guidance document for a rational use of land in the region and better secure the Beni Amrane dam against reservoir siltation.

1. Introduction

Erosion and sediment transport are major constraints to agricultural development and dams' management. They cover 45% of Tell areas, i.e. 12 million cultivable hectares [1]. The volume of materials that collect each year in the reservoirs of the Algerian dams are estimated at 20 million m^3 on average per year [2], which contributes to reduce by 0.3% per year the water storage capacity estimated at 6.2 billion m^3 [3].

Diversity of hydrological and sedimentological characteristics in different parts of the basin involves different but related processes. At the basin mouth, sediment transport closely depends on the sediment load from upstream and on training conditions. The measurement done at this point is only the outcome of all heterogeneity of the basin. Spatial variability of the catchment basin, therefore, requires a fine discretization. For that purpose, the vision of these phenomena is very different depending on space and time scale at which each one places itself. Many works have settled strong relationships linking sediment transport with explanatory parameters in order to understand the complex mechanisms of sediment transport and quantify the volume of sediment transported. Because of the high cost of conservation and the concurrent production targets such as the increase in the population, the development of the infrastructures and the degradation of the grounds, it is necessary to target the solutions and the resources in the high-risk zones, rather than to equitably distribute them between the landscape [4]. The cartography of erosion is a fundamental tool to know the distribution and the geographical extent of the phenomenon as well as its qualitative characterization [5]. Our primordial goal is based on the knowledge and understanding of the environmental dynamics. Thus, a diagnosis of natural environment situation on geographical framework is necessary, especially within a geographic information system (GIS).

The case study of Beni Amrane dam Catchment is located at the bottom waters of Isser River. It extends between longitudes 3° 20' and 3° 60' East and between latitudes 36° 40' and 36° 10' North (CF figure 1). It covers an area of 847 km^2, around more than 20% of the total surface area of Isser basin to which it belongs. The Beni Amrane dam catchment has an annual run off mean of 414 million m^3/an. Its mouth is at 70 km south east of Algiers.

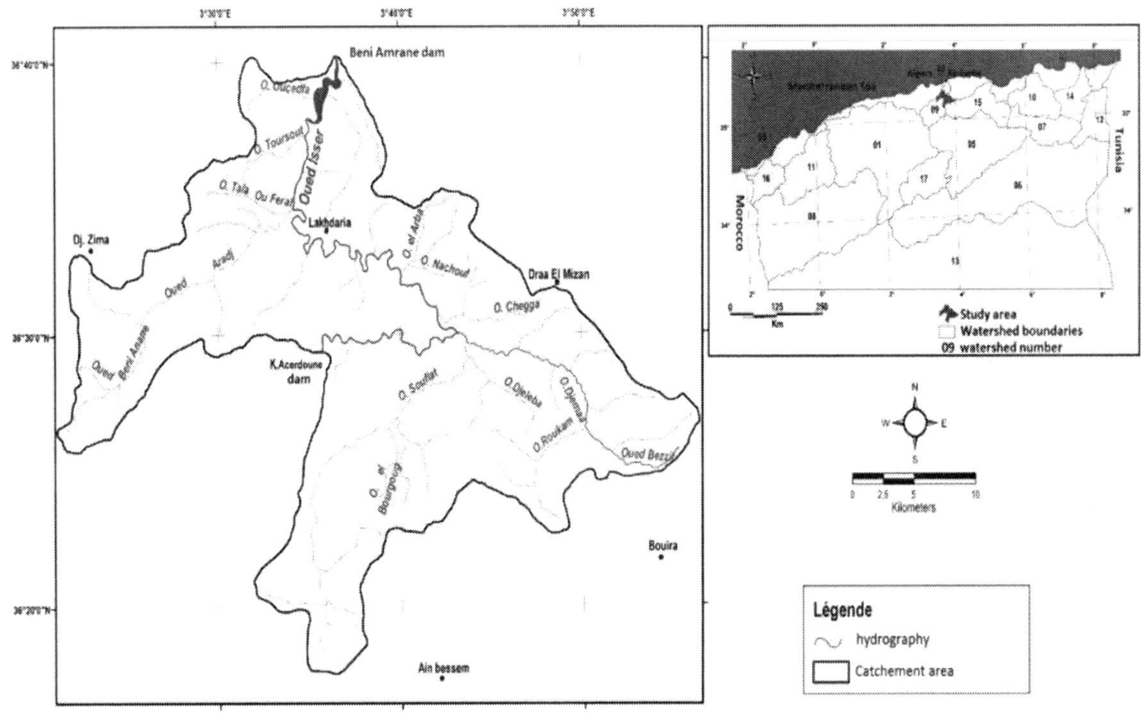

Figure 1. Geographic situation of Beni Amrane dam Catchement area.

2. Materials and methods

Since soil erosion is particularly difficult to measure directly in real time, it is necessary to use models that allow the estimation of the rate or state of this phenomenon. Several erosion models have been developed, some of which have an empirical basis such as the USLE (Universal Soil Loss Equation) [6] and its modified versions, others physically based such as the project of prediction of water erosion, WEPP of Foster and Lane.

The choice of the model depends on the variability of the basin area, punctual data and precipitation [7]. These models require very precise sampling based on punctual measurements and very large scale maps. In the study area, we only have maps at scales less than or equal to 1/500,000, and the lack of data on the region's soil structure and precipitation intensity does not allow us to use a quantitative model of the type USLE. To overcome these drawbacks, we have opted for a method based on a qualitative model of expert system based on a crossing of parameters determining erosion in the form of logical combinations, namely, lithology, rainfall, slope Topography and land use, whose weights have to be balanced [8] based on current knowledge on different types of erosive functioning. Cartographic and descriptive data on the factors influencing the water erosion process will be integrated into a GIS.

The methodology is described according to the flow chart shown in figure 2.

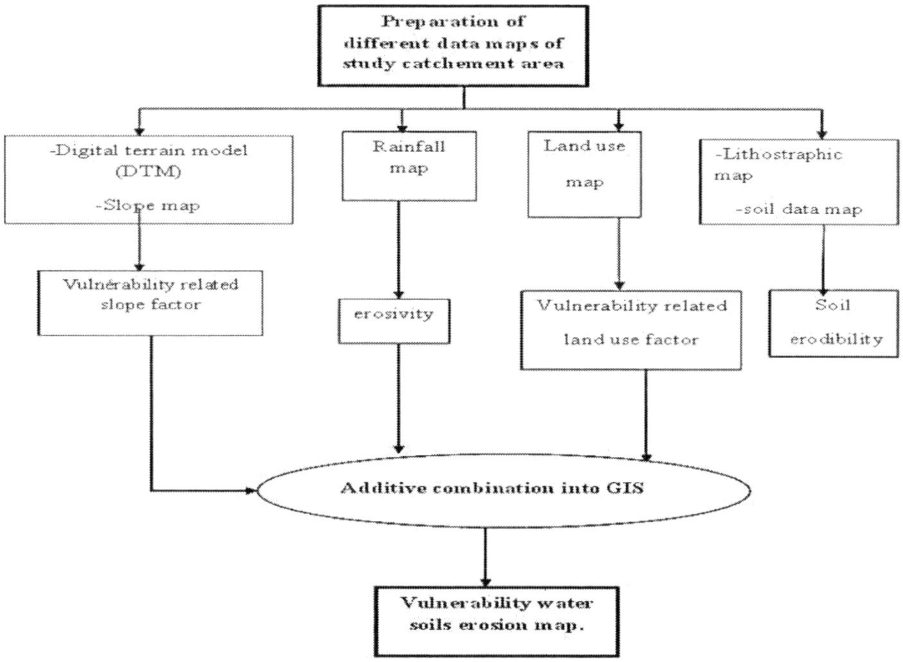

Figure 2. Different processing stages of erosion risk evaluation map.

Data acquisition required a bibliographic approach and electronic data discoveries on the data provided pedologic and geologic service of the National Water Resources Agency [9] and the National Institute of soil, irrigation and drainage [10]. After processing, it was possible to make the main elements and to use the predict approach.

Map data were taken back to ANRH, INSID and the National Meteorology Office (ONM) and were operated. Those data mainly consist of the following:

- A rainfall map of northern Algeria, drawn up on the annual averages of rainfall for the period (1965/1966- 2001/2005), at the scale 1/500000 [9];
- Soil use maps for the following Wilayas: Boumerdes, Bouira, TiziOuzou and Medea at the scale 1/25000: The year 2011 [10];
- Lithostratigraphic map of northern Algeria at the scale 1/200000 [9];
- Hydro-climatological and water quality monitoring map sat the scale 1/500000 [9];
- Shuttle Radar Topography Mission Images With coordinates: N35E003, N35E004, N36E003, N36E004: 23rd September 2014; (source; https://earthexplorer.usgs.gov/).

3. Results

The processing of one-factor vulnerability database was performed with the professional software MapInfo 11.0, and ArcGIS 10.0 (maps digitalization).

3.1. Rainfall map

Rainfall-related erosion parameters are:

The height of precipitation: it has little to do with the amount of erosion, the higher the precipitation height, the greater the volume of water runoff, which promotes water erosion.

Intensity is the main factor in erosion. The greater the intensity, the stronger the threshing effect of the ground.

In the study area only data from the rainfall station of Djebahia (09 04 03) is available, but with a single rainfall station it is impossible to specialize the rainfall intensity.

The classification of annual average rainfall according to the annual average rainfall map of the world is divided into five distinct classes (CF table 1)

Table 1. The world average annual rainfall classification.

World Average annual rainfall (mm)	Classes
< 250	1
250-500	2
500-1000	3
1000-2000	4
>2000	5

This classification was made on a global scale. The average annual rainfall in the study area varies from 350 to 900 m is considered that an average annual rainfall of 1500 mm may favor strong erosion [11]. Referring to these data, discrimination is proposed at the scale of the study area (CF table 2).

Table 2. Average annual rainfall classification of study area.

Average annual rainfall (mm)	Classes
< 500	1
500 – 800	2
> 800	3

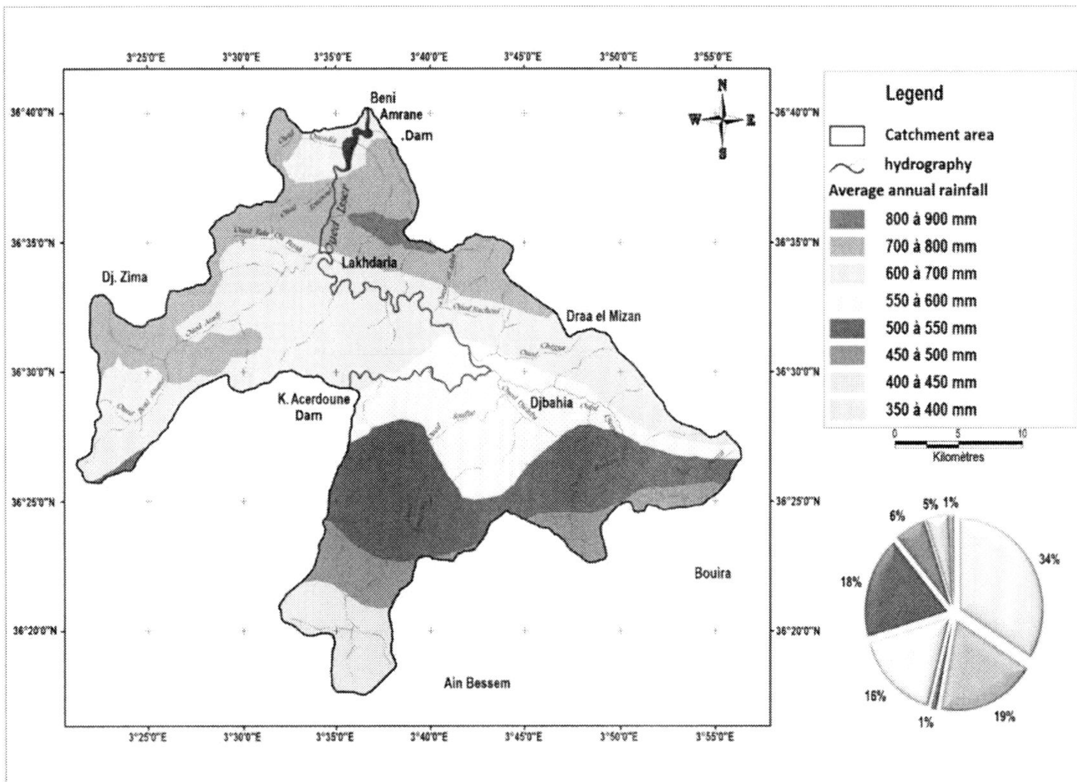

Figure 3. Average annual rainfall map.

From the average annual rainfall map of the Beni Amrane dam basin (CF figure 3), we have classified and then codified the different ranges of average annual rainfall, which resulted in a map of rainfall erosion.

Only low, medium and heavy precipitation is represented on the study site. Low precipitation spreads over an area of 100 km², i.e. 12% of the total area of the basin, and moderate precipitation occupies 87%, while heavy precipitation occupies only 1% of the total area of the basin.

3.2. Slope map

The slope map (CF figure 4) is one of the basic elements for the analysis of the physical characteristics that determine the suitability of the various zones. Indeed, the potential and limits of land use depend to a large extent on the slope since this contributes to the determination of erosion possibilities in relation to other factors of the mechanization of crops, Irrigation, grazing opportunities, installation and development of reforestation vegetation.

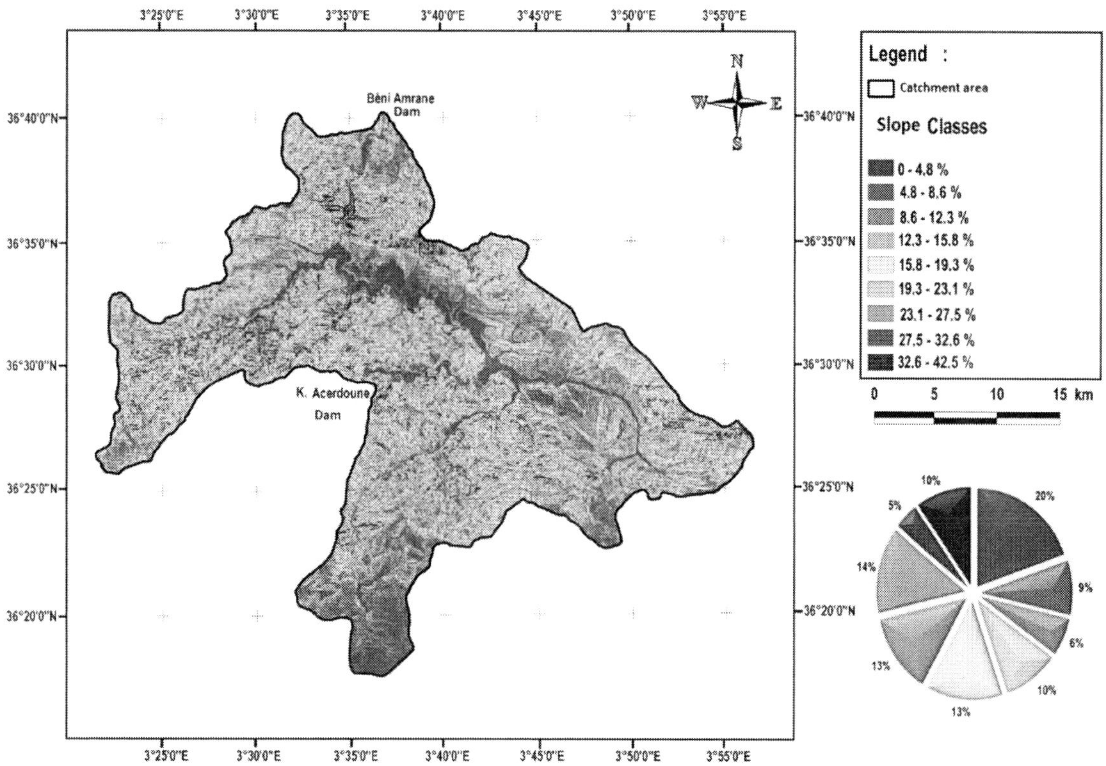

Figure 4. Slope map.

The length of the slope has more uncertain effects [12]. This was not reflected in this study.

The slope map of the study area was reclassified based on the Roose classification (1977) and transformed into a map of vulnerability to soil erosion according to the inclination of the slope.

- 0-5%: weak slope;
- 5-15%: average slope;
- 15 - 25%: steep slope;
- >35%: very steep slope.

For each slope class, an index ranging from 1 to 4 (CF table 3) is assigned, with 1 being assigned to the low slopes (<5°) and 4 to the steep slopes (> 15°).

These four classes cover respectively:

- 13% of the study area for slopes between 0 and 5%;
- 33% for those between 5 and 15%;
- 34% for those between 15 and 35%;
- And 20% for those above 35%.

Table 3. Slopes classes and attributed indices according to Roose, 1977.

Slopes	Code
0 – 5 %	1
5 – 15 %	2
15 – 35 %	3
> 35 %	4

3.3. Land use

From the land use map of the Béni Amrane basin (CF figure 5), we assigned to each class a degree of protection against water erosion.

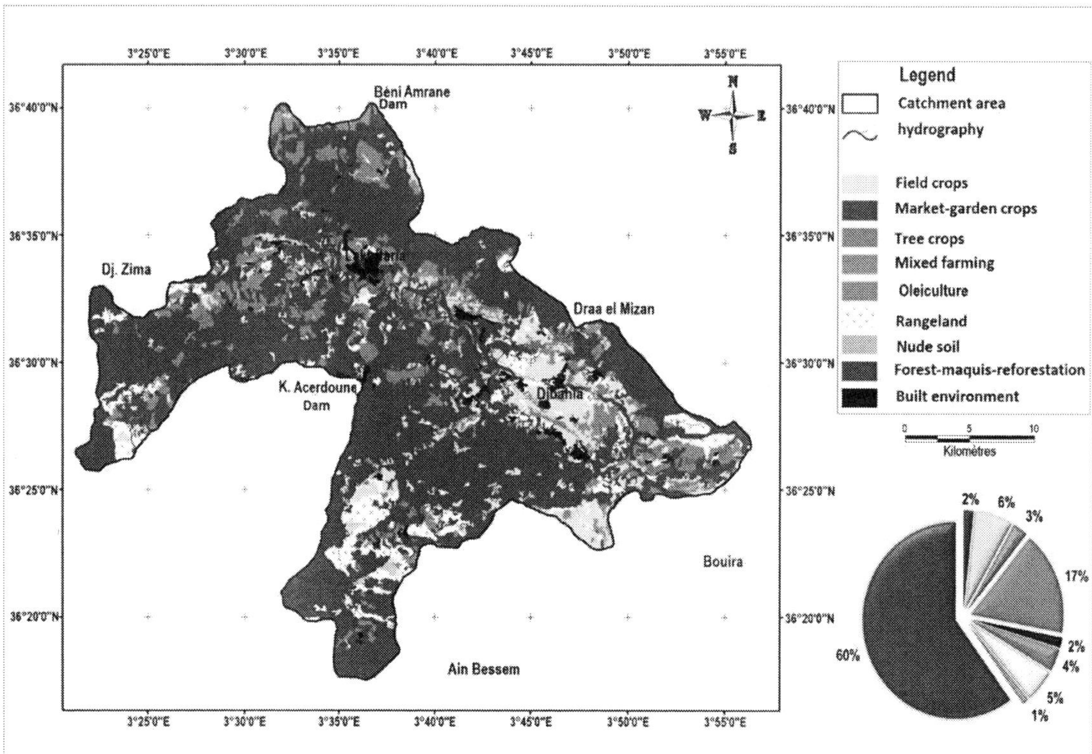

Figure 5. Land use map.

- Degree 1: strongly protective.
- Degree 2: moderately protective.
- Degree 3: slightly protective.

- Degree 4: non-protective.

The classification of Roose (1977) was used to derive an appropriate classification. Roose classifies the vegetable coverings into three groups:

- Permanent vegetation cover.
- Temporary vegetation cover.
- Incomplete vegetation cover

This classification method was detailed by subdividing the incomplete vegetation cover group into two groups: grazing and bare soil.

- Permanent vegetation cover includes:
  - The forest cover
  - Reforestation.
  - Perennial crops.
  - Fruit plantation and fruit trees.
- The temporary vegetation cover includes:
  - Annual crops: Cereals, extensive and semi-intensive agriculture and arable land.
  - Pastures: include very degraded and less degraded pastures

The map established makes it possible to discriminate nine classes of land occupation: 1 / dry field crops; 2 / vegetable crops; 3 / tree crops; 4 / polycultures; 5 / olive growing; 6 / range; 7 / bare soil area; 8 / forest - bush - reforestation; 9 / built space.

The classification is made using a GIS by assigning to each object a value that indicates its degree of sensitivity to erosion. This classification is based on direct observation of land use categories, and has resulted in a land use sensitivity map. Four classes are identified: range lands, bare ground, very dense vegetation, sparse vegetation or medium density.

Each class is assigned a value between 1 and 4, with 1 being assigned to the least vulnerable class and 4 to the most vulnerable class [13]

The dam catchment area is the most affected part by a semi-arid climate regime that determines the vegetal cover. In terms of soil erosion, it is the factor "coverage rate" that occurs more.

The land use map of the dam catchment area (CF figure 5), generated from the land use map prepared by the National Institute of soils, irrigation and drainage (INSID), highlights without a second thought that the vegetal cover mainly consists of trees, shrubs and herbaceous plants, i.e.; 70% of the total area. The rest primarily includes cereal and economic crops, dried vegetable and market gardening in the open fields or under glass.

According to [13] each coverage rate is indexed in table 4 below.

Table 4. Coverage rate classes and assigned indices.

Coverage rate	Index	Assignment
Very dense vegetation	1	Highly protective
Scattered or medium density vegetation	2	Moderately protective
Cultivated land	3	Slightly protective
Bar land	4	Non-protective

According to the map obtained, the distribution of vegetation cover at the level of the catchment area is respectively; 20% of the slightly and non-protective land and 80% are considered protective.

3.4. Soil map

The lithological map of the dam watershed of Beni Amrane reveals a wide variety of surface formations predominantly clay soils derived from marl formations (CF figure 6).

Soil sensitivity classes for each soil type are attributed from their texture, on which the formation and stability of the aggregates depend, and which results in large part from the nature of the origin materials. Some of the soils on massive rocks (granite and limestone) are classified as low and medium erodibility, while soils on friable rocks (sand and molasses) and coarse textured soils, regardless of the origin material, are considered as strongly erodible [14].

Although limestones and sandstones are permeable and hard formations, their association with swelling clays promotes the triggering of runoff from which fine particles are entrained [15].

Taking into account the degree of friability of the substrates., the clay and marly facies have thus been grouped together as brittle, flyshaking as moderately resistant and sandy as very resistant [16].

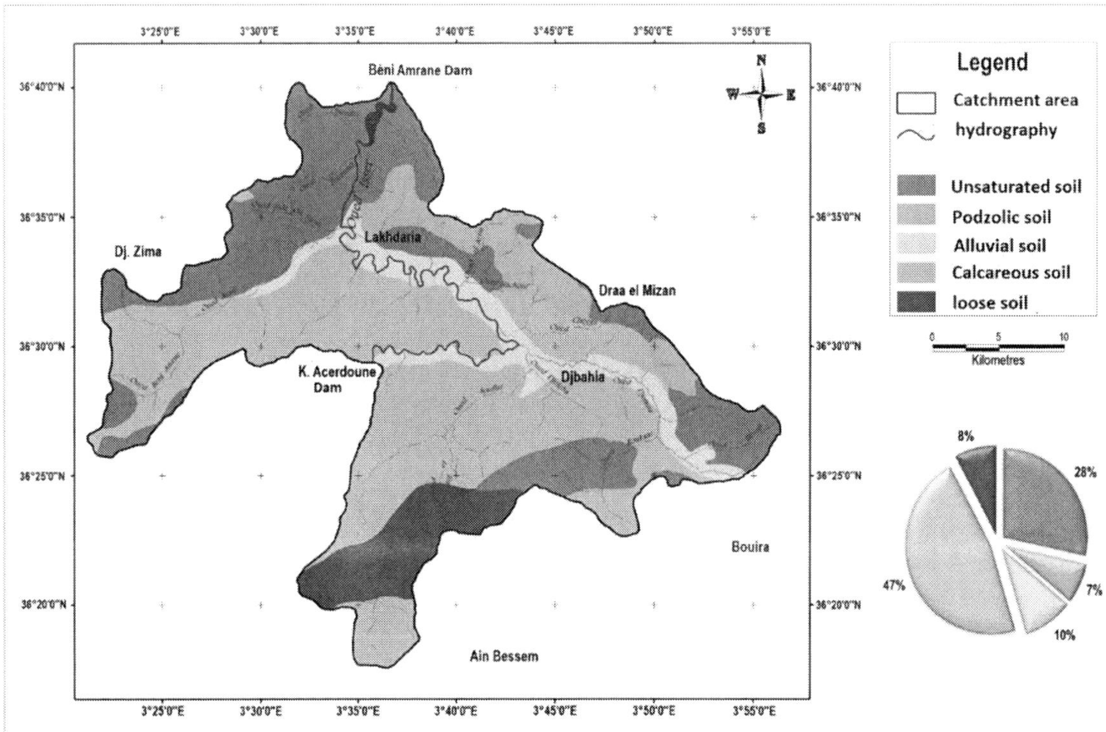

Figure 6. Soil map.

Marly soils, when dry, remain non-erodible, but as soon as they reach certain moisture, their susceptibility to detachability and runoff increases.

The sensitivity classification includes, in ascending order, four classes that have been identified as:

Class 1: Slightly sensitive; Class 2: moderately sensitive; Class 3: highly sensitive; Class 4: highly sensitive.

The lithology map shows that the majority of the basin, more than 75%, consists of weakly or moderately compacted sedimentary rocks or soils and rocks or soils that are not very resistant or strongly altered throughout the basin. We also observe a frequency of 14% of sediments or soft soils, non-cohesive and detritus materials, particularly in the middle of the basin and in areas of low to medium slope. High permeability carbonated rocks and cohesive soils fractured or moderately altered, poorly permeable occupy 11% of the total area of the basin.

The dominance of weakly or moderately compacted sedimentary soils and of soft sediment increases the erosive potential of the watershed.

3.5. Water erosion vulnerability map

The study of the vulnerability of soils to water erosion presupposes, as mentioned above, the logical combination of several factors [17] In a GIS; without forgetting that here we have limited by three external factors, in addition to lithology, at the same time effect and cause.

The problem of weights has been the subject of research by several researchers. The unanimously recognized role of vegetation density and slope against water erosion led us to retain weighting coefficients for each factor.

The crossing of the various thematic maps was carried out by applying weighting coefficients inspired by the general principles used by Le Bissonnais [18]. The highest weighting is attributed to the "vegetation cover" factor, which is considered to be the dominant factor, and the lowest weighting is attributed to precipitation because we do not have rainfall intensities, but only the average annual heights. Factor 2 is attributed to lithology and factor 3 is given to the topography. The general principles used to weight the parameters are as follows [19]:

- Only bare lands and cultivated lands are susceptible to runoff. It has no influence on prairies and dense forests;
- The influence of the slope increases when it is associated with a weak vegetation cover;
- The erodibility factor occurs only in the case of steep slopes;
- The effect of climate increases with susceptibility to erosion, for example in cases of very low susceptibility to erosion (e.g. permanently covered on low slope), the hazard remains very low whatever the rainfall erosion.

We proceed by combining the corresponding attribute tables into the four thematic maps where codes were assigned in order of increasing vulnerability to erosion within each class of factors (tables 2-5). The multifactor vulnerability to erosion of a given sector is evaluated by taking into account the effect of the four parameters by the following formula:

$$I=\sum_{J=1}^{4}\sum_{i=1}^{4} P_i * S_j$$

Where: I is the index of vulnerability to water erosion; Pi is the weighting of parameter i (rainfall, soil, slope, vegetation); Sj is the contribution of the class i for each of the parameters i.

Table 5. Classes of lithofacies and assigned indices.

Lithofacies	**Lithofacies that existant in the study area**	**Code**
Non altered compact rocks, strongly cemented conglomerates, crusts, ferruginous sandstone outcrops (massive limestone, strongly rocky soils, igneous or eruptive rocks, locally encrusted soils).	-	1
Carbonate rocks, fractured or moderately altered cohesive rocks or soils	-Flysch of the Aptian and the Albian -Limestone and marly limestone of Jurassic age	2
Low or moderately compacted sedimentary rocks or soils (slate, shale, marl ...) and rocks or soils that are not very resistant or strongly altered (marl, gypsum, clay slate, etc.)	-Shales and quartzites of the Albian -Clay, marls and sandstones of the Miocene -Paleozoic Micaschists -Marl and Limestone of the Coniacian-Maestrichtien -Marl, sandstones and sandstone limestone of the Upper Eocene	3

IOP Conf. Series: Earth and Environmental Science **82** (2017) 012011 doi:10.1088/1755-1315/82/1/012011

	-Marl and limestone of the Cenomanian.	
Sediment or loose soil, non cohesive and detritus material	-Cailloutis, reddish sandy clay of the Miocene -Quaternary Terraces: eluvium marly limestone -Oligocene series of clays at the base and sandstone at the top -Alluvium of the Quaternary	4

Thus, for classes 1,2,3,4, which represent an area with an average annual rainfall of less than 500 mm with fractured cohesive rocks flushing on a slope varying from 15 to 35% on bare ground, the index is 30 (1 * 1 + 2 * 2 + 3 * 3 + 4 * 4). Four classes of vulnerability to erosion have been drawn: low (indices 4 to 13), average (indices 17 to 24), strong (indices 17 to 32) and very high (32 to 40).

The figure below shows the map of sensitivity to water erosion of the Beni Amrane dam catchment with these four classes.

The soil sensitivity map to water erosion (CF figure 7) is carried out by crossing lithology maps, annual average rainfall, slopes and land cover.

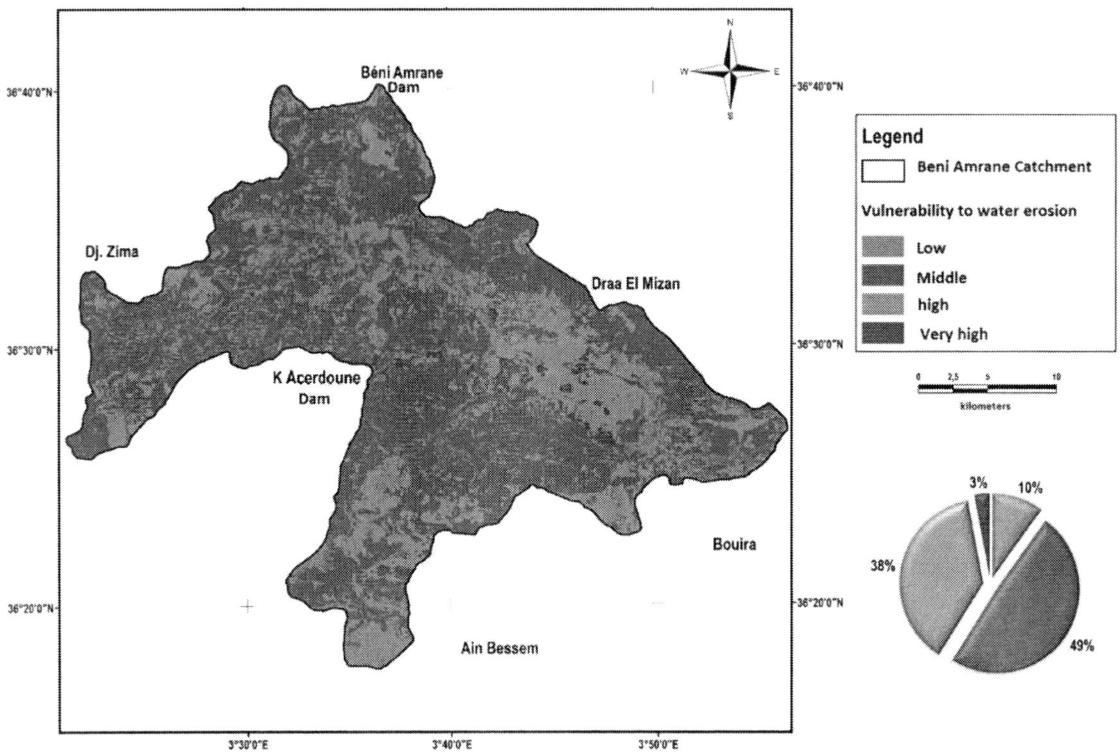

Figure 7. Water erosion vulnerability map.

Four classes appear:

- Class 1: low sensitivity to erosion represents 9% of the total area.

- Class 2: average sensitivity to erosion represents 53% of the total area.
- Class 3: high sensitivity to erosion represents 36% of the total area.
- Class 4: high susceptibility to erosion represents 2% of the total area.

After a general description of the physical environment of our basin, we observed that the part located at the center of the basin and in the West is favorable to the erosive phenomenon. This part is hilly or encountered land steep (> 15°) and a canopy of low density.

There is a dominance of sensitive soils to medium and high water erosion. These affect 89% of the study area, with an almost general distribution in a large sector. A proportion of 36% of the area of the basin is highly sensitive to erosion and 2% of a very high susceptibility. These areas concern land with steep slopes, with crops or matorrals, which possess either herbaceous or degraded vegetation cover, with plowing on sloping land and with slightly resistant surface formation.

The units which are moderately susceptible to erosion occupy 53% of the basin area and correspond to outcrops on low to moderate slopes, and / or are occupied by dense vegetation cover, or soil where cropping practices are appropriate. On the other hand, very low risk soil classes (9% of the basin) are located mainly at the level of tree crops and dense forests, where the well protected soil facilitates runoff and limits the risk of aggressive runoff and erosion; these zones have a slope generally between 0 and 4%. Studies by C.W. Roose (1991) have shown that even on a slope of less than 1%, erosion may occur. This is groundwater erosion, without formation of channels or gullies, with runoff not being able to concentrate.

The mapping of sensitivity to water erosion at the BeniAmrane dam catchment provides information on the extent and current state of soil degradation. They reflect the state of erosion resulting from the predominance of maroon-silty soft formations, steep slopes and aggressive runoff. To refine the results, it will be necessary to update the data, to improve the combined methods implemented and to integrate additional parameters (spatial variability of rainfall intensity, soil infiltrability....).

4. Discussions

Knowing that policy makers are much more interested in the distribution of erosion risk than its absolute value [20], a cognitive qualitative model has been implemented on the watershed of Beni Amrane focusing on simple parameters representing the main factors of water erosion, namely soil inclination, soil occupation, surface soil formations and precipitation.

The classification of sensitivity to water erosion as a function of the inclination of the ground is not refined, as long as it has been made on the basis of a 30 m resolution MNA map, from where the slopes were calculated for each pixel of 30 m by 30 m, which does not always reflect the reality of the terrain. Indeed, it would be more credible to work on 5 m or 10 mb good resolution images even though the MNA map was realized with curves of low equidistance (20 m). Similarly, groundwater erosion whose main cause is the energy of heavy rainfall on bare soils [21] is not considered and occurs in low slope soils, which can generate a high loss of soil.

The sensitivity classification according to the land-use map of the National Institute of Soils, Irrigation and Drainage (INSID) is still reliable, as long as it has been carefully compared with more recent satellite images, where no major changes have been noticed. Nevertheless, consideration should be given to the development of land-use change scenarios [22]. The majority of the surface of our basin is covered by forests and reforestation, that is 60%, where the vegetation cover protects against the phenomenon of rainfall, prolonging the soil permeability and reducing the runoff volume. On the other hand, a large quantity of runoff energy is absorbed by the litter that maintains the mesofauna (influencing the rate of infiltration). However, the state of the vegetation cover directly affects the roughness of the soil. This depends in particular on the number of stems per square meter. A vegetation cover composed of numerous herbaceous plants will have a more effective soil protection action against runoff than trees [20]. Human intervention which accounts for 30% of the surface area accentuates its fragility to groundwater erosion by clearing and cultivating large crops, polycultures and especially crops small and medium-sized slopes.

Although limestones and sandstones are permeable and of hard formations, their association with swelling clays promotes the triggering of runoff from which fine particles are entrained [14]. Marly soils, when dry, remain non-erodible, but as soon as they reach a certain degree of humidity, their susceptibility to detachability and runoff increases. Although sand is inconsistent, it is not certain that sandy areas are always very sensitive to water erosion because water can infiltrate quickly [23].

The spatialization of the erosivity based on the annual mean precipitation map of the period (1965-2005) is not sufficient without taking into account the intensity of the rains during the flood events.

The map that we have realized can be refined by integrating new parameters related to the morphology of the terrain such as the length and direction of the slope and the hydrological behavior such as the variability of rainfall intensity expressed in the dry and humid season.

5. Conclusion

This work is a first contribution in vulnerability to water erosion of Beni Amrane Dam catchment area. This could be a first guidance document for a rational use of land in the region and better secure the dam against reservoir siltation.

The Beni Amrane dam catchment is sensitive to water erosion. The combination of biophysical factors (steep slopes, very erodible soft rock, and very permeable soil) makes the basin a vulnerable zone to water erosion. The results of the study of susceptibility to erosion reveal that 41% of the catchment area is of high to very high erosion susceptibility. The main consequence of this will be the siltation of the Béni Amrane dam, which is an inevitable phenomenon, but which can be reduced through implementation of appropriate technical and biological measures within the basin. We recommend the following means of control: Reforestation, restoration of soils, establishment of benches, creation of small dams (hill reservoirs), planting of crops along the level lines, planting of long-stemmed vegetation in the wadis.

For the silting of dams, several devices have been taken into consideration to combat this phenomenon (methods of hunting, drawing by current density and dredging).

Improvement of pastures: Soils can be protected against erosion only by maintaining the plantations on the slope and shed, at least during the first years, with the aim of ensuring the success of plantations and to achieve a density and even decrease the solid flow. Since breeding is the main resource for the local residents, it is essential to improve the pastures while taking care to avoid overgrazing.

For shoreline protection, the biological fixation method is the most effective, but technical measures (mechanical procedures) are essential in order to reduce the velocity of runoff.

References

[1] Bouanani A 2004 Hydrology, sediment discharge and modeling, study of some basins of Tafna (NW Algeria) (University Aboubeker Belkaid of Tlemcen)

[2] Achit and Touaibia B 2002 Essai regionalization of water erosion in semiarid area of the catchment of Oued Mina (North ouestalgérien) *Actes Int. Water Symp. Quantitative Management Water Resources* (University Hassiba Ben Bouali, Chlef, Algeria, 05-06 February) pp 287-95

[3] ANBT 1990 National Agency Dams *Dam Project Beni Amrane*

[4] Bouchetata A and Bouchetata T 2006 Proposals for the development of the sub-watershed of the oued Fergoug (Algeria) weakened by episodes of drought and submitted to water erosion science and planetary changes/Drought **17(3)** 415-24

[5] Boughalem M 2013 Impact of management systems on the vulnerability of the soil to erosion: the case of the watershed of the Isser - Tlemcen (Algeria) (University Aboubeker Belkaid of Tlemcen)

[6] Cedran O 2001 Analyse and modeling of solid particles on the scale of small cultivated catchments (France: University of Orleans)

[7] Demmak A 1982 Contribution to the study of erosion and sediment transport in northern

Algeria (Paris, France: Engineer University)
[8] Wischmeier W H and Smith D D 1965 Predicting rainfall-erosion losses from cropland east of the Rocky Mountains *Agriculture Handbook* 282 (USDA-ARS)
[9] ANRH 2003 Map of the sub-basins of northern Algeria 2003 (Blida Algeria)
[10] INSID National Institute of soil, irrigation and drainage 2011 Soil use maps for the following wilayas: Boumerdes, Bouira, Tizi Ouzou and Medea at the scale 1/25000
[11] Larfi B and Remini B 2002 Quantification sediment transport in the watershed of Wadi Isser-application siltation dam Beni Amrane *Proc. of the 2nd Mediterranean conference on water and the environment* (Algiers)
[12] Mathys N, Richard D and Grésillon J M 2003 Non-linearity in erosion response of small mountainous and marly basin: the laval in the Draix experimental catchments, south east, France *JEE, Hign National School for Hydraulic* **3** 47-52
[13] Roose E. 1977 *Erosion and runoff in Africa of the West. Twenty years of measures in small parcels.* ed. Orstom, Paris, Series works and Documents 78, p 108
[14] MEDA Program of the European Union 2010 Réalisation discounting Study of the National Water Plan Algérie Tome 3. Siltation of dams (Department of Resources waterfrom in the Republic of Algeria Democratic and Popular)
[15] Meddi M 1998 Quantification of precipitation: Application northwest of algérie *Proc. International Colloquium eau* (Chelef 5-6 February 2002)
[16] Mesrar H, Sadiki A and Jabrane R 2013 *The Risk of Water Erosion "Study and Modeling," If the Downstream Part of the Watershed of Oued Amzaz* (Rif, Morocco)
[17] National Meteorology Office (ONM) 2002 Rainfall map of North Algeria map
[18] Ngoniba P 2011 *Variational Assimilation of Data: An Application in Hydrology* (Germany)
[19] Remmas H 2014 Contribution to mapping of sensibility to water erosion of Saida Wadi Basin (Algeria: Tlemcen University Ecology and Environment)
[20] Roose E, Sabir M, Arabi M, Morsli B and Mazour M 2012 Sixty years of research in cooperation on the'water erosion and the fight erosion control in the Maghreb *Physio-Géo PGE* **6** 43-69
[21] Sadeghi S H 2004 Application of MUSLE in prediction of sediment yield in Iranian conditions *13th Int. Soil Conservation Organization Conf.* (Brisbane ISCO 2004)
[22] Tixeront D J 1960 Solid flow of rivers in Algeria and Tunisia Agricultural Secretariat, Tunis.
[23] Touahir S and Remini B 2007 Comparative study of suspended sediment load between sub basins and wadis zeddine Tikazale (Cheliff) (ENSH. Blida)

IOP Conf. Series: Earth and Environmental Science 82 (2017) 012012 doi:10.1088/1755-1315/82/1/012012

Wave-induced current at Anmok Beach, Korea

H Kim[1], D Park[1,3], S Lee[1] and H-S Lim[2]

[1]Kookmin University, Department of Civil Engineering, Seoul, Korea
[2]Coastal Disaster Research Center, Korea Institute of Ocean Science and Technology, Ansan, Korea

E-mail: qawsedr1991@naver.com

Abstract. Currents were measured at two stations on Anmok Beach, and were reproduced by a numerical model system CST3D-WA. Anmok area is known as wave-dominated coast, and the mean tidal range small. Field data with strong wave-induced current are rare, especially at shallow water depth. Measured currents were strong during three periods. The largest instantaneous current speed at a station during a period was 46 cm/s. CST3D-WA system simulated wave-induced current for the three periods, and compared with measurements. The deviation of the incident peak wave direction from the normal direction to the shoreline strongly affects the wave-induced current speed and direction. When the deviation angle crosses the normal direction, the wave-induced alongshore current direction changes between SE and NW. CST3D-WA is thought to have successfully reproduced the wave-induced current fields on this beach.

1. Introduction

When an artificial structure is to be constructed on a coast, we had better predict the environmental impact caused by the new structure as accurately as possible. If the new structure is planned at shallow water depth, we may need to predict shoreline change around the structure as well. Especially if the purpose of the new structure includes protection of shoreline and onshore zone, the prediction of sediment transport and the consequent morphological change is vital during design stage.

Sediment transport rate has often been described as a function of the bed shear stress, given characteristics of bed material, while the bed shear stress depends on information of waves current and water depth as well. Accurate flow information is the key to accurate prediction of morphological change. Flow information can be prepared from numerical modelling, and should be calibrated with measurements. Measuring meaningful wave-induced current has been hindered by various difficulties at field. In most cases currents include both tidal and wave-induced currents. The wave-induced current develops at high waves, and boats cannot approach both onshore and very shallow depths. Equipments cannot endure wild breaking waves. Thus field data on the wave-induced current are rare, especially at shallow depths.

Anmok Beach is situated on the east coast of Korea, see figure 1. Its shoreline is almost straight, and its normal direction is NE. At the SE end of the beach Kangneung Harbour of about 500 m × 500 m is situated. Another straight Namhangjin Beach runs at the south-east end of the harbour. Anmok beach bed material is mostly sand, and its size is between 180 μm and 560 μm. Crossshore bed slope is around 1/50 up to 30 m depth. Waves of 3 m or bigger height attack the beach, and wave-induced current develops at zone shallower than 10 m. Wave directions are normally from NE, but they deviate from NE during specific typhoons or strong northerly wind period in winter.

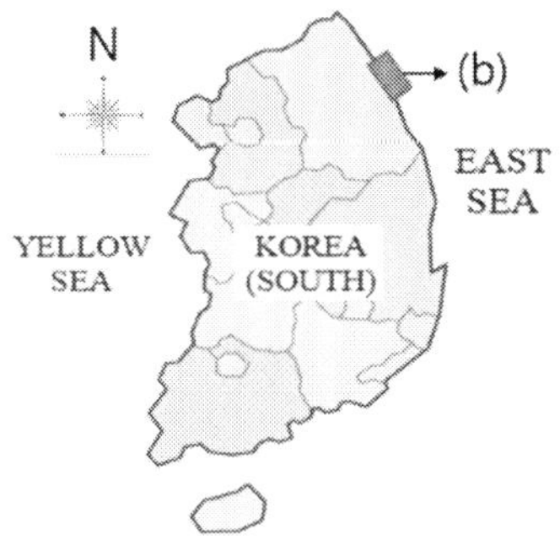

(a) Korea (South)

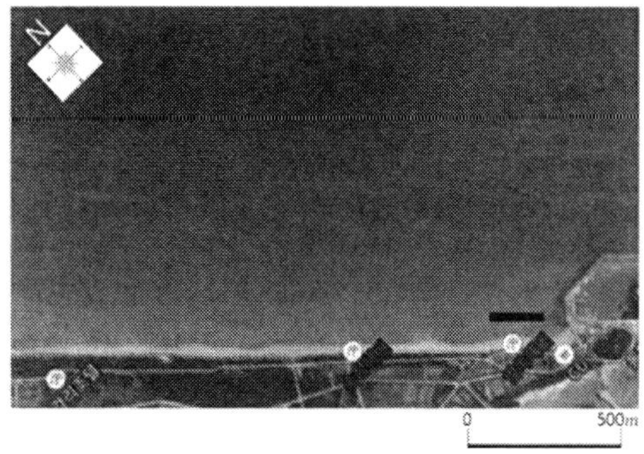

(b) Anmok Beach

Figure 1. Map of Anmok Beach.

Mean tidal range at Anmok Beach is about 14 cm, and the maximum tidal current speeds during mean tidal range in both directions reach up to 10 cm/s at offshore, but are weak near the shoreline. A submerged breakwater was constructed near Kangneung Harbour late 2014 as in figure 1 to protect erosion of beach and the background road. It is important to evaluate its function at this time for prediction of long term future morphological evolution. KIOST [1] has measured the near-coast current at Anmok Beach since early 2015 up to the present. Currents and suspended sediment concentration were measured for a month each time from 12 February-14 February 2015, 27 February-01 March 2015, and 29 August-31 August 2016. Waves were measured continously, and morphology was measured intermittently depending on meteorologic events.

A numerical modelling system CST3D-WA has been developed by Kookmin University with support from Ministry of Ocean and Fisheries, Korea [2-8]. CST3D-WA includes wave, current, sediment transport, and morphological change modules. Existing phase-average wave model SWAN computes wave fields for given boundary conditions [9-11]. Then, wave-induced flow-driving force field is computed from the wave field by making use of Longuet-Higgins and Stewart's [12] radiation stresses and a spreading method of Kim [13]. Flow field is computed for given tidal water level boundary conditions with either additional wave-induced driving force or not depending on situation. Sediment transport is computed with options of suspended load, bed load, or both. CST3D-WA can handle shoreline retreat or advance due to sediment transport around mean water level by adopting the "beachface controlling method" [14].

2. Measurements and reproduction of wave-induced currents at Anmok Beach

Table 1. Details of field measurement at Anmok Beach.

St	Period	X (UTM)	Y (UTM)	Depth(m)	Equipment
W1	2015. 02.12~Date	495269.76	4181915.52	18.26	AWAC
V3	2015.02.12~2015.03.01, 2015.12.08~2015.01.16, 2016.08.25~2016.09.23	494674.25	4181499.83	2.47	Vector 1EA, OBS 1EA
S2	2015.02.12~2015.03.01, 2015.12.08~2015.01.16, 2016.08.25~2016.09.23	494890.07	4181616.19	4.01	ADCP-1200 1EA, Vector 1EA OBS 1EA

Waves and currents were measured at three stations, see figure 2 and table 1. During the third survey period typhoon LIONROCK arrived at the beach, and the biggest significant wave height at the time was 4.01 m, the peak wave period was 11.5 s, and the peak wave direction was about 65.0 ° clockwise from N. Measured current including wave-induced current developed in the alongshore SE direction, and reached 23 cm/s at St. V3 behind the submerged breakwater due to structure-driven breaking and diffraction.

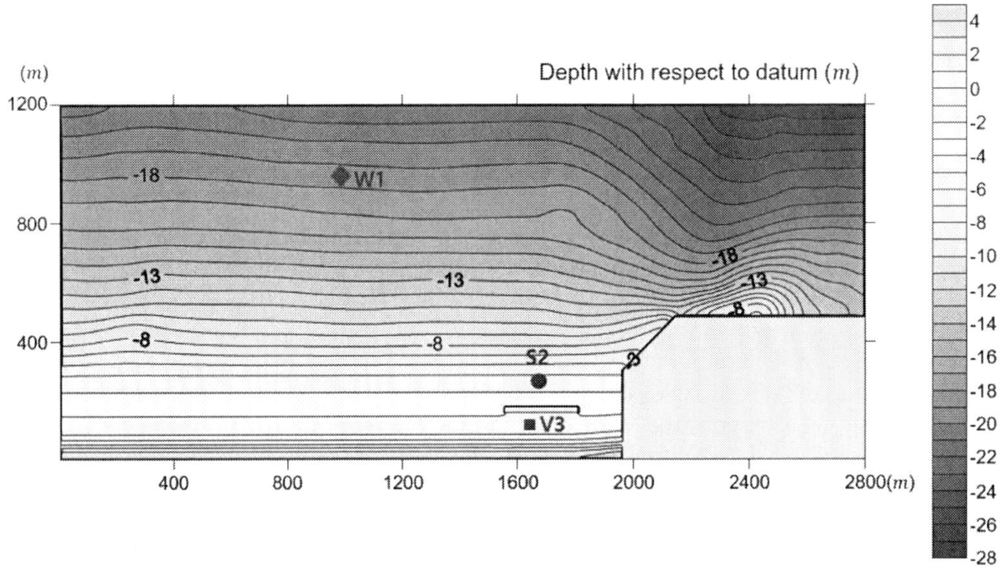

Figure 2. Depth of Anmok Beach.

Several different sources drive coastal currents, e.g. tidal current, wave-induced current, wind-induced current, and density current. Measured current speeds show high correlation with the significant wave height. First, we examine the significance of the tidal contribution to the coastal current at the two stations of this beach.

Simulated residual tidal current field for mean tidal range by CST3D-WA is shown in figure 3, where two circulation cells develop at both sides of Kangneung Harbour. The current speed of the circulation is about 5 cm/s, which is considered small compared to the wave-induced current during typhoons. We try to simulate the wave-induced current separately from the tidal current, considering them negligible during high wave periods.

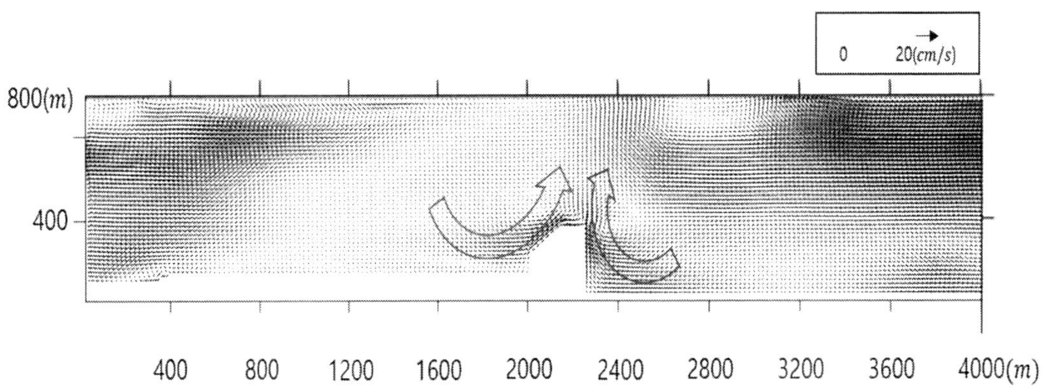

Figure 3. Computed residual mean tidal current field at Anmok and Namhangjin Beaches.

Computed time-series of wave-induced current based on the measured significant wave height, peak wave period, and peak wave direction with the measured current for three terms at the two station are shown in figures 4, 6 and 8, respectively; while the instantaneous wave-induced current fields at four specific instants are shown in figures 5, 7, 9, and 10. Both figures 4 and 6 show south-east alongshore current developed by waves coming from the north, while figure 8 shows transition of the alongshore current direction from north-west to south-east, which reflects well the change of wave direction coming from south-east to north-east. Figures 5, 7, 9, and 10 show local circulations behind the submerged breakwater, which are added to the ambient alongshore current field. Simulated vortices seem stronger than measured ones, see St. V3 measurements in figure 8. Further analysis could explain this deviation between measurements and computation results.

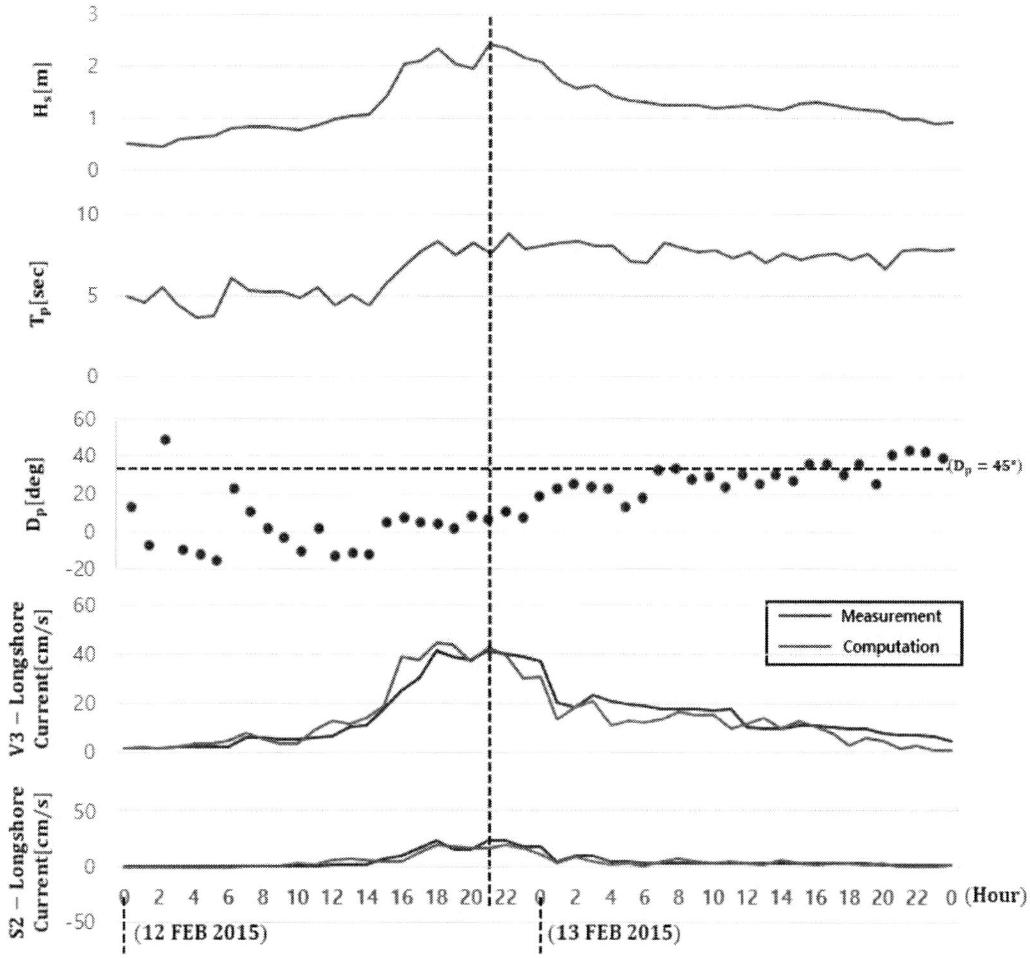

Figure 4. Computed time-series of wave-induced current at two stations from 12 February 2015 to 14 February 2015.

IOP Conf. Series: Earth and Environmental Science 82 (2017) 012012 doi:10.1088/1755-1315/82/1/012012

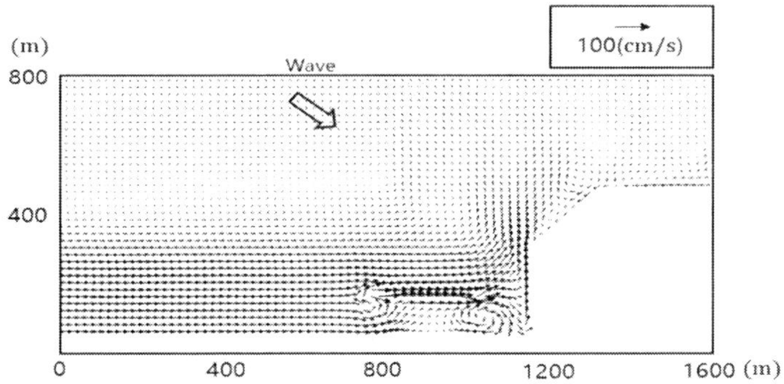

Figure 5. Computed wave-induced current field at Anmok at 21 O'clock, 12 February 2015.

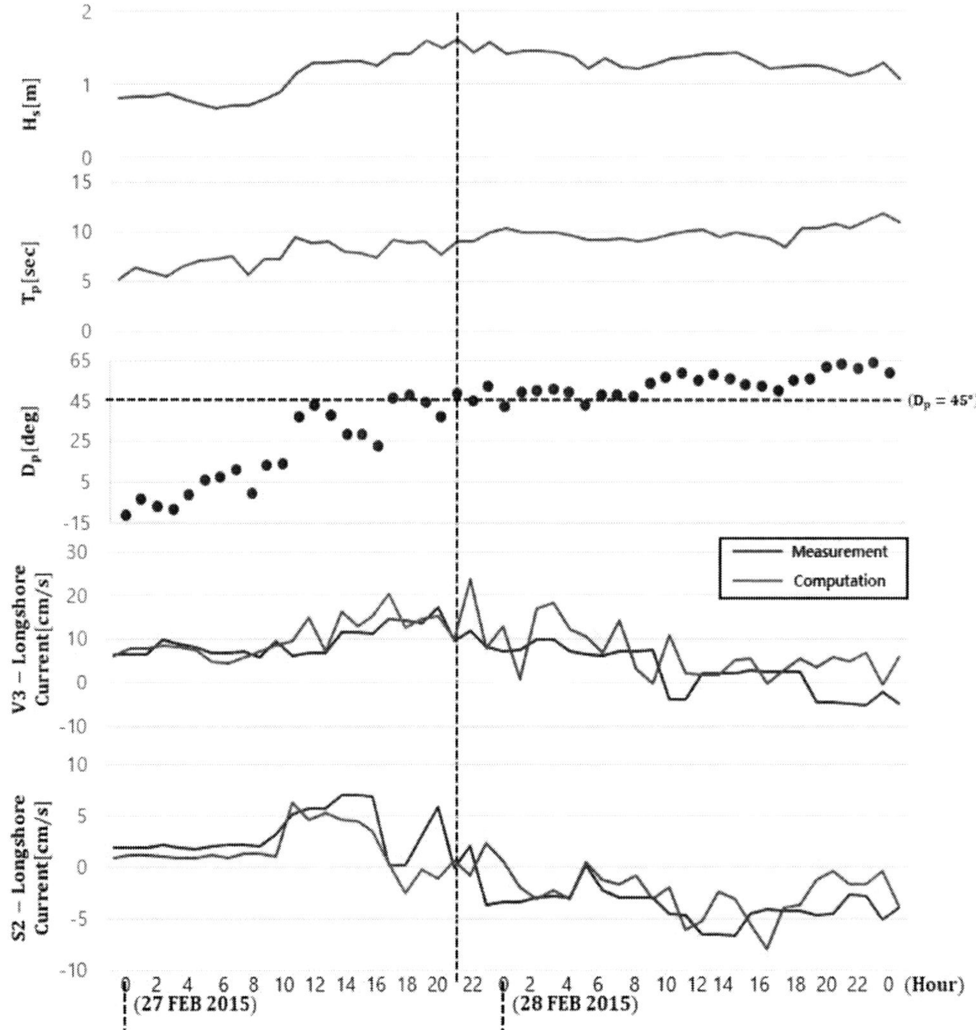

Figure 6. Computed time-series of wave-induced current at two stations from 27 February 2015 to 01 March 2015.

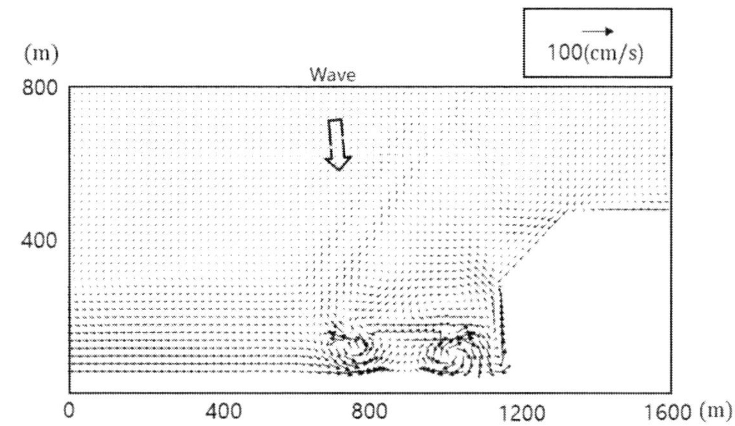

Figure 7. Computed wave-induced current field at Anmok at 21 O'clock, 27 February 2015.

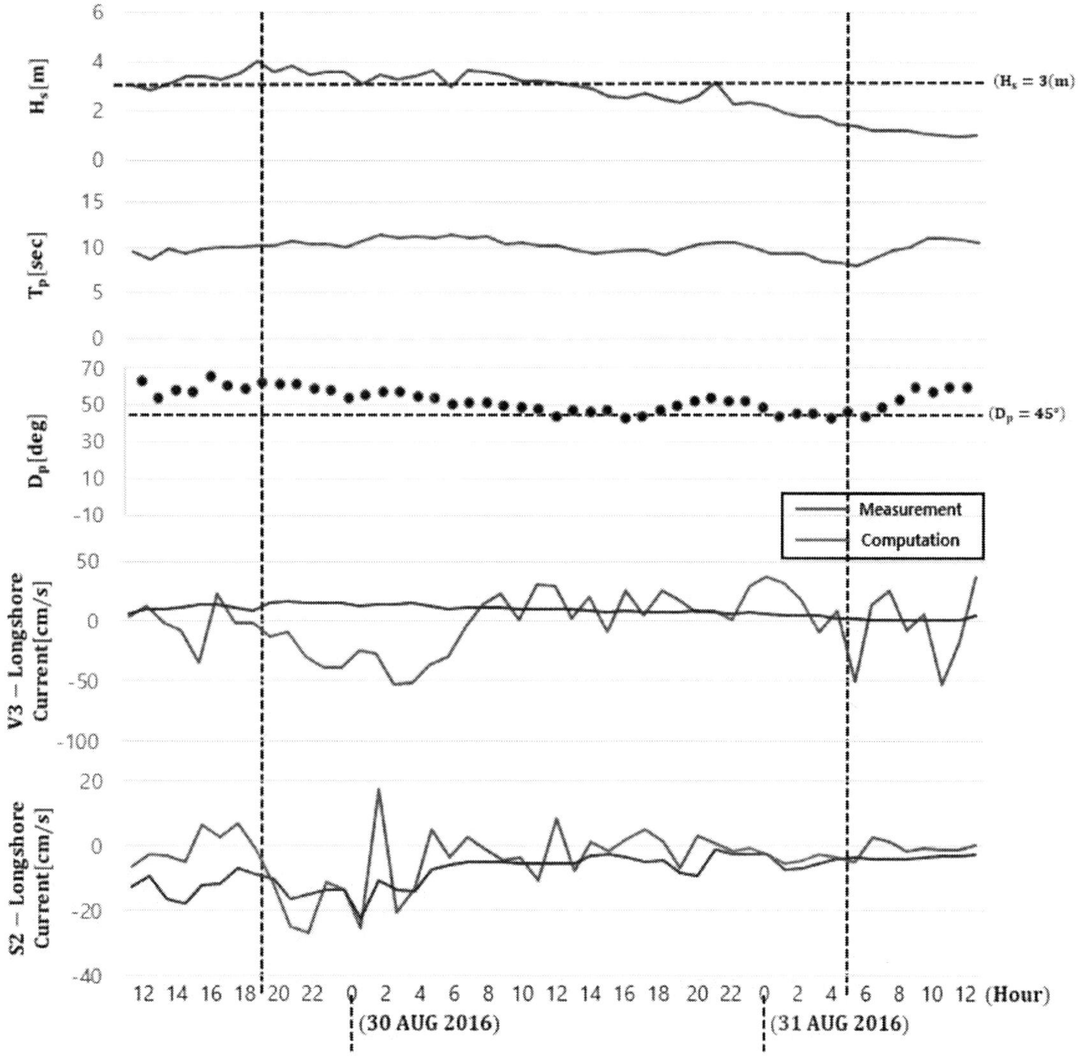

Figure 8. Computed time-series of wave-induced current at two stations from 29 August 2016 to 31 August 2016.

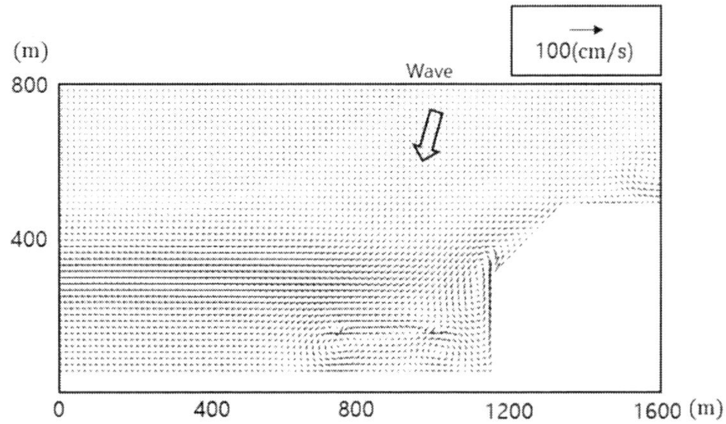

Figure 9. Computed wave-induced current field at Anmok at 19 O'clock, 29 August 2016.

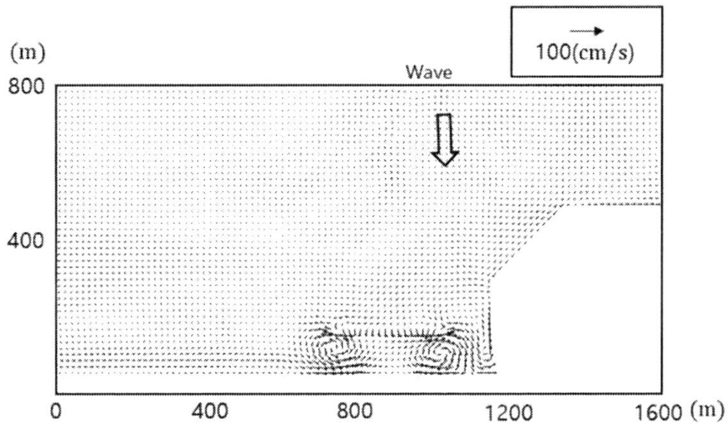

Figure 10. Computed wave-induced current field at Anmok at 05 O'clock, 31 August 2016.

Overall CST3D-WA is believed to have reproduced the wave-induced current fields from the direction and the order of magnitude point of view.

3. Conclusions

KIOST obtained valuable near-coast current data despite various difficulties like deployment of equipment at very shallow water depth, where waves break during typhoons.

CST3D-WA simulated the wave-induced current for three terms when waves were high, and explains correlation between the incident wave direction and the alongshore current direction. The model system also reproduced the current speed reasonably well. It is thought that the CST3D-WA could be applied for simulation of morphological change at the site with high reliability.

Acknowledgments

This is a result of the research project titled "Development of Coastal Erosion Control Technology", funded by the Ministry of Oceans and Fisheries, Korea.

References

[1] KIOST 2016 Development of coastal erosion control technology *Fourth Year Report* (Ministry of Oceans and Fisheries)

[2] Kim H, Ahn J, Jang C, Jin J and Jung B 2013 Extension of piecewise exact solution method for

two-and three-dimensional fluid flows *J. Vibroeng.* **15** 1243-54
[3] Jang C 2009 Development of noncohesive sedimentary environment al prediction system (Korea: Kookmin University)
[4] Yoo H, Kim H, Jang C and Baek S 2012 Considering the effect of reflecting wave on wave-induced current near Sokcho Harbor *The Korean Society for Marine Environment & Energy* 2105-8
[5] Jang C, Kim H and Ihm N 2012 A development of coupled wave-induced current modeling system and its application to the idealized shoreline with detached breakwater *Korean Wetlands Society* 14th **4** 439-55
[6] Kim H 2013 User manual of CST3D-2013-ver1 *Haksan media* Korea
[7] Lee K 2015 Numerical simulation of bathymetric change at Haeundae Beach (Korea: Kookmin University)
[8] Kim H 2017 Costal sediment transport in 3 dimensions available: http://blog.naver.com/ CST3D
[9] Berkhoff J C W 1972 Computation of combind refraction-diffraction *Proc 13th Int. Coastal Eng. Conf. Vancouver* **1972** 471-90
[10] Booij N 1981 Gravity waves on water with non-uniform depth and current Rep 81-1 (Dept. Civil Eng. Delft University of Tech)
[11] The SWAN Team 2008 SWAN user manual (Delft University of Technology Environmental Fluid Mechanics Section)
[12] Longuet-Higgins M S and Stewart R W 1964 1964 Radiation stresses in water waves: A physical discussion with applications *Deep Sea Res.* **11** 529-62
[13] Kim H 2004 Practical spreading of driving forces for wave-driven currents *Ocean Eng.* **31** 435-53
[14] Kim H, Lee K, Lee S, Jin J and Jang C 2016 Understanding tombolo formation via numerical modelling *International Conf. on Coastal Engineering* (Antalya Turkey) p 1369

Numerical simulation on tunnel spillway of Jingping-I hydropower project with four aerators

R Shilpakar[1,3], Z Hua[1], B Manandhar[1], N Shrestha[1], M R Zafar[1], T Iqbal[1] and Z Hussain[2]

[1]School of Renewable Energy, North China Electric Power University, Beijing 102206, China
[2]Pir Mehr Ali Shah-Arid Agriculture University, Rawalpindi, Pakistan

E-mail: razesh@ncepu.edu.cn

Abstract. This Study presents the results of a three dimensional simulation of a tunnel spillway with four aerators of Jingping-I hydropower project as done by using ANSYS. The study uses a combination of realizable k-ε turbulence model and volume of fluid method in order to simulate the characteristics of aeration and cavitation in aerated tunnel spillway. It is possible to yield the water surface profile, pressure distribution, velocity profile and aeration cavity length from the simulation. Results of numerical simulation were compared with experimental results showing that the realizable k-ε turbulence model and volume of fluid model can simulate the aerated tunnel spillway. The aerated cavity zone was found to be stable and within reasonable length. Results from the numerical simulation were found to resemble the experimental result to a great extent, showing the validity of the simulation.

1. Introduction

There is more of cavitation in high velocity and high pressure head in free-flow tunnel spillway. Artificial aerators are generally introduced to reduce cavitation which is very cost-effective and easy measures. Different researches have been done in the study of aeration phenomenon. Chanson [1] has explained the mechanism of aeration by showing that a large amount of air is introduced in the aeration zone along with the air water interface of jet and strong de-aeration tends to occur when undergoing the bottom near nappe impact. In spite of the strong de-aeration occurring in the impact point region, the bottom aerators are found to be very useful for introducing large amount of air over short distance. Pfisher [2] has conducted a hydraulic model test using typical chute aerators with various kinds of flow features including pre-aeration. The study has shown that the pre-aeration in the downstream of the aerators affect the stream-wise development and the bottom air concentration. Chanson [3] has established the method for calculating the cavity length in the bottom and side of sudden enlargement and vertical drop aerator. Both experimental and simulation methods were used to study the aeration in the spillways. Shuai Li et al [4] have presented a study on the diffused downstream flow from a radial sluice and full section aerator where a realizable k-ε turbulent model was combined with a mixture multiphase model. Shuai Li et al [5] have studied the aeration effect of the bottom as well as the side wall of a connective tunnel with high head and large discharge using an experimental test with numerical simulation. Results showed a reduced amount of back water effect in the aeration cavity with the increasing aeration length. Caihuan et al [6] proposed the optimization methods for radial gate with sudden enlargement and offset aerator, Jinhua et al [7] have concluded that the air concentration is affected by the air-cavity length; Piller et al [8] have successfully simulated the turbulent flow in a channel and has obtained the improvement in the results, concluding on the influence of Prandtl number Pr on turbulent transport. Deng et al [9] have revealed the possibility of simulating the complicated flow in the tunnel spillway through the combining k-ε double equations turbulence model and volume of fluid method. Wu J H et al [10] have focused on the necessary of special designs on the different configuration and sizes of the aerators in various tunnel spillways where flow velocity is exceeding 40 m/s. In this study, the realizable k-ε turbulence model was used in a combination with the VOF method in order to determine the characteristics of aerated tunnel spillway in Jingpin-1 Hydropower Station. Jingping-I hydropower project is a tall arch dam located at the Jingping bend of Yalong river, Liangshan, Sichuan province, China. The capacity of this project is 3600 MW and has the tallest 305 m high arch dam. The total length of the flood tunnel spillway is 829 m where the length of the power tunnel is 434 m. The tunnel is free-aerated as air is also provided to the top of tunnel through ventilation shaft. Four aerators are inserted at four different sections. The aerators used here is the combination of deflector and offset type. The designed discharge of the project is 3229 m^3/s and the normal flow at 1880m [11]. The details of the projects were shown in the figure 1.

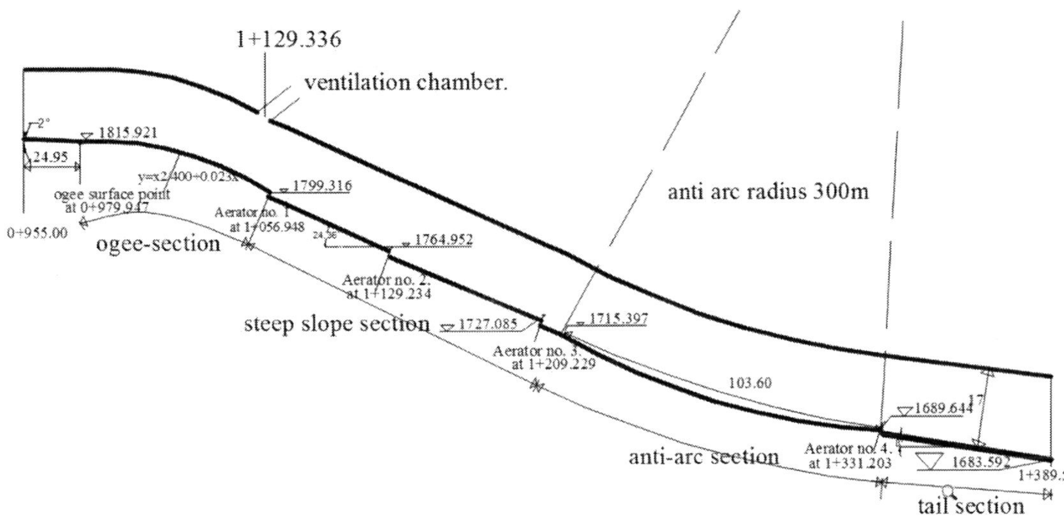

Figure 1. Detailed section of the free-aerated flood tunnel spillway [11].

2. Mathematical model

The realizable k-ε turbulence model has been used to solve the air-water flow .The instantaneous continuity, momentum and energy equations are expressed for a compressible liquid in the following manner.

Continuity equation:

$$\frac{\partial \rho}{\partial t} + \frac{\partial}{\partial x_i}(\rho u_i) = 0 \tag{1}$$

Momentum equation:

$$\frac{\partial}{\partial t}(\rho u_i) + \frac{\partial}{\partial x_j}\left(\rho u_i u_j + p\delta_{ij} - \tau_{ij}\right) = 0 \tag{2}$$

Energy equation:

$$\frac{\partial}{\partial t}(\rho \varepsilon_0) + \frac{\partial}{\partial x_j}\left(\rho u_j \varepsilon_0 + u_j p + q_j - u_i \tau_{ij}\right) = 0 \tag{3}$$

From stokes law, the viscous stress for a Newtonian fluid is

$$\tau_{ij} = 2\mu S_{ij}^{*} \tag{4}$$

Where trace less viscous strain rate is given by:

$$S_{ij}^{*} = \frac{1}{2}\left[\frac{\partial u_i}{\partial x_j} + \frac{\partial u_j}{\partial x_i}\right] - \frac{1}{3}\frac{\partial u_k}{\partial x_k}\delta_{ij} \tag{5}$$

Where u represents velocity, p represents pressure, ρ represents density and μ represents viscosity.

Turbulent flows are characterized by rapid fluctuations of velocity, density, temperature and concentration. The cause for such temporal fluctuations is the non-linear character of the physical-chemical processes. Turbulence is related with random fluctuations of fluid. The most of the engineering flows are turbulent. The transient velocity (being expressed as an instantaneous velocity), which is very difficult to predict due to its random occurrence with time. Instead, velocity is expressed into its steady mean value $\overline{u}_i$ with a fluctuating component u_i' in the following form:

$$u_i = \overline{u}_i + u_i'$$

This process of time averaging is called Reynolds decomposing. This decomposition integrates the instantaneous continuity and momentum equation yielding the Reyonlds Averaged Navier-Strokes (RANS) equations:

$$\frac{\partial}{\partial t}(\rho u_i) + \frac{\partial}{\partial x_j}(\rho u_i u_j) = -\frac{\partial p}{\partial x_j} + \left[\mu\left(\frac{\partial u_i}{\partial x_j} + \frac{\partial u_j}{\partial x_i} - \frac{2}{3}\delta_{ij}\frac{\partial u_k}{\partial x_k}\right)\right] + \frac{\partial}{\partial x_i}(-\rho\overline{u_i'u_j'}) \tag{6}$$

There are many closure model exits but this paper presents only two-equation turbulence models. The Boussinesq hypothesis postulated that effect of turbulence can be represented as an increased viscosity, it is called viscosity model.

$$\left(-\rho\overline{u_i'u_j'}\right) = \mu_t\left(\frac{\partial u_i}{\partial x_j} + \frac{\partial u_j}{\partial x_i} - \frac{2}{3}\delta_{ij}\frac{\partial u_k}{\partial x_k}\right) - \frac{2}{3}\rho k\delta_{ij} \tag{7}$$

RNG k-ε turbulence model and the realizable k-ε turbulence models are two improvement forms of the classical k-ε turbulence model applied in fluent. The realizable k-ε turbulence model guarantees the satisfactory results for rotational flows, flows with adverse gradients within their boundary layers and flows with separation-recirculation effects. In realizable k-ε turbulence model, the transport equation for k and ε are given as:

- For turbulent Kinetic energy (k)

$$\frac{\partial(\rho k)}{\partial t} + \frac{\partial}{\partial x_i}(\rho k u_i) = \frac{\partial}{\partial x_j}\left[\left(\mu + \frac{\mu_t}{\sigma_k}\right)\frac{\partial k}{\partial x_k}\right] + P_k + P_b - \rho_e - Y_m + s_k \tag{8}$$

- For dissipation (ε)

$$\frac{\partial(\rho\varepsilon)}{\partial t} + \frac{\partial}{\partial x_j}(\rho\varepsilon u_j) = \frac{\partial}{\partial x_j}\left[\left(\mu + \frac{\mu_t}{\sigma_\varepsilon}\right)\frac{\partial\varepsilon}{\partial x_j}\right] + \rho C_1 s_\varepsilon - \rho C_2\frac{\varepsilon^2}{k+\sqrt{v\varepsilon}} + C_{1\varepsilon}s_\varepsilon\frac{\varepsilon}{k}C_{3\varepsilon}\rho_b s_\varepsilon \tag{9}$$

Where

$$C_1 = max\left[0.43, \frac{\eta}{\eta+5}\right], \quad \eta = S\frac{k}{\varepsilon}, \quad S = \sqrt{2s_{ij}s_{ij}}$$

μ_t Represents turbulent viscosity calculated by the turbulent kinetic energy (k) and turbulent dissipation rate (ε) as:

$$\mu_t = \rho C_\mu\frac{k^2}{\varepsilon} \tag{10}$$

Where C_μ is no longer a constant which depends on the rate of rotation and strain rate, it is expressed as

$$C_\mu = \frac{1}{A_0 + A_s U^* k/\varepsilon} \tag{11}$$

$A_0=4.0$, $A_S=\sqrt{6}\cos\left\{\frac{1}{3}\cos^{-1}\left[\sqrt{6}\left(\frac{s_{ij}s_{jk}s_{ki}}{\sqrt{s_{ij}s_{ij}}}\right)\right]\right\}$, $U^* = \sqrt{s_{ij}s_{ij} + (\Omega_{ij} - 2\varepsilon_{ij}\,\omega_k)(\Omega_{ij} - 2\varepsilon_{ij}\,\omega_k)}$

The time average rotation rate tensor, $\Omega_{ij} = \frac{1}{2}\left[\frac{\partial u_i}{\partial x_j} + \frac{\partial u_j}{\partial x_i}\right]$.

VOF method is used to track and locate the free contact surfaces between phases. In this study, VOF method was applied for water phase and air phase. The sum of all the volume fractions is equal to one in each control volume. In this research, air is assumed as the first phase and water is assumed as the second phase. For air, the transport equation has the form as:

$$\frac{\partial(\alpha_a)}{\partial t} + \nabla.(\vec{v}\,\alpha_a) = 0 \qquad (12)$$

For the volume fraction α_a of the kth fluid, three conditions are possible; $\alpha_a = 0$ showing; the cell is empty in k^{th} fluid, $\alpha_a = 1$ showing; the cell is full in k^{th} fluid and $0 < \alpha_a < 1$ showing; the cell contain an interface between the fluids.

3. Computational domain and mesh

The flow over the tunnel spillway was modelled in 3-Dimensions in Computational fluid Dynamics (CFD). The geometry of the spillway with different types of aerators was drawn in ANSYS. The meshing of the geometry was done on GAMBIT software. The structural grid meshing was done from the geometry. The grid was made of hexahedral-cells (hex-mesh) on the whole domain. A finer mesh size was considered for better result. The mesh size of 0.4 m was set for hexahedral meshing. It resulted in a mesh volume of 780,000 cells with skewness 0.5. The meshing of the power tunnel was shown in figure 2.

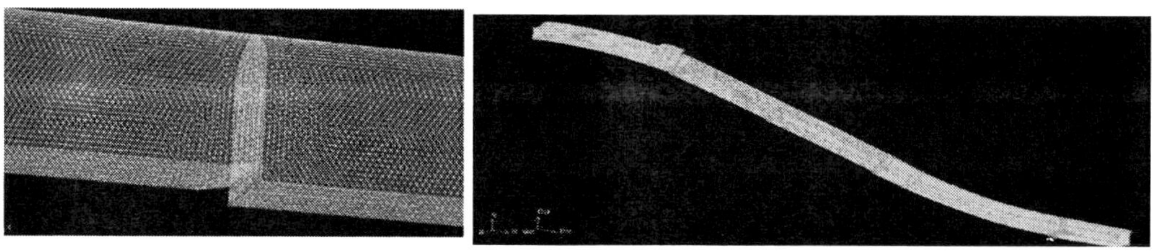

Figure 2. Meshing of the power tunnel.

4. Boundary condition and calculation

Pressure inlet was applied as a boundary condition to the inlet of the water domain where the static pressure developed due to the upstream water level of reservoir of 1880 m. Standard wall functions were used in the bottom surface and the side walls. Pressure boundary was applied on the area above water surface with atmospheric condition for the open channel flow and the outlet boundary was set to a pressure-outlet type with atmosphere. The pressure-based segregated solver was used in ANSYS and the default algorithm. The PISO pressure-velocity coupling scheme was used for the transient flow calculation. PISO can maintain the relatively stable condition when using a large time-step or mesh with a high degree of distortion. An implicit scheme was used with VOF method with default volume fraction cut-off. Realizable k-ε turbulence model with standard wall functions was employed. The Least Square Cell-based method was chosen for interpolation methods. Presto scheme was used as interpolation methods for pressure. The solution was initialized with the standard initialization methods and computed from water-inlet. The patching was done in the assumption that air was initially inside the tunnel before releasing water. The 0.02 time-step size was chosen and the simulation was done for 6000 numbers of time steps for about 120 seconds.

5. Results and discussion

Following are the results obtained from the simulation of the power spillway with four aerators.

5.1. Water surface profile and pressure distribution

Water surface profiles of the power tunnel obtained from the simulation and experimental results [11] for full flow is shown in figure 3. Water surface profile near in all four aerators have small risen due to deflector of aerator. Due to aeration from the upper ventilation pipe and bottom aerators, the water surface at the tail region of tunnel has been more aerated and in milky appearance. The calculated water surface is nearly close to the experimental results [11]. The development of negative pressure inside the tunnel of high velocity region causes cavitation. So, the bottom pressure distribution is important for better understanding. The figure 4 shows the comparison of bottom pressure distributions between simulation and experimental value inside the full flow tunnel spillway.

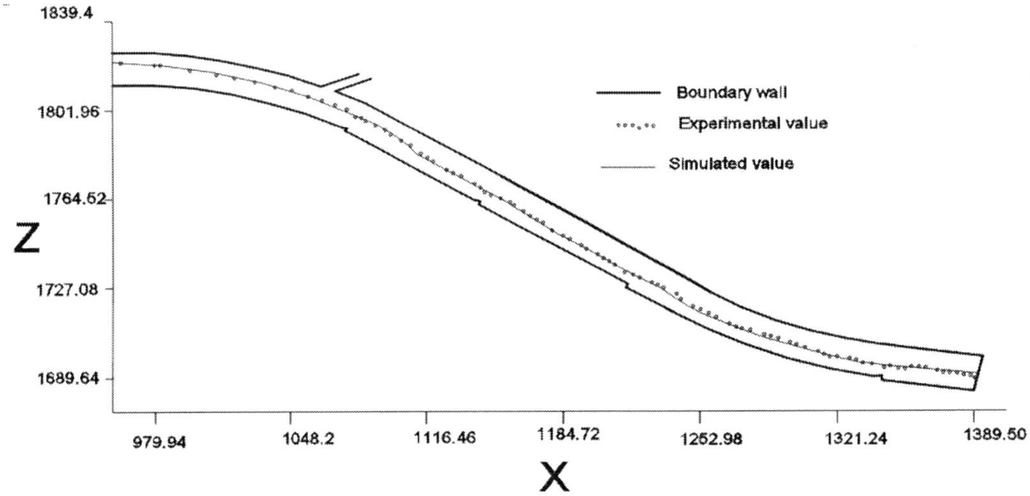

Figure 3. Comparison of water surfaces.

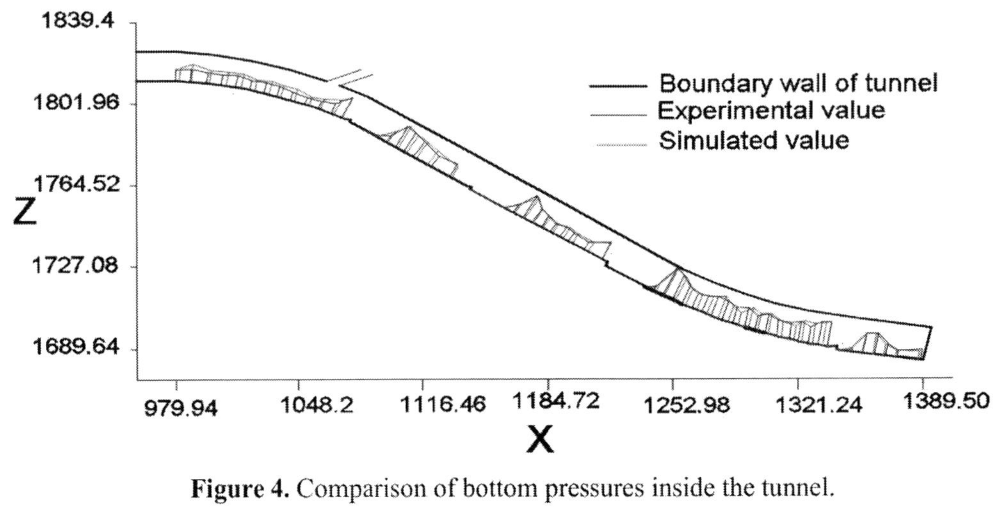

Figure 4. Comparison of bottom pressures inside the tunnel.

The simulated bottom plate pressures distribution on the tunnel spillway is nearly close with that of experimental values [11]. The table 1 clearly shows the bottom pressure developed inside the tunnel for full flow. The simulated bottom plate pressures before aerators 1 & 2 are 120 kPa and 70 kPa which are slightly less than experimental values [11] by 2.75% and 9.88% respectively. The simulated bottom pressure before aerators 3 & 4 are 110 kPa and 135 kPa that are slightly greater than that of experimental values [11] by 7.63% and 0.37% respectively.

Table 1. Bottom pressure development before aerators.

Position	Experimental values [11]	Simulated values	
	Bottom plate pressure (kPa)	Bottom plate pressure (kPa)	Error (%)
Before aeration #1	123.4	120	2.75
Before aeration #2	77.68	70	9.88
Before aeration #3	102.2	110	-7.63
Before aeration #4	134.5	135	-0.37

The figure 5 clearly shows the maximum bottom pressure developed at the floor of the tunnel for fully open gate after each aerator. The aerator jets influenced the pressure distribution in the floor of the spillway where low pressures are obtained near aerators and gradually increases its value to peak value. In the figure 5, blue color shows the low pressure region and red color shows peak value. The calculated peak pressures values in the bottom of spillway after the aerators #1, #2, #3 and #4 are 148 kPa, 135 kPa, 171 kKa and 115 kPa respectively. In aerated cavity, low pressures are obtained which ensured the effective aeration and the minimum pressure of 1.8 kPa,-5.75 kPa,-1.12 kPa and -7 kPa are observed after aerator position's #1, #2, #3 and #4 respectively. The maximum bottom pressure distribution after aerator #2 and # 4 are slightly less than that of experimental results [11]. The peak bottom pressure distribution after aerators #1 and #3 have slightly higher value than that of experimental results [11].

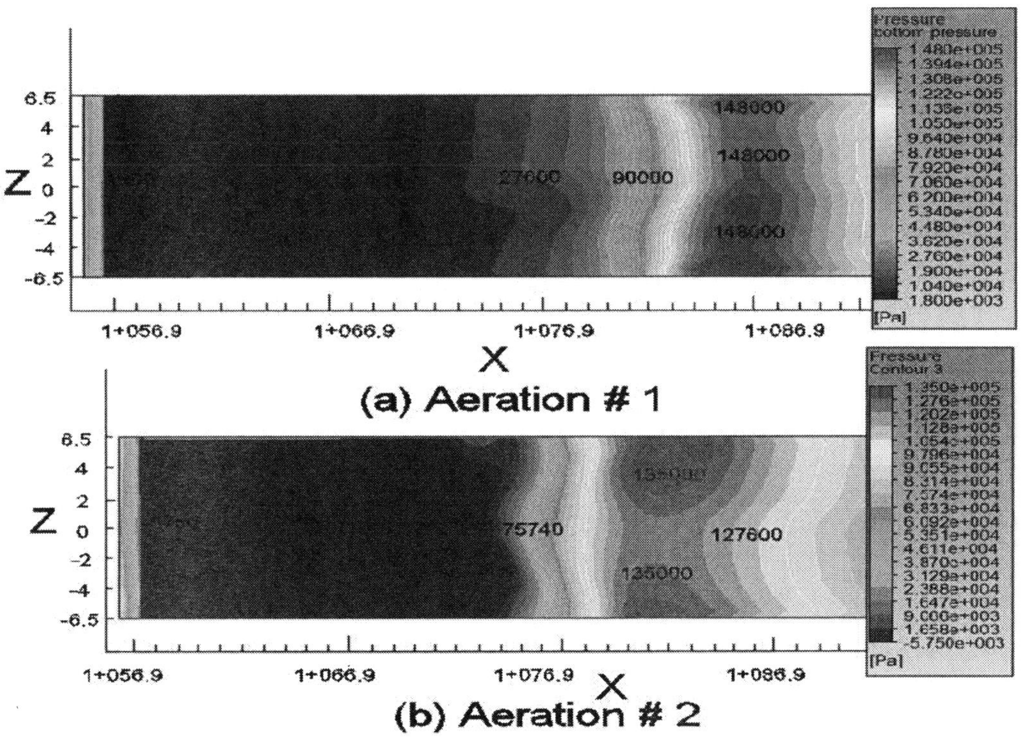

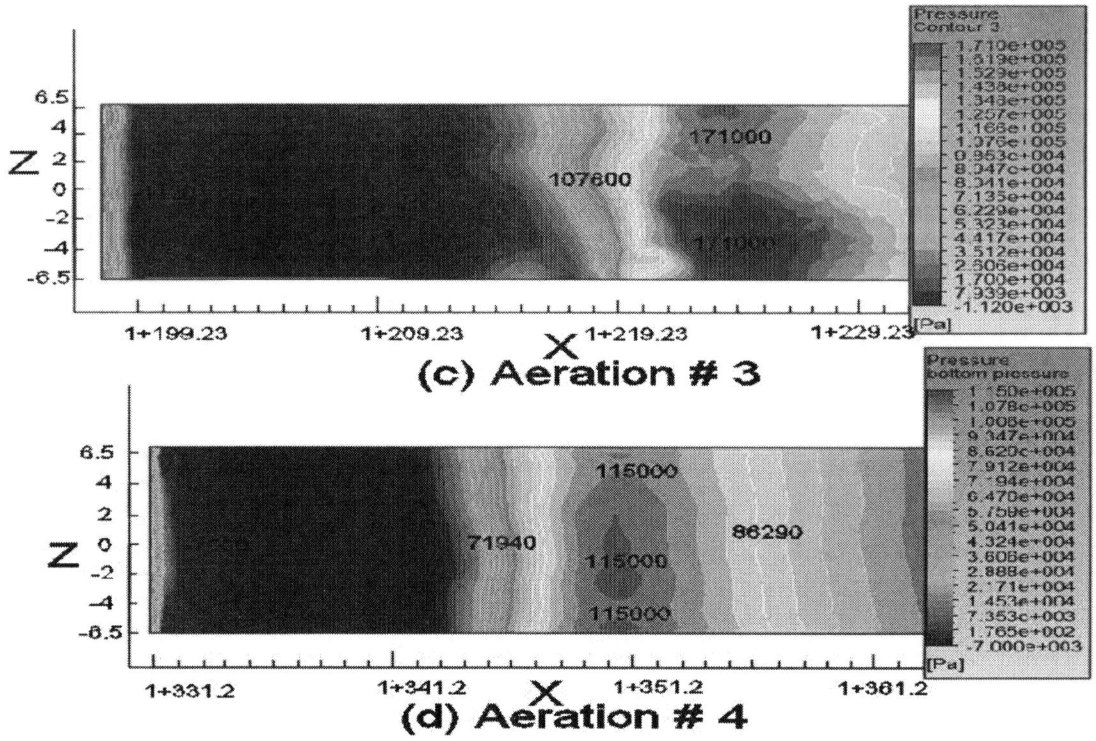

Figure 5. Maximum bottom pressure distributions.

5.2. Velocity distribution

Velocity is one of the most important hydraulic characteristics of the flow. Velocity determines the magnitude of the cavitation erosion. The velocity distribution of the full flow inside tunnel spillway with normal discharge at 1880 m before aerator #1 in the ogee section in the cross section is somehow linear because of no bottom aeration prior aerator #1. The velocity distribution of this ogee section is similar with that of the open channel flow. But the flow before aerator #2 is different from previous, where velocity contour is raised in the middle of the bottom boundary and distributed symmetrically on both sides due to aeration from aerator #1. Similarly, the velocity contour before aerator #3 at the middle of bottom is raised at that time of flow. The velocity distribution is changed due to aeration from the aerator. Water in the tail section of the free-aerated tunnel has been much more aerated and milky in appearance. Following figure 6 gives a detail of velocity development in the cross section of the free-aerated tunnel before aerator.

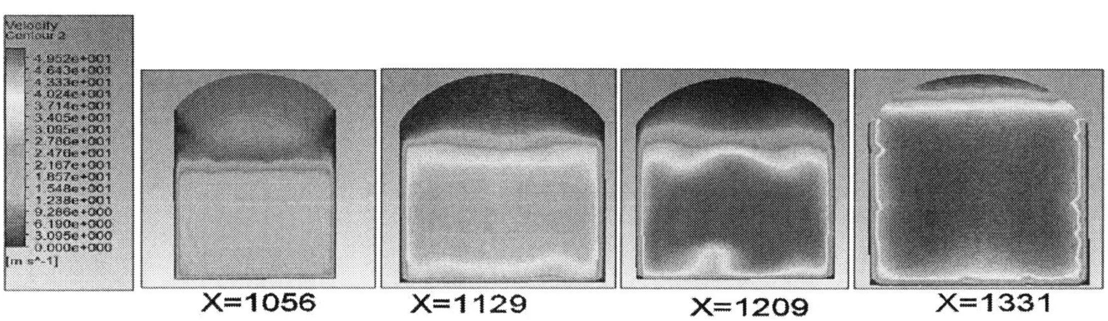

Figure 6. Flow Velocity in the cross section before aerator.

The average velocity and cavitation number of the fully flow in the cross section of the free-aerated tunnel before the aerators is shown in the table 2. The velocity of the flow near aerator #1 is 27.9 m/s, which is less than experimental value [11] with the relative error of 3.46%. The average velocity along the downstream of the tunnel gradually increased up to the end of the tunnel. The average velocity of the flow at the tail section of the tunnel before aerator #4 is obtained to be 46.4 m/s with the relative error of 5.46 % than that of the experimental result [11]. The maximum relative error does not exceed more than 6% which shows that the simulation result agrees well with the experimental result [11]. The cavitation numbers determine the cavitation damages. Aerators are kept in spillway to keep the cavitation index higher than 0.20. The minimum cavitation number obtained from simulation is 0.21 which shows that tunnel surface is free from cavitation damages.

Table 2. Average velocity and cavitation number.

Aerator's Position	Before Stake number	Average velocity (m/s)				
		Experimental Value [11]	Cavitation Number	Simulated value	Error%	Cavitation Number
Aerator # 1	1+056.94	28.9	0.49	27.9	-3.46	0.54
Aerator #2	1+129.23	35.8	0.25	35.7	-0.28	0.26
Aerator #3	1+209.22	41.9	0.21	42.09	0.45	0.23
Aerator # 4	1+331.20	44.0	0.22	46.40	5.45	0.21

5.3. Aeration features

In order to ensure full aeration, there shouldn't be backwater effect after the jump and should possess the stable shape and certain length of aerated cavity. The aeration cavities for fully opening gate inside the power tunnel spillway of four aerators are shown in figure 7. The red and blue color shows water and air respectively in the figure. Low Froude number is obtained near aerator #1 and stable aeration cavity is difficult to form. The impact angle of the jet tip on the slope obtained to be low which has helped in the reducing backwater effect and obtaining stable aeration cavity. The stable aeration cavities are formed after aerator #2, #3 and #4.

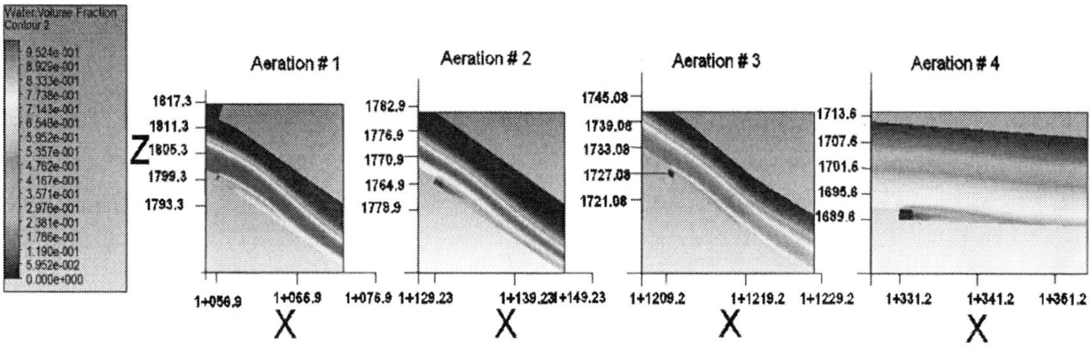

Figure 7. Bottom cavity of four aerators.

The comparative study of aeration cavity for full flow in the power tunnel between the numerical simulation and experimental results is shown in table 3. From the table, it can be easily figured out that the simulated aeration cavity length of 1# aerator is greater than experimental value [11] by 9.8%. But, the aeration cavity length of 2 # aerator is slightly smaller than that of experimental value [11] by 3.8%. The simulated aeration cavity length of 3 # aerator is larger than the experimental results [11] whereas simulated aeration cavity length of 4# aerator is smaller than the experimental results [11]. In aggregate, the simulated results agree with the experimental results [11].

Table 3. The Comparison of aerated cavity length.

Aerator	Aerated cavity length (m)		
	Simulated Value	Experimental Value [11]	Error (%)
1	28	25.5	9.8
2	26	27	-3.8
3	25	24	4.1
4	21.5	22.4	-4

6. Conclusion

The combination of Realizable k-ε turbulence model and VOF method were successfully used to simulate the flow characteristics in aerated tunnel spillway of Jingping-1 Hydropower Project. Other achievements could be presented in following points:

- The numerical simulation results were very similar to that of the experimental results.
- The bottom pressure distribution was reasonably distributed with that of the model test result as the maximum relative error of the pressure is less than 10%.
- Due to the insertion of aerators, there was variation in the velocity of flow. The air entrainment from the both sides of the wall has helped in the symmetrical distribution of the high velocity zones in the cross-section. The relative error of the average velocity didn't exceed 6% which agreed with the result from model test.
- The aerated cavity zone was stable and in reasonable length. The minimum cavitation number (0.21) among calculated was obtained at tail end of tunnel which was greater than 0.2 and justified the necessity of aerators.

Acknowledgments

This paper was a part of master thesis in the field of water resource and hydropower engineering. Authors are indebted to Mr. Sun Shuang Ke of China Institute of Water Resource and Hydropower Research for providing the necessary data.

References

[1] Chanson H 1994 Aeration and de-aeration at bottom aeration devices on spillways *Can. J. Civil. Eng.* **21(3)** 404-9
[2] Pfister M, Lucas J and Hager W H 2011 Chute aerators: Pre-aerated approach flow *J. Hydraul. Eng.* **137(11)** 1452-61
[3] Chanson H 1989 Study of air entrainment and aeration devices *J. Hydraul. Res.* **27(3)** 301-19
[4] Li S, Zhang J, Xu W, Chen J and Peng Y 2016 Evolution of pressure and cavitation on side walls affected by lateral divergence angle and opening of radial gate *J. Hydraul. Eng.* **142(7)** 05016003
[5] Li S, Zhang J, Xu W, Chen J, Peng Y, Li J and He X 2016 Simulation and experiments of aerated flow in curve-connective tunnel with high head and large discharge *Int. J. Civ. Eng.* **14(1)** 23-33
[6] Caihuan Wang, Dongmei Hou, Li Li and Xia Yu 2012 Study of shape design of the aerator with sudden lateral enlargement and bottom drop behind high-head radial gate and its engineering application *J. Hydroelectr. Eng.* **31(5)** 107-13 (in Chinese)
[7] Wu J-H and Ruan S P 2007 Emergence angle of flow over an aerator *J Hydrodynam B* **19.5** 601-6
[8] Piller M, Enrico N and Thomas J H 2002 DNS study of turbulent transport at low Prandtl numbers in a channel flow *J. Fluid Mech.* **458** 419-441
[9] Deng J, Xu W L, Lei J and Diao M J 2005 Numerical simulation of hydraulic characteristics of high spillway tunnel *J. Hydroelectr. Eng.* **36(10)** 1209-18 (in Chinese)
[10] Ruan S P, Wu J H, Wu W W and Xi R Z 2007 Hydraulic research of aerators on tunnel

spillways *J Hydrodynam B* **19.3** 330-4
[11] China Water Conversancy and Research Institute 2011 The Model Test of the Flood Spillway of the Jingping-1Hydropower Station (in Chinese)

Novel approaches in sub-surface parameterisation to calibrate groundwater models

D K S Y Klaas[1,2,4], M A Imteaz[1], I Sudiayem[3], E M E Klaas[3] and E C M Klaas[3]

[1]Department of Civil & Construction Engineering, Swinburne University of Technology, Melbourne, VIC, Australia
[2]Department of Civil Engineering, Politeknik Negeri Kupang, Indonesia
[3]Department of Health, Science and Technology, Klinik Garuda, Malang, Indonesia

E-mail: dklaas@swin.edu.au; duaklaas@yahoo.com

Abstract. Advancement towards improved configuration of pilot point into a model in parameterisation step of groundwater modeling is required to accurately obtain a satisfactory model performance. Nevertheless, the recommendations provided by the majority of current studies on this geostatistical technique are considered too empirical than practical to be applied on real catchment conditions. In this study for the first time a practical approach of using geometric features of pilot point and catchment area is proposed to efficiently configure pilot point distribution in the calibration step of a groundwater model. Three user-friendly ratios, i.e. distance-area (*d/A*), distance-x grid length (*d/Xgrid*), and distance-y grid length (*d/Ygrid*) are introduced in Grid-based (GB) distribution technique. Eight models of eight different pilot point distances (250 m, 500 m, 750 m, 1000 m, 1250 m, 1500 m, 1750 m and 2000 m) were developed using GB technique were constructed on an eogenic karst catchment in Rote Island, Indonesia and their performances were evaluated. Modelers can use the three practical ratios when calibrating groundwater models given the same sub-surface characteristics. This study also provides some insights into the trade-off between maximising and restricting the number of pilot points and offers a relative basis for selecting pilot point properties and distribution method in the development of a physically-based groundwater model.

1. Introduction

Using an accurate spatial conditioning in parameterisation step implies a rigorous effort both numerically and computationally. Nowadays, geostatistical techniques have become important tools in performing a robust model simulation effectively. Among geostatistical techniques, Pilot Points Method (PPM), which has been widely used in different hydrogeological conditions in many parts of the world, is an alternative to the classical zonation method which is conditional upon the number of available field observations and prior knowledge of geological information, such as geological map and stratigraphy [1]. Compared to the zonation method, PPM offers critical advantages when dealing with heterogeneity. Therefore, a number of works have been carried out to enrich the method by introducing stochastic regularization to increase numerical stability when dealing with numerous parameters [2], plausibility term for refining the heterogeneity [3] and singular value decomposition (SVD) to assist a proper selection of pilot points number, hence their locations [4].

In spite of the development on pilot point method, configuring pilot points into a model in parameterisation step is still somehow challenging. Certes and de Marsily have raised a concern on difficulties to properly and practically organise pilot points into a model during parameterisation step to achieve a good fit [5]. They underlined the subjectivity introduced when the number and location of pilot points are self-judged and organised by modellers, and thus brought up a fundamental necessity to develop heuristic recommendations. Some authors then proposed several recommendations to appropriately characterise the pilot point distribution by adjusting the distance between points using correlation length as its relative property [3, 6]. Nevertheless, this theoretical property is considered less practical for real case examples. Motivated by this, in this study we propose a practical approach of using pilot points' properties (i.e. number, distance and distribution method) during calibration step of a groundwater model. For the first time, we introduce the relative distance area ratio (d/A), which refers to distance between points and catchment size.

Thus using a real karst catchment in Rote Island, Indonesia as a study case, the main objectives are to carry out a modelling-oriented analysis of groundwater flow to:

- investigate the impact of pilot points' properties (number, distance and distribution method) on calibration of groundwater models;
- provide recommendation on selection of pilot points' properties (d/A, d/x_{grid}, d/y_{grid}, and zone ratio) to obtain a better calibration result of a groundwater model assessed by statistical measures.

2. Materials and methodology

2.1. Study area

The experimental area is the catchment of Oemau spring located in Rote island, Indonesia, geographically located between latitudes 10°46'42.17"S ~ 10°43'36.91"S and longitudes 123°3'14.84"E ~ 123°9'17.64"E. The catchment is a topographically bounded surface drainage basin area of 20.11 km^2, located around 3 km from Ba'a, the capital of the Rote Island (figure 1).

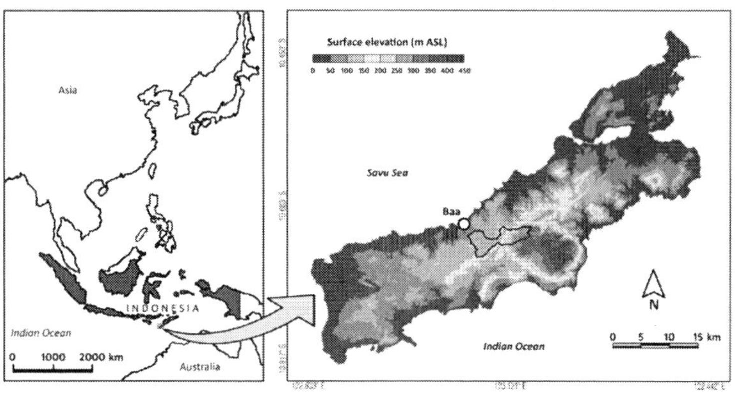

Figure 1. Location of study area and topographical situation of Oemau spring catchment.

The study area has a moonsonal climate characterised by two distinct seasons: dry (May-November) and wet (December-April). Annual rainfall is between 1000 and 2300 mm and humidity is 75-92%. In general, the land surface of the catchment is geologically covered by carbonate formations featured as karst landscape. Hydraulic head and spring discharge data from seven observation wells located within the catchment boundary provide the basis for interpreting the flow system that dictate the subsurface flux. Using modified Boussinesq approach of recession curve, a diffuse flow characterised by low hydraulic conductivity, high storage capacity and laminar flow [7].

2.2. Groundwater model

Eight models were constructed using two pilot point distribution methods (GB). The models are principally differentiated by the distance between points and number of points in each hydraulic gradient zone. A complete explanation is provided in Section 2.4.

The simulations of groundwater flow under both steady-state and transient conditions were carried out using the modular three-dimensional numerical model MODFLOW [8] operated and visualized in the groundwater modeling system, GMS [9] user interface. The conceptual model was constructed on the basis of one horizontal layer which assumes the continuum representation of unconfined carbonate aquifer in the study area. This assumption is considered agreeable to the dominant eogenetic carbonate aquifer in the geologic stratum. The one-layer finite-difference grid is comprised of 106 rows and 210 columns resulting in 8061 active cells (figure 2(a)). The model was discretised in Layer Property Flow, LPF [10] package using a homogenous horizontal model of 50 m by 50 m grid cells. The catchment boundary is set as no-flow boundary assuming no flux flowing through the boundary (figure 2(a)). At downstream, a small stream after the spring was set as specified head boundary using Time-Variant Specified-Head, CHD [11].

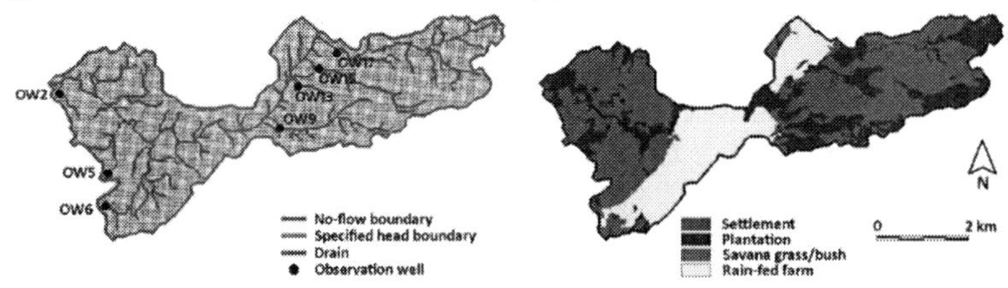

Figure 2. Conceptual model of the model and land use as input for recharge coverage.

The model simulations used prior information of hydraulic conductivity and specific yield values calculated using Boulton method [12] on pumping test data at 3 locations within the catchment. A constant value of 0.3 was set for the porosity to represent the carbonate formation (Johnson 1967). Drain package, DRN (Harbaugh et al. 2000) was used to model the surface drainage network of the recharge area (see figure 2(b) for the drainage network). The streams' surface drain conductance was set between 2,700 and 8,125 m/d correlated with the drain dimension and K values which were derived from local pumping test analysis. To simulate recharge to the aquifer the recharge package, RCH (Harbaugh et al. 2000) was used. The recharge mainly characterised by atmospheric stresses (i.e. precipitation and potential evapotranspiration) and surface (i.e. land use type) was quantified using the water balance method [13]. In this process, SCS-CN [14] and Penman-Monteith [15] methods were applied to calculate surface runoff and evapotranspiration respectively. In the conceptual model, the recharge values are assigned into each land use zonation, illustrated in figure 2(b).

2.3. *Model calibration*

The steady-state calibration was carried to assist a smooth convergence of the model for the transient simulation. Steady-state calibration aimed to determine the distribution of horizontal hydraulic conductivity (K_h) across the model domain during the highest recharge. The K_h values are set to range between 0.1 and 250 m/d to represent karst limestone [16, 17]. Vertical anisotropy was set to 0.2. In steady-state simulation, the maximal rainfall rate of 86.5 mm/day on 19 April 2011 was chosen. The initial K_h values then feed the transient simulation which aims to calibrate the specific yield (Sy) and K_h values. Transient simulation covers the period from January 2011 to August 2011 divided into 8 monthly stress periods with each comprising of 10 time steps. In each step, an iterative matrix solver, Preconditioned Conjugate Gradient 2, PCG2 was used to increase the model's capability to converge. In both steady–state and transient simulations PEST was used. In this automated optimisation code, the pilot points technique [18] was applied for all models by distributing pilot points according to the two distribution methods explained in Section 2.4. In the simulation, Singular Value Decomposition,

SVD (Doherty 2004) and Tikhonov regularisation (Doherty 2003) were used to stabilise the mathematical approximation. Computationally, all models were run on a desktop computer Intel® Core™2 Quad CPU Q6600, clock speed of 2x2.40 GHz using 4 GB RAM.

A calibration criterion for both the steady-state and transient simulations was employed to match simulated heads with observed heads. Daily data collected from 7 observation wells in the catchment for the duration of January, 2011 ~ April, 2012 were used as observed heads. The model calibration was accomplished by analysing models' performance specified by statistical goodness-of-fit measures: Mean Absolute Error (MAE_h), Root Mean Squared Error of head ($RMSE_h$), and Nash-Sutcliffe coefficient (NSE_h) as objective functions.

2.4. *Distribution of pilot points*

GB distribution method is commonly used to deploy pilot points into model domain and is widely used in groundwater modelling [19-21]. The model domain was firstly divided into rectangular uniformity grids. Using orthogonality principle, the pilot points were uniformly distributed with respect to the predefined distinct homogenous distance between points [22]. The grid was rotated perpendicularly to assumed groundwater flow direction (BYU 2014), which is 178.6°. Figure 3 illustrates an example of pilot point assignment using the GB method for our model GB4.

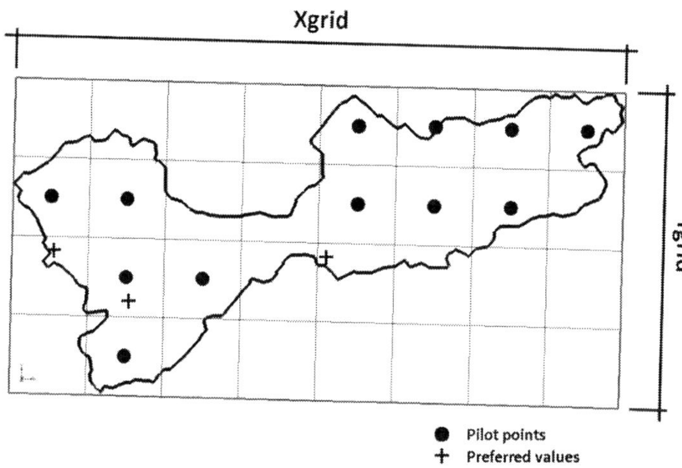

Figure 3. GB spatial distribution of pilot points of model GB4, also showing the incorporation of preferred values into the model domain. *Xgrid* refers to the total length of the horizontal grid and *Ygrid* is the total length of the vertical grid.

Table 1. The model properties, number of pilot points and performances in steady-state simulation.

Model	*d* (m)	Number of pilot points
GB1	2000	9
GB2	1750	12
GB3	1500	13
GB4	1250	15
GB5	1000	26
GB6	750	41
GB7	500	88
GB8	250	322

In this distribution method, eight models of eight different predefined distances were developed (see table 1).

3. Results and discussions

3.1. Model calibration

The complete results of the steady state simulations for the eight models are summarised in table 1. Generally, the error statistics of all models using the two pilot points distribution methods are small in regards to RMSEh values (vary between 0.25 m and 0.81 m), confirming that all models are capable of reproducing hydraulic heads to a satisfactory level. The direction of groundwater flow also confirms the pre-set flow direction, suggesting a well calibrated model (Domenico and Schwartz 1998).

Table 2 summarise the results of the transient simulations for the eight models using GB. The model performances in transient simulation are overall considered satisfactory in regards to $RMSE_h$ values, which vary from 0.29 m to 0.44 m for all the fifteen models. A considerably high NSE_h value (0.81-0.92) is categorised as very good according to the model criteria (Moriasi et al. 2007), which indicate a good simulation achievement.

Table 2. Performances of transient simulations using GB distribution method.

Model	NSE_h	$RMSE_h$ (m)	*d/A* (km^{-1})	*d/Xgrid* (-)	*d/Ygrid* (-)
GB1	0.84	0.42	0.10	0.20	0.40
GB2	0.86	0.38	0.09	0.18	0.35
GB3	0.86	0.35	0.07	0.15	0.30
GB4	0.87	0.32	0.06	0.13	0.25
GB5	0.92	0.29	0.05	0.10	0.20
GB6	0.92	0.32	0.04	0.08	0.15
GB7	0.89	0.36	0.02	0.05	0.10
GB8	0.87	0.35	0.01	0.03	0.05

3.2. Effect of pilot points properties on model performance

Figure 4 shows the model performance in regards to $RMSE_h$ for the eight models as a result of increased distance between pilot points. For the first time, *d/A* ratio was introduced as a practical pilot points property to efficiently classify a grid-guided model. When *d/A* is reduced from 0.10 to 0.05 km^{-1} the model performance increases by 43.18%. The ratio reduction implies that the distances between points become shortened (from 2000 to 1000 m) causing increased pilot point number. In opposition, the model performance drops by 18.27% when *d/A* is reduced from 0.05 to 0.01 km^{-1}, giving a distance shortening from 1000 to 250 m. The figure shows that the maximum model performance is achieved at the value of *d/A* = 0.050. Table 3 shows the comparison of model performances associated with pilot point properties. It shows that the model with the highest performance is GB5 having the values of *d/A*, *d/X*grid and *d/Y*grid as 0.05, 0.10 and 0.20 respectively. This study demonstrates that although pilot points are capable of representing subsurface heterogeneity hence helps to improve model performance, however their capability is influenced by the number of points and the distance between points relative to catchment size. Therefore, an optimal model performance can be reached by appropriate input arrangement of pilot points, which in this study is represented by *d/A* ratio.

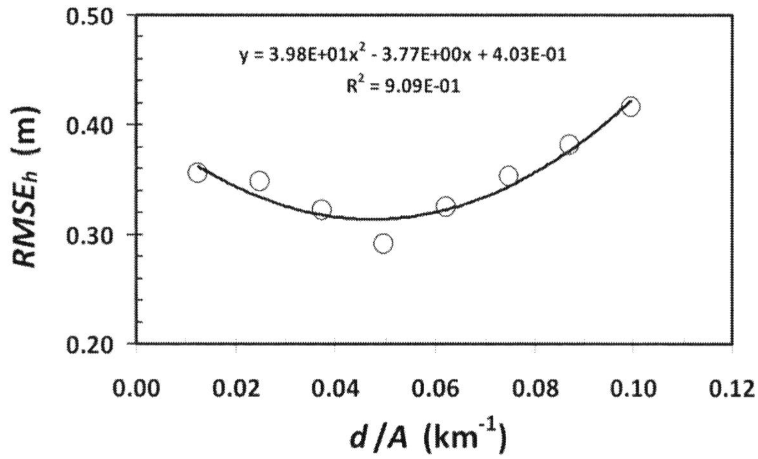

Figure 4. Statistics results ($RMSE_h$) vs. distance-area ratio, which shows the impact of pilot point density in the model domain as a result of distance widening.

The fitting graph demonstrates that adding up the number of pilot points into the models corresponds to the increase in model performance. However, from the figure it is clear that beyond a certain limit adding extra number of pilot points is not recommended, as it results in a drop of model performance. In this study, the reduced model performances after reaching the lowest error values at the "tipping point" might correspond to 'model overfitting', a condition when inaccurate model conditioning including inappropriately assigning location and density of pilot points results is used to force the model output to match the observed values [3]. In our case, the adverse effect of adding more pilot points, which narrows distance between pilot points, beyond a suggested number on model performance may introduce not only redundant measurements [23] but also a possible effect of over-smoothing between points. Therefore, although error values decrease by narrowing distance between pilot points due to increased pilot points number, after the tipping point the pilot points number is considered excessive and results in less-optimal optimisation due to over-parameterisation.

4. Conclusions

Choosing a reasonable pilot point number and appropriate distribution method using a practical guideline is important in the parameterisation step of a groundwater modelling. In this study using the GB method, for the first time distance-area ratio (*d/A*) was introduced. This ratio was found to be a practical tool to represent pilot point property in order to assess the model performance. This study demonstrated that when the distance-area ratio is decreased from 0.10 to 0.05 km^{-1}, implying that the number of pilot points increased due to distance shortening from 2000 m to 1000 m, the model performance increased by 43.18%. On the contrary, when the distance between pilot points was reduced from 1000 m to 250 m, the model performance decreased by 18.27%. The study shows that the maximum model performance is achieved at the value of $d/A = 0.050$.

This study is able to show an alternative practical approach in properly selecting pilot point properties (i.e. distribution method, number and distance). Based on the model simulations developed in this study, the following recommendations are proposed for using GB distribution method:

- The recommended distance-area ratio (*d/A*) is 0.05;
- The recommended distance-x grid length ratio (*d/Xgrid*) is 0.10; and
- The recommended distance-y grid length ratio (*d/Ygrid*) is 0.20.

The outcomes of this study are considered applicable for any modelling analysis with similar soil characteristics, i.e. carbonate dominated aquifer, dominant diffuse flow system and hydro-climate regime, i.e. monsoonal climate. However, although the initial results presented here are promising, it is noteworthy to test the robustness of the result by conducting complementary modeling using the same methodology for other case studies diversified by different and broader model properties, such as catchment size, geological features and climatic setting is required.

References

[1] Klaas D K S Y, Imteaz M A and Arulrajah A 2016 Evaluating the impact of grid cell properties in spatial discretization of groundwater model for a tropical karst catchment in Rote Island, Indonesia *Hydrol. Res.* doi: 10.2166/nh.2016.250
[2] Doherty J 2004 PEST: Model-Independent Parameter Estimation User Manual 5th ed (Brisbane, Australia: Watermark Numerical Computing)
[3] Alcolea A, Carrera J and Medina A 2006 Pilot points method incorporating prior information for solving the groundwater flow inverse problem *Adv. Water Resour.* **29** 1678-89
[4] Christensen S and Doherty J 2008 Predictive error dependencies when using pilot points and singular value decomposition in groundwater model calibration *Adv. Water Resour.* **31** 674-700
[5] Certes C and de Marsily G 1991 Application of the pilot point method to the identification of aquifer transmissivities *Adv. Water Resour.* **14** 284-300
[6] Gómez-Hernánez J J, Sahuquillo A and Capilla J 1997 Stochastic simulation of transmissivity fields conditional to both transmissivity and piezometric data—I. Theory *J. Hydrol.* **203** 162-74
[7] Shuster E T and White W B 1971 Seasonal fluctuations in the chemistry of lime-stone springs: A possible means for characterizing carbonate aquifers *J. Hydrol.* **14** 93-128
[8] McDonald M and Harbaugh A 1988 *A Modular Three-Dimensional Finite-Difference Groundwater Flow Model* (USGS) book 6, Chap A1
[9] BYU 2014 *Groundwater Modeling System* (GMS) version 9.2.9. (Provo, USA: Environmental Modeling Research Laboratory, Brigham Young University)
[10] Harbaugh A W 2005 *MODFLOW-2005*, the U.S. Geological Survey Modular Ground-Water Model, the Ground-Water Flow Process (USGS Techniques and Methods 6-A16, USGS)
[11] Harbaugh A W, Hill M C and McDonald M G 2000 *MODFLOW-2000*, the U.S. Geological Survey Modular Ground-Water Model: User Guide to Modularization Concepts and the Ground-Water Flow Process (USGS Open-File Rep 00-92, USGS)
[12] Boulton N S 1963 Analysis of data from non-equilibrium pumping tests allowing for delayed yield from storage *Proc. Inst. Civil Eng.* (London, Great Britain) **26** 469-82
[13] Bras R L 1990 Hydrology: *An Introduction to Hydrologic Science* (Massachusetts, USA: Addison-Wesley, Reading)
[14] Viessman W and Lewis G L 2003 *Introduction to Hydrology* 5th ed (Upper Saddle River, NJ, USA: Pearson Education)
[15] Allen R G, Pereira L S, Raes D and Smith M 1998 *Crop evapotranspiration-Guidelines for computing crop water requirements* FAO Irrigation and Drainage Paper 56
[16] Davis S N 1969 *Porosity and Permeability of Natural Materials in Flow Through Porous Media* (New York, USA: Academic Press)
[17] Freeze R A and Cherry J A 1979 *Groundwater* (Englewood Cliffs, New Jersey, USA: Prentice-Hall)
[18] de Marsily G 1984 *Spatial Variability of Properties in Porous Media: A Stochastic Approach* ed J Bear and M Y Corapcioglu (Boston, USA: NATO ASI Series E) 719-69
[19] Klaas D K S Y, Imteaz M A and Arulrajah A 2016 Development of a groundwater model for a tropical eogenetic karst aquifer in Rote Island, Indonesia *Proc. Aust. Hydrol. Water Res. Symp.* (Queenstown New Zealand) 228-35
[20] Klaas D K S Y, Imteaz M A and Arulrajah A 2017 Development of groundwater vulnerability zones in a data-scarce eogenetic karst area using Head-Guided Zonation and particle-tracking simulation methods *Water Res.* **122** 17-26
[21] Klaas D K S Y, Imteaz M A and Arulrajah A 2017 Development of groundwater vulnerability

zones in a data-scarce catchment *Proc. 10th World Congress Water Res. Env.* (Athens, Greece)

[22] Doherty J E, Fienen M N and Hunt R J 2010 *Approaches to highly parameterized inversion: Pilot-point theory guidelines, and research directions* USGS Scientific Investigations Report 2010-5168 (USGS)

[23] Klaas Dua K S Y and Imteaz M A 2017 Investigating the impact of the properties of pilot points on calibration of groundwater models: case study of a karst catchment in Rote Island, Indonesia *Hydrogeol. J.* 1-17 doi: DOI 10.1007/s10040-017-1590-4

IOP Conf. Series: Earth and Environmental Science 82 (2017) 012015 doi:10.1088/1755-1315/82/1/012015

A case study on the ecological evaluation of water conservation facilities on the basis of plant diversity

C Y Wu[1], G J Xia[1], T H He[1,2], J Yu[1] and B Qiao[1]

[1]Key Laboratory for Restoration and Reconstruction of Degraded Ecosystem in North-western China of Ministry of Education, Ningxia University, Yinchuan, Ningxia, 750021, China

E-mail: hetonghui@163.com

Abstract. In this experiment, water conservation facilities in the Yinchuan Plains were studied to establish ecological evaluation criterion for water conservation facilities based on plant diversity indexes. Among the tested indexes, the Margalef and the Simpson indexes were found to effectively characterize the ecological degree of water conservation facilities. Natural lake plant diversity has been used to as a metric; the ecological criteria degree of water conservation facilities can be divided into 4 grades. The results showed that the ecological level of the field ditch, primary furrow, and second furrow qualified as class A, which is the highest grade. The branch canal was grade B, which is good. The lateral canal and flood furrow were grade C, which is a medium level. The trunk canal was grade D, which is the poorest. In addition, the results establish an ecologicalization evaluation system based on the plant diversity composite index of the original lake. Metrics such as biomass, carbon and nitrogen fluctuations, temperature, and humidity could be used to comprehensively evaluate the ecologicalization degree of water conservation projects, such as large reservoirs.

1. Introduction

The purpose of ecological hydraulic engineering and water conservation project ecologicalization is to protect the environment. Water resources conservation, space-time allocation, and rational utilization have been carried out for many years in the water industry. Currently, the ecological problems of water conservation facilities are national concerns [1, 2]. Societies must follow basic ecological and economical requirements, and use modern scientific, management, and technological achievements to develop efficient water conservation projects. Water conservation projects should offer a high degree of engineering and ecological benefits. In arid inland areas, water conservation is the lifeblood of both agriculture and ecology. Farmland windbreaks and garden green lands cannot operate without the assistance of water facilities [3]. It is therefore necessary to research and evaluate water conservation project ecologicalization for water conservation project construction. This research could significantly benefit the health and safety of ecological drylands [4].

The scientific community's current understanding of ecological hydraulic engineering lacks a cognitive quantization index; therefore, it is important to explore scientific environmental assessment indicators. Biodiversity is sensitive to the ecological status indicator because the loss of species decreases ecological resilience to disturbance or disruption [5]. Our research of the Yinchuan Plain revealed that the biodiversity index was a better metric for determining the ecology of ditch wetlands [6, 7]. It is commonly known that plant diversity leads to ecological stability, which leads to more complicated energy flows and material cycles [4]. These favorable qualities can increase retention and production, improve the ecology's ability to self-regulate, and ultimately create healthier and more stable ecosystems. Multiple supply and partitioned supply mainly demanded from a tactical point of view infrastructure supply. Multiple supply means that resources and energy are provided uniformly; partitioned supply means that the entire facility is divided into several partitions, and the resources and energy are provided according to each section's needs. The former promotes efficiency and the security of the Foundation Infrastructure Provider. Together, integrity and variety are a representation for foundation infrastructure ecologicalization [8].

Since plant landscapes are not affected by the water deposit status of water facilities, this study evaluates the ecosystems of streams, lakes, and reservoirs in the Yinchuan Plain using plant diversity indexes. The study establishes an evaluation system to judge whether or not water facilities are ecologically sound.

2. The Yinchuan plain and its establishing condition of water facilities

The Yinchuan Plain (105°45′~107°00′, 37°50′~39°20′) is located in the eastern part of northwestern China (figure 1); it belongs to the Ningxia Autonomous Region. It is composed of alluvium and alluvial material from the Yellow River and Helan Mountains; the total area is 7978 km^2. The mean annual precipitation in the Yinchuan Plain is around 200 mm, while the mean annual temperature is 8~10°C. It has a longer winter and a shorter summer, a later spring and an earlier autumn, and a strong wind that carries sand. In LandScope, the Yinchuan Plain is an oasis surrounded by several deserts, which means that water conservation facility ecologicalization is particularly important for this region.

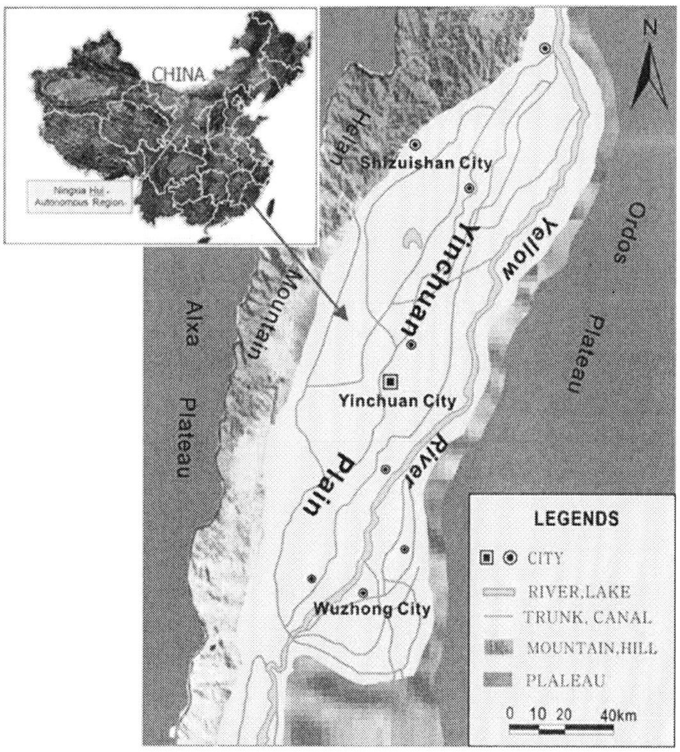

Figure 1. Map of Yinchuan Plain and its location. It is surrounded by the Tengger Desert of Alxa Plateau on the west, the Kubuqi Desert and Mu Us Desert of the Ordos Plateau on the east, and the Wulanbuhe Desert on the north. Thus, the Yinchuan Plain is an oasis formed by Yellow River irrigation.

The Yinchuan Plains were agriculturally developed during the Qin (777~206 BCE) and Han (206 BCE ~ 220 CE) Dynasties; a water system was formed that included a main channel, trunk canal, branch canal, lateral canal, field ditch, and wicking ditch. This system ensured the development of an oasis that irrigated farmlands. By the middle of the 20th century, about 15 canals in total were being used for irrigation, supplying a land area of about 9600 hm^2. In the 1960s, Qingtongxia Reservoir changed control-free irrigation history in the Yinchuan Plain because repairs to the Main channel completely changed the configuration, improved height differences, and expanded the irrigated area. Currently, there are about 18 canals in the Yinchuan plain; their total length is more than 1084 km, their total water flow is 603 m^3/s, and their total irrigated area is 330,000 hm^2. There are about 24 key drains with a total length that exceeds 660 km, a total water flow of 955 m^3/s, and a total irrigated area of 4.2×104 hm^2 [9, 10]. There are 151,800 hm^2 of river and lake wetlands, which amount to about 19.1% of the total area of Ningxia. Some of these wetlands were constructed last year.

3. Materials and methods

3.1. Field investigation and sampling

We carried out a field investigation for water environments in the Yinchuan Plains using a line transect and sample plot. We selected typical lakes, channels, and drainage ditches to set sample plots according to the habitat heterogeneity degree and land use pattern. We chose primordial lakes that had stable ecosystems and displayed without interference. For the main furrow, flood furrow, and trunk canal, we selected sections from each upside, intermediate, and bottom. For general ditches, we selected sections from upstream, midstream, and downstream that were refitted more than three years ago. It was located specifically from the top of one side of the ditch dam to the top of the other side of the dike under the section direction, then lie out samples on the section according to the ecological sequence. In the Yinchuan Plain, the ditches and canals were divided into 5 separate levels according to access relations; the middle stages are often difficult to distinguish in the field. Thus, we classified trunk and branch canals as first stage canals, and lateral and other low-level canals as second stage canals. The quadrat size of herbaceous plants was 1 m × 1 m, but the quadrat size of shrub communities was 4 m × 4 m. Each sample plot had three repetitions. We recorded the name, abundance, height, coverage, and life forms of terrestrial plants, as well as the name and coverage of aquatic plants [11]. We also measured the slope, gradient, slope length, root zone, water depth, flow rate, water flux, and construction status.

3.2. Selecting the diversity index

Biological diversity is a function of biological richness and evenness [12]. There are many relevant calculation models for community biodiversity. However, the various methods weigh richness and evenness differently. This study mainly used the Margalef species richness index, the Simpson species dominance index, the Shannon-Weiner species diversity index, and the Pielou species evenness index [13].

In the below formulas, 'S' indicates the number of species within the samples; 'Pi' indicates the relative important value of a species within the samples (Relative important values = [Relative abundance + Relative frequency + Relative significant]/3).

- Species richness index: Margalef index

$$R=(S-1)/InN \quad (1)$$

This reflects the number of species in the community or habitat. A higher number indicates higher species diversity.

- Simpson species dominance index: Simpson index

$$D = 1 - \sum_{i=1}^{s} p_i^2 \quad (2)$$

This reflects species dominance in the community. A higher number indicates a lower dominant species.

- Species diversity index: Shannon-Weiner index

$$H = -\sum_{i=1}^{s} PiInPi \quad (3)$$

This reflects the species diversity in the community. A higher number indicates more community complexity.

- Pielou species evenness index: Pielou index

$$E = H / InS \tag{4}$$

This reflects the uniformity of spatial distribution for species. A higher number indicates a more homogeneous plant distribution and a greater contribution degree to the ecological environment [12].

Using Microsoft Excel 2007, we calculated the diversity indexes of each lake or canal sample, then concluded that according to its architecture. Finally, we added them separately and averaged them to find the diversity indexes of the lake, canal, ditch, and so on. We often combined the trunk furrows, branch furrow, lateral furrows, field furrow, and little furrows in our calculations because they were too similar to properly distinguish.

4. Results and analysis

4.1. Plant diversity in lakes and ditches

We compared the Margalef index of lakes and ditches in the Yinchuan Plains. The results showed that the flood furrow had the highest number of individual plants, and the trunk canal had the lowest number of individual plants. The difference between the two nearly 1 times; Next to the flood furrow is the second furrow and then the primary furrow, then the lakes, the next are field ditch and trunk canal, species richness of lateral canal are low, close to trunk canal (see figure 2).

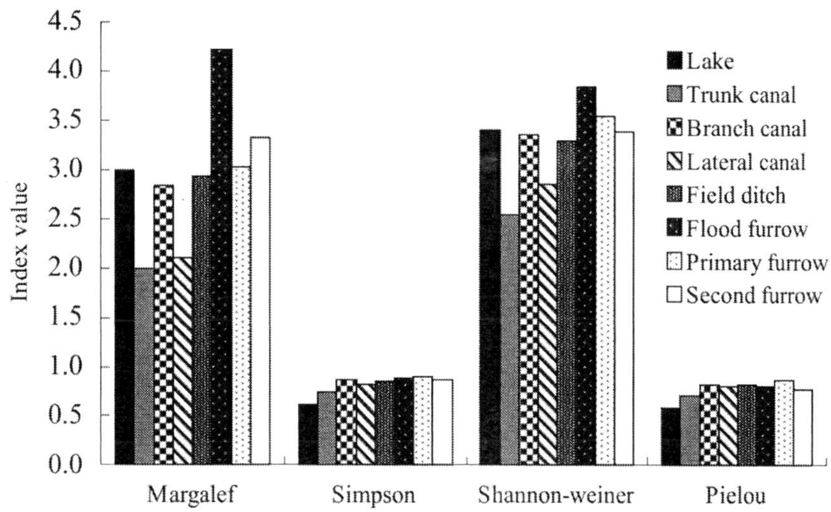

Figure 2. The plant diversity indexes values in the Yinchuan Plain across 4 biological diversity indexes (Margalef, Simpson, Shonn-winner, and Pielou index), across lakes and 6 types of water facilities (trunk canal, branch canal, lateral canal, field ditch, flood furrow, primary furrow and second furrow).

When the species in the level ditches did not display a significant difference in dominance, there were no significant differences between their Simpson indexes. The branch canal and primary furrow had the highest Simpson indexes (0.893, 0.875), while the trunk canal had the lowest (0.743). This illustrates that the dominant species of the trunk canal was relatively more populous, and the dominant species of the branch canal and primary furrow were relatively less populous. The dominant species in other ditches were between these two extremes.

We compared the Shannon-wiener index and Pielou index of each ditch in the Yinchuan Plain (figure 2). The results show that the flood furrow had the highest Shannon-Wiener value at 3.837, but

IOP Conf. Series: Earth and Environmental Science **82** (2017) 012015 doi:10.1088/1755-1315/82/1/012015

its Pielou value was in the middle at 0.798; this means that there were a high number of plants, and the distribution had some aggregation. The trunk canal, lakes, second furrow, branch canal, and field ditch showed middle Shannon-Wiener indexes between 3.298-3.540, but their Pielou index had a large gap between 0.609-0.821; this indicates that the plant distribution was not uniform. The trunk canal and lateral canal had the lowest Shannon-wiener values at 2.541 and 2.861, respectively, and their Pielou indexes were relatively low at 0.709 and 0.798, respectively; this indicates that the plant species distribution was minimal and uneven.

The Simpson and Pielou indexes of different plant communities in the Yinchuan Plain wetlands were very similar. This illustrated that the less dominant species in ditches had large evenness; in contrast, more dominant species in ditches had small uniformity.

4.2. The spatial distribution of plant diversity

The Yinchuan Plains have higher altitudes in the south and west. The 2000 years of oasis development basically lasted the flow direction that from south to north, from the first terrace and shore of the Yellow River Plain to the second terraces of the alluvial plains. In this study, we roughly divided the irrigation area into two stages in chronological order. The oasis was considered an old irrigation area; the oasis subsequently developed into the new irrigation area. According to the survey results, the plant species and community types of the new irrigation areas were richer than that of the old irrigation area; for instance, aquatic plants were extremely rich in the flood furrow of the new irrigation area, except for common species in the old irrigation areas, such as Potamogeton natans L., Nymphoides peltatum (Gmel.) O. Kuntze, Ceratophyllum demersum L., Potamogeton pectinatus L., Utricularia vulgaris L., Salvina natans L., etc. Phragmites australis (Cav.) Trin. ex Steud, Typha angustata Bory & Chaub., Polygonum lapatifolium L., and Polygonum hydropiper L .etc. Typical wetland plant communities grew in the flood furrow and drainage ditch of the new irrigation area, including Comm. Scirpus triqueter, Comm. Scirpus tabernaemontani, Comm. Triglochin palustre, Comm. Acorus calamus, rare species Butomus umbellatus L., Crypsis aculeate (L.) Ait. Wetland vegetation in lakes included Phragmites australis (Cav.) Trin. ex Steud, Scirpus Tabernaemontani, Typha angustifolia L., Scirpus validus Vahl, Sagittaria trifolia L. var. sinensis (Sims.) Makino, Potamogeton distinctus A. Benn., Nymphoides peltatum(Gmel.) O. Kuntze, etc [14].

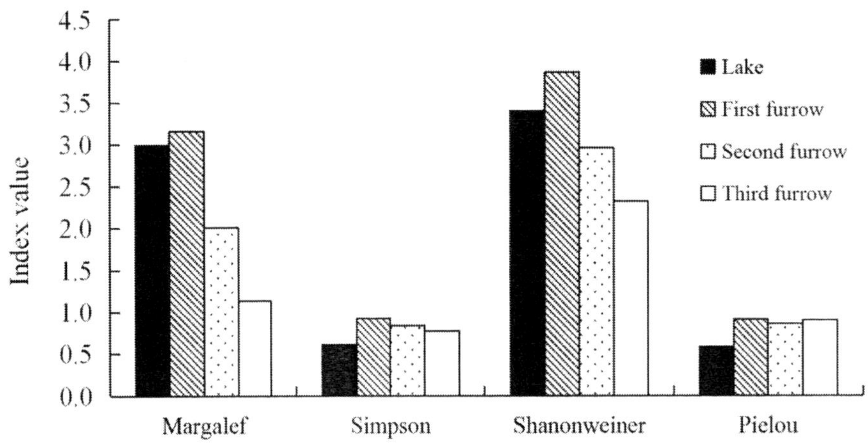

Figure 3. The plant diversity indexes of macro-ditches. There are 5 macro-furrows with lengths greater than 100 km length in the Yinchuan Plain; we selected 3 of them and calculated their biodiversity indexes, which represents the ecological status of water conservation facilities upstream, midstream, and downstream. Additionally, we compared the value differences between them and lakes.

Research has shown that ditch wetland plant diversity in the Yinchuan Plains was high, and it had a strong edge effect [15]. Using lakes and drainage ditches as examples, their species diversity was relatively high, which contrasted with the plant diversity index of the three large drainage ditches of the Yinchuan Plains from south to north in this order: drainage ditch, trunk drainage ditch, primary lake, and drainage ditch. The Margalef and Shannon-Wiener indexes were both the highest in the first drainage ditch (figure 3), and the lowest in the third drainage ditch. Plant species diversity was highest in the first drainage ditch, followed by the lake, and lowest in the trunk drainage ditch. Plant species abundance and complexity decreased from upstream to downstream. The Simpson and Pielou indexes showed few differences in the three drainage ditches, meaning that plant dominance and uniformity did not display many differences between the upstream and downstream areas. The results indicated a significant difference between the drainage ditch and lake [16]. We studied the interspecific relationships between wetland plants in the Yinchuan Plains and concluded that wetland dominant plant populations varied due to annual dynamic changes in wetland large-scale renovation, water contents, and water levels.

5. Evaluating the ecologicalization of lakes and ditches in the Yinchuan Plain

5.1. Ecological index and its effects

To facilitate calculations, the plant diversity indexes of lakes and ditches were carried out in the principal component analysis (table 1). The results showed that the cumulative contribution rate of the Margalef and Simpson indexes were 97.455%, and the Eigenvalue was >1, which indicates that the first two diversity indexes contained more than 85% of the information. Therefore, these two diversity indexes can indicate the ecological degree of lakes and canals in the Yinchuan Plains. The principal component factor loading of each diversity index divided by the Eigenvalue of the arithmetic square root determines the coefficient of each diversity index; the final function expression is:

$$y_1=0.488x_1+0.535x_2+0.507x_3+0.468x_4 \quad (5)$$

$$y_2=0.518x_1-0.438x_2+0.480x_3-0.557x \quad (6)$$

Table 1. Principal component analysis of plant diversity indexes in lakes and ditches.

Ratio/Index	First principal component	Second principal component
Eigenvalue	2.475	1.423
% of Variance	61.884	35.571
Cumulative %	61.884	97.455
Margalef	0.767	0.618
Simpson	0.842	-0.522
Shannon-Weiner	0.797	0.572
Pielou	0.737	-0.665

Table 2. Scores of lakes and ditches on the first and second main component.

Type	First principal component score	Rank	Second principal component score	Rank
Lake	3.786	5	2.596	2
Trunk canal	2.992	8	1.536	8
Branch canal	3.460	6	2.241	6
Lateral canal	3.304	7	1.656	7
Field ditch	3.946	4	2.278	5
Flood furrow	4.847	1	4.422	1
Primary furrow	4.157	3	2.394	4
Second furrow	4.170	2	2.542	3

From table 2, we can find that the scores of the flood furrow on the first main component and second main component were the highest; however, the trunk canal was the lowest on the first and second main component scores. The rank of the branch canal and lateral canal were the same on different main component, but the ranks of other water conservation facilities changed.

5.2. Classification of ecologicalization evaluation

First, plant diversity was generally low in the Ningxia wetland, which was caused by the obvious differentiation and the species single of plant community; for example, reed is the dominant species and constructive species. At the same time, the water environment was relatively homogeneous, which leads to wider ecological reed amplitude with *Typha angustifolia* and other broad nature aquatic plants. It also repulsed other plants, resulting in low biodiversity in a relatively stable lake. In recent years, due to the increase of human disturbance, reed communities in wetlands have been in severe degradation in the Yinchuan Plain [17]. Wetland plant diversity therefore displayed a high discrepancy.

Second, the two main functions of water conservation facilities are transporting and storing water. A facility's ecological stability is just one aspect of this comprehensive function. Ignoring the main functions of water conservation facilities and only paying attention to its ecological construction can lead to wasted resources. In this study, the flood furrow had the highest plant diversity, but the construction of the detention basin and flood frequency has decreased in recent years. The spare problem is conspicuous, which equals to ecological land, but it is not the ideal ecological construction location for water conservation facilities. The Yinchuan Plains are located in a semi-arid area with little rainfall and a strong evaporation ratio. Therefore, we should use water-saving irrigation for wetlands, and consider water resource problems when constructing water conservation facilities [18].

Third, we focused on the ecological evaluation standard of water conservation facilities based on plant diversity. According to the results of the principal component analysis, the Margalef and Simpson indexes contained more than 85% of the information, so their weighted average can be used as the comprehensive evaluation index of ecological water conservation facilities (see table 2). We found that the number of native lakes was lower than the median number. Using the number of native lake as the datum, value 0.25 on both sides and then get an interval. Then, we used a good area when the value was 0.5 on both sides, and so on; table 3 shows the results.

Table 3. Ecological evaluation grading standards of water conservation facilities.

Rank	Interval
CK	3.191
A	2.941~3.441
B	2.691~2.941, 3.441~3.691
C	2.441~2.691, 3.691~3.941
D	<2.441, >3.941

Note: A is best, B is good, C is medium, D is poor, CK is standard.

5.3. Classification of water conservation facilities' ecologicalization in the Yinchuan Plain

According to table 3, water conservation facilities in the Yinchuan Plain can be divided into four levels. Table 4 shows that the ecological level of the field ditch, primary furrow, and second furrow are class A, which is the highest grade. The branch canal is grade B, which is good. The lateral canal and flood furrow are grade C, which is a medium level. The trunk canal is grade D, which is the poorest. Therefore, the field ditch, primary furrow, and second furrow showed high ecological water conservation in the Yinchuan Plains. Even though the ecological evaluation level of the flood furrow was not high at grade C, it displayed the highest plant species diversity (see figure 4) due to high changes in water volume, no seepage prevention, and a slope that produced a variety of moisture gradients. However, these plants are succession pioneer; the species configuration was not stable in water conservation facilities. There was low plant diversity in wetlands such as the trunk canal, which is the main component of the Ningxia canals and important transport canals, and which plays an important role in ecological corridor [19]. Canal beds are typically made of clay or some other type of impenetrable material to minimize loss caused by leakage in the water transport process. Slopes are typically made of cement with a hardening treatment, which severely destroys the water necessary for plants to grow and the soil environment. Some studies have put forward that the main factors influencing vegetation distribution in the Yinchuan Plains are moisture, organic matter, salt, and pH [20], which results in lower plant diversity over a long period of time. These also contribute to wetland vegetation degeneration.

Table 4. Ecological classification of water conservation facilities in the Yinchuan Plain.

Type	Rank	Type	Rank
Lake	A	Field ditch	A
Trunk canal	D	Flood furrow	C
Branch canal	B	Primary furrow	A
Lateral canal	C	Second furrow	A

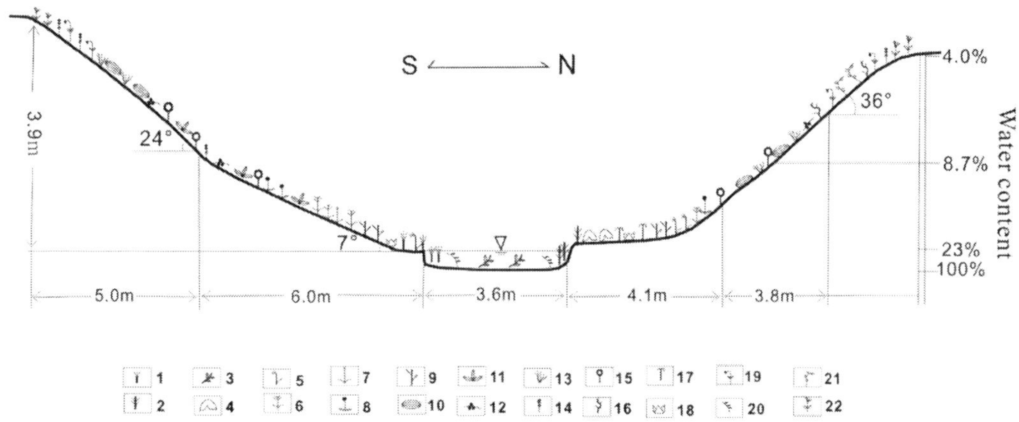

Figure 4. Map of ecological arrays on the slope of the Gaojiazha Ditch. This ditch is a flood furrow stretching from west to east, about 17 km length and 20-30 m width. It provided drainage for flash floods and irrigation water; the slope possessed a significant moisture gradient.

Note: 1 Phragmites australis (Cav.) Trin. ex Steud, 2 Typha orientalis Presl C. Presl, 3 Potamogeton distinctus A. Benn, 4 Sagittaria trifolia L. var. sinensis (Sims.) Makino, 5 Juncellus serotinus (Rottb.) C. B. Clarke, 6 Calamagrostis epigeios (L.) Roth, 7 Setaris viridis (L.) Beauv. Ess. Agrost, 8 Inula japonica Thunb., 9 Equisetum hiemale Desf., 10 Plantago asiatica L., 11 Sonchus brachyotus L. , 12 Xanthium sibiricum Patrin ex Widder, 13 Leymus secalinus (Georgi) Tzvelev, 14 Inula salsoloides (Turcz.) Ostenf, 15 Taraxacum officnala Hand.-Mzt.Monogr.Gothob., 16 Suaeda glauca (Bunge) Bunge, 17 Scirpus triqueter L.Mant., 18 Halerpestes ruthenica(Jacq.) Ovcz., 19 Salsola collina Pall., 20 Ceratophyllum demersum L., 21 Halogeton arachnoideu Moq., 22 Peganum harmala (Maxim.) Bobrov.

6. Conclusion and discussion

This study constructed a practical and reliable grading evaluation criterion for ecological water conservation facilities based on the plant diversity index. The engineering effect of water conservation is usually obvious, and can be constantly adjusted as required. However, the effects of a water conservation project are usually hysteretic and covert. We investigated plant diversity under the influence of water conservation projects, and chose Margalef and Simpson indexes as diversity and evaluation indexes of comprehensive ecological water conservation facilities. This approach considers engineering, ecological, and economic factors.

According to the ecological water resources evaluation standard established in this paper, the eight classes related to water conservation facilities in the Yinchuan Plains can be divided into four grades.

It is possible to use the diversity of native lake composite indexes as a benchmark to evaluate other water conservation facilities in the Yinchuan Plains. The results show that the field ditch, primary furrow, and second furrow had the highest grade (A). The trunk canal had the lowest grade (D).

The plant diversity index for evaluating ecological water conservation is not better when its value is bigger; rather, its purpose is to consider the basic characteristics of the local water ecosystem. Our study revealed that using the average state of native lake wetlands as the standard value to establish an evaluation system is effective. Other studies had proved that factors such as the ecological sequence, hydrological conditions, water conservation project construction, soil, and the water environment can influence plant diversity in the ditches of the Yinchuan Plains [7]. Due to significant differences in the plant diversity index value under different climate and terrain conditions, we think that this is appropriate for evaluating the ecological effect of small and medium-sized water conservation facilities; however, it is not indicative to the material production function of the ecosystem, climate regulation, entertainment, and so on (Jing et al. 2005). If a comprehensive evaluation of an ecological water conservation project is necessary, especially in large lake reservoirs, it is also necessary to introduce other indicators such as biomass, carbon and nitrogen fluctuations, temperature, and humidity.

Acknowledgments
This work was funded by the National Natural Science Foundation of China (41361095). It was also assisted by the Ningxia Wetland Management Center.

References

[1] Fu C and Feng S Y 2000 Study on the principle of the sustainable utilization of water resources (ecological hydraulic engineering) *Adv. Water Sci.* **11** 436-40
[2] Jiang C L and Wang J 2015 Recent advances of ecological water conservation in China *Adv. Sci. Technol. Water Resour.* **35** 168-75
[3] Gallopin G C 1995 The potential of agroecosystem health as a guiding concept for agricultural research *Ecosystem Health* **1** 129-41
[4] Bertollo P 2001 Assessing landscape health: A case study from northeastern Italy *Environ. Manage.* **27** 349-65
[5] Garry P, Craig R A and Holling C S 1998 Ecological resilience, biodiversity, and scale *Ecosystems* 16-18
[6] He T H, Cheng Z and Zhang Y F 2012 Ecological effects based on plant diversity for irrigation-drainage system in Yinchuan Plain *Yellow River* **7** 95-7
[7] He T H, Cheng Z and Zhang Y F 2013 Characteristics of plant diversity of canals and ditches and their influencing factors in Yinchuan Plain *Wetl. Sci.* **11** 352-8
[8] Gu B, Shen Q J and Zheng Z W 2006 Study on ecologicalization of infrastructure - A case on Dongtan of Chongming island Shanghai *Urban Plan. Forum* **4** 20-8
[9] Wu S X and Lu D M 1992 *Ningxia Water Conservation Volunteers* (Yinchuan: Ningxia People Press)
[10] Zhang Y 2005 Benefit evaluation of urban forestry for Yinchuan and the planning of sustainable development (Shanxi: Northwest Agriculture and Forestry University)
[11] Gao Z Z and Dai F H 1988 *Ningxia Vegetation* (Yinchuan: Ningxia People Press)
[12] Curtis J T and Mcintosh R P 1951 An upland forest continuum in the prairie-forest border region of Wisconsin *Eco.* **32** 476-96
[13] Fang J Y, Wang X P and Shen Z H 2009 Methods and protocols for plant community inventory *Biodivers. Sci.* **17** 533-48
[14] Sun S M, He T H and Lou X Q 2012 *Water Ecosystem Rehabilitation and Integrated Management of Yinchuan Lake Wetlands* (Beijing: Heyzer Press)
[15] Zhang J H, He T H and Cheng Z 2014 The comparison research on quantity features of ditch wetland plant communities in Ningxia Plain *J. Arid Land Resour. Environ.* **28** 153-5

[16] Zhao Y Q, He T H and Cheng Z 2013 Interspecific relationships among common plant species in wetlands in the Yinchuan Plain *Arid Zone Res.* **30** 838-44
[17] Zhang Y F, He T H and Zhang J H 2012 Current status of studies in degraded humidogene reed community in China *Wetl. Sci. Manage.* **8** 47-50
[18] Ma J L and He T H 2002 The study on wetlands in Yinchuan Plain *Journal of Ningxia University (Natural Science Edition)* **23** 377-80
[19] Guo L H, He T H and Cheng Z 2011 Progress on the study of ditch wetland ecological environment effects review *J. Water Resour. Res.* **32** 24-7
[20] Zhang J H, He T H and Cheng Z 2013 Analysis of main environmental influencing factors for ditch wetland plant community distribution in Ningxia Plain *Water Saving Irrig.* **3** 35-41

Research of using RF model to drought forecast on Huaihe River

J Wu[1,3] and Y F Chen[2]

[1]Water Resources Department of Hebei Province, Shijiazhuang 050000, China
[2]College of Hydrological and Water Resource, Hohai University, Nanjing 210098, China

E-mail: 805483939@qq.com

Abstract. Random Forest is a combination classification model based on classification and regression tree. RF model can deal with nonlinear problem, therefore, it is able to predict drought. In this paper, RF model is used to forecast drought on the Huaihe River Basin where more and more serious droughts occur. The runoff of twenty-one stations can reflect the overall drought level on Huaihe River. First, the drought has been divided into three levels depending on the SPI criteria. Then, the most important thirty factors are selected as model variables according to the Incnodepurity index after selecting preliminary screening sets. Furthermore, analysis of the drought level of the hydrological stations during the period from 1963 to 2013 is carried out with RF model. The result shows the average accuracy rate of prediction is 73%. The RF drought model is demonstrated as a new effective drought prediction model to reduce or eliminate the loss caused by drought.

1. Introduction

The frequency and intensity of global drought had an increasing tendency under the global climatic changing and human activities. Therefore, taking effective prevention measures to reduce the damage and losses of drought in agriculture, industry and ecology is very important. The most direct way is to predict the drought accurately and effectively. Drought prediction has two mainly method, one based on the physical and the other based on statistical theory. Due to the implicated factors such as climate changes, hydrologic mechanism and so on, leading both of the traditional method which only based on linear smoothly and multivariate model which based on stochastic simulation technology have limitations which are shortcomings on the nonlinear characteristics of forecast of meteorological elements and weather phenomenon. To describe the nonlinearity on forecasting drought climate, Artificial Neural Network (ANN), Grey System (GM) and Support Vector Machine (SVM) have been used to predict drought. Although ANN has strong ability of nonlinear fitting, but its inherent weaknesses are running the prone process, network structure and learning parameters. Moreover, the momentum factor is difficult to select, leading to an over fitting problem and affecting the generalization ability of the network [1]. GM model GM (1, 1) has many advantages, such as small amount of input data, simple principle, moderate calculation and the high prediction precision, but it is only applied to short-term drought forecasting. The prediction accuracy decreases as the predicted time scales increases [2]. In order to improve the accuracy of precipitation forecast, Li [3] combining the BP Neural Network and Gray Model, but it is also applied to short-term drought forecasting. Furthermore, the result of forecast was influenced by the size of the input of the training sample and the way of training screening. G F Fan [4] studied the SVM to predict drought. SVM shovelled the problems of the excessive input varies and over-fitting. However, the optimization of its intrinsic parameters and kernel function is very difficult. It did not satisfy the demand of drought prediction so that the result was only divided into two parts, and we must standardize or normalize the sample and factors in classification and prediction.

Random Forest is an assembly-itemized model that is based on CART. The model can solve nonlinear problem and do not need pre-processing the data. Through the summary on lots of classification tree to improve the precision of the predicted model, it can replace some of traditional machine learning methods such as ANN, GM. In addition, it has been applied more and more in all the aspects of the society also has been used in hydrological analysis and got better effect; Zhao [5] has taken random forests model into the water runoff forecast in Yangtze River.

Random Forests can itemize impact factors and rate the importance of each impact factor that the score as the basis for screening important factors, appling a vote system following the majority rule to determine the final sort. This report is expected by establishing the random forests to classifying drought prediction model. The data during the period from 1963 to 2013 of twenty-one representative station at Huaihe River were used to analysis meteorological drought, and using a meteorological drought index: standardized precipitation index (SPI) analysing drought grade. Firstly preliminary choose 372 factors as the factor set, secondly use RF model selecting 30 importance factors , then use RF drought model to test and forecast at last.

2. Selection of drought classification index

Drought is a natural phenomenon that rainfall, runoff and other natural water in a certain period is less than the normal level in a certain geographic range, water deficit in rivers and lakes. Drought refers to meteorology, hydrology, agriculture and the social and economic disciplines. In addition, its essence is the shortage of water, and what mostly affects the drought is the rainfall. Meteorological drought is associated most closely with precipitation, and is also the most important among the four types of drought. According to the important factors, meteorological drought index is selected to evaluate drought. Meteorological drought index chooses standardized precipitation index (SPI). SPI is considered the quantity of the standard normal distribution, corresponding to the same cumulative frequency of a rainfall value as the standardization of the precipitation rainfall index value (see Appendices).

Because SPI assumes that among all sites where drought and flood happens at the same probability, it is unable to distinguish frequent drought region. This paper respectively divides 21 drought classifications, then we can independently analyse each stand true drought conditions. The scale of analysis used the month, in order to prevent the longer time step indicators being used in region where precipitation is relatively small.

Based on the definition of SPI and analysing of current situation, drought can be divided into three categories based on the threshold of SPI. The SPI value that is corresponding to the frequency of the rainfall is less than 30.9% will be defined as the threshold of drought, rainfall frequency more than 30% of the SPI value will be defined as a threshold of waterlogging, the month's SPI value between two thresholds is normal. Table 1 shows results of drought classification.

Table 1. Result of drought classification based on SPI with 21 stations of Huaihe basin.

Station Name	Level 1 (waterlogging)	Level 2 (normal)	Level 3 (drought)
Wang Jiaba	(~,0.50]	(-0.50,0.50)	[-0.50,~)
Gu Zhen	(~,0.50]	(-0.50,0.50)	[-0.50,~)
Gu Zhen sluice	(~,0.50]	(-0.50,0.50)	[-0.50,~)

Gao Liangjian sluice	(~,0.50]	(-0.50,0.50)	[-0.50,~)
Hua Wu	(~,0.49]	(-0.48,0.49)	[-0.48,~)
Ling Bi	(~,0.50]	(-0.50,0.50)	[-0.50,~)
Ming Guang	(~,0.50]	(-0.46,0.50)	[-0.46,~)
Nan Wan	(~,0.50]	(-0.50,0.50)	[-0.50,~)
Shang Qiu	(~,0.50]	(-0.50,0.50)	[-0.50,~)
She Yangzhen	(~,0.48]	(-0.48,0.48)	[-0.48,~)
Shen Qiu	(~,0.50]	(-0.50,0.50)	[-0.50,~)
Shiliang River Reservoir	(~,0.47]	(-0.50,0.47)	[-0.50,~)
Shi Mantan	(~,0.50]	(-0.50,0.50)	[-0.50,~)
Su Qian sliuce	(~,0.50]	(-0.50,0.50)	[-0.50,~)
Tong Bai	(~,0.50]	(-0.50,0.50)	[-0.50,~)
Xiang Hongdian	(~,0.49]	(-0.52,0.49)	[-0.52,~)
Xin Yanggang sluice	(~,0.50]	(-0.50,0.50)	[-0.50,~)
Xin Zhenf	(~,0.50]	(-0.50,0.50)	[-0.50,~)
Xu Zhou	(~,0.45]	(-0.50,0.45)	[-0.50,~)
Zhang Shan sluice	(~,0.50]	(-0.50,0.50)	[-0.50,~)
Zhu Madian	(~,0.50]	(-0.51,0.50)	[-0.51,~)

Take the March's drought grade of Wang Jiaba station for instance to show the year of drought grade based on SPI classification in table 2.

Table 2. Result of March month drought classification on Wang Jiaba station.

Level	Year
Level 1	1963, 1972, 1977, 1980, 1987, 1991, 1992, 1993, 1997, 1998, 2002, 2003, 2007, 2010
Level 2	1964, 1965, 1966, 1967, 1968, 1969, 1971, 1973, 1974, 1978, 1979, 1982, 1985, 1988, 1990, 1994, 1996, 1999, 2004, 2009, 2012, 2013
Level 3	1970, 1975, 1976, 1981, 1983, 1984, 1986, 1989, 1995, 2000, 2001, 2005, 2006, 2008, 2011

3. Research method

3.1. Principle of random forest

Random Forest (RF) [6] is a kind of statistical learning theory which was put forward by Leo Breiman in 2001. Comparing with other methods, Random Forest's forecasting is more highly accurate and more quick operation which is compared with other machine learning algorithms; it integrates characteristics of split two methods between Bagging and the characteristics of random; it has a good generalization performance and good robustness on abnormal noise and it has achieved good application in different fields. Random forests contain random forest regression and random forest classification.

Definition: A random forest class is a multi-decision classifier consisting of many CART classification models $\{h(X,\theta_k),k=1,2,\cdots\}$, which $\{\theta_k\}$ are the dependent random vectors of identically distribution [7].

The final classification result of the input X is the optimal classification result selected by each decision tree. The basic process of RFC has three steps.

Firstly, take k sample sets from the original training set through the Bootstrap as shown in figure 1, and keep the same size between the original training set and each sample set. Then, train each of k sample sets to set up k decision tree. Finally, the result of RFC is the final classification for all of the models, using the majority rule in the classification vote as shown in figure 2.

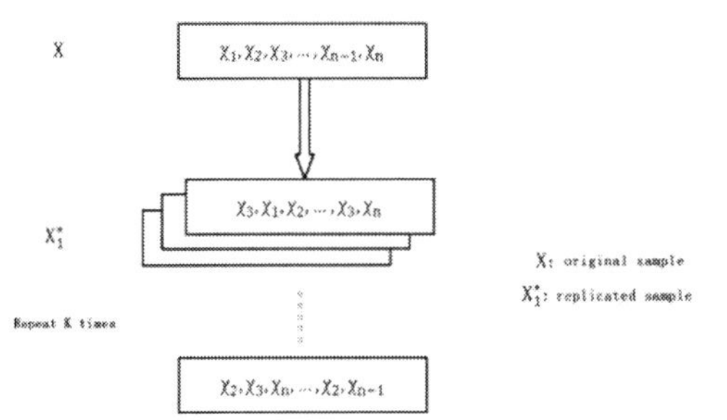

Figure 1. Schematic diagram of bootstrap resampling.

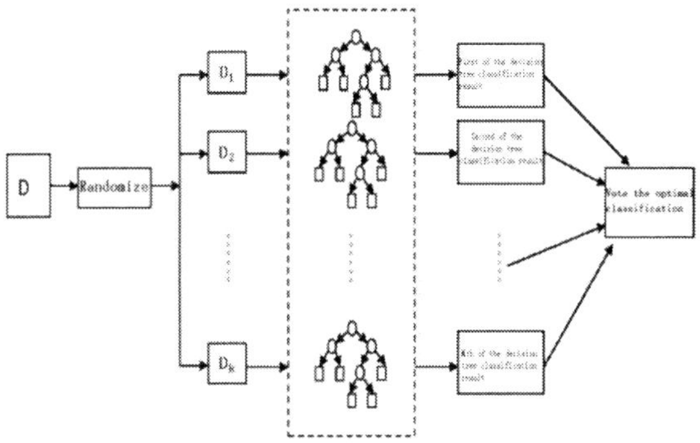

Figure 2. Schematic diagram of random forest structure.

Bagging method is used to form a new training set called RF model, it can divide characteristics randomly, makes random forests be able to tolerate the noise better, and can reduce the correlation between the single tree, not pruning single tree will get lower deviation, guaranteeing classification efficiency on the classification tree. Training set is generated by Bagging methods, the original sample concentration that is close to 37% of the sample will not appear in the training set, the data are called outside the Bag (Out - Of - Bag, OOB) data. Meanwhile, it is available that OOB data can be used to estimate the generalization error. For each decision tree, an error estimation corresponding OOB can be got, by using the random forest OOB error estimates for all the decision trees in average, which will obtain the random forest generalization error estimates. Breiman has shown that OOB error is an unbiased estimation, and relatively to the cross validation by experiments, OOB estimation is effective, and the result is similar to the results of the cross validation. Therefore, the research model is performance evaluation methods ground on generalization error estimation on the OOB estimation method [8].

4. Case study

4.1. Step of the RF model

This report respectively researched and forecasted every month's drought condition of 21 stations in Huaihe River Basin. The model predictor used 12 months 'antecedent precipitation and 74 atmospheric circulation indices that were issued by the National Meteorological Centre. 51 sets of rainfall data from 1963 to 2013 that conducted "three characteristics review" were also used in this research. In the case of the March's drought condition of Wang Jiaba station, the drought prediction is as follows:

- Preliminary selection of forecast factors: Improve the ability of influencing factors selecting and the accuracy of prediction of the model. The selection of explanatory variables should pay attention to the physical or meteorological correlation between the factors and the forecast object. Taking full account of the influence of factors on the spatial and temporal state of the atmosphere and the characteristics, we conducted a preliminary screening of 35 factors from the 74 hydrological and meteorological characteristics, which were related to the formation of meteorological drought.

Table 3. Result of preliminary screening influence factors of random forest model.

List of 24 preliminary screening influence factors		
the area of subtropical high in Northern Hemisphere (5 E-360)	Northern Hemisphere subtropical high ridge (5 E-360)	the intensity of North Hemisphere polar vortex (Region 5, 0-360)
the area of subtropical high in South China Sea (100 E-120 E)	the north boundary of subtropical high on the Northen Hemisphere (5 E-360)	Eurasia zonal circulation index (IZ, 0-150 E)
the area of subtropical high in North America (110 W-20 W)	the north boundary of subtropical high on the South China Ocean (100 E-120 E)	Asia zonal circulation index (IZ, 60 E-150 E)
the area of Subtropical high in Pacific Ocean (110 E-115 W)	the north boundary of subtropical high on the Pacific Ocean (110 E-115 W)	Asia meridional circulation index (IM, 60 E-150 E)
the intensity index of Northern Hemisphere subtropical high (5 E-360)	the area index of Asia Region polar vortex (Region 1,60 E-150 E)	The Tibetan Plateau Index_B (TPI_B) (30 N-40 N, 75 E-105 E)
the intensity index of West Pacific Ocean subtropical high (110 E-180)	the area index of Northern Hemisphere polar vortex (Region 5, 0-360)	India - burma trough (15 N-20 N, 80 E-100 E)
the intensity index of North Africa Atlantic North America subtropical high (110 W-60 E)	the intensity of Asia Region polar vortex (Region 1, 60 E-150 E)	numbered typhoons
the intensity index of South China Sea subtropical high (100 E-120 E)	Southern Oscillation Index	sunspot

- Screening on prediction factors. To reduce the influence of noise added in the evaluation on the accuracy of forecasting random forests, we chose Incnodepurity index to evaluate the importance of the factors, by which we compared the importance of the factor using the impurity reduction value calculated by the segmentation variables of all tree nodes, and the higher values indicated the importance of the factors in the prediction of RF. According to this principle, the 30 factors ordered by the importance of factors on the establishment of the nonlinear relationship can be achieved as shown in table 3. We selected 30 kinds for a total of 372 monthlyobserved hydrology-meteorological data as alternative explanatory variables factor, sequence number of precipitation in the previous 12 months came to 12 and monthly screening of 12 months of 30 atmospheric circulation characteristics arrangement, serial number row to 372. Setting the annual drought level of the Wang Jiaba station as the target variable and all the samples as trained sample sets, we evaluated the importance of the explanatory variables based on the random forest model. According to the explanatory variables importance in descending order, the 30 variables were selected as the final forecasting factor. The impact factor is shown in figure 3, and the prediction of drought in the month is shown in table 4.

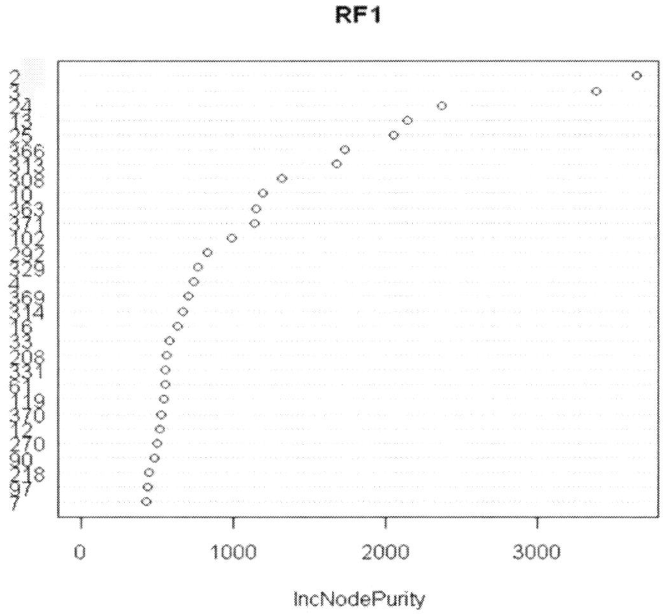

Figure 3. Ranking variable importance that associated with drought classification by random forest.

Table 4. Prediction factors based on RF of March month at Wang Jiaba station.

The forecast factors of March at Wang Jiaba station's RF drought forecast model		
August rainfall in the last year	The Tibetan Plateau Index-B (January)	india-burma trough (March)
December rainfall in the last year	The Tibetan Plateau Index-B (February)	india-burma trough (July)
January rainfall in the current year	the area of subtropical high in North America (January)	the intensity of Pacific Region polar vortex (July)
February rainfall in the current year	the area of subtropical high in North America (September)	Asia meridional circulation index (August)
November rainfall in the last year	the area of Subtropical high in Pacific Ocean (January)	Asia zonal circulation index (March)
February rainfall in the last year	the area of subtropical high in Northern Hemisphere (January)	Asia zonal circulation index (April)
Southern Oscillation Index (October)	the area of subtropical high in Northern Hemisphere (April)	Eurasia zonal circulation index (July)
Southern Oscillation Index (November)	the intensity of North Hemisphere polar vortex (June)	the intensity index of Pacific Ocean subtropical high (June)
Southern Oscillation Index (March)	the area index of Northern Hemisphere polar vortex (February)	the intensity index of North Africa Atlantic North America subtropical high (January)
Southern Oscillation Index (May)	the area index of Asia Region polar vortex (April)	the intensity index of North Africa Atlantic North America subtropical high (June)

4.2. Training and prediction results of RF model

March of Wang Jiaba station of RF drought prediction model took the 1963-2007 data as the model training data, and the error of OOB which was building by model is shown in table 5.

Table 5. The error of OOB at training period of RF model.

Level		Prediction			class rate (%)
		Level 1	Level 2	Level 3	
Real	Level 1	10	2	1	23
	Level 2	2	16	1	16
	Level 3	1	3	9	31

It can be seen from table 5 that the level 1 to 3 of OOB prediction error is 23%, 16% and 31 %, respectively. Average of OOB error of the model is 23%, much better. The result of forecast the grade 2 (normal condition) is the best, because the biggest sample size, or predictors that has not been selected may have little effect on its drought classification. The result of forecast the grade 3 (drought condition) is the worst. It may be related to the sample size and factors that affect the RF model to select the classification level of drought.

Reference climate forecasting business criteria [9], Ts scoring according three grades (Ts = total score/total number of forecasts; see table 6).

Table 6. Ts score standard of 3 level (%).

Level		Prediction		
		Level 1	Level 2	Level 3
Real	Level 1	100	50	0
	Level 2	50	100	50
	Level 3	0	50	100

According to the establishment of the RF model, the results of forecast the drought category from 2008 to 2013 are shown in table 7.

Table 7. Class forecast result of drought from 2008 to 2013 in Wang Jiaba station.

Year	Real Level	Forecasting level	Forecast accuracy
2008	3	2	50
2009	2	2	100
2010	1	2	50
2011	3	2	50
2012	2	2	100
2013	2	2	100

As the table 7 shows, there are 3 years rightly predicted while 3 years predicted differently one-grade. Based on the scoring criteria of table 6, the average forecast accuracy of the random forest model in March is 75%. According to the above prediction process, we made an assay of the drought conditions and obtained the forecast results of drought grade about the 21 stations in Huaihe River Basin as shown in table 8.

Table 8. Drought prediction result of every month of 21stations at the Huaihe River Basin based on RF model (%).

Station Name	January	February	March	April	May	June	July	August	September	October	November	December	Accuracy of station (%)
Wang Jiaba	58.3	66.7	75	83.3	83.3	100	91.7	91.7	75	75	58.3	66.7	77.1
Gu Zhen	58.3	58.3	75	75	66.7	91.7	83.3	75	75	66.7	66.7	58.3	70.8

Gu Zhen sluice	66.7	58.3	66.7	66.7	75	91.7	91.7	75	66.7	66.7	66.7	58.3	70.9
Gao Liangjian sluice	66.7	66.7	75	66.7	75	91.7	83.3	83.3	75	75	66.7	66.7	74.3
Hua Wu	50	58.3	58.3	75	75	83.3	83.3	75	66.7	66.7	58.3	58.3	67.4
Ling Bi	58.3	58.3	66.7	66.7	83.3	91.7	91.7	83.3	75	66.7	58.3	58.3	71.5
Ming Guang	66.7	66.7	66.7	75	66.7	83.3	83.3	83.3	75	75	66.7	66.7	72.9
Nan Wan	66.7	58.3	66.7	75	66.7	83.3	91.7	83.3	75	66.7	66.7	58.3	71.5
Shang Qiu	58.3	66.7	66.7	75	83.3	91.7	83.3	83.3	75	75	66.7	66.7	74.3
She Yangzhen	58.3	58.3	66.7	66.7	75	83.3	83.3	83.3	75	75	66.7	66.7	71.5
Shen Qiu	58.3	66.7	66.7	75	83.3	100	91.7	75	75	75	66.7	58.3	74.3
Shiliang River Reservoir	58.3	58.3	66.7	66.7	75	91.7	91.7	75	75	75	58.3	66.7	71.5
Shi Mantan	50	58.3	66.7	75	75	83.3	83.3	83.3	75	66.7	58.3	58.3	69.4
Su Qian sliuce	58.3	66.7	75	75	66.7	91.7	91.7	83.3	66.7	66.7	66.7	58.3	72.2
Tong Bai	66.7	66.7	66.7	66.7	75	91.7	91.7	91.7	83.3	75	58.3	58.3	74.3
Xiang Hongdian	58.3	58.3	66.7	75	83.3	91.7	91.7	75	75	66.7	58.3	58.3	71.5
Xin Yanggang sluice	58.3	58.3	66.7	75	83.3	83.3	83.3	83.3	75	75	66.7	58.3	72.2
Xin Zhenf	66.7	66.7	75	75	83.3	91.7	91.7	83.3	66.7	75	58.3	66.7	75.0
Xu Zhou	58.3	66.7	66.7	83.3	83.3	91.7	83.3	83.3	75	66.7	58.3	58.3	72.9
Zhang Shan sluice	58.3	66.7	66.7	75	83.3	91.7	91.7	83.3	75	66.7	58.3	66.7	73.6
Zhu Madian	58.3	58.3	66.7	66.7	83.3	83.3	91.7	83.3	75	66.7	66.7	58.3	71.5
Accuracy of month (%)	59.9	62.3	68.3	73.0	77.4	89.7	88.1	81.7	73.8	70.7	62.7	61.5	72.4

The average forecast accuracy of RF drought prediction model for 21 stations in Huaihe River Basin is 72.4%. Among them, the annual average forecast accuracy of Wang Jiaba station is 77.1%, which is the highest while the annual average forecast accuracy of Huawu station is 67.4%, which is the lowest. The highest of the average month forecast accuracy of each station is June with the accuracy rate 89.7%, while the lowest is January with the accuracy rate 59.9%.

Through the analysis, for accurate rate higher sites and months, the influence factors of RF model were selected to have a close relationship with drought in the area, where the cause and effect relationship was larger, so the model could be more accurate to predict the drought level. On the contrary, for accurate rate lower sites and months, the cause of the drought mechanism is affected by more factors, such as the January drought level more affected by winter temperatures, ENSO, etc. It may be also the poor application of drought classification based on SPI in a certain region and in the month.

5. Conclusion

- Basing on the SPI, the paper predicted the drought grade by RF drought prediction model of twenty-one stations during the period from 1963 to 2013 in Huaihe River, choosing 30 factors from the prediction factor set closely related to drought based on physical causes and statistical characteristics as the final prediction factors to inspect and predict by RF model [10].
- The overall average forecast accuracy of RF drought prediction is 72.4%, better than the climate system of the weather forecast accuracy (65%). The RF drought model is demonstrated as a new effective drought prediction model to reduce or eliminate the loss caused by drought.
- The RF model can draw a relatively high prediction accuracy rate, so it can be applied to different areas. However, the classification model based on just for meteorological drought index SPI analysis, in the future, we can consider blending the indexes from hydrologic drought and agriculture drought, so that the drought prediction is more scientific and persuasive from different viewpoints to level the drought [11].

Acknowledgments

Authors wish to acknowledge assistance and encouragement from all of our colleagues, specially Y Kang. We also thank the editors and anonymous reviewers.

Appendices

SPI: its formula is as follows:

$$SPI = S - \frac{t - (c_2 t + c_1)t + c_0}{[(d_3 t + d_2)t + d_1]t + 1.0} \quad (1)$$

t : Cumulative probability function, $t = \sqrt{\ln \frac{1}{F^2}}$, F : Cumulative probability of rainfall; S : Coefficient, when $F > 0.5$, S is the value of positive, otherwise take negative value, c, d are coefficients: $c_0 = 2.515517$; $c_1 = 0.802853$; $c_2 = 0.010328$; $d_1 = 1.432788$; $d_2 = 0.189269$; $d_3 = 0.001308$.

References

[1] Wang C X 2011 *Weather Forecast Method Based on Support Vector Machine (SVM)* (Nanjing University of Information Science & Technology)
[2] Liu D Y, Liang Z M and Zhao W M 2012 Application of grey system theory on drought prediction *Water Power* **38** 10-2
[3] Li X H, Yang Y and Yang H W 2014 Combining BP neural network with gray model to achieve drought predicting *J. Shenyang Agric. Univ.* **45** 253-6
[4] Fan G F, Zhang Y and Yang M 2011 Study of drought prediction based on support vector machine *Chinese J. Agric.* **32** 475-8
[5] Zhao T T G, Yang D W and Cai X M 2012 Predict seasonal low flows in the upper Yangtze River using random forests model *J. Hydroelectric Eng.* **31** 18-24
[6] Li X H 2013 Using "random forest" for classification and regression *Chinese J. Appl. Entomol.* **50** 1190-7
[7] Gislason P O, Benediktsson J A and Sveinsson J R 2006 Random forests for land cover classification *Pattern Recogn. Lett.* **27** 294-300
[8] Li H Y, Xie M and Jiang S 2012 Recognition method for mid-to long-term runoff forecasting factors based on global sensitivity analysis in the Nenjiang River Basin *Hydrol. Process.* **26** 2827-37

[9] Yang J, Lu Z G and Li S P 2009 Hydrological model year forecast based on SVM method *Y R* **31** 29-30
[10] Kang Y, Chen Y F and Gu S H 2014 Assessment of sustainable utilization of regional water resources based on random forest *Water Res. Power* **32** 34-8
[11] Wu J, Chen Y F and Yu S N 2016 Research on drought prediction based on random forest model *China Rural Water Hydropower* **17** 17-22

A new parallelization algorithm of ocean model with explicit scheme

X D Fu

College of Information Science and Engineering, Ocean University of China, Qindao 266100, China

E-mail: fuxiaodan8087@163.com

Abstract. This paper will focus on the parallelization of ocean model with explicit scheme which is one of the most commonly used schemes in the discretization of governing equation of ocean model. The characteristic of explicit schema is that calculation is simple, and that the value of the given grid point of ocean model depends on the grid point at the previous time step, which means that one doesn't need to solve sparse linear equations in the process of solving the governing equation of the ocean model. Aiming at characteristics of the explicit scheme, this paper designs a parallel algorithm named halo cells update with tiny modification of original ocean model and little change of space step and time step of the original ocean model, which can parallelize ocean model by designing transmission module between sub-domains. This paper takes the GRGO for an example to implement the parallelization of GRGO (Global Reduced Gravity Ocean model) with halo update. The result demonstrates that the higher speedup can be achieved at different problem size.

1. Introduction

A branch of modern natural science is oceanic and atmospheric science which involves a lot of ocean-atmosphere coupling models, such as POM (Princeton Ocean Model), ROMS (Regional Ocean Modeling System), FVCOM (Finite Volume Coastal Ocean Model) and CAM (Community Atmosphere Model). With the progress of computer hardware and software architecture, the high parallel performance computing technology has brought opportunities to improve the efficiency of these models; more and more parallel ocean models were developed, such as parallel versions of POM [1, 2], ROMS [3, 4], FVCOM [5] and CAM [6].

There are various software tools and library for developing parallel ocean model so far, such as HPF (high performance fortran), MPI (message passing interface), OpenMP (open multi-processing) and PVM (Parallel Virtual Machine). HPF is an extension of fortran, which has been chosen for the parallelization of FRAME (Fine Resolution AMmonia Exchange) [7]. MPI is a parallel software library and uses message transmission to communicate between parallel processes, which is one of the most commonly used parallel tools for the parallelization of ocean model, for example, the parallelization of FVCOM [5]. OpenMP is a directive-based parallel tool on shared memory platform; and the advantage of OpenMP is that one can quickly translate the serial ocean model to parallel version by means of inserting OpenMP parallel directive into the original serial ocean model [8, 9]. In order to make merit of MPI and OpenMP, Osthoff [10] applied hybrid MPI/OpenMP to parallelize the OLAM (Ocean-land-atmosphere model) model, the results demonstrated that the hybrid version decreases the number of output files and results in better performance for IO operations. PVM groups computing nodes of heterogeneous environment to parallel processors, which masks the differences between computing nodes, and its application in parallel ocean model was demonstrated in reference [11]. However, all these parallel techniques chosen for parallelization are based on developing environment, and are not based on the characteristic of the discretization of the governing equation of ocean models. This paper develops a common parallel algorithm for ocean model with explicit discretization of governing equation.

The time discretization of finite difference scheme of the governing equation of the ocean model is divided into explicit and implicit scheme. The selection of a difference scheme is related to the stability and convergence of the governing equations, and is also related to the selection of parallel technique in the parallelization of ocean model to a certain extent. When the governing equation is discretized with implicit scheme, large sparse linear system is produced, in which case, one can design parallel algorithm aiming at the characteristic of the linear system [12]. For example, when the governing equation is discretized with ADI scheme which produces tri-diagonal linear system, one can parallel design parallel algorithm to parallelize the tri-diagonal linear system [13, 14].

This paper will focus on the parallelization of ocean model with explicit scheme. Aiming at the characteristic of explicit scheme, this paper proposes a new parallel algorithm named halo cell update with tiny modification of the original ocean model. In order to test the performance of the halo cell update algorithm, this paper applies the parallel algorithm to parallelize the Global Reduced Gravity Ocean model under different problem size of ocean model.

2. Halo cell update

Applying the halo cell update to parallelize ocean model, the main concept of the parallel method is "divide and conquer", which contains three steps as follows.

At the first stage, we divide the whole domain involved in the ocean model into sub-domains. Each sub-domain has its own boundary. In this paper, the inner grids which are close to the boundary are called inner boundary. Each sub-domain expands its boundary, which forms outer boundary called halo cells. The halo cells can be set boundary condition used to compute the inner grids.

At the second stage, at each integration step, the inner grids of each sub-domain can be computed according to boundary condition which are saved in halo cells, in which case, the sub-domain can be computed by each processor in a parallel manner, and the results of the inner grids that we get is correct, this is mainly because the inner grids is calculated with the correctly boundary condition saved in the halo cells.

At the third stage, in order to compute correctly at the next time step each sub-domain sends its inner boundary and receives its halo cells from other sub-domains, which is called update step. It is obvious that after the update step the results of the parallel algorithm is identical to the results of the serial algorithm and we can apply the same solver to all sub-domains.

In order to understand the halo cell update correctly, let us consider the horizontal velocity u.

$$u_i^{n+1} = F(u_{i-1}^n, u_i^n, u_{i+1}^n) i=1, 2, \ldots, m, \quad u_0^{n+1}, u_{m+1}^{n+1} = g(t)$$

where u is the horizontal velocity, i is the number of elements in the total domain, n represents the time step, $g(t)$ is the boundary condition and F() is the discretization form of the governing equation of the ocean model. We divide the entire domain into two sub-domains, $u_{1,i}(i=1, \ldots, p)$, and $u_{2,i}(i=p+1, \ldots, m)$. Let us apply the halo cell update algorithm for these two sub-domains. These two sub-domains expand their boundaries, forming halo regions. We set the halo regions with the boundary condition: $u_{1,p+1} = u_{2,p+1}$, $u_{2,p} = u_{1,p}$. Now these two sub-domains have their own boundaries, and can be computed independently by different processors. That is

$$u_{1,i}^{n+1} = F\left(u_{1,i-1}^n, u_{1,i}^n, u_{1,i+1}^n\right), i = 1,2,\cdots,p. \quad u_{1,p+1}^n = u_{2,p+1}^n$$
$$u_{2,i}^{n+1} = F\left(u_{2,i-1}^n, u_{2,i}^n, u_{2,i+1}^n\right), i = p+1,\cdots,m. \quad u_{2,p}^n = u_{1,p}^n$$

At the update stage, the halo region of these two sub-domains must be refreshed; this is because we must use the updated halo region to compute the inner grids at the next time step. In this stage, the update function can be implemented by the basic message passing function:

$$u_{1,p+1}^{n+1} = u_{2,p+1}^{n+1}, \quad u_{2,\ p}^{n+1} = u_{1,p}^{n+1}$$

The update function can be implemented by the use of basic MPI point-to-point communication operations, such as mpi_send and mpi_recv subroutines. The advantage of the halo cell update is as follows:

- This parallel algorithm is simple, i.e., only an update function written in forms of basic message communication subroutines is required to refresh halo cell, and it is clear that the result of the parallel algorithm is identical to the result of the serial algorithm.
- The parallel algorithm has certain extensibility, i.e. we can expand two halo cells to reduce the times of halo cell update, in which case, the times of halo cell update will reduce to a half at the cost of redundant computing of each sub-domain, as is shown in figure 1.

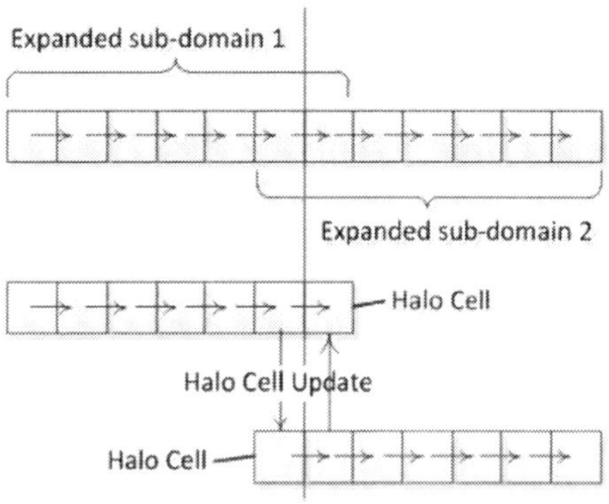

(a) The update of one halo cell.

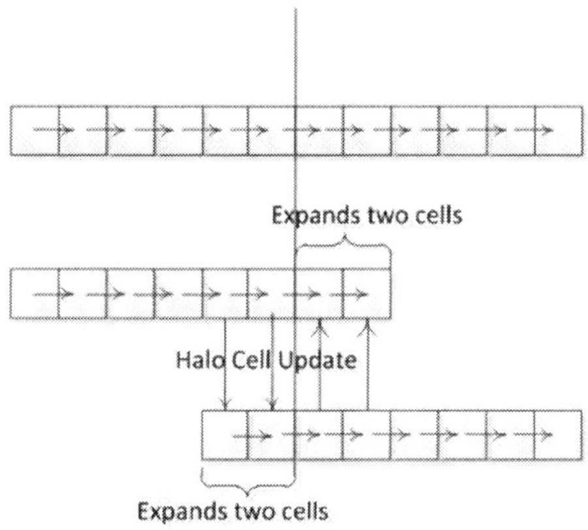

(b) The update of two halo cells.

Figure 1. The update of halo cells.

3. Ocean model

The ocean model used in this paper is an extended 1.5 layer Global Reduced Gravity Ocean model (GRGO) [15, 16], which was first introduced by Cane to study ENSO. In this model, mass, momentum, and heat obey the conservation laws in the upper layer, and the lower layer is assumed to be infinitely deep and motionless to keep the kinetic energy finite, so the momentum, continuity, and thermodynamic equations for the upper can be written as:

$$\frac{\partial U}{\partial t} + U \cdot \nabla U + fK \cdot U - \frac{1}{2} h\nabla b + \nabla(hb) = \frac{\tau}{h} + \mu\nabla^2 u - \frac{1}{h} w_e H(w_e) U \quad (1)$$

$$\frac{\partial h}{\partial t} + \nabla \cdot (hU) = w_e \quad (2)$$

$$\frac{\partial T}{\partial t} + U \cdot \nabla T = \frac{Q}{\rho_0 C_p h} + \kappa\nabla^2 T - \frac{1}{h} w_e H(w_e)(T - T_r) \quad (3)$$

where U=U(x, y, t), h=h(x, y, t), T=T(x, y, t) represent the horizontal velocity vector, thickness of upper layer and surface temperature respectively; x, y represent horizontal coordinates; w_e represents entrainment velocity at the base of mixed layer; $\boldsymbol{\tau}$ represents wind stress vector; b represents buoyancy in the upper layer; Tr represents reference temperature, T represents the surface temperature; Q represents heat flux; κ represents the diffusivity coefficient; ρ_0 is the density; C_p represents thermal expansion coefficient; μrepresents horizontal viscosity coefficient.

The GRGO is about to solve u, v, h, T by the integral of time according to the boundary condition, so in these equations, we mainly solve the $\frac{\partial}{\partial t}$. The main idea of this paper is the domain decomposition in the horizontal directions; we are interested in the discretization of the terms containing $\frac{\partial}{\partial x}, \frac{\partial^2}{\partial x^2}, \frac{\partial}{\partial y}, \frac{\partial^2}{\partial x^2}$ in the above equations.

4. Discretization

This section will deal with the discretization of governing equations of the ocean model. We discretize the governing equations with explicit scheme which is relatively simple to set up and to program in order to find the numerical solutions of the equations. In order to keep the numerical solution stable and avoid generating oscillations, we must choose appropriate time step and spatial step to meet the CFL (Courant–Friedrichs–Lewy) condition. In computational fluid dynamics, the CFL is a necessary condition for the stable while solving certain partial differential equation. In this paper, we choose dt/dx<0.1 (This paper focuses on a two dimensional system). Take the momentum equation (1) and the continuity equation (2) for examples, and the finite difference equations are as follows:

$$\begin{aligned}\frac{u_{i,j}^{n+1}-u_{i,j}^{n}}{\Delta t}+u_{i,j}^{n}\frac{u_{i,j}^{n}-u_{i-1,j}^{n}}{\Delta x}+v_{i-\frac{1}{2},j}^{n}\frac{u_{i,j}^{n}-u_{i,j-1}^{n}}{\Delta y}-fv_{i-\frac{1}{2},j}^{n}-\frac{1}{2}h_{i,j-\frac{1}{2}}^{n}\frac{b_{i,j}^{n}-b_{i-1,j}^{n}}{\Delta x}\\+\frac{b_{i,j}^{n}h_{i,j-\frac{1}{2}}^{n}-b_{i-1,j}^{n}h_{i-1,j-\frac{1}{2}}^{n}}{\Delta x}\\=\frac{\tau_x}{h_{i,j-\frac{1}{2}}^{n}}+\mu\left(\frac{u_{i+1,j}^{n}-2u_{i,j}^{n}+u_{i-1,j}^{n}}{\Delta x^2}+\frac{u_{i,j+1}^{n}-2u_{i,j}^{n}+u_{i,j-1}^{n}}{\Delta y^2}\right)-\frac{1}{h_{i,j-\frac{1}{2}}^{n}}w_{e_{i,j}}^{\ n}H(w_e)u_{i,j}^{n}\end{aligned} \tag{4}$$

$$\frac{h_{i,j-\frac{1}{2}}^{n+1}-h_{i,j-\frac{1}{2}}^{n}}{\Delta t}+\frac{u_{i,j}^{n}h_{i,j-\frac{1}{2}}^{n}-u_{i-1,j}^{n}h_{i-1,j-\frac{1}{2}}^{n}}{\Delta x}+\frac{v_{i-\frac{1}{2},j}^{n}h_{i,j-\frac{1}{2}}^{n}-v_{i-\frac{1}{2},j-1}^{n}h_{i,j-\frac{3}{2}}^{n}}{\Delta y}=w_e \tag{5}$$

Where n represents the time step, i and j represent the number of the elements in the domain. According to the finite difference equations, we can get the relationships of grids shown in figure 2. For example, from equation (4) and figure 2(a), when we compute the u at n + 1 time, we must use u and the four points around u of n time. From equation (5) and figure 2(b), when we compute the h at n + 1 time, we use the $h_{i,j}$, $h_{i-1,j}$ and $h_{i,j-1}$ of n time. We can decide the schema of the domain decomposition according to the relationships of the grids.

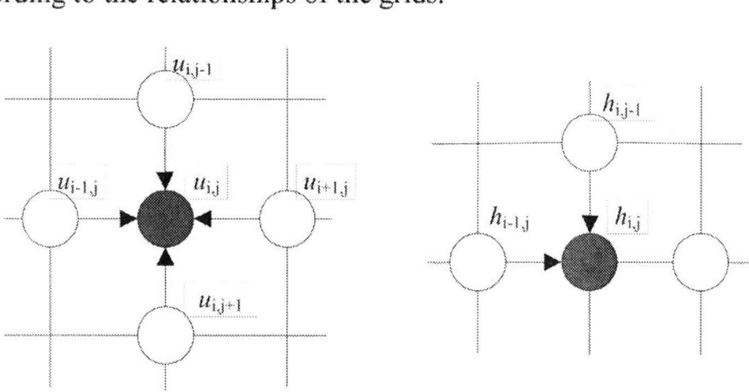

(a). u component of **U** (b). continuity equation

Figure 2. Relationships between grids.

5. Result and analysis

The aim of the parallelization of the ocean model is to reduce the runtime and increase the efficiency of the ocean model. This paper tests the parallel version of GRGO in the following environments:

Platform 1, the number of CPU cores is 4, the memory size is 4 GB, and the operating system is Windows 7.

Platform 2, the number of CPU cores is 8, the memory size is 8 GB, and the operating system is Windows Server.

Platform 3, the number of CPU cores is 16, the memory size is 16 GB, and the operating system is Linux.

This paper use different kinds of resolution of the ocean model to test the performance of the parallel algorithm in the above parallel environment, which means the problem size is different.

The running time of U equation at different problem sizes and different platforms is shown in following tables 1-3.

Table 1. The running time of U equation at different problem size on platform 1.

CPU cores	Different problem sizes (units: s) 1350*6002700*1200	
1	41.32	164.95
2	21.15	84.23
3	14.77	58.04
4	11.52	44.63

Table 2. The running time of U equation at different problem size on platform 2.

CPU cores	Different problem sizes (units: s) 1440*4803000*1000	
1	61.63	268.27
2	32.00	147.40
4	25.53	103.00
6	19.24	80.30
8	16.43	69.20

Table 3. The running time of U equation at different problem size on platform 3.

CPU cores	Different problem sizes (units: s) 1350*6003000*10003600*1600		
1	25.23	94.28	183.17
2	12.86	47.89	94.94
4	6.72	24.07	47.23
10	4.97	17.40	32.51
16	3.56	12.12	22.34

The speedup is an important indicator measuring the performance of parallel algorithm, and is defined as the ratio of serial execution time and parallel execution time. In order to analyse the performance of parallel algorithm, this paper calculates the speedup according to the running time as shown in above tables.

From tables 1-3, the results reveal that the running time reduces as the CPU cores increase on the condition that the problem size is fixed and that the speedup increases, which is shown in figures 3-5. On platform 1, when the number of CPU cores is 4, the speedups of problem size 1350*600 and

2700*1200 are 3.59 and 3.70. On platform 2, when the number of CPU cores is 8, the speedups of problem size 1440*480 and 300*1000 are 3.75 and 3.88. On platform 3, when the number of CPU cores is 16, the speedups of problem size 1350*600, 3000*1000 and 3600*1600 are 7.09, 7.78 and 8.20.

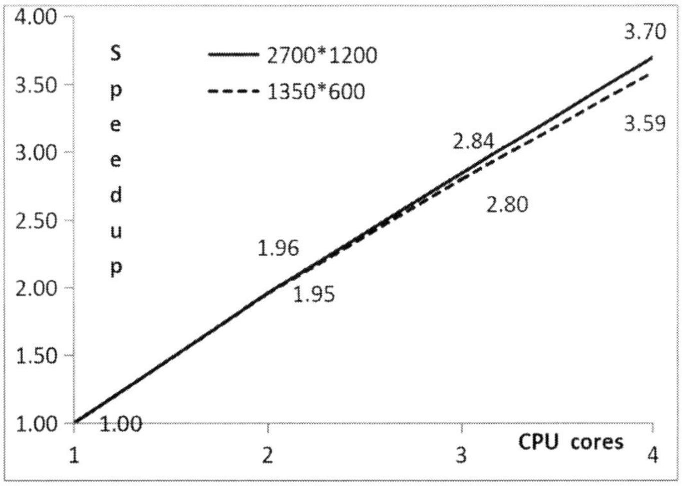

Figure 3. The speedup of different problem size on platform 1.

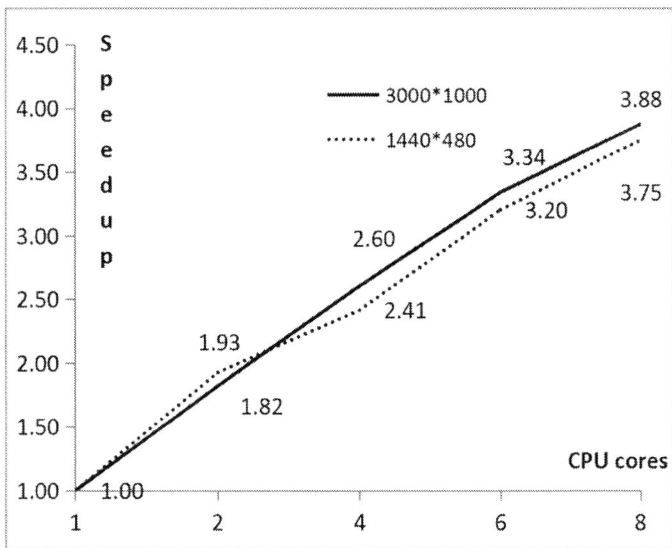

Figure 4. The speedup of different problem size on platform 2.

On platform 1, when the number of CPU cores is fixed, we can get higher speedup as the problem size increases, which is shown in figure 3, i.e., the speedup of problem size 2700*1200 is higher than that of problem size 1350*600. On platform 2 and platform 3, we can get similar results that the higher speedup is gained at the bigger problem size when the number of CPU cores is fixed, as shown in figures 4 and 5, i.e., the speedup of problem size 3000*1000 is higher than that of problem size 1440*480; the speedup of problem size 3600*1600 is the highest one in all of the problem sizes 3600*1600, 300*1000 and 1350*600. The reason of above results is as follows: as the number of CPU cores is fixed, the increased amount of calculation is greater than increased amount of communication when the problem size increases, which demonstrate that the halo cell update has scalability.

From tables 1-3 and figures 3-5, the results reveal that the speedup increases as the number of CPU cores increase and the higher speedup is gained at bigger problem size, which demonstrates that the halo cell update algorithm can get better performance in the parallelization of ocean model.

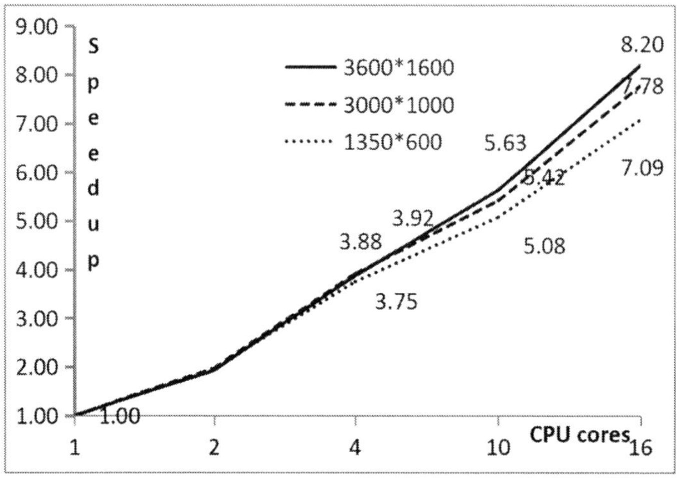

Figure 5. The speedup of different problem size on platform 3.

6. Conclusion

Aiming at the characteristic of the explicit scheme of the ocean model, this paper develops a parallel algorithm used to parallelize ocean circulation model with explicit scheme. In order to measure the performance of the parallel algorithm, this paper applies this parallel algorithm to parallelize GRGO model. The process of parallelizing ocean model is simple, which means that one needs a bit of modification of original ocean model; in addition, the parallel version of ocean model is scalable. This paper tests parallel version of GRGO in three platforms; and the result reveals that the running time reduces as the number of CPU cores increases, and that with fixed number of CPU cores the speedup increases when the problem size increases, which demonstrates that halo cell update is scalability.

References

[1] Giunta G, Mariani P, Montella R, *et al* 2007 pPOM: A nested, scalable, parallel and fortran 90 implementation of the princeton ocean model *Environ. Modell. Softw.* **22** 117-22

[2] Jordi A and Wang D P 2012 sbPOM: A parallel implementation of princenton ocean model *Environ. Modell. Softw.* **38** 59-61

[3] Wang P, Song Y T, Chao Y, *et al* 2005 Parallel computation of the regional ocean modeling system *Int. J. High Perform. C* **19** 375-85

[4] Govett M, Hart L, Henderson T, *et al* 2003 The scalable modeling system: Directive-based code parallelization for distributed and shared memory computers *Parallel Comput.* **29** 995-1020

[5] Cowles G W 2008 Parallelization of the FVCOM coastal ocean model *Int. J. High Perform. C* **22** 177-93

[6] Tseng Y H and Ding C 2008 Efficient parallel i/o in Community Atmosphere Model (CAM) *Int. J. High Perform.* C **22** 206-18

[7] Pais V A, Fournier N, Sutton M A, *et al* 2004 Using high performance fortran to parallelise a multilayer atmospheric transport model *Parallel Comput.* **30** 21-33

[8] Malfetti P 2001 Application of OpenMP to weather, wave and ocean codes *Scientific Programming* **9** 99-107

[9] Wallcraft A J 2000 SPMD OpenMP versus MPI for ocean models *Concurrency: Practice and Experience* **12** 1155-64

[10] Osthoff C, Grunmann P, Boito F, *et al* 2011 Improving performance on atmospheric models

through a hybrid OpenMP/MPI implementation *2011 IEEE 9th Int. Symp. On Parallel and Distributed Processing with Applications* (Busan, South Korea: IEEE) pp 69-74

[11] Beare M I and Stevens D P 1997 Optimization of a parallel ocean general circulation model *Annales Geophysicae* vol 15 (Berlin/Heidelberg: Springer-Verlag) pp 1369-77

[12] Jos D K, Aad V D S, Hakan O and Henk D 2002 A fully implicit parallel ocean model using MUMPS *J. Supercomput.* **23** 167-83

[13] Mendiratta K and Polizzi E 2011 A threaded SPIKE algorithm for solving general banded systems *Parallel Comput.* **37** 733-41

[14] Mattor N, Williams T J and Hewett D W 1995 Algorithm for solving tridiagonal matrix problems in parallel *Parallel Comput* **21** 1769-82

[15] Fang Y 2006 A coupled model study of the remote influence of ENSO on tropical Atlantic SST variability (Texas A & M University)

[16] Cane M A 1979 The response of an equatorial ocean to simple wind stress patterns: I- Model formulation and analytic results; II- Numerical results *J. Mar. Res.* **37** 233-99

Study on hydraulic exciting vibration due to flexible valve in pump system with method of characteristics in the time domain

Y H Yu[1,2], D Liu[1], X F Yang[1] and J Si[1]

[1]College of Water Conservancy and Hydropower Engineering, Hohai University, Nanjing, 210098, China

E-mail: yhyu@hhu.edu.cn

Abstract. To analyse the flow characteristics of leakage as well as the mechanism of self-excited vibration in valves, the method of characteristics was used to assess the effect of flexible valve leakage on the self-excited vibration in water-supply pump system. Piezometric head in upstream of the valve as a function of time was obtained. Two comparative schemes were proposed to simulate the process of self-excited vibration by changing the length, the material of the pipeline and the leakage of valves in the above pump system. It is shown that the length and material of the pipe significantly affect the amplitude and cycle of self-excited vibration as well as the increasing rate of the vibration amplitude. In addition, the leakage of the valve has little influence on the amplitude and cycle of self-excited vibration, but has a significant effect on the increasing rate of vibration amplitude. A pipe explosion accident may occur without the inhibiting of self-excited vibration.

1. Introduction

Hydraulic vibration is a "small fluctuation" or transient flow of pressurized water delivery pipeline system which can be analysed by hydraulic impedance method or transfer matrix method [1]. Suo [2] and Zhou *et al* [3] analysed hydraulic resonance in several hydropower stations and then the methods to reduce vibration were proposed. Yu *et al* [4] demonstrated the possibility of hydraulic resonance of parallel pump system with air surge tank for water hammer protection. Ye [5] and Zhu *et al* [6] analysed the piping vibration accident of a pumped storage power station and concluded that self-excited vibration was caused by defects in seal of flexible valve. It is proposed to prevent or eliminate the vibration by using bypass valve or changing water supply mode of the seal of the ball valve. Zhou *et al* [7, 8] analysed the free vibration of reversible unit operating in unstable region in pumped storage power station, and concluded that the self-excited oscillation was superposed by several oscillation modes whose attenuation factors were positive.

Valve leakage is common in water supply pump system. Valve can be divided into rigid valve and flexible valve according to whether there is a negative slope segment in the characteristic curve of leakage flow of valve. The leakage of flexible valve may cause self-excited vibration and affect the safety of the system [9]. The self-excited vibration caused by leakage of flexible valve in pump system was analyzed by combining with a long-distance water supply project example in this paper. The characteristics of self-excited vibration were obtained to lay a good foundation for the prevention and elimination of self-excited vibration.

IOP Conf. Series: Earth and Environmental Science 82 (2017) 012018 doi:10.1088/1755-1315/82/1/012018

2. The flow characteristics of the leakage of valve

If the leakage flow Q of valve increases with the increase of water pressure difference H between the front and back of valve, the valve is called rigid one. The leakage flow

$$Q = k\sqrt{H} \tag{1}$$

where k is a constant coefficient.

If the leakage area and then the leakage flow Q of valve decrease with the increase of water pressure difference H between the front and back of valve, the valve is called flexible one. The leakage flow

$$Q = k_1\sqrt{H} - k_2 H^{\frac{3}{2}} \tag{2}$$

where $k_1 = \pi D x_0$, $k_2 = \dfrac{\pi D \rho g A}{K}$, D is the diameter of valve. x_0 is the seal gap when the valve pressure difference is 0. A is the area of sealing cover of valve. K is the elastic coefficient of valve seal. ρ is the density of water. g is the acceleration due to gravity.

The leakage characteristics of rigid valve and flexible valve are shown in figure 1. Figure 1 shows that, for the flexible valve, when H=0, Q=0; when H increases to Hc, Q reaches a maximum; when H continues to increase to Hm, Q=0; When $H < H_c$, $\dfrac{dH}{dQ} > 0$, the slope of valve leakage characteristic is positive, the same trend with the leakage characteristics of rigid valve; when $H_c \le H \le H_m$, $\dfrac{dH}{dQ} < 0$, the slope of valve leakage characteristic is negative.

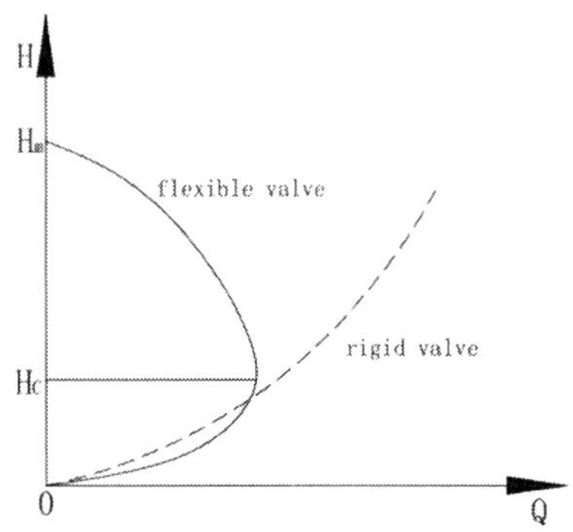

Figure 1. Characteristics of leakage in rigid and flexible valves.

3. Mechanism analysis of self-excited vibration caused by flexible valve

Suppose a simple pipeline system with a reservoir at upstream and a valve at the end, ignoring the hydraulic friction of pipeline, at the valve

$$\frac{h}{q} = -\frac{a}{gA}\tanh\frac{sl}{a} \tag{3}$$

where h and q are the Laplace transforms of hydraulic head and flow pulsation; s is the complex frequency, $s = \sigma + i\omega$; a is the velocity of wave due to water hammer; A is the cross-sectional area of pipe at valve; l is the pipe length andg is the gravitational acceleration.

Linearize the flow characteristic of the valve at the operating point:

$$\frac{h}{q}=\frac{dH}{dQ}=k \tag{4}$$

By summarizing equations (3) and (4), equation (5) is obtained:

$$\sigma=\frac{a}{2l}\ln\left|\frac{1-gAk/a}{1+gAk/a}\right| \tag{5}$$

From equation (5), we can know that when the slope of valve flow characteristic in system is negative, thus $k<0$, $\sigma>0$, the system is unstable and will occur self-excited vibration; when the slope of valve flow characteristic in system is positive, thus $k>0$, $\sigma<0$, the system is stable and will not occur self-excited vibration.

Therefore, when the slope of valve flow characteristic is positive, that is to say, rigid valve or $H<H_c$ segment of flexible valve, the system is stable, self-excited vibration will not happen; when the slope of valve flow characteristic is negative, that is to say, flexible valve section $H_c \le H \le H_m$, the system will occur self-excited vibration.

4. Analysis of self-excited vibration in pump system

Figure 2 is the structure diagram for a parallel pump system of water supply project, where P stands for pump, 1 ~ 11 stand for numbers of each pipe section.

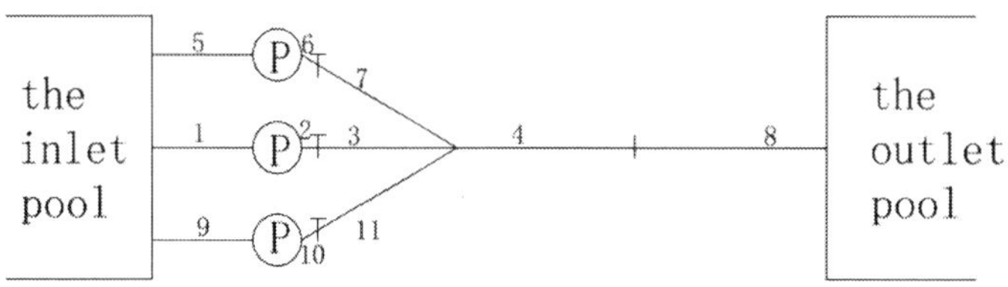

Figure 2. Diagram of pump system.

4.1. The establishment of mathematical mode

Assuming such a condition, the pump system stops running and the three valves are all closed. If the valve of No. 3 pipe is a flexible valve and is leaking, the valves of No. 7 pipe and No. 11 pipe are water-tight, the whole system can be simplified as a series pipeline system composed of No. 8, 4 and 3 pipes which have a reservoir at upstream and a valve at the end. The system can be analyzed by the method of characteristics.

4.1.1. The pipeline. The pipeline is segmented by the space step Δx, the time step is taken as: $\Delta t=\frac{\Delta x}{a}$, so a rectangular grid is formed in the X ~ t plane, and two characteristic line equations about the flow of the point i Q_{pi} and piezometric head H_{pi} can be obtained.

$$C^+:\ H_{pi}=C_P-B_PQ_{pi} \tag{6}$$

$$C^-:\ H_{pi}=C_M+B_MQ_{pi} \tag{7}$$

Where C_P, B_P, C_M, B_M can be calculated by the known flow and piezometric head of nodes i-1 and i+1 at the previous moment. So:

$$C_P = H_{i-1} + BQ_{i-1} \tag{8}$$

$$C_M = H_{i+1} - BQ_{i+1} \tag{9}$$

$$B_P = B + R\left|Q_{i-1}\right| \tag{10}$$

$$B_M = B + R\left|Q_{i+1}\right| \tag{11}$$

where $B = a/(\mathrm{gA})$; $R = f\Delta x/(2gDA^2)$; f the coefficient of friction, a the wave velocity, A the cross-sectional area of pipe, D the diameter of the pipe.

4.1.2. At the outlet pool

$$H_P = H_{P0} \tag{12}$$

$$Q_p = (H_p - C_M)/B_M \tag{13}$$

where H_{P0} is the water level in the outlet pool.

4.1.3. At the flexible valve

$$H_p = C_P - B_P Q_p \tag{14}$$

$$Q_P = k_1\sqrt{H_p - H_w} - k_2(H_p - H_w)^{\frac{3}{2}} \tag{15}$$

where H_P is the piezometric head at the upstream of valve; H_W is the hydraulic head at outlet of valve.

Let $H = H_p - H_w$, by summarizing equations (14) and (15) , equation (16) is obtained:

$$-B_P k_2 H^{\frac{3}{2}} + H + B_P k_1 H^{\frac{1}{2}} + H_W - C_P = 0 \tag{16}$$

Equation (16) takes *H as* variable. The value of *H* can be obtained by the Newton iterative method, and then H_P and Q_P are also obtained.

4.2. Example calculation

The water level of the outlet pool is 1420 m and the elevation of the center of valve Hw is 1260 m. Table 1 lists the pipeline parameters for calculation in the system.

Table 1. Parameters of pipes.

Number of pipe	l (m)	D (m)	f	A (m/s)
3	15	0.219	0.025	1190
4	1300	0.377	0.023	1180
8	1100	0.355	0.021	350

No. 3 & 4 pipes are welded steel pipes.
No. 8 pipe is M - PVC pipe, its wave velocity due to water hammer is small, a=350 m/s.

The characteristic of the flexible is shown in equation (16), letting $k_1 = 1.8\times10^{-4}$, $k_2 = 7.05\times10^{-7}$, H=85 m, $Q_{\max} = 1.1\times10^{-3} m^3/s$. When the pressure difference between the front and back of valve

H=160 m, at this time the valve is in the negative slope section, and will cause the self-excited vibration of system, the leakage flow is $Q = 8.5 \times 10^{-4} m^3 / s$.

Piezometric head in upstream of valve changing with time is obtained by analyzing the self-excited vibration in the system by the method of characteristics. As figure 3 shows:

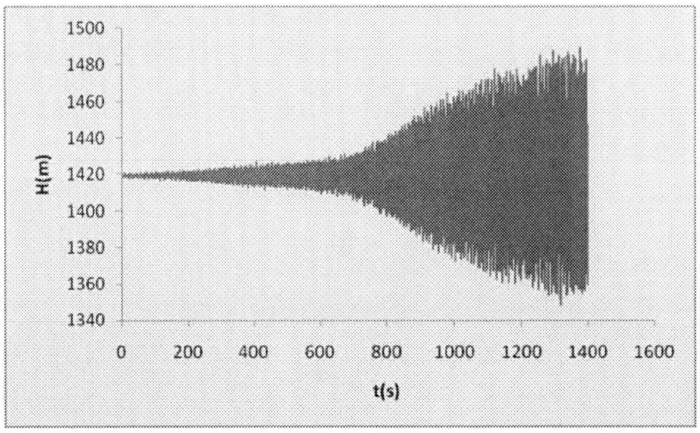

Figure 3. Piezometric head in upstream of valve changing with time.

4.3. Analysis of calculation result

The calculation result shows that: the amplitude of vibration caused by flexible valve leakage in the pipeline is relatively small in the initial phase of the oscillation, as shown in figure 3, the oscillation amplitude between 100 s ~ 200 s is only 5.7 m, as time goes by, the oscillation amplitude at valve increases continuously. The calculation shows that the oscillation amplitude can reach 12.1 m in 500 s, after 800 s, that the amplitude increases sharply to 2000 s, and that the oscillation amplitude will reach 142 m.

Jeager *et al* [10] have used graphical method for the derivation of the hydropower station system: the amplitude of the self-excited oscillation in a simple pipeline is two times the static head. However, the above calculation results show the amplitude of the self-excited vibration of the pump system after a certain period of time has a large gap with the theoretical oscillation amplitude of the hydropower station. Compared with the theoretical hydropower station system, the design flow of the pump system is only 0.043 m^3/s, leakage flow is only $8.5 \times 10^{-4} m^3 / s$, the No. 8 pipe of pump system is made of PVC pipe, its water hammer wave velocity is 350 m/s, far less than 1180 m/s of steel pipe, it makes the vibration cycle increase.

If the material of No. 8 pipe is changed to steel pipe, the length of No. 8 pipe is changed to 275 m, the length of No. 4 pipe is changed to 325 m and other parameters are unchanged, we call it No. 1 scheme. The results of the calculation are shown in figure 4.

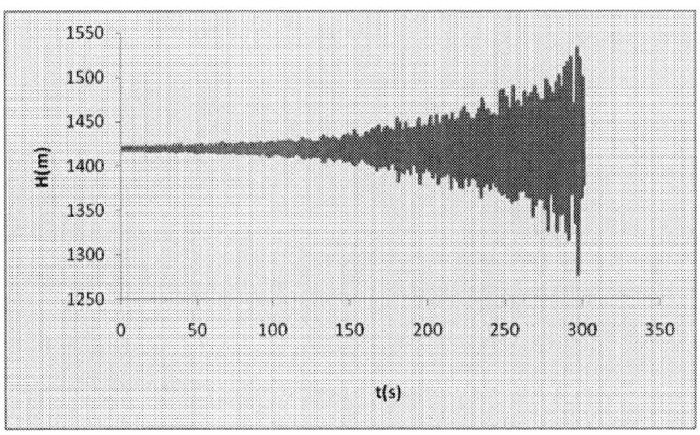

Figure 4. Piezometric head in upstream of valve changing with time in No. 1 scheme.

If the leakage flow of valve two times increases, let $k_1 = 3.6\times10^{-4}$, $k_2 = 1.41\times10^{-6}$ and the remaining parameters are unchanged, we call it No. 2 scheme. The results of the calculation are shown in figure 5.

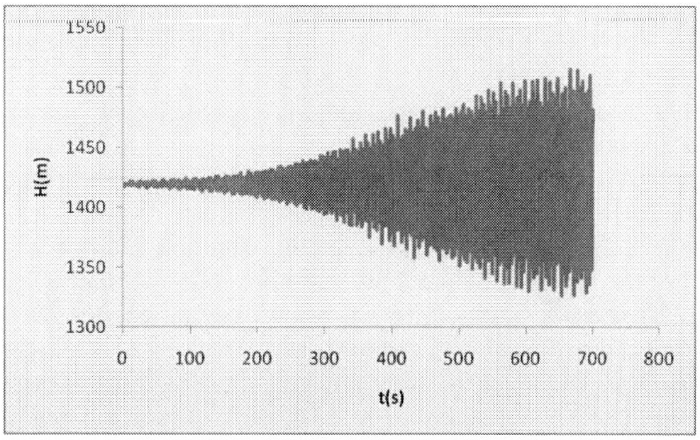

Figure 5. Piezometric head in upstream of valve changing with time in No. 2 scheme.

Figure 4 shows that the amplitude of self-excited vibration increases significantly when the length of pipe is shortened and the material of pipe is changed. The amplitude increases to 255 m, the cycle is shortened and the speed of the development of self-excited vibration increases dramatically.

Figure 5 shows that when the valve leakage increases, the amplitude of the self-excited vibration is 105 m, the amplitude and cycle are slightly changed and the speed of the increasing of amplitude increases obviously.

The comparative analysis mentioned above shows that the length and material of the pipe significantly affect the amplitude and cycle of self-excited vibration as well as the increasing rate of the vibration amplitude. In addition, the leakage of the valve has little influence on the amplitude and cycle of self-excited vibration, but has a significant effect on the increasing rate of vibration amplitude.

Although the self-excited vibration amplitude of water pump system is not as big as simple pipe system, but the self-excited vibration caused by flexible valve leakage will still threaten the safety of pipeline. The water hammer pressure is considered 30% of the pressure in working conditions, and the calculation results show that the amplitude of vibration has exceeded the set value of water hammer pressure in 900 s. If the self-excited vibration continues, vibration amplitude will also continue to increase. The accident of explosion of pipe will occur without the holding back of self-excited vibration.

5. Conclusion

Valve leakage is very common in pressurized water transfer projects. The leakage flow characteristic curve of flexible valve has a negative slope segment. The theoretical analysis indicates that when the slope of valve leakage flow characteristic is negative, the attenuation factor of hydraulic vibration is greater than 0, the system is not stable, and it will occur self-excited vibration.

The length and material of the pipe significantly affect the amplitude and cycle of self-excited vibration as well as the increasing rate of the vibration amplitude. In addition, the leakage of the valve has little influence on the amplitude and cycle of self-excited vibration, but has a significant effect on the increasing rate of vibration amplitude. When the pump system shut down and the valve closed, the self-excited vibration caused by flexible valve water leakage may lead to an accident of explosion without the inhibiting of self-excited vibration.

Only one self-excited vibration condition is analyzed in the dissertation: the pump system stops running, only a flexible valve is leaking. In fact, there are other conditions needing further study, such as when all the pumps of parallel pump system stop, the three flexible valves are leaking and whenonly one pump of the parallel pumps system stops, the flexible valve of its export is leaking, while the rest pumps of the system work.

References

[1] Wylie E B, Streeter V L and Suo L S 1993 *Fluid Transients in Systems* (New Jersey, USA: Prentice Hall Incorporation)
[2] Suo L S, Zhou J X and Liu D Y 1998 Hydraulic resonance in pressurized piping system of hydropower plants *Adv. Sci & Tech Wa Res* **8** 12-4
[3] Zhou J X, Suo L S and Guo R Q 1998 A case study of hydraulic resonance at a hydropower plant *J Hydro Engng* **4** 50-4
[4] Yu Y H, Wang J X and Qin X F 2013 Hydraulic vibration analysis of parallel pumps system with air surge tank for water hammer protection *J Drain Irr Mach Engng* **31** 958-63
[5] Ye F M, Zhu Y Y, Fan H G, *et al* 2007 Research on self-excited oscillations happened in a complex pipe system of a pumped storage power stations *J Hydro Engng* **26** 135-40
[6] Zhu Y Y, Fan H G, Chen N X, *et al* 2006 Prevention of self-excited oscillations in complicated water diversion channels of pumped storage power stations *J Tsinghua Uni: Sci & Tech* **46** 1434-7
[7] Zhou J X, Suo L S and Hu M 2007 Study on self-excited vibration of hydro-mechanical system in pumped storage plants *J Hydr Engng* **9** 1080-4
[8] Zhou J X, Karney B W, Hu M and Xu J C 2011 Analytical study on possible self-excited oscillation in S-shaped regions of pump-turbines *Proc. of the Institution of Mechanical Engineers, Part A: J Power Eng* **225** 1132-42
[9] Zhang S C 1990 Vibration caused by the interaction between fluid and valve mechanism in pipeline *Youth Academic Paper on Water Science* (Beijing: Water Conservancy and Electric Power Press) pp 20-6
[10] Jaeger C, Kerr L S and Wylie E B 1948 Water hammer effects in power conduits *Civ Engng Pub Works Rev* **23** 500-3

Multifactor analysis and simulation of the surface runoff and soil infiltration at different slope gradients

J Huang[1,2,3], Q Kang[1,2], J X Yang[1,2] and P W Jin[1,2]

[1]Soil and Water Conservation Monitoring Center of Pearl River Basin, Guangzhou 510611, China
[2]Pearl River Hydraulic Research Institute, Guangzhou 510611, China

E-mail: hjnwsuaf@outlook.com

Abstract. The surface runoff and soil infiltration exert significant influence on soil erosion. The effects of slope gradient/length (*SG/SL*), individual rainfall amount/intensity (*IRA/IRI*), vegetation cover (*VC*) and antecedent soil moisture (*ASM*) on the runoff depth (*RD*) and soil infiltration (*INF*) were evaluated in a series of natural rainfall experiments in the South of China. *RD* is found to correlate positively with *IRA*, *IRI*, and *ASM* factors and negatively with *SG* and *VC*. *RD* decreased followed by its increase with *SG and ASM*, it increased with a further decrease with *SL*, exhibited a linear growth with *IRA* and *IRI*, and exponential drop with *VC*. Meanwhile, *INF* exhibits a positive correlation with *SL*, *IRA* and *IRI* and *VC*, and a negative one with *SG and ASM*. *INF* was going up and then down with *SG*, linearly rising with *SL*, *IRA* and *IRI*, increasing by a logit function with *VC*, and linearly falling with *ASM*. The *VC* level above 60% can effectively lower the surface runoff and significantly enhance soil infiltration. Two *RD* and *INF* prediction models, accounting for the above six factors, were constructed using the multiple nonlinear regression method. The verification of those models disclosed a high Nash-Sutcliffe coefficient and low root-mean-square error, demonstrating good predictability of both models.

1. Introduction

The surface rainfall and soil infiltration are the key factors affecting the soil erosion [1,2] by a highly nonlinear dynamic mechanism influenced by many factors. Studies on the runoff yield and soil permeability for individual rainfall events are very important for understanding and controlling erosion processes [3].

The rainfall is the dominant factor controlling the surface runoff and soil infiltration. The higher the rainfall amount, the higher the runoff volume [4]. A greater rainfall intensity does not necessarily lead to higher soil infiltration [5]. When the rainfall intensity is relatively small, its increase can induce the better permeability, but an excessive rain intensity may damage the top soil structure and seal the soil, as well as significantly reduce the infiltration rate [6, 7]. The slope gradient and length significantly influence the runoff and infiltration. Fox *et al* [8] proved indirectly that the runoff volume was growing with the slope gradient under laboratory conditions. However, Govers [9] has revealed the opposite trend under field conditions. With other influence factors kept constant, the runoff volume became larger with the slope length, when the latter was relatively small [10]. However, when the slope length exceeds a threshold value, the runoff volume displays a downward trend [11, 12]. Janeau *et al* [13] found that the steady final infiltration rate increased sharply as a result of a growing slope gradient in a series of laboratory experiments with 1.0 m^2 soil bins. Although there is no consensus on the slope gradient/length effects on the runoff and infiltration, the existence of the slope gradient and length threshold values is widely supported in studies [14-16]. The vegetation cover influences the runoff and infiltration changing the rainfall-runoff-infiltration process on the slopes [17, 18]. The vegetation cover can significantly reduce the surface runoff by the interception of the rainfall, and increase the soil infiltration by reducing raindrop impacts [1]. The antecedent soil moisture is also considered the important factor controlling the runoff and infiltration [19-21]. The more the antecedent soil moisture and less the infiltration capacity, the larger the runoff volume [22]. However, Luk [23] indicated that effect of the antecedent moisture was not only confined to the enhanced runoff capacity in the tested soils, having the cohesive structure.

Since most previous studies discussed the effect of only one or two factors on the slope runoff or soil infiltration, it is topical to elucidate the effect of multiple factors on their patterns. Therefore, the object of this study, making use of the field data, is to (1) investigate the quantitative relationships between the slope runoff, soil infiltration and the influence factors, and (2) construct the prediction model of the slope runoff and soil infiltration in the course of individual rainfall events.

2. Materials and methods

2.1. Site description

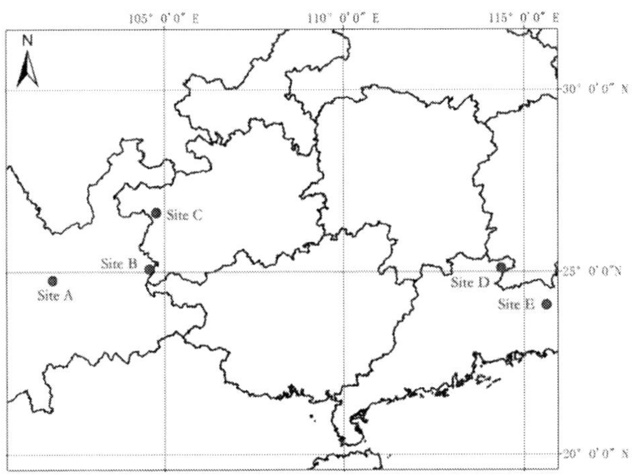

Figure 1. Localities of the five experimental catchments.

The data were obtained in the course of natural rainfall events from 261 runoff plots of five field catchments in the southern China. The localities of these five catchments are depicted in figure 1. Main soil types and vegetation species, as well as several physical soil properties of the soil for these five catchments are shown in table 1. These five catchments (sites A, B, C, D and E) have 10, 12, 12, 13 and 8 runoff plots of different sizes, whose detailed parameters are also given in table 1.

Table 1. Several physical properties of the experimental catchments and runoff plot sizes.

Sites	Main soil types	Main vegetation species	Composition				Runoff plots		
			Clay, %	Silt, %	Sand, %	Organic matter, g Kg^{-1}	Gradient, °	Length, m	Width, m
A	Red soil Purple soil	Orange tree Masson pine	31.8	42.1	26.2	23.1	5	20, 60	5
							15	10, 20,100	
							20	10, 20	
							25	20, 100	
							20	20	
B	Red soil	Pinus yunnanensis Pear tree	27.4	35.3	37.3	26.9	5	20, 60	
							15	10, 20, 100	
							20	10, 20, 100	
							25	20, 100	
							30, 35	20	
C	Yellow soil	Juniperus indica Cryptomeria	53.1	15.7	31.3	42.9	5	20, 60	
							15	10, 20	
							20	10, 20, 100	
							25	20, 100	
							30	20	
							35	20, 60	
D	Red soil Yellow soil	Peach tree White popinac	40.7	44.9	14.3	22.2	5	20, 60	
							15	10, 20, 100	
							20	10, 20, 100	
							25	20, 100	
							30	20	
							35	20, 60	
E	Red soil	Eucalyptus Masson pine	42.6	9.11	48.3	9.5	5	60	
							15	10,20, 100	
							20	20, 100	
							25	20, 100	

2.2. Measurements

The runoff volume for a given period was used to randomly determine the frequency of measurements (ranging from 5 min to 30 min) for the collection of the surface runoff volume with a cylindrical container of a 3000 mL volume. After the collection and clarification of turbid water, the clear water volume was regarded as the runoff volume. The deposited sediment was air-dried and weighted to determine the sediment yield. The rainfall data were measured automatically by an electronic rain gauge.

There was one sample point per 25 m^2 to collect soil samples for determining soil mositure, and moisture levels were measured by the oven drying method. For each plot, the soil sampling points should be uniformly distributed over the slope surface. According to the weather forecast information, soil moisture measurements were carried out before the rainfall and, as soon as possible, after it.

The crown cover and undergrowth were assessed by the photographic method [1]. The specific steps are as follows: firstly, take three to five JPEG format photos with a digital camera to record the data on the current state of the crown cover and undergrowth; secondly, convert the JPEG photos into the ITFF format with a lab color channel using the Photoshop software package; finally, the crown cover and undergrowth are calculated using the Image-J software package. We chose three to five sampling points and repeated measurements 2-3 times per sampling point in each plot. The average value of the crown cover and undergrowth was regarded as the vegetation cover.

For a given individual rainfall event, let *IRA*, *RD*, *INT* represent the rainfall amount, runoff volume and vegetation intercept water, respectively, and assume that the evaporation and water filling of the concave over the slope surface during rainfall are negligible. Based on the water balance equation, the soil infiltration water (*INF*) can be assessed from the following expression:

$$INF=IRA-RD-INT \tag{1}$$

Here the *INT* calculation refers to our previous study, and detailed information can be found elsewhere [24].

3. Results

3.1. Correlation between runoff depth, soil infiltration and influence factors

Results of the Pearson analysis of correlation between the runoff depth (*RD*) and soil infiltration (*INF*) for individual rainfall events and six influence factors are shown in figure 2. All the correlation coefficients reached the significance level ($P \leq 0.01$) except for the one between *RD* and the slope length (*SL*). The coefficient between *RD* and the individual rainfall amount (*IRA*) was positive and the highest. The relationships between *RD* and the individual rainfall intensity (*IRI*) and the antecedent soil moisture (*ASM*) were also positive, and their coefficients were 0.621 and 0.673, respectively. The slope gradient (*SG*) and vegetation cover (*VC*) exhibited a strong negative correlation with *RD*, with their coefficients equal to 0.503 and 0.705, respectively. The coefficient for *IRA*, *IRI* and *INF* was positive and maximum, while that for *ASM* and *INF* was negative and minimum. The results indicate that *INF* increased with *IRA* and *IRI*, but decreased with an increase in *ASM*. The relationships between *SL*, *VC* and *INF* were positive and their correlation coefficients were equal to 0.573 and 0.589, respectively ($P \leq 0.01$). The correlation coefficient for *SG* and *INF* was the least and equaled 0.442.

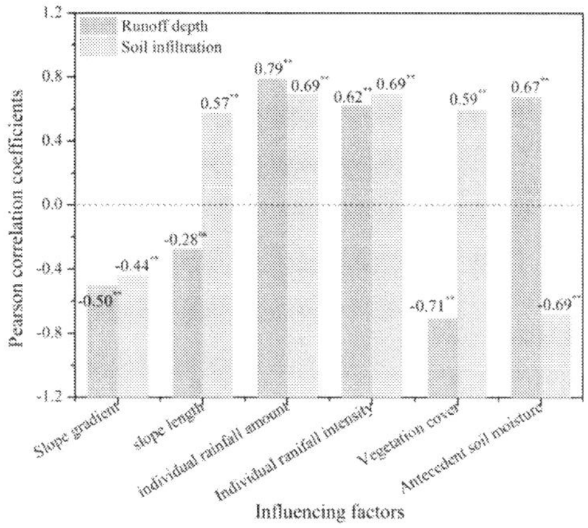

Figure 2. Pearson analysis of correlation between the runoff, infiltration and influence factors (N=363).
(**) significant at P≤0.01.
(ns) insignificant.

3.2. Analysis and simulation of runoff depth

The scatter diagrams for the average values of *RD*, *INF* and six factors were plotted, in order to study the quantitative relationship between *RD* and *INF* and influence factors for individual rainfall events as shown in figures 3 and 4.

RD came down followed by its growth with *SG* increase from 5° to 35°. The fitted equation reached the significance level, and the coefficient of determination R^2 was 0.944. Assuming that the first derivative of the fitted equation is equal to zero, the hump (knee point) on the fitted curve was obtained as (20.35°, 1.20 mm). *RD* was growing followed by a fall with *SL* increase from 10 to 100 m. However, the fitted equation did not reach the significance level, and the knee point was located at (54.22 m, 17.70 mm). *RD* increased consistently as a result of an increase in *IRA* and *IRI*. Both fitted equations reached the significance level with R^2 being equal to 0.902 and 0.857, respectively. The fitted equation for *RD* and the vegetation cover (*VC*) displayed an exponential pattern, acquiring the significance level. *RD* went down sharply with *VC* for *VC*<50%, while a drop of *RD* was slight for *VC*>50%. *RD* decreased with a following rise with *ASM*. The fitted equation reached the significance level and R^2 was 0.9553. The negative hump (knee point) was observed at (10.01%, 1.25 mm), so *RD* fell down consistently with *ASM* for *ASM*<10.01%, while *RD* grew persistently for *ASM*>10.01%.

Based on the fitted relationships between *RD* and influence factors as shown in figure 3, the following multi-parameter nonlinear model for predicting *RD* was constructed by the regression analysis considering 70% of the data:

$$\boldsymbol{RD} = \frac{\boldsymbol{SG}^2 - 23\boldsymbol{SG} - 0.2\boldsymbol{SL}^2 + 18\boldsymbol{SL} + 6\boldsymbol{IRA} + 7\boldsymbol{IRI}}{100} + 4.6e^{-\frac{3VC}{100}} + \frac{\boldsymbol{ASM}^2 - 22\boldsymbol{ASM} - 39}{100} \tag{2}$$

R^2 in equation (2) is equal to 0.9095 (P<0.001), and the residual sum of squares (RSS) is 1.709. All of the regression coefficients in equation (2) reached the significance level (P<0.001) by the T-test.

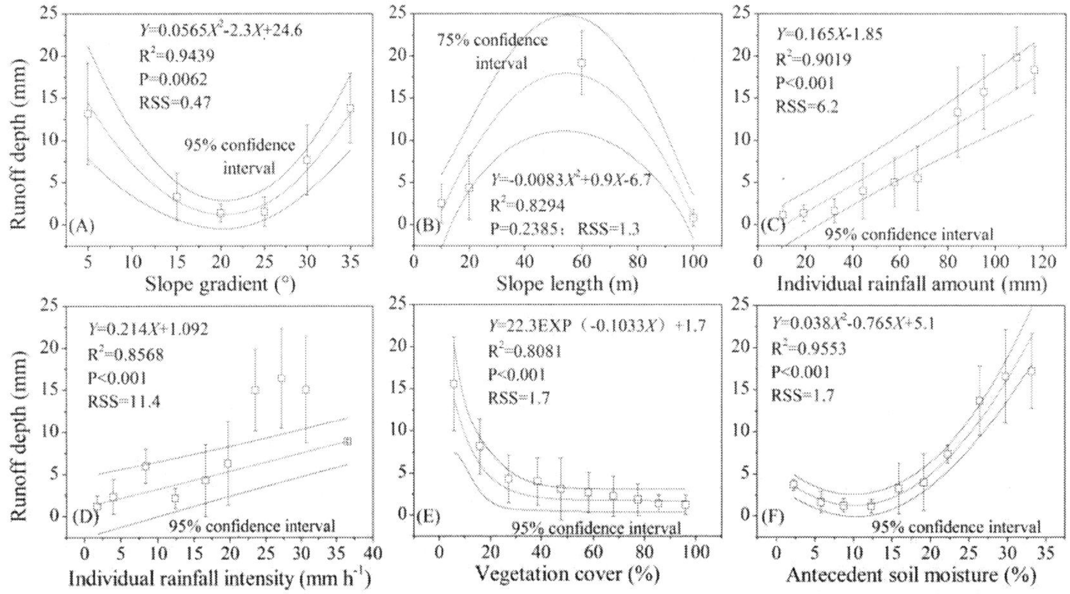

Figure 3. Runoff depth as a function of: (a) slope gradient, (b) slope length, (c) individual rainfall amount, (d) individual rainfall intensity, (e) Vegetation cover, and (f) antecedent soil moisture.
Note: Blue curves represent the confidence intervals at different significance levels.

3.3. Analysis and simulation of soil infiltration

INF displayed an increase followed by its fall with *SG,* but the fitted equation possessed poor correlation. The knee point on the fitted curve was (18.16°, 67.06 mm). In general, *INF* raised as a result of *SL* enlargement, but the fitted equation did not reach the significance level, and R^2 was relatively low. *INF* was growing with *IRA* and *IRI*. The two fitted equations reached the significance level, and R^2 values were 0.9919 and 0.9383, respectively. It is noteworthy that an increment of *INF* rate was relatively small for *IRI*<20 mm h^{-1} but became higher for *IRI*>20 mm h^{-1}. *INF* increased logistically with *VC*. The fitted equation reached the significance level, and R^2 was 0.9154. Obviously, the *INF* variation with *VC* can be negligible when *VC*<60%, but it grew dramatically for *VC*>60%. *INF* was going linearly with *ASM*, and the fitted equation reached the significance level.

The multi-parameter nonlinear model for predicting *INF* was constructed by the regression analysis considering 70% of the data, based on the fitted relationships between *INF* and influence factors as shown in figure 4. The following equation was derived:

$$\boldsymbol{INF} = \frac{\boldsymbol{SG}^2 - 29\boldsymbol{SG} + 2\boldsymbol{SL} + 86\boldsymbol{IRA} - 8IRI}{100} - \frac{368.6}{1 - 495.3\boldsymbol{VC}^{40.2}} - \frac{3\boldsymbol{ASM}}{100} + 3.6 \quad (3)$$

R^2 and RSS are equal to 0.9941 (P<0.05) and 2.1569, respectively. All the regression coefficients in equation (3) reached a significance level (P<0.05) by the T-test.

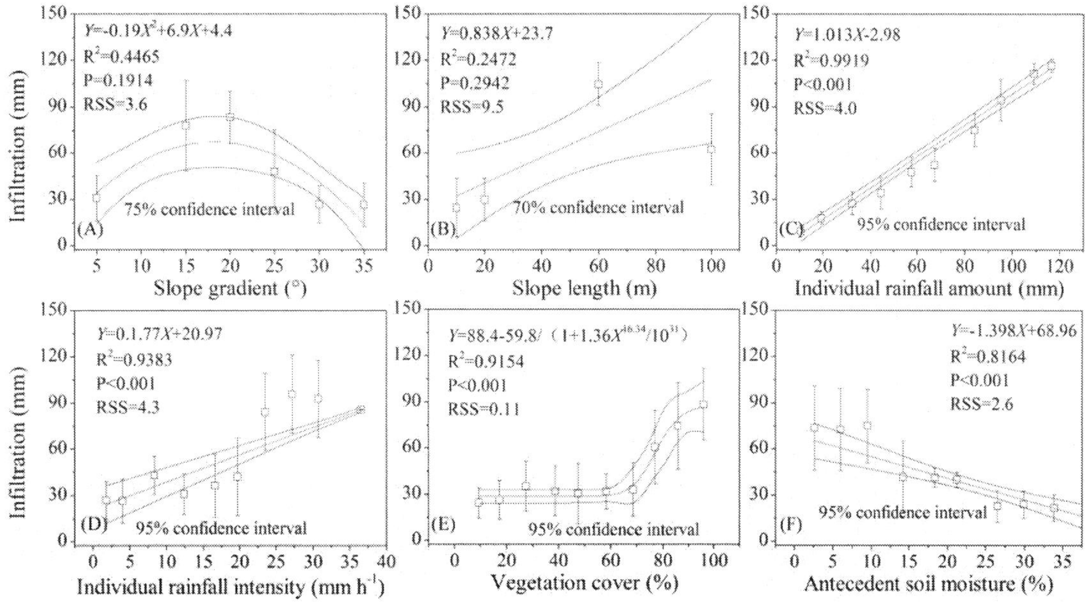

Figure 4. Soil infiltration as a function of: (a) slope gradient, (b) slope length, (c) Individual rainfall amount, (d) Individual rainfall intensity, (e) Vegetation cover, and (f) antecedent soil moisture.
Note: Blue curves represent the confidence intervals at different significance levels.

4. Discussion

The relationship between the six influence factors and the runoff and infiltration for individual rainfall events is systematically analyzed to get a deeper insight into the slope runoff and soil infiltration behavior. The correlation coefficients for *IRA*, *IRI* and *RD*, *INF* were over 0.60 ($P \leq 0.01$), indicating that the rainfall is a very important factor controlling *RD* and *INF* values. This was confirmed by earlier studies of Garcia-Rodeja and Gil-Sotres [25] and Huang *et al* [26, 27]. Although higher infiltration rates are often observed with greater *IRI*, there appears to be no consensus in the literature on the explanation of this phenomenon [28]. Huang *et al* [5] found that the soil infiltration rate grew with *IRI* until a critical threshold was reached, its value ranging from 90 to 150 mm h^{-1}. *VC* can change the rainfall-runoff-infiltration mechanism and then affect the soil infiltration and slope runoff [29, 30]. The results of this study demonstrate that their coefficients were 0.705 and 0.589 ($P \leq 0.01$), respectively. *RD* fell exponentially and *INF* grew logistically with *VC*. *INF* increased sharply and a reduction in *RD* was almost invariant when *VC*>60%. Previous studies by Marston [31] and Zhu *et al* [32] also proved that *VC* of 40-60% or more can significantly reduce runoff and increase infiltration. Karnieli and Ben-Asher [21], Ceballos and Schnabel [19] and Fitzjohn *et al* [20] suggested that *ASM* can also be an important factor controlling the slope runoff and soil infiltration. Zehe *et al* [33] indicated that *ASM* at the forest site can explain 92% of the variability in the runoff coefficients. The higher the *ASM*, the lower the soil permeability and *INF*, and the larger the *RD*. In general, *RD* became higher and *INF* lower as a result of increase in *ASM*. The coefficients of *SG* correlation with *RD* and *INF* were 0.503 and 0.442, respectively ($P \leq 0.01$), indicating that *SG* is an important factor controlling the slope runoff and soil infiltration [8, 9]. In the present study, minimum *RD* and maximum *INF* values were observed for an *SG* value of about

20°. As is known from the literature, *SL* should have a significant impact on the runoff and infiltration [12, 34], but the correlation coefficient for *SL* and *RD* was only 0.276 and, thus, was insignificant. *RD* increased with its further decrease with *SL*, and *SL*=54.22 m was the inflection point of the curve. Ahead of this point, *RD* was continuously growing, while *RD* was going down consistently after this point.

The remaining 30% of the data were used to verify prediction models (2) and (3). The results depicted in figure 5 demonstrate a close of fit of the experimental and predicted results for both *RD* and *INF.* The Nash-Sutcliffe coefficient, root-mean-square error and correlation coefficient between experimental and predicted data were 0.9621, 1.1294, 0.9879 for *RD* and 0.9837, 1.9533, 0.9927 for *INF*, respectively, which indicates that equations (3) and (4) are applicable for predicting *RD* and *INF* values.

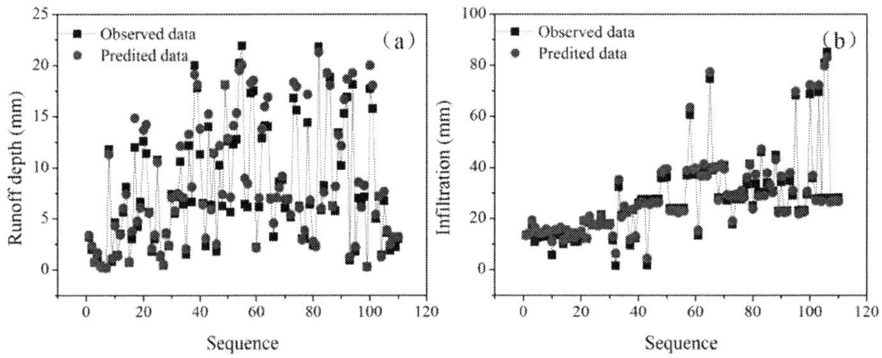

Figure 5. Observed and predicted values of the runoff depth (a) and infiltration (b) considering 30% of the data.

5. Conclusions

This study, based on the data from a series of natural rainfall events in the South of China, generated certain important information for better understanding of the quantitative relationships between the six factors (*SG, SL, IRA, IRI, VC, ASM),* runoff depth *(RD)* and infiltration (*INF).* An empirical model for *RD* and *INF* calculation based on the observation data using the above six factors is proposed. The results obtained are as follows:

The correlation coefficients for *IRA*, *IRI*, *VC* and *RD* were over 0.7 (P≤0.01), while those for *IRA*, *IRI*, *ASM* and *INF* approached 0.7 (P≤0.01). *RD* decreased followed by its increase with *SG* and *ASM*, it grew with a further fall with *SL*, linearly growing with *IRA* and *IRI* and falling exponentially with *VC*. Meanwhile, *RD* was rising consistently with *SG* for *SG*<20.35°, and then falling continuously with *SG* for *SG*>20.35°. *RD* increased persistently with *SL* for *SL*<54.22 m and decreased consistently for *SL*>54.22 m. *INF* became higher and then lower with *SG,* linearly growing persistently with *SL, IRA*, and *IRI,* increased logistically with *VC*, and linearly lowering continuously with *ASM*. *INF* increased continuously with an increase in *SG* for *SG*<18.16° and was falling consistently for *SG*>18.16°. When *VC* was larger than 60%, *RD* was small and its reduction with an increase in *VC* was very slight, while *INF* was sharply growing. Two multi-parameter nonlinear models for predicting *RD* and *INF* were constructed using the regression analysis, and their applicability and feasibility were proved by the validation tests.

Acknowledgments

This work was jointly supported by the CRSRI Open Research Program (CKWV2016389/KY), the National Natural Science Foundation of China (41501019) and the Open Project Program of Key Laboratory of Soil and Water Loss Process and Control on the Loess Plateau of Ministry of Water Resources (2016006).

References

[1] Huang J, Wu P T and Zhao X N 2013 Effects of rainfall intensity, underlying surface and slope gradient on soil infiltration under simulated rainfall experiments *Catena* **104** 93-102

[2] Huang J, Zhao X N and Wu P T 2013 Surface runoff volumes from vegetated slopes during simulated rainfall events *J. Soil Water Conserv.* **68** 283-95

[3] Fu B J, Qiu Y, Wang L and Chen L 2002 Effect simulations of land use change on the runoff and erosion for a gully catchment of the loess Plateau, China *Acta Geographica Sinica* **57** 717-22

[4] Wei L, Zhang B and Wang M 2007 Effects of antecedent soil moisture on runoff and soil erosion in alley cropping systems *Agr. Water Manage.* **94** 54-62

[5] Huang J, Wu P T and Zhao X N 2010 Impact of slope biological regulated measures on soil water infiltration *Tran. of the CSAE* **26** 29-37

[6] Beven K and Germann P 1982 Macropores and water flow in soils *Water Resour. Res.* **18** 1311-25

[7] Romkens M J M, Baumhardt R L, Parlange M B, Whisler F D, Parlange J-Y and Prasad S N 1986 Rain-induced surface seals: Their effect on ponding and infiltration *Ann. Geophysic.* **4** 417-24

[8] Fox D M, Bryan R B and Price A G 1997 The influence of slope angle on final infiltration rate for interrill conditions *Geoderma* **80** 181-94

[9] Govers G 1991 A field study on topographical and topsoil effects on runoff generation *Catena* **18** 91-111

[10] Li X, Liu L, Gao S, Shi P, Zou X and Zhang C 2005 Microcatchment water harvesting for growing *Tamarix ramosissima* in the semiarid loess region of China *Forest Ecol. Manage.* **214** 111-7

[11] De Giesen N V, Stomph T-J, Ajayi A E and Bagayoko F 2011 Scale effects in Hortonian surface runoff on agricultural slopes in West Africa: Field data and models *Agr. Ecosyst. Environ.* **142** 95-101

[12] Smets T, Poesen J and Bochet E 2008 Impact of plot length on the effectiveness of different soil-surface covers in reducing runoff and soil loss by water *Prog. Phys. Geog.* **32** 654-77

[13] Janeau J L, Bricquet J P, Planchon O and Valentin C 2003 Soil crusting and infiltration on steep slopes in northern Thailand *Eur. J. Soil Sci.* **54** 543-54

[14] Liu Q Q and Singh V P 2004 Effect of microtopography, slope length and gradient, and vegetative cover on overland flow through simulation *J. Hydrol. Eng.* **9** 375-82

[15] Parsons A J, Brazier R E, Wainwright J and Powell D M 2006 Scale relationships in hillslope runoff and erosion *Earth Surf. Process Landf.* **31** 1384-93

[16] Rejman J and Brodowski R 2005 Rill characteristics and sediment transport as a function of slope

length during a storm event on loess soil *Earth Surf. Process Landf.* **30** 231-9

[17] Bhark E W and Small E E 2003 Association between plant canopies and the spatial patterns of infiltration in shrubland and grassland of the Chihuahuan Desert, New Mexico *Ecosystems* **6** 185-96

[18] Llorens P and Domingo F 2007 Rainfall partitioning by vegetation under Mediterranean conditions: A review of studies in Europe *J. Hydrol.* **335** 37-54

[19] Ceballos A and Schnabel S 1998 Hydrological behaviour of a small catchment in the *dehesa* landuse system (Extremadura, SW Spain) *J. Hydrol.* **210** 146-60

[20] Fitzjohn C, Ternan J L and Williams A G 1998 Soil moisture variability in a semi-arid gully catchment: Implications for runoff and erosion control *Catena* **32** 55-70

[21] Karnieli A and Ben-Asher J 1993 A daily runoff simulation in semi-arid watersheds based on soil water deficit calculations *J Hydrol.* **149** 9-25

[22] Liu H, Lei T W, Zhao J, Yuan C P, Fan Y T and Qu L Q 2011 Effects of rainfall intensity and antecedent soil water content on soil infiltrability under rainfall conditions using the run off-on-out method *J. Hydrol.* **396** 24-32

[23] Luk S-H 1985 Effect of antecedent soil moisture content on rainwash erosion *Catena* **12** 129-39

[24] Huang J, Jin P W, Li M, Li L B and Jiang X B 2016 Influencing factors of soil infiltration rate and its estimation model at runoff-plot scale *Transactions of the CSAM* **47** 171-8

[25] García-Rodeja I and Gil-Sotres F 1997 Prediction of parameters describing phosphorus-desorption kinetics in soils of Galicia (Northwest Spain) *J. Environ. Qual.* **26** 1363-9

[26] Huang J, Wu P T and Zhao X N 2011 Research on soil infiltration based on BP network and gray correlation analysis *Acta Pedologica Sinica* **48** 1282-6

[27] Huang J, Zhao X N and Wu P T 2012 Factors analysis of runoff and sediment based on path analysis and grey relational analysis *J. Sichuan Univ. (Eng. Sci. Ed.)* **44** 64-70

[28] Foley J L and Silburn D M 2002 Hydraulic properties of rain impact surface seals on three clay soils—influence of raindrop impact frequency and rainfall intensity during steady state *Aust. J. Soil Res.* **40** 1069-83

[29] Kato H, Onda Y, Tanaka Y and Asano M 2009 Field measurement of infiltration rate using an oscillating nozzle rainfall simulator in the cold, semiarid grassland of Mongolia *Catena* **76** 173-81

[30] Molina A, Govers G, Vanacker V, Poesen J, Zeelmaekers E and Cisneros F 2007 Runoff generation in a degraded Andean ecosystem: Interaction of vegetation cover and land use *Catena* **71** 357-70

[31] Marston R B 1952 Ground cover requirements for summer storm runoff control on aspen sites in northern Utah *J. Forest* **50** 303-7

[32] Zhu B B, Li Z B, Li P and You Z 2010 Effect of grass coverage on sediment yield of rain on slope *Acta Pedologica Sinica* **47** 401-7

[33] Zehe E, Graeff T, Morgner M, Bauer A and Bronstert A 2010 Plot and field scale soil moisture dynamics and subsurface wetness control on runoff generation in a headwater in the Ore

Mountains *Hydrol. Earth Syst. Sci.* **14** 873-89

[34] Smets T, Poesen J and Knapen A 2008 Spatial scale effects on the effectiveness of organic mulches in reducing soil erosion by water *Earth-Sci. Rev.* **89** 1-12

IOP Conf. Series: Earth and Environmental Science 82 (2017) 012020 doi:10.1088/1755-1315/82/1/012020

Construction of an evaluation index system of water resources bearing capacity: An empirical study in Xi'an, China

X E Qu[1,2] and L L Zhang[1]

[1]Department of Economy and Financial, University of Xi'an Jiaotong, 74# Yanta Xilu, Xi'an, Shaanxi, 710061, China

E-mail: quxiaoe1227@163.com

Abstract. In this paper, a comprehensive evaluation of the water resources bearing capacity of Xi'an is performed. By constructing a comprehensive evaluation index system of the water resources bearing capacity that included water resources, economy, society, and ecological environment, we empirically studied the dynamic change and regional differences of the water resources bearing capacities of Xi'an districts through the TOPSIS method (Technique for Order Preference by Similarity to an Ideal Solution). Results show that the water resources bearing capacity of Xi'an significantly increased over time, and the contributions of the subsystems from high to low are as follows: water resources subsystem, social subsystem, ecological subsystem, and economic subsystem. Furthermore, there are large differences between the water resources bearing capacities of the different districts in Xi'an. The water resources bearing capacities from high to low are urban areas, Huxian, Zhouzhi, Gaoling, and Lantian. Overall, the water resources bearing capacity of Xi'an is still at a the lower level, which is highly related to the scarcity of water resources, population pressure, insufficient water saving consciousness, irrational industrial structure, low water-use efficiency, and so on.

1. Introduction

The global water crisis has become one of the greatest challenges of the 21st century. Since the 1980s, with the rapid development of China's economy, the water shortage and water environment have become increasingly serious, which has, in turn, created a significant bottleneck that restricts the development of the Chinese economy. In 2002, the Chinese government promulgated and implemented the water law of the People's Republic of China and subsequently developed a series of regulations and policies aimed at addressing water shortage issue and environmental destruction. Such developments have important practical significance and strategic significance to the promotion of sustainable utilization of water resources in China.

In the inland northwest region, the water resources of Xi'an are inadequate and unevenly distributed. Since the west became further developed, the central economic regions have transferred to the west. Furthermore, the population growth of Xi'an is continuing to increase; thus, the economic water consumption and residents' living water consumption increase every year, evidencing the serious connection between urban development and available water supply. Statistical data show that in 2014 the total water resource of Xi'an is 21.92 billion cubic meters, and the per capita water resource is only 268.86 cubic meters, which only accounts for 1/4.2 of the per capita water resources of the Shaanxi Province [1]. At present, both the average annual precipitation and per capita surface water resource of Xi'an are far below the internationally recognized critical value. Inadequate water resources, serious water pollution, inadequate sewage disposal facilities, and other factors cause Xi'an to be a city with one of the worst water scarcities. In order to achieve a sustainable utilization of water resources, the Xi'an government has increased its investment in water conservancy construction and water environmental governance since the 13th five-year plan. In December 2001, the main project of the Heihe Water Diversion was completed. In 2002, the Xi'an government publicized the "Xi'an water resources management approaches", giving priority to the development of water conservancy construction for people's livelihood. Thereafter, Xi'an government successively promulgated a series of local laws and regulations, such as "Xi'an water resources management ordinance" and "Measures for the protection and management of Heihe River".

2. Literature review

The utilization and protection of water resources have attracted the attention of the Xi'an government as well as many scholars. Many scholars have defined water resources bearing capacity; however, water resources carrying capacity is still a concept with a fuzzy extension and connotation. In this paper, with foreseeable technical, economic, and social development levels, based on the principle of sustainable development, we consider water resources carrying capacity to be the maximum supporting capacity of water resources for the social and economic development of the region after a reasonable optimization disposition in a specific historical stage of development.

Internationally, Falkenmark [2] and Bable [3] used the system dynamics method to calculate the usage of water resources in some developing countries and analyzed water resources bearing capacity, thereby laying the foundation for the quantitative study of water resources carrying capacity. With water supply, drainage, sewage, and other indicators, Joardor [4] built a water resources bearing capacity evaluation system and carried out an empirical study on the water resources bearing capacity in India City, believing that the assessment of urban water supply facilities would aid in determining the potential of the city's sustainable growth. Harris and Kennedy [5] focused on the support ability of water resources in agricultural areas worldwide. They consider the water resources bearing capacity and land bearing capacity to be important criterions for regional sustainable development. Rijsberman [6] studied the sustainable development of the Holland urban water system and considered water resource to be an important condition for a city's healthy and rational operation and scientific management. Varis O and Vakkilainen P [7] conducted a comparative analysis of the social economic situation of the Yangtze River Basin and its water environment bearing capacity, thereby revealing the water resources management problems faced by China in the 1980s and 1990s.

Domestically, Wang C [8] used the principal component analysis method to evaluate the water resources bearing capacity of Ordos City and found its water resources bearing capacity to be relatively small. Xie Y [9] studied the water resources bearing capacity of Yulin City through the analytic hierarchy process and "Mode and Sum", and it was found that its water resources bearing capacity has been declining over time. Shao J [10] studied the effective management and sustainable utilization of water resources in Xi'an and claimed that water scarcity has become an important factor restricting the economic development of Xi'an. Based on the natural--artificial two elements model and through the construction of the water resources bearing capacity evaluation system, Li G [11] used the multi-level fuzzy comprehensive evaluation method to study the water resources bearing capacity of Hubei, and it was found that the water resources bearing capacity of Hubei has great potential. Based on the complex system theory, Duan X G and Luan F F [12] used the fuzzy comprehensive evaluation model to evaluate the current situation of the Xinjiang water resources bearing capacity, and they proposed that the development and utilization of water resources in Xinjiang has a certain scale. Wang C J [13] comprehensively evaluated the relative population bearing capacity and the relative economy bearing capacity of resources of the Tarim River Basin and provided some corresponding countermeasures from the aspect of the sustainable utilization of water resources.

3. Study objectives

Based upon the above research, we adopted a compound system theory of "water resources--social economy-ecological environment". By applying the TOPSIS comprehensive evaluation method, we evaluated and analyzed the water resources bearing capacity of Xi'an and explored reasonable methods for the exploitation and utilization of water resources. The results of this paper have great practical significance on the improvement of the water resources bearing capacity of Xi'an. Additionally, this study will aid in the sustainable development of the economy and water resources as well as contribute to the urban ecosystem.

4. Method and material

4.1. Construction of evaluation index system

A comprehensive evaluation of water resources bearing capacity involves taking into consideration the population, economy, society, ecological environment, and so on. Following the guiding ideology and principles in the construction of the evaluation index system, we constructed the evaluation index system of the water resources bearing capacity from three levels: the target layer, criterion layer, and index layer. The target layer is the water resources bearing capacity under the framework of sustainable development and is the ultimate goal of the evaluation index system, which reflects the overall allocation of water resources. The criterion layer, as a subdivision of the target layer, can serve as a criterion for the comprehensive evaluation, and it consists of four aspects: the water resources subsystem, economic subsystem, social subsystem, and ecological subsystem. The index layer is the concrete manifestation of each criterion layer and is the basis of the evaluation index system. The economic meaning and calculation method of the indexes for all levels are shown in table 1.

The original data for the evaluation index are from the "Xi'an statistical yearbook", "Shaanxi statistical yearbook", "Shaanxi Provincial Water Resources Bulletin", "Shaanxi Environmental Statistics Bulletin" and "Shaanxi Province Environmental Status Bulletin.

Table 1. Comprehensive evaluation index system of the water resources bearing capacity in Xi'an.

Target Layer	Criterion Layer	Index Name	Calculation Method	Meaning	Category
water resources bearing capacity (A)	water resources subsystem (B_1)	per capita water resources C_1	total water resources/total population	the situation of per capita water resources	positive
		utilization rate of water resources C_2	total water supply/total water resources	the development and utilization of water resources	negative
		water producing module C_3	total water resources/urban area	the dynamic change of water resources	positive
		annual water supply	statistical data	the situation of urban water supply	positive
		water supply module C_5	total water supply/urban area	water supply capacity of water resources	negative
		surface water resources proportion C_6	surface water resources/total water resources	the distribution of water resources	positive
	economic subsystem (B_2)	GDP C_7	statistical data	the level of urban economic development	positive
		GDP growth rate C_8	GDP growth amount/GDP	the driving forces of urban economic development	positive
		water resource per 10,000 Yuan GDP C_9	total water resources/GDP	economic benefits of water resources utilization	negative
		tertiary industry proportion C_{10}	tertiary industry GDP/GDP	the optimization of economic structure	positive
		water use per 10,000 Yuan industry gross production value C_{11}	total industrial water consumption/industrial value added	the efficiency of industrial water	negative
		repeated use rate of industrial water C_{12}	the repeated industrial water/total industrial water consumption	the efficiency of industrial water reuse	positive
	social subsystem (B_3)	population density C_{13}	total population/region area	the distribution density of population	negative
		total population C_{14}	statistical data	the size of the population	negative

	urbanization rate C_{15}	urban population/total population	the degree of urbanization	positive
	natural population growth rate C_{16}	natural population growth /annual average total population	the natural growth of the population trend	negative
	urban per capita living water consumption C_{17}	urban living water consumption/total urban population	per capita water consumption of urban residents	negative
	urban living water consumption C_{18}	statistical data	the dynamic changes of living water consumption	negative
ecological subsystem (B_4)	forest coverage C_{19}	forest area/city area	the degree of urban greening	positive
	Built-up areas green coverage rate C_{20}	greening area/city area	the degree of built-up areas greening	positive
	per capita public green area C_{21}	total garden green area/ population of non-agriculture	per capita public green area	positive
	per capita garden green area C_{22}	total garden green area/total population	per capita garden green area	positive
	urban wastewater treatment capacity C_{23}	statistical data	city's wastewater treatment capacity	positive
	urban sewage treatment rate C_{24}	wastewater treatment amount/total amount of waste water	the situation of urban wastewater treatment	positive

4.2. Evaluation methods

At present, there are no unified and mature research methods on the water resources bearing capacity either domestically or abroad due to its wide range and complex content. Scholars have adopted different methods for studying the water resources bearing capacity according to their research purpose and data characteristics. Integrated water resources management (IWRM), water impact index, and technique for order preference by similarity to an ideal solution (TOPSIS) can be used to evaluate the bearing capacity of water resources in a particular area. IWRM focuses on the sustainable utilization of water resources, emphasizes the integrated management of water and soil resources and other related resources, and attaches importance to coordination and public participation. It is considered to be an effective and sustainable solution to the problem of water resources, and it is highly praised by the international water community [14]. By setting up the framework of IWRM implementation, we can evaluate and analyze the current situation of water resources management in a specific country or region. However, the evaluation indicators that reflect the key areas of the implement environment, management system, and management tools are primarily abstract and

IOP Conf. Series: Earth and Environmental Science 82 (2017) 012020 doi:10.1088/1755-1315/82/1/012020

qualitative indicators. Thus, it is necessary to combine qualitative analysis with quantitative methods, which is too subjective. The Water Impact Index is a tool developed by Veolia to measure the impact of activities on a local water resource. It integrates the issues of volume, scarcity, and quality into a single indicator that assesses the relative magnitude of potential impacts to freshwater availability generated by a human activity. The Water Impact Index has the potential to serve a number of purposes, including screening and evaluating the usage of water, decision support, and communication. However, the Water Impact Index does not provide a comprehensive assessment of water-related environmental impacts. TOPSIS, first proposed by Hwang C L and Yoon L S [15], is a method of system engineering for multi-objective-decision-making with advantages such as no strict limits to the data distribution, the sample quantity, and the number of indexes. Moreover, TOPSIS has is characterized by scientific evaluation, intuitive geometric meaning, small information distortion, and so on [16]. The basic idea of TOPSIS is to first define the positive ideal solution and the negative ideal solution for the evaluation object and then identify the solution from among the feasible schemes that is closest to the positive ideal solution and farthest away from the negative ideal solution. This solution is the satisfactory solution. Considering the advantages and disadvantages of various methods, we adopted the TOPSIS comprehensive evaluation method, that is, the ideal solution method. The main steps of the TOPSIS method are as follows:

- Construct the evaluation matrix. Assuming there are N evaluation units and k evaluation indexes, the original data matrix is

$$A=(a_{ij})_{n\times k}=\begin{bmatrix} a_{11} & a_{12} & \cdots & a_{1k} \\ a_{21} & a_{22} & \cdots & a_{2k} \\ \cdots & \cdots & \cdots & \cdots \\ a_{n1} & a_{n2} & \cdots & a_{nk} \end{bmatrix}_{n\times k} \tag{1}$$

- Normalize the original data matrix to eliminate the influence of dimension as in equation (2), then construct the normalized decision matrix Z as in equation (3).

$$z_{ij}=a_{ij}\left(\sum_{i=1}^{n}a_{ij}^2\right)^{-\frac{1}{2}},\quad i=1,2,\ldots,n;\ j=1,2,\ldots,k \tag{2}$$

$$Z=(z_{ij})_{n\times k}=\begin{bmatrix} z_{11} & z_{12} & \cdots & z_{1k} \\ z_{21} & z_{22} & \cdots & z_{2k} \\ \cdots & \cdots & \cdots & \cdots \\ z_{n1} & z_{n2} & \cdots & z_{nk} \end{bmatrix}_{n\times k} \tag{3}$$

- Define the positive ideal solution Z+ and the negative ideal solution Z-.

$$z^{+}=(z_1^{+},z_2^{+},\ldots,z_k^{+}),\quad z^{-}=(z-,z_2^{-},\ldots,z_k^{-}) \tag{4}$$

- Calculate the distance between the i evaluation unit and the positive ideal solution D_i^+ and the distance between the i evaluation unit and the negative ideal solution D_i^-.

$$D_i^+ = \sqrt{\sum_{j=1}^{k}(z_{ij} - z_j^+)^2}, \quad D_i^- = \sqrt{\sum_{j=1}^{k}(z_{ij} - z_j^-)^2} \tag{5}$$

- Calculate the relative approach degree between the i evaluation unit and the ideal solution C_i.

$$C_i = \frac{D_i^-}{D_i^+ + D_i^-} \tag{6}$$

Sort the relative approach degree (C_i) by numeric value. The larger the value is, the closer it is to the ideal state. Accordingly, a larger value also corresponds to better evaluation results and a more secure water resources bearing system. Conversely, the smaller the value is, the less secure the water resources bearing system is. When the value is 1, all the indexes of the water resources bearing system are optimal, and the system is in the ideal condition.

5. Results and discussion

5.1. Longitudinal evaluation and analysis of the water resources bearing capacity of Xi'an

Following the principles and procedures of the TOPSIS comprehensive evaluation method, and based on the normalized data, we evaluated the water resources bearing capacity of Xi'an in 2000-2014. The larger the value is, the stronger the water resources bearing capacity is. Conversely, the smaller the value is, the weaker the water resources bearing capacity is. The evaluation results are shown in table 2.

- The target layer evaluation results show that the water resources bearing capacity of Xi'an during the research period increased with fluctuation, and the evaluation value increased from 0.2511 in 2000 to 0.2953 in 2014. More specifically, the comprehensive evaluation value of the water resources bearing capacity decreased from 0.511 in 2000 to 0.1692 in 2008 and then increased to 0.8512 in 2009, before decreasing to 0.2953 in 2014. The water resources bearing capacity reached a maximum value of 0.8512 in 2009 and a minimum value of 0.1490 in 2007. The contribution of the water resources subsystem is the largest with a mean value of 0.4998, followed by the social subsystem, ecological subsystem, and economic subsystem. Overall, the evaluation values of the four subsystems initially increased and then decreased. Since the 20th century, Xi'an has gradually increased its investment in ecological and environmental governance, and the per capita green area has increased year by year. In 2000, the per capita green area was only 5.98%; however, by 2014, it had increased to 23.20% with an average annual increase of 3.9 percentage points [17]. After the promulgation of "The People's Republic of China Water Law" (2002), the Xi'an government published the "Xi'an water resources management approaches, 2002" and revised it in 2004. In May 2002, the Xi'an government established the Xi'an water authority in order to intensify protection of water resources. Since then, the government has successively implemented a series of policy measures, such as the "sustainable utilization plan of water resources in Xi'an" and the "water saving water plan of Xi'an". The government's work has caused the management and protection of water resources to be both a scientific and institutionalized venture. Meanwhile, the government attaches importance to the popularization of new technologies and equipment in industry that are capable of conserving water. Additionally, they have limited high water consumption projects in order to improve wastewater treatment and the reuse rate of industrial water. These policy measures have played a positive role in improving the water resources bearing capacity of Xi'an.

Table 2. Evaluation results of water resources bearing capacity of Xi'an in 2000-2014.

Year	Comprehensive bearing capacity (A)	Water resources subsystem (B_1)	Economic subsystem (B_2)	Social subsystem (B_3)	Ecological subsystem (B_4)
2000	0.2511	0.4514	0.2022	0.8056	0.2624
2001	0.2680	0.4545	0.2001	0.7657	0.2778
2002	0.2360	0.5172	0.1985	0.6448	0.2293
2003	0.2549	0.6150	0.1971	0.8275	0.2001
2004	0.2473	0.5993	0.1914	0.8057	0.3497
2005	0.2308	0.5530	0.1815	0.7409	0.3229
2006	0.1828	0.4455	0.1208	0.7307	0.5485
2007	0.1490	0.3800	0.1016	0.5671	0.4521
2008	0.1692	0.7977	0.0148	0.3388	0.2111
2009	0.8512	0.4325	0.9449	0.3142	0.2408
2010	0.3804	0.3965	0.3841	0.2113	0.1624
2011	0.2625	0.0824	0.2705	0.2251	0.4697
2012	0.2914	0.5391	0.2814	0.0780	0.4652
2013	0.3054	0.7242	0.2765	0.0718	0.4614
2014	0.2953	0.5096	0.2661	0.0450	0.4907
Mean	0.2930	0.4998	0.2554	0.4781	0.3429

- The criterion layer evaluation results (table 1) show that the evaluation value of the water resources subsystem, economic subsystem, and ecological subsystem increased with fluctuation. The bearing capacity of each subsystem steadily increased, as did their contributions to the comprehensive bearing capacity. For the water resources subsystem, its main index water supply module rose from 3.30 in 2000 to 5.33 in 2014, and the annual water supply increased from 30,273 million cubic meters in 2000 to 53,799 million cubic meters in 2014 [17]. For the economic subsystem, the water use per 10,000 Yuan industry gross production value decreased from 76.54 cubic meters in 2000 to 6.54 cubic meters in 2014, and the repeated use rate of industrial water increased from 47% in 2000 to 68.9% in 2014 [18]. For the ecological subsystem, its increasing evaluation value strongly relates to the significant attention that the Xi'an government placed on the protection and management of the ecological environment. Specifically, forest coverage increased from 41.8% in 2000 to 50.4% in 2014, and the urban sewage treatment rate rose from 23.11% in 2000 to 92.71% in 2014 [18]. For the social subsystem, however, its evaluation values show a significant downward trend; it decreased from 0.8056 in 2000 to 0.4781 in 2014. The social subsystem is mainly affected by the total population, population density, rate of urbanization, urban per capita living water consumption, and urban living water consumption. From the changes to these indexes, we found that the urban per capita living water consumption and urban living water consumption increased with the increase in the total population, population density, and rate of urbanization. For example, the total urban living water consumption increased from 10,949 million cubic meters in 2000 to 27,819 million cubic meters in 2014 [18], which put pressure on the social subsystem and reduced its contribution to the water resources bearing capacity.

- The dynamic evolution trend shows that the changing trend of the water resources bearing capacity is similar to those of the economic subsystem and water resources subsystem, indicating that the economic subsystem is an important factor affecting the water resources bearing capacity. This may be because the total demand of water resources increases with the rapid development of economy, which to some extant increases the pressure of the water resources bearing capacity. The total water resource is an important base of the water resources bearing capacity. The average contribution of the water resources subsystem is 4.998, which is the greatest contribution of the four subsystems. However, the uneven time distribution of the water resources leads to a fluctuation in the evaluation value of the water resources subsystem. Therefore, in order to cause the water resources subsystem evaluation value to steadily increase, it is necessary to take the uneven time distribution of water resources into consideration. We can accumulate the water resources of the wet season to make up for the dry season water resources. Although the Xi'an water resources bearing capacity increases over time, the evaluation value is still at a low level of about 0.29. The reasons are as follows: 1) the scarcity of water resources and the low utilization of water resources; 2) the population pressure and residents' insufficient water saving consciousness; and 3) the irrational industrial structure. In Xi'an, the proportion of the second industry and the third industry is high, thereby increasing industrial water and agricultural irrigation water. At present, Xi'an is in an important period of rapid economic and social development, and the gradual improvement of people's living standard has objectively caused pressure on the water resources bearing capacity.

5.2. Transverse evaluation and analysis of the water resources bearing capacity of Xi'an

Through transverse evaluation and analysis, we discovered the regional differences in the water resources bearing capacities of the Xi'an districts. This allows for an exploration of effective ways to improve the bearing capacity of water resources from the perspective of regional coordination. The evaluation results are shown in table 3.

Table 3. The differences in the water resources bearing capacities of the Xi'an districts in 2014.

Region	Comprehensive bearing capacity (A)	Water resources subsystem (B_1)	Economic subsystem (B_2)	Social subsystem (B_3)	Ecological environment subsystem (B_4)
Urban areas[a]	0.7432	0.8150	0.7334	0.8908	0.8341
Lantian	0.2499	0.3750	0.1256	0.1721	0.2337
Zhouzhi	0.3377	0.3700	0.1255	0.3373	0.4142
Huxian	0.3388	0.3656	0.0074	0.3565	0.4293
Gaoling	0.2998	0.0268	0.4081	0.2549	0.1900

[a]Urban areas include the Xincheng District, Beilin District, Lianhu District, Baqiao District, Weiyang District, Yanta District, Yanliang District, Lintong District, and Changan District.

- The target layer evaluation results (table 3) show that in 2014 the comprehensive evaluation value of the water resources bearing capacity of urban areas was 0.7432. The water resources bearing capacities of the Xi'an districts from high to low are as follows: Huxian, Zhouzhi, Gaoling, and Lantian. Lantian's evaluation value was the lowest at 0.2499, with both the evaluation values of the economic subsystem and social subsystem below 0.2. The main reason for the high bearing capacity of Huxian is the great contributions of the ecological subsystem and the water resources subsystem. Zhouzhi is second only to Huxian due to the great contribution of the ecological subsystem. Overall, the ecological subsystem and water resources subsystems contribute the most to the improvement of the water resources bearing capacity in Xi'an, whereas the economic subsystem and social subsystem's contributions are lower. At present, in the transition of rapid social and economic development, the increase in the residents' living water consumption objectively causes pressure on the water resources bearing capacity, which reduces the contributions of the economic subsystem and social subsystem. At the same time, it is difficult to make a breakthrough in technology concerning water resources utilization and water saving.
- The criterion layer evaluation results show that the evaluation value of the water resources subsystem in urban areas was 0.8150 in 2014. The evaluation values of the Xi'an districts from high to low are as follows: Lantian, Zhouzhi, Huxian, and Gaoling. Gaoling's evaluation value was the lowest at 0.0268, which is correlated to its low precipitation and low water supply comprehensive production capacity. For the economic subsystem, the evaluation values from high to low are as follows: urban areas, Gaoling, Lantian, Zhouzhi, and Huxian. The highest value was that of the urban areas at 0.7334, and the lowest value was that of Huxian at 0.0074. In Huxian, the water demand for production and operation increased with the development of economy; however, in this region, water-use efficiency is very low, causing much pressure on its water resources bearing capacity. For the social subsystem, the evaluation values from high to low are as follows: urban areas, Huxian, Zhouzhi, Gaoling, and Lantian. The highest value was that of the urban areas at 0.8908, and the lowest value was that of Lantian at 0.1721, which is strongly related to its expanding population and increasing per

capita daily water consumption. In regards to the ecological subsystem, the evaluation values from high to low are urban areas, Huxian, Zhouzhi, Lantian, and Gaoling. Gaoling needs further attention and support with the lowest value of 0.19.

6. Recommendations

Based on the above results and discussion, several important recommendations are made below:

For the water resources subsystem, we need to rationally develop, utilize, and protect water resources in order to improve the water resources bearing capacity, and the government should increase its investment in water conservancy construction in order to increase the overall supply of water resources. First, new water resources should be found and explored. In the medium to long term, in addition to the Heihe Water Diversion Project and Lijiahe Reservoir Water Diversion project, the Xi'an government should promote the development of the Northeastern Suburbs Water Resources, Jinghe Dongzhuang Reservoir Water Supply, and Two Rivers Water Transfer project. Second, the protection and governance of water resources should be strengthened. The government needs to protect drinking water sources, promote water-saving technology, and make full use of surface water resources. Meanwhile, there should also be efforts to scientifically and rationally mine water resources and actively collect rainwater resources. Third, the efficiency of urban sewage reuse should be improved, and sewage should be transformed into resources. By improving the urban sewage treatment system by promoting mid-water reuse technology, we are able to use water resources more effectively and comprehensively. The treatment of sewage can lead to an achievement of water quality standards, and it can be used in the fields of urban landscape, agricultural irrigation, industrial cooling water, and public facilities (such as flushing).

For the economic subsystem, first, the provincial government should adjust the industrial structure in order to optimize the industrial layout, giving priority to the development of low water consumption projects and moderately-controlled high water consumption projects. Furthermore, the Xi'an government needs to formulate the industrial layout properly according to the water resources in different regions and avoid developing high water consumption industries in water scarcity areas. In addition, the government should eliminate industries with high water consumption, heavy pollution, and out-of-date technology. Second, we should improve water use efficiency and water saving efficiency in industry. As one of the pillar industries in our economy, industry water consumption is second only to agriculture. In Xi'an, industrial water consumption is increasing with the acceleration of industrialization. Therefore, we should accelerate innovations in production technology and equipment modification in industry to improve the utilization efficiency of industrial water resources and perfect the industrial wastewater treatment facilities to improve the water reuse rate and reduce water waster. At the same time, the government needs to strengthen its supervision of the production process and focus on promoting water saving technology in five high water consumption industries: thermal power, textile, petrochemical, paper, and steel.

For the subsystem, first, residents' water saving consciousness should be raised through information and education. The pressure from urban water resources was primarily linked to population because water demand inevitably increases with an expanding population. At present, the population density of Xi'an has increased from 696 people/square kilometers in 2000 to 854 people/square kilometers in 2014, and the total population has increased from 688 million in 2000 to

815 million in 2014, thereby intensifying water supply pressure. Therefore, the government should reasonably control the population growth so as to alleviate the social pressure of water supply and demand. Meanwhile, residents' traditional water habits need to be changed through information and education. For example, the government can promote water-saving appliances to the public in order to improve the utilization efficiency of water resources and conserve water resources. In addition, in order to achieve the optimal allocation or quasi-optimal allocation of water resources, the government should establish standard water prices according to the scarcity degree of water resources, thus improving the pricing mechanisms of water resources.

For the ecological subsystem, it is important to strengthen the protection and governance of the ecological environment in Xi'an. First, the forest cover rate in Xi'an needs to be increased. Although the forest cover rate has increased to around 50% in the past 10 years, we still need to take measures to increase the forestation area in the future. Meanwhile, the government should strengthen the protection and management of the forest ecosystem, giving full attention to the water storage and water production function of forest resources, which is an effective way to improve the water resources bearing capacity. Second, measures should be taken to strengthen the prevention of water pollution. Although the quality of water resources in Xi'an has improved significantly recently, urban sewage discharge still increases every year, of which second industrial wastewater and domestic sewage discharge occupies a great proportion. Therefore, the government should take relevant measures to strengthen the prevention and governance of wastewater. In terms of prevention, the government should strengthen the supervision of wastewater discharge and prohibit the direct discharge of untreated wastewater. In terms of governance, the government should focus on the protection of the water resources ecological environment and increase its investment in water pollution governance in order to improve the sewage treatment capacity.

7. Conclusions

In this paper, by constructing an evaluation index system including the economy, society, water resources, and ecological environment, we empirically evaluated the water resources bearing capacity of Xi'an in 2000-2014. The conclusions are as follow: 1) The water resources bearing capacity of Xi'an reached a maximum in 2007 and a minimum in 2009, which strongly relates to the fact that Xi'an increased its investment in water conservancy construction and developed a series of policies and measures to protect the environment since the 20th century. 2) The water resources bearing capacity varies across districts, and the evaluation value of the water resources bearing capacities from high to low areas as follows: urban areas, Huxian, Zhouzhi, Gaoling, and Lantian. Lantian requires further attention because it had the lowest evaluation. 3) In general, the water resources bearing capacity increases over time. The contributions of the subsystems from high to low are the water resources subsystem, social subsystem, ecological subsystem, and economic subsystem. 4) The water resources bearing capacity of Xi'an is still at a low level. The possible reasons for this are as follows: the scarcity of water resources, population pressure, insufficient water saving consciousness of residents, increasing per capita living water consumption, and irrational industrial structure.

Acknowledgments

Authors would like to thank the anonymous referees for their constructive comments which substantially improved the presentation of the paper. This research was financially supported by the National Social Science Foundation of research on total factor Productivity of Chinese industry based on energy and environmental constrains (Number: 13BJY037), and Xi'an soft science research project of research on water environmental bearing capacity and restoration measures in Xi'an (Number: SF1505), and Xi'an Jiaotong University Humanities and Social Sciences "Cross-disciplinary category" Fund Project: Two types of social construction research under the background of energy saving, low carbon, environmental protection, NO: sk2014033.

References

[1] National Bureau of Statistics 2015 *China Statistical Yearbook* (Beijing: China Statistics Press)
[2] Falkenmark M and Lundqvist J 1998 *Nat. Resour. Forum* **22** 37-51
[3] Babel M S, Gupta A D and Nayak D K 2005 *Water Resour. Manag.* **19** 693-712
[4] Joardar S D 1998 *Habitat Int.* **22** 327-37
[5] Harris J M and Kennedy S 1999 C *Ecol. Econ.* **29** 443-61
[6] Rijsberman M A and Ven F H M V D 2000 *Environ. Impact. Asses.* **20** 333-45
[7] Varis O and Vakkilainen P 2001 *Geomorphology* **41** 93-104
[8] Wang C J, Feng L H and Lu X Q 2012 *J. Water Resour. Water Eng.* **2** 77-80
[9] Xie Y 2011 Study on comprehensive evaluation of water environmental bearing capacity in Yulin city (Xi'an: Northwestern University)
[10] Shao J 2010 Study on water resources management and sustainable utilization in Xi'an city (Xi'an: Northwestern University)
[11] Li G 2012 The evaluation of water resources bearing capacity in Hubei Province (Hubei: China University of Geosciences)
[12] Duan X G and Luan F F 2014 *Chin. Popul. Resou. Environ.* **24** 119-22
[13] Wang C J 2015 *Ecol. J.* **35** 1-19
[14] Savenije H H G and Zaag P V D 2008 *Phys. Chem. Earth* **33** 290-7
[15] Hwang C L and Yoon K S 1981 *Methods for Multiple Attribute Decision Making* (Berlin: Springer Berlin Heidelberg)
[16] Lu Y 2008 Research on comprehensive evaluation model of harmonious society and analysis of regional economic development (Nanjing: Nanjing University of Information Science & Technology)
[17] Xi'an Municipal Bureau of Statistics 2015 *Statistical Yearbook of Xi'an City* (Shaanxi: Shaanxi Statistics Press)
[18] Shaanxi Provincial Bureau of Statistics 2015 *Shaanxi Statistical Yearbook* (Shaanxi: China Statistics Press)

Treatment of slaughter wastewater by coagulation sedimentation- anaerobic biological filter and biological contact oxidation process

M Sun[1], P F Yu[1,2], J X Fu[1], X Q Ji[1] and T Jiang[1]

[1]School of Municipal and Environmental Engineering, Shenyang Construction University, Shenyang110168, China

E-mail: 571453288@qq.com

Abstract. The optimal process parameters and conditions for the treatment of slaughterhouse wastewater by coagulation sedimentation-AF - biological contact oxidation process were studied to solve the problem of high concentration organic wastewater treatment in the production of small and medium sized slaughter plants. The suitable water temperature and the optimum reaction time are determined by the experiment of precipitation to study the effect of filtration rate and reflux ratio on COD and SS in anaerobic biological filter and the effect of biofilm thickness and gas water ratio on NH_3-N and COD in biological contact oxidation tank, and results show that the optimum temperature is 16-24℃, reaction time is 20 min in coagulating sedimentation, the optimum filtration rate is 0.6 m/h, and the optimum reflux ratio is 300% in anaerobic biological filter reactor. The most suitable biological film thickness range of 1.8-2.2 mm and the most suitable gas water ratio is 12:1-14:1 in biological contact oxidation pool. In the coupling process of continuous operation for 80 days, the average effluent's mass concentrations of COD, TP and TN were 15.57 mg/L, 40 mg/L and 0.63 mg/L, the average removal rates were 98.93%, 86.10%, 88.95%, respectively. The coupling process has stable operation effect and good effluent quality, and is suitable for the industrial application.

1. Introduction

Industrial wastewater and domestic sewage directly or substandard's emissions are the main causes of pollution. Among them, slaughter industry is one of the largest organic pollution sources in China [1]. The national slaughter wastewater's discharge accounts for about 6% of the total industrial wastewater discharge and its proportion is still increasing [2]. Slaughter wastewater is a typical high concentration organic wastewater which has characteristics of large displacement and strong volatility [3]. The concentration of COD is generally 1000 ~ 2000 mg/L, BOD/COD is greater than 0.5, and the biodegradability of wastewater is better. The main pollution indicators were COD, suspended solids, organic nitrogen, TN, NH_3-N, NO_3-N, TP, SO_4^{2-}, sulfide, total alkalinity and Escherichia coli [4].

At present, the treatment of this kind of waste water is mainly based on biological process, including aerobic, anaerobic and facultative treatment systems [5-7]. The front end of the biological treatment process is provided with bailing machine, a rack, an oil separating tank, a regulating pool or a sedimentation tank, to reduce the suspended substance and the oil content in the biological treatment structure, and to ensure the normal operation of the structure.

In order to solve the high concentration of organic wastewater from the slaughter house, the treatment effect and influencing factors of slaughter wastewater by coagulation sedimentation -- AF -- biological contact oxidation process were studied and they provide support for the engineering application.

2. Materials and methods

2.1. Experimental device

The AF reactor is made of Polymethyl methacrylate (PMMA), H=2.88 mm, d=100 mm, two filter columns, the total effective volume is 14.1 L, the water flow is from the upper part of the reactor to the bottom, and the reactor water temperature is between 20 and 25 degrees Celsius. The biological contact oxidation device is made of organic glass, the size of the reactor is 0.5 m*0.5 m*0.5 m, and volume is 100 L, moreover, the reaction chamber is filled with composite filler composed of plastic rings and high fiber bundle. The experimental flow chart is shown in figure 1.

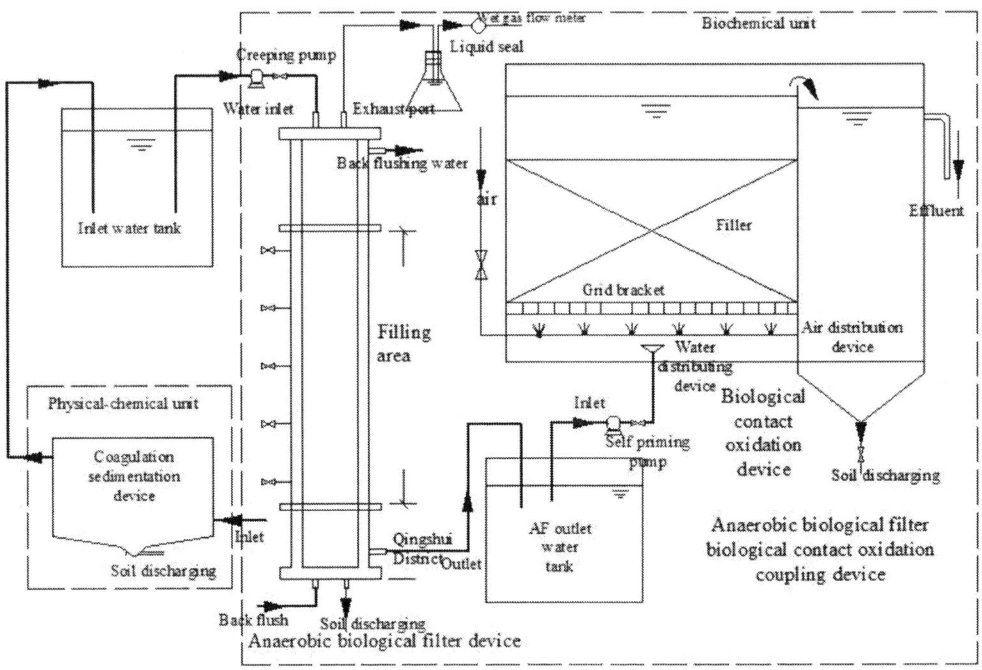

Figure 1. AF -biological contact oxidation combined process.

2.2. Inoculated sludge

The excess sludge of urban sewage treatment plant was inoculated into the sludge of anaerobic biological filter reactor. In the initial stage, adding the flocculant of polyacrylamide to reactor to increase the stability of sludge and the depths of biological film, the concentration of polyacrylamide flocculant is 2-3 mg/L. The bio contact oxidation reactor's sludge was taken from the northern Shenyang wastewater treatment plant, and the mixed liquid suspended solid was 2.539 g/L.

2.3. Experimental water

The experimental water is from a slaughterhouse, and the quality of the water is shown in table 1.

Table 1. Slaughter waste water table.

COD g/L	BOD_5 g/L	NH_3-N mg/L	TN mg/L	PO_4^{3-}-P mg/L	SS mg/L	Grease mg/L	pH	chroma degree	turbidity NTU	coli MPN/L
1.5-2.4	1-1.5	70-95	97-135	8.5-25	500-800	50-150	6.5-7.7	150-250	750-900	2.5×10^4

2.4. Analysis method

Experimental analysis methods are shown in table 2.

Table 2. Analytical methods.

Items	Methods	Instruments
COD	Fast hermetic digestion method (photometric)	WFJ2100 type visible spectrophotometer
NH_4^+-N	Nessler regent spectrophotometric	
TN	Persulfate oxidation-UV spectrophotometry	UV-Vis spectrophotometer
TP	Mo-Sb Anti-spectrophotometry	
chroma	Platinum cobalt standard colorimetric method	WFJ2100 type visible spectrophotometer

3. Results and discussion

3.1. Influencing factors of coagulation sedimentation

3.1.1. Reaction temperature. The removal of COD and turbidity in coagulation sedimentation experiment are shown in figure 2.

From figure 2, the COD and turbidity's removal rate increased with the increase of response temperature [8]. This is because when the water temperature is low, the viscosity of water is larger and the flocculation rate is slow, so it cannot get a better effect in a short time. As the temperature increased rapidly from 8℃ to 16℃, the COD's removal rate increased from 43.2% to 65.8%, and turbidity's removal rate increased from 67.8% to 85.3%, and when the temperature reached 16℃, the removal rate increased slowly. At this time, the influence of water temperature gradually began to weaken. In summary, the suitable water temperature of coagulation sedimentation was 16-24℃, the COD and turbidity's removal rate were stable at 70% and 90%, respectively.

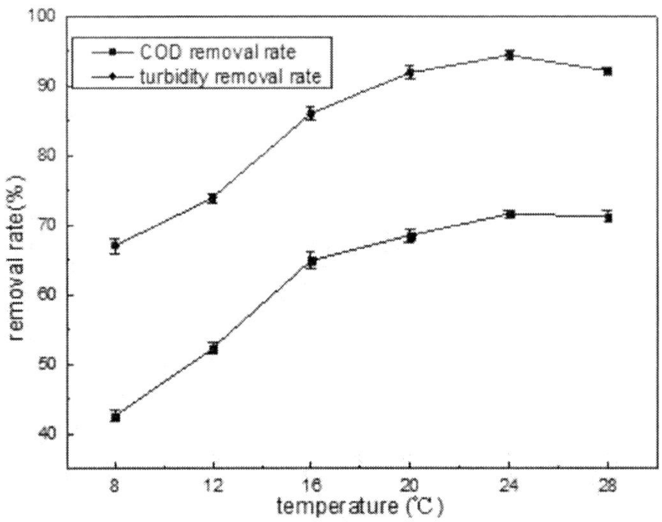

Figure 2. Effect of feed temperature on COD.

3.1.2. Reaction time. The COD and turbidity's removal in coagulation sedimentation experiment are shown in figure 3.

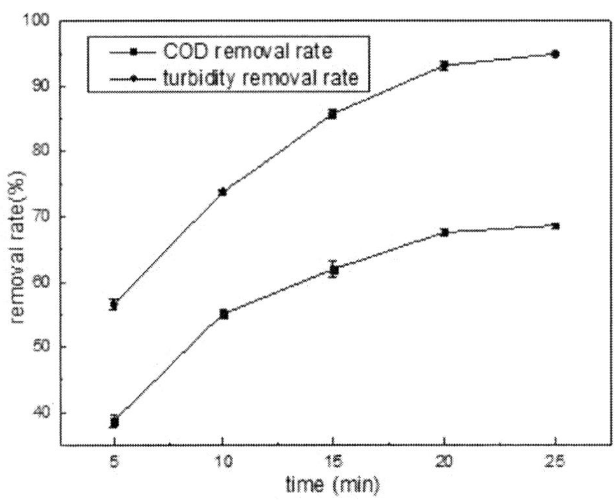

Figure 3. Effect of flocculation time of COD.

From figure 3, the COD and turbidity's removal rate increased with the increase of response time. When the reaction time was 20 min, the COD and turbidity's removal rate were basically stable. So the optimal reaction time was 20 min, and the removal rates were stable at 65.2% and 92.2%, respectively.

3.2. Influencing factors of AF reaction

3.2.1. Effect of filtration rate on the removal of COD and SS. The concentration of influent COD was 850 mg/L, and the influent concentration of SS was controlled by 82 mg/L.

The effect of filtration speed on COD and SS's removals is shown in figure 4.

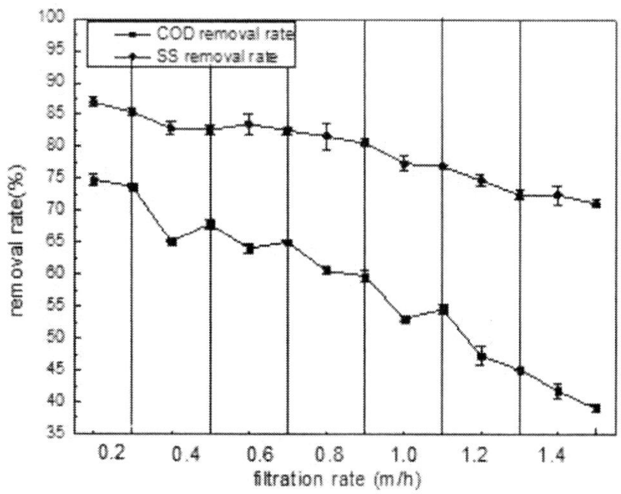

Figure 4. Effect of the filtration rates on COD and SS.

From figure 4, the COD and SS's removal rate decreased with the increase of filtration rate. When the filtration rate ratio was 0.2, best removal effect, but the length of the hydraulic retention time will affect the economic efficiency of sewage treatment. When the filtration rate was 1.4 m/h, the COD's average removal rate was 40.76%, and the removal rate of SS was only 71.84%. The removal rate was mainly due to the interception and adsorption of filter material and microorganism. In summary, the reaction should keep the reaction filter rate of 0.6 m/h, and at this time, the mean value of COD and SS removal rate was 64.42% and 82.55%.

3.2.2. Effect of reflux ratio on the removal of COD and SS. The COD and SS's removals are displayed in figure 5.

From figure 5, COD's removal rate increased with the increase of reflux ratio, and the SS's removal rate decreased with the increase of reflux ratio, and the reason is that the flow of water to make the suspended solid suspended matter has been absorbed into the water body so that the water quality becomes worse. When reflux ratio increased from 300% to 500%, the COD's removal rate increased from 78.4% to 82.5%, and the change of COD removal rate was not obvious. Therefore, select 300% as the test reflux ratio, and the COD and SS's removal rates were 78.40% and 71.22%, respectively.

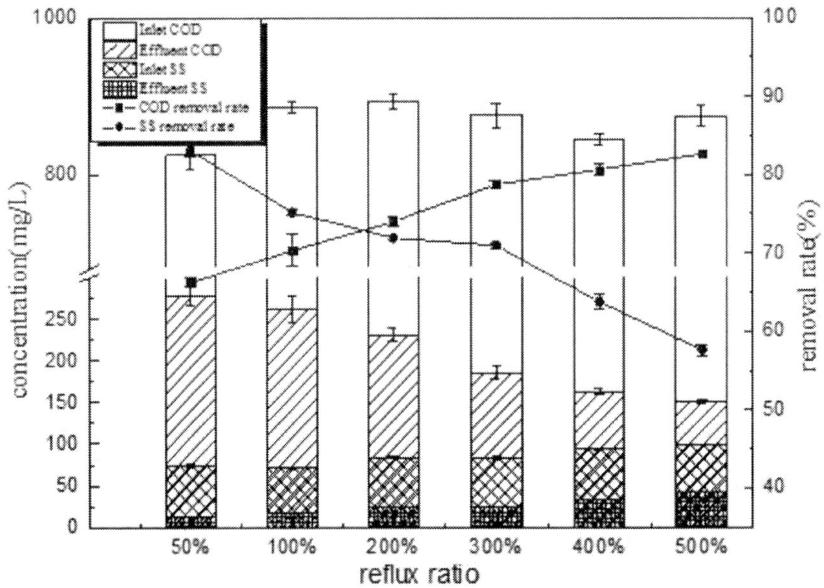

Figure 5. Influence of reflux ratio on SS and COD removal efficiency.

3.3. Influencing factors of biological contact oxidation pool reaction

3.3.1. Effect of biofilm thickness on the removal of COD and NH3-N. The removal efficiency of NH_3-N and COD is shown in figure 6.

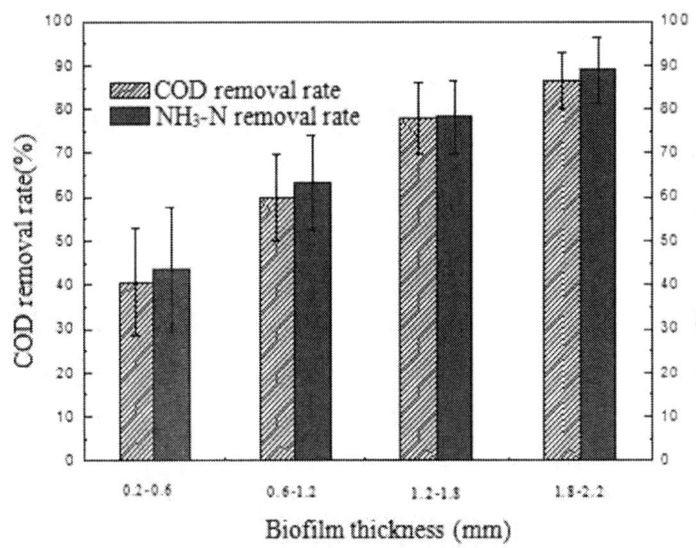

Figure 6. Biofilm characteristics influence on simultaneous nitrification and denitrification.

From figure 6, the COD and NH_3-N's removal rate increased with the increase of biofilm thickness. This is because of the biological membrane with adsorption and carbon storage function. In the early stage of the reaction, some of the organic matter is adsorbed and transferred to the biological membrane, and it is used as a carbon source for denitrification. When the thickness of the biofilm is relatively small, the anoxic environment in the biofilm is not fully formed, and the carbon source is less, which leads to the lack of denitrification ability, and the effect of nitrogen removal is not good. When the biofilm thickness increased from 0.2-0.6 mm to 1.8-2.2 mm, the COD's removal rate increased from 40.67% to 86.45%, and the NH_3-N's removal rate increased from 43.65% to 88.96%. Considering synthetically, the optimum biofilm thickness should be 1.8 mm-2.2 mm, and the removal rates of NH_3-N and COD were 88.96% and 86.45%, respectively.

3.3.2. Effect of gas water ratio on the removal of COD and ammonia nitrogen. Effect of gas water ratio on removal of ammonia nitrogen and COD is shown in figure 7.

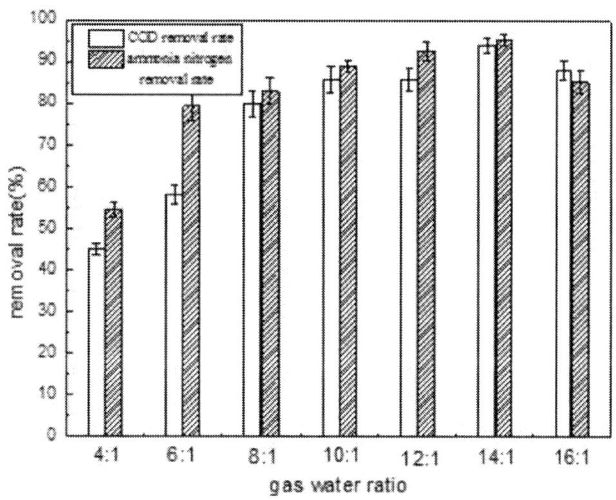

Figure 7. Effect of gas water ratio on ammonia nitrogen and COD.

From figure 7, the average COD's removal rate increased from 44.91% to 93.94% as the gas water ratio increased from 4:1 to 14:1. But after continuing to increase the gas water ratio, the COD's removal rate decreased. The reason is that the gas water ratio is increased, the DO's concentration is gradually increasing in the water, and the removal of organic matter is also gradually strengthened. The nitrification bacteria in the biofilm belong to aerobic bacteria, and the nitrification intensity increases with the increase of the concentration of DO. The biological contact oxidation pool is mainly transmitted through the interface transfer pathway. According to the double film theory [9], the obstruction of the air and hydraulic two-phase membrane decided the oxygen transfer rate, and the concentration of oxygen in aerobic zone will be increased by the increase of the ratio of air to water, and promote the degradation of organic matter by aerobic microorganisms. When the gas water ratio is too large, DO concentration will no longer increase. On the one hand, the amount of aeration is too large, the corresponding power consumption will increase. On the other hand, dissolved oxygen in the water will occur under the strong turbulence, biological membrane on the filler will also occur off, and the water quality of the effluent will be poor [4]. With the increase of gas-water ratio from 4:1 to 14:1, the ammonia nitrogen's removal rate first increased and then decreased. When the gas to water ratio was 12:1--14:1, the concentration of dissolved oxygen was 3.4-4.6 mg/L. At the time, the ammonia nitrogen's removal rate was the highest. This is because the low concentration of dissolved oxygen resulted in the nitration reaction is inhibited, the erosion of biological membranes are small enough, the aging biofilm was not updated, resulting in the decrease of overall activity of biofilm, which is bound to reduce the degradation of biological contact oxidation pool. Therefore, the comprehensive economic and treatment effects, the gas water ratio should not be too high, the most suitable gas water ratio is 12:1-14:1, the COD and ammonia nitrogen's removal rate are 90% and 92.55%.

3.4. Treatment effect of AF- biological contact oxidation process

According to the above experimental conditions, stable operation has lasted for 80 daysto investigate the removal efficiency of COD, TN and TP by AF- biological contact oxidation process.

3.4.1. Removal effect of AF-biological contact oxidation process on COD. The COD's removal by AF-biological contact oxidation process under optimized conditions is shown in figure 8.

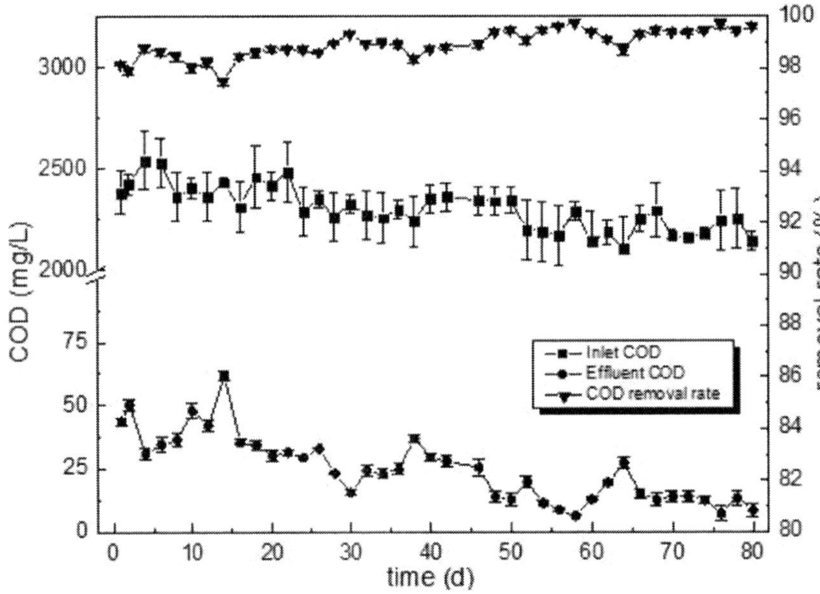

Figure 8. AF - biological contact oxidation coupling process during the stable operation of COD removal efficiency.

From figure 8, we can know that the COD's concentration was about 2400 mg/L in the influent, after physical and chemical treatment and the average value was 836.5 mg/L. After a brief stay in anaerobic biofilter about 3 h-5 h, the treatment of the concentration of COD was about 680 mg/L-800 mg/L, To a certain extent, anaerobic biological filter can improve the biodegradability of the slaughter wastewater, which makes the subsequent biological contact oxidation process be reduced by the impact of the effect of slaughterhouse wastewater treatment [10]. Finally, the removal rate of COD

IOP Conf. Series: Earth and Environmental Science 82 (2017) 012021 doi:10.1088/1755-1315/82/1/012021

was about 98.93%, and the COD was basically stable at about 40 mg/L, the effluent quality was better, and it could meet the discharge standards.

3.4.2. Removal effect of AF-biological contact oxidation process on TN. The TN's removal efficiency by AF- biological contact oxidation process under optimized conditions is shown in figure 9.

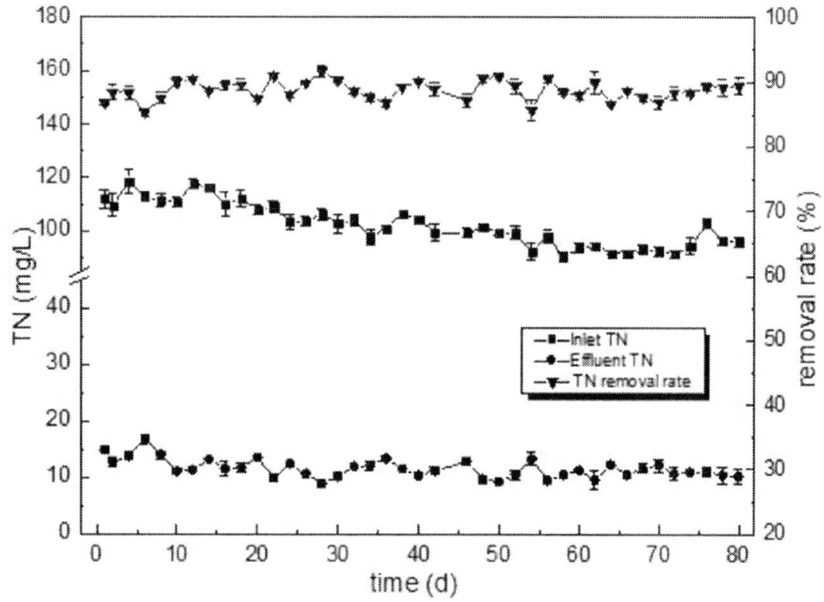

Figure 9. AF - biological contact oxidation coupling process during the stable operation of TN removal efficiency.

From figure 9, we can know that the concentration of TN in the influent was about 110 mg/L, during the anaerobic biological filter, TN's removal rate was low, and the average TN removal rate was about 88.82% and after the process of biological contact oxidation tank, the average TN's concentration in the effluent was about 11.44 mg/L. This is because the anaerobic biological filter process makes the slaughter wastewater biodegradability be further improved, at the same time, it is conducive to the conversion of organic nitrogen in wastewater to NH_3-N, which is conducive to the subsequent nitrification and denitrification.

3.4.3. Removal effect of AF-biological contact oxidation process on TP. The removal efficiency of TP by AF- biological contact oxidation process under optimized conditions is shown in figure 10.

From figure 10, we can see that the average concentration of TP in the waste water is 5.5 mg/L after physical and chemical treatment. Due to the death of their own microorganisms increasing the concentration of TP, anaerobic biological filter's effluent TP removal rate is only about 5%. Therefore, the influent concentration of TP in the biological contact oxidation tank process was essentially flat with the concentration of TP in the slaughter wastewater, and the average TP's concentration was about 5.0 mg/L, but because the anaerobic biological filter can improve the biodegradability of wastewater, and directly affect the operation effect of the process of biological contact oxidation pool, the final average TP's removal rate was about 88.95% and the average effluent TP's concentration was about 0.63 mg/L, so it needs further depth processing to reach the wastewater discharge standards.

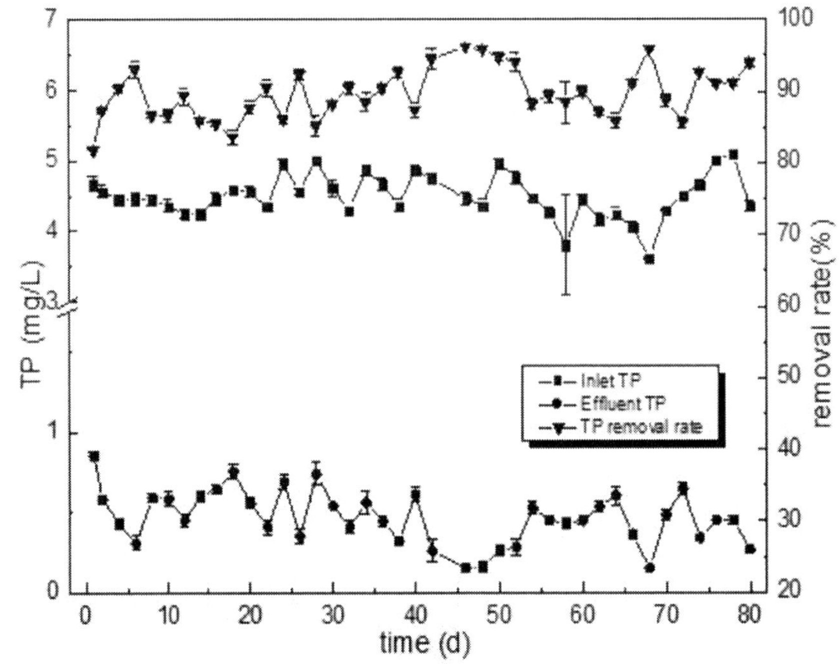

Figure 10. AF - biological contact oxidation coupling process during the stable operation of TP removal efficiency.

4. Conclusion

Coagulation sedimentation experiments determine that the best reaction temperature is 16-24 degrees Celsius and the best reaction time is 20 min. When the reaction pH was about 7, after coagulation and sedimentation treatment, the COD's removal rate and turbidity were 70% and 90%, which reached the ideal physical and chemical treatment effect.

Treatment of slaughterhouse wastewater by AF process, the best reaction filtration rate is 0.6 m/h, after the treatment process by AF, the COD and SS's removal rates were 64.42% and 83.06%. The optimum reflux ratio is 300%; the COD and SS's removal rates were 78.40% and 71.22% under the reaction conditions to achieve the best treatment effect.

When the biological contact oxidation process is treated with separate treatment of slaughter wastewater, the optimum biofilm thickness is 1.8-2.2 mm, the optimum reaction gas-water ratio is 12:1--14:1, and the best treatment effect is achieved under the above conditions.

According to slaughter wastewater treatment under the optimal parameters of AF - biological contact oxidation process, the effluent concentration of COD, TP, TN were 40 mg/L, 0.63 mg/L and 11.44 mg/L. The average removal rates were 98.93%, 88.95%, 88.82% and wastewater treatment effect is very good.

Acknowledgments

The paper is supported by Major Science and Technology Program for Water Pollution Control and Treatment of China (2012ZX07212-001 and 2012ZX07202-011).

References

[1] He Q and Long T Y 1995 Evaluation of slaughter wastewater treatment technology *Chongqing Environ. Sci.* **17(3)** 41-3

[2] Yu F and Chen H B 2005 Treatment technology and application progress of slaughter wastewater *Environ. Sci. Manage.* **30(4)** 84-7

[3] Lin C 2011 *Study on the Production of Slaughter Wastewater Treatment in Suihua Volkswagen* (Harbin: Harbin Institute of Technology)

[4] Wang Q 2013 *Application of Biological Aerated Filter in Advanced Treatment of Slaughter Wastewater* (Jinan: Shandong Jianzhu University)

[5] Deng X Q 2015 *Application of Anaerobic and Aerobic Combined Process in the Treatment of Slaughter Wastewater* (Sichuan: Southwest Jiao Tong University)

[6] Venkata S and Mohan 2006 Low-biodegradable composite chemical wastewater by bioflim configurated sequencing batch reactor (SBBR) *Haz. Mat.* **175(3)** 59-62

[7] Li W G, Zhao Q L and Ma F 2000 Treatment of slaughter wastewater by sequencing batch biofilm reactor *Chin. Wat. Waste* **16(10)** 59-60

[8] Wu Z H 1989 Application and development of low temperature and low turbidity water purification technology *Chin. Wat. Waste* **5** 43-6

[9] Yang C S 2011 Effect of gas/water ratio on SND of biological aerated filter *Anaer Biolog. Filt. Reactor* **8** 63-6

[10] Wu R X, Hu S B and Zhao Y J 2012 Experimental study on the effect of AF+SBR process on slaughterhouse wastewater treatment *J. Agr. Mech. Res.* **7** 173-6

The characteristics and evaluation of water pollution in Ganjiang Tail River

W J Liu[1,2], Z B Li[1], D S Zou[2], C J Ren[3] and Q B Pei[1,3,4]

[1]Xi'an University of Technology 710048, Xi'an, Shannxi Province, China
[2]Planning and Design Institute of Water Resources of Jiangxi Province 330020, Jiangxi Province, China
[3]Nanchang Institute of Technology 330099, Jiangxi Province, China

E-mail: peiqb8310@126.com

Abstract. The water quality in Ganjiang River has an important impact on the ecological environment of Poyang Lake, because Ganjiang River is an important water supply of Poyang Lake. In this paper, the electrical conductivity (ED), turbidity (NTU), suspended solids (SS), total phosphorus (NP), total nitrogen (NT), ammonia nitrogen (NH_4-N), nitrate nitrogen (NO_3-N), and chemical oxygen demand quantity (COD) have been considered as indicators of water quality while performing an assessment of water in Ganjiang River. We evaluated and analyzed comprehensively the quality of surface and underground water by using the Water Quality Identification Index Method. The sample water was retrieved every 50 days from eight monitoring points located in three sections of downstream Ganjiang River in Nanchang city; the study was conducted from September 10, 2015 to June 1, 2016. The results indicate that the pollution index of northern, central, and southern tributaries in Ganjiang River downstream are 3.807, 3.567, and 3.795, respectively; these results were obtained by performing the primary pollutants quality identification index method (PP-WQI); the pollution index for the same tributaries was found to be 3.8077, 3.5003, 3.7465, respectively when we performed comprehensive water quality identification index method (CWQI). The water pollution grades are between level 3 and level 4. The main pollutants are COD, TN, and SS; moreover, there is a linear relationship between the pollution index in groundwater and surface water. The water quality is the best in the central branch, and worst in the south; the water quality is moderate in the north. Furthermore, the water of upstream is better than that of downstream. Finally, the water quality is worst in summer but best in winter.

1. Introduction

There are many evaluation methods to comprehensively assess the quality of river water in our country; water quality assessment methods can be classified into two types: a single factor evaluation method and synthetic evaluation method. Because of their "one-vote veto system", a single factor method can hardly reflect water quality pollution; therefore, it is rarely used by researchers. Water quality is preferably determined using a comprehensive evaluation method [1]. With this method, we can accurately estimate the overall pollution in water that declines its quality. Nemero index method is most frequently used statistical techniques [2], and it based on the combination of weights; this method is integrated with the conventional methods, such as water quality evaluation method [3], comprehensive water quality index method [4] and evaluation based on factor analysis method [5], water quality identification index [6], and other evaluation methods. Among the various evaluation methods used to determine water quality, the most effective one is the water quality identification index method [7, 8]; this method has numerous advantages, such as water quality category, quantitative pollution, water (environmental) functional zone standard, water (environmental) management information; all these parameters can be optimized both qualitatively and quantitatively in order to make a reasonable assessment of water quality. Therefore, the water quality identification index method has been widely used to evaluate the quality of river water [8, 9-11]. Water quality identification index and single-factor identification index method [7], comprehensive water quality identification index method [12, 13], and major pollutants identification index method [14].

In recent years, agricultural and industrial emissions have been increasing with the development of industry and agriculture in Nanchang, resulting in varying degrees of pollution that declines the water quality of Ganjiang River; it is important to note that Ganjiang River of Poyang Lake is an important source of water. It is bound to have a greater impact on the ecology of lower reaches of Poyang Lake. It is important to monitor the changes in Ganjiang River with respect to time and season; with Xu Z X [15] integrated the pollution index for analyzing and evaluating the water quality in Ganjiang Jian River. Then, Hu C [16] combined analytical hierarchy process with comprehensive index method to perform a critical evaluation of Ganjiang River health in Nanchang city of China. Furthermore, Li A P [17] used artificial neural network forecasting method to obtain a better prediction of water quality in Nanchang section of Ganjiang River in China.

In research studies that focus on improving the various prediction methods of water quality, there have been numerous advancements. For example, the grey correlation method is based on the following technologies: computer artificial intelligence [18], analytic hierarchy process [19], artificial neural network method [1], fuzzy mathematics method [20], regression support vector machine, projection pursuit regression method [21], principal component analysis method [13, 15], fuzzy identification method [16], and indicator probability density method [22] All these methods were successfully applied as forecasting methods to predict water quality. Because of this innovative strategy, we can offer ecological protection to Poyang Lake. To analyze the effects of recent changes in Ganjiang River, we used the method of water quality identification index; thus, we analyzed Ganjiang River's water in terms of space and time scales. The results of these analyses helped us compile a scientific report on water quality evolvement and regional ecological management of Poyang Lake.

2. Material and methods

2.1. Regional overview

The Ganjiang River is high in the west and low in the north; it flows northwards when exposed to a mild climate and abundant rainfall; the river flows through a region that can be included in a subtropical monsoon climate zone. Spring and summer are affected by the subtropical monsoon winds, so the air is warm and humid. The air flows in the basin in the South, whereas the Northern cold air often flows over the valley; it is long-lasting current that precipitates as frontal rain over a wide range. In the Ganjiang River basin, precipitation mostly occurs for four consecutive months; however, the monsoon season lasts for about 3–6 months, which is a long period of concentrated rainfall. The average runoff has been constant at 684.2 m^3 since many years (outside the continent mentioned above), whereas the depth of average runoff has been 845.2 mm. Ganjiang Tail river is located in the north of Jiangxi Province and in the southwestern bank of Poyang Lake. This region receives high rainfall in the range 1400–1600 mm, and the average annual runoff is 11.74 billion m^3; however, the average runoff is 941.3 mm. The region is rich in water and soil resources. Besides, it offers excellent services to its citizens, including convenient transportation, strong industrial base, excellent conditions for agricultural production, and an industrial agglomeration degree of significance. The Ganjiang River extends from the south to the north of Nanchang city in the northeast of Nanchang, and it is divided into three branches (North Branch, Middle Branch, and South Branch) that merge with the Poyang Lake to form an excellent network of water resources in the region; therefore, this region is developed in terms of industrial agriculture.

2.2. Sampling program

The surface water quality monitoring and the sampling principle were combined with the actual situation in the North Branch, in the branch, South branch points; thus, a total of 8 sampling points were set up in the section (figure 1). The sampling period extended from September 10, 2015 to June 1, 2016. During this period, five samplings were collected every 50 days. Using a polyethylene bottle, we collected samples from the surface, and the middle and bottom part of fracture surface. Finally, the test was conducted and the weighted average was achieved; the values of test results have been provided in the section of indicator values. In accordance with The Surface Water Environmental Quality Standards (GB3838-2002), we included the following parameters in the index: pH values, conductivity (EC), turbidity, suspended solids (SS), total phosphorus (TP) and total nitrogen (TN), ammonia nitrogen (NH_4–N), nitrate-nitrogen (NO_3–N), and chemical oxygen demand (COD).

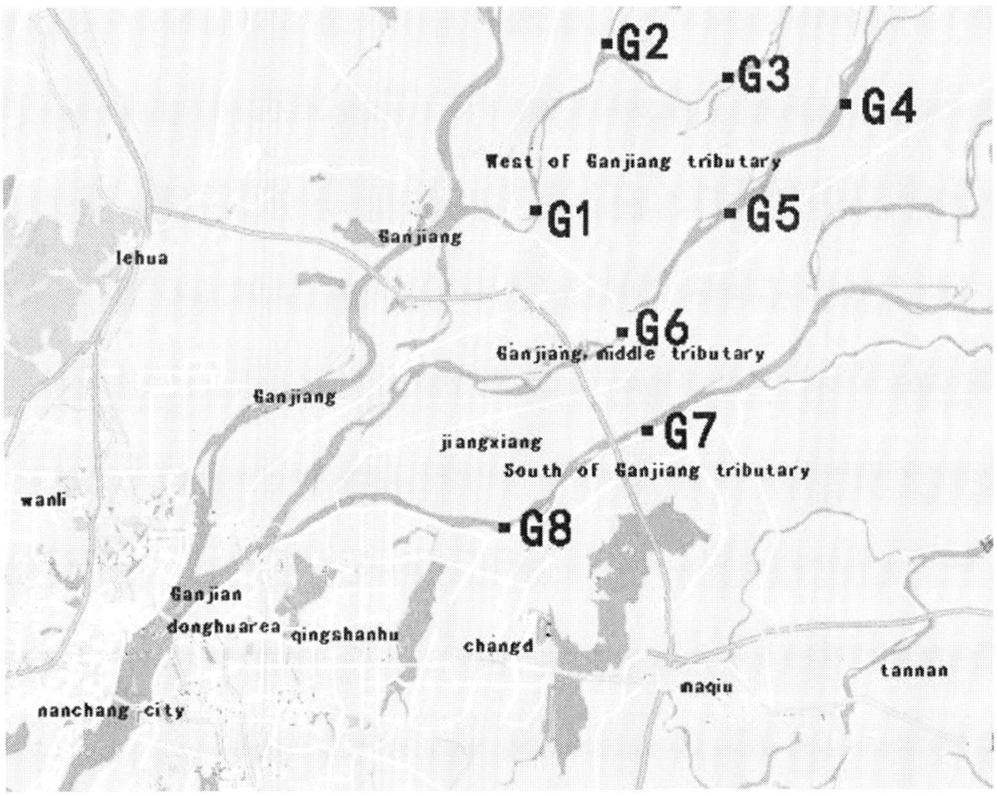

Figure 1. The sample points in the Ganjiang Tail River.

Notes: G1 is North Port station; G-2 is North branch in Tankou village power stations; G-3 is North Hezhou electric power stations of irrigation and drainage; G 4 is Middle branch of Xialou village; G-5 is Middle branch of the South Village; G-6 is Middle of the bridge before the bridge; and G-7 is South Branch of Chubei bridge; G-8 is South Branch of Beiwang bridge.

2.3. Measured values of water quality objectives

The average value of water quality indexes and specific test statistics are shown in table 1: it displays 8 monitoring points and 5 monitoring water quality indicators; both types of points are measured with respect to time and space.

IOP Conf. Series: Earth and Environmental Science 82 (2017) 012022 doi:10.1088/1755-1315/82/1/012022

Table 1. The measured value for water quality monitoring data.

	ED	NTU	SS	TP	TN	NH_4-N	NO_3-N	COD
T1	124.34	23.16	41.83	0.058	0.058	0.709	7.44	103.48
T2	132.78	28.47	47.58	0.302	1.457	1.626	14.90	80.38
T3	108.93	29.22	43.89	0.339	1.599	1.485	15.87	64.29
T4	83.59	27.08	48.90	0.294	0.818	0.972	16.17	61.20
T5	114.89	24.12	67.82	0.375	1.695	1.170	17.92	89.15
G1	99.42	19.90	51.33	0.303	1.019	1.031	13.12	88.58
G2	102.17	30.18	49.67	0.252	0.989	1.032	13.96	89.92
G3	135.83	42.16	67.25	0.278	0.919	0.859	9.13	112.67
G4	97.67	26.89	39.67	0.252	0.993	1.240	13.30	70.42
G5	97.50	24.46	36.42	0.173	0.898	1.343	13.43	67.75
G6	97.92	26.53	41.92	0.221	0.900	1.038	11.75	68.42
G7	116.67	32.86	54.17	0.250	0.910	1.689	13.92	77.17
G8	125.00	28.97	50.00	0.299	0.926	1.312	13.83	64.50
G9	142.88	8.88	16.25	0.200	1.331	1.243	20.72	54.00

Note: G9 is average value of groundwater index of 8 monitoring points. Tables in addition to NTU index is dimensionless units, ED units (uS/cm), but the remaining indicators are in (mg/L). T1 for October 30, 2015, T2 for December 4, 2015, T3 for January 12, 2016, T4 for March 10, 2016, T5 for May 31, 2016.

3. Evaluation methods

According to the method of water quality identification index (WQI) [7], we expressed water quality as a decimal with algebraic properties; they provide a quantitative estimate of water quality at any given point of time. In this method, we first determined the strength of the water monitoring data with respect to all the participating factors using a single-factor water quality identification index. Secondly, we considered the Surface Water Environment Quality Standard (GB3838-2002) and the classification of water quality standards, Water Function Zone Division Standard (GB/T50594-2010) to develop a functional classification standard of water; then, we calculated water quality identification index.

3.1. Single factor water quality identification index method

The method of single-factor water quality identification index (SWQI) [7] reflects the following parameters: water quality categories (X_1), water quality category of interval (X_2), and the relationship between individual indicators and water (environmental) function areas (X_3). Three digits, including an integer, two or three digits after the decimal point could be shown as follows:

$$P_i = X_1.X_2X_3 \tag{1}$$

$\boldsymbol{P_i}$ is the identification index of water quality indicators; $\boldsymbol{X_1}$ is the purpose of water quality indices of water quality categories; $\boldsymbol{X_2}$ is the monitoring data of $\boldsymbol{X_1}$ (Water quality changes in the position, according to the rounding principle); and $\boldsymbol{X_3}$ is the comparison with ribbon category for the water category.

3.2. Primary pollutant-water quality identification index

Primary Pollutant-Water Quality Identification Index (PP-WQI) [14] is a modification of SWQI, and it reflects the major classes of pollutants defined by water quality parameter X_4.

$$PP-WQI = X_1X_2X_3(F) \tag{2}$$

wherein, $PP-WQI$ is the monitoring data that represents the water quality with respect to main

IOP Conf. Series: Earth and Environmental Science **82** (2017) 012022 doi:10.1088/1755-1315/82/1/012022

pollutants in the index of water body identification; X_1 and X_2 are the parameters that have been described above in the single-factor water quality identification index; the difference is X_3 and it is used to provide an overall assessment of water quality indicators, below the number of categories of water quality standards and water quality indicators; For F = the monitoring data of main pollutants in water ($P_i = X_1X_2$ Maximum water quality indicators).

3.3. Comprehensive water quality identification index method

Comprehensive water quality identification indexes (CWQI) [16] are defined as follows: the average of the total representative of water quality and zoning comparison; the participation is inferior to the Ribbon among the overall indicators of water quality, which are identified by the evaluation standard number of water quality objectives.

$$CWQI = X_1X_2X_3X_4 \tag{3}$$

$$X_1X_2 = \frac{1}{n}\sum_{i=1}^{n} X_1X_2 \tag{4}$$

$CWQI$ represents all participants in the evaluation system of comprehensive results in X_1, X_2. With the above equations (3) and (4), we obtained different results after X_1X_2 calculation as compared to those obtained with single factor water quality identification index; this is because the evaluation method of X_1X_2 is different in this case. On an average, X_3 refers to the participants in the comprehensive evaluation of water quality indicators, so it is inferior to the single index number of targets in the Ribbon of water environment; X_4 refers to the comprehensive water quality categories defined after comparing the results of the water of the Ribbon category; the participation emphasizes on the single factor identification index P_i in X_3 Not for 0 To determine the number.

4. Results and discussion

According to Surface Water Environment Quality Standard (GB3838-2002) and the Water Function Zone Division Standard (GB/T 50594-2010), we defined the classification standard of water quality. As per the grading standards, we defined the classification standard values of water quality index in table 2.

Table 2. Classification standard values of water quality index.

Water quality type	I	II	III	IV	V	Bad V
ED	400	1000	1500	2000	3000	3500
NTU	15	20	25	35	50	80
SS	20	25	30	60	150	200
NP	0.02	0.1	0.2	0.3	0.4	0.5
NT	0.2	0.5	1.0	1.5	2.0	2.5
NH_4-N	0.15	0.5	1.0	1.5	2.0	4.0
NO3-N	10	15	20	25	35	50
COD	15	15	20	30	40	80

Table 3 shows that all monitoring points of ED meet class requirements; moreover, NO_3-N is somewhat greater than lever III of water quality standards. In addition, NH_4-N and TN class III meet water quality standards. It is important to note that TP and TN represent light pollution, whereas NTU and SS indicate more serious pollution, while mean COD is an indicator of worst pollution.

Table 3. The water quality of the monitored data of the corresponding category.

	ED	NTU	SS	TP	TN	NH_4-N	NO_3-N	COD
T1	I	IV	IV	II	I	II	I	Bad V
T2	I	IV	IV	IV	IV	IV	II	V
T3	I	IV	IV	IV	V	III	III	IV
T4	I	IV	IV	IV	III	II	III	IV
T5	I	IV	V	V	V	III	III	V
G1	I	III	IV	IV	IV	III	II	V
G2	I	IV	IV	IV	III	III	II	V
G3	I	IV	V	IV	III	II	II	Bad V
G4	I	IV	IV	IV	III	III	II	IV
G5	I	IV	IV	III	III	III	II	IV
G6	I	IV	IV	III	III	III	II	IV
G7	I	IV	IV	IV	III	IV	II	V
G8	I	IV	IV	IV	III	III	II	IV

4.1. Single factor water quality identification method for evaluating results

According to sections 2.1–2.3, we combined water quality identification index calculation method and the results of monitoring data to provide an evaluation of water quality. Table 4 presents a complete evaluation of results by single factor water quality index method.

Table 4. Single factor water quality identification index method for evaluating results.

	ED	NTU	SS	TP	TN	NH_4-N	NO_3-N	COD	WQI	PP-WQI	CWQI
T1	1.30	4.11	4.41	2.50	1.30	2.40	1.90	6.13	3.17	3.03(COD)	3.030
T2	1.30	4.31	4.61	4.51	4.91	4.31	3.00	5.42	4.06	4.06(COD)	4.063
T3	1.30	4.41	4.51	4.91	5.12	4.00	3.10	4.31	3.93	4.05(TN)	3.952
T4	1.20	4.31	4.61	4.41	3.60	2.90	3.10	4.11	3.58	3.54(SS)	3.541
T5	1.30	4.11	5.12	5.22	5.22	3.30	3.60	6.04	4.15	4.25(COD)	4.252
G1	1.20	3.60	4.71	4.51	4.02	3.10	2.70	5.92	3.77	3.74(COD)	3.741
G2	1.30	4.41	4.71	4.02	4.00	3.10	2.90	6.02	3.85	3.84(COD)	3.841
G3	1.30	5.01	5.12	4.31	3.80	2.70	2.20	6.33	3.86	3.84(COD)	3.841
G4	1.20	4.31	4.31	4.02	4.00	3.50	2.80	4.71	3.67	3.64(COD)	3.641
G5	1.20	4.11	4.21	3.50	3.80	3.70	2.80	4.51	3.57	3.53(COD)	3.430
G6	1.20	4.21	4.41	3.80	3.80	3.10	2.50	4.61	3.54	3.53(COD)	3.430
G7	1.30	4.51	4.81	4.02	3.80	4.41	2.80	5.12	3.88	3.85(COD)	3.852
G8	1.30	4.41	4.71	4.51	3.90	3.60	2.80	4.31	3.73	3.74(SS)	3.641

4.2. Comparison of different methods for the evaluation of water quality

To determine how different evaluation methods affected the evaluation of the water quality of Ganjiang River, we applied the method of weighted average pollution index on water (environmental) functional area (Pollution Index method, PIM) [1]. We compared the results obtained by PMI method with those obtained using the following methods: single factor index method (SFCI) [1], Nemero index method (NI) [2], major pollutant index method (MPI) [16], single factor water quality identification index (WQI), main pollutants water quality identification index method (PP-WQI), and comprehensive water quality identification index method (CWQI). As shown in table 5, there are seven ways to evaluate the water quality of Ganjiang River. The relationship between groundwater and surface water pollution index is shown in table 5.

Table 5. Assessment of different methods on the water quality of Ganjiang River.

Sampling points	PIM	SFCI	NI	MPI	WQI	PP-WQI	CWQI
T1	0.94	1.72	1.39	4.00	3.17	3.03(COD)	3.0300
T2	1.08	1.59	1.36	5.38	4.06	4.06(COD)	4.0630
T3	1.06	1.60	1.36	5.27	3.93	4.05(TN)	3.9520
T4	0.92	1.63	1.32	4.70	3.58	3.54(SS)	3.5410
T5	1.17	2.26	1.81	5.63	4.15	4.25(COD)	4.2520
G1	1.02	1.71	1.41	4.95	3.77	3.74(COD)	3.7410
G2	1.04	1.66	1.38	5.07	3.85	3.84(COD)	3.8410
G3	1.17	2.24	1.79	5.12	3.86	3.84(COD)	3.8410
G4	0.97	1.33	1.16	4.80	3.67	3.64(COD)	3.6410
G5	0.91	1.29	1.11	4.63	3.57	3.53(COD)	3.4300
G6	0.92	1.40	1.18	4.60	3.54	3.53(COD)	3.4300
G7	1.10	1.81	1.50	5.12	3.88	3.85(COD)	3.8520
G8	1.03	1.67	1.38	4.92	3.73	3.74(SS)	3.6410
North Branch average	1.08	1.87	1.53	5.04	3.83	3.807	3.8077
In the average	0.93	1.34	1.15	4.68	3.59	3.567	3.5003
South Branch average	1.07	1.74	1.44	5.02	3.80	3.795	3.7465

As shown in table 5, the results of seven evaluation methods of water quality are as follows: three-section North Branch (G1, G2, G3), Medium (G4, G5, G6), and South (G7, G8). Using the weighted average pollution index method (PIM), we obtained various results for class III water and good quality water. In the single-factor category index method (SFCI), the worst water quality indicators define the water quality class according to the evaluation criteria of class III water and moderately polluted water; the results are statistically analyzed by nemero index method (NI). According to these criteria, the river water sample belongs to class III water and is slightly polluted; dyeing index method of main pollutants (MPI) analyzed water of grade IV according to their criteria. In addition, there are three water quality identification index (WQI, PP-WQI, CWQI). The evaluation results are close to each other, because all samples are identified as class III water and the deviation is minor (X2 = 0.7); the samples are very similar to water of grade IV, based on the main pollutants water quality identification index method (PP-WQI). To indicate the main pollutants in water, we first consider COD and then TN and SS. G9 shows that groundwater belonged to class II water quality as it included only pollutants of TN.

The method of single factor water quality identification index was used to evaluate various categories of water quality, because it can also reflect the gap between water quality categories and targets of the Ribbon. Integrated water identifies index evaluation method can fully quantify the integrated water information. Furthermore, PP-WQI method is based on water quality identification index method, and it is introduced to reflect the parameters of main pollutants in river water. Thus, it reflects integrated pollution degree of water, and it also reflects the causes of water pollution with its main pollution index. Thus, both it can provide both qualitative and quantitative evaluations. Moreover, it comprehensively reflects the pollution status in water; therefore, we recommend PP-WQI method as the preferred method of water quality evaluation.

Based on time analysis, we present various evaluation methods in table 5: T5 (2016-5-31 in summer) is the period with the highest pollution, followed by T2 and T3 (12-4 and 1-12 in winter) as the second highest pollution periods, T4 (3-10 III) has moderate pollution levels, and T1 (10-30 autumn) has the least pollution. This is because in the summer season, the rainfall is much greater in Ganjiang River of Nanchang city; with a continuous increase in rainfall, various rainwater streams meet into Ganjiang River. The rainwater brings a lot of pollutants into the river. As the relative concentration of pollutants increases sharply, the water quality becomes poor. Consequently, the pollution index increases tremendously. In contrast, the rainfall is less in autumn; with a steady decline in rainfall, the volume of runoff in Ganjiang River also reduces. Moreover, as the rainwater streams reduce into the Ganjiang River, the pollution content reduces very sharply. Thus, as the relative concentration of pollutants decreases, the pollution index also becomes smaller.

Based on space analysis, we consider WQI method as an example: the North Branch G1 Monitoring pollution indexes (Port) 3.77 are less than the two branches and raw materials. Moreover, G2 (Tankou village power station) and G3 (Hezhou electric power stations of irrigation and drainage) monitoring pollution index is 3.85 and 3.86, respectively; G6 Monitoring points (Bridge before the bridge) Pollution index of 3.54 is less than G4 Monitoring points (Xialou village) Pollution index of 3.67. Furthermore, South Branch G8 Monitoring points (Beiwang Bridge) Pollution index of 3.73 is less than G7 Monitoring points (Chubei Bridge) Pollution index of 3.88. The water quality declines from upstream to downstream in space: North-the best, worst is South Center. Further analysis has shown more dredges and raw materials. Some farmers on the coast breed pension ducks, and usually these ducks move into the river for food. Thus, the water quality becomes poor when these ducks thrive into the river.

5. Conclusions

The main conclusions of this study are as follows:

- Water pollution occurs at different time-periods of the year: pollution is highest in summer, but moderate in winter; however, autumn is the season that experiences minimum pollution; the pollution levels increase from upstream to downstream in space: North is the best with least pollution; central region has moderate pollutions, and South Center has the worst pollution.
- The weighted average results of evaluation were obtained by the pollution index method (PIM); the results of main pollutants dye index (MPI) were poor.
- In the Ganjiang River, the main pollutants that decline water quality in the Lake were accounted by COD; the remaining pollutants were accounted by TN and SS.
- Under the influence of surface water, the pollution index of groundwater and surface water is linear.

Recommendation: By comparing seven methods of water quality assessment, we developed an identification index of major pollutants that precisely reflect the comprehensive water pollution levels and the causes of water pollution, which are also known as the major pollution indicators. Both qualitative and quantitative evaluations can comprehensively reflect the pollution of water provided we perform an integrated assessment of water quality in rivers; PP-WQI method was found to be the preferred method of water quality assessment. The main pollutants of Ganjiang Tail River are quantified by COD and TN. This indicates that the river water contains higher levels of organic matter; therefore, the Government should invest more resources in eliminating these sources of organic pollution: pesticide should be sprayed judiciously and there should strict supervision to control the effluents released by chemical plants in the region.

Acknowledgments

This study was conducted as a project of Science and Technology Support Program of Jiangxi Province Water Conservancy (15B016) and Department of Science and Technology in Jiangxi Province Key Research and Development Program (20151BBF60012).

References

[1] Yin H L and Xu Z X 2008 Comparative study on the comprehensive evaluation method of water quality of the river *Resource and Environment in the Yangtze Basin* **5** 729-33
[2] Wang W J, Gao P C, Li J, *et al* 2016 Tributary of the typical storage evaluation and trend

analysis of water quality of Danjiangkou reservoir *Water Resour. Protect.* **3** 127-30

[3] Jing J L, Huang H M and Wei Y M 2004 Water quality evaluation model based on combination weights *J. Hydro-Electric Power* **3** 13-9

[4] Zhou X L 2012 Comprehensive water quality of Dishui lake water quality evaluation index method *Water Purif. Technol.* **2** 66-71

[5] Li W S 2011 Water quality evaluation based on factor analysis method and its application *J. North Univ. (Nat. Sci. Ed.)* **2** 207-11

[6] Zhang H L, Qi S, Lu Q Q, *et al* 2014 Water function regionalization of agricultural non-point source pollution control in Hubei Province *Soil Water Conser. Sci. China* **2** 1-8

[7] Xu Z X 2005 Study on the evaluation method of the single factor water quality identification index in China *J. Tongji Univ. (Nat. Sci. Ed.)* **3** 321-5

[8] Mao F J, He Y L, Xu Z M, *et al* 2014 Based on single factor water quality identification index for evaluation of water quality of Dongjiang Heyuan *J. Safety Environ.* **5** 327-31

[9] Sun W G, Xing J, Ma Y, *et al* 2010 Single factor water quality identification index evaluation method in the evaluation of water quality in a river basin *Environ. Sci. Manage.* **11** 181-4

[10] Fan Z F, Wang L Q, Chen L X, *et al* 2009 Water quality identification index method used in Dianshan Lake water quality assessment *J. Shanghai Marit. Univ.* **3** 314-20

[11] Fu T Y, Zou Z H and Wang X J 2014 Based on multivariate statistics and Liaoyang Taizi River water quality assessment study of water quality identification index *J. Environ. Sci.* **2** 473-80

[12] Hu C and Su D 2011 Application of comprehensive water quality identification index method in evaluation of water quality of Hunhe River *J. Ecol. Environ.* **1** 186-92

[13] Li G F, Liu X B, Liu Z G, *et al* 2011 Based on principal component analysis and water quality identification index for evaluation of water quality of main rivers in Tianjin area *J. Ecol. Rural Environ.* **4** 27-31

[14] Yang L, Song F, Song B, *et al* 2015 Application of water quality identification index method of main pollutants in river water quality assessment *Environ Sci and Technol* **11** 239-45

[15] Xu Z X 2005 Study on the evaluation method of the comprehensive water quality identification index of rivers in China *J. Tongji Univ. (Nat. Sci. Ed.)* **4** 482-8

[16] Hu C and Su D 2011 Application of comprehensive water quality identification index method in evaluation of water quality of Hunhe River *J. Ecol. Environ.* **1** 186-92

[17] Li A P 2013 Study on assessment and protection countermeasures analysis of water quality of Ji'an Section of Ganjiang River (Jiangxi: Nanchang University)

[18] Jia R 2012 Study on health assessment of Nanchang Section of Ganjiang River (Jiangxi: Nanchang University)

[19] Peng X L 2005 Study on water quality prediction model of Nanchang Section of Ganjiang River Water Source (Jiangxi: Nanchang University)

[20] Liu B, Xiao C L, Ran T H, *et al* 2013 Grey correlation and the application of analytic hierarchy process in evaluation of underground water – A case study of Jilin city *Water-Saving Irrig.* **1** 26-9

[21] Lu W X, Li D, Zhang L, *et al* 2011 Application of fuzzy comprehensive evaluation based on analytic hierarchy process in water quality evaluation *Water-Saving Irrig.* **3** 43-6

[22] Fang C S, Meng H, Dan Y S, *et al* Based on GIS Fuzzy evaluation of underground water quality in Jilin province 2011 *J. Jilin Univ. (Earth Sci.)* **S1** 293-7

IOP Conf. Series: Earth and Environmental Science **82** (2017) 012023 doi:10.1088/1755-1315/82/1/012023

Simple method for quick estimation of aquifer hydro-geological parameters

C Ma[1,2,5] and Y Y Li[3,4]

[1]Powerchina Guiyang Engineering Corporation Limited, Guiyang 550081, China
[2]Hydrochina Guiyang Engineering Corporation Geotechnical Engineering Company Limited, Guiyang 550081, China
[3]School of Environmental Science and Engineering, Chang'an University, Xi'an, Shanxi Province 710054, China
[4]Key Laboratory of Subsurface Hydrology and Ecological Effect in Arid Region of Ministry of Education, Chang'an University, Xi'an, Shanxi Province 710054, China

E-mail: mc9066@163.com

Abstract. Development of simple and accurate methods to determine the aquifer hydro-geological parameters was of importance for groundwater resources assessment and management. Aiming at the present issue of estimating aquifer parameters based on some data of the unsteady pumping test, a fitting function of Theis well function was proposed using fitting optimization method and then a unitary linear regression equation was established. The aquifer parameters could be obtained by solving coefficients of the regression equation. The application of the proposed method was illustrated, using two published data sets. By the error statistics and analysis on the pumping drawdown, it showed that the method proposed in this paper yielded quick and accurate estimates of the aquifer parameters. The proposed method could reliably identify the aquifer parameters from long distance observed drawdowns and early drawdowns. It was hoped that the proposed method in this paper would be helpful for practicing hydrogeologists and hydrologists.

1. Introduction

In many watersheds, groundwater is one of the major sources of water supply for industrial, municipal and agricultural users [1]. However, in many parts of the world, groundwater resources have been overexploited [2]. They have been utilized at excessive rates that lead to various harmful environmental side effects that seriously hinder the development of national economy [3]. It is important for water managers and engineers to assess and calculate groundwater quantity accurately and reliably.

Aquifer hydro-geological parameters are important in groundwater resources assessment and in groundwater dynamic prediction. They are usually obtained from the data of unsteady pumping test, which is the main way to determine the aquifer parameters such as transmissivity and storage coefficient [4-6]. The first study to determine the aquifer parameters was analytically derived by Theis [7]. Later, a lot of methods based on the Theis equation were proposed to estimate the hydro-geological parameters by many scientists [8-11].

At present, in view of the non-steady flow pumping test, the Theis curve method, the straight-line method and the intelligent optimization algorithm are widely used to estimate aquifer parameters.

However, these methods have some deficiencies in practice. The main problem of the Theis curve method is that the optional is larger when the coincident points of the measured curve and the theoretical curve are less. Because drawing line and reading data may have artificial arbitrariness, the straight-line method [12] will have a certain influence on the precision of the calculation results and it only can be used for the variable u is small [13]. With the development of computer technology, the intelligent optimization algorithm [14-16] is widely applied to determine the aquifer parameters. However, the calculation process of the intelligent optimization algorithm is complex and its convergence property cannot be always fully guaranteed.

This paper is organized as follows. First, a fitting function of Theis well function is proposed using fitting optimization method when the independent variable u is in the range of 0.01 to 1.1. Second, a unary linear regression equation is established to estimate the aquifer parameters (transmissivity and storage coefficient). Then two examples are performed and their results are presented to demonstrate the advantages of the proposed method. Finally, conclusions are drawn.

2. Fitting function of the Theis well function

The well-known Theis solution, which describes the observation drawdown under the condition of homogenous, isotropy and infinitely extent aquifer, is

$$s = \frac{Q}{4\pi T} \mathrm{W(u)} \tag{1}$$

where s is groundwater drawdown observed in the well, [L]; Q is constant discharge rate, [L^3T^{-1}]; T is transmissivity coefficient of aquifer, [L^2T^{-1}]; $W(u)$ is dimensionless Theis well function which can be written as

$$W(u) = \int_u^\infty \frac{1}{y} \exp(-y) dy \tag{2}$$

in which u can be written as

$$u = \frac{r^2 S}{4Tt} = \frac{r^2}{4at} \tag{3}$$

where r is the radial distance from pumping well, [L]; S is the storage coefficient of aquifer, dimensionless; t is time, [T]; a is the pressure conduction coefficient of aquifer, [L^2T^{-1}].

The scatter diagram of Theis well function is shown in figure 1. Because the well function can't be integrated directly, we need to find out an approximate function to simplify the calculation. Equation (2) can also be written as

$$W(u) = -0.5772 - \ln u + \sum_{n=1}^{\infty} (-1)^{n+1} \frac{u^n}{n \cdot n!} \tag{4}$$

where after the first two items are alternating series. According to the nature of the alternating series, when u is small (u<0.01), that is to say, for small radial distance, or large time, equation (4) reduces to

$$W(u) \approx -0.5772 - \ln u \tag{5}$$

The relative error involved in adopting equation (5) instead of equation (4) to compute the observation well drawdown by equation (1) is less than 2.03%.

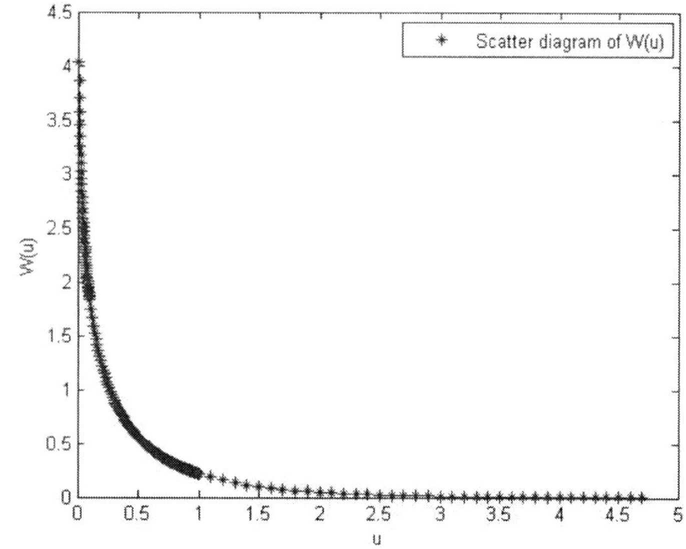

Figure 1. The scatter diagram of the Theis well function.

However, if u is large (u >0.01), equation (5) cannot be used to approximately instead of Theis well function. In this paper, as the independent variable u is within the scope of 0.01 to 1.1, through fitting and optimizing for the Theis well function with the residual sum of squares are minimum, the approximate function of Theis well function is given as

$$W(u) \approx \begin{cases} 11.61282u^{-0.06619373} - 11.69514 & 0.01 \le u \le 0.26 \\ 0.6463112u^{-0.6023161} - 0.4250883 & 0.26 \le u \le 1.1 \end{cases} \tag{6}$$

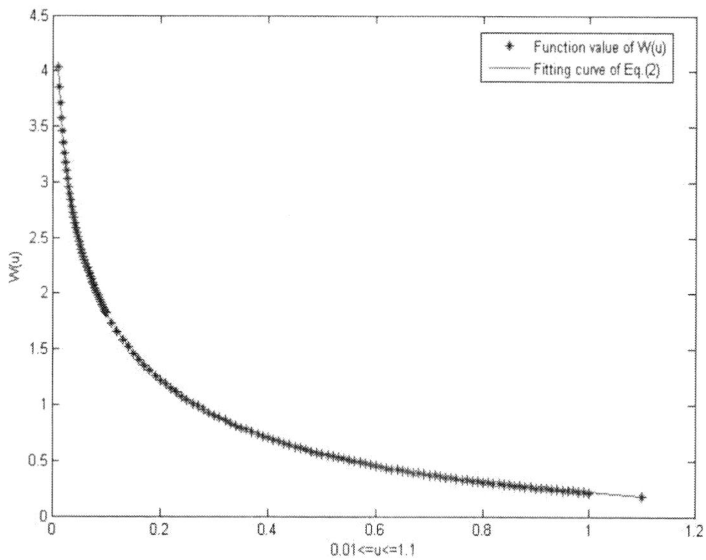

Figure 2. The fitting curve of Theis well function.

As u is within the scope of 0.01 to 1.1, the values of equations (4) and (6) are given in table 1 along with the relative error between Theis well function and the fitting function. As shown in this table, the biggest relative error is less than 1.33%. The fitting curve of Theis well function is shown in figure 2.

It is clear that equation (6) has a high alternative precision and it can satisfy the actual engineering calculation accuracy requirements.

Table 1. Comparison results of equations (4) and (6).

u	The value of equation (4)	The value of equation (6)	Relative error	u	The value of equation (4)	The value of equation (6)	Relative error
0.01	4.0379	4.056475085	0.46%	0.27	0.9849	0.997045456	1.23%
0.012	3.8573	3.867518545	0.26%	0.28	0.9573	0.966232585	0.93%
0.014	3.7054	3.709527591	0.11%	0.29	0.9309	0.937134133	0.67%
…	…	…	…	…	…	…	…
0.24	1.0726	1.068183148	0.41%	0.89	0.2647	0.268217654	1.33%
0.25	1.0443	1.033741162	1.01%	1	0.2194	0.221222956	0.83%
0.26	1.0139	1.000737744	1.30%	1.1	0.186	0.185165148	0.45%

3. Proposed method

In this paper, the fitting function of the Theis well function is used to simply estimate the aquifer parameters. Equation (6) can be transformed into

$$W(u) = Au^{C} + B \tag{7}$$

where A , B and C are constant value. The substitution of equations (3) and (7) into equation (1) gives

$$s = \frac{Q}{4\pi T}(A(\frac{r^2}{4at})^{C} + B) = \frac{AQ}{4\pi T}(\frac{r^2}{4a})^{C}(\frac{1}{t^{C}}) + \frac{BQ}{4\pi T} \tag{8}$$

by defining variables as

$$k = \frac{AQ}{4\pi T}(\frac{r^2}{4a})^{C} \tag{9}$$

$$b = \frac{BQ}{4\pi T} \tag{10}$$

$$x = \frac{1}{t^{C}} \tag{11}$$

Equation (8) can be rewritten succinctly as

$$s = kx + b \tag{12}$$

It is clear that equation (12) is a linear equation that its independent variable is x and dependent variable is s. The aquifer parameters can be estimated by calculating the coefficients of equation (12). Hence, according to the pumping test data $(s_i, x_i)(i = 1, 2, \cdots, n)$, using the least square method, k and b can be estimated by

$$\begin{cases} \hat{k} = \dfrac{\sum_{i=1}^{n}(x_i - \bar{x})(s_i - \bar{s})}{\sum_{i=1}^{n}(x_i - \bar{x})^2} \\ \hat{b} = \bar{s} - \hat{k}\bar{x} \end{cases} \tag{13}$$

where $\bar{x}$ and $\bar{s}$ are average value that can be expressed as

$$\bar{x} = \frac{1}{n}\sum_{i=1}^{n} x_i \tag{14}$$

$$\bar{s} = \frac{1}{n}\sum_{i=1}^{n} s_i \tag{15}$$

Finally, the aquifer parameters can be obtained by substituting equation (13) into equations (9), (10) and (11), leading to

$$T = \frac{BQ}{4\pi b} \tag{16}$$

$$a = \frac{r^2}{4}\left(\frac{Ab}{Bk}\right)^{\frac{1}{C}} \tag{17}$$

$$S = \frac{T}{a} \tag{18}$$

4. Error and reliability criteria

To evaluate the reliability of the calculated aquifer parameters or the goodness-of-fit of the calculated drawdown to the observed drawdown, two different criteria, the mean error (ME) and the standard error of estimate (SEE) criteria, are used to computed the errors between the predicted and the observed drawdown. The ME is given by

$$\mathrm{ME} = \frac{1}{n}\sum_{j=1}^{n} (s - s_e)_j \tag{19}$$

The SEE is given by

$$\mathrm{SEE} = \left(\frac{1}{(n-2)}\sum_{j=1}^{n} (s - s_e)_j^2\right)^{0.5} \tag{20}$$

where n is total drawdown number (nondimension), s is observed drawdown [L], s_e is estimated drawdown [L]. ME gives the normality of distribution about the deviation from the zero mean, and SEE shows the presence of outliers in the data. Low values of these criteria indicate that the calculated aquifer parameters are reliable.

5. Application

In this section, two sets of pumping test data are chosen, one for the radial distance from pumping well is long and the other for the pumping test time is short. The first data are taken from Xue [17] and the second data are from WPRS [18].

5.1. Example 1. Radial distance from pumping well is long

In this example, a complete well discharges a confined aquifer with a pumping rate of 22.60 m^3/h. The distance between pumping well and observation well is 117.85 m. The observation data of groundwater drawdowns and its corresponding time is listed in table 2.

Table 2. The drawdowns and time in pumping test.

Time (h)	Drawdown (m)
8	0.66
10	0.76
12	0.88
15	1.00
20	1.15
25	1.22
30	1.32
35	1.39
40	1.45
50	1.51
60	1.67

The following steps are necessary for an effective application of the presented methodology in the previous sections:

- Calculate the variable x values from equation (11) for each pumping time.

As C= -0.06619373, for each pumping time,
x = (1.1476, 1.1646, 1.1788, 1.1963, 1.2193, 1.2375, 1.2525, 1.2653, 1.2766, 1.2956, 1.3113).
As C = -0.6023161, for each pumping time,
x= (3.499, 4.0024, 4.4669, 5.1095, 6.0762, 6.9503, 7.757, 8.5117, 9.2246, 10.5516, 11.7763).

- Calculate the regression coefficients k and b by the use of the SPSS 19.0 software. When u is within the scope of 0.26 to 1.1, the calculated b is equal to 0.355886, whereas B is equal to -0.4250883 in this condition, from equation (16) we can know that if b and B are opposite, the transmissivity of aquifer is negative, which is impossible. Hence, we can come to a conclusion that u is within the scope of 0.01 to 0.26 during the pumping test. The output of linear regression is shown in table 3, from which we can know that $k = 5.961839$ and $b = -6.159595$.

Table 3. The output of linear regression.

	Coefficients	Standard error	t Stat	P-value	Lower 95%	Upper 95%
Intercept	-6.159595185	0.204048	-30.187	2.35E-10	-6.62118	-5.69801
X Variable 1	5.961839207	0.165556	36.011	4.86E-11	5.587325	6.336353

- Take the significance level $\alpha = 0.05$ and the degree of freedom $n-2=9$, then check the critical value distribution table we conclude that $|t| = 36.01099 > t_{\alpha/2}(9) = 2.262$, which has gone through the t test. Moreover, from table 3 we know that the significance probability value is 4.86×10-11, which is far less than the value of α. Hence we can conclude that under the condition of the significance level $\alpha = 0.05$ the regression is significant.
- Calculate the aquifer transmissivity value from equation (16), which yields, for the data at hand, T =3.414964 m^2/h.
- Calculate the pressure conduction coefficient value from equation (17), which yields a = 2359.64 m^2/h.
- Calculate the storage coefficient value from equation (18). Hence, $S = 1.447\times10^{-3}$.

The calculation results of aquifer parameters by the approach proposed in this paper and the commonly applied method that Theis curve were listed in table 4. From table 4, it is known that the results of aquifer parameters by the approach proposed in this paper are very close to that by the Theis curve method. Hence, it can be said that the approach proposed in this paper is reliable for estimating aquifer parameters from unsteady pumping test data. In addition, the prediction errors are represented in table 4. According to these errors, the values obtained from the proposed method are better than the Theis curve method.

Table 4. The comparison of the method in this paper and the method of Theis curve.

Method	T (m^2/h)	S	ME×10^{-3}	SEE×10^{-3}
Proposed approach	3.415	0.00145	0.004	28.582
Theis curve	3.341	0.00154	15.351	33.727

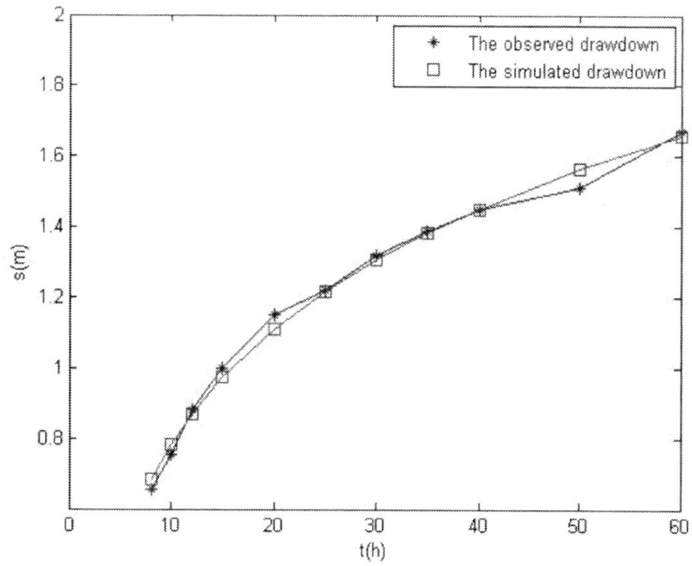

Figure 3. Observed drawdowns and simulated drawdowns by the proposed method.

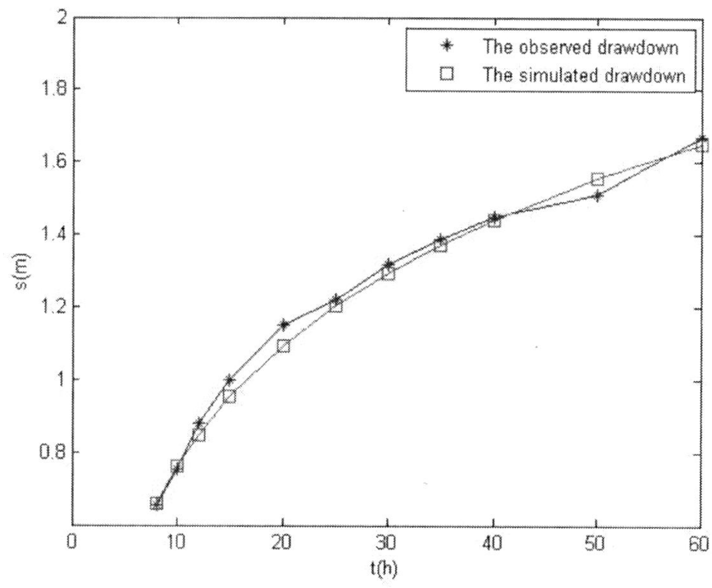

Figure 4. Observed drawdowns and simulated drawdowns by Theis curve method.

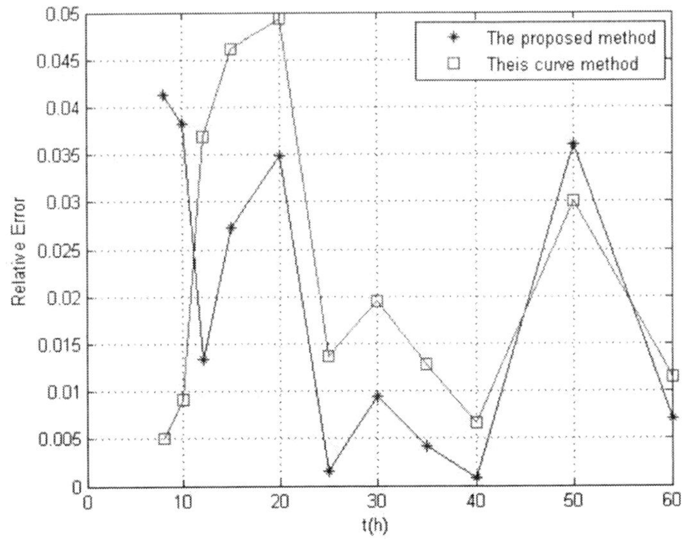

Figure 5. Relative error of simulated drawdowns by the proposed method and Theis curve method.

Taking the solution of aquifer parameters into equation (1), we can get the values of simulated drawdowns. Figures 3 and 4 show the drawdowns simulated by the proposed method and the Theis curve method, respectively. As shown in these figures, both methods are capable of accurately estimating aquifer parameters. Figure 5 shows the relative error of simulated drawdowns between the proposed method and the Theis curve method. It is observed from figure 5 that the accuracy of the proposed method to determine the aquifer parameters is higher than Theis curve method in this example.

5.2. Example 2. Pumping test time is short

In this example, a complete well discharges a confined aquifer with a pumping rate of 4.612 m^3/min. The distance between pumping well and observation well is 30.48 m. The observation data of groundwater drawdowns and its corresponding time is listed in table 5.

Table 5. Pumping test data.

Time (min)	Drawdown (m)
20	0.125
25	0.152
30	0.168
40	0.201
50	0.223
60	0.244
70	0.262
80	0.280
90	0.293
100	0.305
120	0.326
180	0.378
240	0.411
300	0.442
360	0.463
420	0.485

Table 6. The comparison of the method in this paper and the method of Theis curve.

Method	T (m^2/min)	S	ME×10^{-3}	SEE×10^{-3}
Proposed approach	2.893	0.0662	5.023	6.521
Theis curve	2.787	0.0670	9.572	11.205

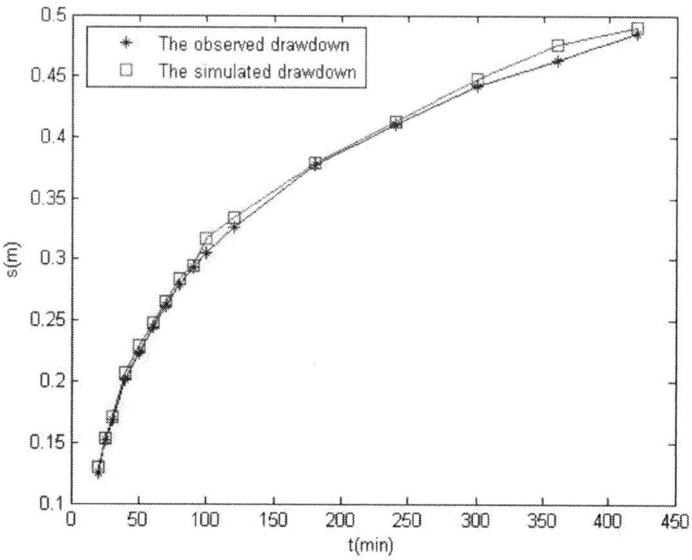

Figure 6. Observed drawdowns and simulated drawdowns by the proposed method.

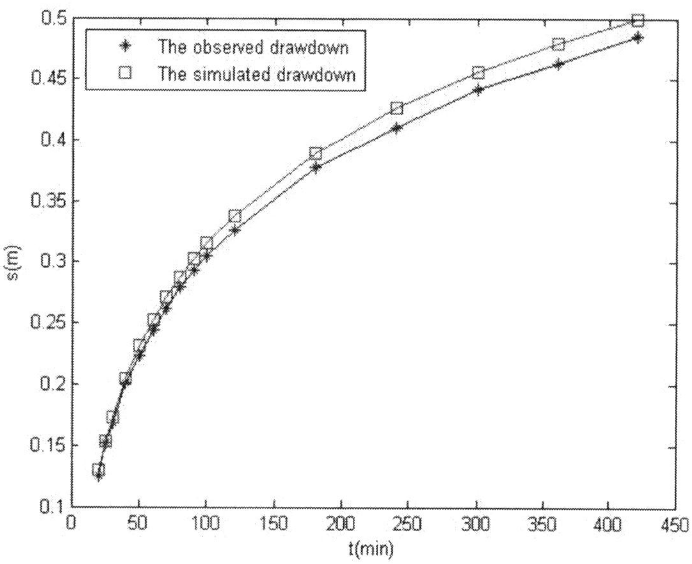

Figure 7. Observed drawdowns and simulated drawdowns by Theis curve method.

It is known that this example is an early pumping test which is another situation that u is large. The application of the proposed method and the Theis curve method is given in table 6. Using the proposed method with the pumping test data results in T and S values of 2.893 m^2/min and 0.0662, the values of T and S obtained by the Theis curve method are 2.787 m^2/min and 0.0670 [18]. The values of ME and SEE in table 6 show that the proposed method has a higher precision than the Theis curve method. Figures 6 and 7 show the drawdowns simulated by the proposed method and the Theis curve method, respectively. Almost all the estimated drawdowns obtained by the proposed method and Theis curve method match the observed drawdowns over the pumping test as shown in figures 6 and 7.

Thus, the proposed method is computationally simple and yields quick and acceptable estimates of aquifer parameters. It has a wider range of application compared with the Cooper-Jacob method. The intelligent optimization algorithm, such as ANN method, GA method and chaotic sequence optimization algorithm (CSOA) which can be used to determine hydro-geological parameters are complicated and can not be operated easily. Theis curve method is a commonly used method to estimate aquifer parameters, but its accuracy is not high. The proposed method can be widely used to the unsteady pumping data.

6. Conclusion

Theis well function is the basis for determination of aquifer hydro-geological parameters. The method of optimum fitting is adopted to approximately estimate the value of Theis well function. Through calculating and contrasting, we know that the maximum relative error between the fitting function and Theis well function is less than 1.33% when the independent variable u is in the range of 0.01 to 1.1. It has been demonstrated that the fitting function which creates the basic conditions for further simply estimating aquifer parameters has a high alternative precision and can satisfy the actual engineering calculation precision requirement.

According to the fitting well function, the aquifer parameters can be estimated by solving the regression coefficients of the unary linear equation. While u is large ($u > 0.01$), Theis curve method is the commonly used method to estimate aquifer parameters. But it has a lot of disadvantages, for example, its calculation process is tedious, its accuracy is not high, and the result of anthropogenic influence is great. As to the comparison between the proposed method and the Theis curve method, two examples of actual time-drawdown data show that the proposed method performs better than the Theis curve method. Observed drawdowns of long distances from pumping well and early pumping test in a confined aquifer can be analyzed for accurately estimate aquifer parameters using the proposed method. The proposed method would be helpful to the field practitioners and engineers.

References

[1] Adamowski J and Chan H F 2011 A wavelet neural network conjunction model for groundwater level forecasting *J. Hydro.* **407** 28-40
[2] Konikow L F and Kendy E 2005 Groundwater depletion: A global problem *Hydro. J.* **13** 317-20
[3] USGS 2010 <http://ga.water.usgs.gov/edu/gwdepletion.html > (accessed 02. 07. 10)
[4] Bear J 1979 *Hydraulics of Groundwater* (London: McGraw-Hill Inc)
[5] Bear J and Cheng A H-D 2010 *Modeling Groundwater Flow and Contaminant Transport* (Dordrecht: Springer)
[6] Fetter C W 1994 *Applied Hydrogeology* (New York: Macmillan College)
[7] Theis C V 1935 The relationship between the lowering of the piezometric surface and the rate and duration of discharge of a well using ground-water storage *Trans. Am. Geophys. Union.* **16** 519-24
[8] Cooper H H and Jacob C E 1946 A generalized graphical method for evaluating formation constants and summarizing well field history *Trans. Am. Geophys. Union.* **27** 526-34
[9] Chow V T 1952 On the determination of transmissivity and storage coefficients from pumping test data *Trans. Am. Geophys. Union.* **33** 397-8
[10] Saleem Z A 1970 A computer method for pumping-test analysis *Ground Water* **8** 21-4
[11] Sen Z 1988 Dimensionless time-drawdown plots of late aquifer test data *Ground Water* **26** 1145-56
[12] Sen Z 2012 Straight-line method generalization for aquifer parameter estimations *J. Irrig. Drain. Eng.* **138** 1082-7

[13] Balkhair K S 2002 Aquifer parameters determination for large diameter wells using neural network approach *J. Hydro.* **265** 118-28
[14] Aziz A R A and Wong K V 1992 A neural-network approach to the determination of aquifer parameters *Ground Water* **30** 164-6
[15] Yeh H D and Huang Y C 2005 Parameter estimation for leaky aquifers using the extended Kalman filter, and considering model and data measurement uncertainties *J. Hydro.* **302** 28-45
[16] Lin G F and Chen G R 2006 An improved neural network approach to the determination of aquifer parameters *J. Hydro.* **316** 281-9
[17] Xue Y Q and Zhu X Y 1979 *Groundwater Dynamics* (Beijing: Geological publishing house)
[18] Water and Power Resources Service (WPRS) 1981 *Ground Water Manual* 2nd ed, US Department

Preliminary investigation on the occurrence of several sulfonamide antibiotics in the Haihe River Basin of China

S L Zhang[1], J Zhang[1,3], Z G Wang[1], Y Z Wang[1], S T Liang[1], C Liu[1,2] and Z Wang[1]

[1]Haihe River Water Conservancy Commission, Ministry of Water Resources, Tianjin, 300170, China
[2]College of Conservancy and Hydropower, Hebei University of Engineering, Handan 056038, Hebei, China

E-mail: zhangjun@hwcc.gov.cn

Abstract. Several samples collected from lakes, rivers and reservoirs in Haihe river basin of China were analyzed for 8 sulfonamide antibiotics by using solid-phase extraction and liquid chromatography with tandem mass spectrometry (HPLC-MS/MS). All water samples were enriched with HLB extraction cartridges. The antibiotics were separated by gradient elution with methanol as the mobile phase adding 0.1% formic acid. The eluate was then analyzed by the mode of multiple reaction monitoring (MRM). The limits of detection (LOD) and quantification (LOQ) were 0.4-1.0 ng/L and 1.0-3.0 ng/L respectively. The method was used for the analysis of 13 samples from Haihe river basin in China. The results showed that sulfamethoxazole was present in all water samples with maximum concentration of 107.59 ng/L. Sulfadiazine was also frequently detected, concentrations ranging from 2.81 ng/L to 85.35 ng/L. Other sulfonamide antibiotics were not detected in most water samples, especially for those samples from drinking water resources.

1. Introduction

Water pollution is one of the most serious environmental problems in northern China. In this area, several industrial pollutants have been monitored for many years. However, some new types of pollutants are still unknown in the surface waters of northern China. Antibiotics are one of the most popular pharmaceuticals. The occurrence of antibiotics in aquatic environments has led to an increasing concern about the environmental risks associated with these compounds [1-3]. Sulfonamide antibiotics (SAs) are widely used antibiotics in modern human and veterinary medical practices. The concentrations of sulfonamide antibiotics measured in surface water samples typically are in the tens of nanograms per liter [4, 5]. Recent studies showed the presence of residual antibiotics in surface waters [6-10]. Although the concentrations are still lower than typical maximum concentrations (in the tens of micrograms per liter) reported before, the effect of continuous exposure to antibiotics on aquatic biota is unknown [11]. In this study, an investigation of 8 sulfonamide antibiotics was carried out in surface waters of lakes, rivers and reservoirs in northern China.

2. Experimental

2.1. Materials and reagents

Symmetry C18 cartridges (150 mm×2.1 mm i.d., 3.5 μm, Waters, USA) and Oasis HLB (500 mg/6 mL, Waters, USA) were used in the study. HPLC grade methanol and formic acid were from J.T. Baker Ltd, USA. Sulfacetamide (SAAM), sulfadiazine (SDZ, 99.5%), sulfamethoxaole (SMZ, 99.5%), sulfathiazole (STZ, 99.5%), sulfamerazine (SMR, 99.2%), sulfisoxazole (SIX, 99.7%), sulfamethazine (SMT, 99.0%) and sulfachloropytidazine (SCP, 99.0%) were obtained from Dr. Ehrenstorfer GhbH, Germany. Ultra-pure water was prepared by a Milli-Q pure water system. Stock solutions of all sulfonamide antibiotics were prepared by dissolving each compound in methanol at a concentration of 100 mg/L. Working solutions of different concentrations were prepared by diluting and mixing the stock solution and stored at 4℃ in the dark.

2.2. Sample location and preparation

Haihe River Basin is located in north of China including Beijing, Tianjin, Hebei Province, Shanxi Province, Shandong Province, Henan Province, Liaoning Province and Inner Mongolia Autonomous Region. Samples in this study were from lakes, rivers (province boundary) and drinking water sources (Panjiakou reservior and Yuecheng reservior) in Haihe river basin, as shown in figure 1.

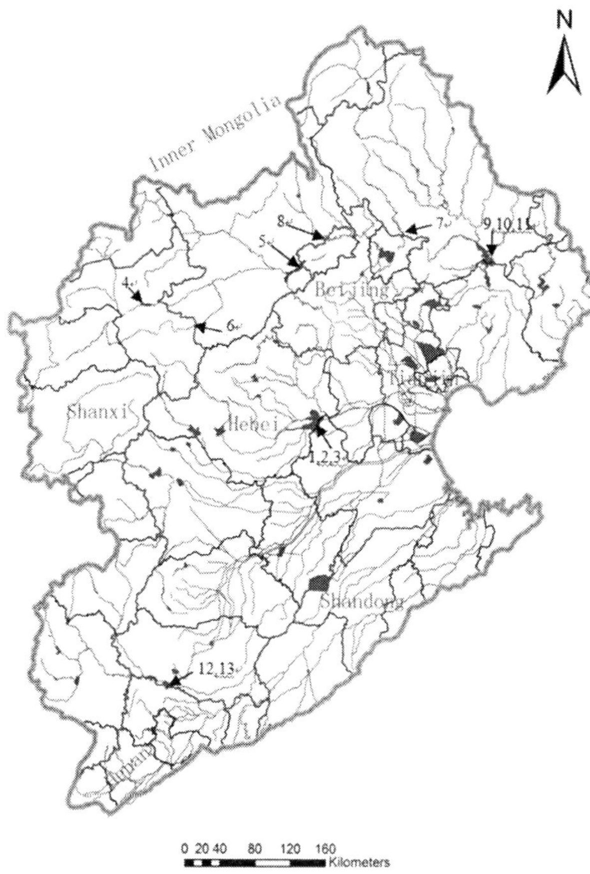

Figure 1. Location for the sampling section in Haihe River basin.

Notes: 1, 2, 3-Baiyang Lake. 4-Cetian reservoir. 5-Guanting reservoir. 6-Huliuhe reservoir. 7-province boundary (Gubeikou). 8-province boundary (Xiabao). 9, 10, 11-Panjiakou reservoir. 12, 13-Yuecheng reservoir.

All water samples were filtered through 0.45 μm glass fiber filters. Five hundred mL of the water samples were acidified to pH=4 by adding formic acid. Oasis HLB cartridges were preconditioned with 5.0 ml of methanol and 5 ml ultra-pure water. Acidified water samples were then passed through Oasis HLB columns at a flow rate of approximately 5 mL/min. The Oasis HLB columns were then rinsed with 5 mL ultra-pure water and dried under vacuum. After drying, sulfonamides were eluted from each cartridge using 2 mL of methanol (3X). Extracts were collected in a glass vial, reduced to a minimum volume, and then dissolved in 20% methanol to a final volume of 1.0 mL.

2.3. Liquid chromatography and mass spectrometry
An Agilent 1260 liquid chromatography system connected to 6410B triplequadrupole mass spectrometer was used in this study. The system was equipped with a vacuum degasser, a binary pump, an autosampler and a temperature controlling system. A Symmetry C18 chromatograph column (Waters, USA, 150 mm×2.1 mm i.d., 3.5 μm) was employed and was operated at 30℃.

The mobile phase consisted of deionized water containing 0.1% formic acid (A) and methanol (B). The gradient was set up as follows: 0-8 min, 20%-80% B; 8-10 min, 80% B; 10-12 min, 80%-20% B; 12-25 min, 20% B. The injection volume was 10 μL and the flow rate was 0.2 mL/min.

The QQQ was operated using an eletrospray ionization (ESI) source in positive mode. High purity liquid nitrogen was used as the nebulizing and drying gas at 350℃ with a flow of 10.0 L/min.

3. Results and discussion

3.1. Chromatography
In this study, we compared two chromatography columns. The results indicate that a Symmetry C18 chromatography column (Waters, USA, 150 mm×2.1 mm i.d., 3.5 μm) achieves better separation of the various sulfonamides than a Zorbax SB C18 chromatograph column (Agilent, USA, 50 mm×2.1 mm). The latter column might be too short to separate the eight sulfonamide antibiotics examined in this study. In addition, it was determined that the solvent used for liquid chromatography is very important. The chromatographic separation results were poor when the sample was resolved in pure methanol. The base line was not flat and there was an obvious solvent peak at 2 min. Also, the peak was not sharp. However, chromatographic separation results were good when the initial mobile phase used was 20% methanol as shown in figure 2.

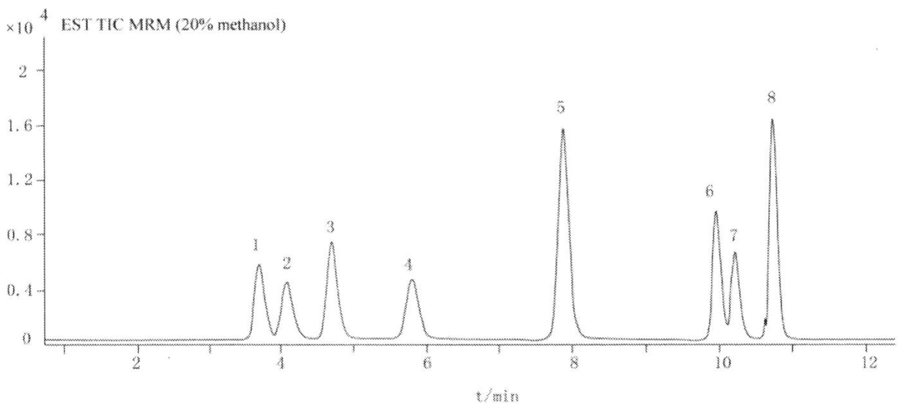

Figure 2. Total ion chromatogram of 8 sulfonamide antibiotics by HPLC-MS/MS with MRM mode.
Notes: 1. SAAM. 2. SDZ. 3. STZ. 4. SMR. 5. SMT. 6.SCP. 7. SMZ. 8. SIX.

3.2. MS/MS-MRM parameters
The mixture of 8 sulfonamide antibiotics was injected for study the initial experiments. The full scan spectra show the presence for protonated molecular ions (mother ions). The mode of multiple reaction monitoring (MRM) was used for detection of all sulfonamide antibiotics. The parameters were set up by following procedures. High purity nitrogen gas was used for

collision-induced dissociation. The fragment ions were formed through the collision with the proper energy. The transition parameters from mother ions to daughter ion were optimized by varying the collision energy. The collision energy (CE) For quantitative and qualitative ions were selected for each compound. Table 1 shows mother ions, daughter ions, optimum fragment voltage and collision energy.

Table 1. MRM Parameters for eight sulphonamides.

Compound	Precursor ion m/z	Quantitative ion m/z	Qualitative ion m/z	Fragment U/V	Collision energy U/V
Sulfacetamide (SAAM)	215.0	156.1	108.1	95	10
Sulfadiazine (SDZ)	251.0	108.1	156.1	105	23
Sulfamethoxazole (SMZ)	254.0	108.1	156.1	105	25
Sulfathiazole (STZ)	256.0	156.1	108.1	95	10
Sulfamerazine (SMR)	265.0	156.1	108.1	115	15
Sulfisoxazole (SIX)	268.0	156.1	108.1	95	10
Sulfamethazine (SMT)	279.0	186.1	124.1	115	15
Sulfachloropyridazine (SCP)	284.9	156.1	108.1	95	10

3.3. Recovery

Rao *et al* [12] reported that the recovery of antibiotics is related to pH and sample volume. Our results also indicate that recovery of the 8 sulfonamide antibiotics is greater at pH 4 than at pH 7. Recovery of 8 sulfonamide antibiotics from the HLB cartridges was measured with 500 mL of tap water spiked at 20 ng/L and 200 ng/L at pH 4. The results indicate that the recoveries for both spiked concentration levels were greater than 80%. The percent recovered was slightly greater at the higher concentrations (200 ng/L) than the lower concentrations (20 ng/L) as shown in table 2.

Table 2. Recoveries for 8 sulfonamide and antibiotics at different spike concentrations.

Compounds	20 ng/L				200 ng/L			
	Recovery (%, n=3)			RSD (%)	Recovery(%, n=3)			RSD (%)
Sulfacetamide (SAAM)	91.5	93.6	95.4	2.1	93.5	95.1	96.8	1.7
Sulfadiazine (SDZ)	95.4	94.8	91.5	2.2	99.0	97.9	95.2	2.0
Sulfamethoxazole (SMZ)	86.0	91.2	93.9	4.4	99.1	93.3	96.2	3.0
Sulfathiazole (STZ)	92.5	97.6	94.5	2.7	92.2	93.6	97.6	3.0
Sulfamerazine (SMR)	89.3	90.6	94.7	3.1	99.7	96.6	93.6	3.6
Sulfisoxazole (SIX)	83.0	83.8	89.9	4.4	98.2	99.6	96.3	1.7
Sulfamethazine (SMT)	96.4	95.1	92.8	1.9	99.0	97.6	95.9	1.6
Sulfachloropyridazine (SCP)	94.5	98.1	98.5	2.3	97.5	99.7	98.9	1.1

IOP Conf. Series: Earth and Environmental Science **82** (2017) 012024 doi:10.1088/1755-1315/82/1/012024

3.4. Calibration, limit of detection and limit of quantification

The method of external calibration was used for the sulfonamide antibiotics with the concentration range of 1-100 μg/L. All sulfonamide antibiotics show good correlation with coefficients R2 of more than 0.9991. Reproducibility of the method was determined by analyzing six replicates of spiked samples at 100 μg/L. The relative standard deviation (RSD) was lower than ±4%. The limits of detection (LOD) and the limits of quantification (LOQ) were calculated based on signal-to-noise ratio of 3 and 10. The result indicates that, LOD and LOQ are 0.2-0.5 μg/L and 0.5-1.5 μg/L, respectively. Considering that the sample volume was 500 ml, the limits of detection (LOD) and the limits of quantification (LOQ) are in the range of 0.4-1.0 ng/L and 1.0-3.0 ng/L under the assumption that the recoveries would be 100%. Results are shown in table 3.

Table 3. Regession analysis of calibration curves, LOD and LOQ of antibiotics.

Compounds	LOD (μg/L)	LOQ (μg/L)	Linear range (μg/L)	Regression equation	R^2
Sulfacetamide (SAAM)	0.2	0.5	1-100	Y=282.51X+82.740	0.9998
Sulfadiazine(SDZ)	0.2	0.7	1-100	Y=57.743X-20.188	0.9995
Sulfamethoxazole(SMZ)	0.3	1.1	1-100	Y=70.531X+48.930	0.9994
Sulfathiazole(STZ)	0.5	1.5	1-100	Y=107.776X+34.305	0.9999
Sulfamerazine(SMR)	0.2	0.7	1-100	Y=42.019X+18.862	0.9991
Sulfisoxazole(SIX)	0.2	0.5	1-100	Y=153.921X+91.560	0.9994
Sulfamethazine(SMT)	0.3	1.1	1-100	Y=137.969X-26.050	0.9993
Sulfachloropyridazine(SCP)	0.2	0.7	1-100	Y=97.017X-56.451	0.9991

Table 4. Concentrations of 8 sulfonamide antibiotics in Haihe river basin.

Compounds	Concentration of antibiotics in different locations (ng/L)												
	1	2	3	4	5	6	7	8	9	10	11	12	13
Sulfacetamide (SAAM)	6.15	ND	ND	ND	ND	ND	ND	ND	ND	ND	ND	ND	ND
Sulfadiazine (SDZ)	85.35	36.24	51.32	6.42	2.81	ND	ND	32.63	ND	ND	ND	ND	ND
Sulfamethoxazole (SMZ)	103.80	75.25	107.59	6.88	29.35	4.56	7.69	67.75	10.25	32.63	10.34	3.10	3.07
Sulfathiazole (STZ)	ND	ND	ND	ND	ND	ND	ND	ND	ND	ND	ND	ND	ND
Sulfamerazine (SMR)	ND	ND	ND	ND	ND	ND	ND	ND	ND	ND	ND	ND	ND
Sulfisoxazole (SIX)	ND	ND	ND	ND	0.91	ND	ND	ND	ND	ND	ND	ND	ND
Sulfamethazine (SMT)	5.88	3.16	4.54	1.74	0.70	ND	1.02	1.19	ND	ND	ND	ND	ND
Sulfachloropyridazine (SCP)	ND	ND	ND	ND	ND	ND	ND	ND	ND	ND	ND	ND	ND

Notes: 1, 2, 3-Baiyang Lake. 4-Cetian reservoir. 5-Guanting reservoir. 6-Huliuhe reservoir. 7- province boundary (Gubeikou). 8-province boundary (Xiabao). 9,10,11-Panjiakou reservoir. 12, 13-Yuecheng reservoir. ND-not detected.

3.5. Analysis of surface water

In this study, an investigation of 8 sulfonamide antibiotics was carried out in surface waters of lakes, rivers and reservoirs in northern China. Thirteen surface water samples were analyzed using the method above as shown in table 4. The results showed that sulfamethoxazole was detected in these surface waters with the maximum concentrations 107.59 ng/L. These results are similar to the previously published research of Cahill *et al* [4]. Sulfadiazine was also frequently detected, at concentrations ranging from 2.81 ng/L to 85.35 ng/L. The other sulfonamide antibiotics were not found in most water samples, especially for those samples from drinking water resources. Our data indicate that the concentration of sulfonamide antibiotics in drinking water resources is low. However, there is an increasing concern about the potential environmental risk.

4. Conclusions

In the study, a method was developed and validated for measuring 8 sulfonamide antibiotics using solid-phase extraction HPLC-MS/MS. The results indicate that it is a simple, rapid and effective method for measuring these antibiotic pollutants in environmental water samples. In the study, sulfamethoxazole was found in every surface water sample with a maximal concentration of 107.59 ng/L. Sulfacetamide, Sulfadiazine and Sulfamethazine were detected in some water samples and other antibiotics were not detected. Though the concentration of most antibiotic is not high, it is a big concern for human health. Therefore, the relationship between antibiotic levels in surface waters and human health risks should be studied further.

Acknowledgments

This work was supported by the International Science & Technology Cooperation Program of China (Grant No. 2013DFA71340) and Program of water quality monitoring in Haihe river basin (Grant No. 126204012002160001).

References

[1] Balakrishnan V K, Terry K A and Toito J 2006 Determination of sulfonamide antibiotics in wastewater: A comparison of solid phase microectraction and solid phase extraction methods *J. Chromatogr. A.* **1131** 1-10

[2] Yang S, Cha J and Carlson K 2005 Simultaneous extraction and analsis of 11 tetracycline and sulfonamide antibiotics in influent and effluent domestic wastewater by solid-phase extraction and liquid chromatography-elerospray ionization tandem mass spectrometry *J. Chromatogr. A.* 1097 40-53

[3] Chang X, Meyer M T, Liu X, Zhao Q, Chen H, Chen J, Qiu Z, Yang L, Cao J and Shu W 2010 Determination of antibiotics in sewage from hospitals, nursery and slaughter house, wastewater treatment plant and source water in Chongqing region of Three Gorge Reservoir in China *Environ. Pollut.* **158** 1444-50

[4] Cahill J, Furlong E T, Burkhardt M R, Kolpin D and Anderson L G 2004 Determination of pharmaceutical compounds in surface- and ground-water samples by solid-phase extraction and high-performance liquid chromatography electrospray ionization mass spectrometry *J. Chromatogr. A.* **1041** 171-80

[5] Xu W H, Zhang G, Zou S C, Li X D and Liu Y C 2007 Determianation of selected antibiotics in the Victoria Harbour and the Pearl River, South China using high-performance liquid chromatography-elerospray ionization tandem mass spectrometry *Environ. Pollut.* **145** 672-9

[6] Watkinson A J, Murby E L, Kolpin D W and Costanzo S D 2009 The occurrence of antibiotics in an urban watershed: From wastewater to drinking water *Sci.Total Environ.* **407** 2711-23

[7] Luo Y, Xu X, Rysz M, Wang Y, Zhang H and Alvarez P J J 2011 Occurrence and transport of tetracycline, sulfonamide, quinolone and macrolide antibiocis in the Haihe River Basin, China *Environ. Sci. Technol.* **45** 1827-33

[8] Jiang L, Chen S, Yang R, Ren Z and Yin D 2008 Occurrence of antibiotics in the aquatic environment of the Changjiang delta, China *Environ. Chem.* **27** 371-4

[9] Gao L, Shi Y, Li W, Liu J and Cai Y 2010 Determ ination of 22 antibiotics in envi ronmental water samples using high performance liquid chro matography electrospray ion ization tandem mass spectrometry *Chinese J. Chromatogr.* **28** 491-7

[10] Tang C, Huang Q, Yu Y and Peng X 2009 Multi residue determination of sulfonamides, macrolides, trimethprim , and chloramphenicolin sewage sludge and sediment using ultra son i c extraction coupled with solid phase extraction and liquid chromatography2 tandem mass spectrometry *Chinese J. Anal. Chem.* **37** 1119-24

[11] Kolpin D W, Furlong E T and Meyer M T 2002 Pharmaceuticals, hormones, and other organic wastewater contaminants in U.S. streams, 1999-2000: A national reconnaissance *Environ. Sci. Technol.* **36** 1202-11

[12] Rao R N, Venkateswarlu N and Narsimha R 2008 Determination of antibiotics in aquatic environment by solid-phase extraction followed by liquid chromatography electrospray ionization mass spectrometry *J. Chromatogr. A.* **1187** 151-64

IOP Conf. Series: Earth and Environmental Science 82 (2017) 012025 doi:10.1088/1755-1315/82/1/012025

A multifactor analysis of the spatial distribution of annual mean extreme precipitation- Taking the Yellow River Basin as an example

B Q Li[1], B D Hou[1], X Y Dong[2,3] and H J Wang[1]

[1]State Key Laboratory of Simulation and Regulation of Water Cycle in River Basin, China Institute of Water Resources and Hydropower Research, Beijing100038, China
[2]Water Conservancy and Hydropower Survey and Design Institute, Xining, 810000, China

E-mail: tongyeti@163.com

Abstract. Because the current interpolation method cannot meet the accuracy specifications, a new method for spatial interpolation of extreme precipitation is developed. It integrates precipitation, longitude and latitude, elevation, and topography data such as slope and aspect, and it is then simulated by the high accuracy surface modelling method and applied in the Yellow River Basin (YRB). Results show that the integrated method affords a very good simulation precision, with its simulation precision error being far below other interpolation methods. In conclusion, the integrated method for spatial interpolation of extreme precipitation can afford a good solution in topographically complex areas.

1. Introduction

The Special Report on Extreme Events (SREX) of the intergovernmental panel on climate change [1] emphasized potentially severe impacts from extreme precipitation. Characteristics of precipitation extremes are important in many practical applications [2]; they include time-domain, frequency domain, and spatial domain. Confirming the spatial domain is extremely important and practical, especially in extreme precipitation.

A large number of authors have studied extreme precipitation in many different regions all over the world. For example, Vahid Rahimpour Golroudbary and Yijian Zeng [3] have shown that the spatial distribution of extreme precipitation clearly reflects the regional differences in the Netherlands. Tao Gao and Lian Xie portrayed a similar result [4]: over the Yangtze River basin, spatiotemporal patterns of precipitation extremes is more uneven and unstable during the years 1981–2011. The spatial and temporal distributions of indices of precipitation extremes in Southeastern Tibet [5] are extremely irregular. The extreme precipitation studies in China [6] showed that the nonstationary GEV distributions performed better than their stationary equivalents. Long-term trends and variability of total and extreme precipitation in Thailand [7] indicate that large-scale climate phenomena in the Pacific Ocean are remote drivers of variability in Thailand's total and extreme precipitation. Elena Voskresenskaya and Elena Vyshkvarkova [8] have indicated that extreme precipitation throughout the Crimean peninsula increases during the winter season and decreases during the summer season. There were no obvious trends shown by the changes in extreme precipitation events in the Loess Plateau

(China) during 1960–2013 under global warming [9]. It is found by Huailiang Wang *et al* [10] that extreme precipitation events over the Yellow Sea western coast make a wetted winter season.

However, most of the studies use traditional methods, such as inverse distance weighting model (IDW), triangle net model (TIN), kreegan model (Kriging), and spline interpolation module (Spline) to confirm the spatial domain of extreme precipitation. These current popular interpolation methods, are theoretical interpolation methods and are not based on physical truth, such as longitude and latitude, elevation, slope, and so on. Therefore, these methods do not have any physical significance and their interpolation results have a higher error. In fact, each of the above methods has theoretical problems. To solve this problem and improve the accuracy of extreme precipitation spatial distribution, the method of high accuracy surface modeling (HASM) is applied, which is based on the theory of differential geometry surface. Due to the spatial distribution of extreme precipitation being closely related to longitude and latitude, elevation, and slope, it would have a better performance and be closer to the actual distribution of extreme precipitation by taking all other influence factors into account when conducting spatial interpolation of precipitation.

2. Materials and methods

2.1. Study area

The Yellow River basin (YRB, figure 1) is situated in the subtropics with latitude between 32° and 42° N and longitude between 96° and119° E, and a length and width at about 1900 km and 110 km, respectively. With a vast basin area, YRB stretches over Qinghai-Tibet Plateau, Inner Mongolian Plateau, Loess Plateau, and Huang-Huai-Hai Plain from west to east. Therefore, it has its own rules and characteristics under the background of climate change. The YRB is located in the arid and semi-arid climate zones; the west of YRB is drought and the east is wet with a drought spring and winter and rainy summer and autumn. Due to concentrated heavy rainfall, sparse vegetation, poor soil erosion resistance, and unreasonable human activities, Loess Plateau, one special landform in YRB, becomes one of the most serious soil erosion areas in China. The special topography Loess Plateau and the growing number of extreme weather events leads to an increasing ecological crisis, such as water loss, soil erosion, and land desertification. Therefore, it is extremely significant to study the temporal variations of extreme precipitation in YRB [9].

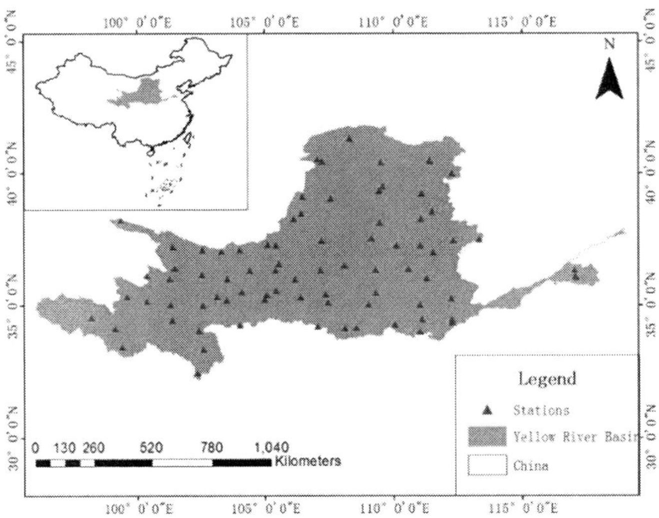

Figure 1. Location of the Yellow River Basin and hydrometeorological stations over the Yellow River Basin.

2.2. Data sources

During 1961–2010, the National Climate Center (NCC) and China Meteorological Administration (CMA) provided daily precipitation dataset from 80 rain gauge stations. Figure 1 shows the distribution of the rain gauge in YRB. In order to guarantee the continuity and unity of all weather station data as much as possible, stations established after the year 1960 are eliminated and the stations whose data are continuous are selected. In the next place, all the data are passed through stringent quality control checks and error value correction.

As to the definition of extreme precipitation events, China Meteorological Administration (CMA) stipulates that the threshold value of extreme precipitation events is 50 mm daily precipitation. However, YRB has special geographical position and climate environment, so it is not accurate to take 50 mm daily precipitation as the threshold value of extreme precipitation events for arid and semi-arid regions in YRB. Based on the world meteorological organization, the international percentile calculation method is used as the threshold value of extreme precipitation events to overcome this deficiency. The exact approach to process precipitation data is: first, sort the daily precipitation data of the selected 80 stations from 1961—2010 in ascending order; second, take the value of the 95th percentile as the threshold value of extreme precipitation events.

2.3. Research methods

Longitude and latitude reflect the influence of the background of climate, circulation, and continentally to some extent. On the other hand, elevation reflects gradient effect of precipitation. Meanwhile, the aspect and slope of the local area in a certain extent represent the influence of prevailing wind-direction to precipitation. Therefore, to improve the space simulation accuracy of extreme precipitation events, longitude and latitude, elevation, aspect, and unobstructed terrain factors are taken into account.

- The coefficient equations of influence on precipitation slope

According to DEM data, the spatial analysis module of ARC/INFO extracts the 1 km grid slope data of YRB. In the extraction results, 0° is on behalf of due north direction, 90° represents due east, 180°represents due south, and 270° represents due west by turning it clockwise. From the view of precipitation distribution affected by the monsoon in YRB, the greatest influence is from the southeast and southwest monsoon, namely warm and wet flow from Pacific Ocean and Indian Ocean. Therefore, the influence of due south on precipitation defines the minimum value 1, and due north -1. Due to flat areas having no influences on precipitation, it is 0. The coefficient equations of slope and its influence on precipitation are performed as:

$$ASP_{index} = -\cos\left(\pi / \left(Aspect / 180^{\circ}\right)\right) \qquad Aspect \neq -1$$
$$ASP_{index} = 0 \qquad Aspect = -1 \tag{1}$$

where ASP_{index} denotes the coefficient equations of slope and its influence on precipitation, $Aspect$ are values of slope.

- The coefficient equations of influence on precipitation of terrain unobstructed factor

In addition to slope, the unobstructed terrain factor, in other words, the maximum elevation rate between one grid cell and the cell around it, has a significant effect on precipitation. The coefficient equation of influence on precipitation of t unobstructed terrain factor is given by:

$$V_{index} = \left[1 + \cos\left(\pi\left(Slope / 180°\right)\right)\right] / 2 \tag{2}$$

Where V_{index} denotes terrain unobstructed factor; $Slope$ is topographic slope.

- The regression equation of extreme precipitation and its influence factor

Every station has precipitation, longitude and latitude, elevation and slope aspect and unobstructed terrain factor calculated by DEM. Annual mean extreme precipitation and its correlation factor have a good correlation. The regression equation is presented as follow:

$$M_{pre} = 2659.733838 + 0.000176X_{cor} - 0.0004791Y_{cor} - \\ 0.084694Z_{elevation} - 40.479061Asp_{index} - 107.599239V_{index} \quad (3)$$

where M_{pre} is the fitted value of annual mean extreme precipitation; X_{cor}, Y_{cor}, $Z_{elevation}$, Asp_{index} and V_{index} represent longitude, latitude, elevation, the coefficient of slope and terrain unobstructed factor, respectively.

The pre-treatment and modifier formulas of annual mean extreme precipitation in YRB can be established according to the above formula:

$$M_{pre_{f0}}(x,y) = M_{pre}(x,y) - 0.000176X_0(x,y) + 0.0004791Y_0(x,y) \\ +0.084694Z_0(x,y) + 40.479061Asp_0(x,y) + 107.599239V_0(x,y) \quad (4)$$

$$M_{pre_{gi}}(x,y) = M_{pre_{f0}}(x,y) + 0.000176X_{fi}(x,y) - 0.0004791Y_{fi}(x,y) \\ -0.084694Z_{fi}(x,y) - 40.479061Asp_{fi}(x,y) - 107.599239V_{fi}(x,y) \quad (5)$$

where $M_{pre_{f0}}(x,y)$, $M_{pre}(x,y)$, $X_0(x,y)$, $Y_0(x,y)$, $Z_0(x,y)$, $Asp_0(x,y)$, $V_0(x,y)$ are the pre-treatment values, annual mean extreme precipitation, longitude, latitude, elevation, the coefficient of slope, and unobstructed terrain factor of every station. $M_{pre_{gi}}(x,y)$, $M_{pre_{f0}}(x,y)$, $X_{fi}(x,y)$, $Y_{fi}(x,y)$, $Z_{fi}(x,y)$, $Asp_{fi}(x,y)$, and $V_{fi}(x,y)$ are the simulation values of every grid cell's annual mean extreme precipitation, the pre-treatment of value of annual mean extreme precipitation, longitude, latitude, elevation, the coefficient of slope, and unobstructed terrain factor.

In the process of simulation, annual mean extreme precipitation of every station can be obtained by analyzing all the 80 stations in YRB, then the $M_{pref0}(x,y)$ of every station can be calculated according to the 4th formula and by using HASM for the interpolation. Finally, use the 4th formula to modify $M_{prefi}(x, y)$, the interpolation value.

3. Results

3.1. The accuracy comparison between the integrated method and other methods

In order to test the accuracy of the interpolation, 8 stations (10% of stations in YRB) are randomly extracted as authentication data, and the other stations (90% of stations in YRB) are interpolation stations. In this paper, four interpolation methods are used: IDW, Kriging, Spline, and HASM. Table 1 presents the statistical results of the different methods.

Table 1. The comparison of simulation accuracy with different methods.

	Error of IDW	Error of Kriging	Error of Spline	Error of Integrated method
simulation precision (mm)	87.01	73.54	137.22	8.88

From table 1, annual mean extreme precipitation obtained with HASM from 1961 to 2010 with mean values shows very good simulation precision with a simulation precision error far lower than other interpolation methods. Spline has the maximum error. We can conclude that the integration of HASM and factors related to extreme precipitation such as longitude, latitude, elevation, the coefficient of slope, and unobstructed terrain factor display a good interpolation precision, and the space analysis function and drawing function of ARC/INFO can perfectly reflect the interpolation results.

3.2. Spatial distribution of annual mean extreme precipitation in YRB
Spatial distribution of annual mean extreme precipitation in YRB is showed by figure 2.

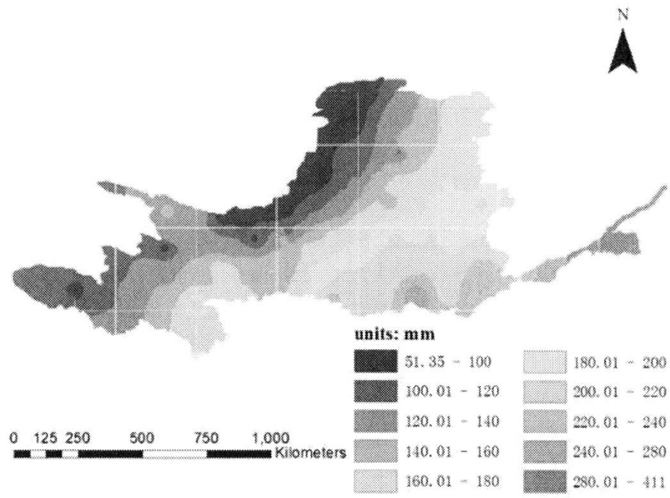

Figure 2. Spatial distribution of annual mean extreme precipitation in YRB.

From figure 2, we can see that the maximum annual mean extreme precipitation is located in the east and southeast parts, and it can be as much as for 280 to 400 mm in one day. The amount of mean annual extreme precipitation in the YRB middle area is from 140 to 280, and the minimum distribute in the northwest part of YRB spreads from 51 to 140.

4. Conclusions and discussions
Because the current interpolation method cannot meet the accuracy specifications, a new method for spatial interpolation of extreme precipitation is developed. It integrates precipitation, longitude and latitude, elevation, and topography data such as slope and aspect. It is then by the high accuracy surface modelling method. The new approach for spatial interpolation of extreme precipitation can greatly improve the accuracy specification.

The YRB is a specific area in China as it stretches over different regions and the spatial distribution of extreme precipitation in it is very important for society, the environment, and the economy. There is a drastic difference in the precipitation in YRB from region to region, and from southeast to northwest the extreme precipitation progressively decreases. This is related to longitude, latitude, elevation, slope, and aspect. The new method for spatial interpolation of extreme precipitation in topographically complex areas such as YRB is a good solution.

Acknowledgments
Special thanks for Fan Zemeng from institute of geographic sciences and natural resources research, Chinese Academic of Sciences, China. His paper called "Downscaling simulation for the scenarios of precipitation in China" gives the authors inspiration to write this paper. We are heartily grateful to him and his work.

This study is jointly supported by The National Science Fund for Distinguished Young Scholars (No. 5612590), Social Water Cycle and High-efficiency Utilization of Water Resources, Hydrology Technology Project of Heilingjiang Province (201305) and National Nature Science Fund Project (51679252).

References

[1] IPCC, 2012. Managing the risks of extreme events and disasters to advance climate change adaptation *A Special Report of Working Groups I and II of the Intergovernmental Panel on Climate Change* (Cambridge, UK; New York, USA: Cambridge University Press) p 582

[2] Rulfová Z, Buishand A, Roth M and Kysely J 2016 A two-component generalized extreme value distribution for precipitation frequency analysis *J. Hydrol,* **534** 659-68

[3] Rahimpour V, Zeng Y, Mannaerts C M and Su Z B 2016 Attributing seasonal variation of daily extreme precipitation events across The Netherlands *Wea. Climate Extremes* **11** 1-11

[4] Gao T and Xie L 2016 Spatiotemporal changes in precipitation extremes over Yangtze River basin, China, considering the rainfall shift in the late 1970s *Global Planet Change* **147** 106-24

[5] Zhang J, Shen X and Wang B 2015 Changes in precipitation extremes in Southeastern Tibet, China *Quaternary Int.* **380-381** 49-59

[6] Gao M, Mo D and Wu X 2016 Nonstationary modeling of extreme precipitation in China *Atmos. Res.* **182** 1-9

[7] Limsakul A and Singhruck P 2016 Long-term trends and variability of total and extreme precipitation in Thailand *Atmos. Res.* **169** 301-17

[8] Voskresenskaya E and Vyshkvarkova E 2016 Extreme precipitation over the Crimean peninsula *Quaternary Int.* **409** 75-80

[9] Sun W, Mu X, Song X, Wu D, Cheng A and Qiu B 2016 Changes in extreme temperature and precipitation events in the Loess Plateau (China) during 1960–2013 under global warming *Atmos. Res.* **168** 33-48

[10] Wang H, Shao Z, Gao T, Zou T, Liu J and Yuan H B 2016 Extreme precipitation event over the Yellow Sea western coast: Is there a trend? *Quaternary Int.* **441** 1-17

IOP Conf. Series: Earth and Environmental Science 82 (2017) 012026 doi:10.1088/1755-1315/82/1/012026

Oxygen, pH, and Eh microprofiles around submerged macrophyte Vallisneria natans response to growing stages

B Dong[1,3], G X Wang[2] and H G Yu[1]

[1]Shandong Provincial Key Laboratory of Water and Soil Conservation & Environmental Protection, College of Resources and Environment, Linyi University, 276000, China
[2]School of Environment, Nanjing Normal University, Nanjing, 210023, China

E-mail: dongbin@lyu.edu.cn

Abstract. The periphyton, attached to the surfaces of submerged plants, has important effects on plant growth and development in eutrophic waters. Periphyton complicates the microenvironment of diffusive boundary layer around submerged plants. We researched periphyton characteristics, oxygen (O_2), pH, and Eh microprofiles at various growing stages of *Vallisneria natans*. The results suggested that during the growing period of *V. natans*, O_2 concentration and pH decreased from 0 to 2 mm above the leaf surface, whereas the Eh increased. As *V. natans* grew, O_2 and pH gradually increased until they peaked during stable growing stages, while the Eh decreased. However, during the decline stage, O_2 and pH gradually decreased, and Eh increased. To summarise, O_2 and pH showed a unimodal pattern in response to the life cycle of *V. natans*, with the maximum levels during the stable growth stage and the minimum levels during the rapid growth and decline stages. Our study demonstrated that *V. natans* growth induced steep gradients in O_2 concentrations, pH, and Eh at the DBL by increasing the layer's thickness, macrophyte photosynthetic capacity, and periphyton biomass in eutrophic waters.

1. Introduction

Macrophytes provide the primary accessible surface area for the periphyton in eutrophic lakes. The periphyton attached on submerged macrophytes is an assemblage of microorganisms, inorganic particles, and organic detritus, forming a diffusive boundary layer (DBL) of periphyton–macrophyte association [1]. The DBL is important for macrophyte growth, nutrient transformation and biogeochemical cycles in eutrophic ecosystems [2, 3]. Oxygen (O_2) concentration, pH, Eh, and dissolved substances in the DBL differ considerably from the surrounding water, exhibiting steep gradients around the macrophyte surfaces [1, 4]. Characteristics of this DBL, including O_2, pH, and Eh levels, are affected by factors such as hydrodynamics, surface flux, concentration gradients, and periphyton surface structure. And light, temperature, plant growth stages and species of the associated macrophyte, also affect diffusive boundary layers indirectly, by affecting the flux of oxygen into and out of the leaf surface [2].

Of these factors, the effect of aquatic plant age on the DBL has been relatively understudied. We previously found that submerged plant *Potamogeton malaianus* leaf age had significant influences on the distribution of O_2, pH, and Eh at microscale, which exhibited the steepest gradient on adult leaves and the shallowest on old leaves [1]. Other studies have supported this pattern; for example, O_2

concentrations in the microprofiles of *Littorellauni flora* and *Potamogeton crispus* were significantly higher in early June than in late August. Moreover, as host plants grow, the periphyton is reported to increase in size, density, composition, structure, and species diversity, with senescent plants being the most densely colonised [5, 6]. However, we currently know little about how the characteristics of microenvironmental variables in the DBL may vary across different macrophyte growth stages. In fact, no study exists that investigates periphyton–macrophyte association across growth stages simultaneously with microprofile characteristics (specifically, measurements of O_2, pH, and Eh) around submerged plant leaves. This lack of data hinders a comprehensive understanding of the DBL's ecological function in eutrophic waters. We therefore require a detailed investigation of how plant growth affects microprofile dynamics and the structure of the periphyton–macrophyte association.

In this study, we used microsensors to measure the O_2, pH, and Eh microprofiles of submerged macrophytes at various growing stages. The aims of this study were to track the characteristics of O_2, pH, and Eh microprofiles, as well as the fine structure of the periphyton–macrophyte association, across the life cycle of submerged plants in eutrophic lakes. The consequent research results might help to make clear the processes and mechanisms of the DBL affecting nutrient cycle in eutrophic lakes and to fully understand the ecological remediation function and role of macrophytes in eutrophic waters.

2. Materials and methods

Vallisneria natans rhizomes were collected from Weishanhu District (34°46′ N, 117°08′ E) of Nansi Lake in China during November 2015. Sediment from the top 20 cm at the same site was gathered using a sediment corer. After removing large particles and plant/animal residue, the sediment was mixed uniformly and then stored until early March 2016, when it was used to plant *V. natans* rhizomes in the glass greenhouse (Linyi University, Linyi, China). Once the seedlings were 20 cm in height, during late April 2016, they were transplanted into high-density polyethylene bucket (Φ100 cm, H 120 cm) with 10-cm deep sediment. 260 seedlings each were planted in four such plant-sediment systems and incubated in the greenhouse. Water from the Nansi lake was injected slowly into the buckets, and water quality parameters (total nitrogen [TN], NH_4^+-N, NO_3^--N, NO_2^--N, total phosphorus [TP], chlorophyll a [Chl-a]) were measured after one week. We incubated *V. natans*in water that matched the quality of Weishanhu District. Nutrients, plankton and water were replenished regularly whenever depleted through sampling or evaporation. Monthly until the macrophytes died and decayed, we measured O_2, pH, and Eh microprofiles, periphyton, rapid light curves (RLCs), and the plant growth index of *V. natans*.

We analysed O_2, pH, and Eh microprofiles using microsensors (Unisense, Science Park Aarhus, Denmark), following methods modified from literature [1]. All experiments and calibrations were conducted at 20°C. To prepare the plants for measurements, we fastened macrophyte leaves on a plastic holder in a glass beaker. The experiments were carried out under controlled light from a fibre-optic halogen lamp (100 μmol photons·m^{-2}·s^{-1}) in laboratory. O_2 microelectrodes of 10 μm tip diameters were used to study O_2 microprofiles, and the O_2 microelectrode was mounted on a motor-driven micromanipulator (Unisense, Denmark) and connected via a Microsensor Multimeter to a personal computer. Linear calibration was performed using an O_2-free solution and air-saturated distilled water. The pH and Eh microelectrodes were calibrated and positioned according to the literature [1]. Measurements occurred before there was sufficient time for a gradient to develop. The gradients were measured with the periphyton intact.

Periphyton was scraped from a 10-g fresh weight sample of *V. natans* with a fine brush, and leaves were rinsed with sterile water. Subsequently, the water containing the periphyton was diluted with distilled water to 500 mL. Dry weight (DW), ash weight (AW), ash-free dry weight (AFDW), and Chl-a were measured according to previously described methods [1]. We also measured the rapid light curves (RLC) of *Vallisneria natans* with the pulse amplitude modulated fluorometer (Diving-PAM) following methods from literature [1]. Data were analysed using Wincontrol v2.0 (Walz GmbH, Effeltrich, Germany).

Before statistical analyses, all data were tested for homogeneity and normal distributions of variance. The data of DW, AW, AFDW, and the Chl-a content of the periphyton were analysed with One-way ANOVA, with aquatic plant growth in months as the independent variable. If these results were statistically significant, the differences are experimented by a Tukey's honest significant difference (HSD) test with a confidence interval of 95%.

3. Results

The rapid growth stage for *V. natans* in this study spanned June, July, and August; the stable growth stage, September and October; and the decline stage, November and December. During the growing period of *V. natans*, the O_2 concentration and pH decreased from 0–2 mm above the leaf surface, whereas the Eh increased (figures 1a, 1b and 1c). The O_2 concentration and pH experienced a gradual increase as macrophytes entered the vigorously growing stage (figures 1a and 1b). This gradual increase continued with *V. natans* growth until both variables peaked during the stable growing stages. In contrast, during the decline stage, O_2 concentration and pH decreased as the plant grew older. Overall, O_2 and pH increases peaked in September, respectively reaching 552.7 $\mu mol \cdot L^{-1}$ and 9.4. This period coincided with the point at which macrophytes entered reproductive growth and growth slowed (figures 1a and 1b). Next, starting from October, O_2 and pH decreased gradually, dropping to their minimum values of 348.2 $\mu mol \cdot L^{-1}$ and 7.8, respectively, during December. Correspondingly, the spatial characteristics of O_2 and pH differed in June compared with December.

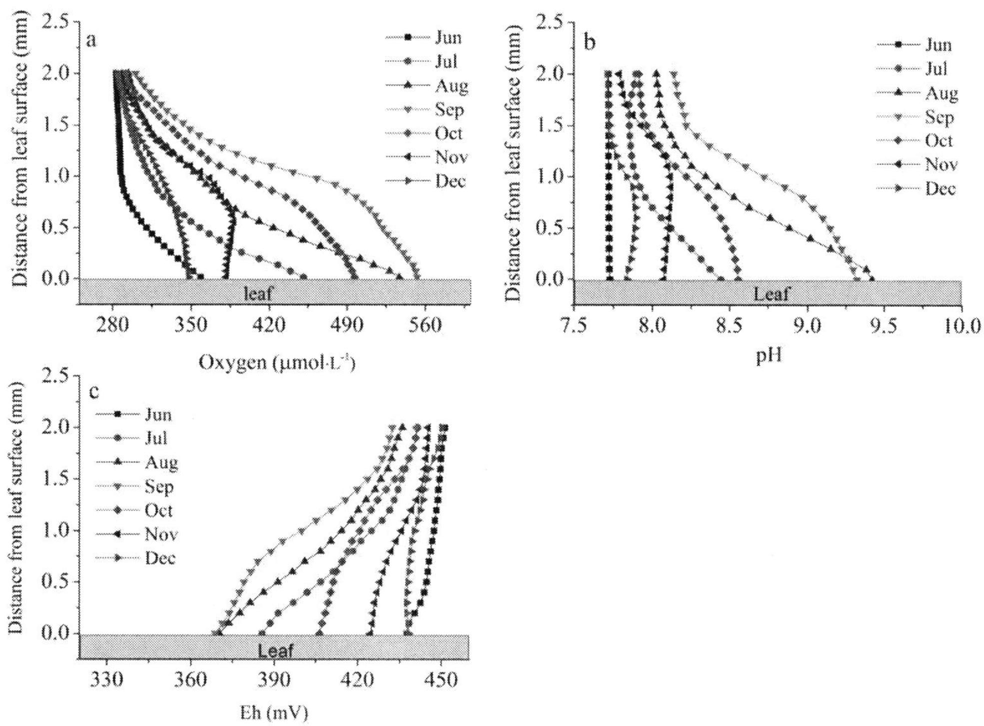

Figure 1. O_2, pH, and Eh microprofiles around *Vallisneria natans* leaves across months. O_2 microprofiles (a), pH microprofiles (b), and Eh microprofiles (c) were measured with microsensors as the macrophytes grew. Data points represent the means of triplicates.

Contrary to the microprofiles of O_2 and pH, the Eh markedly increased from 0–2 mm above the leaf surface, reaching its minimum value at the leaf surface. The decrease in Eh accompanied the entry of macrophytes into their rapid growth stage (figure 1c). Overall, Eh decreased gradually from June to September, reaching the lowest point (368.81 mV) in September (stable growing stage). Next, Eh increased gradually from October to December (decline stage). Neither December nor June experienced significant changes in Eh microprofiles, but the spatial characteristics of Eh differed in December compared with June.

The periphyton exhibited clear changes on the *V. natans* leaf surface during the macrophytes' life cycle (table 1). All measured variables except Chl-a (i.e. DW, AFDW, AW, and thickness) increased gradually from June to December, peaking in the latter month. Compared with their values in June, the DW, AFDW, AW, and thickness in December were 13.67, 14.17, and 13.52, 13.62 times higher, respectively. The proportion of AFDW in the periphyton increased gradually from June (18.81%), peaking (23.63%) in October, and then declining.

Table 1. Characteristics of periphyton on the *Vallisneria natans* across months.

Month	DW mg·cm^{-2}	AFDW mg·cm^{-2}	AW mg·cm^{-2}	Chl-a μg·cm^{-2}	Thickness μm
Jun	0.31 ±0.02 a	0.06 ± 0.01 a	0.25 ±0.01 a	1.03 ± 0.10 a	40 ± 3 a
Jul	0.59 ± 0.04 a	0.12 ± 0.01 a	0.47 ± 0.03 a	2.39 ± 0.22 b	77 ± 5 b
Aug	0.99 ± 0.07 b	0.21 ± 0.01 b	0.78 ± 0.01 b	3.26 ± 0.30 b	128 ± 9 c
Sep	1.67 ±0.08 c	0.37 ± 0.02 c	1.30 ± 0.07 c	6.31 ± 0.62 c	217 ± 15 d
Oct	2.32 ± 0.22 d	0.55 ± 0.04 d	1.77 ± 0.15 d	10.65 ± 0.83 d	330 ± 20 e
Nov	3.32± 0.27 e	0.71 ± 0.05 e	2.60 ± 0.22 e	15.13 ± 1.24 e	430 ± 30 f
Dec	4.24 ± 0.36 f	0.85 ± 0.06 f	3.38 ± 0.25 f	15.97 ± 1.27 e	550 ± 33 g

Notes: Different lowercase letters indicate that there is a significant difference.

The RLC, the curve of electron transfer rate changing with light intensity, contains information on physiological flexibility with which a plant can adapt its photosynthetic apparatus to rapid changes of light intensity. It allows insight into characteristics of actual photosynthesis of plant. Hence, RLC can appraise the photosynthesis capability of plant leaves. The RLC of *V. natans* leaves differed significantly across various months (figure 2), indicating that the leaves' photosynthetic capacity experienced monthly fluctuation Photosynthetic capacity increased gradually as *V. natans* grew, and reached its maximum in September (the stable growth stage), with a maximum potential rate of electron transport (ETR_{max}) of 34.85 μmol electrons·m^{-2}·s^{-1}. Starting from October, photosynthetic capacity then gradually decreased, reaching the lowest point in December (the decline stage), with an ETR_{max} that was 25.82% of the stable growth stage. The RLC patterns of variation were similar to patterns observed for O_2 and pH (figures 1a, 1b and 2).

IOP Conf. Series: Earth and Environmental Science 82 (2017) 012026 doi:10.1088/1755-1315/82/1/012026

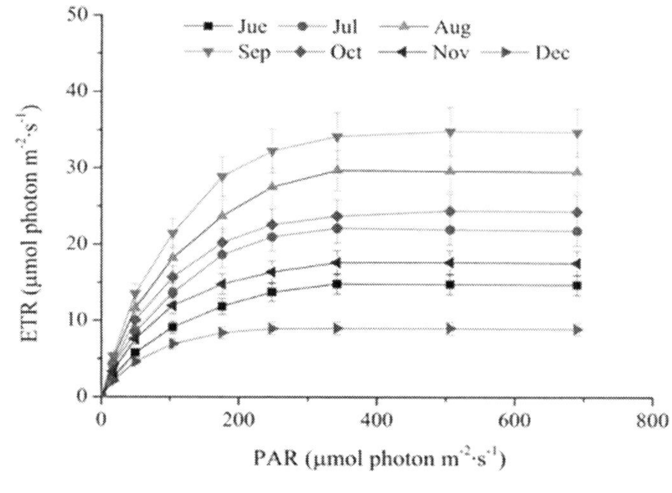

Figure 2. Rapid light curves of *Vallisneria natans* across months. Data points and vertical bars represent the mean of triplicates and standard errors, respectively. (photosynthetically available radiation, PAR; electron transport rate, ETR)

4. Discussion

Our results demonstrated that stages in the macrophyte life cycle could diversify O_2, pH, and Eh microprofiles (figures 1 and 3). The patterns of O_2 and pH values presented a unimodal pattern, being lower during the plant's rapid growth and decline stages, and higher during the stable growing stage (figure 3). Variation in the Eh was the opposite of the O_2 and pH patterns. These outcomes can be explained by the observation that while cell division and metabolism were quite fast during the rapid growth stage, photosynthetic capacity was also relatively weak due to low chlorophyll content in the young plant. Thus, the macrophyte exerted little effect on O_2, pH, and Eh microprofiles during this time.

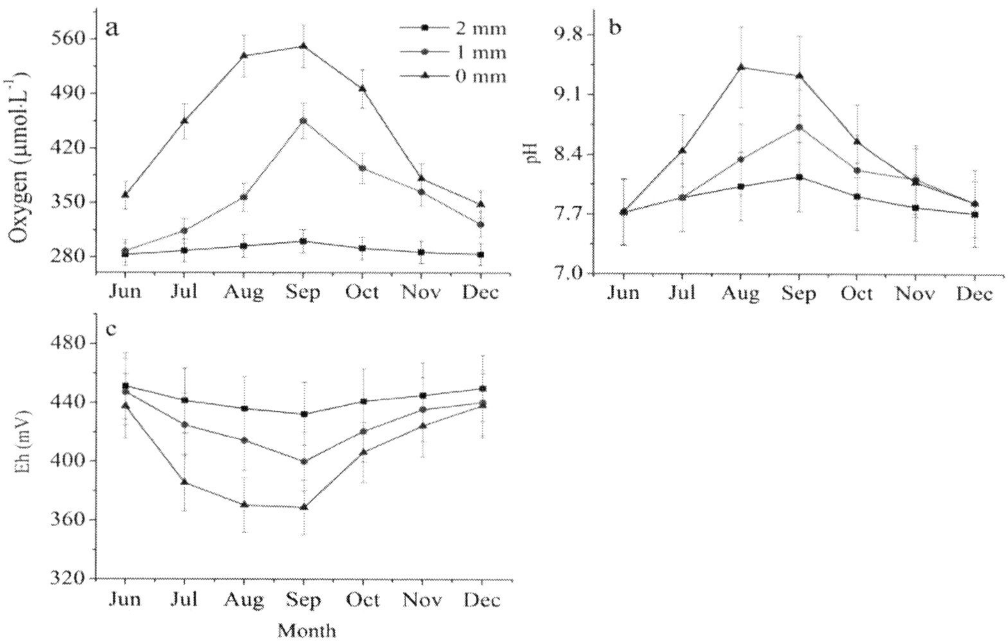

Figure 3. Magnitude of O_2, pH, and Eh at different distances from the *Vallisneria natans* leaf across months. O_2 concentration (a), pH (b), and Eh (c) values were measured with microsensors at distances of 0, 1, 2 mm above the macrophyte leaf surface. Data points and vertical bars represent the means of triplicates and standard errors, respectively.

However, once the maximum photosynthetic capacity was reached at the stable growth stage, *V. natans* was capable of exerting a powerful influence on O_2, pH, and Eh microprofiles. We therefore propose that the macrophyte life cycle affects O_2, pH, and Eh microprofiles via altering photosynthetic capacity. Different growth stages may also affect the distributions of CO_2, phosphates, nitrates, and microbes in the DBL. If these connections can be supported with future empirical research, it should provide new clues for the control mechanism of hydrophytes on nutrient cycle in eutrophic lakes.

Our data also support the idea of the periphyton being a major factor affecting O_2, pH, and Eh microprofiles, although the strength of the effect is strongly linked with the macrophyte life cycle. This link exists because periphyton accumulation and loss naturally accompanies macrophyte growth. During the rapid growth stage, *V. natans* was reported to continuously produce new tissue and slough off old tissue, which results in the easy detachment of the periphyton [6]. Additionally, previous research has also shown that the periphyton was inhibited by the phenolic acid secretions of *V. natans* during the rapid growth stage. Therefore, the periphyton was very sparse during this period in the macrophyte life cycle and did not exert any obvious effect on O_2, pH, and Eh microprofiles (table 1).

However, during the stable growth stage (September–October), *V. natans* biomass and physiological activity peaked, increasing the amount of microorganisms attached to the leaf surface. In turn, these microbes continuously secreted extracellular polymeric substance, which allowed more microorganisms and inorganic particles to adhere. As a result, the periphyton grew thicker and more biologically complex (table 1), with microbes actively modifying the colonised plant microenvironment. Thus, O_2, pH, and Eh gradients became significantly steeper due to the thicker periphyton and the high physiological activity of *V. natans* (figure 1). These changes in O_2, pH, and Eh increased the suitability of the external environment, allowing small microbial colonies to grow and eventually merge, supporting the dense periphyton [7].

Finally, during the decline stage, *V. natans* physiological activity decreased substantially, leading to a reduction in the fluctuation intensity of O_2, pH, and Eh. As a result, the distribution of O_2, pH, and Eh became dominated by the thick periphyton, rather than by the macrophyte.

To summarise, the O_2, pH, and Eh microprofiles measured in this study were the combined outcome of macrophyte growth stages and changes to the periphyton. We note that our results are supported by data from *Potamogeton crispus* in Almind Lake of Denmark, which demonstrated variations in O_2 microprofiles that are consistent with this study [2].

We found that the DBL microenvironment became more complex in response to submerged macrophyte growth and periphyton increases. The thickness of the DBL is positively correlated with highly eutrophic water. Moreover, the DBL can hinder soluble substance transport to the macrophyte, thus slowing macrophyte growth or even accelerating macrophyte decline. Therefore, research on O_2, pH, and Eh microprofiles around submerged macrophytes at various stages in their life cycle reveal more information about the DBL microenvironment, providing significant insights into macrophyte management and nutrient cycling in eutrophic waters.

5. Conclusion

Our results provide evidence that macrophyte growth stages induced steep gradients inO_2 concentrations, pH, and Eh of eutrophic waters, via increasing DBL thickness, macrophyte photosynthetic capacity, and periphyton biomass. We advocate more studies on how soluble substances and microbes are distributed or transformed in microprofiles, because the resultant data are necessary to understand how the DBL surrounding macrophyte leaves regulates nutrient cycling in eutrophic, polluted waters.

Acknowledgments

This research was supported by National Natural Science Foundation of China (41603071 and 41173078), Introduction of Talent Research Funding (LYDX2016BS074) and National Undergraduate Innovation and Entrepreneurship Training Program (201610452018).

References

[1] Dong B, Han RM, Wang GX and Cao X 2014 O_2, pH, and redox potential microprofiles around *potamogetonmalaianus* measured using microsensors *PLoS One* **9** e101825
[2] Sand-Jensen K, Revsbech NP and Barker Jörgensen B1985 Microprofiles of oxygen in epiphyte communities on submerged macrophytes *Mar Biol* **89** 55-62
[3] Eriksson PG 2001 Interaction effects of flow velocity and oxygen metabolism on nitrification and denitrification in biofilms on submersed macrophytes *Biogeochemistry* **55** 29-44
[4] Dong B, Han RM, Wang GX, Mao LN, Ji F and Ma J2015Micro-optode analysis of the spatio-temporal dynamics of O_2 in the micro-boundary layer around *Potamogetoncrispus* leaves and stems *Acta Ecol Sin* **35** 8202-10
[5] Romanów M andWitek Z2011 Periphyton dry mass, ash content, and chlorophyll content on natural substrata in three water bodies of different trophy *Oceanol Hydrobiol St* **40** 64-70
[6] Dong B, Lu QP, Wang GX, Mao LN, Lin Hand Zhou F2013 The impact of periphyton attached on *Potamogetoncrispus*on nitrogen and phosphorus loads in water *J Lake Sci* **25** 359-65
[7] Pereira T A, Felisberto S A and Fernandes VO 2013 Spatial variation of periphyton structural attributes on *Eichhorniacrassipes* (Mart.) Solms in a tropical lotic ecosystem *Acta Sci: Biol Sci* **35** 319-26

IOP Conf. Series: Earth and Environmental Science 82 (2017) 012027 doi:10.1088/1755-1315/82/1/012027

Urban permeable pavement system design based on "sponge city" concept

M M Yu[1,2], J W Zhu[1], W F Gao[1], D P Xu[1] and M Zhao[1]

[1]Xi'an University of Technology, Xi'an, China

E-mail: ymm_fish@163.com

Abstract. Based on the "sponge city" concept, to implement the goal of building a city within the city to solve the sponge waterlogging, rational utilization of water resources, reduce water pollution this paper, combined with the city planning level in China, establishes the design system of city road flooding from the macro, medium and micro level, explore the design method of city water permeable pavement system, and has a practical significance the lower flood risk water ecological problems. On the macro level, we established an urban pavement sponge system under the regional ecological pattern by "spot permeable open space – low impact developing rain water road system – catchment area and catchment wetland". On a medium level, this paper proposed the permeable suitability of pavement and the planning control indicators when combined with urban functional districts to conduct permeable pavement roads plans and controls. On micro level, the paper studied sponge technology design of permeable pavement from road structure, surface material, and other aspects aimed at the pavement permeability requirements.

1. Introduction

The urbanization accelerated in recent years while the speed of the municipal facility construction in many cities is much further behind, causing many urban diseases. The waterlogging problem shave especially become normal after heavy rain, seriously constraining urban development.

According to the 2010 research of Chinese 351 cities for three years during 2008-2010, 62% of the city incurred waterlogging events [1] and a series of problems, such as loss of rainwater resources, runoff pollution, flood risk, and so on. The city's hard impervious surface is one of the primary reasons causing waterlogging. Now the average area of impervious is 50% in our country, while in the megacities Beijing and Shanghai, the average impervious area reaches 71.3% [2], pavement waterproof, blocking the rain adding to groundwater, rainwater exclude poor aggravate the automobile exhaust pollution, rainwater drainage, evaporation increased dust pollution in the city, and hard surface absorbs heat to increase urban heat island effect.

Searching for a new method for storm water management is the fundamental approach to solving these problems [3]. In the 1990s, America proposed low impact development model (LID) based on urban ecology environment to realize the comprehensive management of storm water runoff control and rainwater exploitation. Based on this, China proposed the "Sponge city" theory in 2009, and the Ministry of Housing and Urban-Rural Development issued "Sponge City Guide-Low Impact Development System Construction" and ruled sponge city construction, which indicated permeable pavement being an important method. The permeable pavement refers to the use of good permeability materials on ground pavement, enhancing ground air and water permeability, and the natural rainwater infiltration and purification. It has various functions to reduce surface runoff, relieve urban waterlogging, purify rainwater runoff pollution, improve river water quality, effectively supply groundwater, cooling, humidification, reduce noise, improve urban environment, restore soil ecological environment, promote biodiversity and more [2]; the permeable pavement is very significant for solving the urban waterlogging problem, improving urban environments, and promoting urbanization. At present, our country's primary study of permeable pavement focused on design, construction and other technical aspects [4-8]. It has less relation with urban planning, which led to the lack of systematic theoretical guidance and operational methods for permeable pavement construction. This paper is based on "Sponge city" theory combined with urban planning, and established urban permeable pavement design system. It has crucial realistic significance to implement the sponge city constructing target and solve waterlogging and ecological problems.

2. Sponge city concept

"Sponge city" theory was proposed based on low impact development rainwater exploitation model and water sensitive city planning theory; its essence is to realize the urban development coordinates and resources. By controlling rainwater runoff, natural accumulation, and peak regulation, the sponge city proposes low impact development to play regulation function to promote rainwater's natural infiltration, repair water circulation system, restore water ecology, and purify water quality by nature to solve water storage problem [9]. By infiltration, storage, purification and reuse, water circulation ecological balance between ecology and human is achieved.

Water ecological problems relate to basin problems, so sponge city should not only focus on inner city's water system regulation, rainwater processing, and site measures, but should also focus on basin and region issues and combine urban planning's scale and stage to study cross regional, scale urban planning, and construction based on water ecology (figure 1).

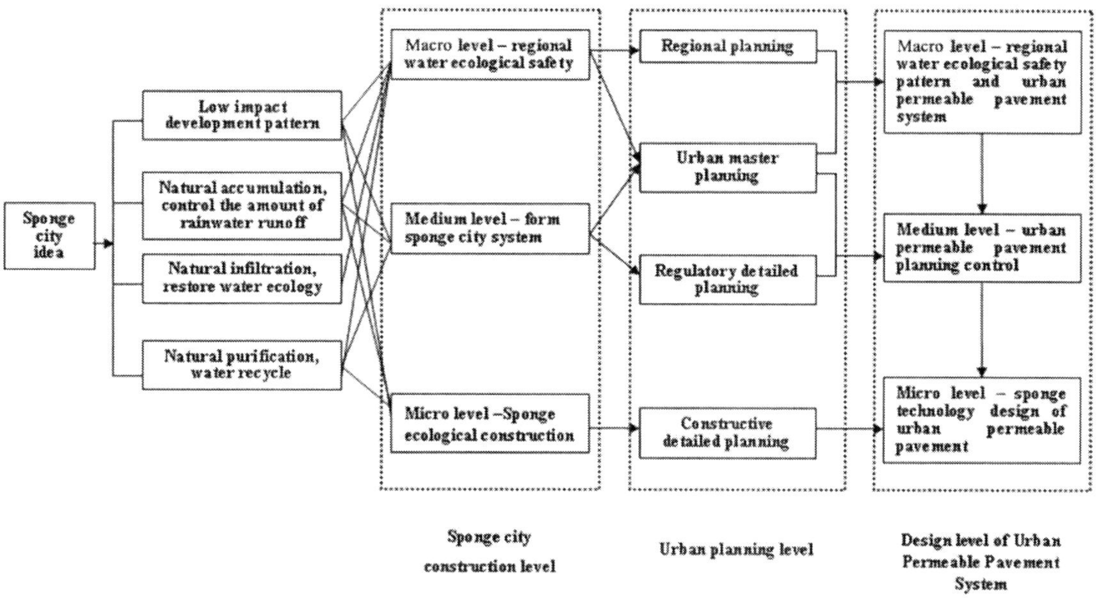

Figure 1. Sponge city concept combined with urban planning work.

On the macro level, sponge city mainly refers to the urban water system. The construction key is to study regional water ecological safety, establish safe patterns for the water ecological system by analyzing important locations and relationships of the system, implementing it into regional and urban planning, and gradually realize water ecology by controlling the space administration of region and urban construction. The system provides control objectives for urban annual runoff, gives clear low impact development strategies and construction processes, and brings relative requirements into municipal projects planning.

On a medium level, sponge city primarily refers to urban areas where the construction key is effectively used in regional water systems. The urban master and regulatory planning is combined with urban pavement patterns to set watershed and catchment areas, form the sponge city system with spots, lines, and surface combinations, and satisfy functions like regional rainwater storage and regulations, purify the water quality, and protect water ecological wetlands.

On the micro level, sponge city primarily refers to the construction of some water catchment units like urban road, square, parking, and garden sites. The key problem of construction is the technical implementation of water ecological infrastructure, including the urban rainwater administration green sponge technique, human wetland water purification technique, water ecological restoration technique, permeable pavement design, construction technology, and so on [10]. By the construction of detailed planning and design and the use of modern technology, maximum develop the water ecological system's function to realize construction aims of sponge city.

Permeable pavement is used as the sponge city's measure of construction, which should integrate the concept into the entire urban planning and each step throughout the construction. It should follow the guidance of "regional systematic technology" and combine the permeable pavement design with urban planning working steps to establish an urban permeable pavement design system, and clear every steps' content of urban permeable pavement: on the macro level, systematic planning is based on regional water ecological patterns. On the medium level, planning control is combined with urban functional distracts. On the micro level of permeable pavement, the sponge technology is designed.

3. Urban permeable pavement systematic composition

Urban pavement consists of road, square, parking lot, and urban site pavements. Each type has different water requirements due to different traffic quantities, and road loads directly impact the permeable pavement design (table 1). The quality of runoff road rainfall water has a relationship with geographical environment, nature of surrounding land, road grade, rainfall intensity, and rainfall duration [4], and the permeable requirements are also based on runoff water quality. According to China's current urban road classification standards and road characteristics, expressway, trunk road, logistics, and warehouse and car traffic, the industrial sites have a large amount of traffic volume, and rainwater quality is poor with serious pollution, therefore the permeable requirement is low. Secondary roads and parking lots have a large traffic volume, and the rainwater quality is general, so the road permeable requirement is high. Garden and green land have good rainwater quality, so the permeable requirement is high.

Table 1. Urban permeable pavement construction analysis.

Permeable pavement construction		Characteristics	Quality of rainwater analysis	Permeable requirement
Urban road surface	Express way	Large amount of traffic volume, high demands on road load	SS[a]: >1000 mg/L COD: >800 mg/L	Relatively low
	Main road	Large amount of traffic volume, high demands on road load	SS>1000 mg/L COD>800 mg/L	Low
	Secondary road	Large amount of traffic volume, high demands on road load	SS: 500-1000 mg/L COD[b]: 400-800 mg/L	General
	Branches	Small amount of traffic volume, low demands on road load	SS: 100-400 mg/L COD: 100-400 mg/L	Relatively high
Urban square,	Urban square	Pedestrian traffic, low demands on road load	SS: 100-200 mg/L COD: 100-200 mg/L	High

parking lot surface	Parking lot	Car traffic, high demands on road load	SS: 100-400 mg/L COD: 100-400 mg/L	General
Urban site surface	Urban garden, green space	Pedestrian traffic, low demands on road load	SS: ＜100 mg/L COD: ＜100 mg/L	High
	Other sites	Pedestrian and car traffic, general demands on road load (high demands on logistics warehouse, industry site)	SS: 100-200 mg/L COD: 100-200 mg/L	Relatively high (low demands on logistics warehouse, industry site)

[a]SS (Suspended Solids) refers to the content of suspended solids in water, including insoluble in water, inorganic matter, organic matter and mud sand, clay, microorganisms, etc., is one of the indicators to measure the degree of water pollution.
[b]Under certain conditions, COD (Chemical Oxygen Demand) refers to the chemical oxygen demand to take a certain strong oxidant treatment of water samples consumed by the amount of oxidation, often used as a measure of the amount of organic matter in water indicators. The bigger the COD, the more serious the pollution of organic matter.

4. Multi scale urban permeable pavement system design on sponge city theory

4.1. Macro level – regional water ecological safety pattern and urban permeable pavement system

On the aspect of regional and urban master planning, we analyzed some key water ecological safety issues using the ecological safety pattern method to regulate flood storage, water quality and safety, water environment protection, and to establish comprehensive water ecological safety pattern surrounding ecosystem services.

The water ecological safety pattern uses low impact development rainwater exploited mode to plan urban road system, and to enhance roads and surrounding green land construction. The permeable pavement method is used for road green space and rainwater tree pools to improve road capacity of absorbing water and reducing road rainwater runoff quantity. The continuous and open space system is formed with high permeable requirements of urban square, garden, and green lands, combined with rainwater runoff pipe system and exceeding rainwater runoff discharge system to enhance the rainwater absorbing and discharging capability of urban square, green land, and garden, for the support of the urban water system. By using nature, traffic volume, geological conditions, and climate conditions, urban sites determine the comprehensive permeable pavement catchments. The urban pavement sponge system is finally formed by "spot permeable open space – low impact developing rain water road system – catchment area and catchment wetland" (figure 2). It proposed the requirements and goals of all constitutes under the low impact development, then translated them into special planning design requirements. Urban green space planning should combine spot permeable space to reasonably reserve space, help square, garden, green land, and their surrounding hardening areas' rainwater runoff infiltrate and regulate, purify, and should connect the rainwater drainage system. Combined with road and green water retention facilities, traffic planning should coordinate the land and space use and vertical design inside and outside of the road boundaries to achieve low impact road development targets.

4.2. Medium level – urban permeable pavement planning control

Urban regulatory detailed planning stage decomposes and refines the low impact development aims and requirements, which is proposed by urban master planning on the aspects of spot infiltrate open space, low impact development rainwater road system, catchment area, and wetland in urban permeable pavement system, and brings land planning points into focus, converts specific lot control indicators, overall coordinates, and is systematically designed and constructed.

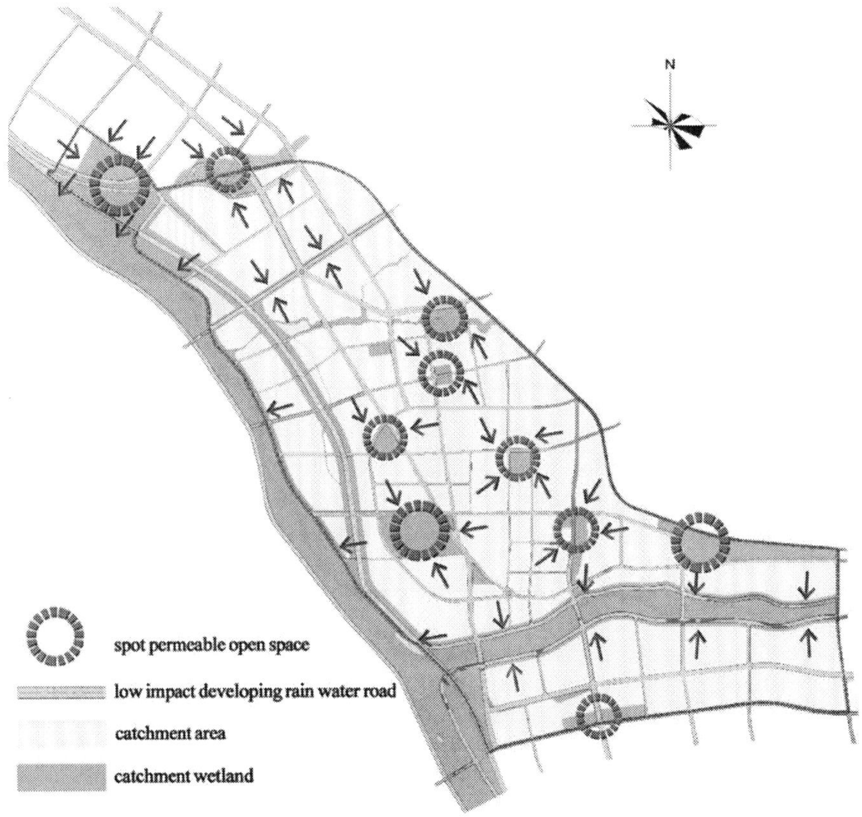

Figure 2. Urban pavement sponge system.

Combined with existing research and practice [11-15], and some related technology standards and requirements, such as "Outdoor Drainage Design Specifications" and "Green Building Evaluation Criteria", the effective control indicators of urban permeable pavement mainly includes the permeable ground proportion, runoff coefficient, and low elevation green space ratio. The permeable ground proportion and runoff coefficient are important indicators to measure rainwater infiltration capability which directly affects pavement permeable effects. The low elevation green space is an effective measure to maximize the permeability of pervious pavement and rainwater storage and achieve water ecological system regulating balance. Because the permeable pavement is different from city lake water system the site aspects, land using characteristics, service objects, water quality, and landscape demand, its control indicators should be determined by urban functional distract and related requirements in order to provide future urban road surface technical designs.

Urban comprehensive functional distract generally includes green leisure area, residential area, public service area, logistics and warehouse area, and industrial area. Based on the analysis of traffic volume and road surface load, the green leisure area is primarily for walking as its air condition and rainwater quality is good and its landscape requirements are relatively high. The residential area has traffic volume, but its number and road load are small, and the air condition and rainwater are good. The public service area has a large traffic volume and road load, and a normal air condition and rainwater quality. The logistics warehouse area and industrial area have large traffic volume and road load and its air condition and rainwater quality are poor, therefore the high permeability will cause pollutants to seep with rain, causing contamination to the soil and underground water. Based on the above analysis, this paper holds suitable classification on urban permeable pavement and proposes corresponding planning control indicators recommendations. (tables 2 and 3 and figure 3)

Table 2. Suitability partition of urban permeable pavement.

Urban functional distract	Traffic volume and road load	Rainwater quality	Landscape demands	Water permeable suitability grade
Green leisure area	Pedestrian traffic, low demands on road load	Good	high	I
Residential area	Small volume of traffic, low demands on road load(refer to branches)	Relatively	Relatively high	II
Public service area	Large volume of traffic, high demands on road load(refer to secondary road)	General	Relatively high	III
Logistics and warehouse area, industry area	Large amount of traffic volume, high demands on road load(refer to main road and secondary road)	Bad	General	IV

Table 3. Urban permeable pavement planning control.

Urban permeable pavement grade	Indicator control demands
Water permeable grade I	Proportion of permeable pavement ⩾ 70%, runoff coefficient ⩽ 0.15, proportion of low elevation green space⩾80%
Water permeable grade II	Proportion of permeable pavement ⩾ 50%, runoff coefficient ⩽ 0.45, proportion of low elevation green space⩾70%
Water permeable grade III	Proportion of permeable pavement ⩾ 50%, runoff coefficient ⩽ 0.45, proportion of low elevation green space⩾50%
Water permeable grade IV	Proportion of permeable pavement ⩾ 40%, runoff coefficient ⩽ 0.6, proportion of low elevation green space⩾30%

The above indicators are combined with recommended values of practice experience and research. Because the differences between nature conditions, development status, and construction speed, as well as the problems between new and old cities, to take reference and adjustments, it is important to start from the actual situation during the specific construction and planning controls of urban permeable pavement. Old city area control indicators could be moderately reduced and large rainfall area of urban control indicators could be moderately improved.

4.3. Micro level – sponge technology design of urban permeable pavement

Based on the LID technology, urban permeable pavement design must construct new types of permeable pavement (table 4). On the one hand, to enhance the capability of rainfall runoff to infiltrate into urban sponge system. On the other hand, permeable pavement should combine low elevation green space, rain garden, and ecological grass site to form the sponge system, collect the rainwater, then store, filter, and purify it.

Figure 3. Urban permeable pavement planning control classification.

Table 4. Urban permeable pavement technical design.

Permeable pavement type		Pavement structure	Surface material
Urban road	Expressway	Surface layer uses permeable materials, base layer and cushion layer use poor permeable and impermeable materials	Sidewalk use permeable brick; roadway use permeable concrete, permeable asphalt mixture
	Main road	Surface layer uses permeable materials, base layer and cushion layer use poor permeable and impermeable materials	Sidewalk use permeable brick; roadway use permeable concrete, permeable asphalt mixture
	Secondary road	Surface layer and basement have good permeability, cushion layer has low permeability	Sidewalk use permeable brick; roadway use permeable concrete, permeable asphalt mixture
	Branches	Surface layer and basement have good permeability, cushion layer has low permeability	Sidewalk use permeable brick; roadway use permeable concrete, permeable asphalt mixture
Urban pavement	square	Surface layer, base layer, cushion layer have good permeability and water storage capability	Permeable brick, color permeable concrete
Parking lot		Surface layer and base layer have good permeability, cushion has low	Surface use permeable concrete, permeable asphalt mixture

	permeability	
Garden, green space	Surface layer, base layer, cushion layer have good permeability and water storage capability	Permeable brick, color permeable concrete
Logistics and warehouse, industrial area pavement	Surface layer uses permeable materials, base layer and cushion layer use poor permeable and impermeable materials	Site road surface use permeable concrete, permeable asphalt mixture materials; other pavement use permeable brick or color permeable concrete
Other sites	Surface layer and base layer have good permeability, cushion has low permeability	Site road surface use permeable concrete, permeable asphalt mixture materials; other pavement use permeable brick or color permeable concrete

For the road types, permeable demands are different for each area, and permeable pavement has relative difference on road structure and surface materials. By the current practice, water permeable brick, pervious concrete, and permeable asphalt mixture are good permeable pavement materials.

High permeable required pavement: pavement surface layer, base layer, and cushion layer must have good water permeability and water storage capability. The surface layer usually uses larger pore permeable brick, pervious concrete, and pervious asphalt mixture. The cushion layer could use pervious sand cushion to enhance the pervious effect. It is mainly applicable to the road of urban square, garden and green space, and the pavement permeability could reach 70%.

Medium permeable required pavement: surface layer and base layer must have good water permeability. The surface layer is usually use pervious brick, pervious concrete, and pervious asphalt mixture. The base layer could use good permeable grading gravel, and the cushion layer has low permeability and could use impervious materials, such as asphalt sand. Anti-infiltration earthwork cloth could be put on the ground to protect the subgrade. It is suitable for the urban parking lot road, sub roads, and branches, and the pavement permeability could reach 50%.

Low permeable required pavement: the pervious surface layer must use pervious materials, and the base and cushion layers must use impervious or poor permeability materials, such as asphalt. Rainwater passes through the surface layer, then flows directly out of the roadbed and protected roadbed. It is suitable for the urban expressway, main roads, logistics and warehouse area, industrial area, and so on, and its pavement permeability could reach 40%.

5. Conclusion

Urban permeable pavement is the most effective way to solve urban rainwater drainage, flood regulation, and water purification, but partial construction and spot repair could not achieve all of the demands. This paper established a system design fame combined with urban planning work stages. To ensure its operability, such aspects of study need to be continued to truly achieve the goal of sponge city construction:

On the technique research aspect, we should strengthen the research on permeable pavement of urban control, choosing suitable technology to take long-term monitor on permeable pavement, and collect continuous data to evaluate different road drainage and storage effect by different functional distracts. To make permeable pavement more adaptable, the permeable pavement design research of sponge technology should be enhanced, especially on the aspects of road surface layer pervious performance, material diversification, drainage mode, and its reasonable research.

The technique operation aspect should be based on and suit the regional condition. To reduce the safety hazard of groundwater pollution, it is unsuitable to use the high permeable road type in unfavorable geological conditions, unfavorable soil permeability, and highly polluted areas. The construction of legal system should be improved as soon as possible to make the legal protection on permeable pavement construction. To simultaneously conduct the new and old area, the urban permeable pavement system steps should be reasonably created.

References

[1] Li X Z and Li N 2012 "Hard" pavement and permeable pavement *Enc. Knowl.* **16** 26-7
[2] Shu P and Zhang F X 2016 Permeable pavement application of old quarter in North under the background of sponge City *Mod. Dec. (Theory)* **10** 202
[3] Feng B 2016 Discussion on comprehensive utilization of low impact development rainwater model *Chin. High-tech. Ent.* **33** 109-10
[4] Zhu M and Dang Q P 2014 Design and utilization of urban green road rainwater control *Sichuan Arch.* **34(1)** 213-5
[5] Dong X and Shen Z 2010 Analysis on choice of engineering site and type in Chinese urban pervious pavement construction *J. Gansu Agric. Univ.* **6** 145-50
[6] Li M L, Chen J, Wang Z H and Di S G 2015 Evaluation of ecological effect of permeable slow-traffic system based on low-carbon perspective *J. Highway Tran. Res. Dev.* **4** 40-4, 51
[7] Xu D K and Lv W Y 2012 Low impact development model application-pervious pavement development review *N. Build. Mater.* **3** 31-4
[8] Chu T, Han B and Xu Y J 2014 A discussion on city pedestrian pervious pavement and its pervious surface layer materials *Chin. Build Mater. Sci. Tech.* **4** 68-70
[9] Qiu B X 2015 Connotation, approaches and prospect of LID *Constr. Sci. Tech.* **1** 1-6
[10] Yu K J, Li D H, Yuan H, Fu W and Wang S S 2015 "Sponge city": Theory and practice *City Plan. Rev.* **39(6)** 26-35
[11] Wang Q, Gu L R, Li W Y and Liu Z 2016 The sponge city practice of the Gui'an innovation park *City & House* **7** 9-14
[12] Zheng D Y and Zhou X Y 2016 A study on technical guideline of sponge city regulatory planning *Planners* **32(5)** 17-22
[13] Zhang L 2016 Path of sponge city construction in northwestern China: An empirical study on Xixian new area *City Plan. Rev.* **40(3)** 108-12
[14] Li K 2014 Indicator analysis of integrated rain water management in low-carbon eco-city *Green Build.* **2** 17-20
[15] Su Z Y and Guo T 2016 Study on runoff control zoning and LID measures quantification in sponge city planning: A case study of Kunming *Shanghai Urban Plan. Rev.* **4** 115-9

SD simulation study on degraded farmland policy on farming-pastoral area under the constrains of water resources--Taking Tongliao City of Inner Mongolia as example

D P Xu[1,2,4], B Zhao[3], T S Li[4,5], J W Zhu[1,2] and M M Yu[1]

[1]School of Civil Engineering and Architecture, Xi'an University of Technology, Xi'an 710048, China
[2]State Key Laboratory of Ecological Water Conservancy in the Northwest Arid Area, Xi'an 710048, China
[3]Academic Affairs Office, Xi'an University of Technology, Xi'an 710048, China
[4]College of Urban and Environment Science, Northwest University, Xi'an 710127, China

E-mail: 12551727@qq.com

Abstract. Water resources are the primary factor in restricting the sustainable development of farming-pastoral regions. To support the sustainable development of water resources, whether or not the land uses patterns of farming-pastoral areas is a reasonably important issue. This paper takes Tongliao city as example for the purpose of sustainably developing the farming-pastoral area in the north. Several scientific productions and evaluations were conducted to study the farming-pastoral landuse pattern, which is the key problem that effects sustainable development of farming-pastoral areas. The paper then proposes that 1:7 landuse pattern is suitable for the sustainable development of farming-pastotal area. Based on the analysis of the research findings on sustainable development of farming-pastoral area, the paper established a suitability evaluation indicators system of degraded farmland policies in Tongliao city, and used an Analytical Hierarchy Process (AHP) method to determine the weight to run system dynamic (SD) model. The simulation results were then obtained on social economic ecological development in Tongliao city under different degraded farmland policies, and used the comprehensive evaluation model to optimize the results. It is concluded that stabilizing the policy of degraded farmland policy is the preferential policy in Tongliao, which provides useful theoretical research for the sustainable development of farming-pastoral area.

1. Introduction

The northern farming-pastoral area is the typical ecological fragile area, which is the crucial ecological safety barrier in China [1]. Since the 1980s, the developing model of irrigated agriculture being the leading industry made the regional landuse pattern turn to an unsustainable state, resulting in a series of problems, such as an imbalance in development of agriculture and husbandry, depletion of surface water, decline of groundwater level, and serious desertification. On the one hand, the imbalanced farming and pastoral landuse ratio led to the waste of too much water resources (average of 75%) in low productive irrigating production sectors, restraining the development of secondary and tertiary industry. On the other hand, the regional ecological environment was seriously damaged [2]. In this case, it becomes promoting regional sustainable development that retionally determines the landuse proportion of farming-pastoral and husbandry areas become the important scientific issue, as well as the water resources allocation in all sectors of social economy.

The sustainable development of farming-pastoral area is the current hot topic, and some scholars hold research on the relationship between the vulnerability of farming-pastoral area and its sustainable development. Hu Binghui *et al* [3] state that there is a close relationship between ecological vulnerability and social construction of sustainable development. Xu Erqi *et al* [4] think that over exploration on agriculture is the root of ecological degradation on farming-pastoral area. Wang Xiuhong *et al* [5] believe ecological restoration is the most effective way to enhance sustainability, and the primary factor effecting sustainable development of farming-pastoral area is resource capability. Zhang Juntao *et al* [6] think water resources and environmental conditions are the key factors effecting regional sustainable development. Long Yinhui *et al* [7] believe that strength of water resource capability is the important factor of ecosystem sustainable development in grassland area, and the suitable use of land resource is another key factor to control the sustainable development of farming-pastoral area. Qi Yuepu [8] and Jin Xin *et al* [9] research the rural sustainable development and the strength and potential of farmland use. Ma Jiwei *et al* [10] studied the complex adaptive configuration model of soil and water resources. Zheng Jiuyu *et al* [11] constructed the matching calculation model on agricultural water and land resources, and evaluated the matching degree on agricultural water and land resources in Hetao irrigation area.

Many research results show the importance of the sustainable development of farming-pastoral area. Furthermore, the harmonia development in farming-pastoral area relates to the ecological safety and economic development in local and large parts of the country. The comprehensive study discovered that farming-pastoral landuse pattern is rarely involved in present studies. Therefore, this paper uses Tongliao city as the typical farming-pastoral area to study sustainable developing pattern under different farming-pastoral areas, and uses system dynamics to simulate the effect of the policy, which is degraded farmland to grassland. The paper provides theoretical and practice research on enhancing the sustainable developing capability in Tongliao and larger farming-pastoral areas.

2. Research Region

The farming-pastoral regions in northern China mainly spread in the southern edge of Inner Mongolia plateau and the area along the Great Wall that has an annual rainfall of around 200~400 mm and degreeness of 1~2. It is the ecological barrier connecting the east and west region of China. Table 1 shows that the urbanization level, industrial structure, industry growth rate, water use efficiency, and the proportion of ecological water use in Tongliao are similar to those in the northern region of farming-pastoral area. Tongliao city can be used as a typical representative of the northern farming-pastoral region. It is worth mentioning that the proportion of the primary industry output in the

Table 1. Comparison table of social and economic development between Tongliao and Northern Agro-pastoral transitional zone.

region	urbanization (%)	Capita GDP (10^4yuan/p)	Industry structure ——	Proportion of agriculture water (%)	Agriculture water-use efficiency (yuan/m^3)	Total water-use efficiency (yuan/m^3)	Proportion of ecological water (%)
The northern farming-pastoral area	38	5.7	14:58:28	85	10.8	63.4	36
Tongliao City	39	2.54	13:59:28	76	9.8	67.5	32

industrial structure of Tongliao and Northern China farming pastoral region is greater than 10%. According to Chenery's theory of standard industrial structure, which means that the primary industry occupy a more important position, the farming and animal husbandry is the main part of the northern farming-pastoral areas. Therefore, it is very important to study the structure of the proportion of agriculture and animal husbandry in the primary industry in the northern farming-pastoral areas.

Tongliao city is located in the eastern part of Inner Mongolia 43°22'~43°58'N 121°42'~123°02'E. The average rainfall is 350~400 mm and the evaporation is about five times the rainfall. In recent years, the Liaohe River, Xinkai River, and Jiaolai River in Tongliao dried up throughout the year due to the arid climate and human development; more than 70 small water reservoirs dried, and local underground water level declined, resulting in a lack of water.

Water shortage, imbalance of water use structure, low water use efficiency, and deterioration of water environment are found to be the important factors restricting the sustainable development of Tongliao. The reason lies in the unbalanced development of agriculture and animal husbandry in Tongliao. The long-term excessive development of agriculture led to the high ratio of farmland. In 2014 the ratio of farmland and grazing grassland is 1:1.4, which is much higher than the water resources carrying capability in farming-pastoral areas, resulting in agricultural irrigation water reaching 85% of the total use of water in Tongliao city, and the output being only 8%. With no more than 15% of usable water resources left, it is hard to allocate into other production departments and urban rural sectors. The way of this inefficient and imbalanced water resources allocation is dangerous for the social economic sustainable development in Tongliao city. Comprehensive analysis shows that the water resources exploration is close to the limit, but the social economic developing status is still on a lower level. Regional water resources will exhaust fast if the maintenance of present farming-pastoral development status is continued, leading irreversible ecological disaster. Therefore, it is crucial to implement the policy of degraded farmland to grassland and only improve the situation of use water resources. It is also necessary to study the suitable ratio of farmland and pastoral land.

3. Evaluation on farming and pastoral landuse pattern

The primary industry in farming-pastoral areas plays an important role on the development of society and economy. This tuition is quite typical in Tongliao city. The primary industral sectors in Tongliao city mainly consist of agriculture and animal husbandry, and the forestry and fisheries only account for 4%. Because agriculture has higher efficiency than husbandry, a large-scale land reclamation began in Tongliao city since the 1980s, resulting in a drastic shrinkage in husbandry and prosperous agriculture. However, contrary to rain-rise husbandary, agriculture needs many water resources, which is the short board in farming-pastoral areas. Through the comprehensive study and interpretation of remote sensing data and long-term observation of underground water data, in the 20 years since 1995, the agriculture land increased 35% and the husbandary land decreased 28% in Tongliao city. At present the surface water resources nearly dried up and the water resources are all from under the ground; the underground water level declines 0.1m each year. It declined from 3.5 m to 5.5 m in the Horqin area, which formed an underground funnel that was 2- meters deep. With the decline of the underground water level, a series of ecological problems followed as the lakes and wetlands disappeared, rivers cutoff, lands deserted, and biodiversity diminished. The development of water and soil resources in this uncontrolled farming and grazing system poses a great threat to the sustainable development of Tongliao's economy and society.

Based on the comprehensive study on the northern farming-pastoral areas and investigation of Tongliao city, the paper constructed different evaluation tables of agriculture and husbandry developing patterns (table 2). Through interpretation of remote sensing data, the investigation shows that the ratio of agriculture and husbandry lands in Tongliao city is 1:1.4. According to the 60% irrigation rate and average water amount, agriculture irrigation water is about 42.81×10^8 m^3, accounting for 85% of total water consumption at 50.36×10^8 m^3. According to "the most stringent water resources administration of water resources department in Inner Mongolia", the average annual available water resources is only 31.54×10^8 m^3, and use of the high water will lead to the overdraft of

59.67% underground water; long-term persistence will lead to ecological disaster. If maintaining the development of promoting agriculture and degraded husbandry, there may be a 1:1 ratio of agriculture and husbandry lands. Although the increased farmlands are slope or barren lands, the popularity of irrigation rate will be reduced to 50%, its total annual water demands will still reach 52.88×10^8 m^3, and the overdraft rate will reach 67.66%. This is the dangerous state of comprehensive development of economy society ecology. To seriously prevent the ecological disasters, it should be strictly controlled on the aspect of the continued expansion of farmland. In the sense of possible ecological disasters, the suitable and reliable ratio of agriculture and husbandry lands is 1:3, which reduced the scale of agriculture development and increased hasbandry landusing. Although the degraded farmlands are mainly the low-yielding fields whose irrigating technology is hard to cover, total irrigation rate increased to 60%, the irrigation technology would improve, and the water consumption would greatly reduce to about 28.47×10^8 m^3.

Using the situation to comprehensively consider the domestic and foreign farming-pastoral areas' agriculture water, the proportion of agriculture irrigation water will reduce to 80%, and the total assumption of water will be 35.58×10^8 m^3. At this time, at a groundwater overdraft rate of 12.82%, it still in the long-term overdrafted situation and is not safe. Considering it is still in an insecure state, keep degraded farmland scale, and reduce the ratio of agriculture and husbandry lands to 1:5, and the overdrafted rate of underground water is -13.14%; the underground water would get a better supplementary. Long-term maintainance would make the economic, social, and ecological development a much safer situation. It is the sustainable situation that the proportion of landuse for agriculture and husbandry is 1:7, and the overdrafted rate is -25.84%. It is ideal, but hard to realize that the proportion of landuse for agriculture and husbandry is 1:9 and the overdrafted rate is -33.18.

Table 2. State evaluation table of development pattern of different agriculture and husbandary in Tongliao.

Ratio of agriculture and husbandary	Farmland area 1×10^4 km^2	Agriculture irrigation rate %	Irrigation water demands 1×10^8 m^3	Total water demands 1×10^8 m^3	Annual water resources available 1×10^8 m^3/a	Groundwater overdraft rate %	State evaluation
1:1	2.14	50	44.95	52.88	31.54	67.66	Super dangerous
1:1.4	1.78	60	42.81	50.36	31.54	59.67	Present state
1:3	1.07	70	28.47	35.58	31.54	12.82	Unsafe state
1:5	0.71	80	20.55	27.40	31.54	-13.14	Safe state
1:7	0.54	90	16.37	23.39	31.54	-25.84	Sustainable state
1:9	0.43	100	13.7	21.07	31.54	-33.18	Ideal state

4. System dynamics simulation

4.1. System dynamics method

Contrasting with other methods, the system dynamics method has time-varing, nonlinear and multiple feedback characteristics, such as analytic hierarchy process [12] information entropy [13] and principle component analysis [14]. It has its own advantage [15] when dealing with the problem of complex and dynamic social economic ecological sustainable development. This paper takes water resource as an important prerequisite to constrain Tongliao's sustainable development. Based on the present farming-pastoral landuse pattern, the simulation model of system dynamics was established to carry on the simulation study of "degraded farmland to grassland" strategy. The evaluating process of farming-pastoral landuse pattern was searched in order to provide useful theoretical exploration of the sustainable development of farming-pastoral areas in northern China.

4.2. System dynamics simulation

This paper uses Vensim as a simulation platform to establish a system dynamics model of the sustainable development in Tongliao city with regional, dynamic, and limited characteristics, which include social development subsystem, economic development subsystem, water resources support subsystem, ecological environmental subsystem, and so on. Figure 1 describes the structure of the system dynamics model in detail.

It took Tongliao city around 30 years to change from being advantaged in husbandry production to agricultural production. During this period, uncontrolled development of agriculture industry made an accumulated destruction on Tongliao's ecological environment. This situation could not be reversed in one day; long-term strategies are needed to reach a sustainable situation in Tongliao city. The fast or slow implementation of degraded farmland has different degrees of impact on Tongliao's economy development due to its fragile ecology, and the normal prerequisite is to take "emergency degraded farmland policy", "stable degraded farmland policy", and "relieve degraded farmland policy". In this paper, the definition of "emergency degraded farmland policy" refers to reaching a 1:7 ratio of agriculture and husbandary lands by 2030, and "stable degraded farmland policy" and "relieve degraded farmland policy" refers to reaching a 1:7 ratio by 2040 and 2050. This model has the ability to consider the ecological protection in operation processes, as well as economic development by allocating the water saved by degrading farmland to the secondary and tertiary industry sectors to ensure regional ecological protection and smooth development of the economy. In order to provide useful theoretical reference for the development of farming-pastoral policies in Tongliao city, this paper uses system dynamics to simulate the social economic ecological developing situations under three types of policies, to enrich the sustainable development theory of northern farming-pastoral areas.

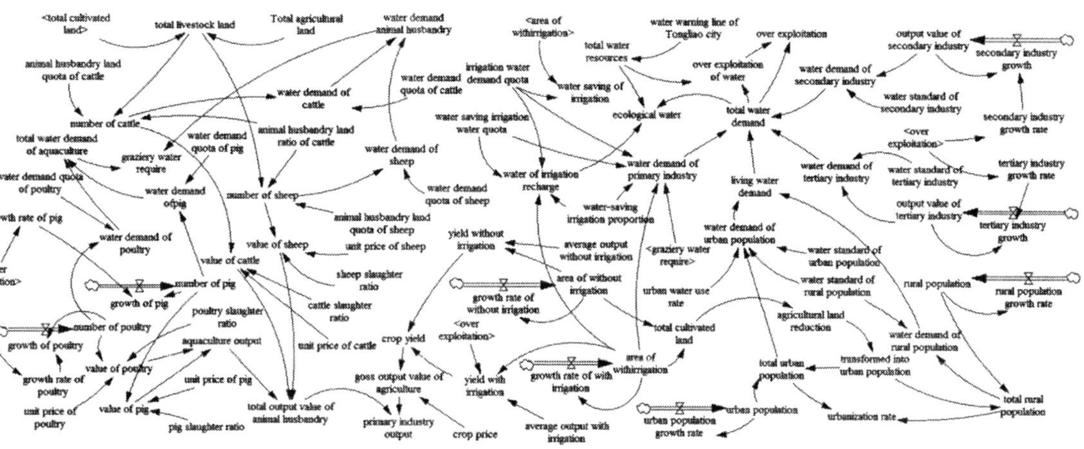

Figure 1. System dynamics simulation model of sustainable development in Tongliao.

4.3. Simulation results and scheme optimization

4.3.1. Simulation resutes. By running the system dynamics model from 2013 to 2050, we obtained the different simulation results separately under fast and slow policies in Tongliao city. Table 3 shows the typical index simulation results, such as rate of urbanization, total GDP, industrial structure, capita GDP, water demands, groundwater overdraft rate, comprehensive water efficiency, intensity of conversion to grassland, and so on during key years.

Table 3. Dynamic simulation results of different policies of degraded farmland of Tongliao city.

degraded farmland policy	indicator	2020	2030	2040	2050
Emergency	Urbanization %	54.73	74.80	77.85	80.77
Stable		50.35	64.91	77.85	80.77
Relieve		48.33	60.36	71.23	80.77
Emergency	Industrial structure	18:54:28	16:52:32	10:51:39	7:50:43
Stable		18:54:28	20:51:29	13:52:35	7:52:41
Relieve		18:54:28	20:51:29	20:50:30	10:52:38
Emergency	Total water demands 1×10^8 m^3	40.6	29.0	31.2	31.5
Stable		44.0	36.0	28.2	30.8
Relieve		45.5	39.7	30.8	27.6
Emergency	Wateruse efficiency yuan/m^3	54.27	125.21	257.02	457.1
table		49.29	81.88	204.49	437.8
Relieve		47.23	71.86	126.42	359.7
Emergency	GDP 1×10^8 yuan	2202.88	3627.94	8017.23	14392.5
Stable		2167.86	2950.64	5762.85	13496.1
Relieve		2150.59	2853.17	3894.65	9923.43
Emergency	groundwater overdraft rate %	28.29	-7.84	-0.15	1.43
table		39.00	14.61	-9.79	-0.69
Relieve		43.92	26.28	-1.38	-11.11
Emergency	Intensity of degraded farmland %	5.78	13.69	13.69	13.69
Stable		3.16	4.63	8.62	8.62
Relieve		2.18	2.79	3.86	6.29

4.3.2. Evaluation method. Based on the obtainable possibility principle of the output results of the system dynamic model and the relationship of reason and result impact, the paper established the suitable evaluation index system of degraded farmland policy in Tongliao city. The system biases on the water resource's support ability and farming-pastoral developing pattern, therefore, the primary indicators are urbanization rate, total GDP, industrial structure, capita GDP, water requirement, groundwater overdraft rate, comprehensive water efficiency, intensity of conversion to grassland, and so on, which are all listed in table 4. According to the impact mechanism of degraded farmland policy and some principles like the comprehensive and conclusive characteristics of index system, the system and hierarchy, dynamic and static, and paper divided the suitable evaluation indicators into the level of evaluation target, level of evaluation subsystem, and level of evaluation indicators by assigning the weight of each index using AHP. This specific method is referenced in [16].

Table 4. Suitable evaluation system of degraded farmland policy in Tongliao city.

Evaluation target	Weight	Evaluation subsystem	Weight	Evaluation indicator	unit	Weight	direction
Suitability of degraded farmland policy	1	Social development subsystem	0.0435	urbanization	%	0.0435	+
		Economic development subsystem	0.2472	GDP	108 yuan	0.0817	+
				Industry	--	0.0724	+
				Capita GDP	104 yuan/ person	0.0931	+

Water resource subsystem	0.2114	Total water demands	$108m^3$	0.0805	-
		Groundwater over production rate	%	0.0501	-
		Total water-use efficiency	$yuan/m^3$	0.0808	+
Degraded farmland excutive subsystem	0.4979	Intensity of degraded farmland to grassland	km^2/a	0.4979	-

4.3.3. Scheme optimization. By the method in reference [16], this paper took comprehensive evaluation on system dynamics simulation results. As shown in table 5, the dynamic evaluation results were obtained from 2020-50 in Tongliao city.

Table 5. Comprehensive evaluation of 3 degraded farmland policies of Tongliao city 2020~2050.

Degraded farmland policy	2020	2025	2030	2035	2040	2045	2050
Emergency	0.47797	0.47797	0.48087	0.48350	0.37150	0.37185	0.37198
Stable	0.52825	0.54547	0.51472	0.54653	0.59408	0.63249	0.66055
Relieve	0.52203	0.52203	0.52203	0.50246	0.51462	0.62850	0.62589

The dynamic evaluation results show advantages and disadvantages under three policies of degraded farmland during 2015~2050: "emergency degraded farmland policy" is weaker than "stable degraded farmland policy", which is weaker than "relieve degraded farmland policy". The simulation result of emergency degraded farmland policy appeared best, but evaluated lowest, because Tongliao is located in northeast inland area of China and lags in economic development. It still has a middle stage of social economic development when using the Chenery standard industrial structure and industrialization stage theory, which means the industrial foundation is relatively weak, causing a large number of landless peasants and unemployed if the "strong and emergency degraded farmland policy" is used turning this period. The husbandry that has limit capacity could not involve large amounts of agricultural labor, and the urban development could not provide sufficient employment to absorb unemployed farmers, therefore causing a wide range of social contradictions. The "relieve degraded farmland policy" slowly reduced the ratio of agriculture and left enough time and space for Tongliao's social economic development. It could also better absorb labors that came from the degraded farmland policy and better relieve the social contradictions. However, the simulation shows that the "relieve degraded farmland policy" has a long developing period, which leads to poor societal and economic performances in various indicators, resulting in a slightly weak comprehensive evaluation. The "stable degraded farmland policy" could absorb surplus agriculture labors better and also has a better societal and economic development performance, making it the ideal degraded farmland policy in future of Tongliao city.

5. Conclusion and discussion

- The present farming-pastoral developing pattern could not support the future sustainable development. According to the comprehensive evaluation, the farming-pastoral landuse pattern is between a safe and dangerous situation. If developing agriculture industry and expanding agriculture scales is continued, it will cause Tongliao city to have a dangerous ecological safety status and will cause ecological disaster. Therefore, improving the landuse pattern of agriculture and husbandry has become Tongliao's urgent task.

- The farming-pastoral pattern is the key factor impacting the sustainable development. The farming-pastoral pattern has great influence on Tongliao, which has a lack of water resource. The best way to realize Tongliao city's sustainable development is to implement the degraded farmland to a grassland policy, reducing the ratio of agriculture land and forming the 1:7 ratio of agriculture and husbandry in order to reverse the situation that agriculture production consumes large amount of groundwater resources.
- Reasonable carrying out the degraded farmland policy is the key of sustainable development. We established a system dynamic model, imitated three different types of policies: "emergency degraded farmland policy", "stable degraded farmland policy", and "relieve degraded farmland policy", and established a degraded farmland suitable evaluation indicator system. We then conducted a comprehensive evaluation based on the simulation results. The results show that "emergency degraded farmland policy" may obtain good results, but its intensity is great and easy to cause social contradiction, making it inappropriate to adopt. The "relieve degraded farmland policy" could sufficiently relive social contradiction, but its effects are too slow to obtain the recognition, so it is the suboptimal selection. The "stable degraded farmland policy" could better resolve social problems and quickly develop social economy, so is the best choice.
- The paper enriched the theory and practice of sustainable development in the northern farming-pastoral areas, which Tongliao city is the representative of. By conducting simulation research on different degraded farmland policies, this paper obtained different proactive effects based on different degraded policies of sustainable development, provided good practice guidance for the implementation of the degraded farmland policy, and enriched the sustainable developing theory for farming-pastoral areas.

Acknowledgments
This paper is founded by Public Welfare Project of Ministry of Water Resources (201501031): Study on key technologies of water cycle and water ecological security in semiarid area and NSFC (51479160): Study on the change mechanism and adaptive regulation of soil and water resources utilization pattern for eco city.

References

[1] Liu H L, Wan Y M and Dou X 2009 Progress and perspective of ago-pasturage Ecotone *Acta Ecol. Sinica* **29(8)** 4420-5
[2] Liu M 2015 Agricultural industrialization: Theoretical basis and analysis of sustainable development *Product Res.* **5** 38-45
[3] Hu B H, Liao Y C, Wang K Q, *et al* 2011 The comparison and evolvement of agroecosystem structure in representative counties of the farming-pastoral ecotones in north China *Res. Soil Water Conserv.* **18(2)** 35-9
[4] Xu E Q and Zhang H Q 2012 Research on the evaluation of land use sustainability in ecologically fargile region in China *Chin. J. Agric. Res. Reg. Plan.* **33(3)** 1-6
[5] Wang X H and Shen J X 2013 Suatainability analysis of cropland use in cropping-pastoral ecotone before and after the grain-for-green policy: A case study from Yanchi county in Ningxia *Agric. Sci. Tech.* **14(1)** 88-93
[6] Zhang J T and Fu X F 2005 Research on the sustainable development of the vulnerable eco-regions *Chin. Pop. Res. Environ.* **15(5)** 58-62
[7] Long Y H, Guo Z X, Liao Z L, *et al* 2014 Water resource carrying capacity evaluation method in typical grassland *Res. Soil Water Conserv.* **21(3)** 222-6
[8] Qi Y P, Cui Y Q and Li X W 2010 Analysis of cultivated land intensity and potential in the farming-pastoral fragile ecological area of northern Tongliao *J. Nat. Resour.* **25(12)** 2072-80

[9] Ji X 2013 Restriction factors analysis of sustainable development of rural areas in the agriculture and animal husbandry staggered ecological fragile region—Take Tongliao City as an example *For. Pos.* **22** 77-9
[10] Ma J W, Guo X Y, Fu Q, *et al* 2014 Study on complex adaptive allocation of regional agricultural water and soil resources —Taking Sanjiang plain for example *Res. Soil Water Conserv.* **21(3)** 256-60
[11] Zheng J Y, Zhao X N, Cao X C, *et al* 2015 Study on spatiotemporal matching pattern of agricultural water and land resources in Hetao irrigation district *Res. Soil Water Conserv.* **22(3)** 132-6
[12] Ding R N and Liu J 2014 Study of coal mine sustainable development based on AHP method *Coal Technol.* **33(9)** 31-3
[13] Liu X J and Fu H L 2015 Evaluation on water resources carrying capacity based on variable-weight and entropy-weight methods with improved TOPSIS *Bull. Soil Water Conserv.* **35(6)** 34-9
[14] Heng Y, Zhang W J and Wang N 2014 Assessment of the carrying capacity of agricultural water resources in western region of Jilin based on principal component analysis *J. Yangtze River Sci. Res. Inst.* **31(9)** 11-6
[15] De W M and Crookes D J 2013 Improved decision-making on irrigation farming in arid zones using a system dynamics model *South African J. Sci.* **109(11-12)** 1-8
[16] Yang D and Yang X Q 2001 A method of quantitative evaluating of regional sustainable development and its application *J. Northwest Normal Univ. (Nat Sci Ed)* **37(1)** 83-8

IOP Conf. Series: Earth and Environmental Science 82 (2017) 012029 doi:10.1088/1755-1315/82/1/012029

A novel water poverty index model for evaluation of Chinese regional water security

L Gong[1,2], C L Jin[1], Y X Li[1] and Z L Zhou[1]

[1]School of Civil Engineering, Lanzhou Jiaotong University, Lanzhou, 730070, China

E-mail: gongl@mail.lzjtu.cn

Abstract. This study proposed an improved Water Poverty Index (WPI) model employed in evaluating Chinese regional water security. Firstly, the Chinese WPI index system was constructed, in which the indicators were obtained according to China River reality. A new mathematical model was then established for WPI values calculation on the basis of Center for Ecology and Hydrology (CEH) model. Furthermore, this new model was applied in Shiyanghe River (located in western China). It turned out that the Chinese index system could clearly reflect the indicators threatening security of river water and the Chinese WPI model is feasible. This work has also developed a Water Security Degree (WSD) standard which is able to be regarded as a scientific basis for further water resources utilization and water security warning mechanism formulation.

1. Introduction

Water security is a water-centered complex huge system, which has a significant influence on social, economic and ecological aspects. To date, the unscientific water utilization and management in many regions resulted in the water shortage and pollution. It leads to not only water imbalance, but also reduction of the water quality. As a result, the water function was being weakened and even lost. Once the water body loses its economic or social value, it leads to water crisis for basic human demands, preventing human economy and society from moving forward [1].

Water security evaluation is the basis to formulate the regional water security planning and strategy for better water management. It is also the premise for regional water security decision-making. At present, researches on regional water security evaluation are still on the preliminary stage, and only some discusses about the qualitative conception, connotation and countermeasures.

For researchers in foreign countries, Falkenmark and Widstrand proposed that per capita water resources can measure regional water resources scarcity level [2]. In moderately developed countries, the thresholds of per capita water resources for arid regions were obtained, shown as follows:

- The water is on the shortage level when the per capita water resource is blow 1700 m^3 but higher than 1000 m^3.
- The water is on the severe shortage level when the per capita water resource is blow 1000 m^3 but higher than 500 m^3.
- The water is on the extreme shortage level when the per capita water resource is below 500 m^3.

Falkenmark and Widstrand improved these indicators to discriminate the water sources scarcity levels and security status in various countries and regions.

In China, Hong Y *et al* [3] suggested that the water security should be analyzed from external environment and conditions. Jia S-F *et al* [4] considered that the water was safe as long as the regional water supply can meet the needs of long-term economy and social progress. Zhang X *et al* [5] defined water security as the non-threatened and controlled influence of water content and activity on economy and society stability and development. Chang M-Q and Huang Q [6] considered that the water security was the status that human existence and development would not be threatened by water problems.

The above discussions show that there were very few quantitative analyses on regional water security evaluation. Therefore, particular emphasis was put on the water security evaluation index system for quantitative processing in this study. It introduced WPI into water security evaluation model, and explored index system in WPI, especially, several components composed of detailed indicators. Meanwhile, it proposed water security evaluation index system adapted to Chinese water. In advanced sustainable use of regional water resources, this study assures the regional water resources security, so it accommodates both theoretical significances and practical significances, and drive sustainable society and economy development.

2. Materials and methods

2.1. WPI

The Water Poverty Index (WPI) method was proposed in 2002 by the researchers in the Center for Ecology and Hydrology (CEH) to monitor the water status [7]. Based on the interdisciplinary theory, the WPI provides a standard framework for evaluation of regional water exploitation and utilization. It particularly evaluates the effects of water scarcity on society and economy. The WPI model has simple calculation, low cost and understandable property. Hence, the WPI has been drawn great concern of the researchers in water resource field all over the world, especially for the 3rd World Water Forum in Kyoto.

The WPI is a standard framework to guide how to rationally manage water. It is appropriate for different regions and countries on different levels. Meanwhile, it is easy to obtain data of the index system in WPI model for a specific region or country. In WPI index system, Resource (R), Access (A), Use (U), Capacity (C) are the five components (indicators of them are shown in table 1), the values of which vary from 0 to 100. The higher the value is, the better the performance is.

Table 1. Indicators in the five WPI components.

WPI components	Indicators
R	Ground water, surface water, water quality value, soil moisture, reliability data.
A	Household water supply, facilities for irrigation, foodstuffs import, state of health.
U	Water usage of agriculture, water usage of industry, water usage of life, water usage of livestock, water utilization efficiency in different industrial sectors, total water usage
C	GDP per capita, levels of household consumption
E	Living species, ecological environment, water pollution, flood, water and soil erosion.

Table 1 shows that the index system includes not only water and environment indicators, but also economic and social indictors. Water security problems in the early human civilization are exhibited by natural disasters, such as arid, flood, river diversion. With the human civilization progress development, the contents of water security in economy and society are enriched and extended.

The equation of the WPI model given by CEH is shown as follow:

$$WPI = \frac{\sum_{i=1}^{N} w_{x,i} X}{\sum_{i=1}^{N} X_{x,i}} \quad (1)$$

Equation (1) shows WPI value for a certain region. N denotes the number of WPI components. X is the components values in this region, including R, A, U, C or E. $w_{x,i}$ denotes weights of the ith indicator in the components X . It is obvious that the WPI is obtained by the weighted average value of all five components. The following is the rearranged equation:

$$WPI = \frac{w_r R + w_a A + w_c C + w_u U + w_e E}{w_r + + w_a + w_c + w_u + w_e} \quad (2)$$

2.2. *Chinese WPI index system*

In general, the WPI index system could reflect the water and water industry state in different regions. So WPI theory was spread to 147 countries by the Center for Ecology & Hydrology (CEH), while WPI value varied from 48 to 55.9 in China. However, the 5 components in the WPI index system were determined according to Europe realities. In other regions or countries, some indicators of the 5 components are difficult to be obtained. For instance, the indicators related to household water and household welfare are important in water management decision making in Europe and America. While some of them are hard to collect because of large population. Meanwhile, it is lack of some indicators which could better reflect the Chinese water situation in the index system.

Typically, this study firstly proposed the Chinese indicators of the 5 components in the WPI index system on basis of the CEH [7] and realities in China, which is shown in table 2. On account of the natural attribute, social attribute and humanity attribute of the water security, the related indicators should be considered in the WPI index system. Besides the regional water conditions, the factors related to environment, society and economy in China are also included.

In order to get the indicator values, a large number of reliable data were obtained from the relevant agencies and departments termed as step one. After that, the data were processed in different ways. There is no doubt that it was a complicated and time-consuming process. For the qualitative description, it is necessary to consult experts to get much more accurate results. For indictors in different regions, the calculation bases are different. So how to choose suitable variables to establish the mathematical model is an important work.

The WPI index system in China was formed according to situation of China, related economic development, social insurance, and water supply. It could be better to reflect the water shortage on influencing human life. The Chinese WPI system provides conveniences for water managers to check the water availability and social economic factors changes. It provides comprehensive information about water resources and social economic status to further establish the water security system.

2.2.1. Resource (R) in China. The Resource is the component showing physical availability of water resources in the study area. The larger the component value is, the more the water availability is considering internal-annual and seasonal variability. It includes 4 indicators:

- Surface water amount.
- Overdraft rate of groundwater.
- Water supply / demand ratio.
- Quality compliance rate of drinking water.

Table 2. Chinese WPI index system.

WPI components	General indicators	Indicators in China	Dimension	Nature
R	Evaluation of surface water with hydrological technology	Surface water amount	m^3	↑
	Evaluation of ground water evaluation with hydrogeological technology	Groundwater overdraft rate	%	↓
	Resource reliability evaluation about quantitative and qualitative evaluation	Supply/demand ratio of water	%	↑
	Water quality reliability values evaluation about quantitative and qualitative	Drinking water quality compliance rate	%	↑
A	Acquisition approaches of clean water	Percentage of running water users	%	↑
	Proportion of human with health facilities	Urbanization rate	%	↑
	Water collection time, including waiting time	Water resources utilization ratio	%	↑
	Irrigation way	Irrigated area ratio	%	↑
U	Industrial water consumption per capita	Water consumption based on industrial output (/10^4Yuan)	$m^3/10^4$Yuan	↓
	Household water consumption per capita	Daily per capita water consumption	/ day	↑
	Agricultural water consumption per capita	Agricultural water consumption per unit area	m^3/ha.	↓
C	GDP per capita	GDP water consumption per 10,000 Yuan	$m^3/10^4$Yuan	↓
	Education level	Number of university students per 10,000 persons	/10^4 persons	↑
	Household consumption level	Farmers' annual net income	Yuan/ year	↑
	Drink coefficient	Town Engel coefficient	Dimensionless	↓
	Child under 5 mortality rate (Health Indicators)	Mortality of infant and child	%	↓
E	Observations of water quality	Integrated pollution index of surface drinking water	Dimensionless	↓
	Environmental management and governance	Urban sewage treatment rate	%	↑
	Water pollution load	Industrial wastewater specific pollutants emissions	Kg	↓
	Biodiversity	Green coverage	%	↑

Notes: "↑"denotes the performance increases with the increase of the value.
"↓"denotes performance decreases with decreasing value.

2.2.2. Access (A) in China. The access is the component that considers population with reasonable access to sanitation for health and much well-being, and to adequate safe drinking water. It includes 4 indicators:

- Percentage of running water users.
- Urbanization rate.
- Water resources utilization ratio.
- Irrigated area ratio.

2.2.3. Use (U) in China. Water use is one of the most important things in human activity, and the water consumption increases with the society and economic development. Therefore, Use is the component that indicates the ability of human use water effectively and the water to be contributed to society and economy. It includes 3 indicators:

- Daily per capita water consumption.
- Water consumption based on industrial output (/104Yuan).
- Agricultural water consumption per unit area.

2.2.4. Capacity (C) in China. The Capacity is the components that refer to the current potential to manage water. It includes 5 indicators:

- GDP water consumption per 10,000 Yuan.
- Famers' annual net income.
- Town Engel coefficient.
- Number of university students per 10,000 persons.
- Infant and child mortality.

2.2.5. Environment (E) in China. The environment is the component that shows the ability that the ecosystem continuously provides benefits to society and the environment influence water quantity and quantity. It includes 4 indicators:

- Integrated pollution index of surface drinking water.
- Urban sewage treatment rate.
- Industrial wastewater with specific pollutants emissions.
- Green coverage.

2.2.6. WPI Mathematical model in this study. Take Capacity(C) for example, its calculation steps were shown as follows:

- Divide each of the indicators of Capacity (C) in a certain region by its corresponding national average value to get the ratios. For Capacity (C), the indicators are GDP water consumption per 104 Yuan (C1), urban worker wage level (C2), number of university students per 104 persons (C3), Town Engel coefficient (C4), infant and child mortality (C5).
- Average all of these ratios. The result reflects Capacity (C) of this region compared to the national level.
- Obtain the Capacity(C) value by multiplying the above average value with Chinese Capacity(C) value given by the CEH. Thus, the component value and WPI reflect the relative level of the region in one country.

Based on the CEH model and the above steps, the WPI mathematical model given by this study is concluded as follow:

$$X = \frac{\sum_{1}^{N} \frac{X_i}{X_{i-CN}}}{N} \times X_{CEH} \tag{3}$$

IOP Conf. Series: Earth and Environmental Science **82** (2017) 012029 doi:10.1088/1755-1315/82/1/012029

Suppose X_i denotes the ith indicator in the WPI component X . X_{i-CN} denotes the indicator's national average value. N denotes the indicator numbers in the component. X_{CEH} is the component value given by CEH. Then the component value can be calculated as follow:

In situations where it is difficult to obtain the components weights, the equal weighted method was adopted to ensure the evaluation transparency. The WPI value can be calculated by equation (4) and all of the components weights to 1.

$$WPI = \frac{R + A + C + U + E}{5} \tag{4}$$

2.3. Water security degree (WSD)

The study firstly proposed the Water Security Degree (WSD) in China in order to give direct water security levels. So the thresholds and ranges of each degree should be confirmed. With the purpose of getting accurate thresholds, 20 indicators which can better describe water security status [8] are chosen, and all of the indicators values are quantized and classified. After that, the international standards, and the standards and development plan in China were collected as the basis and standards. The WSD thresholds and ranges were obtained by Expert Consultation Method. 5 degrees of WSD are shown as follows:

- WSD1 - *Very Safe* degree (WPI > 62): Development of society and economy is **highly satisfied** with the water resources and water environment.
- WSD2 - *Safe* degree (62 > WPI > 56): The society and economy development is **satisfied** with the water resources and water environment.
- *WSD3- Generally Safe* degree (56 > WPI > 48): The society and economy development is **generally satisfied** with the water resources and water environment.
- *WSD4- Unsafe* degree (48 > WPI > 35): The society and economy development is **dissatisfied** with the water resources and water environment.
- *WSD5- Extremely Unsafe* degree (WPI < 35): Entirely degraded water and water environment which seriously hinder the society and economy sustainable development.

2.4. Calculation examples

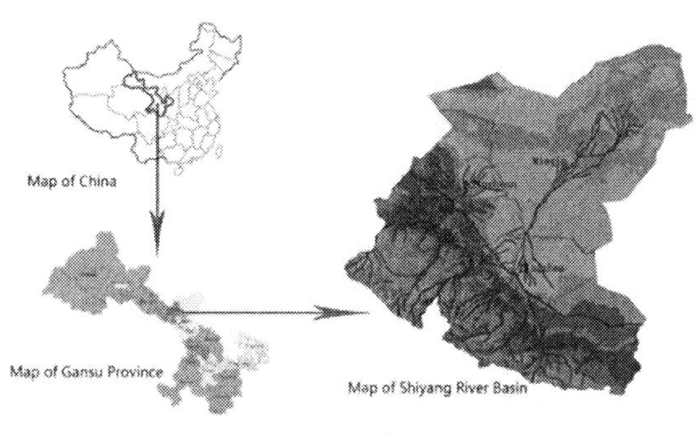

Figure 1. Shiyanghe River Basin.

The Shiyanghe river basin lies to the eastern of the Hexi Corridor, which is on the western of Wushaoling Mountain and the northern of Qilian Mountain. It encompasses 41,600 km^2 area, lying between 101°41′ to 104°16′ E and 36°29′ to 39°27′ N. The Shiyanghe River is in China hinterland with the continental temperate arid climate. The characteristics are abundant sunshine, large temperature difference, strong solar radiation, rapid evaporation, less rainfalls and the dry air. The river basin from south to north is divided to 3 climatic regions. The southern part is along Qilian Mountain with climate of alpine, semi-arid and semi-humid. The middle part along the Hexi corridor plain is the cool and arid climate. The southern part is the warm and arid area. The Shiyanghe River is composed of 8 rivers and many streams. The 8 rivers are Dajing River, Gulang River, Huangyang River, Zamu River, Jinta River, Xiying River, Dong River and Xi River from south to north. The river is feed by snow melt water in the mountain and atmospheric precipitation. The Shiyanghe River runoff area is 11,100 km^3 and average annual runoff is 1.56 trillion m^3 [9, 10]. The Shiyanghe River is shown in figure 1. The statistical data of Jinchang City, Wuwei City and Minqin County in Gansu province of China were used in the calculation examples.

3. Results

Based on the Chinese WPI index system, the WPI values, components values and the water security levels of Jinchang City, Wuwei City, Minqin County and Gansu Province in Shiyanghe River Basin were calculated, the results are shown as follows.

Table 3. WPI and components values of the Shiyanghe River.

River secondary region	R	A	U	C	E	WPI value	Degree of Water Security
Jinchang (J)	21.5	78.1	55.2	59.7	23.0	49.1	WSD3
Wuwei (W)	25.3	44.1	39.5	45.1	22.3	35.3	WSD4
Minqin (M)	2.7	14.3	15.1	13.4	3.1	9.7	WSD5
Gansu Province (G)	45.9	42.1	53.8	50.1	54.6	49.3	WSD3

Table 3 shows that the highest WPI value is in Jinchang City, while the lowest one is in Minqin County. Overall, the water resource in Shiyanghe River is on WSD4- *Unsafe* degree. It seemed that the security level in Jinchang City is on WSD3- *Generally Safe* degree, while its WPI value 49.1 was close to 48 which is the marginal value of WSD4. Wuwei City is on WSD4 with the WPI value of 35.3, which is close to the marginal value of WSD5- *Extremely Unsafe* degree. The WPI value of Minqin County is 9.7, indicating water and water environment system were degraded entirely, which had seriously hindered the society and economy sustainable development.

4. Discussion

The Radar Chart is usually applied to describe WPI value. Figures 2-5 are the Radar Charts showing the WPI and its components values in regions of Shiyanghe River. Jinchang City lacks water resources and good environment. While it has good performances in urbanization, resident income and water use,

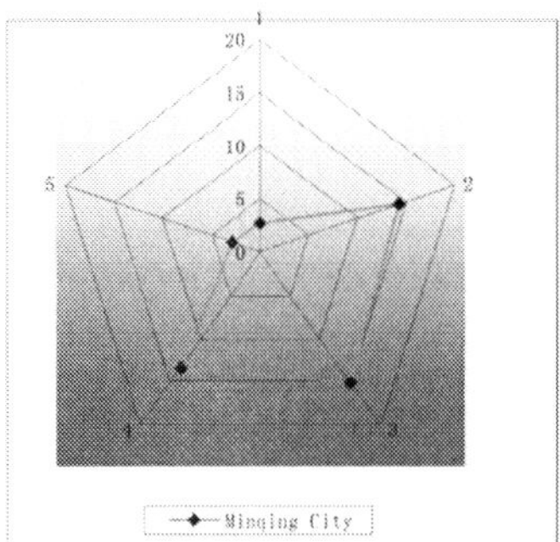

Figure 2. Radar chart- WPI values in Minqin: WPI (M) =9.7.

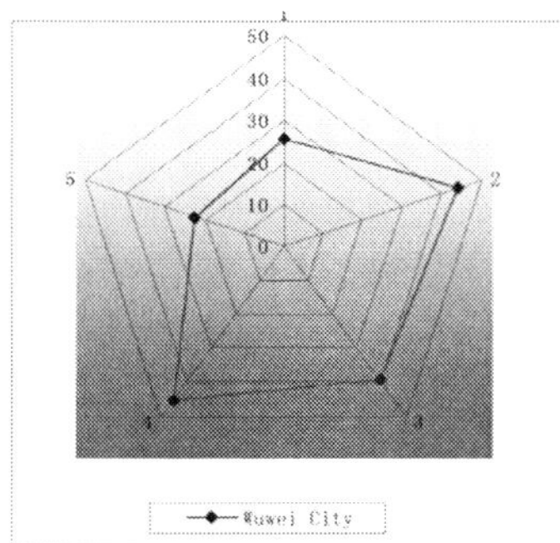

Figure 3. Radar chart- WPI values in Wuwei: WPI (W) =35.3.

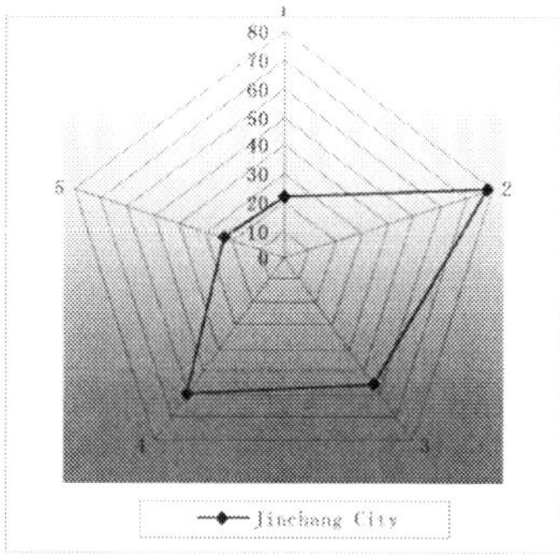

Figure 4. Radar chart- WPI components values in Jinchang: WPI (J) = 49.1.

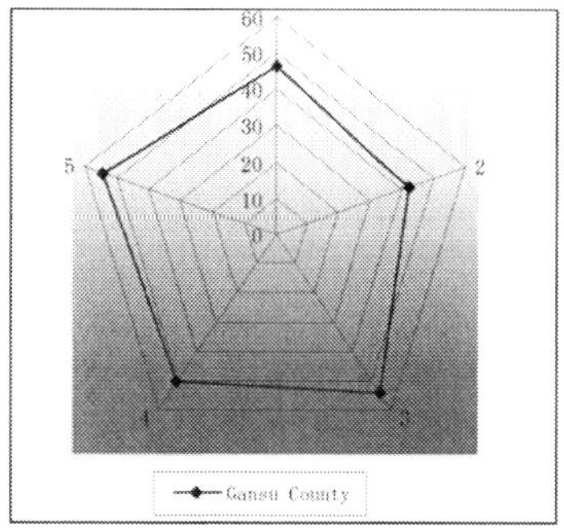

Figure 5. Radar chart- WPI components values in Gansu: WPI (G) =49.3.

making the WPI value in the lower-middle level. So Jinchang City is on WSD3- *Generally Safe* degree. Wuwei City has scarce water resource and degraded environment. Meanwhile, most social indicators in the index system were below the national level in China, causing the WPI value lower than average. So Wuwei City is on WSD4- *Unsafe* degree.

The extremely low WPI value shows that the water security in Minqin County is on WSD5- *Extremely Unsafe* degree. In the upper and middle reaches of Minqin, the water resource use increases, so the surface water amount flowing in Minqin County has declined from the 590 million m^3 in 1950s to less than 100 million m^3 in 2005. With the increasing of the water requirement, the ground water exploitation for Minqin Basin has been up to 520 million m^3, overexploiting 300 million m^3. The ecological deterioration of Minqin County was very serious. The serious overexploitation of water resource and the environmental degradation have formed a vicious circle. With conditions going out of control, Minqin County would probably become the second Lop Nor, which not only threatens the local livelihoods, but also brings bad effect on development of the entire region.

The composite WPI value of cities in Gansu Province was 49.3, which was on WSD3- *Generally Safe* degree, close to WSD4- *Unsafe* degree. It should be paid more attentions to increase water utilization efficiency and improve water security situation.

5. Conclusions

- WPI model was proposed to construct river water security index system in China, on the basis of its traditional applications on different environments and ecologies comprehensive evaluation. It indicates the path to establish universal river water security index system.
- The Chinese WPI index system was applied in the water security evaluation of Jinchang City, Wuwei City, Minqin County and Gansu Province. By analyzing the 5 components in the Chinese WPI system, the main factors affecting water security in this region. The result shows that the system provides theoretic basis for coordination between regional water security and economic development.
- Certain indicators in the WPI index system can be appropriately adjusted for different regions. Therefore, the evaluation results could objectively reflect the water security characteristics in different regions. It is because the complex water types and different water security problems in different regions.
- The application of the Chinese WPI index system in Shiyanghe River proved that this system could clearly indicate the factors threatening water security. The Water Security Degree (WSD) proposed in this study provides evidence for security warning mechanism of water resources.

Acknowledgments
The authors are grateful to National Natural Science Foundation of China: 51669010, 51541902 and

National Natural Science Foundation of Gansu Province: 1506RJZA082.

References

[1] Gong L and Jin C L 2004 Problems about ecological environment construction and water resources sustainable utilization in northwest China *J. Des. Res.* **24** 513-7
[2] Falkenmark M and Widstrand C 1992 Population and water resources: A delicate balance *Popul. Bull.* **47(3)** 1-36
[3] Hong Y 1999 China's water safety in the 21st century *Environ. Prot.* **10** 30-1
[4] Jia S F, Zhang J Y and Zhang S F 2002 Regional water resources stress and water resources security appraisement indicators *Prog. Geog.* **21** 538-45
[5] Zhang X, Xia J and Jia S F 2005 Definition of water security and its assessment using water poverty index *Resour. Sci.* **27** 145-9
[6] Chang M Q and Huang Q 2006 The theory and method of water resources security (Shaanxi: Xi'an Polytechnic University)
[7] Lawrence P, Meigh J and Sullivan C 2003 The water poverty index: An international comparisons *Keel Econ. Res. Pap.* **19(19)**
[8] Wang J S, Li Y Q and Ding J 2007 Evaluation method of water security based on indicator system *Chin. Rur. Water Hydrop.* **2** 116-9
[9] Song D M, Xiao W N and Zhang Z C 2004 Environmental characteristics of sandstorm in Minqin county in recent 50 years *J. Des. Res.* **24** 257-60
[10] Wang Y L and Li C L 2006 Problems existing in water resources utilization and water-economizing ways in the Minqin oasis of Gansu province *J. Des. Res.* **26** 103-7

Performance assessment and application of the DSSAT-CERES-Maize model for simulating maize yield under water stress conditions

S M Geng[1], D H Yan[2,5], Z B Zhang[2], Z L Wang[3] and A Girmad[4]

[1]College of Civil and Hydraulic Engineering, Tsinghua University, Beijing 100084, China
[2]State Key Laboratory of Simulation and Regulation of Water Cycle in River Basin, China Institute of Water Resource and Hydropower Research, Beijing 100038, China
[3]Anhui institute of hydraulic research, Bengbu, Anhui 233000, China
[4]College of Environmental Science and Engineering, Donghua University, Shanghai 201620, China

E-mail: simin0818@126.com

Abstract. Summer maize is a major crop in the Northern Huaihe Plain, where is one of the most important grain producing areas in Anhui Province, while the maize yield has been seriously affected by drought stress for a long period of time. To maximize maize yield and improve water use efficiency, the rule of yield loss should be clarified which was caused by drought stress in the Northern Huaihe Plain. DSSAT-CERES-Maize model was used to simulate maize yield losses caused by water stress in Bengbu Station of the Northern Huaihe Plain, which was calibrated and validated with drought stress experiments conducted in 2013. Comparing the simulated yield with measured yield under different drought stresses at different growth stages, we found that simulated losses under mild drought were higher than that of measured in all growing stages, on the contrary, simulated data under moderate and severe drought were lower than measured ones in each stage. There were 8 years that loss percentage was higher than 20% from 1961 to 2015. Results implied that maize yield under rain-fed environment was higher than that of suitable water condition, which might benefit from overcompensation of rehydration after appropriate drought stress.

1. Introduction

Summer maize is a main crop in the Northern Huaihe Plain, which is known as one of the most important grain producing areas in Anhui Province. Generally, its supply water for summer maize is precipitation or/and irrigation during the growing periods and the yield has been seriously affected by drought stresses for a long time [1]. Quantitative analysis of yield loss caused by water stresses is the basis for rational allocation of water resources and it is important criterion for rational water use efficiency.

However, there are a wide variety of component factors for yield loss such as droughts, floods, plant diseases and insect pests, insufficient fertility, heat waves, therefore, there is a complex mechanism among them. In addition, synergistic effects of factors are also common which including drought and locusts. Considering the above-mentioned factors, there are a lot of difficulties in determining the effects of drought on maize yield. Field experiment is the most simple and traditional

IOP Conf. Series: Earth and Environmental Science **82** (2017) 012030 doi:10.1088/1755-1315/82/1/012030

method to understand the influencing mechanism. Therefore, in order to study the effect of different drought stresses during different growth stages on the final product, it needs doing large-scale and long-term experiments. However, these require a lot of manpower, material and financial resources and it is also time-consuming [2]. Furthermore, drought effects on maize yield varies with strength and lasting time of drought stress, growing stage, as well as the compensation effect after rehydration [3-6]. Considering all these factors, there were great limitations in field experiments.

With the development and application of crop model, this problem is being solved step by step. Decision Support System for Agrotechnology Transfer (DSSAT) [7] is one of the extensive application and highly praised crop model. The CERES (Crop Estimation through Resource and Environment Synthesis) model simulates the growth and development of crops including maize and wheat et al in response to meteorological and management factors. DSSAT-CERES has been used world wise under a wide range of climatic and soil conditions and it has been proven as a useful tool for determining management practices due to its outstanding performance in simulating the growth, biomass accumulation, yield and water resource use efficiency in response to various environmental factors and management. Lots of researchers simulated winter wheat, rice and maize response to irrigation management, such as deficit irrigation, water efficient utilization, conservation agriculture practices [8-14]. DSSAT models have been used in many regions of China in recent years. However, they have not been used in the Northern Huaihe Plain. In addition, the soil in the Plain is mainly lime concretion black soil, which has never been simulated by DSSAT models.

2. Materials and methods

2.1. DSSAT CERES-Maize model

DSSAT-CERES model could simulate water balance, crop growing process, biomass and yield of maize, wheat, soybean, rice, millet and sorghum [10]. DSSAT-CERES-Maize model is one modular of DSSAT-CERES, which could simulate the effects of soil water, fertilizer conditions, field managements (sowing date, irrigation, different cultivation way, and fertilization et al.) and external environmental factors like temperature, water stress, wind speed and solar radiation on maize growth and yield. With the comprehensive consideration of crop genetic characteristics, soil, climate, and field management, DSSAT-CERES-Maize model has obvious advantages of simulating the influence of climate change, geographical condition change and management change on maize growth and yield. It can be used in all geographical areas, soil and climate conditions due to its less restrictions in its application.

The model was constituted by data module, simulation module, analysis module and tool module. This study focused on soil water and the yield, and computing method of crop yield is introduced below.

$$dW/dt = W^* - S_L - S_S - S_R \tag{1}$$

$$W^* = E \times \left(P_g - R_m\right) \tag{2}$$

$$P_g = K_p \times PGMAX\left(PAR\right) \times f_L \times f_O \times f_N \times f_T \tag{3}$$

$$R_m = LEAF \times 0.03 + STEM \times 0.015 + ROOT \times 0.01 + SO \times 0.01 \tag{4}$$

Where t represents the time, SL, SS and SR represent aging pare of leaf, stem and root, E is energy transfer, Pg is gross photosynthesis, Rm is maintenance respiration, which is calculated by Penning de Vries-SUCROS model [15], Kp represents adaptability of soil fertility, PGMAX (PAR) means response on photosynthetic active radiation (PAR=400-700 nm), fL is the response on LAI (0-1), fO is

IOP Conf. Series: Earth and Environmental Science **82** (2017) 012030 doi:10.1088/1755-1315/82/1/012030

response on water supply (0-1), *fN* is response on nitrogen sink of leaf (0-1), *fT* is response on daily temperature (0-1), *LEAF*, *STEM*, *ROOT* and *SO* represent protein activity of leaf, stem, root and storage organ, respectively.

DSSAT CERES-Maize model required different kinds of data, such as genotype coefficients of maize, soil data, daily meteorological data and field management data. The general input of the model and the data sources are shown in table 1, where *P1*, *P2* are vegetatively accumulated temperature and photoperiod sensitivity coefficient, respectively, *P5* is accumulated temperature from tasseling to mature period, and *G2* and *G3* represent grain number per plant corn and grain filling rate.

Table 1. General input of DSSAT CRESE-Maize model and data sources.

Input	Elements	Data source
genotype coefficients of maize	*P1*, *P2*, *P5*, *G2*, *G3*	obtained from fitting the model simulation results against the measured data
physical and hydrological properties of lime concretion black soil	bulk density, drained upper and lower limit, saturated water content, field capacity, pH, particle composition, fertility	observations obtained from 2013 field experiment
daily meteorological data at Bengbu Meteorological Station for the period of 2013	total solar irradiance, minimum and maximum temperatures, precipitation, wind speed, air humidity	China meteorological data sharing service system http://data.cma.cn/data/index.html
field management data	seeding date, fertilization, irrigation	consistent with field experiment

2.2. Site and field experiment

2.2.1. Field. We first studied summer maize yield response to water stresses in different growth stages through field experiments, which were carried out in Xinmaqiao experimental station located in Bengbu, where the climate and soil type (lime concretion black soil) are both typical in Huaibei plain. Then we calibrated DSSAT-CERES-Maize model and evaluated the performance of the model in simulating maize yield under water stresses using field experiment datasets. Based on this, we simulated yield losses caused by water stress in Bengbu Station from 1961 to 2015. Clarifying the water stress as the unique influencing factor on maize yield and identifying the risk of losses caused by drought in Bengbu, rational evaluation and planning of irrigation and water conservancy facilities construction would be provided.

Field experiments were conducted at Xinmaqiao experimental station (Lat. 33°09′ N, Long. 117°22′ E, elevation 19.7 m) in Anhui institute of hydraulic research, which is located in the Northern Huaihe Plain of Anhui Province. The area located in warm temperate semi humid monsoon climate, with mean annual precipitation of 911 mm (more than 60% is concentrated in June to September) and evaporation of 916 mm, and the mean annual temperature is 14.2 ℃. The main crops are summer maize and winter wheat. The main soil type is lime concretion black soil, a typical mid-low yield soil in the Northern Huaihe Plain, which is characterized by relatively high content of clay with poor drainage and water holding capacity. Under this climate and soil background, the area is prone to drought and flood during summer maize growth season.

The soil pH is 7.4 with middle fertility (table 2). The field water-holding capacity and saturated water content is 0.298 and 0.355, with the wilting point of 0.164 and effective water content of 0.134, and the bulk density is 1.42. The tested maize is Yedan 13, which belongs to late maturing varieties with high fertilizer requirement.

Table 2. Initial fertility of tested soil (g/kg).

type	organic matter	total nitrogen	hydrolytic nitrogen	total phosphorus	available phosphorus	available potassium
content	11.95	0.973	0.065	1.6	0.036	0.093

2.2.2. Experimental design. Experiments were carried out in 6.75 m^2 (2.7 m×2.5 m) test-pits under rain-proof shelter. Soil depth was 180 cm with filtering layer and water supply/drainage pipeline system at the soil bottoms. In order to monitor soil moisture at any time, moisture detectors were placed at 40 cm depth of the soil. Summer maize was planted by dry seed method on June 10, 2013. In each test-pit 6 rows of maize seeds were sowed and the space between the rows was 50 cm. 9 maize seeds were sown in each row and the space between each maize crop was 30 cm, which is equivalent to 80000 plants per hectare. The maize crop was harvested on September 15, 2013 and the growth season lasts 98 days. Nitrogen, Phosphorus and Potassium Fertilizers were applied at the rate of 180 Kg N ha-1, 90 Kg P ha-1 and 90 Kg K ha-1. During the whole growth season the rain-proof shelter was opened in sunny days and closed in rainy days.

The growing season was divided into four stages-seeding stage, such as seeding, jointing, heading-filling and matured stage, which were indexed as Ⅰ, Ⅱ, Ⅲ and Ⅳ, respectively. The control factor was soil water content in different growing stages of maize. Long term water control tests indicated that lower limits of the suitable water content at each stage were 65%, 70%, 75% and 70% of soil water holding capacity in lime concretion black soil. The water was available for maize which was generally the effective water content, which was 0.134 in the tested soil. Therefore, in order to better reflect the difficulty for the crop to use water and the water stress grades, the effective water content was separated into different sections. The grade classification of available water with soil moisture percentage in field capacity was shown in table 3. According to the table 3, it can be seen that the lower limits of four stages are 22%, 33%, 44% and 33% available water.

Table 3. Water grade classification and comparison of available water with percentage of soil moisture in field water capacity.

Soil water grade	Available water (%)	Soil water content (cm^3/s)	Percentage of soil moisture in field water capacity (%)
1	(40～50]	(0.218, 0.231]	(73.02, 77.52]
2	(30～40]	(0.204, 0.218]	(68.52, 73.02]
3	(20～30]	(0.191, 0.204]	(64.03, 68.52]
4	(10～20]	(0.177, 0.191]	(59.53, 64.03]
5	(0～10]	(0.164, 0.177]	(55.03, 59.53]
6	(-10～0]	(0.151, 0.164]	(50.54, 55.03]
7	(-20～-10]	(0.137, 0.151]	(46.04, 50.54]
8	(-30～-20]	(0.124, 0.137]	(41.54, 46.04]
9	<-30	<0.124	<41.54

Each test-pit was irrigated up to 50% of available water on June 1 in order to keep water content in each test-pits coincident at the beginning of the experiments. According to the lower limits of suitable water content (θ_{Li}, i= Ⅰ, Ⅱ, Ⅲ, Ⅳ), three water stress scenarios were set, mild drought, 1 grade lower than the grade of θ_{Li} (5% lower than θ_{Li}); moderate drought, 3 grades lower than the grade of θ_{Li} (15%

lower than θ_{Li}); severe drought, 6 grades lower than the grade of θ_{Li} (25% lower than θ_{Li}). The present study examined 12 water stress (shown in table 4) treatments and 1 CK (soil water content was suitable at each growth stage). Each treatment had 2 replications and 39 sets of test-pits.

Water content at moisture-control growth stage in each treatment maintains within the water grade which corresponds to the drought scenario-when water content was less than the lower limit of the water grade (G_{Li}, i=1, 2, 3, ...9), it would be irrigated to the upper limit (G_{Ui}, i=1, 2, 3, ...9), with sprinkling irrigation method and the water use efficiency was 0.8. While water content at the other stages maintain within the θ_{Li} and the corresponding G_{Ui}, when water content was less than the θ_{Li}, it would be irrigated to the G_{Ui}. Taking the treatment of moderate drought at jointing stage as an example, θ_{L2} was 0.21 (70% of water capacity), with the available water of 33% and the suitable water grade 2. Due to drought classification standard, water grade of moderate drought at jointing stage was 5 and water content should be controlled at (0.164, 0.177], when water content was less than 0.164, it would be irrigated to 0.177. As θ_{L1}, θ_{L2}, and θ_{L3}, were 0.194, 0.224 and 0.209, with the water grades of 3, 1 and 2, the moister at seeding stage, heading-filling stage and mature stage should be maintain at (22.2, 30], (44.4, 50] and (33.3, 40], respectively.

Table 4. The 12 treatments and CK of field experiments (%).

Stage / No.	treatment	Stage Ⅰ	stage Ⅱ	Stage Ⅲ	Stage Ⅳ
1	Mild drought in stage Ⅰ	(10, 20]	(33, 40]	(44, 50]	(33, 40]
2	Moderate drought in stage Ⅰ	(-10, 0]	(33, 40]	(44, 50]	(33, 40]
3	Severe drought in stage Ⅰ	<-30	(33, 40]	(44, 50]	(33, 40]
4	Mild drought in stage Ⅱ	(22, 30]	(20, 30]	(44, 50]	(33, 40]
5	Moderate drought in stage Ⅱ	(22, 30]	(0, 10]	(44, 50]	(33, 40]
6	Severe drought in stage Ⅱ	(22, 30]	(-30, -20]	(44, 50]	(33, 40]
7	Mild drought in stage Ⅲ	(22, 30]	(33, 40]	(30, 40]	(33, 40]
8	Moderate drought in stage Ⅲ	(22, 30]	(33, 40]	(10, 20]	(33, 40]
9	Severe drought in stage Ⅲ	(22, 30]	(33, 40]	(-20, -10]	(33, 40]
10	Mild drought in stage Ⅳ	(22, 30]	(33, 40]	(44, 50]	(20, 30]
11	Moderate drought in stage Ⅳ	(22, 30]	(33, 40]	(44, 50]	(0, 10]
12	Severe drought in stage Ⅳ	(22, 30]	(33, 40]	(44, 50]	(-30, -20]
CK	Keep suitable water condition in the whole growing season	(22, 30]	(33, 40]	(44, 50]	(33, 40]

2.3. Model calibration and validation

In order to improve the model simulation performance, model calibration is considered as an important process. To evaluate the model performance, the simulated values of grain yields was compared with the measured values. All samples were collected from our field experiments. In order to analyze model fitting performance, we compared the results of simulated with these of measured grain yields under different drought treatments.

Based on the above-mentioned data, we choose some indices to evaluate the fitting degree of model, such as the coefficient of determination (r^2), normalized Root Mean Square Error (*nRMSE*), coefficient of residual mass (*CRM*) [16], index of agreement (*D-index*) [17], and Nash-Sutcliffe coefficient (*E*). All indices measures the fitting degree (the mean of measured data, *M*) between simulated (E_i) and measured (M_i) data. For *nRMSE*, the value of the *nRMSE* was less than 10%, the fitting degree is very good. The *nRMSE* was greater than 10% but less than 20% which was considered as good. When the *nRMSE* was greater than 20%, the fitting degree of model was general or bad [18].

$$nRMSE = \sqrt{\sum_{i=1}^{n} \frac{(E_i - M_i)}{n}} \times \frac{100}{\overline{M}} \tag{5}$$

The value of *CRM* was a minus representing that simulated value was larger than measured value, and the value of *CRM* was more than zero representing that simulated value was less than measured value.

$$CRM = 1 - \sum_{i=1}^{n} E_i \Big/ \sum_{i=1}^{n} M_i \tag{6}$$

The value of *D-index* was 1 representing s good fitting degree of model between measured and simulated data. The value of *D-index* values was close to 0.0 indicating that the simulated values are close to the average of measured data, i.e., $E_i \sim M$, indicating bad fitting degree of model.

$$D - index = 1 - \sum_{i=1}^{n} (M_i - E_i)^2 \Big/ \sum_{i=1}^{n} \left(\left|E_i - \overline{M}\right| + \left|M_i - \overline{M}\right|\right)^2 \tag{7}$$

The value of Nash-Sutcliffe efficiencies (*E*) was 1 representing s perfect fitting degree of model between measured and simulated data. When *E* was less than 0, the mean value of measured data was considered as a better predicator than simulated data.

$$E = 1 - \sum_{i=1}^{n} (M_i - E_i)^2 \Big/ \sum_{i=1}^{n} \left(M_i - \overline{M}\right)^2 \tag{8}$$

2.4. Model simulations of the percentage of loss in yield caused by water stress using long-term meteorological data

Precipitation during four growth stage and the whole growing period from 1961 to 2015 in Bengbu Station was shown in figure 1. Mean precipitation during the whole growing period was 513 mm, with the mean rainfall of 141 mm, 172 mm, 99 mm and 101 mm in each growth stage. Precipitation varied greatly between different years and growth periods.

Yield was simulated by DSSAT-CERES-Maize model and the loss percentage in yield caused by water stress was calculated as follow.

$$L_i = \frac{Y_{si} - Y_{ri}}{Y_{si}} \times 100\% \tag{9}$$

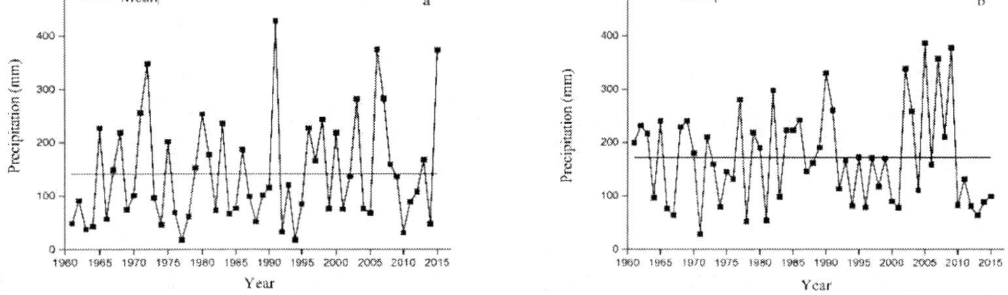

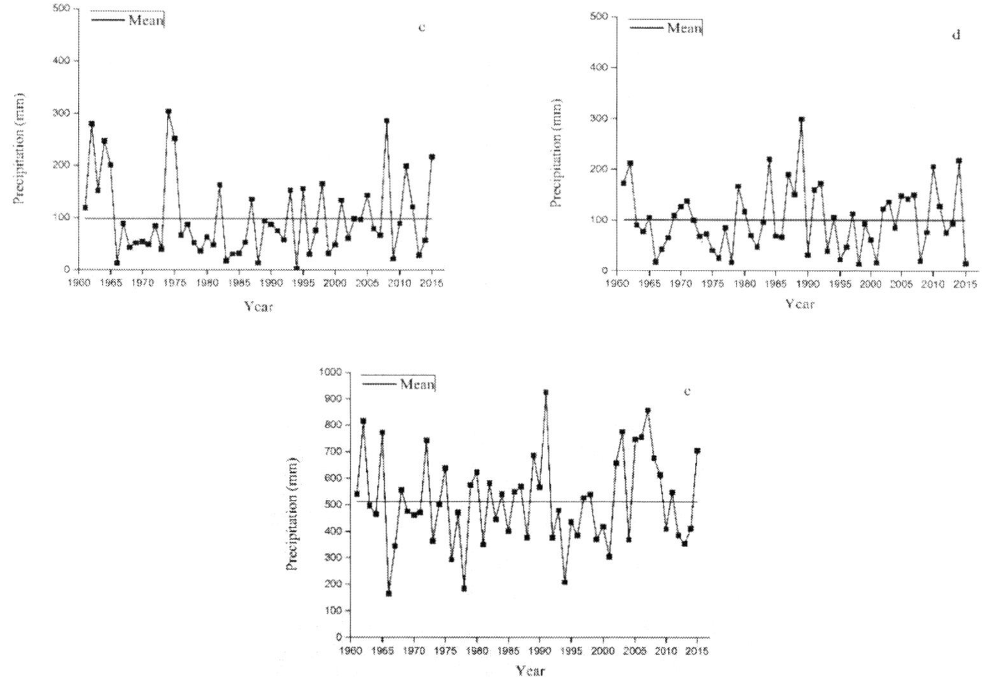

Figure 1. Precipitation of Bengbu Station from 1961 to 2015 during each of the 4 growth stage (growth stage Ⅰ(a), stage Ⅱ(b), stage Ⅲ(c), and stage Ⅳ(d)) and the whole growing period (e).

Where L_i is the loss percentage in yield caused by water stress in the year (1961-2015), Y_{si} is the yield maintaining suitable water content under climate background of the year, Y_{ri} is the yield under rain-fed agriculture of the year, respectively.

3. Materials and methods

3.1. Model calibration and validation

Based on experimental datasets, we applied Generalized Likelihood Uncertainty Estimation (GLUE) method to calibrate the model. The GLUE method was considered as a separate tool by R programming language embed into DSSAT model. We simulated 2000, 4000, 6000, 8000 and 10000 times to estimate the values of the five genetic coefficients by using GLUE method after determining the parameter of crop (Maize) and experiment processing (drought). Many times of experiment showed that the results were good by simulating 6000 times, and we obtained the values of the five genetic coefficients for maize as showed in table 5.

The simulated model was evaluated using field experiments data of summer maize yield in response to different drought treatments. Based on experimental datasets, we compared the variable of measured and simulated grain yields to validate the calibrated DSSAT CERES-Maize model. In order to validate summer maize yield performance of the validated model, we calculated the value of above statistic operation parameters based on the grain yields as shown in table 6. For the validation, most of *nRMSE* were less 20%, which indicated that the simulated and measured grain yields fitted each other good. With the *D-index* and *E* close to 1, the fitting degree of the calibrated model was perfect. The regression of simulated and measured grain yield indicated a good agreement with most of r^2 higher than 0.98. In addition, the *CRM* test showed that there were some negative values and some positive values but all values were close to 0, results indicated slight overestimation or underestimation of grain yields.

Table 5. Empirical ranges and calibrated values of model parameters.

	P1	*P2*	*P5*	*G2*	*G3*
Parameter name	Thermal time above 8 ℃ from emergence to the end of the juvenile	photoperiod sensitivity coefficient	Thermal time (above 8 ℃) from silking to physiological maturity	Maximum number of kernels per plant	Kernel filling rate under optimum conditions
Unit	℃·d	D/h	℃·d	number/plant	mg/day
Empirical range	100-400	0.1-0.8	600-1000	560-850	5-12
Calibrated value	260	0.3	850	800	9.6

Table 6. Comparison of measured (*MEA*) and simulated (*SIM*) grain yield predicated by the calibrated DSSAT CERES-Maize model.

Drought treatments		*MAX*	*MIN*	*MEAN*	*SE*	*N*	R^2	*P-VALUE*	*nRMSE*	*CRM*	*D-INDEX*	*E*
1	SIM	8313	1742	5448.6	2717.4	10	0.98	0.0011	11.09	0.02	0.99	0.97
	MEA	8573	1859	5562	2995.3							
2	SIM	7563	1411	4597.8	2452.9	10	0.99	0.0004	14.94	-0.11	0.99	0.94
	MEA	6834	1327	4155.6	2314.5							
3	SIM	8386	1709	5403.8	2749.3	10	0.99	0.0003	10.85	0.01	0.99	0.98
	MEA	8803	1391	5450.8	3112.1							
4	SIM	7283	1324	4423.6	2477.2	10	0.98	0.0016	25.79	-0.17	0.97	0.87
	MEA	6203	1125	3774.2	2414							
5	SIM	4712	983	3046	1519.1	10	0.99	0.0004	22.12	-0.14	0.98	0.92
	MEA	4299	654	2678	1595.3							
6	SIM	7975	1585	5063.2	2617.7	10	0.98	0.0017	12.07	-0.001	0.99	0.97
	MEA	8276	1423	5057	2843.6							
7	SIM	6217	1014	3615	2088.4	10	0.97	0.0018	14.54	-0.04	0.99	0.93
	MEA	5509	955	3492.2	1785.2							
8	SIM	3581	438.5	1931.6	1290.7	10	0.94	0.0065	24.48	-0.04	0.97	0.84
	MEA	3020	386	1859.2	1015.6							
9	SIM	9050	1847.3	5780.6	2900.8	10	0.98	0.0013	15.50	-0.08	0.99	0.95
	MEA	9117	1358	5354.8	3034.7							
10	SIM	8190	1847.3	5570.6	2636.8	10	0.98	0.0014	14.33	-0.05	0.99	0.97
	MEA	7838	1231	5303.6	2667.4							
11	SIM	7328	1847.3	5257	2297.9	10	0.99	0.0007	17.18	-0.12	0.98	0.89
	MEA	6928	1527	4700	2226.0							
12	SIM	9435	1824	5899.2	3092.8	10	0.98	0.0011	15.71	-0.08	0.99	0.96
	MEA	9349	1391	5457.8	3222.5							

In addition, data of simulated and measured yield were plotted for comparison as shown in figure 2 during the growing conditions of different drought treatments, respectively. Results indicated that the calibrated model simulated grain yields and measured grain yields fitted each other quite well in response to different drought scenarios.

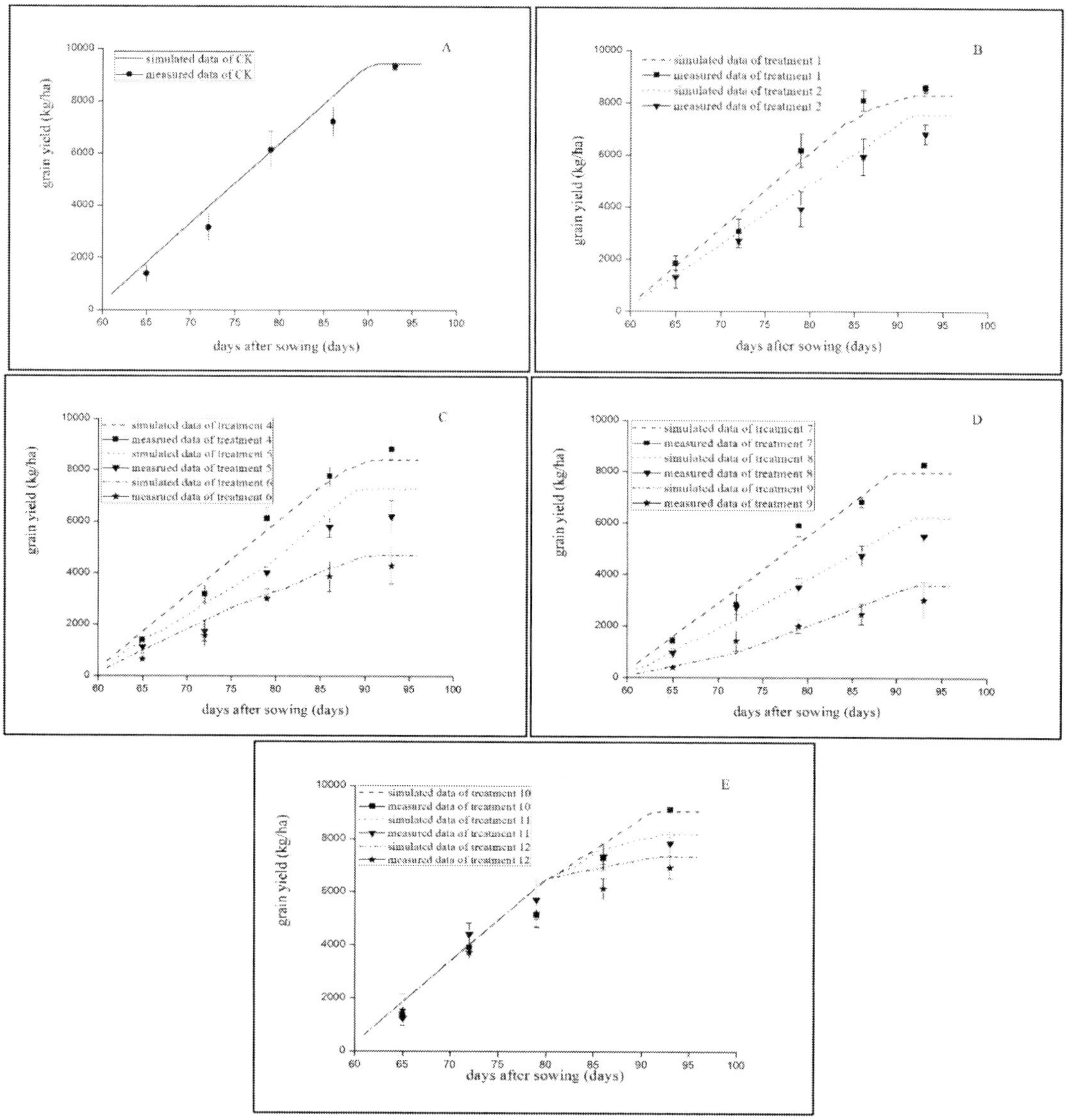

Figure 2. Simulated and measured yield of summer maize under CK (A) and mild, moderate and severe drought at stage Ⅰ (B), stage Ⅱ (C), stage Ⅲ(D), and stage Ⅳ(E).

YLP in maize yield caused by drought under different drought scenarios in different growing stages were shown in figure 3. Excepting failure of emergence under severe drought at seeding stage, YLP of the other treatments ranged from 2% to 70%. YLP in seven out of twelve treatments were higher than 20%, and four of them were higher than 40%, treatment 3, 9, 6 and 8. After suffering from consistent severe drought at stage Ⅰ, YLP would reach 100% due to the failure of emergence. Besides, the biggest YLP occurred in severe drought at stage Ⅱ and Ⅲ, which was around 70% and 50%. Some studies also showed that stage Ⅲ was the critical water requirement stage for summer maize [1, 19-24], which was the period for corn storage capacity building and expanding.

Simulated losses percentages under mild drought were higher than these of measured in all the growing stages, on the contrary, simulated data under moderate drought were lower than these of measured in each stage. Under severe drought, simulate results were lower than these of measured at stage Ⅱ, stage Ⅲ and stage Ⅳ, while both simulated and measured YLP at stage Ⅰ was 100%.

Higher simulate yield loss under mild drought might be caused by the ill-conceived of DSSAT-CERES-Maize model about the yield promotion of mild drought stress and the compensation effect after rehydration. On the other hand, lower simulate yield loss under moderate and severe drought indicated that the model might underestimate the negative effect of drought on maize yield.

Many researchers studied the reasons of yield loss caused by drought stress [4, 25-28]. Hao and Mei [6] emphasized that drought stress cut down photosynthesis rate (PR) and transpiration rate (TR) and then cause decreasing in summer maize yield. In addition, the reduction degree varied with growing stage, drought stress and rehydration timing. The PR reduced 38.8-45.7% and 11.2-36.2% at stage Ⅲ and Ⅱ, with the reduction in TR of 45.6-55.9% and 29.7-52.1%. After rewatering, PR recovered to 62.0-80.0% and 75.6-89.6% at stage Ⅲ and Ⅱ, while TR recovered to 78.1-84.2% and 82.1-92.1%. Beyond that, Geng and Yan [29] found that recovery of soil microorganism, which played an important role in maintaining soil ecosystem productivity [30], varied according to the drought stress. Briefly, the reduction of PR, TR and soil productivity caused drought stress and their insufficient recovery after rehydration resulted in the loss of maize yield, while the differences between them led to difference between yield losses.

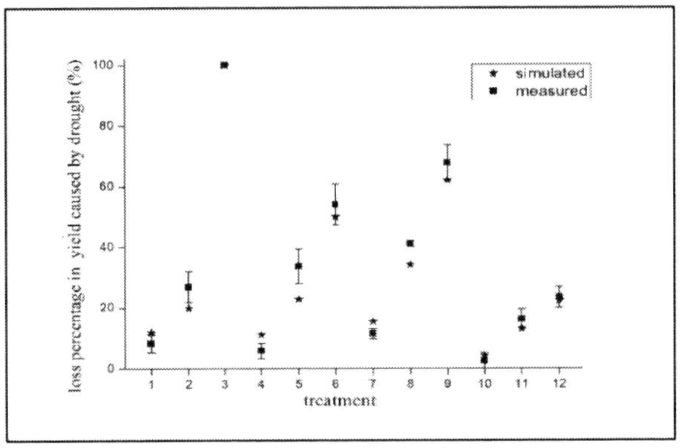

Figure 3. Simulated and measured loss percentage in maize yield caused by drought of the 12 treatments.

3.2. Model simulations of maize yield under rain-fed agriculture in 1961-2015

After calibrated and validated, DSSAT-CERES-Maize model was applied to simulate maize yield under rain-fed and suitable water content. Water stress caused yield loss percentages of Bengbu Station in Huaibei Plain as shown in figure 4. Loss percentage fluctuates according to the precipitation during the growth periods and the change rate ranged from -8.7% to 61.0%. There were 4, 4, 1, 0, 2 and 1 years that loss percentage ranges from 10-20, 20-30, 30-40, 40-50, 50-60 and 60-70, respectively.

The precipitations of the first three growth stages were significantly lower than the mean value in both 1966 and 1994, with very low rainfall at the fourth stage in 1966 and a much higher rainfall in 1994. However, there was no significant difference between two years' yield loss. We came to the conclusion that compensation of rehydration at the fourth stage after long-term severe drought stress was very little, due to the excessive damage of vegetative growth and storage capacity building and expanding at the first three stages. Precipitation at stage Ⅰ, Ⅱ and Ⅳ were very little in 1964, but precipitation in stage Ⅲ was much larger than the mean value with large yield compensation effect, therefore, the loss percentage was not significant, which was consistent with the findings of many other researches. Cakir [31] indicated that compensation effects on the yield of rehydration at flowering stage after experiencing drought stress at vegetative growth stage was high, while yield compensation effects of rehydration after encountering drought stress at silking and filling stage was very low. Hao [6] found that re-watering after short period of time (mild) drought could replenish soil moisture in time and promote utilization of deep soil water, while re-hydration after long time (severe) drought could not completely restore deep soil water because of over consumption during drought, which causes a negative water balance.

Precipitations at all stages in both 1978 and 1966 were much less than the mean value, while at stage III in 1966 rainfall was four times higher than that of 1978, resulting in more than double yield loss percentage of 1978, which also indicated that stage III was the most critical period for yield formatting.

There were 8 years out of all years that the loss percentage was negative, which means that the yield under rain-fed was higher than that of under suitable water content. In both 2005 and 1961, rainfall in stage I was lower than the mean value, while precipitation in stage II, III and IV was larger than the mean. The yield increased might gained from the overcompensation of in time rehydration after appropriate drought stress in seeding stage. The conception of crop overcompensation originated from the leaves being picked and increased in grain yield [32, 33], which implied increasing in the amount of biomass and yield after suffering from harsh conditions like water stress, insect pest, disease, etc.

Besides, the determination of suitable soil water content was unreasonable. Suitable water content was affected by meteorological elements such as temperature and solar radiation. The present study used the same suitable water content to simulate maize yield from 1961 to 2015, due to climate backgrounds were different, results might not be in conformity with the actual.

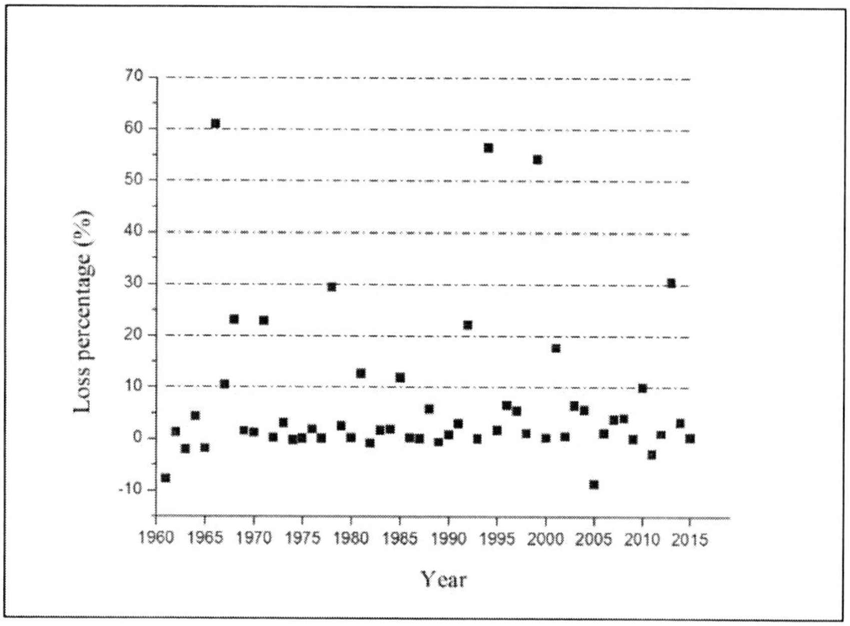

Figure 4. Loss percentages in maize yield from 1961 to 2015 and their distribution.

4. Conclusions

Maize yield losses caused by water stress in Bengbu Station in the Northern Huaihe Plain from 1961 to 2015 were simulated using DSSAT-CERES-Maize model. First, the model was calibrated and validated with drought stress experiments conducted in 2013. Then the model was evaluated by comparing the simulated yield with measured yield under different drought stresses at different growth stages. Subsequently, yield losses caused by water stress from 1961 to 2015 in Bengbu were simulated. The main findings are as follows:

- In view of parameters of summer maize growth and yield validated model, the performance of simulated model was considered to be perfect. Results indicated that the calibrated model simulated grain yields and measured grain yields fitted each other quite well in response to different drought scenarios.

- Simulated losses percentage under mild drought were higher than these of measured in all the growing stages, on the contrary, simulated data under moderate drought were lower than these of measured in each stage. Under severe drought, simulate data were lower than these of measured in stage II, stage III and stage IV, while both simulated and measured YLP in stage I were 100%.
- Yield loss percentages caused by water stress from 1961 to 2015 in Bengbu Station were simulated. There were 4, 4, 1, 0, 2 and 1 years that loss percentage ranges from 10-20, 20-30, 30-40, 40-50, 50-60 and 60-70, respectively.
- There were 8 years out of all years that the loss percentage was negative, which means that the yield under rain-fed condition was higher than that of suitable water content. Higher maize yield might be gained from the overcompensation of rehydration after appropriate drought stress.

Under the conditions of this study, maize yield losses caused by drought of the other meteorological station in Northern Huaihe Plain could be simulated and the risk distribution of the Plain would be mapped. Based on this, rational allocation of water resources, which would assist the Northern Huaihe Plain with highest maize yield and water use efficiency, could be established.

Acknowledgments
Authors are grateful to the Chinese National Natural Science Foundation (No. 51279207, No. 51409266), and the National Key Research and Development Project (No. 2016YFA0601503) for supporting the research. Last but not the least, we appreciate the assistance of Anhui institute of hydraulic research for supplying experimental test sites, and wish to acknowledge assistance or encouragement from colleagues, and special work by technical staff or financial support from organizations should do so in an unnumbered

References
[1] Liu S, et al 2010 Crop yield responses to climate change in the Huang-Huai-Hai Plain of China *Agr. Water. Manage* **97(8)** 1195-209
[2] Bhatia V S, *et al* 2008 Analysis of potential yields and yield gaps of rainfed soybean in India using CROPGRO-Soybean model *Agr Forest Meteorol* **148(8-9)** 1252-65
[3] Song F and Dai J 2000 Effect of drought stress on growth and development of female inflorecence and yield of maize *Journal of Jilin Agricultural University* **22** 18-22
[4] Bai L, *et al* 2004 Effects of soil water stress on morphological development and yield of maize *Acta Entomol Sin.* **24(7)** 1556-60
[5] Igbadun H E, *et al* 2008 Effects of deficit irrigation scheduling on yields and soil water balance of irrigated maize *Irrigation Sci* **27(1)** 11-23
[6] Hao W P 2013 *Influence of Water Stress and Rewatering on Maize WUE and Compensation Effects* (Beijing: Chinese Academy of Agricultural Sciences) pp 1-3
[7] Drury C F and Hoogenboom G 2012 Optimizing parameters of CSM-CERES-maize model to improve simulation performance of maize growth and nitrogen uptake in northeast china *J Intger Agr* **11(11)** 1898-913
[8] Tong F and Guo P 2013 Simulation and optimization for crop water allocation based on crop water production functions and climate factor under uncertainty *Appl Math Model* **37(14-15)** 7708-16
[9] Dai A, Trenberth K E and Karl T R 1998 Global variations in droughts and wet spells: 1900-1995 *Geophys Res Lett* **25(17)** 3367-70
[10] Yang Y, *et al* 2006 Estimation of groundwater use by crop production simulated by DSSAT-wheat and DSSAT-maize models in the piedmont region of the North China Plain *Hydrol Process* **20(13)** 2787-802

[11] Qi Z, *et al* 2013 Simulating dryland water availability and spring wheat production in the northern great plains *Agron J* **105(1)** 37
[12] Devkota K P, *et al* 2015 Simulating the impact of water saving irrigation and conservation agriculture practices for rice-wheat systems in the irrigated semi-arid drylands of Central Asia *Agr Forest Meteorol* **214** 266-80
[13] Attia A, *et al* 2016 Application of DSSAT-CERES-Wheat model to simulate winter wheat response to irrigation management in the Texas High Plains *Agr Water Manage* **165** 50-60
[14] Jiang Y, *et al* 2016 Modeling irrigation management for water conservation by DSSAT-maize model in arid northwestern China *Agr Water Manage* **177** 37-45
[15] Keulen V H 1982 Crop production under semi-arid conditions, as determined by moisture availability *Simulation of Plant Growth & Crop Production* 234-251
[16] Loague K 1991 The impact of land-use on estimates of pesticide leaching potential-assessments and uncertainties *J Contam Hydrol* **8(2)** 157-75
[17] Willmott C J 1982 Some comments on the evaluation of model performance *B Am Meteorol Soc* **63(11)** 1309-13
[18] Bannayan M and Hoogenboom G 2009 Using pattern recognition for estimating cultivar coefficients of a crop simulation model *Field Crop Res* **111(3)** 290-302
[19] Singh B R and Singh D P 1995 Agronomic and physiological-responses of sorghum, maize and pearl-millet to irrigation *Field Crop Res* **42(2-3)** 57-67
[20] Stone L R, *et al* 1996 Response of corn, grain sorghum, and sunflower to irrigation in the High Plains of Kansas *Agr Water Manage* **30(3)** 251-9
[21] Stone P J, *et al* 2001 Water deficit effects on sweet corn. I. Water use, radiation use efficiency, growth, and yield *Austra J Agr Res* **52(1)** 103-13
[22] Zhang J, Mu Q and Huang J 2016 Assessing the remotely sensed drought severity index for agricultural drought monitoring and impact analysis in North China *Ecol Indic* **63** 296-309
[23] Zhang Z H and Cai H J 2001 Influences of water deficit on growth, development and yield in plastic-mulched spring corn *Irrig Drain* **20(2)** 13-16
[24] Anderson M C, *et al* 2011 Evaluation of drought indices based on thermal remote sensing of evapotranspiration over the continental United States *J Climate* **24(8)** 2025-44
[25] Jamieson P D, Martin R J and Francis G S 1995 Drought influences on grain-yield of barley, wheat, and maize *New Zeal ournal of Crop and Hort* **23(1)** 55-66
[26] Schussler J R and Westgate M E 1995 Assimilate flux determines kernel set at low water potential in maize *Crop Sci* **35(4)** 1074-80
[27] Fan Z X, *et al* 2001 Study on grain filling characteristics of transplanted summer maize J *Maize Sci* **9(2)** 47-49
[28] Zhang J, *et al* 2010 Effect of moisture and mulching on filling characteristics and yield of summer maize *Chinese Journal of Eco-Agriculture* **18(3)** 501-6
[29] Geng S M, *et al* 2015 Effects of drought stress on agriculture soil *Nat Hazards* **75(2)** 1-15
[30] Biederbeck V O, *et al* 1994 Labile soil organic matter as influenced by cropping practices in an arid environment *Soil Biol Biochem* **26(12)** 1647-56
[31] Cakir R 2004 Effect of water stress at different development stages on vegetative and reproductive growth of corn *Field Crop Res* **89(1)** 1-16
[32] Belsky A J 1986 Does herbivory benefit plants - A review of the evidence *Am Nat* **127(6)** 870-92
[33] Maschinski J and Whitham T G 1989 The continuum of plant responses to herbivory: The influence of plant association, nutrient availability, and timing *Am Nat* **134(1)** 1-19

IOP Conf. Series: Earth and Environmental Science 82 (2017) 012031 doi:10.1088/1755-1315/82/1/012031

Higher moments method for generalized Pareto distribution in flood frequency analysis

C R Zhou[1], Y F Chen[1,3], Q Huang[1] and S H Gu[2]

[1]College of Hydrology and Water Resources Hohai University, No. 1 Xikang Road, Gulou District in Nanjing, Jiangsu Province, China
[2]Shanghai Hydrological General Station, No. 241 East Longhua Road, District Xuhui in Shanghai, China

E-mail: 19870056@hhu.edu.cn

Abstract. The generalized Pareto distribution (GPD) has proven to be the ideal distribution in fitting with the peak over threshold series in flood frequency analysis. Several moments -based estimators are applied to estimating the parameters of GPD. Higher linear moments (LH moments) and higher probability weighted moments (HPWM) are the linear combinations of Probability Weighted Moments (PWM). In this study, the relationship between them will be explored. A series of statistical experiments and a case study are used to compare their performances. The results show that if the same PWM are used in LH moments and HPWM methods, the parameter estimated by these two methods is unbiased. Particularly, when the same PWM are used, the PWM method (or the HPWM method when the order equals 0) shows identical results in parameter estimation with the linear Moments (L-Moments) method. Additionally, this phenomenon is significant when $r \geq 1$ that the same order PWM are used in HPWM and LH moments method.

1. Introduction

Peak Over Threshold (POT) series is an alternative to the annual maximum series when studying characteristics of extreme events. Several distributions can be used to describe extreme events of discharge observations (annual floods). For POT series, the generalized Pareto distribution (GPD) has been proven to be a perfect candidate for modeling POT samples [1, 2]. The GPD can describe the partial duration series (or POT series) which are obtained by letting a threshold select the highest values from an ordinary hydrological time series [3].

However, the POT approach remains under-employed recently, mainly due to the complexities associated with its usage [4], such as the choice of thresholds, the selection of criteria for retaining floods peaks, and so on.

Parameter estimation is a crucial part of flood frequency analysis (FFA). Many types of estimators have been developed to improve the efficiency and robustness of the fitting techniques. Moment methods, including the method of moments (MOM), the probability weighted moments (PWM) method, and the linear moments (L-Moments) method, use sample moments to estimate population moments. The law of large numbers is the basic discipline of these methods. The PWM method showed advantages for small datasets due to its smaller uncertainty than ordinary moments [3]. This characteristic suggested that the PWM method was a more suitable method for relatively short length of flood series. The L-Moments method was first put forward by Hosking [5] and it has been widely applied to estimate various distributions. L-Moments, which are linear combinations of the PWM, tend to share similar characteristics with PWM. Additionally, Wang [6] developed LH moments which are generalizations of L-Moments. Higher L-Moments (LH moments) rely on the linear combinations of higher-order statistics with the capability of a more detailed analysis of annual flood peak data [7].

Previous researchers have concluded that higher moments gave more consideration to significant flood peaks and extracted information existing in the POT series to the greatest possible extent. Several studies have attempted to deduce calculation procedures for ordinary moments, such as the emergence of , LH moments or higher PWM for common distributions (generalized Extreme Value, generalized Pareto, Pearson Type-III, Gumbel and Weibull distributions [8], to name a few).

This paper is structured as follows: in the first section, a brief introduction to GPD and estimators based on PWM will be presented; in section 2, statistical experiments were given to investigate the performances of all methods studied in this paper. In section 3, these methods were applied to a case study of the POT series of the Yangtze River, China.

2. Backgrounds

The Pareto distribution (PD) was first put forward by Vilfredo Pareto (an Italian civil engineer, economist, and sociologist). Pickands [9] extended it to the generalized Pareto distribution (GPD), and revealed its great importance to extreme value theory. In this study, GPD with three parameters, μ,σ,ζ, which refer to the location, scale and shape of GPD respectively, was analyzed. The shape of GPD differs with the value of shape parameter ζ. The definition of GPD is given by:

$$G(\mu,\sigma,\zeta)=\begin{cases}1-(1+\zeta\dfrac{x-\mu}{\sigma})^{-\frac{1}{\zeta}},\zeta\neq 0\\ 1-\exp(-\dfrac{x-\mu}{\sigma}),\zeta=0\end{cases} \tag{1}$$

Considering practical application, adequate flood quantile estimates are crucial. The quantile function is given by equation (2).

$$x_{(G)}=\begin{cases}\mu-\dfrac{\sigma}{\zeta}\left[1-(1-G)^{-\zeta}\right],\zeta\neq 0\\ \mu+\sigma\left[-\ln(1-G)\right],\zeta=0\end{cases} \tag{2}$$

The probability weighted moments (PWM) method is a widely used estimator in parameter estimation for hydrological distributions, including GPD. However, one should note that the PWM method is only applicable when $\zeta<0.5$. Considering X is the random variable and its distribution is $F(x)$, the PWM of X is given by:

$$M_{p,r,s}=\int_0^1[x_{(F)}]^p F^r(1-F)^s dF \tag{3}$$

Where $x_{(F)}$ is the quantile function; p,r,s are real numbers; and r refers to the order of PWM. To obtain excellent statistics characteristics of GPD, $p=1,s=0$ were used. The PWM of X shown as equation (4).

$$\beta_r=M_{1,r,0}=\int_0^1 x_{(F)}F^r dF;(r=0,1,2,...,n) \tag{4}$$

Let $\{x_1,x_2,x_3,...x_n\}$ be random samples selected from X. Sort the series ascending, then it will become a new series $\{x_{(1)},x_{(2)},...,x_{(n)}\}$ (note $x_{(1)}\leq x_{(2)}\leq...\leq x_{(n)}$). Therefore, PWM can be calculated by equation (5). The b_r of samples are estimators to the β_r of population.

$$b_r = \frac{1}{n}\sum_{i=1}^{n}\frac{(i-1)(i-2)...(i-r)}{(n-1)(n-2)...(n-r)}x_{(i)} \tag{5}$$

If $\zeta \neq 0$, the quantile function of X is $x_{(G)} = \mu - \frac{\sigma}{\zeta}[1-(1-G)^{-\zeta}]$. Thus, β_r is given by:

$$\beta_r = \int_0^1 x_{(G)}G^r dG = \int_0^1 [\mu - \frac{\sigma}{\zeta}(1-(1-G)^{-\zeta})]G^r dG = \frac{\mu}{r+1} - \frac{\sigma}{\zeta(r+1)} + \frac{\sigma}{\zeta}\int_0^1 G^r(1-G)^{-\zeta} dG \tag{6}$$

According to the GPD-based method of moments (MOM), the equations below (see equations (7) to (9)) put a constraint condition on the use of moment methods. Consequently, they are theoretically applicable only when $\zeta \neq \frac{1}{3}$, and ζ must not exceed 0.5 for the reason that when $\zeta \to 0.5$, the variance will tend to be infinite. However, researchers have previously suggested that moments' estimators for ζ <0.25.

$$E(X) = \mu + \frac{\sigma}{1-\zeta} \tag{7}$$

$$Var(X) = \frac{\sigma^2}{(1-\zeta)^2(1-2\zeta)} \tag{8}$$

$$Skew(X) = \frac{2(1+\zeta)\sqrt{1-2\zeta}}{1-3\zeta} \tag{9}$$

If $\zeta < 1$, then $\int_0^1 G^r(1-G)^{-\zeta} dG = B(r+1,1-\zeta)$, where $B(r+1,1-\zeta)$ is the Beta-function which can be expressed as $B(p,q) = \int_0^1 x^{p-1}(1-x)^{q-1}dx$. Equation (6) can be changed into equation (10).

$$\beta_r = \frac{\mu}{r+1} - \frac{\sigma}{(r+1)\zeta} + \frac{\sigma}{\zeta}B(r+1,1-\zeta) \tag{10}$$

3. Higher moments methods

3.1. *Higher probability weighted moments method*

Higher PWM (HPWM) give more weights to larger flows [10] because the power function F^r increases with the values of samples. The HPWM method has been developed to be an estimator for many hydrological distributions [11, 12]. These applications have further proven the availability and robustness of HPWM method. In this paper, the HPWM method was developed for GPD, and the estimator to each parameter was listed.

According to the descriptions above, if $\zeta < 1, \zeta \neq 0$, the PWM for GPD can be calculated as equation (11).

$$(r+1)\beta_r = \mu - \frac{\sigma}{\zeta} + (r+1)\frac{\sigma}{\zeta}B(r+1,1-\zeta) \tag{11}$$

If $\zeta < 1$, one can calculate as follows according to the HPWM method based on GPD.

IOP Conf. Series: Earth and Environmental Science **82** (2017) 012031 doi:10.1088/1755-1315/82/1/012031

$$(r+2)\beta_r = \mu - \frac{\sigma}{\zeta} + (r+2)\frac{\sigma}{\zeta}B(r+2,1-\zeta) \tag{12}$$

$$(r+2)\beta_{r+1} - (r+1)\beta_r = (r+2)\frac{\sigma}{\zeta}B(r+2,1-\zeta) - (r+1)\frac{\sigma}{\zeta}B(r+1,1-\zeta) \tag{13}$$

$$\frac{(r+2)\beta_{r+1} - (r+1)\beta_r}{(r+3)\beta_{r+2} - (r+2)\beta_{r+1}} = \frac{(r+2)B(r+2,1-\zeta) - (r+1)B(r+1,1-\zeta)}{(r+3)B(r+3,1-\zeta) - (r+2)B(r+2,1-\zeta)} \tag{14}$$

$$\sigma = \frac{\zeta[(r+2)\beta_{r+1} - (r+1)\beta_r]}{(r+2)B(r+2,1-\zeta) - (r+1)B(r+1,1-\zeta)} \tag{15}$$

$$\mu = (r+1)\beta_r + \frac{\sigma}{\zeta} - (r+1)\frac{\sigma}{\zeta}B(r+1,1-\zeta) \tag{16}$$

The key, clearly, is to establish a method to estimate ζ. To calculate the estimator $\hat{\zeta}$ of ζ, let the left side of equation (14) be Z; then, its estimator can be expressed by $\hat{Z} = \frac{(r+2)b_{r+1} - (r+1)b_r}{(r+3)b_{r+2} - (r+2)b_{r+1}}$. Then, the relationship between Z and ζ can be investigated. Considering that ζ is a random number within 0~1, so equation (17) is given below.

$$Z = \frac{(r+2)B(r+2,1-\zeta) - (r+1)B(r+1,1-\zeta)}{(r+3)B(r+3,1-\zeta) - (r+2)B(r+2,1-\zeta)} \tag{17}$$

If r and ζ are given, Z in equation (17) can be easily calculated. Relationship between Z and ζ were drawn, and then these curves were fitted by polynomial. The results apparently showed linear relationships between Z and ζ (see figure 1). The linear fitting coefficients and linear fitting formulas were shown in table 1.

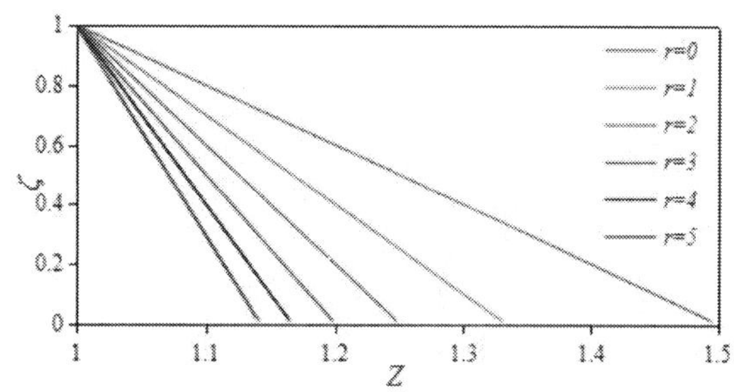

Figure 1. Relationship diagram of $Z \sim \zeta$.

Table 1. Linear fitting coefficients and linear fitting formulas.

Order r	Fitting coefficients and correlation coefficients			Linear fitting formulas
	α_1	α_0	R^2	
0	-2.0000	3.0000	0.9999990	$\zeta = -2.0000Z + 3.0000$
1	-3.0000	4.0000	0.9999998	$\zeta = -3.0000Z + 4.0000$
2	-4.0000	5.0000	0.9999451	$\zeta = -4.0000Z + 5.0000$

3	-5.0001	6.0001	0.9999975	$\zeta = -5.0001Z + 6.0001$
4	-6.0000	7.0000	0.9999994	$\zeta = -6.0000Z + 7.0000$
5	-6.9999	7.9999	0.9999996	$\zeta = -6.9999Z + 7.9999$

In conclusion, when $\zeta < 1, \zeta \neq 0$, HPWM can be used to estimate the parameters of GPD according to the following steps:

- ●Step 1: calculate PWM of the samples (b_r, b_{r+1}, b_{r+2}), and apply them to compute $\hat{Z}$ as the estimator of Z;
- ●Step 2: substitute $\hat{Z}$ into the linear fitting formulas in table 1 to calculate $\hat{\zeta}$, such as we can calculate the estimate of $\hat{\zeta}$ when the order of the HPWM equals 0: $\hat{\zeta} = -2.0000\hat{Z} + 3.0000$.
- ●Step 3: substitute $\hat{\zeta}$ into equation (15) to compute $\hat{\sigma}$;
- ●Step 4: use $\hat{\zeta}, \hat{\sigma}$ to calculate $\hat{\mu}$ by equation (16).

3.2. L-Moments

Hosking introduced L-Moments as an alternative to Moments Method for characterizing distributions and data [10]. They are linear combinations of PWM and have shown excellent performances as estimators [13-15]. PWM are essential components of L-Moments, and PWM (β_r with $r = 0,1,2$) can be calculated by equation (18).

$$\beta_0 = \mu - \frac{\sigma}{\zeta} + \frac{\sigma}{\zeta} B(1,1-\zeta)$$

$$\beta_1 = \frac{\mu}{2} - \frac{\sigma}{2\zeta} + \frac{\sigma}{\zeta} B(2,1-\zeta) \tag{18}$$

$$\beta_2 = \frac{\mu}{3} - \frac{\sigma}{3\zeta} + \frac{\sigma}{\zeta} B(3,1-\zeta)$$

Given the relationship between L-Moments and PWM, L-Moments can be expressed by equation (19).

$$\lambda_1 = \beta_0$$

$$\lambda_2 = 2\beta_1 - \beta_0 \tag{19}$$

$$\lambda_3 = 6\beta_2 - 6\beta_1 + \beta_0$$

Referring to the introduction of PWM and HPWM, the L-moments of samples can be estimated by its PWM as $\hat{\lambda}_1 = b_0; \hat{\lambda}_2 = 2b_1 - b_0; \hat{\lambda}_3 = 6b_2 - 6b_1 + b_0$.

Therefore, according to equations (18) and (19), L-M of the GPD samples with $r = 1,2,3$ can easily lead to the expression of $\lambda_1, \lambda_2, \lambda_3$ (see equation (20)).

$$\lambda_1 = \mu - \frac{\sigma}{\zeta} + \frac{\sigma}{\zeta} B(1,1-\zeta) = \mu + \frac{\sigma}{1-\zeta}$$

$$\lambda_2 = \frac{\sigma}{\zeta}[2B(2,1-\zeta) - B(1,1-\zeta)] = \frac{\sigma}{(1-\zeta)(2-\zeta)} \tag{20}$$

$$\lambda_3 = \frac{\sigma}{\zeta}[6B(3,1-\zeta) - 6B(2,1-\zeta) + B(1,1-\zeta)] = \frac{\sigma(1+\zeta)}{(3-\zeta)(2-\zeta)(1-\zeta)}$$

Thus, equations (21) to (23) can be used to calculate the estimation of each parameter.

$$\hat{\zeta} = \frac{3\hat{\tau}_3 - 1}{1+\hat{\tau}_3}; \hat{\tau}_3 = \frac{\hat{\lambda}_3}{\hat{\lambda}_2} \quad (21)$$

$$\hat{\sigma} = (1-\hat{\zeta})(2-\hat{\zeta})\hat{\lambda}_2 \quad (22)$$

$$\hat{\mu} = \hat{\lambda}_1 - (2-\hat{\zeta})\hat{\lambda}_2 \quad (23)$$

3.3. LH Moments

Wang [10] was the first to suggest LH moments as the generalization of L-Moments. Previous researches have applied it into fitting with hydrological distributions [16-18], which have revealed the limitation in the usage of the proposed LH moments. Many studies introduced the LH moments, but there are still rare studies about its corresponding calculation methods.

Similar to L-Moments, the first three orders of LH moments can be defined by equations (24) and (25).

$$\lambda_1^r = (r+1)\beta_r$$

$$\lambda_2^r = \frac{r+2}{2!}[(r+2)\beta_{r+1} - (r+1)\beta_r] \quad (24)$$

$$\lambda_3^r = \frac{r+3}{3!}[(r+4)(r+3)\beta_{r+2} - 2(r+3)(r+2)\beta_{r+1} + (r+1)(r+2)\beta_r]$$

$$\tau_3^r = \frac{r+3}{3(r+2)} \frac{(r+4)(r+3)\beta_{r+2} - 2(r+3)(r+2)\beta_{r+1} + (r+2)(r+1)\beta_r}{(r+2)\beta_{r+1} - (r+1)\beta_r} \quad (25)$$

According to the GPD-based PWM, β_r in equation (25) above can be easily obtained. So equations above can be transformed into equations (26) and (27).

$$\lambda_1^r = (r+1)\beta_r = \mu - \frac{\sigma}{\zeta} + \frac{\sigma}{\zeta}(r+1)B(r+1,1-\zeta)$$

$$\lambda_2^r = \frac{r+2}{2}\frac{\sigma}{\zeta}[(r+2)B(r+2,1-\zeta) - (r+1)B(r+1,1-\zeta)] \quad (26)$$

$$\lambda_3^r = \frac{r+3}{6}\frac{\sigma}{\zeta}[(r+4)(r+3)B(r+3,1-\zeta) - 2(r+3)(r+2)B(r+2,1-\zeta) + (r+2)(r+1)B(r+1,1-\zeta)]$$

$$\tau_3^r = \frac{\lambda_3^r}{\lambda_2^r} = \frac{r+3}{3(r+2)} * \frac{(r+4)(r+3)B(r+3,1-\zeta) - 2(r+3)(r+2)B(r+2,1-\zeta) + (r+2)(r+1)B(r+1,1-\zeta)}{(r+2)B(r+2,1-\zeta) - (r+1)B(r+1,1-\zeta)} \quad (27)$$

Referring to equation (27), ζ is the only unknown parameter corresponding to a given order r.

Because the calculation expression of ζ is not that ready to get, it leads one to seek the numerical methods to it, and find the approximate relationship between τ_3^r and ζ.

Let the value of ζ change within the range of [-2, 1), and one can calculate each τ_3^r corresponding to different ζ. Then draw the points (τ_i, ζ_i) corresponding to each given order r. Fit these points with the polynomial functions. Finally, cubic polynomial performs well in fitting with the points. So the approximate relationship between τ_3^r and ζ can be expressed as by equation (28).

$$\zeta = \alpha_3(\tau_3^r)^3 + \alpha_2(\tau_3^r)^2 + \alpha_1\tau_3^r + \alpha_0 \tag{28}$$

Figure 2 shows the results of polynomial fitting and the relationship between τ_3^r and ζ ; Table 2 gives out the coefficients of polynomial fitting and correlation coefficients.

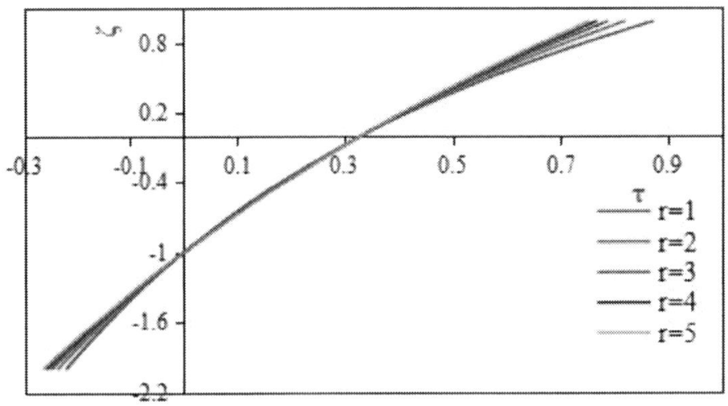

Figure 2. Approximate relationship between τ and ζ.

Table 2. The coefficients of polynomial fitting and correlation coefficients.

Order r	**The coefficients of polynomial fitting** α_3	α_2	α_1	α_0	**Correlation coefficients** R^2
1	1.0552	-2.6431	3.7896	-1.004	0.999999
2	0.7668	-2.0926	3.624	-10018	0.999999
3	0.5745	-1.7192	3.5152	-1.0009	0.999999
4	0.4437	-1.4539	3.4387	-1.0005	0.999999
5	0.3518	-1.2574	3.3821	-1.0003	0.999999

The value of τ_3^r can be estimated by the PWM of samples (see equation (29)). Thus the estimation of ζ can be obtained by the polynomial relationship between them (equation (28)).

$$\hat{\tau}_3^r = \frac{r+3}{3(r+2)}\frac{(r+4)(r+3)b_{r+2} - 2(r+3)(r+2)b_{r+1} + (r+2)(r+1)b_r}{(r+2)b_{r+1} - (r+1)b_r} \tag{29}$$

When the estimation of ζ has been calculated, the other two unknown parameters σ, μ would be estimated by equations (30) to (31).

$$\hat{\sigma} = \frac{2\hat{\zeta}\lambda_2^r}{(r+2)[(r+2)B(r+2,1-\hat{\zeta}) - (r+1)B(r+1,1-\hat{\zeta})} \tag{30}$$

$$\hat{\mu} = \lambda_1^r + \frac{\hat{\sigma}}{\hat{\zeta}} - \frac{\hat{\sigma}}{\hat{\zeta}}(r+1)B(r+1,1-\hat{\zeta}) \tag{31}$$

4. Statistical experiments and results

Statistical experiments were used to evaluate and compare the methods of PWM, HPWM, L-Moments and LH moments. Cunnane [19] demonstrated that the unbiased plotting position, which was the mean of the order statistics in samples from the reduced variety population, had the smallest mean square error among all such estimates. In order to reduce sampling errors as much as possible, a random-number generator was used in Matlab, to generate random numbers to follow the GPD. Two random datasets, each with a size of 1000*1000 (parameters: $\zeta=0.2$, $\sigma=0.1$, $\mu=0.1$, $\zeta=-0.1$, $\sigma=1$, $\mu=1$) are generated. The data was ascendinglysorted.

According to theory of order statistics, the mean of each order statistics was calculated as an unbiased plotting position. The parameters of the samples were estimated by the method given or deduced above. All results are given in table 3.

From table 3, it was clear that if the first three PWM were used, the results of different methods were identical to each other. This was clear to see from the results of PWM, HPWM ($r=0$), L-Moments and LH moments ($r=0$). Moreover, when the same PWMs were used, LH moments methods and HPWM method (when $r\geq 1$) showed almost the same performances.

Table 3. Estimators and their performances.

Distribution Function: GPD;		Parameters: $\zeta=0.2, \sigma=1, \mu=1$					
Parameters Estimators	PWM	HPWM					
		$r=0$	$r=1$	$r=2$	$r=3$	$r=4$	$r=5$
ζ	0.2005	0.2005	0.2001	0.1998	0.1995	0.1992	0.1991
σ	0.9993	0.9993	1.0000	1.0008	1.0016	1.0023	1.0027
μ	1.0000	1.0000	0.9995	0.9988	0.9980	0.9972	0.9967
Parameters Estimators	L-M	LH moments					
		$r=0$	$r=1$	$r=2$	$r=3$	$r=4$	$r=5$
ζ	0.2005	0.2005	0.2019	0.2002	0.1995	0.1992	0.1989
σ	0.9993	0.9993	0.9961	0.9999	1.0016	1.0026	1.0034
μ	1.0000	1.0000	1.0002	0.9995	0.9998	0.9969	0.9959
Distribution Function: GPD;		Parameters: $\zeta=-0.1, \sigma=1, \mu=1$					
Parameters Estimators	PWM	HPWM					
		$r=0$	$r=1$	$r=2$	$r=3$	$r=4$	$r=5$
ζ	-0.1002	-0.1002	-0.1002	-0.1002	-0.1002	-0.1002	-0.1000
σ	1.0006	1.0006	1.0006	1.0005	1.0006	1.0004	0.9997
μ	1.0000	1.0000	1.0000	1.0000	1.0000	1.0001	1.0005
Parameters Estimators	L-M	LH moments					
		$r=0$	$r=1$	$r=2$	$r=3$	$r=4$	$r=5$
ζ	-0.1002	-0.1002	-0.0950	-0.0977	-0.0988	-0.0993	-0.0996
σ	1.0000	1.0000	0.9916	0.9955	0.9974	0.9984	0.9989
μ	1.0006	1.0006	1.0045	1.0032	1.0023	1.0018	1.0014

5. Case study and results

Daily runoff series of Cuntan and Hankou gauged stations (Yangtze River in China) was selected for further study to investigate the applicability of GPD and HPWM to the POTS observations and test the performance of HPWM and LH moments method in parameter estimation. Cuntan gauged station was set up in the year of 1893. It monitors the fluxes and sand from the upper stream of Yangtze River, and this makes it be an important station to control the inflow of the Three George Project [20]. Hankou gauged station is set up in the year of 1865, but the daily runoff time series from 1952 to 2005 is included in this research due to they can be taken as stationary(pass the statistical test).

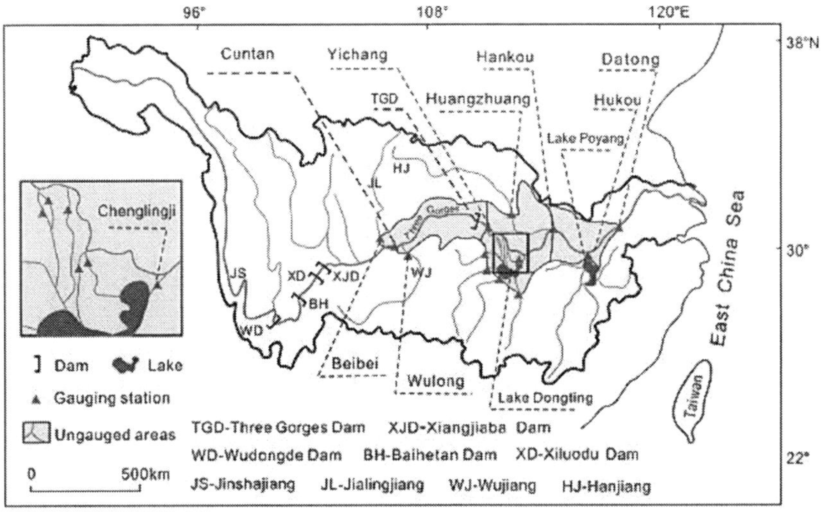

Figure 3. The location of Cuntan Station (Download from the paper of Yang, et al. 2007).

As a result of anthropogenic processes as well as the construction and operation of hydroelectrolical projects, the hydrologic properties of Cuntan station have dramatically changed. Thus, a consistency test must be studied before applying it to the FFA series.

Therefore, the runoff data from 1979 to 2014 was selected in this study. According to the results of statistical test (significance level=0.05), the series is stationary.

Previous studies [21] showed more reasonable results will be obtained when the number of annual average peaks to be 2 to 3. So the threshold value was determined to be 30000 m^3/s in this study. There are 95 peaks occurred in the series, so approximately 2.63 peaks per year. The same work was done to Hankou gauged station, the threshold is settled as 40000 m^3/s, and 133 peaks can be obtained in total, or 2.463 peaks each year (within 2 to 3)

All methods mentioned in this paper are used to fit with the peak runoff points. The results were listed in table 4, and all theoretical curves and empirical points are drawn in figure 4.

Table 4 shows that all of the methods performed well for parameter estimation. Particularly, the HPWM method and LH moments method seem to provide a broad range of selections because they can give out different results that correspond to different orders. Thus, one can select a better set of parameters to describe GPD in fitting with empirical points. The distributions used to fit with peak runoff points were drawn where the best fit was achieved. In order to describe the goodness of fit, sum relative square variation (WLS) is used in this study:

$$WLS = \sum_{i=1}^{n} \left(\frac{x_{ip} - x_p}{x_p}\right)^2 \times 100\% \tag{32}$$

Where WLS is the sum relative square deviation and x_{ip} is the estimator of x_p corresponding to the frequency p.

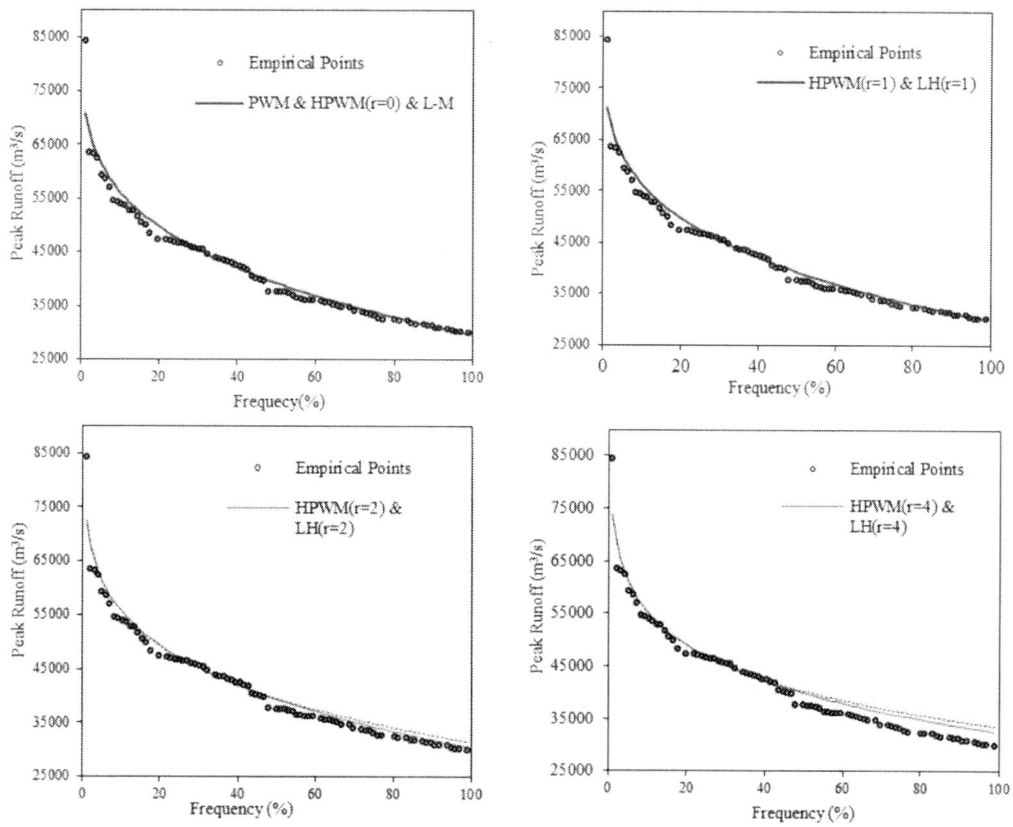

Figure 4. Relative diagrams of empirical points and theoretical curves, for which parameters estimate by all methods mentioned in this paper, take Cuntan gauged station as an example.

Table 4. Results of parameter estimation by several methods.

	PWM	Cuntan gauged station: HPWM					
Parameters Estimators		$r=0$	$r=1$	$r=2$	$r=3$	$r=4$	$r=5$
ζ	-0.2547	-0.2547	-0.2414	-0.201	-0.1507	-0.101	-0.0575
σ	15280	15280	14966	13909	12540	11198	10051
μ	29566	29566	29710	30335	31305	32409	33478
	L-M	Cuntan gauged station: LH moments					
Parameters Estimators		$r=0$	$r=1$	$r=2$	$r=3$	$r=4$	$r=5$
ζ	-0.2547	-0.2547	-0.2363	-0.1983	-0.1491	-0.1002	-0.0571
σ	15280	15280	14847	13838	12499	11175	10042
μ	29566	29566	29765	30377	31335	32428	33486
	PWM	Hnakou gauged station: HPWM					
Parameters Estimators		$r=0$	$r=1$	$r=2$	$r=3$	$r=4$	$r=5$
ζ	-0.3297	-0.3297	-0.3025	-0.3065	-0.3146	-0.3235	-0.3325

σ	13921	13921	13367	13460	13669	13925	14203
μ	40874	40874	41118	41066	40932	40752	40542
Parameters Estimators	L-M	Hankou gauged station: LH moments					
		$r=0$	$r=1$	$r=2$	$r=3$	$r=4$	$r=5$
ζ	-0.3297	-0.3297	-0.2977	-0.3038	-0.3129	-0.3224	-0.3320
σ	13921	13921	13271	13397	13625	13894	14186
μ	40874	40874	41160	41101	40960	40773	40554

Table 5 lists the means of WLS between theoretical curves (for which the parameters are estimated by different methods) and empirical points, especially large empirical points (for which the frequency is not over 50% ($P(X) \le 50\%$)). According to the information shown in table 5 and figure 4, the performance of method does not significantly improve as higher PWM are used. It is clear that when the order ranges from 0 to 4, WLS decreases; however, it increases after the order is more than 4. The relative WLS shows the percentage of WLS of $P(X) \le 50\%$ takes in the sum WLS. Clearly to see that with the order increases, the relative WLS decreases dramatically. So the theoretical curves which parameters are estimated by higher moments' methods can fit with the large flood points better.

Table 5. Sum relative square variations (WLS) between flood peaks and theoretical values.

WLS Methods	PWM	Cuntan gauged station: HPWM					
		$r=0$	$r=1$	$r=2$	$r=3$	$r=4$	$r=5$
WLS ($P \le 50\%$)	0.058726	0.058726	0.056927	0.052042	0.04736	0.044855	0.045356
WLS (Summary)	0.077692	0.077692	0.077314	0.088834	0.137434	0.238788	0.38688
Relative WLS	75.59%	75.59%	73.63%	58.58%	34.46%	18.78%	11.72%
WLS Methods	L-M	Cuntan gauged station: LH moments					
		$r=0$	$r=1$	$r=2$	$r=3$	$r=4$	$r=5$
WLS ($P \le 50\%$)	0.058726	0.058726	0.056262	0.05173	0.046637	0.044823	0.045373
WLS (Summary)	0.077692	0.077692	0.0773178	0.090045	0.13554	0.240924	0.388189
Relative WLS	75.59%	75.59%	72.77%	57.45%	34.41%	18.6%	11.69%
WLS Methods	PWM	Hankou gauged station: HPWM					
		$r=0$	$r=1$	$r=2$	$r=3$	$r=4$	$r=5$
WLS ($P \le 50\%$)	0.02999	0.02999	0.02713	0.02742	0.02774	0.02776	0.02746
WLS (Summary)	0.03513	0.03513	0.03377	0.03357	0.03277	0.03184	0.03134
Relative WLS	85.37%	85.37%	80.34%	81.68%	84.68%	87.19%	87.62%
WLS Methods	L-M	Hankou gauged station: LH moments					
		$r=0$	$r=1$	$r=2$	$r=3$	$r=4$	$r=5$
WLS ($P \le 50\%$)	0.02999	0.02999	0.02667	0.02723	0.02649	0.02776	0.02746
WLS (Summary)	0.03513	0.03513	0.03367	0.03370	0.03137	0.03192	0.03137
Relative WLS	85.37%	85.37%	79.21%	80.80%	84.44%	86.97%	87.54%

Note: Relative WLS = WLS ($P \le 50\%$) / WLS (summary).

6. Conclusions

Several estimators, like PWM and L-Moments, were first introduced in this paper. In addition, the GPD-based approximate calculation methods of HPWM and LH moments were mainly deduced. A statistical experiment and a case study were applied to investigate and the WLS was the feasible measurement of these methods. Following conclusions can be drawn from the results of this study.

- Both the statistical experiment and case study proved that the GPD-based HPWM and LH moments deduced in this paper were feasible and applicable in parameter estimation.
- When the same order PWM were used in HPWM and LH moments methods, the results were identical. For example, the HPWM method (when $r = 0$) showed similar performance to the methods of PWM, L-Moments and the LH moments (when $r = 0$); the LH moments method (when $r \geq 1$) performed as well as the HPWM method (when $r \geq 1$) when the same PWM were used in calculation.
- However, from the standpoint of WLS which was used to measure the goodness-of-fit of each method, the order of the PWM affected the performances of the higher moments' methods. For example, when the order r increased from 0 to 4, the performances of the HPWM method and LH moments method became better. But when r exceeded 4, the performances of either the HPWM method or LH moments method became noticeably worse.
- The methods with higher moments tended to show more flexibility in fitting flood series on the up tail for example, the theoretical curves fit well with the empirical points which frequency were less than 50%.
- Comparing all of the methods outlined in this paper, it became clear that the HPWM method showed the same performance as the LH moments method and they used much more information in the flood series than the PWM method and L-Moments method. However, unlike LH moments method, the HPWM method was much easier to calculate. So HPWM method is a more preferable one.

References

[1] Wang J and Song S 2010 Application of generalized Pareto distribution in flood series frequency analysis *J. Northwest A. & F. Univ. (Nat. Sci. Ed.)* **38** 191-6 (in Chinese)
[2] Bermudez P D Z and Kotz S 2010 Parameter estimation of the generalized Pareto distribution part - II *J. Stat. Plan. Infer.* **140** 1374-88
[3] Diedda R and Puliga M 2009 Performances of some parameter estimators of the generalized Pareto distribution over round-off samples *Phys. Chem. Earth* **34** 626-34
[4] Lang M, Ouarda T B M J and Bobée B 1999 Towards operational guidelines for over- threshold modeling *J. Hydrol.* **225** 103-17
[5] Hosking J R M 1990 L-moments: Analysis and estimation of distributions using linear combinations of order statistics *J. Roy. Stat. Soc. B.* **52** 105-24
[6] Wang Q J 1997 LH moments for statistical analysis of extreme events *Water Resour. Res.* **33** 2841-8
[7] Meshgi A and Khalili D 2009 Comprehensive evaluation of regional flood frequency analysis by L- and LH moments I. A re-visit to regional homogeneity *Stoch. Environ. Res. Risk Assess.* **23** 119-35
[8] Busababodhin P, Seo Y A, Park J, *et al* 2016 LH-moment estimation of Wakeby distribution with hydrological applications *Stoch. Environ. Res. Risk Assess.* **30** 1757-67
[9] Pickands J 1975 Statistical inference using extreme order statistics Ann Stat **3** 119-31
[10] Wang Q J 1997 Using higher probability weighted moments for flood frequency analysis *J. Hydrol.* **194:** 95-106
[11] Song S, Li Y and Kang Y 2012 Calculation higher probability weighted moments for generalized Extreme value in hydrology *Adv. Mater. Res.* **518-523** 4015-21 (in Chinese)
[12] Li Y and Song S 2012 Application of higher probability weighted moments of P-III distribution to flood frequency analysis *J. Hydroele. Eng.* **33** 10-8 (in Chinese)

[13] Pilon P J, Adamowski K and Alila Y 1991 Regional analysis of annual maxima precipitation using L-moments *Atmos. Res.* **27** 81-92
[14] Hosking J R M and Wallis J R 1993 Some statistics useful in regional frequency analysis *Water Resour. Res.* **29** 271-81
[15] Chen Y, Wang Q, Sha Z, *et al* 2002 Application of L-moments based regional flood frequency analysis method to middle and lower Yangtze River *J. Hohai Univ. (Nat. Sci. Ed.)* **31** 207-11 (in Chinese)
[16] Bhuyan A, Borah M and Kumar R 2010 Regional flood frequency analysis of North-bank of the river Brahmaputra by using LH-Moments *Water Resour. Manag.* **24** 1779-90
[17] Hewa G A, Wang Q J, McMahon T A, *et al* 2007 Generalized extreme value distribution fitted by LH moments for low-flow frequency analysis *Water Resour. Res.* **430** 1-10
[18] Murshed M S, Park B J, Jeong B Y, *et al* 2009 LH-Moments of some distributions useful in hydrology *Communications of the Korean Statistical Society* **16** 647-58
[19] Cannane C 1978 Unbiased plotting positions - a review *J. Hydrol.* **37** 205-22
[20] Yang S L, Zhang J and Xu X J 2007 Influence of the Three Gorged Dam on downstream delivery of sediment and its environmental implications, Yangtze River *Geographs. Res. Lett.* **34** Doi: 10.1029/2007GL029472
[21] Fang B, Guo S L, Chai X L, *et al* 2005 Flood frequency analysis based on FPOT method *Water Power* **31** 9-12 (in Chinese)

IOP Conf. Series: Earth and Environmental Science **82** (2017) 012032 doi:10.1088/1755-1315/82/1/012032

The distribution characteristics of polycyclic aromatic hydrocarbons in typical water system of North China

Z Wang[1,4], L H Zong[2], X Z Meng[1], S L Zhang[1], W Y Cui[1], C Liu[3] and J Zhang[1]

[1]Haihe River Water Conservancy Commission, Ministry of Water Resources, Tianjin 300170, China
[2]Haihe River Water Conservancy Commission, Tianjin 300170, China
[3]Hebei University of Engineering, Handan 056000, China

E-mail: 896928483@qq.com

Abstract. The temporal and spatial distribution of PAHs in surface-water from the typical rivers of North China was investigated. PAHs in typical rivers ranged from n.d. to 1.98 ng/L. LMW PAHs (2 and 3-ring) were abundant in surface water with high detection frequencies in Beisanhe River, while HMW PAHs (4 and 6-ring) were abundant in Luanhe River. Specific PAHs with its ratios were measured in order to deduce the possible sources of PAHs contaminant. The results indicate that the PAHs were from combustion processes in Beisanhe River, while PAHs were from petroleum products in Luanhe River.

1. Introduction

Polycyclic aromatic hydrocarbons (PAHs) are series of typical persistent organic pollutants originating from fossil fuels or imperfect combustion, pyrolysis of other organic matter. PAHs are made up of two or more than two benzene rings arranging in linear, angular, or clumps of hydrocarbons. With the development of industry, the cost of coal and oil has risen sharply, which is the one of main reasons for PAHs [1]. PAHs have high octanol water partition coefficient, sediment/water partition coefficient, carcinogenicity, teratogenesis and mutagenicity [2, 3]. Therefore, PAHs can produce great harm to human health and ecological environment. In recent years, 16 PAHs have been enrolled in the items of priority-control pollutants [4, 5], and the issue of PAHs have been a great concern for scientist and the public

Heihe River, one of the largest rivers in China, Luanhe River and Beisanhe River, two of the largest rivers in North China, flow across Hebei province and Tianjin. Previous studies focused on distribution, composition and sources of PAHs in water and sediments of upper reach of Huaihe River [6, 7]. However, there was little information about PAHs in Haihe River

In this work, the distribution characteristics and possible sources of PAHs were detected to confirm the pollution levels in Luanhe River and Beisanhe River of North China. Furthermore, this work could provide the reference for the investigation of occurrence and trend of pollutants in river system.

2. Materials and methods

2.1. Instrument and mehtod

Agilent HPLC 1260, Exlipse PAHs (3.5 μm, 2.1 × 100 mm), Oasis HLB (500 mg/ 6 mL, Waters, USA) was used in the study. PAHs are purchased from AccuStandard, 2 mg/mL. HPLC grade acetonitrile, methanol, dichloromethane and formic acid were purchased from J.T. Baker Ltd, USA.

Equipment Condition: The mobile consist of water (A) and acetonitrile (B). The gradient was set up. The temperature of column is 35°C, flow rate is 0.3 mL/min. The UV detection wavelengt is 254 nm.

2.2. Sample pretreatment

The water samples gathered from Beisanhe River and Luanhe River in North China, including: 1 - SiDaoDian 2 - DACaoPing, 3 - GuBeiKou; LuanHe River water system, 4 - PuHeKou, 5 - WuLongJi, 6 - LiuHeKou, 7 -KuanChengQu 8 - TingBaShang, 9 - TingWangXiang, 10 - XiaChi, 11-PanBaShang12 - PanJiaKou , 13 – ShiHeShuiKu, were chosen as research objects. 1 L of water samples of each section were kept in brown reagent bottle at 0°C.

Methanol and NaCl with 5 mL, respectively, solution (1 g/mL) were added into 500 mL water sample. The Oasis HLB was preconditioned with 5.0 mL of methanol, 5 mL dichloromethane and 5 mL ultra-pure water, successively. Thereafter, the water samples pass the Oasis HLB columns at 10 mL/min. The Oasis HLB column was then rinsed with 5 mL ultrapure water and dried under vacuum. After drying, each cartridge was eluted with 2 mL × 3 dichloromethanel. The extracts were collected in a glass vial, reduced to minimum volume, and then dissolved in 40 % acetonitrile to a final volume of 1.0 mL.

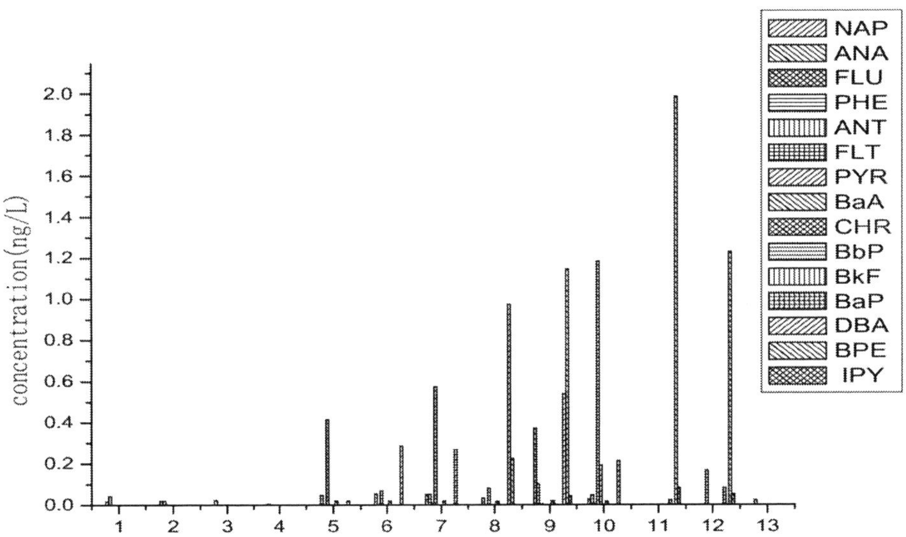

Figure 1. The concentration of PAHs in water sample from several rivers in north China.

3. Results and discussion

The spatial and temporal distribution of PAHs in surface-water from the Beisanhe River and Luanhe River were investigated in this work. As shown in figure 1, 15 kinds of priority control of PAHs recommended by EPA of America including naphthalene (NAP), acenaphthene (ANA), benzo [a] anthracene (BaA), flexion (FLT), benzo [b] fluoranthene (BbP), benzo [k] fluoranthene (BKF) were not detected. However, fluorine (FLU), phenanthrene (PHE), anthracene (ANT), fluoranthene (FlU), Pyrene (PYR), benzo [a] pyrene (BAP), dibenzo (a, h) anthracene (DBA), Indeno (1, 2, 3 - cd) pyrene (IPY), Benzo (g, hi) perylene (BPE) were detected at all sampling sites. In addition, the PAHs in Beisanhe and Luanhe ranged from n.d. to 0.0581 ng/L, n.d. to 1.98 ng/L, respectively. The United States has released carcinogenic PAHs concentration in groundwater was 0.2 ~ 6.9 ng/L, and the concentration in the surface-water was 0.1 ~ 800 ng/L, most in 2 ~ 50 ng/L [9]. Thus, the concentration of PAHs in Beisanhe and Luanhe River are not cause of cancer.

IOP Conf. Series: Earth and Environmental Science **82** (2017) 012032 doi:10.1088/1755-1315/82/1/012032

3.1. Source of PAHs

PAHs were primarily derived from Chemical sources (e.g. oil spills) and pyrogenic processes, including the combustion of organic matter such as coal and wood combustion. According to research, the relative abundance of the PAHs reflects its origin [8]. Main sources of PAHs are surface-water with atmospheric precipitation, city life sewage, industrial emissions (waste water and waste), surface runoff. PAHs in the groundwater mainly come from polluted surface-water leakage or supply, sewage irrigation, the soil pollution by solid waste disposal site and the infiltration [9, 10]. The United States has released carcinogenic PAHs concentration in groundwater which was 0.2 ~ 6.9 ng/L, and the concentration in the surface-water was 0.1 ~ 800 ng/L, most in 2 ~ 50 ng/L [11].

The low molecular weight (LMW) PAHs (2~3 rings, LPAHs) mainly originated from the pollution of petroleum and the PAHs (4 rings and above, PAHs) were derived from the fossil fuel combustion. The detected percentages of LPAHs (mainly including naphthalene acenaphthene, ANA, PHE, ANT and FLU and PAHs, PYR and BaA, BkF, BaP, DBA, IPY, BPE) were shown in figure 2. As can be seen, the distribution characteristics of PAHs changed with monitoring section. Considering the similar sampling period, the low ring PAHs were important component in Beisanhe River. While in Luanhe River the percentage of low ring PAHs was lower than that of middle ring PAHs.

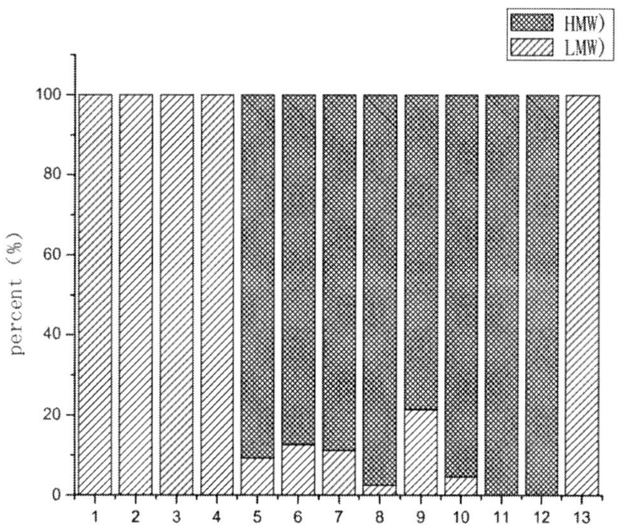

Figure 2. The percentage of low and high ring PAHs in water sample from the section in north China.

The sources of PAHs could be diagnosed by the ratio of isomers [12, 13]. For example, when the ratio of Ant/(Ant + Phe) was above 0.1, [14] the PAHs was originated from combustion processes. When the ratio was below 0.1, the sources were fossil fuels burning [15]. When the Fla/(Fla + Pyr) [16] among 0.4 and 0.5, PAHs was originated from petroleum combustion. When ratio was below 0.4, the sources were petroleum combustion. The sources were from grass combustion, when the ratio was above 0.5.The Ant/ (Ant + Phe) in BeiSanHe samples were above 0.1, while in LuanHe water samples, the Ant/(Ant + Phe) were below 0.1. In all water samples, Pyr was not detected. All the results indicated PAHs sources of BeiSanHe River or LuanHe River were from combustion processes and fossil fuels burning, respectively.

4. Conclusions

The spatial and temporal distribution of PAHs in surface-water in Luanhe River and Beisanhe River has been studied and the results are as follows:

- Concentration of PAHs in several rivers in north China ranges was from n.d.to 1.98 ng/L

- Considering the same sampling period, the low ring PAHs were main component in Beisanhe River. While in Luanhe River the percentage of low ring PAHs was lower than middle ring PAHs.
- The ratios of specific PAHs indicated that the source of PAHs in Beisanhe River and Luanhe River were from combustion processes and fossil fuels burning, respectively.

References

[1] Cui B W, Luo S F and Xue J 2005 Determination of polycyclic aromatic hydrocarbon in water course soil by high performance liquid chromatography Chemical Analysis and Meterage **14(5)** 30c

[2] Liu H Y, Meng F B, Tong Y D and Chi J 2014 Effect of plant density onphytoremediation of polycyclic aromatic hydrocarbons contaminated sediments with Vallisneria spiralis *Ecol. Eng.* **73** 380-5

[3] Xing W Q, Luo Y G, Wu L H, Song J, Qian W and Christie P 2006 Spatialdistribution of PAHs in a contaminated valley in Southeast China *Environ. Geochem. Health* **28** 89-96

[4] Deutsch-Wenzel R P, Brune H, Grimmer G, *et al* 1983 Experimental studies in rat lungs on the carcinogenictity and doserespose relationships of eight frequently occurring environmental polycyclic aromatic hydrocarbons J. Natl Cancer Inst **71(3)** 539-44

[5] Thyssen J, Althoff J, Kimmerle G, *et al* 1981 Inhalation studies with benzo [a] pyrene in syrian golden hamsters J. Natl Cancer Inst **66(3)** 575-7

[6] Feng J L, Zhai M X, Sun J H and Liu Q 2012 Distribution and sources of polycyclicaromatic hydrocarbons (PAHs) in sediment from the upper reach of HuaiheRiver, East China *Environ. Sci. Pollut. Res.* **19** 1097-106

[7] Liu Q 2009 The study of persistent toxic pollutants (PTS) in water from the upperreach of the Huaihe River (Henan: Henan Normal University) (in Chinese).

[8] Baek S, Field R A and Goldstone M E 1991 A review of atmospheric polycyclic aromatic hydrocarbons: Sources, fate and behavior Water, Air & Soil Poll. **60** 279-300

[9] Guo Y, Meng X Z, Tang H L, *et al* 2008 Distribution of polybrominated diphenyl ethers in fish tissues from the Pearl River Delta, China: Levels, compositions, and potential sources Environ Toxicol Chem **27** 576-82

[10] Harmann P C, Quinn J G, Caims R W, *et al* 2004 The distribution and source of polycyclic aromatic hydrocarbons in NarragansettBay surface sediments Mar Pollut Bull **48** 359-70

[11] Manoli E and Samara C 1999 Polycyclic aromatic hydrocarbons in natural waters: Sources, occurrence and analysis Trac-Trend Anal Chem **18(6)** 417-28

[12] Lumer M 1976 Polycyclic aromatic hydrocarbons in nature *Sci. Am* **234** 34- 5

[13] Liu M L, Feng J L, Hu P T, *et al* 2016 Spatial-temporal distributions, sources of polycyclic aromatic hydrocarbons (PAHs) in surface water and suspended particular matter from the upper reach of Huaihe River China Ecol Eng **95** 143-51

[14] Yim U H, Hong S H, Shim W J, Oh J R and Chang M 2005 Spatio-temporaldistribution and characteristics of PAHs in sediments from Masan Bay, Korea Mar Pollut Bull **50(3)** 319-26

[15] Unker M B, Snowdon L R, Macdonald R W, *et al* 1996 Polycyclic aromatic hydrocarbons composition and optional source for sediments samples from the Beaufort and Barents Seas Environ Sci Technol **225** 181-205

[16] Dickhut R M, Canuel E A, Gustafson K E, Liu K, Arzayus K M, Walker S E, Edgecombe G, Gaylor M O and MacDonald E H 2000 Automotive sources ofcarcinogenic polycyclic aromatic hydrocarbons associated with particulatematter in the Chesapeake Bay region Environ. *Sci. Technol.* **34(21)** 4635-40

Occurrence, composition and ecological restoration of organic pollutants in water environment of South Canal, China

Y Z Wang[1,2], C Lin[3,5], X S Zhou[1], Y Zhang[4] and C G Han[4]

[1]Haihe River Water Environmental Monitoring Center, Tianjin 300170, China
[2]Key Laboratory of Pollution Processes and Environmental Criteria (Nankai University), Ministry of Education, Tianjin 300071, China
[3]Haihe Water Resources Protection Bureau, Tianjin 300170, China
[4]Zhangweinan Canal Management Bureau, Haihe River Water Conservancy Commission, Dezhou 253000, China

E-mail: linchao@hwcc.gov.cn

Abstract. Ecological restoration of polluted river water was carried out in South Canal by adding microbial water purifying agents and biological compound enzymes. The objective of present study was to investigate the ecological restoration effect of organic pollutants by this efficient immobilized microbial technologies, analysis the occurrence and composition of organic pollutants including fifteen persistent organochlorine pesticides (OCPs), seventeen polycyclic aromatic hydrocarbons (PAHs) and eighteen organophosphorus pesticides (OPPs) both in natural water environment and ecological restoration area of South Canal, China. Results showed that the total concentrations of OCPs ranged from 1.11 to 1.78 $ng \cdot L^{-1}$, PAHs from 52.76 to 60.28 $ng \cdot L^{-1}$, and OPPs from 6.51 to 17.50 $ng \cdot L^{-1}$. Microbial water purifying agents and biological compound enzymes essentially had no effects on biological degradation of OCPs and PAHs in the river, but could remove OPPs with degradation rates ranging from 19.6% to 62.8% (35.2% in average). Degradation mechanisms of microbial water purifying agents and biological compound enzymes on OCPs, PAHs and OPPs remained to be further studied. This technology has a certain value in practical ecological restoration of organic pollutants in rivers and lakes.

1. Introduction

With the rapid development of social economy, a large number of domestic sewage and industrial wastewater were discharged into river directly without treatment, which caused serious water pollutions and destroyed the ecological environment of the river, and finally weakened the purification capacity of the river. Highly effective microorganism technologies are a new kind of environmental biotechnologies combined by bioecological remediation and water restoration [1]. It can effectively improve concentrations and activities of microorganisms in water environment by adding exogenous microbial strains and biological enzymes [2]. Organic materials, nitrogen, phosphorus and other pollutants are degraded, transformed and absorbed through the assimilation and dissimilation by microorganisms. At present, conditions of water environment had been improved apparently using high efficient microbial technology by a large number of researchers, while they carried out indoor water purification experiments or practical applications by adding high efficient microbial agents into the water [3-6]. However, previous studies mainly focused on conventional indicators concentrations of water environment including ammonia nitrogen, chemical oxygen demand, total nitrogen and total phosphorus degraded effectively by adding microbial agents to pollutants. Practical applications of microbial technology for organic pollutants restoration in polluted river water were relatively less, and more experiments for specific organic pollutants degradation were carried out in the stage of laboratory research through screening microbial strains by single or multiple strains [7-10]. Faced with complex system of organic pollution in the river, little literature had been reported organic pollutants treatment using high efficient microbial technology, such as persistent organic pollutants (POPs), organophosphorus pesticides (OPPs).

Owing to the characteristics of stable physical and chemical properties, POPs have been detected in various environmental media such as atmosphere, plants, water, sediments and soils. POPs have strong toxicity, can translate with long chain migration, accumulate along the food chain and cause serious harm to human health. Organochlorine Pesticides (OCPs) such as Hexachlorocyclohexanes (HCHs) and Dichlorodiphenyltrichloroethanes (DDTs) were still widely existed in various environmental media because of its persistence, bioaccumulation and toxicity even though it had been banned since 1980s. Polycyclic aromatic hydrocarbons (PAHs) are a wide group of organic contaminants composed of fused aromatic rings, known to be ubiquitous in environment and of great concern due to their toxic, mutagenic and carcinogenic potentials to the ecosystem and human beings. PAHs are widely distributed in the river water through atmospheric deposition, surface runoff, municipal and industrial effluents, oil spillage or leakage [11, 12]. Its main sources are divided into two categories including fuel combustion and petrochemical products [13]. OPPs are kinds of pesticides occupied the largest pesticide market share in China and especially widely used in agricultural activities. A large number of OPPs pesticides applied in the agricultural production period could enter rivers, lakes and other water bodies through rainfall, runoff and other means, causing surface water pollution and toxic effects on microorganisms, fish, human and other mammals [14]. Although OPPs are thought to be easily degraded and low residues in environment, some studies have shown that it can be transformed into some kinds of persistent pollutants [15]. Pollution status, composition and distribution characteristics of OCPs, PAHs and OPPs have been extensive reported in multi medium environment such as surface water [16-19], soil [20, 21], sediment [22], and estuary [23, 24] in Haihe River Basin. However, little attention has been paid attention to the pollution characteristics of OCPs, PAHs and other persistent organic pollutants in surface water of Zhangweinan subwatershed, except Cao *et al* [25] studied the pollution characteristics and risk assessment of dissolved PAHs. In addition, OPPs have not been reported in various environmental media related to the Zhangweinan subwatershed. The widespread existences of persistent OCPs and PAHs, as well as the extensive use of OPPs in agricultural activities, posed a potential risk and threat to the ecological environment and human health.

As a part of the Grand Canal from Beijing to Hangzhou, South canal is one of the main flood discharge channel in Zhangweinan River, Haihe River Basin. South Canal carried a large number of domestic sewage and industrial sewage from upstream and surrounding without treatment, and immediately created extremely serious water pollution. Contaminated river water have threatened human health through drinking water and ingesting food around South Canal basin, which showed that comprehensive management and restoration of water pollution become essential. In order to deal with the water pollution situation of the South Canal, efficient immobilized microbial technologies coupled with efficient microbial purifying agents and biological compound enzymes were used in ecological restoration experiment carried out in South Canal water. Pollution loads including concentrations of COD, ammonia and other pollution indicators in South canal were gradually reduced, and also the water ecosystems were reconstructed and gradually recovered in South canal. The objective of present study was to investigate compositions, distributions and concentrations of fifteen persistent OCPs, seventeen PAHs and eighteen OPPs in surface water of South Canal, Zhangweinan River Basin, obtain the pollution characteristics of organic pollutants in ecological restoration water environment of the South Canal and discuss the ecological restoration effect of organic pollutants coupled by efficient microbial water purifying agents and biological compound enzymes.

2. Materials and methods

2.1. Study area

The study area was located in the Dezhou section of the South Canal, Dezhou city, Shandong province, China. The distance from Sinvsi Gate to the lower reaches of the South Canal were 665 meters and this sections were selected as the experimental area (as shown in figure 1). Dam with 1 meters height was built in the downstream to maintain the river water level at an average depth of 2.5 m and ensure river water stability overflows. The average width of water surface was 60 m and storage capacity of this area was about 10×10^4 m^3. Velocity of the inflow river water from Sinvsi Gate was about 1×10^4 m^3 every day and water quantity was in stable.

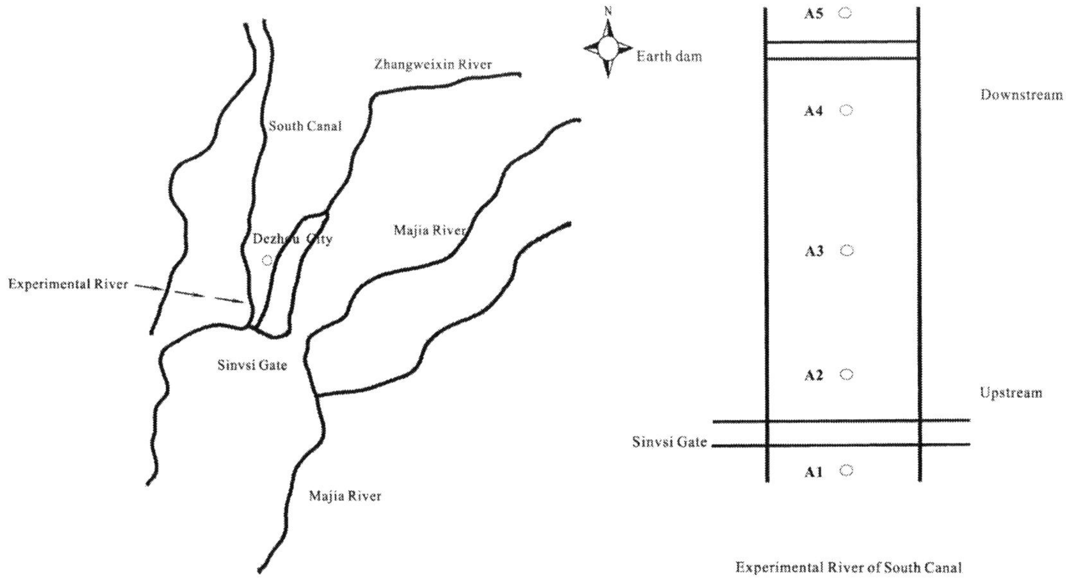

Figure 1. Experimental sites of South Canal and sampling sections.

2.2. Chemicals and instruments

Agilent 7890A/7000B (Agilent, USA), Automatic solid phase extraction instrument (GL Science, Japan); Bond Elut C18 Solid phase extraction column (500 mg/6 mL, Agilent).

Standard mixture solution for OCPs contained fifteen components, including *α*-HCH, *β*-HCH, *γ*-HCH, *δ*-HCH, *p, p'*-DDE, *p, p'*-DDD, *o, p'*-DDT and *p, p'*-DDT from research institute of standard sample, ministry of environmental protection. Heptachlor, Heptachlor epoxide, Aldrin, Dieldrin, Isodrin, *α*-Endosulfan and *β*-Endosulfan were obtained from research Institute of environmental protection, Ministry of Agriculture.

Standard mixture solution for PAHs contained seventeen components, including Naphthalene (Nap), Acenaphthene (AcP), Acenaphthylene (AcPy), Fluorene (FL), Phenanthrene (Phe), Anthracene (Ant), Carbazole (Car), Fluoranthene (Flu), Pyrene (Pyr), Benz(a)anthracene (BaA), Chrysene (Chr), Benz(b)fluoranthene (BbFL), Benz(k)fluoranthene (BkFL), Benzo(a)pyrene (BaP), Indeno (1,2,3,-cd)pyrene (InP), Benzo(g,h,i)pylene (BghiP) and Drbenz(a,h)anbhracene (DBA) were obtained from accustandard company, USA.

Standard mixture solution for OPPs contained eighteen components, including Methamidophos (MTP), (Dichlorvos, DDVP), Acephate (ACP), Omethoate (OMT), Ethoprophos (ETP), Monocrotophos (MOP), Phorate (PHO), Dimethoate (DIT), Isazophos (ISP), Diammonium phosphate (DAP), Methyl parathion (MEP), Malathion (MAT), Chlorpyrifos (CHP), Parathion (PAT), Bromophos (BRP), Isofenphos-methyl (IPM), Methidathion (MED) and Triazophos (TZP) were obtained from research institute of standard sample, ministry of environmental protection.

Ethyl acetate and dichloromethane were chromatography pure and from Duksan, Korea. Methanol was HPLC grade and from Baker company, USA.

Bpa-1017 series microbial water purifying agents and Bacto-Zyme 1011 series of biological enzymes were purchased from Beijing Bangyuan environmental protection technology Co., Ltd.

2.3. Experiment methods

2.3.1. Field experiments design

Experiments were carried out from April to July in 2014. A certain proportion of the microbial water purifying agents (Bpa-1017 series) and biological compound enzymes (Bacto-Zyme 1011 series) were added to water body of the experimental rivers. One aerator was installed in upstream of experiment section to provide sufficient oxygen for the water body and increase the flow of river water. Several artificial floating beds were built in surface water of the experimental area and a large number of artificial forests were fixed at the bottom of the water as efficient microbial carriers, which provided attachment and growth conditions for microorganisms, protozoa, small pelagic animal that participated in water purification.

2.3.2. Sample collection

As shown in figure 1, five sampling sections were set up, include one background section above Sinvsi Gate (A1) to reflect the pollution status of organic pollutants in the South Canal, and four sections in experimental area of South Canal to reflect ecological restoration effect of organic pollutants, which located in upstream (A2), midstream (A3), downstream (A4) of the river and under the river dam (A5) of the south canal respectively. Glass bottles were used to collect the surface water (0.5 m) samples of 1 L for laboratory analysis after ecological restoration experiments finished. Concentrations of dissolved oxygen in surface water were measured by portable multi parameter detector (YSI EXO2).

2.4. Analysis of organic materials

2.4.1. Sample extraction. Water samples were filtered and enriched by Bond Elut C18 solid-phase extraction cartridges. Volume of the samples were 500 mL and methanol was added according to the rate of V (water samples): V (methanol) = 100:1. Solid-phase extraction columns were placed in automatic solid-phase extraction apparatus, washed and purified by 5 mL dichloromethane and ethyl acetate at a flow rate of 3 mL·min^{-1} respectively, and then activated by 10 mL methanol and ultra water at a flow rate of 5 mL·min^{-1}. Water samples were passed through the cartridges at about the rate of 15 mL·min^{-1} under vacuum. After extraction, the cartridges were dried and then eluted three times by 2 mL ethyl acetate, 2 mL dichloromethane and 2 mL dichloromethane with the rate of 1 mL·min^{-1}, 1 mL·min^{-1} and 2 mL·min^{-1} respectively. Elutes were collected and concentrated to 1 mL by dichloromethane for GC/MS/MS analysis.

2.4.2. Gas chromatograph and tandem mass spectrometry analysis. HP-5MS column (30 m × 0.25 mm × 0.25 μm, Agilent) was used to detect the concentrations of OCPs in water samples. High purity nitrogen was used as carrier gas at 1.2 mL/min with the pressure of 11.681 psi under a constant flow mode. Transmission line temperature was maintained at 280°C. Injection volume was 1.0 μL with splitless inject mode.

Oven program for OCPs was programmed to keep the initial temperature at 80℃ for 1 minute, then raised to 230℃ at the rate of 20℃·min^{-1} and held for 6 minutes. The total analysis time was 14.5 minute. Injector temperature was 270℃. For PAHs was programmed to keep the initial temperature at 80℃ for 1 minute, then raised to 250℃ at the rate of 20℃·min^{-1}, and then raised to 300℃ at the rate of 10℃·min^{-1} and held for 5.5 minute. The total analysis time was 20.0 minute. Injector temperature was 270℃. And for OPPs was programmed to keep the initial temperature at 80℃ for 1 minute, then raised to 200℃ at the rate of 15℃·min^{-1} and held for 3 minute, and then raised to 250℃ at the rate of 20℃·min^{-1} and held for 2 minute. The total analysis time was 17.5 minute. Injector temperature was

240℃.

Mass spectrometer conditions were selected as ionization mode: EI, electron energy 70 eV, interface temperature 280℃, ion source temperature 230℃, quadropole temperature 150℃, solvent delay 7.5 min. The parameters of chromatography and mass spectrometry were optimized by SCAN and productor mode and detection mode were in multiple reaction monitoring (MRM).

2.4.3. Quality control and quality assurance. Quality control and quality assurance in sample analysis refer to the requirements of USEPA and QA/QC in water environment standard of China (SL 219-2013). The quality of the monitoring results can be guaranteed by blank, standard added recovery and parallel sampling. In this study, recoveries of OCPs in water samples ranged from 70.3% to 115.1%, and the relative standard deviations (RSD) ranged from 0.6% to 12.2%, and the detection limits from 0.08 to 0.38 ng·L^{-1}. Recoveries of PAHs ranged from 69.8% to 111.6%, relative standard deviations (RSD) from 4.9% to 12.6%, and detection limits from 0.11% to 0.40 ng·L^{-1}. Recoveries of OPPs ranged from 86.1% to 117.2%, relative standard deviations (RSD) from 1.1% to 15.8%, and detection limits from 0.09% to 0.82 ng·L^{-1}. Fifteen OCPs, seventeen PAHs and eighteen OPPs were quantitatively analyzed by Mass Hunter data acquisition software of Agilent 7890A/7000B gas chromatography-tandem mass spectrometry instrument.

2.5. Statistical analysis
SPSS17.0 statistical software was used for data analysis.

3. Results and discussion

3.1. Occurrence and compositions of organic pollutants in natural water environment of South Canal
Several components of OCPs, PAHs and OPPs were detected in natural water environment of South Canal reflected by A1 sampling sites. As shown in table 1, total concentrations of OCPs were 1.45 ng·L^{-1}, which stands for three OCPs including α-HCH, β-HCH and γ-HCH examined in this study. Eleven categories of PAHs including Nap, AcP, AcPy, FL, Phe, Ant, Car, Flu, Pyr, BaA and Chr were detected and total concentrations were 54.70 ng·L^{-1}. Two kinds of OPPs including DDVP and IPM were detected and total concentrations were 17.50 ng·L^{-1}.

Pollutions of OCPs and PAHs in this study were lower than that of the other rivers and lakes in China. According to literatures, surface water were polluted by OCPs with varying degrees, such as HCHs, DDTs, Heptachlor, Heptachlor epoxide, Aldrin in China's seven river basin including Yellow River, Yangtze River, Pearl River, Hai River, Liao River, Songhua River and Huai River [26, 27]. For example, concentrations of HCHs ranged from 9.27 to 10.51 ng·L^{-1} in water environment of Nanjing section of Yangtze River, and DDTs from 1.57 to 1.79 ng·L^{-1} [28], while HCHs in Haihe River ranged from 3.95 to 20.15 ng·L^{-1}, and DDTs from 6.88 to 35.23 ng·L^{-1} [29]. For the concentrations of PAHs, they were 321.80 to 6234.90 ng·L^{-1} in Yangtze River [30], 987.13 to 2878.50 ng·L^{-1} in Pearl River [31], as well as 1800 to 35000 ng·L^{-1} in Hai River [32]. In addition, total concentrations of dissolved PAHs in surface water of Sinvsi Gate ranged from 74.50 to 99.00 ng·L^{-1} in the year of 2008 in South Canal according to the survey of pollution characteristics by Cao Zhiguo [25]. Results of this study indicated that low pollution levels of PAHs currently relative to that of in history.

Most parts of OPPs in agricultural production activities could not be efficiently used by crops, and conversely entered into rivers, lakes and other water bodies through rain, sedimentation and runoff. Total concentrations of OPPs in the Pearl River estuary area ranged from 4.44 to 635 ng·L^{-1} [33]. Eight kinds of OPPs were found in three lakes of Beijing, with total concentration ranged from 0.18 to 0.63 ng·L^{-1} [34], among them four categories including dichlorvos, methamidophos, dimethoate and methyl parathion were detected. It can be seen that the concentrations of OPPs in rivers and lakes in some areas of China were much higher than that of in South canal.

Table 1. Occurrence and compositions of organic pollutants in natural water environment of South Canal/ng·L^{-1}.

Pollutants	Concentrations	Pollutants	Concentrations	Pollutants	Concentrations
α-HCH	0.30	Nap	9.50	MTP	ND
β-HCH	0.78	AcP	4.94	DDVP	15.31
γ-HCH	0.37	AcPy	0.68	ACP	ND
δ-HCH	ND	FL	2.09	OMT	ND
p,p'-DDE	ND	Phe	10.29	ETP	ND
p,p'-DDD	ND	Ant	23.00	MOP	ND
o,p'-DDT	ND	Car	0.52	PHO	ND
p,p'-DDT	ND	Flu	1.80	DIT	ND
Heptachlor	ND	Pyr	1.44	ISP	ND
Heptachlor epoxide	ND	BaA	0.32	DAP	ND
Aldrin	ND	Chr	0.11	MEP	ND
Dieldrin	ND	BbFL	ND	MAT	ND
Isodrin	ND	BkFL	ND	CHP	ND
α-Endosulfan	ND	BaP	ND	PAT	ND
β-Endosulfan	ND	InP	ND	BRP	ND
—	—	BghiP	ND	IPM	2.19
—	—	DBA	ND	MED	ND
—	—	—	—	TZP	ND
∑OCPs	1.45	∑PAHs	54.70	∑OPPs	17.50

Note: ND stands for not detected. "—"stands for no data.

Table 2. Human health ambient water quality criteria and aquatic life criteria by EPA/μg·L^{-1}.

OCPs	Water quality criteria for human health		Water quality criteria for aquatic organisms (fresh water)	
	Consumption of water and organism	Consumption of organism only	Acute toxicity	Chronic toxicity
α-HCH	0.00036	0.00039	—	—
β-HCH	0.0080	0.014	—	—
γ-HCH	4.2	4.4	0.95	—
p,p'-DDD	0.00012	0.00012	—	—
p,p'-DDE	0.000018	0.000018	—	—
p,p'-DDT	0.000030	0.000030	1.1	0.001
Heptachlor	0.0000059	0.0000059	0.52	0.0038
Heptachlor epoxide	0.000032	0.000032	0.52	0.0038
Aldrin	0.00000077	0.00000077	3.0	—
Dieldrin	0.0000012	0.0000012	0.24	0.056
Isodrin	0.03	0.03	0.086	0.036
α-Endosulfan	20	30	0.22	0.056
β-Endosulfan	20	40	0.22	0.056

Note: "—"stands for no data.

Environmental water quality criteria is the basis for establishing the limits of water quality standards, which is of great significance for the prediction, evaluation, control and treatment of pollutants, maintain rivers or lakes in a good ecological environment. As shown in table 2, according to the toxicological data and numerical calculation, EPA formulated OCPs water quality criteria for human health and aquatic organisms. Human health quality criteria represent the maximum acceptable concentrations of pollutants that do not produce harmful effects on the human health through drinking water or by eating aquatic organisms. This criterion indicates that the lowest concentration of specific pollutants in water will not pose a significant risk for most aquatic species, including the values of acute toxicity and chronic toxicity. According to human health standard of OCPs for water quality and aquatic water quality criteria, the south canal water OCPs concentration was less than water quality criteria for human health and freshwater aquatic organisms.

Because of high hydrophobic and lip solubility, PAHs could be enriched and enlarged in the organism through food chain, and then entered into the human body by fish and shellfish intake. Some countries and international organizations have established the safety standards for aquatic organisms exposed to PAHs contaminated water through a large number of available toxicological data and a series of extrapolation methods [35] (as shown in table 3). Data in safety standard indicated the concentration of PAHs that could constitute threat to human health by bioaccumulation of aquatic organisms through food chain, especially mussels and oysters filter feeding animal. As reflected in table 3, single components of PAHs in surface water of the South Canal did not exceed the safety standards of each countries or organizations referring to the international safety standards, but the total concentration of PAHs exceeded the quality standard of the United States and environmental maximum allowable concentration limits in European Union.

Table 3. Safety guidelines for aquatic organisms exposured to PAHs in water/$\mu g \cdot L^{-1}$.

PAHs	Maximum allowable concentration limits in Ireland	Water quality standards draft in Canada	Ecotoxicological evaluation criteria by Khosrow and Paris Committee	Water quality assessment criteria in Danmark	Environmental quality standards in United State	maximum allowable concentration limits in European Union
Nap	—	11.0	1.0~10.0	1.0	—	—
Phe	2.0	0.8	—	—	4.6	—
Ant	—	0.12	0.005~0.05	0.01	—	—
Flu	0.5	—	0.05~0.5	—	—	—
BaA	0.2	—	—	—	—	—
BkFL	0.1	—	—	—	—	—
BaP	0.1	0.008	0.01~0.1	—	—	—
BghiP	0.02	—	—	—	—	—
∑PAHs	—	—	—	—	0.03	0.2

Note: "—"stands for no data.

For the safety standards for aquatic organisms exposed to OPPs shown in table 4, some information could be collected from different countries and organizations except from water quality standards draft in Canada, water environmental quality standards in China and Environmental quality standards in United State. According to existed safety guidelines, it was obvious that each component of OPPs in surface water of the South Canal did not exceed the safety standards of Canada, China and United State.

Table 4. Safety guidelines for aquatic organisms exposured to OPPs in water/μg·L^{-1}.

PAHs	Water quality standards draft in Canada	Water environmental quality standards in China	Environmental quality standards in United State
DDVP	—	50	—
DIT	6.20	80	—
MEP	—	2	—
MAT	—	50	0.10
CHP	0.02	—	—
PAT	—	3	0.013

Note: "—"stands for no data.

3.2. Occurrence and compositions of OCPs in experimental area

Water quality of South Canal had been greatly improved from worse than grade V to grade IV by adding high efficient microbial water purifying agents and biological compound enzymes. As shown in figure 2, *α*-HCH, *β*-HCH and *γ*-HCH in fifteen OCPs were detected in surface water sampling sites of A1, A2, A3, A4 and A5, and the other twelve OCPs were not detected. *α*-HCH and *β*-HCH were detected in each sampling sites and the detection rate was 100%, and *γ*-HCH was only detected in sites of A1, A3 and A4 with detection rate 60%. Total concentrations of HCHs ranged from 1.11 to 1.78 ng·L^{-1} (1.43 ng·L^{-1} in average), and *β*-HCH was the highest concentrations of HCHs in each sampling sites.

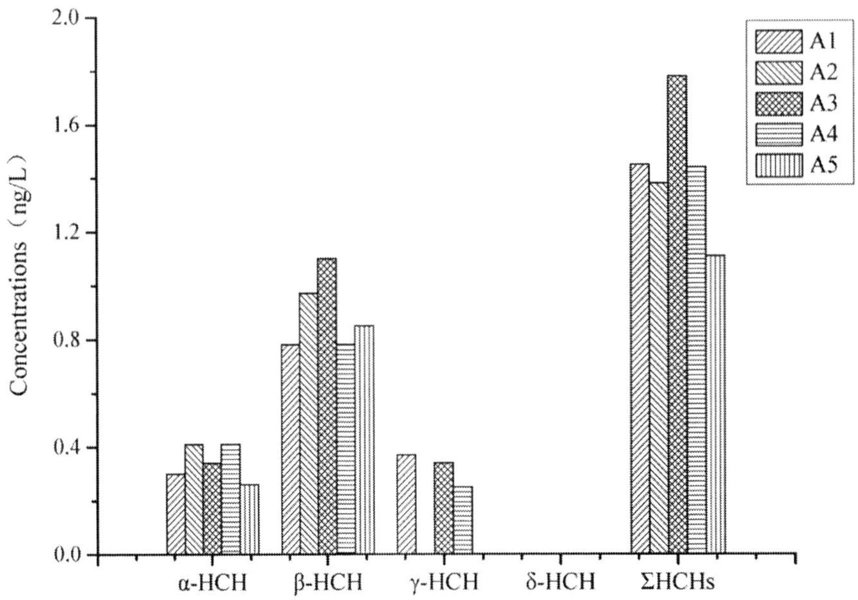

Figure 2. Concentrations of organochlorine pesticides in experimental water body of South Canal.

In this study, concentrations of organic pollutants in A1 sampling sites were used to reflect the background organic pollution status of water environment in South Canal, and microbial water purifying agents and biological enzymes were not added into this area. Concentrations of organic pollutants in A2-A5 sampling sites were considered as ecological restoration area and degradation efficiency could be observed when concentrations of organic pollutants in which were compared with A1 sampling sites. Total concentrations of OCPs and its isomers showed no significant difference in different sampling sites between background and ecological restoration area (P>0.05), which could explain that the microbial water purifying agents and biological compound enzymes used in our research could not degrade OCPs in river water environment obviously. Studies have shown that microorganisms could degrade OCPs, and microbial coupled with the enzyme morphology on the degradation of OCPs are closely related to dehydrochlorinase, hydrolase and dehydrogenase. Degradation processes of OCPs were completed by the ways of co-metabolism, metabolism or mineralization synergism during different species [36]. At present preliminary microorganisms proved to degrade OCPs were pseudomonas that contained sphingomonas UT26 [37], B90 [38, 39] and brown rot fungi [40], and also including serial genuses such as clostridium, bacillus, corynebacterium, micrococcus and hydrogenomonas. Meanwhile, persistent OCPs degradation in river water environment was affected by many environmental factors, such as organic carbon source [41], pH [42], humic acid [42], and so on.

In addition, according to the existing literature, degradations of chlorinated organic contaminants by microorganisms were carried out in the laboratory conditions with high initial concentration. While as organochlorine pollutants reached to a certain low concentrations, the degradation rates changed slowly and degradation process became time consuming, because extremely low concentrations of chlorinated aromatic compounds could not be used by microorganisms [43]. Currently, pollutions of organochlorine pollutants in surface water environments such as rivers and lakes are relatively in a low level, and the concentration of pollutants is oligotrophic for degrading bacteria. How to improve the degradation efficiencies of chlorinated organic contaminants by microorganisms in oligotrophic environment was the key to solve water pollution problems [43]. OCPs were not degraded by microorganisms and enzyme technology in South Canal, and how to use this technology to treat organochlorine pollutants in actual water ecological restoration process need to be further studied.

3.3. Occurrence and compositions of PAHs in experimental area

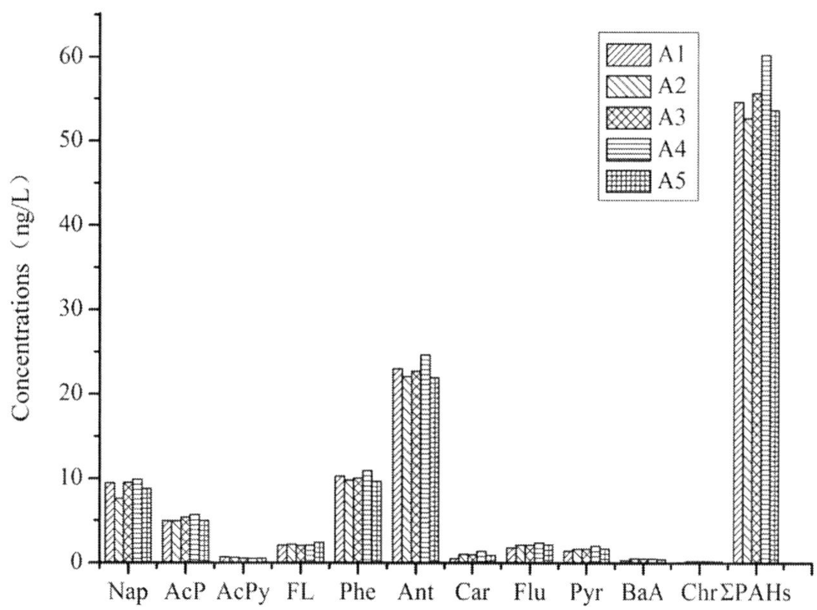

Figure 3. Concentrations of PAHs in experimental water body of South Canal.

As shown in figure 3, eleven kinds of PAHs including naphthalene, acenaphthylene, acenaphthene, fluorene, phenanthrene, anthracene, carbazole, fluoranthene, pyrene, benz(a)anthracene and chrysene were detected in experimental area in South Canal. Other six kinds of PAHs including benzo[b]fluoranthene, benzo[k]fluoranthene, benzo[a]pyrene, Indeno[1,2,3-cd]pyrene, benzo(g,h,i)pyrene and drbenz(a,h)anbhracene were not detected. Concentrations of anthracene is the highest in all PAHs components with 22.87 ng·L^{-1} in average, and followed by phenanthrene and naphthalene, the average concentrations were 10.18 ng·L^{-1} and 9.07 ng·L^{-1} especially. Total concentrations of PAHs ranged from 52.76 to 60.28 ng·L^{-1} with 55.44 ng·L^{-1} in average. While for the composition of PAHs in surface water of South Canal, 2 to 4-rings PAHs were in dominant and compositions above the 5-rings were not detected. Among them, concentrations of 3-rings PAHs accounted for the largest proportion in each sampling sites with the range from 72.8% to 75.1%, and followed by the 2-rings PAHs, the proportion of which ranged from 16.3% to 18.9%. 4-rings PAHs were occupied the least proportion and ranged from 3.7% to 5.1%.

Comparison of PAHs concentrations in surface waters in A1 sampling sites and ecological restoration experiment sections from A2 to A5 sampling sites, concentrations of eleven PAHs showed no significant differences ($P>0.05$). Microbial water purifying agents and biological compound enzymes had no obvious effect on the degradation of PAHs, which showed similar degradation effects of OCPs. Studies had shown that treatments of PAHs in water environment were mainly contributed by microbial degradations. Naphthalene and other low molecular weight of PAHs were relatively easy to be degraded, but for the high molecular weight PAHs above 4-rings or 4 rings, the characteristics of hydrophobicity and stability increased with increasing molecular weight, such as the components of chrysene, benzo[a]pyrene were persistent in environment and were difficult to be degraded by microorganisms [44]. It was generally believed that monooxygenase and dioxygenase were key enzymes for microbial degradation of PAHs [45]. The main reason that PAHs were not easy to be degraded by microorganisms was that no suitable efficient catalytic enzymes existed to improve degradation efficiency of microbial [46]. In our study, low molecular weight components accounted for most of the PAHs, and microbial water purifying agents and biological compound enzymes showed no degradation effect, which indicated that the added microbial water purifying agents and biological compound enzymes were not suitable for PAHs degradations. According to the pollution conditions of PAHs in natural river water, it was necessary to find a highly effective degradation enzyme to improve the biodegradation efficiency.

3.4. Occurrence and compositions of OPPs in experimental area

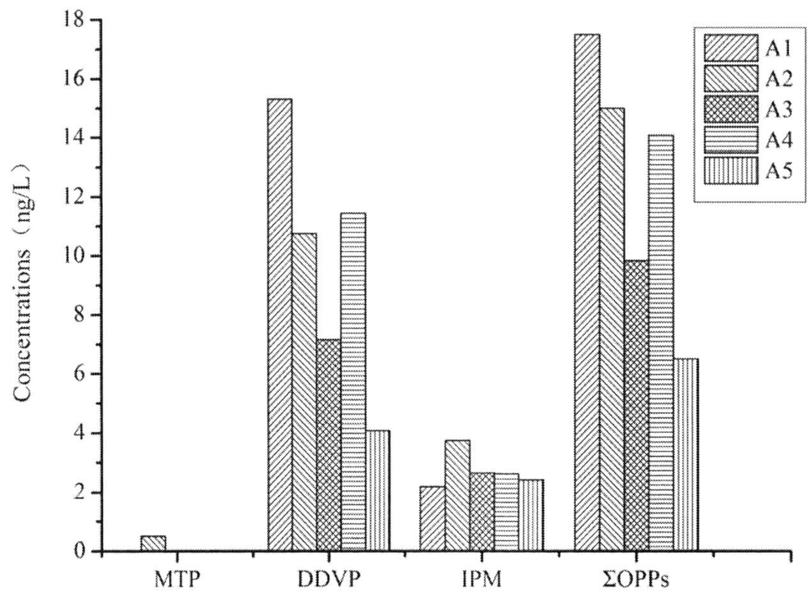

Figure 4. Concentrations of organophosphorus pesticides in experimental water body of South Canal.

As shown in figure 4, methamidophos, dichlorvos and isofenphos methyl were detected in surface water South Canal respectively, and other fifteen kinds of OPPs were not detected. Among them, dichlorvos and isofenphos-methyl were detected in each sampling sites and the detection rates were 100%; methamidophos was only detected in A2 sites and the detection rate was 20%. It can be also seen from figure 4 that concentration of dichlorvos was highest among three kinds of OPPs. Total concentrations of OPPs in experimental area ranged from 6.51 to 17.50 ng·L^{-1}, and the average value was 12.58 ng·L^{-1}.

As shown in figure 4, total concentrations of OPPs in A1 sites were higher than that of in A2-A5 sites and showed significant difference ($P<0.05$), which indicated that microbial water purifying agents and biological compound enzymes had a certain effect on the degradation of OPPs in water environment of South Canal. According to the analysis of monitoring data, the degradation rate of OPPs ranged from 19.6% to 62.8% with 35.2% in average. Water in experimental area was in an aerobic environment with sufficient dissolved oxygen under the action of aeration device. Concentrations of dissolved oxygen ranged from 10.80 to 14.00 mg·L^{-1} in each sampling sites. Adequate concentrations of oxygen ensured that aerobic microorganisms had sufficient activity to promote the degradation of organic pollutants. Mechanisms of microbial degradation of OPPs were similar to that of OCPs, and mainly comprised three types including co-metabolism, communal synergistic metabolisms and mineralization [47]. A variety of microorganisms were separated to degrade OPPs residues including bacteria, actinomycetes, fungi and algae, especially bacteria and fungi were studied thoroughly [47, 48]. P-N, P-O and P-S chemical bonds of OPPs were broken in different degrees and conducted its degradation [49]. Many kinds of enzymes could degrade OPPs including oxygenase, dehydrogenase, azo reductase and peroxidase, and microbial degradations of OPPs were accomplished by the synergistic action of multiple enzymes in many cases. Factors affecting microbial to degrade OPPs mainly included pH, temperature, water types and properties, physical and chemical factors such as oxidation-reduction potential, substrate type and its chemical structure, substrate concentration, biochemical factors such as activity and types of microbial, factors of substrate concentration [50]. A variety of bacteria, fungi and algae were probably existed in the water environment of South Canal and influencing factors of OPPs degradation consisted of the degradation reasons in a certain extent. According to the situations of generally relatively high concentrations of OPPs in water bodies of actual ecological restoration demonstration project, application this ecological restoration technology coupled with microorganisms and enzymes to degrade OPPs have a future prospect. Next steps of the research should consider how to improve degradation efficiency of OPPs in actual ecological restoration process.

4. Conclusion

Occurrence, composition and ecological restoration of organic pollutants in water environment of South Canal were studied by monitoring thirty-two persistent organic pollutants including fifteen OCPs and seventeen PAHs. Results showed that microbial water purifying agents and biological compound enzymes had not obvious effects on degradation of OCPs and PAHs in water body, while provided a certain effect on OPPs. The degradation rate of microbial water purifying agents and biological compound enzymes to total OPPs ranged from 19.6% to 62.8% with 35.2% in average. Degradation mechanisms of OCPs, PAHs and OPPs by high efficient microbial water purifying agents and biological enzymes in the river water environment remains to be further studied. Ecological restoration technologies coupled with microorganisms and enzymes to treat organic pollutants used in our study had a certain reference to practical application on the rivers and lakes.

Acknowledgments

This paper was supported by Key Laboratory of Pollution Processes and Environmental Criteria (Nankai University), Ministry of Education (KL-PPEC-2015-3), the Science & Technology Promotion Projects of Ministry of Water Resources of China (TG1408), International Science & Technology Cooperation Program of China (2013DFA71340) and Major National Science & Technology project of Water Pollution Control and Management of China (2012ZX07203-002).

References

[1] Wen Y, Zhao G Z, Zhou C B, *et al* 2011 Research progress of microbial agents in ecological engineering *Acta Ecol. Sin.* **31** 6287-94 (in Chinese)
[2] Luo Y, Jin S Q and Zhou J B 2013 A comparison study on purifying ability of different types of miroorganism water purification agent *Environ. Pollut. Control* **35** 41-5 (in Chinese)
[3] Cheng Z W, Lu L C, Kennes C, *et al* 2016 A composite microbial agent containing bacterial and fungal species: Optimization of the preparation process, analysis of characteristics, and use in the purification for volatile organic compounds *Bioresour. Technol.* **218** 751-60
[4] Chen F, Ke Z D, Zhou J P, *et al* 2007 Control of lake eutrophication by compound bacterium *Environ. Sci. Technol.* **30** 29-33 (in Chinese)
[5] Yang L, Lin F K, Xu Z, *et al* 2005 Study on in-situ bioremediation technology of municipal eutrophication river using multiple enzymes *Environ. Pollut. Control* **27** 607-10 (in Chinese)
[6] Mcnamara C J, Anastasious C C, O'flaherty V, *et al* 2008 Bioremediation of olive mill wastewater *Int. Biodeterior. Biodegrad.* **61** 127-34
[7] Alessandrello M J, Tomas M S J T, Isasc P, *et al* 2017 PAH removal by immobilized bacterial cells-support systems using low-cost culture media for biomass production *Int. Biodeterior. Biodegrad.* **120** 6-14
[8] Meng C, Chngchun S, Guo Y H, *et al* 2004 Study on characteristics of biocometabolic removal of omethoate by *Aspergillus sp. Water Res.* **38** 1139-46
[9] Sari A A and Tachibana S 2012 Determination of co-metabolism for 1,1,1-trichloro-2,2-bis(4-chlorophenyl) ethane (DDT) degradation with enzymes from Trametes versicolor U97 *J. Biosci. Bioeng.* **114** 176-81
[10] Baccellaa S, Cerichellib G, Chiarini M, *et al* 2000 Biological treatment of alkaline industrial waste waters *Process Biochem.* **35** 595-602
[11] Guo J Y, Liang Z, Liao H Q, *et al* 2011 Sedimentary record of polycyclic aromatic hydrocarbons in Lake Erhai, Southwest China *J. Environ. Sci.* **23** 1308-15
[12] Zheng B H, Wang L P, Lei K, *et al* 2016 Distribution and ecological risk assessment of polycyclic aromatic hydrocarbons in water, suspended particulate matter and sediment from Daliao River estuary and the adjacent area, China *Chemosphere* **149** 91-100
[13] Wu Y L, Wang H X, Li Y Y, *et al* 2011 Occurrence of polycyclic aromatic hydrocarbons (PAHs) in seawater from the Western Taiwan Strait, China *Mar. Pollut. Bull.* **63** 459-63
[14] Sultatosl G 1994 Mammalian toxicology of organophosphorus pesticides *J. Toxicol. Environ. Health.* **43** 271-89
[15] Macdonald R W, Barrie L A, Bidleman T F, *et al* 2000 Contaminant in Canadian Arctic: 5 years of progress in understanding sources, occurrence and pathways *Sci. Total Environ.* **254** 93-234
[16] Halecky M, Spackova R, Paca J, *et al* 2014 Biodegradation of nitroglycerin and ethylene glycol dinitrate by free and immobilized mixed cultures original research article *Water Res.* **48** 529-37
[17] Chi J 2009 Distribution of organochlorine pesticides in water and potamogeton crispus L. from the Haihe River *Res. Environ. Sci.* **22** 1008-12 (in Chinese)
[18] Cao Z G, Liu J L, Luan Y, *et al* 2010 Distribution and ecosystem risk assessment of polycyclic aromatic hydrocarbons in the Luan River, China *Ecotoxicology* **19** 827-37
[19] Wang Y Z, Zhang J, Zhou X S, *et al* 2017 Seasonal pollution characteristics and source identification of polycyclic aromatic hydrocarbons and organochlorine pesticides in surface water of Baiyangdian Lake *Environ. Sci.* **38** 964-78 (in Chinese)
[20] Ren N Q, Qi H, *et al* 2009 Levels, distributions and source-acceptor relationship of organochlorine pesticides in the top soils in Northeastern China *J. Environ. Sci.* **21** 1541-6.
[21] Zhu Y Y, Tian J, Wei E Q, *et al* 2014 Characteristics, sources apportionment and ecological risks assessment of polycyclic aromatic hydrocarbons in soils of Tianjin, China *Environ. Chem.* **33** 248-55. (in Chinese)

[22] Ma M, Wang Z J and Anders S 2001 Contamination of PCBs and organochlorinated pesticides in the sediment samples of Guanting Reservoir and Yongding River *Environ. Chem.* **20** 238-43 (in Chinese)
[23] Qin Y W, Zhang L, Zheng B H, *et al* 2010 Distribution characteristics and sources of OCPs in estuary sediments of main runoff flowed into Bohai Bay, China *J. Agro-Environ. Sci.* **29** 1900-6 (in Chinese)
[24] Zhao L, Hou H, Guo P, *et al* 2009 Distribution of organochlorine pesticides in soils in Haihe River and Haihe Estuary area, China Chinese *J. Environ. Sci.* **30** 543-9 (in Chinese)
[25] Cao Z G, Liu J L, Wang X M, *et al* 2010 Pollution characteristics, ecological risk assessment and sources of polycyclic aromatic hydrocarbons (PAHs) in surface water from the Zhangweinan River *Acta Sci. Circumst.* **30** 254-60 (in Chinese)
[26] Gao J J, Liu L H, Liu X R, *et al* 2008 Occurrence and distribution of organochlorine pesticides-lindane, p,p'-DDT, and heptachlor epoxide-in surface water of China *Environ. Int.* **34** 1097-103
[27] Yuan X Y, Yang Y Y, Li Q X, *et al* 2013 Present situation and distribution characteristics of persistent organic pollutants in freshwater in China *Environ. Chem.* **32** 2072-81 (in Chinese)
[28] Jiang X, Xu S F, Martens D, *et al* 2000 Polychlorinated organic contaminants in waters, suspended solids and sediments of the Nanjing section, Yangtze River China *Environ. Sci.* **20** 193-7 (in Chinese)
[29] Chi J 2009 Distribution of organochlorine pesticides in water and *Potamogeton crispus L.* from the Haihe River *Res. Environ. Sci.* **29** 1008-12 (in Chinese)
[30] Feng C L, Xia X H, Shen Z Y, *et al* 2007 Distribution and sources of polycyclic aromatic hydrocarbons in Wuhan section of the Yangtze River, China Environ. Monit. Assess. **133** 447-58.
[31] Luo X J, Mai B X and Yang Q S 2004 Polycyclic aromatic hydrocarbons (PAHs) and organochlorine pesticides in water columns from the Pearl River and the Macao harbor in the Pearl River Delta in South China *Mar. Pollut. Bullet.* **48** 1102-15
[32] Cao Z H, Wang Y Q, Ma Y M, *et al* 2005 Occurrence and distribution of polycyclic aromatic hydrocarbons in reclaimed water and surface water of Tianjin, China *J. Hazard. Mater.* **122** 51-9
[33] Zhang Z, Dai M and Hong H 2002 Dissolved insect icides and polychlorinat ed biphenyls in the Pearl River Estuary and South China Sea *J. Environ. Monit.* **4** 922-8
[34] Wan Y W, Kang T F and Zhang Y 2008 Determination of organic phosphorous pesticides in surface water by solid-phase extaction gas chromatography *Chi. Environ. Monit.* **24** 40-3 (in Chinese).
[35] Law R J, Dawes V J, Woodhead R J, *et al* 1997 Polycyclic aromatic hydrocarbons (PAH) in seawater around England and Wales *Mar. Pollut. Bullet.* **34** 306-22
[36] Lu X M and Chen P P 2012 Progress in degradation of organochlorinated pesticides by microorganism *Environ. Sci. Technol.* **35** 89-93 (in Chinese)
[37] Nagata Y, Miyauchi K and Takagi M 1999 Complete analysis of genes and enzymes for γ-hexachlorocyclohexane degradation in *Sphin-gamonas paucimobilis* UT26 *J. Ind. Microbiol. Biotechnol.* **23** 380-90
[38] Kumari R, Subudhi S, Suar M, *et al* 2002 Cloning and characterization of Lin genes responsible for the degradation of hexachlorocyclohexane isomers by *spin-gamonas paucimobilis* strain B90 *Appl. Environ. Microbiol.* **68** 6021-8
[39] Sharma P, Raina V, Kumari R, *et al* 2006 Haloalkane dehalogenase LinB is responsible for β- and δ- hexachlorocyclohexane transformation in *Sphingobium indicum* B90A *Appl. Environ. Microbiol.* **72** 5720-7
[40] Adi S P, Ichiro K and Ryuichiro K 2008 Degradation of 1,1,1-trichloro-2,2-bis(4-chlorophenyl) ethane(DDT) by brown-rot fungi *J. Biosci. Bioeng.* **105** 614-21
[41] Zhao H M, Quan K, Yang F L, *et al* 2002 Basic research on the bioremediation of p,p'-DDT

contaminated sediment by adding short chain organic substrate *Acta Sci. Circumst.*, **22** 51-5 (in Chinese)
[42] Du L Y, Zhang G Y and Jin W 2005 Influence of microorganism and humic acid on organochlorine pesticides under different pH conditions *Jiangsu J. Agric.Sci.* **5** 120-3 (in Chinese)
[43] Du L Y, Zhang G Y and Jin W 2006 Effects of soil water content and humic acid on degradation of organochlorine pesticides *Acta Pedol. Sin.* **43** 332-6 (in Chinese)
[44] Wu H Z, Feng C H, Yu X B, *et al* 2012 Formation of persistent organic chlorides in aquatic environments and their control strategies *J. South China Univ. Technol. (Nat. Sci. Ed.)* **42** 29-36 (in Chinese)
[45] Albert L J and Ravendra N 2000 Bioremediation of high molecular weight polycyclic aromatic hydrocarbons: A review of the microbial degradation of benzo[a]pyrene *Int. Biodeterior. Biodegrad.* **45** 57-88
[46] Wen H Y, Liao Y Z and Li X D 2005 Advance of biodegradation of Polycyclic aromatic hydrocarbons (PAHs) by microorganisms *J. Microbiol.* **25** 73-5 (in Chinese)
[47] Liu J L 2010 Present situation and development of microbial degradation of organphosphorus pesticides residues in fruits and vegetables *Guangdong Agric. Sci.* **37** 107-10 (in Chinese)
[48] Bo W Q, He F Q and Qiu X H 2004 Advances in biodegradation of organophosphates Chinese *J. Appl. Environ. Biol.* **10** 57-66 (in Chinese)
[49] Lu Y, Ni F, Xu P X, *et al* 2005 ^{31}P NMR and HPLC to screen and trace the degradation of methamidophos (MAP) by bacteria *Environ. Chem.* **24** 41-3 (in Chinese)
[50] Wu Y P and Zheng F C 2007 Advance in microbial degradation of organophosphate pesticides *Guangxi Agric. Sci.* **38** 637-42 (in Chinese)

Stability mechanisms of barchan dunes: A case study of Hexi Desert area in Gansu Province

Z F Chang[1,2], J H Zhang[1], Q Wang[1], D K Zhang[1], J N Tang[1], J H Zhang[1] and Q Q Wang[1]

[1]Gansu Desert Control Research Institute, Lanzhou 730070, China

E-mail: czf123@sina.com

Abstract. Barchan dunes, an important type of dunes, are distributed on the edge of oases. Research of stability mechanisms of barchan dunes is of great importance to the determination of regulation of aeolian sand movement on the edge of oases. However, the stability of barchan dunes remains an unresolved scientific issue. In this paper, the stability mechanisms of the top of barchan dunes in Hexi desert area of Gansu Province were analyzed. The results showed that: (1) in Hexi desert area of Gansu Province, patches of barchan dunes were distributed downwind on the edge of desert. The distribution areas were flat sandy-clay or sandy-gravel beaches. The lowlands among the dunes were wide and had the prominent prevailing wind. (2) The prevailing northwestern wind caused the highest point and sand ridge line of the dunes to coincide, and the dunes to move forward and become lower. The southeastern wind (opposite to the prevailing wind) caused the highest point and sand ridge line of the dunes to separate, the leeward slope to be eroded by wind and the dunes to rise. (3) The higher the barchan dunes were, the longer the two flanks were, and the smaller the angle was. (4) The wide flat beaches were vital to the formation of barchan dunes and a chain of barchan dunes. The largely prevailing wind was the key factor for the stability of barchan dunes and a chain of barchan dunes. The opposite wind and the corridor effect of the two flanks were key factors for the increase of barchan dunes in size and height. The speed and distribution frequency of the prevailing and opposite winds were factors restraining the height of barchan dunes and a chain of barchan dunes here.

1. Introduction

In most desert areas of Hexi region in Gansu Province, there are obvious prevailing wind directions, and there are no more than three groups of winds with similar power. That is, there is no repeated shaping process of contrary winds or gales in other directions. However, these high barchan dunes and a chain of barchan dunes can exist stably, and their top can not be scraped by wind. There are lots of reports on barchan dunes at home and abroad, but most reports focus on their morphological characteristics [1-4], flow field [5-7] and surface grain size [8-10] instead of stability of their top.

Xu Zhenwen *et al* [11] have analyzed the quantity, scale and sand granularity of barchan dunes in the Kumtag Desert and the Ulan Buh Desert as well as along Changli gold coast and Emerald Isles of Beidaihe. Bishop [12] and Durán *et al* [13] have suggested that the distribution of barchan dunes undergoes a process of self-regulation, and their size and spacing tend to be consistent with their movement and merging [14]. Hersen *et al* [3] have pointed out that the size and spacing of barchan dunes on each sand are consistent, and the bigger the barchan dunes are, the quicker the barchan dunes breed. Elbelrhiti *et al* [15] have suggested that big barchan dunes change into small barchan dunes after the loss of sand materials on both sides, thereby maintaining the development pattern and size distribution of barchan dunes. During the interaction process of barchan dunes, their height distribution will not change [16, 17].

Airflow is the dynamic factor of formation of barchan dunes, so it is the focal point of research on barchan dunes [18]. Bourke [4] and Yang Yanyan *et al* [18] Zhang Cunlai *et al* [19] have pointed out that deposition is more intense than erosion in locations where the windward slope of barchan dunes becomes gentle, while erosion is more intense than deposition in other locations, and the sand transport rate and erosion intensity are the highest at their top. Based on the observation and research of moving dunes in the Mu Us Sandy Land, Yao Honglin *et al* [20] have suggested that when wind velocity increases to the threshold leading to sand movement, the top of the dues is being eroded by wind; as wind velocity increases gradually, sand is accumulating at their top. Belrhiti *et al* [21] have pointed out that only when a stable polygon is formed on both flanks and at the top of a barchan dune can the barchan dune be invariant in shape during the process of movement.

To sum up, the top stability of barchan dunes remains an unresolved scientific issue in the field of desert ecology. Research on the top stability of barchan dunes not only has important academic value to the revelation of sand movement rules, invasion of sand flow, and expansion of desert on the edge of oases, but also has considerable practical significance to the construction of sand defense and stabilization, resource development and protection of ecological environment in desert areas.

2. General situation of the study area and research methods

2.1. The study area

Hexi region (92°45'-104°15' E, 36°35'-42°45' N) in Gansu Province includes Wuwei, Jinchang, Zhangye, Jiuquan and Jiayuguan cities on the west of the Yellow River. The region is located on the north of the Qilian Mountains and is surrounded by the Tengger Desert in the east, the Badain Jaran Desert in the north, and the Kumtag Desert in teh west. In the region, the area of desert and scattered sand is 754 km^2, and there are large numbers of tall barchan dunes and a chain of barchan dunes on the edge of oases. Minqin County is situated on the western edge of the Tengger Desert and on the northwest of Hexi Corridor, and barchan dunes and a chain of barchan dunes concentrate most densely here; most barchan dunes and a chain of barchan dunes are distributed on the NW edge of oases, namely the upwind edge of desert (figure 1).

2.2. Research methods

Based on the complete investigation of barchan dunes and a chain of barchan dunes on the edge of oases in Hexi region of Gansu Province, the flow field, wind erosion and accumulated sand of typical barchan dunes were observed. That is, two tall barchan dunes were chosen from Minxin County, and observation recorders of wind speed and direction were put in their key positions (figure 2). Wind speed at No. 1 dune was observed from April to September in 2014, and the highest point and sand ridge line of the dune separated at the initial stage of observation; wind speed at No. 2 dune was observed from April to September in 2015, and the highest point and sand ridge line of the dune coincided at the initial stage of observation. Meanwhile, wind erosion poles were put in the key positions of the dunes to measure the depth of wind erosion and thickness of accumulated sand. Moreover, sand sources around the dunes and vegetation before and after the dunes were observed. Wind speed and direction were observed at a height of 50 cm, and sampling was conducted one time every 20 min. Every 2 m/s was as a grade of wind speed, and the maximum wind speed was 14.9 m/s. To observe the depth of wind erosion and thickness of accumulated sand, one end of a No. 8 steel wire was buried underground, and the height of the steel wire above the ground was 30 cm. The height of the steel wire above the ground was observed. In this paper, the prevailing wind was NW wind, and wind opposite to the prevailing wind was SE wind. NW slope was defined as windward slope. The apex of the dunes referred to the highest point of the dunes, and the chord length was the distance between the endpoints of their two flanks; their length-width ratio was the ratio of their total length to total width; the thickness of back of a bow was the distance from the bottom of windward slope to the bottom of leeward slope (figure 3). Climatic data were from local meteorological stations and China meteorological science data sharing service network.

2.3. Data analysis

Every 2 m/s was as a grade of wind speed of the prevailing wind, and the wind speed S was divided into five grades, including $S < 2$ m/s, 4 m/s>S≥2 m/s, 6 m/s>S≥4 m/s, 8 m/s>S≥6 m/s, and S≥8 m/s. The speed of wind opposite to the prevailing wind was small, so every 1 m/s was as a grade of wind speed. Data of wind speed were analyzed by using SPSS 13.0.

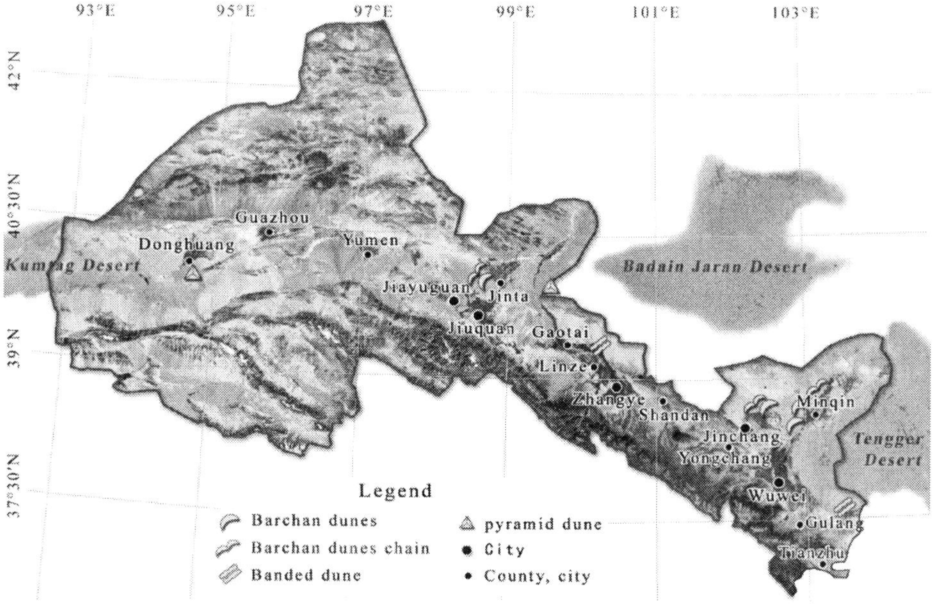

Figure 1. Distribution of barchan dunes in Hexi desert area of Gansu area.

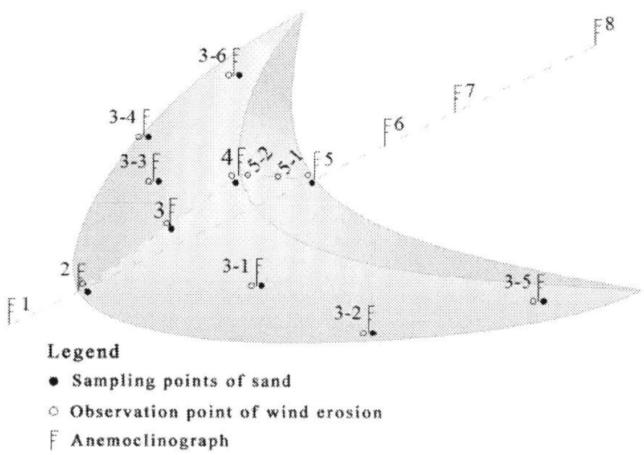

Figure 2. Observation sampling points of barchan dunes.

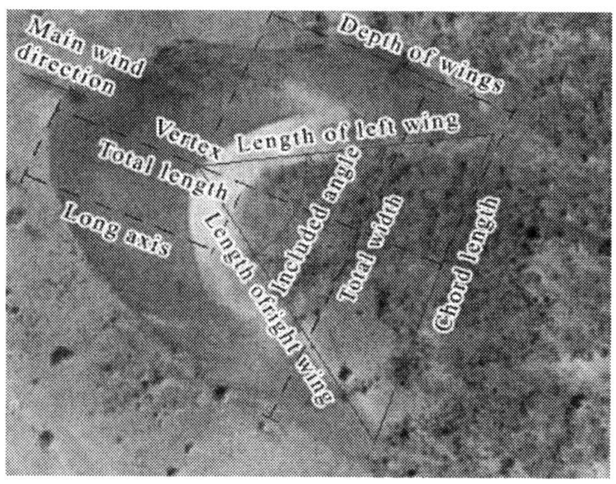

Figure 3. Feature representation of barchan dunes.

3. Results and analysis

3.1. Distribution environment of barchan dunes

Barchan dunes are often formed open sandy-clay or sandy-gravel beaches. Barchan dunes develop from patches of sand and sand piles, that is, after encountering obstacles such as bushes and gravel, sand flow piles up to form patches of sand firstly and then peltate sand piles and prototype of barchan dunes [18], thereby developing into barchan dunes.

In the desert areas of Hexi region in Gansu Province, barchan dunes were the most on the edge of oases in Minqin in the east, followed by the edge of oases in Jinta in the west, while there were few barchan dunes in the middle of Hexi Corridor. The basic characteristics of barchan dunes are shown as follows: their average height was 8.1 m, and the maximum was up to 11.6 m, while the minimum was 3.6 m; the thickness of back of a bow averaged 101.1 m, and the maximum reached 155.7 m, while the minimum was 64.1 m; their total length averaged 162.3 m, and the maximum was 294.4 m, while the minimum was only 86.5 m; their total width averaged 188.9 m, and the maximum reached 227.4 m, while the minimum was 57.3 m; the slope of leeward slopes averaged 31.8°, and the slope of the steepest slope was 32.9°, while the slope of the gentlest slope was 30.1° (table 1)

Distribution environmental characteristics of barchan dunes are shown as follows: barchan dunes were located on the downwind edge of desert or scattered sand, namely on the upwind edge of oases. For instance, barchan dunes in Minqin desert area were distributed on the upwind edge of oases in western Minqin; barchan dunes in Jinchang desert area were distributed on the edge of desert and oases in northern Jinchang; barchan dunes in Jinta desert area were located on the edge of scattered desert on the west of oases in Jinta (figure 1). A chain of barchan dunes could be found in regions where barchan dunes appeared. Open and flat gravel beaches were distributed in front and at the back of barchan dunes; the average width of upwind beaches was 224.1 m, while the maximum was up to 662.9 m; the average width of downwind beaches was 359.9 m, while the maximum reached 697.3 m. There was a single main wind direction. For example, the main wind direction was NW in Minqin and Jinchang desert area and NW and WNW in Jinta desert area. The forward direction of barchan dunes changed from N45°W—N48°W in Minqin desert area to N63°W in Jinta desert area. All barchan dunes were moving dunes, that is, there was nearly no vegetation on their windward slopes and leeward slopes; vegetation was sparse in beaches in front and at the back of barchan dunes, and vegetation coverage ranged from 0.5% to 4.0 %.

Table 1. Morphological characteristics of barchan dunes in Hexi desert area of Gansu.

No. of dunes	Plots	Height (m)	Thickness of back of a bow (m)	Length (m)	Width (m)	Slope of a leeward slope (°)	Forward direction (°)	Length of upwind beaches (m)	Length of downwind beaches (m)
1	Minqin 1	9.8	123.3	214.6	202.5	32.6	N48°W	662.9	378.0
2	Minqin 2	11.2	155.7	294.4	227.4	32.2	N48°W	163.3	491.6
3	Minqin 3	9.5	120.5	197.5	209.8	32.1	N48°W	134.2	176.6
4	Minqin 4	3.7	78.7	86.5	81.8	31.8	N48°W	251.8	367.8
5	Minqin 9	7.9	90.8	112.5	81.5	32.9	N45°W	214.3	253.3
6	Minqin 10	7.6	86.7	136.8	64.7	31.5	N46°W	137.9	430.1
7	Minqin 11	11.6	104.4	183.6	57.3	31.6	N46°W	179.7	243.1
8	Minqin 12	8.6	64.1	131.4	77.9	30.1	N46°W	135.2	697.3
9	Jinchang 1	8.4	130.3	197.6	123.2	31.7	N55°W	227.4	372.1
10	Jinta 1	6	83.7	125.3	104.8	31.5	N63°W	250.4	129.5
11	Jinta 2	5.3	73.5	104.8	76.6	31.7	N63°W	108.2	419.8

3.2. Main and anti-main wind direction process

3.2.1. Forward movement and drop of the apex of barchan dunes and its coinciding with a sand ridge line in the main wind direction. In Minqin desert area, the main wind direction was NW and the forward direction of barchan dunes was NW-SE, while the wind speed causing sand movement ranged 4.5—5.0 m/s [22]. The highest point and sand ridge line of No. 1 barchan dune separated before observation. The observation results of wind speed show that when wind speed was ≥8 m/s at the upwind reference point, wind speed between the highest point and sand ridge line of barchan dunes was smaller than the wind speed at the upwind reference point, that is, there was a weak wind zone between the highest point and sand ridge line of barchan dunes (Figure 4(a)). As the highest point and sand ridge line of No. 2 barchan dune coincided, wind speed was the largest at the top of the barchan dune (Figure 4(b)).

According to the observation results of wind erosion from March 31 to May 8 in 2014, under the effect of prevailing wind (NW wind), various parts of barchan dunes on windward slopes were being eroded by wind to a degree, and wind erosion depth was 6.7 cm at the bottom of windward slopes, 13.0 cm in the middle of windward slopes, 16.8 cm between the highest point and sand ridge line of barchan dunes, 6.0 cm in the middle of the western flank of barchan dunes, 0.0 cm on the edge of the western flank, 11.0 cm in the middle of the eastern flank of barchan dunes, and 0.5 cm on the edge of the eastern flank respectively. That is, wind erosion was the strongest at the top of barchan dunes, and the height of barchan dunes dropped.

The highest point and sand ridge line of barchan dunes were often separate in spring. When gale and sandstorm happened, sand flow rose along the windward slopes of barchan dunes to the highest point, and then load flow continued to move along the windward slopes under the effect of inertia force (figures 5(a) and 5(b)). The height between the highest point and sand ridge line of barchan dunes was lower than the highest point of barchan dunes, so there was a weak wind zone between the highest point and sand ridge line of barchan dunes.

When sand flow passed the highest point of barchan dunes, a large quantity of sand settled (figure 5(b)), so the highest point of barchan dunes moved forward until it coincided with the sand ridge line. Afterwards, a large quantity of sand fell on the upside of the leeward slope and then slipped down freely, so barchan dunes moved forward (Figure 5(c)). From March 31 to May 8, the bottom of the leeward slope moved forward by 1.34 m.

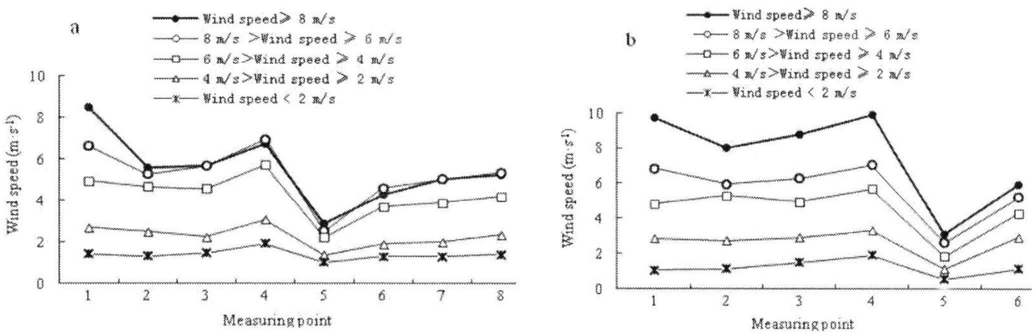

Figure 4. Change trends of wind speed in various parts of the dunes in the main direction.

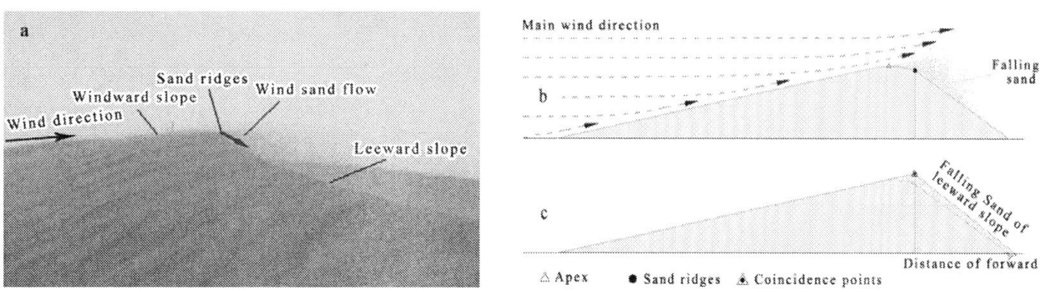

Figure 5. Schematic diagram of the main wind direction process.

3.2.2. Rise of the apex of barchan dunes and its separation from a sand ridge line in the anti-main wind direction. After spring, the highest point and sand ridge line of barchan dunes in Minqin desert area coincided. In summer, SE wind would appear here. As its wind speed reached the speed causing sand movement, wind in a SE direction (or ESE and SSE direction) blew across a large quantity of sand on their leeward slopes, and the sand passed sand ridge lines and continued to move upward along their leeward slopes. The larger the wind speed was, the higher the sand moved. Their leeward slopes were steep, and sand was fine, so the sand moved by sand flow upward was high; coarse sand fell and deposited on the upside of their windward slopes, while fine sand fell at the bottom of their windward slopes or even far (Figures 6(a) and 6(b)).

If SE winds with wind speed high than wind speed causing sand movement were frequent in summer, wind erosion appeared on the upside of their leeward slopes, and the leeward slopes became gentle; the the highest point of barchan dunes separated from the sand ridge lines and moved toward the windward slopes; the height of barchan dunes increased at the same time (Figure 6(c)).

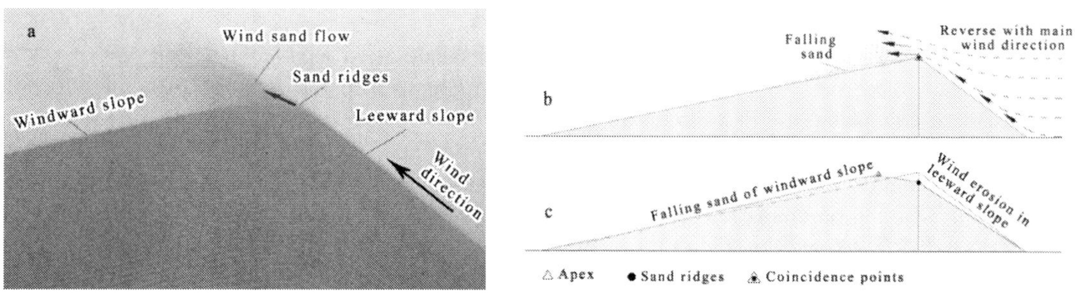

Figure 6. Schematic diagram of the anti-main wind direction process.

3.3. Functions of their two flanks as galleries

3.3.1. Features of wind speed in the anti-main wind direction. In Minqin desert area, wind speed was high in NW (including WNW and W) direction but slow in SE (including ESE and E) direction (figure 7). Therefore, the forward direction of most barchan dunes was NW—SE, that is, NW was the main wind direction of barchan dunes, while SE was their anti-main wind direction. According to table 1, the observation results of wind speed in NW direction at No. 1 barchan dune are shown as follows: wind speed reduced from the open upwind ground (No. 1 observation point) to the bottom of its windward slope (No. 2 observation point); wind speed increased from the bottom of its windward slope to its top (No. 4 observation point); wind speed was the smallest at the bottom of its leeward slope (No. 5 observation point), and then it rose gradually (Figure 4(a)).

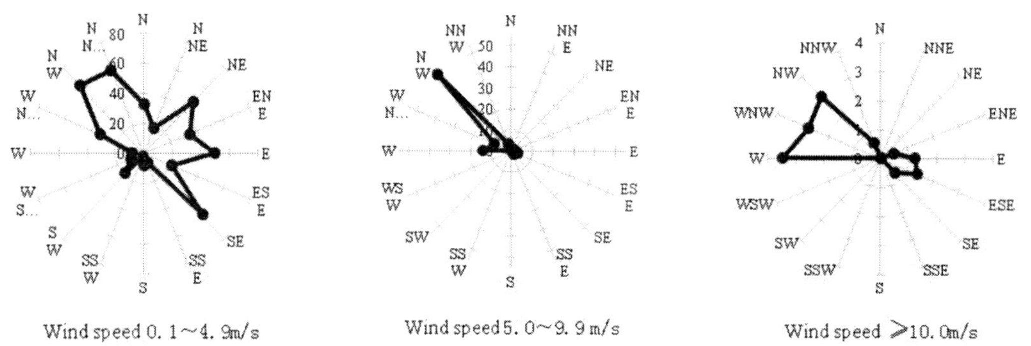

Figure 7. Distribution of wind speed and wind direction in Minqin desert area from 2000 to 2014.

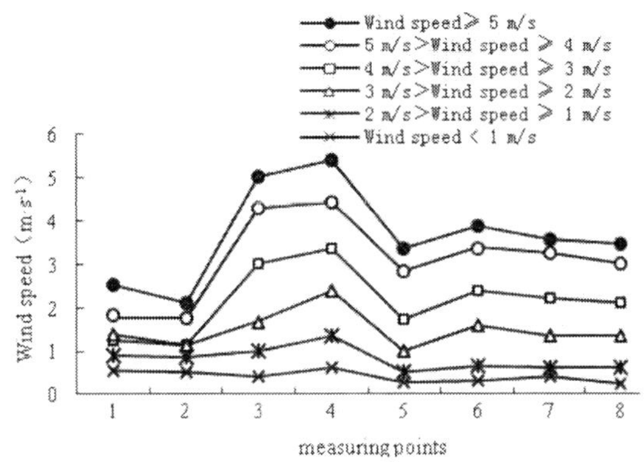

Figure 8. Change trends of wind speed in various parts of the dunes in the opposite direction.

The observation results of wind speed in SE direction at No. 1 barchan dune (figure 8) were contrary to that in NW direction. That is, on the one side of its leeward slope, wind speed increased gradually from the far observation point (No. 8 observation point) to the near observation point (No. 6 observation point) and reduced at the bottom of its leeward slope (No. 5 observation point). The acceleration of wind speed in SE direction on the one side of its leeward slope correlated positively with wind speed. For instance, from No. 8 observation point No. 6 observation point, ≥ 5 m/s, ≥4 m/s—>5m/s, ≥3 m/s—> 4m/s, ≥2 m/s—>3m/s, ≥1 m/s—>2m/s wind speed increased by 0.417, 0.366, 0.272, 0.268, and only 0.031 m/s respectively. The variation of wind speed was the smallest at the bottom of its windward slope but the most obvious at the top of the dune. In the main wind direction, the variation of wind speed was the smallest at the bottom of its leeward slope (No. 5 observation point) but the most obvious on the smooth upwind ground of its windward slope (No. 1 observation point). The decrease of wind speed from the middle to the bottom of its windward slope was far larger that from the top of the dune to the middle of its windward slope. In the main wind direction, the increase of wind speed from the bottom to the middle of its windward slope was close to that from the middle of its windward slope to the top of the dune.

3.3.2. Characteristics of sand accumulation in the anti-main wind direction. According to the observation results of wind erosion at No. 1 barchan dune in table 1, under the effect of the main wind (NW wind), the height of the dune dropped by 37.7 cm from March 31 to May 8 in 2014; the depth of wind erosion at the bottom and in the middle of the windward slope was 6.7 and 13.0 cm respectively; the depth of wind erosion in the middle and on the edge of its western flank was 6.0 and 0.0 cm respectively; the depth of wind erosion in the middle and on the edge of its eastern flank was 11.0 and 0.5 cm respectively. That is, wind erosion was the strongest at the top of the dune.

On September 21, 2015 after the anti-main wind process, the distance between the highest point and sand ridge line of No. 1 barchan dune was 8.3 m, and its height rose by 43 cm compared with that on May 22, 2015.

4. Conclusions

- The most obvious characteristics of distribution environment of barchan dunes and a chain of barchan dunes are shown as follows: firstly, they were distributed on sandy-clay or sandy-gravel beaches; secondly, they were distributed in flat regions (especially in ancient riverbeds), and the differentiation between barchan dunes and beaches as underlying surface was obvious; thirdly, upwind and downwind beaches around barchan dunes were open. On the flat and open underlying surface, sand flow ran smoothly. On the one hand, sand carried by sand flow rose along barchan dunes and accumulated on them; on the other hand, sand on both sides of barchan dunes was moved forward by gale to form two flanks of barchan dunes [6]. In Hexi

Corridor, patches of barchan dunes were distributed downwind on the edge of desert, and the average width of upwind beaches around barchan dunes was 246.4 m, while the average width of downwind beaches was 374.4 m,

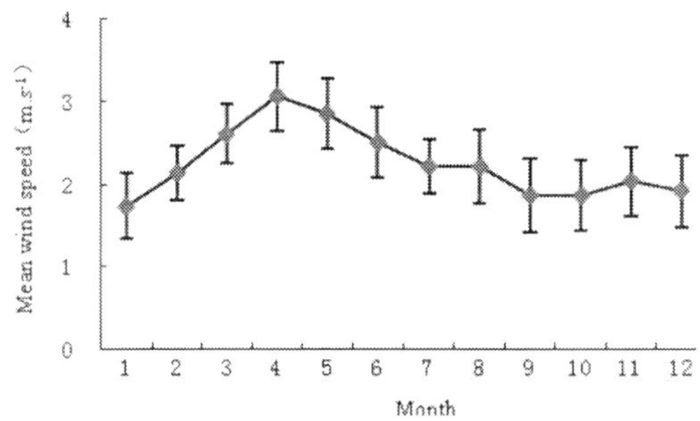

Figure 9. Average wind speed of each month in Minqin desert area, 1961-2016.

- Minqin desert area is mostly NW wind in spring and winter and the wind speed is larger. Mostly SE wind in summer and autumn and the wind speed is smaller (figure 9). Generally speaking, the separation of the highest point and sand ridge line of barchan dunes alternated with their overlap one time every year, and their overlap was frequent in summer, while their separation was common in autumn and winter. In special years, SE wind speed was small, or there was no separation in local regions sometimes; when NW wind speed was small in spring in a year, their separation appeared in the whole year. Existence and disappearance of the weak wind zone between the highest point and sand ridge line of barchan dunes took place by turn. Therefore, the height of barchan dunes fluctuated in a year. The apex of a barchan dune moved forward or back alternately within a year, and the movement distance mainly depended on the speed and frequency of NW and SE wind in the year. During the process, even if the speed and frequency of SE wind were high, the wind could erode the upside of a leeward slope to make its gentle. However, it could not make the base of a leeward slope move back, that is, the base of a leeward slope could move forward. Hence, the forward movement of a barchan dune should depend on the base of its leeward slope (No. 5 observation point).
- According to the observation results of wind erosion in various parts of barchan dunes, their top was the most unstable, namely being in the dynamic process of moving forward—moving back and rising—dropping. NW and SE wind were distributed alternatively in various seasons to maintain the basic shape of barchan dunes. The stability of barchan dunes was dynamic. If there was no SE wind causing sand movement, the highest point and sand ridge line of a barchan dune would not separate; the top of a barchan dune would be relatively stable; barchan dunes would move forward more quickly.
- Some scholars suggest that coarse sand changes into fine sand from the bottom of a windward slope to the top of a dune [23, 24], while other scholars have obtained the opposite conclusion [8, 25], which are related to the observation and sampling season. In fact, it is caused by seasonal alternation between NW and SE wind. In Minqin desert area, middle sand increased but fine sand decreased in quantity from the bottom of a windward slope to the top of a barchan dune, while fine sand tended to increase in the two flanks in early May. If observation was conducted in the late period of NW wind before the appearance of SE gale in late spring and early summer, similar results could be obtained. However, opposite results might be obtained if observation was conducted in autumn and winter.
- In Minqin desert area, the height of tall barchan dunes reached above 10 m (table 1), and barchan dunes lower than 4-5 m in height were small barchan dunes that were developing. The change of patches of sand and sand piles into peltate sand piles and prototype of barchan dunes started from the increase in the thickness of back of a bow [18]. When sand flow met patches of

sand and sand piles, airflow rose or its direction changed, so wind speed dropped, and sand accumulated. Small archan dunes were short and their windward slopes were short, and the wind aggregation role of the prevailing wind on windward slopes was smaller than that of tall barchan dunes, so the highest point and sand ridge line of small barchan dunes could not coincide mostly. As the back of a bow of small barchan dunes continued to thicken, the length of their windward slopes increased gradually, and their height also increased. Due to the increase of barchan dunes in height, sand flow changed into reflection eddy current on their leeward slopes, and sand falling slopes and sand ridge lines appeared on their leeward slopes gradually. With the formation and extension of their two flanks, the role of anti-main wind increased, and the increase of dunes in height accelerated. In specific environment, the accumulation and erosion of the main wind and anti-main wind were close, the height of dunes tended to be stable. In Minqin desert area, the height of barchan dunes increased in years when the power and frequency of the anti-main wind were high but decreased in years when they were low.

5. Conclusions

In Hexi desert area of Gansu Province, patches of barchan dunes and a chain of barchan dunes were distributed downwind on the edge of desert. The distribution areas were flat sandy-clay or sandy-gravel beaches. The beaches among the dunes were wide, and downwind beaches were more open than upwind beaches. There was the prominent prevailing wind direction in their distribution areas, and wind speed in other direction was small, or there was a gale blowing once in a while.

The prevailing northwestern wind caused the highest point and sand ridge line of the dunes to coincide, and the dunes to move forward and become lower. The southeastern wind (opposite to the prevailing wind) caused the highest point and sand ridge line of the dunes to separate, the leeward slope to be eroded by wind and the dunes to rise. NW wind was dominant here, so barchan dunes moved forward from NW to SE, while they were eroded by wind and the slope of their leeward slopes decreased under the effect of SE wind.

The length and angle between two flanks of barchan dunes were closely related to their height. That is, the higher the barchan dunes were, the longer the two flanks were, and the smaller the angle was. Two flanks of barchan dunes are one of important morphological characteristics to maintain the stability of barchan dunes.

The wide flat beaches were vital to the formation of barchan dunes and a chain of barchan dunes. The largely prevailing wind was the key factor for the stability of barchan dunes and a chain of barchan dunes. The opposite wind and the corridor effect of the two flanks were key factors for the increase of barchan dunes in size and height. The speed and distribution frequency of the prevailing and opposite winds were factors restraining the height of barchan dunes and a chain of barchan dunes here.

Acknowledgments

Sponsored by National Natural Science Foundation of "Ecological effects on prevention and control of sand encroachment by developing photovoltaic industry in the desert and gobi area" (41671528) & National Natural Science Foundation of "study on the surface crust Eco-hydrological effects of Nitrariasand-pile in desert-oasis ecotone" (41361004).

References

[1] Jiao Y X, Mu Y R, Zhang W S, *et al* 2013 Morphologial characteristics of Kumtagh sand in Beishan, Xinjiang Uygur autonomous region *Remote Sensing for Land & Resources* **25(2)** 138-42

[2] Sauermann G, Rognon P, Poliakov A and Herrmann H J 2000 The shape of the barchan dunes of Southern Morocco *Geomorphology* **36(1-2)** 47-62

[3] Hersen P 2004 On the crescentic shape of barchan dunes The European Physical *Journal B-Condensed Matter and Complex Systems* **37(4)** 507-54

[4] Bourke M C 2010 Barchan dune asymmetry: Observations from Mars and Earth *Icarus* **205(1)**

183-97

[5] Han Z W, Gou Qi Q, Du H Q, *et al* 2012 The piecewise fitting of sand flux vertical distribution of wind-sand flow within 100-cm height above the barchan dune surface *Scientia Geographica Sinica* **32(7)** 892-7

[6] Jiang L J and Ma G S 2010 Numerical modeling of flow over an isolated barchans dune *Journal of Lanzhou University (Natural Sciences)* **46(1)** 48-52, 58-8

[7] Sauermann G, Andrade J S Jr, Maia L P, Costa U M S, Araújo A D and Herrmann H J 2003 Wind velocity and sand transport on a barchan dune *Geomorphology* **54(3-4)** 245-55

[8] Zhou L, Zhang C L and Liu Y G 2011 Variation of grain size on surface of barchans in Mainling Great Valley, Yarlung Zangbo River *Scientia Geographica Sinica* **31(8)** 958-63

[9] Durán O, Schwämmle V, Lind P G and Herrmann H J 2009 The dune size distribution and scaling relations of barchan dune fields *Granular Matter* **11(1)** 7-11

[10] Palmer J A, Mejia-Alvarez R, Best J L and Christensen K T 2012 Particle-image velocimetry measurements of flow over interacting barchan dunes *Exp. Fluids* **52(3)** 809-29

[11] Xu Z W, Wang X D, Wang G J, *et al* 2013 Comparative analysis of morphology and formation mechanism of barchans dune of northwest inlands and coastal areas *Journal of Changchun Normal University* 32(4) 63-6

[12] Bishop M A 2007 Point pattern analysis of north polar crescentic dunes, Mars: A geography of dune self-organization *Icarus* **191(1)** 151-7.

[13] Durán O, Claudin P and Andreotti B 2011 On aeolian transport: Grain-scale interactions, dynamical mechanisms and scaling laws *Aeolian Res.* **3(3)** 243-70

[14] Kocurek G and Ewing R C 2005 Aeolian dune field self-organization-implications for the formation of simple versus complex dune-field patterns *Geomorphology* **72(1-4)** 94-105

[15] Elbelrhiti H, Claudin P and Andreotti B 2005 Field evidence for surface-wave-induced instability of sand dunes *Nature* **437(7059)** 720-3

[16] Endo N, Taniguchi K and Katsuki A 2004 Observation of the whole process of interaction between barchans by flume experiments *Geophys. Res. Lett.* **31(12)** L12503.1-L12503.3

[17] Katsuki A, Nishimori H, Endo N and Taniguchi K 2005 Collision dynamics of two barchan dunes simulated using a simple model *J. Phys. Soc. Jpn.* **74(2)** 538-41

[18] Yang Y Y, Liu L Y, Qu Z Q, *et al* 2014 A review of barchans dunes *Scientia Geographica Sinica* **34(1)** 76-83

[19] Zhang C L, Hao Q Z, Zou X Y, *et al* 1999 Response of morphology and deposits to surface flow on windward slope of barchans dune *J. Desert Res.* **19(4)** 359-63

[20] Yao H L, Yan D, Hu X L, *et al* 2001 Study on rules wind erosion and accumulated sand for moving dune in Maowusu Sandy land *Inner Mongolia Forestry Science & Technology* **1** 3-9

[21] Belrhiti H E and Douady S 2011 Equilibrium versus disequilibrium of barchan dunes *Geomorphology* **125(4)** 558-68

[22] Minqin Desert Control Experiment Station 1975 *Gansu Desert and Its Control* (Lanzhou: Gansu People's Publishing House)

[23] Li Z Z and Guan Y Z 1996 Experimental study on imitative flow pattern of longitudinal dunes and transverse dunes *J. Desert Res.* **16(4)** 360-3

[24] McLean S R, Nelson J M and Wolfe S R 1994 Turbulence structure over two-dimensional bed forms: Implications for sediment transport *J. Geophys. Res.* **99(C6)** 12729-47

[25] Wang X M, Dong Z B and Zhao A G 2004 Airflow and particle-size distributions and their significance on the dynamic process of a simple transverse dune *J. Arid Land Resour. Environ.* **8(4)** 29-33

IOP Conf. Series: Earth and Environmental Science **82** (2017) 012035 doi:10.1088/1755-1315/82/1/012035

Soil nutrient concentration and distribution at riverbanks undergoing different land management practices: Implications for riverbank management

X H Xue[1], S Chang[1] and L Y Yuan[2,3]

[1]College of Bioscience and Technology, Hubei University for Nationalities, Enshi, Hubei 445000, China

[2]College of Horticulture and Landscape Architecture, Yangtze University, Jingzhou, Hubei 434025, China

E-mail: yzq29@sina.com

Abstract. Riverbanks are important boundaries for the nutrient cycling between lands and freshwaters. This research aimed to explore effects of different land management methods on the soil nutrient concentration and distribution at riverbanks. Soils from the reed-covered riverbanks of middle Yangtze River were studied, including the soils respectively undergoing systematic agriculture (gathering young tender shoots, reaping reed straws, and burning residual straws), fires and no disturbances. Results showed that the agricultural activities sharply decreased the contents of soil organic matter (SOM), N, P and K in subsurface soils but less decreased the surface SOM, N and K contents, whereas phosphorus were evidently decreased at both surface and subsurface layers. In contrast, the single application of fires caused a marked increase of SOM, N, P and K contents in both surface and subsurface soils but had little impacts on soil nutrient distributions. Soils under all the three conditions showed a relative increase of soil nutrients at riverbank foot. This comparative study indicated that the different or even contrary effects of riverbank management practices on soil nutrient statuses should be carefully taken into account when assessing the ecological effects of management practices.

1. Introduction

Riverbanks are important ecotones between aquatic and terrestrial ecosystems. The nutrient inputs from riverbank soils have been considered as an important nutrient source of freshwater ecosystems [1] and play a vital role in the nutrient cycle and eutrophication of water bodies [2]. However, it is notable that riverbanks are often suffering from various management practices for human benefits. Soil nutrients have been taken as an important part when assessing the ecological effects of those varied riverbank management methods or when evaluating the efficiency of riverbank restorations [3]. Understanding the nutrient status of riverbank soils under different land management practices can provide meaningful insights for riverbank restoration and management.

Changes of soil properties are always accompanied with land use change. Studies have attempted to connect soil nutrient stocks and forms with land management [4]. Most of these studies were focused on large-scale landscapes. Studies on the influences of land management on the nutrient inputs into rivers were carried out at a watershed scale [5], but the freshwater–land transition zones were often seen as homogeneous transport passages. More recently, the importance of resources movement across ecosystems has being increasingly recognized [2]. The heterogeneity and complexity of management practices imposed on riverbanks are worthy of careful consideration.

Recent studies reveal that riverbanks represent a considerable nutrient source rather than a nutrient pool [1], and the nutrient inputs from riverbank soils into rivers were considered as an unneglectable contributor of aquatic eutrophication [6, 7]. Nutrients in riverbank soils are transferred into rivers not only by the way of surface flow but by ways of subsurface flow [8] and riverbank erosion [1, 9], and an obviously higher phosphorus concentration in suspended sediments and bed sediments than in soils were observed [7, 10]. Studies have also attempted to connect the nutrient dynamics of riverbank soils with hydrogeomorphological attributes such as geomorphic patterns [8], riverbank erosion [1, 9], sedimentation [6], and hydrologic connectivity [11]. Notably, the close relation of land use with water quality and aquatic ecological processes [10] requires a better understanding of the nutrient status in riverbank soils within a human-activity-related context. Studies have evaluated the role of riverbanks in the nutrients reduction [12] and examined the validity of riparian zone restorations [3], yet knowledge about the differences in the soil nutrients at riverbanks similar in land cover but undergoing different management practices is limited. More care should be taken on restoration practices in consideration of the potential eco-obstacles resulted from riparian restoration [6, 7]. Riverbank soils have been used as a management tool to mitigate the nutrient inputs into rivers, but the reduction efficiency is not necessarily creditable [8]. As researchers have pointed out [6], conservation and restoration of floodplain ecosystems often suffer from a deficient knowledge on the basic information in soils. Improper riverbank management methods would produce unpredictable ecological consequences such as aggravating nutrient inputs [9].

Therefore, an evaluation of the soil nutrient status under different land management practices is needed for sustainable riverbank management and restoration. We consider the nutrient concentration and distribution in riverbank soils would be a tangible variable as a response to riverbank management practices. Using the reed-covered riverbanks of middle Yangtze River under different land management practices as an example, we aimed (1) to investigate the soil organic matter, nitrogen, phosphorus and potassium contents and pH value at these riverbanks, (2) and to evaluate the nutrient status of the riverbank soils having similar vegetation cover but suffering from different management practices.

2. Study area and method

Yangtze River, the world's third largest river in the world and the largest river in China, is an important ecological and economic corridor of China. Its middle reach ranges from Yichang to Hukou (figure 1). The floodplains along the middle Yangtze River are utilized at a high level of intensity for agriculture. The sloping riverbank lands are commonly covered by reeds (*Phragmites australis*), which is a perennial and tall herbaceous plant.

Three typical locations of the reed-covered riverbanks along the middle Yangtze River (figure 1) were selected to represent three different statuses of land management: the Soils with Systemic Agricultural activities, SSA; the Soils with Single application of Fires, SSF; and the Soils with No Disturbances, SND. The SND soils were used as a control for understanding the soil nutrient status after the application of management methods. The contents and distributions of soil nutrients in the SSA and SSF soils were compared with those in the SND soils, and the significance of difference were statistically tested.

The slopes of these three riverbanks for study have a length of about 130 m at a gradient of about 4.5-5.6%. The floodplains adjacent to these riverbanks are used as farmland for planting oilseed rape or wheat. At the SSA riverbanks, in spring, local people gather the young tender shoots, which is a widely-used vegetable in China; in autumn, people reap the reed straw for paper making and burn the residual straws in situ. These sequential agricultural activities have been imposed on SSA riverbanks over six decades since the land development in 1950s, but there are no other cultivation activities such as fertilization, spraying pesticides or ploughing. The SSF riverbanks are always adjacent to farming lands, and the annually burning of reeds at autumn is just a supplementary method for the cultivations in the adjacent farm lands. The SND riverbanks are not suffering from any direct disturbances.

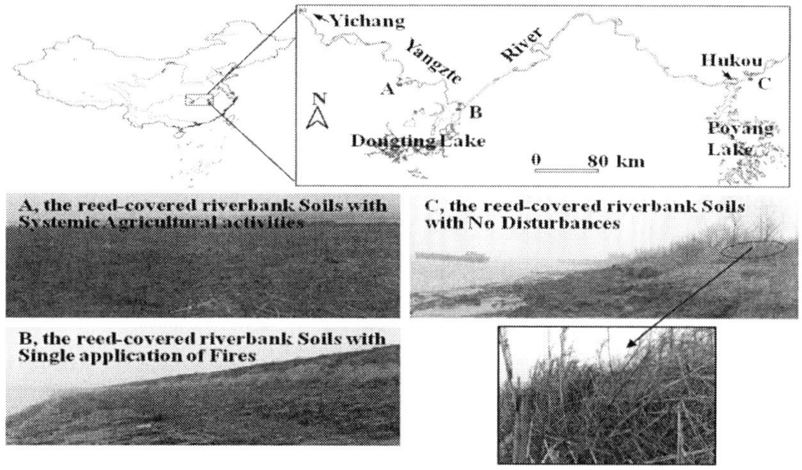

Figure 1. Study area and the three locations for soils sampling.

Soil sampling was carried out in early spring, and there had been about ten days without rains before sampling. At each location, we selected three sampling transects at 5-m intervals, and set seven sampling points at equidistant intervals along each transect, covering the gradient from the top to the bottom of the riverbank. A total of 126 soil samples at depths 0–10 cm and 10–20 cm were collected.

Analysis methods for soil pH and nutrient content followed the related Chinese national standards. Specifically, soil pH was determined by using a glass electrode in a 1 mol/L potassium chloride solution, in accordance with the NY/T 1377-2007 standard published by Ministry of Agriculture of the People's Republic of China. Determination of soil organic matter content (SOM, g/kg) was based on the standard NY/T 121.6-2006, using the potassium dichromate volumetric method. In the soil alkali-hydrolyzable nitrogen content (N) determination based on the Chinese standard LY/T 1229-1999, the air-dried soils were hydrolyzed with 1.8 mol/L sodium hydroxide solution; the alkali-hydrolyzable nitrogen was led in a boric acid receiving solution of 20 g/L; the receiving solution was titrated using hydrochloric acid; then the alkali-hydrolyzable nitrogen content in the soil sample was calculated. The soil available phosphorus content (P) was measured based on the Chinese standard HJ 704-2014 using the sodium hydrogen carbonate solution-Mo-Sb anti spectrophotometric method. Contents of available potassium (K) in soil samples were determined with ammonium acetate extraction using flame photometer, according to the Chinese standard NY/T 889-2004.

3. Results

3.1. The nutrient levels of riverbank soils under different land management practices

The average and variation of soil nutrient contents along the riverbank slope are shown in table 1. At soil surface layer (0–10 cm), the SOM contents in the SSA and SSF soils were higher than that in the SND soils; the nitrogen content was highest in the SSF soils followed by the SND soils; the SSA soils was remarkably lower in available P content than the SSF and SND soils; and the available K content was highest in the SSF soils followed by SND soils. At soil subsurface layer (10–20 cm), the SOM, N, P and K contents in the SSA soils were evidently less than those in the SSF and SND soils, and the SSF soils had the highest nutrient contents. Although all the riverbank soils had higher contents of soil nutrients in the surface layer than in the subsurface layer, the difference of soil nutrient contents between the surface and the subsurface layers was largest in the SSA soils and was smallest in the SND soils.

Table 1. The soil nutrient contents in the riverbank soils undergoing different land management practices.

Riverbank soils		SOM /g.kg^{-1}		N /mg.kg^{-1}		P /mg.kg^{-1}		K /mg.kg^{-1}		pH	
		Avg	SD	Avg	SD	Avg	SD	Avg	SD	Avg	SD
0–10cm	SSA	13.85	5.58	56.98	27.84	6.74	1.80	61.17	22.03	8.55	0.09
	SSF	14.58	3.22	77.99	15.88	19.10	8.96	99.85	54.41	8.18	0.17
	SND	11.84	4.46	61.27	17.00	21.26	7.38	63.35	31.67	8.48	0.06
10–20cm	SSA	10.51	2.49	43.56	10.21	4.19	1.35	31.15	11.54	8.62	0.07
	SSF	13.49	3.10	69.96	13.93	21.41	8.04	74.28	19.52	8.27	0.16
	SND	11.16	2.61	64.24	19.62	15.13	5.29	55.38	35.01	8.48	0.05

Note: SOM, Soil Organic Matter; N, alkali-hydrolyzable Nitrogen; P, available Phosphorus; K, available potassium; Avg, average of soil nutrient contents; SD, the standard deviation of soil nutrient contents along soil transect; SSA, the Soils with Systemic Agricultural activities; SSF, the Soils with Single application of Fires; SND, the Soils with No Disturbances.

All soils especially the SSA soils showed that the variation of soil nutrient contents along the riverbank slope was larger at the surface layer than at the subsurface layer (table 1). Specifically, surface SOM and N concentrations exhibited a larger degree of variation along the SSA soil transect than along the SSF and SND soil transects. In contrary, the variations of subsurface SOM and N contents along the SSA soil transect were smaller. The soil P and K contents at either surface or subsurface layer had the smallest variations along the SSA soil transect.

3.2. Soil nutrient distribution along the riverbank slopes under different land management practices

The distribution of soil N contents was generally similar to that of SOM contents (figure 2). Along the riverbank slope in downward direction, there was a decreasing trend of SOM and N contents especially for the SSF and SND soils. It is noteworthy that a marked increase of SOM and N contents at the riverbank foot was observed, and that the SOM and N contents at the SSA riverbank foot were obviously lower than those at the SSF and SND riverbank foots.

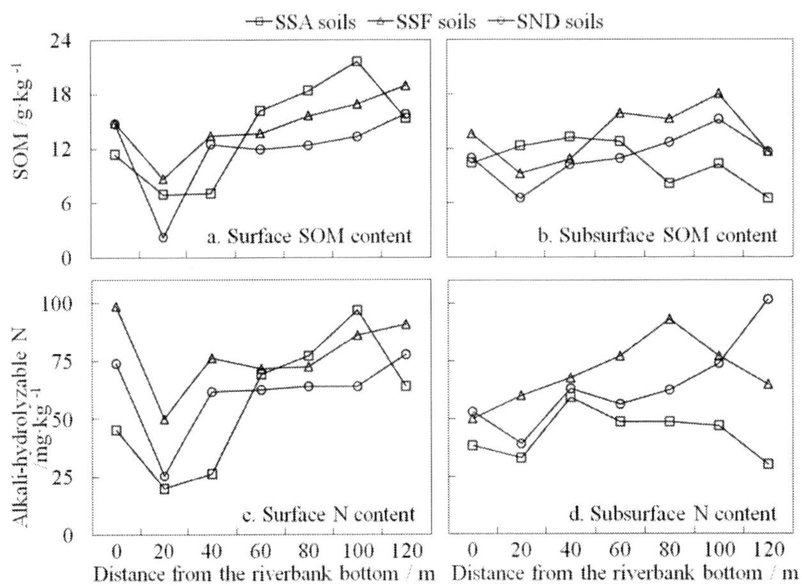

Figure 2. The distribution of SOM and N content along the riverbank slope (SSA, the Soils with Systemic Agricultural activities; SSF, the Soils with Single application of Fires; SND, the Soils with No Disturbances).

In either the SSF or SND soils, the nutrient distribution patterns at surface and subsurface layers were similar. In contrast, the surface and subsurface layers of SSA soils presented different nutrient distribution patterns, and the subsurface SOM and N contents varied little along the riverbank slope.

Both the surface and subsurface P contents in the SSA soils varied little along the riverbank slope, and the vertical difference of P contents was small (figure 3). Contrary, a more complex distribution pattern of P contents along the SSF and SND soil transects were observed. At the middle part of SSF and SND transects the soil P content decreased along the riverbank slope in downward direction. But there were an obvious increasing trend of P content at the bottom part and a decreasing trend of P content at the top part. Similar to SOM and N, the P content in all the soils obviously increased at the riverbank foot.

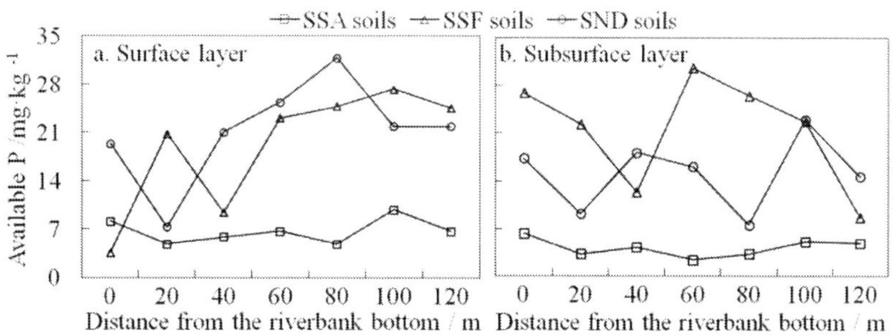

Figure 3. The distribution of soil available P content along the riverbank slope (SSA, the Soils with Systemic Agricultural activities; SSF, the Soils with Single application of Fires; SND, the Soils with No Disturbances).

Along the riverbank slope in downward direction, on the whole, the surface K contents in all the soils presented a decreasing trend except an increase of K content occurring at the riverbank foot (figure 4). In the SSF and SND soils, the distribution patterns of surface and subsurface K contents were similar. But the K contents in the SSA soils had different distribution patterns at the surface and subsurface layers, and the subsurface K contents varied little along the riverbank slope.

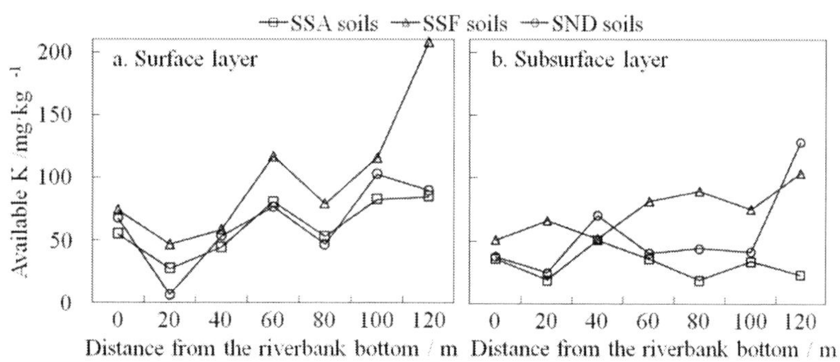

Figure 4. The distribution of soil available K content along the riverbank slope (SSA, the Soils with Systemic Agricultural activities; SSF, the Soils with Single application of Fires; SND, the Soils with No Disturbances)

3.3. The pH of riverbank soils under different land management practices

Table 1 and figure 5 show that either at surface or subsurface layer, the soil pH value was the highest in the SSA soils, lower in the SND soils, and lowest in the SSF soils. The average pH value of the surface SSA soils was 8.55 (range: 8.43–8.70), and the average pH values in the surface SND and SSF soils were respectively 8.48 (range: 8.43–8.57) and 8.18 (range: 7.92–8.42). At subsurface layer, the average pH values of the SSA, SND and SSF soils were 8.62 (range: 8.54–8.72), 8.48 (range: 8.43–8.56), and 8.27 (range: 8.06–8.53). All soils showed a higher pH value at the subsurface layer than at the surface layer (figure 5), and the vertical difference of soil pH was largest in the SSF soils. Both SSA and SND soils showed an increasing trend of soil pH value along the downward riverbank slope, while a clear decreasing trend of pH value was observed in the SSF soils.

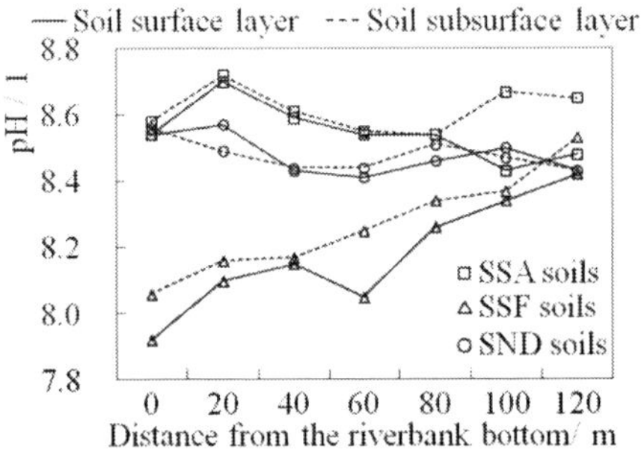

Figure 5. The distribution of soil pH along the riverbank slope (SSA, the Soils with Systemic Agricultural activities; SSF, the Soils with Single application of Fires; SND, the Soils with No Disturbances).

4. Discussions

4.1. Impacts of riverbank management practices on the nutrient levels in riverbank soils

The comparison between the SSA and SND soils indicates that the agricultural activities imposed on the reed-covered riverbank have led to the significant decrease of subsurface soil nutrient contents (table 2). The subsurface N, P and K contents in the SSA soils were significant lower than those in the SND soils ($p < 0.05$, table 2). Notably, the agricultural activities have resulted in the significant decrease of soil P contents at either surface or subsurface layer ($p < 0.05$). Impacts of agricultural activities on the SOM, N and K contents in surface soil layer were not significant.

Table 2. The nonparametric-test statistical significance of the differences in soil nutrient content and pH between the riverbank soils undergoing different land management practices.

Paired samples		SOM	N	P	K	pH
SSA-SND	0–10 cm [a]	0.310 [b]	0.612 [c]	0.018 [c, *]	0.612 [c]	0.046 [b, *]
	10–20 cm [a]	0.735 c	0.018 [c, *]	0.018 [c, *]	0.018 [c, *]	0.018 [b, *]
	0–20 cm [a]	0.612 [b]	0.063 [c]	0.018 [c, *]	0.063 [c]	0.018 [b, *]
	VD [d]	0.048 [*]	0.040 [*]	0.024 [*]	0.655	0.847
SSF-SND [e]	0–10 cm [a]	0.018 [b, *]	0.018 [b, *]	0.398 [c]	0.018 [b, *]	0.018 [c, *]
	10–20 cm [a]	0.028 [b, *]	0.310 [b]	0.063 [b]	0.128 [b]	0.034 [c, *]
	0–20 cm [a]	0.018 [b, *]	0.018 [b, *]	0.310 [b]	0.028 [b, *]	0.028 [c, *]
	VD [d]	0.565	0.405	0.798	0.949	0.046 [*]
SSA-SSF	0–10 cm [a]	0.612 [c]	0.128 [c]	0.028 [c, *]	0.018 [c, *]	0.018 [b,*]
	10–20 cm [a]	0.063 [c]	0.018 [c, *]	0.018 [c, *]	0.018 [c, *]	0.018 [b, *]
	0–20 cm [a]	0.043 [c, *]	0.018 [c, *]	0.018 [c, *]	0.018 [c, *]	0.018 [b, *]
	VD [d]	0.035 [*]	0.201	0.038 [*]	0.949	0.275

[a] The Wilcoxon two-sample test is used (2-tailed).
[b] Based on positive ranks.
[c] Based on negative ranks.
[d] *VD* is the vertical difference of soil nutrient content and pH between the surface and subsurface layers, and the Mann-Whitney two-sample test is used (2-tailed).
[e] SSA, the Soils with Systemic Agricultural activities; SSF, the Soils with Single application of Fires; SND, the Soils with No Disturbances.
'*', Significant at p=0.05 level.

In contrast, the single application of fires has remarkably increased the soil nutrient content at the reed-covered riverbanks. At the depth 0–20 cm, the SOM, N and K contents in the SSF soils were significant higher than those in the SND soils ($p < 0.05$, table 2). Specifically, the increases in surface SOM, N and K contents and subsurface SOM content were significant at $p = 0.05$ level. The subsurface SSF soils had higher contents of N, P and K than the subsurface SND soils, but the difference was not statistically significant.

Furthermore, there are evident differences in soil nutrient content between the riverbank soils suffering from fires and the soils impacted by agricultural activities. At 0–20 cm soil layer, the SSF soils had significantly higher contents of SOM, N, P and K than the SSA soils ($p < 0.05$, table 2).

For the SSA soils, the marked decrease in subsurface nutrient contents as well as the decrease in surface P content is connected with the annual reaping of reeds, which means the direct removal of soil nutrient source. The unnoticeable change in the SOM, N and K contents in surface SSA soils is an integrated result of the soil-nutrient source removal and the accelerated nutrient-transformation rate by fires.

The black carbon produced by fires may play an important role in the increase of SOM content in SSA and SSF soils. Studies have reported that black carbons can strongly influence the organic carbon dynamics in soils [12]. The effect of fires on soil nitrogen status is complex. Some studies have demonstrated that fires can decrease the nitrogen content in forest soils because of the poor quality litter produced by incomplete burning [13], but others claimed that fires can increase the soil N availability [14]. In contrast to trees, reeds are easy burned and the unburned litter is of much lower quantity, and more organic carbon and nitrogen are released. The higher SOM and N contents in the SSF soils in this study indicated that fires can accelerate the releasing of carbon and nitrogen. It has long been recognized that agricultural activities reduce the phosphorus content of soils [15], while fires are generally considered to benefit the accumulation of available phosphorus in soils [16].

On the whole, agricultural activities can lead to a remarkable decrease of nutrient contents in the subsurface riverbank soils, and have a stronger impact on mineral nutrients (P and K) than on the SOM and N. Fires can sharply increase the nutrient contents especially the mineral nutrient content in soils.

4.2. Impacts of riverbank management practices on the pH value of riverbank soils

Agricultural activities can result in a higher soil pH value, which may be associated with the burning of residual straws (figure 5 and table 2). The single application of fires, however, has resulted in a significant decrease of soil pH value ($p < 0.05$, table 2) and has created a different distribution pattern of soil pH along the riverbank slope. This may be in connection with the uneven burning of reeds across the riverbanks. Different from the prescribed burning of residual straws, burning reeds at the wild-reeds covered riverbanks is just a supplementary method for the cultivation at the adjacent farm land. And the burning severity decreases with the distance from the riverbank top. Therefore, the soil pH decreases along the riverbank slope at the downward direction. Additionally, the relatively higher content of SOM favors the lowering of soil pH.

4.3. Impacts of riverbank management practices on the soil nutrient distribution along the riverbank slope

As shown in figures 2–4, the soil nutrient distribution pattern along riverbank slope was quite similar between the soils with single application of fires (SSF) and the soils with no disturbances (SND), and it means that the single application of fires has little impacts on the soil nutrient distribution at riverbanks. Along the riverbank slope in downward direction, the nutrient content in both SSF and SND soils decrease on the whole, and the distribution patterns at surface and subsurface layers differ little. But there is a marked increase of soil nutrient content at the riverbank foot, implying that the nutrient accumulation at riverbank foot should be carefully taken into account when assessing the filtration effects of riverbanks.

In contrast, the agricultural activities showed remarkable influences on soil nutrient distribution. Specifically, the subsurface nutrient content in the SSA soils vary little along the riverbank slope, and the vertical differences of SOM and N contents were significantly larger in the SSA soils than in the SND soils ($p < 0.05$, table 2). That is, agricultural activities have resulted in the relative enrichment of soil nutrients at the soil surface layer, and variations of both surface and subsurface P contents along the SSA riverbank slope are evidently smaller. Additionally, the soil nutrient contents at the SSA riverbank foot were lower than those at the SND riverbank foot.

4.4. The potential impacts of riverbank management practices on the nutrient inputs into the river

The comparison between the SND and SSF soils has shown that fires have caused a remarkable increase of SOM, N, P and K contents in riverbank soils. Furthermore, the soil nutrient content increase at riverbank foot was larger for the SSF soils than for the SND soils. That is, the SSF soils have a higher accumulation of soil nutrients at riverbank foot. The single application of fires can accelerate the soil nutrients cycling at riverbanks and potentially increase the nutrient inputs into the river.

Although there was no significant difference in surface SOM, N and K contents between the SSA and SND soils, the subsurface N and K contents were remarkably lower in the SSA soils. Notably, both the surface and subsurface P contents were much lower in the SSA soils than in the SND soils. Furthermore, the increase of soil nutrient contents at the SSA riverbank foot was fainter. It can be seen that the agricultural activities may decrease the nutrient inputs (especially N, P and K) into the river. Additionally, the agricultural activities have enhanced the relative enrichment of soil nutrient at the surface layer and would make the surface soil be the major passage for the nutrient inputs into the river.

5. Conclusions

This study at the reed-covered riverbanks of the middle Yangtze River indicates that different management practices have significant and different impacts on the concentration and distribution of soil nutrients. The subsurface SOM, N, and K contents were decreased in the riverbank soils with systemic agriculture, but the surface SOM, N, and K contents were less decreased. The agricultural practices applied to the studied riverbanks evidently decreased the P contents at both surface and subsurface layers. Soil nutrient distribution was also affected by these agricultural practices, that is, the vertical and horizontal variations of soil nutrient contents were enlarged. In addition, soil alkalinity was increased in the soils with systemic agriculture. In contrast to the agricultural practices, the single application of fires evidently increased contents of SOM, N, P and K, and the soil pH level, but impacted little on the soil nutrient distribution. In all the studied circumstances, there is a generally decreasing trend in soil nutrient content toward the river except a relative increase at riverbank foot.

These results indicate that the agricultural activities – gathering young tender shoots, reaping mature reeds, and burning residual straws – potentially decrease the nutrient inputs (especially P and K) into the river waters, whereas fires might largely increase the nutrient inputs into the river. The agricultural practices with a moderate degree of disturbance may be a management method providing ecological and economic benefits for the reed-covered riverbanks of middle Yangtze River, but fires are not advisable.

Acknowledgments

This project is supported by National Natural Science Foundation of China (Grant No. 31460132 & 31170400) and the PhD Research Start-up Foundation of Hubei University for Nationalities (Grant No. 2014196).

References

[1] Ishee E R, Ross D S, Garvey K M, Bourgault R R and Ford C R 2015. Phosphorus characterization and contribution from eroding streambank soils of Vermont's Lake

IOP Conf. Series: Earth and Environmental Science 82 (2017) 012035 doi:10.1088/1755-1315/82/1/012035

Champlain Basin *J. Environ. Qual.* **44** 1745-53

[2] Bartels P, Cucherousset J, Steger K, Eklöv P, Tranvik L J and Hillebrand H 2012 Reciprocal subsidies between freshwater and terrestrial ecosystems structure consumer resource dynamics *Ecology M.* **93** 1173-82

[3] Heimann D C, Morris D M and Gemeinhardt T R 2015 Nutrient contributions from alluvial soils associated with the restoration of shallow water habitat in the lower Missouri River *River Res. Appl.* **31** 323-34

[4] Bahr E, Zaragocin D C and Makeschin F 2014 Soil nutrient stock dynamics and land-use management of annuals, perennials and pastures after slash-and-burn in the Southern Ecuadorian Andes *Agric. Ecosyst. Environ.* **188** 275-88

[5] Clutterbuck B and Yallop A R 2010 Land management as a factor controlling dissolved organic carbon release from upland peat soils 2: Changes in DOC productivity over four decades *Sci. Total Environ.* **408** 6179-91

[6] Klaus V H, Sintermann J, Kleinebecker T and Hölzel N 2011 Sedimentation-induced eutrophication in large river floodplains–An obstacle to restoration? *Biol. Conserv.* **144** 451-8

[7] Ekholm P and Lehtoranta J 2012 Does control of soil erosion inhibit aquatic eutrophication? *J. Environ. Manag.* **93** 140-6

[8] Liu X, Vidon P, Jacinthe P A, Fisher K and Baker M 2014 Seasonal and geomorphic controls on N and P removal in riparian zones of the US Midwest *Biogeochemistry* **119** 245-57

[9] Tufekcioglu M, Isenhart T M, Schultz R C, Bear D A, Kovar J L and Russell J R 2012 Stream bank erosion as a source of sediment and phosphorus in grazed pastures of the Rathbun Lake Watershed in southern Iowa, United States *J. Soil Water Conserv.* **67** 545-55

[10] Iglesias M L, Devesa-Rey R, Pérez-Moreira R, Díaz-Fierros F and Barral M T 2011 Phosphorus transfer across boundaries: From basin soils to river bed sediments *J. Soil Sediment.* **11** 1125-34

[11] Wolf K L, Noe G B and Ahn C 2013 Hydrologic connectivity to streams increases nitrogen and phosphorus inputs and cycling in soils of created and natural floodplain wetlands *J. Environ. Qual.* **42** 1245-55

[12] Rumpel C, Alexis M, Chabbi A, Chaplot V, Rasse D P, Valentin C and Mariotti A 2006 Black carbon contribution to soil organic matter composition in tropical sloping land under slash and burn agriculture *Geoderma* **130** 35-46

[13] Wilson C A, Mitchell R J, Boring L R and Hendricks J J 2002 Soil nitrogen dynamics in a fire-maintained forest ecosystem: Results over a 3-year burn interval *Soil Biol. Biochem.* **34** 679-89

[14] Schafer J L and Mack M C 2010 Short-term effects of fire on soil and plant nutrients in palmetto flatwoods *Plant Soil* **334** 433-47

[15] Hedley M J, Stewart J W B and Chauhan B S 1982 Changes in inorganic and organic soil phosphorus fractions induced by cultivation practices and by laboratory incubations *Soil Sci. Soc. Am. J.* **46** 970-6

[16] Turrion M B Lafuente F, Aroca M J, López O, Mulas R and Ruipérez C 2010 Characterization of soil phosphorus in a fire-affected forest Cambisol by chemical extractions and 31 P-NMR spectroscopy analysis *Sci. Total Environ.* **408** 3342-8

Performance and mechanism of ferric tannate in nitrogen removal from wastewater

R N Zhang[1], Y T Guan[1,3], J X Liu[2] and Z Liang[2]

[1]Graduate School at Shenzhen Tsinghua University, Shenzhen, Guangdong 518055, China
[2]Research Center for Eco-Environmental Sciences, Chinese Academy of Sciences, Beijing 100085, China

E-mail: guanyt@tsinghua.edu.cn

Abstract. Ferric tannate (FT) has capacities on simultaneous adsorption of NH_4^+-N and NO_2^--N. So far, no study has been reported on the mechanism of FT in the removal of NH_4^+-N and NO_2^--N. In this work, FT was prepared for nitrogen removal, and the potential of FT in treating NH_4^+-N and NO_2^--N from aqueous solution was investigated. The influence parameters e.g. substrate concentration and FT dosage on removing nitrogen from aqueous solution were optimized. The results showed that more N_2 generated when FT was added to the oxidation-reduction reaction between NH_4^+-N and NO_2^--N. The concentration of NO_2^--N was higher than that of NH_4^+-N would be conductive for the N_2 generation when adding FT. The Fourier Transform Infrared (FTIR) Spectrometer results showed that the structure of the FT (Fe-O) changed in the redox between NH_4^+-N and NO_2^--N. The X-ray photoelectron spectroscopy (XPS) results showed that Fe^{3+} and Fe^{2+} coexisted in FT.

1. Introduction

At present, more and more contaminated wastewater which contained NH_4^+-N and NO_2^--N are generated with the rapid development of industries, especially catalyst, chemical, and steel industries [1, 2]. Although nitrogen is a very important nutrient for algae, the excessive discharge of nitrogen will lead to eutrophication of receiving water bodies [3, 4]. The eutrophication of water will not only harm the environment, but also endanger the human health. And it has been reported that nitrite nitrogen causes blue baby syndrome, and is a precursor to the carcinogenic nitrosoamine [5]. Therefore, NH_4^+-N and NO_2^--N removal from wastewater has attracted more interests recently.

Traditional methods, including biological treatment [6], ion exchange [7], adsorption [8-10] and catalysis etc., [11-14] have been applied to remove NH_4^+-N and NO_2^--N from wastewater. Currently, long periods on bacteria acclimation for biological treatment and low total nitrogen (TN) removal when the influent biodegradable COD to TN ratio is low, which presents a problem in practical use [15]. Compared with the biological treatment, physical-chemical treatment has become a promising technology due to its high safety, high nitrogen removal, with practicable application [16]. Through the adsorbent and ion exchange resin can quickly remove nitrogen from wastewater, but they must be regenerated after adsorption nitrogen saturation, and the regenerated solution needs treatment and disposal. If the absorbent can be regenerated in situ, the adsorption will be an attractive alternative nitrogen removal technology.

Theoretically, the redox can occur between NO_2^- and NH_4^+ according to their standard redox potential [17], which its products are N_2 and H_2O. To enhance NO_2^- and NH_4^+ removal from wastewater simultaneously, a new material which had the ability of both NO_2^- and NH_4^+ adsorption is needed. Previous a patent pointed that the colloial particles of $Fe(OH)_3$, Ferric tannate (FT) and Fe_2O_3 could be used as catalysts during the process of NH_4^+-N reacted with NO_2^--N at room temperature [18]. However, to date, no research has been focused on factors of catalytic denitrification performance and mechanism on FT to remove NH_4^+-N and NO_2^--N.

Thus, the potential catalytic denitrification of FT to remove NH_4^+-N and NO_2^--N from aqueous solution simultaneously is investigated. The influence of various parameters (substrate concentration and FT dosage) on capacity and performance of the FT to remove nitrogen efficiently from aqueous solution were optimized. And before and after nitrogen removal, the FT were characterized through Fourier Transform Infrared (FTIR) Spectrometer and X-ray photoelectron spectroscopy (XPS).

2. Materials and methods

2.1. FT preparation

In this study, FT was prepared from the reaction of tannic acid and ferric chloride at ratio of 1:20 according to the reference [19].

2.2. Experiment procedue

The solution used in the test was mixture of ammonium chloride (≥99.5%, Sinopharm Chemical reagent Co., China) and sodium nitrite (≥99.0%, Sinopharm Chemical reagent Co., China). Ammonium (1000 mg/L) and nitrite (1000 mg/L) stock solution were prepared by dissolved ammonium chloride and sodium nitrite in deionized water, respectively. In these experiments, the concentrations of N in total were varied from 0 to 500 mg/L.

The adsorption-catalytic test was carried out in a 250 mL glass flask which had a working volume of 100 mL with the ratio of FT/ NH_4^+-N and FT/ NO_2^--N was 100 mg/mg and 50 mg/mg, respectively. There were two holes on the plug. The diameter of each hole was 50-70 mm. Two glass tubes connected with the flask through the holes for collecting samples and purging air. The gas of He was pumped into the flask to discharge all the air. Then the sealed flask was shocked at 200 rpm for 48 hours under ambient temperature, take the gas sample and the water sample from the flask every 2 hour. An additional flask that without ultrasound treatment was used as a control.

After filtration (microfiltration membrane pore size 0.45 μm, Beijing river the integrity of membrane technology development center, China), the concentrations of NH_4^+-N and NO_2^--N were determined.

2.3. Analysis and calculation methods

The concentrations of NH_4^+-N were measured using Nessler'S reagents spectrophotometer methods according to standard methods [20]. The concentrations of NO_2^--N were measured using hydrochloride naphthodiamide spectrophotometer methods on the basis of standard methods [20].

The content of N_2 was determined using the Agilent 6890N Gas Chromatograph equipped with TCD detector, and the gas of He was used as a carrier. During the procedure, temperature of the oven and the detector was 100℃ and 250℃, respectively.

The functional group of the tannin acid and FT were measured using Fourier Transform Infrared Spectrometer (FTIR Spectrometer, Tensor 27, Germany) over the range of 400～4000 cm^{-1} region [21]. In the test, the dried tannin acid and FTs were mixed potassium bromide (KBr) in the mass ratio of 1:20, respectively.

The binding energy of Fe and O on the surface of FT were obtained by XPS analysis. The sample of FT was packed carefully on XPS sampling template under the anaerobic condition to avoid surface oxidation. The XPS analysis was carried out using a Mutilab 2000 (ESCALAB250Xi, PANalytical co, Holland) with an Mg Kα X-ray (1253.6 eV) having a source power of 200 W. In the procedure of XPS

IOP Conf. Series: Earth and Environmental Science **82** (2017) 012036 doi:10.1088/1755-1315/82/1/012036

measurements, the sample chamber was evacuated (the base pressure was 3×10^{-9} mbar) before changing the gas atmosphere. The flood gun for charge compensation was used and calibrated the binding energy on the carbon 1s peak at 284.6 eV.

Fe content of FT was determined by spectrophotometric method using a Reflectoquant RQflex plus 10 (Merck chemical (Shanghai) Co., Ltd, China). The measuring principle was as follows. In acid solution (pH 2.5-9), a red-violet complex formed when Fe^{2+} ions react with a triazine derivative that is determined reflectometrically. Samples must be reduced by ascorbic acid before Fe^{3+} ions and total iron (sum of Fe^{2+} and Fe^{3+}) can be measured. The procedure of the sample pretreatment was as follows. At first, 0.01 g of the dried FT was dissolved in 10 mL HCl (10%), the solution was divided into two equal parts. 5 mL solution was measured by a Reflectoquant RQflex plus 10 to determine the content of Fe^{2+}. In order to measure the content of total Fe, the other 5 mL solution was dissolved in 5 mL ascorbic acid (100 mg/L).

The content of Fe^{2+} was calculated according to equation (1).

$$Fe^{3+}\ (mg/L) = Total\ Fe - Fe^{2+} \tag{1}$$

Where Fe^{2+} was measured by measuring the $Fe(phen)_3^{2+}$ (phen=1,10-phenanthroline) complex formed during acid condition , total Fe was measured by converting any Fe^{3+} of FT to Fe^{2+} using ascorbic acid as a reducer.

The adsorption performances were evaluated according to the amount of NH_4^+-N and NO_2^--N absorbed on per mass of FT (q_e in mg-N/g), which could be calculated according to equation (2).

$$q_e = (C_0 - C_e)/m \tag{2}$$

Where q_e was the equilibrium contaminant adsorbed by FT (mg/g). Co was the initial concentrations (mg N/L). Ce was the equilibrium concentration (mg N/L). V was solution volume (L) and m was the mass of FT used (g).

3. Results and discussion

3.1. The adsorption-catalytic denitrogenation performance on FT

The long reaction time (48 h) for FT in the study was chose in the study, which might be due to the reaction time for NH_4^+-N and NO_2^--N was unknown when FT added. To be on the safe side, the long reaction time (48 h) was chose for FT in the initial study.

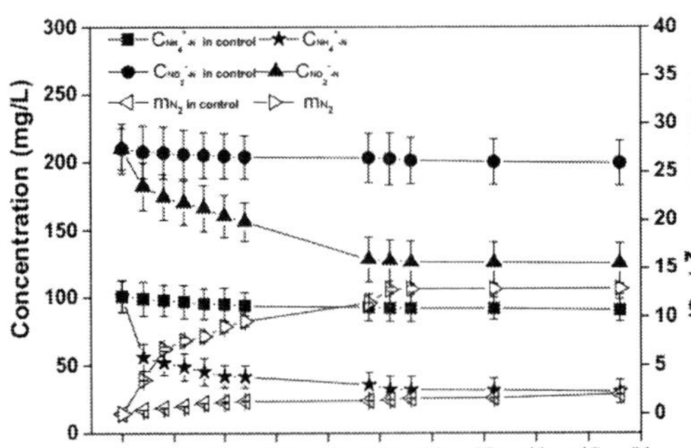

Figure 1. The removal of NH_4^+-N and NO_2^--N and N_2 production with the addition of FT.

In the present study, dosage of ferric tanate, concentration of NH_4^+-N and NO_2^--N was 10.0 g/L, 100 mg/L and 200 mg/L, respectively. The results showed that N_2 increased when the concentration of

NH_4^+-N and NO_2^--N decreased as time goes on (figure 1). In the procedure, NH_4^+-N reacted with NO_2^--N as the equation (3).

$$NH_4^+ + NO_2^- \longrightarrow N_2 + H_2O \quad (3)$$

After 24 hours, the concentration of NH_4^+-N and NO_2^--N reduced to 30-35 mg/L and 125-128 mg/L, respectively. N_2 as product increased to 11.5-12.8 mg. And the removal rate of NH_4^+-N and NO_2^--N was 64.6% and 39.0%. Meanwhile, the removal of NH_4^+-N and NO_2^--N was 8.14% and 3.29%, respectively. And only 1.34 g N_2 generated in the control group. The results showed that more N_2 generated when FT was added to the oxidation-reduction reaction between NH_4^+-N and NO_2^--N. Furthermore, according the result analysis, we confirmed the reaction time (10 h) for FT in the following study.

3.2. Adsorption of NH_4^+-N or NO_2^--N on the FT

In order to investigate the nitrogen removal mechanism of FT, sequencing batch experiments for adsorbing NH_4^+-N and NO_2^--N were conducted separately.

The adsorption capacity could describe well the distribution between liquid and solid phases at equilibrium [22, 23]. So the adsorption capacity of NH_4^+-N and NO_2^--N on FT was measured in this work (figure 2). The maximum adsorption capacity of NH_4^+-N and NO_2^--N were 13.4 mg/g and 10.0 mg/g (figure 2) under ambient temperature, respectively. And in the same experimental condition, the adsorption capacity of NH_4^+-N on FT was higher than NO_2^--N. Thus, the results suggested that the FT showed better selectivity adsorption towards NH_4^+-N compared with NO_2^--N. In addition, there was no N_2 generated in the processor. Hence NH_4^+-N and NO_2^--N were removed by the adsorption of FT.

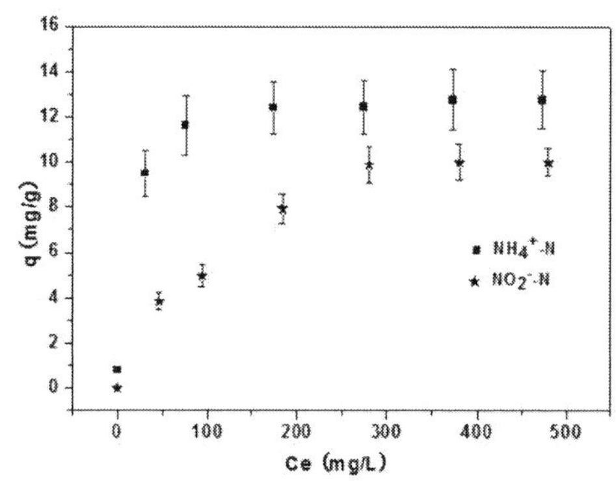

Figure 2. Adsorption isotherms of NH_4^+-N and NO_2^--N on FT.

There were ferric ions and negative oxygen ions existed in FT, thus which can enhance NH_4^+-N and NO_2^--N adsorption through the electrostatic interaction. The other important point is FT has the capacity of speeding this reaction up under ambient temperature. So the FT has the ability of regenerated in-situ when it used in NH_4^+-N and NO_2^--N coexisted wastewater.

Therefore, in view of the cheaper cost and the simpler operation, FT is a good choice. More N_2 generated when FT was added to the reaction between NH_4^+-N and NO_2^--N. Thus, the FT may be used as a promising material to remove nitrogen from wastewater in the future.

3.3. Factors influencing the reaction between NH_4^+-N and NO_2^--N

The initial substrate concentration and the dosage of FT have important influence on the reaction of NH_4^+-N and NO_2^--N. Therefore, it was necessary to investigate these significant factors. The experimental condition was listed as table 1, and the result of different initial substrate concentrations and different dosages of FT were illustrated as figures 3-6, respectively.

Table 1. The relative data in the experiments.

Test	1	2	3	4	5	6	7	8
NH_4^+-N (mg/L)	25	50	100	100	100	200	100	400
NO_2^--N (mg/L)	100	100	25	100	200	100	400	100
NH_4^+-N/NO_2^--N	1:4	1:2	4:1	1:1	1:2	2:1	1:4	4:1
FT (g/L)	0.5	0.5	0.5	0.5	0.5	0.5	0.5	0.5
V (mL)	200	200	200	200	200	200	200	200

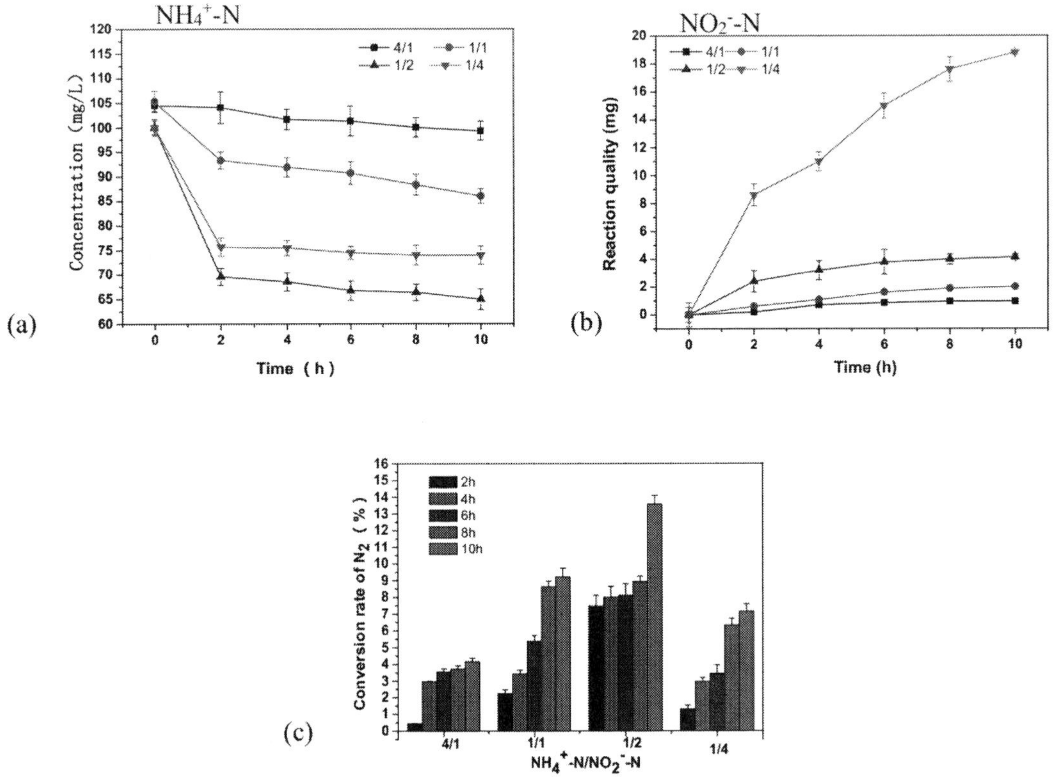

Figure 3. The variation of NH_4^+-N (a) and NO_2^--N (b) and conversion rate of N_2 (c) with the addition of FT.

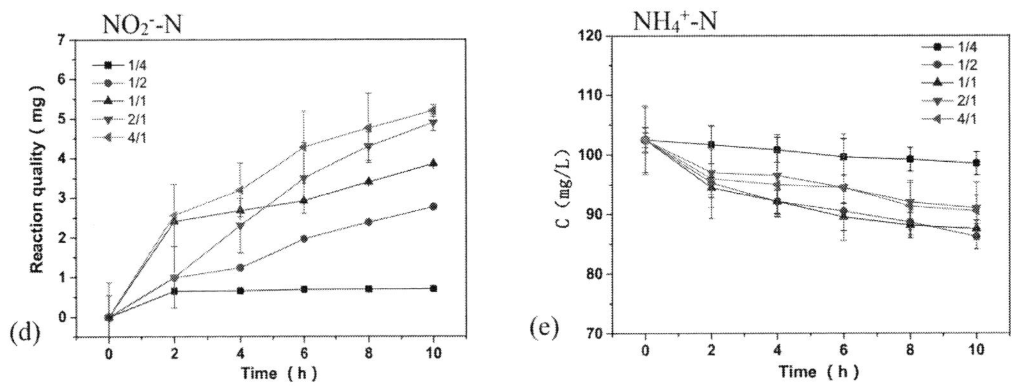

IOP Conf. Series: Earth and Environmental Science **82** (2017) 012036 doi:10.1088/1755-1315/82/1/012036

(f)
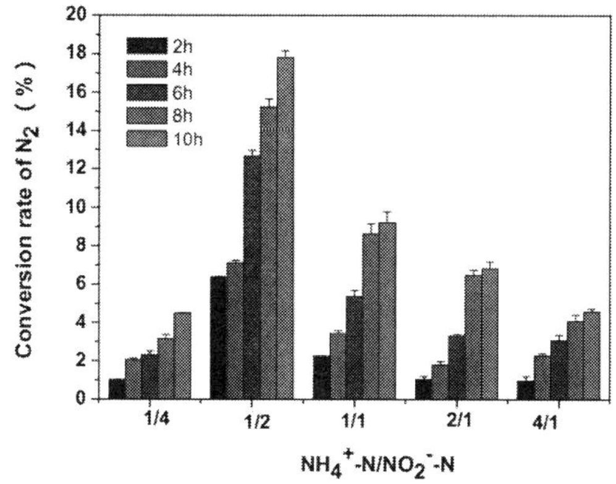

Figure 4. The variation of NH_4^+-N (d) and NO_2^--N (e) concentration and conversion rate of N_2 (f) with the addition of FT.

(g)
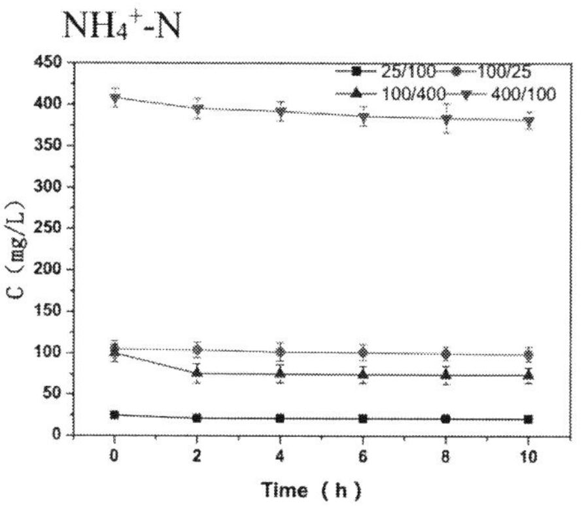

(h)
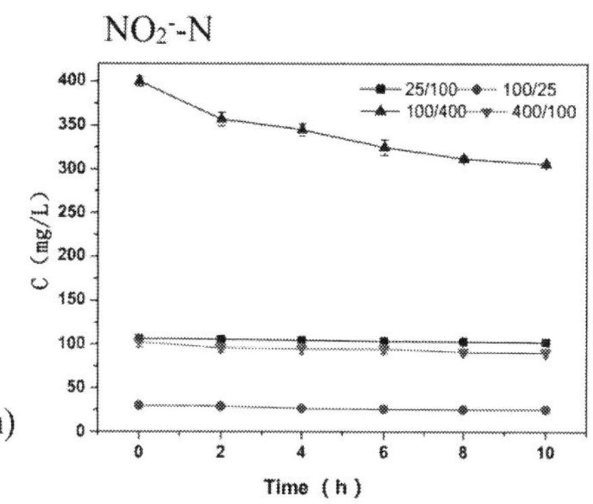

(i)
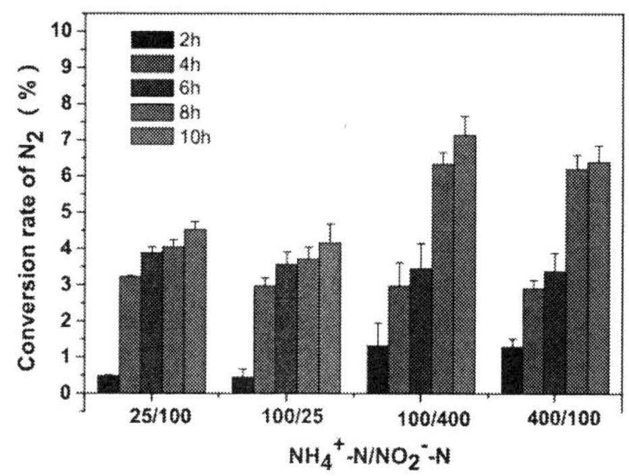

Figure 5. The variation of NH_4^+-N (g) and NO_2^--N (h) concentration and conversion rate of N_2 (i) with the addition of FT.

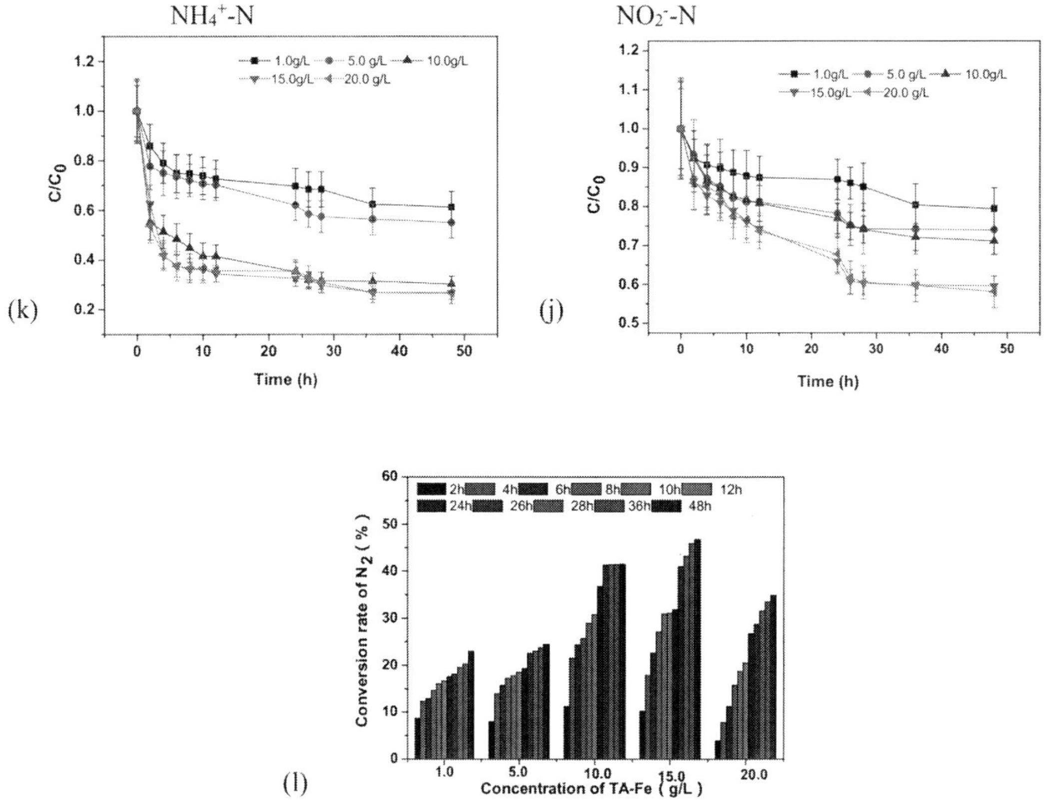

Figure 6. The variation of NH_4^+-N (j) and NO_2^--N (k) removal and conversion rate of N_2 (l) with the addition of FT.

3.3.1. Effect of the substrate concentration.

- The initial NH_4^+-N concentration was 100 mg/l, effect of the NO_2^--N concentration on the reaction.

The results showed that the concentration of H_4^+-N decreased to 99.3-64.3 mg/L and the removal quality of NO_2^--N from 0.98 mg to 18.8 mg with NH_4^+-N to NO_2^--N ratio decreased after 10 hours (figure 3). In the reaction, the production of N_2 was from 1.31 mg to 7.34 mg. And the most amount of N_2 (8.42 mg) generated at 1:2. The result indicated that the highest conversion rate of N_2 was achieved at 1:2 when the initial NH_4^+-N concentration was 100 mg/L.

- The initial NO_2^--N concentration was 100 mg/l, effect of the NH_4^+-N concentration on the reaction.

The results showed that the concentration of NO_2^--N decreased to 98.5-87.5 mg/L and the removal quality of NH_4^+-N from 0.70 mg to 5.20 mg with NH_4^+-N to NO_2^--N ratio increased after 10 hours (figure 4). In the procedure, the production of N_2 was from 1.00 mg to 5.43 mg. And the most amount of N_2 (5.43 mg) generated at 1:2. The result indicated that the highest conversion rate of N_2 was attained at 1:2 when the initial NO_2^--N concentration was 100 mg/L. While a closer look showed that the highest conversion rate of N_2 was obtained at NH_4^+-N to NO_2^--N ratio 1:2 when the NO_2^--N concentration kept constant.

- The initial amount of TN kept constant, effect of the NH_4^+-N to NO_2^--N ratio on the reaction.

The initial content of TN in the group one (Test 1 and Test 3), group two (Test 7 and Test 8) were identical, respectively. After 10 hours reaction, the production of N_2 was 1.21 mg when the NH_4^+-N to NO_2^--N ratio of 25:100, which was higher than the ratio of 100:25 (1.01 mg) when the initial TN concentration was 125 mg/L. The production of N_2 was 7.14 mg when the NH_4^+-N to NO_2^--N ratio of 100:400, which was higher than the ratio of 400:100 (4.66 mg) (figure 5) When the initial TN concentration was 500 mg/L. The results indicated that concentration of NO_2^--N was higher than NH_4^+-N would be conductive to the N_2 generation with FT addition.

Results of the substrate concentration indicated that the concentration of NO_2^--N was higher than NH_4^+-N would be conductive to the N_2 generation with FT addition. And in the present study, the optimum NH_4^+-N to NO_2^--N ratio was 1:2, the removal rate of NH_4^+-N and NO_2^--N was 64.3% and 7.0%, respectively. Meanwhile, the N_2 conversion ratio was 14.0%.

3.3.2. FT dosage. The addition of FT accelerated the reaction of NH_4^+-N and NO_2^--N, and more N_2 generated. The effect of the dosage of FT on the redox between NH_4^+-N and NO_2^--N was investigated.

In the study, FT dosages ranging from 0 to 20 g/L was optimized under optimum water conditions of NH_4^+-N 100 mg/L and NO_2^--N 200 mg/L. The results showed that When the FT dosage was below 15 g/L, the production of N_2 increased as the increased addition of FT (figure 6). When the FT dosage continue increased (15-20 g/L), the production of N_2 and the removal of NH_4^+-N and NO_2^--N no longer increased. Therefore, FT dosage was 15 g/L preferred for the further studies. And when the concentration of FT was 15 g/L, the highest removal rate of 68.0% for NH_4^+-N and 32.7% for NO_2^--N could be reached. Meanwhile, the production of N_2 was 12.8 mg after 26 hours reaction.

3.4. The surface chemistry structure variation of FT

After the adsorption and adsorption-catalytic nitrogen removal reactions in the present study, FT were collected, and dried in a freezer dryer at minus 50℃. Then the structure analysis of FT was measured using FTIR spectroscopy. FTIR analysis permitted spectrophotometric observation of the material surface in the range 400-4000 cm^{-1}. The characteristic FTIR adsorption peaks could indicate some surface functional groups exist on the FT.

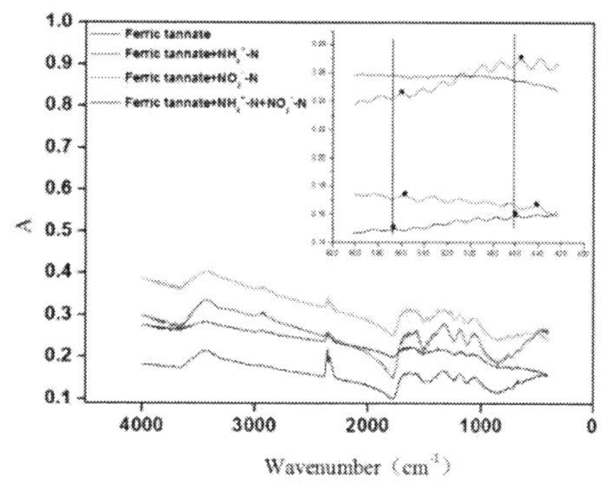

Figure 7. FTIR spectra of FT before and after reaction.

The smooth and broad adsorption peak at around 3500 cm^{-1} appeared both the fresh and the used FT (figure 7), which was the characteristic FTIR adsorption peaks of adsorbed water or hydroxyl groups [24]. The broad Fe-O bands at around 567 cm^{-1} and 459 cm^{-1} appeared in the fresh FT, and the bands shifted to lower wavenumbes after either NH_4^+-N (559 cm-1 and 455 cm-1) or NO_2^--N (557 cm-1 and 440 cm-1) adsorption. After experiments of adsorption-catalysis, the bands disappeared. The results implied that the structure of the FT (Fe-O) changed after it was added to the system where NH_4^+-N and NO_2^--N coexisted.

3.5. The binding energy of O and Fe analysis in FT

The effects of chemical changes on the photoelectron peak shape using XPS have been of particular interest [25]. The technology of XPS was used to determine the chemical composition and the valence state on the surface of substances [26]. In this study, the electron binding energy of O and Fe on the surface of FT were measured.

The O1s spectra were given as figure 8. In the fresh FT, the electron binding energy of O1s was 531.43 eV. After the FT was added to the system with NH_4^+-N or NO_2^--N, or NH_4^+-N and NO_2^--N coexisted, the electron binding energy of O1s was changed to 531.28 eV, 531.33 eV and 531.38 eV. The Fe 2p spectra were given as figure 8. In the fresh FT, the electron binding energy of Fe 2p was 710.93 eV. After the FT was added to the system with NH_4^+-N or NO_2^--N, or NH_4^+-N and NO_2^--N coexisted, the electron binding energy of Fe 2p was changed to 710.93 eV, 710.88 eV eV and 711.53 eV.

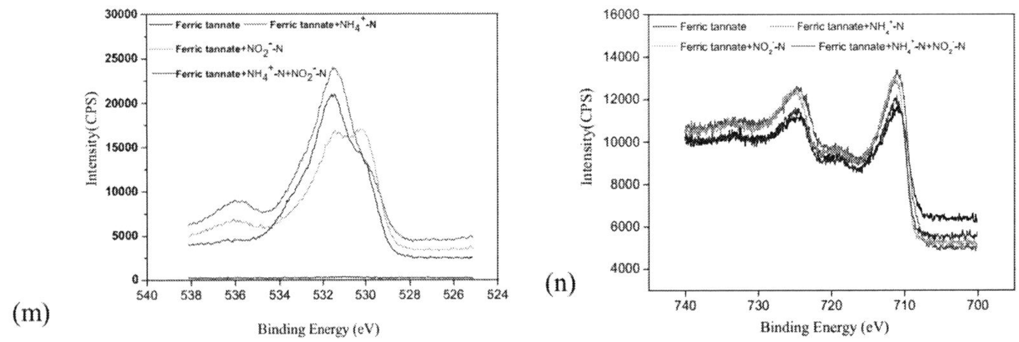

Figure 8. XPS spectra of FT surface (m) O1s (n) Fe2p before and after reaction.

There are four primary factors that affected the element's electron binding energy as follows. (1) The spatial structure of the substance (2) The electronegativity of the substance (3) The oxidative state of the substance's central ion (4) chemical environment of the element [27]. In this work, the electron binding energy of O1s on FT decreased. It might be the chemical environment of O1s had changed in the reaction, which led to the electron binding energy of O1s on the FT varied [28]. It was therefore suggested that the –O Fe was the core of the FT with the nitrogen removed function. And the difference in Fe 2p binding energy might be attributed to the oxidative state of Fe changed, or chemical environment of Fe changed [29]. The system with NO_2^--N existed, NO_2^--N would be combined with Fe ions of the FT through the electrostatic interaction. So the electronic cloud density of Fe ions on the FT increased, which could be observed from the binding energy of Fe ions decreased.

Furthermore, after and before adsorption-catalytic, the binding energy of Fe ions of FT were around 710 eV and 725 eV. Combined with the literature investigation and the standard XPS analysis method, it was found that Fe existed in FT as a form of Fe_3O_4 [30]. It might be concluded that the Fe^{3+} and Fe^{2+} coexisted in the FT, which could be determined in the next work.

3.6. The determination of the content of Fe in FT

FT that prepared in the study was difficult to dissolve in water. In the interest of analyzing the mechanism of the nitrogen removal using FT, before and after the reaction, the quantity change of FT was measured using the content of Fe (Fe^{3+} and Fe^{2+}) in the substance with phenanthroline spectrophotometric method. The total amount of Fe in 1.5 g fresh FT was 0.1365 g, including a large quantity of Fe^{3+} (92.3%) and a small quantity of Fe^{2+} (7.70%) (figure 9).The result was consistent to the prediction of the XPS characterization with the Fe^{3+} and Fe^{2+} coexisted in the FT. And after the FT was added to the system which existed NH_4^+-N or NO_2^--N, or NH_4^+-N and NO_2^--N coexisted, the percentage composition of Fe^{3+} and Fe^{2+} was varied, while the total amount of Fe (Fe^{3+} and Fe^{2+}) remained constant.

The percentage composition of Fe^{3+} in FT decreased quickly, and the percentage composition of Fe^{2+} increased obviously simultaneously in the first 2 hours in the system of NH_4^+-N and NO_2^--N.

Later, the quantity of Fe^{3+} and Fe^{2+} in FT transformed each other to a certain degree with time prolonging (figure 9). While in the system with NH_4^+-N or NO_2^--N existed dependently, the percentage composition of Fe3+ in FT increased, and the percentage composition of Fe^{2+} decreased simultaneously with the time prolonging. And Fe^{3+} and Fe^{2+} in FT transformed each other without observed in the procedure. Therefore, only the redox reaction of NH_4^+-N and NO_2^--N was coupled with the redox reaction between Fe^{3+} and Fe^{2+}. In the procedure, FT has the ability to improve the redox between NH_4^+-N and NO_2^--N. And Fe ions which contained in the FT represented as the electron transfer.

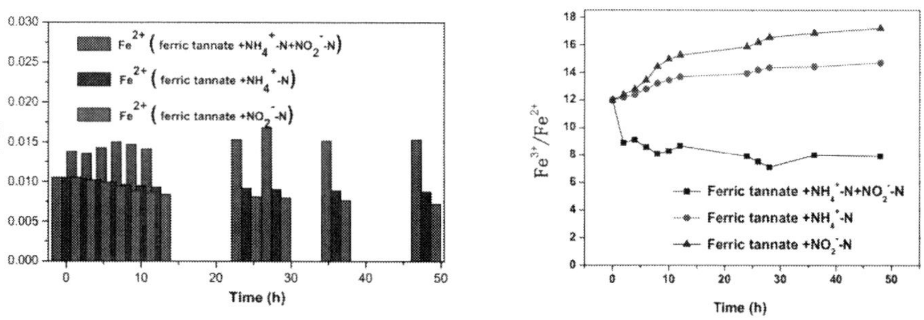

Figure 9. The content of Fe^{3+} and Fe^{2+} of FT before and after reaction.

4. Conclusions

More N_2 generated when FT was added to the oxidation-reduction reaction between NH_4^+-N and NO_2^--N. Furthermore, the FT has good performance for ammonium and nitrite removal with up to 68.0% and 32.7% removal efficiency, respectively. The concentration of NO_2^--N was higher than NH_4^+-N would be conductive to the N_2 generation with FT addition. When the FT concentration was 15 g/L, the production of N_2 was 12.8 mg, meanwhile the removal of NH_4^+-N and NO_2^--N was 68.0% and 32.7%, respectively.

The FTIR results implied that the structure of the FT (Fe-O) changed after it be added to the system where NH_4^+-N and NO_2^--N coexisted. XPS analysis found that Fe existed in FT as a form of Fe3O4, which meant the Fe^{3+} and Fe^{2+} coexisted in it. Results of the phenanthroline spectrophotometric method showed that in the system with NH_4^+-N and NO_2^--N existed dependently, without observed Fe^{3+} and Fe^{2+} in FT transformed each other. And only the redox between NH_4^+-N and NO_2^--N was coupled with the redox reaction between Fe^{3+} and Fe^{2+}. In the procedure, FT has the ability to improve the redox reaction of NH_4^+-N and NO_2^--N, thus more N_2 generated. And Fe ions of the FT represented as the electron transfer.

Acknowledgments

This research was supported by the Major Science and Technology Program for Water Pollution Control and Treatment, National Water Grant (No. 2012ZX07301-001), Shenzhen Science and Innovation Commission (No. JSGG20160428181710653).

References

[1] Zhao Y X, Yang Y N, Yang S J and Zhang Z Y 2013 Adsorption of high ammonium nitrogen from wastewater using a novel ceramic adsorbent and the evaluation of the ammonium-adsorbed-ceramic as fertilizer *J. Colloid. Interf. Sci.* **393** 264-70

[2] Li M, Feng C, Yang S and Sugiura N 2010 Treatment of nitrate contaminated water using an electrochemical method *Bioresource Technol.* **101** 6553-7

[3] Bassin J P, Pronk M, Kraan R, Kleerebezem R and Loosdrecht M C 2011 Ammonium adsorption in aerobic granular sludge, activated sludge and anammox granules *Water Res.* **45** 5257-65

[4] Widiastuti N, Wu H W, Ang H M and Zhang D 2011 Removal of ammonium from greywater using natural zeolite *Desalination* **277** 15-23
[5] Jellali S, Wahab M A, Riahi K and Jedidi N 2011 Biosorption characteristics of ammonium from aqueous solutions onto Posidonia oceanica (L.) fibers *Desalination* **270** 40-9
[6] Carrera J, Baeza J A, Vicent T and Lafuente J 2003 Biological nitrogen removal of high-strength ammonium industrial wastewater with two-sludge system *Water Res.* **37** 4211-21
[7] Thornton A, Pearce P and Parsons S A 2007 Ammonium removal from solution using ion exchange on to MesoLite, an equilibrium study *J. Hazard. Mater.* **147** 883-9
[8] Babou Kammoe R B and Hamoudi S 2014 Investigation of ammonium ion removal from aqueous solutions using arene- and propylsulfonic acid functionalized mesoporous silica adsorbents *J. Environ. Qual.* **43** 1032-42
[9] Zheng H, Han L, Ma H, Zheng Y, Zhang H, Liu D and Liang S 2008 Adsorption characteristics of ammonium ion by zeolite 13X *J. Hazard. Mater.* **158** 577-84
[10] Ye Z, Wang J, Sun L, Zhang D and Zhang H 2014 Removal of ammonium from municipal landfill leachate using natural zeolites *Environ. Technol.* **36** 2919-23
[11] Sune D E, Barbara L M and Leon L 2008 In situ attenuated reflection infrared (ATR-IR) study of the adsorption of NO^{2-}, NH_2OH, and NH^{4+} on Pd/Al_2O_3 and Pt/Al_2O_3 *Langmuir* **2** 869-79
[12] El-Kabbany F, Badr Y and Tosson M 2006 Tosson Infrared study of the high-temperature phase transition phase transition III- I of Na_2SO_4 *Phys. Status. Solidi (a)* **63** 699-704
[13] Reddy K J and Lin J P 2000 Nitrate removal from groundwater using catalytic reduction *Water Res.* **34** 995-1001
[14] Ou H H, Hoffmann M R, Liao C H, Hong J H and Lo S L 2010 Photocatalytic oxidation of aqueous ammonia over platinized microwave-assisted titanate nanotubes *Appl. Catal. B: Environ.* **99** 74-80
[15] Dey A 2010 Modeling simultaneous nitrification–denitrification process in an activated sludge continuous flow stirred-tank reactor: system optimization and sensitivity analysis *Environ. Eng. Sci.* **27** 757-65
[16] Terry P A 2009 Removal of nitrates and phosphates by ion exchange with hydrotalcite *Environ. Eng. Sci.* **26** 691-6
[17] Fanning J C 2000 The chemical reduction of nitrate in aqueous solution *Coordin Chem Rev* **199** 159-79
[18] Yuasa Y and Kakai J P 1987 197 196[87 197196] 31 Aug.
[19] Zhang R N, Li L and Liu J X 2015 Synthesis and characterization of FT as anovel porous adsorptive-catalyst for nitrogen removal from wastewater *RSC Adv.* **5** 40785-91
[20] Mccrady M H 1966 *Am. J. Public Health and the Nations Health* **56** 684-5
[21] Liu W G, Xu Y M, Zhou W, Zhang X F, Cheng X L, Zhao H, Gao S and Huo L H 2017 A facile synthesis of hierarchically porous TiO microspheres with carbonaceous species for visible-light photocatalysis *J. Mater. Sci. Technol.* **33** 39-46
[22] Awwad A M and Salem N M 2014 Kinetics and thermodynamics of Cd(II) biosorption onto loquat (Eriobotrya japonica) leaves *J. Saudi Chem. Soc.* **18** 486-93
[23] Farhan A M, Al-Dujaili A H and Awward A M 2013 Equilibrium and kinetic studies of cadmium (II) and lead (II) ions biosorption onto ficus carcia leaves *Int. J. Ind. Chem.* **4** 1-8
[24] Demirbas E, Dizge N, Sulak M T and Kobya M 2009 Adsorption kinetics and equilibrium of copper from aqueous solutions using hazelnut shell activated carbon *Chem. Eng. J.* **148** 480-7
[25] Grosvenor A P, Kobe B A, Biesinger M C and Mcintyre N S 2004 Investigation of multiplet splitting of Fe 2p XPS spectra and bonding in iron compounds *Surf. Interface Anal.* **36** 1564-74
[26] Cerrato J M, Jr H M, Knocke W R, Dietrich A M and Cromer T F 2010 Use of XPS to identify the oxidation state of Mn in solid surfaces of filtration media oxide samples from drinking water treatment plants *Environ. Sci. Technol.* **44** 5881-6

[27] Li C, Hamasaki M, Manaka H and Obara K 2014 Charge transfer mechanism and spatial density correlation of electronic states of excited Zinc (3d9) film *Open J Phys Chem* **4** 44-51

[28] Dimitrov V and Komatsu T 2003 Correlation of optical basicity and O1s chemical shift in XPS spectra of oxide glasses *Phy Chem Glasses* **44** 357-64

[29] Biesinger M C, Payne B P, Grosvenor A P, Lau L W M, Gerson A R, Smart RC *et al* 2011 Resolving surface chemical states in XPS analysis of first row transition metals, oxides and hydroxides: Cr, Mn, Fe, Co and Ni *Appl Surf Sci* **257** 2717-30

[30] Reddya G K, Boolchandb P and Smirniotis P G 2011 Sulfur tolerant metal doped Fe/Ce catalysts for high temperature WGS reaction at low steam to CO ratios – XPS and Mössbauer spectroscopic study *J Catal* **282** 258-69

Cd pollution and ecological risk assessment for mining activity zone in Karst Area

B Yang[1], J L He[1], X M Wen[1] and H Tan[1,2]

[1]Guizhou Academy of Testing and Analysis, Guiyang, China

E-mail: 357688237@qq.com

Abstract. The monitored soil samples were collected from farmland in the area with mining activity in Karst area in Liupanshui. In this article, moss bag technology and TSP were used simultaneously for Cd transportation and deposition in the study area. Geostatistics and GIS were then used for the spatial distribution of Cd in the soil. Afterwards, Cd pollution to the soil environment and human health was studied by using the geo-accumulation index and potential ecological risk index methods. The results indicated that atmospheric deposition is the major route of Cd pollution. A moderate to strong pollution of Cd in the area and the degree of potential ecological risk was in a high level in the study area. Furthermore, Cd pollution in Liupanshui may originate from mining activity and atmospheric deposition.

1. Introduction

Cadmium (Cd) is one of air pollutant which is harmful to human beings [1], and it plays a key role in the agricultural soil quality which has been a hot topic in the field of environmental research in recent years [2].

One of the important environment issues today is the pollution due to atmospheric deposition. Atmosphere works as a part of environment and has a significant effect on the ecosystem. Atmospheric deposition is a kind of circulation from earth to atmosphere through the earth's surface. Atmospheric deposition helps the transportation of trace metal in environment, while it's the main external origin of bio-available heavy metal in the environment [3]. There are reports on atmospheric deposition which is the main route for exchange of substances between the land surface and atmosphere and the atmospheric Cd content in urban area and rural areas and industrial district. Nriagu [4] reported atmospheric deposition made crucial contribution to Cd content in soil as external source. Atmospheric deposition has been one of global environmental problems and has led to the effect of agricultural soil interface with potential ecological risk.

A series of ecological environmental problems (e.g. water and soil loss) in Guizhou Province due to Karst environment is fragile and vulnerable [5]. Mining activities in Karst area as rich mineral resource regions cause heavy metal emission from atmosphere to the soil [6].

Guizhou Province has abundant lead and zinc resources. During the process of zinc or lead smelting, Cd is obtained as a by-product and can be emitted into atmosphere [7, 8]. The variation of metallic element ratio and equilibrium concentration in soil pose is a potential ecological risk to the local field. The study is focused on determining the atmospheric deposition of Cd and concentration level of agricultural soil. The potential ecological risk of surrounding agricultural soil is assessed as well.

2. Material and methods

2.1. Sampling collection

2.1.1. Study site. In the study, the effect of gradient atmospheric sediment of Cd to soil was evaluated in the primitive zinc production area.

300 samples were collected within an area of 50 km^2 in Laoyingshan town (N 26°30', E 105°03'), which is located in Shuicheng (a town in Liupanshui area), Guizhou Province (the sampling points showed in figure 1). The samples were collected at a depth of 20 cm at different distances.

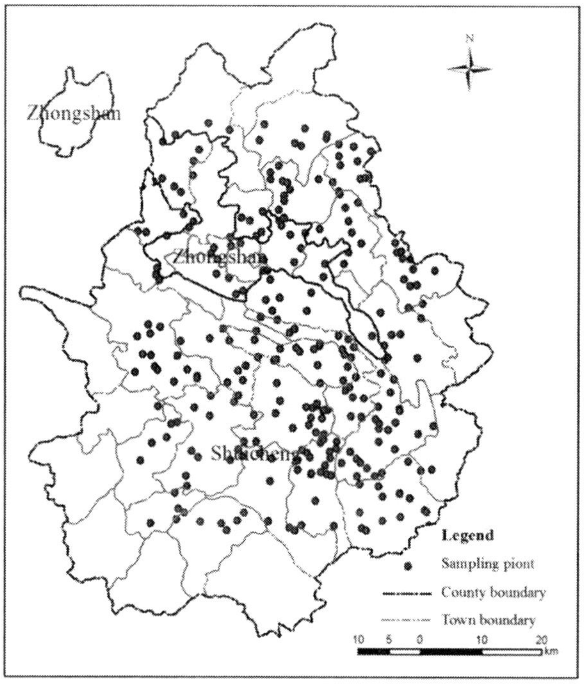

Figure 1. Location of the soil samples.

2.1.2. Sampling and analytical methods. Moss-bag techniques were used for domestic investigation of deposition of mercury in mercury and zinc production area [9, 10].

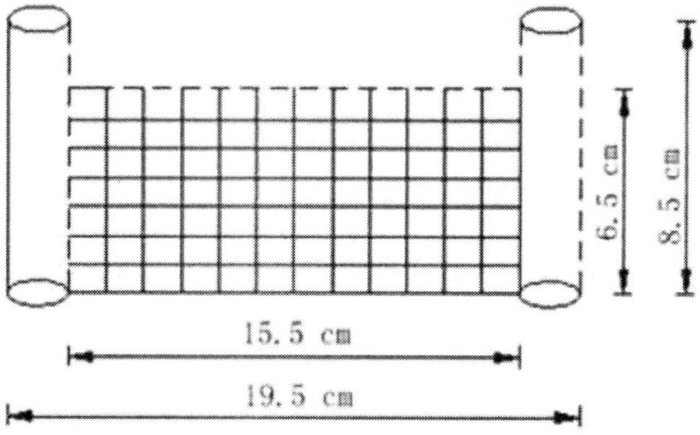

Figure 2. The geometry dimension of moss bag.

Fabrication Method of Moss Bag *Hypnum plumaeforme* Wils (with > 6 cm height) from Fanjing Mountain with good environment conditions was collected for packing moss bag. The chosen *H. plumaeforme* was cleaned with distilled water and immersed in 0.5 mol/L HCl for 24 hours and rinsed thoroughly with distilled water before drying. Dry moss (~3.2 g) was packed in a small nylon net with a mesh of 1.5 × 2.0 mm [11]. The installed bag with 100 cm^2 surface area is shown in figure 2. In addition, the concentration of Cd in moss was measured for eight times, and the average blank value of Cd was 0.077 mg kg^{-1}.

2.1.3. Monitoring of Moss Bag. The bags were exposed to the air by hanging 3-3.5 m off the ground and were away from the main road at least 100 m. Samples were collected from downwind direction. Each sample was collected with covered and uncovered moss bag, while total deposition of Cd was collected for 30 days under normal climate in 2005. All sites were dried at 60±2℃ and weighed with loss <5%.

2.2. Geo-accumulation index

The geo-accumulation index (I_{geo}) for Cd concentrations was calculated by using the following formula [12, 13].

$$I_{geo}=Log_2(C_n/(k*B_i)$$

where, C_n = measured concentration of Cd (mg/kg) in soil; Bi = geochemical background value (mg/kg) of the element in the background sample; k = factor introduced to minimize the effects of possible due to lithogenic effect [14, 15]. The Cd background value in the primitive zinc production area was chosen as B_i. The classification of pollution degree is shown in table 1. In the study, B_i is 0.31 mg· kg^{-1} [16].

Table 1. Geo-accumulation index and classification of pollution degree.

Igeo value	Igeo Class	Pollution level
≤0	0	Unpolluted
0-1		Unpolluted to moderately polluted
1-2	2	Moderately polluted
2-3	3	Moderately to strongly polluted
3-4	4	Strongly polluted
4-5	5	Strongly to very strongly polluted
>5	6	Very strongly polluted

2.3. Potential ecological risk assessment

The methods for heavy metal potential ecological risk assessment including Geo-accumulation Index (GI), Sediment Enrichment Factor Method (SEF), Toxicity Characteristic Leaching Procedure (TCLP), Potential Ecological Risk Index (PERI) etc. were used [17-22]. In the study, PERI method was to evaluate heavy metal contamination from the perspective of sedimentology. PERI was calculated by using the following formula:

$$RI = \sum E_i = \sum \left[T_i\left(C_\delta^i / C_n^i\right)\right]$$

where Ei = potential ecological risk index of single element; T_i =toxic-response factor for the given element (i.e. Cd=30); C_δ^i = concentration of given element in the soil; C_n^i = reference value of given element.

Table 2. Classfication of PERI parameters.

E_i	**RI**	**Potential ecological risk grade**
<40	<150	Low
40-80	150-300	Moderate
80-160	300-600	Considerable
160-320	-	High
>320	>600	Very high

2.4. Geostatistical analysis
Semi-variogram was used for analyzing the spatial variability in a region. The formula is:

$$\gamma(h)=\frac{1}{2}Var\left[Z(x+h)-Z(x)\right]$$

Where γ(h) measures the mean variability between two points x and x+h, as a function of their distance h [23]. The geostatistical analysis was carried out using GIS 9.3. Based on the fitted semi-variogram models, the ordinary Kriging provided by the software Arc/GIS 9.3 was used to map the distribution of heavy metals.

3. Results and discussion

3.1. Cd concentration in soil
In order to study the extent of metal contamination in this area, investigation of Cd concentration in soil at various distance to the zinc working industries are recorded. Table 3 shows the distribution of Cd from selected sites of the zinc production area. Exponential curve of the trend of Cd pollution in soil of zinc-smelting area is shown in figure 3. The average concentration decreased along the distance to the zinc working industries.

Table 3. Concentration of Cd in different distance of soil samples in farmland.

Distance (km)	Average (mg kg^{-1})	Max (mg kg^{-1})	Min (mg kg^{-1})	SD	CV (%)
5	3.68	3.92	3.37	0.16	4.3%
10	2.75	3.06	2.44	0.17	6.2%
15	2.32	2.76	1.89	0.25	10.8%
20	2.01	2.44	1.63	0.22	10.9%
25	2.53	2.89	2.02	0.21	8.1%
30	1.54	1.82	1.25	0.17	11.3%
35	1.34	1.66	0.98	0.22	16.4%
40	0.91	1.14	0.76	0.11	12.0%
45	0.68	1.04	0.33	0.22	31.8%
50	0.52	0.84	0.22	0.21	40.4%

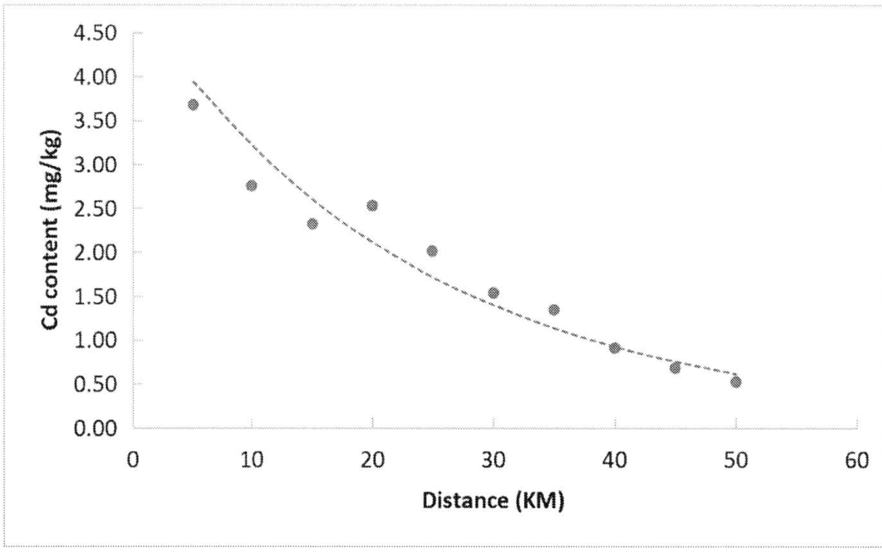

Figure 3. The trend of Cd pollution in soil along the wind direction in the area.

3.2. Cd deposition rate by Moss-bag techniques

The Cd content in 3.2 g moss indicates the Cd deposition rate during 30 days in the investigated area. The ratio of dry deposition rate ranged from 58% to 96%, with an average of 79%. It showed that dry deposition accounts a large proportion of total deposition rate. Therefore, the process of Cd to the mark from the atmosphere is a gravitation-induced mechanism which corresponds to the report.

Table 4. Deposition rate of Cd by monitoring of moss bag.

Distance/ (km)	Total Depostion Rate/($mg \cdot m^{-2} \cdot mon^{-1}$)	Dry Depostion Rate/($mg \cdot m^{-2} \cdot mon^{-1}$)	Dry/Total (%)
5	8.14	4.72	58%
10	5.18	3.62	70%
15	3.59	2.12	59%
20	1.43	1.35	94%
25	0.91	0.86	95%
30	0.43	0.31	72%
35	0.24	0.23	96%
40	0.16	0.12	75%
45	0.09	0.08	89%
50	0.05	0.04	80%

3.3. Estimation of pollutant indicators

The Geo-accumulation index value of Cd from 2.25 to 3.66 indicates a range of Geo-accumulation index class from 3 to 4. It showed all the sediments in these stations were moderately contaminated to be strongly polluted by Cd.

3.4. Potential ecological risk assessment

Potential ecological risk indexes for Cd in the topsoil were computed, and the value ranged between 143 and 380. Mean value of Cd was 241 and represents high level of potential ecological risk. The graph showed the ecological risk assessment of the study area. The primitive zone area along the wind direction indicated a very high level of potential ecological risk as shown in figure 4.

IOP Conf. Series: Earth and Environmental Science **82** (2017) 012037 doi:10.1088/1755-1315/82/1/012037

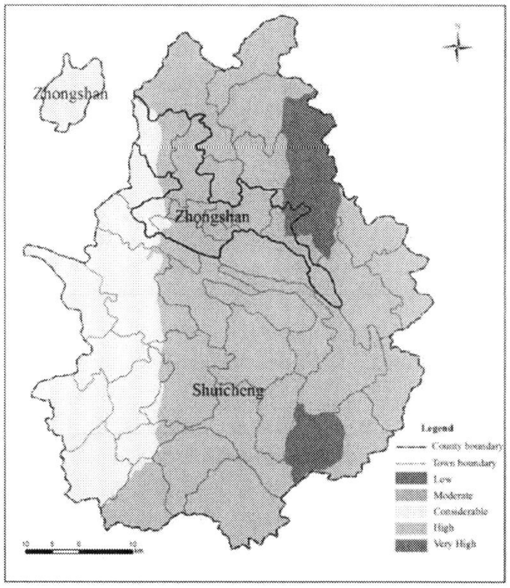

Figure 4. Spatial distribution of potential ecological risk index of Cd.

3.5. Geostatistical analysis

The results of the best-fit spatial model and model parameters are shown in table 5. In table 5, C_o is nugget, C_1 is sill, and A is range. The nugget/sill ratios were <25%, suggesting that the variable had strong spatial dependence. The ratios were between 25% and 75%, suggesting that the variable had moderate spatial dependence. The ratios were >75%, suggesting the variable had weak spatial dependence [24]. The ratio of Cd was 44.6%, showing that the factors had moderate spatial dependence.

Table 5. Semivariance function model and parameters of Cd content.

Sample Size	Model	Nugget	Sill	Range	Nugget/sill
300	Gaussian	0.0906	0.2031	0.0291	0.446

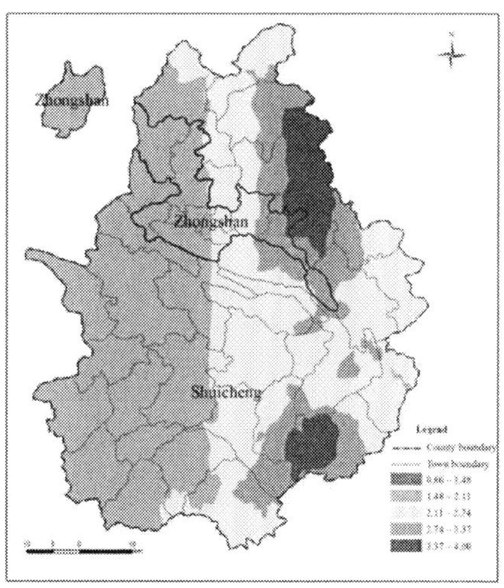

Figure 5. Spatial distribution of Cd content in topsoil.

Karst Area in Guizhou showed a high Cd background level due to mining activity. The characterization (rich in calcium and magnesium, alkaline) of Karst ecosystem indicated soil heterogeneity. Because the karst environment is sensitive and fragile, it is impossible to be restored once destroyed. The present repair mechanism of heavy metal pollution in other area is not applicable for Karst environment. Therefore, a series of researches about heavy metal pollution in karst area can provide a different theoretical system for Cd pollution. The topsoil in the study site has a high-level risk of Cd pollution which caused by mining activity as showed in the result of PERI and pollutant indicators as shown in figure 5.

4. Conclusion

Cd content in TSP, soil and the deposition rate of Cd decreased with the distance away from the pollutant source. Eighty percent dry deposition is the dominated process of Cd to the mark from the atmosphere. It suggested that dry deposition determination is better due to the loss of moss bag derived from the effect of precipitation to the total deposition rate. Furthermore, heavy metal potential ecological risk assessment was done based on Geo-accumulation Index (GI) and potential ecological risk index (PERI) method. The calculation indicates all the sediments in these stations were from moderately to strongly pollute by Cd and a high level of potential ecological risk to the study area by Cd. Nugget/sill ratio implied that Cd pollution in Liupanshui may have originated from mining activity and atmospheric deposition.

In the polluted central zone, the usage of soil is classified based on legislation "Soil Pollution Control Plan" for release the impact on Cd pollution [25]. In order to protect this area, more risk assessment and agricultural quality investigation should be done in this area and we need an integrated control system for water-soil-agricultural products. On the other hand, several measures helped decrease the impact on Cd pollution, including culturing microbial communities, spraying organic fertilizer, increasing the pH value of soil and planting less solanaceae.

Acknowledgments
This work was supported by the National Natural Science Foundation of China (Grant No. 21667009).

References

[1] He Z L, Zhou Q X and Xie Z M 1998 *Balance of Soil Beneficial and Harmful Elements* (Beijing: China Environmental Science Press)
[2] Egwurugwu J N, Ufearo C S, Abanobi O C, *et al* 2007 Effects of ginger (Zingiber officinale) on Cd toxicity *Afr. J. Biotechnol.* **6(18)** 2078-82
[3] Quan Q, Ying C, Qingwei M A, Wang F, Xi M and Bo W 2016 The impact of atmospheric deposition of cadmium on dominant algal species in the east china sea *J. Ocean Univ. Chin.* 15(2) 271-82
[4] Nriagu J O 1984 *Changing Metal Cycles and Human Health* (Berlin: Springer)
[5] Xiao D, Zhang W, Xiong K, *et al* 2009 Study on Karst landscape spatial pattern based on remote sensing and GIS *Proc. Spie* **7498** 74981D-7
[6] Lin W J, Xiao T F, Zhou W·C, Ao Z Q and Zhang J F 2009 Environmental concerns on geochemical mobility of lead, zinc and cadmium from zinc smelting areas: Western Guizhou, China *Environ. Sci.* **30(7)** 2065-70
[7] Zhou Q X 1994 Cd contaminated soil ecosystem and its theoretical model release mechanism *Journal of Zhejiang University (Natural Science)* **28** 283-44
[8] Sun T H, Zhou Q X and Li P J 2001 *Pollut. Ecol.* (Beijing: Science Publisher)
[9] Tan H, He J L and He T T 1997 Study of Hg deposition near a Hg mine by moss bag adsorption *Environ. Sci.* **18(6)** 71-2
[10] Xie F, Tan H, Yang B, *et al* 2014 The study of atmospheric transport and deposition of Cd emitted from primitive Zinc production area *Water Air Soil Pollut.* **225(11)** 1-7
[11] Temple P J 1981 Moss bag as monitors of atmospheric deposition *J. Air Pollut. Control Assoc.*

IOP Conf. Series: Earth and Environmental Science **82** (2017) 012037 doi:10.1088/1755-1315/82/1/012037

31(6) 668-70
[12] Wan J B, Wang J Y and Wu D 2008 Current assessment on heavy metal pollution in the sediment on Le'an River *Environ. Sci. Technol.* **31(11)** 130-3
[13] Hanif N, Ali S M, Cincinelli A, Ali N, Katsoyiannis I A and Tanveer Z I, *et al* 2016 Geo-accumulation and enrichment of trace metals in sediments and their associated risks in the chenab river, Pakistan *J. Geochem. Explor.* **165** 62-70
[14] Wang L, Yu R, Hu G and Tu X 2009 Speciation and assessment of heavy metals in surface sediments of Jinjiang River tidal reach, southeast of china *Environ. Monit. Assess.* **165(1-4)** 491-9
[15] He S L, Long C L, Liu Y Z, *et al* 2004 Geochemistry and environment of the Cd in the soil and sediments at the surface of Guizhou Province *Guizhou Geol.* **21(4)** 245-50
[16] Yi X, Gu X J, Hou Y Q, *et al* 2010 Assessment on soil heavy metals pollution by geoaccumulation index in Jinghuiqu irrigation district of Shanxi Province *J. Earth Sci. Environ.* **32(3)** 288-91
[17] Müller G 1979 Schwermetalle in den sedimenten des Rheins-Vera E'nderungenseit 1971' *Umschau* **79** 778-83
[18] Nobi E P, Dilipan E, Thangaradjou T, *et al* 2010 Geochemical and geo-statistical assessment of heavy metal concentration in the sediments of different coastal ecosystems of Andaman Islands, India *Estuar. Coast. Shelf. S.* **87(2)** 253-64
[19] Gopinath A, Nair S M, Kumar N C, *et al* 2010 A baseline study of trace metals in a coral reef sedimentary environment, Lakshadweep Archipelago *Earth Sci. Environ.* **59(6)** 1245-66
[20] Guo W, Liu X, Liu Z, *et al* 2010 Pollution and potential ecological risk evaluation of heavy metals in the sediments around Dongjiang Harbor, Tianjin *Procedia Environ. Sci.* **2(1)** 729-36
[21] Gargouri D, Azri C, Serbaji M M, *et al* 2011 Heavy metal concentrations in the surface marine sediments of Sfax Coast, Tunisia *Environ. Monit. Assess.* **175(1-4)** 519-30
[22] Nemr A E, El-Said G F, Khaled A, *et al* 2016 Distribution and ecological risk assessment of some heavy metals in coastal surface sediments along the Red Sea, Egypt *Int. J. Sediment Res.* **31(2)** 164-72
[23] Wang Z Q 1999 *Geo-statistics and Its Application in Ecology (In Chinese)* (Beijing: Science Press) pp 25-31
[24] Journel A G 1993 *Quantitative Geology and Geostatistics//Quantitative geology and geostatistics* (U.S.A.: D. Reidel Pub. Co.; Canada: Kluwer Academic Publishers; Kluwer Academic Publishers Group) pp 213-24
[25] Shen G Y 2016 The State Council issues soil pollution control action plan *Fine and Specialised Chemicals* **6** 12

IOP Conf. Series: Earth and Environmental Science **82** (2017) 012038 doi:10.1088/1755-1315/82/1/012038

Radioactivity of seawater nearby Daya Bay Nuclear Power Station

F Deng[1,2], Q Lin[1], R D Zhou[1], W L Chen[1] and Z D Chen[1]

[1]Guangdong Environment Radiation Monitoring Center, Guangzhou, 510300, China

E-mail: 843456548@qq.com

Abstract. The radioactivity of seawater nearby Daya Bay NPPs was introduced in this paper. The average annual concentrations of ^{90}Sr and ^{137}Cs ranged from 0.55~2.38 and 0.81~4.68 Bq·m^{-3} respectively during 1993~2015. The average annual concentration range of ^{3}H is 0.8~7 Bq·L^{-1}. The concentration range of ^{14}C is 7.6~8.6 Bq·m^{-3}, and the concentration of ^{3}H and ^{14}C is higher than that of the reference location or the background, but with no cumulative effect. The concentrations of ^{90}Sr and ^{137}Cs in seawater have no obvious relationship with the nuclear liquid effluent discharging. The average concentrations of ^{238}U and ^{232}Th were 24.6±10.6 and 18.1±7.5 Bq·m^{-3} respectively during 2000~2008.

1. Introduction

Natural nuclides ^{40}K, U, Th and their decay products, cosmogenic nuclides ^{14}C and ^{3}H, released nuclides ^{90}Sr and ^{137}Cs due to nuclear test or accident in last century; moreover, these radionuclides appear in the seawater if without the influence of artificial radionuclides sources.

Dayawan Nuclear Power Station (GNPS), Lingao Nuclear Power Station I (LNPS I) and Lingao Nuclear Power Station II (LNPS II) co-locate in the Daya Bay, 6 REP NPPs started to generate electricity in 1994, 2002 and 2010 respectively. In normal operating state, liquid effluent rarely contains ^{90}Sr, ^{137}Cs, ^{124}Sb, ^{58}Co and ^{54}Mn, and the activity concentrations of ^{124}Sb, ^{58}Co and ^{54}Mn are usually below the detection limit in West Daya Bay. ^{110m}Ag is rarely detected since 2009 when silver flange was replaced by other materials, while because of the influence of nuclear explosion test and accident such as chernobyl, ^{90}Sr, ^{137}Cs could still be detected. ^{3}H and ^{14}C are accounted for a great proportion in liquid effluent, in the same time they naturally present in the environment.

In order to evaluate the effect of liquid effluent from nuclear power plants, ^{90}Sr, ^{3}H, ^{14}C, γ nuclides (such as ^{137}Cs, ^{124}Sb, ^{60}Co, ^{58}Co, ^{54}Mn, ^{110m}Ag) and ^{238}U, ^{232}Th were monitored in seawater in Daya Bay.

In this paper, the activity concentrations of ^{90}Sr and ^{137}Cs are accounted and analyzed during different stages of 6 REP reactors in Daya Bay from 1993 to 2015. Also, the activity concentration of ^{3}H in 1993-2014, the concentration of U and Th, the activity of ^{14}C in 2015 were presented.

2. Experimental part

2.1. Monitoring scheme

The radioactivity of seawater in West Daya Bay has been surveyed since 1993. The emission location was moved towards the east for about 1 km in order to reduce the liquid effluent discharge influence in 2000. So the sampling point location amount varied from the original 18 to 10 for better optimizing monitoring of seawater in West Daya Bay, which also strengthened the monitoring of Dapeng'ao area which has the weakest ability of dilution.

Before 2002, we set 18 sampling point locations which numbered from W1 to W18. The sampling frequency was 3~4 times per year. ^3H was analyzed in all samples, while ^{90}Sr and γ nuclides were only analyzed in the locations of W2, W4, W9, W11, W12, W13 and W14. The distribution of the monitoring point diagram is shown in the left panel of figure 1.

We optimized the sampling point location amount to 10 in 2002, which were marked as L1~L10. Every year we took samples 2 or 4 times. ^3H was analyzed in all samples. But ^{90}Sr and γ nuclides were only analyzed in L1, L4, L9 and L10. The right panel of figure 1 gives the distribution of the monitoring point location diagram.

The monitoring of ^{14}C began at 2015, for which the monitoring points locate in L1, L4, L10 and the liquid effluent discharge channel.

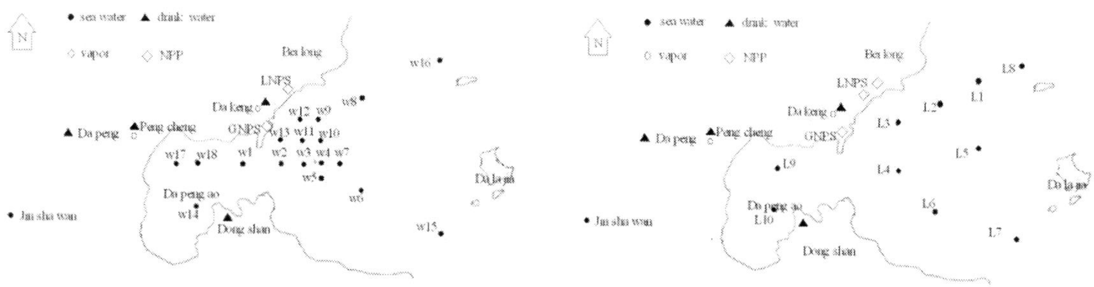

Figure 1. Sampling point locations for tritium monitoring nearby NPPs before 2001 (left panel) and after 2002 (right panel).

2.2. Analysis method
In this paper, ^3H, ^{14}C, ^{90}Sr, ^{137}Cs, U and Th have been monitored using different methods follow state standard or method published in paper. Table 1 showed the methods used in this paper.

Table 1. Analyzed nuclides and their relative method.

Radionuclide	Method	Reference
^3H	analytical method of tritium in water	GB12375-90
^{14}C	liquid scintillation	Guo Guiyin's approach [1]
^{90}Sr	Extraction chromatography by di-(2-ethylhexyl) phosphoric acid	GB6764-86
^{137}Cs, U and Th	adsorption-coprecipitation method	Den Fei's approach [2]

GB12375-90 analytical method of tritium in water was used to detect ^3H. The samples used for ^3H measurement are not concentrated by electrolysis. For monitoring ^3H, the low background liquid scintillation equipment Quantulus 1220 was used. For this equipment, the background turns to be lower than 0.7 cpm and the detection efficiency is greater than 60%. The detection limit was about 1.2 $Bq \cdot L^{-1}$ if the measurement time was set 24 h.

The monitoring method for ^{14}C followed Guo Guiyin's approach [1]: take 20 L pellucid samples to the four neck flask, stirring with magnetic stirrer when adding sulfuric acid. CO_2 origins from inorganic carbon were carried into the collecting bottle by N_2; $K_2S_2O_8$, $FeSO_4$ and H_2O_2 were added respectively and then collect CO_2 from organic carbon. Finally, the sample was measured by method of liquid scintillation with Carb-Sorb E absorption CO_2.

Radiochemical analysis of strontium-90 in water precipitation by fuming nitric acid GB6764-86 and Radiochemical analysis of strontium-90 in water extraction chromatography by di-(2-ethylhexyl) phosphoric acid GB6766-86 were used for the detect of ^{90}Sr, "GB6764-86"and "GB6766-86" were updated by "HJ815-2016" Radiochemical analysis of strontium-90 in water and ash of biological samples.

MnO_2, produced by chemical reaction between H_2O_2 and $KMnO_4$, along with $K_2CoFe(CN)_6$, were used for the precessing of γ radionuclides like ^{137}Cs, U and Th in seawater samples by adsorption-coprecipitation method [2].

When processing statistical analysis, sample activity is taken into consideration by 1/2 of the detection limit in case where activity is smaller than detection limit. The number of calculating samples is m/n, while n represents total number of samples and m refers to the amount of the samples which possess an activity higher than detection limit. The activity would be provided in the form of A±X, where A is means activity and X represents to the uncertainty caused by the counting error of the instrument during the measuring process.

3. Results and discussions

3.1. Artificial radionuclides ^{90}Sr, ^{137}Cs, et al

The activity concentration of ^{90}Sr and ^{137}Cs in the seawater of West Daya Bay among 1993~2016 is illustrated in table 2, which gives a value varies between 0.55~2.38, 0.81~4.68 Bq•m^{-3} respectively, while the maximum value in one sample turns out to be 4.70 and 7.05 Bq•m^{-3} respectively. These maximums appeared between 1993~1996 when NPPs was in pilot run. According to the NPP operation record, abnormal emissions took place during 1995~1997 which were mainly caused by maintenance and other activities. During this period, in some patches of seawater a trace of ^{124}Sb,

Table 2. Concentration of 90Sr and 137Cs in seawater nearby Daya Bay Nuclear Power Station.

Date	^{90}Sr (Bq·m^{-3})			^{137}Cs (Bq·m^{-3})		
(year)	samples (m/n)	range	average	samples (m/n)	range	average
1993	4/4	1.80-2.70	2.38 ±0.40	4/5	<1.0-2.00	1.30±0.55
1994	10/10	1.20-4.70	2.06 ±0.99	9/10	<1.0-5.13	3.06±1.67
1995	18/18	0.45-3.00	1.44±0.76	21/21	2.90-7.05	4.68±1.27
1996	16/16	0.36-1.84	1.32±0.46	16/16	2.50-6.10	4.23±1.12
1997	10/10	0.61-1.51	1.12±0.26	11/15	<1.0-6.00	2.65±1.90
1998	12/12	0.69-3.20	1.55±0.61	19/20	<1.0-5.50	2.97±1.31
1999	19/19	0.49-2.10	1.16±0.40	19/19	1.48-3.09	2.43±0.39
2000	16/16	1.0-3.1	1.54±0.56	16/16	1.66-3.07	2.48±0.44
2001	12/13	<0.51-1.98	1.17±0.56	11/13	<1.9-2.84	2.13±0.58
2002	5/5	0.75-0.99	0.85±0.10	14/14	1.16-3.20	2.23±0.56
2003	10/10	0.44-2.02	0.79±0.46	11/15	<1.20-3.80	1.86±0.88
2004	4/4	0.75-1.00	0.87±0.12	14/16	<1.5-2.71	1.82±0.55
2005	4/4	0.91-1.40	1.10±0.23	8/16	<1.4-2.80	1.44±0.77
2006	4/4	0.96-1.50	1.32±0.24	6/16	<1.3-2.31	1.19±0.62
2007	12/12	0.73-2.00	1.32±0.42	7/16	<1.2-2.96	1.26±0.66
2008	16/16	0.66-2.05	1.29±0.39	9/16	<1.3-2.30	1.09±0.50
2009	4/4	0.82-1.89	1.22±0.51	2/12	<1.1-1.7	0.81±0.37
2010	9/12	<0.4-3.45	1.16±0.97	6/16	<1.1-2.50	1.10±0.65
2011	5/13	<0.17-2.28	0.74±0.59	6/16	<1.1-1.79	0.98±0.36
2012	13/16	<0.41-1.30	0.55±0.31	13/16	<1.1-1.99	1.42±0.47
2013	15/15	0.31-2.09	0.90±0.54	7/16	<1.3-2.18	1.08±0.51
2014	16/16	0.35-2.50	1.18±0.54	13/16	<1.0-2.90	1.52±0.62
2015	6/8	<0.6-1.15	0.73±0.32	6/8	<1.2-2.1	1.49±0.52
2016	8/8	0.36-1.22	0.90±0.29	8/8	1.65-2.30	1.99±0.22

^{110m}Ag and ^{58}Co was detected. As for example, the activity of ^{124}Sb and ^{58}Co in four samples in July 1994 ranged 0.55~2.38 and 0.81~4.68 respectively, the ^{110m}Ag activity of four water samples in April 1995 and three samples in 1997 varied between 1.1~23.3 and 1.52~1.67 respectively [3], while for the rest samples, all the activities were below detection limit. In conclusion, the phenomenon that ^{137}Cs activity turned out to be much higher during this period may relate to the abnormal emissions of NPPs. Also, the liquid effluent emissions of NPPs show no significant influence in ^{90}Sr and ^{137}Cs concentrations through figure 2, which well satisfied the natural decay law, in West Daya Bay.

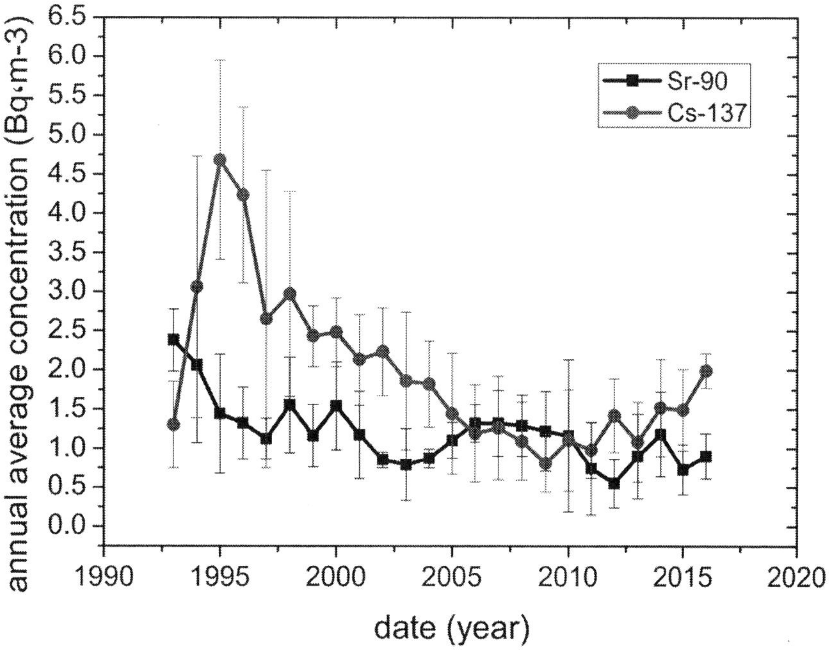

Figure 2. Annual average concentration of ^{137}Cs and ^{90}Sr in seawater nearby Daya Bay NPPs.

Table 3 and figure 3 give the ^{90}Sr and ^{137}Cs activities for the effluent that has not been well diluted. The results in tables 3 and 2 do not tell much difference, which indicates that ^{90}Sr and ^{137}Cs give a trace amount in the NPPs liquid effluent. Figure 4 showed the relationship between annual emission and specific activity of ^{137}Cs in West Daya Bay seawater samples. When comparing data between tables 2 and 3 a conclusion could be made that the operation of LNPS I and LNPS II does not emit a detectable amount of ^{90}Sr and ^{137}Cs.

Table 3. Activity of ^{90}Sr and ^{137}Cs in the effluent of NPPs.

Date	^{90}Sr ($Bq \cdot m^{-3}$)			^{137}Cs ($Bq \cdot m^{-3}$)		
(year)	samples m/n	range	average	samples m/n	range	average
1999	3/3	1.00-1.80	1.40±0.40	3/3	1.59-3.36	2.54±0.89
2000	3/3	0.98-1.70	1.39±0.37	3/3	2.14-2.53	2.37±0.20
2001	3/3	0.46-2.82	1.46±1.22	3/3	1.92-3.60	2.48±0.97
2002	1/1	1.2	1.2	3/3	1.29-2.90	2.23±0.84
2003	2/2	0.92-1.41	1.17±0.35	2/3	<1.6-2.08	1.60±0.70
2007	3/3	0.87-1.20	1.01±0.17	0/3	<1.6	<1.6
2008	3/3	0.66-0.88	0.80±0.12	3/4	<1.0-1.87	1.26±0.54

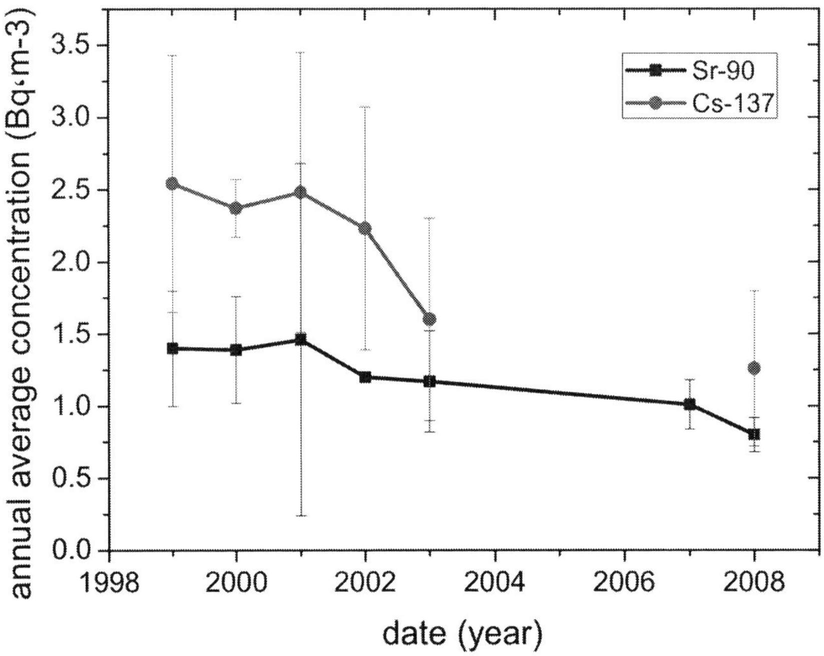

Figure 3. Annual average concentration of ^{137}Cs and ^{90}Sr in the effluent of NPPs.

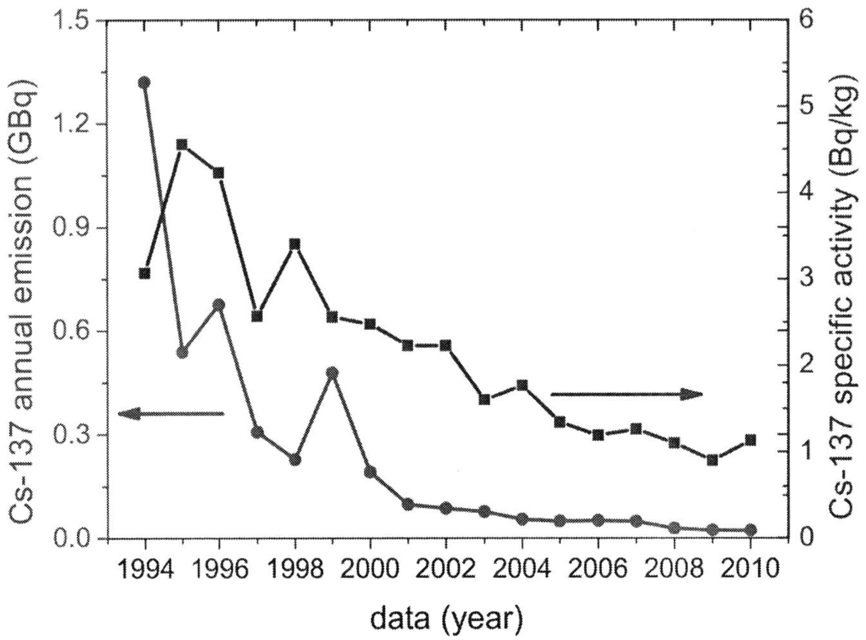

Figure 4. Annual emission and specific activity of ^{137}Cs in West Daya Bay seawater sample.

3.2. ^{3}H and ^{14}C in West Daya Bay seawater

Due to the lack of relevant method and standard, the ^{14}C activity monitoring in the seawater nearby Daya Bay Nuclear Power Station was launched until 2015. The result shows that the activity concentrations of ^{14}C were 15.6, 7.6, 8.5, 8.6 Bq·m^{-3} for seawater monitoring point locations discharge channel, L1, L4, L10 respectively, as a comparison, the value turns to be 5.5 Bq·m^{-3} for seawater in reference location Maoming, which locates several hundred kilometers from Daya Bay [internal information]. The ^{14}C activity is higher in no well diluted discharge channel and in L4/L10 which have a weak dilution capacity, than in L1 which possesses a strong dilution capacity. Activity concentration of ^{14}C in West Daya Bay seawater is more than 50% higher than that in the reference location Maoming.

The activity of ^{3}H in West Daya Bay seawater ranges from a value below 0.6 to 70.7 $Bq \cdot L^{-1}$, while the annual average value ranges 0.8~7 $Bq \cdot L^{-1}$. The annual average activities of ^{3}H are three to ten times higher than the background activities, which might be caused by the emission of liquid effluent from NPPs, the difference between sampling time and discharge time. No concentration effect of ^{3}H appears facing the operations of the 6 NPPs [4], which indicates that the dilution-exchange capacity of West Daya Bay seawater satisfies the requirements on the discharge of liquid effluent.

3.3. Concentrations of ^{238}U and ^{232}Th

The method used to monitor ^{238}U and ^{232}Th is introduced in the literature [2]. The average value of ^{238}U and ^{232}Th were 24.6±10.6, 18.1±7.5 $Bq \cdot m^{-3}$ in 177 samples sampled in West Daya Bay during 2000-2008 respectively. Convert to mass concentration of U and Th were 2.0±0.86, 4.47±1.85$\mu g \cdot L^{-1}$ respectively (The activity of ^{238}U in natural uranium is 12.35 $kBq \cdot g^{-1}$, the activity of ^{232}Th in natural thorium is 4.046 $kBq \cdot g^{-1}$).The concentration of U is 3.1-3.9 $g \cdot L^{-1}$ reported in the literature [5, 6], which is higher than the previous values. The main reason is that the samples were monitored before the decay balance between ^{238}U and its progeny ^{234}Th, but take the activities of ^{238}U deduced from ^{234}Th. The concentration of U was 2.82 $\mu g \cdot L^{-1}$ in seawater from West Daya Bay detected by using sodium polyphosphate as fluorescence enhancement agent while the verified value was 2.83 $\mu g \cdot L^{-1}$ by ICP-MS [7]. This value was a little lower than that in the literature [6].

4. Conclusion

The Average annual concentration of ^{90}Sr, ^{137}Cs and ^{3}H ranged 0.55~2.38, 0.81~4.68 and 0.8-7 $Bq \cdot L^{-1}$ respectively during 1993~2015, the concentration ranges of ^{14}C was 7.6-8.6 $Bq \cdot m^{-3}$ in 2015 in seawater nearby Daya Bay Nuclear Power Station. The concentrations of ^{3}H and ^{14}C were higher than that in the background or in the reference location, but with no cumulative effect. The trend of ^{90}Sr and ^{137}Cs concentrations was consistent with the natural decay law, with no relationship when comparing the discharging of liquid effluent from the NPPs. The Average concentration of ^{238}U and ^{232}Th were 24.6±10.6 and 18.1±7.5 $Bq \cdot m^{-3}$ respectively during 2000 – 2008.

Artificial radionuclides could not be detected except for part of the samples sampled during 1995-1997 during which there was abnormal emission or nuclear power accident. Trace amounts of ^{124}Sb, ^{60}Co and ^{110m}Ag were detected in seawater from West Daya Bay.

References

[1] Guo G Y, Huang Y J. ZHANG B, *et al* 2015 Analysis of ^{14}C in environmental water based on wet-oxidation pretreatment *Radiat. Prot.* **35(6)** 339-44 (in Chinese)
[2] Deng F, Chen Z D, Lin Q, *et al* 2006 Source of ^{3}H in monitored well water at Daya Bay Nuclear Power Station *Radiat. Prot. Bull.* **26(2)** 10-4 (in Chinese)
[3] Lin Q, Deng F, Chen Z D, *et al* 2003 Concentrations of nuclide ^{110m}Ag in marine samples around Daya Bay nuclear power station *Radiat. Prot. Bull.* **23(1)** 35-7 (in Chinese)
[4] Deng F, Lin Q, Wang J Y, *et al* 2016 Tritium discharged by Daya Bay nuclear power plant and its environmental impact *Environ. Chem.* **35(5)** 956-63 (in Chinese)
[5] Zhang Y A, Li Q Z and Zeng Z W 1983 Rapid determination of trace uranium in seawater by laser fluorescence *Acta Oceanol. Sin.* **5(4)** 467
[6] Chen J H, Edwards R L and Wasserburg G J 1986 ^{238}U, ^{234}U and ^{232}Th in Seawater *Earth and Planet Sci. Lett.* **80(3-4)** 241
[7] Qin L J and Li M L 2016 Determination of trace uranium in seawater with sodium polyphosphate as fluorescence intensifier *Radiat. Prot. Bull.* **36(2)** 39-42 (in Chinese)

Numerical studies on groundwater-grassland relations in an inland arid region in China

J R Wang[1,2], L T Hu[1,2,3], K N Sun[1,2] and X M Liu[1,2]

[1]College of Water Sciences, Beijing Normal University, Beijing 100875, China
[2]Engineering Research Centre of Groundwater Pollution Control and Remediation, Ministry of Education, Beijing Normal University, Beijing 100875, China

E-mail: litanghu@bnu.edu.cn

Abstract. In this study, a 2-D numerical model was developed to assess the impacts of groundwater on grassland ecology in the Hulun Lake Basin. An extreme dry climate scenario and water resource management scenario and their interactions in the Hulun Lake Basin were designed, and their influence on groundwater was evaluated. The results show that the grassland ecology is heavily dependent on groundwater, and a distribution of groundwater with a depth of 8 m correlates well with the distribution of grassland. Under the water resource management scenario, the groundwater level will increase to a maximum value of 2.5 m after 15 years around Hulun Lake. The groundwater level will decrease dramatically under the extreme dry climate scenario, thus affecting the environment.

1. Introduction

Covering approximately 3.98 million km^2 in China, grasslands play an important role in economic development and the environment [1]. Grasslands are usually groundwater-dependent all over the world, especially in arid or semi-arid regions where rainfall is seasonally variable and significantly lower than the evaporation rate [2]. Groundwater-dependent grassland will be at risk if groundwater levels are reduced. An understanding of groundwater-grassland relations is of great importance to maintaining a healthy grassland ecology [3]. Available and adequate water supplies are essential for maintaining grasslands, and thus water resources management is critical to meet the water demands of grasslands, especially during record drought conditions. As the fifth largest lake in China, the Hulun Lake has suffered from the drawdown of groundwater levels and degradation of grasslands around the lake in the past few decades due to excessive groundwater exploitation [4]. To protect and restore grassland ecosystem function, water resources management projects, including water transfer projects within the basins and from adjacent basins, and groundwater extraction control measures should be drafted by the local government. The objectives of this paper are to develop a numerical model using the FEFLOW software [5], a commonly used groundwater modeling system, to quantitatively assess the impacts of groundwater on grasslands and explore the extent to which the potential water resource management project may impact the grassland. In this study, the groundwater depth across the Hulun Lake Basin will be forecasted so that a risk assessment can be performed, aiming to assess the sensitivity of grasslands to groundwater.

IOP Conf. Series: Earth and Environmental Science **82** (2017) 012039 doi:10.1088/1755-1315/82/1/012039

2. Methods

2.1. Study area

Located in the northeastern part of the Inner-Mongolia Autonomous Region of China, the Hulun Lake Basin (115°30'-115°30' E, 47°40'-49°50' N) has an approximate drainage area of 34299 km^2. The study area can be divided into two parts, mountainous in the west and high plains in the east. The elevation ranges from 560 m in the east to 900 m in the west (figure 1). The main rocks types are Paleozoic metamorphic rocks, Mesozoic volcanic rocks and the Yanshan Period granites in the west, and whereas they are Cainozoic sandstone or clayey breccia in the east. The rock type in the central area is mostly sandy loam derived from sedimentary river and lake environments. Study area has a mean minimum yearly temperature of -42.7℃ and a mean maximum yearly temperature of 40.1℃, whereas the mean annual rainfalls are approximately 275 mm.

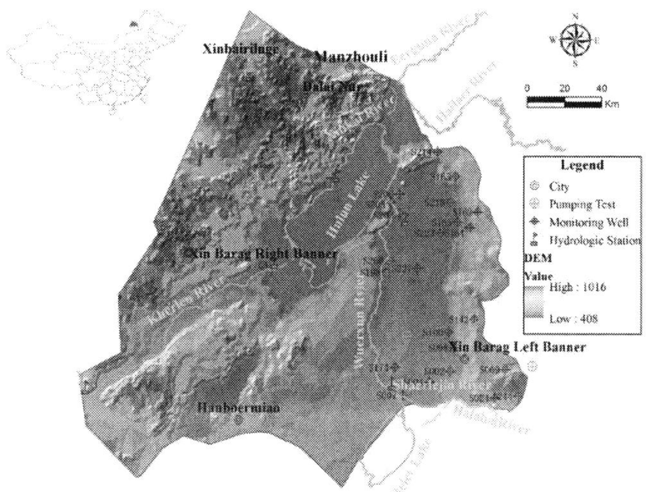

Figure 1. Map of the study area.

2.2. Conceptual model

The study area is chosen to be the model area. According to the characteristics of the aquifer formation and the hydrological conditions in the Hulun Lake Basin, hydrogeological models were developed. Considering the complex rock types in the phreatic aquifers in the study area, the local media are assumed to be heterogeneous. Due to a scarcity of data, interpolated Digital Elevation Model (DEM) values can be acquired using the Kriging method in the SURFER software. Both the western boundary and southern boundary are assumed as Neumann boundary conditions, and the flux of water flow through the boundary is estimated to be 10^{-5} m/d. The northern boundary is regarded as a no-flux boundary. The eastern boundary is represented by a flux boundary with very low flow rate. The relation between the Hulun Lake and groundwater is a very important part in this model. The Hulun Lake will recharge the groundwater when the lake level is greater than the groundwater level, and vice versa. Therefore, the lake is represented by a transfer boundary, and the exchange capacity can be calculated by using Darcy's Law. The initial conditions of the groundwater system are estimated from a steady-state groundwater flow simulation with the average recharge rate. The model is two-dimensional and was developed using the FEFLOW software.

2.3. Data collection

2.3.1. Grassland coverage. Remote sensing imagery of the Hulun Lake area from the Landsat TM dataset for the year 2000 was used to identify the grassland coverage. Land use maps were collected for assisting the interpretation of the remote sensing images. The land use types in the study area can divided into three types, (1) high coverage grassland, (2) moderate coverage grassland, and (3) low coverage grassland, by using ERDAS IMAGINE 8.4 and ArcGIS 9.2 software in combination with the field observations of the spatial distribution of land-use in the Hulun Lake Basin. The high coverage grassland is closely related to the water body and mainly distributed near Hulun Lake, the Wuerxun River, the Kherlen River and the marshland that located to the east of Hunlun Lake. The moderate coverage grassland is widely distributed over highland regions. The low coverage grassland overlays the western hilly region. Saline-alkali land is mostly distributed in the high plain area.

2.3.2. Hydrogeology and observational data. 23 monitoring wells are distributed in the study area (figure 1) and the water-level data from the observation wells are listed in table 1. Based on a geological map of China at a scale of 1:2500000 (http://ngac.org.cn/onemap/index.html), the study area is divided into 20 zones. From hydrogeological reports and pumping tests, the hydraulic conductivities vary from 0.1 m/d to 12.5 m/d, and the porosity ranges from 0.03 to 0.15. The Kherlen River and Wuerxun River are the two main rivers that interact with groundwater in study area. The exchange flux can be calculated by FEFLOW automatically based on the corresponding parameters (e.g., stream stage and groundwater level). Statistical data show that the average level of groundwater exploitation and the return flow are approximately 68.01 million m^3 per year and 1.39 million m^3 per year, respectively.

Table 1. Water-level data from observation wells in 2009.

Month	Water level for each observation well (m)							
	S002	S004	S007	S024	S069	S098	S100	S142
May.	582.39	576.58	575.46	621.72	651.59	574.91	568.48	643.04
Jun.	582.37	576.46	575.42	621.67	651.49	574.99	568.29	642.96
Jul.	582.53	576.60	575.80	621.72	651.55	574.97	568.27	642.94
Aug.	582.77	576.86	576.23	621.88	651.66	574.73	568.52	643.18
Sep.	583.09	577.23	576.33	622.01	651.60	574.61	568.70	643.22
Month	Water level for each observation well (m)							
	S154	S155	S162	S165	S174	S198	S200	S203
May.	554.44	536.80	626.38	566.28	570.20	534.24	537.49	533.60
Jun.	554.42	536.75	626.32	566.26	570.13	534.20	537.41	533.51
Jul.	554.50	536.81	626.24	566.30	570.13	534.23	537.47	533.52
Aug.	554.59	536.95	626.40	566.40	570.24	534.32	537.70	533.72
Sep.	554.58	536.95	626.50	566.41	570.28	534.37	537.77	533.68
Month	Water level for each observation well (m)							
	S205	S206	S214	S218	S223	S227	S244	
May.	544.92	547.20	570.08	547.12				
Jun.	544.85	547.14	570.00	547.07	537.41	541.46	697.90	
Jul.	544.79	547.11	570.01	547.12	537.36	541.38	697.77	
Aug.	545.02	547.33	570.20	547.31	537.59	541.66	697.91	
Sep.	545.10	547.70	570.23	547.26	537.63	541.69	698.01	

2.4. Evaluation of groundwater-grassland relations

Groundwater-grassland relations are evaluated by analyzing the relation between the depth to groundwater and the area of the grassland coverage. The depth to groundwater is simulated using the FEFLOW software, and the suitable depth for grass growth was assumed as ecology water level.

3. Results of model evaluation

The model area is discretized into 3374 nodes and 6482 elements in total. The sedimentary period is set from January to December of 2009. The calibration targets are observed groundwater levels. The period for the observation is from May to September of 2009. The comparison of observed and simulated hydraulic head is shown in figure 2. Relative error is defined as the ratio of the absolute difference between the simulated and observed water level to the observed results. The correlation between observed and simulated values illustrates that the relative errors are less than 2%, which means that the simulated results are basically coincident with the observed results.

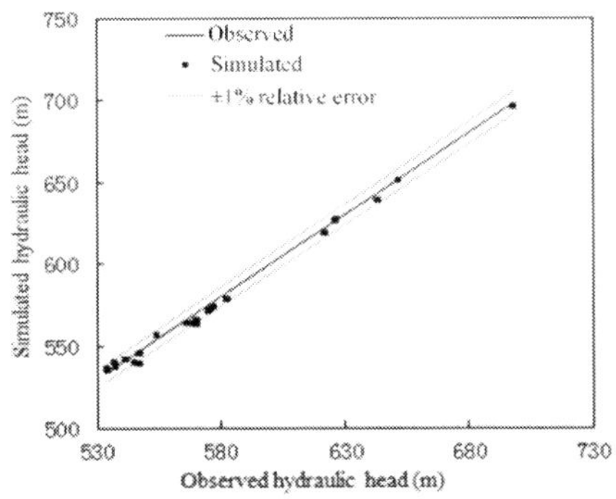

Figure 2. Comparison of observed and simulated hydraulic head.

3.1. Groundwater balance analysis

The results reveal that the annual precipitation infiltration is 0.45 billion m^3, which accounts for 98.70% of the total recharge. The annual groundwater evaporation is approximately 0.38 billion m^3, accounting for approximately 82% of the total discharge. Groundwater abstraction is approximately 15% of the total discharge. The yearly water supply is 0.46 billion m^3.

3.2. Distribution of groundwater depth and its relation with the grassland

Figure 3(a) shows the map of depth to groundwater, where a depth of less than 4.1 m (the darkest color), less than 8 m (the darker color) and less than 40 m (the gray color) are represented by different colors. The depth to groundwater is within 8 m near the Wulannuoer and the Wuerxun Rivers, where the distribution essentially coincides with the remote-sensing images of land-use and shrub growth. Thereinafter, a groundwater depth of 8 m is assumed to be the ecology water level.

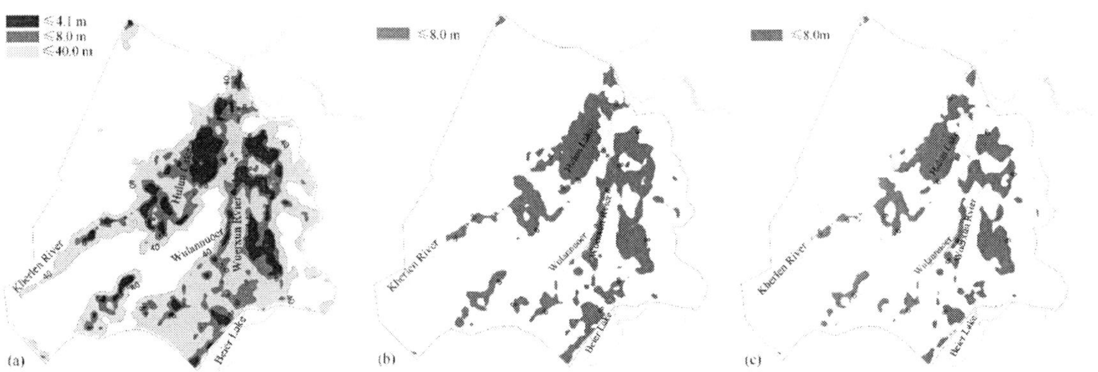

Figure 3. Distribution of the groundwater depth for the base scenario (a), the change in the groundwater depth (b) between the base scenario and the management scenario, and (c) that between the base scenario and the extreme dry climate scenario.

4. Discussions

For fulfilling environmental protection in the study area, a water resources management project for the Hulun Lake Basin is proposed. Two scenario are set: a base scenario and a management scenario. In the base scenario, the water level of Hulun Lake is 541.87 m, and the streamflow rate of the Kunduleng and Ganzhuhua Rivers is 0.679 and 0.735 billion m^3/a, respectively; whereas in the management scenario, the water level of the lake is 544.51 m, and the streamflow rate at the two stations is 0.552 and 0.606 billion m^3/a, respectively. Based on the groundwater balance analysis, precipitation infiltration is the major recharge to groundwater and thus has an important influence on the groundwater level. An additional scenario is added with an extreme dry climate with an annual precipitation of 247.5 mm under the management scenario. This scenario is aimed at revealing the sensitivity of grasslands to precipitation infiltration. The simulation duration is set as 15 years.

The simulated results under the management scenario and extreme dry climate scenario show the change in the groundwater depth (figures 3(b) and 3(c)). The numbers denote the change of groundwater depth between the base scenario and the management scenario, as well as that between the base scenario and the extreme dry climate scenario, and positive values mean that the groundwater table rises. The results show that the maximum increase in groundwater level around Hulun Lake is 2.5 m, mainly due to the rising of Hulun Lake water levels. Thus, the lake will recharge the groundwater, and the grassland ecological function will be further restored. The results show that, with the decline of precipitation infiltration, the depth to groundwater will decrease at a rate of 5 to 10 m over 15 years in the high plain zone compared with the management scenario, thus affecting the ecological environment to a great extent (figure 3(c)).

5. Conclusions

Groundwater-dependent grasslands are very important to the environment. A numerical model was developed to assess groundwater-grassland relations and their changes under a water resource management scenario and an extreme dry climate scenario in the Hulun Lake Basin. Climate, hydrology, digital elevation maps and geological maps, in conjunction with boreholes and monitoring wells, were used to construct the numerical model. Model calibration was carried out and the simulated results have a strong correlation with the observations. The results also show that precipitation infiltration is the main recharge source for the groundwater system and that a distribution of groundwater with a depth of 8 m correlates well with the distribution of grasslands. The results under the management scenario and the extreme dry climate scenario show that groundwater levels will be increased to a maximum value of 2.5 m after 15 years around Hulun Lake after the rising of the Hulun Lake level and that extreme dry weather will cause the decline of groundwater levels and the degradation of grasslands.

Acknowledgments

This work was supported by the National Natural Science Foundation of China (Grant number: 41572220) and the Beijing Natural Science Foundation (J150002).

References

[1] Jiang H 2005 Grassland management and views of nature in China since 1949: Regional policies and local changes in Uxin Ju, inner Mongolia *Geoforum* **36** 641-53

[2] Jolly I, Mcewan K and Holland K 2008 A review of groundwater-surface water interactions in arid/semi-arid wetlands and the consequences of salinity for wetland ecology *Ecohydro.* **1** 43-58

[3] Zhao C, Wang Y, Chen X and Li B 2005 Simulation of the effects of groundwater level on vegetation change by combining FEFLOW software *Ecol. Model.* **187** 341-51

[4] Wang G, Song J, Xue B, Xu X and Otsuki K 2012 Land use and land cover change of Hulun Lake nature reserve in Inner Mongolia, China: A modeling analysis *J. Fac. Agr. Kyushu. U.* **57** 219-25

[5] Trefry M G and Muffels C 2007 FEFLOW: A finite-element ground water flow and transport modeling tool *Ground Water* **45** 525-8

IOP Conf. Series: Earth and Environmental Science **82** (2017) 012040 doi:10.1088/1755-1315/82/1/012040

Assessment of polycyclic aromatic hydrocarbons (PAHs) ecological risk in surface seawater from the west Bohai Bay, China

X M Jia[1], J Y Li[1] and S Y Tian[1,2,3]

[1]College of Marine and Environmental Sciences, Tianjin University of Science and Technology, Tianjin 300457, China
[2]Tianjin Key Laboratory of Marine Resources and Chemistry, Tianjin 300457, China

E-mail: tiansy@tust.edu.cn

Abstract. In order to understand the pollution characteristics of polycyclic aromatic hydrocarbons (PAHs) in west Bohai Bay, China, the sixteen PAHs listed by US EPA in surface seawater were investigated in the Qinhuangdao (QHD), Caofeidian (CFD) and Huanghuagang (HHG) areas. Distribution and composition of PAHs in the dissolved phase (DP) and the suspended particulate associated phase (SP) were detected by gas chromatography and mass spectrometry (GC-MS), respectively. An ecological risk assessment was evaluated using risk quotients (RQ). The results showed that the 16 PAHs were prevalent in west Bohai Bay and the 2-3 rings PAHs were the predominant chemicals. The QHD area had the highest concentration of total PAHs (TPAHs), which ranged from 186.3 to 531.3 ng/L in DP and 222.7 to 715.7 ng/L in SPM. In all three areas, the PAHs levels in SPM were higher than those in DP. The risk assessment showed that there was a relatively low potential ecological risk of PAHs in the coastal regions of Bohai Bay.

1. Introduction

Polycyclic aromatic hydrocarbons (PAHs), which have the potential to be toxic, mutagenic and carcinogenic, can enter the marine environment through a variety of ways, such as river runoff, atmospheric precipitation, sewage outfalls and maritime transport [1]. PAHs are hydrophobic organic compounds that are prone to combine with particles in the water column and settle into the bottom sediment. Therefore in aquatic environments, water is considered a pollution transfer media and sediments are regarded as a reservoir for PAHs [2]. Due to their persistence, PAHs can exist in the environment for a long time and accumulate in organisms through the food chain [3]. However, there are only limited data about the level of PAHs in the marine environment and their ecological risk [4,5].

Bohai Bay is a typical semiclosed shallow sea in northeast China surrounded by the industrialized coastal regions including Tianjin and Hebei Province. Therefore, the marine environment in Bohai Bay has become one of the most severely polluted coastal areas in China [6]. PAHs contamination from the petroleum industry, shipping industry and other human activities cannot be ignored in this area. Because of the potential impact of PAHs on organisms in the marine environment, it is necessary to study the ecological risk of PAHs in Bohai Bay. However, there is not enough current research about the state of PAHs pollution in the seawater of Bohai Bay to make a risk assessment [7].

The aim of the present study is to evaluate the ecological risk of PAHs in surface seawater by investigating the pollution status of PAHs. The three typical coastal regions, QHD, CFD and HHG, selected all have a developed industry or port. This research could provide the basic dates for contamination control of PAHs and coastal environment management.

2. Materials and methods

2.1. Study area and sampling

A total of 36 samples were analysed to detect the PAHs concentrations in surface seawater in the west Bohai Bay. The investigation was carried out in three areas: QHD (8 samples), CFD (15 samples) and HHG (13 samples). The sampling sites are shown in figure 1. The samples were collected by a surface water sampler and were stored in pre-cleaned brown glass bottles at -4℃ until analysis.

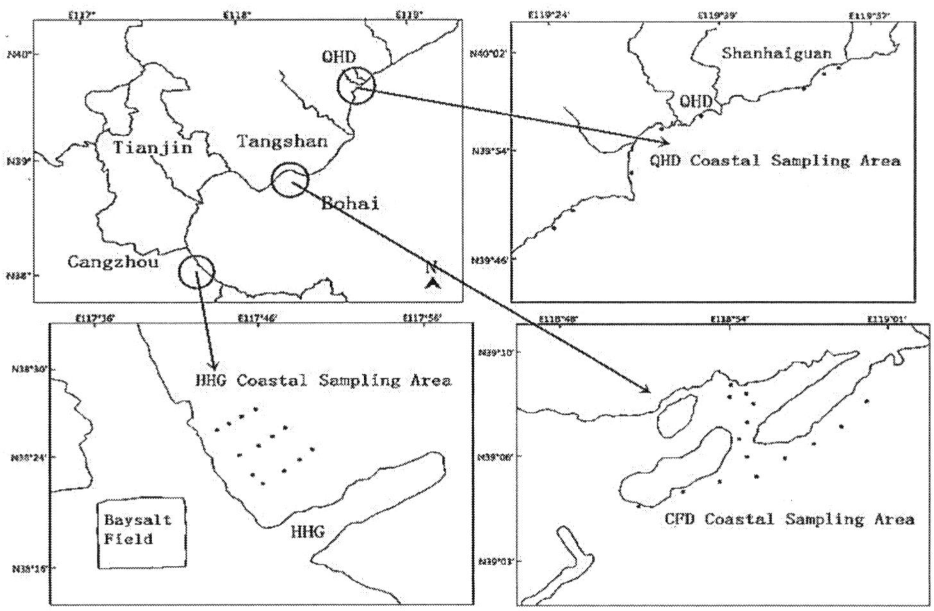

Figure 1. Location of sampling areas in QHD, CFD and HHG of the west Bohai Bay, China.

2.2. Reagents and materials

US EPA priority control 16 PAHs (purity > 99%) mixed standards and phenanthrene-d10 were purchased from AccuStandard in USA. Methanol and dichloromethane (DCM) (purity ≥ 99.9%) were purchased from Biologic Reagent. Anhydrous sodium sulfate was heated at 600℃ for 3 hrs in a muffle furnace and placed in a desiccator. C18 columns were obtained from Simon Aldrich in Germany.

2.3. Sample extraction and analysis

Samples were filtered by glass fiber filters (Whatman GF/F, pore size is 0.7 μm, Φ47 mm). The filtered seawater was used to analyze the dissolved phase (DP) PAHs in seawater. The solid materials in the filter membrane were analyzed for suspended particulate associated phase (SP) PAHs in seawater. The analysis was done with the following procedure. First, each water sample was gathered by solid-phase extraction (SPE) with a C18 column (preconditioned with 10 ml DCM, 10 ml methanol and 10 ml ultrapure water two times) at a flow rate of 5 ml·min^{-1}. Then, PAHs were eluted from the column with 12 ml DCM. The extraction was concentrated to 0.5 ml using Termovap Sample Concentrator in a water bath (45℃) and analyzed with GC-MS. Second, the solid materials separated by GF/F glass fiber filters were freeze-dried and weighed. The dried filters were extracted with 150 ml of the mixed solvent n-hexane and DCM (1:1, v/v) for 24 hrs using a soxhlet apparatus. Extractions were purified by a self-packet silica-alumina column (from top to bottom: anhydrous sodium sulphate, 1 cm; the mixture of silicone and n-hexane, 10 cm; anhydrous sodium sulphate, 1 cm; a few glassfiber). Finally, the PAHs eluent was concentrated to 0.5 ml for GC-MS analysis.

The quantification of PAHs was performed by gas chromatography and mass spectrometry (GC-MS: Agilent 7890A GC-5975 C MS). The GC was equipped with a DB-5 MS capillary column (30 m×0.25 mm×0.25 μm). GC operating conditions were programmed from 70℃ (initial time, 1 min) to 260℃ at a rate of 10℃·min^{-1}, 120℃ (initial time, 4 min) to 300℃ at a rate of 5℃·min^{-1}, and held at 300℃ for 4 min. PAHs were then quantified by selected ion monitoring (SIM).

2.4. *Quality assurance/quality control*

Some quality assurance measures were taken to vouch for the reliability of the results. Phenanthrene-d10 was added as a recovery indicator to each sample before extraction and the method recovery of 16 PAHs in DP and in SP was between 80% and 110%. Pyrene-d10 was applied as the internal standard to calculate the concentration of 16 PAHs.

2.5. *Risk assessment*

Risk quotients (RQ) were proposed by Kalf [8] to calculate the ecological risk of PAHs and also used to evaluate risk of PAHs on aquatic organisms [4,9]. RQ are commonly determined as follow:

$$RQ_{NCs} = C_{PAHs} / C_{NCs} \tag{1}$$

$$RQ_{MPCs} = C_{PAHs} / C_{MPCs} \tag{2}$$

Where C_{NCs} is the negligible concentrations (NCs) of PAHs and C_{MPCs} is the maximum permissible concentrations (MPCs) of PAHs in water [8]. RQ_{NCs} and RQ_{MPCs} are risk quotients of the NCs and MPCs of PAHs in water, respectively. In addition, the above method was further optimized to judge the ecological risk of 16 PAHs [10]. The risk grade of individual PAH and total PAHs (TPAHs) are listed in table 1.

Table 1. Risk grade of individual PAH and TPAHs [10].

individual PAHs			TPAHs		
Risk Grade	RQ_{NCs}	RQ_{MPCs}	Risk Grade	RQ_{NCs}	RQ_{MPCs}
Lower risk	0		Lower risk	=0	
			Low risk	≥1&<800	=0
Medium risk	≥1	<1	Medium risk 1	≥800	=0
			Medium risk 2	<800	≥1
High risk		≥1	High risk	≥800	≥1

3. Results and discussion

3.1. *PAHs distribution in surface water*

The DP and SP PAHs concentrations in the three investigated areas are listed in table 2. The SP PAHs concentrations were higher than DP PAHs concentrations, which indicate that PAHs are prone to being absorbed by particles in water column. The highest concentration of TPAHs was detected in QHD. This may be due to the rivers nearby the QHD sampling sites such as the Shi and Xinkai Rivers. These rivers bring terrigenous pollution into the coastal area. The concentration of TPAHs in CFD was the lowest among the three survey areas. The variation of PAHs concentrations between the three areas may be due to differences in industrial development. However, there is no significant difference of the PAHs concentrations between the three areas (p>0.05, one-way anova).

Table 2. PAH concentrations in surface seawater from the three coastal areas of the west Bohai Bay (ng/L).

PAHs	QHD		CFD		HHG	
	DP	SP	DP	SP	DP	SP
naphthalene (Nap)	50.8-321 (194.4)	24.5-253.6 (163.8)	8.84-437 (119.3)	31.62-264.57 (172.61)	10.2-418 (111.0)	22.79-519.61 (229.88)
acenaphthylene (Acy)	0.69-2.1 (1.50)	0.37-5.08 (2.8)	0.10-1.91 (1.01)	1.25-4.70 (3.32)	0.27-2.15 (1.2)	1.87-8.63 (4.10)
acenaphthene (Ace)	1.54-25.0 (7.94)	1.47-9.90 (6.1)	0.72-11.9 (5.22)	2.81-9.43 (6.32)	1.06-24.3 (8.6)	3.00-65.02 (10.65)
fluorene (Flu)	11.9-49.6 (23.21)	12.59-32.56 (26.1)	3.49-36.9 (19.90)	14.74-33.23 (23.45)	2.7-53.2 (24.31)	17.75-45.07 (26.12)
phenanthrene (Phe)	26.1-87.8 (45.24)	43.87-98.93 (73.8)	22.6-93.1 (50.97)	32.68-77.56 (54.67)	5.2-108 (51.8)	40.83-98.65 (59.49)
anthracene (Ant)	3.38-7.09 (5.02)	8.85-12.45 (9.5)	2.49-7.90 (5.48)	4.66-12.61 (8.17)	0.91-11.5 (5.5)	5.65-15.49 (9.32)
fluoranthene (Flua)	6.62-17.8 (11.62)	18.83-43.78 (34.5)	5.45-27.6 (15.68)	13.77-33.64 (23.85)	2.09-25.0 (14.9)	18.81-43.51 (29.10)
pyrene (Pyr)	22.0-35.0 (27.00)	31.74-93.36 (70.1)	10.8-35.7 (28.29)	38.99-99.16 (60.38)	5.1-45.0 (26.9)	46.13-114.05 (77.94)
benzo[a]-anthracene (BaA)	0.31-1.42 (0.83)	2.13-19.29 (5.6)	0.26-1.67 (0.96)	1.09-3.04 (2.03)	0.11-1.77 (0.92)	1.60-4.28 (2.56)
chrysene (Chr)	0.69-2.01 (0.98)	2.44-18.48 (5.8)	0.27-2.47 (1.14)	0.78-3.13 (2.15)	0.17-4.37 (1.28)	1.60-5.38 (2.89)
benzo[b]-fluoranthene (BbF)	0.74-2.31 (1.34)	3.73-32.49 (10.5)	0.47-2.40 (1.47)	1.83-6.80 (3.71)	0.52-2.11 (1.27)	2.37-8.04 (4.14)
benzo[k]-fluoranthene (BkF)	0-0.91 (0.55)	0.70-7.39 (2.7)	0.24-1.06 (0.66)	0.71-2.33 (1.32)	0.34-0.94 (0.61)	0.74-2.31 (1.23)
benzo[a]pyrene (BaP)	0.43-0.86 (0.65)	1.67-12.54 (4.3)	0.18-1.22 (0.71)	1.91-4.86 (2.27)	0.36-1.51 (0.94)	1.44-4.17 (2.11)
indeno[1,2,3-cd]-pyrene (InP)	0-0.96 (0.60)	1.59-15.05 (4.8)	0.15-1.59 (0.70)	0.65-2.96 (1.70)	0.29-1.14 (0.64)	1.35-3.65 (2.12)
dibenzo[a,h]-	0-0.81	1.06-	0.03-1.09	0.81-1.53	0.05-	0.66-1.71

anthracene (DahA)	(0.35)	196.79 (29.4)	(0.44)	(1.29)	0.91 (0.44)	(1,12)
benzo[ghi]perylene (BghiP)	0-1.03 (0.55)	1.81-53.23 (10.7)	0.14-1.02 (0.67)	1.00-3.44 (2.04)	0-1.26 (0.62)	1.71-4.38 (2.28)
TPAHs	186.3-531.3 (321.8)	222.7-715.7 (460.5)	124.1-581.9 (252.6)	198.4-528.3 (369.3)	48-607 (251)	283-934.99 (465.1)

The data shows the range (mean).

Compared to reports on other coastal areas, the PAHs concentrations in west Bohai Bay are at a medium pollution level. For example, TPAHs concentrations in west Bohai Bay were higher than in Xiamen coastal area [5], Western Taiwan Straits [11] and Northwestern Mediterranean Sea in France [12]. However, the PAHs concentrations in west Bohai Bay were lower than that in the Persian Gulf [13]. The deteriorating PAHs contamination condition in the Persian Gulf may be due to the Gulf War which led to a large amount of oil pollution [13]. According to the criteria distinguishing the PAHs pollution level [4], the three investigated areas in the present research can be classified as a mild or moderate PAHs pollution status.

3.2. *Composition of PAHs*

The composition patterns and relatively abundant DP and SP PAHs individuals in the investigation areas are shown in figure 2. The compositions of DP PAHs were similar as those in SP PAHs except for the QHD area. For the CFD and HHG areas, Nap was the most dominant chemical and comprised almost 50% TPAHs for both DP and SP PAHs. In contrast, Nap comprised more DP PAHs (about 60%) than SP PAHs (about 40%) in the QHD area.

Furthermore, the proportion of 2-ring and 3-ring low molecular weight (LMW) PAHs ranged from 83.4% to 88.9% in DP PAHs and from 69.6% to 78.3% in SP PAHs. This was higher than the proportion of 4-6 rings PAHs in DP, which are defined as high molecular weight (HMW) PAHs. This phenomenon may be because HMW PAHs have low water solubility and are prone to being absorbed by the particles [4] and deposit in sediments [13].

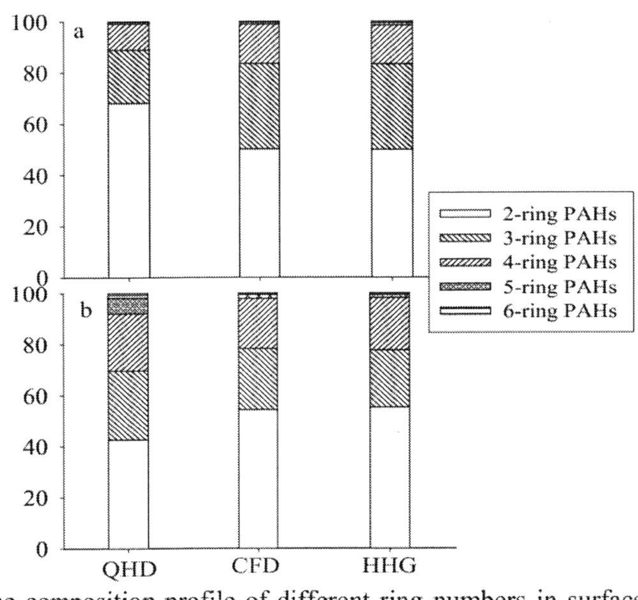

Figure 2. The composition profile of different ring numbers in surface seawater from QHD, CFD and HHG in DP (a) and SP (b) PAHs.

3.3. Ecological risk assessment

The mean values of RQ_{NCs} and RQ_{MPCs} of the PAHs of QHD, CFD and HHG in the west Bohai Bay are listed in table 3. For the three investigated areas, the RQ_{MPCs} of individual PAHs ranged from 0 to 0.50 and the RQ_{NCs} of almost all of the 16 PAHs except for Chr and DahA were higher than 1.0, demonstrating a medium risk level of all individual PAHs. For 2-4 rings PAHs, e.g. Nap, Flu, Phe and Pyr, the RQ_{NCs} values were high especially for Flu and Pyr. This suggests that there is a higher ecological risk for these PAHs individuals. In particular, the RQ_{NCs} values of Nap and Acy in QHD were higher than those in CFD and HHG. This suggests that the ecological risk for these chemicals is higher in QHD. However for some PAHs individuals, including Ace, Flu, Phe, Ant, Chr, BaP and DahA, the RQ_{NCs} values in CFD and QHD were higher than those in HHG, indicating a higher ecological risk in the two areas. Generally, the ecological risk for TPAHs in the surface waters of the three investigated areas was low. The $RQ_{TPAHs(NCs)}$ value was between 1.0-800 and the $RQ_{TPAHs(MPCs)}$ value was <1.0. The values of $RQ_{TPAHs(NCs)}$ and $RQ_{TPAHs(MPCs)}$ for the three areas were similar. Thus, the ecological risk of TPAHs for the three areas is similar as well. The results indicate that TPAHs in the surface water of west Bohai Bay have mild potential risk on the aquatic biota. The management for PAHs pollutions in west Bohai Bay should be considered, especially for Flu and Pyr.

Table 3. Mean values of RQ_{NCs} and RQ_{MPCs} for PAHs in the surface water of west Bohai Bay, China.

		Water (ng/L)		QHD		CFD		HHG	
PAHs	TEFs[a]	NCs[b]	MPCs[c]	RQ_{NCs}	RQ_{MPCs}	RQ_{NCs}	RQ_{MPCs}	RQ_{NCs}	RQ_{MPCs}
Nap	0.001	12.0	1200	16.20	0.16	9.94	0.10	9.25	0.09
Acy	0.001	0.7	70	2.14	0.02	1.45	0.01	1.77	0.02
Ace	0.001	0.7	70	11.34	0.11	7.45	0.07	12.35	0.12
Flu	0.001	0.7	70	33.16	0.33	28.43	0.28	34.73	0.35
Phe	0.001	3.0	300	15.08	0.15	16.99	0.17	17.25	0.17
Ant	0.001	0.7	70	7.17	0.07	7.82	0.08	7.92	0.08
Flua	0.01	3.0	300	3.87	0.04	5.23	0.05	4.96	0.05
Pyr	0.001	0.7	70	38.57	0.39	40.42	0.40	38.44	0.38
BaA	0.1	0.1	10	8.30	0.08	9.62	0.10	9.23	0.09
Chr	0.01	3.4	340	0.29	0.00	0.34	0.00	0.38	0.00
BbF	0.1	0.1	10	13.40	0.13	14.70	0.15	12.72	0.13
BkF	0.1	0.4	40	1.38	0.01	1.64	0.02	1.53	0.02
BaP	1	0.5	50	1.30	0.01	1.42	0.01	1.87	0.02
InP	0.1	0.4	40	1.50	0.02	1.76	0.02	1.60	0.02
DahA	1	0.5	50	0.70	0.01	0.87	0.01	0.89	0.01
BghiP	0.01	0.3	30	1.83	0.02	2.23	0.02	2.07	0.02
TPAHs		27.2	2720	11.83	0.12	9.29	0.09	9.23	0.09

[a] The values of toxic equivalency factors (TEFs) were from references [4,14].
[b, c] The values were from reference [10].

4. Conclusions

The concentrations of DP TPAHs were lower than those of SPPAHs for the three investigated areas. The TPAHs in QHD were higher than those in CFD and HHG. For all three areas, the PAHs contamination was mild to moderate.

Acknowledgments

This study was supported by National Science Foundation of China (Funding Numbers: 21307045, 41303070).

References

[1] Maher W A and Aislabie J 1992 Polycyclic aromatic hydrocarbons in nearshore marine sediments of Australia *Sci. Total Environ.* **112** 143-64

[2] Budzinski H, Jones I and Bellocq J 1997 Evaluation of sediment contamination by polycyclicaromatic hydrocarbons in the Gironde Estuary *Mar. Chem.* **58** 85-97

[3] Huang W, Wang Z and Yan W 2012 Distribution and sources of polycyclic aromatic hydrocarbons (PAHs) in sediments from Zhanjiang Bay and Leizhou Bay, South China *Mar. Pollut. Bull.* **64** 1962-9

[4] Li P, Cao J, Diao X, Wang B, Zhou H, Han Q, Zheng P and Li Y 2015 Spatial distribution, sources and ecological risk assessment of polycyclic aromatic hydrocarbons in surface seawater from Yangpu Bay, China *Mar. Pollut. Bull.* **93** 53-60

[5] Ya M L, Wang X H, Wu Y L, Ye C X and Li Y-Y 2014 Enrichment and partitioning of polycyclic aromatic hydrocarbons in the sea surface microlayer and subsurface water along the coast of Xiamen Island, China *Mar. Pollut. Bull.* **78** 110-7

[6] Qian X, Liang B, Fu W, Liu X and Cui B 2016 Polycyclic aromatic hydrocarbons (PAHs) in surface sediments from the intertidal zone of Bohai Bay, Northeast China: Spatial distribution, composition, sources and ecological risk assessment *Mar. Pollut. Bull.* **112**349-58

[7] Zhang Y D, Tian S Y, Liu X B, Li G F, Zhu W J, Jia R. 2013 Distribution, source and ecological risk assessment of polycyclic aromatic hydrocarbons in surface water from the northwest Bohai sea *Oceanologia Limnologia Sinica* **44** 255-61

[8] Kalf D F, Crommentuijn T and van de Plassche E J 1997 Environmental quality objectives for 10 polycyclic aromatic hydrocarbons (PAHs) *Ecotoxicol. Environ. Saf.* **36** 89-97

[9] Wang Y, Wang J, Mu J, Wang Z, Yao Z and Lin Z 2014 Aquatic predicted no-effect concentration for three polycyclic aromatic hydrocarbons and probabilistic ecological risk assessment in Liaodong Bay of the Bohai Sea *Environ. Sci. Poll. Res.* **21** 148-58

[10] Cao Z G, Liu J L, Wang X M and Jie X 2010 Pollution characteristics, ecological risk assessment and sources of polycyclic aromatic hydrocarbons (PAHs) in surface water from the Zhangweinan River *Acta Scien. Circum.* **30** 254-60

[11] Wu Y-L, Wang X-H, Li Y-Y and Hong H-S 2011 Occurrence of polycyclic aromatic hydrocarbons (PAHs) in seawater from the Western Taiwan Strait, China *Mar. Pollut. Bull.***63** 459-63.

[12] Guigue C, Tedetti M, Ferretto N, Garcia N, Mejanelle L and Goutx M 2014 Spatial and seasonal variabilities of dissolved hydrocarbons in surface waters from the Northwestern Mediterranean Sea: Results from one year intensive sampling *Sci. Total Environ.* **466** 650-62

[13] Sinaei M and Mashinchian A 2014 Polycyclic aromatic hydrocarbons in the coastal sea water, the surface sediment and Mudskipper Boleophthalmus dussumieri from coastal areas of the Persian Gulf: Source investigation, composition pattern and spatial distribution *J. Environ. Health Sci. Eng.* **12** 1-11

[14] Nisbet I C T and Lagoy P K 1992. Toxic equivalency factors (TEFs) for polycyclic aromatic hydrocarbons (PAHs) *Regul. Toxicol. Pharm.* **16** 290-300

IOP Conf. Series: Earth and Environmental Science 82 (2017) 012041 doi:10.1088/1755-1315/82/1/012041

Regulation research on flow backward of Poyang Lake

B Xu[1,4], G C Chen[1,2] and X N Li[3]

[1]Water Resources Department, Yangtze River Scientific Research Institute, Wuhan 430010, China
[2]Planning Bureau of Changjiang Water Resources Commission, Wuhan 430010, China
[3]Guizhou Survey and Design Research Institute for Water Resources and Hydropower, Guiyang 550000, China

E-mail: taibaidugu@163.com

Abstract. The main factors affecting the water level of Poyang Lake include the inflow, outflow, precipitation and flow backward, etc. The Hydrological Alteration Diagnose System was used to analyze the water level series of Hukou Station to uncover the alteration point, and the alteration point was taken as the basis to divide the time series as before the alteration point and after the alteration point, in order to study the alteration regulation of flow backward. The results show that, compared with the flow backward series before the alteration point, the water level and discharge of flow backward after the alteration point were much more intensity, and the tendency of flow backward discharge was decreasing, the times and lasting period of flow backward were much more reduced at the same time. The regulation appeared above may be one of the reasons that cause the water level decreasing, and dry season advancing of Poyang Lake.

1. Introduction

River water backward flow is a typical feature of lakes connected to rivers, providing the function of regulating and storing flood from the Yangtze River, maintaining lake water level, ensuring safe utilization of water resources at lake areas and promoting the ecological evolution of marshlands and wetlands. Due to the impact of the changing relations between rivers and lakes, river water backward flow also exhibits a number of new change rules. Current research on the relations between rivers and lakes is mostly concentrated on the correlation between lake water level and silt in the changing environment. Guo Hua, *et al* [1], based on data analysis for 2004-2008, believe that the large storage of water in the Three Gorges Reservoir in October substantially reduces the frequency of action of the Yangtze River on Poyang Lake, resulting in reduced let-down flow of the river and affecting the water level of the lake to some extent. Wang Yingchun, *et al* [2] use the coupling hydrodynamic model of rivers and lakes at the middle reaches of the Yangtze River for simulation, which indicates that the water storage of the Three Gorges Reservoir will result in reduced water level at Duchang station of Poyang Lake by 0.09-1.11 m. Fang Chunming, *et al* [3] predict that due to the application of the Three Gorges Reservoir and under the combined action of river channel erosion and water storage, the dry season of Poyang Lake will occur about one month earlier; and they also propose simplified conditions to determine the occurrence of backward flow at Hukou station, i.e. daily increased amount of the flow of Jiujiang River exceeds the flow of Hukou of the previous day. Hu Chunhong [4], Zhu

Lingling, *et al* [5] carried out a research on the change of sediment accumulation of mainstream channels of the Yangtze River and Poyang Lake after the Three Gorges Project was put into operation and they pointed out that the operation of the project reduced the speed and volume of sediment accumulation of the lake and increased the river channel erosion.

To sum up, a number of scholars regard the operation of the Three Gorges Reservoir as the node of change regarding the river-lake relations of Poyang Lake but they often fail to carry out further hydrological alteration analysis and the selection of the node lacks theoretical basis; secondly, river water backward flow plays a very obvious role in the replenishment of water for lakes connected to rivers and is susceptible to environmental change; and the backward flow volume and duration are also changing quietly but currently there is a lack of research on the change rule. In view of the above problems, this paper takes backward flow of river water with regard to lakes connected to rivers as the research object and adopts the hydrological alteration diagnosis system to carry out alteration diagnosis of the water level at Hukou station of Poyang Lake; and, based on the diagnosis result, the paper conducts a systematic sorting and analysis regarding the annual change rule of river water backward flow prior to and after the alteration.

2. Database and hydrological alteration diagnose system

2.1. Data sources

Water level of lakes is one of the key factors that affect lake ecological safety. It is the comprehensive reflection of the integrity of the structure and function of lake ecological system and it should reflect the river-lake relations between Poyang Lake and the Yangtze River. Therefore, this paper takes the water level sequence of Hukou station of Poyang Lake as the river-lake relations change node analysis sequence; at it uses the actual runoff sequence of the station to analyze the annual change rule of the river water backward flow prior to and after the alteration.

2.2. Hydrological alteration diagnosis system

The hydrological series composed by definite composition and stochastic composition [6], the Hydrological Alteration Diagnosis System (HADS) was put forward by XIE to diagnose the alteration of jump and trend, by the investigation of the actual conditions, the alteration style will be confirmed at last.

3. Water level series of Hukou station alteration

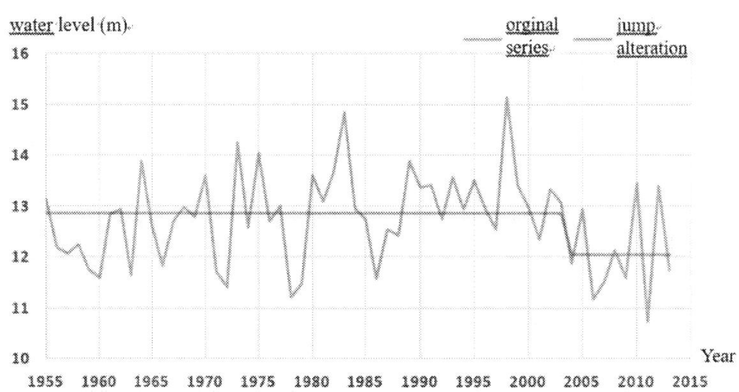

Figure 1. Annual average water level series and jump alteration figure.

With the first degree and second degree of confidence: α=0.05 and β=0.01, the water level series from the year of 1955 to 2013 of Hukou station was diagnosed by HADS. The result shows that the water level of Hukou station experienced a moderate alteration in a leap-down manner in 2003. The mean value of the water level for 1955-2003 before the alteration is 12.85 m and the mean value for 2004-2013 is 12.05 m. The annual water level decrease of Hukou station after the alteration is 0.8 m as shown in the figure 1.

4. Alteration regulation analysis of flow backward

On the basis of the result of the alteration diagnosis of the water level sequence of Hukou station, the river water backward flow sequence of the station for 1958-2012 is divided into two parts, i.e. 1958-2003 before alteration (hereinafter referred to as before alteration,) and 2004-2012 after alteration (hereinafter referred to as after alteration), and the annual change rule of river water backward flow for Poyang Lake is analyzed.

4.1. Water level and discharge alteration regulation of flow backward

With the number of days of backward flow as the sample, analysis is carried out regarding the overall sample water level and flow in the first place.

Based on the measured data of Hukou Station, the time of flow backward was 574 days before the alteration, and 82 days after the alteration, shown in figures 2 and 3 below.

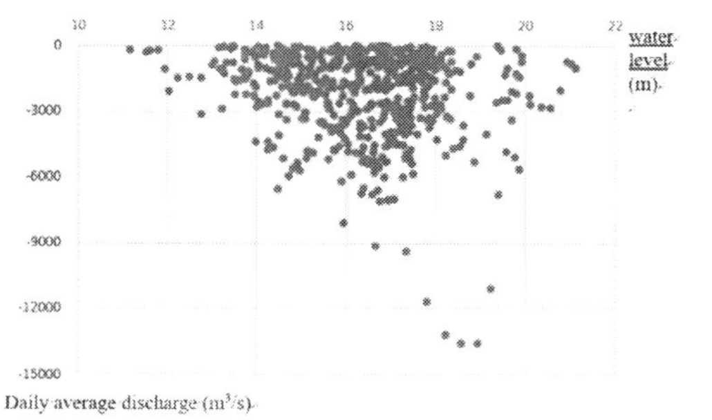

Figure 2. Daily average discharge of Hukou Station before jump alteration.

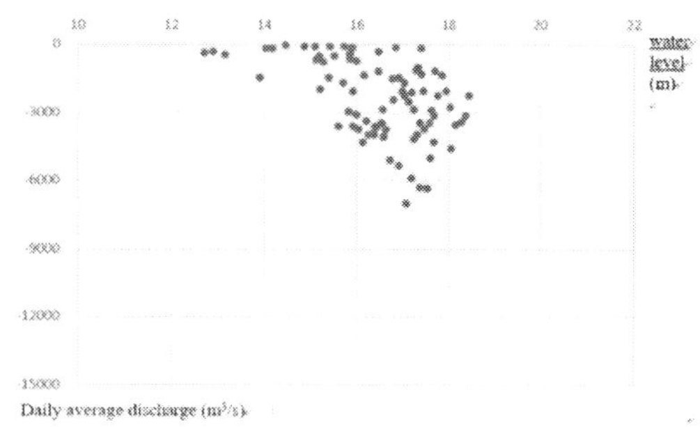

Figure 3. Daily average discharge of Hukou Station after jump alteration.

Table 1. Flow backward water level and discharge border variation before and after the alteration.

Item	Minimium Waterlevel (m)	Maxium Waterlevel (m)	Maxium daily average discharge (m^3/s)
BA	11.16	21.12	13600
AA	12.73	18.46	7000

Notes: "BA" means Before Alteration; "AA" means After Alteration; same as follows.

Table 2. Flow backward water level distribution zones variation before and after the alteration.

Water level of flow backward (m)	BA		AA	
	Times	Probability	Times	Probability
Less than 12	5	0.87	0	0.00
12-14	39	6.79	4	4.88
14.01-16	175	30.49	21	25.61
16.01-18	288	50.17	51	62.20
18.01-20	59	10.28	6	7.32
Bigger than 20	8	1.39	0	0.00
Sum	574	100	82	100

The extreme value of water level and discharge of Hukou Station when the flow backward happened before and after the alteration were shown in table 1.

From table 2, it can be seen that the water level and maximum daily average flow in case of backward flow of river water experiences substantial changes before and after water level alteration of Hukou station. First, the water level distribution interval in case of river water backward flow after alteration is more concentrated, only at the interval of 12.73-18.46 m, a decrease of 4.23 m compared to the value of 11.16- 21.12 m before alteration; secondly, the maximum value of daily average backward flow decreases remarkably from 13,600 m^3/s to 7,000 m^3/s, representing a decrease rate of 48.5%.

The total flow backward water level samples were satisfied, shown as the table 2.

From table 3, it can be seen that the water level distribution interval in case of river water backward flow of Hukou station experiences some change. First, the water level distribution after alteration is more concentrated, the probability of occurrence at 16.01-18 m increases to some extent compared to that before alteration and the water level interval decreases; secondly, at the interval of water level less than 12 m and more than 20 m after alteration, there is no river water backward flow.

The total flow backward discharge samples were satisfied, shown as table 3.

Table 3. Flow backward discharge distribution zones variation before and after the alteration.

Discharge of flow backward (m^3/s)	BA		AA	
	Times	Probability	Times	Probability
Less than 1500	271	47.21	28	34.15
1500-3000	152	26.48	22	26.83
3001-4500	77	13.41	24	29.27
4501-6000	54	9.41	5	6.10
6001-7500	12	2.09	3	3.66
Bigger than 7500	8	1.39	0	0.00
Sum	574	100	82	100

From table 4, it can be seen that the river water backward flow distribution interval before and after water level alteration experiences some change. Before alteration, about half of the backward flow is distributed within the range less than 1,500 m^3/s and, with the increase of backward flow, the probability of occurrence decrease accordingly, showing substantial regularity; after alteration, only about one third of the backward flow is distributed within the range less than 1,500 m^3/s and the distribution probability in the interval of 1,500-3,000 m^3/s and 4,501-6,000 m^3/s is very close, showing substantial decentrality of backward flow.

4.2. Times of flow backward alteration regulation

Table 4. Flow backward annual times variation before and after the alteration.

annual times	BA		AA	
	Years	Probability	Years	Probability
1	7	20.00	2	28.57
2	5	14.29	3	42.86
3	11	31.43	1	14.29
4	7	20.00	1	14.29
5	3	8.57	0	0.00
6	1	2.86	0	0.00
7	1	2.86	0	0.00
Sum	35	100	7	100

IOP Conf. Series: Earth and Environmental Science 82 (2017) 012041 doi:10.1088/1755-1315/82/1/012041

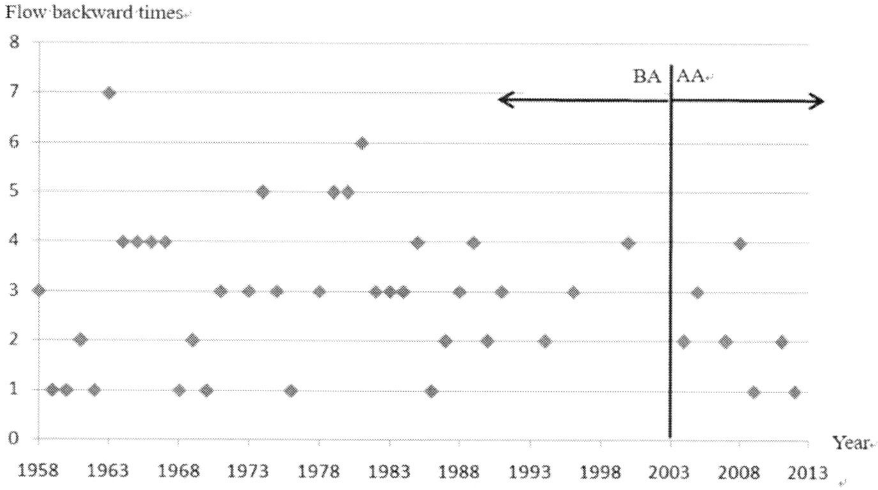

Figure 4. Flow backward annual times distribution before and after the alteration.

With the frequency of backward flow as the sample, analysis is made regarding the annual change rule of the sample water level and flow.

According to the actual backward flow data of Hukou station, river water backward flow occurs in 35 years before alteration, totalling 106 times; and seven years after alteration, totalling 15 times. The number of times of backward flow each year before and after alteration is shown in figure 4 and table 4.

Table 5. Flow backward lasting days variation before and after the alteration.

Lasting days	BA		AA	
	Years	Probability	Years	Probability
1-5	5	14.29	3	42.86
6-10	6	17.14	0	0.00
11-15	7	20.00	1	14.29
16-20	7	20.00	1	14.29
21-25	2	5.71	2	28.57
26-30	6	17.14	0	0.00
31-40	1	2.86	0	0.00
>40	1	2.86	0	0.00
Sum	35	100	7	100

From table 5, it can be seen that when backward flow occurs before alteration, the largest probability is three times per year, totaling 11 years; the probability within four times a year is 85.72%, totaling 30 times. When backward flow occurs after alteration, the largest probability is two times a year, totaling three years and the probability within four times a year is 100%; the number of times of backward flow after alteration decreases substantially compared to that before alteration.

The annual flow backward duration samples before and after the alteration were satisfied, shown as the table 5 and figure 5.

From table 5 it can be seen that when backward flow occurs before alteration, the probability of various durations is relevantly average, with the largest probability not more than 20%. After alteration, the backward flow duration is reduced significantly, with the largest probability within five days and it is distributed in a concentrated manner; in addition, there is no backward flow that lasts for more than 25 days.

IOP Conf. Series: Earth and Environmental Science **82** (2017) 012041 doi:10.1088/1755-1315/82/1/012041

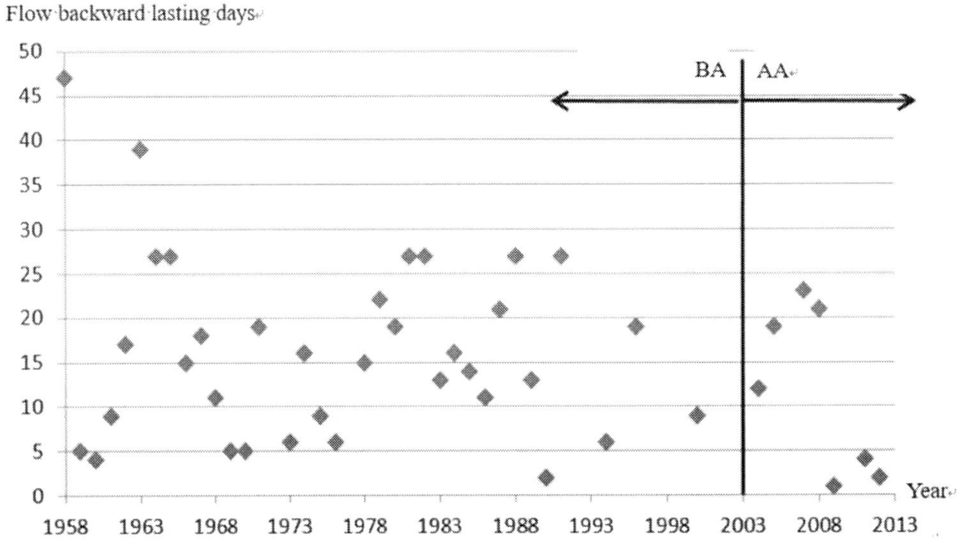

Figure 5. Flow backward annual lasting days distribution before and after the alteration.

5. Conclusions and discussions

5.1. Conclusions

Hydrological alteration diagnose system was used to diagnose the series of measured water level of Hukou Station, and based on the alteration results, the year of 2003 was taken as the point to divide the series as before alteration and after alteration, therefore, the alteration regulation could be studied, the main concusions as follows:

- The water level of Hukou station experienced a moderate alteration in a leap-down manner in 2003. The mean value of the water level for 1955-2003 before the alteration is 12.85 m and the mean value for 2004-2013 is 12.05 m. The annual water level decrease of Hukou station after the alteration is 0.8 m.
- Based on the measured data of Hukou Station, the time of flow backward was 574 days before the alteration, and 82 days after the alteration. The water level became much more concentrate after the alteration, only at the interval of 12.73-18.46 m, a decrease of 4.23 m compared to the value of 11.16- 21.12 m before alteration. The maximum discharge decreasing apparently, from 13,600 m^3/s to 7,000 m^3/s, representing a decrease rate of 48.5%. The backward flow distribution interval before and after water level alteration experiences some change. Before alteration, about half of the backward flow is distributed within the range less than 1,500 m^3/s and, with the increase of backward flow, the probability of occurrence decrease accordingly, showing substantial regularity; after alteration, only about one third of the backward flow is distributed within the range less than 1,500 m^3/s and the distribution probability in the interval of 1,500-3,000 m^3/s and 4,501-6,000 m^3/s is very close, showing substantial decentrality of backward flow.
- According to the actual backward flow data of Hukou station, river water backward flow occurs in 35 years before alteration, totaling 106 times; and seven years after alteration, totaling 15 times.

When backward flow occurs before alteration, the largest probability is three times per year, totaling 11 years; the probability within four times a year is 85.72%, totaling 30 times. When backward flow occurs after alteration, the largest probability is two times a year, totaling three years and the probability within four times a year is 100%; the number of times of backward flow after alteration decreases substantially compared to that before alteration.

When backward flow occurs before alteration, the probability of various durations is relevantly average, with the largest probability not more than 20%. After alteration, the backward flow duration is reduced significantly, with the largest probability within five days and it is distributed in a concentrated manner; in addition, there is no backward flow that lasts for more than 25 days.

5.2. Discussion

There still exist some problems needing to be studied in the next step.

- Although the alteration happened in the year of 2003 at Hukou Station, the attribution degree of the Three Gorges needs to be studied in the next step.
- Compared with the series before the alteration point, the flow backward times, duration, and the water quantity are all decreasing after the alteration, that maybe one of important reasons that result in the low water level in the dry season, but the attribution degree and countermeasures according need to be studied in the next step.

Acknowledgments

This study was financially supported by the Basic Research Fund of Central Public-interest Scientific Institution (NO. 20150914091245); also supported by the National Natural Sciences Foundation of China (NO. 51409014) and by Guizhou Science and Technology Project (Planning of Technology in Guizhou [2016]2903).

References

[1] Guo H, Hu Q and Zhang Q 2011 *Acta Geogr. Sin.* **66** L609
[2] Wang Y C, Lai X J and Jiang J H 2011 *J. Lake Sci.* **23** L191
[3] Fang C M, Cao W H and Mao J X 2012 *J. Hydraul. Eng.* **43** L175
[4] Hu C H and Wang Y G 2014 *J. Yangtze River Sci. Res. Inst.* **31** L107
[5] Zhu L L, Chen J C and Yuan J. 2014 *Adv. Water Sci.* **25** L348
[6] Xie P, Chen G C and Lei H F 2009 *J. Basic Sci. and Eng.* **17** L1

Characteristics of precipitation in the rainy season of Koxkar Glacier, Tianshan Mountain, China

C C Zhao[1,2], S X Yao[2], H D Han[3,4] and Q D Zhao[3]

[1]School of Geography and Planning, Lanzhou City University, Lanzhou 730070, China
[2]School of Mathematics, Lanzhou City University, Lanzhou 730070, China
[3]State Key Laboratory of Cryospheric Science, Cold and Arid Regions of Environmental and Engineering Research Institute, Chinese Academy of Sciences, 320 Donggang West Road, Lanzhou 730000, China

E-mail: hhd@lzb.ac.cn

Abstract. Precipitation is widely recognized as the major driving force of hydrologic system, and changes of its pattern could have direct impacts on water resources, especially in mountains. The total precipitation and rain days based on observation were analysed in Koxkar glacier catchment in rainy season during the period of 2009 - 2016. The results have shown that the total precipitation of rainy season fluctuates randomly with no significant trend of increasing or decreasing. The rain days of different precipitation classes differ noticeably, and are inconsistent with the contribution of total precipitation in rainy season. The maximum daily precipitation influences the monthly precipitation. These characteristics of rainy season precipitation display the important factor for the study of glacier changes.

1. Introduction

Precipitation is an important part of fundamental components of the climate and hydrological cycle in understanding the water balance and the complex process of energy interactions among the small- and large-scale [1, 2]. In general, precipitation data can be obtained from three observational techniques, 1) measurement directed by rain gauge, 2) ground based radar networks, 3) space-borne satellite instruments [3, 4]. Although the weather radar and satellite remote sensing techniques are provided high spatial-temporal resolution with complete area coverage and available in real-time [5, 6], the rain gauge observations are considered relatively accurate in situ measurements in spite of suffering from various problems, such as poor spatial sampling, undercatchment due to wind and evaporation, especially snowfall [7, 8]. The observation has been widely applied in climate analysis, climate diagnostic, drought relief, and flood forecasting [9-11].

Tianshan Mountain is one of the largest mountains in the northwest of China, which is a typical arid and semi-arid region. It stretches from west to east for about 1700 km, and from south to north for about 250-300 km. The average elevation is approximately 4000 m.s.l. [12, 13]. Because it locates in the hinterland of Eurasian continent, which is far away from the ocean, precipitation is the main source of surface water to supply for river runoff, farm irrigation and drinking. Since it is located in the prevailing westerlies, the major water vapor is come from west wind current, and the dry, cold current of Arctic Ocean is the minor resource which accounts for 25-33% of that of the west wind current. Therefore, Tianshan Mountain originates several important inland rivers in northwest China and

Central Asia, and has become the mainly water resources of downstream oasis. The condition and variability of precipitation over Tianshan Mountain play an important role for the ecosystems and environment, and regional sustainable development [14, 15].

2. Materials and methods

2.1. Study area

The Koxkar glacier catchment (41°42' N-41°53' N and 79°59' E-80°10' E) is located in the southwest Tianshan, Wensu County in Xinjiang Province, China. The whole catchment covers an area of 116.5 km^2, out of which 62.8% is glacierized area, with an average elevation of approximately 4000 m.a.s.l. [16]. The Koxkar Glacier extends 25.1 km in length and covers an area of 73.2 km^2, and 93.6% of discharge of the Koxkar Glacier occurs in the ablation season [17]. The mean annual air temperature observed near the terminus of the glacier is 0.778℃, while the mean temperature of summer is as high as 7.748℃. Mean annual precipitation is near the terminus is 608 mm, of which >80% occurs during the ablation period [18]. A belt of low rainfall is around approximately 3700 m, while greater in the terminus and accumulation zones of the glacier [19].

2.2. Methods

A manual observation station is located near the glacier terminus at an elevation of 3009 m from 2008 to present in rainy season. Precipitation observation carried out at GMT 8:00 am and 8:00 pm in every day respectively. The amount of day precipitation is recorded from GMT 20:00 to 20:00 of next day. Though, there are seven automatic weather stations installed in the catchment which observed the air temperature and the precipitation automatically recently. Due to the short period of these automatic weather stations, analyzed the characteristics of precipitation is only used the data observed by manual. The precipitation was missed in September 2008; the period of analysing is from May to September when the major rainfall occurred in the period of 2009-2016.

2.3. Precipitation classes

During rainy season, the number of days with precipitation is determined in particular classes according to Elżbieta [20] criterion as shown in table 1:

Table 1. Precipitation classes.

Amount (mm/d)	Classes
<10	Light
10~24.9	Moderate
⩾25	High

3. Results

3.1. Total precipitation in rainy season

The variation of rainy season precipitation is large in Koxkar glacier catchment over the period of 2009 – 2016 (figure 1). The average precipitation is 453.5 mm during the observation period. The maximum of rainy season is 546.4 mm occurred in 2013, while the minimum is 335.8 mm occurred in 2014. The difference exceeds 200 mm. The rainy season precipitation decreased dramatically over the period of 2010 – 2012, but increased significantly over the period of 2014 – 2016.

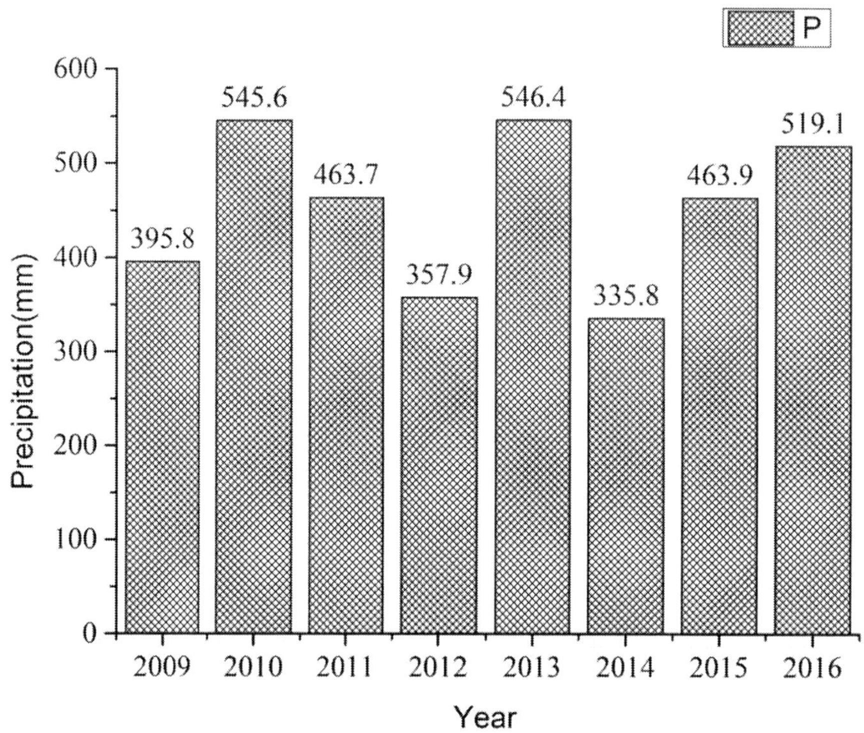

Figure 1. Total rainy season precipitation over the period of 2009 – 2016.

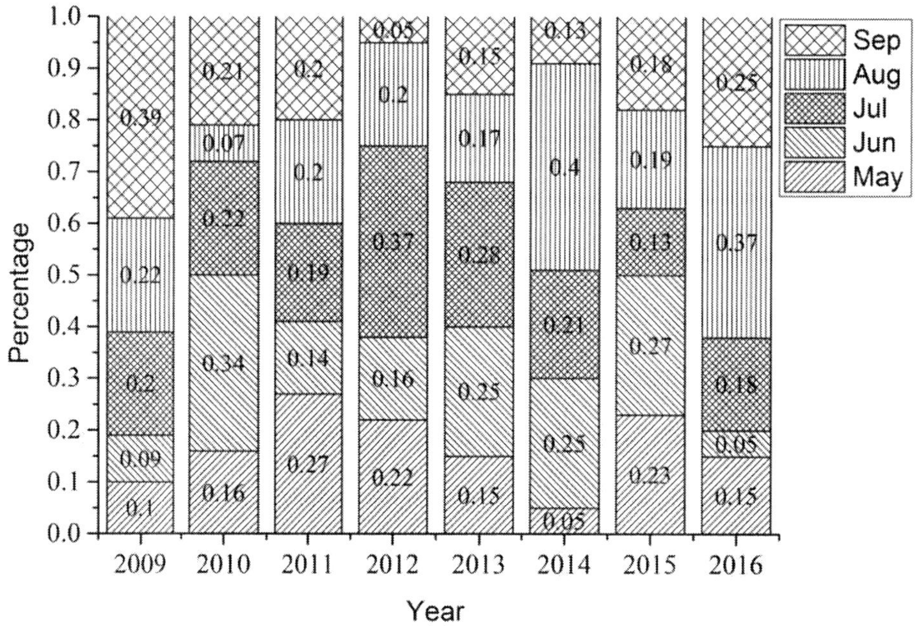

Figure 2. The monthly percentage in rainy season over the period of 2009 – 2016.

In order to better analyse the characteristic of rainy season, we calculated the monthly percentage in rainy season during the observation period (figure 2). The variation of monthly percentage is similar to that of the rainy season, undulated significantly during the observation period. In May, the percentage of rainy season precipitation ranges from 0.1 to 0.2, and the mean value is 0.18 except 2014. The maximum value 0.27 occurred in 2011, but the minimum is only 0.1 occurred in 2014 in which the rainy season precipitation is minimum. In June, the percentage of rainy season precipitation ranges from 0.14 to 0.34, the mean value is 0.24 except in 2009 and 2016. In July, the percentage of rainy season precipitation is the range 0.13 to 0.37, the mean value is 0.22. The percentage of July is above the 0.2 from 2009 to 2014, below 0.2 nearly two year. In August, the percentage of rainy season precipitation ranges from 0.17 to 0.4, the mean value is 0.25 except in 2010. In September, the percentage of rainy season precipitation ranges from 0.13 to 0.39, the mean value is 0.22 except in 2012. The percentage of September decreased first, and then increased.

3.2. Precipitation days in rainy season

On a rainy season basis, the number of rainy days is insufficient half of observation days except 2016 (table 2). The maximum value of rainy days occurred in 2016, while the minimum occurred in 2014 which the rainy season precipitation is the minimum. It is obvious that the days of light rain are major in rainy season accounted for 70% above. However, the amount of light rain is fall short of half total precipitation in rainy season. The number of moderate rain days is below 15, there are few days of high rain. The amount of moderate rain accounts for 40% to 50 % of total precipitation in rainy season. Though the fewer days of high are occurred, destructive flood can still form due to large precipitation intensity.

Table 2. Precipitation days in different classes.

Year	Days			Amount (mm)		
	Light	Moderate	High	Light	Moderate	High
2009	53	11	2	145.2	174.9	75.7
2010	53	13	4	212.3	223	110.3
2011	59	14	1	244.2	193.4	26.1
2012	64	8	2	182.5	115.5	59.9
2013	54	15	4	188.8	222.6	135
2014	52	15	1	158.2	149.4	28.2
2015	62	15	1	182.9	237.5	43.5
2016	67	13	3	220.5	214.8	83.8

It is shown that the variation of light rainy days has direct and strong impact on the variation of rain days, and the variation of moderate rainy days and high days has weaker influence than that of light rainy days. In order to deeply analyse the effect of different precipitation classed on total precipitation of rainy season, we investigate the relationship between amount of different precipitation classes and total precipitation in rainy season (figure 3). It is easy to see that the correlation is the best between amount of moderate rain and total precipitation, the correlation coefficient is up to 0.86. However, the amount of high rain is few, the correlation is well between high rain and total precipitation, with the correlation coefficient is up to 0.68. The correlation is weak between light rain and total precipitation in spite of the maximum of rainy days.

3.3. Maximum daily precipitation in rainy season

Table 3 shows that the maximum daily precipitation in rainy season over the period of 2009 – 2016. Although the maximum amount of monthly precipitation did not occur in September, the maximum daily precipitation is the largest in September compared to that in other months. The maximum daily precipitation occurred twice, in May and July respectively, only once in August, and none in June. The amount of maximum daily precipitation accounts for more than 30% of the monthly precipitation.

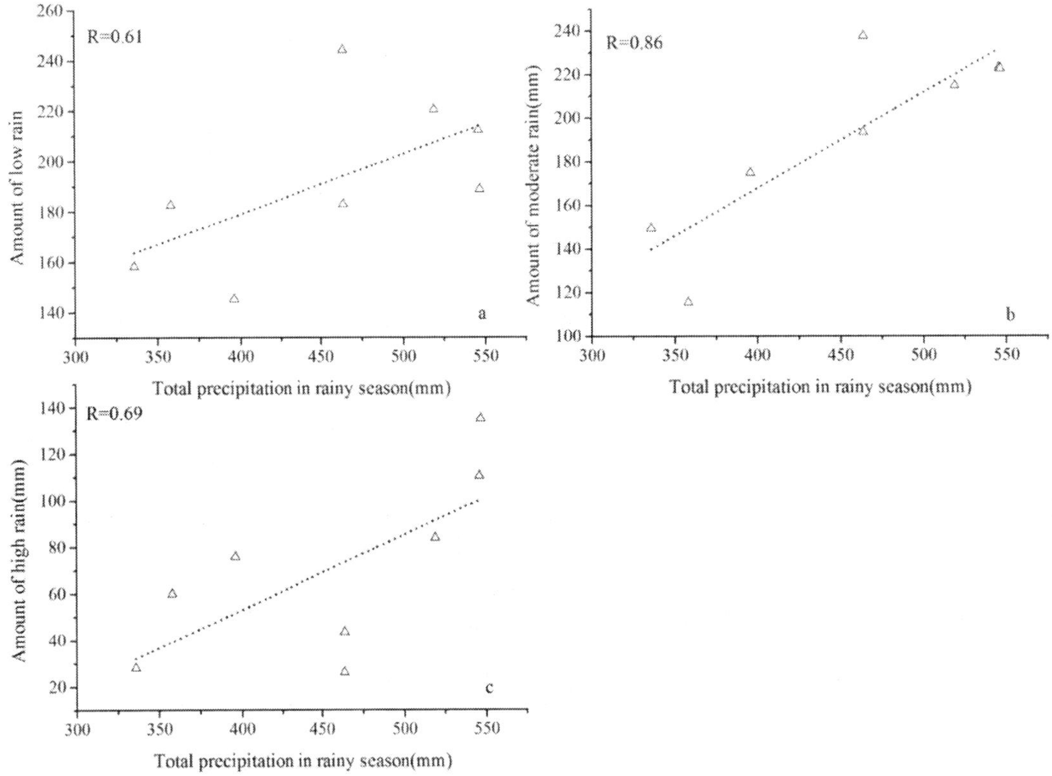

Figure 3. The relationship between amount of light, moderate, high and total precipitation in rainy season.

Table 3. Maximum daily precipitation in rainy season over the period of 2009 – 2016.

Maximum daily precipitation	Amount of precipitation (mm)
5 Sep, 2009	38.2
17 Sep, 2010	29.4
11 Aug, 2011	26.1
12 Jul, 2012	31.9
16 Sep, 2013	45.9
30 Jul, 2014	28.2
18 May, 2015	43.5
12 May, 2016	29.8

4. Conclusion

The total precipitation, precipitation days and maximum daily precipitation in rainy season have been exhibited in this paper during the period of 2009 – 2016 over Koxkar glacier catchment. The main conclusions of the study are summarized as follows:

- In study area, due to the large variability of precipitation in rainy season, trends of significant increase or decrease did not exist. The difference of rainy season precipitation is significant during the period of observation. The variability of monthly precipitation is similar to that of total precipitation. It is critical to the study of glacier changes.

- The days of precipitation are basically above 70 days except in individual years, which account for 40% over the period of observation. The light rain days take up the majority, and the high rain days are very few. The moderate rain days are below 15 days, but the amount of moderate rain affected the total precipitation significantly in rainy season.
- The maximum daily precipitation occurs at different times during the period of observation, and its amount has contributed to the monthly precipitation.

Acknowledgments
This study was supported by the National Natural Science Foundation of China (No. 41361013; No. 41471060; No. 31300388). The authors would like to express their gratitude to Dr. Q F Li and S H Guo, State Key Laboratory of Cryospheric Science, Cold and Arid Regions of Environmental and Engineering Research Institute, Chinese Academy of Sciences, for their valuable suggestions and cooperation.

References

[1] Adler R F, Huffman G J, Chang A, Ferraro R, Xie P P, Janowiak J, Rudolf B, Schneider U, Curtis S, Bolvin D, Gruber A, Susskind J, Arkin P and Nelkin E 2003 The version-2 global precipitation climatology project (GPCP) monthly precipitation analysis (1979-present) *J. Hydrometeorol.* **4(6)** 1147-67
[2] Kidd C and Huffman G 2011 Global precipitation measuremen *Meteorol. Appl.* **18(3)** 334-53
[3] Rubel F, Brugger K and Michaelides S 2009 3-hourly quantitative precipitation estimation over central and northern europe from rain gauge and radar data *Atmos. Res.* **94(4)** 544-54
[4] Tapiador F J, *et al* 2012 Global precipitation measurement: methods, datasets and applications *Atmos. Res.* **104-105(1)** 70-97
[5] Chen M, Xie P, Janowiak J E and Arkin P A 1997 Global land precipitation: A 50-yr monthly analysis based on gauge observations *B. Am. Meteorol. Soc.* **78(11)** 2539-58
[6] Symeonakis E, Bonifaçio R and Drake N 2009 A comparison of rainfall estimation techniques for sub-saharan Africa *Int. J. Appl. Earth Obs.* **11(1)** 15-26
[7] Ye B, Yang D, Ding Y, Han T and Koike T 2007 A bias-corrected precipitation climatology for China *J. Hydrometeorol.* **5(6)** 1147
[8] Yang D Q, Wang N L, Ye B S and Ma L J 2009 Recent advances in precipitation-bias correction and application *Sciences in Cold and Arid Regions* **1(3)** 193-8
[9] Curtis S and Adler R 2000 Enso indices based on patterns of satellite-derived precipitation *J. Climate* **13(15)** 2786-93
[10] Stephenson D B, Chauvin F and Royer J F 1998 Simulation of the asian summer monsoon and its dependence on model horizontal resolution *J. Meteorol. Soc. Jpn.Ser.II* **76(2)** 237-65
[11] Dai A 2001 Global precipitation and thunderstorm frequencies. Part I: Seasonal and Interannual vAriations *J. Climate* **14** 1092-111
[12] Wang T, Ren G, Chen F and Yuan Y 2014 An analysis of precipitation variations in the west-central tianshan mountains over the last 300years *Quatern Int.* **358** 48-57
[13] Zhao C, Yao S, Zhang S, Han H, Zhao Q and Yi S 2015 Validation of the accuracy of different precipitation datasets over tianshan mountainous area *Adv. Meteorol.* **2015(3)** 1-10
[14] Huang B and Gong Z 2005 Geochemical barriers and element retention in soils in different landscapes of the tianshan mountain area, xinjiang, China *Geoderma* **126(3-4)** 337-51
[15] Zhang M J, Li R X, Jia W X and Wang X F 2009 Temporal and spatial changes of potential evaporation in Tianshan Mountains form 1960 to 2006 *Acta Geogr. Sin.* **64** 798-806
[16] Han H D, Ding Y J, Liu S Y and Wang J 2015 Regimes of runoff components on the debris-covered koxkar glacier in western China *J. Mt. Sci.* **12(2)** 313-29
[17] Han H, Liu S, Jian W, Qiang W and Xie C 2010 Glacial runoff characteristics of the koxkar glacier, tuomuer-khan tengri mountain ranges, China *Environ. Earth. Sci.* **61(4)** 665-74
[18] Han H D, Liu S Y, Ding Y J, *et al* 2008 Near-surface meteorological characteristics on the

koxkar baxi glacier,tianshan *J. Glaciol. Geocryol.* **30(6)** 967-75
[19] Wang X, Liu S, Han H, Wang J and Liu Q 2012 Thermal regime of a supraglacial lake on the debris-covered koxkar glacier, southwest tianshan, China *Environ. Earth. Sci.* **67(1)** 175-83
[20] Elżbieta R 2014 Classification of precipitation intensity during vegetation season in central-eastern poland (1971–2005) *J. Ecol. Eng.* **15(3)** 51-5

Statistical analysis of soil moisture at grassland in the Horqin Sand Land

S X Yao[1], C C Zhao[1,2] and S Y Wang[1]

[1]Lanzhou City University, Lanzhou 730070, China

E-mail: zhao_chch1978@163.com

Abstract. Firstly, based on the observed data of soil moisture (SM) at grassland from May to September in 2009~2012, we analyzed the variation and statistical characteristics of SM from 0 to 150 cm in growing seasons (from May to September) at grassland of Horqin Sand Land by statistical analysis and the test method of normal distribution, and the normality of SM in different soil layer and different month were tested by Skewness, Kurtosis test method. Secondly, based on the SM of obeying normal distribution, we obtained the estimates and the confidence interval of the mean and variance for the normal distribution. The results indicated that: (1) The SM of grassland at the same soil layer and same month passed the hypothesis testing of normal distribution at the significance level of $\alpha = 0.1$, but the SM of grassland at different months did not pass this hypothesis testing; (2) The value of SM at grassland ranged from 2%~10%; The standard deviation of SM at grassland was 1~5; The variation coefficient of SM at grassland was between 0~1; (3) Influenced by precipitation, SM for grassland was higher in July, and significant different from other months in the growing season (from May to September); The estimates and confidence interval for the mean and variance of SM at grassland in July were significantly different from other months of the growing season; The variation of SM at grassland with month increased firstly and then decreased. (4) SM of grassland was lower at 0~10 cm and different from other soil layer; With the increasing of soil depth, SM of grassland presented a "adverse S" type.

1. Introduction

Soil moisture (SM) is the main source of plant water consumption, and has a significant impact on the physiological activities of plants [1]. Moreover, SM is not only an environmental factor sensitive to climate change, but also an important link of water cycle [2]. Therefore, the study of SM in soil-plant-atmosphere continuum (SPAC) is one of the hot topics of research in the current international academic circle [3]. However, for semi-arid sandy land, water is a key factor determining the structure and function of ecosystem in the region, and the changes in its hydrological cycle process are often a direct driving force for desertification and all other ecological problems in the region [3]. Therefore, the SM in semi-arid sandy land is an important parameter for the study of drought characteristics [4], and hence has become a hot spot of research on ecological environment and a basic scientific issue to be addressed first for the protection of ecological environment and vegetation restoration and reconstruction in sandy land [3]. At present, there have been some studies about SM of sandy land [5, 6], but there are few reports on the statistical analysis on the SM in grassland in Horqin Sand Land.

Horqin Sand Land is one of China's four major sandy lands. The ecological environment here is fragile, and SM is a key factor for its ecosystem stability, structure and normal functioning [7]. Due to moisture restraints, the restoration of ecosystem in arid areas has become a worldwide problem [8]. Therefore, it is necessary to carry out statistical analysis on the SM in grassland, which was one of the main types of sandy land in the study area. With the help of the SM data on grassland by means of positioning and monitoring in 4 growing seasons from 2009 to 2012 of Naiman Station for Desertification Experiment and Research of the Chinese Ecosystem Research Network, in this paper, we used Skewness and kurtosis test method in statistics to analyze the normality of SM in the grassland, and obtained the mean value and point estimation of variance and the confidence interval at the confidence level of 0.95, aiming to provide a theoretical basis and guidance to improve the utilization rate of SM in grassland in this region.

2. Materials and methods

2.1. Study area description

The study area is located in the southern part of the the Horqin Sand Land in eastern Inner Mongolia, China (42°55′ N, 120°42′ E, 345 m a.s.l.). The Horqin Sand Land is one of four well-known sandy areas in northern China, which are generally thought to originate from sand and dust storms that occur frequently in the arid and semi-arid regions of northern China [9]. This area has a temperate continental semi-arid monsoonal climate. Annual precipitation is 366 mm. Mean annual potential pan-evaporation is approximately 1935 mm.

2.2. Soil properties of the site

Soil saturated hydraulic conductivity, bulk density and soil organic matter content were listed in table 1.

Table 1. Soil physical and chemical properties.

Soil layer (cm)	Soil saturated hydraulic conductivity (mm/ d)	Bulk density (g/cm^3)	Organic matter content (%)
0~10	403.2	1.51	3.20
10~30	376.4	1.43	2.74
30~60	275.7	1.41	3.43
60~90	201.4	1.34	3.36
90~120	171.3	1.31	3.31
120~150	132.6	1.27	3.43
Mean	260.1	1.38	3.25

2.3. Methods

CNC100 (Beijing) neutron moisture meter was employed for determination of SM. Three neutron moisture meters at the depth of 2 m were buried at an equal interval on each sample land for regular observation of SM at the depth of 0~150 cm in grassland. Every 10 cm constituted a new soil layer. From early May to the end of September every year, SM was determined once every 10d without rain, and one more time after each rainfall for 4 consecutive years (2009-2012), and the determination was repeated three times every time. The measurement values of the 3 neutron moisture meters were calibrated by the traditional drying method to calculate the average moisture for analysis.

"Skewness and Kurtosis test method" in statistics was employed for normality test of SM [10]. The basic idea is as follows: If the population X is a normal variable, assume that X_i is a sample from the population X, and propose the testing hypothesis that H_0: X is a normal population, let

$$\sigma_1 = \sqrt{\frac{6(n-2)}{(n+1)(n+3)}};\sigma_2 = \sqrt{\frac{24(n-2)(n-3)}{(n+1)^2(n+3)(n+5)}},$$

IOP Conf. Series: Earth and Environmental Science **82** (2017) 012043 doi:10.1088/1755-1315/82/1/012043

$$U_1 = G_1 / \sigma_1; U_2 = (G_2 - \mu_2) / \sigma_1; \mu_2 = 3 - 6/(n+1).$$

If H_0 is true, and n is sufficiently large, the observed values of $|U_1| = |G_1/\sigma_1|$ and $|U_2| = |G_2 - \mu_2/\sigma_2|$ should not be too large, that is, $|u_1| = |g_1/\sigma_1|$ and $|u_2| = |g_2 - \mu_2/\sigma_2|$ should be smaller. Where, g_i is the observed value of G_i, $i = 1, 2$.

In addition, when n is sufficiently large,

$$G_1 \sim N(0, \frac{6(n-2)}{(n+1)(n+3)}), G_2 \sim N(3 - \frac{6}{n+1}, \frac{24(n-2)(n-3)}{(n+1)^2(n+3)(n+5)}).$$

Where $G_1 = B_3 / B_2^{3/2}$ and $G_2 = B_4 / B_2^2$ were the third-order origin moment and the fourth-order origin moment of the moment estimators for X, respectively [10].

Data processing was completed by using 2003 software. Normality test and parameter estimation of SM as well as analysis on the differences of SM at different months and at different soil layers were all completed using SPSS 17.0 statistical analysis software [11].

3. Results

"Skewness and kurtosis test method" was used for testing of 0~150 cm SM in grassland on Horqin Sand Land between May and September during 2009-2012. The results showed that SM at the same soil layer in the same month passed the normal distribution test with significance level $\alpha = 0.1$, while the SM at the same soil layer in different months failed to pass the normality test.

Let $\overline{X}$ and S^2 was the unbiased estimated value of sample mean value (μ) and sample variance (σ^2), respectively. The confidence interval of $\overline{X}$ and S^2 at the confidence levels of 0.95 obtained from SM which have passed normality test in grassland was shown in table 2.

As shown in table 2, the range of the unbiased estimated value of sample mean value in grassland was 2%~10%, mainly concentrated within 5%~8%, accounting for 76.7%. Due to the heaviest precipitation in July in the study area [12], and given the fact that the main source of SM in the region is recharge from precipitation [13], the SM in July is significantly higher than that in other months. The confidence interval of μ was mainly between 2% and 10%.

From table 2 we also seen that both the sample mean value and the sample variance estimated value of SM in grassland in July were greater than those in other months, and the confidence interval significantly shifted to the right compared with other months. The statistical characteristics of SM in grassland (table 2) showed, the standard deviation of SM in grassland of different depths in different months was basically 1~5; the coefficients of variation were 0~1, that was, it was belonged to the moderate variability. In addition, the unbiased estimated of σ^2 in grassland was 1~29, and the confidence interval of σ^2 was mainly between 1 and 47. The range of the confidence interval of σ^2 was wider than μ.

By significant test (table 2), we found that there were significant differences ($P<0.05$) between SM in July and other months at the same soil layer, and SM at 0~10 cm, 10~30 cm and 30~60 cm were significantly different from each other.

By further analysis the variation of SM in grassland with month and soil layer (figure 1) we discovered that, influenced by precipitation, the SM at different soil layers was highest in July (figure 1(a)). After analysing the various SMs with soil layer (figure 1(b)), we found that SM at 0~10 cm was the lowest, and with increasing of soil depth, SM increased firstly, and decreased after then, and increased finally.

Table 2. Statistical characteristics, unbiased estimated and confidence interval at 0.95 levels of μ (%) and σ^2 of SM at grassland.

Soil layers/ (cm)	Month	Unbiased estimated of μ	Unbiased estimated of σ^2	Confidence interval of μ	Confidence interval of σ^2	Statistical Characteristics of SM	
		$\overline{X}$	S^2			SD	CV
0 ~ 10	5	3.64^{ab}	5.17	(2.93, 4.34)	(3.50, 8.41)	2.25	0.62
	6	3.86^{ab}	5.97	(3.13, 4.59)	(4.09, 9.53)	2.42	0.63
	7	5.47^{a}	25.67	(4.08, 6.86)	(18.10, 40.20)	5.02	0.92
	8	3.27^{b}	10.50	(2.41, 4.14)	(7.47, 16.20)	3.21	0.98
	9	2.71^{b}	2.38	(2.16, 3.26)	(1.54, 4.17)	1.52	0.56
	Mean	3.79^{C}	9.94	(2.94, 4.64)	(6.94, 15.70)	2.88	0.74
10 ~ 30	5	7.27^{ab}	20.01	(5.88, 8.67)	(13.50, 32.50)	4.42	0.61
	6	7.25^{ab}	8.47	(6.38, 8.13)	(5.81, 13.50)	2.88	0.40
	7	9.69^{a}	23.99	(8.35, 10.0)	(16.90, 37.60)	4.85	0.50
	8	6.58^{b}	10.74	(5.70,7.45)	(7.64, 16.60)	3.25	0.49
	9	5.61^{b}	2.54	(5.05, 6.18)	(1.64, 4.45)	1.57	0.28
	Mean	7.28^{B}	13.15	(6.27, 8.09)	(9.10, 20.93)	3.39	0.46
30 ~ 60	5	7.87^{ab}	29.04	(6.19, 9.55)	(19.70, 47.20)	5.32	0.68
	6	7.62^{ab}	17.09	(6.38, 8.87)	(11.70, 27.30)	4.09	0.54
	7	9.5^{a}	17.97	(8.41, 10.70)	(12.70, 28.10)	4.20	0.44
	8	6.24^{b}	4.21	(5.70, 6.79)	(3.00, 6.50)	2.03	0.33
	9	6.68^{b}	7.19	(5.73, 7.63)	(4.65, 12.60)	2.64	0.40
	Mean	7.60^{A}	15.10	(6.48, 8.71)	(10.35, 24.34)	3.66	0.48
60 ~ 90	5	5.63^{b}	3.67	(5.03, 6.23)	(2.49, 5.97)	1.89	0.34
	6	5.59^{b}	2.20	(5.14, 6.04)	(1.51, 3.52)	1.47	0.26
	7	8.07^{a}	13.04	(7.08,9 .06)	(9.19, 20.40)	3.58	0.44
	8	6.01^{b}	5.80	(5.37, 6.65)	(4.13, 8.96)	2.39	0.40
	9	6.80^{ab}	9.12	(5.73, 7.87)	(5.90, 16.00)	2.97	0.44
	Mean	6.42^{AB}	6.77	(5.67, 6.70)	(4.64, 10.97)	2.46	0.38
90 ~ 120	5	5.77^{b}	3.26	(5.21, 6.33)	(2.21, 5.30)	1.78	0.31
	6	5.58^{b}	2.16	(5.13, 6.02)	(1.48, 3.45)	1.45	0.26
	7	7.30^{a}	10.02	(6.43, 8.17)	(7.06, 15.70)	3.13	0.43
	8	5.88^{b}	4.63	(5.31, 6.45)	(3.30, 7.15)	2.13	0.36
	9	6.82^{ab}	8.61	(5.78, 7.86)	(5.57, 15.10)	2.89	0.42
	Mean	6.27^{B}	5.74	(5.57, 7.00)	(3.92, 9.34)	2.28	0.36
120~ 150	5	6.82^{ab}	4.34	(6.17, 7.46)	(2.94, 7.06)	2.06	0.30
	6	6.56^{b}	2.87	(6.05, 7.07)	(1.97, 4.58)	1.68	0.26
	7	7.87^{a}	9.67	(7.01, 8.72)	(6.82, 15.20)	3.08	0.39
	8	6.08^{b}	1.61	(5.75, 6.42)	(1.15, 2.49)	1.26	0.21
	9	7.44^{ab}	7.76	(6.45, 8.43)	(5.02, 13.60)	2.74	0.37
	Mean	6.95^{B}	5.25	(6.29, 7.62)	(3.58, 8.59)	2.16	0.31

Note: Different letters for average values indicate significant differences at $P < 0.05$.

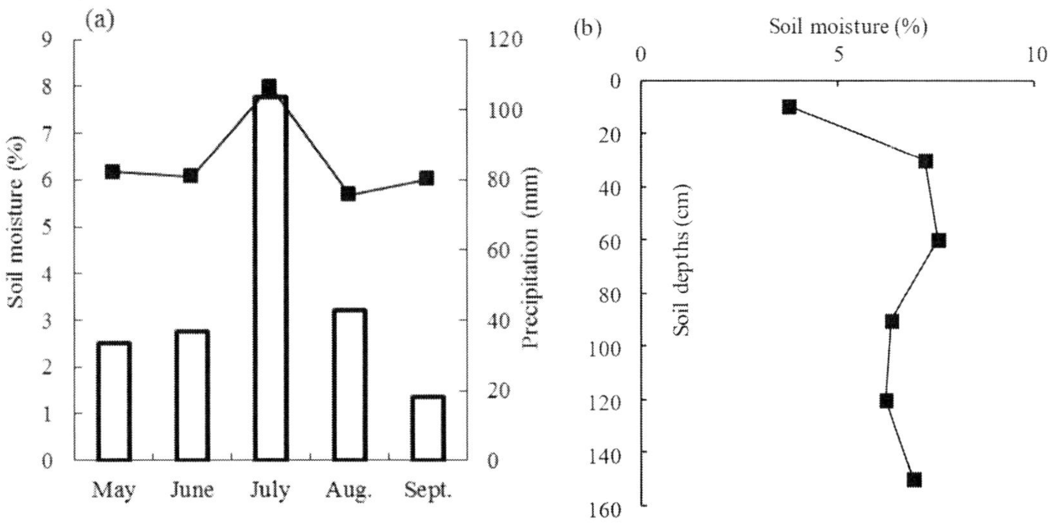

Figure 1. Variation of SM along with (a) month, (b) soil depth at grassland.

4. Conclusions

- The average values of SM in grassland vary within the range of 2%~10%; SM in grassland on Horqin Sandy Land at the same soil layer and same month obeys normal distribution;
- SM in grassland in July is significantly higher than that in other months, and significantly different from other months, and the confidence interval of SM in July shows a trend of significantly rightward shifting; From May to September every year, SM shows an "increasing firstly, and then decreasing" trend, and peaks in July;
- SM in grassland at the depth of 0~10cm is considerably lower than other soil layers, and significantly different from other soil layers; SM in grassland shows an "increased firstly, then decreased and eventually increased again" trend with changes in soil depth.

Acknowledgments
This study was supported by the National Natural Science Foundation of China (No. 31300388; No. 41361013).

References

[1] Zhang Z F 2006 Quality control schemes for hourly soil moisture from automatic weather stations in China *Arid Land Geogr.* **36(1)** 101-8
[2] Zhang H F, Wang J S and Huang B 2006 Vertical distribution of soil moisture in Xifeng wheat field of Loess Plateau *Chin. J. Soil Sci.* **37(6)** 1081-5
[3] Liu X P 2008 *Water Dynamics and Water Resources Evaluation in Horqin Sandy Land* (Chinese Academy of Sciences: Cold and Arid Regions Environmental and Engineering Research Institute)
[4] Wang M C, Yang S T, Dong G T, *et al* 2012 Estimating soil water in Northern China based on vegetation temperature index (VTCI) *Arid Land Geogr.* **35(3)** 446-55
[5] Li H J, Wang M B and Chai B F 2003 Spatial and temporal characteristics of soil moisture dynamics in Loess Plateau *Chin. J. Appl. Ecol.* **14(4)** 515-9
[6] Yao S X, Zhang T H, Zhao C C, *et al* 2012 Spatio-temporal variability of soil moisture in different dunes of Horqin Sandy Land *J. Soil Water Conserv.* **26(1)** 251-4
[7] Yao S X, Zhang T H and Zhao C C 2013 A comparison of soil saturated hydraulic conductivity (KFS) in different Horqin Sand Land *Acta Pedologica Sin.* **50(3)** 45-53

[8] Li H S, Wang W F, Liu B L, *et al* 2013 Applying isolation method on study of soil water source in an extremely dry area *Arid Land Geogr* **36(1)** 92-100
[9] Wang T 2000 Land use and sandy desertification in the North China *J Desert Res* **20(2)** 103-7
[10] Sheng J, Xie S Q and Pan C Y 2000 Probability Theory and Mathematical Statistics (Beijing: Higher Education Press)
[11] Liu Z, Wu G, Ding W D, et al 2011 SPSS Statistical Analysis and Application (Beijing: Electronic Industry Press)
[12] Yao S X, Zhao C C and Zhang T H 2014 Multiple-time-scale analysis on precipitation in Naiman Banner of Horqin Sandy Land in 1970-2010 *J Desert Res* **34(2)** 542-9
[13] Li Y Q, Zhang T H, Liu X P, et al 2010 Change pattern of soil water content in different dunes and grassland in Horqin Sandy Land *Bull Soil Water Conserv* **30(3)** 31-5

IOP Conf. Series: Earth and Environmental Science 82 (2017) 012044 doi:10.1088/1755-1315/82/1/012044

Drought-waterlog encounter probability research between the diversion region and benefited region in the Middle Route of South-to-North Water Transfer Project

Z Yuan[1,2], J J Xu[1], J J Huo[1], L Q Yao[1], Y R Lin[1] and X F Hong[1]

[1]Changjiang River Scientific Research Institute, Wuhan, China

E-mail: yuanzhe_0116@126.com

Abstract. One of the major risk factors in managing water resources in the South to North Water Transfer Project will be due to spatial and temporal variations in precipitation. This is because the synchronous encounter probability of drought and waterlog can be different at the diversion region and benefited region. This study analyzed the statistical characteristic of drought and waterlog in the diversion region and benefited region and the wetness-dryness encountering probability among the diversion region and benefited region. Besides, the wetness-dryness encountering probability and the surface water resource in the diversion region were also projected based on the outputs of GCMs. The results showed that: (1) The concurrence of drought or waterlog in diversion region and benefited region is uncommon and it benefits the water transfer. (2) In the future, the synchrony of drought/waterlog will increase in wet season while it will decrease in dry season. But the synchrony of drought/waterlog between the north of Haihe River Basin and diversion region will decrease in the late spring, early summer and autumn. (3) The surface water resource is projected to be increased by 8.1~11.5% compared to the period from 1961 to 2011. The climate change in the future would benefit the water transfer.

1. Introduction

The water stress has been rising in large cities of northern China since the 1980s. Even though some temporary solutions have been taken, the water scarcity problem is still serious [1]. To make balance of water availability in China, the South to North Water Transfer Project (SNWTP) was launched in 2002. About 44.8×10^9 m^3 of water will be diverted from the wetter regions of southern China to northern China annually. This project involves three routes, namely, the western, eastern and central routes, to the drier northern regions [2]. The main channel is so long that it spans several climate zones. The Diversion region and benefited region are affected by the climate condition, hydrologic characteristics and human activities. The precipitation and runoff possess the characteristic of wetness-dryness alternation and have great variation [3]. The wetness-dryness encountering means the combined event of dry and wet in different regions that is the result of the imbalance spatial-temporal distribution of water resources. The wetness-dryness alternation in diversion region may have influence on the transferable water and it may affect the water requirement and guarantee rate of water supply. However, the wetness-dryness encountering will affect the economic benefits of SNWTP [4, 5]. The worst condition of the project is that suffer from continuing drought in both diversion region and benefited region. There would be no enough water in diversion region and the water demands would increase in benefited region. Many methods have been used to quantify the characteristic of wetness-dryness encountering probability in diversion region and benefited region. Feng *et al* took the Xianshui River, Zumuzu River, Zhuosijia River and the Yellow River as study areas. With the 4-dimensional joint distribution model, the wetness-dryness encountering probability among these regions was calculated. It can be found that the favorable probability for water transfer is 43.8-77.8% [6]. Zhang *et al* analyzed the drought-waterlogging encounter probability of water source area and Hai River receiving area and the results showed that the profitable frequency of the drought-waterlogging encounter for water transfer was 56.7%, and the disadvantageous frequency was 24.9% [7]. Kang *et al* (2011) found that the risk probability is less than 25% for water transfer to receiving areas (Tangbaihe River Basin, Huaihe River Basin, South of Haihe River Basin and North of Haihe River Basn) based on a risk analysis model [8]. However, there are three problems in the previous studies. More specifically, (1) most studies focus on the wetness-dryness encountering probability in terms of inter-annual variation. The study of variation within the year is scanty. (2) Previous researches only analysis the characteristic of wetness-dryness encountering through meteorological factor (eg. precipitation). Limited research considered the water source quantity in diversion region; (3) Most of the studies only focused on the historical characteristic of wetness-dryness encountering. In fact, climate change potential effects in future can be assessed with the outputs from GCMs.

The main argument presented here is analyzing drought-waterlogging encounter probability of water source area (Dangjiangkou Catchment) and Haihe River Basin in the history and future. The results of this study will benefit the planning and management of the SNWTP.

2. Materials and methods

2.1. Study area

The water diversion of SNWTP is in Danjiangkou reservoir which is located in Hanjiang River. This water diversion project went through Fangchengyakou that is the divide of Yangtze River Basin and Huaihe River Basin. The diversion canal is excavated along the edge of the west of Huanghuaihai Plain and cuts through the Yellow River in Gubaiju by tunnel or aqueduct. Then the canal heads towards the north along the west side of Beijing-Guangzhou railway. The water can flow automatically to Beijing and Tianjin (figure 1). The area of benefited region is 1.5×10^6 km^2. The water-transmission line from the Dangjiangkou reservoir to TuanCheng Lake in Beijing is as long as 1267 km. The lengths of sections in south of Yellow River, crossing the Yellow River and in the north of Yellow River are 477 km, 10 km and 780 km. The length of section from Xushui, Heibei province to Tianjing is 154 km. This study focuses on the benefited region in Haihe River Basin, which is divided into three parts. The information of each part is listed in table 1.

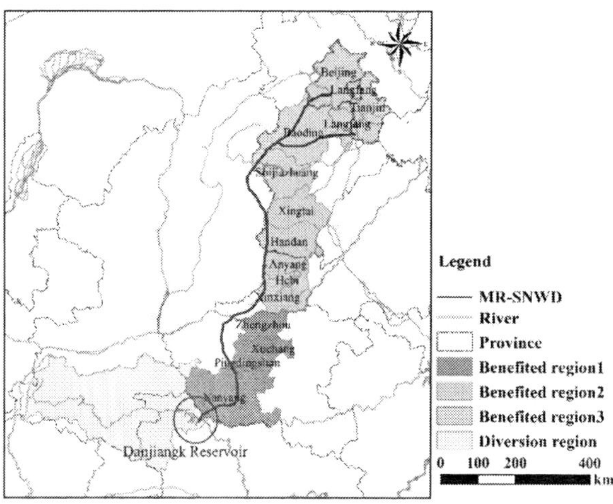

Figure 1. The route of SNWTP.

Table 1. The cities in benefited region.

Region	City
Benefited region 1	Nanyang, Pingdingshan, Xuchang, Zhengzhou
Benefited region 2	Xinxiang, Hebi, Anyang, Handan, Xingtai
Benefited region 3	Shijiazhuang, Baoding, Beijing, Tianjin, Langfang

2.2. Data

The data used in this study were the daily precipitation data (from 1961 to 2011) of meteorological stations which are located in the study area (the stations near the boundary are also included). The study period 1961 to 2010 was chosen because it has good representativeness in the station data. Pre-1961 records of precipitation were less consistent than that during the 1961 to 2011 period. For some stations which contain partial or no data, they were deleted from the analysis of precipitation. In order to improve the precision of spatial analysis, stations with less than 80% temporal coverage during the 51-year period were omitted. Finally, data of 331 meteorological stations were used in this research (figure 2).

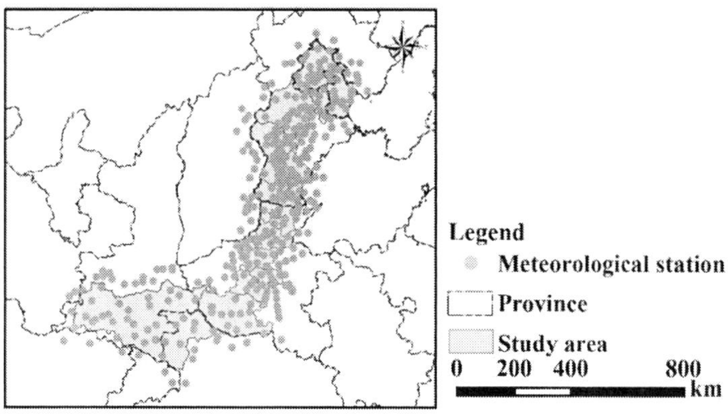

Figure 2. Meteorological stations of study area.

Runoff data used in this research come from the hydrological station of Dangjiangkou reservoir. The discharge in Dangjiangkou reservoir from 1961 to 2011 and the annual mean monthly discharge in Dangjiangkou reservoir are shown as figure 3.

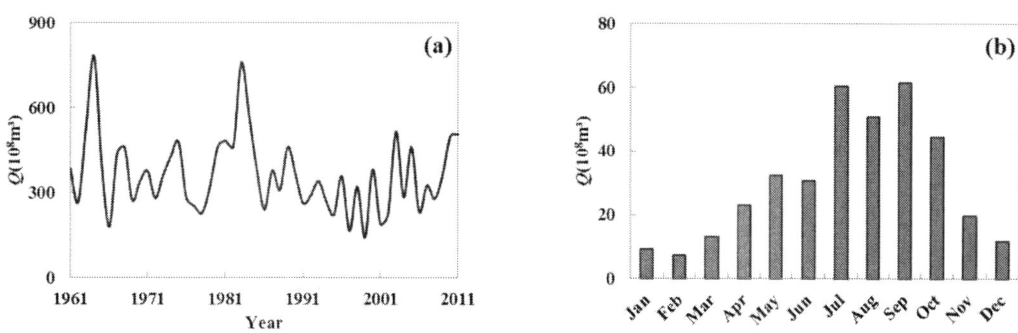

Figure 3. (a) The discharge in Dangjiangkou reservoir from 1961 to 2011 and (b) The annual mean monthly discharge in Dangjiangkou reservoir.

Data of the regional climate change scenarios comes from ISI-MIP (The Inter-Sectoral Impact Model Intercomparison Project, http://www.isi-mip.org) [9, 10]. The general circulation models are list as table 2. The scenarios of RCP2.6, RCP4.5 and RCP8.5 are used in this study.

Table 2. List of general circulation models (GCMs) used in this study.

Centre	Country	Name	Properties
Geophysical Fluid Dynamics Laboratory(GFDL)	United States	GFDL-ESM2M	1951/1/1 -2050/12/31
Hadley Centre for Climate Prediction and Research, Met Office	United Kingdom	HADGEM2-ES	70.25° E-140.25° E, 15.25° N-55.25° N 0.5°×0.5°
L'Institut Pierre-Simon Laplace (IPSL)	France	IPSL-CM5A-LR	
Technology, Atmosphere and Ocean Research Institute, and National Institute for Environmental Studies	Japan	MIROC-ESM-CHEM	
Norwegian Climate Centre	Norway	NORESM1-M	

2.3. Assessment of drought/waterlog magnitude

In this study, the standard deviation method was used to assess the magnitude of drought/waterlog. The method assumes that the precipitation obeys the normal distribution. Then the standard deviation of precipitation is used to assess the magnitude. The formula is shown as the follow.

$$K = \frac{P_i - \overline{P}}{\sigma} \tag{1}$$

P_i is the monthly precipitation, $\overline{P}$ is the annual mean monthly precipitation, σ is mean square error of precipitation. Standard is as follows (table 3).

Table 3. The Flood and Drought magnitude of K.

K	Magnitude
$Z \geqslant 2$	Severe Waterlog
$1 < Z < 2$	Moderate Waterlog
$-1 < Z \leqslant 1$	Normal
$-2 < Z \leqslant -1$	Moderate Drought
$Z < -2$	Severe Drought

2.4. Set pair analysis

Set pair analysis (SPA) method [11] was adopted for the dividing of alternate drought /waterlog-prone region in the study area. The key of the method is to bring two correlated groups from the system for set pair (A, B) and analyze the identity (I), discrepancy (D) and contradiction (C) of the two. After these, the connection degree (μ) can be calculated which can reflect the relationship of set A and set B. The connection degree is defined as

$$\mu_{A-B} = \frac{s}{n} + \frac{f}{n} i + \frac{p}{n} j \tag{2}$$

where s, f and p are the number of the identical terms, the discrepant terms and the contradictory terms, respectively. i means the uncertainty coefficient of discrepancy, a numeric value between 0 and 1; j means the uncertainty coefficient of contradictory, taking -1 usually in study. $I = s/n$, $D = f/n$ and $C = p/n$ are the identity degree, the discrepancy degree and the contradictory degree, respectively. So, equation (7) can be redefined as

$$\mu_{A-B} = I + Di + Cj \tag{3}$$

3. Results

3.1. The drought and waterlog features in diversion region and benefited region from 1961 to 2011

The frequency of drought and waterlog is high from the 1960s. From the figure 4 we can conclude that the frequency of drought and waterlog is higher in the period from May to September. The drought occurred frequently in May and July. The frequencies of drought in these months from 1961 to 2011 are 11.8-21.6% and 13.7-19.6% respectively. However, the waterlog occurred frequently in September. It is inconformity of the drought and waterlog in terms of temporal dimension. Figure 5 shows the connection degree of drought/waterlog in the diversion region and benefited region of Haihe River Basin. The degree is lower in the period from May to September. It means that concurrence of drought or waterlog in diversion region and benefited region is uncommon. Besides, this period is drought-prone time in Haihe River Basin. This feature is benefits the water transfer.

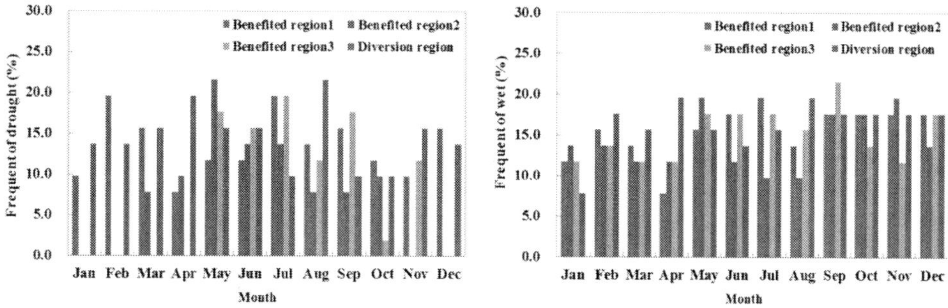

Figure 4. The frequency of drought and waterlog in diversion region and benefited region.

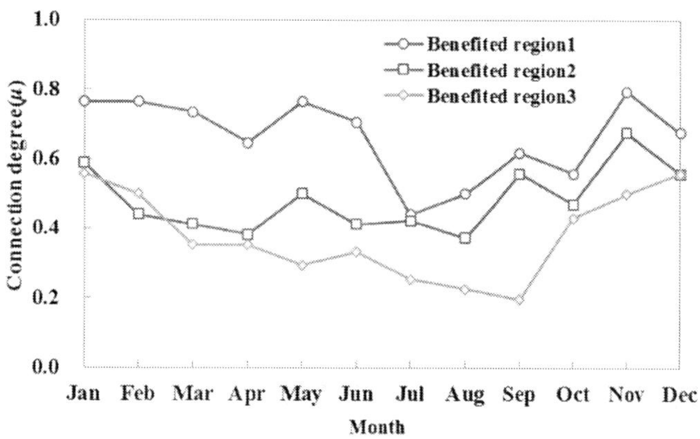

Figure 5. The connection degree of drought/waterlog in the diversion region and benefited region.

3.2. The drought-waterlogging encounter in diversion region and benefited region from 2011 to 2050

Based on the monthly precipitation of diversion region and benefited region from 2011 to 2050 and equations (2) and (3), the connection degree of diversion region and benefited region in the future can be calculated (figure 6). Take RCP4.5 as example, the connection degree will increase significantly in in the period of July to September. In benefited region 1, the degrees will increase by 29.8%, 20.0% and 14.5% in July, August and September respectively in terms of Muti-average of model outputs, while the degrees will decrease by 17.3%, 13.4%, 11.4% and 22.4% in January, February, May and

June. In Benefited region 2, the connection degree will increase significantly in February, March, April and August, especially in March and April, rate of increase is more than 20%. However, the connection degree will decrease significantly in May, June, September and November. In Benefited region 3, the change situation is in accordance with benefited region 2. The rate of increase in March, May, July, August and September is more than 20%. But it will be decreased by more than 10% in October and November. In conclusion, the synchrony of drought/waterlog between the south of Haihe River Basin and diversion region will increase in wet season while it will decrease in dry season. But the synchrony of drought/waterlog between the north of Haihe River Basin and diversion region will decrease in the late spring, early summer and autumn.

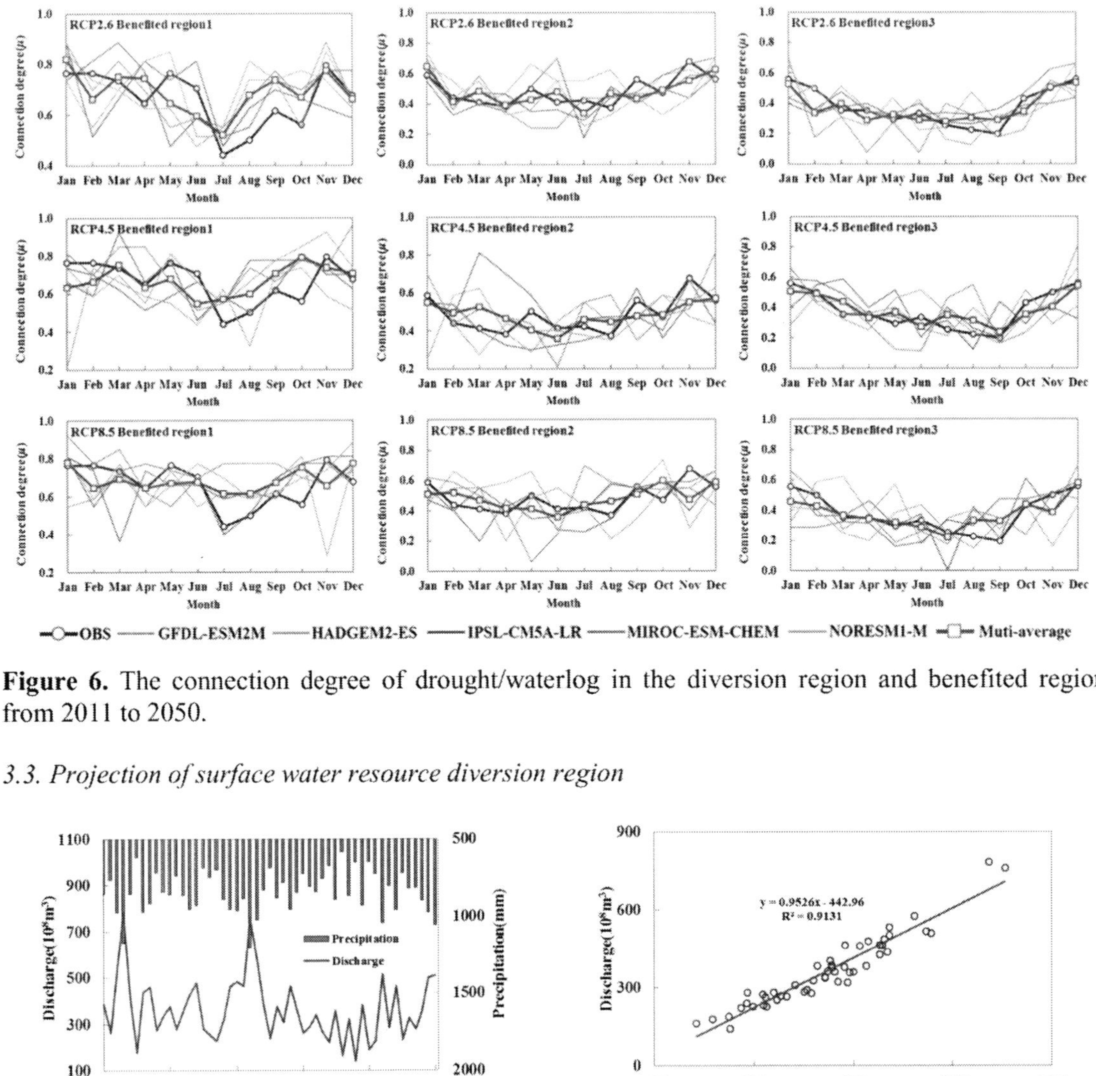

Figure 6. The connection degree of drought/waterlog in the diversion region and benefited region from 2011 to 2050.

3.3. Projection of surface water resource diversion region

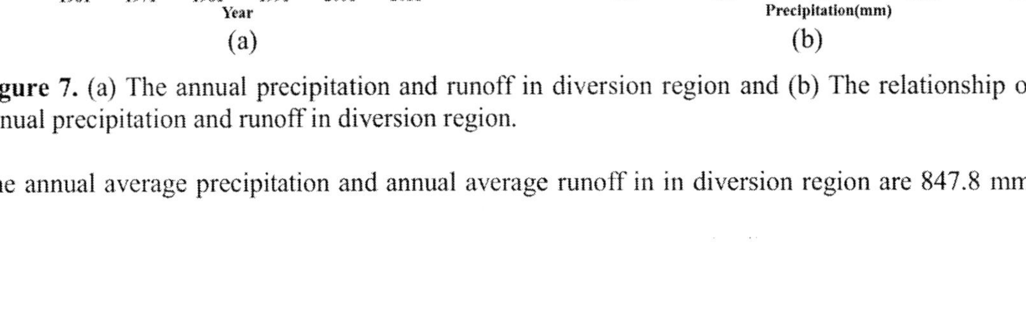

Figure 7. (a) The annual precipitation and runoff in diversion region and (b) The relationship of the annual precipitation and runoff in diversion region.

The annual average precipitation and annual average runoff in in diversion region are 847.8 mm and

364.7×10^8 m^3, respectively. Both the precipitation and runoff had decreased since the 1980s. But the precipitation and runoff have been increasing since early 21st century (figure 7(a)). The runoff has a linear dependence relation with the precipitation factor (R=0.9526P+442.96). Correlation coefficient (R^2) is more than 0.9 (figure 7(b)). So, we can project the future surface water resource by the precipitation that comes from the outputs of GCMs.

Based on the relationship of the annual precipitation and runoff in diversion region, we can calculate the future surface water resource from 2011 to 2050 (figure 8). It could be found that there was a decreasing trend of surface water resource in diversion region during the last 50 years. It had decreased 22.0×10^8 m^3 per 10-year from 1961 to 2011. However, the surface water resource will increase in the future (from 2011 to 2050). It will increase by 19.1×10^8 m^3/10a, 11.9×10^8 m^3/10a and 6.8×10^{8}m^3/10a under RCP2.6, RCP4.5 and RCP8.5 scenarios respectively. Increase rate of surface water resource is larger in low emission scenarios than high emissions scenario. The muti-average results show that the annual mean surface water resource would increase by 8.1% (-1.1-18.7%), 11.5% (-5.8-26.1%) and 4.8% (-11.0-17.4%) compared to the period from 1961 to 2011. That means the climate change in the future would benefit the water transfer.

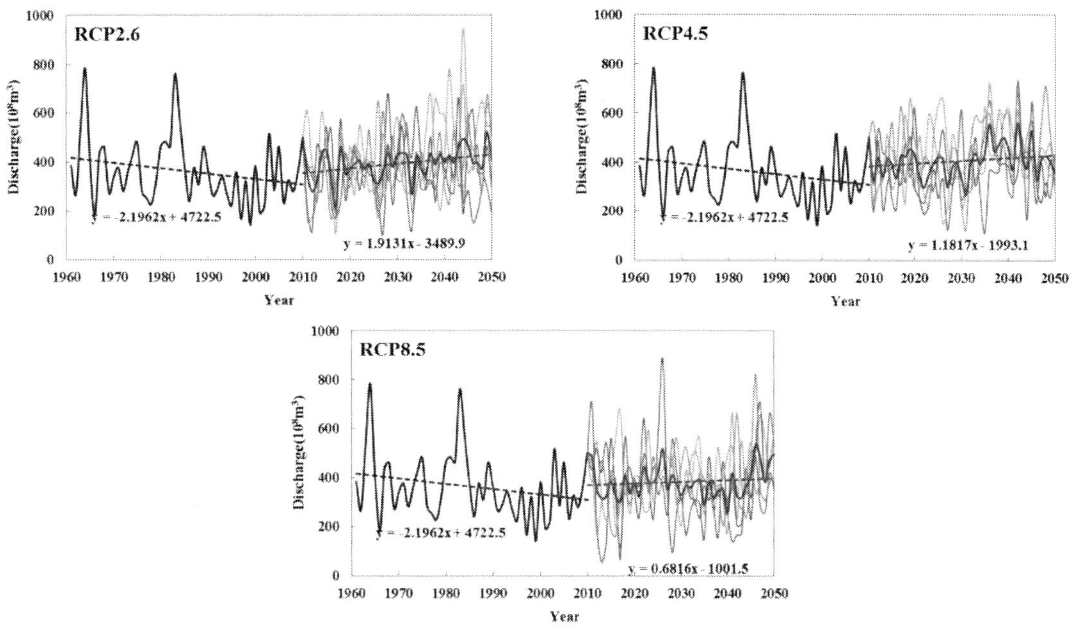

Figure 8. The surface water resource in the diversion region from 1961 to 2050.

4. Conclusions and discussion

The SNWTP is one of important projects to relieve the conflict of water supply and demand in the North China. This study analyzed the statistical characteristic of drought and waterlog in the diversion region and benefited region in SNWTP and the wetness-dryness encountering probability among the diversion region and benefited region. Besides, the wetness-dryness encountering probability and the surface water resource in the diversion region were also projected based on the outputs of GCMs.

In the benefited region in Haihe River Basin, the frequency of drought and waterlog is higher in the period from May to September. The drought occurred frequently in May and July, while the waterlog occurred frequently in September. In the diversion region, the drought often happened in July and September, while the August was waterlog-prone time. The connection degree is lower in the period from May to September. It means that concurrence of drought or waterlog in diversion region and benefited region is uncommon. So, this feature benefits the water transfer. In the future, the synchrony of drought/waterlog between the south of Haihe River Basin and diversion region will increase in wet season while it will decrease in dry season. But the synchrony of drought/waterlog between the north of Haihe River Basin and diversion region will decrease in the late spring, early summer and autumn. The surface water resource is projected to be increased by 8.1% (RCP2.6), 11.5% (RCP4.5) and 4.8% (RCP8.5) compared to the period from 1961 to 2011. The climate change in the future would benefit the water transfer.

Acknowledgments

This research was financially supported by International science & Technology Cooperation Program of China "Integrated Risk Management of Water Resources for Adaptation to Climate Change in Jinsha River Basin" (No. 2014DFA71910), National Public Research Institutes for Basic R & D Operating Expenses Special Project (CKSF2017029/SZ, CKSF2017061/SZ), National Natural Science Foundation of China (41601043), Major Competitive Project of Hydraulic Science and Technology (KT201514) and Natural Science Foundation of Hubei Province(2013CFB400).

References

[1] Xia J, Zhang L, Liu C and Yu J 2007 *Water Resour. Manag.* **21** 233-47
[2] Li L, Zhang L, Xia J, Gippel C J, Wang R and Zeng S 2015 *Water Resour. Manag.* **29** 2563-79
[3] Ren G, Jiang T, Li W, Zhai P, Luo Y and Ma Z 2008 *Adv. Water Sci.* **19** 772-9 (in Chinese)
[4] Yan B, Guo S and Xiao Y 2007 *J. Hydraul. Eng.* **38** 1178-85 (in Chinese)
[5] Sun P and Chen J 2009 *J. Environ. Sci.* **30** 47-51 (in Chinese)
[6] Feng P, Niu J, Zhang Y and Hou H 2010 *J. Hydraul. Eng.* **41** 900-7 (in Chinese)
[7] Zhang L, Qin L, Zhang D and Zeng S 2010 *Resour. Environ. Yangtze Basin* **19** 940-5 (in Chinese)
[8] Kang L and He X 2011 *Adv. Water Sci.* **22** 44-50 (in Chinese)
[9] Piani C, Weedon G P, Best M, Gomes S M, Viterbo P, Hagemann S and Haerter J O 2010 *J. Hydrol.* **395** 199-215
[10] Hagemann S, Chen C, Haerter J O, Heinke J, Gerten D and Piani C 2011 *J. Hydrometeorol.* **12** 556-78
[11] Wang W, Jin J, Ding J and Li Y 2009 *Sci. Chin. E. Technol. Sci.* **52** 3017-23

Spatio-temporal dynamics of NDVI and its response to climate factors in the Heihe River Basin, China

C X Cheng[1,2], Z H Wang[1,4], P Q Xiao[1], Z X Xu[2], P Jiao[1], G T Dong[1] and G J Wei[3]

[1]Yellow River Institute of Hydraulic Research, Zhengzhou 450003, China
[2]College of Water Sciences, Beijing Normal University, Beijing 100875, China
[3]College of Water Resources and Electric Engineering, Qinghai University, Xining, China

E-mail: wzh8588@aliyun.com

Abstract. Ecosystem in the Heihe River Basin (HRB) is very vulnerable and sensitive to variations in the climatic conditions and human activities. In this study, The SPOT and MODIS NDVI time series data from 1998 to 2012 were used to analyze NDVI variation characteristics and its response to climatic factors in the HRB based on Mann–Kendall test, relative Sen's slope, linear regression and partial correlation analysis. The results indicate that: (1) There is an obvious decreasing trend during 1998-2001 and a significantly increasing trend from 2001 to 2012. The vegetation in the HRB was improved by 1.5% of annual average level from 1998 to 2012, and rate of change is greatest in the lower reaches. (2) During 1998-2001, the rainfall is proved to be the most significant factor influencing NDVI for each reaches, and temperature and radiation are also related significantly with NDVI in the middle and lower reaches respectively. However, during 2001-2012, all climatic factors are not significantly related with NDVI for each reaches, excepting rainfall in the middle reaches. This demonstrates that the climatic variation cannot totally explain the significant improvement of vegetation during 2001-2012 in the HRB, whereas ecological restoration projects implemented in this period might be a reasonable factor for the improvement of ecological environment in the HRB.

1. Introduction

Vegetation is an important component of global ecosystems, and plays important roles in the interception of rainfall, alleviation of runoff, prevention of desertification, and conservation of soil and water for terrestrial ecosystems, especially for arid and semi-arid zones [1-3]. Vegetation cover and density are commonly considered as the sensitive indicators for ecological environment change [4-7]. Numerous studies have been reported that vegetation cover dynamics and its driving factors, including climatic change, land use transformation, and the fertilization effect of CO_2, in various ecosystems in China [8-12].

In the past two decades, with the development of earth observation system and the long-term remote sensing data accumulation, most previous studies have employed time series of indicator derived from remote sensing data exhibiting vegetation coverage, primarily the Normalized Difference Vegetation Index (NDVI), and meteorological data measured from ground stations to analyze the vegetation response to climate change and human activities [13]. The understanding of spatio-temporal characteristics of vegetation variation induced by climate change and human activities at region scale is very helpful for local decision makers to make more scientific and effective eco-management planning.

Vegetation dynamics in arid and semi-arid areas have been proved to be more sensitive to climate change when compared with those in other regions [12, 13]. The vulnerable terrestrial ecosystems are generally thought to be highly susceptible to degradation and desertification with climatic fluctuations. In addition, vegetation deterioration and inappropriate land use practices by humans can also lead to severe land degradation and other serious problems to the ecology and environment. The response of arid ecosystems to water supply variation, climate change and human activities has been widely concerned because of the economic, hydrological and ecologic significance of most arid and semi-arid areas around the world.

The Heihe River Basin (HRB) located in northwestern China is the second largest inland river basin, and is characterized by an arid and semi-arid climate. Glacier/snow meltwater from the Qilian Mountain in the upper reaches and precipitation can provide water resources for the Heihe River, and spatial distribution pattern of vegetation types in the HRB is totally affected by the site-specific climatic conditions. The HRB has experienced a significant vegetation degradation and desertification as a result of long-term limited water availability [14, 15]. By the end of the 20th century, a series of ecological restoration projects, including water diversion project (EWDP), were implemented by the central government of China to address these serious ecological and environmental problems [1]. The variation of vegetation in the HRB inevitably would affect the regional hydrological, ecological process and local economic development. Therefore, long-term and large-scale assessments of vegetation dynamics and their response to hydro-climatic change and human activities during recent period in the HRB are urgently needed to implement a more effective policy.

In this study, spatio-temporal variation characteristics and their responses to climatic and non-climatic factors in the HRB from 1998 to 2012 are investigated. Firstly, the SPOT and MODIS NDVI time-series data are used to analyze the spatial distribution characteristics, temporal fluctuation and variation trend for vegetated areas. Secondly, the responses of vegetation to hydro-climatic and non-climatic factors at pixel-level and regional scale in the upper, middle and lower reaches with significantly different hydro-climatic conditions are analyzed to explore the driving factors leading to vegetation variation in the HRB.

2. Study area

HRB lies between 96°30' and 102°0' E, and 337°30' and 43°0' N and covers an area of approximately 130,000 km^2 (figure 1). The HRB is the second largest inland river basin of China, extending over the provinces of Qinghai, Gansu and Inner Mongolia. Its upper, middle and lower reaches stretch from the middle of the Hexi Corridor in Gansu to western Inner Mongolia.

The southern Qilian Mountains covered by nature grassland and forests, with remarkable vertical zonality, is the river source area of Heihe River, with the elevation range from 2,000m to 5,500 m and a mean annual precipitation of 250~500 mm.

The middle Hexi Corridor covered by cropland and grassland, located between the Qilian Mountains and the Beishan Mountains, is characterized with a elevation range from 1,000 m to 2,000 m and mean annual precipitation of 100~200 mm. The northern Alxa High-plain is mainly covered by the sparse shrubs, bare Gobi, and *populus diversifolia*, with mean elevation of 1000 m and a mean annual precipitation of <50 mm. Therefore, the characteristics of geographic distribution in the HRB result in significantly distinctive climatic conditions and associated vegetation types in the upper, middle and lower reaches.

By the end of the 20th century, an ecological restoration including EWDP and grass protective project was successfully implemented by the government of China, consequently, the severe deterioration of ecological environment in the HRB, especially for the lower reaches, has been significantly alleviated [1].

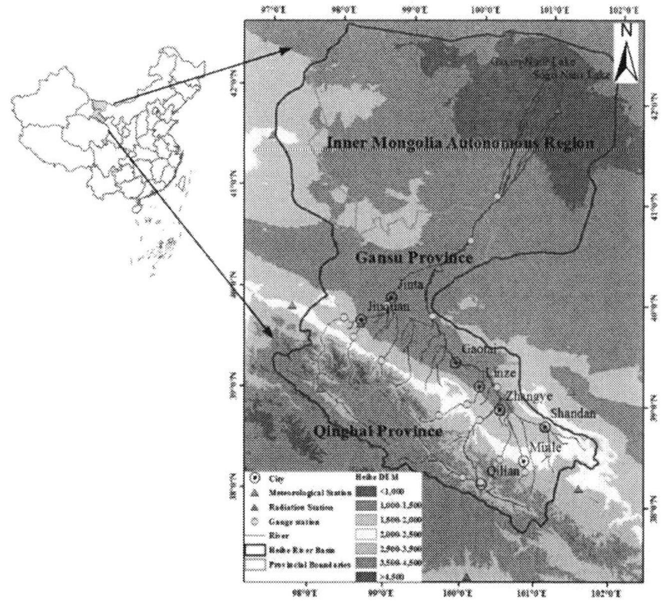

Figure 1. The locations of the Heihe River Basin and associated meteorological, radiation and gauge stations.

3. Materials and methods

3.1. Materials

3.1.1. Remote sensing data. NDVI was considered as an effective indicator of greenness and coverage of vegetation [4, 11]. In this study, NDVI dataset derived from satellite remote sensing data was employed to reflect spatial-temporal dynamics of vegetation. Both SPOT/VEGETATION NDVI product and MODIS Terra NDVI product were used to obtain the NDVI time-series from 1998 to 2012 in the HRB. The SPOT/VEGETATION NDVI time-series data at 10-days period temporal resolution and 1-kilometer spatial resolution was downloaded freely from the Environmental and Ecological Science Data Center for West China (http://westdc.westgis.ac.cn), covering the period from April 1998 to December 2007. The MODIS Terra monthly NDVI time-series (MOD13A3) at 1-kilometer spatial resolution was obtained from National Aeronautics and Space Administration (http://ladsweb.nascom.nasa.gov), covering the period from January 2008 to December 2012. Firstly, the SPOT/VEGETATION 10-day-period NDVI data was processed into monthly NDVI time-series using the maximum value composites method to match the temporal resolution of MODIS product, and then NDVI monthly time-series from 1998 to 2012 was obtained by combing those two NDVI products.

3.1.2. Meteorological data. In this study, monthly precipitation, temperature and sunshine hours at 16 meteorological stations and monthly solar radiation at 6 radiation stations within and nearby the HRB (figure 1) from 1998 to 2012 were downloaded from the China Meteorological Administration Data Sharing Service System (http://cdc.nmic.cn/home.do). The monthly gridded precipitation, temperature and sunshine hours at a spatial resolution of 1-km were produced based on in-situ measurements from meteorological sites and gridded digital elevation model (DEM) at 1-km spatial resolution using the ANUSPLIN software package version 4.3 where the original and partial thin plate smoothing spline surface fitting technique was considered to be greatly suitable for spatially interpolating meteorological data [16]. The monthly gridded radiation data were then produced based on in-situ data from radiation sites and gridded sunshine hours at 1-km spatial resolution using ANUSPLIN.

3.2. Reconstruction of NDVI time-series

Although the composite NDVI used in this study is the maximum value during one month, there are always some noise effects caused by cloud and sensor in original NDVI time-series [4, 7]. Consequently, NDVI is not able to reflect the real situation of vegetation. To solve this problem, in this study, Savitzky-Golay smoothing filter method was employed to smooth contaminated pixel in the NDVI time-series, consequently, the effects from cloud and sensor on NDVI value could be eliminated to some extent.

Savitzky and Golay [17] proposed a smoothing method using simplified least-squares-fit convolution for a noise curve. In this study, the Savitzky–Golay filter was employed to smooth NDVI time series with some noise signal, and general equation of the Savitzky–Golay filter can be given as follows:

$$Y_j^* = \frac{\sum_{i=-m}^{i=m} C_i Y_{j+i}}{N} \quad (1)$$

Where Y is the original NDVI, Y_j^* is the smoothed NDVI, C_i is the coefficient for the ith NDVI in the filter window, and N is the number of convoluting integers and is equal to the filter window size (2m+1), where m is half-width of the filter window. In this study, m=2 was used to obtain high-quality monthly NDVI time-series according to repeated trials. The method of calculating coefficients C_i in the previous study of Steinier et al [18] was employed in this study.

3.3. Temporal trend analysis

The Mann–Kendall (MK) test was generally used to examine the significance of temporal trends (monotonic upward or downward) of time series data. This statistical method has been widely used to evaluate variation trend and its significance of hydrological and climate time series since 1945 [19, 20]. MK test is a non-parametric test, which means that the test is not influenced by the actual distribution of the data and is less sensitive to outliers [21].

The statistic Z of the MK test is expressed as follows:

$$Z_c = \begin{cases} \frac{S-1}{\sqrt{\mathrm{var}(s)}}, S > 0 \\ 0, \ S = 0 \\ \frac{S+1}{\sqrt{var(s)}}, \ S < 0 \end{cases} \quad (2)$$

where,

$$S = \sum_{i=1}^{n-1} \sum_{j=i+1}^{n} sgn(x_j - x_i) \quad (1 \le k < j \le n)$$

$$sgn(x_j - x_i) = \begin{cases} 1, x_j - x_i > 0 \\ -1, x_j - x_i < 0 \\ 0, x_j - x_i = 0 \end{cases}$$

$$\mathrm{Var}(s) = \sum_{i=1}^{n} n(n-1)(2n+5)/18$$

where, x_i and x_j are annual observations, and n is the length of sample data.

For MK trend analysis, H_0 is null hypothesis of no trend for an independent and identically distributed data samples X={x_1, x_2 . . .,x_n}. Alternative hypothesis H_1 represents that there is a monotonic trend for the data set X. For the null hypothesis H_0, if $|Z_c| > Z_{(1-\alpha)/2}$, where $Z_{(1-\alpha)/2}$ is the standard normal distribution, and α is the significance level for the test, the null hypothesis H_0 is rejected, which means the confidence level of this trend is less than α, and trend variation is significant under the significance level of α. Conversely, the trend is not significant. In this study, when $|Z_c| < Z_{(1-0.1)/2}$ =1.645, there is an insignificant trend, when $|Z_c| > Z_{(1-0.1)/2}$ =1.645, there is a significant trend, and when $|Z_c| > Z_{(1-0.01)/2}$=2.567, there is a highly significant trend. Meanwhile, Z>0 reflects the increasing trend of time series. Conversely, Z<0 indicates the decay trend.

Sen's slope is usually regarded as the absolute magnitude of trend variation, it is expressed as follows:

$$\beta = \text{Median}\left(\frac{x_j - x_i}{j-i}\right) \quad (1 \le k < j \le n) \tag{3}$$

$$R = \frac{\beta}{\frac{1}{n}\sum_{i=1}^{n} x_i} \times 100\% \tag{4}$$

$$\overline{R} = \frac{\sum_{i=1}^{M} R_i}{M} \tag{5}$$

where, β is the Sen's slope, n is the total number of temporal epochs in time-series. A positive β represents a rising trend, whereas a negative β represents a decreasing trend. R is the relative Sen's slope, which makes the variation magnitudes of time series of different pixels comparable. The higher the R is, the higher the relative magnitude of trend variation for this pixel will be. $\overline{R}$ is the spatial average of R across the whole basin, M is the total number of the pixels which have significant variation trend based on MK test in the HRB. $\overline{R}$ is able to exhibit the overall variation trend and magnitude at watershed scale, taking into account area ratio and relative variation magnitude for areas with statistically significant change trend.

3.4. *Partial correlation analysis*

Correlation analysis is one of the most commonly used statistical methods to explore the influence of one variable on another. Pearson Correlation calculated using equation (6) is applied to measure the relationship between only two variables. When the variable was comprehensively influenced by multiple factors, partial correlation coefficient (equation (7)) is a more scientific indicator to exhibit the relationship between two variables while removing the effects of other variables.

$$r = \frac{\sum_{i=1}^{n}(x_i - \overline{x})(y_i - \overline{y})}{\sqrt{\sum_{i=1}^{n}(x_i - \overline{x})^2 \cdot (y_i - \overline{y})^2}} \tag{6}$$

$$r_{12,34\ldots p} = \frac{r_{12,34\ldots(p-1)} - r_{1p,34\ldots(p-1)} \cdot r_{2p,34\ldots(p-1)}}{\sqrt{1 - r_{1p,34\ldots(p-1)} \cdot {}^2}\sqrt{1 - r_{2p,34\ldots(p-1)}{}^2}} \quad (P \ge 3) \tag{7}$$

where, r is the Pearson correlation coefficient which is also called as 0-order correlation coefficient. $r_{12,34\ldots p}$ is (p-2)-order partial correlation coefficient, which was calculated by (p-1)-order partial correlation coefficient ($r_{12,34\ldots(p-1)}$).

In general, the t-test statistic is calculated to evaluate the significance of the correlation coefficient, as illustrated in equation (8), in this study, each partial correlation coefficient is tested using t-test value at the significant level of 0.05.

$$t = \frac{r\sqrt{n-q-2}}{\sqrt{1-r^2}} \tag{8}$$

where, t is the t-test statistic, r is the partial correlation coefficient, n is the sample number, q is order of partial correlation coefficient, and (n-q-2) is the degrees of freedom.

In this study, pixel-level partial correlation analysis was employed to analyze the relations between NDVI and climatic factors (temperature, precipitation and radiation) and time respectively during 1998 to 2012. The partial correlation coefficient between NDVI and time was considered as the influence of non-climatic factors on vegetation. In addition, region-level partial correlation analysis was used to explore the correlations between NDVI and climatic factors including temperature, precipitation, and radiation at different reaches with different dominant vegetation types during different periods.

4. Result and discussion

4.1. Vegetation spatial characteristics in the HRB

The annual maximum NDVI represent the best status of vegetation during vegetation growing season, which could effectively characterize the discrepancy for different vegetation species in different areas in HRB, in this study, the multi-year average of annual maximum NDVI was calculated to describe the overall spatial distribution of vegetation during a fifteen-year period. It is well known there is an ambiguous ecological meaning for NDVI variation in unvegetated areas [22]. Thus, all NDVI pixels of vegetated areas in the HRB were used to exhibit the local vegetation situation in this study. Firstly, the vegetated area in HRB was identified using annual maximum NDVI time-series, if the NDVI value was greater than 0.1 at least two consecutive years, the pixel was classified as the vegetation. This treatment not only guaranteed that occasionally vegetated areas were taken into account but also NDVI variations not caused by vegetation change were excluded from further analysis [11].

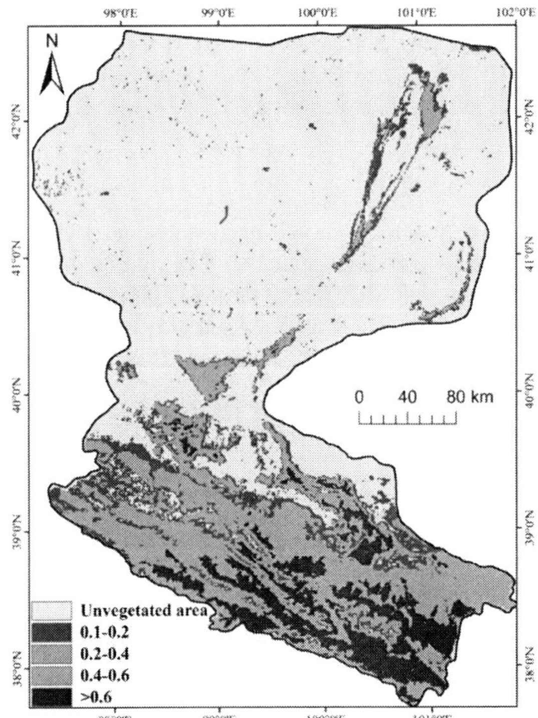

Figure 2. Spatial distribution of multi-year average of annual NDVI maximum time-series for vegetated areas in the HRB.

The spatial distribution characteristic of the average annual maximum NDVI of vegetation area from 1998 to 2012 was shown in figure 2. According to figure 2, it is illustrated that the vegetated area and unvegetated area account for 27.9% and 73.1% in HRB respectively, which demonstrates that the HRB is a typical arid and semi-arid inland river basin where the overall vegetation cover is very low. For different typical areas in the HRB, vegetated areas account for 90%, 56.5% and 4.9% in the upper reaches, middle reaches and lower reaches respectively, which illustrates an obvious gradient change of vegetation cover from upper reaches to lower reaches. This could be explained by the local climate condition that more precipitation and glacier/snow meltwater in the mountainous region, whereas there is extremely low rainfall and high evapotranspiration in the desert.

It is shown that the areas with NDVI less than 0.2 are mainly distributed in the desert, transit zone between oasis and desert, transit zone between mountains and desert and Gurinai basin which is mainly due to dwarf and sparse shrubs are the dominant vegetation species in these regions where there are very limited available water resources. The NDVI of Ejin oasis dominated by *Euphrates poplar* forests and shrubs is lowest in all oases in HRB, with NDVI value between 0.2-0.4. The artificial oases in the middle reaches (Jinta, Jiuquan and Zhangye) have relatively higher NDVI because the cropland is the dominant vegetation in this region where there are above 90% population and production activities in the HRB, hence water resources supply is guaranteed by EWDP for necessarily economic development. Ganzhou irrigated area has the highest NDVI higher than 0.6 among artificial oases. For the Qilian monotonous region dominated by nature forests and grassland, it is observed that almost one has the NDVI higher than 0.6, and there are higher NDVIs in the eastern region of Qilian mountain than the western region (Taolaihe sub-basin and Hongshuihe sub-basin), last but not the least, the NDVI in the middle elevation is higher than that in the higher and lower elevation in Qilian mountain.

In this study, the histogram of NDVI was also constructed from all the pixels for vegetated area in the HRB to assess the overall situation of vegetation across the whole basin, as illustrated in figure 3. It is concluded that the distribution characteristic of NDVI for vegetated area in HRB is not the unimodal distribution but bimodal distribution, the two peaks of NDVI appears between 0.15-0.25 and between 0.59-0.68, which means a large number of vegetation with low cover and high cover coexist in the HRB.

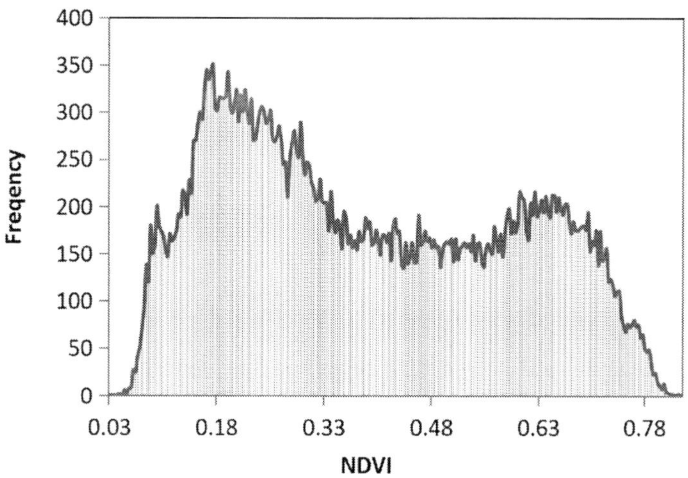

Figure 3. The histogram of NDVI for vegetated areas in the HRB.

4.2. NDVI temporal and spatial variation from 1998 to 2012 in the HRB

4.2.1. Variation trends of pixel-level NDVI in the HRB. The leaves of vegetation only keep green in annual growing season in this arid and semi-arid zone, so in this study, the average NDVI during the growing season (April to November for this region) was employed to represent the overall condition and status of vegetation during one year. The MK test and relative Sen's slope were employed to explore the temporal variation trend of average NDVI during the growing season time-series from 1998 to 2012. The spatial pattern of statistic Z for NDVI in HRB was obtained by calculating statistic Z of NDVI time series for each pixel. The degrees of vegetation change were divided into five levels (figure 4(a)), consisting of highly significant decrease ($Z \leq -2.567$), significant decrease ($-2.567 < Z \leq$-

1.645), insignificance change (-1.645<Z<1.645), significant increase (1.645≤Z<2.567) and highly significant increase (Z≥2.567). For the areas with significant change trend, four levels were divided, including -23.2%~-3%, -3%~-0.1%, 0.1%~3% and 3%~15.8%, based on relative Sen's slope of variation trend in order to demonstrate the spatial distribution of variation magnitude, as illustrated in figure 4(b).

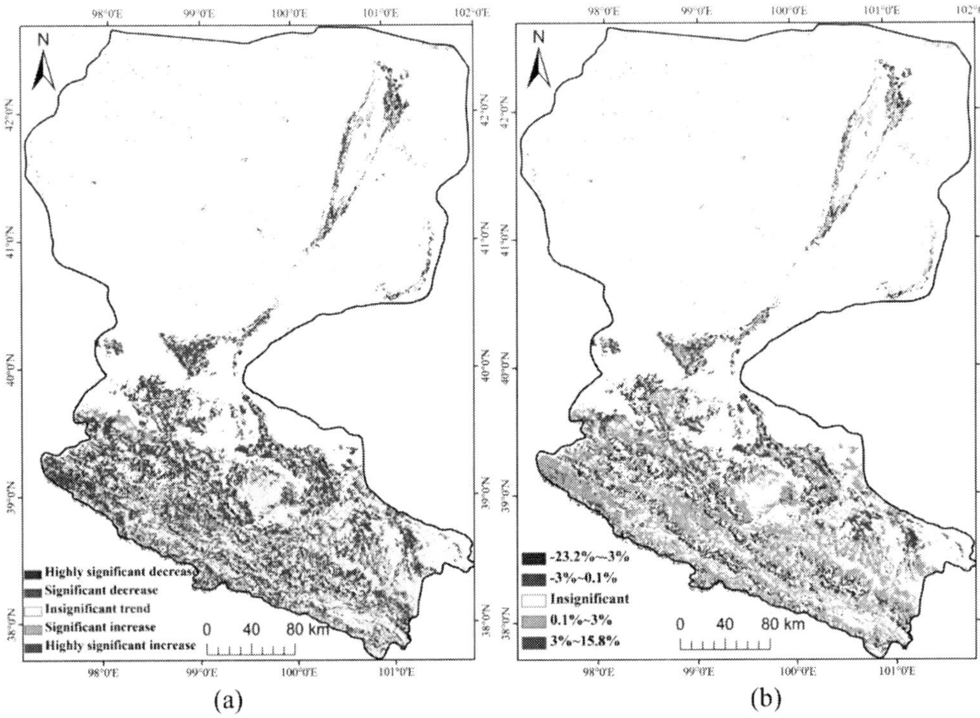

Figure 4. Spatial distributions of variation trend and significance (a) and relative Sen's slope of NDVI time-series.

MK test result shows that the variation trends of NDVI pixels are not significant in the area, accounting for 44.9% in the total vegetated area, located at Sandan, Minle, Ganzhou and transit zone between mountain and oasis, and parts of grassland in the upper reaches. The variation trend is significantly increasing for the majority of vegetated area, accounting for 46.3%, and there is only 8.8% of the vegetation covered area shows a significantly decreasing trend, furthermore, the highly significantly decreasing ($P<0.01$) areas where water resources supply is very limited only account for 3% in vegetated area, however, the highly significantly increasing ($P<0.01$) areas account for 27.4%, which demonstrates that the overall area of ecologically improved vegetated area overweight that of degraded vegetated area during 1998-2012.

The spatial distribution of NDVI pixels with significant variation has distinct regional characteristics. According to the figure 4(a), the significantly deteriorated areas in the upper reaches are mainly located in the mountainous region where the elevation is higher than 4200 m in the Qilian Mountain, which proves that the alpine vegetation is experiencing degradation. For the oasis plains, the areas with a declining trend of NDVI are located at the northern part of Hexi Corridor and southern part of Linze. For the desert in the lower reaches which is traditional area characterized with vegetation degradation, a significant decreasing trend is still observed in the areas discretely distributed near the downstream channel and northern part of Ejin oasis. The areas with significant and highly significant improvement in the upper reaches are mainly located at the Taolaihe sub-basin and Hongshuihe sub-basin in the western parts of Qilian Mountain, some scattered regions with middle elevation in Qilian Mountain, and the foot of the Northern mountain. For the oasis plains, the majority of areas with increasing trend are characterized with highly significant increase, especially for the Jinta, Jiuquan, Gaotai, Linze and Ganzhou, which demonstrates that the condition of cropland in this area is experiencing an increasing trend. The herbs and shrubs on both sides of downstream river channel and parts of Ejin oasis exhibit the highly significant improving trend according to the Figure 4(a).

The pixels showing significant change trend based on MK test (figure 4(a)) were selected to further analyze the relative change magnitude which reflects the changing degree. In this study, relative Sen's slope (R) of variation trend was proposed to represent the variation magnitude from 1998 to 2012. The spatial distribution of relative Sen's slope in NDVI was shown in figure 4(b). For the significantly increasing vegetated area, the areas with R less than -3% and higher than -3% account for 42.1% and 57.9% respectively; for the significantly decreasing vegetated area, the areas with R less than 3% and higher than 3% account for 74.1% and 25.9%.

For the spatial distribution characteristic of R, the pixels exhibiting decreasing trend, with R less than -0.01%, are scatted in the whole basin. The pixels with R higher than 3% are mainly distributed in oasis plain, Minle and Sandan grassland in the middle reaches, the both sides of downstream river and the Ejin oasis. The relative magnitude of pixels characterized with increasing trend in the upper reaches is mainly between 0.1%-3%. It is illustrated that the lowest R (-23.2%) appears in the western part of Qilian mountain, and the highest R (15.8%) appears in the Ejin oasis. In this study, the $\overline{R}$ was proposed to represent the overall variation trend and magnitude at watershed scale. The $\overline{R}$ calculated using equation (5) for the HRB during 1998 to 2012 is 1.5%, which illustrates that vegetation in the HRB was improved by 1.5% of annual average level from 1998 to 2012.

4.2.2. Variation trends of region-level NDVI for different reaches in the HRB. There are different dominant vegetation species in the upper, middle and lower reaches with significantly different climatic conditions in the HRB, the variation trend of region-level NDVI time-series was also analyzed to explore the change characteristics of vegetation under different environment. The average NDVI values were calculated within the upper, middle and lower reaches respectively in each year, and region-level NDVI time-series in three areas were shown in figure 5.

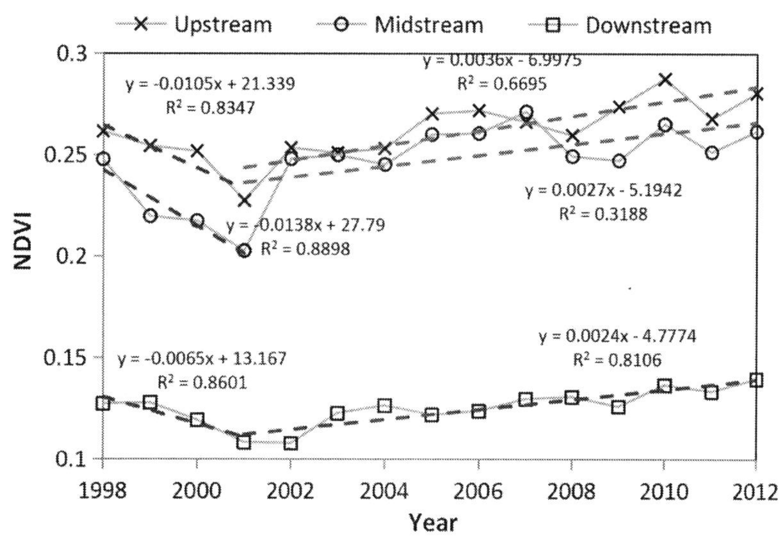

Figure 5. Variation trends of region-level NDVI in the upper, middle and lower reaches respectively.

The MK test was employed to analyze the time-series variation trend, and the same varying patterns were found in different regions: 1) the year of 2001 is the abrupt point, which coincides with the implement time of EWDP proposed by Chinese central government; 2) there is an obvious decreasing trend during 1998-2001; 3) a significantly increasing trend was observed from 2001 to 2012, which could explain that the relative variation magnitude is very low in the HRB during 1998 to 2012, and the variation degree is expected to be greater during 2001 to 2012. For the upper reaches, there is a highly significant increase ($P<0.01$), with a relative increasing magnitude of 0.8%; for the middle reaches, there is a significant increase ($P<0.1$), with a relative increasing magnitude of 0.86%; for the lower reaches, there is a highly significant increase ($P<0.01$), with a relative increasing magnitude of 1.06%. It is observed according to comparison that there is a highest NDVI value but lowest increasing magnitude for the upper reaches, however, there is a lowest NDVI value but highest increasing magnitude in the lower reaches. It is concluded that the changing degree of vegetation is greatest for the sparse shrubs in the lower reaches, and is smallest for the forest and grassland in the upper reaches.

4.3. Correlation between pixel-level NDVI and climatic and non-climatic factors

Precipitation, solar radiation and temperature are the key climatic factors constraining the vegetation growth [23, 24]. Meanwhile, the effect from human activities on vegetation is also non-ignorable, especially in the human living areas. The partial correlation analysis was used to determine the influence of one factor on vegetation, taking away the influence from other factors. In this study, the partial correlation analysis between NDVI and four variables including precipitation (P), temperature (T), Radiation (R), time (t) at pixel scale was performed in the HRB according to equation (7), and the significance of partial correlation coefficients between NDVI and variables were determined based on t-test (equation (8)) at the significant level of 0.05. If the partial correlations between NDVI and multiple variables are significant, it is demonstrated that the variation of NDVI is affected by the multiple variables. It should be noted that partial correlation coefficient between NDVI and time, taking away the impacts from all climatic factors, was considered to be the influence from non-climatic factor mainly including human activities on vegetation.

In this study, overlay analysis of pixel-level NDVI variation trend produced in the 4.2.1 section and pixel-level partial correlations between NDVI and four variables were performed to determine which factor(s) result in an increase or decrease of NDVI during 1998 to 2012, as illustrated in the figure 6. It is observed that there are no significant partial correlation between NDVI and meteorological factors involving temperature and radiation, which demonstrates that temperature and radiation are not the driving factors constraining vegetation dynamics in the HRB. According to the figure 6, the precipitation and time are the two main driving factors affecting vegetation variation. Spatially, the significant vegetation degradation in the northern part of Heixi Corridor and some southern parts in the Linze is caused by decreasing precipitation. Although the transit zone between desert and southern part of the Ganzhou oasis and Minle are not characterized with significant NDVI change trend, the precipitation and time are still the driving factors influencing the fluctuate up and down of vegetation. It is very obvious that the time is the most important factor which leads to NDVI increase significantly. The NDVI increasing trend in the foot of west-end of Qilian Mountain and east-end of Taolaihe is mainly caused by increasing precipitation. It is shown that NDVI increase at the transit zones between mountain and oasis in the middle reaches as a result of combined effects from precipitation and time. More importantly, the vegetation improvement in the areas with middle elevation in the Qilian Mountain, the oasis plains involving Jinta, Jiuquan, Gaotai, Linze in the middle reaches, the areas along downstream river channel, Ejin oasis and Gurinai basin was only influenced by non-climatic factor. In the middle reaches and lower reaches, the influence of non-climatic factor is reasonably considered as effects from EWDP implementation, which indicates that the ecological and economic effects of EWDP implementation are positive for the HRB during past fifteen years.

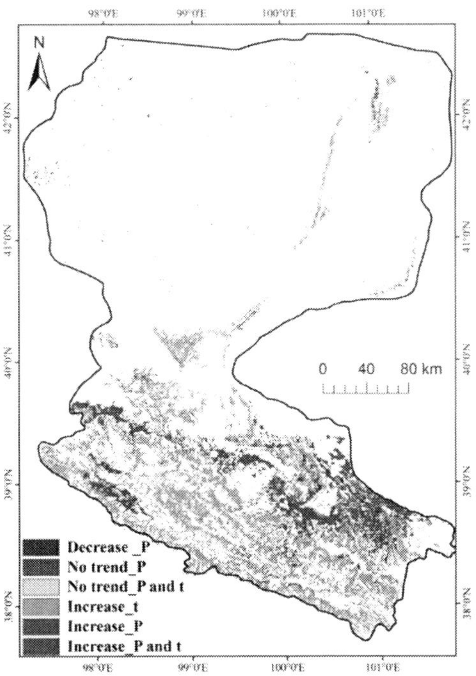

Figure 6. The spatial distribution of NDVI variation trends caused by precipitation, non-climatic factors and the combination of precipitation and non-climatic factors.

4.4. Effects of climatic factors on region-level NDVI for different periods

To explore the driving factor affecting region-level NDVI variation, in this study, the spatial average of annual average temperature (°C), annual total precipitation (mm) and annual total radiation (MJ/m^2) were calculated based on gridded meteorological dataset within the upper, middle and lower reaches respectively.

The linear regression analysis between NDVI and climatic factors were used to derive determination coefficient for different reaches during different periods, as illustrated in the figure 7. The result shows that 1) for the period from 1998 to 2001, rainfall is an important climatic factor influencing NDVI in the upper reaches (R^2=0.54) and middle reaches (R^2=0.83), and radiation is a significant factor influencing NDVI in the lower reaches (R^2=0.96), and the higher the radiation is, the lower the NDVI in the lower reaches will be; 2) for the period from 2001 to 2012, the influence of rainfall on NDVI is still significant, however determination coefficient has reduced to 0.42, and influence of other climatic factors on NDVI in each reaches were not so significant as the period from 1998 to 2001.

For determining the driving factor leading to the vegetation variation in different reaches for different , the partial correlations between NDVI and three climatic variables at region-level were calculated and significance testing was performed using t-test statistic method, if the significance level is greater than 0.1, there is no statistical significance for this correlation. The significance levels of partial correlation between NDVI and all factors were illustrated in the table 1. It is indicated that effects of climatic factors on NDVI varied with intervention of intensive human activities since 2001. The rainfall is proved to be the most significant factor relating with NDVI variation in the upper reaches, and rainfall and temperature are significant factors influencing NDVI variation in the middle reaches, and rainfall and radiation are significant factors influencing NDVI variation in the lower reaches during the period (1998-2001) when there are relative small amount of anthropogenic activities in the HRB. However, there is only a significant climatic factor (rainfall) relating NDVI variation in the middle reaches, and all climatic factors are not significantly related with NDVI in the upper reaches and lower reaches during the period (2001-2012) when lots of ecological restoration projects were implemented. This could demonstrate that due to implementation of lots of ecological restoration projects since 2001, the effect of climatic factors on vegetation change in the HRB is getting weak from 2001 to 2012, and influence of human activities is getting strong.

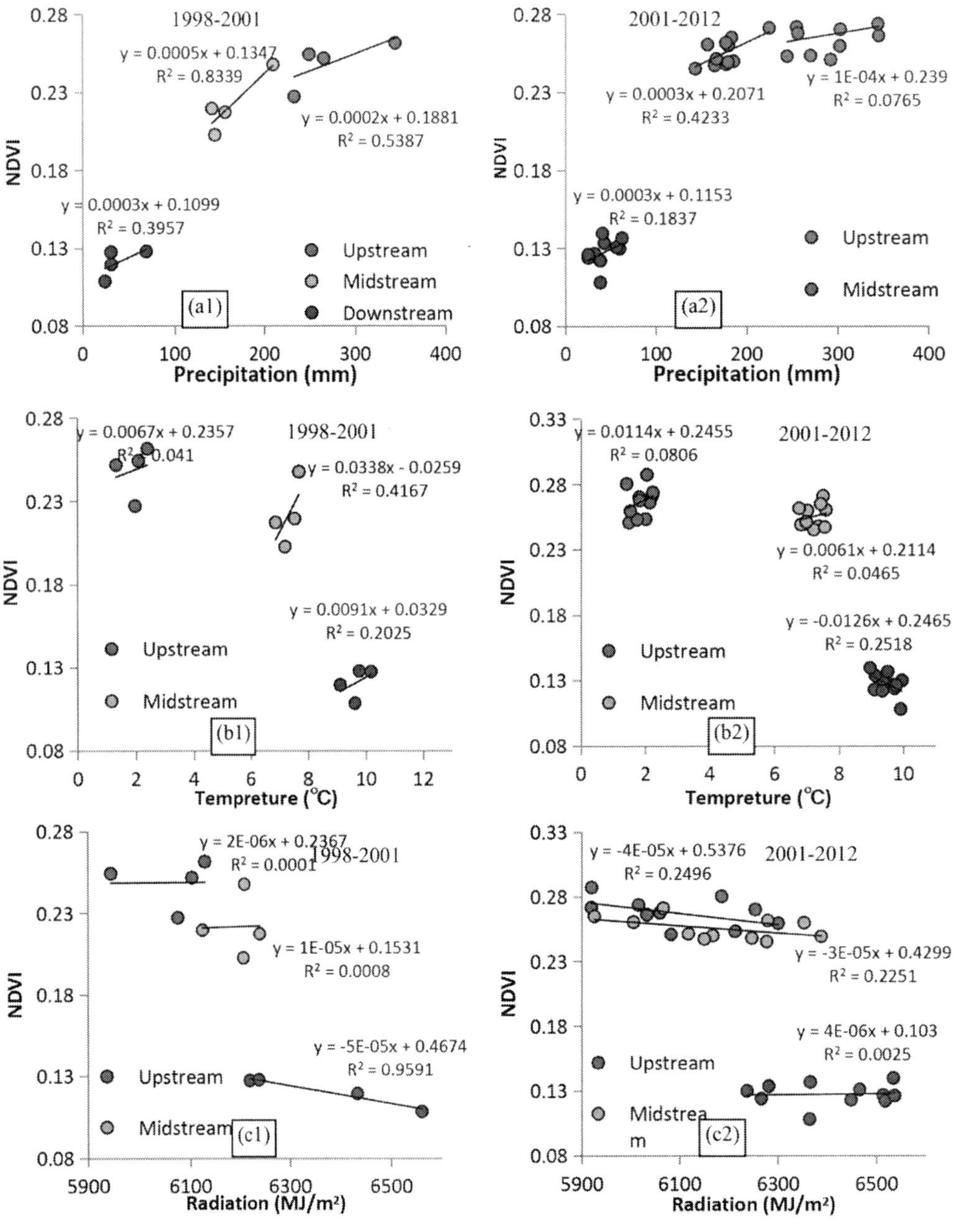

Figure 7. Linear regression analysis between NDVI and climatic factors in the different reaches for different periods in the HRB.

Table 1. The partial correlation coefficients between region-level NDVI and climatic factors (precipitation, temperature and radiation) based on t-test for different periods.

Periods	1998-2001			2001-2012		
Regions / Factors	Upper reaches	Middle reaches	Lower reaches	Upper reaches	Middle reaches	Lower reaches
Precipitation	0.72*	0.91*	0.71*	0.07	0.64*	0.41
Tempreture	0.20	0.64*	0.4	0.27	0.31	0.50
Radiation	0	0	0.96*	0.48	0.46	0

* P<0.05

5. Conclusion

In this study, spatial-temporal dynamics of vegetation and its response to climatic change in the HRB, a typical arid region with vulnerable ecosystems were explored. The Mann–Kendall test was used in combination with relative Sen's slope to analyze the NDVI time-series at pixel and regional scale respectively. The linear regression and partial correlation analysis were then employed for exploring the driving forces affecting the vegetation changes at pixel and regional scale respectively. According to the analysis and discussion of results presented, we may come to the following conclusions:

- It is concluded that the vegetated area and unvegetated area account for 27.9% and 73.1% in the HRB respectively, and vegetated area account for 90%, 56.5% and 4.9% in the upper reaches, middle reaches and lower reaches respectively. The NDVI of Ejin oasis dominated by *Euphrates poplar* forests and shrubs is lowest in all oases in the HRB, and Ganzhou irrigated district has the highest NDVI higher than 0.6 among artificial oases. For the Qilian monotonous region dominated by nature forests and grassland, it is observed that almost one-third of this area has the NDVI higher than 0.6, and the NDVI in the middle elevation is higher than that in the higher and lower elevation in Qilian mountain. It is shown that the distribution characteristic of NDVI histogram for vegetated area in the HRB is not the unimodal distribution but bimodal distribution.
- The area where the variation trend of vegetation is significantly increasing accounting for 46.3% in the vegetated area in the HRB during 1998-2012, and there is only 8.8% of the vegetation covered area shows a significantly decreasing trend. The relative Sen's slope (R) of variation trend was proposed to represent the variation magnitude from 1998 to 2012. For the significantly increasing vegetated area, the areas with R less than -3% and higher than -3% account for 42.1% and 57.9% respectively; for the significantly decreasing vegetated area, the areas with R less than 3% and higher than 3% account for 74.1% and 25.9%. The $\overline{R}$ was proposed to represent the overall variation trend and magnitude at watershed scale, which illustrates that vegetation in the HRB was improved by 1.5% of annual average level from 1998 to 2012.
- On the regional scale, same variation trajectories of vegetation were found in different regions: 1) the year of 2001 is the catastrophe point, which coincides with the implement time of EWDP proposed by Chinese central government; 2) there is an obvious decreasing trend during 1998-2001; 3) a significantly increasing trend was observed from 2001 to 2012. The relative Sen's slope is 0.8%, 0.86% and 1.06% for the upper, middle and lower reaches respectively. It is concluded that the changing degree of vegetation is greatest for the sparse shrubs in the lower reaches, and is smallest for the forest and grassland in the upper reaches.
- Effects of climatic factors on NDVI varied with intervention of intensive human activities since 2001. The rainfall is proved to be the most significant factor relating with NDVI variation in the upper reaches, and rainfall and temperature are significant factors influencing NDVI variation in the middle reaches, and rainfall and radiation are significant factors influencing NDVI variation in the lower reaches during the period (1998-2001) when there are relative small amount of anthropogenic activities in the HRB. However, there is only a

significant climatic factor (rainfall) relating NDVI variation in the middle reaches, and all climatic factors are not significantly influence the NDVI change in the upper reaches and lower reaches during the period (2001-2012) when lots of ecological restoration projects were implemented to improve ecological effects of Heihe River.

Acknowledgments

The authors gratefully acknowledge the financial support provided for this research by Fund of development on science and technology Yellow River Institute of Hydraulic Research Science and Technology Foundation Programme (HKF201602), National Key Research Priorities Program of China (2016YFC0402402).

References

[1] Cheng G, Xin L, Zhao W, *et al* 2014 Integrated study of the water–ecosystem–economy in the Heihe River Basin *Natl. Sci. Rev.* **1** 413-28
[2] Mohammad A G and Adam M A 2010 The impact of vegetative cover type on runoff and soil erosion under different land uses *Catena* **81** 97-103
[3] Mata-González R, Martin D W, Mclendon T, *et al* 2012 Invasive plants and plant diversity as affected by groundwater depth and microtopography in the Great Basin *Ecohydrology* **5** 648-55
[4] Cao L, Xu J, Chen Y, *et al* 2013 Understanding the dynamic coupling between vegetation cover and climatic factors in a semiarid region—a case study of Inner Mongolia, China *Ecohydrology* **6** 917-26
[5] Xin Z B, Xu J X and Wei Z 2008 Spatiotemporal variations of vegetation cover on the Chinese Loess Plateau (1981-2006): Impacts of climate changes and human activities *Sci. Chin. Earth Sci.* **51** 67-78
[6] Wang J, Wang K, Zhang M, *et al* 2015 Impacts of climate change and human activities on vegetation cover in hilly southern China *Ecol. Eng.* **81** 451-61
[7] Yu K and Hu C 2013 Changes in vegetative coverage of the Hongze Lake national wetland nature reserve: A decade-long assessment using MODIS medium-resolution data *J. Appl. Remote Sens.* **7** 3589
[8] Zhang J, Zhang Q, Yang L and Li D 2006 Seasonal characters of regional vegetation activity in response to climate change in West China in recent 20 years *J. Geogr. Sci.* **16** 78-86
[9] Duan H, Yan C, Tsunekawa A, *et al* 2011 Assessing vegetation dynamics in the Three-North Shelter Forest region of China using AVHRR NDVI data *Environ. Earth Sci.* **64** 1011-20
[10] Li A, Wu J and Huang J 2012 Distinguishing between human-induced and climate-driven vegetation changes: A critical application of RESTREND in inner Mongolia Landscape *Ecol.* **27** 969-82
[11] Wang Y, Shen Y, Chen Y, *et al* 2013 Vegetation dynamics and their response to hydroclimatic factors in the Tarim River Basin, China *Ecohydrology* **6** 927-36
[12] Wang J, Meng J J and Cai Y L 2008 Assessing vegetation dynamics impacted by climate change in the southwestern karst region of China with AVHRR NDVI and AVHRR NPP time-series *Environ. Geol.* **54** 1185-95
[13] Zhang Y, Yu J, Wang P, *et al* 2011 Vegetation responses to integrated water management in the Ejina basin, northwest China *Hydrol. Process* **25** 3448-61
[14] Qi S Z and Fang L 2005 Water environmental degradation of the Heihe River Basin in arid northwestern China *Environ. Monit. Asses.* **108** 205-215
[15] Qi S and Cai Y 2007 Mapping and assessment of degraded land in the Heihe River Basin, arid northwestern China *Sensors* **7** 2565-78
[16] Hijmans R J, Cameron S E, Parra J L, *et al* 2005 Very high resolution interpolated climate surfaces for global land areas *Int. J. Climatol.* **25** 1965-78
[17] Savitzky A and Golay M J E 1964 Smoothing and differentiation of data by simplified least

IOP Conf. Series: Earth and Environmental Science **82** (2017) 012045 doi:10.1088/1755-1315/82/1/012045

squares procedures *Anal. Chem.* **36** 1627-39
[18] Steinier J, Termonia Y and Deltour J 1972 Comments on smoothing and differentiation of data by simplified least squares procedure *Anal. Chem.* **44** 1906-9
[19] Yue S and Wang C Y 2002 Applicability of prewhitening to eliminate the influence of serial correlation on the Mann–Kendall test *Water Resour. Res.* **38** 41-47
[20] Xu J H, Chen Y N, Li W H, Ji M H, Dong S and Hong Y L 2009 Wavelet analysis and nonparametric test for climate change in Tarim River Basin of Xinjiang during 1959–2006 *Chin. Geogr. Sci.* **19** 306-13.
[21] Hamed K H 2008 Trend detection in hydrologic data: The Mann–Kendall trend test under the scaling hypothesis *J. Hydrol.* **349** 350-63
[22] Zeng B and Yang T B 2008 Impacts of climate warming on vegetation in Qaidam Area from 1990 to 2003 *Environ. Monit. Asses.* **144** 403-17
[23] Churkina G and Running S W 1998 Contrasting climatic controls on the estimated productivity of global terrestrial biomes *Ecosystems* **1** 206-15
[24] Nemani R R, Keeling C D, Hashimoto H, *et al* 2003 Climate-driven increases in global terrestrial net primary production from 1982 to 1999 *Science* **300** 1560-3

IOP Conf. Series: Earth and Environmental Science 82 (2017) 012046 doi:10.1088/1755-1315/82/1/012046

Estimation of pumpage from a riverside well at the lower reaches of the Songhua River

K N Sun[1,2], L T Hu[1,2,3], M L Zhang[1,2] and X M Liu[1,2]

[1]College of Water Sciences, Beijing Normal University, Beijing 100875, China
[2]Engineering Research Centre of Groundwater Pollution Control and Remediation of Ministry of Education, Beijing Normal University, Beijing 100875, China

E-mail: litanghu@bnu.edu.cn

Abstract. As a kind of vital water resources exploitation mode, riverside pumping has the important advantage of maintaining the stability of water supply and purifying surface water quality. It is of great significance to estimate the pumpage from a riverside well for the sustainable utilization of water resources in the area. In this article, the method of image and Girinskii's potential function are used to derive the pumpage. A case study in the Jamusi City shows that the pumpage from a riverside well varies from 9000 to 15000 m^3/d. The sensitivity of factors (h_0, K, S_{max} and r_w) is analysed, indicating that the influence degree from large to small is h_0, K, S_{max}, and r_w. Besides, h_0, K, S_{max} are far greater than r_w. It is hoped that this paper can provide some theoretical reference for the rational utilization of groundwater in the areas near the rivers.

1. Introduction

Compared with direct exploitation of surface water, riverside pumping is advantageous because of the better quality of riverside groundwater due to water treatment through the underground media, and reliable quantities from river recharge [1]. So riverside pumping has been practiced in many areas over the world for a long time. However, how to determine the appropriate pumpage and ensure the quality of groundwater, is the focus of domestic and foreign scholars in recent years. Up to now, many scholars have paid much attention to the estimation of pumpage from a riverside well. The principle of water balance is firstly used to estimate the pumpage, where the method is simple, but it cannot accurately calculate the pumping rate from a riverside well. Numerical simulation is an effective method for analyzing surface water-groundwater interactions, but it will cost much time and data to construct a model. In most cases, analytical solutions provide a quick and draft estimation of pumpage from a riverside well [2].

Concerning with the analytical solutions, Theis obtained the first unsteady solution about riverside pumping problem in the form of an integral in 1941 [3]. In 1954, this same solution was rewritten by Glover and Balmer using the complimentary error function [4]. Then in 1965, Hantush [5] studied a problem which differed from the problem considered by Theis in 1941 only by the inclusion of a vertically semi-pervious material along the stream edge. In 1999, Hunt [6] studied an analytical model that considered the effects of streambed clogging and partial stream penetration. Then many scholars, such as Chen [7], Singh [8], Singh [9] and Zhang [10] studied the river-aquifer interaction. These solutions provide a theoretical basis for the calculation of pumpage and are applicable to the ideal or simple structure aquifer. Up to now, there are still some problems to be addressed, such as partially penetrated river, influences of semi-pervious beds, presence of unsaturated zone between river and groundwater. Therefore, the premise of applying analytical method is to make a reasonable generalization and comprehensive consideration about hydrogeological conditions

The objective of this paper is to propose a new method for estimating pumpage using the method of image and Girinskii's potential function, and then analyze the sensitivity of factors that impact the pumpage. A case study in the Jiamusi City, downstream of the Songhua River, is given to demonstrate the application of the analytical solutions.

2. Method

In most riverside areas, groundwater system may be under a dynamic steady state, due to the influences of river recharge, pumping well and rainfall infiltration. As shown in figure 1(a), the river can be considered as a constant-head linear boundary. It is also assumed that the phreatic aquifer is horizontal and has a uniform thickness, and the aquifer medium is homogeneous and isotropic. So the Girinskii's potential function can be used to analyze the problem of groundwater flow and the potential of the unconfined aquifer is written as

$$\varphi = Kh^2/2 \quad (1)$$

where φ represents the Girinskii's potential; K denotes the hydraulic conductivity; h means the thickness of the saturated area in the unconfined aquifer.

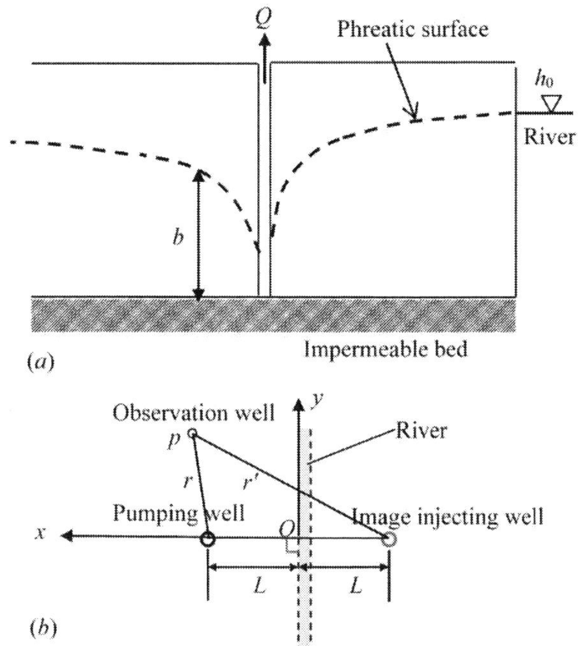

Figure 1. Schematic diagram of riverside pumping problem.

For an unconfined aquifer, assume that the steady radial groundwater flow to a pumping well is horizontal, and the aquifer has uniform thickness. Then Girinskii's potential can be described as follows

$$Q = 2\pi r\, d\varphi/dr \quad (2)$$

The general solution of equation (2) is

$$\varphi = \frac{Q}{2\pi} ln\, r + C \quad (3)$$

where Q is pumpage rate for the single well; C is a constant determined by boundary conditions; r is the distance away from any point to the pumping well.

The solution of pumping flow near a riverside can be obtained by the method of image. The method is to consider the influences of an image injection well and then substitute the actual effect influenced by the river [11] (figures 1(a) and 1(b)). The effects of pumping well and river can be superposed. Girinskii's potential satisfies the Laplace equation. So the superposition rule can be used. Considering equations (1) and (3), one gets the solution

$$h^2 = h_0^2 - \frac{Q}{\pi K} ln\left(\frac{r'}{r}\right) \quad (4)$$

$$r' = \sqrt{(x + L)^2 + y^2} \quad (5)$$

$$r = \sqrt{(x - L)^2 + y^2} \quad (6)$$

where r is the distance from the observation well $p(x, y)$ to the pumping well; r' is the distance from the observation well $p(x, y)$ to the image well; L is the distance between the river and well.

Equation (4) is the solution for a riverside well. When the observed location is in the pumping well, one can get the following equation (7).

$$Q = \frac{\pi K}{\ln\left(\frac{2L}{r_w}\right)}\left(h_0^2 - (h_0 - s_{max})^2\right) \quad (7)$$

where h_0 is the thickness between the river surface and impervious base; r_w is the radius of the well; S_{max} is the maximum allowable drawdown.

3. Case studies and discussions

3.1. Study area

The Jiamusi City (129°55'~135°04' E, 46°56'~48°30' N) (figure 2) is located at the lower reaches of the Songhua River flood plain. The terrain is zonal and slightly inclined from the southwest to the northeast. Groundwater is mainly distributed in the Quaternary strata of the alluvial plain. The main groundwater type is pore phreatic water. The main components of groundwater recharge are rainfall infiltration and surface water recharge. The main discharge items are the exploitation of mining wells and discharge to the rivers. According to the hydrogeological characteristics and water abundance, the study area can be divided into two hydrogeological zones (I and II): (1) Zone I is mainly distributed in the area along the Songhua River, where the thickness of the aquifer in this area is 50~105 m, and the

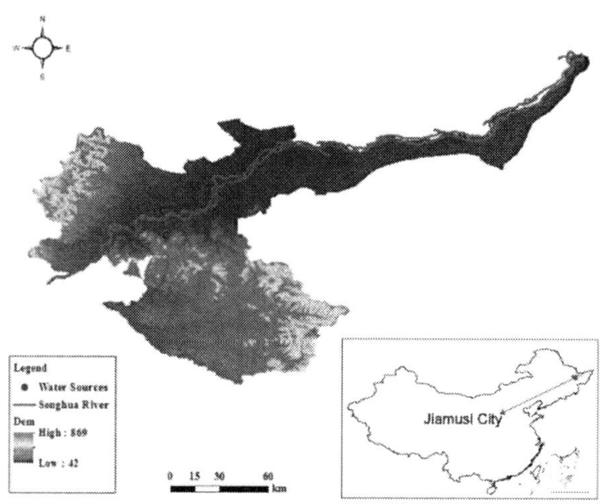

Figure 2. Map of study area.

hydraulic conductivity is about 39~144 m/d. (2) Zone II is mainly distributed in the northwest and the thickness of the aquifer is 80~89 m. The hydraulic conductivity in Zone II is about 24~26 m/d.

In 2005, the total water consumption in the area was about 3.50×10^8 m^3, whereas annual groundwater supply was about 0.70×10^8 m^3, accounting for about 20% of the total water consumption. Meanwhile, the urban water supply sources in the Jiamusi city are from groundwater. Because of the natural supply source of the Songhua River, riverside pumping becomes the main form of groundwater exploration in the Jiamusi city. The two major groundwater sources (figure 2) in the Jiamusi city are located in the lower part of the Songhua River.

3.2. Change of pumpage with the distance from the river

According to the previous studies, the average aquifer thickness in this area is about 70 m, and the average hydraulic conductivity is about 30 m/d. In China, the allowable drawdown for a pumping well is usually set as 7 m. Most pumping wells are set in a range of about 3000 m away from the river. Based on equation (7), the change of pumpage with the distance from the river is shown in figure 3. It can be clearly seen that the pumpage from a well varies from 9000 to 15000 m^3/d. With the increase of L, Q will gradually reduce. Moreover, with the increase of L, the degree of the influence of L on Q will decrease, mainly because groundwater system may obtain less recharge from the river when the distance away from the river is increased.

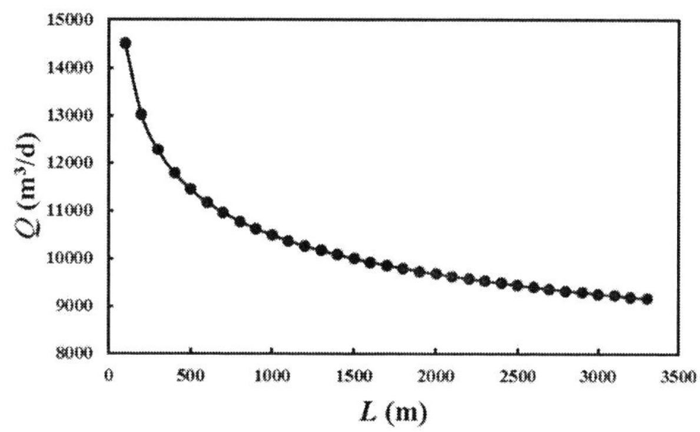

Figure 3. Schematic diagram of the relationship between pumpage and the distance between well and river.

For better understanding change of water table under the influences of different pumpage, figure 4 is drawn, which is obtained when L is 1000 m and the cross-section is set along the line from the pumping well to the imaging well. It can be seen that when water table increases inversely with the pumpage (Q) varying from 9000 m^3/d to 15000 m^3/d. The maximum difference of water table under the different pumpage is 3.47 m, and water table near the river is higher than that at the opposite side and the same distance from the well.

3.3. Sensitivity analysis of key factors on pumpage

From the hydrogeological conditions, we know the range of K is 24~144 m/d and the range of h_0 is 50~105 m. According to empirical value, S_{max} and r_w should not be too large. The appropriate ranges of S_{max} and r_w are 1~22 m and 0.1~2.3 m. Based on equation (7), the relationships between Q and S_{max}, Q and K , Q and h_0, Q and r_w can be obtained and are shown in figures 5(a)-5(d). It can be easily seen from the following curves that with the increase of S_{max}, K, h_0 and r_w, Q will also increase. K and h_0 are both positively proportional to Q; S_{max} and Q present the relationship of quadratic function; Logarithmic function relationship exists between r_w and Q.

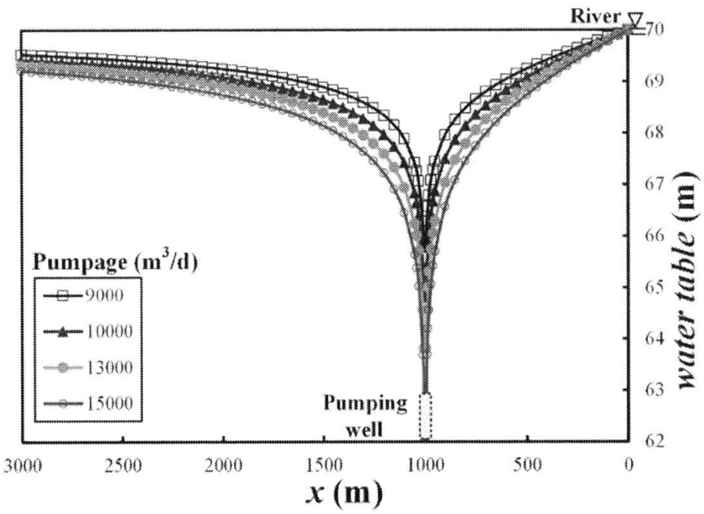

Figure 4. Change of water table along x axis under different pumpage conditions.

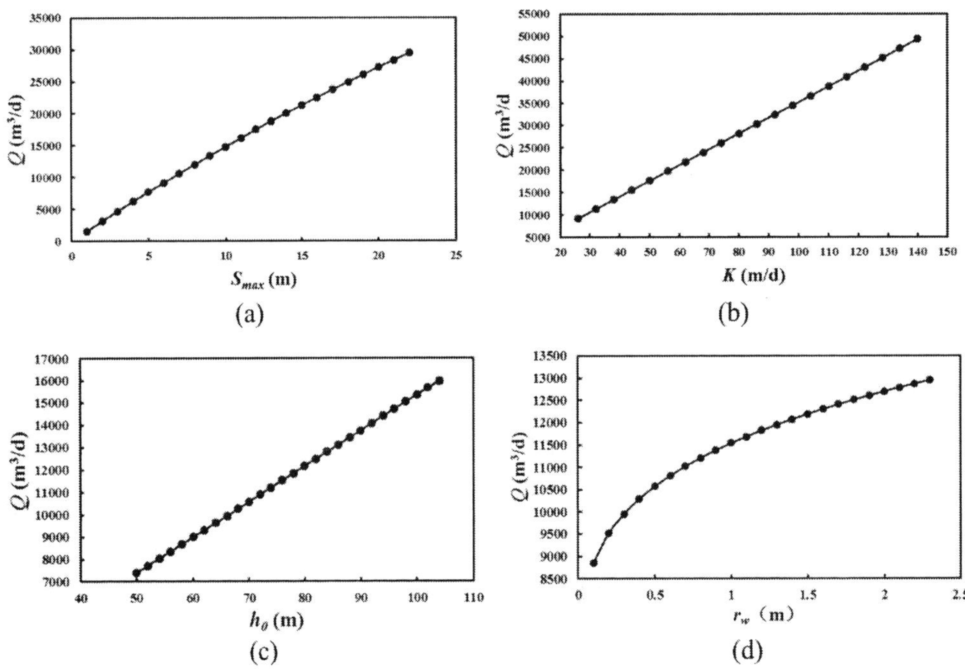

Figure 5. Schematic diagram of the relationship between Q and the maximum allowable drawdown (a), the hydraulic conductivity (b), the thickness between the river surface and impervious base (c), the radius of the well (d).

The sensitivity of different parameters is analyzed when the distance between well and river is 1000 m. The sensitivity of each factor on pumpage is calculated using equation (8).

$$J_i = \partial y / \partial \alpha_i \quad (8)$$

Where J_i is the sensitivity coefficient which indicates the degree of influence on the output y when a_i changes; i is the parameter serial number; Because of the different units of different parameters, the formula 9 is obtained after the formula 8 is standardized.

$$J_i = (\partial y/y)/(\partial \alpha_i/\alpha_i) \approx ((y_{(\alpha_i+\Delta\alpha_i)} - y_{(\alpha_i)})/y_{(\alpha_i)})/(\Delta\alpha_i/\alpha_i) \quad (9)$$

The sensitivity coefficients of S_{max}, K, h_0 and r_w are obtained according to formula 9 and the range of values is 0.08~1.07. Among them, the sensitivity coefficient of r_w is the smallest (0.08~0.13). The sensitivity coefficient of the other three parameters is basically same (J_{h_0} =1.07~1.04; J_K=1; $J_{S_{max}}$=0.82~0.99). So we know that the influence degree of h_0 is greatest followed by K, S_{max} and r_w and the effect of h_0, K and S_{max} on Q is much higher than that of r_w.

4. Conclusions

The estimation of pumpage from a riverside well is very important for the sustainable development and utilization of regional groundwater in the area of riverside pumping. This study proposes an analytical solution of estimating pumpage using the method of image and Girinskii's potential. And then a case study in the Jiamusi City is used to analyze the single well pumpage. The result shows that the pumpage from a riverside well may vary from 9000 to 15000 m^3/d (S_{max}=7 m, r_w=0.5 m). The sensitivity of factors that impact the pumpage is analyzed. The influence degree of h_0 is greatest and r_w is smallest. But the effect of h_0, K and S_{max} on Q are basically equal and are much higher than that of r_w.

The solutions can be well used to quickly estimate the pumpage at the downstream of the Songhua River. However, some assumptions are made in the derived analytical solutions, including fully-penetrate river, the horizontal, homogeneous and isotropic aquifer with a uniform thickness. So the method is only applicable for the draft and quick estimation of pumpage in an area. There is an important relationship between the maximum pumpage, the hydrogeological parameters and the design of the well, so it is suggested to carry out detailed investigation to identify the value of each hydrogeology properties to further refine the calculation in the field cases.

Acknowledgments
This work was supported by the National Natural Science Foundation of China (Grant number: 41572220) and Beijing Natural Science Foundation (J150002).

References

[1] Hu B, Teng Y, Zhai Y, Zuo R, Li J and Chen H 2016 Riverbank filtration in china: A review and perspective *J. Hydrol.* **541** 914-27
[2] Chen X and Shu L 2006 Groundwater evapotranspiration captured by seasonally pumped wells in river valleys *J. Hydrol.* **318** 334-47
[3] Theis C V 1941 The effect of a well on the flow of a nearby stream *Eos Trans. Am. Geophys. Union* **22** 734-8
[4] Glover R E and Balmer G G 1954 River depletion resulting from pumping a well near a river *Eos Trans. Am. Geophys. Union* **35** 468
[5] Hantush M S 1965 Wells near stream with semipervious beds *J. Geophys. Res.* **70** 2829-38
[6] Hunt B 1999 Unsteady stream depletion from ground water pumping *Ground Water* **37** 98-102
[7] Chen X 2003 Analysis of pumping-induced stream–aquifer interactions for gaining streams *J. Hydrol.* **275** 1-11
[8] Singh S K 2003 Flow depletion of semipervious streams due to pumping *J. Irrig. Drain. Eng.* **129** 449-53
[9] Singh S K 2005 Rate and volume of stream flow depletion due to unsteady pumping *J. Irrig. Drain. Eng.* **131** 539-45
[10] Zhang Y, Hubbard S and Finsterle S 2011 Factors governing sustainable groundwater pumping near a river *Ground Water* **49** 432-44
[11] Chen C, Hu L and Wang X 2006 Analysis of steady ground water flow toward wells in a confined-unconfined aquifer *Ground Water* **44** 609-12

IOP Conf. Series: Earth and Environmental Science 82 (2017) 012047 doi:10.1088/1755-1315/82/1/012047

The effect of time series length on runoff characteristics analysis

S Zhang[1], D F Liu[1,4], Q Huang[1], X Wei[1], J L Feng[2] and Mu Lin[3]

[1]State Key Laboratory Base of Eco-hydraulic in Arid Area, School of Water Resources and Hydropower, Xi'an University of Technology, Xi'an 710048, China
[2]Shanxi Provincal Water and Soil Conservation and Ecological Environment Construction Center, Taiyuan Shanxi 030002, China
[3]School of Statistics and Mathematics, Central University of Finance and Economics, Beijing 100081, China

E-mail: liudf@xaut.edu.cn

Abstract. The length of time series is of great importance, which is an uncertain factor in runoff characteristics analysis, especially in the analysis of trend and period. This paper investigates the tendency and period using runoff observation data of 107 years from 1909 to 2015 at Monticello hydrological station in the Upper Sangamon River in USA. The data series of 30 years, 40 years, 50 years and so on until 100 years and 107 years are selected to test their trend and period, respectively, and then are compared with each other. The results show that, with the increase of time series length, the tendency and period tend to be stable, and the annual runoff begins to show a significant increase trend once the length of time series is more than 80 years. The period tends to a dynamic stability value of 22 years when the time series length is more than 70 years. When the series length is larger than 70 years, the results of runoff characteristics analysis are much more reliable.

1. Introduction

In recent years, most of river runoff has changed to a great extent due to the impact of climate change and human activities [1]. Hence, runoff characteristics have been widely analysed to provide a basis for the rational development and utilization of water resources [2]. In the last few decades, scholars have been much more mature on the research of trend and period analysis of hydrological series. Sang et al [3] analysed the determined period by using empirical mode decomposition (EMD) and maximum entropy spectral analysis (MESA) which can efficiently avoid the impacts of noise on trend and period analysis. Burn and Hag Elnur [4] utilized the Mann-Kendall non-parametric test to detect trends and found out that a greater number of trends are observed than are expected to occur by chance. Hamed [5] improved Mann-Kendall test to study the significant trends in annual maximum flow of a group of 57 worldwide total annual runoff time series. Dixon et al [6] used the non-parametric Mann-Kendall test with bootstrap resampling to calculate annual and seasonal trends and then concluded that a marked east-west gradient of streamflow trend emerged in the mountainous west. Labat [7] gave a review of concepts of Wavelet analyses and the recent advances, which can be of great importance to identify period components. Gaucherel [8] exploited the potential of continuous wavelet transform on flow curves to detect new periodicities or time annual features. Sang et al [9] developed a discrete wavelet spectrum (DWS) approach to identify non-monotonic trend patterns in hydrological time series and evaluate their statistical significance.

The length of time series plays an important role in hydrological research. Xu and Xie [10] established the Autoregressive Integrated Moving Average (ARIMA) model to analyse the influence of the length of time series for the results of monthly runoff forecast. However, up to now, few people have investigated the effects of time series length on the analysis of runoff characteristics. Therefore, in this paper, we preliminarily discuss the optimum length of time series using runoff observation data of 107 years from 1909 to 2015 at Monticello hydrological station in the Upper Sangamon River in USA.

2. Study area and data

The Upper Sangamon River Basin (USRB) is located in the Upper Illinois River in Illinois, USA, with a watershed area of about 3150 km^2. The United States Geological Survey (USGS) sets up a water level gage at Monticello (N40°01′51″, W88°35′20″) as shown in figure 1, with a catchment area of about 1430 km^2. Monticello water level gage has collected daily runoff observation of 107 years from 1909 to 2015, which provides important basic data for analysing the long-time characteristics of runoff in this watershed [11].

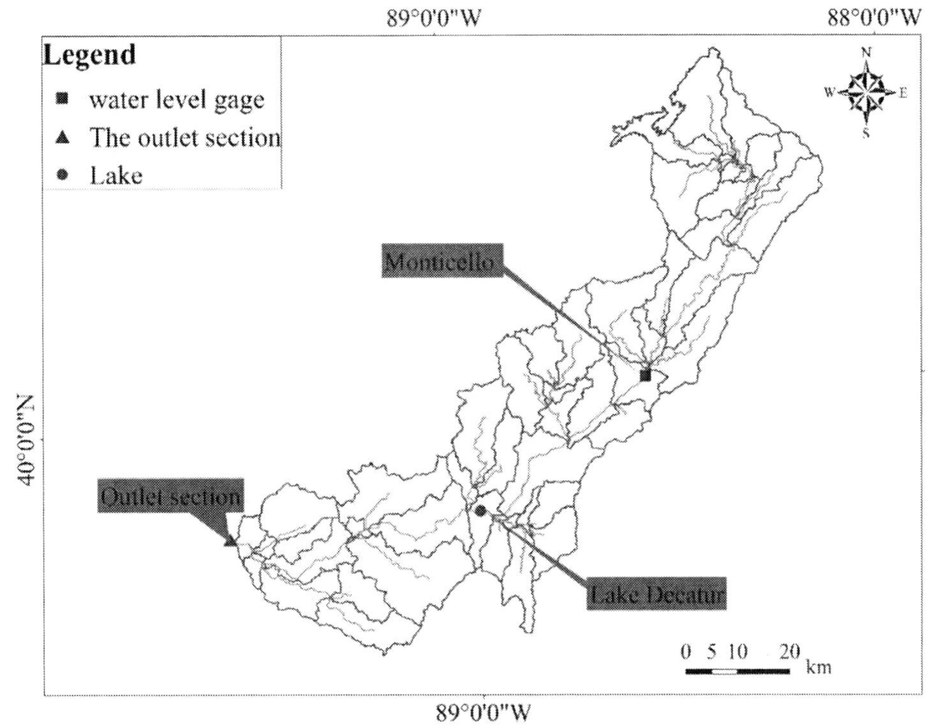

Figure 1. Location of the Monticello hydrological station in the USRB.

USRB has been affected by human activities including construction of railway and drainage systems, excavation of drains, laying drainage pipe underground and other activities since the 19th century. In the midstream of USRB, there is a manmade lake named Decatur, where a reservoir lies. Because of the lack of reservoir's water supply data, the water level gage is finally selected to make runoff characteristics analysis, which is located in the upstream of Decatur Lake [12].

The runoff data are obtained from USGS. They are divided into 9 groups with a fixed year of 1909 and a step of 10 years. In other words, we take the runoff observation of 30 years from 1909 to 1938, 40 years from 1909 to 1948, …, 100 years from 1909 to 2008 and 107 years from 1909 to 2015. And then their tendency and period are analyzed, respectively and are compared.

3. Methods

In this paper, three methods are applied to analyse the runoff characteristics, among which Correlation Coefficient Analysis (CCA) and Linear Regression Trend Test (LRTT) for tendency analysis and Wavelet analysis (WA) for period analysis.

3.1. Correlation coefficient analysis (CCA) and linear regression trend test (LRTT)

It is assumed that there is a linear trend component in a hydrologic sequence x_t, and the formula is as follows:

$$x_t = T_t + \eta_t = a + bt + \eta_t \tag{1}$$

where, T_t represents the trend component; η_t represents other components; a and b represents the parameters of the linear equation in the trend component respectively; t is time. The expressions of each parameter are as follows [13]:

$$b = \frac{\sum_{t=1}^{n}(t-\bar{t})(x_t-\bar{x})}{\sum_{t=1}^{n}(t-\bar{t})^2} \tag{2}$$

$$a = \bar{x} - b\bar{t} \tag{3}$$

$$S_b^{\ 2} = \frac{S^2}{\sum_{t=1}^{n}(t-\bar{t})^2}, \qquad S^2 = \frac{\sum_{t=1}^{n}\eta_t^{\ 2}}{n-2} \tag{4}$$

$$\sum_{t=1}^{n}\eta_t^{\ 2} = \sum_{t=1}^{n}(x_t-\bar{x})^2 - b^2\sum_{t=1}^{n}(t-\bar{t})^2 \tag{5}$$

where, S_b represents the standard deviation of b ; $\sum_{t=1}^{n}\eta_t^{\ 2}$ represents the sum of square of deviation; $\bar{x}$ and $\bar{t}$ represent the mean of x_t and t, respectively; n represents the capacity of sample.

r is taken as the correlation coefficient between x_t and t, and T is defined as the statistic value, there calculation formulas are as follows:

$$r = \frac{\sum_{t=1}^{n}(x_t-\bar{x})(t-\bar{t})}{\sqrt{\sum_{t=1}^{n}(x_t-\bar{x})^2 \cdot \sum_{t=1}^{n}(t-\bar{t})^2}} \tag{6}$$

$$T = b / S_b \tag{7}$$

In this paper, there are two steps for trend analysis:

- r is calculated first, and then find the correlation coefficient checklist to identify whether the relationship between x_t and t is significant or not. In other words, we want to find out whether the time series have a specific change or not with the influence of time. If it is, continue to the second step to obtain what change it is; otherwise, stop it to get a conclusion that there is no relationship between x_t and t, hence there is no tendency there.

- T is calculated using equation (7). At the beginning, b is assumed to be equal to 0, and T is subject to t-distribution with $D = n-2$. Giving α, $t_{\alpha/2}$ can be checked in the checklist of t. Then compare $|T|$ and $t_{\alpha/2}$, if $|T| > t_{\alpha/2}$, the initial assumption of b will be rejected, that is to say that the trend component is significant; otherwise, the trend component can be ignored.

3.2. Wavelet analysis (WA)

WA, called the 'mathematic microscope' [14], is an important break-through in the development history of Fourier transform (FT), which can identify very weak period signal compared to FT and has a powerful local time-frequency resolution for description of local signal characteristics [15]. In this paper, WA is used to test period component of different length of time series, it is as follows [11]:

$$W_f(a) = \int_{-\infty}^{\infty} \left| W_f(a,b) \right|^2 db \tag{8}$$

where, $W_f(a)$ represents wavelet variance; a represents scale parameter; b represents translation parameter; $W_f(a,b)$ represents wavelet coefficient.

Using WA, the figure of wavelet variance is shown to identify the period change of different time scales with the development of distribution of a. The peak in each figure is defined to be the major period of different time scales.

4. Results

4.1. Tendency analysis

Table 1. Comparison chart of correlation coefficient.

n /years	30	40	50	60	70	80	90	100	107
r	0.138	0.0964	0.0173	0.0387	0.0721	0.1204	0.2156	0.208	0.2287
$r_{0.05}$[a]	0.361	0.312	0.270	0.254	0.235	0.220	0.207	0.197	0.190
$r_{0.01}$[b]	0.463	0.403	0.361	0.330	0.306	0.286	0.270	0.256	0.248

[a] The correlation coefficient of each sequence length when the probability level is 5%.
[b] The correlation coefficient of each sequence length when the probability level is 1%.

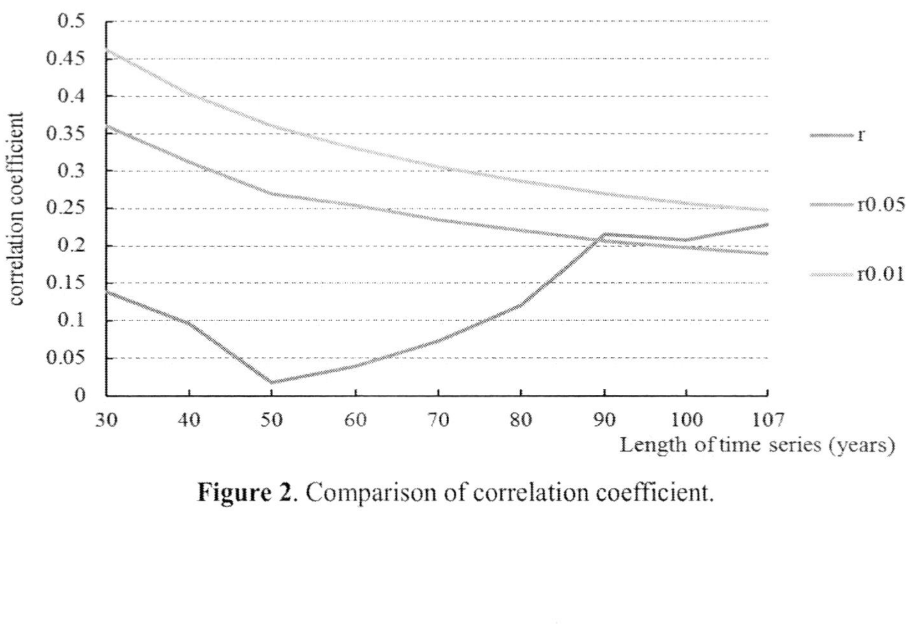

Figure 2. Comparison of correlation coefficient.

IOP Conf. Series: Earth and Environmental Science **82** (2017) 012047 doi:10.1088/1755-1315/82/1/012047

Through CCA, the correlation coefficient of each group is calculated in table 1. Compared with the critical value of the correlation coefficient from the correlation coefficient checklist, i.e. $r_{0.05}$ and $r_{0.01}$, the results are shown in table 1 and figure 2. The correlation coefficient is between $r_{0.05}$ and $r_{0.01}$ when n is more than 80 years. In other words, when the length of hydrological sequence exceeds 80 years, it exhibits a specific change with the influence of time and the trend component begins to appear. Then we adopt LRTT to test the significance of tendency for the hydrological sequence whose length is more than 80 years.

The statistic value T is calculated using equation (7), and listed in table 2 along with the t-test value $t_{\alpha/2}$ where α is defined to be equal to 0.05. When n is greater than 80 years, $|T|$ is always larger than $t_{\alpha/2}$. So the increasing tendency is significant. Hence, when the length of time series is more than 80 years, the effect of trend becomes significant.

Table 2. Comparison chart of the statistic value T and the t-test value $t_{\alpha/2}$.

n /years	90	100	107
T	93.15	109.86	138.05
$t_{\alpha/2}$ (α =0.05)	1.987	1.984	1.98

4.2. Period analysis

The curves of wavelet variance have been shown in figure 3. It is obvious that the wavelet variance curves of each group are similar in shape. They all have 2 or 3 peak values, and the first, second, third peak value represents the first, second, third main period respectively [16]. And the results are shown in table 3 and figure 4. From figure 3 and table 3, the first main period is generally expressed as a half of the length of time series, which has a great possibility to be the pseudo period. The first and second main period is approximately equal to the multiple of the third main period, and therefore the third main period can be ignored. Examining the second main period, we can conclude that the period of USRB tends to a dynamic stability when the length of hydrological sequence is more than or equal to 80 years. Therefore, we recommend that the length of hydrological sequence should be more than or equal to 80 years when analyzing period component in order to get a better result.

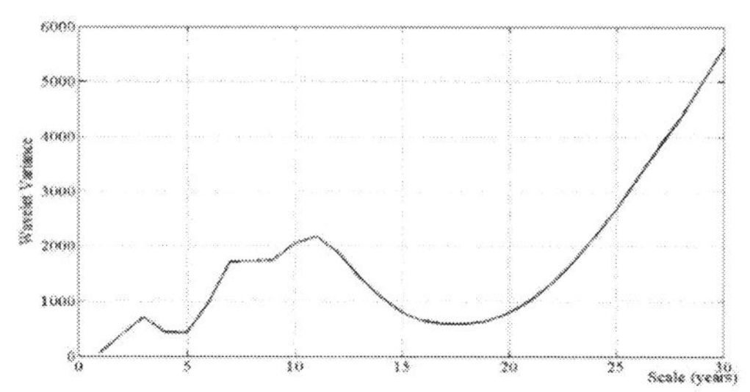

(a) 30 years

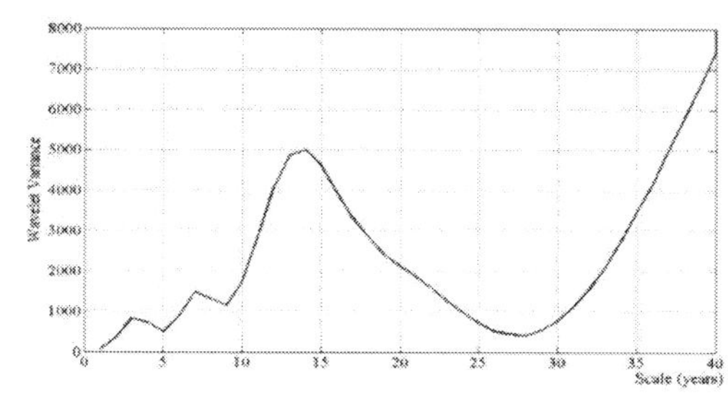

(b) 40 years

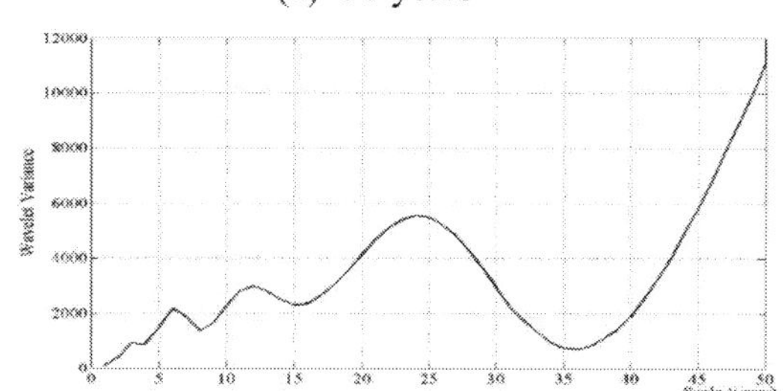

(c) 50 years

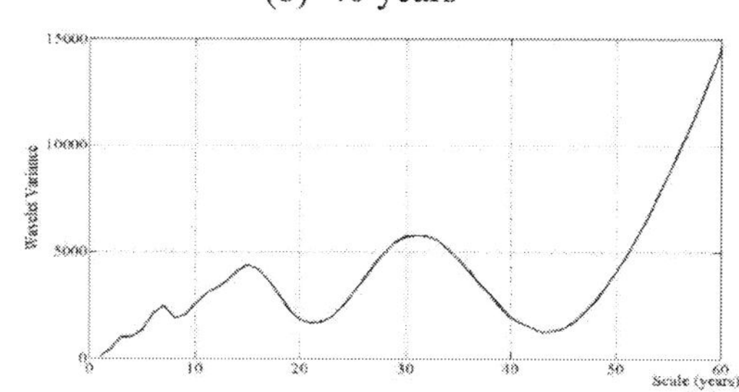

(d) 60 years

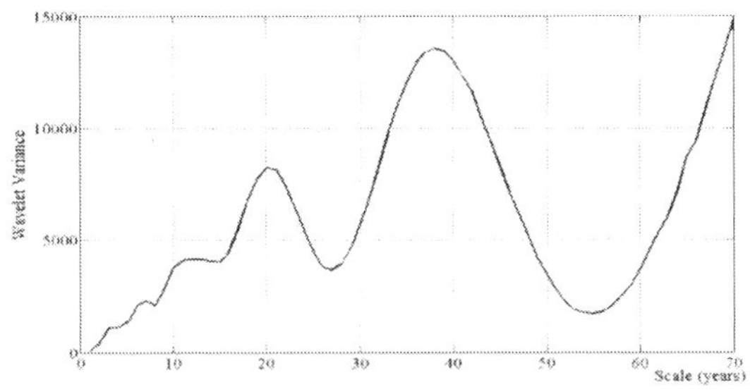

(e) 70 years

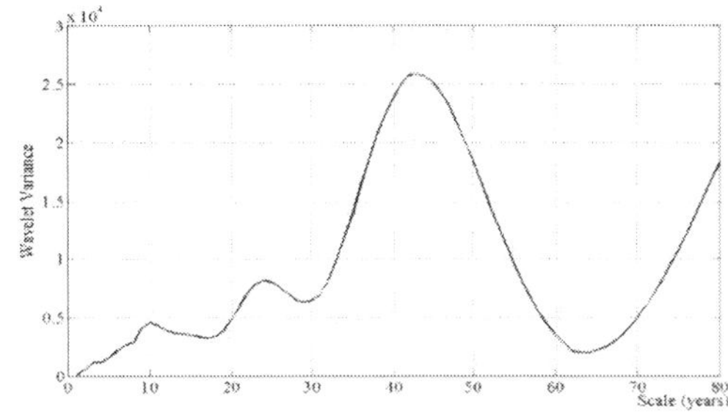

(f) 80 years

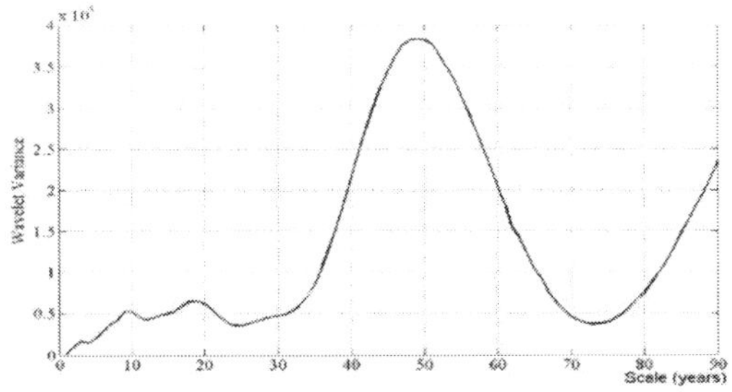

(g) 90 years

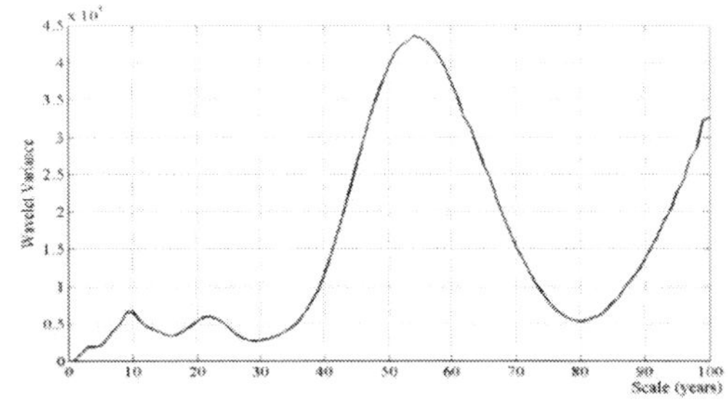

(h) 100 years

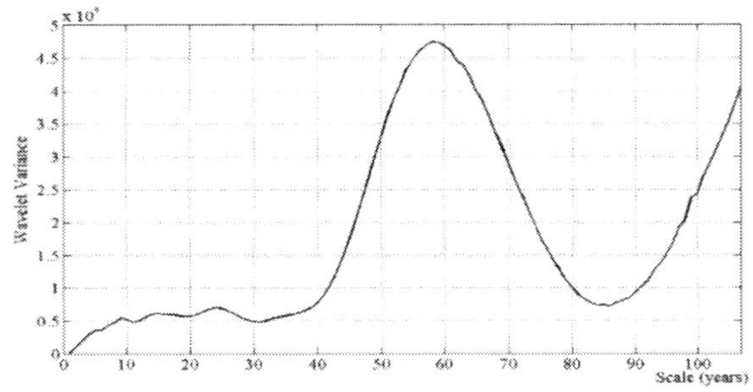

(i) 107 years

Figure 3. The curves of wavelet variance of different groups.

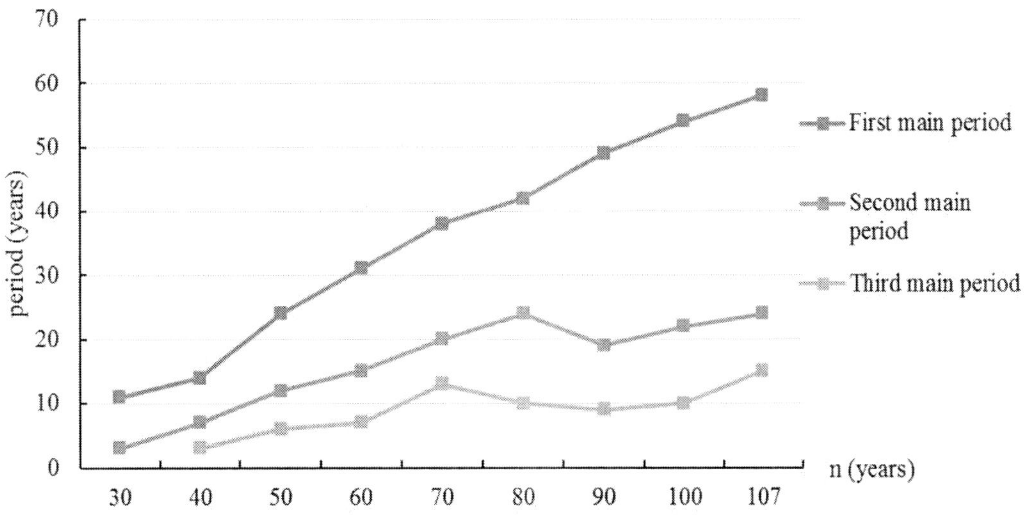

Figure 4. The periods and the length of the data

Table 3. Comparison chart of different main period in each group.

n /years	30	40	50	60	70	80	90	100	107
First main period	11	14	24	31	38	42	49	54	58
Second main period	3	7	12	15	20	24	19	22	24
Third main period		3	6	7	13	10	9	10	15

5. Conclusion

Most of the runoff observation data of the watershed in China is about 60 years, which limits our research on runoff characteristics analysis. This paper discusses trend and period components using runoff observation data of 107 years from 1909 to 2015 at Monticello hydrological station in the Upper Sangamon River in USA, which are divided into 9 groups.

The results show that, in USRB, we can detect a significant increasing tendency when the length of series is more than 80 years; based on WA, and the curves of wavelet variance are used to identify the periods of different group. We find out that the length of series should be more than or equal to 80 years when analyzing period component in order to get a better result. Therefore, it is recommended that the length of hydrological series should not be less than 80 years when analyzing runoff characteristics. But future runoff can be completely different, that should be taken into account with the actual situation. This study can provide the reference for future research.

Acknowledgments

This work was partially supported by the National Natural Science Foundation of China (NSFC) (Grant No. 51309188, 51609270). Funding was also provided by the Basic Research Plan of Natural Science of Shaanxi Province (2016JQ5105), SRFDP (20136118120021) and National Department Public Benefit Research Foundation of Ministry of Water Resources (201501058).

References

[1] Huang S Z, Huang Q, Wang Y M and Chen Y T 2014 The study on the runoff change based on the heuristic segmentation algorithm and approximate entropy *ASNUS* **53** 154-60

[2] Huang J X, Xu Z X and Gong T L 2007 Characteristics and driving factors of the runoff variations in the Yarlung Zangbo river *J. China Hydrol.* **27** 31-5

[3] Sang Y F, Wang Z G and Liu C M 2012 Period identification in hydrologic time series using empirical mode decomposition and maximum entropy spectral analysis *J. Hydrol.* **424-425** 154-64

[4] Burn D H and Hag Elnur M A 2002 Detection of hydrologic trends and variability *J. Hydrol.* **255** 107-22

[5] Hamed K H 2008 Trend detection in hydrologic data: The Mann-Kendall trend test under the scaling hypothesis *J. Hydrol.* **349** 350-63

[6] Dixon H, Lawler D M and Shamseldin A Y 2006 Streamflow trends in western Britain *Geophys. Res. Lett.* **33** L19406

[7] Labat D 2005 Recent advances in wavelet analyses: Part 1. A review of concepts *J. Hydrol.* **314** 275-88

[8] Gaucherel C 2002 Use of wavelet transform for temporal characterisation of remote watersheds *J. Hydrol.* **269** 101-21

[9] Sang Y F, Sun F B, Vijay P S, Xie P and Sun J 2017 A discrete wavelet spectrum approach to identifying non-monotonic trend pattern of hydroclimate data *Hydrol. Earth Syst. Sci . Discuss.* doi:10.5194/hess-2017-6

[10] Xu M and Xie Q Q 2014 Analysis of the effect of time series length on monthly runoff forecast based on ARIMA model *J. Yangtze Univ. (Nat. Sci. Edit.)* **11** 6-10

[11] Liu D F, Lei J Q, Huang Q and Lin M 2014 Analysis of runoff evolution characteristics with

observation data for nearly hundred-year *J. Water Resour. Water Eng.* **25** 24-8

[12] Liu D F, Tian F Q and Hu H P 2009 Sediment simulation at Upper Sangamon River Basin using THREW model *IAHS Publication* **335** 187-95

[13] Huang Q and Zhao X H 2008 *Theory and Method of Time Series Analysis and Forecast for River Runoff* (Zhengzhou, Henan Province: The Yellow River Water Conservancy Press) 18-9

[14] Xie Y Y, Huang Q, Chang J X, Liu S Y and Wang Y M 2016 Period analysis of hudrologic series through moving-window correlation analysis method *J. Hydrol.* **538** 278-92

[15] Yang Y, Shu H P, Ma J Z and Yang Z F 2017 Small and medium scale of climate variation characteristics of Baiyin city, Gansu Province *J. Arid Land Resour. Environ.* **31** 126-31

[16] Shao X M, Xu Y Q and Yan C R 2006 Wavelet analysis of rainfall variation in the Yellow River Basin *ASNUP* **42** 503-9

IOP Conf. Series: Earth and Environmental Science 82 (2017) 012048 doi:10.1088/1755-1315/82/1/012048

Study of risk and early warning index of rainstorm waterlogging in Wuhan City

Z Y Huang[1,2,3], T Peng[1], H Y Zhang[3] and W L Yao[4,5]

[1]Institute of Heavy Rain, Wuhan, CMA 430205, China
[2]China University of Geosciences, Wuhan 430074, China
[3]South-central University for Nationalities, Wuhan 430074, China
[4]Wuhan Meteorological Bureau, 430000, China

E-mail: whqxywl1121@sina.com

Abstract. In the present study, rainfall thresholds creating waterlogging and a waterlogging risk map were established by analyzing waterlogging monitoring and rainfall observation data in Wuhan City from 2011 to 2016. The study found that 1) Wuhan City rainstorm waterlogging occurred during the summer months, mostly in June and July, accounting for 90% of the total waterlogging; 2) Waterlogging generally occurred along the rivers and lakes near low-lying areas; 3) The longest continuous waterlogging time was greater than 10 h and the deepest water depth was 2 m; 4) The greater the cumulative rainfall over 24 h, 12 h, 6 h, or 3 h, the more severe the waterlogging was; 5) The thresholds of continuous 24 h rainfall of mild, moderate, and severe waterlogging in Wuhan City were 123, 163, and 196 mm, respectively, while 12 h cumulative rainfall thresholds were 86, 145, and 187 mm, respectively. The 6 h cumulative rainfall thresholds were 68, 113, and 135 mm, respectively, and 3 h cumulative rainfall thresholds were 40, 74, and 94 mm, respectively. If the city's drainage capacity increases, the cumulative rainfall of waterlogging warning indicators will also need to be adjusted.

1. Introduction

Rainstorms are the major meteorological disaster in China because heavy rainfall can lead to flooding, urban waterlogging, landslides, and other derivative disasters [1]. Wuhan City is located in the middle reaches of the Yangtze River, Yangtze River, and Han Jiang River, as well as near dozens of large and small lakes. In Wuhan City the average annual rainfall is approximately 1200 mm, which is mainly concentrated during the summer months [2]. Owing to the concentration of rainfall and low-lying aspect of the city, there is frequent flooding in Wuhan City, which has a great impact on the population's production and everyday life. City waterlogging is in China, India and other emerging developing countries in the process of urbanization problems [3].

At present, there are three main methods for assessing the risk of urban waterlogging disasters. The first method is studying the history of disaster statistics [3]. Using mathematical statistics, analysis of historical disaster data can be undertaken to determine the patterns of disaster development, establish the statistical model of disaster occurrence probability and its influencing factors, and then estimate the possible losses caused by future disasters [4-6]. This method is simple to calculate and does not require detailed geographical background data. However, at present, because of limited historical observation data, it has limitations in its application. The second method is the index system method. Based on the characteristics of the disaster system, a certain index system is selected, the indicators are established, and the process of regional disaster risk can be obtained. The accuracy of this method for calculating results is low and, therefore, has fewer applications [7-12]. The third method is the hydrological hydraulics model and simulation method. By setting the different frequency rainstorm scenarios, using the numerical simulation of the basin production and flooding model and the flood evolution model, the flooding range, flooding depth, and flooding of urban floods caused by heavy rain in different scenarios can be predicted. This method requires a large amount of data, the calculation is more complex, and therefore the practical application is less. The establishment of an urban waterlogging disaster index system, as well as a risk assessment and early warning system is required from an urban waterlogging disaster research point of view [13-15].

In the present study, the waterlogging monitoring data and rainfall observation data for Wuhan City between 2011 and 2016 are utilized to study the temporal and spatial distribution characteristics of waterlogging in Wuhan City, and the relationship between waterlogging and rainfall is analyzed. Based on this rainfall threshold and waterlogging risk map, this city waterlogging warning can be utilized to provide a reference. On this basis, it is possible for the city to have a reference for a future waterlogging warning system.

2. Data and methodology

2.1. Data

The data used in the present study are from waterlogging monitoring points and rainfall observations in Wuhan City from 2011 to 2016. The spatial layout of the waterlogging monitoring points and the multi-element automatic meteorological observation stations are shown in figure 1.

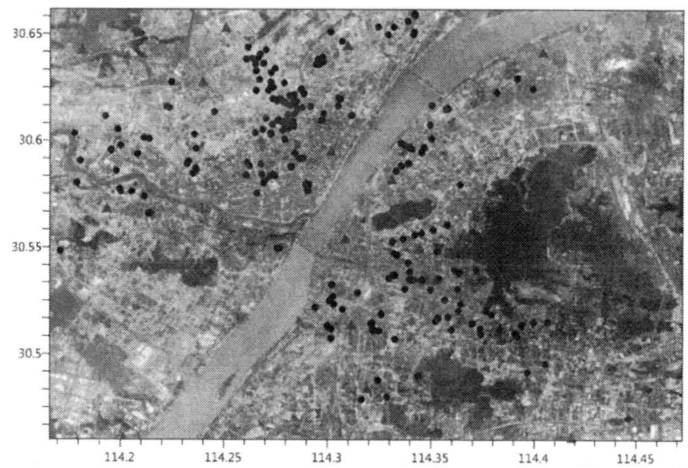

Figure 1. Spatial distribution of waterlogging monitoring points (blue dots) and meteorological observation stations in Wuhan (red triangles).

Waterlogging monitoring data were obtained from the Wuhan Municipal Water Affairs Bureau and included the location of waterlogging points, waterlogging times and the depths of waterlogging. The waterlogging point is not an isolated point in space, rather it represents a community, or a culvert, or it might even be a crossing. In Wuhan City, there are 258 waterlogging monitoring points. The time of waterlogging is not accurate to the minute; it is only a rough record of the occurrence of waterlogging and the approximate end time. The depth of the water is accurate to 0.1 m and only the maximum depth of a waterlogging process is recorded.

The rainfall observation data was obtained from the Wuhan Meteorological Bureau, which records the hourly rainfall and is accurate to 0.1 mm. In Wuhan City, there are 12 multi-element automatic meteorological observation stations.

2.2. *Analysis methods*
First, based on the waterlogging monitoring data during the period from 2011 to 2016, the waterlogging distribution map of Wuhan city was established. Second, the cumulative rainfall and precipitation intensity characteristics of typical waterlogging were analyzed, and the corresponding relationship between waterlogging and rainfall established, and the threshold value of rainfall determined. Finally, a comprehensive analysis of waterlogging distribution and rainfall threshold was completed, which resulted in the establishment of a waterlogging risk map for Wuhan City.

3. Results and discussion

3.1. *Waterlogging characteristics*
Statistics showed that during 2011–2016 Wuhan City had a total of 32 waterlogging processes caused by rainfall, which equated to an average of 5.3 times a year. Because of the East Asian monsoon climate present in Wuhan City, the heavy rainfall occurred mainly during the summer months; therefore, the waterlogging caused by this precipitation occurred in June and July, which accounted for 90% of the total number of waterloggings.

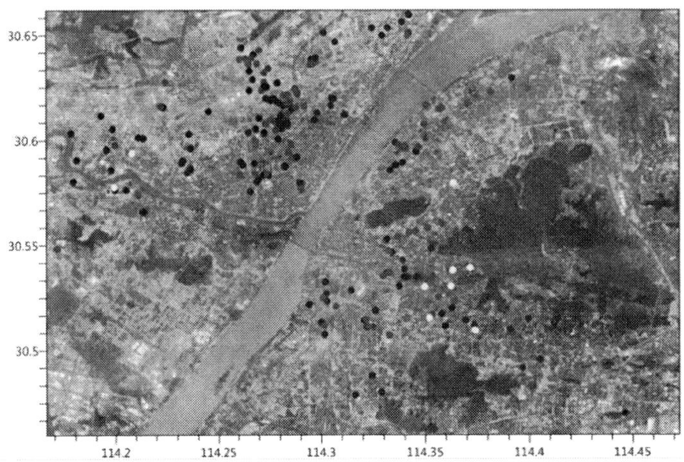

Figure 2. The waterlogging situation synthesis map in Wuhan City (yellow dots: the maximum water depth ≥ 1.0 m, blue dots: the maximum water depth < 0.5 m, and red dots: the maximum water depth ≥ 0.5 m and <1.0 m).

Figure 2 is based on the 32 waterlogging processes obtained from the Wuhan City waterlogging map and shows the location of possible waterlogging in Wuhan City and the maximum possible waterlogging depth. The different colored dots represent the positions of the waterlogged points, which are relatively fixed and are divided into three grades based on the waterlogging depth (i.e., > 1.0 m, 0.5–1.0 m, and <0.5 m). For each waterlogging point, the maximum waterlogging depth was selected from the 32 waterlogging processes and classified to the corresponding level. Figure 2 demonstrates that waterlogging in Wuhan City generally occurs along the rivers and lakes near low-lying areas. In addition, the intersection of two or more streets, culverts, and overpasses are also prone to waterlogging and the lower the terrain, the greater the possible waterlogging depth.

3.2. *The relationship between waterlogging and precipitation*
The present study revealed that if the actual occurrence of waterlogging monitoring points that accounted for the proportion of total waterlogging monitoring points reached 40% (103 waterlogging points), as a large range of waterlogging processes, then from 2011 to 2016 waterlogging processes occurred a total of seven times over a wide range in Wuhan City (table 1).

Table 1. Waterlogging duration, maximum depth, and number of waterlogged spots in Wuhan City from 2011 to 2016 ("d" represents the depth of water, unit: m).

Case	Date	Waterlogging duration (h)	Maximum depth (m)	Number of waterlogged spots			
				total	d≥1.0	0.5≤d<1.0	d<0.5
1	June 8, 2011	4	1.6	236	17	107	112
2	June 27, 2012	1	0.6	105	0	42	63
3	July 12, 2012	4	1.6	234	17	104	113
4	June 6, 2013	2	0.7	167	0	76	91
5	July 6, 2013	12	2.0	258	23	141	94
6	July 23, 2015	11	2.0	258	21	144	93
7	July 5, 2016	11	2.0	258	22	143	93

It can be seen from table 1 that the longest waterlogging time of the seven waterlogging processes lasted approximately 12 h, the minimum waterlogging time was only 1 h, and the maximum waterlogging depth exceeded 2 m. In addition, case 5 had the widest range of waterlogging and the most serious waterlogging, with all 258 monitoring points experiencing waterlogging, and 64% of the maximum waterlogging points greater than 0.5 m depth. Twenty-three waterlogging points had a maximum water depth of greater than 1.0 m. The waterlogging situation for case 6 was similar to case 5. Case 2 had the least number of waterlogging spots, with less than half the monitoring spots experiencing waterlogging, and the maximum waterlogging depth was less than 1 m. The same was true for the waterlogging situation in case 4. Case 1 and case 3 had similar waterlogging situations, with 92% of monitoring points experiencing waterlogging. The number of monitoring points with water depth of greater than 1.0 m was less than that of case 5, case 6, and case 7.

3.2.1. Rainfall characteristics during waterlogging. It can be seen from figure 1 that there is not a one-to-one relationship between the automatic weather station and waterlogging monitoring point. In the present study, the most recent rainfall observation data from the location of the stained water point represents the amount of rainfall at that waterlogging point. If there are several automatic weather stations closer to a point of waterlogging, then the maximum rainfall observed was taken from these automatic weather stations to represent the amount of rainfall at this waterlogging point. Finally, the rainfall per hour takes the maximum value of the observed rainfall at all weather stations.

Table 2 shows the cumulative rainfall for different hourly periods for the seven cases. The cumulative rainfall for 24 h is the sum of the hourly rainfall from 00:00 (UTC) to 23:00 (UTC). The accumulated rainfall of 12 h is the maximum of the sum of the hour-by-hour rainfall for 12 consecutive hours from 00:00 (UTC) to 23:00 (UTC), and so on. From table 2 it can be seen that the minimum value of the cumulative rainfall of the seven cases is 122.9 mm and the maximum value is 251.9 mm over a 24 h period.

Comparing the cumulative rainfall from 12 h to 24 h, the ratios of the former to the latter are 85%, 70%, 98%, 89%, 94%, 95%, and 86% for the seven cases. The maximum cumulative rainfall of 12 h is the main contributor to the 24 h total rainfall. Similarly, the maximum cumulative rainfall of 6 h is the main contributor to the maximum accumulated rainfall of 12 h, and the maximum accumulated rainfall of 3 h is the main contributor to the 6 h maximum accumulated rainfall. This indicates that the rainfall in these seven cases is extremely uneven in time distribution.

Table 2. Hourly cumulative rainfall data of waterlogging in Wuhan City from 2011 to 2016.

Case	Date	Maximum cumulative rainfall (mm)			
		24 h	12 h	6 h	3 h
1	June 8, 2011	170.7	145.2	112.6	74.0
2	June 27, 2012	122.9	85.9	68.4	39.7
3	July 12, 2012	162.8	159.0	138.8	91.0
4	June 6, 2013	124.3	111.2	76.5	61.2
5	July 6, 2013	251.9	237.8	156.0	126.3
6	July 23, 2015	195.5	186.5	144.6	108.8
7	July 5, 2016	241.5	206.8	134.7	93.6

However, in case 2 and case 4, although the cumulative rainfall after 24 h is very similar, the maximum cumulative rainfall in other periods is relatively dissimilar, especially for the maximum cumulative rainfall at 3 h. This shows that the rainfall time of case 4 is more concentrated.

3.2.2. The relationship between waterlogging and rainfall intensity. In the present study, we selected the smallest, medium, and largest of the continuous 24 h cumulative rainfall from table 2 as typical examples to analyze the relationship between the degree of waterlogging and rainfall intensity. These were case 2, case 1, and case 5, respectively. Figure 3 shows the hourly rainfall distribution for these three typical cases.

It can be seen from figure 3 that the 24 h rainfall of the three typical cases was reflected in the multi-peak form. Case 1 had four peak rainfall values (figure 3(a)), which appeared at 02:00, 09:00, 14:00, and 16:00. The rainfall was very small within 6 h after the first peak (1 h rainfall <5 mm). From the second peak to the third peak within 8 h, the rainfall did not stop and gradually increased, with the cumulative rainfall of 8 h being 134.2 mm (accounting for 79% of the total 24 h rainfall). This shows that the rainfall at this stage was not only relatively large, but was also very concentrated. Waterlogging occurred at 13:00, with the third and fourth peak rainfall, at which point the waterlogging increased, the depth of waterlogging increased, and led to a waterlogging time of 6 h.

Case 2 had three peaks of rainfall (figure 3(b)), which appeared at 6:00, 10:00, and 23:00. The first and second peaks occurred after a total of 5 h and a cumulative rainfall of 55.9 mm (45% of the total 24 h rainfall) and resulted in waterlogging. However, since the rainfall was very small within approximately 10 h after the second rainfall peak, the duration of waterlogging was only 1 h. The third rain peak did not cause waterlogging. Comparing case 1 and case 2, it was found that case 2 did not only have small cumulative rainfall but it was also scattered, with 30% of the rainfall (3rd rainfall peak) not contributing to waterlogging. Therefore, the waterlogging point of case 2 is less than that of case 1, and the waterlogging time is much shorter than that of case 1.

Case 5 (figure 3(c)) had three peak rainfall values, which appeared at 12:00, 17:00, and 23:00. The three peaks lasted for 11 h and the cumulative rainfall was 237.8 mm (94% of the total 24 h rainfall). During this 11 h, the rainfall was constant, and 1 h maximum rainfall reached 58.6 mm. This shows not only that the cumulative rainfall was very large and the rainfall was very concentrated, but the rainfall intensity was also very large. Waterlogging was at a peak after the first rainfall. With the second and third peak rainfalls, the waterlogging point increased rapidly, the depth of waterlogging significantly increased, the maximum water depth was 2 m, and the waterlogging time was more than 12 h.

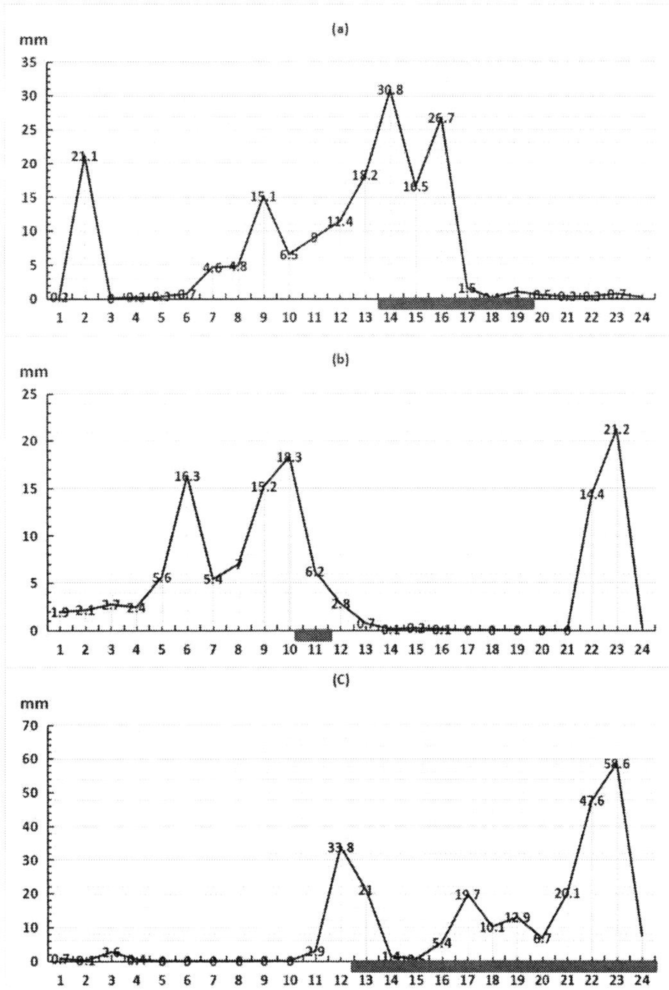

Figure 3. Typical case of hourly rainfall distribution, a: June 8, 2011(case 1); b: June 27, 2012 (case 2); and c: July 6, 2013 (case 5) (X-axis is time, unit: h; Y-axis is rainfall, unit: mm. The red rectangle represents the duration of waterlogging).

In summary, as the cumulative rainfall for 24 h, 12 h, 6 h, and 3 h increased, the more serious was the waterlogging. Waterlogging occurred after the cumulative rainfall reached a certain level because the runoff and confluence takes some time. In addition, the duration of waterlogging was not only related to the cumulative rainfall, but also to the urban drainage capacity.

3.3. The rainfall threshold for waterlogging

Based on results in table 1, the seven cases were divided into three grades (mild, moderate, and severe) according to the waterlogging condition. Case 2 and case 4 were mild waterlogging; case 1 and case 3 were moderate waterlogging; and case 5, case 6, and case 7 were severe waterlogging. For each level of waterlogging, first, the minimum value of the maximum cumulative rainfall for 24 h, 12 h, 6 h, and 3 h was selected for the threshold of rainfall that might lead to waterlogging (table 3). Then, the place where waterlogging occurred in each of the cases was recorded as was the maximum waterlogging depth of each waterlogging site. Finally, the actual waterlogged spots and their maximum waterlogging depth were plotted on the city map, resulting in a waterlogging risk map (figure 4).

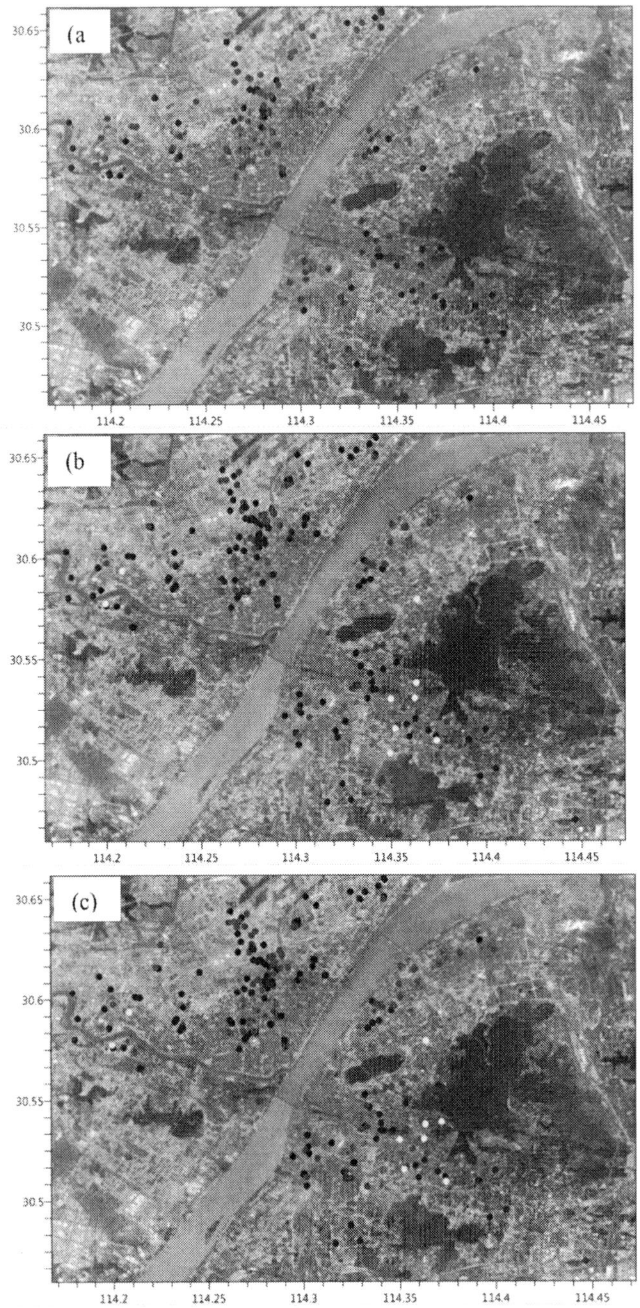

Figure 4. Wuhan City waterlogging risk map. a, Mild waterlogging; b, Moderate waterlogging; and c, Severe waterlogging. (Yellow dots: maximum water depth ≥ 1.0 m, blue dots: maximum water depth <0.5 m, and red dots: maximum water depth ≥ 0.5 m and <1.0 m).

If the accumulated rainfall in a given period reaches the thresholds in table 3, the possible waterlogging situation can be seen from figure 4. For example, if the maximum cumulative rainfall of 3 h is 38 mm, it can be seen from table 3 that there might be mild waterlogging in the future, the longest water duration would be 2 h, a maximum water depth of 0.7 m, and the specific waterlogging point and waterlogging depth can be seen in figure 4(a). Based on the future possible rainfall obtained from table 3 the corresponding waterlogging level can be determined and then, from figure 4, an estimated possible waterlogging point and waterlogging depth.

Table 3. Precipitation threshold of waterlogging in Wuhan City.

Waterlogging grade	Cumulative rainfall index (mm)				Maximum waterlogging depth (m)	Longest duration of waterlogging (h)
	24 h	12 h	6 h	3 h		
Mild	123	86	68	40	0.7	2
Moderate	163	145	113	74	1.5	5
Severe	196	187	135	94	2.0	10

4. Conclusions

In the present study, using data from waterlogging monitoring points and rainfall observations in Wuhan City from 2011 to 2016 and based on the relationship between waterlogging characteristics and rainfall, the threshold value of rainfall caused by waterlogging was established. The main conclusions are as follows:

- Rainfall during the summer months causes waterlogging in Wuhan City, on average 5.3 times a year with June and July having the most rainfall. This accounts for 90% of the total number of waterloggings. Waterlogging occurs generally along rivers and lakes near low-lying areas, the longest continuous waterlogging time is greater than 10 h, and the deepest water depth is 2 m.
- The degree of waterlogging is not only related to the cumulative rainfall, but also to the time distribution of rainfallalong with rainfall intensity. In general, as the cumulative rainfall for 24 h, 12 h, 6 h, and 3 h increases, the more severe the waterlogging is.
- The thresholds of continuous 24 h rainfall for mild, moderate, and severe waterlogging in Wuhan City are 123, 163, and 196 mm respectively, 12 h cumulative rainfall thresholds are 86, 145, and 187 mm, respectively, 6 h cumulative rainfall thresholds are 68, 113, and 135mm respectively, and 3 h cumulative rainfall thresholds are 40, 74, 94 mm, respectively.

Since there are only six years of data, the sample size is small, which means it is not possible to run further correlation analysis on this data. Therefore, the above conclusions need to continue to be complemented and improved upon in future studies. In addition, if the city's drainage capacity increases, the cumulative rainfall of waterlogging warning indicators also needs to be adjusted.

Acknowledgments

This work is supported by the Research and Operation Project of the Institute of Heavy Rain, CMA, Wuhan (JK201707, JK201704 and JK201711) and the Public welfare industry special funds (grant GYHY201306012) and the National Natural Science Foundation of China (51379149).

References

[1] Lin J and Yang G M 2014 Spatio-temporal characteristics of rainstorm in China during 1981-2010 *Meteorol. Mon.* **40(7)** 816-26
[2] Tang Y L, Xu G R, Wan R, *et al* 2016 Analysis of climate change of Meiyu precipitation in Hubei Province in recent 52 years *Torrential Rain and Disasters* **35(6)** 511-20.
[3] Zhang D D, Yan D H, Wang Y C, *et al* 2014 Research progress on risk assessment and integrated strategies for urban pluvial flooding *Journal of Catastrophology* **29(1)** 144-9
[4] Benito G, Lang M, Barriendos M, *et al* 2004 Use of systematic, palaeoflood and historical data for the improvement of flood risk estimation, review of scientific methods *Nat. Hazard* **31(3)** 623-43
[5] Nott J 2006 *Extreme Event: A Physical Reconstruction and Risk Assessment* (London: Cambridge University Press)
[6] Hans de M, Jeroen C, Aerts J H, *et al* 2011 Development of flood exposure in the Netherlands

during the 20th and 21st century *Global Environ. Chang.* **21** 620-7
[7] Cutter S L, Boruff B J and Shirley W L 2003 Social vulnerability to environmental hazards *Soc. Sci. Q.* **84** 242-61
[8] Cutter S L, Mitchell J T and Scott M S 2000 Revealing the vulnerability of people and places: A case study of Georgetown County, South Carolina *Ann. Assoc. Am. Geogr.* **90** 713-37
[9] Clark G, Moser S, Ratick S, *et al* 1998 Assessing the vulnerability of coastal communities to extreme storms: The case of Revere, MA, USA *Mitig. Adapt .Strat. Gl.* **3** 59-82
[10] Rygel L, O' Sullivan D and Yarnal B 2006 A method for constructing a social vulnerability index: An application to hurricane storm surges in a developed country *Mitig. Adapt .Strat. Gl.* **11** 741-64
[11] Aleksandra K Z and Gina C 2011 Surface water flooding risk to urban communities: Analysis of vulnerability, hazard and exposure *Landscape Urban Plan.* **103** 185-97
[12] Christopher Z 2001 Review of urban storm water models *Environ. Modell. Softw.* **16** 195-231
[13] Ceng G P 1990 A model to simulate stormwater runoff in urban area *J. Hydraul. Eng.* **10** 68-75
[14] Qiu J W, Li N, Cheng X T and Xia X G 2000 The simulation system for heavy rainfall in Tianjin City *Xueli Xuebao* **11** 34-42
[15] Li N 2002 Study on simulation system of rainstorm waterlogging in Tianjin City *J. Nat. Disast.* **11(2)** 112-8

IOP Conf. Series: Earth and Environmental Science 82 (2017) 012049 doi:10.1088/1755-1315/82/1/012049

Evaluation of water resources system vulnerability based on co-operative co-evolutionary genetic algorithm and projection pursuit model under the DPSIR framework

Y Zhao[1,4], X H Su[1], M H Wang[2], Z Y Li[1], E K Li[1] and X Xu[3]

[1]Yellow River Institute of Hydraulic Research, China
[2]Water Resources Management and Regulation Bureau, YRCC, China
[3]Liaocheng Water Conservancy Bureau, China

E-mail: wssf9303@163.com

Abstract. Water resources vulnerability control management is essential because it is related to the benign evolution of socio-economic, environmental and water resources system. Research on water resources system vulnerability is helpful to realization of water resources sustainable utilization. In this study, the DPSIR framework of driving forces-pressure–state–impact -response was adopted to construct the evaluation index system of water resources system vulnerability. Then the co-evolutionary genetic algorithm and projection pursuit were used to establish evaluation model of water resources system vulnerability. Tengzhou City in Shandong Province was selected as a study area. The system vulnerability was analyzed in terms of driving forces, pressure, state, impact and response on the basis of the projection value calculated by the model. The results show that the five components all belong to vulnerability Grade II, the vulnerability degree of impact and state were higher than other components due to the fierce imbalance in supply-demand and the unsatisfied condition of water resources utilization. It is indicated that the influence of high speed socio-economic development and the overuse of the pesticides have already disturbed the benign development of water environment to some extents. While the indexes in response represented lower vulnerability degree than the other components. The results of the evaluation model are coincident with the status of water resources system in the study area, which indicates that the model is feasible and effective.

1. Introduction

As the combined influence of climate change and human activities, the shortage of local water resources is becoming an increasingly serious problem. In order to realize the water resources system sustainable utilization, it is necessary to make a reasonable evaluation of the water resources system vulnerability. At present, the research on water resources system vulnerability is mainly carried out from qualitative to quantitative gradually. In general the water resources system is always regarded as a whole to conduct the research, while it is seldom to analyse the influence of the different components to the whole system.

The purpose of this study is to evaluate the vulnerability of regional water resource system. An index system based on the DPSIR framework and the Projection Pursuit model are applied to synthesizing the socio-economic and environmental factors which may impact the vulnerability of water resources and understanding the distinct cause-and-effect relationships in regional water resources system.

2. Construction of the evaluation index system of water resources system vulnerability

2.1. Connotations of water resources system vulnerability

In the process of water resources utilization in domestic living, production and environment, water resources system vulnerability, which is derived from the temporal-spatial distribution characteristic of water quality and quantity in particular area, has easily damaged property showed up when system is threatened by climate change, natural disaster, human activity and so on. If the damage is allowed to continue, the forces will be accumulated to a certain extent which causes severely damage to water resources system [1-3].

Combined with the characteristics of water resources system, the connotations of water resources system vulnerability in this study are as follows: (1) Water resources system vulnerability is a state which reflects the vulnerability degree acted by the external forces. (2) Water resources system vulnerability has the temporal-spatial properties. (3) Natural condition is the determinant of the intrinsic vulnerability of the water resources system, while the human activity plays a determined role in the extrinsic influences and it is the driving factor. (4) Time delay characteristics: When acted by the external forces, water resources system will try to keep the original state, thereby the system will show up time delay features from being acted by the external forces to showing up the vulnerability. (5) Recessive characteristics: when the system hasn't show up the vulnerability clearly, it cannot be accounted for the inexistence. Only by being forced to withstand enough external threatens, the system vulnerability characteristics could be shown up clearly. (6) Reversibility: System vulnerability can be reduced by effective measures implementation.

2.2. Construction of the evaluation index system of water resources system vulnerability based on DPSIR framework

A number of studies in water resources system problems analyzing were used the DPSIR framework model [4, 5]. Bowen and Riley used the framework for analyzing the interrelationships of socio-economic and coastal environmental dynamics. The crucial driving forces which impacted on the changes in groundwater ecosystems status due to human activities were analyzed [6, 7]. Combined with other methodologies, DPSIR framework was applied to research the correlative pressures and influences of changes in coastal area water quality for evaluating the feasibility of the European Water Framework Directive [8]. Considered with environmental and socio-economic sciences, the original DPSIR framework was used for addressing the primary issue of eutrophication [9]. Bianet Jago-on et al. integrated the existing data of subsurface environment to understand the interrelationships among subsurface environmental issues with DPSIR framework [10]. River environmental conditions in Balkans were analyzed with DPSIR framework [11]. Correlation studies have demonstrated that the DPSIR framework is an effective method to assess the sustainability in agriculture and water management with the core of "cause-effect relationship" [12, 13]. The DPSIR framework was utilized widely in water resources system sustainability and vulnerability evaluation through combining with data mining and intelligent algorithm technology [14, 15]. Thus, the DPSIR framework is a helpful approach in analyzing the relationships between water resources and socio-economic systems because of its effect on addressing the causes, impacts and responses to changes in water resources system [16, 17].

For evaluating the water resources system vulnerability, socio-economic factors, ecological environment factors and intrinsic factors of water resources system should be synthesized. At present, the widely used DPSIR framework can reflect the system's integrity, causal association and co-evolution characteristics. Also it can represent the continuous feedback mechanism of the different kinds of factors in water resources system, which provides a good research approach to analysis the water resources system vulnerability. By consulting the existing framework of the index system of water resources system and the correlational researches of the water resources system vulnerability, 31 indexes were chosen in this study to construct the evaluation index system of water resources system vulnerability based on the DPSIR framework (table 1). In the progress of choosing the indexes, one important consideration was the overlap of the indexes. If some indexes had similar meaning, the more commonly used index would be chosen.

In this study, the DPSIR framework was applied to understanding the linkages and interdependencies of socio-economic, eco-environment and water resources system. Causal relationships of water resources system issues are analyzed using information from water resources bulletins, statistical communiques on national economic and social development, and results from previous studies in order to create links and indicators of the components of DPSIR.

Based on the review of different studies, researches in vulnerability evaluation field (Kyung, et al. 2011), the cause and effect relationship between socio-economic system and water resources system in the DPSIR framework are shown in figure 1.

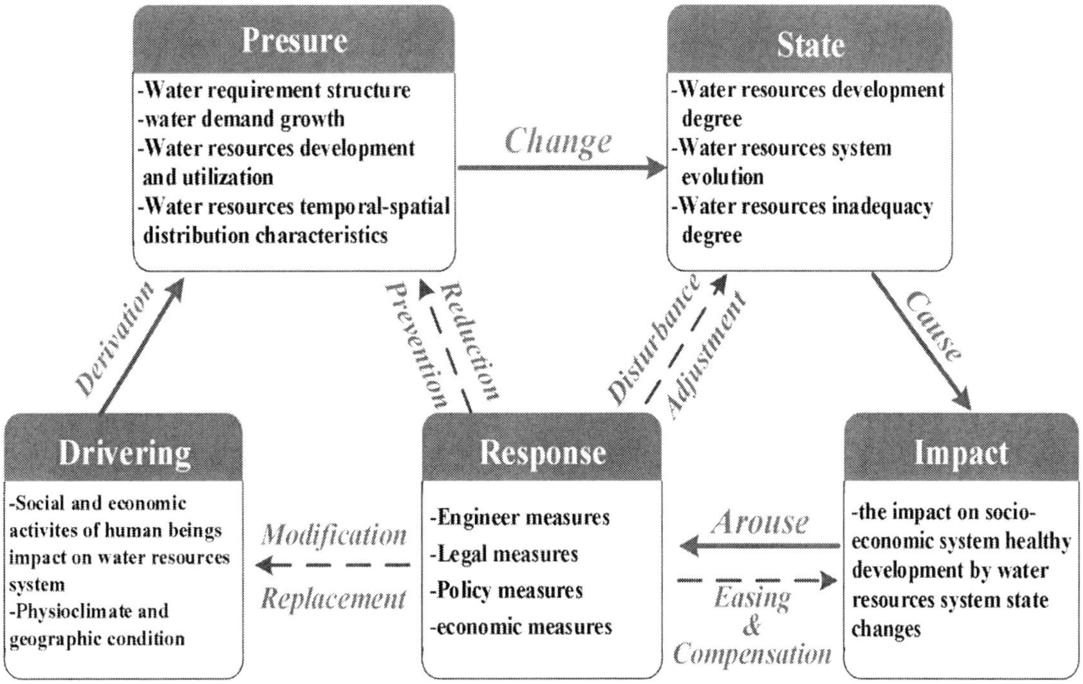

Figure 1. The cause and effect relationships of DPSIR Framework in this study.

2.3. Standard of the evaluation index

For evaluating the vulnerability scientifically and reasonably, the grade standard is essential to measure the degree of vulnerability. Based on the review of related researches (Wang and Zhang, 2010), the hierarchical evaluation grades dividing is a critical and difficult step. If the number of grades is too few, the accuracy of the evaluation results would be affected; otherwise it is too cumbersome with too many grades. In this study, the evaluation criterion of water resources system vulnerability was divided into 4 grades: grade Ⅰ -weak vulnerability, grade Ⅱ -medium vulnerability, gradeⅢ-strong vulnerability, gradeⅣ- very strong vulnerability (table 1).

Table 1. Evaluation index system and evaluation criterion of water resources system vulnerability based on DPSIR.

	Indicators	Units	Description	Grade Ⅰ	Grade Ⅱ	GradeⅢ	Grade Ⅳ
Driving forces	Drought index D_1		annual evaporative power	<1	1~2	2~4	>4

			divided by annual precipitation				
	Precipitation D_2	mm	water depth that was converted from liquid and solid water falling down to the earth from the atmosphere	>800	500~800	200~500	<200
	Natural population growth rate D_3	‰	trend of natural growth of population	<2	2~15	15~20	>20
	Primary industry proportion D_4	%	Ratios of primary industry to GDP	<3	3~15	15~30	>30
	Tertiary industry proportion D_5	%	Ratios of tertiary industry to GDP	>60	45~60	30~45	<30
	Urbanization rate D_6	%	Ratios of urban population to total population	>80	50~80	20~50	<20
Pressure	Average per capita water resources P_1	m^3	multi-year average per capita water resources	>1700	1000~1700	500~1000	<500
	Average per mu water resourcesP_2	($m^3.hm^{-2}$)	multi-year average per mu water resources	>1500	700~1500	400~700	<400
	Cropland acreage per capita P_3	hm^2	Cropland acreage per capita in Tengzhou	>1	0.6~1	0.1~0.6	<0.1
	Industrial COD emission concentration P_4	($mg.L^{-1}$)	COD emission concentration in industrial sewage in the region	<60	60~100	100~150	>150
	Urban per capita green area P_5	m^2	urban public green area per capita in the region	>10	8~10	7~8	<7
	Per capita water consumption P_6	($m^3.a^{-1}$)	annual per capita water consumption in the region	<500	500~800	800~1000	>1000
	Urban per capita domestic water consumption P_7	($L.d^{-1}$)	per capita domestic water consumption in the urban area	<75	75~150	150~200	>200
State	Water consumption for industrial added value of ten thousand yuan S_1	(m^3. ten thousand $yuan^{-1}$)	Ratios of industrial water supply to gross industrial production in the region	<30	30~90	90~200	>200
	Exploitation and utilization	%	Ratios of exploited	<30	30~60	60~100	>100

	efficiency of underground water S_2		groundwater to exploitable groundwater				
	Agricultural water proportion S_3	%	Ratios of agricultural water to total water consumption in the region	<40	40~55	55~75	>75
	Leakage rate of pipe network S_4	/%	Ratios of water leakage to water consumption leaving the factory in the region	<5	5~15	15~30	>30
	Unit area irrigation water consumption S_5	(m^3. hm^{-2})	Average irrigation water consumption in per mu of farmland	<3000	3000~5200	5200~7200	>7200
	Water consumption per unit of GDP S_6	(m^3. ten thousand yuan $^{-1}$)	Ratios of water supply in the region to Gross Domestic Product	<100	100~200	200~400	>400
Impacts	Forest coverage rate I_1	/%	Percentage of forestry area in the region to area of land	>60	40~60	10~40	<10
	Ratio up to the standard of water quality I_2	%	Ratios of amounts that is up to the standard of water quality to total amounts of water sources in the city	100	98~100	90~98	<90
	Green coverage rate I_3	%	Percentage of green coverage area to total area of the land in the city	>60	30~60	10~30	<10
	Proportion of eco-environmental water consumption I_4	%	Ratios of off-stream eco-environmental water consumption to total water consumption	>5	3~5	2~3	<2
	Grain yield per cubic meters of water I_5	(kg.m^{-3})	Ratios of agricultural production water consumption to grain yield	>3	1.5~3	0.6~1.5	<0.6
Responses	Utilization rate of the recycled water	%	Ratios of utilized recycled water	>20	10~20	5~10	<5

IOP Conf. Series: Earth and Environmental Science 82 (2017) 012049 doi:10.1088/1755-1315/82/1/012049

R_1		consumption to total recycled water consumption				
Water-saving irrigation area proportion R_2	/%	Ratios of water-saving irrigation area to effective irrigation area	>60	30~60	10~30	<10
Wter saving devices pervasion in city R_3	%	Ratios of amounts of water saving devices to amounts of water devices	>90	60~90	30~60	<30
Tap water pervasion in city R_4	%	Ratios of numbers of people that using tap water to total numbers of people in the city	>80	60~80	40~60	<40
Utilization coefficient of irrigation water R_5		Ratios of irrigation water consumption to total water consumption of head works	>0.7	0.5~0.7	0.4~0.5	<0.4
Repeating utilization rate of industrial water use R_6/	%	Ratios of repeating utilization water consumption for industry to total water consumption	>80	60~80	40~60	<40
Contribute rate to economic increasing by scientific and technological progress R_7	/%	Percentage of factors which contribute to economic increasing by scientific and technological progress in the GDP growth rate	>70	55~70	40~55	<40

3. Establishment of evaluation model of water resources system vulnerability based on co-operative co-evolutionary genetic algorithm (CCGA) and projection pursuit

3.1. Co-operative co-evolutionary genetic algorithm

CCGA is a kind of GA based on multi-population's concurrent evolution, which is more suitable for solving the complex optimization problems because multi-population can better correspond to objects of different nature.

CCGA provides a feasible approach in which it is relatively convenient to integrate a variety of evolutionary methods, while still containing the fundamental evolutionary behaviors of iterative

IOP Conf. Series: Earth and Environmental Science 82 (2017) 012049 doi:10.1088/1755-1315/82/1/012049

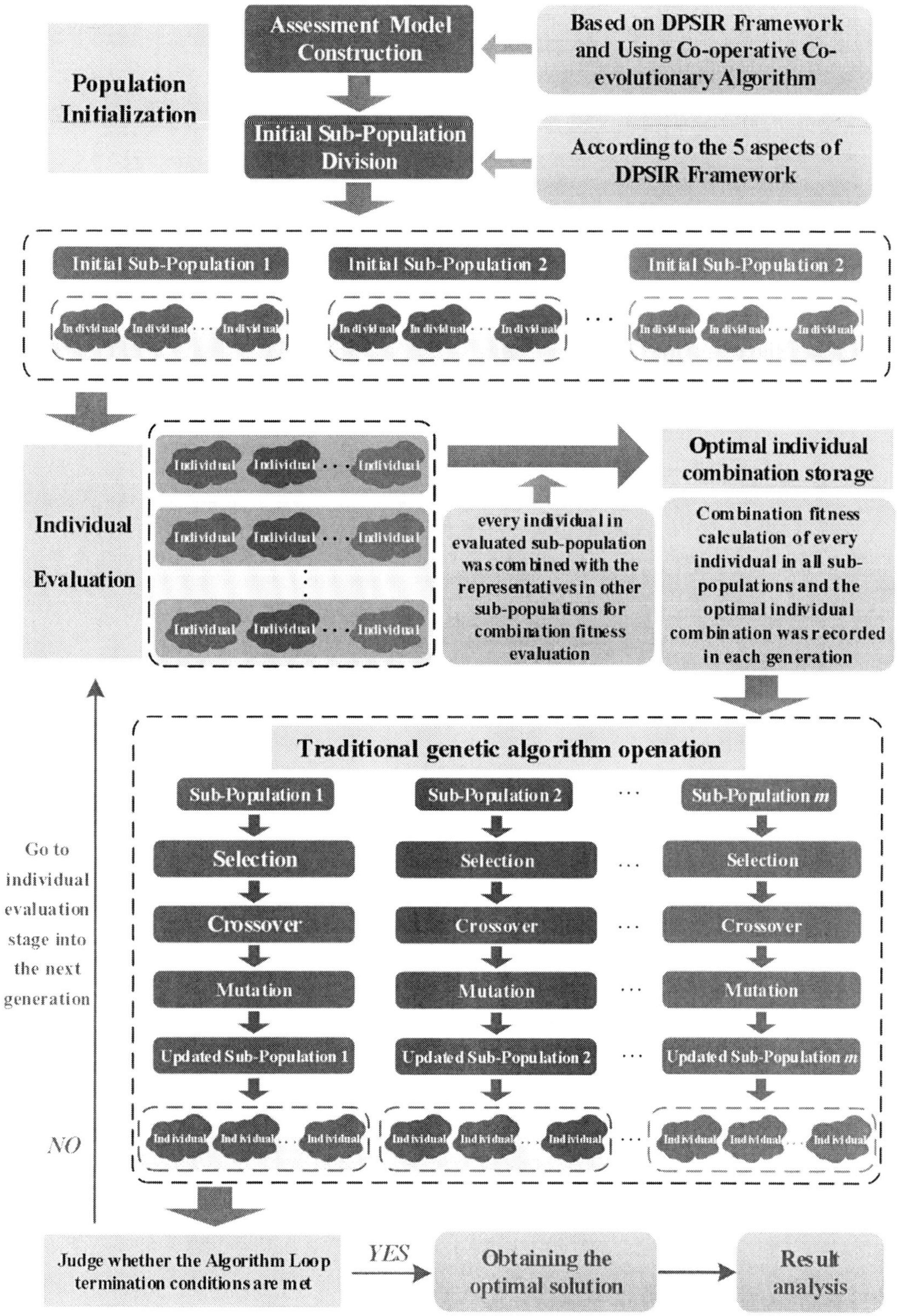

Figure 2. The Calculation procedures of CCGA according to DPSIR framework [22].

procedures in the algorithm [18-20]. Considering the five components of DPSIR framework, CCGA was applied by dividing five sub-populations as a suitable approach to calculate the optimal projection value of Projection Pursuit Model for evaluating the vulnerability of water resources system. The CCGA computing process is shown in figure 2 and the main procedures are as follows [21]:

Main procedures:

```
gen =0
for each sub-population s do begin
    Pop_s(gen) = randomly initialized population (0~1)
    evaluate fitness of each individual in Pop_s(gen)
end for
while termination condition = false do begin
    gen = gen +1
    for each sub-population s do begin
        select Pop_s(gen) from Pop_s (gen -1) based on fitness evaluation
        apply genetic operators to Pop_s(gen)
        evaluate fitness of each individual in Pop_s(gen)
    end for
end while
```

First, based on the research contents and the review of related researches, the evaluation model of water resources system vulnerability was established. And the initial population was divided into five sub-populations according to the structure of DPSIR framework. In this way the complex multivariable problem could be converted into five simple problems by sub-population genetic algorithm operating. Each sub-population evolved independently and only exchanged information when it was time to evaluate the individuals with other sub-populations. Secondly, an optimal individual and a representative individual in each other sub-population were selected to form be-evaluated combinations with the individuals in the be-evaluated sub-population, each be-evaluated individual was participated in two combinations forming, one combination was formed with the optimal individual in each other sub-population, and the other combination was with the representative individual in each other sub-population. Thirdly, every individual in be-evaluated sub-population was evaluated by comparing the fitnesses of two combinations in which the be-evaluated individual had been participated the forming. And the combination with the better fitness was preserved (figure 3). The fitness of each individual in each sub-population was calculated and the optimal combination of each generation was preserved after comparing procedure. Then all the optimal combinations of each generation were compared for finding the global optimal combination as the optimal projection direction to calculate the projection values.

3.2. Projection pursuit

In this study, projection pursuit was selected due to its fundamental which high dimensional data can be projected to low dimensional space. By optimizing projection index function, projected vector, which can reflect the characteristics of high dimensional data can be found. Without giving the weight of the evaluation index in advance, the artificial disturbance can be avoided by analyzing data in the low-dimensional space. p was defined as the amount of indexes in water resources system vulnerability evaluation. n was defined as the amount of samples. $\left\{x_{ij}^{*} \middle| i = 1,2,...,n; j = 1,2,...,p\right\}$ was defined as sample value of the evaluation index system of water resources system vulnerability. $\left\{x_{ij} \middle| i = 1,2,...,n; j = 1,2,...,p\right\}$ was defined as standardized index value. z_i was defined as projection eigenvalue of sample i in one-dimensional linear space [23, 24].

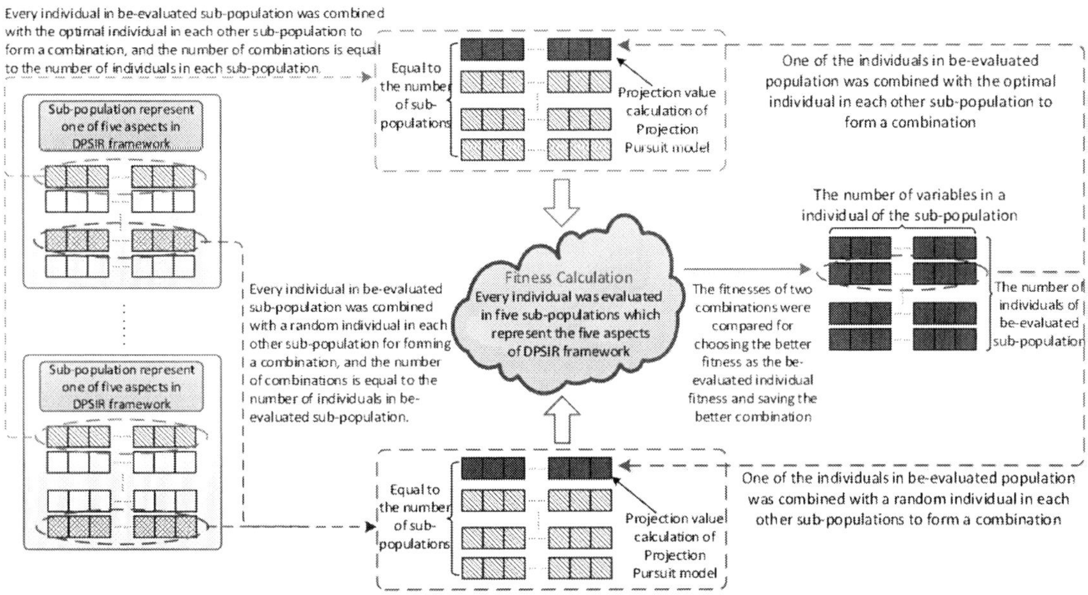

Figure 3. Fitness calculation and evaluation in CCGA [22].

3.3. Evaluation model of water resources system vulnerability based on co-operative co-evolutionary genetic algorithm and projection pursuit

The steps of the establishment of evaluation model of water resources system vulnerability based on co-evolutionary genetic algorithm and projection pursuit are as follows.

Step 1: standardization of indexes. Since different units of basic indexes were utilized in socio-economic, water resources and environmental system of DPSIR framework, a trade-off operate was required to normalize the actual values of the indexes. The following equation was used on the indexes which had the opposite change tendency compared with the water resources system vulnerability.

$$x_{ij} = (x_{ij}^* - x_{min,j})/(x_{max,j} - x_{min,j}) \tag{1}$$

In a similar way the following equation was used on the indexes which had the same change tendency compared with the water resources system vulnerability.

$$x_{ij} = (x_{max,j} - x_{ij}^*)/(x_{max,j} - x_{min,j}) \tag{2}$$

In equation (2), $x_{max,j}$ and $x_{min,j}$ were defined as the maximum and minimum of the index j of the index system respectively. The values of $x_{i,j}$ which were calculated by equations (1) and (2) were evaluation indexes in the region of [0, 1].

Step 2: index function construction of projection pursuit. The essence of projection pursuit was to find the optimal projection direction which could reflect the data characteristics to maximum extent and contain the data information fully. And z_i was the one-dimensional projection value by multiplying $x_{i,j}$ by a = $(a_1, a_2, ..., a_p)$.

$$z_i = \sum_{j=1}^{p} a_j \cdot x_{ij} \tag{3}$$

In equation (3), a_j was defined as vector of unit length, and $\sum_{j=1}^{p} a_j^2 = 1$.

If a = $(a_1,a_2,...,a_p)$ was the optimal projection direction, put it into the equation (3) and the projection values of water resources system vulnerability could be calculated, and then water resources system vulnerability could be analyzed quantitatively. When projection values were synthesized, more information of z_i was expected to extract. So projection indicator function *Q (a)* was constructed as the basis to choose the optimal projection direction. *Q (a)* could be calculated by the following equation:

$$Q(a) = S_z D_z \tag{4}$$

$$S_z = \sqrt{\frac{\sum_{i=1}^{n}(z_i - \bar{z})^2}{n-1}} \tag{5}$$

$$D_z = \sum_{i=1}^{n}\sum_{k=1}^{n}(R - r_{ik}) \cdot f(R - r_{ik}) \tag{6}$$

In equations (4) and (5), S_z was defined as standard deviation of projection value z_i. D_z was defined as local density of projection value z_i. $\bar{z}$ was defined as average value of projection value $z_1, z_2,...,z_m$. *R* was defined as window radius of local density, which could be got by experiment and the value was 0.1 times of S_z in general. $f(R - r_{ik})$ was defined as unit step function. When $R > r_{ik}$, $f(R - r_{ik}) = 1$; otherwise $f(R - r_{ik}) = 0$. The bigger D_z was, the more obvious the classification was.

When projection indicator function *Q (a)* reached the maximum, the optimal projection direction could be found.

Step 3: projection indicators function optimization using CCGA model. As the scheme set of samples was confirmed, the change of projection indicator function *Q (a)* was only concerned with the projection direction *a*. When *Q (a)* reached the maximum, *a* was the optimal projection direction. The maximum objective function and constraint condition were as follows:

$$\begin{cases} \max Q(a) \\ \|a\| = 1 \end{cases} \tag{7}$$

The maximum objective function was a complex non-linear optimization problem, in which $\{a_j | j = 1,2,...,p\}$ was optimization variable. Then CCGA was chosen to optimal the results, there were five sub-populations, and the variables amount in Driving forces sub-population was 6, in Pressure sub-population was 7, in State sub-population was 6, in Impact sub-population was 5, in Response sub-population was 7.

Step 4: classification of evaluation standards and the evaluation model application in study area. First, according to the index value corresponded with the vulnerability evaluation grade, put the optimal projection direction a^* calculated by standardization into equation (3) to get the range of the projection value corresponded with each grades, so that the grade division standard of the system vulnerability could be determined. Secondly, put $\{x_{ij} | i = 1,2,...,n; j = 1,2,...,p\}$ into equation (3) to calculate the projection value *y* of water resources system vulnerability in the study area. The vulnerability range could be confirmed and the evaluation of water resources system vulnerability in study area could be measured comparably with the projection values of evaluation standard grades.

4. Application example

4.1. General situation in study area

The data used in this study mainly includes meteorological, socio-economic, and water utilization data. The socio-economic data were taken from the "Tengzhou Statistical Yearbook". The hydrologic data were provided by the water authority in Tengzhou (Water Authority in Tengzhou, 2009) and obtained from Tengzhou City Water Resources Sustainable Utilization Planning Report.

Tengzhou City in Shandong Province was selected as the example to conduct the evaluation of water resources system vulnerability in this study. Tengzhou City is located at 116°49′E ~ 117°24′E and 34°50′N ~ 35°17′N in the south of Shandong Province (figure 4), with plenty of sunshine and good rainfall. the multi-year (1956 ~ 2008) mean precipitation is 753.9 mm. The historical maximum precipitation is 1216.8 mm (2003), and the historical minimum precipitation is 433.0 mm (1981). The area belongs to Nansi Lake in Huaihe River Basin. There are hundreds of rivers in the city and five of them are main rivers which had more than 100 km^2 drainage area. There are 2 large and medium-sized reservoirs, 4 small reservoirs of which the capacity is more than 1 million but less than 10 million cubic meters, 22 small reservoirs of which the capacity is more than 0.1 million but less than 1 million cubic meters, and 227 pond and retaining dams. Tengzhou City has edged itself into the top one hundred countries of the economic competition for years. The industrial structure proportion is 9.6:58.4:32, and per capita GDP has reached 30460 yuan (2008). Called the granary in the south of Shandong Province, it is listed as the commodity grain base and high quality vegetable base in the country. The industry of coal, electricity, chemical industry, construction materials, light industry, cottonocracy and others are all benignly developed in the city. Because of the important occupation in economic and social development, Tengzhou City has a large demand of water. The evaluation of water system vulnerability of Tengzhou City is helpful to make a further understanding about the state of water resources in this region, so that water resources can be taken more advantages and be protected , which will promote the sustainable development of water resources system.

Figure 4. Location of the study area.

4.2. Water resources vulnerability evaluation

According to the collected data, the data in 2008 wwere more informative, comprehensive and accurate. In order to ensure the reliability of the research, the data in 2008 was chosen to evaluate water resources system vulnerability in Tengzhou City. Taken collected data, related reports and statistic yearbook into consideration, 31 indicator values of driving forces-pressure-state-impact-response would be achieved. After standardizing the indices by equations (1) and (2) in accordance with each characteristics, and then the standardized indices were optimized by equations (3) to (7). The optimal projection direction of each index was listed in table 2.

Table 2. Optimal projection direction and the value.

serial number	index value	projection direction	Serial number	index value	projection direction	Serial number	index value	projection direction
D_1	1.36	0.136	P_6	241.4	0.146	I_4	2.55	0.185
D_2	757.1	0.207	P_7	61.6	0.215	I_5	3.8	0.167
D_3	3.75	0.153	S_1	30.1	0.16	R_1	16.7	0.162
D_4	6.63	0.132	S_2	96.88	0.141	R_2	46.13	0.152
D_5	32	0.187	S_3	66.05	0.164	R_3	75	0.195
D_6	25.08	0.181	S_4	22	0.187	R_4	100	0.21
P_1	407	0.21	S_5	249.78	0.193	R_5	0.59	0.171
P_2	563	0.161	S_6	83.9	0.217	R_6	71.1	0.181
P_3	0.7	0.186	I_1	26.8	0.205	R_7	56	0.185
P_4	68.78	0.171	I_2	100	0.166			
P_5	9.03	0.196	I_3	17.52	0.198			

According to tables 1 and 2, the degree of system vulnerability could be calculated by equation (3) (table 3). After standardized index value of Tengzhou City was put into equation (5), the projection value of vulnerability was 3.087. The result showed that the water resources system vulnerability in Tengzhou City belonged to grade II in the medium degree.

Table 3. Degree of water resources System Vulnerability.

Value of Z	Grade	Degree of vulnerability
$z \geq 3.929$	Grade I	Weak vulnerability
$2.738 \leq z < 3.929$	Grade II	Medium vulnerability
$1.565 \leq z < 2.738$	GradeIII	Strong vulnerability
$z < 1.565$	GradeIV	Very strong vulnerability

According to the collected information of water resources development and utilization in 2008, the agricultural water had occupied a large proportion in Tengzhou City and most was adopted broad irrigation. According to Surface Water Environmental Quality Standard (GB 3838-98), water quality of Cross river in the upper and middle basin was ranked III and water quality of other places was ranked IV or V, so the water environment condition were not going well. There is an increasing trend of regional water resources system vulnerability due to the combined influence of water consumption and pollution. Auxiliary construction of project was on the low levels. Main irrigation canal didn't match and the water efficiency in canal system of irrigation was very low in our country. Although the study area belongs to the water-affluent area in Shandong Province, water resources development and protection is also of significance. According to the results of evaluation of the model, water resources system vulnerability in Tengzhou City was belonged to the Grade II in the medium degree. It is showed that the result was conformed with the water resources situation to some extent, which indicated that the model was feasible and effective.

Water resources system vulnerability is the comprehensive performance which reflects the effect produced by different kinds of factors to water resources system. The same or similar evaluation results of water resources vulnerability may be caused by various reasons and the reflected problems are multifarious. In order to analyze the impact of various development state in different components of index framework to the vulnerability, first of all the index system of Driving forces-Pressures-States-Impacts-Responses was put into the model to calculate the optimal projection direction, then the projection values of z^*_{cd} (c=1,2,...5 was defined as the different departments of

DPSIR, *d*= Ⅰ, Ⅱ, Ⅲ, Ⅳ was defined as the different grades of the standard) corresponded with the grade and the projection values y_c could be calculated. Secondly the projection values of z^*_{cd} and y_c were adjusted to integer between 0 and 4 for comparing and analyzing effectively. For example, the scope of projection values [1.475, 2.066) in grade Ⅱ of driving type was adjusted to [2, 3) and the y_c=1.58 was adjusted to 2.18. The other projection values were adjusted in the similar way (table 4). Lastly, radar map was drew to show the projection values after the standard transformation (figure 5). From figure 5, results could be concluded as the followings: (1) the five components of the DPSIR framework are all belonged to grade Ⅱ in the medium degree and the vulnerability degree of indexes of impacts type and states type was higher than others, which indicated that the influence of high speed economic and social development and the overuse of the pesticide had already destroyed the water environment to some extent. (2) the water resources system vulnerability has been mainly influenced by the fierce imbalance between supply-demand of water resources and the weakness condition of water resources utilization. (3) the indexes of responses type had less impact which indicated that by measures, such as improving the utilization ratio of reuse water and saving more water, adaptive capacity of water resources system towards social-economic development could be improved accordingly relying on scientific and technological progress.

Table 4. The contrast of the projection values before and after the standard transformation.

Index type	Projection values of different grades $z^*(i)$				Projection values y	y projection values after the integer normalization
	Ⅰ	Ⅱ	Ⅲ	Ⅳ		
D	≥2.066	[1.475, 2.066)	[0.861, 1.475)	<0.861	1.58	2.18
P	≥1.501	[1.001, 1.501)	[0.667, 1.001)	<0.667	1.24	2.48
S	≥2.086	[1.635, 2.086)	[0.999, 1.635)	<0.999	1.67	2.09
I	≥1.208	[0.821, 1.208)	[0.358, 0.821)	<0.358	0.83	2.02
R	≥1.83	[1.12,1.83)	[0.547, 1.12)	<0.547	1.62	2.70

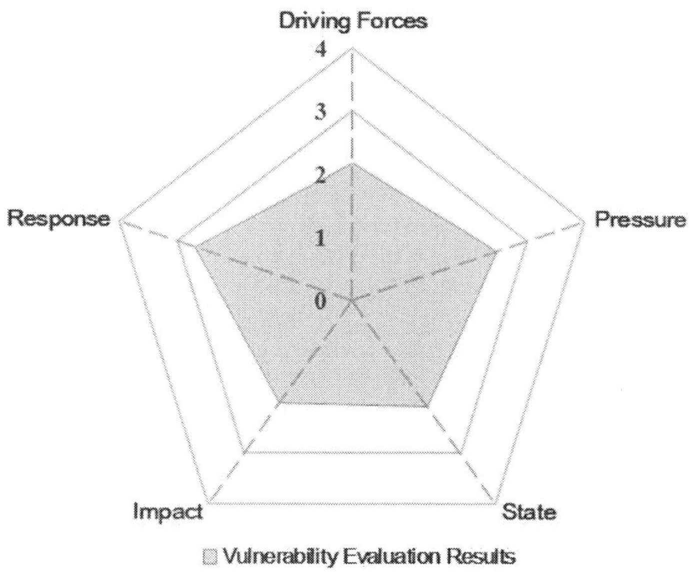

Figure 5. The structure of the water resources system vulnerability in study area.

5. Conclusions

The DPSIR framework was adopted to construct the evaluation index system of water resources system vulnerability. And then the co-evolutionary genetic algorithm and projection pursuit was used to establish evaluation model of water resources system vulnerability. The system vulnerability in Tengzhou City was analyzed in terms of Driving forces, Pressure, State, Impact and Response on the basis of the projection value calculated by the model. The results are coincident with the state of water resources in study area, while indicate that the model is feasible and effective.

Due to the limitations of data, this study only evaluated the vulnerability in 2008. On condition that enough data are available, multi-sample analysis of water resources system vulnerability could provide more basises for the local water resources management. Water resources system vulnerability can be reduced by improving water usage efficiency, capital investment, pollution control, total water consumption control and water quota management.

Acknowledgments

This study is supported by the Basic Scientific Research Business Expense of Yellow River Institute of Hydraulic Research (HKY-JBYW-2016-25), and the National Natural Science Foundation of China (51409091).

References

[1] Wang X, Ma F B and Li J Y 2012 Water resources vulnerability assessment based on the parametric-system method: A case study of the Zhangjiakou Region of Guanting Reservoir Basin, North China *The 18th Biennial Conf. of the Int. Society for Ecological Modelling* **13** 1204-12

[2] Liu L L 2002 Concept and quantitative assessment of vulnerability of water resource *Bull. Soil Water Conserv.* **2** 41-4

[3] Liu L L, Xu H M and Ma S M 2014 Understanding of climate change impacts, adaptation and vulnerability in urban areas and rural areas *Adv. Climate Change* **10** 254-9

[4] Walmsley J J 2002 Framework for measuring sustainable development in catchment systems *Environ. Manage.* **29** 195-206

[5] Giupponi C 2002 From the DPSIR reporting framework to a system for a dynamic and integrated decision making process *Conf. on European Policy and Tools for Sustainable Water Management* (Italy, Venice)

[6] Bowen R E and Riley C 2003 Socio-economic indicators and integrated coastal management *Ocean Coast Manage.* **46** 299-312

[7] Danielopol D L, Griebler C, Gunatilaka A and Notenboom J 2003. Present state and future prospects for groundwater ecosystems *Environ Conserv.* **30** 104-30

[8] Borja A, Galparsoro I, Solaun O, Muxika I, Tello E M, Uriarte A and Valencia V 2006 The European water framework directive and the DPSIR, a methodological approach to assess the risk of failing to achieve good ecological status *Estuarine Coastal Shelf Sci.* **66** 84-96

[9] Pirrone N, Trombino G, Cinnirella S, Algieri A, Bendoricchio G and Palmeri L 2005 The driver-pressure-state-impact-response approach for integrated catchment-coastal zone management: Preliminary application to the Po catchment-Adriatic Sea coastal zone system *Reg. Environ. Change* **5** 111-37

[10] Bianet K A, Kaneko S, Fujikura R and Fujiwara A 2009 Urbanization and subsurface environmental issues: An attempt at DPSIR model application in Asian cities *Sci. Total Environ.* **407** 3089-104

[11] Skoulikidis N T 2009 The environmental state of rivers in the Balkans – A review within the DPSIR framework *Sci. Total Environ.* **407** 2501-16

[12] Gleick P H and Palaniappan M 2010 Peak water limits to freshwater withdrawal and use *Proc. Natl. Acad. Sci. U.S.A.* 107-11155

[13] Timmerman I G, Beinat E and Termeer C I 2011 Developing transboundary river basin monitoring programs using the DPSIR indicator framework *I Environ. Monitor.* **13** 2808-18

[14] Dong S F, Dong Z C and Chen K N 2010 Analysis of water resources system vulnerability

based on DPSIR conceptual model *Water Resour. Prot.* **25** 1-3
[15] Wang S J and Zhang X L 2010 Projection pursuit dynamic cluster model and its application to water resources carrying capacity evaluation *J. Water Resour. Prot.* **2** 449-54
[16] Fernando S M, Berta M L, Marina G L 2013 Unraveling the relationships between ecosystems and human wellbeing in Spain *PLoS ONE* **8** e73249
[17] Zhou S D, Muller F, Burkhard B 2014 Socioeconomic influences on biodiversity, ecosystem services and human well-being: a quantitative application of the DPSIR model in Jiangsu, China *Sci. Total Environ.* **490** 1012-28.
[18] Potter M A and De Jong K A 1994 A cooperative co-evolutionary approach to function optimization *Problem Solving from Nature PPSN-III* ed Y Davidor, H-P Schwefel and R Manner R Parallel *Proc. of the Int. Conf. on Evolutionary Computation, Lecture Notes in Computer Science* **866** 249-57
[19] Sun S K, Wang Y B, Liu J, Cai H J, Wu P, Geng Q L and Xu L J 2016 Sustainability evaluation of regional water resources under the DPSIR framework *J. Hydrol.* **532** 140-8
[20] Gu J W, Gu M Z, Cao C W and Gu X S 2010 A novel competitive co-evolutionary quantum genetic algorithm for stochastic job shop scheduling problem *Comput. Oper. Res.* **37** 927-37
[21] Iorio A and Li X 2002 Parameter control within a co-operative co-evolutionary genetic algorithm *Lect. Notes Comput. Sci.* **2439** 247-256
[22] Zhao Y, Wang M H, Li H B, Du K, Su X H and Xiao H 2017 *Study on Synergetic Development of Complex Water Resources System* (Zhengzhou: The Yellow River Water Conservancy Press)
[23] Zhao Y, Dong Z C and Guo H F 2011 Projection-pursuit model based on shuffled frog leaping algorithm and its application *Proc. of 2011 Int. Conf. on Transportation, Mechanical and Electrical Engineering* **1** 610-4
[24] Wang M H, Dong Z C and Ma H L 2014 Assessment of water resources system vulnerability based on shuffled frog leaping algorithm and projection pursuit model *Int. J. Hydroelectric Energy* **9** 31-5

IOP Conf. Series: Earth and Environmental Science 82 (2017) 012050 doi:10.1088/1755-1315/82/1/012050

Extreme water level and wave estimation for nearshore of Ningde City

Y D Jin[1,2], E K Wang[1] and G Q Xu[1]

[1]First Institute of Oceanography, SOA, Qingdao 266061, China

E-mail: jyd@fio.org.cn

Abstract. The high and low design water levels are calculated by observation tidal data in sea areas of Ningde offshore wind power project from September 2010 to August 2011, with the value 318 cm and -246 cm, respectively. The extreme high and low levels are also calculated using synchronous difference ratio method based on station data from 1973 to 2005 at Sansha station. The value is 431 cm and -378 cm respectively. The design wave elements are estimated using the wave data from Beishuang Station and Pingtan station. On this basis, the SWAN wave model is applied to calculating the design wave elements in the engineering sea areas. The results show that the southern sea area is mainly affected by the wave effect on ESE, and the northern is mainly affected by the E waves. This paper is helpful and useful for design and construction of offshore and coastal engineering.

1. Introduction

The design of water level and wave elements is an important part of offshore engineering design. Teng and Xiu [1] used several station observations to study the extreme waves in the south of Bohai Sea. Since the advent of satellites, altimeter measurements of wave heights have been available. Wang *et al* [2] used satellite wave data to estimate extreme wave in the Bohai Sea.

In the practical engineering design, due to the lack of long-term wave observation data, the wind wave data are usually used to calculate the offshore wave elements in the engineering area. The numerical simulation method (hindcast and forecast) has become more important for wave investigation since the appearance of models. Lei *et al* [3] used a wave model to simulate the wave elements in the the Bohai Sea. Yin *et al* [4] studied interaction of waves and tide surge by a joint numerical model. Qiao *et al* [5] gave the extreme wave parameters by wave simulation under tropical cyclones wind forcing of Tonkin Gulf. Wang *et al* [6, 7] studied waves in the Bohai Sea and South China Sea respectively such as wave characteristics and extreme parameters.

Ningde City is located in the northeast of Fujian Province, with a coastline of 1046 km and sea area of 600 km^2. There are 344 islands and the coastline accounts for about 1/3 in the province of Fujian. However, there is lack of studies in this sea area. The object of this paper is to calculate the design water level and design wave element in the engineering sea area as shown in figure 1 according to simulation and observation data analysis. There 4 parts in this paper. The second part is to estimate the engineering water level. The design waves in the open sea and nearshore are calculated in the third part. Finally the conclusion is given in the fourth part.

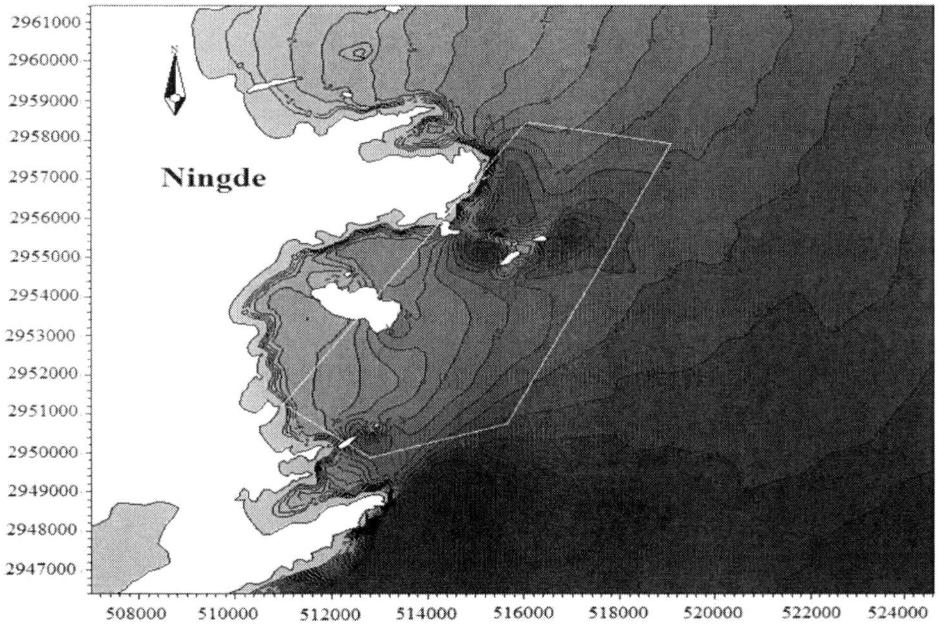

Figure 1. The depth and points for wave calculation.

2. Engineering water level

In the aspect of engineering water level, Dong *et al* [8] calculated the return values of typhoon significant wave height by a Poisson-maximum entropy distribution. Wu *et al* [9] studied storm surge during typhoon progress in the sea area around Shandong Peninsula and got the extreme parameters of sea level by numerical simulation and statistical analysis. Based on 20-year numerical hindcast data, Wang *et al* [10] studied currents and sea levels in the Bohai Sea such as long-term characteristics and extreme parameters.

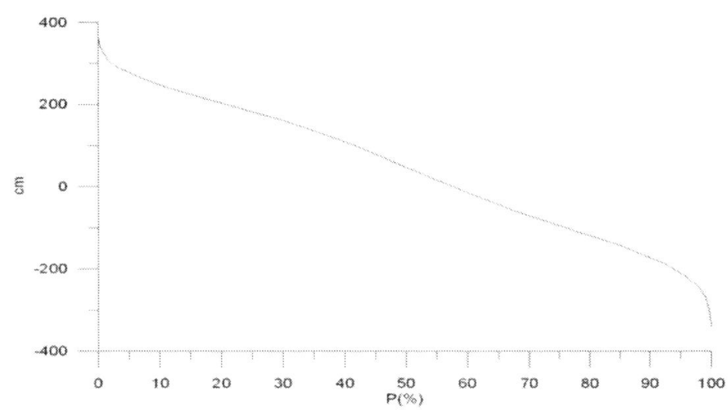

Figure 2. The cumulative frequency curve of design water level.

There should be one or more years of measured tidal data used to determine the design of high and low water level by the method of the high tide and low tide cumulative frequency statistics. The cumulative frequency analysis was performed by using the annual hourly tide data from September 2010 to August 2011. The design high water level is taken as the cumulative frequency of 1% tide and the design low water level is taken as the cumulative frequency of 98% tide. The results are shown in figure 2. The cumulative design high water level is 318 cm and the low water level is -246 cm respectively.

The highest and lowest tidal level data in the Sansha station from 1973 to 2005 year by year are used to obtain the extreme high and low water level, according to the Gumbel distribution method. The results are shown in table 1. The water surface depended on the 1985 national height datum.

Table 1. Extreme water level at Sansha station.

Return period	Sansha station		Engineering sea area	
	High level	Low level	High level	Low level
50a	441	-381	431	-378
20a	422	-374	413	-371
5a	392	-363	383	-360
1a	328	-340	320	-337

The relationship of high tide and low tide between engineering sea area and Sansha station is established by statistical analysis of 1 month synchronous tide observation. The high tide regression equation is

$$Y=0.987X+2.226 \tag{1}$$

The correlation coefficient is 0.997 (figure 3(a)). The low tide regression equation is

$$Y=0.992X+3.496 \tag{2}$$

The correlation coefficient is also 0.997 (figure 3(b)).

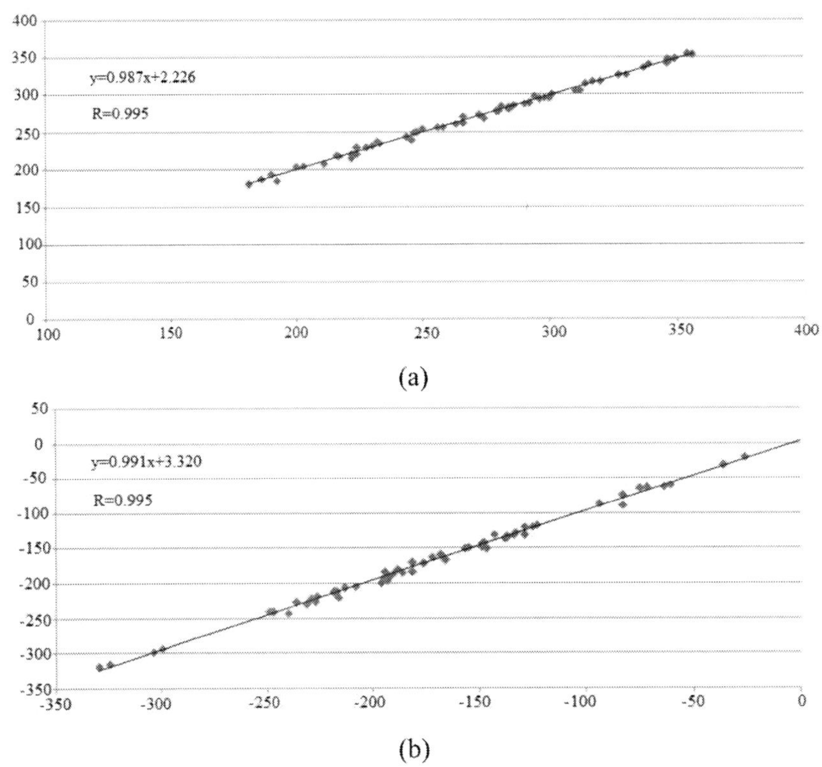

Figure 3. (a) Regression analysis of high tide of engineering sea area and Sansha station and (b) Regression analysis of low tide of engineering sea area and Sansha station.

According to the regression equation of high and low tide of two stations, the extreme water level of different return periods is calculated. The results are shown in table 1. The extreme high level is 431 cm and the extreme low level is -378 cm in the engineering sea area.

3. Extreme wave parameter

3.1. Extreme wave in the open sea

There are many islands in the engineering sea area with East peninsula to the west and open sea to the east. East is the dominant wave direction. The huge waves are mainly from NE to SE direction. There are two wave stations which are Beishuang and Pingtan located to the east of the engineering. The observation data from 1981 to 2000 at Beishuang station and from 1960 to 1982 at Pingtan station is used to extreme wave estimation. The results show the wave in the ESE direction is not hazardous at Beishuang station as it is under cover of Dongshuang Island to the SE. On the other hand, the strong wave directions at Pingtan station are from E to SE. Considering the engineering safety, waves for extreme estimation in the direction of NE an E are taken from Beishuang station while those in the direction of ESE are taken from Pingtan station. The P-III distribution is applied to estimate wave height $H_{1/10}$ of different return period. The results are shown in table 2. The max value is in the E direction with the wave height $H_{1/10}$ 11.45 m of 50 year return period.

Table 2. Design wave height $H_{1/10}$ of different return period (m).

Return period	NE	E	ESE
50a	8.54	11.45	10.11
20a	7.10	10.25	8.17
10a	6.07	9.23	6.75
1a	3.88	1.28	2.95

3.2. Extreme wave in the shallow water

The wave model Simulating Waves Nearshore (SWAN) model is used in this study. The action density $N(\sigma, \theta)$ is applied to calculating in SWAN model, which is equal to variance density divided by relative frequency,

$$N(\sigma,\theta) = E(\sigma,\theta) / \sigma \tag{3}$$

where σ is the relative frequency and θ is the wave direction.

The evolution of the wave spectrum is described by the spectral action balance equation which for Cartesian coordinates is

$$\frac{\partial N}{\partial t} + \frac{\partial}{\partial x} c_x N + \frac{\partial}{\partial y} c_y N + \frac{\partial}{\partial \sigma} c_\sigma N + \frac{\partial}{\partial \theta} c_\theta N = \frac{S}{\sigma} \tag{4}$$

The first term represents the local rate of change of action density in time. The propagations of action in geographical space with propagation velocities c_x and c_y in the x and y direction are the second term and the third term, respectively. The fourth term represents shifting of the relative frequency due to variations in depths and currents. The fifth term represents depth-induced and current-induced refraction. S is the source term which represents the effects of nonlinear wave-wave interactions, wind generation and dissipation. The dissipation mainly includes bottom friction, white capping and depth-induced wave breaking.

There were 30 frequencies logarithmically spaced from 0.05 Hz to 1.0 Hz. The number of iterations was set from 1, which is the default value, to 4 in order to enhance the numerical accuracy as the model integrated to the next time step. The linear wind input and growth method is used by Cavaleri and Malanotte-Rizzoli [11]. The exponential wind input and growth method is applied from Komen *et al* [12]. The whitecapping dissipation method is from Komen *et al* [12]. The non-linear interaction method is used by Hasselmann *et al* [13].

The extreme wave parameters are taken as the boundary conditions. The triple triangular mesh is set to the model. The range is from 119°32′ E to 120°40′ E at longitude and 26°13′ N to 27°10′ N at latitude with the resolution 140 near the engineering. The bathymetric details for the SWAN are shown in figure 4.

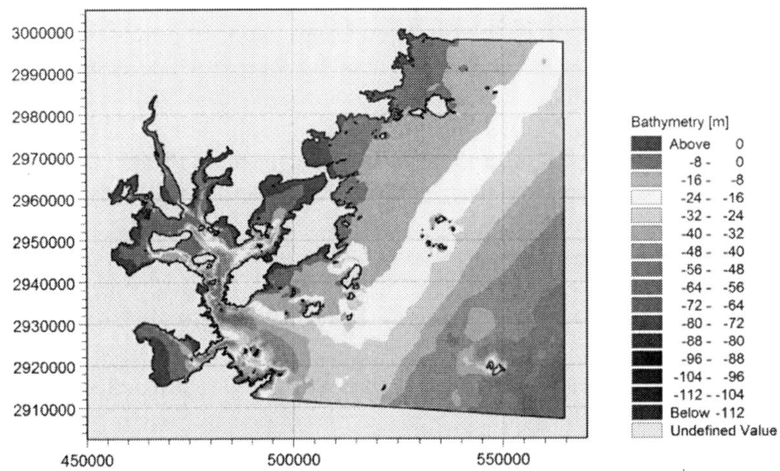

Figure 4. The bathymetric details for wave simulation.

The extreme wave parameters are estimated under condition of 4 water levels and 3 wave directions. The design water levels include design high and low water levels and extreme high and low water levels. The wave directions are NE, E and ESE which influence the engineering sea area most. The H_s fields of 50 year return period under 4 water levels and 3 wave directions are shown in figures 5 to 7. Furthermore, 12 points are selected as the representative points for statistical analysis. The point locations are shown in figure 1. The design wave of 50 year return period at 12 points is given in table 3. The results show that the design wave parameters are higher in the E direction than ESE direction. The south of engineering is mainly influenced by wave from ESE, while the north is mainly influenced by wave from E.

Table 3. Extreme parameters of 50 year return period in 12 points.

	NE		E		ESE	
	H_s(m)	T_m(s)	H_s(m)	T_m(s)	H_s(m)	T_m(s)
A1	4.98	10.5	5.98	12.1	5.72	11.4
A2	5.13	10.5	6.21	12.1	5.93	11.4
A3	5.34	10.5	6.42	12.1	6.08	11.4
A4	5.28	10.5	5.98	12.1	5.70	11.4
A5	5.47	10.5	5.61	12.1	5.83	11.4
A6	5.56	10.5	6.25	12.1	6.46	11.4
A7	5.42	10.5	6.38	12.1	6.78	11.4
A8	5.25	10.5	6.25	12.1	6.84	11.4
A9	5.07	10.5	5.61	12.1	5.84	11.4
B1	5.37	10.5	6.15	12.1	6.18	11.4
B2	4.61	10.5	5.93	12.1	5.98	11.4
B3	5.08	10.5	5.93	12.1	5.41	11.4

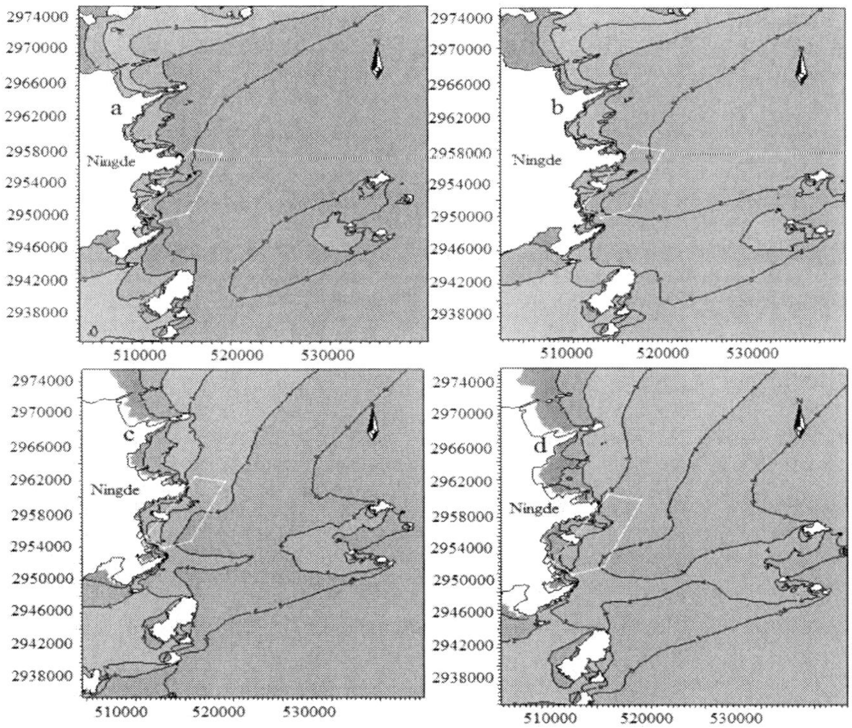

Figure 5. Hs of 50 year return period in NE (a: extreme high water, b: design high water c: design low water d: extreme low water).

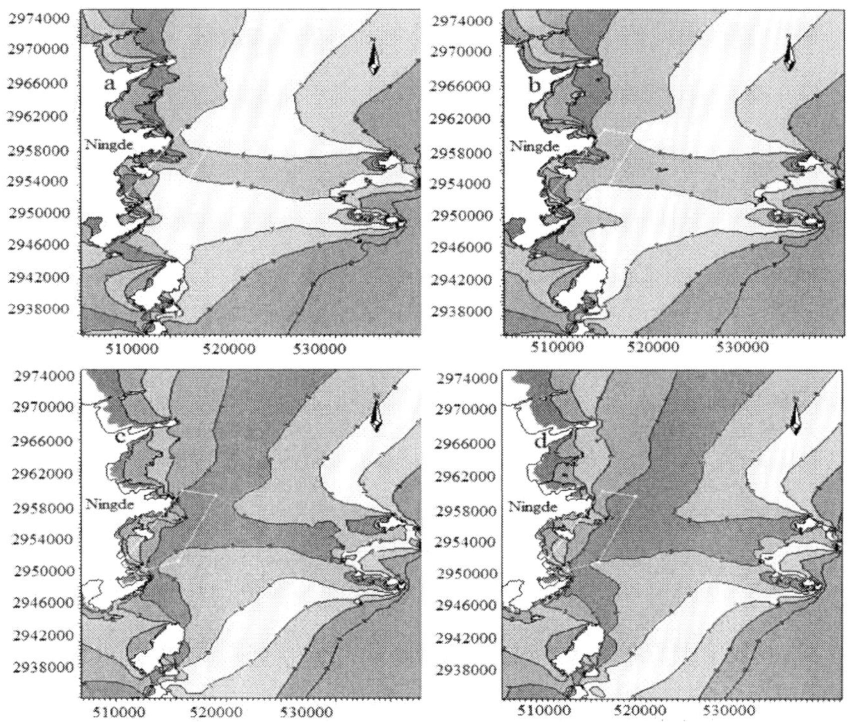

Figure 6. H_s of 50 year return period in E (a: extreme high water; b: design high water; c: design low water; d: extreme low water).

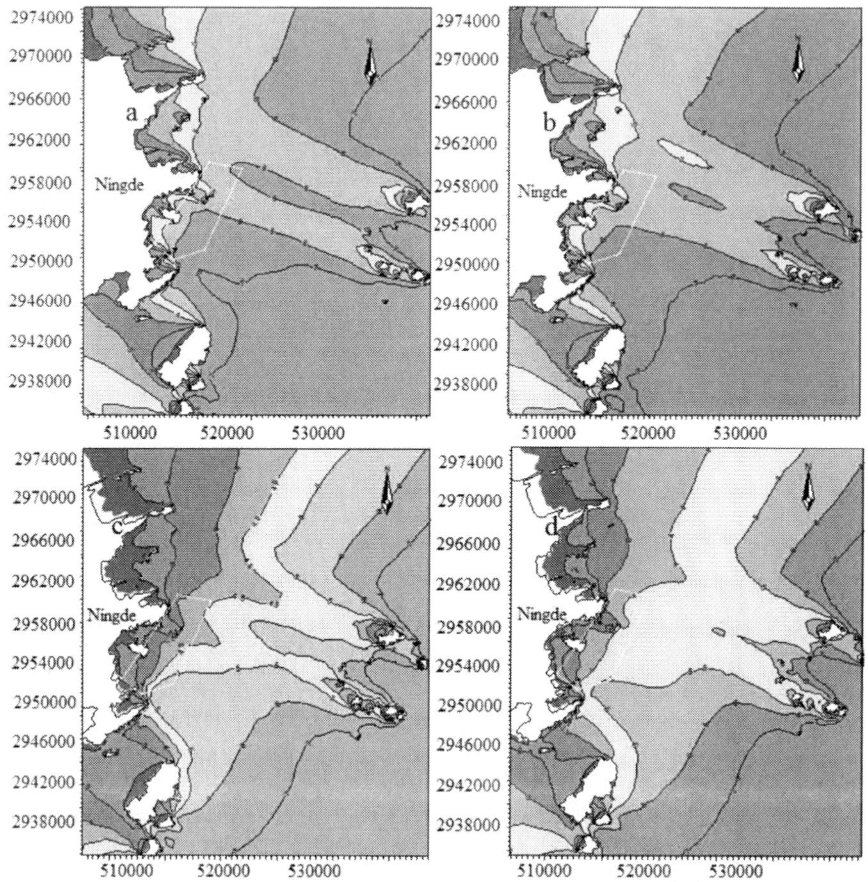

Figure 7. H_s of 50 year return period in ESE (a: extreme high water, b: design high water c: design low water d: extreme low water).

4. Conclusion

In this paper, the water level and the design elements of offshore wave are calculated using the measured data, and then the SWAN wave model is applied to simulating the wave propagation in shallow water.

The design high and low water level is 318 cm and -246 cm respectively, and the extreme high and low water level is 431 cm and -378 cm with the 1985 national height datum as the water surface. The engineering sea area is mainly influenced by wave from E direction. The wave height $H_{1/10}$ of 50 year return period is 11.45 m.

On the whole, this paper is useful and helpful for marine disaster and coastal engineering nearshore of Ningde.

References

[1] Teng X C and Wu X J 1994 A study on the characteristics of annual extreme and design waves in the South Region of the Bohai Sea *Journal of Oceanograpgy of Huanghai & Bohai Seas* **12** 3-12 (in Chinese)

[2] Wang G Y, Zhao J P and Song R S 2000 Sea-wave extreme estimation using satellite macro-wave remote-sense data *Port Eng. Technol.* **1** 11-8 (in Chinese)

[3] Lei F H, Wang J Z and Chen B H 1992 Wave elements simulation of petroleum development areas in the Bohai Sea *China Offshore Oil and Gas (Engineering)* **4** 47-51 (in Chinese)

[4] Yin B S, Wang T, Hou Y J, Cheng M H, Fan S T and Su J Z 2001 Numerical study of the

influence of waves and tide surge interaction on waves in Bohai Sea *Oceanologia Et Limnol Sinica* **32** 109-16 (in Chinese)

[5] Qiao F L, Yu W D and Pan Z D 1997 Study on the wind and wave extreme parameters of Tonkin Gulf in the South China Sea-the Applications of LAGFD Numerical *Models J. Hydrodyn.* **12** 75-86 (in Chinese)

[6] Wang Z F, Wu K J, Zhou L M and Wu L Y 2012 Wave characteristics and extreme parameters in the Bohai Sea *Chin. Ocean Eng.* **26** 341-50

[7] Wang Z F, Zhou L M, Dong S, Wu L Y, Li Z B, Mou L and Wang A F 2014 Wind wave characteristics and engineering environment of the South China *Sea J. Ocean Univ. Chin.* **13** 893-900

[8] Dong S, Liu W, Zhang L and Guedes Soares C 2013 Return value estimation of significant wave heights with maximum entropy distribution *J. Offshore Mech. Arc. Eng.* **135** 031103-1-10

[9] Wu Y N, Wang Z F, Dong S and Li J J 2014 Numerical simulation and statistic analysis of typhoon storm surge in the sea area around Shandong *Peninsula Proc. of the 11th Pacific/Asia Offshore Mechanics Symposium* (Shanghai, China) pp 235-8

[10] Wang Z F, Dong S, Chen C C and Soares C G 2015 Long-term characteristics and extreme parameters of currents and sea levels in the Bohai Sea based on 20-year numerical hindcast data *Nat. Hazards* **76** 1603-24

[11] Cavaleri L and Malanotte-Rizzoli P 1981 Wind wave prediction in shallow water: Theory and applications *J. Geophys. Res.* **86** 10961-73

[12] Komen G J, Hasselmann K and Hasselmann K 1984 On the existence of a fully developed wind-sea spectrum *J. Phys. Oceanogr.* **14** 1271-85

[13] Hasselmann S, Hasselmann K, Allender J H and Barnett T P 1985 Computations and parameterizations of the nonlinear energy transfer in a gravity wave spectrum Part II: Parameterizations of the nonlinear transfer for application in wave models *J. Phys. Oceanogr.* **15** 1378-91

Multi-timescale analysis of rainfall in Karst in Guizhou, China

X N Li[1,2], X J Zhao[1,2,5], B Xu[3] and G C Chen[4]

[1]Guizhou Water Conservancy and Hydroelectric Power Investigation, Design and Research Institute, Guiyang 550002, China
[2]Guizhou Engineering Technology Research Center for Exploitation and Utilization of Water Resources in Karst region, Guiyang 550002, China
[3]Water Resources Department, Yangtze River Scientific Research Institute, Wuhan 430010, China
[4]Planning Bureau of Changjiang Water Resources Commission, Wuhan 430010, China

E-mail: tzhaoxj@vip.sina.com

Abstract. The rainfall data are evaluated by performing multi-timescale analysis. Weixin station was considered as a representative location for collecting rainfall data in Karst area of Guizhou Province in China. The results are as follows: (1) there is an increasing trend in annual precipitation, whereas the spring, summer, and winter seasons exhibit a positive contribution to the growth of annual rainfall; however, autumn shows a negative contribution; (2) the annual precipitation series is strongly volatile in the period extending from 2005–2016; summer and autumn show the highest volatility in recorded rainfall, whereas the change in rainfall is not that significant in spring and winter; (3) the main factor contributing to a change in rainfall is not the frequency of rainfalls, but the condition of a single rainfall in a certain time.

1. Introduction

Guizhou mountainous area has numerous sources of water. As a result, water resources per capita value of this region is higher than the national average in China. However, the region has karst topography, so there are very scarce sources of surface water. On the other hand, surface water readily transforms into groundwater; therefore, it is difficult to harvest water resources in this region. Thus, a paradoxical situation created in that the region faces the problem of drought frequently [1-3]. In recent years, soil erosion has increased massively, and flood disasters have occurred frequently. Consequently, there has been an adverse impact on social development. Furthermore, there is significant ecological imbalance in this region, which is accentuated with climate change and special geological conditions inherent in karst topography [4-6]. In the 12th five-year plan drafted by policy makers of Guizhou province, there were strategies proposed to tackle ecological imbalance and to improve social development in the region. In the 13th five-year plan drafted by the government of Guizhou province, the objectives addressed were as follows: healthcare, big data, ecology, and tourism. To address these objectives, a common minimum program for development was presented in the 13th five-year plan in Guizhou province, China. In accordance with this plan, policy makers felt it is necessary to conduct a multi-timescale analysis of rainfall in Guizhou province having Karst topography. The results of the analysis can be used for optimum development of water resources, ecological management and protection, and sustainable social structure in Guizhou province, China.

In the Weixin rainfall station, a key water-control project was developed by Bijie city in Guizhou province, China. Under this project, the study example was the rainfall data: this was evaluated to analyze the climate changes according to the rules presented in multi-timescale method. Based on the analysis, we deciphered the trends in climate change. In addition, we also elucidated the relationship between different times and rainfall in a region with typical karst topography. The methodology is an analytic method with which we evaluate rainfall data on a multi-timescale level (annual, monthly, and four seasons), and we further determine rainfall frequency in every timescale. After obtaining the results of the study, we would like to reiterate that it is a preliminary study based on multi-timescale analysis; so the results not only explain the relationship of various rainfall times at a single station but they also provide an investigative account of how the intensity of water problem varies with seasonal climatic changes in the mountainous region with karst topography.

2. Study area and data

Weixin station is located in the Weixin township of Nayong County, Guizhou Province, China. The geographical (latitude and longitude) coordinates of this station are 105°11′ E, 27°01′ N. The station is located in a subtropical region with a typical monsoon climate. Because it is located in the middle of Guizhou plateau, it has distinctly cold mountainous areas with the following geographical characteristics: high altitude and low temperature. The location at which Weixin station is situated has four distinct seasons: i) monsoon with abundant rainfall, ii) a long winter season, and iii) a short summer season. The temperature difference is distinct between day and night; the temperatures vary appreciably in wet and dry seasons. Meanwhile, there are widespread rivers and complex geological conditions in this region with karst topography.

The rainfall data were collected on a daily basis from 2005 to 2016. The data collection was done by Guizhou Bureau of Hydrology and Water Resources in China.

Table 1. Monthly rainfall in study station.

year/month	1	2	3	4	5	6	7	8	9	10	11	12
2005	20.1	24.1	17.6	48.1	141.6	124.3	140.5	192.8	83.1	40.6	34.5	37.5
2006	12.6	24.7	36.7	24.1	141.6	142.5	99.8	81.4	135.3	98.8	30.8	15.0
2007	6.0	21.5	31.1	50.1	159.1	187.0	236.0	181.6	110.4	62.7	16.5	30.5
2008	29.0	28.2	33.8	57.5	126.6	278.8	170.3	188.9	255.4	134.2	96.2	22.3
2009	22.3	14.3	39.5	88.5	101.5	101.9	95.6	179.6	19.6	93.9	14.4	35.3
2010	1.6	17.6	8.3	48.9	62.2	159.3	196.5	72.1	148.4	101.8	33.2	19.8
2011	11.0	19.5	22.5	55.5	45.5	127.5	83.0	90.5	84.5	52.5	35.5	36.5
2012	26.0	21.0	21.5	46.0	134.0	190.5	399.0	107.5	153.0	55.0	26.5	22.5
2013	16.5	12.5	55.0	50.0	73.0	155.0	61.5	160.0	99.0	45.5	26.5	29.0
2014	17.5	33.0	93.0	54.5	105.5	142.5	217.0	122.5	152.0	64.5	52.5	22.0
2015	43.0	23.5	43.0	98.5	189.0	140.5	125.0	190.5	154.5	120.0	9.5	29.5
2016	24.5	18.0	50.5	127.5	118.5	384.5	105.5	118.0	104.5	66.5	40.5	16.0
Mean value	19.2	21.5	37.7	62.4	116.5	177.9	160.8	140.4	125.0	78.0	34.7	26.3
Slope γ_i	1.26	-0.19	3.21	4.95	-1.36	7.51	-0.136	-2.09	0.89	-0.42	-0.64	-0.54
η_i /%	10.1	-1.5	25.8	39.8	-11.0	60.4	-1.1	-16.8	7.1	-3.4	-5.2	-4.3

3. Monthly rainfall analysis

The slope of the series representing rainfall data illustrates the trends of decrease or increase in rainfall at different points of time; negative and positive values indicate trends of decrease and increase in rainfall, respectively. In the months of February, May, July, August, October, November, and

December, the slope of the rainfall data series is negative; therefore, we infer that the monthly rainfall decreases in these months. The month of August has the least value of negative slope at –2.09. In the remaining months, we were able to obtain series with positive slope; the positive slope had a maximum value of 7.51 in the month of June. The rate of contribution represents the contribution degree of monthly rainfall in a year; positive value represents positive contribution, while negative value represents negative contribution. Please see formula 1. As shown in table 1, the rates of contribution are 10.1%, –1.5%, 25.8%, 39.8%, –11.0%, 60.4%, –1.1%, –16.8%, 7.1%, –3.4%, –5.2%, and –4.3% for the months ranging from January to December, respectively. This indicates that August was the month that witnessed maximum negative contribution at –16.8%, whereas June is the month that witnessed maximum positive contribution at 60.4%.

$$\eta_i = \gamma_i / \sum \gamma_i \times 100 \quad (1)$$

Where η_i is the rate of contribution in time i, γ_i is the slope of time i.

4. Rainfall analysis for annual and seasonal periods

As shown in table 2 and figure 1, the mean values of rainfall are 216.6 mm, 479.1 mm, 237.7 mm, and 67.0 mm in spring, summer, autumn and winter; the annual rainfall in the region is 1000.5 mm. This indicates that rainfall mostly occurs in summer; the amount of rainfall decreases gradually as seasons change from spring and autumn; the minimum rainfall is recorded in winter season. In autumn, the slope of rainfall series has a negative value, indicating a decreasing trend in rainfall. In contrast, the slopes of rainfall series have positive values in the remaining seasons, with spring season having a maximum positive slope value of 6.8. The slope of annual rainfall series is 12.44; the rainfall shows an increasing trend in Weixin station. Meanwhile, the rates of contribution are 54.6%, 42.5%, –1.4%, and 4.3% in the period progressing from spring to winter; the rate of contribution is negative in autumn, whereas the rate of contribution is positive in spring, summer, and winter seasons.

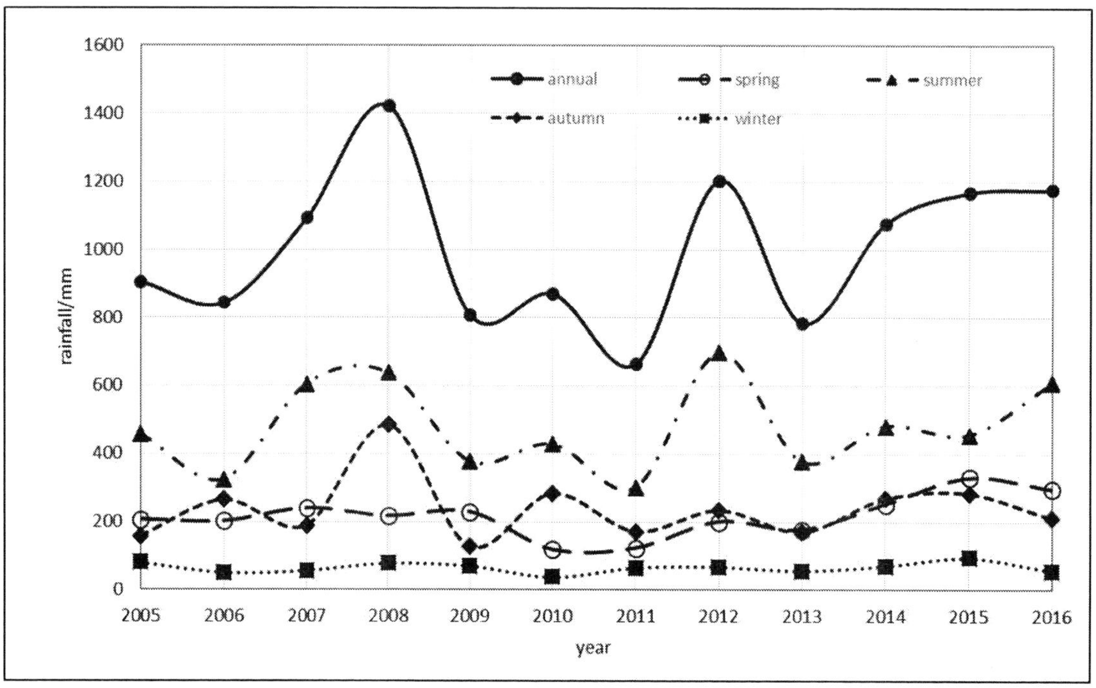

Figure 1. The scatter diagram of seasons and annual rainfall.

Table 2. Seasons and annual rainfall.

year/seasons	Spring	Summer	Autumn	Winter	annual
2005	207.3	457.6	158.2	81.7	904.8
2006	202.4	323.7	264.9	52.3	843.3
2007	240.3	604.6	189.6	58.0	1092.5
2008	217.9	638.0	485.8	79.5	1421.2
2009	229.5	377.1	127.9	71.9	806.4
2010	119.4	427.9	283.4	39.0	869.7
2011	123.5	301.0	172.5	67.0	664.0
2012	201.5	697.0	234.5	69.5	1202.5
2013	178.0	376.5	171.0	58.0	783.5
2014	253.0	482.0	269.0	72.5	1076.5
2015	330.5	456.0	284.0	96.0	1166.5
2016	296.5	608.0	211.5	58.5	1174.5
Mean value	216.6	479.1	237.7	67.0	1000.5
Slope γ_i	6.80	5.29	-0.18	0.53	12.44
η_i /%	54.6	42.5	-1.4	4.3	/

As shown in figure 1, the scatter diagram presents the trends in rainfall series. The annual rainfall series changed dramatically from 2005 to 2016; the year 2008 received maximum rainfall of 1421.2 mm, while lowest rainfall of 664.0 mm was recorded in 2011. The changes in summer rainfall series was a consistent trend that complied with that witnessed in annual rainfall series. There was a drastic change in rainfall series in autumn; however, the change in rainfall series was almost steady in spring and winter.

5. Analysis of rainfall times

As shown in figure 2, the monthly rainfall frequency is similar from January to December. The month of June had the maximum rainfall frequency of 18.6 times, while the minimum rainfall frequency of 12.3 times was recorded in February. As shown in figure 3, the annual rainfall frequency is 178.8 times. The rainfall frequency in spring, summer, autumn, and winter seasons are 46.1 times, 46.1 times, 45.7 times, and 41.3 times, respectively. However, rainfall times are similar in all the four seasons. Therefore, rainfall frequency is not the main factor that influences the distribution of rainfall.

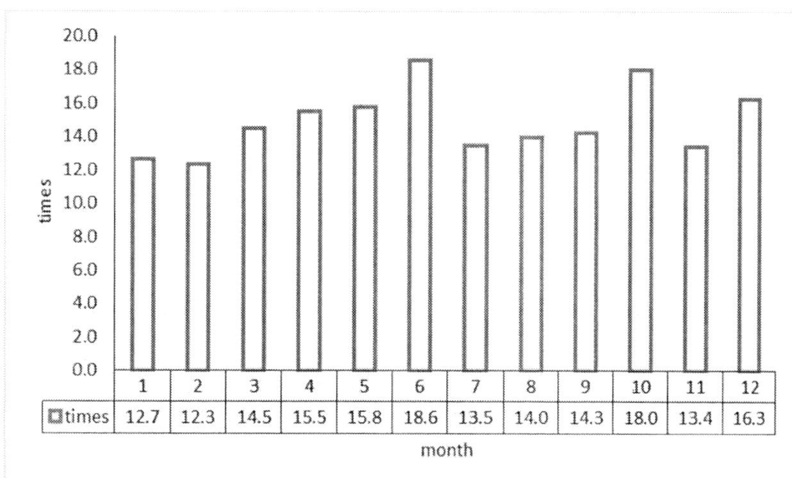

Figure 2. Times of monthly rainfall per year.

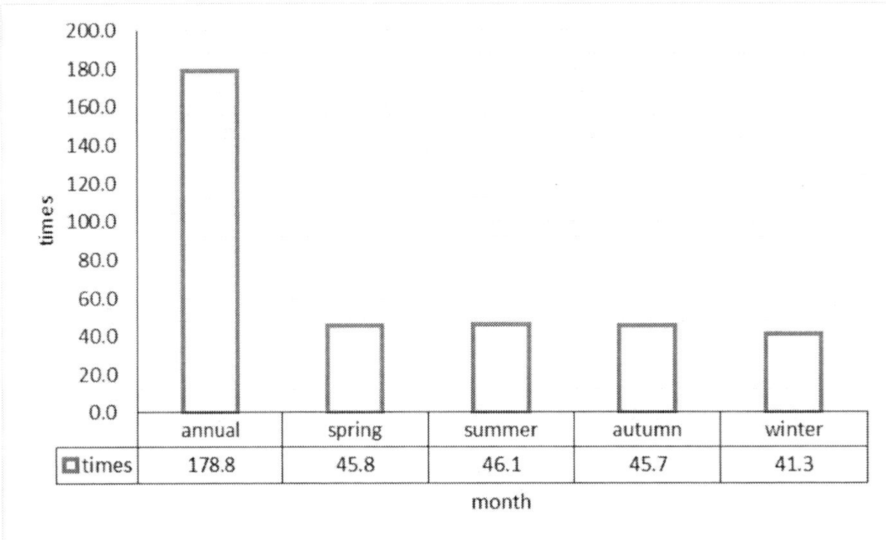

Figure 3. The times of seasons and annual rainfall.

6. Results and discussion

The trend analysis approach is a simple and useful way to investigate the rainfall series in a multi-timescale parameter. Therefore, we analyzed the trend in rainfall distribution in this paper.

The annual rainfall has an increasing trend, with the slope of its trend line being 12.44. The series of spring, summer, and winter witness an increasing trend; the slopes of trend lines for the series of spring, summer, and winter are 6.80, 5.29, and 0.53, respectively. The main contribution of rainfall mainly occurred from spring and summer, whereas the contribution of rainfall is negative in autumn.

Based on the changes witnessed in the rainfall series recorded from 2005–2016 , we infer that the change in rainfall series is a steady state in spring and winter, with the slopes of rainfall series in spring and winter being 6.8 and 0.53. The main factor responsible for the growth in annual rainfall is the growth of rainfall in spring. Meanwhile, the mean value of summer rainfall is maximum: the mean value of summer rainfall is 485.8mm, whereas the slope of summer rainfall series is 5.29. The rate of contribution of summer rainfall is 42.5%. All these results indicate that the main factor responsible for the growth in annual rainfall is also summer rainfall.

Because the rainfall frequency is similar in most months, the four seasons also do not show any drastic change in rainfall frequency. Hence, "rainfall times" is not the main factor influencing the annual rainfall; however, the single rainfall precipitation is one of the main factors influencing annual rainfall growth.

7. Conclusions

The meteorological condition is very unique in karst mountainous area; it adversely affects the development and utilization of water resources. It also sharply influences the regional climatic conditions.

Rainfall is one of the main factors governing the meteorological condition in karst area, so it is important to know the development of different climatic conditions and their impact in karst area.

Due to the availability of limited rainfall data and a significant rise in water conservancy in Guizhou region, both the climatic conditions and human activities equally influence the region. This article is a preliminary study conducted by performing multiple time-scale analysis; its objective was to analyze recent climatic conditions, so only the recent daily rainfall data was selected for research.

Weixin station is a representative karst area in Guizhou Province, China. The results explain the relationship between rainfall and times at a single station. The results of this study can serve as reference standards for future research studies intending to investigate the changes in climatic conditions in karst mountainous area.

Acknowledgments
This study was financially supported by Guizhou Science and Technology Project (Planning of Technology in Guizhou [2016] 2903 and [2015] 3006); also supported by the Basic Research Fund of Central Public-interest Scientific Institution (NO.20150914091245) and by the National Natural Sciences Foundation of China (NO. 51409014).

References

[1] Chu W H, Yan G Q and Su W C 2008 *J. of Hydraul. Eng.* **39** 753-7
[2] Wang L C and Shi Y L 2006 *Sci. Geogr. Sin.* **26** 2173-8
[3] Zhang X Q, Hu M C and Jiang R 2014 *Water Resour. Hydropower Eng.* **45** 40-3
[4] Peng J T, Zhang Q and Liu C L 2012 *Pearl River* **33** 13-7
[5] Tong H F and Ye A Z 2012 *Pearl River* **33** 24-8
[6] Tang S S, Liao W H and Song W Z 2016 *Pearl River* **37** 22-8

Storm surge characteristics and extreme parameters in the Chengshantou sea area

Y D Jin[1,2], X Zhao[1] and L N Ma[1]

[1]First Institute of Oceanography, SOA, Qingdao 266061, China

E-mail: jyd@fio.org.cn

Abstract. Storm surge in the Chengshantou sea area is simulated by the advanced circulation (ADCIRC) model. The Regional Atmospheric Modeling System (RAMS) wind model is used to simulate wind parameters used to be wind force of storm surge model. The comparison results of storm surges show good agreement between simulations and observations. The storm surge is caused by typhoon in summer but extratropical cyclone in winter. By statistical analysis of storm surge in 20 years, it is found that the annual max negative storm surge are all caused by extratropical cyclone while the frequency of annual max positive storm surge caused by tropical cyclone is about 20%. The extreme parameters are estimated by Gumbel distribution. The positive extreme storm surge of 100 year return period is determined as 95 cm and the negative is -124 cm.

1. Introduction

Recent advances have provided new tools for modeling storm surges in estuarine and coastal sea areas. The advanced circulation (ADCIRC) model incorporates a good combination of these three elements, making storm surge simulation one of its main applications [1]. Androulidakis *et al* [2] explored the trends of storm surge extremes in the Mediterranean Sea for a period of 150 years (1951–2100) by using a high-resolution storm surge model. Tomkratoke *et al* [3] used a numerical simulation to determine the effective resonance period, quality factor Q and linear friction coefficient and mechanism of tide and storm surge in the Gulf of Thailand. Lawler *et al* [4] investigated the impact of spatial scaling, mesh resolution, storm characteristics and bottom friction on storm surge in wetland areas in the barrier island system of the Delmarva Peninsula using the coupled hydrodynamic-wave model (ADCIRC+SWAN). Soomere & Pindsoo [5] addressed the possibilities of a separation of the overall increasing trend in maximum water levels of semi-enclosed water bodies into associated trends in the heights of local storm surges and basin-scale components of the water level based on recorded and modeled local water level time series. Mo *et al* [6] developed a three-dimensional numerical model to study storm surges induced by cold waves. Yin *et al* [7] investigated the effects of potential sea level rise as well as typhoon intensification (TI) on storm surges through the use of coupled ADCIRC+SWAN models. Lewis *et al* [8] presented the first assessment of the potential impact of storm surges on tidal-range power. Krien *et al* [9] made use of a state-of-the art numerical modeling system with improved bathymetric and topographic data to identify the strengths, weaknesses, and to suggest areas for improvement of current storm surge models in northern Bay of Bengal.

Chengshan is located in the coastal area of the east of Shandong Peninsula (figure 1), facing the Yellow Sea. It is the storm more serious areas in the north of China. The purpose of this paper is to study the characteristics of storm surge in Chengshantou Sea by means of numerical simulation, and to provide scientific support for disaster prevention and mitigation in this area. The paper is divided into 4 parts. Numerical simulation is applied in section 2. Section 3 is the results and discussion. Finally, the conclusion is given in section 4.

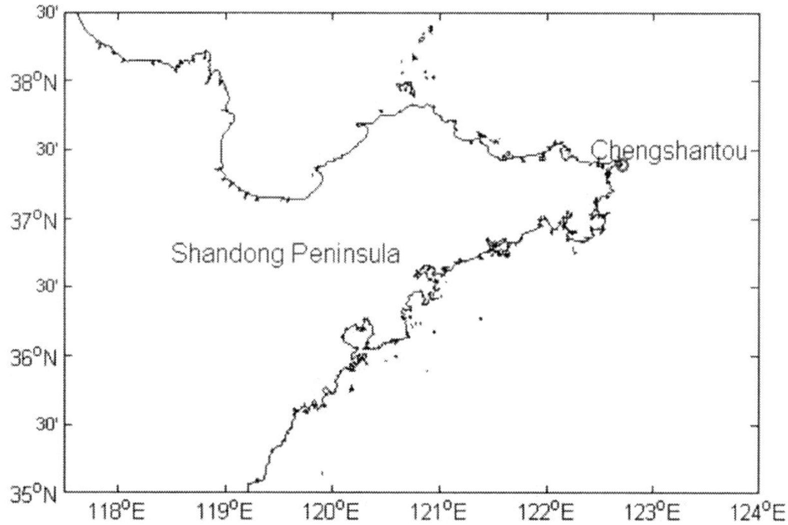

Figure 1. Location of Chengshantou.

2. Numerical simulation

2.1. Model description

The advanced circulation (ADCIRC) model is a highly-developed computer program for oceanic, coastal, and estuarine waters. All nonlinear terms have been retained in these equations and ADCIRC can be run with either a spherical coordinate or Cartesian system, both of which make ADCIRC preferred to some other storm surge models. The two-dimensional governing equations for storm surge in Cartesian coordinates are the Primitive continuity equation:

$$\frac{\partial \zeta}{\partial t}+\frac{\partial UH}{\partial x}+\frac{\partial VH}{\partial y}=0 \tag{1}$$

and the Primitive momentum equations:

$$\frac{\partial U}{\partial t}+U\frac{\partial U}{\partial x}+V\frac{\partial U}{\partial y}-fV=-\frac{\partial}{\partial x}\left[\frac{p_s}{\rho_0}+g\,\zeta-g\,(\eta+\gamma)\right]+\frac{\tau_{sx}}{\rho_0 H}-\frac{\tau_{bx}}{\rho_0 H}+D_x-B_x \tag{2}$$

$$\frac{\partial V}{\partial t}+U\frac{\partial V}{\partial x}+V\frac{\partial V}{\partial y}+fU=-\frac{\partial}{\partial y}\left[\frac{p_s}{\rho_0}+g\zeta-g(\eta+\gamma)\right]+\frac{\tau_{sy}}{\rho_0 H}-\frac{\tau_{by}}{\rho_0 H}+D_y-B_y \tag{3}$$

where, x and y are the Cartesian coordinates, with east and north having positive values; u and v are the depth-averaged components of velocity in the x and y direction, respectively; t is the time; f is the Coriolis parameter; ρ_0 is water density; H is the total water depth; p_s is the atmospheric pressure on the sea surface; g is acceleration due to gravity; ζ is the elevation of the sea surface above the undisturbed water level; τ_{sx} and τ_{sy} represent the components of wind stress on the sea surface; τ_{bx} and τ_{by} represent the components of sea floor stress; and $(\eta+\gamma)$ represents the Newtonian tidal potential, earth tide, self-attraction, and load tide.

$$B_x = \frac{g}{H}\int_{-h}^{\zeta}\left[\frac{\partial}{\partial x}\int_{z}^{\zeta}\left(\frac{\rho-\rho_0}{\rho_0}\right)dz\right]dz \quad B_y = \frac{g}{H}\int_{-h}^{\zeta}\left[\frac{\partial}{\partial y}\int_{z}^{\zeta}\left(\frac{\rho-\rho_0}{\rho_0}\right)dz\right]dz \tag{4}$$

$$D_x = \frac{E_h}{H}\left[\frac{\partial^2 UH}{\partial x^2} + \frac{\partial^2 UH}{\partial y^2}\right] D_y = \frac{E_h}{H}\left[\frac{\partial^2 VH}{\partial x^2} + \frac{\partial^2 VH}{\partial y^2}\right] \tag{5}$$

$$\tau_{bx} = U\tau_*; \tau_{by} = V\tau_*; \tau_* = \frac{C_f(U^2+V^2)^{1/2}}{H} \tag{6}$$

$$\frac{\partial^2\zeta}{\partial t^2} + \tau_0\frac{\partial\zeta}{\partial t} + \frac{\partial A_x}{\partial x} + \frac{\partial A_y}{\partial y} - UH\frac{\partial\tau_0}{\partial x} - VH\frac{\partial\tau_0}{\partial y} = 0 \tag{7}$$

where:

$$A_x = U\frac{\partial H}{\partial t} + H\{-U\frac{\partial U}{\partial x} - V\frac{\partial U}{\partial y} + fV - \frac{\partial}{\partial x}\left[\frac{p_s}{\rho_0} + g\zeta - g(\eta+\gamma)\right] + \frac{\tau_{sx}}{\rho_0 H} - \frac{\tau_{bx}}{\rho_0 H} + \tau_0 U + D_x - B_x\} \tag{8}$$

$$A_y = V\frac{\partial H}{\partial t} + H\{-U\frac{\partial V}{\partial x} - V\frac{\partial V}{\partial y} - fU - \frac{\partial}{\partial y}\left[\frac{p_s}{\rho_0} + g\zeta - g(\eta+\gamma)\right] + \frac{\tau_{sy}}{\rho_0 H} - \frac{\tau_{by}}{\rho_0 H} + \tau_0 V + D_y - B_y\} \tag{9}$$

2.2. Wind force and bathymetry data

The wind parameters used to simulate storm surge are simulated by the Regional Atmospheric Modeling System (RAMS [10]). The equations of RAMS are as follows:

$$\frac{\partial u}{\partial t} = -u\frac{\partial u}{\partial x} - v\frac{\partial u}{\partial y} - w\frac{\partial u}{\partial z} - \theta\frac{\partial \pi'}{\partial x} + fv + \frac{\partial}{\partial x}(k_m\frac{\partial u}{\partial x}) + \frac{\partial}{\partial y}(k_m\frac{\partial u}{\partial y}) + \frac{\partial}{\partial z}(k_m\frac{\partial u}{\partial z}) \tag{10}$$

$$\frac{\partial v}{\partial t} = -u\frac{\partial v}{\partial x} - v\frac{\partial v}{\partial y} - w\frac{\partial v}{\partial z} - \theta\frac{\partial \pi'}{\partial y} - fu + \frac{\partial}{\partial x}(k_m\frac{\partial v}{\partial x}) + \frac{\partial}{\partial y}(k_m\frac{\partial v}{\partial y}) + \frac{\partial}{\partial z}(k_m\frac{\partial v}{\partial z}) \tag{11}$$

$$\frac{\partial w}{\partial t} = -u\frac{\partial w}{\partial x} - v\frac{\partial w}{\partial y} - w\frac{\partial w}{\partial z} - \theta\frac{\partial \pi'}{\partial z} - \frac{g\theta_v'}{\theta_0} + \frac{\partial}{\partial x}(k_m\frac{\partial w}{\partial x}) + \frac{\partial}{\partial y}(k_m\frac{\partial w}{\partial y}) + \frac{\partial}{\partial z}(k_m\frac{\partial w}{\partial z}) \tag{12}$$

$$\frac{\partial \theta_{il}}{\partial t} = -u\frac{\partial \theta_{il}}{\partial x} - v\frac{\partial \theta_{il}}{\partial y} - w\frac{\partial \theta_{il}}{\partial z} + \frac{\partial}{\partial x}(k_h\frac{\partial \theta_{il}}{\partial x}) + \frac{\partial}{\partial y}(k_h\frac{\partial \theta_{il}}{\partial y}) + \frac{\partial}{\partial z}(k_h\frac{\partial \theta_{il}}{\partial z}) + (\frac{\partial \theta_{il}}{\partial t})_{rad} \tag{13}$$

$$\frac{\partial r_n}{\partial t} = -u\frac{\partial r_n}{\partial x} - v\frac{\partial r_n}{\partial y} - w\frac{\partial r_n}{\partial z} + \frac{\partial}{\partial x}(k_h\frac{\partial r_n}{\partial x}) + \frac{\partial}{\partial y}(k_h\frac{\partial r_n}{\partial y}) + \frac{\partial}{\partial z}(k_h\frac{\partial r_n}{\partial z}) \tag{14}$$

$$\frac{\partial \pi'}{\partial t} = -\frac{R\pi_0}{C_V\rho_0\theta_0}(\frac{\partial \rho_0\theta_0 u}{\partial x} + \frac{\partial \rho_0\theta_0 v}{\partial y} + \frac{\partial \rho_0\theta_0 w}{\partial z}) \tag{15}$$

where, u and v represent the zonal wind vector and meridional wind vector respectively; w represents the vertical wind vector; k_m and k_h represent eddy viscosity coefficient; f represents Coriolis parameter; θ_v represents virtual potential temperature; θ_{il} represents ice-liquid water potential temperature; r_v is water vapor mixing ratio; g represents the gravity coefficient; ρ represents density; R is the gas constant; C_v is specific heat capacity and π represents total Exner function.

The RAMS model is run every hour in this study. The time range is from 1987 to 2006. The horizontal resolution of RAMS is set to 0.1° by 0.1°. There are 16 layers in the vertical directions. The Klemp/Wilhelmson scheme is used in the lateral boundary. The initial fields are NCEP reanalysis wind data which the horizontal resolution of that is 0.25° by 0.25° and the time resolution is 6 hours.

The topography datasets used in this paper are shown in figure 2. The topography datasets are derived from admiralty chart in the Bohai Sea and ETOPO5 in the Yellow Sea [11].

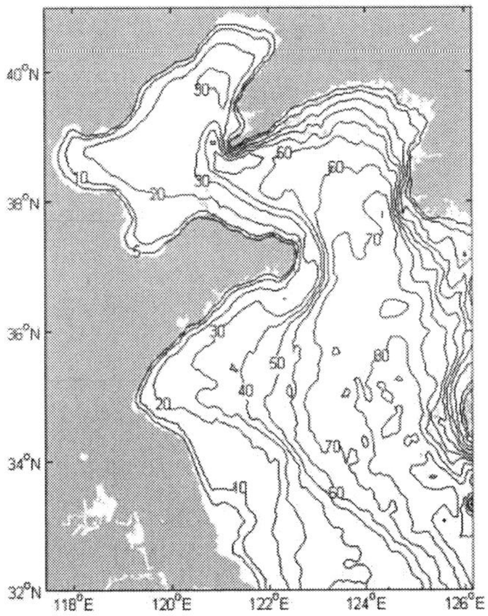

Figure 2. Depth of computational domain.

2.3. Model set-up and validation
The ADCIRC-2DDI model adopts the finite element (FE) method using triangular mesh, to improve description of the complex coastal and island boundaries. The computational domain covers approximately 8.75° longitude and 9° latitude which is from 117.4167° E to 126.1667° E and from 32° N to 41° N.

The initial condition state is set to zero including currents and surface elevation, $\zeta = U = V = 0$. The lateral boundary conditions are driven by four main astronomical tidal constituents (K_1, O_1, S_2, and M_2) along the open boundary, and are assumed to be zero when there is normal flow to the solid boundary. The sea floor stress has the parameters of a quadratic law in terms of the horizontal velocity vector at the bottom sea layer. In addition, the dry and wet method is adopted to satisfy the moving boundary; allowing veritable simulation of the process of storm surge across floodplains.

The storm surge during typhoon and extratropical cyclone are both validated. Two typhoons 8509 and 9216 are selected for data validation and the progress of extratropical cyclone is in April 1987 and February 1994 respectively. The validation results are shown in figures 3 and 4. The dotted line is the observation data in Chengshantou station and the solid line is the simulation data by ADCIRC. It could be seen that ADCIRC can simulate the storm surge well in the Chengshantou sea area.

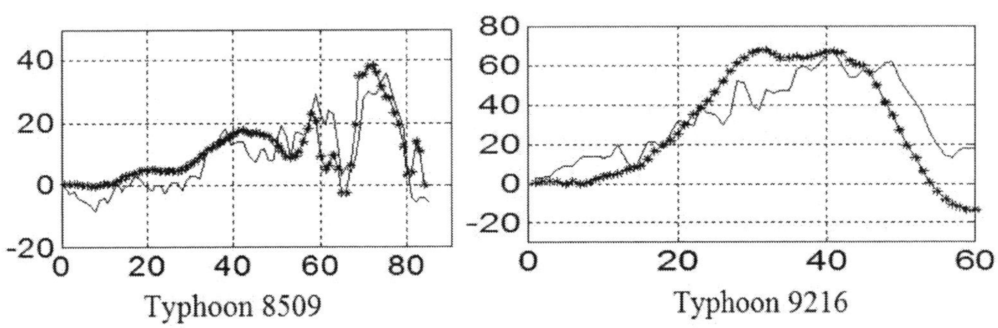

Figure 3. Storm surge validation duirng typhoon 8509 and 9216 (The dotted line is the observation data and the solid line is the simulation data).

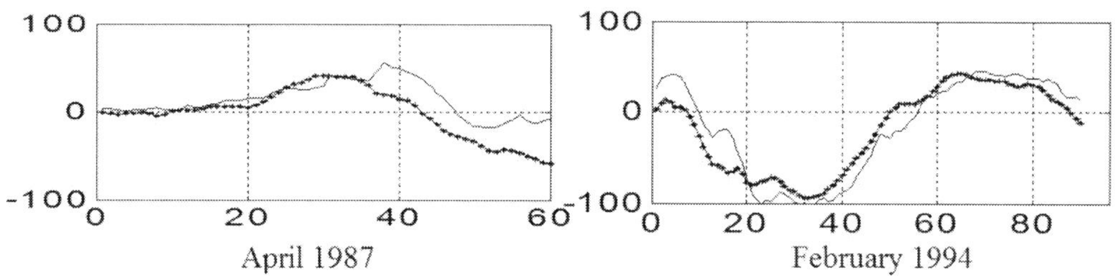

Figure 4. Storm surge validation during extratropical cyclone in 1987 and 1994 (The dotted line is the observation data and the solid line is the simulation data).

3. Results and discussion

3.1. Statistical analysis

The results show that the storm surge is caused by typhoon in summer but extratropical cyclone in winter by statistical analysis of storm surge in 20 years between 1987 and 2006. The processes of typhoon occur in late July, August and September each year while the extratropical cyclone usually happens in January, February and December. Frontal system is the most important weather system in eastern Shandong. Storm surge caused by frontal system generally appears in the spring, autumn and winter. Among those, the max value of negative storm surge caused by frontal system usually happens from January to April.

The annual extreme values of positive and negative storm surge between 1987 and 2006 are shown in figure 5. The red circle indicates the storm caused by the tropical cyclone. There are four years which the extreme values of positive storm surge are caused by tropical cyclone. The frequency is about 20%. On the other hand, the annual extreme values of negative storm surge are all caused by extratropical cyclone.

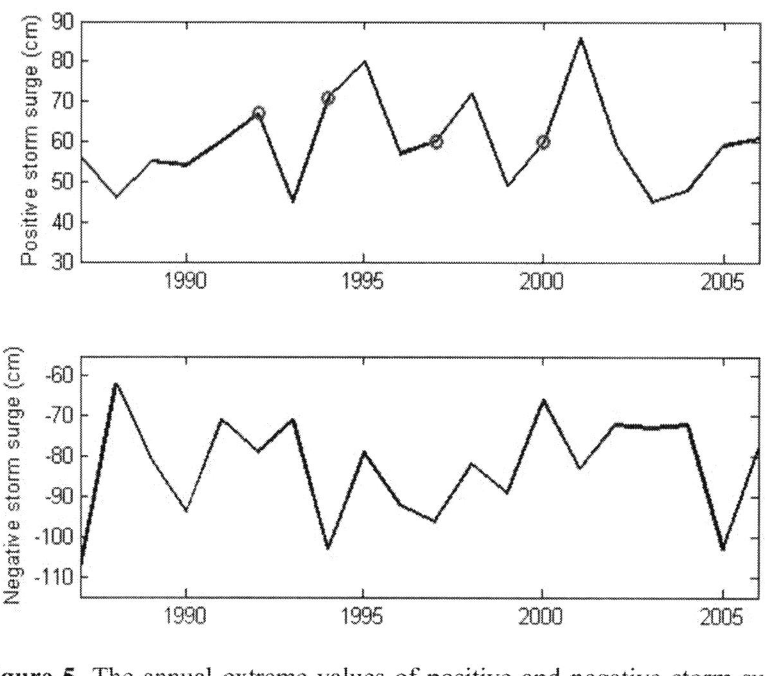

Figure 5. The annual extreme values of positive and negative storm surge between 1987 and 2006.

3.2. Extreme parameter

The extreme parameters of storm surge are estimated by three methods which are Weibull, Gumbel and P-III distribution. The biggest advantage of P-III distribution is its high elasticity. The theoretical curve and empirical frequency points could be fit well by curve fitting or adjust the deviation coefficient and mean value. The disadvantage is the optional arbitrarily increase when the data sequence is short especially there are several same values in the statistical sample. The probability density function of P-III distribution is as follows,

$$f(x)=\frac{\beta^{\alpha}}{\Gamma(\alpha)}(x-a_0)^{\alpha-1}e^{-\beta(x-a_0)},\ x\geq a_0,\quad \alpha>0 \tag{16}$$

Where a_0 is the location parameter; α is the shape parameter and β is the scale parameter.

The Weibull distribution is derived from the combination of theory and empirical data. There are three (or two) dependent parameters of the sample in its distribution. The application of Weibull distribution is more flexible. The slope and intercept of those parameters can be directly from the probability map, or could be got by least square method. The location parameter is usually obtained by the successive approximation method. The probability density function of P-III distribution is as follows,

$$f(x)=\frac{\alpha}{\beta}(x-a_0)^{\alpha-1}\exp\left[-\frac{(x-a_0)^{\alpha}}{\beta}\right]\quad x\geq a_0 \tag{17}$$

The Gumbel distribution is the first type of distribution derived from the extreme distribution theory. The greatest advantage of this distribution is that the theory is based on fullness, no arbitrariness, and the degree of fit near the median is better. The disadvantage is it is difficult to treat the problem of singular extreme value. When the max value of sample Xmax is less than the specific value x, the distribution function of Gumbel distribution is as follows,

$$F(x)=P\{X_{\max}<x\}=\exp\left[-\exp(-\frac{x-\mu}{\sigma})\right] \tag{18}$$

The extreme parameters of storm surge are estimated by the above statistical methods using the annual extreme positive and negative data. The results of 100 year return period are shown in table 1.

Table 1. The positive and negative storm surge of 100 year return period.

	Weibull	Gumbel	PIII
Positive storm surge	86	95	94
Negative storm surge	-113	-124	-121

It could be found that the positive storm surge estimated by Gumbel distribution is the largest in these three methods with the value 95 cm. The nest is that estimated by P-III distribution with the value 1 cm less than Gumbel distribution. On the other hand, the negative storm surge estimated by Gumbel distribution is the largest in these three methods with the value -124 cm. The nest is that estimated by P-III distribution with the value 3cm less than Gumbel distribution. The results of the Weibull distribution are generally smaller than those of the other two methods. From a conservative and security perspective, the positive extreme storm surge of 100 year return period is determined as 95 cm and the negative is -124 cm. All the values are estimated by Gumbel distribution.

4. Conclusion

This study presents a study on storm surge characteristics and extreme parameters in the Chengshantou sea area. The storm surges are simulated by ADCIRC from 1987 to 2006 with model input of wind force from RAMS. Comparisons of storm surge between simulations and observations show good agreement.

Results show that there are four years which the annual max positive storm surges are caused by tropical cyclone. The frequency is about 20%. The annual max negative storm surges are all caused by extratropical cyclone. The positive extreme storm surge of 100 year return period is determined as 95cm and the negative is -124 cm. All the values are estimated by Gumbel distribution.

In brief, this paper increases the understanding of storm surge in the Chengshantou. It is useful and helpful for marine engineering and disaster prevention and mitigation.

References

[1] Wu Y N, Wang Z F, Dong S, Li J J 2014 Numerical simulation and statistic analysis of typhoon storm surge in the sea area around Shandong Peninsula *Proc. of the Eleventh (2014) Pacific/Asia Offshore Mechanics Symposium* (Shanghai, China) 10 12-6

[2] Androulidakis Y S, Kombiadou K D, Makris C V, Baltikas V N and Krestenitis Y N 2015 Storm surges in the mediterranean sea: Variability and trends under future climatic conditions *Dynam. Atmos. Oceans* **71** 56-82

[3] Tomkratoke S, Sirisup S, Udomchoke V and Kanasut J 2015 Influence of resonance on tide and storm surge in the gulf of thailand *Cont. Shelf Res.* **109** 112-26

[4] Lawler S, Haddad J and Ferreira C M 2016 Sensitivity considerations and the impact of spatial scaling for storm surge modeling in wetlands of the mid-atlantic region *Ocean Coastal Manage.* **134** 226-38

[5] Soomere T and Pindsoo K 2016 Spatial variability in the trends in extreme storm surges and weekly-scale high water levels in the eastern baltic sea *Cont. Shelf Res.* **115** 53-64

[6] Mo D, Hou Y, Li J and Liu Y 2016 Study on the storm surges induced by cold waves in the northern east china sea *J. Marine Syst.* **160** 26-39

[7] Yin K, Xu S D, Huang W R and Xie Y 2017 Effects of sea level rise and typhoon intensity on storm surge and waves in Pearl River Estuary *Ocean Eng.* **136** 80-93

[8] Lewis M J, Angeloudis A, Robins P E, Evans P S and Neill S P 2017 Influence of storm surge on tidal range energy *Energy* **122** 25-36

[9] Krien Y, Testut L, Islam A K M S, Bertin X, Durand F, Mayet C and Ballu V 2017 Towards improved storm surge models in the northern Bay of Bengal *Cont. Shelf Res.* **135** 58-73

[10] Wang Z F, Dong S, Li X and Guedes S C 2016 Assessments of wave energy in the Bohai Sea China *Renew. Energ.* **90** 145-56

[11] Wang Z F, Dong S, Chen C C and Guedes S C 2015 Long-term characteristics and extreme parameters of currents and sea levels in the Bohai Sea based on 20-year numerical hindcast data *Nat. Hazards* **76** 1603-24

Bioaccumulation of polybrominated diphenyl ethers (PBDEs) in sediment aged for 2 years to carps (*Cyprinus carpio*)

S Y Tian[1,2,3], J Y Li[1] and X M Jia[1]

[1]College of Marine and Environmental Sciences, Tianjin University of Science and Technology, Tianjin 300457, China
[2]Tianjin Key Laboratory of Marine Resources and Chemistry, Tianjin 300457, China

E-mail: tiansy@tust.edu.cn

Abstract. In order to understand the risk of polybrominated diphenyl ethers (PBDEs) existing in sediment for years, the accumulation of PBDEs in sediment aged for 2 years to fish was investigated. Simulated aquatic system microcosms were conducted with PBDE contaminated sediment aged for 2 years and carps were cultured in the microcosms for 20 days. PBDE concentrations in carp tissues were analyzed to estimate the bioavailability of aged PBDEs in carps. The main spiked PBDE congeners were detected in sediment even though the contaminated sediment was aged for 2 years. Similarly, the five PBDE (BDE-28, 47, 100, 153 and 154) congeners which probably were bioaccumulated by carp were detected in fish tissues, indicating that PBDEs could be bioaccumulated after aging for 2 years. The PBDEs distribution revealed that the concentrations of polybrominated diphenyl ethers in tissues of *Cyprinus carpio* is in this order of magnitude: gut > liver > gill > fillet. The PBDEs concentrations in fillet were as high as 67.9 ng/g dry wt, in which BDE-47 contributed almost 50% in profile.

1. Introduction

Polybrominated diphenyl ethers (PBDEs) are a kind of brominated flame retardants thay are widely used in human livings and industry. They have been detected in a variety of environments due to their big production and usage since they came out [1-5]. Because of their persistence, toxicity, bioavailability and long distance transportation. Some PBDE congeners were identified as persistent organic pollutants since 2009.

PBDEs are very hydrophobic, thus are prone to attachment on solid particles. Therefore, sediment becomes the important sink of PBDEs and could impact the transport and fate of PBDEs in environment importantly. PBDEs deposited in sediment are potentially accumulated by aquatic organisms. Several studies have reported the accumulation of PBDEs beimg deposited in sediment by benthic organisms [6-8]. The longer PBDEs are deposited in the sediment, the less of the rapid desorbing part for the sediment-associated PBDEs. Similarly, the proportion of the slowly desorbing part was increased gradually [9]. Consequently, the bioaccumulation of PBDEs deposited in sediment decreases with increase in depositing time. However, little study has been conducted to detect the bioavailable degree of PBDEs deposited in sediment for years through bioaccumulation experiments. The evaluation of the available characteritics of PBDEs to organisms after deposited in sediment for years is helpful for assessment of the PBDEs environmental health risk.

In the present study, simulated pond system microcosms were built up with PBDE contaminated sediment aged for 2 years and carps were cultured in the microcosms for 20 days. PBDE concentrations in carp tissues were analyzed to evaluate the bioaccumulation of PBDEs in carps. The results would help us to evaluate the risk of sediment-associated PBDEs already existing in environment and nanoparticles.

2. Materials and methods

2.1. Chemicals
The PBDEs standard solution were bought from AccuStandard company (New Haven, CT, USA). Commercial products were purchased from Great Lakes Chemical Corporation (West Lafayette, IN, USA). All solvents used for the analysis were chromatographic and obtained from Concord Science and Technology Co. (Tianjin, China).

2.2. Sediment preparation
The clean sediment was from an ecological preservation area. The procedures to prepare and contaminate the sediment have been reported in details [8]. The spiked sediment was aged for 2 years in the microcosms before the experiment.

2.3. Experimental setup
Microcosms were set up using 6 clean rectangular glass aquaria. There were experimental group and control test. The control test was conducted with unspiked sediment, free of PBDEs. The experimental group was built up with spiked sediment aging for 2 year. The experimental aquaria was built up by depositing the sediment at the bottom, then, water was added to overlay the sediment to obtain a simulated lake system. The microcosms were equilibrated for 1 month before 20 carps (*Cyprinus carpio*) were added. The exposure lasted for 20 days and all fish were sampled for analysis.

2.4. Analysis of PBDEs
Organism tissues and the sediment in experimental microcosms were lyophilized and weighed. The concent of PBDEs in fish tissues and sediment was analyzed by the method reported by Zhu *et al* [10]. The measurement was carried out using gas chromatograph (Agilent 7890A) mass spectrometer (Agilent 5975C). GCMS ranned in the mode of negative chemical ionization. The column mode was DB-5 capillary column.

2.5. Quality assurance and quality control
The sample treatment and clean up procedure were carried out in a clean room. The procedural blank was done to ensure the accuracy of the treatment. The qualitative and quantitive of PBDEs contents were conducted by the mixed standards containing known amounts of PBDEs congeners. BDE-77 was used as recovery indicator and the recovery of the ananlysis was between 80% and 120%. The reported data were not corrected by recovery.

3. Results and discussions

3.1. PBDE concentrations variation after aging
Three kinds of PBDEs commercial products were used to contaminate the experimental sediment. They were penta-, octa- and deca-BDE. All the main PBDE congeners in the commercial products could be still detected even though the contaminated sediment were aged for 2 years. The detected PBDE congeners contents decreased after aging for 2 years except for BDE-28 and BDE-47 as shown in figure 1. The main reason for the decreasing concentration may be the transformation of PBDEs. The ratio of BDE-47 to BDE-99 was about 0.65 in the primal spiked sediment. But the ratio of BDE-47 to BDE-99 increased to 1.1 when the contaminated sediment aged for 2 years. Even higher ratios of BDE-47 to BDE-99 were reported in field sediments and biota. For example, it was reported

that the ratio of BDE-47 to BDE-99 were 2.45 ± 0.95 in sediment and even higher ratio of 5.08 ± 2.07 in mussel from Bo Sea [11]. There were two reasons for the difference of the ratios between sediment and organism. On the one hand, BDE-99 is more liabel to be debrominated than BDE-47. On the other hand, BDE-47 is easier accumulated by organism than BDE-99. Furthermore, the concentrations of BDE-183 and BDE-206 in aged sediment were only half of those in primal spiked sediment. And the concentrations decreased more for BDE-207, BDE-208 and BDE-209. It has been verified that BDE-209 would transform to hexa-BDE congeners through nona-BDE congeners. It was reported that meta-position and para-position debromination are preferred pathway for BDE-209, which produced BDE-207 and BDE-208 [12]. The increasing concentration for BDE-47 and BDE-28 could be attributed to the transformation of high brominated congeners to low brominated congeners.

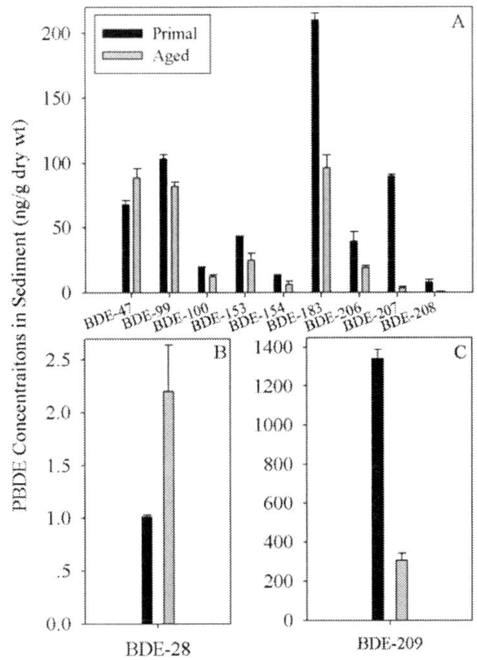

Figure 1. PBDE concentrations in the primal spiked sediment and sediment aged for 2 years. A: the concentrations for BDE-47, BDE-99, BDE-100, BDE-153, BDE-154, BDE-183, BDE-206, BDE-207and BDE-208; B: the concentrations for BDE-28; C: the concentrations for BDE-28.

3.2. PBDEs concentrations in tissues of Cyprinus carpio

After exposure in the microcosms for 20 days, several PBDE congeners were detected in fillet, liver, gill and gut tissues of *Cyprinus carpio*. The variation of PBDEs in the fish tissues are shown in figure 2. Five PBDE (BDE-28, 47, 100, 153 and 154) congeners were detected in the fish tissues. The PBDEs distribution shows that the concentrations in different tissues is in this order of magnitude: gut > liver > gill > fillet. Several studies have found that PBDEs are prone to be accumulated in liver tissues [13, 14]. The PBDEs concentrations in fillet were even up to 67.9 ng/g dry wt, indicating that PBDEs could be bioaccumulated after aging for 2 years.

As showed in figure 3, the first congener in all the tissues was BDE-47, which accounted for 50% to 70% of the total amount of the detected congerners. BDE-154 was the second congener in the five congeners. This result was in agreement with other studies. BDE-47 always accounts for the largest propotion in organism tissues whether in field research or in lab experiment. In the commercial PBDE products of Penta and Octa, BDE-99 and BDE-183 is the important congeners, respectively. And the contents of the both congeners were relatively high in spiked sediment, However, they were not detected in any of the fish tissues. This finding was also aggress with other reports [13]. The reason for the phenomenon that BDE-99 and 183 were not detected in fish tissues was because for the biotransformantion in biota tissues. Roberts *et al* [15] reported that some congeners, such as BDE-153 and BDE-183, are liable to be transformed into some lower brominated congeners by common carp through *in vitro* experiment, while other congeners, such as BDE-47 and BDE-154 were resistant to debrominating metabolism.

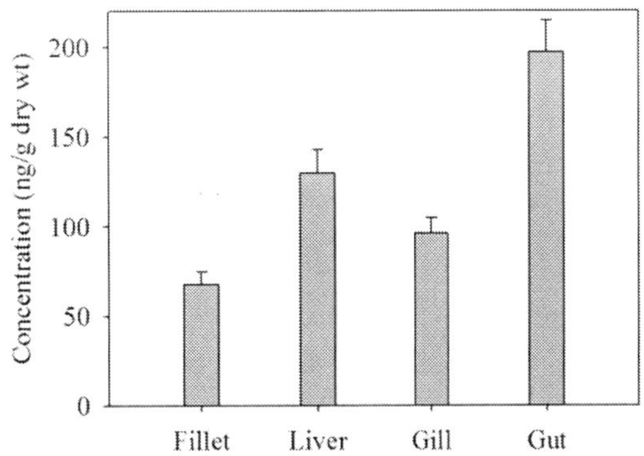

Figure 2. PBDE concentrations (ng/g dry wt) in *Cyprinus carpio* tissues.

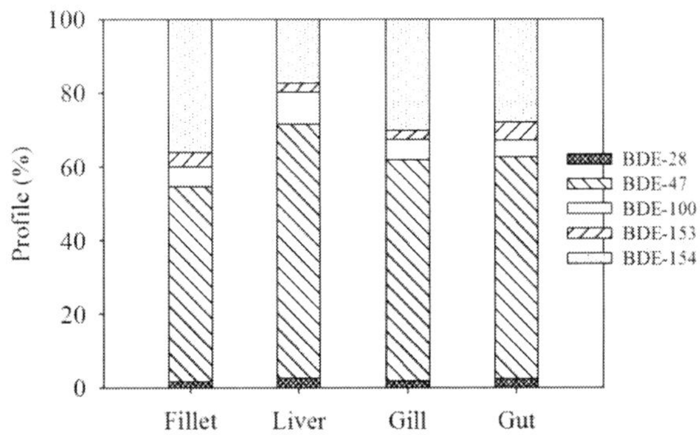

Figure 3. PBDE congener profiles in *Cyprinus carpio* tissues.

4. Conclusions

The sediment-associated PBDEs could be bioaccumulated by fish even aged for 2 years in a pond simulated microcosm.

Acknowledgments

This research was supported by a project grant from the Chinese Natural Science Foundation (Grand No. 41303070) and China Postdoctoral Science Foundation (Grand No. 2013T60256).

References

[1] Kwan C S, Takada H, Mizukawa K, *et al* 2013 Sedimentary PBDEs in urban areas of tropical Asian countries *Mar. Pollut. Bull.* **76** 95-105

[2] Ondarza P M, Gonzalez M, Fillmann G, *et al* 2014 PBDEs, PCBs and organochlorine pesticides distribution in edible fish from Negro River basin, Argentinean Patagonia *Chemosphere* **94** 135-42

[3] Wang H-S, Jiang G-M, Chen Z-J, *et al* 2013 Concentrations and congener profilesof polybrominated diphenyl ethers (PBDEs) in blood plasma from Hong Kong: Implications for sources and exposure route *J. Hazard Mater.* **261** 253-9

[4] Ma J, Qiu X, Zhang J, *et al* 2012 State of polybrominated diphenyl ethers in China: An overview *Chemosphere* **88** 769-78

[5] Tian S, Zhu L and Liu M 2010 Bioaccmulation and distribution of polybrominated diphenyl ethers in marine species from Bohai bay China *Environ. Toxicol. Chem.* **29** 2278-85

[6] Ciparis S and Hale R C 2005 Bioavailability of polybrominated diphenyl ether flame retardants in biosolids and spiked sediment to the aquatic oligochaete, Lumbriculus variegatus *Environ. Toxicol. Chem.* **24** 916-25

[7] Leppänen M T and Kukkonen J V K 2004 Toxicokinetics of sediment-associated polybrominated diphenylethers (flame retardants) in benthic invertebrates (Lumbriculusvariegatus, oligochaeta) *Environ. Toxicol. Chem.* **23** 166-72

[8] Tian S and Zhu L 2011 Bioaccumulation kinetics of sediment-associated DE-83 in benthic invertebrates (Nereis succinea, polychaete) *Chemosphere* **84** 160-5
[9] Liu M, Tian S, Chen P, *et al* 2011 Predicting the bioavailability of sediment-associated polybrominated diphenyl ethers using a 45-d sequential Tenax extraction *Chemosphere* **85** 424-31
[10] Zhu L Ma B and Liang X 2008 Quantitative analysis of polybrominated diphenyl ethers in earthworms and soil by gas chromatography coupled to ion-trap tandemmass spectrometry *Rapid Commun. Mass Spectrom.* **22** 394-400
[11] Wang Z, Ma X D, Lin Z S, *et al* 2009 Congener specific distributions of polybrominated diphenyl ethers (PBDEs) in sediment and mussel (Mytilus edulis) of the Bo Sea, China *Chemosphere* **74** 896-901
[12] Holden A, Park J S, Chu V, *et al* 2009 Unusual hepta- and octa-brominated diphenyl ethers and nonabrominated diphenyl ether profile in California, USA, Peregrine Falcons (Falco Peregrius): More Evidence For Brominated Diphenyl Ether-209 Debromination *Environ. Toxicol. Chem.* **28** 1906-11
[13] Tian S, Zhu L, Bian J, *et al* 2012 Bioaccumulation and metabolism of polybrominated diphenyl ethers in carp (Cyprinus carpio) in a water/sediment microcosm: Important role of particulate matter exposure *Environ. Sci. Technol.* **46** 2951-8
[14] Stapleton H M, Alaee M, Letcher R J, *et al* 2004 Debromination of the flameretardant decabromodiphenyl ether by juvenile carp (*Cyprinus carpio*) following dietary exposure *Environ. Sci. Technol.* **38** 112-9
[15] Roberts S C, Noyes P D, Gallagher E P, *et al* 2011 Species-specific differences and structure-activity relationships in the debromination of PBDE congeners in three fish species *Environ. Sci. Technol.* **45** 1999-2005

IOP Conf. Series: Earth and Environmental Science **82** (2017) 012054 doi:10.1088/1755-1315/82/1/012054

Differences in methods of quantifying the vertical hyporheic flow for streambank flow field

C P Lu[1,2,5], C C Yao[1], B N Shu[3], X D Huang[4] and H Lv[1]

[1]Hohai University, State Key Laboratory of Hydrology-Water Resources and Hydraulic Engineering, Nanjing, 210098, China
[2]China Institute of Water Resources and Hydropower Research, State Key Laboratory of Simulation and Regulation of Water Cycle in River Basin, Beijing, 100038, China
[3]Jining Hydrology Bureau of Shandong Province, Jining, 272000, China
[4]Qingdao Hydrology Bureau of Shandong Province, Qingdao, 266000, China

E-mail: luchengpeng@hhu.edu.cn

Abstract. One-dimensional analytical methods are generally used to estimate the vertical hyporheic velocity, and inevitably yield values that differ from the true values when assumptions are not satisfied. In this study, numerical simulations were used to determine the amount of errors in the vertical hyporheic flow from several 1D analytical methods, including the solutions from amplitude ratio, phase shift of temperature, and hydraulic gradient with depth, under gaining and losing conditions. Results demonstrate that the velocity at the shallow layer is generally higher than the velocity at the deeper layer. Horizontal flow dominates the hyporheic exchange in the streambank-aquifer continuum at distances further away from the central channel. The amplitude ratio method and gradient method estimate the vertical velocity well when the vertical flow dominates the hyporheic flow and the flow is upwelling; otherwise, these two methods yield very inaccurate results. The amplitude ratio method and phase shift method yield better results compared with the gradient method for the streambank area. Our simulations indicate that two or more 1D analytical methods should be used in conjunction, and that the multi-dimensional hydraulic heads should be measured via an appropriate monitoring network to more accurately understand the hyporheic flow of a streambank.

1. Introduction

Heat transport is an important component in a variety of hydrologic and environmental processes. Diurnal and annual variations in hyporheic exchange, which means the water that flows from streams or rivers into shallow subsurface and then returns to the stream [1], have been directly linked to diurnal and annual variations in stream water temperature [2, 3]. Heat tracing is useful to depict the hyporheic flow path and quantify the associated flow [4]. Constantz impressed that advancements of temperature - based methods could produce a new "streambed science" [3].

Analytical solutions of the heat transport equation [5-8] have been built to evaluate surface water-groundwater interactions. However, the analytical methods are limited to environments with vertical 1D flow and sinusoidal temperature fluctuations [9-11]. In fact, the flow field in a natural stream is rather complicated, especially in the downstream reach which generally exhibits meanders. However, the stream channel within the mountainous area is rather straight and simple compared to the downstream reach. That is why 1D analytical model was widely used.

A generally simplification of the groundwater flow governing equations was originated from the assumption of the strictly vertical flow into or out of the streambed [2, 12-15].The heat tracing method for hyporheic exchange was in a large part derived by Stallman [6]. A special solution of a 1D vertical hyporheic exchange was developed and applied [7, 8, 16, 17].

The hyporheic flow path is apt to tend toward a vertical direction near the riverbed interface and along a horizontal direction beneath the riverbed. For such cases, more sophisticated 2D numerical models have been used [18-20], but 2D numerical models generally require more information and therefore, can be difficult to apply. Analytical 1D models solving the heat transport equation have been popular for interpreting temperature data at the scale of a streambed due to their ease of use and cost-effectiveness, and because the heat flow parameters vary much less across a range of sediment textures than, for example, hydraulic conductivity [21].

Notably, the natural hyporheic exchange is more complicated than the assumed 1D flow solutions. These 1D solutions are based on simplified assumptions that always do not conform to the laws of nature. A non-vertical flow field, which is common in natural streams, can result in wrong result. Lautz [10] indicated that the deviation between the estimated and natural flux is mainly caused by non-vertical flow.

Analysis of non-vertical flow through heat tracing has been well documented by Jensen and Engesgaard [4], Cardenas and Wilson [9], Lautz [10], and Roshan et al [22]. Jensen and Engesgaard, Lautz, Shanafield et al and Swanson and Cardenas [4, 10, 19, 23,] documented that 1D temperature - based solutions inevitable make errors when estimating the vertical velocity with the flow dominated by horizontal flow. Cardenas and Wilson [9, 24] investigated the non-vertical flow induced by current-bed forms generated complex flow patterns when heat is used as a tracer. Additionally, Lautz [10] reported the deviations of the 1D vertical flow detailedly. Cuthbert and Mackay [25] inferred that convergent or divergent flow is occurring and that a 1-D analysis is inappropriate. Irvine et al [26] reported the best heat tracing method for a heterogeneous field.

Roshan et al [22] noted that horizontal flow is likely to occur in natural streams for not wide streams. In fact, a degree of horizontal flow always occurs beneath the streambed and especially near the stream banks. Different zones with identical hydraulics conditions have been recognized along the cross-section direction. The field works by Jensen and Engesgaard [4], and Shanafield et al [19] approved that temperature measurements are well dependent with its horizontal location. Moreover, Shanafield et al [19] also stated that 1D analytical solution can only fit the field temperature measurements at one side of the channel.

Shanafield et al [19] emphasized that horizontal movement of water through stream banks, as opposed to the streambed, may play a significant role in total infiltration; however, this idea has received little attention. To our knowledge, the impacts of non-vertical flow subject to a bank-aquifer continuum on thermal-based and gradient-based hyporheic flux estimates have not been studied in any detail. This type of hyporheic process could yield a highly variable flow field in the conditions of losing and gaining, which would result in severe deviations from the 1D field. A riverbank case is used herein to investigate the applicability of Heat-based analytical solutions and the Darcy-based gradient method.

The aim of this study is twofold: to characterize the flow paths within a riverbank under losing and gaining conditions and to assess the ability of analytical heat transport models and the gradient method for quantifying vertical fluid fluxes within a riverbank. For this, virtual experiments with numerical simulations were used. First, the 2D heat-mass transport equation was solved using VS2DH to generate hyporheic flow fields under both gaining and losing conditions and associated temperature distributions for a riverbank system. Secondly, the estimated streambed temperature was used to estimate the vertical velocity via these 1D analytical solutions. In addition, the vertical velocities were also calculated via the Darcy-based gradient method. Finally, comparison between these estimates and the simulated velocities was performed to identify the impact of horizontal flow component from these analytic methods.

2. Methods

2.1. Mass and heat transfer equation

2.1.1. Energy transport equation. The energy transport equation, which is actually a form of the advection-dispersion equation, is derived by balancing the changes in energy stored within a volume of porous media. Such changes occur due to the flow of ambient water with different temperature into the volume, thermal conduction into or out of the volume, and energy dispersion into or out of the volume. The unsaturated zone is an essential part of the hydrologic continuum in a streambank-aquifer system. The advection-diffusion equation is shown as following:

$$\frac{\partial}{\partial t}[\theta C_w + (1-\phi)C_s]T = \nabla K_T(\theta)\nabla T + \nabla \theta C_w D_h \nabla T - \nabla \theta C_w vT + qC_w T^* \quad (1)$$

where t refers to time, T refers to temperature, T^* is temperature of fluid source, ϕ is the sediment porosity, θ is the percentage of volume water content, K_T is the thermal conductivity of the bulk sediments, D_h is the thermal mechanical dispersion, v is the flow velocity, and q is the rate of flow source. C_s and C_w are the thermal capacities of the sediment and water, respectively. Further details on equation (1) can be obtained from Voss [27] and Kipp [28] .

The thermal mechanical dispersion is a tensor defined as [29]

$$D_h = \alpha_t |v| \delta_{i,j} + \frac{(\alpha_1 - \alpha_t) v_i v_j}{|v|} \quad (2)$$

where α_l and α_t are the longitudinal and transverse dispersivities, respectively, $\delta_{i,j}$ is the Kronecker delta function and equals 1 if i=j and otherwise is equal to 0, v_i and v_j are the ith and j^{th} component of the velocity vector, respectively, and $|v|$ is the magnitude of the velocity vector. Rau et al [30] and Roshan [22] indicated that the thermal dispersion induced less change compared to the hydraulic conductivity in the heat transport results using a range of sandy field velocities.

2.1.2. Fluid Flow Equation. The flow velocity within variably saturated material from equation (1) is generally described by Richards' equation. The hyporheic flow considered in this study is a 2D flow, including horizontal (x) and vertical (z) dimensions. The fluid flow is governed by equation (3) as follows:

$$C(h)\frac{\partial h}{\partial t} = \frac{\partial}{\partial x}\left[K(h)\frac{\partial h}{\partial x}\right] + \frac{\partial}{\partial z}\left[K(h)\frac{\partial h}{\partial z}\right] + \frac{\partial K(h)}{\partial z} \quad (3)$$

where $C(h)$ is the specific moisture capacity, which is the slope of the retention curve = $d\theta/dh$, h is the pressure head, t is time, and K is the unsaturated hydraulic conductivity and is calculated by van Genuchten model.

2.2. Numerical model description

2.2.1. Model domain, initial condition (IC), and boundary conditions (BCs). The synthetic temperature series generated by numerical model were used to evaluate the impact of horizontal flow on vertical water velocity estimates. A streambank-aquifer continuum was conceptually modeled, in which hyporheic exchange occurs perpendicular to the channel. Exchange may occur in the 3D domain; however, the scenarios along the stream flow have been discussed by Lautz [10] , which are regarded as constant in this case. Therefore, our numerical model was developed according to the 2D flow field. Simplifying the model extent to a streambank 2D cross-section in a lateral flow direction helps reduce model complexity for defining the boundary conditions and recognizing the hyporheic flow paths.

The solution of equations (1) and (3) requires a boundary condition within the model domain combined with the pressure head and temperature distribution:

$$\begin{cases} T(x,y,t) = T_0(x,y) \\ h(x,y,t) = h_0(x,y) \end{cases} \quad t=0 \tag{4}$$

where the initial pressure head and temperature are set to 0 m and 15℃, respectively. The top of the domain as a stream was set by Dirichlet-type boundary condition for temperature,

$$T(x,0,t) = T_0 + \Delta T \cdot \cos\left[\frac{2\pi t}{\mathrm{Pr}}\right] \tag{5}$$

where T is of given function on the specific boundary, and Pr refers to the period of temperature variation (Pr = 1 day). T_0 and ΔT refer to the mean temperature and the amplitude ratio on the surface ($T_0 = 15°C$ and $\Delta T = 5°C$), respectively. A constant head on the top stream boundary was set at 0 m. A water head with positive (0.3 m, 0.2 m, and 0.1 m) and negative values (-0.1 m, -0.2 m, and 0.1 m) was also applied on the left boundary for losing and gaining condition, respectively. The left side of the domain was head boundary, where different heads generated different average gradients (-0.1, -0.067, -0.033, 0.33, 0.67, and 0.1) as shown in table 1.

Table 1. Numerical scenarios of the flow and heat transfer model.

Code of scenario	Total head of left side boundary	Total head of stream	Average gradient	Type of stream
L3	-0.3	*0*	-0.1	*Losing*
L2	-0.2		-0.067	
L1	-0.1		-0.033	
G1	0.1		0.033	*Gaining*
G2	0.2		0.067	
G3	0.3		0.1	

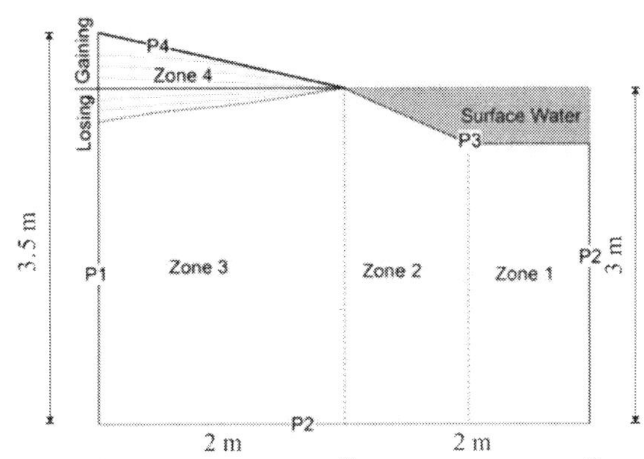

Figure 1. Schematic of the model domain including the boundary conditions using VS2DH.
The losing and gaining condition was achieved by modifying the pressure of the left boundary P1. P1 and P3 are total head boundary conditions, and P2 and P4 are the no-flow boundaries.

IOP Conf. Series: Earth and Environmental Science **82** (2017) 012054 doi:10.1088/1755-1315/82/1/012054

Ronan et al [31] showed that the center vertical line can be regarded as a no-flow boundary for both gaining and losing streams at the steady state. Therefore, the right side and the bottom of the domain were modeled as a no-flow boundary. The settings of the model domain setting are shown in figure 1, which depicts the cross section of streambank-aquifer continuum. According to a general scale of mountainous stream, the depth of the domain was set at 3 m. And the widths of the surface water and model domain shown in figure 1 were set at 2 and 4 m, respectively. Moreover, the 3m depth was determined because there is not apparent changes in heat transfer when the depth surpassed 3m in a bank-aquifer flow system with the dimensions in this case.

For the flow system, P1 (different pressure at each simulation) and P3 (representative of surface water) are the total head boundary conditions, and P2 and P4 are the no-flow boundaries. A simple total head boundary of surface water was used for P3. In terms of energy transport, P2 is the no-energy flux boundary, and P1, P3, and P4 are specific temperature boundary conditions according to the temperatures of groundwater, surface water, and the ground, respectively.

Four zones are marked in figure 1. Zone 1 is the area beneath the central streambed, and zone 2 is the area beneath the side slope of the streambed. Zones 3 and 4 are beneath the river bank, and represent the saturated zone and vadose zone, respectively. The unsaturated zone was also included in the simulations because changes in the ground temperature could have an impact on the ground water temperature in a relatively shallow system. However, the unsaturated flow and energy transport are notably different in the saturated zone and this is inappropriate for heat tracing methods [7,8] .Therefore, the calculated temperature series data collected from zone 4 was not used to analyze the discrimination in 1D analytical solution. The thermal and hydraulic properties of water and sediment adopted in the numerical simulations are listed in table 2.

Table 2. Physicalproperties of water and sediments used in numerical model.

Parameter	Unit	Symbol	Value
Solid Thermal Conductivity	$W(mC)^{-1}$	k_s	2.4
Water Thermal Conductivity	$W(mC)^{-1}$	k_f	0.6
Solid Specific Heat Capacity	$J(kgC)^{-1}$	c_s	865
Water Specific Heat Capacity	$J(kgC)^{-1}$	c_f	4201
Solid Density	kg m^{-3}	ρ_s	999.7
Water Density	kg m^{-3}	ρ_f	2650
Porosity	-	n	0.375
Longitudinal Thermal Dispersivity	m	α_l	0.1
Transverse Thermal Dispersivity	m	α_t	0.1
Saturated Hydraulic Conductivity	m/s	K	4.05×10^{-5}

2.2.2. Change in Peclet number and seepage velocity. The Peclet number (Pe) is a dimensionless number relevant in the study of transport phenomena in fluid flows. It is defined as the ratio of the rate of advection of a physical quantity by a flow to the rate of dispersion of the same quantity driven by an appropriate gradient. The hydraulic gradient changed in every simulation scenario, resulting in different seepage velocities and the associated Peclet number.

In the context of the transport of heat, the Peclet number is equivalent to the product of the Reynolds number (Re) and the Prandtl number (Pr).

For diffusion of heat (thermal diffusion), the Peclet number is defined as:

$$Pe = \mathrm{Re} \cdot \mathrm{Pr} = \frac{LU}{\alpha} \tag{6}$$

where L[m] is the characteristic length (considered as constant grid cell length), U[m/s] is the seepage velocity [m/s], and α[m^2/s] is the thermal diffusivity,

$$\alpha = \frac{k}{\rho c_P} \tag{7}$$

where k [W/(m·K)] is the thermal conductivity, ρ [kg/m3] is the density, and c_P [J/(kg·K)] is the heat capacity.

2.2.3. Numerical code and simulation time. The USGS variably saturated 2D hydraulic model, VS2DH [29, 32, 33], is a 2D finite difference model that simulates heat and water flow in the hyporheic zone beneath a river. VS2DH has been successfully used to describe heat transport in variably saturated material at several sites near streams [31, 34]. The program provides numerical solutions of variably saturated porous media flow and heat transport problems, and is applicable to the study of unsaturated-zone riparian hyporheic exchange problems. Heat transport is simulated using the advection-diffusion equation and the flow of water is described by the Richards' equation, shown in equations (1) and (3).

VS2DH was used herein to generate synthetic time series temperature data for the streambed. The number of grids is 140×160, with the size of each grid equals to 25 mm × 25 mm. For each hydro-scenario, a three days period of the numerical model was simulated when a steady state flow field was attained. Each scenario was simulated for a period of three days. During the first two days (48 h) of each simulation, the flow field was influenced by the initial conditions, and therefore, the results of these two days were excluded from the analysis and only information from the last day was analyzed, as recommended by Lautz [10].

2.3. 1D analytical modeling of streambed heat transport and vertical gradient

Vertical velocities of hyporheic flow were estimated from the vertical pairs of modeled time series temperature based on analytical solutions to the 1D heat transfer equation and several assumptions [6-8]. This heat tracing method is advantageous because hyporheic flow can be obtained over time periods without calibrating two- or three-dimensional models. More details on the analytical solution can be found in Hatch et al. and Keery et al [7, 8].

The MATLAB-based VFLUX program solves analytical approximations [7, 8] of the 1D advection-diffusion equation to calculate vertical fluid flux in saturated porous media. First, VFLUX re-samples the measured data to achieve optimal sampling frequency (12-24 samples/d) for power-spectral analysis, which is used to identify the dominant diurnal component of the signal. This spectral analysis and subsequent extraction of the diurnal signal from each time series is performed using dynamic harmonic regression (DHR), as implemented by the Captain Toolbox [35, 36]. VFLUX then reports the amplitude ratio and phase shift of each temperature time series at the reduced sampling frequency (here, 2-hr time steps), and calculates the fluid flux based on the Ar and Δϕ between the upper and lower diurnal signals. The same measured thermal parameters used in VS2DH were employed for the analytical models.

The analytical solutions to the 1D advection-diffusion equation, as presented by Hatch et al [7], predict Ar and Δϕfor any given rate of vertical flux (q) based on vt:

$$q = \frac{c}{c_w}\left(\frac{2k_e}{\Delta z}\ln A_r - \sqrt{\frac{\alpha + v_t^2}{2}}\right) \tag{8}$$

$$|q| = \frac{c}{c_w}\sqrt{\alpha - 2\left(\frac{4\pi\Delta\phi k_e}{P\Delta z}\right)^2} \tag{9}$$

where C_w and C are the heat capacity of the water and sediment-water matrix, respectively; note that only absolute values of vertical fluid flux can be determined with equation (9) due to similar $\Delta\phi$ reduction with both upwelling and downwelling. An upward velocity (gaining condition) and a downward velocity (losing condition) are defined as positive and negative, respectively.

Probe arrays T1, T2, and T3 were positioned at the central line of zone1, zone 2, and zone 3, respectively. Roshan et al [22] indicated that deeper probes were more probably to lead to significant errors. Therefore, two probes in every probe array were placed close to the sediment water interface (SWI) at depths of 0.1 and 0.3 m below the SWI. The derived flux can be apparently compared with the seepage velocity at a cell 0.2 m below the SWI. The average velocity from each zone was also compared with the estimated velocity via analytical solutions.

The numerical model listed the hydraulic head for each cell, which was used to calculate the vertical velocity through the Darcy-based gradient method. Under the general assumption of strictly vertical flow under the streambed, the vertical velocity can be estimated using Darcy's Law, the hydraulic heads, the distance between the two points, and the average hydraulic conductivity between the two points. The estimated vertical velocities from the gradient method and the heat transport method (Ar and $\Delta\phi$ methods) were compared with the 'true' velocity generated by the numerical model. The points located at the depths of 0.1 m and 0.3 m below the SWI were analyzed using the heat transport method and the gradient method.

3. Results

In general, the estimated velocity errors of stream bank-aquifer system is greater compared to the purely vertical flow. To generate a wider range of perturbations in the propagation of surface heat signals into the streambed, several horizontal gradients were applied in this study. These simulations could be used to reveal the effects of non-vertical flow on the vertical flow estimates using 1D analytical heat tracing and Darcy-based solutions for the riverbank cross-section.

3.1. Point and averaged Vz and Vx under different simulation scenarios

The flow velocity of each zone defined in figure 1 has its own spatial distribution. In the numerical model, the domain of zone 3 was changed and easily determined according to the saturation of each grid. Then, the average velocity of each zone was obtained. The observed points were assigned at the central line of each zone, and located 20 cm beneath the upper boundary of each zone. The temperatures in the shallow area, i.e. T1, T2, and T3, were used to analyze the vertical velocity through 1D analytic solution. Therefore, first the relationship between the point observed velocity and the average velocity was analyzed.

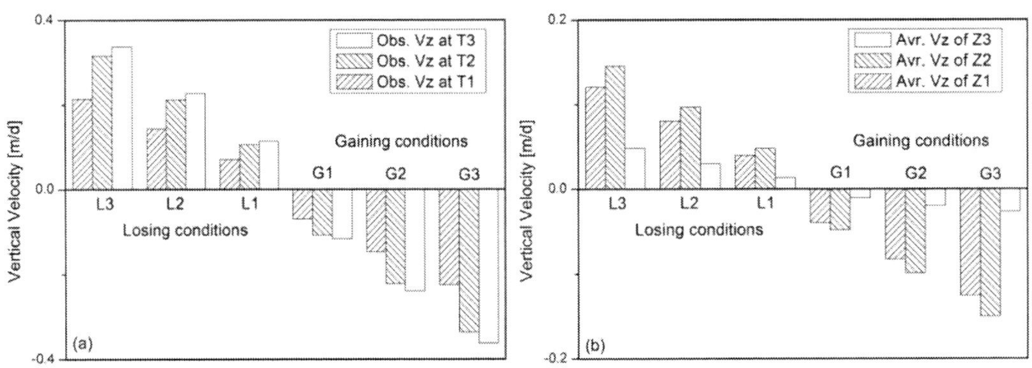

Figure 2. Measured point vertical velocities (a) and average vertical velocities (b) under different hydro-scenarios.
Positive and negative Vz indicate potential flow in the upwelling and downwelling directions, respectively.

Figure 2 illustrates the point vertical Vz (figure 2(a)) as well as the average Vz for each zone (figure 2(b)) under six different conditions; L1, L2, and L3 for losing, which means the surface water recharge the groundwater. G1, G2, and G3 for gaining conditions, which indicates the groundwater recharge the surface water. The relationship between the point vertical velocities at the three zones can be explained using figure 2(a). For the point vertical observation, the velocities observed at T3 were the largest among the three points. The T1 point velocities were the smallest, and T2 had medium-level point velocities. Similar patterns were observed under gaining and losing conditions.

Interestingly, the average velocities show a different pattern (figure 2(b)). With respect to the average Vz values, zone 2 had the largest value, while zones 1 and 3 had the second largest and smallest values, respectively. This pattern is different from that shown in figure 2(a).

The point Vz values at zones 1 and 2 are around 1.77 times and 2.22 times larger than the corresponding average Vz values. However, the point Vz values at zone 3 ranged from 6.91 to 13.75 times greater than the average values of the six simulation scenarios. The hydraulic gradient of these scenarios decreased from L3 (or G3) to L1 (or G1). Both the point Vz and average Vz showed a consistent decreasing trend with the hydraulic gradient.

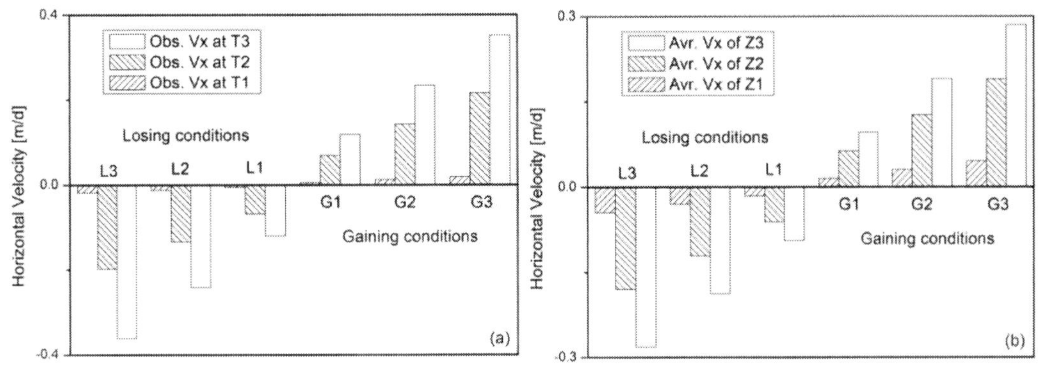

Figure 3. Measured point horizontal velocities (a) and average horizontal velocities (b) under different hydro-scenarios.

Although only the vertical velocity can be estimated using equations (8) and (9), the natural movement of flow is 2D even within a 3D flow field. In the cross-sectional numerical model in this study, the horizontal velocity can be easily obtained. The point horizontal velocity and average Vx are shown in figure 3 under different scenarios. The relationship among the average Vx of the three zones is similar to that of the point Vx among the three observed points, and is relatively more straightforward as compared with the Vz pattern. The order of magnitude of the Vx of the three points was T1>T2>T3. Similarly, the order of average Vx magnitude was Zone 1 > Zone 2 > Zone 3. The point Vx shows the same pattern as the average values, but the magnitude of the point Vx is significantly different than the zonal Vx. The Vx at T1 for the L3 scenario is -0.018 m/d, and the average Vx of Zone1 for L3 is -0.045 m/d. The ratio of the average Vx to the point value for zone 1 is generally around 2.46. This ratio decreases to 0.90 and 0.80 for zone2 and zone 3, respectively. This shows that the superficial horizontal velocity is much higher than the horizontal velocity at deeper locations in zone 1, and that this pattern is opposite for zones 2 and 3.

Both the point Vx and the average Vx show an increasing trend with the general horizontal gradient from L1 (G1) to L3 (G3). This trend is the same as that of the Vz in figure 2. This indicates that both the Vx and Vz are closely controlled by the hydraulic gradient. However, the ratio of Vx to Vz at each location did not vary significantly with the hydraulic gradient in our results. Therefore, the hydraulic gradient between groundwater and surface water varied the magnitude of Vx and Vz spatially, but the flow path and direction did not change significantly with the hydraulic gradient.

The 1D heat tracing analytic method requires diurnal temperature variations. The observed diurnal temperature variations at the points were used to estimate the hyporheic flux. The 2D bank-aquifer flow fields, which is obtained by the numerical simulations under losing (L3) and gaining (G3) conditions, are presented in figures 4(a) and 4(b), respectively. Figures 4(a) and 4(b) show that the velocity over the cross-section is not consistent; the velocity vector directions are almost opposite, and the magnitude for any point within zones 1, 2, and 3 are almost equal.

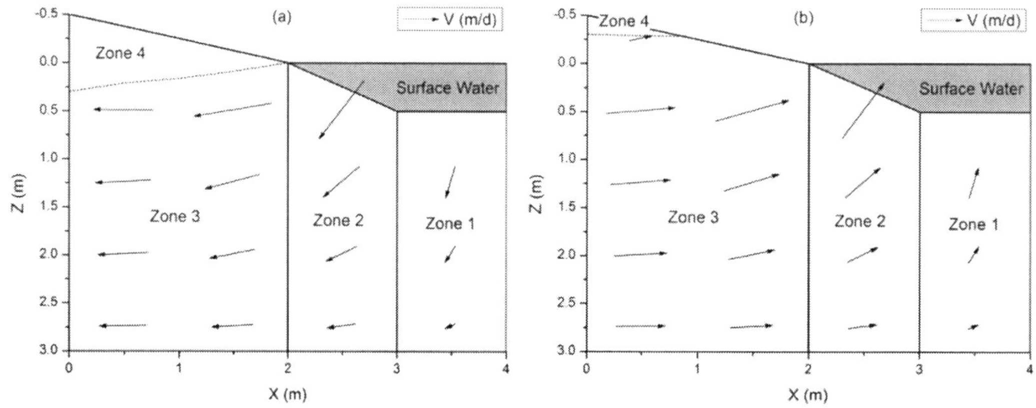

Figure 4. Flow vectors in the cross-section on the third day of scenarios L3 (a) and G3 (b).

The vertical velocity in real is not always constant with depth for most field settings [22]. The simulation results show that the velocity distribution can be described by a function, which depends on depth and location if the hydraulic heads of the boundaries are unchanged. For zone 1, which is beneath the center of the streambed, the shallower part has a larger velocity, and the horizontal component is relatively smaller than the vertical component. With increasing depth, the magnitude becomes smaller whereas the ratio ofthe horizontal component to the vertical component of flow velocity increases to become larger than one.

Jensen and Engesgaard [4] suggested that a higher degree of hyporheic flow is promoted at the middle of a stream than at the two banks due to the presence of more loose and coarse-grained sediments in the middle of the stream. This finding was not verified in this study, but the area near the SWI has greater velocity. The spatial pattern of velocity of zone 2 beneath the riverbank is similar to that of zone 1, although the magnitude of the velocity at the same specific elevation in zone 2 is greater than the velocity in zone 1. The ratio of the horizontal component to the vertical component is slightly different between the two zones. The zone 3 is the farthest one from the steam among the three zones. The total velocity at the shallow part of zone 3 was relatively larger than that at the deeper part and the horizontal component of velocity was always larger the vertical component, indicating that horizontal flow dominated the hyporheic flow in zone 3.

3.2. Deviations of the Vz estimated by the 1D heat tracing and gradient methods from the numerical model's values

The velocity, head, and temperature of any grid can be easily extracted from numerical results. Although the velocity varies spatially, the values from the shallow observation points are usually used to calculate the vertical velocity via the heat tracing and gradient methods. In this study, shallow observation points, at a depth of 20 cm beneath the SWI, were positioned for the three zones. Since only vertical velocity can be obtained using the heat tracing methods, the vertical velocity from numerical models, and the estimated velocities from the phase shift, amplitude ratio, and gradient methods were compared and are shown in figure 5.

For zone 1 (figure 5(a)), the amplitude ratio method estimates the downwelling vertical velocity well. The maximum error in the three scenarios (L1, L2, and L3) is <0.96% in L1. When the flow is upwelling, the amplitude ratio method gives the worst results; the estimated velocity has the correct direction, but the relative error is up to 90%. The error in the estimated velocity from the phase shift method is generally considerable in the six scenarios. The estimated error diminished with the magnitude of the velocity. The results of this study demonstrate that the error is no more than 20% when the true velocity is larger than 0.1 m/d. The gradient method also estimates the velocity under both losing and gaining conditions well. The relative errors range from 12.75% to 13.07% in the six scenarios.

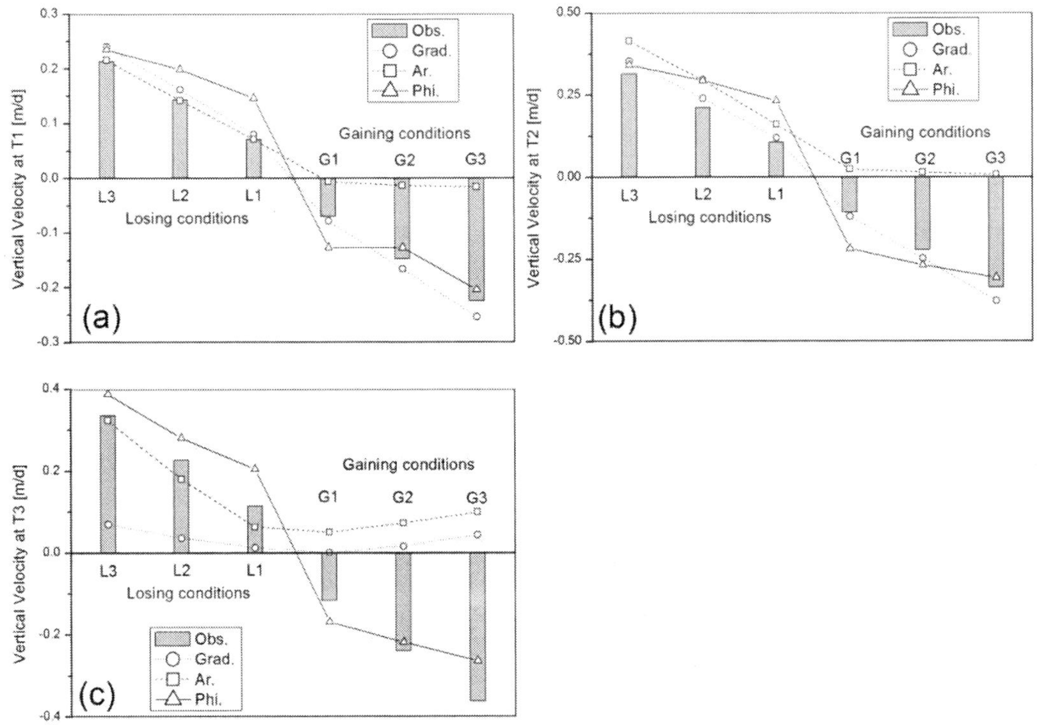

Figure 5. Comparison of the vertical velocity estimated from the 1D heat transfer analytical solution, the Darcy-based gradient method, and the numerical solution.

Horizontal flow dominated the flow field in zone 3 (figure 4), resulting in quite different estimated velocity results for zone 1 (figure 5(c)). Roshan et al [22] suggests that large variations from one side of the stream to the other exists in velocity estimates, which could be an indicator of potential errors resulted from significant horizontal flow components. For zone 3, the relative error of the downward estimated velocity from the amplitude ratio method was no more than 50%, and the relative error decreased with increasing velocity. For upward flow, to a certain extent, the direction and magnitude of the velocity could not be precisely estimated via the amplitude ratio method. In terms of accuracy, the phase shift method produced results comparable to that for zone 1. Figure 5(c) shows that the gradient method provided a low-quality estimation of the velocity. The estimated velocities from the gradient method have relative errors of -79.39 % to -89.02% and -100.75 to -112.07% for losing and gaining conditions, respectively. At the moment, it is not possible to explain why the methods provided an inaccurate flux except that there was probably a significant horizontal flow component that caused errors in both the amplitude ratio and gradient methods.

The flow field of zone 2 is a transition area between zone 1 and zone 3. The true velocity and the estimated values from the three methods are shown in figure 5(b). The estimated errors for this zone lie in between those for zone 1 and zone 3. No more detailed descriptions for zone 2 are provided in the remainder of this paper.

Analysis of the methods provides some key points. When the hyporheic flow is downward and vertical, the amplitude ratio method generates the most accurate vertical velocity estimates followed by the gradient and the phase shift methods. In all, the three methods estimate the vertical velocity well when the true velocity is almost vertical and downward. When the flow is vertical but upward, the gradient method shows the highest accuracy followed by the phase shift method. However, the amplitude ratio method has poor accuracy under gaining conditions. The error from the phase shift method did not vary significant with the flow condition. Although the phase shift method can yield reasonable estimates with a small error in all types of flow conditions, a disadvantage is that the flow direction is unknown and needs to be determined in advance. Moreover, the gradient method performs inadequately under non-vertical flow conditions, as does the amplitude ratio method.

The results of Roshan et al [22] also indicate that the amplitude ratio method do not perform well under gaining conditions, while the phase shift method performs well. For a losing stream, both methods can be applied but the amplitude ratio method achieves a better estimation than the phase shift method. Lautz [10] concluded that analytical methods using the amplitude ratio to derive flux are less prone to error than methods using phase shift under non-ideal field conditions. Swanson and Cardenas [23] indicated that flux estimation methods using a temperature time-series amplitude ratio analysis more closely matched field measurements than the phase shift methods at their study site. In fact, the differences between the velocity estimates from the amplitude ratio and phase shift methods only roughly indicate probable errors in the velocity estimates and it seems as if there is no obvious answer with respect to the extent to which the field velocity data can be suffered the impact of the horizontal flow. In this study, horizontal flow became more dominant when going from zone 1 to zone 3, and the error in the estimation increased significantly.

Lautz [10] stated that in the case of non-vertical flow fields (losing condition), the estimated vertical velocity from a 1D analytical solution is always higher than the true vertical velocity. However, the study of Roshan et al. do not support this conclusion [22]. In fact, our results (figure 5) show that the relationship between the estimated vertical velocity and the true velocity changes among the three zones.

3.3. Effect of Pe number on the error of the estimated flux

Jesen et al. [4] emphasized that the Pe number is key to the temperature distribution of a steady-state hyporheic exchange obtained via the analytical solution of Bredehoeft and Papaopulos [37] . Cardenas and Wilson [24] used a global Pe number reflecting the relative dominance of heat conduction versus advection in a dune geometry-induced hyporheic heat transfer. Moret [38] also stated that thermal Pe affects the numerical solution of the advection-diffusion heat transport significantly.

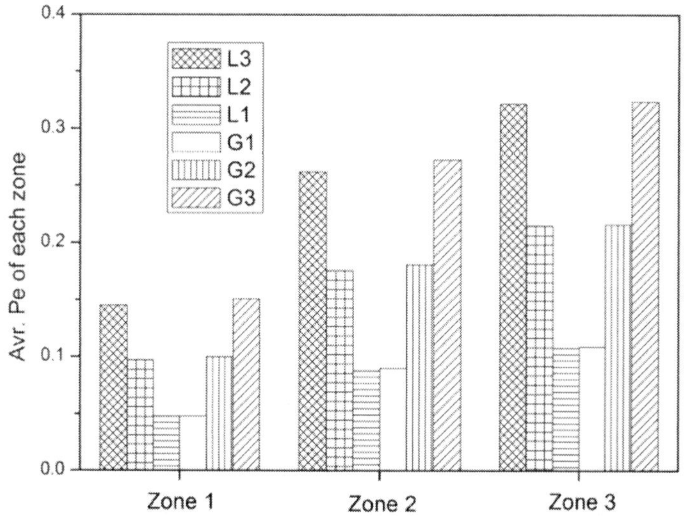

Figure 6. The Pe numbers of T1, T2, and T3 under different hydro-scenarios.

The Pe can be calculated using equations (6) and (7). For any location in this study, the Pe only changed with the seepage velocity while the other parameters were constant during each simulation. The hydraulic gradient varied in the six different scenarios due to the change in the hydraulic head at the left boundary as discussed previously. The losing and gaining conditions were also altered by the change in the hydraulic head at the boundary. With the decrease in the hydraulic gradient from L3 (G3) to L1 (G1), the corresponding Pe (figure 6) showed a linear correlating decrease with the decrease in the boundary head.

Though the average hydraulic gradient of L1 (L2, L3) was designed to be equal to the gradient of G1 (G2, G3), the calculated Pe for the same location shows little difference with the simulated velocity, and the gaining scenarios generally have a slightly larger Pe value, as shown in figure 6. The Pe also varied among the three zones. The Pe of zone 3 (river bank) always had the greatest value, and that of zone 1 (streambed below the center stream) the least. As the 2D flow field in figure 4 shows, zone 1 has the greatest vertical velocity component, but the overall velocity is the least among the three zones. Since water flows from zone 1 to zone 3, the horizontal velocity component gradually increases and the vertical velocity finally approximates to zero; in addition, the overall velocity becomes higher.

Ronan et al [31] studied the relationship between the FSD (full-scale deflection) error and Pe, and concluded that the FSD error in the vertical velocity estimates decreases with increasing Pe. The parameter that varies with a constant Peclet number is the stream width in Ronan et al [31] . However, the seepage velocity was the variable in this study. The results from the different zones show that the effects of Pe on the error of the estimates varied among the zones.

Because the Pe number is significantly affected by the boundary conditions of the numerical model, this spatial pattern of Pe is not explained further. The Pe induced error of the vertical velocity estimation is shown in figure 7. For the losing condition (figure 7(a)), the estimated error from the phase shift method exhibited a dramatic decreasing trend with increasing Pe; the amplitude ratio method gave relatively stable and accurate estimates even with changing Pe. The gradient method provided very close estimates of the vertical velocity except for the horizontal flow dominated zone 3.

For the gaining condition (figure 7(b)), the results are somewhat different from that under the losing condition. The error in the vertical velocity estimates increases with decreasing Pe in the phase shift method. The velocity estimation from the phase shift method also shows a decreasing trend with the Pe. The estimated velocity from the phase shift method overestimates the velocity at a low Pe. With the increasing of Pe, the estimated error decreases and then remains at a relatively low level. As shown in figure 7(b), the amplitude ratio method is highly inaccurate under gaining conditions. The gradient method provides quite good results when the horizontal flow is weaker than the vertical flow, but otherwise, provides quite bad results when the Pe becomes larger.

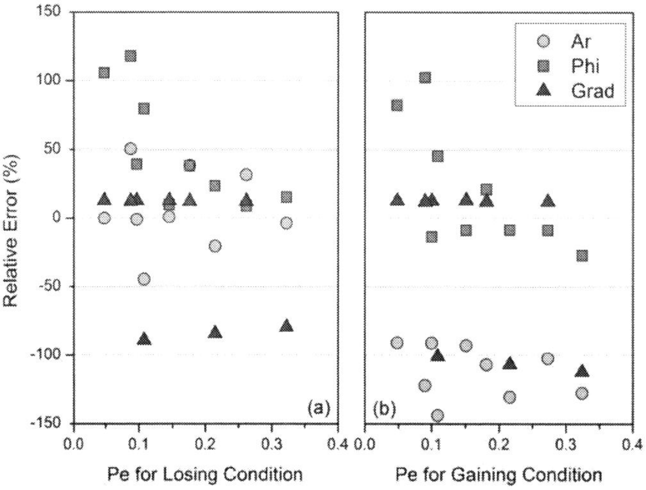

Figure 7. Estimated error against Peclet number in the three analytical methods.

4. Discussion

Since temperature is a primary variable for many physical, chemical, and biological processes occurring in streams, lakes, estuaries, and oceans, and in their bottom sediments, the ramifications of our simulation results on choosing an appropriate method for describing a hyporheic field and the use of the gradient method are briefly discussed here.

Generally, the impacts of horizontal flow on vertical velocity estimates are not straightforward and require comprehensive analyses of the flow field. Nevertheless, our current analysis provides some guidance, and these observations are overall agreed with and extended from the findings of Jensen and Engesgaard, Lautz, Shanafield et al, and Ronan et al [4, 10, 11, 31]. The 1D analytic solution of the heat advection-diffusion equation is obtained under the assumption of vertical flow. Therefore, both the amplitude ratio and the phase shift methods basically inherit the advantages of the analytic solution for the hyporheic zone right beneath the streambed. In particular, these methods estimate the vertical velocity for a losing stream quite well. On the other hand, for a gaining stream, the phase shift method also yields fair results, but the amplitude ratio method's results are very poor. The Darcy-based gradient method estimates the velocity quite well under both losing and gaining conditions.

In our simulations, horizontal flow became dominant with increasing distance from the center of the stream. The estimated error in the velocity from the three methods differed by a large amount, whose maximum value can reach nearly 200%. As discussed in section 3.2, the results of the three methods for the hyporheic zone right below the streambed are similar. However, the differences became larger for the hyporheic zone at a greater distance from the stream (zone 3 in this study). The phase shift method can be used for a gaining stream, and the amplitude ratio method can only be used for a losing stream. The gradient method provides very error-prone results for the hyporheic zone at a greater distance from the stream; a possible explanation for this is that the flow here is not strictly vertical. Lautz [10] stated that the use of temperature data to quantify flux across the streambed is a promising alternative to the more commonly used approaches, such as the Darcy flux calculations. However, it is known that the amplitude ratio method and phase shift method have their own advantages and perform better than the gradient method for areas far from the center of a stream.

Roshan et al [22] also noticed that using only one analytic heat transport solution could cause a large uncertainty in the estimation, and that using both methods in combination was preferable. However, one shortcoming of the phase shift method is that it does not provide the flow path and the amplitude ratio method cannot capture the direction and magnitude of the vertical velocity for a gaining steam reasonably. A combination of the two methods may not result in a credible estimate as shown by Kalbus et al [39], who demonstrated that the heat-tracing methods should not be applied along. Information on the hydraulics of a system in the flow direction is required to guide the choice of analytic methods. The simple measurement of hydraulic heads at different depths for a single location will yield erroneous flow direction, and the Darcy-based gradient method will provide inaccurate estimates of the vertical velocity.

5. Conclusions

Through numerical simulation of streambank-aquifer flow and heat transport, a general flow pattern was determined. The heat transport provided information on the spatial and temporal variations in temperature, which were used to estimate the vertical velocity via the amplitude ratio and phase lag methods. The vertical hydraulic gradient was also used to obtain the vertical velocity. The true vertical velocity obtained from numerical simulation was compared with the indirect methods. Certain indicative points were determined in our study.

For a streambank-aquifer continuum, the velocity at the shallower part is generally higher than the velocity at the deeper part. The vertical component dominates the velocity vector beneath the center of a stream. At large distances from the center of the stream, horizontal flow dominates the hyporheic exchange in the stream bank-aquifer continuum. Although the hydraulic gradient can affect the magnitude of the vertical and horizontal velocities, the ratio of vertical velocity to horizontal velocity did not change significantly with the hydraulic gradient in both losing and gaining conditions in the present study.

Our results indicate that the relationship between the estimated vertical velocity and the true velocity is location and method dependent. The amplitude ratio method and gradient method estimate the vertical velocity well when vertical flow dominates the hyporheic flow and the flow is upwelling. For the other hydro-scenarios, these two methods provide inaccurate results. The phase shift method estimates the vertical velocity relatively well whatever the location and flow conditions, but requires the direction of the hyporheic flow to be known beforehand. The amplitude ratio method and phase shift method yield better results compared with the gradient method for the streambank area.

The Peclet number was affected by the hydraulic gradient in our simulations, with losing and gaining conditions affecting the average Pe insignificantly. The estimated error in the phase shift method showed a dramatic decreasing trend with increasing Pe. However, the amplitude ratio method was not sensitive to the Pe. The relationship between the estimates from the gradient method and the Pe is location-dependent, but at each location, no increasing or decreasing trend was found.

As discussed above, non-vertical flow could be the factor most responsible for the error in the estimates by the amplitude ratio method, phase method, and gradient method. Each analytic method has its own advantages and two or more methods should be used in conjunction to more accurately understand the true flow field; the hydraulic heads should be obtained from the appropriate monitoring network.

Acknowledgments

The researchers would like to extend their thanks to the National Natural Science Foundation of China grants (41201029and 51279208). This study was also supported by the Open Research Fund of State Key Laboratory of Simulation and Regulation of Water Cycle in River Basin (China Institute of Water Resources and Hydropower Research), Grant NO: IWHR-SKL-201502, Fundamental Research Funds for the Central Universities (2015B14414), and Open Foundation of State Key Laboratory of Hydrology-Water Resources and Hydraulic Engineering (2014490711).

References

[1] Boano F, Harvey J W, Marion A, Packman A I, Revelli R, Ridolfi L and Wörman A 2014 Hyporheic flow and transport processes: Mechanisms, models, and biogeochemical implications *Rev. Geophys.* **52** 603-79

[2] Lapham W W 1989 Use of temperature profiles beneath streams to determine rates of vertical ground-water flow and vertical hydraulic conductivity *US Geol. Surv. Water Suppl. Pap.* **2337** 1-35
[3] Constantz J, Thomas C L and Zellweger G 1994 Influence of diurnal variations in stream temperature on streamflow loss and groundwater recharge *Water Resour. Res.* **30** 3253-64
[4] Jensen J K and Engesgaard P 2011 Nonuniform groundwater discharge across a streambed: Heat as a tracer *Vadose Zone J.* **10** 98-109
[5] Suzuki S 1960 Percolation measurements based on heat flow through soil with special reference to paddy fields *J. Geophys. Res.* **65** 2883-5
[6] Stallman R W 1965 Steady one-dimensional fluid flow in a semi-infinite porous medium with sinusoidal surface temperature *J. Geophys. Res.* **70** 2821-7
[7] Hatch C E, Fisher A T, Revenaugh J S, Constantz J and Ruehl C 2006 Quantifying surface water & ndash; groundwater interactions using time series analysis of streambed thermal records: Method development *Water Resour. Res.* **42** 2405-11
[8] Keery J, Binley A, Crook N and Smith J W N 2007 Temporal and spatial variability of groundwater-surface water fluxes: Development and application of an analytical method using temperature time series *J. Hydrol.* **336** 1-16
[9] Cardenas M B and Wilson J L 2007a Effects of current–bed form induced fluid flow on the thermal regime of sediments *Water Resour. Res.* **43** 1050-6
[10] Lautz L K 2010 Impacts of nonideal field conditions on vertical water velocity estimates from streambed temperature time series *Water Resour. Res.* **46** 3473-5
[11] Shanafield M, Hatch C and Pohll G 2011 Uncertainty in thermal time series analysis estimates of streambed water flux *Water Resour. Res.* **47** 341-51
[12] Schmidt C, Bayerraich M and Schirmer M 2006 Characterization of spatial heterogeneity of groundwater-stream water interactions using multiple depth streambed temperature measurements at the reach scale *Hydrol. Earth Syst. Sci.* **10** 849-59
[13] Conant B 2004 Delineating and quantifying ground water discharge zones using streambed temperatures *Ground Water* **42** 243-57
[14] Fryar A E, Wallin E J and Brown D L 2000 Spatial and temporal variability in seepage between a contaminated aquifer and tributaries to the Ohio River *Ground Water Monit. Rem.* **20** 129-46
[15] Silliman S E and Booth D F 1993 Analysis of time-series measurements of sediment temperature for identification of gaining vs. losing portions of Juday Creek, Indiana *J. Hydrol.* **146** 131-48
[16] Luce C H, Daniele T, Frank G and Ralph A 2013 Solutions for the diurnally forced advection-diffusion equation to estimate bulk fluid velocity and diffusivity in streambeds from temperature time series *Water Resour. Res.* **49** 488-506
[17] Goto S, Yamano M and Kinoshita M 2005 Thermal response of sediment with vertical fluid flow to periodic temperature variation at the surface *Phys. Fluids* **17** 211-26
[18] Constantz J 2008 Heat as a tracer to determine streambed water exchanges *Water Resour. Res.* **44** 0-10
[19] Shanafield M, Pohll G and Susfalk R 2010 Use of heat-based vertical fluxes to approximate total flux in simple channels *Water Resour. Res.* **46** 742-50
[20] Naranjo R C, Niswonger R G, Mark S, Clinton D and Alan M 2012 The use of multiobjective calibration and regional sensitivity analysis in simulating hyporheic exchange *Water Resour. Res.* **48** 273-9
[21] Constantz J and Stonestrom D A 2003 Heat as a tracer of water movement near streams *U.S. Geol. Surv. Circ.* **1260** 1-6
[22] Roshan H, Rau G C, Andersen M S and Acworth I R 2012 Use of heat as tracer to quantify vertical streambed flow in a two-dimensional flow field *Water Resour. Res.* **48** 197-205
[23] Swanson T E and Cardenas M B 2010 Diel heat transport within the hyporheic zone of a

pool-riffle-pool sequence of a losing stream and evaluation of models for fluid flux estimation using heat *Limnol. Oceanogr.* **5** 1741-54

[24] Cardenas M B and Wilson J L 2007b Thermal regime of dune-covered sediments under gaining and losing water bodies *J. Geophys. Res.* **112** 497-507

[25] Cuthbert M O and Mackay R 2013 Impacts of nonuniform flow on estimates of vertical streambed flux *Water Resour. Res.* **49** 19-28

[26] Irvine D J, Cranswick R H, Simmons C T, Shanafield M A and Lautz L K 2015 The effect of streambed heterogeneity on groundwater-surface water exchange fluxes inferred from temperature time series *Water Resour. Res.* **51** 198-212

[27] Voss C I 1984 SUTRA, A finite-element simulation model for saturated-unsaturated, fluid-density-dependent ground-water flow with energy transport or chemically-reactive single-species solute transport *Water Resour. Invest. Rep.* **84-4369** 1-260

[28] Sheffield D, Smith T, Broadwell S, Polk D and Lepore S 1987 HST3D: A computer code for simulation of heat and solute transport in three-dimensional ground-water flow systems *Water Resour. Invest. Rep. (USA)* **86-4095** 127

[29] Healy R W 1990 Simulation of solute transport in variably saturated porous media with supplemental information on modifications to the U.S. Geological Survey's computer program VS2D *U.S. Geol. Surv. Water Resour. Invest. Rep.* **90-4025** 125 pp

[30] Rau G C, Andersen M S and Acworth R I 2012 Experimental investigation of the thermal dispersivity term and its significance in the heat transport equation for flow in sediments *Water Resour. Res.* **48** 1346

[31] Ronan A D, Prudic D E, Thodal C E and Constantz J 1998 Field study and simulation of diurnal temperature effects on infiltration and variably saturated flow beneath an ephemeral stream *Water Resour. Res.* **34** 2137-53

[32] Healy R W and Ronan A D 1996 Documentation of computer program VS2Dh for simulation of energy transport in variably saturated porous media; modification of the US Geological Survey's Computer Program VS2DT *U.S. Geol. Surv. Water Resour. Invest. Rep.* **96-4230** 36

[33] Hsieh P, Wingle W and Healy R 2000 VS2DI—A graphical software package for simulating fluid flow and solute or energy transport in variably saturated porous media *U.S. Geol. Surv. Water Resour. Invest. Rep.* **99-4130** 16

[34] Constantz J, Stewart A E, Niswonger R and Sarma L 2002 Analysis of temperature profiles for investigating stream losses beneath ephemeral channels *Water Resour. Res.* **38** 1316

[35] Young P C, Pedregal D J and Tych W 1999 Dynamic harmonic regression *J. Forecast.* **18** 369-94

[36] Taylor C J, Pedregal D J, Young P C and Tych W 2007 Environmental time series analysis and forecasting with the Captain toolbox *Environ. Modell. Softw.* **22** 797-814

[37] Bredehoeft J D and Papaopulos I S 1965 Rates of vertical groundwater movement estimated from the Earth's thermal profile *Water Resour. Res.* **1** 325-8

[38] Moret G 2007 Annual variations in ground-water temperature as a tracer of river -aquifer interactions *Diss. Theses Gradworks*

[39] Kalbus E, Reinstorf F and Schirmer M 2006 Measuring methods for groundwater - surface water interactions: A review *Hydrol. Earth Syst. Sci.* **10** 873-87

The effects of a low-permeability lens on hyporheic exchange intensity

X R Su[1], L C Shu[1], G Zhao[2], M M Wang[2], C P Lu[1,3] and C C Yao[1]

[1]State Key Laboratory of Hydrology-Water Resources and Hydraulic Engineering, Nanjing, 210098, China
[2]Hydraulic Research Institute of Jiangsu Province, Nanjing, 210017, China

E-mail: luchengpeng@hhu.edu.cn

Abstract. Hyporheic exchange induced by dunes is a key process controlling water fluxes and biogeochemical process in river network, which has gained significant advances. Owing to the limitation of instrumental detection at small spatial scales, previous studies mainly focused on dune-induced hyporheic exchange in homogeneous systems and the impacts of a low-permeability lens on hyporheic process is still unknown. 2-D laboratory flume experiments were conducted in this study to quantitatively analyze the response mechanism of hyporheic exchange to the spatial locations of a low-permeability lens. Results indicated that the lens has hindering effects on the hyporheic exchange process and when located at the center of the horizontal location, the hindering effects were the weakest. The effect weakens when the vertical locations of the lens getting deeper. With the lens moving from the locations closer to the upstream face to that closer to the downstream face of the dunes, or moving from a shallow depth to a deep depth, the flowing path of hyporheic exchange become less influenced by the lens and the area of HZ is getting larger.

1. Introduction

Hyporheic exchange is an important component of groundwater-surface water (GW-SW) interactions, which means the water that flows from streams or rivers into shallow subsurface and then returns to the stream [1]. This kind of GW-SW exchange has significant influences on contaminant transport and remediation, water resources management, and the functioning of aquatic ecosystems [2]. Investigating the mechanisms that drive hyporheic exchange is very important to the further understanding of the physical, biogeochemical and thermal processes in streams, hyporheic zones and aquifers [3].

One fundamental mechanism of hyporheic exchange is the existence of pressure gradients at the streambed surface generated by surface water flowing over dunes in streambed surface [4]. Previous studies have made significant advances in the understanding of dune-induced hyporheic exchange by conducting laboratory experiments and numerical simulations. Elliott and Brooks [5, 6] provided analytical approximations of velocity fields, residence times and net mass exchange in the hyporheic zone generated by dunes. Jin *et al* [7] studied solute transport process in a streambed with dunes by combining laboratory experiments and numerical simulations.

Despite significant advances, most dune-induced hyporheic process studies have focused on homogeneous systems. However, the heterogeneity of streambed sediments is recognized to have significant effects on subsurface flow process and solute transport [8]. Especially, hydrological and ecological processes can be significantly influenced by the presence of fine-grained sediments with low permeability [9]. To better analyze the effect of low-permeability sediments on hyporheic exchange, we focus on low-permeability lens features in this study. Such features are widespread in natural streambeds [10]. Research has suggested that in heterogeneous sediments, the transportation of aqueous solute [11] and nonaqueous solute [12] are both controlled by the presence of low-permeability lenses. Gomez-Velez *et al* [13] studied the impacts of low-permeability layers on spatial distributions of hyporheic flow. However, the mechanism of hyporheic flow under the control of a low-permeability lens is still unknown. No studies have revealed the impact a low-permeability lens with varying sizes and locations on hyporheic process.

In this study, we focus on hyporheic process at the scale of a bedform, aiming to uncover the impact of a low-permeability lens on dune-induced hyporheic exchange. The response mechanism of hyporheic exchange to a low-permeability lens was quantitatively analyzed by conducting laboratory flume experiments. One single lens with varying spatial location was imbedded in the streambed to study its effects on the hyporheic process. How hyporheic exchange intensity and hyporheic flow paths respond to the presence of a low-permeability lens was revealed in this study.

2. Methodology

2.1. Flume Setup

The dune-induced hyporheic processes was researched by a rectangular recirculating flume, shown as figure 1. The recirculating flume consisted of a 200 cm long, 10 cm wide and 120 cm high sandbox (located in the center of the flume), two rectangular water tank (located on the left and right sides of the sandbox, respectively) and a recirculating pipe system. The sandbox was filled with sand which has a mean diameter that equals to 0.5 mm and a porosity that equals 0.3. The hydraulic conductivity was estimated by the in-situ permeability testing method which equals 0.075 cm/s [14]. The sand was made as dune shaped, which had the length of 58 cm and the height of 17 cm. To keep the triangular dunes remain stationary, wet method in Jin *et al* [7] was applied when making dunes. The dunes sizes were kept unchanged along the cross section of the flume. The mean water depth was 13 cm in this study, which equals to the distance from the water surface to the average bed level. The left and right water tank represented upstream and downstream of the surface water, respectively. It can be seen from figure 1 that a centrifugal pump drives the water flow in the flume. The overlying surface water velocity was kept to be 10 cm/s, under which the dunes was found to remain stationary.

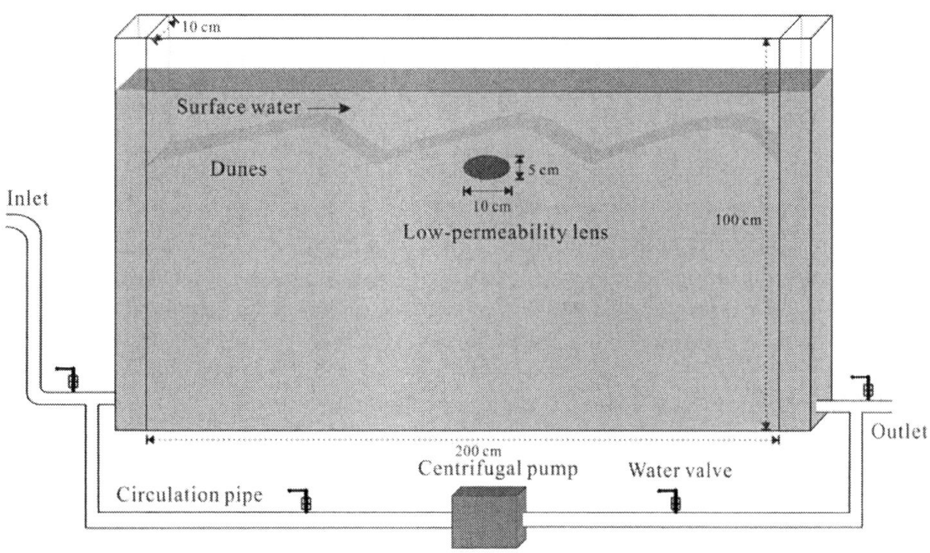

Figure 1. Schematic illustration of the experimental recirculating flume.

IOP Conf. Series: Earth and Environmental Science 82 (2017) 012055 doi:10.1088/1755-1315/82/1/012055

The low-permeability lens was made by kaoline, which could be treated as impermeable compared with its surrounding sand in the sandbox. The length and width of the lens was set to be 10 cm and 5 cm, respectively (figure 1). The thickness of the lens was 10 cm, which was equal to the width of the sandbox. The lens was placed into the sandbox with its surfaces stick to the wall, thus the water flowing and solute transportation in the sandbox can be treated as two-dimensional processes.

2.2. Hyporheic exchange

According to Fox *et al* [15], hyporheic exchange, or q_H, can be estimated by mass balance equations.

For a nonreactive tracer, the mass balance equations were shown as following:

$$W\frac{\mathrm{d}C}{\mathrm{d}t} = Aq_H(C_s - C) \tag{1}$$

$$W_s\frac{\mathrm{d}C_s}{\mathrm{d}t} =- Aq_H(C_s - C) \tag{2}$$

$$C(t = 0) = C_0 \tag{3}$$

$$C_s(t = 0) = 0 \tag{4}$$

where t represents the time after thorough blending of the tracer in the surface water, A represents the streambed area, W and W_s are the total volume of surface water and the water volume in the pore of sand, respectively. $C(t)$ and $C_s(t)$ represent the mean tracer concentrations in surface water and pore water, respectively.

According to Grant *et al* [16], during the initial periods of each experiment, the concentration within sand could be negligible ($C_s \approx 0$). Thus the concentration in stream can be simply given by:

$$\frac{\mathrm{d}C}{\mathrm{d}t} =- \frac{q_H}{d}C \tag{5}$$

where $d = W/A$ is the average depth of surface water. According to the equations (5) and (3):

$$\mathrm{C(t) = C_0 exp}\left(-\frac{\mathrm{q_H}}{\mathrm{d}}\mathrm{t}\right) \tag{6}$$

Equation (6) can be simplified into

$$C^*(t) = \frac{C(t)}{C_0} = \exp\left(-\frac{q_H}{d}t\right) = \exp\,(St) \tag{7}$$

where:

$$S =- \frac{q_H}{d} \tag{8}$$

Thus q_H can be estimated as:

$$q_H =- Sd \tag{9}$$

The conservative tracer in this study was dissolved NaCl, whose concentration was measured by a conductivity meter with an interval of 1 minutes. Water temperature was maintained at 19℃.

The red dye was injected into the dunes at multiple locations to study the flow path of hyporheic exchange (figure 2). The injected locations were equally distributed in dunes. With the centrifugal pump unopened, dissolved red dye was injected to the interior of the dunes by a transfer pipette with external diameter equals to 0.5 mm, which was smaller enough to avoid changing the structure of the dunes. After the overlying water started to flow, the flowing process of the dye was monitored by photographs taken every 10 minutes.

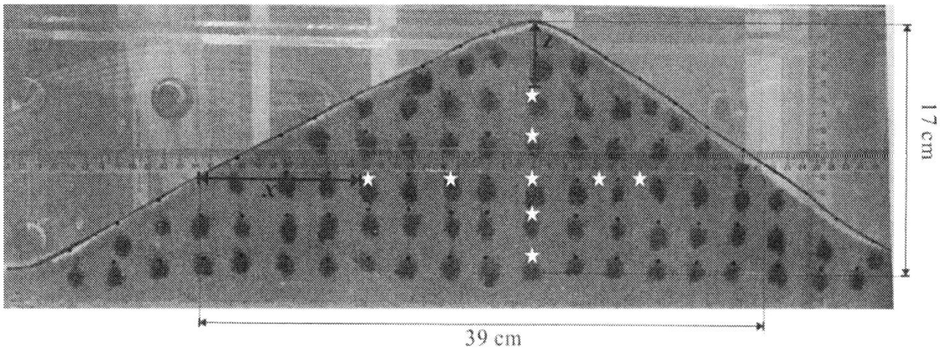

Figure 2. Schematic illustration of dye injections and the locations of the low-permeability lens.

3. Results and discussion

To research the impacts of spatial locations of lens on hyporheic exchange, the lens was imbedded at five different horizontal locations and depths (shown as white five-pointed star in figure 2). Firstly, the vertical distance of the lens to the peak point of the dunes (shown as z in figure 2) was kept to be stable and moved the lens from the upstream face of the dunes to that of the downstream face. The ratio of the horizontal distance of the lens to the upstream face of the dunes (shown as x in figure 2) to the length of dunes at this depth (39 cm here) was recorded to be the relative horizontal location of the lens to the dunes. Similarly, when changing the depth of the lens, the horizontal locations of the lens was kept at the center of the dunes (beneath the peak point) and the relative depth of the lens refers to the ratio of z to the height of the dunes (17 cm here). The relative horizontal locations and depth of the low-permeability lens is shown in table 1.

Table 1. Relative horizontal locations and depth of the low-permeability lens (dimensionless).

Relative horizontal locations	0.286	0.429	0.571	0.714	0.786
	Relative depth = 0.632				
Relative depth	0.316	0.474	0.632	0.779	0.947
	Relative horizontal locations = 0.571				

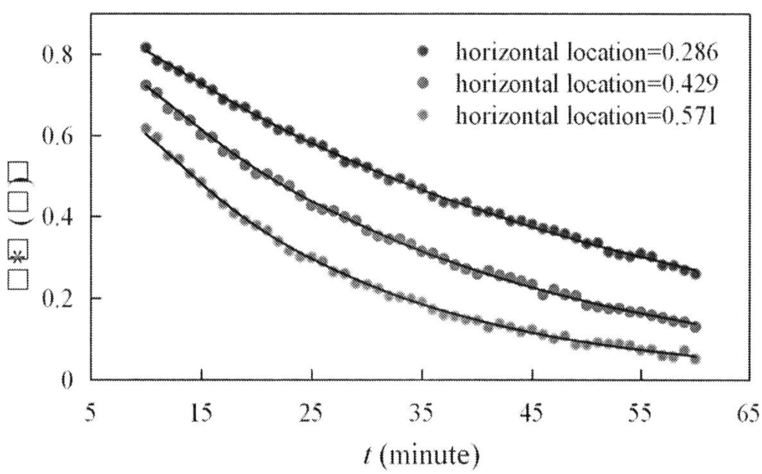

Figure 3. $C^*(t)$ curve at various horizontal locations.

The hyporheic exchange was estimated by equations (7) and (9). The variation of $C^*(t)$ with time was shown in figure 3. Relative horizontal locations of 0.286, 0.429 and 0.571 were selected as representatives. The $C^*(t)$ curve was fitted by the exponential function, shown as black line in figure 3. It can be seen from figure 3 that the fitted results were acceptable.

How the hyporheic exchange intensity q_H varied with the horizontal locations and depth of the lens were shown in figures 4 and 5, respectively. At each spatial location, the nonreactive tracer experiments were conducted repeatedly by 3 times. Figures 4 and 5 displayed the variation of the mean hyporheic exchange intensity and the errors bars at each spatial location. The hyporheic exchange intensity without a lens existing in the flume equals to 0.847 cm/min, which is higher than that when there is a lens in the dunes (figures 4 and 5), which means the existence of the low-permeability lens weakens the hyporheic exchange intensity. That is to say, the lens has hindering effects on the hyporheic exchange process.

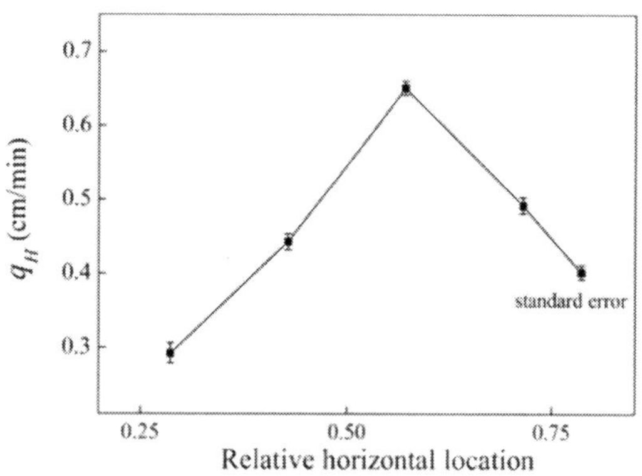

Figure 4. Variation of q_H and error bars with relative horizontal locations of the low-permeability lens.

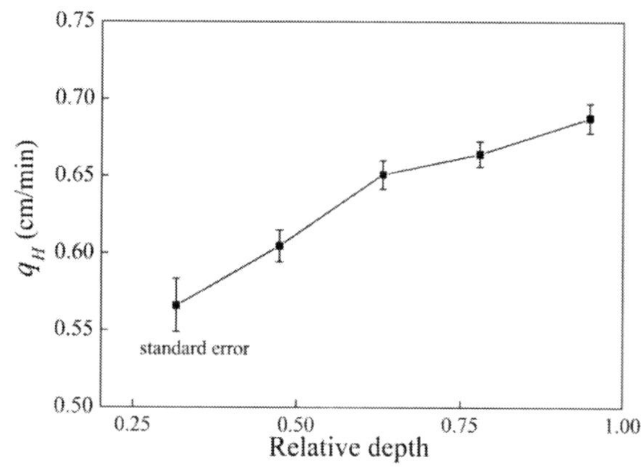

Figure 5. Variation of q_H and error bars with relative depth of the low-permeability lens.

It can be seen from figure 4 that with the lens moving from the upstream face of the dunes to the upstream face, the q_H (shown as rectangle points in figure 4) increases first until reaching its maximum value at the horizontal location of 0.571, after that, q_H decreases gradually. That is to say, the low-permeability lens has the weakest hindering effects on hyporheic exchange when located at the center of the horizontal location. The hindering effects strengthen when the lens getting closer to the upstream or downstream face of the dunes. The relative locations of 0.429 and 0.714 have the same horizontal distance to the central point (the horizontal location of 0.571), while q_H has a smaller value when the lens is closer to the upstream face of the dunes. Besides, q_H has its minimum value when the lens located at horizontal location of 0.286. Thus it can be inferred that the hindering effects of a lens located closer to the upstream face of the dunes is higher than that when closer to the downstream face of a dunes. The error bars were calculated by the stand error, which can be estimated by dividing the standard deviation by the square root of number of measurements (3 here) [16]. The error bars reflect how reliable that the mean value represents the true value. The wider the error bars, the less reliable of the results [16]. It can be seen from figure 3 that all the error bars were relatively narrow, which means the results of the physical experiments are reliable and acceptable.

The error bars are also narrow when the lens located at different depth thus the results in figure 5 are also acceptable. With the low-permeability lens moving from the shallowest depth to the deepest depth, the hyporheic exchange intensity shows a gradual increasing trend. That is to say, the hindering effect of on hyporheic process weakens with the vertical locations of the lens getting deeper.

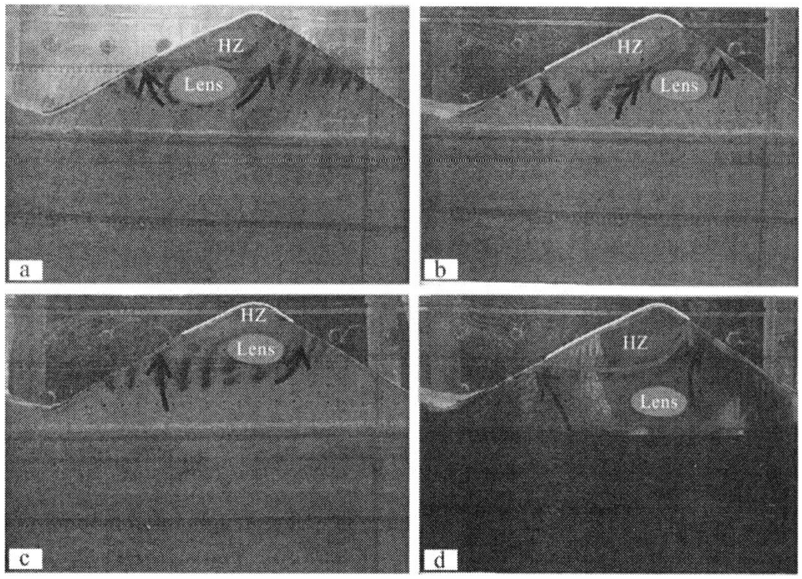

Figure 6. Hyporheic flow paths under the impact of dunes with varied spatial locations. The lines with arrows refer to the streamlines of water in the dunes. The blue lines show the deepest streamline of hyporheic exchange. The zones between the yellow lines and the blue lines refer to the hyporheic zones (HZ).

After one hour, the flow path of hyporheic exchange under the impact of dunes with varied horizontal locations (a and b) and depth (c and d) of the lens are shown in figure 6. Hyporheic exchange refers to the water that flows from the streams into the subsurface and then returns to the surface. The black lines in figure 6 only reflect the process of water in the sand bed flowing to the surface water while do not present the process of surface water flowing into the dunes, which do not belong to hyporheic exchange. Thus the blue lines in figure 6 show the deepest streamline of hyporheic exchange. It can be seen from figures 6a and b that with the lens moving from the locations closer to the upstream face to that closer to the downstream face of the dunes, the flowing path of hyporheic exchange become less influenced by the lens and the area of HZ is getting larger. Similarly, the hyporheic flow is hindered significantly by a lens that located at shallow depth while getting almost un-affected when the lens locating at deeper depth (figures 6c and d). The area of HZ also increased when the lens getting deeper. These results were consistent with the results of hyporheic exchange intensity, which have revealed that the hindering effect of the lens weakens when the horizontal locations change from upstream to downstream and the vertical locations become deeper.

4. Conclusion

This study focuses on dune-induced hyporheic processes, uncovers the impact of a low-permeability lenses on hyporheic exchange, which promotes better understanding of the impact of heterogeneity on hyporheic process at small spatial scales to some extent. The following is a summary of this study:

- The existence of the low-permeability lens weakens the hyporheic exchange intensity. The lens has hindering effects on the hyporheic exchange process.
- The low-permeability lens has the weakest hindering effects on hyporheic exchange when located at the center of the horizontal location. Meanwhile, the hindering effects weaken when the lens getting closer to the upstream or downstream face of the dunes. The hindering effect of a lens located closer to the upstream face of the dunes is higher than that when closer to the downstream face.
- The hindering effect of a lens on hyporheic process weakens when the vertical locations of the lens getting deeper.

- With the lens moving from the locations closer to the upstream face to that closer to the downstream face of the dunes, or moving from a shallow depth to a deep depth, the flowing paths of hyporheic exchange become less influenced by the lens and the area of HZ is getting larger.

Acknowledgements
This research was supported by the program of China Scholarships Council (No. 201706710023), the National Natural Science Foundation of China (41201029, 41572210), Fundamental Research Funds for the Central Universities of China (2015B14414, 2016B40114), and the Open Research Fund of State Key Laboratory of Simulation and Regulation of Water Cycle in River Basin (IWHR-SKL-201502).

References
[1] Winter T C, Harvey J W, Franke O L and Alley W M 1998 Ground water and surface water: A single resource *Usgs U.S. Geol. Surv.* 1139-79
[2] Boulton A, Datry T, Kasahara T, Mutz M and Stanford J 2010 Ecology and management of the hyporheic zone: Stream-groundwater interactions of running waters and their floodplains *J. N. Amer. Benthol. Soc.* **29(1)** 26-40
[3] Jin G, Tang H, Li L and Barry D A 2011 Hyporheic flow under periodic bed forms influenced by low-density gradients *Geophys. Res. Lett.* **38(22)** L22401.1-6
[4] Boano F, Harvey J W, Marion A, Packman A I, Revelli R, Ridolfi L and Wörman A 2014 Hyporheic flow and transport processes: Mechanisms, models, and biogeochemical implications *Rev. Geophys.* **52(4)** 603-79
[5] Elliott A H and Brooks N H 1997 Transfer of nonsorbing solutes to a streambed with bed forms: Theory *Water Resour. Res. A* **33(33)** 123-36
[6] Elliott A H and Brooks N H 1997 Transfer of nonsorbing solutes to a streambed with bed forms: Laboratory experiments *Water Resour. Res. B* **33(1)** 137-51
[7] Jin G, Tang H, Gibbes B, Li L and Barry D A 2010 Transport of nonsorbing solutes in a streambed with periodic bedforms *Adv. Water Resour.* **33(11)** 1402-16
[8] Freitas J G, Rivett M O, Roche R S, Durrant M, Walker C and Tellam J H 2015 Heterogeneous hyporheic zone dechlorination of a TCE groundwater plume discharging to an urban river reach *Sci. Total Environ.* **505(505)** 236-52
[9] Karwan D L and Saiers J E 2012 Hyporheic exchange and streambed filtration of suspended particles *Water Resour. Res.* **48(1)** 273-9
[10] Wu G, Shu L, Lu C and Chen X 2016 The heterogeneity of 3-D vertical hydraulic conductivity in a streambed *Hydrol. Res.* **47(1)**15-26.
[11] Milosevic N, Thomsen N I, Juhler R K, Albrechtsen H J and Bjerg P L 2012 Identification of discharge zones and quantification of contaminant mass discharges into a local stream from a landfill in a heterogeneous geologic setting *J. Hydrol.* **446-447** 13-23
[12] Faisal A A H, Kassim W M S and Hussein T K 2011 Influence of clay lens on migration of light nonaqueous phase liquid in unsaturated zone *J. Environ. Eng.* **137(1)** 9-14
[13] Gomez-Velez J D, Krause S and Wilson J L 2014 Effect of low-permeability layers on spatial patterns of hyporheic exchange and groundwater upwelling *Water Resour. Res.* **50(6)** 5196-215
[14] Hvorslev M J 1951 Time lag and soil permeability in ground water obsevations *U.S. Army Crops Eng. Waterw. Exp. Sta., Vicksburg Miss.* **36**
[15] Fox A, Boano F and Arnon S 2014 Impact of losing and gaining streamflow conditions on hyporheic exchange fluxes induced by dune-shaped bed forms *Water Resour. Res.* **50(3)** 1895-907
[16] Zhang X, Shasha D E, Song Y and Wang J T L 2012 Fast elastic peak detection for mass spectrometry data mining *IEEE Trans. Knowl. Data Eng.* **24(4)** 634-48

Effect of hydrophilic foamed copper on dynamic membrane formation in dynamic membrane bioreactor (DMBR)

C T Li[1], H J Liu[1,2], X Z Qian[1] and H H Yang[1]

[1]School of Space and Environment, Beihang University, 37 Xueyuan Road, Haidian District, Beijing 100191, China

E-mail: liuhj@buaa.edu.cn

Abstract. The effect of hydrophilic membrane material on the formation of dynamic membrane (DM) was investigated, by using a membrane module combined with a hydrophilic modified foam copper with 0-10 mm of thickness, 300 μm of pore size and an activated carbon sponge with 10 mm of thickness. The appropriate thickness of the hydrophilic membrane was selected to treat the refining wastewater. Hydrophilic modification and membrane formation experiments showed that water molecules were easier to spread on the surface of the foam copper, which could reduce the accumulation of sludge layer on the surface of the membrane module and increase the effluent's stability of the combined membrane module. Under experimental conditions, the proper thickness of the foam copper was 5 mm, the sludge layer thickness on the membrane surface was close to 0mm, and the effluent flux was stabled as 44 $L/(m^2 \cdot h)$, which was higher than that of the activated carbon sponge without using the hydrophilic foam copper, increased by 107%. Wastewater treatment results showed that the removal rate of COD in the DMBR system was 95.6%, and the removal rate of ammonia nitrogen was 98.5%.

1. Introduction

In the development of wastewater treatment technology, the use of membrane bioreactor (MBR) has become more and more extensive; however, membrane cost, high energy consumption and membrane fouling have become the main factors that limit the further promotion of MBR [1]. Thus, the dynamic membrane bioreactor (DMBR), with low energy consumption and low film cost [2], is paid more and more attention by researchers. Dynamic membrane bioreactor, with large pore size membrane substrate, is a new MBR process instead of conventional MBR, in which microporous membrane and ultrafiltration membrane are used [3]. DMBR's filtration process, intercepting the sludge and flocs, which formed the "dynamic film" for effective filtering process, is the biggest difference from traditional MBR [4]. While maintaining advantages of high filtration efficiency and small covering area, DMBR system significantly reduces the cost of the membrane, greatly improve the effluent flux [5].

Large-diameter membrane module was used as the substrate of Dynamic Membrane Bioreactor [6], microorganisms, suspended particulates and flocs of mixture in the reactor, is partially trapped in the grid, when passes through the membrane module [7], due to the suction pressure, moreover, partly attached to the mesh due to the adsorption of the material, after a series of complex processes, a dense layered structure formed [8]. Dense layer structure has a perfect filter performance for the suspended particulates in reactor [9], which could intercept most of particulate matter, like MLSS and dissolved

EPS, and maintain the stability of the effluent quality [10]. However, with the continuous operation of the experiment, the lamellar structure, accumulated on the surface of the membrane module, will collapse under the action of suction pressure, forming a thicker and more dense structure, thus, the effluent flux of the membrane reactor will gradually decrease [11]. Fan *et al* [12] reported that, the structure of dynamic membrane could mainly be divided into two parts, the mud cake layer and the gel layer. The structure of the cake layer, which contributes to the main filtration resistance, filters the large particle substances, the gel layer mainly adsorbed small molecules and dissolved substances. Zhang Jian *et al* [13] investigated the removal efficiency of COD, ammonia nitrogen, total nitrogen and other pollutants, the results showed that, dynamic membrane had some ability to remove pollutants, which may be related to the adsorption and filtration ability of the gel layer. Nowadays, industrial filter cloth, activated carbon sponge, nylon mesh, polyester filter cloth and silk screen are commonly used as membrane substrate [14], which is always hydrophobic materials, especially. You *et al* [15] reported that, using of hydrophobicity material, leaded to the accumulation of organic matter which was always negatively charged , and formed a relatively solid dense cake layer structure, because negatively charged substance could be easy adsorbed to hydrophobic membrane. The cake layer could not be removed by normal aeration scour, thus, the effluent flux and membrane lifecycle reduced, the degree of backwash affected and system operating costs increased [16]. To solve these problems, Xiong *et al* [17] used an elastic membrane material, this membrane could be retracted when sucked, and formed a thick and dense sludge layer, while it could protrude in the backwash process, conducive to the scour for sludge. Meng *et al* [18] reported that, hydrophobic solute, like EPS, produced by activated sludge could approach to the membrane surface, which was an entropy spontaneous process, because no hydrogen bonding between the hydrophobic membrane surface and H_2O, and the membrane filtration pressure (TMP) increased due to the surface adsorption and accumulation of hydrophobic solute.

In this study, a hydrophilic foamed copper material was attached to the activated carbon sponge. A composite membrane module was made up to investigate the mechanism of dynamic membrane formation, and the effects on effluent was also analyzed. The microorganism, suspended particles and flocs could be more evenly spread to the inner layer of activated carbon sponge, which was a hydrophobic material [19], due to the hydrophilic foam copper on the surface. This spatial structure could increase the stability of the effluent, reduce cake layer of the membrane surface, increase the flux of effluent, and improve treatment efficiency of DMBR system.

2. Materials and methods

In this study, a foam copper with 300 μm pore size was selected and modified by surface controlled oxidation method [20], attached to an activated carbon sponge with 10 mm thickness, as dynamic membrane module. After the domestication of the sludge, we started investigating the effects of membrane substrate in DMBR system.

2.1. Materials

The activated sludge was taken from Qinghe Sewage Treatment Plant, Beijing, China. Activated sludge, mixed with the refinery wastewater, was domesticated to adapt the quality of refinery wastewater. The MLSS concentration of domesticated activated sludge was 7.5 g/L, 5.5 g/L of MLVSS concentration. 10 mm thickness of activated carbon sponge was chosen, the hydrophilic copper foam components was 240×110 mm, 300 um diameter, and its' thickness was 1 mm, 2 mm, 5 mm, 10 mm, respectively. Membrane substrate materials attached to both sides of ABS material, chosen for external support frame, sealed with glass glue. A set of MBR systems from Japan was used, and its' basic structure shown in figure 1.

Refinery wastewater was pumped to the reactor, by the peristaltic pump. Mud and water separation continued after organic matter degradation in the mixture. The suction power was provided by the suction pump, and oxygen was provided by a quartz sand microporous head underneath the membrane module.

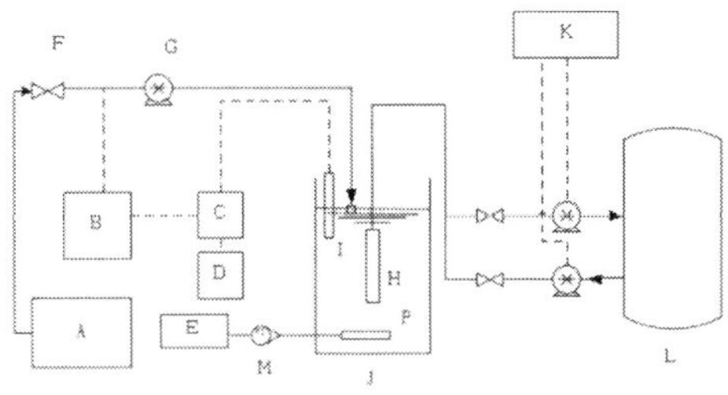

Figure 1. Experimental setup and experiment process.
A. waste water tank; B. transformer; C. level control controller; D. power supply; E. air compressor; F. valve; G. suction pump; H. DM module; I. level meter; J. reactor; K. PLC; P. aeration head; L. collection tank; M. flowmeter.

Pollutants in the refinery wastewater are shown in table 1.

Table 1. Water quality index of refinery wastewater.

	COD (mg/L)	NH_3-N (mg/L)	pH	NTU
parameter	633.23-812.67	32.9-33.4	11.2	77-863

2.2. Methods

2.2.1. Membrane hydrophilic modification. The surface properties of foamed copper were modified by "surface controlled oxidation", the foamed copper component was immersed in hydrochloric acid of 4.0 mol/L for 15 min, then washed with deionized water. 0.065 mol/L of potassium persulfate ($K_2S_2O_8$) and 2.5 mol/L of potassium hydroxide mixed aqueous solution were prepared. The washed foamed copper component was immersed in the above solution, heated at 60°C for 1 hour, then washed with deionized water, air dried, finally heated at 180°C for 2 hours to obtain a CuO film material.

The surface properties of the material were characterized by the German DataPhysics OCA20 contact angle tester, and the wettability of the material was analysed.

2.2.2. Membrane component selection. According to previous research [21], appropriate material mesh pore size of DMBR is between 100 um to 500 um. Thus, a pore size of 300 μm foam copper material was selected in our experiment, with thickness of 1 mm, 2 mm, 5 mm and 10 mm, respectively. Membrane components was made up by activated carbon sponge, which attached with hydrophilic foamed copper component. Under the same operating conditions, with a speed of 15 r/min, peristaltic pump pumped for 24 hours, measuring the thickness of cake layer and effluent flux.

2.2.3. Wastewater analysis. After the DM formed, wastewater was treated by the DMBR system. COD as well as ammonia nitrogen of influent, supernatant and effluent, were analyzed during the system running process. COD was measured by fast confined catalytic digestions, while ammonia nitrogen was measured by Nessler reagent spectrophotometry [22].

3. Results and discussion

3.1. Surface properties of membrane
German DataPhysics OCA20 contact angle tester was used to measure the contact angle of membrane material. The contact angle could not be determined accurately. As the foam copper was a spatial mode structure, only its wettability could be roughly analyzed. Results of unmodified foam copper's

wettability were shown in figure 2.

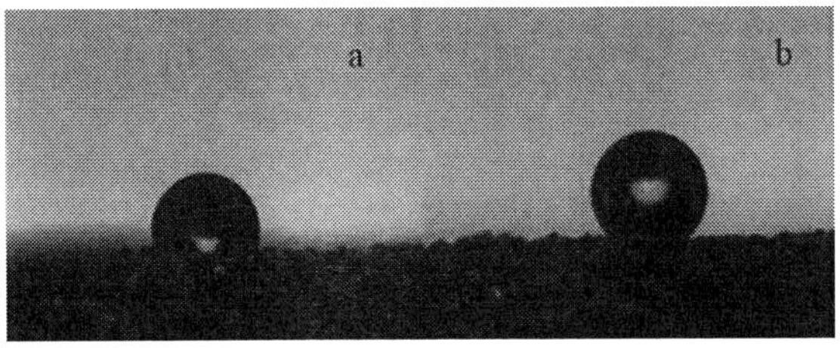

Figure 2. Results of unmodified foam copper's contact angle.

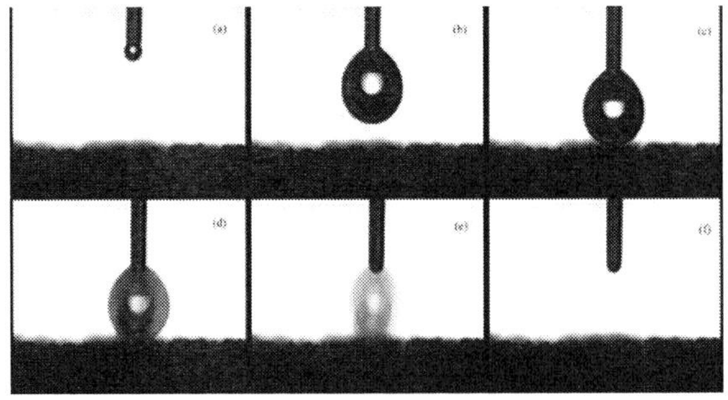

Figure 3. Results of modified foam copper's contact angle.

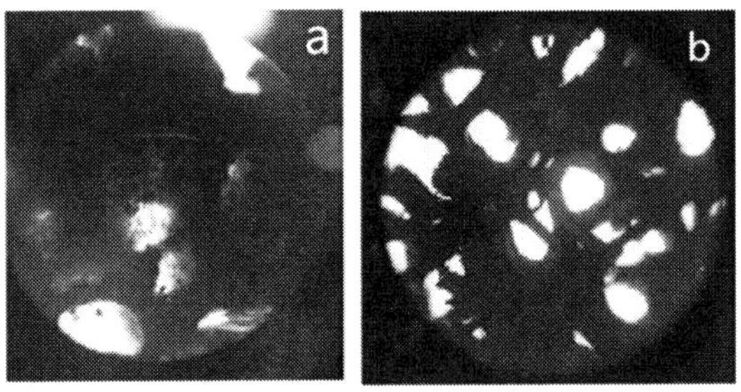

Figure 4. Pictures of flocs and microorganism on membrane module. (a), activated carbon sponge; (b), modified foam copper.

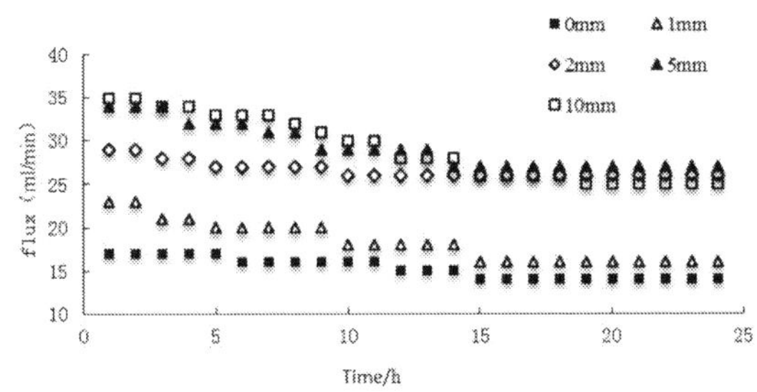

Figure 5. The effluent flux in membrane formation period.

Results of modified foam copper's wettability were shown in figure 3.

Pictures of flocs and microorganism, gathered on modified foam copper and activated carbon sponge after backwashing, were shown in figure 4.

It can be seen from figure 2, the macroscopic surface of the modified foam copper was not planar, and was composed of a porous structure, consisting of copper wires and gaps. The average contact angle of macroscopic surface was 123.5°, this material was in a hydrophobic state, thus, water molecules could not spread out on the surface of the foam copper.

As can be seen from figure 3, d and e, when a 5 μL of droplet approached to the surface of the material, the droplet was sucked into the material. Figure 4 indicated that, flocs and microbial particles combined fast with hydrophobic material, while formed a loose structure in modified foamed copper. Compared with unmodified foam copper, hydrophilicity of modified foam copper improved obviously, water droplets could be completely spread out on the modified foam copper, and the solid particles in water could attach to the surface of the copper mesh or infiltrate to the internal of modified foam copper.

3.2. Membrane formation analysis

The modified foamed copper membranes, with the thickness of 1 mm, 2 mm, 5 mm and 10 mm, were attached to the activated carbon sponge, respectively. Under a condition of 1.5 L/min aeration, a peristaltic pump was used, with the speed of 15 r/min, to pump the mixture for 24 h, then measured thickness of cake layer and effluent flux. Flux changed as shown in figure 5.

Thickness of cake layer changed as shown in figure 6.

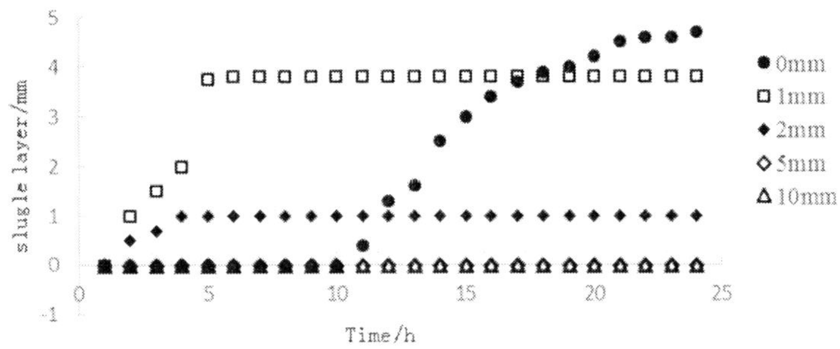

Figure 6. The thickness of sludge layer in membrane formation period.

It can be seen from figures 5 and 6, in a certain range, the flux of the dynamic membrane increased with thicker modified foamed copper. Flux of the membrane module, with a large thickness of the modified foamed copper, tended to be stable, and gradually approached. This might be because, at the beginning of the operation cycle, the sludge particles in the mixture could not accumulate on the surface of the foamed copper when passed through the thicker foam copper, which was evenly dispersed, partially redropped into the reactor. Part of the sludge flocs evenly tiled in the activated carbon sponge spatial structure. Flux of the membrane module increased, because it formed a sludge layer with thin and loose structure. While a thin modified foamed copper used, the sludge particles quickly gathered in the activated carbon sponge spatial structure, and then compressed, forming a thick and dense cake layer, making the membrane flux reduced. It can be seen from figures 4 and 5, appropriate thickness of the modified foamed copper was 5 mm, with which the stability flux was 29 mL/min, no cake layer remained on the surface, and the running cost was less than that of 10mm modified foamed copper moudle.

From the changes of foam copper material's wettability, we can see that, there was a difference in a degree of spreading of water molecules on foam copper material's surface, due to the difference of wettability. Research showed that the sludge layer and the gel layer in the DMBR system, both have the filtering effect [12]. Most of the sludge particles were filtered out by the sludge layer's mechanical filtration, and the small molecules were adsorbed by the gel layer. Water molecules were more difficult to spread inside the hydrophobic material, and air bubbles formed often in the hydrophobic copper mesh or activated carbon sponge pore, forming an unstable factors, which could affect the stability of effluent quality. While water molecules could easily spread out in the hydrophilic material, so that the sludge particles could spread out into the activated carbon sponge spatial structure, forming a stable, dense sludge layer and gel layer, to keep a more stable effluent quality. While sludge could not be firmly bonded to the surface of the copper substrate, due to the spread of water molecules, and the sludge on the surface could naturally fall off through the water erosion, reducing the sludge resistance, increasing the flow of effluent.

3.3. Analysis of effluent quality
A 5 mm thick foamed copper membrane module was selected, and attached to the activated carbon sponge, making a composite membrane module. The membrane module was connected to the MBR system, COD and ammonia nitrogen were measured.

3.3.1. COD analysis. After completion of the sludge domestication, the dynamic membrane module was placed into the reactor, and the COD concentration of the refining wastewater was 743 mg/L on average. During the operation cycle of the DMBR system, effluent and the reactor supernatant COD concentration were measured every two days. The 60-day experimental data were analyzed and the results are shown in figure 7.

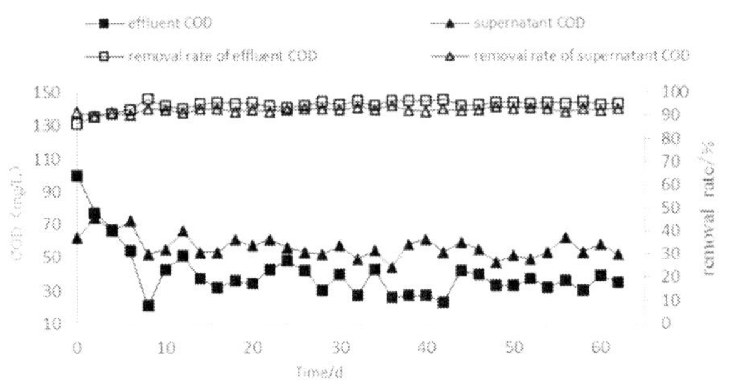

Figure 7. COD changes with time.

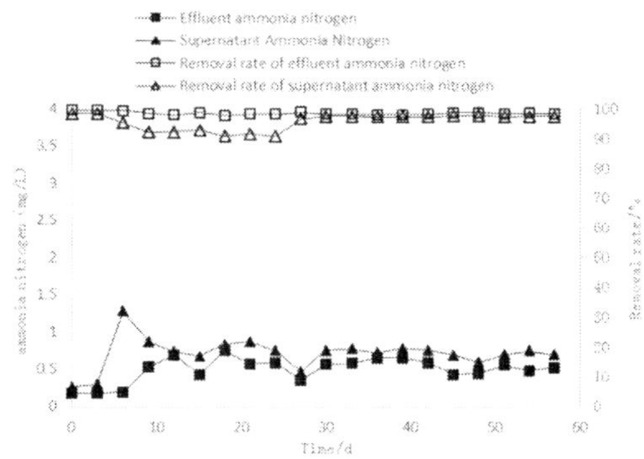

Figure 8. Ammonia nitrogen changes with time.

It can be seen from figure 7 that, the average effluent COD concentration was 34.10 mg/L, the average removal rate was 95.56%. The COD concentration of the supernatant was 56.50 mg/L, and the average removal rate was 92.25%. The dynamic membrane contributed little to the COD removal rate of raw water. The experimental results showed that, the contribution of membrane to COD removal was only 3.5%. In the early stage of the membrane system operation cycle, the COD of effluent dropped sharply, and 10 days later of operation, the COD of effluent became stable. According to the mechanism analysis of dynamic membrane formation, the dynamic membrane was formed by large particle size particles, proteins, polysaccharides and other EPS or SMP, originally, became a cake layer. The surface of the cake layer was gradually deformed, under the action of suction pressure, the pore size became smaller, the effluent quality getting better, and transmembrane pressure difference on the membrane module became larger. When the pressure on activated carbon sponge exceeded a certain degree, the activated carbon sponge would be deformed, part of the original filter layer would be destroyed, and a small part of the small particles leaked out into the effluent, leaded to the slightly addition of COD value. As the new particles continued to fill into the spatial structure, filter layer inside activated carbon sponge became normal, and then formed a complete dynamic membrane. This process continued to cycle and eventually achieved dynamic balance.

3.3.2. Ammonia nitrogen analysis. Ammonia nitrogen removal rate changes with time in figure 8.
It can be seen from figure 8 that, the average removal rate of ammonia nitrogen in the membrane module was 98.50%, the average removal rate of NH_3-N in supernatant was 97.84%, with a larger thickness of the membrane module, some of the ammonia nitrogen, which could not be removed inside the reactor, was partially absorbed and removed in the sludge layer, thus, the ammonia nitrogen removal rate of effluent was higher than that in supernatant. The removal efficiency of NH_3-N by sponge substrate dynamic membrane bioreactor was better than that of polyester nonwoven fabric. Under the same operating conditions, the aerobic and anaerobic regions formed by the sponge membrane substrate, which strengthen the combination of aerobic and anaerobic, achieving a better nitrification effect.

3.4. Discussion

Recently, the researchers of DMBR field mainly focused on the critical flow, the choice of different membrane materials, wastewater treatment effect and the improvement of the DMBR control system. However, there were few scholars focused on the surface performance of the membrane material, the spatial structure of membrane materials, to reduce the membrane fouling problems and enhance the system reliability. Therefore, this research used a hydrophilic state, spatial structure copper foam, as the DMBR membrane module, and investigated its characteristic. Experimental results showed that, the sludge on the surface of the membrane module was obviously reduced, membrane formation time sharply reduced, the flow was increased, and the removal rate of COD was significantly promoted, which fully meet the technological requirements of wastewater treatment. These indicate that, The DMBR system has a good ability for sewage treatment, which could degrade the conventional pollutants such as COD. Meanwhile, this study first used hydrophilic copper foam material as the membrane module, the results indicated that it could effectively reduce membrane fouling, this was because hydrophilic groups on the surface of the membrane module and water molecules combines, forming a stable layer of water film, thus, substances such as sludge cannot combine with membrane components surface. Under the action of water flow, sludge was rinsed clean, thus, the effluent flow of the system was also increased, and the production efficiency of the system improved.

In the backwash process, sludge structure in the hydrophobic material was dense and fast, and big flushing pressure was needed to meet the backwashing requirements, meanwhile, more pressure could damage material's spatial structure, reduce the service life of the membrane. While membrane structure in hydrophilic material was relatively loose, which could be washed under a low flow rate, thus, more energy could be saved. In this case, the choice of the membrane material could tend to be a hydrophilic material of self-cleaning ability, low energy consumption, and long service life.

4. Conclusion

In the research field of dynamic membrane bioreactor, the use of hydrophilic material deserved more attention, because it can not only improve DMBR's defects, but also be a foundation for DMBR mechanism research. For example, the connection between structure characters and the formation of filter layer on the surface, and the relationship between surface structure character and the stability of spatial structure of filter layer, are worthy researching. In a nutshell, the ultimate aim of DMBR research is to improve the filtering stability of the membrane module, eventually to promote the DMBR process.

References

[1] Kiso Y, Jung Y J, Ichinari T, Park M, Kitao T, Nishimura K and Min K S 2000 Wastewater treatment performance of a filtration bio-reactor equipped with a mesh as a filter material *Water Res.* **34(17)** 4143-50

[2] Liang S, Qu L, Meng F, Han X and Zhang J 2013 Effect of sludge properties on the filtration characteristics of self-forming dynamic membranes (SFDMs) in aerobic bioreactors: Formation time, filtration resistance, and fouling propensity *J. Membrane Sci.* **436** 186-94

[3] Zhang Y J 2012 Research progress of dynamic membrane bioreactor *J. Membrane Sci. Technol.* **2032(2013)** 2117-23

[4] Chu H, Zhang Y, Zhou X, Zhao Y, Dong B and Zhang H 2014 Dynamic membrane bioreactor for wastewater treatment: Operation, critical flux, and dynamic membrane structure *J. Membrane Sci.* **450** 265-71

[5] Yu K 2006 Development and prospects on dynamic membrane bioreactor (DMBR) *J. Water Purif. Technol.* **2025(2002)** 2014-8

[6] Wu Y X 2004 Operating characteristics of self-forming dynamic membrane bioreactor at high flux *J. Chin. Water Wastewater* **2020(2002)** 2005-7

[7] Ye M S, Zhang H M and Yang F L 2008 Experimental study on application of the boundary layer theory for estimating steady aeration intensity of precoated dynamic membrane bioreactors Desalination *Environ. Sci.* **230(1-3)** 100-12

[8] Xiong J, Fu D F, Singh R P and Ducoste J J 2016 Structural characteristics and development of the cake layer in a dynamic membrane bioreactor *Sep. Purif. Technol.* **167** 88-96

[9] Ersahin M E, Ozgun H, Tao Y and van Lier J B 2014 Applicability of dynamic membrane technology in anaerobic membrane bioreactors *Water Res.* **48** 420-9

[10] Liu H, Yang C, Pu W and Zhang J 2009 Formation mechanism and structure of dynamic membrane in the dynamic membrane bioreactor *Chem. Eng. J.* **148(2–3)** 290-5

[11] Hu Y, Wang X C, Tian W, Ngo H H and Chen R 2016 Towards stable operation of a dynamic membrane bioreactor (DMBR): Operational process, behavior and retention effect of

dynamic membrane *J. Membrane Sci.* **498** 20-9
[12] Fan B 2003 Filtration capability of the bio-dynamic membrane *J. Environ. Sci.* **2024(2001)** 2091-7
[13] Zhang J 2007 Structure and function of dynamic membrane in the dynamic membrane bio-reacto *J. Environ. Sci.* **2028(2001)** 2147-51
[14] Hong J M 2011 Study on performance of dynamic membrane bioreactor with different membrane materials *Chin. J. Environ. Eng.* **2005(2012)** 2311-4
[15] You C Y 2005 Study on mechanism and control of gel layer forming on membrane surface *J. Jiangsu Environ. Sci. and Technol.* **18(14)** 11-3
[16] Lu J D 2006 A bench-scale dynamic membrane bioreactor for purification of wastewater *Environ Sci. Technol.* **2029(2005)** 2093-4
[17] Xiong J D, Fu D F and Singh R P 2014 Self-adaptive dynamic membrane module with a high flux and stable operation for the municipal wastewater treatment *J. Membrane Sci.* **471** 308-18
[18] Meng F G 2006 Effect of activated sludge properties on short-term membrane fouling in submerged membrane bioreactor based on statistical analysis *Chin. J. Environ. Eng.* **2027(2007)** 1348-52
[19] Qu M N 2016 Research and development of functional superhydrophobic materials *Chem. Prog.* **2016(2012)** 1774-87
[20] Qian B T 2006 Super-hydrophobic CuO nanoflowers by controlled surface oxidation on copper *J. Inorg. Mater.* **21(23)** 747-52
[21] Li X B 2007 Development of research on the dynamic membrane separation technology *J. Membrane Sci. Technol.* **27(24)** 91-5
[22] China National Environmental Monitoring Centre 2002 Water and wastewater monitoring and analysis methods *China Environmental Science Press* **2002** 2210-20 2276-81

IOP Conf. Series: Earth and Environmental Science **82** (2017) 012057 doi:10.1088/1755-1315/82/1/012057

Annual runoff prediction of the upstream of Heihe River Basin, China

Y Y Xie[1], S Y Liu[1,2], Q Huang[1] and Y M Wang[1]

[1]State Key Laboratory Base of Eco-Hydraulic Engineering in Arid Area, Xi'an University of Technology, Xi'an city, 710048, China

E-mail: 719685861@qq.com

Abstract. A prediction method of annual runoff is proposed for the upstream Heihe River Basin. The Mann-Kendall test, order cluster analysis, variance analysis and R/S analysis methods were employed to investigate trends, mutations, periods and persistence of annual runoff time series, respectively. Based on various components, the linear superposition method was applied to forecast annual runoff. The Nash-Sutcliffe efficiency coefficient (*NSE*), the mean absolute percentage error (*MAPE*) and the percentage of pass (*POP*) were utilized to evaluate the simulation and prediction effects of annual runoff. Moreover, the calibration period was between 1958 and 2010, while the validation period was from 2011 to 2015. The results show that i) the annual runoff time series had a significant increasing trend (0.65×10^8 $m^3/10a$), a probable mutation at 2006 and two significant periods (6 years and 22 years) in the calibration period; ii) the randomization of annual runoff time series was improved after removing the trend, mutation and periodic components, with the Hurst exponents 0.52; and iii) the simulation effects of calibration period (*NSE*=0.74, *MAPE*=6.6% and *POP*=90.6%) and the prediction effects of validation period (*MAPE*=6.5%, *POP*=80%) jointly indicate that the annual runoff prediction method is suitable in this river basin.

1. Introduction

Heihe River, the second largest inland river of China, is located in the northwest of China with the arid or semi-arid climate. It supplies the important water resource to the socioeconomic development and ecological environmental protection of northwestern China. The upstream of the river basin is the main runoff-yield region which is fed by rainfall and ice-snow melt water [1]. In the middle of the basin, there distribute many agricultural irrigation districts which consume a large amount of water each year [2]. The downstream of the river flows through desert and Gobi area, where the evaporation is intensive and the ecosystem is fragile [3].

In Heihe River Basin, it is very important to correctly handle the relations among irrigation, ecologic environment and hydroelectric generation under the condition of limited water resources. From 1960th, the annual runoff in the downstream river basin reduced sharply because of the rapid population growth and extensive irrigation area in the middle river basin, causing a serious degradation of ecological environment of the middle and lower river basin [4-6]. Since 2000, the government of China has taken many water allocation measures to relieve the water contradiction between irrigation and ecological environment in the river basin, achieving some success [7-9]. However, the water contradiction is still serious currently, and the downstream ecological environment has not yet restored to the level of the 1980th [10, 11]. Besides, the upstream of the river has rich waterpower resources, and many hydropower plants were built there in the first decade of 21st century. These hydropower stations provide reliable electric power for the local socioeconomic development, so the requirements of the hydropower plants should be properly taken into account when allocating the surface water of Heihe River [12].

The annual runoff at the upper stream could directly influence the water allocation between different users of Heihe River Basin, since the upstream basin is the main runoff generation area. In order to better realize the comprehensive benefits of irrigation, ecologic environment and hydroelectric generation, the decision makers always make the water distribution schemes based on the next year's runoff volume in advance. Therefore, it is of great significance to predict the annual runoff at the upstream of the river basin.

The prediction of annual runoff belongs to long-term runoff forecast, which could be solved by three kinds of common methods. The first one is the time series analysis method, which acquires the future runoff based on the internal relations (autocorrelation and relations with time) of the historical runoff series [13, 14]. The second one is the relevant factor analysis method, which forecasts the runoff based on the statistical relations between runoff and the other factors such as precipitation, atmospheric temperature and wind speed [15-17]. The last one is runoff generation and concentration model, which predicts the runoff according to the physical mechanism of runoff formation [18-20]. Among the three methods, the first one is the simplest because it only needs the historical runoff data, but it lacks physical mechanism and excessively depends on the constant statistical characteristics of runoff series. The second one could demonstrate the explicit causes and course of runoff formation, but it is the most complicated approach which needs a great mass of data (hydrologic data, meteorological data, vegetation data, soil data, land use data, etc.). As to the third one, its advantages and disadvantages are between the first and second methods. On the whole, it is difficult to prove which way is best for forecasting the annual runoff due to many reasons (e.g. assumed conditions, empiric formulas, data deficiencies and error propagation) without considering specific river basins.

The time series analysis method is often employed to forecast the annual runoff of Heihe River because of the insufficient observation data and the unrevealed ice-snow melting mechanism of the river basin. Chen *et al* [17] adopted grey model to the extract tendency part of runoff time series, employed the wavelet analysis method to extract the period part, utilized the autoregressive method to deal with the stochastic part, and then built the runoff prediction model of Heihe River. Jiang and Liu [21] combined the wavelet analysis method with radial basis function network to forecast the annual runoff of Heihe River. Chu *et al* [22] applied Mann-Kendall method to study the trend of runoff time series of Heihe River, used the coupling method of periodic wave extension and stepwise regression to determine the period part. Zhang and Ding [23] employed the grey topological method to predict runoff volume at the valley outlet of Heihe River. Hou *et al* [24] established the annual runoff forecast model of Heihe River through the periodic mean superposition method. Li *et al* [25] proposed a new runoff prediction model of Hiehe River based on R/S analysis method and grey system theory. It can be seen that the applications of the time series analysis method to runoff prediction of Heihe River in the past mainly focused on three components of runoff time series (trend, periodic and stochastic components), and merely provided the deterministic forecasting results to the decision-makers. However, the annual runoff series of the upstream Heihe River probably has mutation and becomes more uncertain in the background of regional climate change [26, 27], which may bring to the decision of water allocation certain risks in future.

Specific to the upstream Heihe River, this study still adopts the time series analysis method to build the annual runoff prediction model, but it is different from the prior studies. Firstly, the mutation of annual runoff series will be investigated before establishing the model. Secondly, the model not only provides the deterministic prediction results but shows the main variation range of runoff, which could help the decision-makers make adequate preparation against the risks of uncertainty. Moreover, the effect of the extraction of deterministic components on annual runoff series will be discussed. Therefore, this study is organized as follows: Section 2 introduces the general situation of the Heihe River; Section 3 displays the data and methods for trend analysis, mutation test, period detection, persistence analysis and prediction of annual runoff; in Section 4, the results are presented and discussed; and the conclusions are drawn in Section 5.

2. Study area

Heihe River is a typical inland river of China, flowing through Qinghai, Gansu and Inner Mongolia provinces from south to north (figure 1). The river originates from the north foot of Qilian Mountain and feeds into the West Juyan Lake and the East Juyan Lake, with a core drainage area of 130,000 km^2 and a mainstream length of 821 km. The upstream Heihe River (UHR), the study area, is between the river sources and Yingluoxia canyon with the drainage area of 10,009 km^2, the river length of 303 km and the average channel gradient of 9.1‰. UHR is characterized by the mountainous terrains with the average elevation of 3737.7 m. The climate of UHR is cold and moist, with the annual average air temperature of 2°C and the annual average precipitation of 350 mm. Many mountain areas of UHR with high elevation (more than 4,000 m) are annually snow-covered. Moreover, a number of modern glaciers distribute in UHR, which is the natural reservoir of northwestern China. Yingluoxia canyon is the outlet of UHR with the annual average runoff volume of 1.6 billion m^3, which will be mainly used for irrigating crops, recharging groundwater, improving ecological environment in the middle and lower Heihe River.

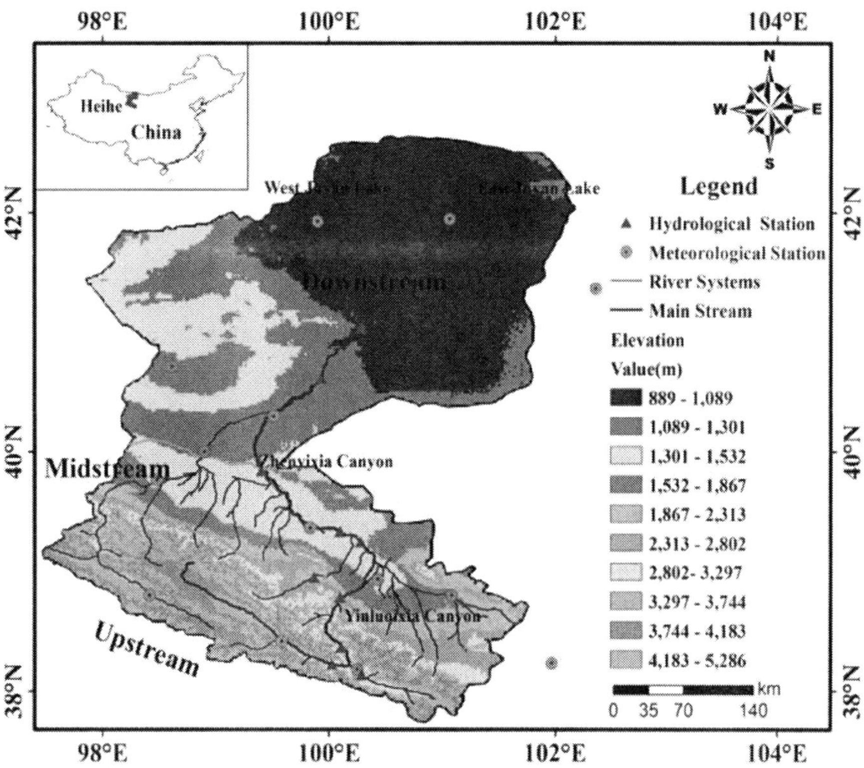

Figure 1. Map of Heihe River basin.

3. Materials and methods

3.1. Materials

The average daily flow data of Yingluoxia canyon is observed at Yingluoxia hydrologic station (figure 1), which was built in 1943. Due to the poor field monitoring condition and the backward data reorganization technology, the reliability of average daily flow data before 1958 is much lower.

Therefore, the annual runoff data of Yingluoxia station between 1958 and 2015 are employed in this study. Moreover, UHR annual runoff data between 1958 and 2010 are applied to calibrate the prediction model, and the remaining data (2011-2015) are utilized to validate the model.

3.2. Methods

3.2.1. Trend analysis method. Mann-Kendall test is widely applied to detect the trends of hydrologic time series because it is a nonparametric test and is not sensitive to the mutations of nonstationary time series. The statistic (U) is built as below [28, 29]:

$$U = \frac{12K - 3N(N-1)}{\sqrt{2N(N-1)(2N+5)}} \tag{1}$$

$$K = \sum_{i=1}^{N-1} \sum_{j=i+1}^{N} k_{ij} \tag{2}$$

$$k_{ij} = \begin{cases} 1 \text{ if } X_j > X_i \\ 0 \text{ if } X_j \le X_i \end{cases}, 1 \le i < j \le N \tag{3}$$

where N is the length of hydrologic time series; X_i (i=1, 2, …, N) is the ith hydrologic variable (e.g. annual runoff); k_{ij} is the paired value between X_i and X_j; and K is the summation of all the paired values.

When N is large enough ($N \geq 10$), the statistic U approximately obeys the standard normal distribution. The original hypothesis is that the hydrologic time series has not any obvious trends. The significance level α is given before the trend test. If $|U| \leq U_{\alpha/2}$, the original hypothesis is accepted, i.e. the series has no obvious trends; otherwise, the original hypothesis is refused, which shows that there exists evident tendency in the series. Besides, the sign of U represents the trend direction ('+' represents the growth trend while '-' indicates the decreasing tendency), and the absolute value of U shows the obvious degree of trend (the degree will become more and more obvious with the increasing of $|U|$).

In addition to the overall trend analysis, the local trends are also investigated in the trend analysis of hydrologic time series. The turning points and stages of various local trends are distinguished by the trend variation index (T), which is defined as follows:

$$T_\tau = | t_{\tau,1} - t_{\tau,2} | \tag{4}$$

where τ ($10 < \tau \leq N-10$) is the division point of series; $t_{\tau,1}$ and $t_{\tau,2}$ represent the linear trend slopes of the former and latter subseries divided by τ, respectively.

In equation (4), when T_τ reaches the maximum value, the division point τ is considered as one turning point, which divides the series into two obvious different stages of local trends. Moreover, if a hydrologic time series has l (l=1, 2, …) turning points, it has l+1 different trend stages.

3.2.2. Mutation analysis method. The mutation represents an irreversible jump of hydrologic time series at a certain time. The order cluster analysis method could effectively find the most probable mutation point of hydrologic time series [30]. Its principle is to find out an optimal segment point dividing the series into two subsequences, where the variable difference in the same subsequence is the minimum and the variable difference between two subsequences is the maximized.

The mutation index (Z) of hydrologic time series is computed as follows [30]:

$$Z_\tau = \frac{(\overline{X_\tau} - \overline{X_{N-\tau}})^2}{S_\tau^2 + S_{N-\tau}^2}, 2 \le \tau < N-1 \tag{5}$$

where τ is possible mutation point of series; $\overline{X_\tau}$ and S_τ are the mean value and the standard deviation of the subsequence before τ, respectively; $\overline{X_{N-\tau}}$ and $S_{N-\tau}$ represent the mean and the

IOP Conf. Series: Earth and Environmental Science **82** (2017) 012057 doi:10.1088/1755-1315/82/1/012057

standard deviation of the subsequence after τ, respectively.

In equation (5), the square of mean difference (the numerator part) represents the variable difference between two subsequences, and the variance sum (the denominator part) reflects the variable difference in the same subsequence. According to the principle of the order cluster analysis method, the point where Z reaches maximum is the most probable mutation point of time series.

After identifying the most probable mutation point, the next necessary step is to test its significance. Rank sum test is also a common nonparametric method in examining the obvious degree of mutation of hydrologic time series [30]. Firstly, the figures of the original series are sorted from small (large) to large (small). Secondly, the rank of each figure of the short subsequence in the sorted series is recorded. Finally, the summation of these the ranks (rank sum) is calculated and employed to establish the statistic of rank sum test.

The statistic (U) of rank sum test is built as below [30]:

$$U=\frac{W-n_1(n_1+n_2+1)/2}{\sqrt{n_1 n_2(n_1+n_2+1)/12}} \tag{6}$$

where W is the rank sum of short subsequence; n_1 is the length of the short subsequence; and n_2 is the length of the long subsequence.

When n_1 is large enough ($n_1 \geq 10$), the statistic U approximately obeys the standard normal distribution. The original hypothesis is that the series has no mutation points, and the significance level α is given before rank test. If $|U| \leq U_{\alpha/2}$, the original hypothesis is accepted, i.e. the series has no mutation; otherwise, the original hypothesis is refused, i.e. there is a remarkable mutation in the series. Moreover, the difference between the means of the series before mutation and after mutation is the jump height, which could also be understood as the mutation component of series.

3.2.3. Period analysis method. Variance analysis is a very useful period identification method, and its obvious advantage is that it is not confined to the exact form of periodic wave function [31, 32]. The period detection process of variance analysis is described in the following [32].

- Divide the time series into an equal number *of* groups based on the length of the assumed period.

$$m_j=\begin{cases} N/P, & \text{if} r=0 \\ (N-r)/P+1, & \text{if} r\neq 0 \& j\leq r \\ (N-r)/P, & \text{if} r\neq 0 \& j>r \end{cases} \tag{7}$$

$$r=\text{mod}(N,P), r<P \tag{8}$$

where P ($2 \leq P \leq N/2$) is the assumed period; r is the remainder when dividing N by P; and m_j (j=1, 2, …, P) is the size of the jth group.

- Calculate the mean values of original time series and each group.

$$\overline{X}=\sum_{i=1}^{N} X_i \tag{9}$$

$$\overline{X_j}=\frac{1}{m_j}\sum_{k=1}^{m_j} X_{kj} \tag{10}$$

$$X_{kj}=\begin{cases} X_{(j-1)m_1+k}, & \text{if} r=0 \\ X_{(j-1)m_1+k}, & \text{if} r\neq 0 \& j\leq r \\ X_{rm_1+(j-r-1)m_{r+1}+k}, & \text{if} r\neq 0 \& j>r \end{cases} \tag{11}$$

where X_i (i=1, 2, …, N) is the ith figure of original time series; $\overline{X}$ is the mean of original series; $\overline{X_j}$ (j=1, 2, …, P) is the mean of the jth group; and X_{kj} (k=1, 2, …, m_j) is the kth figure of the jth group.

- Compute the intra-group and inter-group sum of deviation square.

$$S_1^2=\sum_{j=1}^{P}\sum_{k=1}^{m_j}(X_{kj}-\overline{X_j})^2 \tag{12}$$

$$S_2^2=\sum_{j=1}^{P}m_j(\overline{X_j}-\overline{X})^2 \tag{13}$$

where S_1^2 is the intra-group sum of deviation square; and S_2^2 is the inter-group sum of deviation square.

- Calculate the value of statistic F, give the significance level α and test the significance of the supposed period.

$$F=\frac{S_2^2}{P-1}\bigg/\frac{S_1^2}{N-P} \tag{14}$$

In the definition formula (14), the statistic F follows F distribution with the degree of freedom (P-1, N-P). The significance level α is given before F test. If $F>F_\alpha(P-1, N-P)$, the assumed period P is significant; otherwise, it is not obvious. After finding the first significant period, its periodic wave could be constructed according to the mean values of corresponding groups. Generally, there often exist more than one obvious period in the hydrologic time series. Hence, it is essential to continue to search an additional number of significant periods after extracting the first periodic process. Meanwhile, the number of prominent periods of series is not more than three considering the negative impact of some pseudo periods [32].

3.2.4. Persistence analysis method. The persistence analysis of runoff series has been initiated since the middle of 20th century, which was rapidly expanded to many other geophysical records including temperature and precipitation data [33]. Currently, it is proved that the time series with long range dependence property (persistence) widely exist in the natural world [34]. Hurst exponent H ($0<H<1$), presented by Hurst in 1951, could quantitatively describe the persistence of hydrologic time series very well [35]. Specific to Hurst exponent, there are three kinds of persistence condition in the following.

- If H=0.5, the hydrologic time series is an independent random sequence with a limited variance.
- If $0.5<H<1$, the series has long range dependence property (persistence), which indicates that the future tendency is consistent with the past trend. Moreover, the persistence of series will endlessly strengthen when H approaches 1.
- If $0<H<0.5$, the series still has long run memory property (anti-persistence), which shows that the future tendency is contrary to the past trend. Besides, the anti-persistence of series will continually enhance when H is close to 0.

At present, there are many methods for estimating Hurst exponent [34]. Among them, R/S analysis is the most universal method because of its robustness and nonparametric analysis feature (i.e. it has no special demands on the distribution type of time series). The calculation steps of Hurst exponent by R/S analysis method are expressed as below [33, 34].

- Create an accumulated deviation series (ADS)

$$Y_\tau=\frac{1}{\tau}\sum_{i=1}^{\tau}X_i, 4\le\tau\le N \tag{15}$$

$$A(i,\tau)=\sum_{j=1}^{i}[X_j-Y_\tau], 1\le i\le\tau \tag{16}$$

where τ is the length of ADS; Y_τ is the mean of time series $\{X_1, X_2, \ldots, X_\tau\}$; and $A(i, \tau)$ is the ith variable of ADS.

- Calculate the range and the standard deviation of ADS

$$R(\tau) = \max[A(i,\tau)] - \min[A(i,\tau)] \tag{17}$$

$$S(\tau) = \sqrt{\frac{1}{\tau}\sum_{i=1}^{\tau}[X_i - Y_\tau]^2} \tag{18}$$

where $R(\tau)$ is the range of ADS; and $S(\tau)$ is the standard deviation of ADS.

- Estimate Hurst exponent

$$\lg\frac{R(\tau)}{S(\tau)} = H\lg\tau + C \tag{19}$$

where C is a constant.

In equation (19), a series of data points $\left(\lg\tau, \lg\frac{R(\tau)}{S(\tau)}\right)$ could be obtained when τ ($4\leq\tau\leq N$) is assigned different values. Finally, Hurst exponent H is estimated through the least square method.

3.2.5. Prediction and evaluation method. The linear supposition method is an important time series analysis method [36]. In analyzing hydrologic time series X, it is assumed that the series could be linearly decomposed to five parts including the constant component X_C, the trend component X_T, the mutation component X_M, the period component X_P and the stochastic component X_S. Except for X_S, the other components are collectively called deterministic components. The linear supposition method is expressed as follows [36]:

$$X = X_C + X_T + X_M + X_P + X_S \tag{20}$$

In equation (20), the different components of X series should be estimated and extracted in turn. X_C is considered as the mean value X'_C of X series; X_T is estimated by the linear trend part X'_T of X-X'_C series based on Mann-Kendall test result; X_M is estimated by the jump height X'_M at the mutation point of X-X'_C-X'_T series according to the consequence of mutation analysis; X_P is estimated by the periodic part X'_P of X-X'_C-X'_T-X'_M series through the variance analysis method; and the remaining part X-X'_C-X'_T-X'_M-X'_P (X'_S) is thought as the estimation value of X_S. Moreover, X'_S could be further subdivided into a dependent random component X'_{S1} and an independent random component X'_{S2}. Hurst exponent H of X'_S calculated by R/S analysis method determines the proportions of two components in X'_S series. If $H\leq0.4$ or $H\geq0.6$, the proportion of X'_{S1} could not be neglected. On this occasion, X'_{S1} should be estimated through autoregressive model [17] and stripped from X'_S. Then, the remaining part of X'_S is X'_{S2}. If $0.4<H<0.6$, X'_S could be regarded as an independent random series, i.e. $X'_{S2}= X'_S$. Without taking X'_{S2} into account, the medium estimation value X^M of the X series could be calculated as below.

$$X^{M} = X'_C + X'_T + X'_M + X'_P + X'_{S1} \tag{21}$$

In order to reduce the uncertainty of prediction, it is essential to master the main variation range of X'_{S2}. According to the probability distribution type of X'_{S2}, its lower and upper bounds could be defined as

$$p = p[X'_{S2} \geq X'_{S2}(p)] \tag{22}$$

$$X^{L} = X^{M} + X'_{S2}(q) \tag{23}$$

$$X^{U} = X^{M} + X'_{S2}(1\text{-}q) \tag{24}$$

where p is the occurrence probability of event $X'_{S2}\geq X'_{S2}(p)$; X^L and X^U are the lower and upper estimation values of X, respectively; q ($0.5<q<1$) is the appointed probability; $X'_{S2}(q)$ and $X'_{S2}(1\text{-}q)$ represent the lower and upper bounds of X'_{S2}, respectively.

In this study, Nash-Sutcliffe efficiency coefficient (*NSEC*) and the mean absolute percentage error

(*MAPE*) are employed to evaluate hydrological forecast effects [37, 38]. *NSE* ($NSE \leq 1$) and *MAPE* ($0 \leq MAPE \leq 1$) are two commonly used indexes in evaluating hydrologic prediction model. The forecast accuracy improves with the increasing of *NSE* and the reducing of *MAPE*. Instead of measuring the forecast accuracy, and the percentage of pass (*POP*) is defined to evaluate the reliability of prediction variation range.

$$NSE = 1 - \frac{\sum_{i=1}^{N}(X_i^{\mathrm{R}} - X_i^{\mathrm{M}})^2}{\sum_{i=1}^{N}(X_i^{\mathrm{R}} - \overline{X^{\mathrm{R}}})^2} \tag{25}$$

$$MAPE = \frac{1}{N}\sum_{i=1}^{N}\left|\frac{X_i^{\mathrm{R}} - X_i^{\mathrm{M}}}{X_i^{\mathrm{R}}}\right| \times 100\% \tag{26}$$

$$POP = \frac{M}{N} \times 100\% \tag{27}$$

where X_i^{R} and X_i^{M} are the ith (i=1, 2, …, N) observed and medium estimation values of hydrologic series, respectively; $\overline{X^{\mathrm{R}}}$ is the mean value of observed hydrologic series; and M is the number of events that X^R is located between X^L and X^U.

4. Results and discussion

The average annual runoff of Yingluoxia station was 16.04×10^8 m^3 between 1958 and 2010 (calibration phase). Thus, the constant component X'_{C} of UHR annual runoff series X was equal to 16.04×10^8 m^3, which would be stripped from X before extracting the other components.

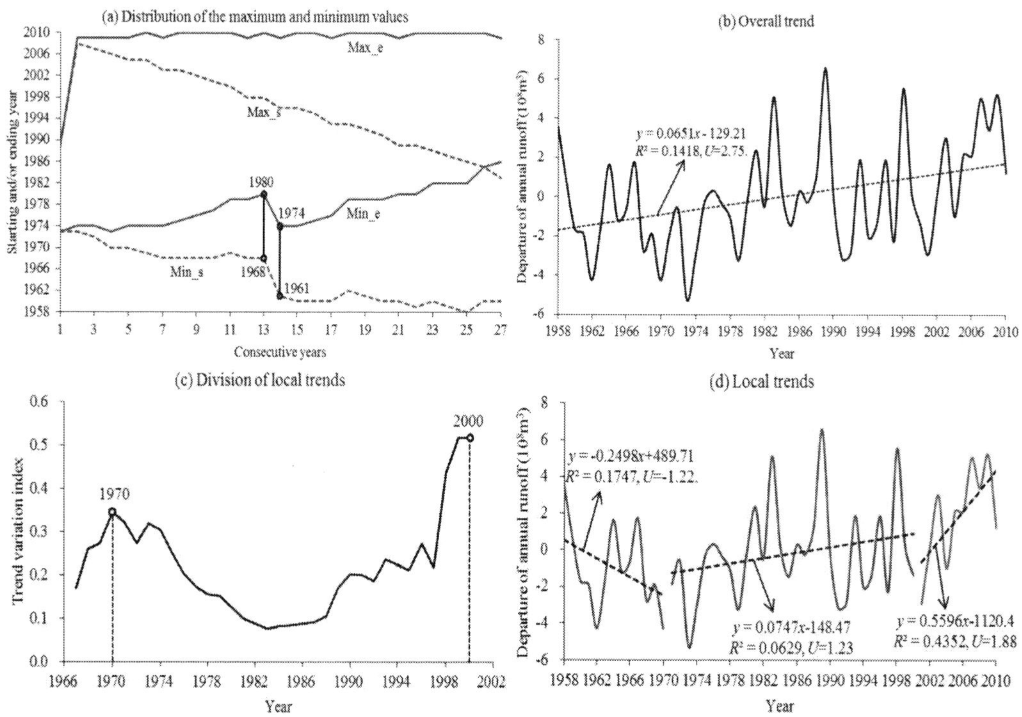

Figure 2. Trend test results.

In graph (a), Max_s and Max_e represent the start years and the end years of n (n=1, 2, …, 27) consecutive years with the maximum runoff volume, respectively; Min_s and Min_e are the start years and the end years of n consecutive years with the minimum runoff volume respectively. In graph (b) and (d), the solid curve is X–X'_{C}

series, and the dotted line represents linear trend.

4.1. Trend component
Based on Mann-Kendall test with the significant level α=0.05 ($U_{\alpha/2}$=1.96), the trend test results of X–X'_C series were shown in figure 2. In graph (a), Max_s curve was above Min_e curve from 1 to 26 consecutive years, which directly indicated that X–X'_C series had a clear increasing tendency between 1958 and 2010. In addition, Min_s and Min_e curves had obvious abrupt turns between 13 and 14 consecutive years, indicating a trough (1968-1974) of X–X'_C series. In graph (b), the overall trend of X–X'_C series was significant with a linear growth speed 0.65×10^8 m^3/10a. The expression of X'_T was the trend equation of X–X'_C series. In graph (c), it showed that X–X'_C series had two remarkable turning points (1970 and 2000). Thus, the series had three stages of local trends, which were displayed in graph (d). The first stage was between 1958 and 1970, with a decreasing speed 2.50×10^8 m^3/10a; the second stage ranged from 1971 to 2000, with a slow growth speed 0.75×10^8 m^3/10a; and the third stage started from 2001 and ends at 2010, with an increasing speed 5.60×10^8 m^3/10a. However, none of the three local trends were significant through Mann-Kendall test. The overall trend and local trends of X–X'_C series were closely related to the variation processes of precipitation and atmospheric temperature in UHR during the past 50 years before 2010 [1, 3, 39].

4.2. Mutation component
The significant overall trend component of hydrologic time series easily misled the mutation analysis, taking the middle year of time series as its mean value mutation year. Hence, it is necessary to strip the trend component from X-X'_C series at first. Through the order cluster analysis method, the mutation analysis result was displayed in figure 3. It could be seen that the most probable mutation point of X-X'_C-X'_T series occurred at 2006, and the jump height at 2006 was 2.0×10^8 m^3 occupying 12.5 percent of the mean annual runoff of UHR. However, as far as the sample size of calibration phase was concerned, the length of annual runoff series after 2005 was so short (only 5 years) that the current mutation test methods (including the rank sum test method) were not suitable to examine the significance of the most probable mutation point. Thus, it could not be confirmed whether the mean value of X-X'_C-X'_T series had a significant mutation at 2006. When estimating the annual runoff series in the validation phase, it was conservatively thought that X-X'_C-X'_T series had no mutation or mutation of mean value at 2006, i.e. X'_M=0 m^3.

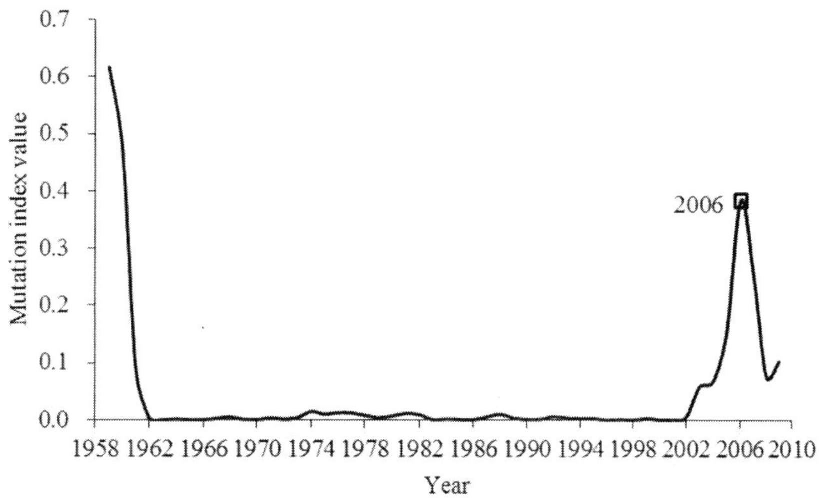

Figure 3. Mutation analysis result.

4.3. Periodic components

The period analysis results were shown in figure 4. The left graph displayed two significant simple periods (P_1=22 years and P_2=6 years) of X-X'_C-X'_T-X'_M series, and the right one presents their respective variation processes. Through the variance analysis method, the 22 years' period was found in the first F test when given the significance level α=0.05. After extracting the 22 years' period component, there was no obvious period component in the remaining part in the second F test when α=0.05. Generally, the potential periods in the second test were less significant than the periods detected in the first test. Based on this, it was imperative to lower the significant level to search the potential periods in the second test. When given the significance level α=0.1, the 6 year' period was found in the second test, but no remarkable periods were found in the third test. So far, the period detection work of X-X'_C-X'_T-X'_M series by variance analysis method was finished. The two significant simple period components of UHR annual runoff series mainly result from the sun activity [17]. At last, the synthesized period component X'_P was acquired through linear superposition of two simple period components.

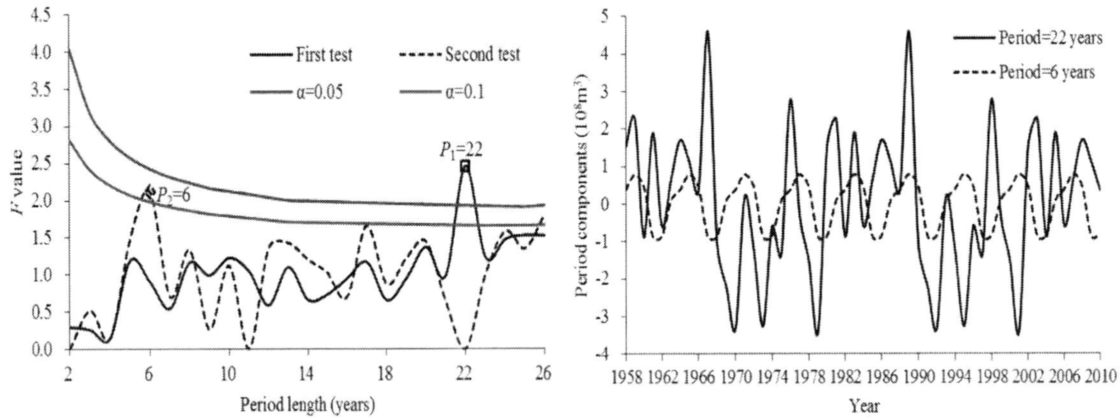

Figure 4. Period detection result (left) and period components (right).

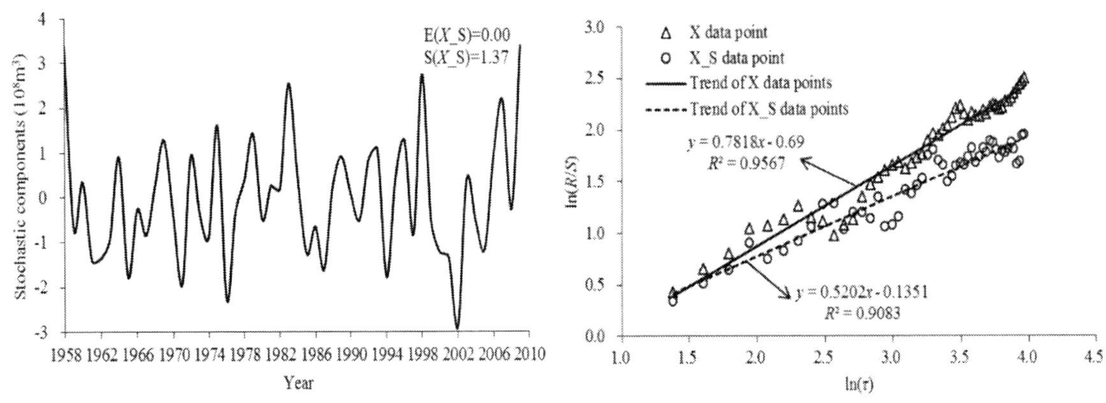

Figure 5. Stochastic component (left) and Hurst exponent variation (right).

4.4. Stochastic component

After extracting all the deterministic components from the UHR annual runoff series X, the remaining part was the stochastic component X'_S which was displayed in the left graph of figure 5. Further, Hurst exponents of X and X'_S were also calculated through R/S analysis method, and the results were presented in the right graph of figure 5. Seen from the right graph, Hurst exponents of X and X'_S were 0.78 and 0.52. The two Hurst exponents showed that X had a remarkable long run dependence property and the randomness of X'_S was very strong. Meanwhile, the great change of Hurst exponent also proved that the persistence of UHR annual runoff series was mainly caused by its deterministic components. Since Hurst exponent of X'_S was very close to 0.5, it was believed that X'_S did not contain any dependent random components. Hence, the following work was to investigate the main distribution range of X'_S as an independent random series.

Numerous studies had shown that the hydrologic time series of various rivers in China could be well fitted by Pearson III (P-III) type distribution [40]. Considering many negative values in X'_S series, P-III distribution type was employed to fit $X'_C+X'_S$ series. The theoretical curve-fitting result was shown in the left graph of figure 6, and it can be seen that the curve fitting (R^2=0.9888) was very successful. Further, the distribution range of X'_S was also obtained based on theoretical frequency curve. In the right graph, according to equations (22)-(24), the accuracy rate of prediction was equal to q-(1-q) (0.5<q<1), and the forecast range is equal to $X'_S(1-q)-X'_S(q)$. The forecast range rose faster and faster with the growth of accuracy rate, which exactly brought a dilemma to the prediction of annual runoff. Even so, the accuracy rate of prediction still attracted more attention. In this study, the appointed probability q was assigned 0.95, and the forecast range was 4.66×10^8 m^3 with the accuracy rate 90%.

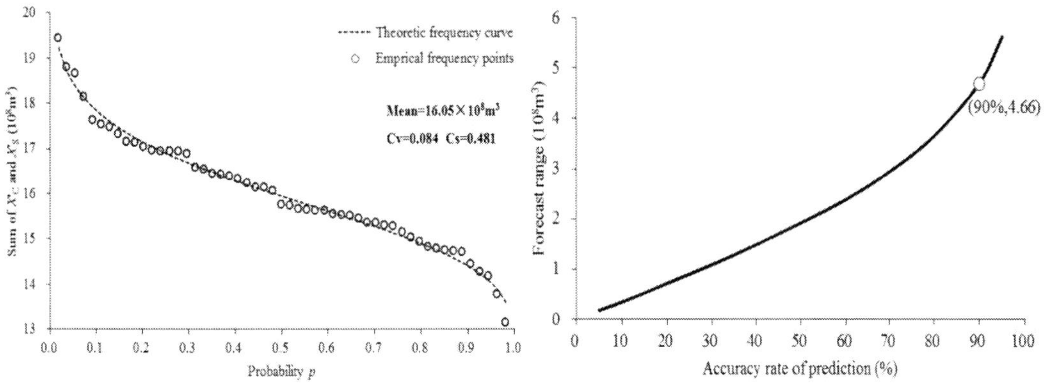

Figure 6. Theoretical curve-fitting result of probability distribution of $X'_C+X'_S$ series (left) and relation between forecast range and accuracy rate of prediction (right).

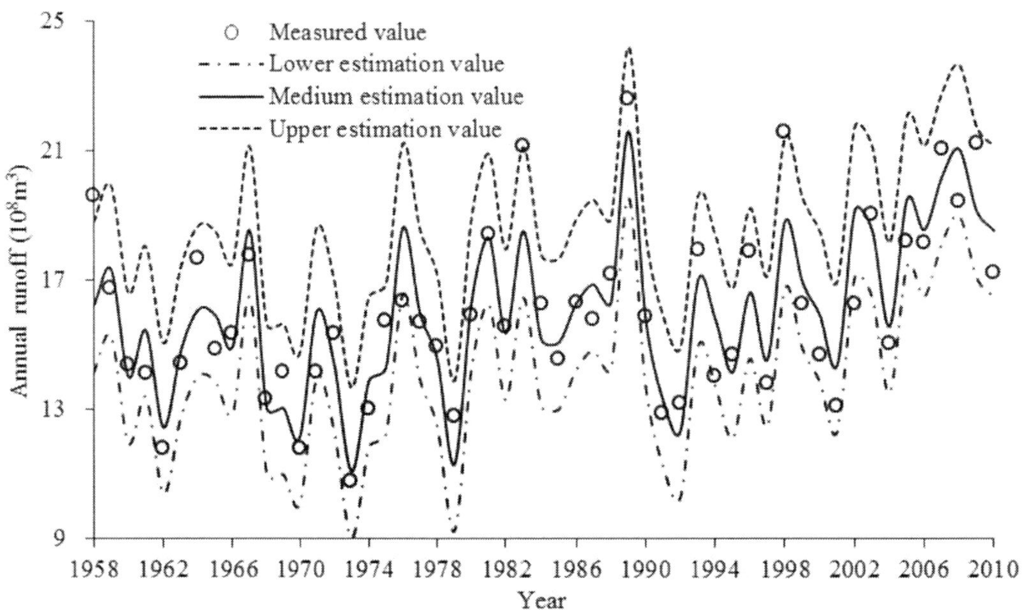

Figure 7. Forecast results of annual runoff of UHR in calibration phase.

4.5. Prediction and evaluation

Based on the deterministic and stochastic components, the estimation results of UHR annual runoff series in the calibration phase (1958-2010) were exhibited in figure 7. The evaluation indexes (NSE=0.74, $MAPE$=6.6%, POP=90.6%) showed that the simulation effects were satisfactory in calibration phase. Moreover, table 1 displayed the medium, lower and upper estimation values of UHR annual runoff series in the validation phase (2011-2015). Seen from this table, the estimation range failed to cover the measured value at 2011, and the medium estimation values were always smaller than the measured values in four consecutive years (2012-2015). Nonetheless, the estimation effects in validation phase were still acceptable on the basis of the evaluation indexes ($MAPE$=6.5%, POP=80%). Therefore, the model based on the time series analysis method was feasible in the prediction of UHR annual runoff.

Table 1. Estimation results of UHR annual runoff in validation phase. X is the measured value of annual runoff; X^M, X^L, and X^U are the medium, lower, and upper estimation values of X, respectively.

Year	X (10^8m^3)	X^M (10^8m^3)	RE (%)	X^L (10^8m^3)	X^U (10^8m^3)
2011	18.95	21.62	14.1	19.56	24.22
2012	19.34	19.15	-1.0	17.09	21.75
2013	19.52	18.40	-5.7	16.34	21.00
2014	18.37	17.12	-6.8	15.06	19.72
2015	20.24	19.28	-4.7	17.22	21.88

Note: RE is the relative error of X^M, i.e. $RE=(X^M-X)/X\times100\%$.

In order to provide the basis for the water allocation of Heihe river basin, the annual runoff series in future five years (2016-2020) were calculated and shown in table 2. In addition, the future UHR annual runoff series probably had some unexpected variations under the double impact of regional climate change and human activities. Hence, it was of great significance to update the current UHR annual runoff prediction model utilizing the latest observation data regularly.

Table 2. Prediction results of UHR annual runoff between 2016 and 2020.

Year	X^M (10^8 m^3)	X^L (10^8 m^3)	X^U (10^8 m^3)
2016	18.02	15.96	20.62
2017	17.17	15.11	19.77
2018	20.02	17.96	22.62
2019	19.66	17.60	22.26
2020	23.61	21.55	26.21

5. Conclusions

An annual runoff prediction method was adopted to forecast annual runoff of UHR based on the linear superposition method. In other words, various components were extracted and investigated by decomposing the historical annual runoff series, hence constructing the forecast model through the linear superposition of these components. The main conclusions are drawn as below:

- During 1958 and 2010, the annual runoff of the upstream Heihe River had an obvious increasing trend with a speed 0.65×10^8 m^3/10a, a probable mutation at 2006, two significant periods (22 years and 6 years) and a persistence property.
- The adopted method based on the linear superposition method is suitable to forecast the annual runoff in the Heihe River.

Acknowledgments

This study is funded by the Natural Science Foundation of China (91325201) and China Scholarship Council.

References

[1] Wang J, Li H, and Hao X 2010 Responses of snowmelt runoff to climatic change in an inland river basin, Northwestern China, over the past 50 years *Hydro. Earth Sys. Sci. Discuss.* **14** 1979-87

[2] Ge Y, Li X, Huang C and Nan Z 2013 A Decision Support System for irrigation water allocation along the middle reaches of the Heihe River Basin, Northwest China *Environ. Modell. Softw.* **47** 182-92

[3] Zhang A, Zheng C, Wang S and Yao Y 2015 Analysis of streamflow variations in the Heihe River Basin, northwest China: Trends, abrupt changes, driving factors and ecological influences *J. Hydrol-Regional Studies* **3** 106-24

[4] Lu L, Li X and Cheng G 2003 Landscape evolution in the middle Heihe River Basin of north-west China during the last decade *J. Arid Environ.* **53** 395-408

[5] Qi S and Luo F 2005 Water environmental degradation of the Heihe River Basin in arid northwestern China *Environ. Monit. Assess.* **108** 205-15

[6] Wang Y, Feng Q, Chen L and Yu T 2013 Significance and effect of ecological rehabilitation project in inland river basins in northwest China *Environ. Manag.* **52** 209-20

[7] Guo Q, Feng Q and Li J 2009 Environmental changes after ecological water conveyance in the lower reaches of Heihe River, northwest China *Environ. Geol.* **58** 1387-96

[8] Zhang Y, Yu J, Qiao M and Yang H 2011 Effects of eco-water transfer on changes of vegetation in the lower Heihe River basin *J. Hydraul. Eng.* **42** 757-65 (in Chinese)

[9] Cheng G, Li X, Zhao W, Xu Z, Feng Q, Xiao S and Xiao H 2014 Integrated study of the water–ecosystem–economy in the Heihe River Basin *Nat. Sci. Rev.* **1** 413-28

[10] Wang G, Cheng G and Shen Y 2002 Dynamic tendency of arid oasis under the influence of water resources decrease-a case study of Ejina oasis in Heihe River Basin *J. Appl. Ecol.* **13** 564-8 (in Chinese)

[11] Qin D, Zhao Z, Han L, Qian Y, Ou Z and Wang M 2012 Determination of groundwater

recharge regime and flowpath in the Lower Heihe River Basin in an arid area of Northwest China by using environmental tracers: Implications for vegetation degradation in the Ejina Oasis *Appl. Geochem.* **27** 1133-45
[12] Zhang Y, Huang Q, Zhang H, Sun X and Gao F 2010 Research on joint operation of cascade reservoirs in the upstream Heihe River *J. Hydroelec. Eng.* **29** 52-6 (in Chinese)
[13] Chou C 2014 Application of set pair analysis-based similarity forecast model and wavelet denoising for runoff forecasting *Water* **6** 912-28
[14] Zhao X and Chen X 2015 Auto regressive and ensemble empirical mode decomposition hybrid model for annual runoff forecasting *Water Resour. Manag.* **29** 2913-26
[15] Qu Y, Zhou J, Liu F, Yang J and Li Y 2006 Medium-and long-term runoff forecasting based on improved elman neural network *J. China Hydrol.* **26** 45-50 (in Chinese)
[16] Shoaib M, Shamseldin A, Melville B and Khan M 2016 A comparison between wavelet based static and dynamic neural network approaches for runoff prediction *J. Hydrol.* **535** 211-25
[17] Chen R, Kang E, Yang J and Zhang J 2002 Nonlinear features of the runoff from mountain areas of the Heihe River, Qilian Mountains *J. Glaciol. Geocryol.* **24** 292-7 (in Chinese)
[18] Demirel M, Venancio A and Kahya E 2009 Flow forecast by SWAT model and ANN in Pracana basin, Portugal *Adv. Eng. Softw.* **40** 467-73
[19] Lü H, Hou T, Horton R, Zhu Y, Chen X, Jia Y, Wang W and Fu X 2013 The streamflow estimation using the Xinanjiang rainfall runoff model and dual state-parameter estimation method *J. Hydrol.* **480** 102-14
[20] Suliman A, Jajarmizadeh M, Harun S and Darus I 2015 Comparison of semi-distributed, GIS-based hydrological models for the prediction of streamflow in a large catchment *Water Resour. Manag.* **29** 3095-110
[21] Jiang X and Liu C 2004 Radial basis function networks based on wavelet analysis for the annual flow forecast *J. Appl. Sci.* **22** 411-4 (in Chinese)
[22] Chu Y, Lan Y, Li X, Zhang Y, Hu X and Liu Q 2005 Models for long-term forecast of annual runoff at Yinluoxia station of Heihe River *J. Des. Res.* **25** 869-73 (in Chinese)
[23] Zhang J and Ding H 2005 Application of the gray topological method in predicting runoff volume at the debouchure of the Heihe River *Arid Land Geogr.* **28** 751-5 (in Chinese)
[24] Hou H, Yang L, Li F and Wang L 2010 Annual runoff forecast of the mainstream of Heihe river based on time series analysis method *Yellow River* **32** 49-53 (in Chinese)
[25] Li J, Zan M and Li B 2014 Grey prediction of out-mountainous annual runoff of Heihe River based on R/S analysis *Areal Res. Dev.* **33** 127-31 (in Chinese)
[26] Shi Y, Shen Y, Kang E, Li D, Ding Y, Zhang G and Hu R 2007 Recent and future climate change in northwest China *Climatic Change* **80** 379-93
[27] Wang L and Zhang X 2010 Effect of the recent climate change on water resource in Heihe River basin *J. Arid Land Res. Environ.* **24** 60-65 (in Chinese)
[28] Yue S and Wang C 2004 The Mann-Kendall test modified by effective sample size to detect trend in serially correlated hydrological series *Water Res. Manag.* **18** 201-8
[29] Tabari H, Marofi S, Aeini A, Talaee P and Mohammadi K 2011 Trend analysis of reference evapotranspiration in the western half of Iran *Agr. Forest Meteorol.* **151** 128-36
[30] Chen Y, Lu B, Zhang Y and Zhou X 2011 Improvement of sequential cluster analysis method for extracting turning point of time series *J. China Hydrol.* **31** 41-4 (in Chinese)
[31] Sang Y, Wang Z and Liu C 2012 Period identification in hydrologic time series using empirical mode decomposition and maximum entropy spectral analysis *J. Hydrol.* **424** 154-64
[32] Tang C, Guo L and Wang X 2007 Application of periodic mean superposition in the middle-to-long term hydrological prediction in Cuntan station *China Rur. Water Hydrop.* **8** 15-7 (in Chinese)
[33] Kantelhardt J, Bunde E, Rybski D, Braun P, Bunde A and Havlin S 2006 Long-term persistence and multifractality of precipitation and river runoff records *J. Geophys. Res-Atmospheres* **111** D1

[34] Jiang T and Deng 2004 Some problems in estimating a Hurst exponent-a case study of applicatings to climatic change *Scient. Geog. Sin.* **24** 177-82 (in Chinese)
[35] Hurst H 1951 Long-term storage capacity of reservoirs *T. Am. Soc. Civil Eng.* **116** 770–99
[36] Xie Y, Huang Q, Chang J, Liu S and Wang Y 2016 Period analysis of hydrologic series through moving-window correlation analysis method *J. Hydrol.* **538** 278-92
[37] Yang T, Yu P and Chen C 2005 Long-term runoff forecasting by combining hydrological models and meteorological records *Hydrol. Process.* **19** 1967-81
[38] Zhang K, Yang Y, Qin Z and Meng Z 2013 Analysis and prediction of annual runoff for Fen river basin *Res. J. Appl. Sci. Eng. Technol.* **6** 872-7
[39] Wang J and Meng J 2008 Characteristics and tendencies of annual runoff variations in the Heihe River Basin during the past 60 years *Scient. Geog. Sin.* **28** 83-8 (in Chinese)
[40] Huang Z, Wang C and Ma J 2002 Adaptability of Pearson III type distribution and analysis for the error of hydrologic design value *Hydrology* **22** 21-4 (in Chinese)

IOP Conf. Series: Earth and Environmental Science 82 (2017) 012058 doi:10.1088/1755-1315/82/1/012058

Carbon tetrachloride pollution pathway of groundwater a typical contaminated site in the east of the city

P Jiang[1,2], Z M Ma[1,2,4], W W Yu[3] and M Wen[1,2]

[1]School of Resources and Environment, University of Jinan, Jinan 250022, China
[2]Research Center of Groundwater Numerical Simulation and Pollution Control Engineering in Shandong Province, Jinan 250022, China
[3]Best Justicial Testing Technology (Qing Dao) Co., Ltd, Qingdao 266500, China

E-mail: stu_mazm@ujn.edu.cn

Abstract. Determine 40 sampling points basing on a comprehensive monitoring. Determine the spatial distribution characteristics of the carbon tetrachloride by using the software of ArcGIS. Determine the location of the pollution sources by using MT3DMS program and Hook-Jeeves arithmetic to simulate, and connecting with the actual situation of carbon tetrachloride to analyze pollution causes. The results show that the source of carbon tetrachloride is located in the northeast near a chemical plant in the study area, whose pollutant concentration is diminishing from northeast to southwest. The main reasons to the pollution are that factories discharge waste water at random, leakage of open channel and culvert, sewage irrigation and the vulnerability of geological conditions in this area.

1. Introduction

The production and use of CCl_4 (carbon tetrachloride) is the main source of CCl_4 pollution of groundwater environment. Groundwater contamination with carbon tetrachloride was common in the industrial complexes [1, 2]. It is commonly found in groundwater and surface water at concentrations exceeding the maximum contaminant level of 5 μg/L [3, 4]. Carbon tetrachloride data from Cape Grim and other strategic sites around the globe have been used to derive global and regional CCl_4 emissions (Asia, about 75% of global emissions; Africa about 10%; North and South America about 10%; Europe about 5%; Australia and New Zealand lest than 1%) [5]. China's contribution to global emissions increases from 7.5% to 19.5% during 1992–2009, but the contribution is reduced to 9.9% and 8.0% in 2010 and 2011 [6]. The shallow groundwater along the Xiaoqing River in Shandong Province has been polluted by CCl_4 with a maximum concentration of 380 μg/L and a pollution area of 80 km^2 [7], based on the investigation and every day monitoring data of the CCl_4 pollution in the groundwater. The contaminated area at the initial stage reached the max contaminant concentration of 249.7 μg/L [8]. Jiangsu Xuzhou, Shandong Zoucheng, Hebei Handan, Beijing suburbs and other areas have found groundwater CCl_4 pollution phenomenon [9]. Through an analysis of shallow groundwater in a certain area show that the detection rate of CCl_4 was 26% in May 2000, 32.5% in December 2000, in May 2001 was 58.06%, 32.69% in July 2001 [10].

Groundwater pollution source identification problem is the inverse problem of the groundwater and optimization method is the best method for solving this problem [11]. The commonly used optimization methods have Newton Method, Simplex Method, Conjugate Gradient Method, Hooke-Jeeves Method, Genetic Algorithm and Simulated Annealing [12]. Ayvaz used Harmony

IOP Conf. Series: Earth and Environmental Science 82 (2017) 012058 doi:10.1088/1755-1315/82/1/012058

Search Algorithm to invert the location and intensity of groundwater pollution source [13]. Ayvaz used a new simulation–optimization approach where a binary genetic algorithm and a generalized reduced gradient method are mutually used that is proposed for solving the areal groundwater pollution source identification problems [14]. Mirghani used Parallel Evolutionary Algorithm to identify groundwater pollution source [15]. Jiang Simin made use of Hybrid Hooke-Jeeves and Attractive Repulsive Particle Swarm Optimization Method, combining MT3DMS program, to invert the location and intensity of groundwater pollution source [16]. Bashi-Azghadi presents a new regret-based optimization model which can rapidly identify an unknown pollution source [17]. Jiang Simin adopted Hybrid Simplex Method of Simulated Annealing to calculate the intensity of groundwater source [18]. Datta presents a new source identification methodology that uses a classical nonlinear optimization model linked to a flow and transport simulation model [19].

Karst water samples sampled at 40 monitoring sites will be analyzed to find out the characteristic of groundwater CCl_4 pollution. Groundwater CCl_4 pollution source identification model will be build, which will be simulate through using MT3DMS program and Hook-Jeeves arithmetic to identify pollution sources. Thinking about pollution sources of CCl_4, hydrogeological condition and pollution way of CCl_4, causes of groundwater CCl_4 pollution will be analyzed.

The study area located in the east of the city with a total area about 120 km^2. Study area belongs to the warm temperate continental monsoon climate. The study area has obvious seasonal precipitation, the average rainfall 660.3mm for many years, more concentrated in the summer months of June to September, accounting for 75% of total annual precipitation. The annual evaporation of study area is 1717.5 mm, which is 2.6 times of the annual precipitation, the month of evaporation of the largest amount is July, the average evaporation is up to 254.8 mm. The formation of a specific area of hydrogeological conditions and distribution will be affected by climate, regional geological structure, stratum lithology, topography factors and other factors, which the geological structure is the decisive factor. The main types of groundwater are loose rock pore water and fracture-karst water in the study area. The study area is located in the north of mounttai block uplift, which belongs to the category of luxi vortex structure system on the geological structure, relatively complete formation development. The formation from south to north from the old to the new is outcrops of Cambrian, Ordovician, Carboniferous-Permian, Mesozoic and Cenozoic Quaternary, Yanshanian igneous rock. Loose rock types of pore water in the study area are mainly distributed in northern piedmont inclined plain to the Yellow River alluvial plain, covering most of the area in the study area of quaternary strata thickness increases gradually from south to north, whose thickness range between 20 to 120 m. Fissure karst water in the study area are mainly distributed in the south of the study area of low hilly land and buried deep under Quaternary pore water aquifer. The depth of shallow ground water in the study area is in commonly 7 meters to 30 meters. The recharge of shallow groundwater mainly comes from atmospheric precipitation in the south. The main supply sources of shallow groundwater in northern of the study area is the recharge of deep groundwater in addition to the atmospheric precipitation and surface infiltration recharge rivers. A large number of shallow groundwater in the study area is used in industrial and agricultural production and the supply of spring water. In addition, the rest run off into the Xiaoqing River.

2. Methods

2.1. Data source

In view of the hydrogeological conditions in the studied area and the distribution characteristics of different enterprises, the sampling points are arranged as follows: make the sampling points distribution on all kinds of hydrogeological conditions, in order to reduce the unnecessary loss because of the concentrated distribution of the sample points; try best to make sure all kinds of industrial waste water pollution of groundwater can be analyzed. According to the above principles of sampling point arrangement, there are 40 different sample points in the studied area as shown in figure 1.

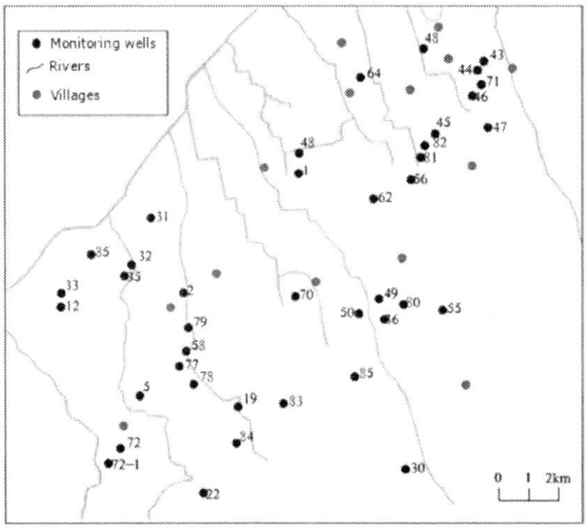

Figure 1. The sample points in the study area.

2.2. Distribution characteristics

Based on the China national standard method and the EPA method, GC/MS method was used to analyze the concentration of CCl_4 to each sample. Get the detection rate and exceeding rate of CCl_4 in the study area as well as the statistical data based on the statistical analysis to the concentration of CCl_4 that is measured. The results show that the detection rate of CCl_4 is 100%, the exceeding rate is 94.83%, and the minimum value is 1.50 μg/L, the maximum is 23.31 μg/L. Choose the ordinary kriging method to do the interpolation of the concentration of the pollution that was measured with the software of ArcGIS. The spatial distribution of CCl_4 pollutants in the study area is showed in figure 2.

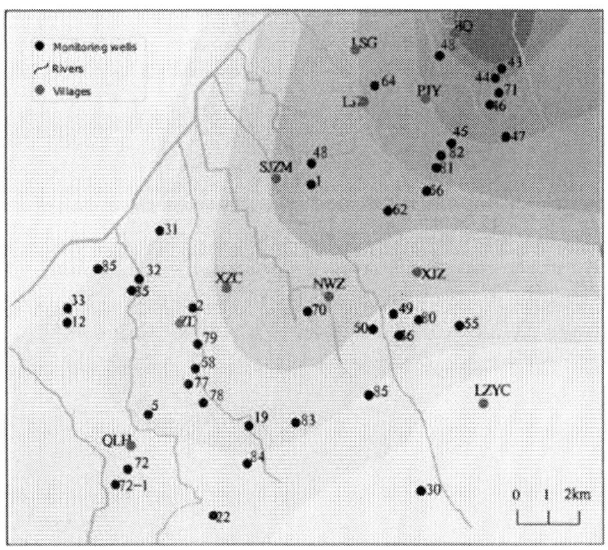

Figure 2. Space distribution of CCl_4 in the study area.

2.3. Source apportionment

Groundwater pollution source identification problem is the inverse problem of the groundwater and is the inversion of the pollution source location and concentration on the basis of the concentration of CCl_4 at sampling sites. Groundwater pollution source inversion can be converted into solving optimization model that regards pollution source location and concentration as the decision variables. In the process of solving optimization model, groundwater flow model and groundwater solute migration model need to be solved continually. So, proper groundwater flow model and groundwater solute migration model are the basis of identifying groundwater pollution source. In this paper, MODFLOW and MT3DMS program are selected to simulate groundwater flow model and groundwater solute migration model, respectively.

Groundwater flow equation of MODFLOW:

$$\frac{\partial}{\partial x}(K_x \frac{\partial h}{\partial x}) + \frac{\partial}{\partial y}(K_y \frac{\partial h}{\partial y}) + \frac{\partial}{\partial z}(K_z \frac{\partial h}{\partial z}) + W = \mu \frac{\partial h}{\partial t} \quad (1)$$

Where, h is groundwater level, K_x, K_y, K_z are hydraulic conductivity in x, y, z direction(m·d^{-1}), W is source and sink (1/d), μ is specific storage (1/d) and t is time (d).

Groundwater solute transport equation of MT3DMS:

$$\frac{\partial(\theta c)}{\partial t} = \frac{\partial}{\partial x_i}(\theta D_{ij} \frac{\partial c}{\partial x_j}) - \frac{\partial}{\partial x_i}(\theta v_i c) + q_s c_s \quad (2)$$

Where, θ is porosity (dimensionless), c is solute concentration (mg·L-1), t is time(d), D_{ij} is coefficient of dispersion (m2·d-1), v_i is the actual water flow rate (m·d-1), q_s is value of recharge and discharge of unit volume aquifer (m3·d-1) and c_s is solution concentration in source and sink (mg·L^{-1}).

Objective function

$$E(x, y, z, q) = \sqrt{\frac{1}{n_g . n_t} \sum_{i=1}^{n_p} \sum_{j=1}^{n_t} \left[c_{i,m}^j - c_{i,g}^j\right]^2} \quad (3)$$

Where, $c_{i,m}^j$ is value of simulation of monitoring site of i in term of j(mg • L^{-1}), $c_{i,g}^j$ is observed value of monitoring site of i in term of j(mg • L^{-1}), n_g is number of observation well, n_t is number of observation.

2.4. Constraint condition:

Except groundwater flow model and groundwater solute migration model mentioned above, the value range of the location of pollution source and pollution intensity should be taken into account. The value range of the location of pollution source is entering study area and the location of pollution source is expressed in layer, row and column of finite difference grid. So, the coordinate of the location should be integer. The pollution intensity should have upper and lower limit in order to find out proper intensity fast. The pollution intensity isn't too maximum but also too minimum.

According to above processing, the problem of groundwater pollution source identification will be changed into optimization model. Based on MT3DMS program, using Hybrid Hooke-Jeeves and Attractive Repulsive Particle Swarm Optimization Method [16], the optimization model of groundwater pollution sources will be simulated and calculated.

3. Results and discussion

3.1. Distribution characteristics of CCl_4

As can see from figure 2, the high values area of concentration appears in the northeast of the study area around the sample point 46,47,48,71. And the concentration was decreasing from the center with high value area to the southwest in the study area, low concentration area is concentrated in the study area in the northwest and south parts. The concentration of CCl_4 near the 48 point is as high as 23.00 μg/L, and near 30 sampling point is only 1.80 μg/L. The standard contamination of CCl_4 in groundwater is 2 μg/L [20]. The concentration of sample that was checked out was 1.50 ~ 23.31 μg/L. The figure shows that the detected value has been exceeded in many areas, and it shows that the pollution of CCl_4 is very serious. The concentration of CCl_4 is as high as 20.87 μg/L near a chemical factory. Thus, it shows that the factory is the main pollution source of CCl_4 in the study area. The highest concentration of CCl_4 is 23.31 μg/L in the study area, which is 11.65 times higher than the standard value. It shows that the pollution of CCl_4 in groundwater is very serious.

3.2. The law of CCl_4 migration

Affecting the groundwater CCl_4 migration mainly factors include convective dispersion, volatilization, adsorption and biodegradation.

Based on the results of the first survey, the change of the concentration of CCl_4 in the aquifers of the study area is simulated and the results are shown in figure 3.

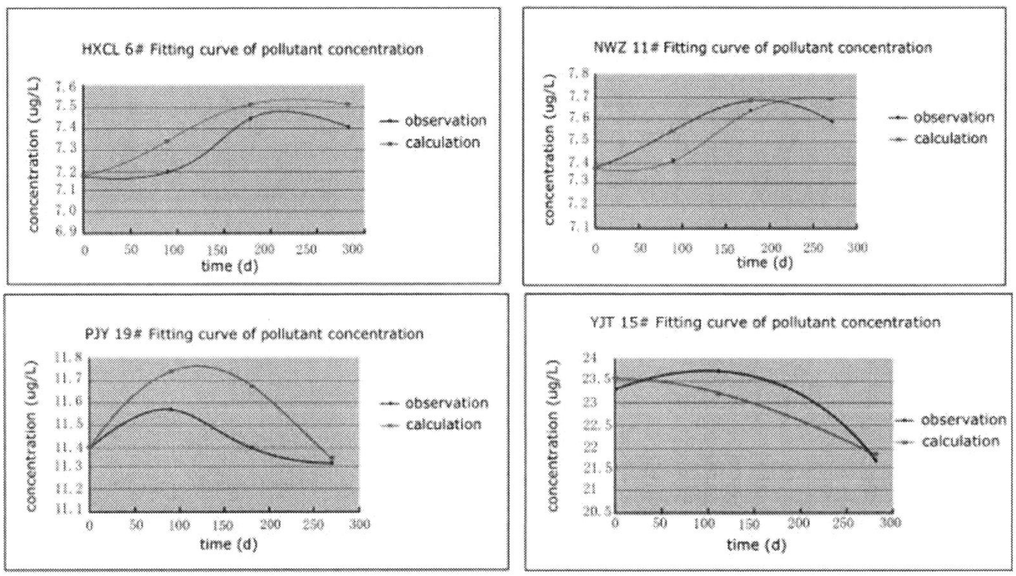

Figure 3. Simulation value and actual value.

Figure 3 shows that the trends of the observation curve and the calculated curve are basically the same in the simulation. The observed curve rises and falls once after the phenomenon, but the range is not large. The speculation may be due to changes in the season or temperature of the study area, leading to changes in the groundwater system under redox conditions. The model can basically reflect the migration and transformation of CCl_4 in the study area.

3.3. Pollution ways of CCl_4

The position of sources is shown in figure 4. It is consistent with spatial distribution of CCl_4 by ArcGIS, showing a high concentration of northeast and gradually decreasing trend to the southwest. There is a chemical factory near the pollution sources through field visits, which discharged large number of CCl_4 by open channel and culvert. The results show that the chemical factory is the pollution source of CCl_4. Pollution ways of CCl_4 of groundwater in the study area are as follows:

- As important industrial and mining enterprises in the study area, in the process of each enterprise production operation, soils in the upper layer are usually polluted first. Pollutants retention infiltrates through leaching. These contaminants run into the underground through rainfall, resulting in widespread contamination of groundwater.

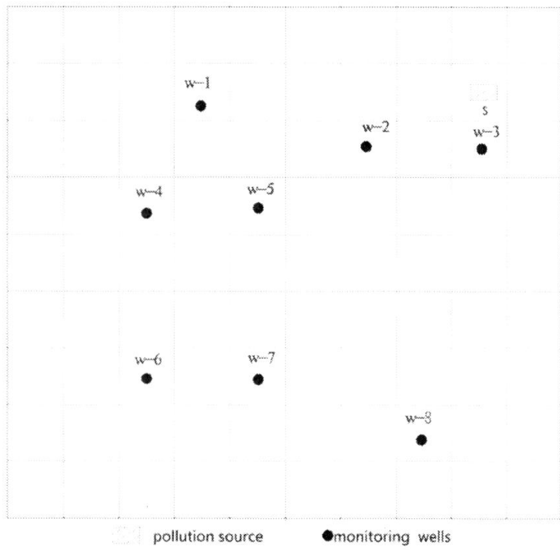

Figure 4. Location of pollution source.

- Research in northern area, the lithology of vadose zone is given priority to with powder sand belt, loose structure, penetration ability is stronger, permeability coefficient is 0.30 ~ 0.35, water depth is general less than 3 m, pollutants directly run into underground or seep into the ground through rainfall. Thus, pollute the groundwater. Moreover, due to the presence of a chemical factory in the northeast of the study area, make the pollution of CCl_4 is most serious nearby. Research in southern area, the lithology of vadose zone is given priority to with sand-loam and cohesive soil gravel, penetration ability is poorer, permeability coefficient is 0.15 ~ 0.20, water depth is general more than 10 m. However, due to the loess with vertical joints, gravel sandwich has some porosity, which leads to poor vulnerability and makes it easy for CCl_4 to infiltrate into ground, pollute groundwater.
- The discharge way of sewage of industrial and mining enterprises in the study area is complex, which have open channel and culvert. The sewage ditches are on the top of the normal water level, the bottom is given priority to with clay and silty clay layer, CCl_4 pollutants seep into the vadose zone though ditches, polluting the groundwater. Most of the sewage channels in the study area are in north-south direction, basically consistent with the groundwater flow direction, but some sewage channels are in east- west direction, the drainage ditches transverse cutting of groundwater flow direction and caused more serious groundwater pollution. The closer distance to the channel is, the more serious the pollution is.
- Most industrial enterprises in the study area don't have a complete management system, and in all aspects of operation or production process there is a serious phenomenon such as run, drip, and leak. And the pollutant of production of these enterprises is most CCl_4, which is difficult to avoid seeping into groundwater and polluting it. In addition, most of irrigation wells in the study area are scrapped due to the reduction of groundwater level, and the farmers have to refer to enterprises waste water or river water for irrigation. It is easy to cause a large area of groundwater pollution.

4. Conclusion

Determine 40 sampling points basing on a comprehensive monitoring. Determine the spatial distribution characteristics of the CCl_4 by using the software of ArcGIS. Direct supply of atmospheric precipitation and surface water to ground water, complementary relationship between loose rock pore water and fracture-karst water all lead to hydrogeological conditions vulnerable in the study area, which makes it easy for surface pollutants to infiltrate into ground and thus pollute groundwater. It is found that through research and analysis, the ability of the aquifer to remove CCl_4 is not prominent. By analyzing industrial layout and hydrogeological conditions, there are mainly two causes of carbon tetrachloride pollution: the vulnerability of geological conditions in this area; the other being the serious pollution in this area caused by remaining pollution sources of the industrial.

Acknowledgment

This work was supported by Shandong provincial key research project (no. 2015GSF117025) and Shandong provincial natural science foundation, China (no. ZR2014DM011). The authors thank anonymous reviewers for their constructive comments.

References

[1] Han B, Wang X, Zhu X and He K L 2004 Groundwater contamination by carbon tetrachloride in Karstic area in China *Acta Sci Circumstantiae* **24(6)** 982-8
[2] Yu S, Lee P K and Hwang S I 2015 Groundwater contamination with volatile organic compounds in urban and industrial areas: Analysis of co-occurrence and land use effects *Environ. Earth Sci.* **74(4)** 3661-77
[3] Reynolds G W, Hoff J T and Gillham R W 1990 Sampling bias caused by materials used to monitor halocarbons in groundwater *Environ. Sci. Technol.* **24:1(1)** 135-42
[4] Xu M, Gu X, Lu S, Qiu Z, *et al* 2015 Degradation of carbon tetrachloride in aqueous solution in the thermally activated persulfate system *J. Hazardous Mater.* **286(3)** 7-14
[5] Fraser P J, Dunse B L, Manning A J, *et al* 2014 Australian carbon tetrachloride emissions in a global context *Environ. Chem.* **11(1)** 77-88
[6] Bie P, Fang X, Li Z, Wang Z and Hu J 2017 Emissions estimates of carbon tetrachloride for 1992-2014 in China *Environ. Pollut.* 1-9
[7] Fang S and Chen X L 2001 Environmental problems caused by groundwater development and management *Groundwater* **23(1)** 8-11
[8] Guo Y, Tian X Z, Zhang L B, Wei H E and Shan Q 2011 Characteristics of carbon tetrachloride Pollution in groundwater in a water source *South-to-North Water Diversion and Water Sci. Technol.* **9(3)** 109-13
[9] Pei Z P 2009 The carbon tetrachloride pollution mechanism of a karst groundwater supply source in one city (Xuzhou: China University of Mining and Technology)
[10] Zhang D Z, Chen H H, Li H M, Zou S Z and Liu L C 2002 Halogenated hydrocarbon contaminants in shallow groundwater Chinese *Geology* **29(3)** 326-9
[11] Sun N 1999 *Inverse Problems in Groundwater Modeling* (Netherlands: Kluwer Academic Publishers)
[12] Jiang S M, Cai Y, Wang M, *et al* 2012 Simultaneous identification of groundwater contaminant source and aquifer parameters by harmony search algorithm *Shuili Xuebao* **43(12)** 1470-7
[13] Ayvaz T M 2010 A linked simulation-optimization model for solving the unknown groundwater pollution source identification problem *J. Contam. Hydrol.* **117** 46-59
[14] Ayvaz T M 2016 A hybrid simulation–optimization approach for solving the areal groundwater pollution source identification problems *J. Hydrol.* **538** 161-76
[15] Mirghani Y B, Mahinthakumar K G, Tryby M E, *et al* 2009 A parallel evolutionary strategy based simulation-optimization approach for solving groundwater source identification problem *Adv. Water Resour.* **32** 1373-85
[16] Jiang S M, Wang P, Shi X Q, *et al* 2012 Groundwater contaminant source identification by hybrid Hooke-Jeeves and attractive repulsive particle swarm optimization method *J. Jilin Univ. (Earth Sci. Edit.)* **42(6)** 1866-72
[17] Bashi-Azghadi S N, Kerachian R, Bazargan-Lari M R and Nikoo M R 2016 Pollution source identification in groundwater systems: Application of regret theory and Bayesian networks *IJSTC* **40(3)** 1-9
[18] Jiang S M, Zhang Y L, Cai Y, *et al* 2013 Groundwater contaminant identification by hybrid

simplex method of simulated annealing *J. Tongji Univ.* **41(2)** 253-7
[19] Datta B, Chakrabarty D and Dhar A 2011 Identification of unknown groundwater pollution sources using classical optimization with linked simulation *J. Hydro-Environ. Res.* **5(1)** 25-36
[20] GB5749-2006 *Standards for Drinking Water Quality* (China)

IOP Conf. Series: Earth and Environmental Science 82 (2017) 012059 doi:10.1088/1755-1315/82/1/012059

Nitrate-nitrogen contamination in groundwater: Spatiotemporal variation and driving factors under cropland in Shandong Province, China

J Liu[1], L H Jiang[2], C J Zhang[1], P Li[1] and T K Zhao[1,3]

[1]Institute of Plant Nutrition and Resources, Beijing Academy of Agriculture and Forestry Sciences, Beijing 100091, China
[2]Institute of Plant Nutrition and Resources, Shandong Academy of Agriculture Sciences, Jinan 250100, China

E-mail: tkzhao@163.com

Abstract. High groundwater nitrate-N is a serious problem especially in highly active agricultural areas. In study, the concentration and spatialtemporal distribution of groundwater nitrate-N under cropland in Shandong province were assessed by statistical and geostatistical techniques. Nitrate-N concentration reached a maximum of 184.60 mg L^{-1} and 29.5% of samples had levels in excess of safety threshold concentration (20 mg L^{-1}). The median nitrate-N contents after rainy season were significantly higher than those before rainy season, and decreased with increasing groundwater depth. Nitrate-N under vegetable and orchard area are significantly higher than ones under grain. The kriging map shows that groundwater nitrate-N has a strong spatial variability. Many districts, such as Weifang, Linyi in Shandong province are heavily contaminated with nitrate-N. However, there are no significant trends of NO_3^--N for most cities. Stepwise regression analysis showed influencing factors are different for the groundwater in different depth. But overall, vegetable yield per unit area, percentages of orchard area, per capita agricultural production, unit-area nitrogen fertilizer, livestock per unit area, percentages of irrigation areas, population per unit area and annual mean temperature are significant variables for groundwater nitrate-N variation.

1. Introduction

Natural groundwater resource provides drinking water for nearly half the population in China and other areas in world [1-3]. However, groundwater is unavoidable to be polluted by industry and municipal wastewater discharges and excessive use of fertilizer in agriculture [4, 5]. In intensive cropland areas, nitrate-N in groundwater commonly exceed natural levels, which may produce potential hazards to human health, such as birth defects, digestive cancers and leading to increase in the occurrence of an immune-system impairment [6-10].

Nitrate-N pollution in agriculture has been an environmental issue in world [11]. Higher concentrations in shallow groundwater in intensive agricultural regions are generally believed to come from agricultural activities [12]. Many efforts have been made to understand how factors affect nitrate concentration in groundwater, such as soil properties, iron, dissolved oxygen and redox conditions in groundwater and so on [13-15]. However, for mitigating nitrate pollution, it is difficult to alter these factors, such as redox conditions [16, 17]. Therefore, it is necessary to analyze the driving factors for regulating groundwater nitrate contamination. Many studies have showed that fertilizer and irrigation in agriculture as the main factor for increasing nitrate-N concentration in groundwater [18-20]. Furthermore, manure applied in agriculture has impacted nitrate-N concentration in subsurface drainage water. Excess fertilizer in agriculture resulted in greater residual nitrate in soil and cause increased NO_3^- leaching to the groundwater [21-23]. Thus, it is imperative to grasp the correlation between agriculture and groundwater quality and develop good agricultural practices by farmers for agricultural sustainable development. Additionally, it is also necessary to understand non-artificial reasons for groundwater nitrate contamination, such as climate factors [24].

Shongdong province was one of the intensive agricultural areas in the North China Plain, which called "China's granary" [25]. Monitoring and understanding trends of groundwater nitrate concentration in this area are imperative for protection of groundwater quality. Zhang W L *et al* (1996) found nitrate pollution in groundwater caused by agriculture in the North China Plain [26]. Following, many districts have been reported heavy contamination of groundwater nitrate [27-29]. Zhao T K *et al* conducted the survey of groundwater nitrate-N pollution in seven provinces of the North China, including Shandong province [30]. In recent years, many scientists carried out research from contamination evaluation, tracing source, nitrogen balance and effecting factor analysis on groundwater pollution in Shandong [29, 31-34]. However, the spatialtemporal distribution, variation trend and key factors for nitrate pollution under intensive croplands in this area are inadequate, which is conducive to develop suitable agricultural practices to alleviate the groundwater nitrate pollution. Therefore, this study aimed at protection groundwater, a strategic resource in Shandong, by analysing above problems, especially from aspects of groundwater depth, time series and comprehensive agricultural practices.

2. Materials and methods

2.1. Study area

Shandong province is located in eastern China and occupies a total area of 156,700 km^2. The Yellow River, running through southwest and northern part of the province, form large area of flood alluvial plain. The main crops are wheat, maize, greenhouse vegetables, cotton and orchard, and the wheat-maize rotation vegetable and fruit planting account for multitude cultivated area. The typical monsoonal climate results in 70–80% of annual precipitation concentrating from June to September in study area. The annual average temperature ranges from 11℃ to 14℃, with greater differences in east-west region than in north-south. Annual precipitation ranges from 550 mm in the northwest to 950 mm in the southeast.

2.2. Samplings and data collection

About 2297 groundwater samples from 2006 to 2012 were collected, covering 17 cities in Shandong province (figure 1). The sampling conducted in May (before rainy season) and October (after rainy season) for every site, and mainly involved in intensive cropland, including grain, vegetable and orchard. Samples were collected in polyethylene bottles after pumping the wells for about 30 min. The pumping ensured samples not coming from the water in the storage within the wells. First, the polyethylene bottles were rinsed with the water samples to be collected to reduce the probable contamination. All samples were transported to the laboratory with an ice-cooled cabin and they were stored in a cooler until analysis. Nitrate-N concentration of water samples were analyzed in laboratory using a colorimetric method with an UV-vis spectrophotometer [35].

During sampling, basic information about well location, groundwater depth, planting pattern and surrounding land-use were collected. According to depth, sampling wells were divided into three types, shallow (0-30 m), middle-deep (30-100 m) and deep (>100 m). Furthermore, fourteen possible variables, including livestock per unit area, annual precipitation, annual mean temperature, population per unit area, percentages of crop area, percentages of grain crop area, grain yield per unit area, percentages of vegetable area, vegetable yield per unit area, percentages of orchard area, fruit yield per unit area, per capita agricultural production, percentages of irrigation areas, nitrogen fertilizer per unit area, were assembled at the municipal level from statistical data and literatures [36, 37].

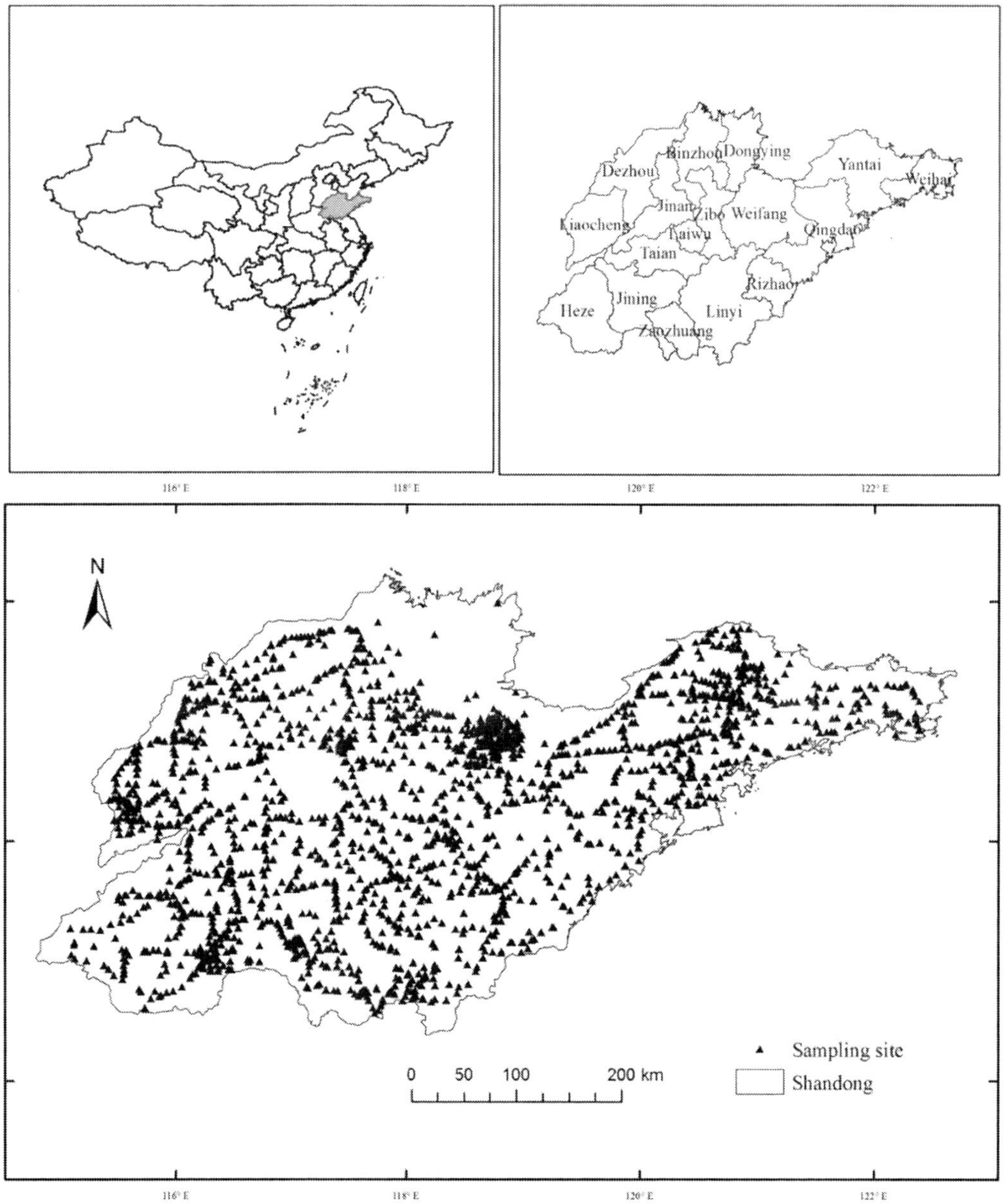

Figure 1. Sampling sites for groundwater nitrate contamination in Shandong, China from 2006 to 2012.

2.3. Statistical and geostatistical analyses

Analyzed with the Shapiro–Wilk test, the data regarding NO_3-N concentration were not normally distributed. Then, the nonparametric Mann-Whitney U test or Kruskal-Wallis test were used to compare differences of groundwater nitrate-N concentration among depth, sampling seasons (before and after a rainy season) and cropland uses.

The significant factors and the regression model for nitrate-N variation were then completed by stepwise regression with stepwise Akaike's Information Criterion (AIC). In statistical analysis, square-root transformation was applied to all data to transform non-normal data to a data set that is reasonably normal. All statistical analyses were conducted using R software (version 3.0.3).

3. Results and analysis

3.1. Nitrate-N content in groundwater

Nitrate-N concentration collected in May from 2006 to 2012 run up to 181.60 mg L^{-1} with a median value of 9.47 mg L^{-1}, except for some sites being not detected. Correspondingly, nitrate-N concentrations in October peak at 184.60 mg L^{-1} with a median value of 10.96 mg L^{-1}, higher than those in May. According to Quality standard for groundwater in China (GB/T 14848-1993) [38], NO_3-N contents in 70.56% of samples in May are below safety threshold concentration (20 mg L^{-1}), while 70.44% within the threshold in October (figure 2). In sum, there were statistically significant differences before and after rainy season (May and October, respectively) with the Mann-Whitney U test ($p<0.05$).

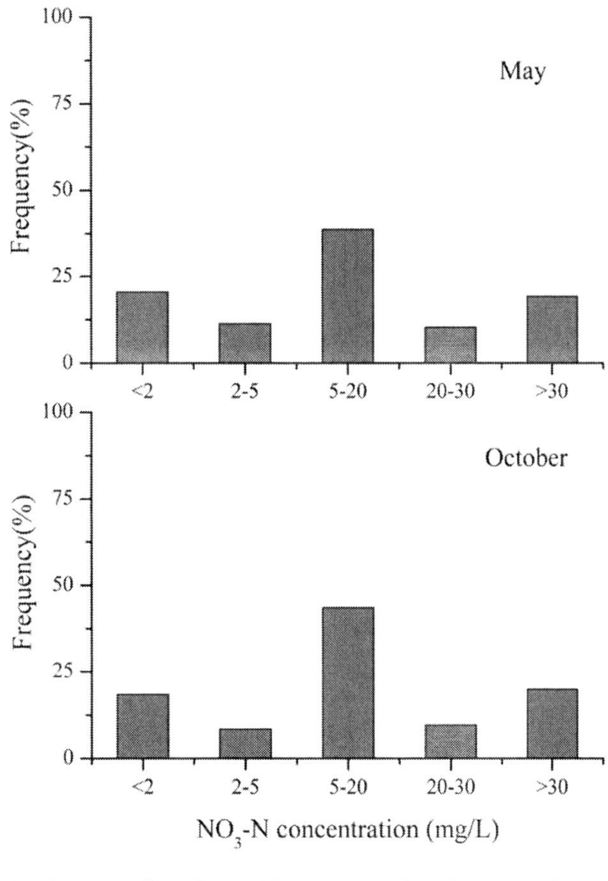

Figure 2. Frequency distribution for nitrate-N concentration in groundwater samples in May and October.

Generally, NO_3^--N content decreased with groundwater depth (figure 3). In study, groundwater NO_3^--N content in shallow wells (range=0-184.6 mg L^{-1}, mean=19.2 mg L^{-1}) are significantly higher than ones in middle-deep (range=0-171.4 mg L^{-1}, mean =16.18 mg L^{-1}) and deep wells (range=0-181.6 mg L^{-1}, mean =15.66 mg L^{-1}), analyzed with Kruskal-Wallis test ($p<0.05$). However, there are no difference between middle-deep and deep wells ($p>0.05$).

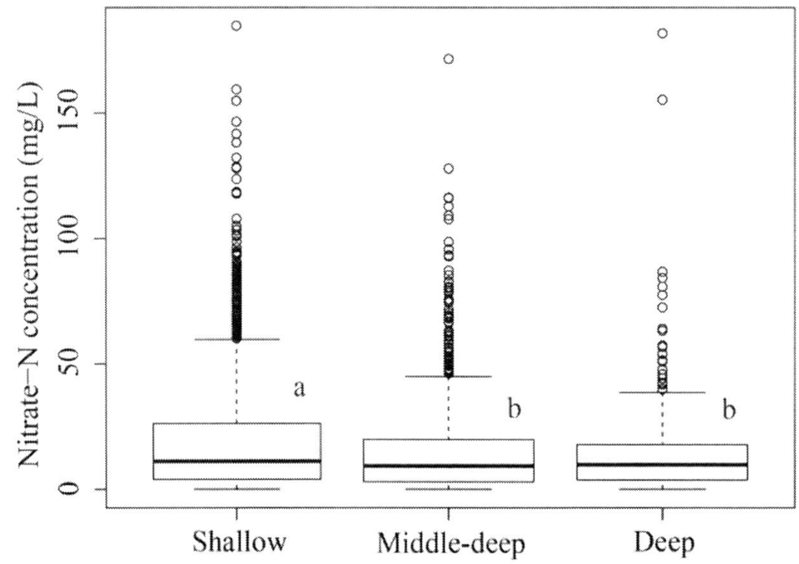

Figure 3. Comparison of measured groundwater nitrate-N concentration among three depths of wells in study area. (Different letters on each box represent a significant difference across cropland use type ($p<0.05$).

A non-parametric test showed there are significant differences in nitrate-N concentration among grain, vegetable and orchard with a 1% significance level (figure 4). The highest content of NO_3^--N was under vegetable area (range=0-171.4 mg L^{-1}, mean=27.40 mg L^{-1}) followed by orchards (range=0-184.6 mg L^{-1}, mean=24.6 mg L^{-1}) and grains (range=0-155.1 mg L^{-1}, mean=13.53 mg L^{-1}). However, there are no difference between vegetable and orchard land ($p>0.05$).

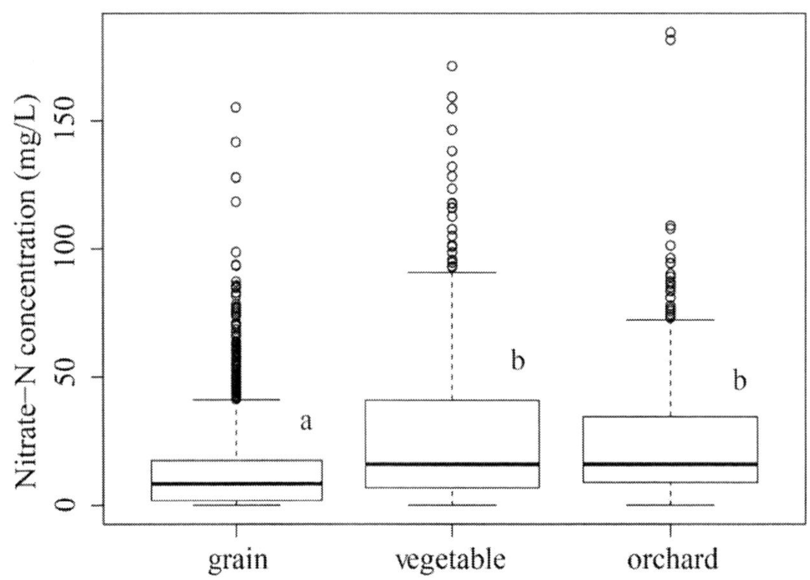

Figure 4. Comparison of measured groundwater nitrate concentration among different cropland use types in shallow wells in study area. (Different letters on each box represent a significant difference across cropland use type ($p<0.05$).

3.2. Spatiotemporal variability of nitrate-N

Distribution of nitrate-N contamination in study area was estimated with spatial statistics. The computed model parameters are shown in table 1. The spatial semivariogram may be affected by intrinsic (physical, chemical, and biological characteristics of hydraulic and geographic conditions) and/or extrinsic (agricultural management practices, such as fertilization, irrigation and animal wastes) factors and can be classified into three categories according to a nugget-to-sill ratio (%). A ratio of <25% indicates a strong spatial dependence, a ratio of 25%-75% indicating a moderate spatial dependence and a ratio of >75% indicating a weak spatial dependence [39]. In study, the ratios of nugget-to-sill for NO_3^--N ranged from 0.609 to 0.742 in 2007, 2008, 2010 and 2011, showing a moderate spatial dependence, which showed NO_3^--N might be determined by extrinsic and intrinsic factors. However, the ratios of nugget-to-sill for NO_3^--N ranged from 0.880 to 0.999 in 2006, 2009 and 2012, indicating a weak spatial dependence and NO_3^--N might be affected by extrinsic factors.

Table 1. Model parameters for semivariograms of logit-transformed NO_3^--N concentration in shallow wells in Shandong from 2006 to 2012.

year	Model	Nugget	Sill	Range (m)	Nugget/Sill	r^2	RSS
2006	Exponential	0.001	1.779	43000	0.999	0.167	0.355
2007	Spherical	0.495	1.373	332000	0.639	0.949	0.0327
2008	Exponential	0.767	2.609	284000	0.706	0.921	0.138
2009	Spherical	0.192	1.596	684000	0.88	0.964	0.0466
2010	Exponential	0.222	0.86	2110000	0.742	0.817	0.0031
2011	Exponential	0.322	0.823	73000	0.609	0.671	0.04
2012	Spherical	0.229	2.042	1270000	0.888	0.93	0.0598

Spatial distribution of nitrate-N concentrations was shown in figure 5. The location and areal coverages of five NO_3^- -N concentration classes were different from 2006 to 2012. According to Quality standard for groundwater in China [38], the safety threshold interval of NO_3^--N (< 20 mg L^{-1}) generally occupy the majority area in study, especially in 2008 and 2009. On the contrary, areas with the higher interval of NO_3^- -N (> 30 mg L^{-1}) is relatively more in 2006. Kriging prediction maps indicated that higher nitrate-N concentration mainly encountered in the middle-eastern region of Shandong, and Weifang and Linyi district have been hotspot areas for nitrate-N contamination. However, NO_3^- -N concentration of groundwater is relatively low in western area in Shandong.

Temporal variability of groundwater nitrate-N in study area, as depicted in figure 5, had a certain trend over the years. Generally, mean nitrate-N concentration decreased from 2006 to 2009, and increased from 2009 to 2013. Additionally, the mean measured nitrate-N concentrations in 17 cities of Shandong province from 2006-2012 in shallow, middle-deep and deep wells are all listed in table 2. However, the Mann-Kendall trend test showed there are no significant trends for most cities for the 7-year period (p>0.05), except Rizhao city. The latter had positive significant Kendall's Tau values in deep groundwater, consistent with an upward trend of nitrate-N concentrations over the 7-year period. In contrast, other cities had no significant Kendall's Tau values, probably indicating there is no clear trend in nitrate-N concentration over the years.

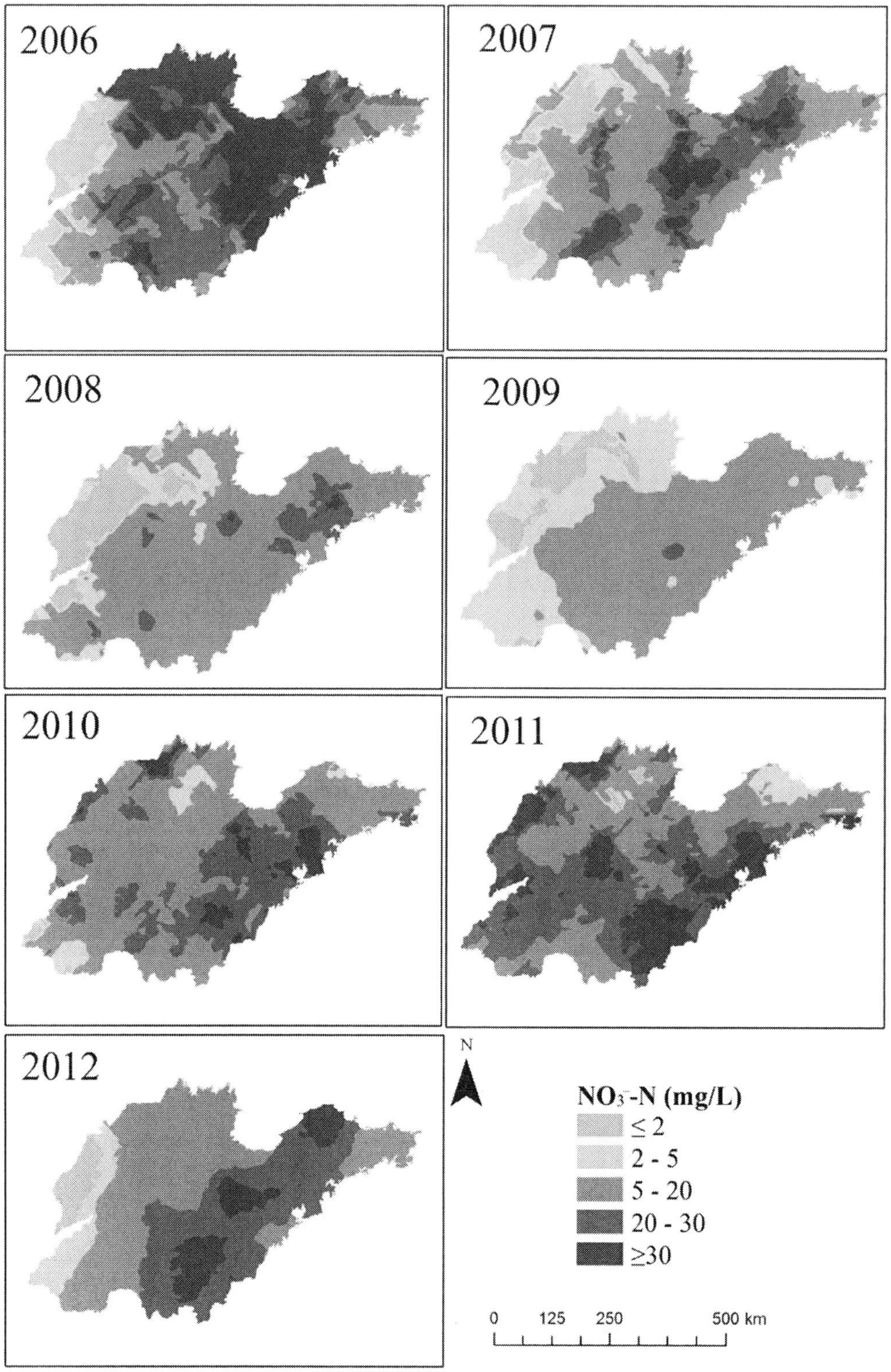

Figure 5. Spatial and spatiotemporal nitrate variability in groundwater resources of Shandong regions, China, 2006-2012.

Table 2. Mean nitrate-N (mg L[-1]) and the Mann-Kendall trend test results for nitrate concentrations in the time series of wells during 2006-2012 in Shandong province, China.

City	well deep	NO3-N concentration 2006	2007	2008	2009	2010	2011	2012	Mean	Kendall's tau	*p*
Binzhou	1	21.17	9.38	4.42	2.11	23.35	16.22	7.04	11.96	-0.143	0.764
	2	0.17	0.56	1.14	--	6.05	0.81	8.91	2.94	0.733	0.060
	3	--	3.04	7.73	0.03	4.89	--	9.39	5.02	0.40	0.462
Dezhou	1	24.33	2.46	3.64	1.03	26.41	42.40	8.55	15.54	0.238	0.548
	2	0.56	2.68	1.05	1.74	4.58	6.70	1.64	2.71	0.429	0.230
	3	--	--	0.12	0.20	5.92	--	--	2.08	1.0	1.0
Dongying	2	--	0.18	0.46	0.00	--	--	--	0.21	-0.333	1.0
	3	--	0.00	--	1.72	3.87	--	--	1.86	1.0	1.0
Heze	1	13.35	0.61	4.64	1.95	13.93	30.12	2.36	9.56	0.238	0.548
	2	--	0.18	0.06	7.36	4.18	22.57	0.49	5.81	0.333	0.452
Jinan	1	13.72	10.74	6.08	4.92	22.56	8.35	21.75	12.59	0.048	1.000
	2	18.79	31.36	21.93	--	8.70	11.29	13.05	17.52	-0.333	0.452
	3	1.58	7.31	2.13	1.05	9.30	50.37	10.50	11.75	0.524	0.133
Jining	1	18.87	10.53	16.13	3.90	10.97	13.28	11.39	12.15	-0.143	0.764
	2	41.35	9.17	7.48	4.93	9.18	48.77	5.65	18.07	-0.143	0.764
	3	0.19	28.38	5.54	4.21	11.75	16.53	17.92	12.08	0.429	0.230
Laiwu	1	28.66	4.29	9.47	6.78	6.99	33.70	19.06	15.56	0.238	0.548
	2	9.63	8.27	8.76	--	--	--	--	8.88	-0.333	1.0
	3	--	2.08	7.11	--	--	--	--	4.60	--	--
Liaocheng	1	0.28	2.80	0.34	2.04	21.41	27.44	0.72	7.86	0.429	0.230
	2	0.85	0.74	0.30	1.86	6.30	4.95	0.42	2.20	0.143	0.764
	3	--	0.02	--	5.76	3.87	--	--	3.22	0.333	1.000
Linyi	1	20.12	21.51	10.96	10.60	25.18	37.87	32.17	22.63	0.429	0.230
	2	14.79	11.13	9.80	10.54	19.88	27.08	25.83	17.01	0.429	0.230
	3	13.60	20.67	15.81	--	18.04	--	--	17.03	0.333	0.734
Qingdao	1	61.95	26.44	23.34	8.31	29.92	35.75	23.02	29.82	-0.238	0.548
	2	45.42	14.14	21.17	6.01	31.61	36.11	25.32	25.68	0.048	1.0
Rizhao	1	--	9.86	7.13	6.57	--	--	--	7.85	1.0	1.0
	2	68.94	28.88	9.17	--	36.23	35.23	16.43	32.48	-0.333	0.452
	3	--	--	2.66	3.96	13.88	19.65	22.85	12.60	**1.0**	**0.027**
Taian	1	25.67	10.81	8.80	8.24	17.10	27.30	19.88	16.83	0.143	0.764
	2	4.72	17.29	21.91	8.02	18.27	35.55	25.88	18.81	0.619	0.072
	3	--	10.05	5.54	7.06	14.55	47.03	17.60	16.97	0.60	0.133
Weifang	1	40.29	30.28	13.84	9.53	27.69	22.27	26.46	24.34	-0.333	0.368
	2	22.83	27.43	17.74	9.22	24.58	23.88	23.80	21.35	-0.048	1.000
	3	0.83	32.94	5.12	11.20	14.18	14.71	24.49	14.78	0.524	0.133
Weihai	1	--	17.23	11.42	5.96	--	--	--	11.54	1.0	1.0
	2	--	5.59	8.35	8.73	25.37	32.02	7.53	14.60	0.467	0.260
Yantai	1	36.40	29.24	21.91	7.69	13.54	8.97	30.35	21.16	-0.333	0.368
	2	13.23	28.21	8.01	8.39	9.58	5.96	29.57	14.71	-0.048	1.000
	3	--	11.23	23.14	11.17	14.69	9.67	14.15	14.01	-0.20	0.707
Zaozhuang	1	45.65	43.31	12.48	7.91	20.66	16.93	32.28	25.60	-0.238	0.548
	2	21.27	21.66	20.60	6.77	3.70	1.50	24.26	14.25	-0.333	0.368
	3	--	14.56	9.85	20.04	--	--	--	14.82	0.333	1.000
Zibo	1	--	10.03	7.85	7.57	4.62	0.77	18.42	8.21	-0.333	0.452
	2	4.44	8.53	9.73	8.88	3.34	0.22	10.19	6.48	0.048	1.0
	3	6.06	77.38	1.76	3.43	0.72	--	--	17.87	-0.60	0.221

Note: 1, shallow wells; 2, middle-deep wells; 3, deep wells.

[a] Significant trends are in bold (p=0.05). Note that cities with positive significant values had an upward trend in

nitrate concentrations over the 7-year period while those with significant negative values had a downward trend.

3.3. Analysis for effective variables

Groundwater nitrate-N varied significantly in their spatial behaviors due to many factors. To select significant variables for nitrate-N variation in three depth groundwater, a stepwise regression method with step AIC was used. Regression results showed livestock per unit area, annual mean temperature, vegetable yield per unit area, and percentages of irrigation areas are significant variables for variance of NO_3^--N in shallow groundwater, and in regression equation the interception is 14.71 and coefficient are 0.665, -2.7475, 1.034 and -30.36, respectively. For variance of NO_3^--N in middle-deep groundwater, annual mean temperature, population per unit area, vegetable yield per unit area, percentages of orchard area, per capita agricultural production and percentages of irrigation areas are significant variables, and the interception is 25.92 and coefficient of these variables are -5.78, 2.38, 0.85,-0.86, -1.50 and -33.3, respectively. Additionally, livestock per unit area, annual mean temperature, per capita agricultural production, percentages of irrigation areas and unit-area nitrogen fertilizer are significant variables for variance of NO_3^--N in deep groundwater, and the interception is -5.81 and their coefficient are -0.51, 2.44, 4.54, -25.63 and -3.32 , respectively.

4. Discussion

According to Quality standard for groundwater in China [38], about 29.5% of groundwater samples have been in excess of NO_3^--N safety threshold concentration (20 mg L^{-1}), which showed groundwater nitrate-N pollution in Shandong province cannot be ignored. Long-term studies on groundwater quality monitoring are needed, which provide scientifically needed data on relationship between groundwater quality and land use management and understand if agriculture practices are effective in protecting groundwater resources.

Difference of NO_3^--N concentration before and after the rainy season in study is consistent with many reports, which showed larger amounts of precipitation can cause rapid infiltration (3.8–5.8 mm/min). The contamination or excess fertilizer application might have entered the groundwater via heavy rainfall after the rainy season, especially from June to August [40].

Difference of groundwater nitrate-N contamination among cropland rooted in agriculture management practices such as cropping pattern, fertilizer input and irrigation amounts and methods [41]. Many reports showed excessive nitrate-N in groundwater is closely related to vegetation planting, for its higher loads of nitrogen [26, 30, 32, 42] with no exception in this research. Geospatial variation of nitrate-N concentration in figure 5 shows areas with high nitrate-N concentration locates in Weifang district, where vegetable cultivation area is larger as we all know. Additionally, with multivariate statistical analysis in study, the variable of vegetable yield per unit area is significant factor for variation of groundwater nitrate-N especially in shallow and middle-deep wells.

Besides vegetation variable, the percentage of orchard area also is significant variable for variation of nitrate-N in middle-deep groundwater. Zhao T K *et al* show groundwater nitrate-N concentration is higher under orchard area than in grain crop area [30], which is consistent with results in study. The variable of livestock per unit area is significant variable for variation of nitrate-N in shallow and deep groundwater. Previous reports showed animal wastes are considered as highly concentrated pollutants that may reach the water table [43, 44]. For sites with similar aquifer features and groundwater depth, contamination must exist in some sites located closely to a septic tank [45]. Pasten-Zapata *et al* also showed that livestock activities in a 1000-m radius contributed significantly to NO_3^--N in shallow groundwater in Mexico [46].

Gu *et al* reported that nitrate leakage to groundwater was significantly related to population density in 2013 [47]. The variable is significant variable for variation of groundwater nitrate-N in deep wells. The effects of natural factors (mean annual temperature and precipitation) on groundwater nitrate-N variation were different in many reports [45, 47]. In this study, variables of annual mean temperature will significantly affect nitrate-N content in three types of depth groundwater. However, precipitation is not significant variable for variation of NO_3^--N in study, probably for their no significant difference among regions.

It is noteworthy that nitrogen fertilizer per unit area is only included in regression equation for nitrate-N in deep wells. Hu *et al* reported groundwater pollution by NO_3^--N leaching occurred due to irrigation with wastewater and excessive application of fertilizers in 2005 [48]. Especially in shallow groundwater systems, frequent and excessive use of fertilizers resulted in a significant nitrate pollution [49]. During the last two decades, groundwater contaminations from extensive fertilizer applications have been reported by many studies [48-52]. Nevertheless, in this study, variables about nitrogen fertilizer and crop areas were based on unit area, and excess fertilizer in agriculture activities is normal in Shandong [53], which probably provides the reasons for having no significant difference among districts.

It has been found that unreasonable management of water and N led to nitrate pollution in groundwater and surface water [54]. Deficit irrigation has been proposed as an alternative to reduce nitrate leaching. However, agriculture in Shandong located in zones of high water demand, and flood irrigation is the common way, which is the reason that the variable of percentages of irrigation area is significant factor for variation of groundwater nitrate-N. In all, statistical analysis showed there are many influencing factors for nitrate-N variation in space and time. Perhaps, this is the reason why there is no significant trend for NO_3-N for many cities in Shandong.

5. Conclusions

High NO_3^--N in groundwater is a serious problem especially in highly active agricultural areas, worldwide. In this work, nitrate-N concentration in groundwater amounts to 184.60 mg L^{-1}, and 29.5% of samples had levels in excess of nitrate-N safety threshold concentration (20 mg L^{-1}), which could result in a large risk to the health of rural populations. The median nitrate-N concentrations in October (after rainy season) were significantly higher than those in May (before rainy season). Nitrate-N content decreased with increasing groundwater depth ($p<0.05$). And nitrate-N under vegetable and orchard area are significantly higher than ones under grain ($p<0.05$). The kriging map of groundwater nitrate-N shows that groundwater nitrate-N has a strong spatial-temporal variability. Higher nitrate-N concentration mainly encountered in the middle-eastern region of Shandong, and Weifang and Linyi district have been hotspot areas. Generally, mean nitrate-N concentration decreased from 2006 to 2009, and increased from 2009 to 2013. Nevertheless, there are no significant trends for most cities for the 7-year period with results of the Mann-Kendall trend test.

The significant factors and the regression model for nitrate-N variation were then completed by stepwise regression, which showed livestock per unit area, annual mean temperature, vegetable yield per unit area, and percentages of irrigation areas in shallow wells, annual mean temperature, population per unit area, vegetable yield per unit area, percentages of orchard area, per capita agricultural production and percentages of irrigation areas in middle wells and livestock per unit area, annual mean temperature, per capita agricultural production, percentages of irrigation areas and unit-area nitrogen fertilizer in deep wells are significant variables for variance of NO_3^--N.

Nitrate contamination in groundwater should be attached importance for developing and implementing strategies especially in rural areas. Measures must be taken to alleviate nitrogen pollution, such as changing the cropping pattern, adjusting the irrigation and fertilization programs and applying limited water and nitrogen fertilizers split application and reducing pollution from livestock and wastewater discharge. Results in present study may be more applicable in regions with similar climate, topography, agriculture model and urbanization in China.

Acknowledgments

This work was supported by agricultural ecological environmental protection project from Ministry of Agriculture of the People's Republic of China from 2006 to 2012 and the National Science and Technology Projects in the 12th Five-Year Program (Grant No. 2012BAD15B02-1) and Postdoctoral foundation of Beijing Academy of agriculture and Forestry Sciences (2015-2017). Additional support for fieldwork was provided by Shandong Academy of Agricultural Sciences.

References

[1] Lu Q Y, Li C L and Li T 2006 General situation of groundwater pollution in China (in Chinese) *Journal of Occupational Health and Occupational Diseases in China* **24(5)** 317-20

[2] Nolan BT and Hitt K J 2006 Vulnerability of shallow groundwater and drinking-water wells to nitrate in the United States *Environ. Sci. Technol.* **40(24)** 7834-40

[3] Kaown D, Hyun Y, Bae G O and Lee K K 2007 Factors affecting the spatial pattern of nitrate contamination in shallow groundwater *J. Environ. Qual.* **36(5)** 1479-87

[4] Zhao J C, Li Y Z, Yamashita I, Xu C Y, Li Q Z and Jia X F 2010 Summary on deduction and trace the source methods for ground water nitrate contamination (in Chinese) *Chinese Agr. Sci. Bull.* **26(18)** 374-8

[5] Rekha P N, Kanwar R S, Nayak A K, Hoang C K and Pederson C H 2011 Nitrate leaching to shallow groundwater systems from agricultural fields with different management practices *J. Environ. Monitor.* **13(9)** 2550-8

[6] Gelberg K H, Church L, Casey G, London M, Roerig D S, Boyd J and Hill M 1999 Nitrate levels in drinking water in rural New York State *Environ. Res. Sect.* **A80** 30-40

[7] Gulis G, Czompolyova M and Cerhan J R 2002 An ecologic study of nitrate in municipal drinking water and cancer incidence in Trnava District, Slovakia *Environ. Res. Sect.* **A88** 182-7

[8] Knobeloch L, Krenz K, Anderson H and Hovel C 1992 Methemoglobinemia in an infant—Wisconsin *MMWR-Morbid. Mortal. W.* **42(12)** 217-9

[9] Fan A M and Steinberg V E 1996 Health implications of nitrate and nitrite in drinking water An update on methemoglobinemia occurrence and reproductive and developmental toxicity *Regul. Toxicol. Pharmacol.* **23** 35-43

[10] Johnson P T J, Townsend A R, Cleveland C C, Patricia M G, Robert W H, Valerie J M, Eliska R and Ward M H 2010 Linking environmental nutrient enrichment and disease emergence in humans and wildlife *Ecol. Appl.* **20** 16-29

[11] Spalding R F and Exner M E 1993 Occurrence of nitrate in groundwater—A review *J. Environ. Qual.* **33** 392-402

[12] Fabro A Y R, Ávila J G P, Alberich M V E, Sansores S A C and Camargo-Valero M A 2015 Spatial distribution of nitrate health risk associated with groundwater use as drinking water in Merida, Mexico *Appl. Geogr.* **65** 49-57

[13] Burow K R, Nolan B T, Rupert M G and Dubrovsky N M 2010 Nitrate in groundwater of the United States, 1991–2003 *Environ. Sci. Technol.* **44** 4988-97

[14] Li S L, Liu C Q, Lang Y C, Zhao Z Q and Zhou Z H 2010 Tracing the sources of nitrate in karstic groundwater in Zunyi, Southwest China: A combined nitrogen isotope and water chemistry approach *Environ. Earth Sci.* **60** 1415–23

[15] Kaushal S S, Groffman P M, Band L E, Elliott E M, Shields C A and Kendall C 2011 Tracking nonpoint source nitrogen pollution in human-impacted watersheds *Environ. Sci. Technol.* **45** 8225-32

[16] Stigter T Y, Carvalho D A M M and Ribeiro L 2011 Major issues regarding the efficiency of monitoring programs for nitrate contaminated groundwater *Environ. Sci. Technol.* **45** 8674-82

[17] Bryan N S and Loscalzo J 2011 *Nitrite and Nitrate in Human Health and Disease* (New York: Humana Press)

[18] Delgado J A and Bausch W 2005 Potential use of precision conservation techniques to reduce nitrate leaching in irrigated crops *J. Soil Water Conserv.* **60** 379-87

[19] Jaynes D B, Hatfield J L and Meek D W 1999 Water quality in Walnut Creek watershed: Herbicides and nitrate in surface waters *J. Environ. Qual.* **28(1)** 45-59

[20] Stites W and Kraft G J 2001 Nitrate and chloride loading to groundwater from an irrigated north-central U.S. sand-plain vegetable field *J. Environ. Qual.* **30** 1176-84

[21] Kanwar R S, Karlen D L, Cambardella C A and Cruse M R 1995 Swine manure and

N-management systems: impact on groundwater quality Clean water, clean environment-21st century *Conf. Proc. ASAE* (Kansas City, MO. 5–8 Mar) vol 2 pp 91-4
[22] Karlen D L, Cambardella C A and Kanwar R S 2004 Challenges of managing liquid swine manure *Appl. Eng. Agric.* **20(5)** 693-9
[23] Bakhsh A, Kanwar R S and Karlen D L 2005 Effects of liquid swine manure applications on NO3-N leaching losses to subsurface drainage water *Agric. Ecosyst. Environ.* **109** 118-28
[24] Randall G W and Mulla D J 2001 Nitrate nitrogen in surface waters as influenced by climatic conditions and agricultural practices *J. Environ. Qual.* **30** 337-44
[25] Zhang Y, Liu X, Zhang F S, Ju X T, Zou G Y and Hu K L 2006 Spatial and temporal variation of atmospheric nitrogen deposition in the North China *Plain Acta Ecologica Sinica* (in Chinese) **26** 1633-8
[26] Zhang W L, Tian Z L, Zhang N and Li X Q 1995 Investigation of nitrate pollution in ground water due to nitrogen fertilization in agriculture in North China (in Chinese) *Plant Nutrition and Fertilizer Sciences* **1(2)** 80-7
[27] Wang L, Zhang G Y, Sun S Y, Ru S H and Geng N 2008 Effect of fertilization on nitration concentration of groundwater in high-yielding vegetable land of Heibei Province (in Chinese) *J. Heibei Agric. Sci.* **12(10)** 75-7
[28] Liu X Q, Xu J Y, Jiang L H, Huang J X, Wang L M, Liu J and Zou J Q 2010 Spatial variability and distribution pattern of groundwater nitrate pollution in farming regions of Shandong Province, China (in Chinese) *J. Agro-Environ. Sci.* **29(6)** 1172-9
[29] Kou C L, Guo Z L, Ma Z H, Li L D, Wang S G, Shen A L and Zhao L J 2013 Assessment of groundwater vulnerability to nitrate in Henan Province of China based on principal component regression (in Chinese) *Chinese J. Appl. Ecol.* **24(10)** 2912-6
[30] Zhao T K, Zhang C J, Du L F, Liu B C and An Z Z 2007 Investigation on nitrate concentration in groundwater in seven provinces (City) surrounding the Bohai Sea (in Chinese) *J. Agro-Environ. Sci.* **26(2)** 779-83
[31] Lin H T, Jiang L H, Song X Z, Zheng F L, Tan D S, Gao X H and Liu Z H 2011 Nitrate concentration of groundwater and its affecting factors in Shandong Province, China (in Chinese) *J. Agro-Environ. Sci.* **30(2)** 353-7
[32] Kou C L, Ju X T and Zhang F S 2005 Nitrogen balance and its effects on nitrate-N concentration of groundwater in three intensive cropping systems of North China (in Chinese) *China J. Appl. Ecol.* **16(4)** 660-7
[33] Gao X H, Jiang L H, Liu Z H, Li X L, Lin H T, Tan D S, Song X Z and Zheng F L 2011 Investigation and evaluatioin on nitrate contamination status of groundwater in rural areas of Shandong Province (in Chinese) *Chinese J. Agrometeorology* **32(1)** 89-93
[34] Li Y Z, Jia X F, Xu C Y, Wang Q S and Li Q Z 2013 A study on the source tracing of groundwater nitrate in Shandong province (in Chinese) *Ecol. Environ. Sci.* **22(8)** 1401-7
[35] Rowell D L 1994 *Soil Science: Methods and Applications* (Harlow: Longman Group)
[36] CMA 2013 *C.M.A. Dataset of Surface Climate Datain China* China Meteorological Data Sharing Service System, http://cdc.cma.-gov.cn/home.do..
[37] NBS 2013 *N.B.o.S., China Statistical Yearbook for Regional Economy 2013 B* (Beijing: China Financial and Economic Publishing House)
[38] Bureau N T M 1993 *Quality Standard for Groundwater in China (GB/T 14848-1993)*
[39] Cambardella C A, Moorman T B, Parkin T B, Karlen D L, Novak J M, Turco R F and Konopka A E 1994 Field-scale variability of soil properties in central Iowa soils *Soil Sci. Soc. Am. J.* **58** 1501-11
[40] Min L L, Yu J J, Song R L, Hu S S, Hu K and Zhang G Y 2010 Infiltration characteristics under simulated rainfall over platycladus orientalis land in Taihang mountainous region (in Chinese) *J. Soil Water Conserv.* **24(1)** 28-32, 68
[41] Kurunc A, Ersahin S, Sonmez N K, Kaman H, Uz I, Uz B Y and Aslan G E 2016 Seasonal changes of spatial variation of some groundwater quality variables in a large irrigated coastal

Mediterranean region of Turkey *Sci. Total Environ.* **554** 53-63

[42] Wang C H, Huang Q W, Zhang Y Z, Zhou W J, Huang Y X and Wang K Y 2004 Evaluation on status of nitrate pollution in vegetables—soils—groundwater system in the open location of vegetables (in Chinese) *J. Hunan Agric. Univ. (Nat. Sci.)* **30(4)** 374-7

[43] Adriano D C, Pratt P E and Bishop S E 1971 Nitrate and salt in soils and groud waters from land disposal of dairy manure *Soil Sci. Soc. Am. J.* **35** 759-62

[44] Salman A, Al-Qinna M and Al Kuisi M 2014 Spatial analysis of soil and shallow groundwater physicochemical parameters in El-Mujib Basin-central Jordan *J. Asian Earth Sci.* **79** 366-81

[45] Wang S Q, Tang C Y and Song X F 2015 Factors contributing to nitrate contamination in a groundwater recharge area of the North China Plain *Hydrol. Proc.* **30** 2271-85

[46] Pasten-Zapata E, Ledesma-Ruiz R, Harter T, Ramirez A I and Mahlknecht J 2014 Assessment of sources and fate of nitrate in shallow groundwater of an agricultural area by using a multi-tracer approach *Sci. Total Environ.* **470-471** 855-64

[47] Gu B J, Ge Y, Chang S X, Luo W and Chang J 2013 Nitrate in groundwater of China: Sources and driving forces *Global Environ. Change* **23** 1112-21

[48] Hu K L, Huang Y F, Li H, Li B, Chen D and White R E 2005 Spatial variability of shallow groundwater level, electrical conductivity and nitrate concentration, and risk assessment of nitrate contamination in North China Plain *Environ. Int.* **31(6)** 896-903

[49] Kurunc A, Ersahin S, Uz B Y, Sonmez N K, Uz I, Kaman H, Bacalan G E and Emekli Y 2011 Identification of nitrate leaching hot spots in a large area with contrasting soil texture and management *Agric. Water Manag.* **98** 1013-9

[50] Frind E O, Duynisveld W H M, Strebel O and Boettcher J 1990 Modeling of multicomponent transport with microbial transformation in groundwater - the Fuhrberg case *Water Resour. Res.* **26** 1707-19

[51] Jalali M 2005 Nitrates leaching from agricultural land in Hamadan, western Iran *Agric. Ecosyst. Environ.* **110** 210-8

[52] Sonmez I, Kaplan M and Sonmez S 2007 Investigation of seasonal changes in nitrate contents of soils and irrigation waters in greenhouses located in Antalya-Demre region *Asian J. Chem.* **19** 5639-46

[53] Ju X T, Kou C L, Christie P, Dou Z X and Zhang F S 2007 Changes in the soil environment from excessive application of fertilizers and manures to two contrasting intensive cropping systems on the North China Plain *Environ. Pollut.* **145** 497-506

[54] Zhu Z L and Chen D L 2002 Nitrogen fertilizer use in China – Contributions to food production, impacts on the environment and best management strategies *Nutr. Cycl. Agroecosys.* **63(2)** 117-27

IOP Conf. Series: Earth and Environmental Science 82 (2017) 012060 doi:10.1088/1755-1315/82/1/012060

Comparative analysis of year-end water level determining methods for cascade carryover storage reservoirs

J Liu[1,2], C H Huang[1] and G Z Zeng[1]

[1]Key Laboratory of the Pearl River Estuarine Dynamics & Associated Process Regulation, MWR, Guangzhou 510611, China

E-mail: 13560030408@163.com

Abstract. Based on the analysis of the determining methods of year-end water level for the cascade carryover storage reservoir, two prediction methods have been used to study the year-end water level: Multi-objective decision model and statistical regression predictive function. The advantages and applicable conditions of each of them have been compared and discussed.

1. Introduction

The issue of determining year-end water level of carryover storage reservoir is very important in the operation of cascade reservoirs. It not only affects the operation status and economic benefit of current year, but also the following many years and the whole cascade reservoirs [1-3]. At present, research activities about year-end water level of carryover storage reservoir are still not deep going, and the solution methods are not good enough. Two methods have been widely applied: multi-objective decision model considering the max energy of cascade power [4, 5] and statistical regression predictive function based on the formerly operation experience. The two methods are used to analyze the year-end water level of two carryover storage reservoirs, which are Hongjiadu and Goupitan reservoirs in the WuJiang River basin (figure 1), the advantage and applicable condition have been compared with the two methods.

2. Study on year-end water level based on multi-objective decision model

The storage capacity of the carryover storage reservoir can be divided into yearly regulation and overyear regulation [6]. The yearly regulation storage is used for generating in current year, and overyear regulation storage is used for generating in years after. The energy storage in the overyear regulation reservoir contains hydropower generation in current year and power storage. Consequently, the energy storage in the yearly regulation reservoir or runoff hydropower station is just the hydropower generation in current year for the lack of regulating ability. Actually, the objective function should include hydropower generation and power storage which are relative in the cascade reservoirs. More water must be expended and year-end water storage must be reduced to increase hydropower generation of cascade stations in current year. On the contrary, less water will be expended and hydropower generation will be restricted to increase the year-end energy storage. Therefore, how to coordinate the hydropower generation of cascade stations with the year-end water storage is the key to solve the conflict [7, 8].

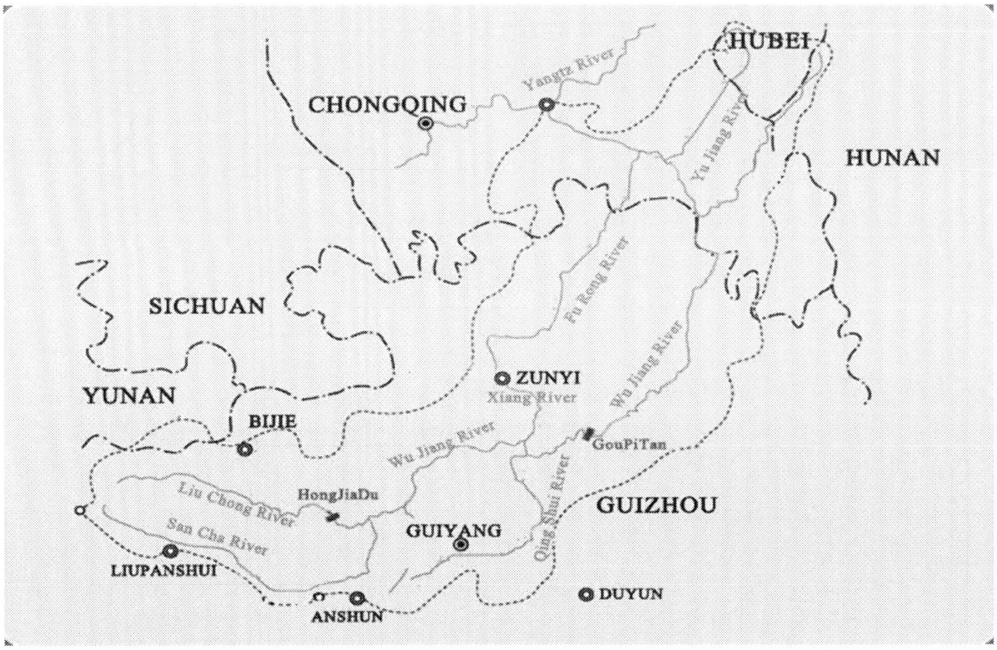

Figure 1. The map of WuJiang River cascade hydropower station.

In this paper, the year-end water level prediction model of carryover storage reservoir was established, whose objective function is energy maximization of the cascade stations [9], which is

$$\mathrm{MaxE} = \mathrm{Max}(\mathrm{E_d} + \mathrm{E_s}) \tag{1}$$

Where E is the total energy storage of cascade stations; E_d is hydropower generation in current year; and E_s is year-end power storage.

In equation (1):

$$E_d = \sum_{i=1}^{n}\sum_{t=1}^{T} K_i Q_{i,t} H_{i,t} \tag{2}$$

$$E_s = \sum_{j=1}^{m} A_j V_{nj} \overline{H}_j \tag{3}$$

Therefore, equation (1) can be re-written as

$$\mathrm{MaxE} = \mathrm{Max}[\sum_{i=1}^{n}\sum_{t=1}^{T} \mathrm{K_i Q_{i,t} H_{i,t}} + \sum_{j=1}^{m} \mathrm{A_j V_{nj} \overline{H}_j}] \tag{4}$$

where n is the total number of cascade reservoirs; m is the total number of overyear regulation reservoirs; T is the total number of operating time period; K_i is the integrated output power coefficient of hydroelectric station i; $Q_{i,t}$ is the Power Generating Flow at a time period t at a hydroelectric station i; $H_{i,t}$ is the power generating net head at a time period t at a hydroelectric station i; $A_j = K_j/3600$; V_{nj} is the year-end water storage of overyear regulation reservoir j;

$\overline{H}_j$ is the average power generating net head in year-end water storage of overyear regulation reservoir j; and $\overline{H}_j$ is the sum of average power generating water head at a downstream hydroelectric station, if there is any.

The objective function (equation (1)) is subjected to the following system constraints:

- Mass balance equation. The reservoirs system follows the mass balance principle which is given by the following equation:

$$V(i,t+1) = V(i,t) + (Q_I(i,t) - Q_O(i,t)) \times \Delta t \tag{5}$$

where $V(i,t), V(i,t+1)$ are the starting and ending reservoir storage volume (Mm^3) at a time period t at a hydroelectric station i; and $Q_I(i,t)$、$Q_O(i,t)$ are inflow and outflow at a time period t at a hydroelectric station i.

- Flow balance equation

$$Q_I(i+1,t) = Q_O(i,t) + q(i,t) \tag{6}$$

where $q(i,t)$ is the local inflow during time period t between hydroelectric station i and $i+1$.

- Water level bounds

$$Z_{\min}(i,t) \le Z(i,t) \le Z_{\max}(i,t) \tag{7}$$

where $Z_{\max}(i,t), Z_{\min}(i,t)$ are the minimum and maximum limits of water level during time period t in a reservoir i; $Z_{\min}(i,t)$ generally is the dead water level and $Z_{\max}(i,t)$ must be considered with the flood control.

- Release bounds

$$Q_{O\min}(i,t) \le Q_O(i,t) \le Q_{O\max}(i,t) \tag{8}$$

where $Q_{O\max}(i,t), Q_{O\min}(i,t)$ are the minimum and maximum limits of water release during time period t at the ~~of~~ reservoir i; $Q_{O\min}(i,t)$ is considered with water use and navigation; and $Q_{O\max}(i,t)$ must be considered with the flood control.

- Power production constraint

$$N_{\min}(i,t) \le N(i,t) \le N_{\max}(i,t) \tag{9}$$

where $N_{\min}(i,t)$、$N_{\max}(i,t)$ are the minimum and maximum power production limits. Depending upon the power plant capacity and reliability, the minimum and maximum production has to be fixed for any power plants.

Based on the energy maximization of the cascade stations model above, the influence of inflow and year-start water level for year-end water level in the overyear regulation reservoir has been analyzed. The model was applied in the Wu-Jiang River, which has two overyear regulation reservoirs (Hongjiadu and Goupitan). Based on the inflow record (from 1951 to 2014), the scheme set of year-end water level of overyear regulation reservoirs (Hongjiadu and Goupitan) has been developed. The process of simulated calculating is as follows: first, according to the inflow frequency and the year-start water level in Hongjiadu and Goupitan, the year-end water level can be determined by the scheme set described above. Second, the hydropower generation maximization model (equation (4)) has been applied to optimal operation of the cascade stations. Third, the results of hydropower generation and firm power of the stations have been compared to the design values. This comparison indicates that the model is reliable.

Table 1. Comparison of conventional and optimization scheduling for key parameters.

Reservoirs		Hongjiadu	Goupitan
Average annual power generation (10^8 KW·h)	Conventional Scheduling	15.59	96.67
	Optimization Scheduling	16.89	98.08
	Increasing rate (%)	1.3	1.41
Firm power (MW)	Conventional Scheduling	159.1	746.4
	Optimization Scheduling	155.1	841.3
	Increasing rate (%)	-4	95.9
Assurance rate (%)	Conventional Scheduling	95	95
	Optimization Scheduling	94	96
	Increasing rate (%)	-1	1

From table 1 we can see that the optimization operation can get the maximum average annual power generation and firm power. Therefore, prediction method of year-end water level based on Multi-objective decision model is reasonable and reliable.

3. Study on year-end water level based on statistical regression analysis

The conflict of hydropower generation and energy storage of carryover storage reservoir can be resolved by applying the multi-objective decision model to determine the year-end water level. This model assumes that either the hydropower generation in current year or the power storage in the subsequent years is optimal. Due to uncertainty of inflow condition, the influence of inflow in subsequent years on cascade reservoirs operation cannot be considered properly in the optimal model. So, a statistical regression function model was developed which describes by the relationship between the year-end water level and related factors.

The statistical regression function model for determining the year-end water level is widely used. As well known, the reservoir operation chart is based on the classification and summary of reservoir operation history. Like the method for obtaining operation chart, the statistical regression function for prediction of the year-end water level can be developed as follows. First, the deterministic model of cascade reservoirs operation must be established. Second, the long-series actual flow records must be used for the simulation calculation. At last, the long-term optimal operation rules and the variation range of year-end water level in the carryover storage reservoir can be obtained.

Table 2. Regression effect analyzing with different degrees of confidence.

Reservoirs	Hongjiadu	Hongjiadu	Goupitan	Goupitan
Degrees of confidence	0.1	0.001	0.1	0.001
Amount of forecast factor	4	3	2	2
Sum of squares	16288.69	16288.69	1723.84	1723.84
Residual sum of squares	1984.54	2203.66	1059.82	1316.03
Regression square sum	14304.16	14085.03	664.02	407.81
Multiple correlation coefficient	0.94	0.91	0.62	0.46
Residual standard deviation	6.57	6.85	4.70	5.18
Mean relative errors	0.39%	0.42%	0.53%	0.58%

IOP Conf. Series: Earth and Environmental Science 82 (2017) 012060 doi:10.1088/1755-1315/82/1/012060

The cascade hydropower generation maximization model (equation (2)) can be used for the long-series simulation operation. Based on the inflow record (from 1951 to 2014), the change low of year-end water level in overyear regulation reservoirs (Hongjiadu and Goupitan) has been established. According to the long-series results, the relationship between year-end water level and the related factors, such as year-start water level, inflow, power output and power generation can be determined. In order to analyze the accuracy of the simulation results, the statistical regression data were calculated with different degrees of confidence (0.1 and 0.001). The results are shown in table 2.

The Stepwise Regression predicting model for Hongjiadu is

$$Z_{year\text{-}end}=611.3808+3.34*10^{-4}*(Z_1)^2+0.9243*(Q_1)-2.4119*10^{-3}*(Q_1)^2+0.0643*(Q_2) \quad (10)$$

The Stepwise Regression predicting model for Goupitan is

$$Z_{year\text{-}end}=578.2711+0.1085*(Q_1)-5.94*10^{-5}(Q_1)^2 \quad (11)$$

where, $Z_{year\text{-}end}$ is the year-end water level of overyear regulation reservoirs, Z_1 is the year-start water level, Q_1 and Q_2 is the predicting inflow in the current and subsequent years.

The year-end water level predicting model results were compared with the actual data (figures 2 and 3). According to the T-Test and F-Test results, the regression effect is significant in theory.

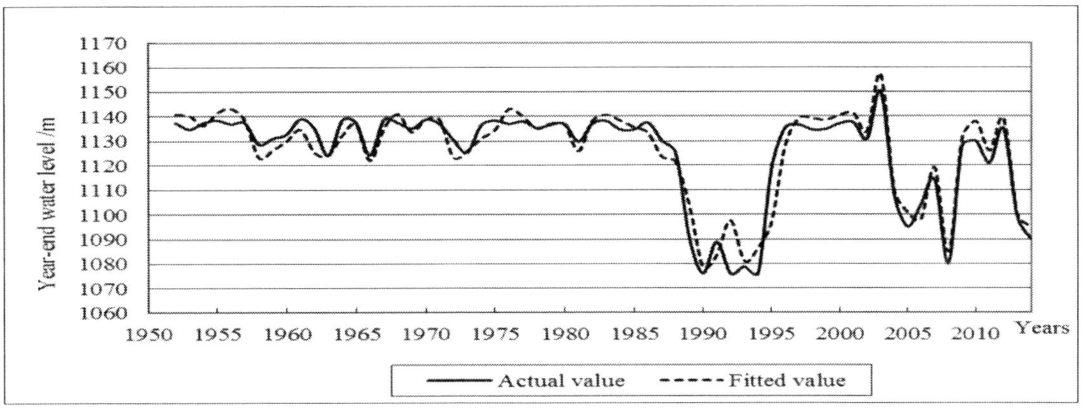

Figure 2. The actual and calculated values of the year-end water level in the Hongjiadu reservoir.

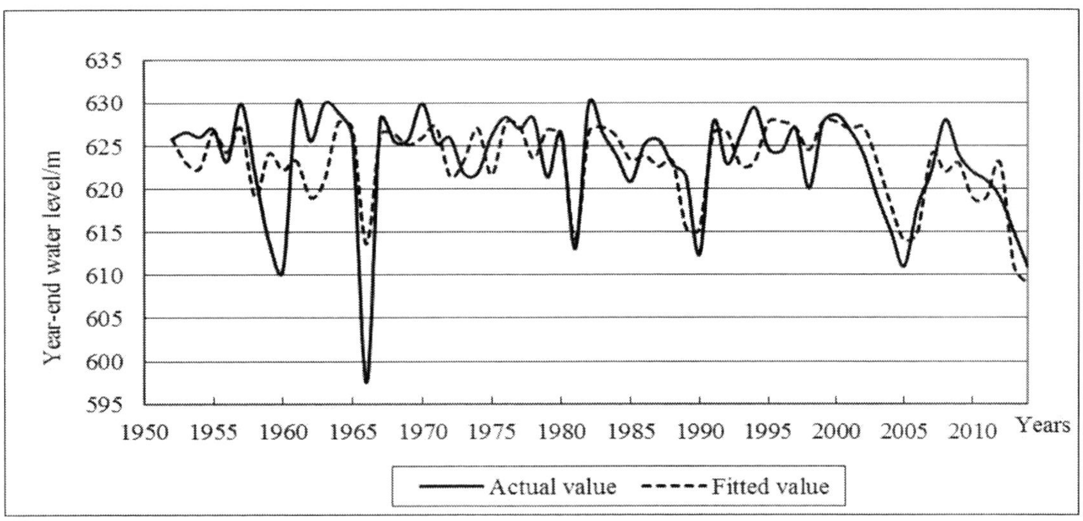

Figure 3. The actual and calculated values of the year-end water level in the Goupitan reservoir.

4. Compared analysis of the two year-end water level prediction model

Based on the study of year-end water level of overyear regulation reservoirs, two methods for determining the optimal year-end water level were established, namely the Multi-objective decision model and the Statistical regression function model. Two methods are applied in the actual overyear regulation reservoirs which are Hongjiadu and Goupitan in Wujiang River basin. According to the long series simulation calculation results of cascade reservoirs in Wujiang River basin, the two methods for determining the optimal year-end water level are reasonable and reliable, and suitable for cascade reservoirs operation. The compared results of two different year-end water level prediction models are presented in table 3, figures 4 and 5.

Table 3. Compared results of different year-end water level prediction model.

Reservoirs		Hongjiadu	Goupitan
Average annual power generation (10^8KW·h)	Design value	15.59	96.67
	Multi-objective decision model	16.89	98.08
	Statistical regression function	17.12	100.55
Firm power (MW)	Design value	159.1	746.4
	Multi-objective decision model	155.1	841.3
	Statistical regression function	156.3	851.9
Assurance rate (%)	Design value	95	95
	Multi-objective decision model	94	96
	Statistical regression function	95	98

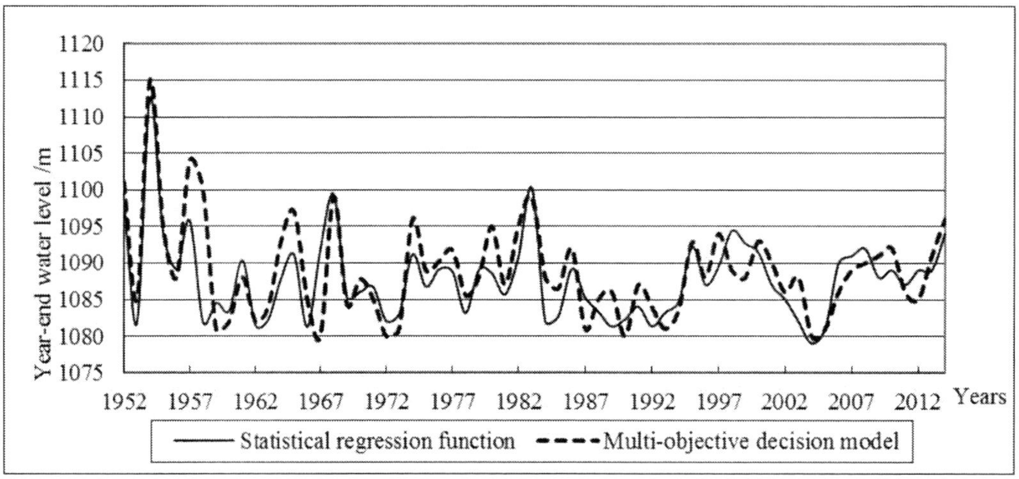

Figure 4. The year-end water level in the Hongjiadu reservoir derived from different models.

A comparison of two year-end water level prediction models in figures 4 and 5 shows that the two methods have the same trend, which is consistent with the long series optimal operation. However, some parts of the two curves show differences, which are caused by the following reasons: first, the year-end water level determined by statistical regression function is based on long series optimal operation results, which require long enough sample series. In the paper, there is only the 63-year inflow record, which cause the statistical regression function model only shows the prediction trend, but the extremal problem cannot be solved. Second, due to the error accumulation of year-end water level, the simulation model has a disadvantage as uncertainly of the function type and relatively large prediction error, which impact the prediction accuracy. Third, the year-end water level has a great influence on the reservoir operational mode, while the multi-objective decision model for determining year-end water level was insensitive to that influence. In other words, the multi-objective decision model has a great effect on the different inflow frequencies, which causes the error of the extreme value at the edge zone of decision interval.

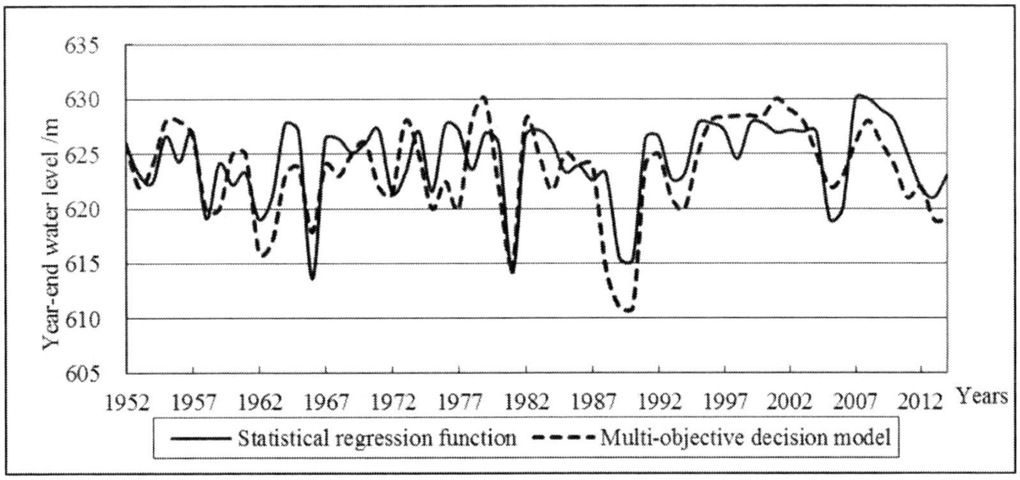

Figure 5. The year-end water level in the Goupitan reservoir derived from different models.

Because of the Goupitan reservoir is in the middle of the Wujiang River basin, its year-end water level is influenced by a lot of factors. Therefore, the statistical regression function model cannot be completely applicable, thus the multi-objective decision model could be used more successfully. From the figure 5, we can see that, the two methods of determining year-end water level of Hongjiadu reservoir give prediction results without big differences. However, in dry years, the prediction of year-end water levels from the statistical regression model is relatively low, which causes the increase in hydropower generation at cascade stations. Considering the good regulation performance of the Goupitan reservoir, the statistical regression method should be given priority for the Hongjiadu reservoir in dry years. Two determining methods are suitable for Hongjiadu reservoir in wet or average water years. In practical use, the year-end water level of overyear regulation reservoirs can be calculated by two methods separately, and finally the best option could be determined by the actual situation and power plant dispatcher's operation experience.

5. Conclusions

The issue of determining year-end water level at the carryover storage reservoir has always been one of the most complicated decisions in water resources management. Two methods for determining the optimal year-end water level are presented in this paper, which are the multi-objective decision model and the statistical regression function model. The two methods have their own advantages and disadvantages. In general, the statistical regression function is suitable for the conditions of steady inflow and the long enough series record. The multi-objective decision model is suitable for the conditions with serious contradiction in reservoir function. From the long term joint operation of cascade reservoirs, the two year-end water level determining methods are reasonable and effective.

References

[1] Wang Y M, Chang J X, Huang Q, *et al* 2005 Water level control of Longyangxia reservoir based on BP at the end of year *J. Northwest Sci.-Tech. Univ. Agric. For. (Nat. Sci. Ed)* **33**

IOP Conf. Series: Earth and Environmental Science **82** (2017) 012060 doi:10.1088/1755-1315/82/1/012060

68-72
[2] Yeh W W-G 1985 Reservoir management and operations models: A state-of-the-art review *Water Resour. Res.* **21** 1797-818
[3] Oliveira R and Loucks D P 1997 Operating rules for multi-reservoir systems *Water Resour. Res.* **33** 839-52
[4] Yuan W L and Wang F L 2012 Study on year-end water level of multi-year regulating storage reservoir in power market *J. Hydroelectric Eng.* **31** 95-8
[5] Zou J and He S H 2006 Fuzzy multi-objective decision-making model for year-end dwawdown level of multi-year regulating storage reservoirs in cascade reservoirs *Hydro.-Sci. Eng.* **2** 18-23.
[6] Zhang S H, Zhang Z B, Xu W H and Wang P 2012 Study on year-end level of carryover storage reservoir based on decision tree *J. Hydroelectric Eng.* **31** 44-8
[7] Shi P, Ji C M, Li J W, *et al* 2014 Forecast model of year-end level for multi-year storage regelation reservoirs based on operation trends *J. Hydroelectric Eng.* **33** 58-64
[8] Wang J L, Huang W B, Ma G W, *et al* 2015 Study on multi-objective prediction model for year-end water level of carryover storage reservoir *J. Hydroelectric Eng.* **34** 28-34
[9] Xie Y Y, Wang Y M and Huang Q 2014 Year-end water level and water discharge in flood season for Longyagnxia reservoir *J. Northwest Sci.-Tech. Univ. Agric. For. (Nat. Sci. Ed.)* **42** 223-7

IOP Conf. Series: Earth and Environmental Science 82 (2017) 012061 doi:10.1088/1755-1315/82/1/012061

Chemical water shutoff profile research status and development trends

L T Xu

College of Chemistry and Chemical Engineering Lab of Fine Chemical Engineering, Northeast Petroleum University, Heilongjiang, Daqing, 163318, China

E-mail: xulitingxuliting@163.com

Abstract. Excess water production is now a common problem encountered in almost every water flooding mature oilfield. The exploitation of oil field is faced with great challenge because of the decrease of oil field production. For the development of high water cut rare the status quo chemical water shutoff profile control technology is an important solution to solve this problem. Oilfield chemical water shutoff has important application prospects. This paper analyzes the water shutoff profile control and water shutoff profile control agent currently oilfield applications, moreover the use and development of blocking agent profile technology is to improve reservoir recovery and propose solutions. With the constant increase in water cut, profile technology should be simple, efficient, practical and profile control agent of development should be economic, environmental, and long period

1. Introduction

The status quo of oil fields in China, an injection profile along with the increasing of water injection, is increasingly uneven, so the well water yield increases/decreases. For example, In Daqing oil field, the average water cut oil wells are more than 95%. The good physical conditions of remaining reserves depend on deep mining so as to reduce reservoir. High temperature conditions of reservoir heterogeneity have become the focus of the development of oilfield forcing the water shutoff technology continue developing, in order to improve the recovery rate to a good development effect. According to the construction of different objects, water shutoff technology can be divided into water injection profile control and water shutoff. But these two kinds of operation modes are used on a large number of chemical agents, so these can also be used by chemical means for water plugging, and currently the most be used by control.

2. Water shutoff technology status

Water shutoff technology is composed of the near well profile control technology and the deep profile control technology. Near wellbore profile control technology mainly solves the reservoir heterogeneity between the contradictions, as strong or stronger gel plugging agent; In the late stage of oilfield development, water injection wells in the reservoir margin under the condition of less use of deep profile control technology; The plugging agent for particles or gel. The near wellbore area weakened with the oilfield development, and the increasingly deep profile control [1], including deep weak gel profile control system, has more application.

Water shutoff profile control technology application research in China generally has the following four stages of development, as described in table 1.

Table 1. The development of water shutoff profile control technology application in China.

Years	Stage of oil field	Technical type	Plugging agent
50s -70s	Early stage	Mechanical water plugging profile control technology	Cement, resin, heavy oil, calcium chloride, etc.
70s-80s	Early stage	Physical barrier blocking	Strong gel plugging agent based polymer and crosslinking agent
90s	high water cut stage	Polymer water shutoff and deep water plugging and profile control	Applied in different system of nearly 100 kinds of plugging agent
After2000	Extra high water cut stage	Deep water shutoff and the deep fluid diversion	Deep water flooding technology supporting water shutoff agent

Flooding problems in high water cut period are increasingly complex, so requirement of water shutoff technology is becoming higher, driven by its rapid development. In recent years, water shutoff technique and plugging agent research in China has made many advances in the international leading position, R & D, containing weak gel dispersion gel, swellable particle profile control technology, the high water cut oilfield water flooding development effect is improved, and improved the recovery ratio [2, 3].

3. Field application of chemical plugging agent type

Non selective and selective water shutoff agents are currently two kinds of commonly used chemical plugging agents.

3.1. Non selective water shutoff agent

Non selective water shutoff technology is used to seal single or high aquifer. The plugging agent has no selectivity to oil and water, so it can be blocked. Before the profile, make sure the water layer section, plugging agent injection water layer, use the appropriate method to separate oil and water layer, injected water plugging agent formation blockage can be achieved. Main plugging agents are cement, calcium silicate gel, resin, gel, etc [4].

3.1.1. Weak gel plugging agent. The weak gel is a kind of polymer which is close to the polymer flooding, formed by adding a small amount of delayed crosslinking agent [5]. Usually use the concentration of 800 ~ 3000 mg/L of high molecular weight polyacrylamide to advocate agent for crosslinking. There are many different kinds of crosslinking agents, like resin, dialdehyde, polyvalent metal ion, etc. Moreover acetic acid chromium, aluminium citrate and glyoxal are commonly used abroad [6], while phenolic composite of resin, poly lactic acid, acetic acid, chromium, chromium etc. are commonly used in Chinese oil fields.

3.1.2. Oily sludge plugging agent. Oily sludge is a kind of industrial waste due to its complex chemical properties. It is one of the main by-products of crude oil production. The main ingredients are water, mud, colloid, asphalt, wax etc. In general, the oil sludge is a stable suspension emulsion, which is formed by the stabilization of the system, but it is difficult to realize multiphase separation. As a particle plugging agent, it has characteristics of high yield, high oil, viscosity, fine grain, hard to dehydration, etc. Compared with other chemical agents, it has better resistance to salt and high temperature and shear resistance. It has obvious advantages in high dose injection, low cost and good effect. Furthermore, it can also solve the problem of oily sludge treatment and emission, reduce the environmental pollution and the oil sludge solidification cost, and deal with the harmful components in time.

West GuDong in Shengli oilfield in 1992 used the HPAM/chromium acetate system to test three well group, which was the earliest application of this technology in our country. A total of 155 thousand cubic meters of profile control agent, the use of 3000 mg/LHPAM and 500 mg/L chromium acetate system, water injection wells after the injection pressure profile increased by about 3 MPa on average, the cumulative increase of oil 9800 t [7]. Liaohe Oilfield used weak gel deep profile control and the effect was the best overall, the temperature is about 55℃, reservoir depth is 1550 ~ 1700 m, 2200 mg/L mineralization, average porosity is 20%, the permeability is 1.13 D, which has significantly improved the efficiency of water flooding, valid for 3 years [8].

3.1.3. Cement plugging agent. As one of the early use of plugging agents, cement plugging agent is of low prices, big intensity, easy to adapt to the temperature environment, etc. After the solidification of cement impervious to the use of this mechanism to plug the hole, for example, the lower water, channeling water, or squeeze into the water blocking, Due to the large size of cement particles, it is difficult to enter the middle and low permeability formation, so the plugging strength and success rate are low and the validity period is short, so the application scope is limited [9], In recent years, superfine cement and cement additive agent have been developed successfully and have broad prospects for cement plugging.

3.1.4. Resin plugging agent. The resin plugging agent is a polymer material produced by condensation of low molecular substances, with the body structure is difficult to soluble fusion. After heated by the change of nature, it can be divided into thermosetting resin and thermoplastic. Thermosetting resins such as phenolic resin, epoxy resin, sugar alcohol resin and so on are often used as non-selective plugging agents. Phenolic resins are commonly used in oil fields [10]. Will preshrinking liquid mixed with thermosetting phenolic resin curing agent into the water, suitable for formation temperature and curing agent, the crosslinking, within a certain time to form difficult to soluble phenolic resin melt, can block the formation of the layer. The advantages of Resin Plugging Agent: high strength, many kinds of pore. The cured resin is neutral, and the chemical stability is not easy to react with the underground liquid. However, before the curing of resin is sensitive to pollution, such as water, watch live agent, acid and alkali, must pay attention to detect horizon and isolation before use.

3.2. Selective water shutoff agent

Selective plugging agent is used to separate water and profile control by using the difference between oil and water, oil layer and water layer. With the rapid development, there are many methods of water shutoff, such as water shutoff agent, oil based plugging agent and alcohol based plugging agent type of different dispersion medium, water, oil and alcohol solvent, respectively.

3.2.1. Oil-based plugging agent. Oil-based plugging agent is mainly composed of thickener BCI, amphoteric polymer, diesel and other components [11]. High pressure plugging agent is injected into the oil well stratum, polymer by oil and oil flow channel of high water content and low degree of crosslinking and hydrolysis characteristics is difficult, and can avoid the emergence of gel which did not affect the flow of oil, at the same time, it can be mined with the flow of oil, into the water channel into the stratum.

3.2.2. Water plugging agent. Water plugging agent due to the variety and the lowest cost is now widely used in China's oil production. One of the most commonly used water-soluble polymers [12]. Its aqueous solution will be priority into containing high water-bearing formation, through the adsorption effect of hydrogen bonding in water layer at the grass-roots level surface to make further reserves, based on polypropylene acyl ammonia as the main raw material of the application principle of plugging agent.

3.2.3. Alcohol-based plugging agent. Alcohol-based plugging agent with resin dimer, alcohol based compound plugging agent, etc., not often used in practical production. It is commonly used to explore the possibility of selective plugging of reservoir water under the conditions of high temperature. Alcohol-based plugging agent is mainly composed of water-soluble polymer and sodium silicate aqueous ethanol solution [13].

4. Classification of oilfield water shutoff profile control technology

4.1. Mechanized water shutoff profile control technology

Mechanical water shutoff profile control technology is mainly based on the water pipe column and the relevant packer. The separator water layer will prevent water from entering the septum in the wellbore in the well. In the field practice, after entering high water cut period in mining mechanization technology is gradually to the adjustable layer technology and hydraulic adjustable two aspects of technology development, water shutoff profile control with the help of this kind of mechanization technology development will also improve the efficiency of water plugging in the actual work, also can achieve the larger goal in control water stabilizing oil production process [14].

4.2. Fine chemical profile control technology

Fine chemical profile control technology is mainly targeted according to the injection Wells in high water absorption layer on the longitudinal distribution of specific characteristics to carry on the fine chemical profile control, usually used to supplement mechanical profile control technology, to improve the remaining reserves of various types of reservoir the mining efficiency through the fine adjustment of injection profile [7]. This technology has a positive effect on improving the stability of oil and water control and the potential exploitation of oil reservoirs, improving the efficiency of oilfield development and the economic benefits etc. all have positive role, and oilfield development residual oil reservoir at the high water-cut stage is effective means of profile control.

4.3. Deep profile control technology

Deep profile control technology is injected the plugging agent deep into reservoirs. In the oil field actual mining for deep reservoir area, requires a lot of chemicals to injection, a prajnaparamita normal can use the block water cut 1/4 or 1/3 of the spacing of the actual processing. For water injection well in large pores or obvious cracks is required by 1/2 and above the well spacing, in order to reach the deeper parts of the high permeability reservoir plugging water layer, the water absorption layer is reduced and the actual permeability affected area to expand, improve the recovery factor [15].

4.4. Block water shutoff in oil field

The water injection wells with more than 1/3 are selected in the oil field blocks that need to be dealt with by the reasonable profile control and water shutoff measures. According to the characteristics of the reservoir and the overall requirements of the oil field development, the auxiliary measures such as adjusting the parameters, acidizing, filling holes and other auxiliary measures are adopted to achieve the purpose of improving the overall development effect.

5. The development trend of profile control water plugging technology

At this stage the profile control technology applied in the actual production of the oil field in our country has the following trends:

- Based on the principle of molecule and the principle of design, this paper studies the water absorbing delayed expansion agent, and has good research results;
- Application in horizontal well water shutoff profile control technology research has clear progress;
- Physical simulation technology, simulation of profile control technology should pay more attention to practicality, simplicity.

- The development of new technology of environmental protection water shutoff and profile control, which is based on the selectivity of nano molecular materials, and it will be the main research field in the future.

5.1. Microbial technology applied in the profile control water plugging
Microbial profile control technology is injected into formation microorganisms or used to activate the formation source. Microbial metabolism, growth and reproduction will produce gas or bio polymer, inorganic salts, the precipitation of these substances can be used to adjust the formation of the biofilm water injection profile. For water flow channel, the breeding of microorganism in the porous media, can reduce the permeability of high permeable zone formed hyphae community, to improve recovery factor [16]. Microbial profile control technology is applied in the exploitation of oilfield remaining reserves, has brought good economic benefit, environmental protection, safety, less material consumption, wide raw material sources, and the advantages of less equipment, good application prospect.

5.2. CO_2 deposition technology applied in the profile control water plugging
Adding CO_2 in the foaming agent can increase the resistance factors of foam system, the ability of the seal steam enhanced; The resistance factor decreased with the increase of temperature, and the foam stability of CO_2 foam was good under high temperature and high pressure; The resistance factor of CO_2 foam increases with the decrease of oil saturation and the increase of formation permeability. It can be seen that the CO_2 foam can be used to block the steam channeling channel and the large pore channel [17]. CO_2 foam flooding has good adaptability and good environmental protection, but there are few applications and large space for development, which is technically feasible and economically feasible still needs further validation. CO_2 water shutoff profile control, as a new technology with cationic reaction precipitation formation water, the purpose of the change of formation permeability to water shutoff profile control, Also it is the development direction of the future.

6. Conclusion
Water shutoff profile control technology has experienced decades of development, has formed a more mature water control oil stabilization technology. It can improve water flooding development effect according to different reservoir conditions, and it should be toward the low input, high-yield, multi-function, the direction of good efficiency in the future, and for different fields, practical solutions for the reservoir in high water cut period should be developed to achieve efficient development.

The technical research and development should be carried out for different conditions of the reservoir:

- According to the specific situation of the oilfield development economy long-term new water plugging material. It is important to consider the economic feasibility of water shutoff profile control in high water cut reservoir.
- Study on how to improve the mechanism of water flooding with water plugging and profile control technology. According to the current situation of oilfield development, the establishment of physical model and the corresponding evaluation, combined with the test, correlation of the real conditions of reservoir diversion agent materials and oilfield, study new control technology for different reservoir conditions.
- Precise numerical simulation and design software research and development. Most of the oil fields in our country have been greatly changed due to the long-time of strong injection and high water cut reservoir conditions. The numerical simulation and optimization design method of the traditional seepage theory is not enough to improve the present situation of water plugging and profile control technology. In combination with the practical situation of reservoir, research and development of the corresponding numerical simulation and optimization design software, make full use of the plugging agent, and achieve the best effect of development.

References

[1] Chen T L, Xin X, Zhang Y, *et al* 2008 Study on AMPS salt resistant weak gel *J. Southwest Petrol. Univ.* **30(1)** 126-8

[2] Wang P M, Luo J H, Bai F L, *et al* 2001 Research progress of water plugging technology at home and abroad well *Drill. Prod. Technol.* **24(4)** 28-30

[3] Field R 1994 In-depth drive fluid diversion using an evolution of colloidal dispersion gels and new bulk Gels-an operational case history of north rainbow ranch unit *SPE* **27773** doi: 10.2118/27773-MS

[4] Zhang X P, Zhao X T, *et al* Matching relationship between inorganic particle plugging agent and formation permeability *Drill. Prod. Technol.* **18(4)** 80-2

[5] Zaitoun A, Bertin H and Lasseux D 1998 Two-phase flow property modifications by polymer absprption *SPE* **39631** 285-92

[6] Seright R S 1996 Use of preformed gel s for conformance control infractured systems *SPE* **35351** 56-7

[7] Liu Y Z and Lv X H 1994 Exploration and practice of improving oil recovery by chemical method in Shengli Oilfield *Oil Gas Field Recov. Technol.* **1(1)** 25-8

[8] Tang X F, Wu Q, Liu G H, *et al* 2003 Field test and effect of block weak gel flooding *J. Petrol.* **24(4)** 58-61

[9] Omari A, Rabary R and Rebard M 2000 Controlling gelation time and microgel water shutoff *SPE* **59317** 55-8

[10] Hu Y X, Wang X S and Qin T 2002 Development and application of composite resin grain plugging agent *Adv. Fine Petrochem.* **3(2)** 19-21

[11] Wang L, Zhao L Q, *et al* 2005 Indoor evaluation of acid resistant high strength modified resin plugging agent *Geol. Sci. Technol. Inform.* **2(4)** 109-12

[12] Chen C 2016 Application of chemical plugging agent in oil field production *Technol. Platform* **1** 22

[13] Song Y G, Niu J, He H and Wang W 2008 Research progress of chemical water plugging and profile control agent in Oilfield *Adv. Fine Petrochem.* **12(50)** 5-11

[14] Seright R S, Lane R H and Sydansk R D 2003 A strategy for attacking excess water production *SPE* **84966** 158-69

[15] Chen Y 2008 Technology for improving the effect of deep water plugging in oil wells *Spec. Oil Gas Reserv.* **15(6)** 79-82

[16] Jiang Y, Xu D T, Chen J B, *et al* 2005 Single well treatment technology and its application effect analysis *Petrol. Explor. Develop.* **32(2)** 104-6

[17] Wu Y B, Hu D D, Chang Y W, *et al* 2010 Application status of CO_2 flooding to improve recovery of low permeability reservoir *Xinjiang Oil & Gas* **6(1)** 36-54

IOP Conf. Series: Earth and Environmental Science **82** (2017) 012062 doi:10.1088/1755-1315/82/1/012062

The response of drought in Beiluo River Basin of China based on the comprehensive method of Pa, SPI and fuzzy

L P Zhang[1], D F Liu[1,2], H X Zhang[1], Q Huang[1] and J X Chang[1]

[1]State Key Laboratory Base of Eco-hydraulic Engineering in Arid Area, School of Water Resources and Hydropower, Xi'an University of Technology, Xi'an 710048, China

E-mail: liudf@xaut.edu.cn

Abstract. The meteorological drought is threatening the agricultural economic development with the change of the climate. In order to analyze the characteristics of drought spatio-temporal change, the precipitation data of eight meteorological stations in the Beiluo River Basin of Shaanxi Province of China have been collected, and the drought index of Pa, SPI and FSE have been selected to analyze the drought in Shaanxi Province for the last 55 years. The results of Pa, SPI and FSE test show that the droughts happened in the Beiluo River Basin are 149, 215 and 203 times in the past 55 years, respectively. Overall, the Beiluo River has a tendency to dry out. The main type of drought is low-grade drought, followed by the medium-grade drought, and the specially-grade drought happened least. The average rainfall decreases in the Beiluo River Basin from the southeast to the northwest, and the change of the number of drought is just opposite to that of precipitation trend, which increases from southeast to northwest. The results will provide the scientific basis for the monitoring, evaluation, early warning and drought relief.

1. Introduction

Drought is one of the most common natural disasters in the world and is regarded as one of the most serious types of natural disasters in the world [1]. In the context of global warming, the frequency and intensity of extreme weather events, such as severe drought and high temperatures, are increasing [2, 3]. China is one of the countries with the largest number of natural disasters, the most frequent disasters and the most serious disasters. More than 70% of the annual economic losses of national average are caused by natural disasters because of the meteorological disasters. In the context of climate warming, drought disasters have been intensified. In the 20th century, there were many serious droughts in the northwest, the interval of drought occurred was shorter and shorter, the drought duration was longer and longer, and the drought intensity was bigger and bigger. The drought assessment, monitoring and forecasting research of the government and academia attract great importance to the hot issues, which has great practical significance. For a long time, a large number of research works have been carried out in the related industries all over the world, and many quantitative evaluation indexes have been established from climate, hydrology and agriculture [4]. Drought indicators are the basis for droughts study and the key to measuring the degree of drought. Over the years, many scholars have studied drought changes in different research areas based on different drought indicators [5-7], but it is rare to use a variety of indicators to compare and study at the same area.

Climate is the basis cause of drought, but also the main cause of drought heterogeneity. Among these climatic factors, precipitation is an important basis for the quantitative study of regional dry and wet succession. The standardized precipitation index (SPI) and the percentage of precipitation anomalies (Pa) are common indicators for the assessment of drought trends based on precipitation changes [8, 9]. SPI is recommended by the World Meteorological Organization (WMO) as a useful indicator of drought and is widely used around the world. Pa is more extensive used in China. Huang *et al* used the SPI and Copula functions to calculate the joint recurrence period in two typical drought scenarios of the Weihe River Basin [2, 10]. Wei *et al* used the Pa to analyze the spatial distribution of drought in Shaanxi Province [3]. Zhao used the Pa, SPI and SPEI to compare and analyze the eastern Qinghai-Tibet Plateau alpine grassland drought [4].

The Weihe River Basin is located in the semi-humid and semi-arid areas in the northwest of China. The ecological environment is fragile and the natural disasters are frequent. The drought is the most frequent, the heaviest and the most harmful disaster. It has great influence on the regional economic structure adjustment and rational exploitation and utilization of water resources [11]. The Beiluo River Basin is located in the Weihe River Basin. The climate and environment are complex and the drought disaster threatens the development of agriculture. The identification of the drought index is a powerful tool for monitoring drought and is the basis for studying the drought. Therefore, this paper chooses the comprehensive method of SPI, Pa and fuzzy to analyze the temporal and spatial characteristics of arid in the Beiluo River Basin, aiming at adjusting the agricultural planting structure, disaster prevention and mitigation, and comparing the applicability of SPI and Pa in drought monitoring in Shaanxi Province, providing scientific advices for water resources assessment and drought monitoring in the area.

In recent years, great progress has been made in the development and application of drought indexes. Drought index, as quantitative indicators of drought degree, plays an important role in drought monitoring, forecasting and management of water resources, especially drought index provides method and basis for the application of regional resource environment and interdisciplinary fields. At the same time, drought index is also an effective tool for hydrology and water resources research. It can provide a basis for slowing down and preventing the adverse effects of drought on agriculture and making scientific policy.

2. Data source

In this study, the daily precipitation, daily temperature, sunshine hours and other meteorological data from eight meteorological stations (figure 1) in the Beiluo River Basin (1960-2015) were collected from the Chinese Meteorological Science Data Sharing Service Network. The distribution of each meteorological station is uniform, which can be used to reflect the spatial changes of hydrological meteorology in the basin. For the missing meteorological data, linear interpolation was applied in the station. The precipitation and surface evaporation are calculated from the precipitation of each station. The evaporation data are calculated by Tyson polygon method in ArcGIS.

Beiluo River is first tributary of the Weihe River, the second tributary of the Yellow River, a total length of 680.3 km, and is the longest river in Shaanxi Province. The average runoff of the Beiluo River Basin is 943 million m^3, and in Shaanxi Province is 873 million m^3, and the inter-annual variation is large [2, 3]. Annual average sediment transmission is 0.833×10^8 t. The basin is semi-arid continental monsoon climate, the annual average precipitation is 548mm, the distribution of precipitation in the Beiluo River is uneven during the year, and the precipitation in June to September accounts for about 70% of the total annual precipitation, during which rainstorm happens frequently, the 3 months contain 60% of the annual runoff, and more than 90% of the flood occurred from July to August. Climate is semi-arid and semi-humid to the dry climate of the transition zone, with obvious continental monsoon climate characteristics, cold and dry in winter, in spring dry and windy, cool and humid in autumn. The average annual temperature of 7.8℃, the extreme maximum temperature is 37.4℃, and the extreme minimum temperature is -28.0℃.

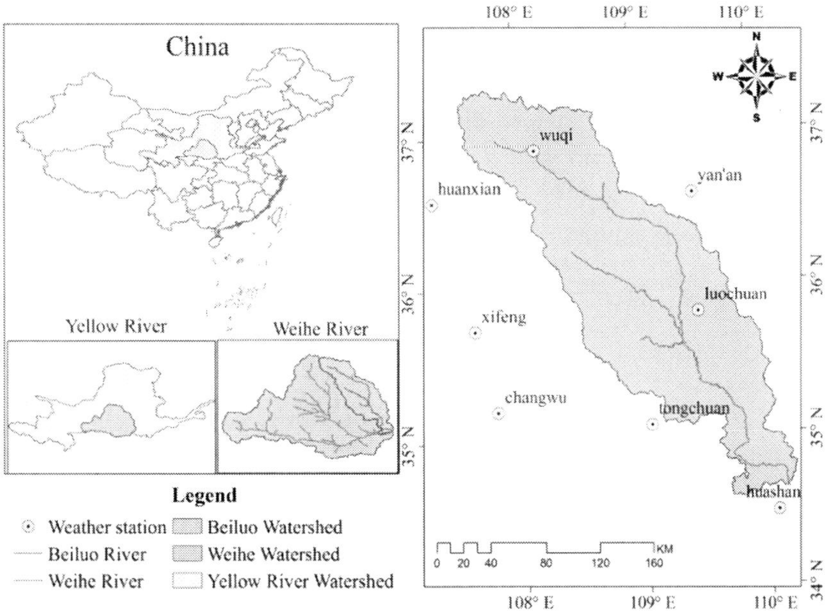

Figure 1. Location of meteorological stations in the Beiluo River Basin.

3. Methodology

3.1. Pa

The percentage of precipitation anomalies can be used to reflect the degree of deviation from the average state of precipitation over a period of time and to directly reflect the drought caused by precipitation anomalies [4, 9]. The formula is:

$$Pa = \frac{P - \overline{P}}{P} \times 100\%$$

where, P is the precipitation in a certain period of time, $\overline{P}$ is the average monthly precipitation in the same period. In this study, P is the monthly precipitation, which has been the average precipitation for the past 55 years. The advantage of the Pa is that the calculation method is simple and intuitive, but the disadvantage is that the response is relatively slow. In China, Pa index is used in the animal husbandry drought test, and the China Meteorological Administration has determined the semi-arid, semi-humid areas of drought criteria (table 1).

Table 1. The classification of drought magnitude.

Drought magnitude	Non drought	Low-grade drought	Medium-grade drought	Heavy-grade drought	Specially-grade drought
Pa	-40<Pa	-60<Pa<-40	-80<Pa<-60	-95<Pa<-80	Pa<-95
SPI	-0.5<SPI	-1<SPI<-0.5	-1.5<SPI<-1	-2<SPI<-1.5	SPI<-2
FSE	FSE<0.7	0.7<FSE<1.5	1.5<FSE<2	2<FSE<2.6	2.6<FSE<4

3.2. SPI

The SPI index uses mathematical methods to convert the cumulative frequency distribution of precipitation into a standard normal distribution, which has dimensionless and standardized characteristics and is applicable in different regions [4, 8, 10]. Suppose the precipitation in some time is x, assuming that the precipitation sequence obeys the Gamma distribution, then the probability density function is:

$$P(x)=\frac{1}{\beta^{\alpha}\Gamma(x)}x^{\alpha-1}e^{-x/\beta},x>0$$

$$\alpha=\frac{1+\sqrt{1+4A/3}}{4A}$$

$$\beta=\bar{x}/\alpha$$

$$A=ln\left(\bar{x}\right)-\frac{\sum_{i=1}^{n}ln\,x_i}{m}$$

where, β and α are the scale parameters and the shape parameters respectively, greater than zero, are obtained by the maximum likelihood estimation method, n is the length of the precipitation sequence, where the number of items for zero is m, $\bar{x}$ is the average of non-zero precipitation sequence.

Let $q=m/n$, the cumulative probability of a certain time scale is:

$$H(x)=q+(1-q)G(x)$$

$$G(x)=\int_0^x p(x)dx$$

The cumulative probability distribution is converted to the standard normal distribution, that is, the corresponding SPI value.

When $0<H(x)\leq 0.5$, and let $k=\sqrt{ln\frac{1}{H(x)^2}}$, then

$$SPI=-\left(k-\frac{c_0+c_1k+c_2k^2}{1+d_1k+d_2k^2+d_3k^3}\right)$$

When $0.5<H(x)<1$, and let $k=\sqrt{ln\frac{1}{\left[1\text{-}H(x)\right]^2}}$, then

$$SPI=k-\frac{c_0+c_1k+c_2k^2}{1+d_1k+d_2k^2+d_3k^3}$$

where, $c_0=2.515517, c_1=0.802853, c_2=0.010328,$ $d_1=1.432788, d_2=0.189269$
$d_3=0.001308$

The advantage of SPI index is that the calculation stability is great, the time scale is flexible, and the indication of spatiotemporal change is effective and can reflect the drought and flood conditions effectively. This paper chose the time scale of 6 months SPI as the research object, and the drought magnitude standard is shown in table 1.

3.3. Fuzzy synthetic evaluation based on entropy method

Entropy weigh method is an objective empowerment method, and the weight of each evaluation index can be determined according to the difference degree of each index value in drought evaluation. The fuzzy comprehensive evaluation method is to quantify the fuzzy indexes of the object by constructing the hierarchical fuzzy subsets (ie, determine the membership), and then use the fuzzy transformation principle to evaluate the indexes [12, 13]. When the fuzzy comprehensive evaluation method is used to evaluate some results, the evaluation results are intuitionistic and can be solved satisfactorily. The fuzzy evaluation can objectively solve the problems of uncertainty existed in the evaluated things, and can deal with the fuzzy evaluation object exactly by digital means, and give a comprehensive, accurate, and realistic quantification results. The generally follow the steps below:

Step 1: Standardize raw data matrix and Define entropy, count and record the Entropy Weight.

Step 2: divide the study area D into k small drought evaluation areas, that is, $D=|D_1,D_2,...,D_k|$. Determine U, the set of factors for comprehensive drought assessment, and determine the m drought evaluation indexes, $U=|u_1,u_2,...,u_m|$. Determine the set of possible assessment results in the drought evaluation, the set of drought magnitude set $B=|B_1,B_2,...,B_n|$.

Step 3: Determine the membership matrix, starting from a single index, determine the evaluation of the degree of membership of the drought magnitude. Membership can be obtained by membership function:

$$r_{i1}=\begin{cases}1,x<x_1\\(x_2-x)/(x_2-x_1),x_1<x<x_2\\0,x>x_2\end{cases}$$

where, $1<k<5$, x_1,x_2,x_3,x_4,x_5 are the critical values of drought index at the different drought magnitude, mild drought, medium drought, severe drought and extreme drought, respectively.

Step 4: Determine the membership weight of the evaluation index. Use the entropy method to establish the mathematical model to calculate the weight of the index. $A=(\alpha_1,\alpha_2,...,\alpha_m)$, and α_i is the drought evaluation index membership weight coefficient.

Step 5: The membership matrix R is multiplied by the evaluation index membership weight A, and the matrix W is obtained. W shows all the effects on different drought magnitude, and calculates the fuzzy comprehensive drought evaluation index DI.

$$w_i=\sum_{j=1}^{m}(r_{ji}\alpha_i)$$

$$DI=\sum_{i=1}^{m}(w_iB_i)$$

where, B_i is the weight coefficient for different drought magnitude. Drought is divided into five magnitudes: drought, mild drought, moderate drought, severe drought and extreme drought. The standard of classification is shown in table 1, using the classification method of "meteorological drought magnitude" (GB/T 20481-2006).

4. Results and discussion

4.1. Time scale analysis of drought

The Pa and SPI6 indexes of each station were calculated by the meteorological data from 1961 to 2015. Pa and SPI of each month in recent 55 years in the Beiluo River Basin were shown in the following figures 2(a) and 2(b), and at the same time according to the drought magnitude (table 1) the number of Pa and SPI in the Beiluo River was tallied (table 2). It can be seen from figure 2(a) that the overall Pa of the Beiluo River Basin shows a decreasing trend, and the Mann-Kendall test has a statistical value of -0.812, and its absolute value is less than 1.96, failing to pass the 95% confidence test, which indicates that the Beiluo River Basin has no obvious dry-out trend. It can be seen from table 2 that the results of Pa test show that the number of drought in the Beiluo River Basin was 149, and 72 (10.91%) and 41 (6.21%) occurred in Low-grade drought and Medium-grade drought), while the number of Heavy-grade and Specially-grade drought was relatively small, which occurred for 18 (2.73%) times. It can be seen from figure 2(b) that the overall SPI6 exhibits a decreasing trend, which is statistically -3.165 and an absolute value larger than 1.96, which pass the Mann-Kendall test. The 95% confidence test shows that the Beiluo River Basin has a dry-out trend. From table 2, the results of SPI test showed that the number of drought in the Beiluo River Basin was 215 times in the past 55 years, among which 102 (15.45%) and 74 (11.21%) occurred in Low-grade drought and Medium-grade drought), while the number of Heavy-grade drought and Specially-grade drought was relatively small, which occurred for 25 (3.79%) and 14 (2.12%) times, respectively.

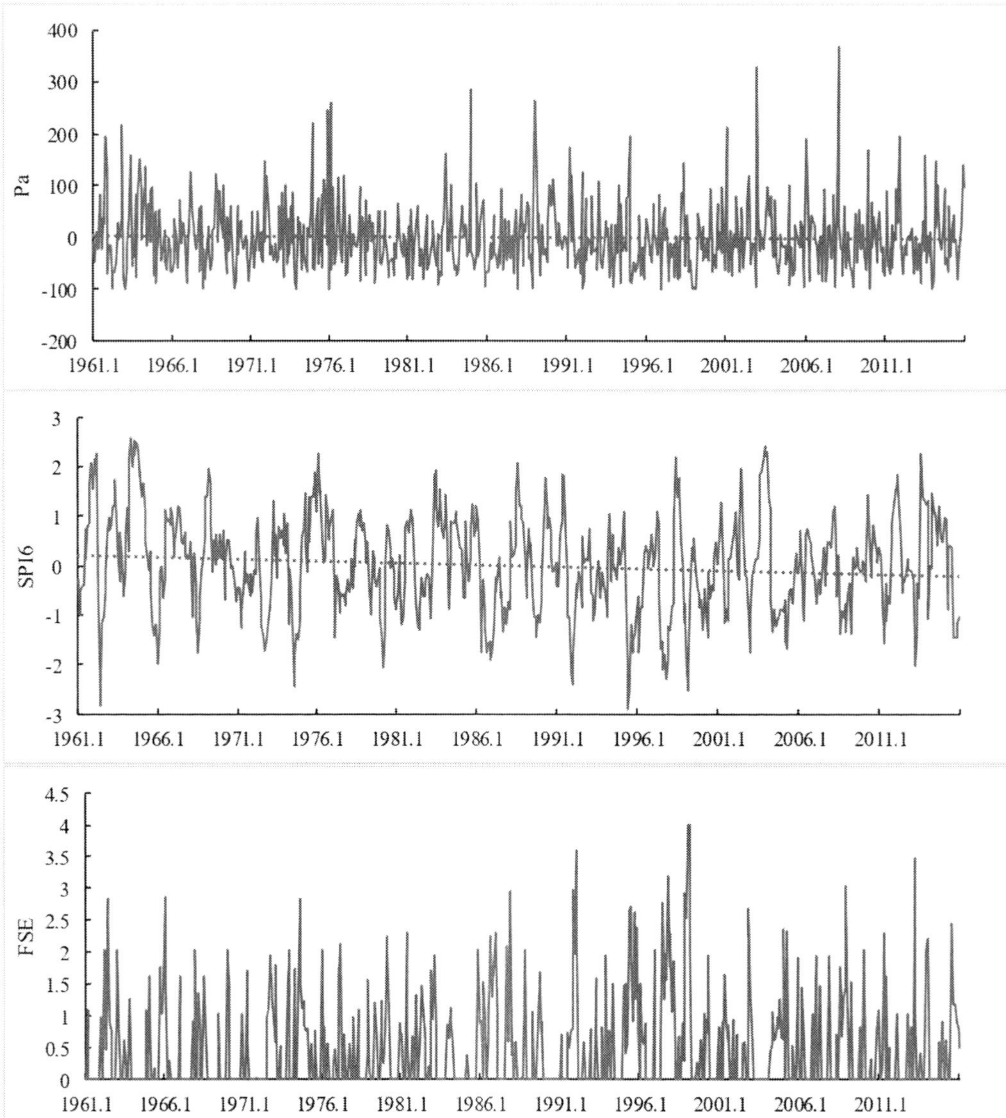

Figure 2. The monthly changes of Pa (a), SPI (b) and FSE (c) index.

Table 2. The quantitative statistics of drought occurrence on Beiluo River Basin.

Drought index	Non drought	Low-grade drought	Medium-grade drought	Heavy-grade drought	Specially-grade drought
Pa	511	72	41	18	18
SPI	445	102	74	25	14
FSE	457	118	50	18	17

In order to reflect the drought condition in the Beiluo River better, this paper uses the fuzzy synthetic evaluation method based on the entropy method to evaluate the wet and dry conditions of the basin (figure 2(c)) in case of avoiding the adverse effects of a single indicator. The results of FSE method showed that the number of month that drought happened was 203 times, and there were 118 (17.88%) and 50 (7.58%) times occurred in Low-grade drought and Medium-grade drought in the Beiluo River, respectively, and in Heavy-grade drought and Specially-grade drought was 18 (2.73%) and 17 (2.58%) times, respectively. Overall, the Beiluo River Basin has a dry trend, the main type of drought was low-grade drought, followed by Medium-grade drought, and the least was specially-grade drought.

The number of drought in different grades happened indifferent seasons was tallied according to the FSE by month. It can be seen from the figure 3 that the each grade drought happened most in winter, and the numbers are 41, 23, 7, and 6. The autumn was in the second place, and drought happened least in summer. Over all, there were large differences among the four seasons in which the drought happened in Beiluo River Basin, and the number that drought happened in the four seasons were 42, 32, 52, and 77, respectively, among which winter was the drought most likely happened. The Beiluo River Basin is a region that needs high attention and further study, autumn and winter are the most drought-prone seasons, high frequency of drought.

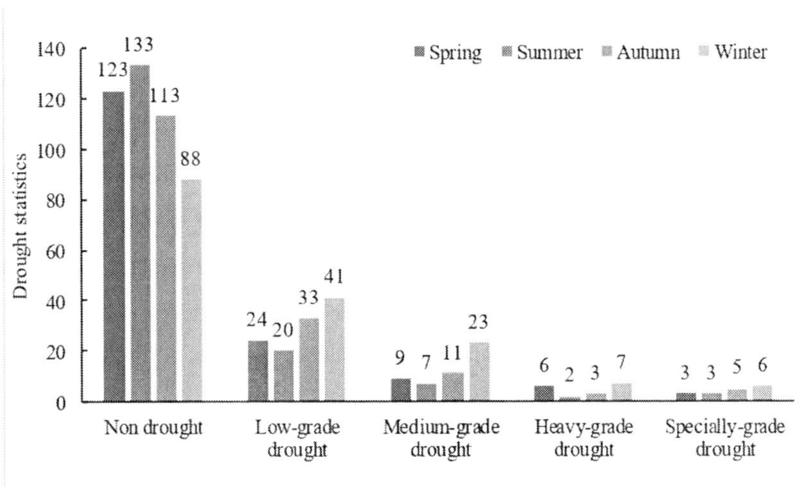

Figure 3. The frequency statistics of seasonal drought.

4.2. Spatial scale analysis of drought

The number of drought occurred years in each meteorological station in the Beiluo River Basin is shown in table 3, and the number of years of drought in each station is 177 ~ 251 times. As shown in figure 4(a), the average rainfall in the Beiluo River Basin gradually decreased from the southeast to the northwest, the precipitation of each station is between 430-810 mm, among which Huashan has the highest precipitation (810.33 mm),while the Huanxian station is the least (432.27 mm). The spatial distribution of drought situation on the Beiluo River Basin was shown in figure 4(b), and the times of drought occurred has the opposite trend of precipitation, increase from southeast to northwest, the most frequent occurrence in Low-grade drought is the Wuqi station, and in the Xifeng station the Medium-grade drought and Heavy-grade drought occurred most, and the most frequent occurrence in Specially-grade drought is Huanxian station.

Table 3. The quantitative statistics of drought occurrence with eight weather stations.

Weather station	Non drought	Low-grade drought	Medium-grade drought	Heavy-grade drought	Specially-grade drought
Huashan	483	79	61	25	12
Huanxian	412	85	75	38	50
Luochuan	444	80	64	39	33
Tongchuan	442	89	61	42	26
Wuqi	409	98	75	31	47
Xifeng	443	68	78	43	28
Yan'an	417	90	65	42	46
Changwu	455	70	68	39	28

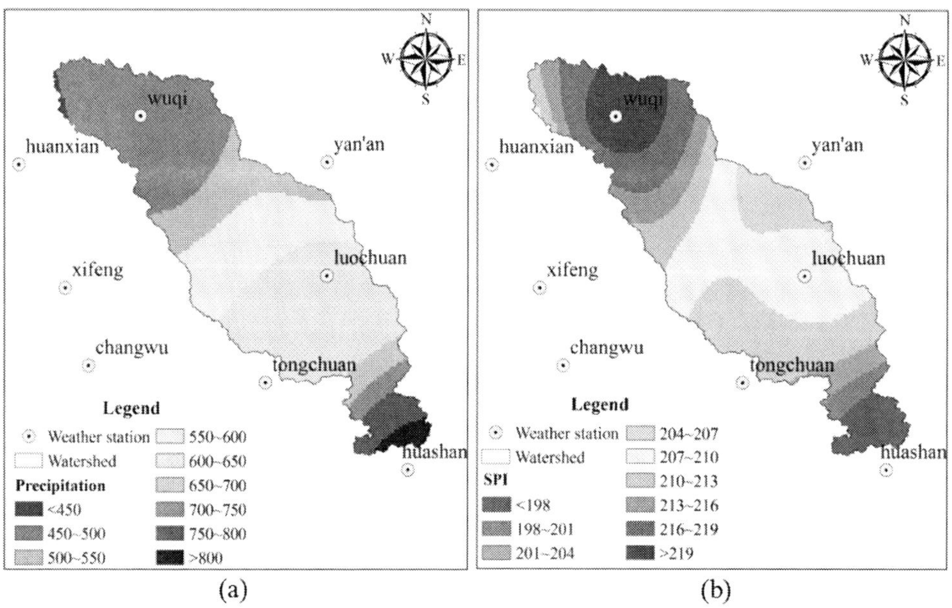

Figure 4. Spatial distribution of annual mean precipitation (a) and drought degree (b) in the Beiluo River Basin.

5. Conclusions

The spatial and temporal changes of the drought degree in the Beiluo River Basin were analyzed by Pa, SPI and FSE. The results are as follows:

- Pa and SPI test results showed that, numbers of years of drought occurred in the Beiluo River Basin in the past 55 years were 149 and 215, among which Low-grade drought occurred 72 times and 102 times, Medium-grade drought occurred, respectively, 41 times and 74 times. The occurrence of Heavy-grade drought and Specially-grade drought was relatively less, among which the number from Pa method test was 18 times. Heavy-grade drought and Specially-grade drought from SPI method was 25 times and 14 times, respectively.

- The result of FSE method showed that the number of drought happened in Beiluo River Basin was 203 times, among which the four grades drought happened for 118 (17.88%), 50 (7.58%), 18 (2.73%) and 17 (2.58%), respectively. Overall, the Beiluo River Basin has a dry-out trend, the main type of drought is low-grade drought, followed by the medium-grade drought, and the specially-grade drought happened least.
- The average annual precipitation in the Beiluo River Basin is decreasing from the southeast to the northwest, and the number of arid occurrences is opposite to that of the precipitation. And the most frequent occurrence of Low-grade drought is at the Wuqi station, Heavy-grade drought is the most at the Xifeng station, and the most frequent occurrence of Specially-grade drought is at Huanxian station.

Acknowledgments
This work was supported in part by the National Natural Science Foundation of China (NSFC) (Grant No. 41330858, 51309188). Funding was also provided by the Basic Research Plan of Natural Science of Shaanxi Province (2016JQ5105), National Department Public Benefit Research Foundation of Ministry of Water Resources (201501058) and Project of School of water resources and hydropower of Xi'an University of Technology (2016ZZKT-15).

References
[1] Zhao M and Running S W 2010 Drought-induced reduction in global terrestrial net primary production from 2000 through 2009 *Science* **329(5994)** 940
[2] Huang S Z, Huang Q, Wang Y M, *et al* 2015 Evolution of drought characteristics in the Weihe river basin based on standardized precipitation index *J. Nat. Disaster* **24(1)** 15-22 (in Chinese)
[3] Wei K, Wang Q J, Zhou B B, *et al* 2017 Analysis of drought characteristics in Shaanxi province based on precipitation anomaly percentage *J. Soil Water Conserv.* **31(1)** 318-22
[4] Zhao X L, Li W L, Guo X L, *et al* 2017 The responses of Pa, SPI and SPEI to dry climate in alpine meadows of eastern Qinghai-Tibet Plateau *Pratacult. Sci.* **34(20)** 273-82 (in Chinese)
[5] Chen Y T, Chang J X, Huang S Z, *et al* 2014 Variation characteristics of drought in Weihe River Basin based on Palmer drought severity index *J. Nat. Disaster* **23(5)** 29-37 (in Chinese)
[6] Dai A, Trenberth K E and Qian T 2009 A global data set of palmer drought severity index for 1870–2002: Relationship with soil moisture and effects of surface warming *J. Hydrometeorol* **5(6)** 1117-30
[7] Tsakiris G, Pangalou D and Vangelis H 2007 Regional drought assessment based on the Reconnaissance Drought Index (RDI) *Water Resour. Manag.* **21(5)** 821-33
[8] Amirataee B and Montaseri M 2017 The performance of SPI and PNPI in analyzing the spatial and temporal trend of dry and wet periods over Iran *Nat. Hazards* **86** 1-18
[9] Chanda K and Maity R 2015 Meteorological drought quantification with standardized precipitation anomaly index for the regions with strongly seasonal and periodic precipitation *J. Hydrol. Eng.* **20(12)** 06015007
[10] Huang S, Chang J, Leng G, *et al* 2015 Integrated index for drought assessment based on variable fuzzy set theory: A case study in the Yellow River Basin, China *J. Hydrol.* **527(527)** 608-18
[11] Chang J, Li Y, Ren Y, *et al* 2016 Assessment of precipitation and drought variability in the Weihe River Basin, China Arabian *J. Geosci-czech* **9(14)** 633
[12] Chowdhury S and Champagne P 2008 Selecting water disinfection processes using fuzzy synthetic evaluation (FSE) technique *Water Qual. Res. J. Can.* **43(1)** 1-10
[13] Xu Y, Du P and Wang J 2017 Research and application of a hybrid model based on dynamic fuzzy synthetic evaluation for establishing air quality forecasting and early warning system: A case study in China *Environ. Pollut.* **223** 435-48

IOP Conf. Series: Earth and Environmental Science 82 (2017) 012063 doi:10.1088/1755-1315/82/1/012063

Impacts of climate change and human activities on runoff in Weihe Basin based on Budyko hypothesis

H S Wu[1], D F Liu[1], J X Chang[1,2], H X Zhang[1] and Q Huang[1]

[1]State Key Laboratory Base of Eco-hydraulic Engineering in Arid Area, Xi'an University of Technology, Xi'an 710048, China

E-mail: chxiang@xaut.edu.cn

Abstract. The Weihe River Basin (WRB) is the largest tributary of the Yellow River and plays an irreplaceable role in the Shaanxi-Gansu-Ningxia area. In recent years, owing to the human activities and climate change, the runoff of the WRB has reduced, wherefore, it is necessary to analyze the impact on runoff quantitatively. By using the data of Huaxian and Zhuangtou stations, we can respectively calculate the changes in runoff for climate change and human activities via Budyko hypothesis. The trend of runoff, precipitation, temperature, potential evapotranspiration and the break points are examined by Mann-Kendall test (M-K method), cumulative anomaly method and ordered cluster analysis. The results show that the break points of runoff series in WRB are 1970 and 1989, so that the runoff series can be divided into the baseline period and the changed period. Based on the data of potential evapotranspiration and Budyko formula, the contribution rates of climate change and human activities to runoff are 41% and 59% in 1970-1989. From 1990 to 2010, the contribution rates of climate change and human activities are 37% and 63%, respectively.

1. Introduction

Weihe River is one of the important rivers in the northwest region. It is the largest tributary of the Yellow River, flows through Shaanxi, Gansu and Ningxia provinces (autonomous regions). The WRB is located in the northwest of China, whose ecosystem is fragile and the occurrence of natural disaster is frequent. Among them, the frequency of drought is the highest, which is the main factor restricting the further development of the national economy. The WRB plays an important role in the production, living and ecology in the northwest area. The change of the runoff will directly affect the regional water resources and destroy the local water structure. In recent years, because of the dual role of climate change and human activities, the runoff in WRB has been decreasing year by year, and thus produces a series of problems [1, 2]. The main performances are as follows: (1) the water quantity in the watershed has been reducing; (2) the water quality has been polluted seriously; (3) water and sand balance has been destroyed, resulting in the sediment deposition, and the flood discharge capacity has been reducing.

The dramatic changes in climate and the increasing human activities have had a tremendous impact on the river basin in the world scope, and one such river basin is the WRB. However, this is also one of the areas where lacks water, water shortages have become a bottleneck restricting local economic development. During the last several decades, we have built a large number of water conservancy projects in the WRB, such as reservoir and hydropower station et al. Climate change surveys in the Yellow River basin suggested that trends at a rate of 1.28 ℃ over the past half century, while the average precipitation dropped by approximately 8.8 % over the second half of the 20th century [3]. The combination of these effects resulted in a reduction in runoff [4, 5]. Climate change affects runoff, chiefly through the variability of potential evapotranspiration and precipitation [6-9]. Human activities can cause changes in the underlying conditions. Moreover, unrestrained exploitation of surface water and groundwater will have an unpredictable impact on the runoff generation and confluence process. It is of great practical significance to quantitatively separate the impacts of climate change and human activities on runoff so that they can be applied to land management, water resource management and so on. Over the past few decades, hydrologists had been working to observe changes under the impact of climate change and human activities on runoff [10-13]. Based on the Budyko hypothesis [14], this study quantitatively analyzes the contribution rates of climate change and human activities to the reduction of runoff, which is of great reference value to watershed management and ecological sustainable development in the WRB.

2. Study area and data

2.1. Study area

The WRB (figure 1) is located in E 104° 25 '~ 110° 25', N 33° 75 '~ 37° 25', which originates in Gansu Province Weiyuan County NiaoShu mountain, and mainly follows through the Shaanxi-Gansu-Ningxia region, and finally through the Tongguan County, Shaanxi Province, into the Yellow River. The Weihe River flows across Shaanxi and Gansu provinces with a total length of 818 km and a total area of 134700 km^2. The average annual runoff is 72.7 × 10^8 m^3. Jinghe River, the largest tributary in the WRB, is located in the central Loess Plateau, and it has a total length of 455 km, and watershed area of 45400 km^2; the Beiluo River is the second largest tributary in Weihe, and it has a total length of 680 km, and watershed area of 26900 km^2.

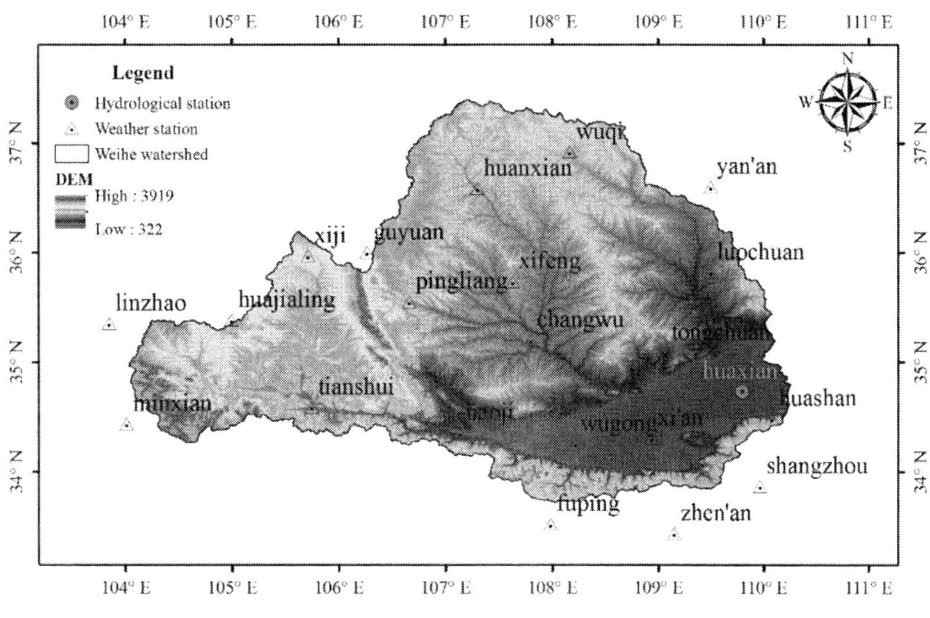

Figure 1. Meteorological stations and their control basins in WRB.

The WRB belongs to the typical temperate continental monsoon climate. The local climate is dry in winter and hot in summer, and the precipitation is concentrated in the July to September. The average annual precipitation is 636 mm, and the average annual temperature is 9.42℃. In addition, the extreme maximum and minimum temperatures are 42.8℃ and -28.1℃. The average annual evapotranspiration is 848.78 mm, and gradually increased from south to north. The eastern part of Guanzhong is a high evapotranspiration area, where the annual evapotranspiration is more than 1200 mm. The western part of Guanzhong and Qinling Mountains are the areas with low evapotranspiration, which are less than 800 mm. The minimum monthly evapotranspiration usually occurs in January or December, and the maximum one appears in July.

2.2. Data

The meteorological data of the 21 meteorological stations of the WRB in this study are provided by the China Meteorological Data Sharing Network. The measured runoff data is selected from the Huaxian and Zhuangtou station, which control the study area [15]. The Penman-Menteith formula is used to calculate the potential evapotranspiration [16]. For the missing precipitation and meteorological data, we use the linear interpolation method of adjacent meteorological stations to interpolate the extension. By Utilizing ArcGIS 9.3, we can convert precipitation and potential evapotranspiration into watershed data.

3. Study methods

3.1. Budyko hypothesis

Based on the global water and energy balance, Budyko pointed out that the watershed evapotranspiration was controlled by both the ability of precipitation and evaporation. The precipitation P is used to express the moisture supply condition, and the potential evapotranspiration E_0 is used to denote the energy supply condition. Thus, in the calculation of land surface evapotranspiration, there are the following boundary conditions [17]:

In extreme arid regions, all of the precipitation will be converted into evapotranspiration:

$$\frac{E_0}{P} \to \infty, \frac{E}{P} \to 1 \tag{1}$$

In extremely humid regions, all of the evapotranspiration energy will be converted to latent heat:

$$\frac{E_0}{P} \to 0, \frac{E}{E_0} \to 1 \tag{2}$$

Thus, we can obtain the coupled equilibrium equation of water and heat:

$$\frac{E}{P} = \mathrm{f}(\frac{E_0}{P}) = \mathrm{f}\ (\emptyset) \tag{3}$$

$$\mathrm{f}(\emptyset) = [\emptyset \tan h\,(\frac{1}{\emptyset})(1 - e^{-\emptyset})]^{1/2} \tag{4}$$

$\emptyset$ (the degree of dryness), which can be used as an indicator of water and heat to divide climate and vegetation zones. $\tan h$ is a hyperbolic tangent function [17]. f is a universal function [18]. Based on the hypothesis of Budyko, Fu, a meteorologist in China, put forward the analytical expression of the actual evapotranspiration [19]:

$$\mathrm{E} = \mathrm{P} + E_0 - [P^{\omega} + E_0^{\omega}]^{1/\omega} \tag{5}$$

ω is a parameter that indicates the efficiency of evapotranspiration, and its range is $(1,+\infty)$. ω mainly reflects the underlying surface conditions, such as topography, vegetation and other conditions. According to Fubao Sun, Dawen Yang and other data calculated in 2007, in this study ω take 3.22 [20].

For a certain closure area, the expression of its multi-year water balance is:

$$\mathrm{P} = \mathrm{E} + \mathrm{Q} + \Delta\mathrm{S} \tag{6}$$

The Δ S is the watershed storage variables, in the years of average scale, it can be neglected [21].

The changes in runoff caused by climate change are quantitatively calculated according to the method proposed by Milly and Dunne *et al* [22]:

$$\Delta Q_c = \frac{\partial Q}{\partial P} \Delta P + \frac{\partial Q}{\partial E_0} \Delta E_0 \quad (7)$$

Where ΔP and ΔE_0 represent the changes of precipitation and potential evapotranspiration, $\frac{\partial Q}{\partial P}$ and $\frac{\partial Q}{\partial E_0}$ respectively represent the sensitive coefficient of runoff to precipitation and potential evapotranspiration. Based on Fu's formula, the sensitive coefficient can be expressed as:

$$\frac{\partial Q}{\partial P} = (1 + x^{\omega})^{(\frac{1}{\omega})} - x^{\omega}(1 + x^{\omega})^{(\frac{1}{\omega}-1)} \quad (8)$$

$$\frac{\partial Q}{\partial E_0} = x^{(\omega-1)} (1 + x^{\omega})^{(\frac{1}{\omega}-1)} - 1 \quad (9)$$

3.2. Cumulative anomaly method [23]

The cumulative anomaly method is also a widely used method to judge the trend of series change from the curve. It can reflect the trend of the change of the elements. The cumulative anomaly curve shows an upward trend, which indicates that the anomaly value increases, otherwise the anomaly value decreases. Through the ups and downs of the curve can also determine the break points of the series.

For a given sequence X, the cumulative anomaly value at a given moment is expressed as

$$X = \sum_{i=1}^{t} x_i - x(t=1,\ 2,\ \ldots,\ n) \quad (10)$$

Among $x = \frac{1}{n} * \sum_{i=1}^{t} x_i$

3.3. Mann-Kendall Test (M-K) [24]

Mann-Kendall nonparametric test method is widely used for its advantages that it is not affected by sample value and the type of distribution. So it is commonly used in hydrological, meteorological and other non-normal distribution of the sequence .The basic principles are as follows:

- Given a time sequence X ($X_1, X_2, \ldots X_n$), construct a rank sequence M_i, which indicates the cumulative number of samples($X_i > X_j$ ($1 \le j \le i$)). Define d_k as:

$$d_k = \sum_{1}^{k} M_i \quad (2 \le k \le n) \quad (11)$$

- The mean and variance of d_k:

$$E(d_k) = \frac{k(k-1)}{4} \quad (12)$$

$$var(d_k) = \frac{k(k-1)(2k+5)}{72} (2 \le k \le n) \quad (13)$$

- In time sequence, the statistics are defined independently:

$$UF_K = \frac{d_k - E(d_k)}{\sqrt{var(d_k)}} (k = 1,2, \ldots, n) \quad (14)$$

Firstly, given the confidence level α, if the $|UF_K| > UF\alpha/2$, indicates that the sequence has significant trend. Then, the time sequence is arranged in reverse order. According to the equation (14) calculation, while making

$$UB_k = -UF_k \quad (15)$$

$$K=n+1-k \quad (16)$$

Finally, UB_k and UF_k are drawn as UB and UF curve. If there is an intersection between the two curves, the intersection is the beginning of the mutation.

3.4. Ordered cluster analysis [25]

With the "like attracts like" to describe the cluster analysis, it can express the idea of cluster analysis vividly. If the order of classification cannot be disrupted, such a classification is called ordered classification. In order to estimate the most likely break point τ in orderly classification, the essence is to find the optimal segmentation point which make the same classes ' sum of deviation square is smaller, while the different classes is larger. For the hydrological sequence X_1, X_2 ..., X_n, the optimal two-segment method is as follows:

Suppose that the possible break point is τ, then the square sum of the deviations before and after the mutation is

$$V_\tau = \sum_{i=1}^{\tau}(x_i - x_\tau)^2 \tag{17}$$

$$V_{n-\tau} = \sum_{i=\tau+1}^{n}(x_i - x_{n-\tau})^2 \tag{18}$$

In this equation, x_τ and $x_{n-\tau}$ are the mean of the two parts before and after the τ.

The sum of squared deviations

$$S_n(\tau) = V_\tau + V_{n-\tau} \tag{19}$$

Then, when the $S = \min_{2 \le \tau \le n-1}\{S_n(\tau)\}$, τ is the optimal two-segment point and it can be speculated as a break point.

4. Calculation results and analysis

4.1. The analysis of variability and trend

The break points of the precipitation, temperature and runoff in the WRB were determined by the M-K method. The break points of precipitation appeared in the range from 1971 to 1985, and further through the cumulative anomaly and ordered cluster analysis to determine the break point was 1985 (figure 2). At the same time, the break point of temperature was determined in 1991 (figure 3). In addition, with the M-K test of runoff, the break point was determined in 1989, further through the cumulative anomaly to determine the break points were 1970 and 1989 (figure 4). The results of the break point test are shown in figures 2-4:

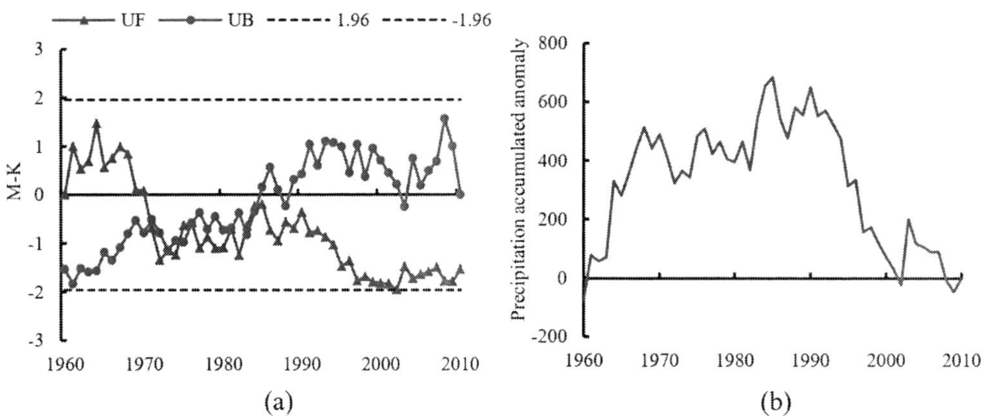

Figure 2. The break point test of annual precipitation with M-K (a) and cumulative anomaly (b).

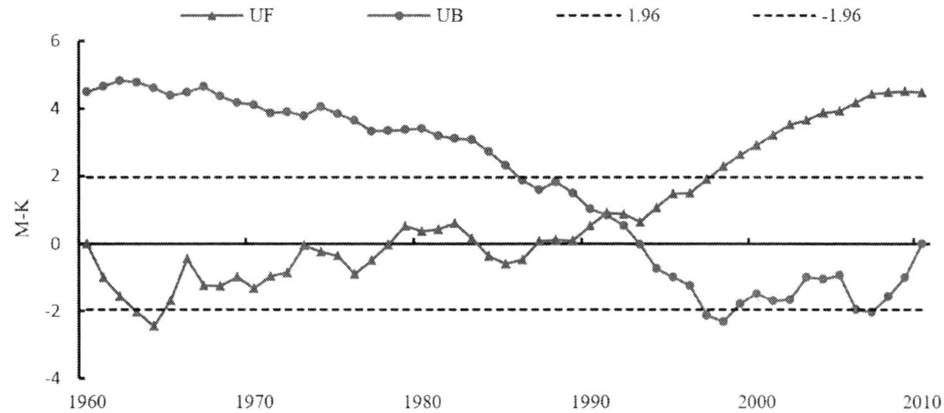

Figure 3. The break point test of annual temperature with M-K.

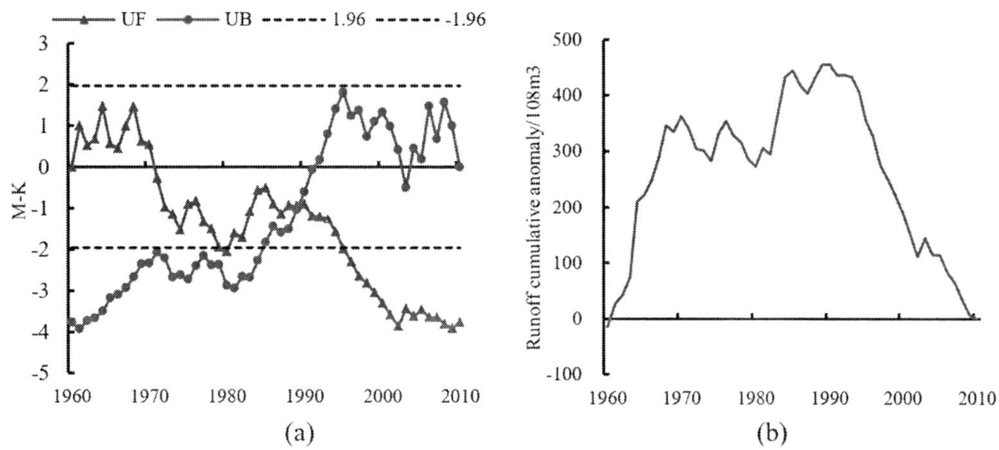

Figure 4. The break point test of annual runoff with M-K (a) and cumulative anomaly (b).

Figure 5 shows that the annual runoff and precipitation of the WRB have a decreasing trend, while the temperature and potential evapotranspiration are increasing. The maximum annual precipitation appears in 1964 (808.7 mm), and the minimum annual precipitation occurs in 1997 (372.4 mm), the precipitation series decreased by 12.7% (figure 5 (a)). In addition, the maximum annual runoff appears in 1964 (207.8×10^8 m^3), and the minimum annual runoff occurs in 1995 (21×10^8 m^3), and the runoff series decreased by 58.9% (figure 5 (b)). The maximum annual temperature appears in 1998 (10.6℃), and the minimum annual temperature occurs in 1967 (8.2℃), the temperature series increased by 14.4% (figure 5 (c)). Besides, the maximum potential evapotranspiration occurs in 1997 (934.8 mm), and the minimum potential evapotranspiration appears in 1964 (721.7 mm), the evapotranspiration series increased by only 1.2% (figure 5 (d)).

4.2. Calculation of contribution rate of climate change and human activities on runoff

According to the above results, the break points of the runoff are determined in 1970 and 1989, so the measured runoff series can be divided into three time periods. The baseline period is from 1960 to 1970, and the 1971-1989, 1990-2010 are the changed periods. The information for each parameter of each period can be seen from table 1, moreover, the results are shown in table 2.

IOP Conf. Series: Earth and Environmental Science 82 (2017) 012063 doi:10.1088/1755-1315/82/1/012063

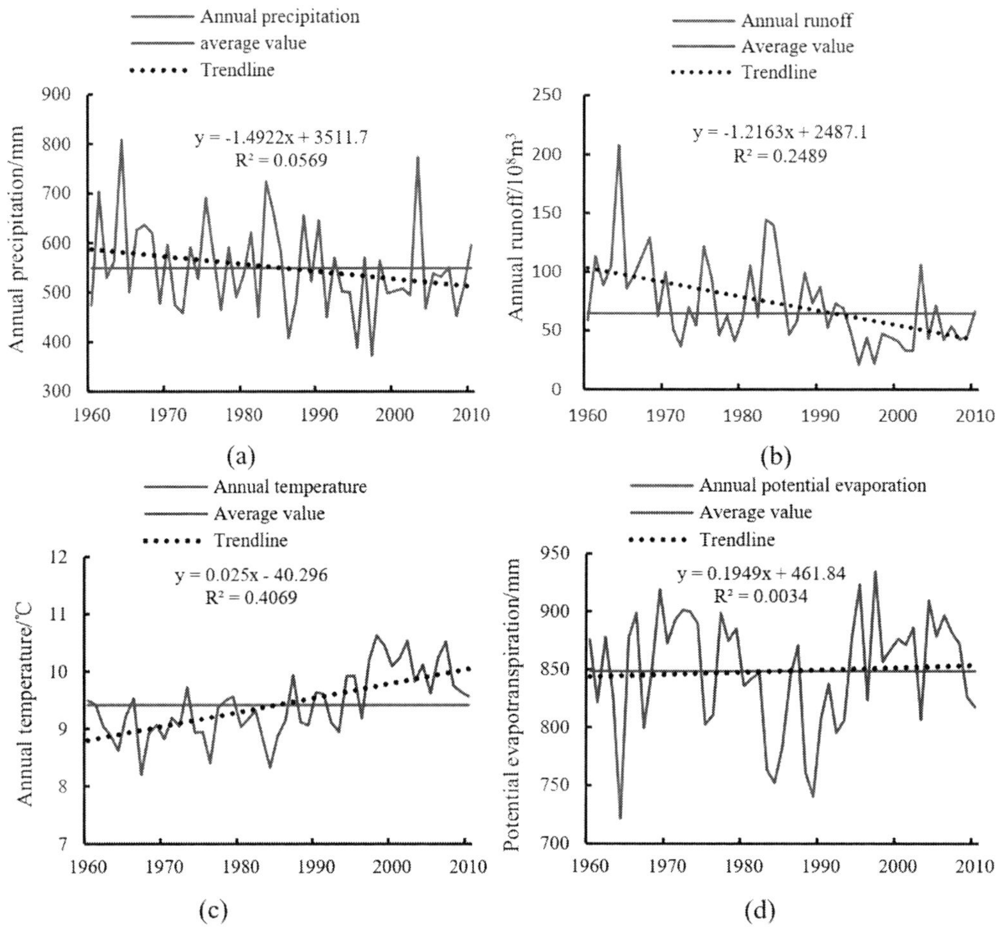

Figure 5. Trend test for precipitation (a), runoff (b), temperature (c) and evapotranspiration (d).

Table 1. Parameter information for each period.

Time periods	Multi-year average runoff Q (10^8m^3)	Multi-year average evapotranspiration E_0 (mm)	Multi-year average precipitation P (mm)	$x=E_0/P$	Amount of runoff change ΔQ (10^8m^3)
1960-1970	105.71	849.15	594.25	1.43	
1971-1989	76.92	836.60	563.18	1.49	-28.79
1990-2010	51.78	859.61	553.32	1.55	-53.93

Table 2. Contribution of climate change and human activities on runoff based on Budyko hypothesis.

Time periods	$\partial Q/\partial P$	$\partial Q/\partial E_0$	ΔQc (mm)	Climate change		Human activities	
				Qc (10^8m^3)	η_c	Q_H (10^8m^3)	η_H
1960-1970							
1971-1989	0.35	-0.16	-8.99	11.84	0.41	16.95	0.59
1990-2010	0.33	-0.14	-14.97	19.71	0.37	34.22	0.63

Compared with the baseline period, the annual runoff decreased by 28.79×10^8 m^3 in 1971-1989, accounting for about 27.2% of the baseline period. In addition, the annual precipitation decreased by 31.07 mm (5.2%) compared with the baseline period. The potential evapotranspiration decreased by 12.55 mm (1.5%). According to equation (7), the contribution rate of climate change to runoff is 41%, and the human activity is 59%. In 1990-2010, the annual runoff decreased by 53.93×10^8 m^3 compared with the baseline period, accounting for about 51% of the baseline period. The annual precipitation decreased by 40.93 mm (6.9%). But the annual potential evapotranspiration is 10.46mm higher than the baseline period, accounting for 1.2% of the baseline period. The contribution rate of climate change to runoff change is 37%, and the human activity is 63%.

The results show that the runoff demonstrates a significant reduction from 1960 to 2010, and the 1970 and 1989 are the break points. At the same time, the descending trend of the precipitation series after 1985 is not obvious. It is shown that human activities rather than climate change are the main reason for the reduction of runoff. Human activities change the underlying conditions of the basin, directly or indirectly affecting the quality, quantity and processes of runoff. Human activities have both positive and negative effects on runoff. Human beings can build water conservancy projects such as reservoirs, silt dams and water tanks, which can reduce the peak flow, adjust the runoff process. The principle of the soil conservation measures is to reduce the slope of the original ground, truncate slope length, and increases the roughness of the ground, thereby increasing the amount of infiltration, extending the confluence time and reducing the peak, so that the flow process line becomes flat. In addition, humans can also play a role in regulating runoff by planting trees and increasing the forest coverage.

However, unreasonable human activities, such as excessive deforestation, uncontrolled exploitation can cause serious soil erosion. Besides, factory waste and chemical pesticides, not only undermine the regulation of soil runoff, but also pollute the water seriously. Meanwhile, the acceleration of the urbanization process is also one of the factors affecting the reduction of runoff [26-28].

5. Conclusion

Based on the Budyko hypothesis, this study analyzes the impact of climate change and human activities on runoff of Weihe River from 1960 to 2010. And the conclusions are as follows:

- With the analysis of the runoff, precipitation, potential evapotranspiration and temperature in the WRB, the results show that: the annual precipitation and runoff both show a decreasing trend, while the decreasing trend of runoff is more obvious. Conversely, the potential evapotranspiration and temperature present an increasing trend, but the increasing trend of the former is not significant.
- Compared with the baseline period, the runoff in the WRB has greatly reduced over the past 20 years, and the annual runoff (51.78×10^8 m^3) is 51% lower than the baseline period (105.71×10^8 m^3).
- Utilizing the M-K method and the cumulative anomaly method, we can determine the break points of runoff in WRB are 1970 and 1989, and the break point of the precipitation is 1985. From 1971 to 1989, the contribution rates of climate change and human activities to runoff are 41% and 59% respectively. And the contribution rates of climate change and human activities to runoff are 37% and 63% in 1990-2010. The results that the impact of human activities on runoff is increasing are in agreement with the findings of other scholars [29-32].

Acknowledgments

This study was supported by the National Natural Science Foundation of China (51679187 and 91647112) and the National Key Research and Development Program of China (2016YFC0400906). Funding was also provided by the Basic Research Plan of Natural Science of Shaanxi Province (2016JQ5105). The authors thank the editor and the reviewers for their valuable comments and suggestions.

References

[1] Yao W G 1994 Comprehensive treatment of Weihe river basin is eyebrow cilium *Water Resour. Hydraul. Eng.* **06** 45-7 (in Chinese)

[2] Wang Y L, Wang W-K and Yang Z-Y 2004 Discussion on eco-environmental water demand in Weihe River Basin of Shaanxi Province *J. Nat. Resour.* **19** 69-78 (in Chinese)

[3] Yang D, Li C, Hu H, Lei Z, Yang S, Kusuda T, Koike T and Musiake K 2004 Analysis of water resources variability in the Yellow River of China during the last half century using historical data *Water Resour. Res.* **40** 308-22

[4] Ye X C, Zhang Q, *et al* 2009 Impacts of climate change and human activities on runoff of Poyang Lake Catchment *J. Glaciol. Geocryol.* **31** 835-42 (in Chinese)

[5] Chang J X, Wang Y, Istanbulluoglu E, Bai T, Huang Q, Yang D and Huang S 2015 Impact of climate change and human activities on runoff in the Weihe River Basin, China *Quatern. Int.* 380-1/169–79

[6] Ding Y J, Ye B S and Liu S Y 2000 *J.* Impact of climate change on the Alpine streamflow during the past 40 a in the middle part of the QiIian Mountains, Northwestern China *J. Glaciol. Geocryol.* **3** 193-9 (in Chinese)

[7] Zhang G H 2006 Analysis on potential effects of global climate change on natural runoff in the Yellow River Basin *Geogr. Res.* **2** 268-75 (in Chinese)

[8] Ward P J, van Balen R T, Verstraeten G, Renssen H and Vandenberghe J 2009 The impact of land use and climate change on late Holocene and future suspended sediment yield of the Meuse catchment *Geomorphology* **103** 389-400

[9] Chang F J, Chang L C, Kao H S and Wu G R 2009 Assessing the effort of meteorological variables for evaporation estimation by self-organizing map neural network *J. Hydrol.* **384** 118-29

[10] Wang G Q, Zhang J Y, *et al* 2008 Quantitative assessment for climate change and human activities impact on river runoff *Chin. Water Resour.* **2** 55-8 (in Chinese)

[11] Ding Y H 2008 Human activity and the global climate change and its impact on water resources *Chin. Water Resour.* **2** 20-7 (in Chinese)

[12] Roderick M L and Farquhar G D 2011 A simple framework for relating variations in runoff to variations in climatic conditions and catchment properties *Water Resour. Res.* **47** W00G07

[13] Destouni G, Jaramillo F and Prieto C 2013 Hydroclimatic shifts driven by human water use for food and energy production *Nat. Climate Change* **3** 213-7

[14] Budyko 1974 *M I. Climate and Life* (San Diego, CA: Academic)

[15] Su X L, Kang S Z, Wei X M, *et al* 2007 Impact of climate change and human activity on the runoff of Wei River basin to the Yellow River *J. Northwest Sci.-Tech. Univ. Agric. For. (Nat. Sci. Ed.)* **35** 154-9 (in Chinese)

[16] Sun J, Ruan B Q and Jiang R F 2006 Study on the crop reference evapotraspiration and the meteorological variables in the Yellow River irrigation area of Ningxia Region *J. Irrig. Drain.* **25** 54-61 (in Chinese)

[17] Guo S L, Guo J Y, Hou Y K, *et al* 2015 Prediction of future runoff change based on Budyko hypothesis in Yangtze River basin *Adv. Water Sci.* **26** 151-60 (in Chinese)

[18] Sun F B, Yang D W, Liu Z Y, *et al* 2007 Validation of coupled water- energy balance in the Haihe River Basin and Inland River Basins *Hydrology* **4** 7-10 (in Chinese)

[19] Fu B P 1996 On the calculation of the evaporation from land surface in mountainous areas *Scientia Meteorol. Sin.* **16** 328-35 (in Chinese)

[20] Sun F B, Yang D W, Liu Z Y, *et al* 2007 Study on coupled water-energy balance in Yellow River basin based on Budyko Hypothesis *J. Hydraul. Eng.* **38** 409-16 (in Chinese)

[21] Li B, Li L J, Qin Y C, *et al* 2011 Impacts of climate change on streamflow in the upper and middle reaches of the Taoer River based on the Budyko Hypothesis *Resour. Sci.* **33** 70-6 (in Chinese)

[22] Milly P C D and Dunne K A 2002 Macroscale water fluxes 2 water and energy supply control

of their interannual variability *Water Resour. Res.* **38** 21-4
[23] Wei F Y 1999 *Modern Climate Statistical Diagnosis and Prediction Technology* (Beijing: Meteorological Press)
[24] Cao J P, Chi D C, Wu L Q, *et al* 2008 Mann-Kendall examination and application in the analysis of precipitation trend *Agricultural Science & Technology and Equipment* **5** 35-40 (in Chinese)
[25] Fan B W, Mu Y H and Qiu W J 2008 Consistency and variation analysis of runoff time series in Taohe River Basin *Hydrology* **3** 73-6 (in Chinese)
[26] Wang L, Zhu C B, *et al* 2002 Evaluation of the influence of urbanization on runoff producing by artificial neural network *Yangtze River* **33** 21-4 (in Chinese)
[27] Chen D C, Li X P, Yang J S, *et al* 2002 The evolution of river networks in the process of urbanization in Shanghai *Urban Problems* **5** 31-5 (in Chinese)
[28] Zou X and Liu J M 2016 Research on urban rainfall runoff model and simulation comparison *Chin. Rural Water Hydro.* **12** 101-5 (in Chinese)
[29] Fan J J, Huang Q, Liu D F, *et al* 2016 Impacts of climate change and human activity on runoff variation of Beiluo River *J. Northwest Sci.-Tech. Univ. Agric. For. (Nat. Sci. Ed.)* **44** 221-34 (in Chinese)
[30] Zhan C S, Jiang S S, Sun F B, *et al* 2014 Quantitative contribution of climate change and human activities to runoff changes in the Wei River Basin, China *Hydrol. Earth Syst. Sci.* **18** 3069-77
[31] Zhang L P, Liu D F, Zhang H X, *et al* 2016 Impact of climate change and human activities on runoff variation in Beiluo River Basin *J. Hydroelectric Eng.* **35** 55-66 (in Chinese)
[32] Guo A J, Chang J X, Huang Q, *et al* 2014 Quantitative analysis of the impacts of climate change and human activities on runoff change in Weihe Basin *J. Northwest Sci.-Tech. Univ. Agric. For. (Nat. Sci. Ed.)* **42** 212-20 (in Chinese)

Evaluation of mangrove ecosystem service functions of Ximen Island Marine Specially Protected Areas in Yueqing Bay, China

D G Wang[1,2,4], L Sun[1], Y H Tan[1], A Q Shi[3] and J Cheng[1]

[1]Second Institute of Oceanography, SOA, Hangzhou, 310012, China
[2]Nanjing University, Nanjing, 210000, China
[3]State Key Laboratory of Satellite Ocean Environment Dynamic, Second Institute of Oceanography, SOA, Hangzhou, 310012, China

E-mail: Degang_w@sio.org.cn

Abstract. Taking the mangrove ecosystem of Ximen Island National Marine Specially Protected Areas as the research object, the ecological service value of the mangrove forest was evaluated and analyzed using a market value method, an ecological value method and a carbon tax method. The results showed that the ecosystem service value of the mangrove forest on Ximen Island is worth a total of 16,104,000 CNY/a. Among the value of individual ecosystem services, the direct value of material production function and leisure function reached 1,385,000 CNY/a, with a ratio of 8.6%. The indirect value of disturbance regulation, gas regulation, water purification, habitat function and culture research reached 14,719,000 CNY/a, with a ratio of 91.4%. Among the above sub-items, the proportion of disturbance regulation value, habitat function value and cultural research function value reached 78.8%, which reflects the important scientific value and ecological value of the Ximen Island mangrove ecosystem, especially its vital importance in providing a habitat for birds and playing a role in disaster prevention and mitigation.

1. Introduction

Mangrove ecosystems refer to systems consisting of producers (mangroves, semi-mangroves, etc.), consumers (fish, benthic animals, birds, etc.), decomposers and inorganic elements [1]. Mangroves are mainly distributed in river estuaries and bays. They are important ecological systems in regions where land to ocean transition occurs. These systems have high primary productivity, which supports land or sea ecosystems. In addition, these systems have high ecological value and aesthetic value, as well as research and education value [2]. Due to the important ecological function and high economic value of mangrove forests, mangrove ecosystem services have received increasing attention in recent years, and the evaluation of mangrove ecological service value has gradually increased.

Since the publication of the global ecosystem services paper by Constanz in 1997 [3], the United Nations implemented the Millennium Ecosystem Assessment in 2003 [4], national ecologists have established a multi-level assessment system of ecosystem service value which strongly contributed to the scientific assessment of ecosystem services.

Studies investigating the value of mangrove ecosystem services have advanced in method and theory after the long-term accumulation of scientific information. The research results are mainly reflected in the definition and quantification of mangrove ecosystem values, the identification of mangrove production and maintenance of biodiversity, and the accumulation of mangrove nutrients [5-8]. China's mangrove ecosystem services research area is mainly located in the south of Fujian, including Fujian province, Guangdong province, Guangxi province and Hainan province, where is in the south latitude area of Ximen Island.

Zhejiang Ximen Island mangrove forest is the most northern end of the mangrove forest, composed of only Kandelia candel, whose height is approximately 2-3 m. The Ximen mangrove forest has adapted to the local climate and environment and has natural reproductive capacity. The introduction of the mangrove forest to Ximen Island has broken through the traditional growth of the Chinese mangrove forest. It not only has high ecological and tourism value but also has high scientific research value and needs to be protected and cultivated. In Ximen Island, the China-Italy climate change cooperation project "ecosystem capacity building adapt to climate change in the coastal areas" get implemented. At the same time, the United Nations Development Program (UNDP)/Global Environment Facility (GEF)/Small Grants Program Project (SGP)" Environmental Protection and Integrated Management of Mangroves Ecosystems in Ximen Island, Zhejiang Province" get implemented. Therefore, the study of the ecological value of the mangrove forest on Ximen Island can supplement the research results of high-latitude mangroves in China to a certain extent. Due to the importance of the Ximen Island mangrove, we put forward the hypothesis of "There are ecological service value difference between the natural growth mangrove at low latitude and artificial introduction mangrove at high latitude".

2. Study area

2.1. The location of Ximen Island

Ximen Island is located in the south of Zhejiang, China (figure 1). The area of this island is approximately 7 km^2, with a coastal wetland area of 19.2 km^2. In 2005, Ximen Island Marine Specially Protected Areas were established, including Ximen Island and its coastal wetlands, which mainly protected the mangrove communities in the northern latitudes of China, as well as benthic organisms and wetland birds. The total protected area is 30.80 km^2. There are 37 rocky reef organisms and 92 muddy beach organisms in the protected areas. This is an area of high biological value along the Zhejiang coastal area that has considerable biological diversity.

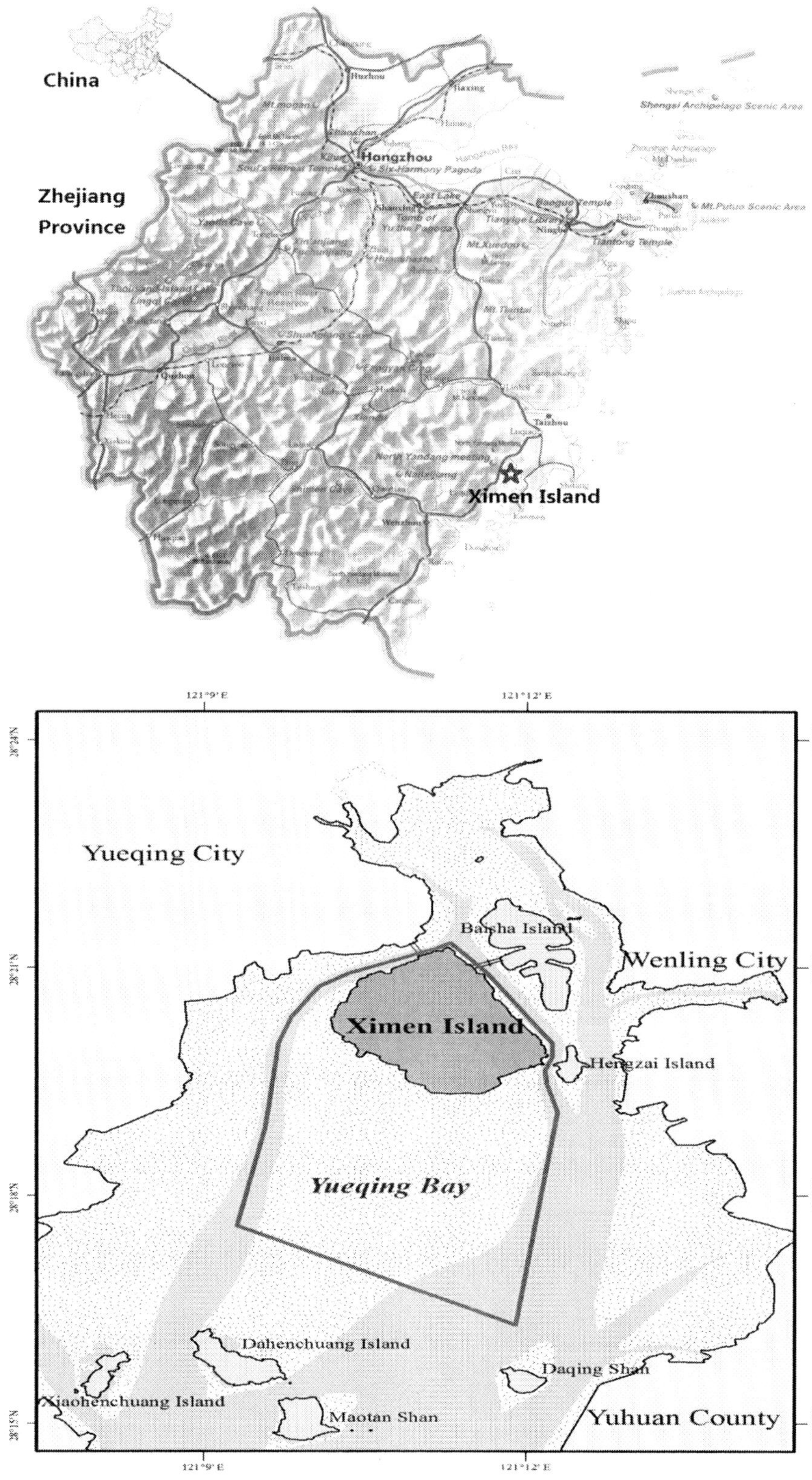

Figure 1. The location of Ximen Island Marine Specially Protected Areas.

2.2. Distribution of mangroves
Using the WorldView-2 high-resolution satellite image interpretation, the distribution of mangrove communities in the Ximen Island Marine Specially Protected Areas was obtained in 2016. The mangrove communities in the protected areas are mainly distributed in the northern part of the island (Area B) and the northwestern (Area A) and western part of the island (Area C) which can be seen in figure 2. As of 2016, the total area of the mangrove community in the Ximen Island Marine Specially Protected Areas was approximately 26.68 Ha.

Figure 2. Distribution and live photo of mangrove communities in Ximen Island in 2016.

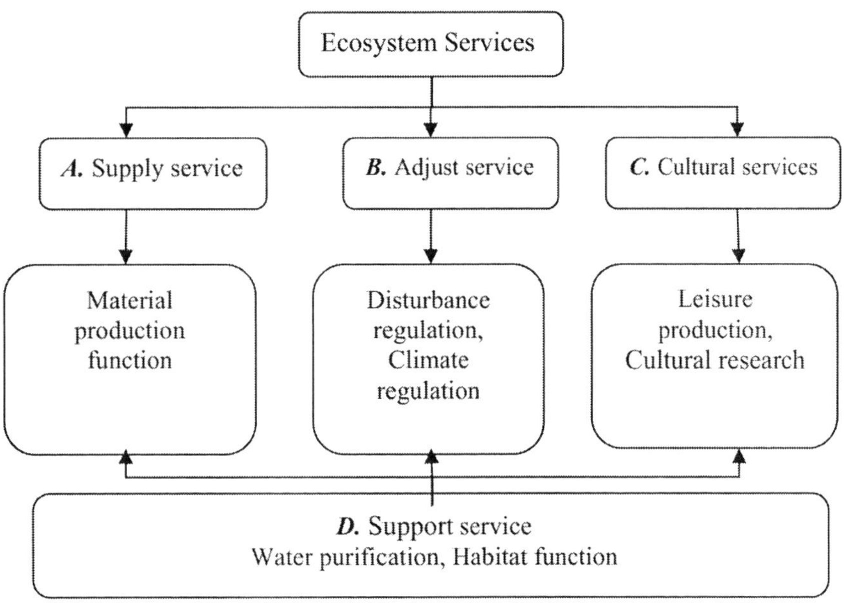

Figure 3. The function classification system of mangrove ecosystem in Ximen Island.

3. Evaluation of ecological service value of the mangrove forest

3.1. The function and value of the mangrove ecosystem

The classification system of the United Nations' Millennium Ecosystem Assessment divides ecosystem services into four categories: supply services, regulatory services, cultural services and support services [9].

On the basis of MA (Millennium Ecosystem Assessment) general classification system, we selected the main ecological service value of mangrove forest system in Ximen Island. According to the characteristics of the Ximen Island Marine Specially Protected Areas, the mangrove ecosystem function is divided into material production function (MP), disturbance regulation (DR), climate regulation (GR), water purification (WP), habitat function (HF), leisure production (LP) and cultural research (CR) which can be seen in figure 3.

3.2. Ecological value assessment method

Mangrove wetlands have many functions that are important to human beings. The ecosystem service function is multidimensional and therefore has widespread value, which brings considerable social and economic benefits to mankind. According to the benefits, the economic value can be divided into direct use value, indirect use value, non-use value, presence value and genetic value. The value of ecosystem services is different, and the corresponding assessment methods are different (table 1). The selection of the assessment method should be based on feasibility and maneuverability, and the most appropriate method should be selected. Because genetic values are often difficult to assess, they are not evaluated in the paper. And the presence value is reflected in the indirect value and non-indirect value, in order to avoid duplication of calculation, so in the paper has not been assessed. According to the actual situation of the Ximen Island Marine Specially Protected Areas, the evaluation method of the mangrove ecosystem service function value was selected.

Table 1. Evaluation methods of ecosystem value.

Value type	Ecosystem service value	Assessment method
Direct use value	Material production value	Market value method
	Leisure production value	Travel fee method
Indirect use value	Disturbance regulation value	Shadow engineering method
	Climate regulation value	Carbon tax law
	Water purification value	Fuzzy mathematics
	Habitat function value	Ecological value method
	Cultural research value	Travel fee method
Non-use value	Presence value	Willingness to pay
	Genetic value	Willingness to pay

3.3. Calculation of ecological service value of mangrove forests

3.3.1. Material production value (MP). Mangrove ecosystems are among the most productive ecosystems in the world, and their material production function is mainly embodied in living trees and fallen material (e.g., leaves and twigs). According to the results presented by Han Weidong (2009) [10], Chinese mangrove forest production volume is 4.4 $m^3/(ha *a)$, and the log market price is 1,200 CNY/m^3. In the current study, the market value method was used to estimate the value (MP_1) of the mangrove forest on Ximen Island.

$$MP_1=S*Y_1*P_1=26.68*4.4*1200=14.1*10^4\ CNY/a \quad (1)$$

Where S is the mangrove area; Y_1 is the annual mangrove growth of Chinese mangroves; and P_1 is the price of Chinese logs.

The Mangrove fallen leaves in the Ximen Island Protected Area are an important source of food for winter migratory birds. The mangrove annual dislocations are approximately 9.4t / (ha *a), and the market price of various types of waterfowl feed is 2,800 CNY/t. The paper uses the market value method to estimate the dislocation value (MP_2) in the protected area of Ximen Island.

$$MP_2=S*Y_2*P_2=26.68*9.4*2800=70.2*10^4\ CNY/a \quad (2)$$

Where S is the mangrove area; Y_2 is the mangrove annual dislocation; and P_2 is the waterfowl feed price.

Thus, the total value (MP) of the mangrove material production for Ximen Island is obtained by the sum of the above two values.

$$MP= MP1+ MP2=84.3*10^4\ CNY/a \quad (3)$$

3.3.2. Leisure production value (LP). The Ximen Island Marine Specially Protected Areas do not charge for entry tickets and have not released the statistics pertaining to annual traffic. Due to proximity to the national 5A-level Yandang Mountain Scenic Area, Ximen Island has many tourists. With reference to the Leizhou mangrove wetland tourism value assessment conducted by Han Weidong [11], the unit area annual tourism value of the Ximen Island mangrove is $2.03 * 10^4$ CNY/(ha*a). By multiplying the per unit area annual tourism value by the mangrove area, the mangrove leisure travel value (LP) of Ximen Island can be obtained.

$$LP=L*S=2.03*26.68*10^4 =54.2*10^4\ CNY/a \quad (4)$$

Where S is the mangrove area, and L is the mangrove area per annual tourism value.

3.3.3. Disturbance regulation value (DR). Mangroves adapt to the tide and flood impacts, forming a unique ventilation organization, canopy and other morphological features that have a strong anti-wind and anti-wave function. In addition, the mangrove-dense roots function as berms by slowing the velocity of the water, settling the suspended particles in the water, and promoting the formation of soil. Therefore, the disturbance regulation value of the mangroves on Ximen Island mainly includes the function of the coastal flood control and berms.

The mangrove belt on Ximen Island is approximately 3 km long. According to research results presented by Lu Changyi, the mangrove coastline can provide typhoon disaster protection benefits of approximately $8.0*10^4$ CNY/(km*a) [12], and the ecological conservation benefits of the embankment are $64.7 * 10^4$ CNY (km*a) [13]. Combined with the length of the mangrove forest, the value of the Ximen Island protected area (DR_1) is $24 * 10^4$ CNY/a, and the embankment value (DR_2) is $194.1 * 10^4$ CNY/a.

The total disturbance regulation value of the mangroves on Ximen Island is the sum of the wave value and the berm value.

$$DR= DR_1+ DR_2=218.1*10^4\ CNY/a \quad (5)$$

3.3.4. Climate regulation value (GR). According to Kang Wenxing *et al* [14], the mangrove ecosystem absorbs 21.48 t/ha of CO_2 per year. The total amount of CO_2 in the mangrove forest was estimated by calculation.

$$C=GP*S=21.48*26.68=573.1(t/a) \quad (6)$$

Where C is the amount of CO_2 absorbed; GP is the amount of CO_2 absorbed per unit area of mangroves per year; and S is the mangrove area.

The amount of CO_2 absorbed and the amount of O_2 released can be obtained through the photosynthesis equation, that is, the absorption of 1 g CO_2 by plants can result in the release of 0.75 g O_2. Therefore, the CO_2 absorbed by the mangrove forest on Ximen Island is 573.1 t/a, and the amount of O_2 released is 429.8 t/a.

Using the internationally recognized carbon tax rate that is, 150 US dollars/t (in accordance with the exchange rate, which is 1,050 CNY/t), and the value of fixed CO_2 (GR_1) is $60.2*10^4$ CNY/a. Using the alternative market method, the release value of O_2 can be calculated. The industrial oxygen price is 1,000 CNY/t; therefore, the value (GR_2) of O_2 released is $43.0 * 10^4$ CNY/a.

The climate regulation value of the Ximen Island mangrove (GR) is the sum of the above two terms.

$$GR = GR_1 + GR_2 = 103.2*10^4 \text{ CNY/a} \quad (7)$$

3.3.5. Water purification value (WP). There are four villages on the island of Ximen, and considerable domestic sewage flows through the mangrove area into Yueqing Bay. According to the results presented by Zhang and Yu *et al*, the annual adsorption capacity of mangroves with respect to nitrogen, phosphorus and heavy metals is $3.77*10^4$ CNY/(ha*a) [15]. Combined with the area of Simmental mangroves, the value of purified water (WP) can be calculated.

$$WP = W*S = 100.6*10^4 \text{ CNY/a} \quad (8)$$

Where W is the water purification value of the mangroves per unit area per year; S is the mangrove area.

3.3.6. Habitat function value (HF). The Ximen Island Marine Specially Protected Areas are an important habitat for migratory wetland waterfowl, which are included in the Agreement on the Conservation of Migratory Birds and Their Habitats in China and Japan and the Agreement on the Conservation of Migratory Birds and Their Habitats in China and Australia. The habitat function value of the protected area is the sum of the habitat value of important and ordinary birds. In the calculation, the conservation costs of important and ordinary birds can be multiplied by the number of birds, and then the results are added to obtain the habitat function value of the Ximen Island protected area.

According to the "Ximen Island Marine Specially Protected Areas General Plan", the number of world-class endangered Black Mouth Gulls along the Wenzhou coastal region is approximately 3,000 in winter, whereas Egrets and other ordinary birds account for approximately $2.5 * 10^4$. The number of birds in the mangrove area of Ximen Island was calculated as 1/50, that is, 60 important birds and 500 ordinary birds. Significant bird protection costs are $10 * 10^4$ CNY /a, and ordinary bird protection costs are 1,000 CNY/a [16]. The habitat function value of Ximen Island mangroves is the sum of these two amounts.

$$F = A_1*N_1 + A_2*N_2 = 650*10^4 \text{ CNY/a} \quad (9)$$

Where A_1 is the protection cost of important birds; N_1 is the number of important birds; A_2 is the protection cost of ordinary birds; and N_2 is the number of ordinary birds.

3.3.7. Cultural research value (CR). According to the work summary of the Ximen Island Marine Specially Protected Areas presented by the Yueqing Marine and Fishery Bureau in 2016, the investment cost of research and education in 2005-2015 was $8,000 * 10^4$ CNY, and the average annual investment was $800 * 10^4$ CNY. Research funding for the mangrove forest was calculated as 50% of the total funds, resulting in $400 * 10^4$ CNY/a, i.e., CR=$400 * 10^4$ CNY/ a.

3.3.8. Non-use value. Because non-use value depends on human subjectivity, it is relatively difficult to calculate this value; therefore, no quantitative assessment was performed.

4. Discussion and conclusion

4.1. Discussion

In this paper, mangroves in different latitudes in China were selected from the north to the south, which were mangrove forests in Ximen Island, Zhejiang Province, Zhangjiangkou mangrove forest in Fujian Province, Futian mangrove forest in Shenzhen Bay, Guangdong Province and Dongzhai port mangrove forest in Hainan Province. Based on the research of this paper, the research of mangrove forest in Zhangjiangkou of Fujian Province by Zhang [15], research of Futian mangrove forest in Shenzhen, Guangdong Province by Li [17] and research of mangrove forests in Dongzhai port, Hainan Province by Ding [6], the comparison of the mangrove species, height, category, latitude, the main ecological service value type and the ecological service value per unit area were analyzed (table 2).

Table 2. Comparison of mangrove in Ximen Island and other mangroves in different latitudes.

Area	The main species	Height (m)	Category	Latitude	Main ecological service value type and proportion	Ecological service value per unit area (CNY/ha*a)
Ximen Island, Zhejiang Province	Kandelia candel	2-3	Artificial cultivation	28°21′-28°16′	Habitat function value (40.4%), Cultural research value (24.8%)	23,078
Zhangjiangkou , Fujian Province	Avicennia marina, Bruguiera gymnorrhiza, Kandelia candel	2-5	Natural growth	23°53′-23°56′	Disturbance Regulation value (46.9%), Water purification value (20.9%)	59, 991
Futian, Shenzhen Bay, Guangdong Province	Kandelia candel, Aegiceras corniculata, Avicennia marina	2-6	Natural growth	22°30′-22°32′	Disturbance Regulation value (23.4%), Water purification value (14.2%)	107,400
Dongzhai port, Hainan Province	Sonneratia caseolaris, Avicennia marina	3-10	Natural growth	19°51′-20°01′	Climate regulation value (32.6%), Disturbance Regulation value (22.3%)	225,500

Through the comparison of the ecological services value per unit area, it can be concluded that, Hainan Province> Guangdong Province > Fujian Province > Zhejiang Province. This indicates that the value of ecological services per unit area of mangrove forests is closely related to the latitude, that is, the growth of mangrove forests in the low latitudes is lusher and the value of ecological services is higher.

Secondly, it also has some differences between natural growth and artificial introduction mangroves in the ecological service value type composition. The main ecological service value of the natural growth mangrove forest is the value of disturbance Regulation, Water purification or climate regulation, while the artificial introduction forest of Ximen Island is mainly the value of habitat and cultural research. This conclusion provides a solid theoretical basis for the introduction project of China's high latitude mangrove forest by State Oceanic Administration.

4.2. Conclusion

4.2.1. Ecological value of the mangrove forest on Ximen Island. Through the evaluation of the mangrove ecosystem service value for the Ximen Island Marine Specially Protected Areas, the total service function value, each individual function value and their proportion were obtained.

The results in table 3 show that the total value of the mangrove ecosystem service for the Ximen Island Marine Specially Protected Areas is 16,104,000 CNY/a, the material production value is 843,000 CNY/a, the value of leisure is 542,000 CNY/a, the disturbance regulation value is 2,181,000 CNY/a, the value of climate regulation is 1,032,000 CNY/a, the value of water purification is 1,006,000 CNY/a, the habitat function value is 6,500,000 CNY/a, and the cultural research value is 4,000,000 CNY/a.

Table 3. Components of the ecosystem value of mangrove on Ximen Island.

Ecosystem Service Value	Value (CNY/a)	Proportion (%)
Material production value (MP)	843,000	5.2
Leisure production value (LP)	542,000	3.4
Disturbance Regulation value (DR)	2,181,000	13.6
Climate regulation value (GR)	1,032,000	6.4
Water purification value (WP)	1,006,000	6.2
Habitat function value (HF)	6,500,000	40.4
Cultural research value (CR)	4,000,000	24.8
Total	16,104,000	100

From the value composition analysis, HF> CR> DR> GR> WP> MP> LP. Among them, the habitat value, cultural research value and disturbance regulation value dominate, accounting for 78.8% of the total value. The composition of the value well reflects the practical role of the Ximen mangrove forest in the region, especially reflecting the significant scientific research value of this forest as the most northern latitude distribution of China. At the same time, it also reflects the importance of Yueqing Bay as an important international migratory habitat, as well as the important disaster prevention and mitigation effects of the mangrove forest in Wenzhou.

4.2.2. Analysis of the ecological service value of mangroves on Ximen Island. The value of mangrove ecosystem services for Ximen Island is mainly composed of direct use value (material production value and leisure value) and indirect use value (disturbance regulation value, climate regulation value, water purification value, habitat function value, cultural research value). The results show that the proportion of the direct use value relative to the total value is 8.6%, and the proportion of the indirect use value is 91.4%. This indicates that material production is not the main ecological service function of the mangrove forests on Ximen Island, and the invisible ecological service function strongly indicates its ecological value, such as the value of habitat, the value of cultural research and the value of disturbance regulation. At the same time, the results also show that the ecological value and scientific research value of the Ximen Island mangrove are extremely high.

Therefore, in the protection of mangroves on Ximen Island, we should pay special attention to the protection of ecological processes. At the same time, the management agency should strengthen the research investment in Ximen Island mangroves in order to enhance the cultural and scientific value of Chinese high-latitude mangrove forests. In the later construction of the Ximen Island Marine Specially Protected Areas, we should expand the mangroves associated with the "South Mangrove and North Chinese Tamarisk Twig" program of the State Oceanic Administration. The expansion plan will enhance the service value of the mangrove ecosystem and realize the sustainable development of the Ximen Island Marine Specially Protected Areas.

Acknowledgments

The Zhejiang Provincial Natural Science Foundation of China (Grant No. LY15D060003) and the Scientific Research Fund of the Second Institute of Oceanography, SOA (Grant No. JT1204) supported this research.

References

[1] Wang W Q 2007 *Chinese Mangrove Forest* (Beijing: Science Press)
[2] Jiang J and Hu B Q 2013 The services function and value of mangrove ecosystem *J. Anhui Agric. Sci.* **41** 3958-60
[3] Costanza R, Agre R and De G 1997 The value of the world's ecosystem and natural capital *Nature* **387** 253-60
[4] Zhao S-D and Zhang Y-M 2004 Concepts, contents and challenges of ecosystem assessment—Introduction to "ecosystems and human well-being: A framework for assessment" *Adv. Earth Sci.* **19** 42-6
[5] Li W H 2002 *Research on Ecosystem Service Function* (Beijing: Meteorological Press)
[6] Ding D J, Liao B W and Guan W 2016 Evaluation on service value of coastal wetland ecosystem in Dongzhai harbor mangrove nature reserve *Ecol. Sci.* **35** 182-90
[7] He H J, Wen J S and Zhang J W 2015 Ecological functions assessment of mangrove wetland in Hainan Province *Ecol. Econo.* **31** 145-9
[8] Zhao S, Hong H-S and Zhang L-P 2007 Emergy value of mangrove ecosystem services in China *Resour. Sci.* **29** 147-54
[9] Millennium Ecosystem Assessment (MA) 2003 *Ecosystems and Human Well-being: A Framework for Assessment* (Washington DC: Island Press)
[10] Han W-D, Gao X-M and Lu C-Y 2000 The ecological values of mangrove ecosystems in China *Ecol. Sci.* **19** 40-6
[11] Han W D 2009 *Mangrove Ecosystem of Leizhou Peninsula and Its Protection Strategy* (Guangzhou: South China University of Technology Press)
[12] Lu C Y 1991 The development of mangrove forest is an effective countermeasure to reduce the tide of typhoon *J. Nanjing Univ. (Nat. Sci.)* **1** 70-5
[13] Fan H Q 1995 An ecological pattern of sea dyke maintenance by mangroves and assessment of its benefits along Guangxi Coast *Guangxi Sci.* **2** 48-52
[14] Kang W X, Zhao Z H and Tian D L 2008 CO_2 exchanges between mangrove and shoal wetland ecosystems and atmosphere in Guangzhou *Chin. J. Appl. Ecol.* **19** 2605-10
[15] Zhang H-Y, Chen C-M and Zheng H-Y 2013 Evaluation of value of wetland ecosystem services of Zhangjiang estuary mangrove national nature reserve *Wetl. Sci.* **11** 108-13
[16] Dai Y B 2004 Preliminary exploration on forest ecological function and value estimate of Daxueshan nature reserve *Forest Invent. Plan.* **29** 83-7
[17] Li Y-L, Ning T-Z and Xu H-L 2011 Evaluation of mangrove ecosystem service functions for Futian Mangrove Reserve in Shenzhen Bay, Southern China *J. Cent. South Univ. For. Technol.* **31** 41-8

IOP Conf. Series: Earth and Environmental Science 82 (2017) 012065 doi:10.1088/1755-1315/82/1/012065

Transformation rules and degradation of CAHs by Fenton-like oxidation in growth ring of water distribution network-A review

D Zhong[1], W C Ma[1,4], X Q Jiang[1], Y X Yuan[1,2,4], Y Yuan[3], Z Q Wang[1], T T Fang[1] and W Y Huang[1]

[1]School of Municipal and Environmental Engineering, Harbin Institute of Technology, Harbin 150090, China
[2]State Key Laboratory of Urban Water Resource and Environment, Harbin Institute of Technology, Harbin 150090, China
[3]School of Biological Engineering, Beijing Polytechnic, Beijing 100176, China

E-mail: damahit@163.com; yyx1957@163.com

Abstract. Chlorinated hydrocarbons are widely used as organic solvent and chemical raw materials. After treatment, water polluted with trichloroethylene (TCE)/tetrachloroethylene (PCE) can reach the water quality requirements, while water with trace amounts of TCE/PCE is still harmful to humans, which will cause cancers. Water distribution network is an extremely complicated system, in which adsorption, desorption, flocculation, movement, transformation and reduction will occur, leading to changes of TCE/PCE concentrations and products. Therefore, it is important to investigate the transformation rules of TCE/PCE in water distribution network. What's more, growth-ring, including drinking water pipes deposits, can act as catalysts in Fenton-like reagent (H_2O_2). This review summarizes the status of transformation rules of CAHs in water distribution network. It also evaluates the effectiveness and fruit of CAHs degradation by Fenton-like reagent based on growth-ring. This review is important in solving the potential safety problems caused by TCE/PCE in water distribution network.

1. Introduction

Chlorinated hydrocarbons (CAHs) are widely used as organic solvent and chemical raw materials in chemical, leather, medicine and electronic industries. CAHs contaminated soil and groundwater has resulted due to past disposal directly onto land, accidental spills, and leaking storage tanks, posing a threat to human health and environment [1]. Atmosphere, water body, soil and refuse landfill have been reported of pollution of CAHs [2-4]. CHAs are common persistent organic pollutants and they are easy to be enriched in living organisms. CHAs are characterized by high stability, toxicity and "mutagenesis, carcinogenesis, teratogenesis". Besides, CHAs are especially troubling since their high densities as compared to water allow them to move downward through the water table and collect in hard to reach regions. Here they continue to contribute to contamination [5, 6]. Due to their toxicity and refractory, halogenated organic solvents are listed as priority control environmental contaminants. In 1979, United State Environmental Protection Agency (US EPA) released 129 kinds of environmental priority control pollutants and organic halogenated hydrocarbons and their derivatives accounted for more than 60. In the "blacklist" released by European Communities, CAHs and chlorinated hydrocarbon compounds that can be formed in the environment are listed first [7]. Besides, TCE and PCE are among the most common pollutants in CAHs, and they are high-toxic substances and widely used. Both of them draw utmost attention, and the both are included in 129 kinds of environmental priority control pollutants released by US EPA in 1976 and "blacklist of priority pollutants" put forward in China [8-10].

World Health Organization (WHO)'s Drinking Water Quality Indicators (2005 year, the third edition) regulates that TCE and PCE in drinking water are not more than 0.02 mg/L and 0.04 mg/L, respectively. US EPA regulates that Maximum Contaminant Levels (MCL) of both of TCE and PCE are 5 μg/L and Superfund sites polluted by TCE amount to 336 [11]. According to the survey of shallow underground water of cities, the maximum concentrations of TCE and PCE are as high as 6000 μg/L and 200 μg/L separately [12]. PCE has become one of five kinds of the most widely distributed pollutants in the groundwater [13, 14]. 61% leach liquor samples of the toxic waste collected from 1984 to 1991 year in America were detected to contain PCE [15]. Tap Water Quality Standards promulgated in Japan regulates that the maximum concentrations of TCE and PCE are 0.03 mg/L and 0.01 mg/L separately. According to the research conducted by environmental protection bureau, a wide range of groundwater was polluted by TCE, and that amount of groundwater was polluted by high concentration of TCE and the number of polluted wells was still increasing, especially in Morioka which had a population of 300000. In Morioka, 15% tap water come from polluted groundwater [16]. After testing 5900 water samples in Japan nationwide, we found that PCE concentrations of 2.3% samples did not reach the standards of drinking water [17]. Standards for Drinking Water Quality (GB 5749—2006) promulgated in China regulates that the maximum concentrations of TCE and PCE are 0.07 mg/L and 0.04 mg/L separately. An investigation on groundwater of a city in north China indicates that TCE exceed ratio and concentration are both the highest, followed by PCE. The maximum TCE concentration reaches 1000 μg/L, 200 times higher than the standard of US EPA, with the exceed ratio 44.4% [18]. Another investigation of organic pollution in shallow groundwater of north China indicates that local areas are seriously polluted by CAHs, and the pollution of TCE and PCE is much more serious than trichloromethane (CF) and tetrachloromethane (CT). In this investigation, TCE and PCE are the main pollutants and the highest concentrations are 487.55 μg/ L and 63.74 μg/L, respectively [19].

TCE is volatile organic compounds (VOCs), and it is denser and less viscous than water resulting in rapid rates of subsurface migration and eventually in the formation of non-aqueous phase liquid (NAPL) [20], which can be extremely difficult to remediate due to their inaccessibility and insolubility, potentially leading to long term contamination of aquifers. TCE and its degradation products, dichloroethylene (DCEs) and vinyl chloride (VC), are toxic carcinogenic substances, posing a great threat to human health [21]. PCE usually exists in the forms of immiscible residual liquid, dissolved phase, gas phase or adsorbed phase in vadose zone and underground water [22]. PCE gathers near the water table first, and the height provides enough pressure to overcome capillary forces at the oil-water interface and force the PCE to move downwards. Finally, PCE moves through the entire aquifer, and collects at the bottom of the aquifer or in low areas [23], leading to long term contamination [24].

With the development of relevant industries, TCE and PCE contamination will aggravate, and the contradiction between drinking water security and industrial developments will be more acute. Therefore, in order to secure water quality, it is urgent to know and control transformation rules of TCE/PCE in water treatment and conveyance processes, and it is necessary to construct TCE/PCE degradation system by advanced technology.

2. Growth-ring in water distribution network

Water distribution network is an important part of water supply system. Treated water is transported to consumers through huge complex network. Water distribution pipelines runs for years, and irregular annular mixture, "growth-ring", forms gradually along the inner wall of pipeline [25, 26]. It is the mixture of deposits, corrosions, slimes and biofilms [27-29]. Water flows through pipelines for tens of or even hundreds of kilometers, and the residence time can reach as high as several hours or even several days. In the process of transportation, water contacts with growth-ring. If water is polluted during the transportation, it will lead to water deterioration and a secondary pollution occurs. Therefore, the effect of growth-ring on water quality draws more and more attention.

In the early stage of the formation, "growth-ring" is loose, so it is easily washed away. When flow direction or flow rate of water suddenly changes, sediments are stripped from pipe wall, impacting water quality of consumers. Sediments stay on pipe wall deposit and harden gradually, providing a haven for bacteria. The formation mechanism of "growth-ring" is complicated, including physical deposition, electrochemical corrosion, influences of chemical and biological stability. Besides, for metallic pipe, ferrous iron forms under the function of iron bacteria and sulfate-reducing bacteria; after further oxidation, metallic pipes release ferric iron, and its volume is hundreds of times larger than before, blocking pipelines and leading to microbial corrosion. Micro-flocs of aluminum salt and Fe/Mn oxides and hydroxides attach to pipe surfaces, and a slime layer forms, providing protection for bacteria and they can regrow and reproduce. Accompany with the reproduction of iron bacteria, electrochemical corrosion occurs, leading to formation of rust tumor. In the process, impurities in water continuously adhere to the inner wall of pipes. "Growth-ring" is formed along the inner wall of pipes with time and the "growth-ring" has a thicker bottom and a thinner roof [30].

3. Analysis of growth-ring components and microbial species

Scholars make a research on the composition of corrosion products and microbial species of water supply pipelines [31-33].

Yuan Y X and Li Xin et al tested and analyzed growth-ring in cast iron pipes of actual water supply network. The results showed that growth-ring had a large surface area, which is up to 41.35~132.3 m^2/g. The main components of growth-ring included Fe, O, S and Si. Besides, it also contained Mg, Al, P, Cl, Ca, Mn and K etc.. Growth-ring existed mainly in the form of goethite (α-FeOOH), lepidocrocite (γ-FeOOH) and FeS, and Fe existed mainly in the form of ferrous iron and ferric iron. Escherichia coli, iron bacteria and sulfate-reducing bacteria were also detected in growth-ring [34].

Wu J J and Jiang W C et al investigated growth-ring in grey cast iron pipes and steel pipes without coating in the actual pipelines of a mountainous city. The results showed that tuberculations of grey cast iron pipes and steel pipes had similar chemical compositions, mainly Fe and O, followed by C, Al, Si, Cl, Ca, S, Mn, P and K etc.. The main component of tuberculations was iron oxide. Corrosion products of cast iron pipe mainly contained α-FeOOH, Fe_3O_4, γ-FeOOH, γ-Fe_2O_3 and a little $Fe(OH)_3$. The surface layer of corrosion products of steel pipes mainly contained $CaCO_3$, α-FeOOH and Fe_3O_4. The hard crust mainly contained α-FeOOH, Fe_3O_4, γ-FeOOH and γ-Fe_2O_3. The kernel layer mainly contained α-FeOOH, γ-Fe_2O_3 and γ-FeOOH [31, 35].

Zhang X Y, Yuan Y X and Li X investigated growth-ring in cast iron pipes of actual water supply network. The results showed that the main elements of growth ring are O and Fe, followed by organic functional groups and a little Si, C and S. The main components of growth-ring were goethite, lepidocrocite and FeS. Isolated bacteria included bacillus, Escherichia coli, Enterococcus faecalis, paenibacillus elgii, pseudomonas and mycobacterium [36].

Devet et al investigated bacteria diversity in underground water and filtered water. The results showed that similar to Cenarchaeum and Nitrosopumilus, ancient nitrifying bacteria largely existed in nitrated filtered water. Grenathrix polyspora, iron oxidizing bacteria, existed in groundwater which contained carbon dioxide [37]. The corrosion process of iron release process is shown in figure 1.

Li Y Q, Yuan Yi X, et al investigated biofilm community structure and biodiversity in PE, stainless steel and cast iron pipes. They got 59 strains of bacteria and found that bacillus accounted for a large proportion. They identified 47 strains of bacteria successfully, which belonged to 18 genera and 13 families, and Sphingobium, appeared most frequently. The degree of biological diversity of cast iron pipes was the highest and the bacteria in cast irons belonged to 14 genera. The bacteria in stainless steel pipes and PE belonged to 7 and 3 genera respectively. Bacteria genera isolated from those three pipes were largely different and the only genus those three kinds of pipes shared was only

Sphingomonas [38].

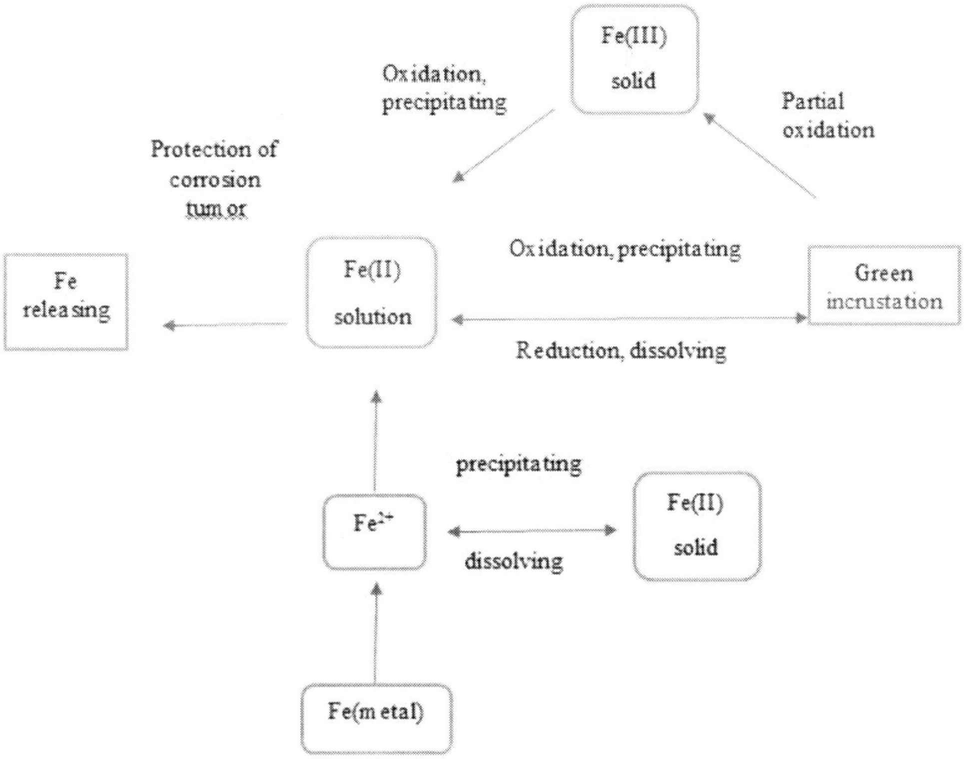

Figure 1. Picture of corrosion process of iron release process.

Yu J X, et al investigated bacteria on biofilm in actual water supply network and got 7 strains of bacteria including Micrococcus roseus, Micrococcus uarians, white bacillus pumilus, Brevibacterium linens, arthrobacter flavescent, Microbacterium lactium and Alkalophilic Bacillus [39]. Bai X H, et al investigated biofilm in a main pipeline of actual water supply network and the relevant adjusting valve wall. The results showed that Bacillus accounted for a large part of microorganisms in pipelines. In addition, they got actinomycetes and fungus [40]. Liu X L, Liu W J, et al investigated biofilm in cement-lined steel pipes and steel pipes of actual water supply network in Beijing. They found five kinds of bacteria including Bacillus cereus, pseudomonas, Bacillus thuringiensis, Acinetobacter haemolyticus and uncultured bacteria [41]. Zhang X Y, Liu W J et al. investigated biofilm in cement-lined steel pipes and steel pipes of actual water supply network and detected 15 species of bacteria, 1 species of Streptomyces, 4 species of fungi and 1 species of yeast. Gram-positive aerobic bacillus accounted for a large part of bacteria, and Gram-negative bacteria and cocci accounted for a small part [42]. Li Jun, Zhao Xin hua et al. investigated biofilm in simulated network and they got 32 strains of bacteria belonging to 10 genera including Pseudomonas, Chryseobacterium, Xanthomonas, Bordetella, Stenotrophomonas, Burkholderia cepacia, Chromobacterium, Aeromonas and non-fermenting bacteria [43]. Other researchers also have similar results [44-46].

4. Degradation mechanism of chlorinated hydrocarbon

It is thus obvious that iron is the main element in growth-ring. Iron is active metal and has reducing capacity. So iron can displace metals which rank behind iron in Metal Active Order Form and the reduced metals deposit on the surface of iron. Iron can also reduce ions, compounds and some organics with strong oxidability. Fe^{2+} has reducing capacity, too. Fe^{2+} can be oxidized to Fe^{3+} when oxidizing agent exists in water. When cast/pure iron with impurities is mixed with carbon particles in water, countless tiny primary cells will be formed between iron and carbon or other elements. When the solution is alkaline and oxygen exists, Fe^{2+} and Fe^{3+}can be transformed into flocculent precipitates, $Fe(OH)_2$ and $Fe(OH)_3$, respectively. $Fe(OH)_3$ may also hydrolyze, generating $Fe(OH)^{2+}$, $Fe(OH)^{2+}$ and other complex ions. Those complex ions have strong flocculation performance. It is thus obvious that zero-valent iron can remove pollutants through reduction, microelectrolysis, coagulation and adsorption and combined effect [47]. Research shows that under the condition of normal temperature and pressure, zero-valent iron can reduce a variety of chlorinated organic compounds [47, 48]. After a number of studies, Gillham et al. found that zero-valent iron can dechlorinate TCE quickly.

There are three ways for TCE dechlorination: ① the metal react directly, that is, the electron on the surface of zero-valent iron is transferred to chlorinated hydrocarbon: $Fe+RCl+H+\rightarrow RH+Fe^{2+}+Cl-$; ② ferrous iron, the direct product of iron corrosion has reducing capacity, and it can dechlorinate a part of chlorinated hydrocarbon. But this reaction proceeds very slowly; ③ hydrogen produced by the reaction of iron can reduce chlorinated hydrocarbon. Under anaerobic condition, H_2O can act as electron acceptor, and the following reactions occur: $2H_2O+2e\rightarrow H_2+2OH-$, $Fe+2H_2O\rightarrow Fe^{2+}+H_2+2OH^-$, thus $H_2+RCl\rightarrow RH+H^++Cl-$. Among the reactions, the most important reaction is the first one [49, 50]. With suitable catalysts, all of the three ways are effective. The catalysts can be provided by surface of iron, impurities in iron and other solid phases in system [47].

Other researchers found that Fe/FeS nanoparticles were able to remove TCE in water quickly [51]. There are mainly two kinds of bacteria that are capable of reductive dechlorination. The one is a strain such as dehalobacter restrictus, dehalospirillum multivorans etc. that can transform PCE and TCE into dichloroethylene (DCE). The other one is a strain that can continually transform DCE and vinyl chloride (VC) into ethylene(ETH). Indigenous microorganisms that can degrade TCE include Sphaerotilus natans and Crenothrix of iron bacteria, Thiothrix in Filamentous sulfur bacteria, Rhodococcus TA1, Pseudomonas putida TA2 and Nocardia AR1, two kinds of microorganisms: P.putida DC1 and Onion pseudomonas genus GR3 [1, 21, 52].

5. Research status of Fenton-like

Similar to the traditional Fenton reagent, "Fenton-like reagent" consists of H_2O_2 and Fe^{3+} salt solution, soluble iron or iron oxide minerals such as goethite, lepidocrocite, magnetite, pyrrhotite, hematite, siderite etc. [53-55]. In the system, H_2O_2 is catalyzed and decomposed into HO-, then degrade organic compound [56].

Fenton-like reaction equations are as follows:

$$Fe^{2+}+H_2O_2 \rightarrow Fe^{3+}+OH\text{-}+\text{-}OH$$

$$Fe^{3+}+H_2O_2 \rightarrow Fe^{2+}+HO_2\cdot + H^+$$

$$HO\cdot+Fe^{2+} \rightarrow Fe^{3+}+OH\text{-}$$

$$Fe^{3+}+HO_2\cdot \rightarrow Fe^{2+}+O_2+H^+$$

$$HO\cdot+H_2O_2 \rightarrow HO_2\cdot+H_2O$$

The mechanism of the degradation of TCE under the catalysis of siderite, hematite and magnetite is that hydrogen peroxide produces OH under the catalysis of ferrous iron and produces $HO_2\cdot$, O_2, $O\cdot$- under the catalysis of ferric iron. The OH is added to the TCE double bond. And the chlorine of the TCE molecule are nucleophilic substituted by the O_2 and $O\cdot$-, resulting in the final conversion of TCE to CO_2, H_2O and release Cl^- [54].

In the case of siderite-catalyzed TCE degradation, the specific reaction equation are as follows:

$$OH\cdot+CCl_2{=}CClH\rightarrow CCl_2OH\text{-}CClHOH$$

$$2OH\cdot+CCl_2OH\text{-}CClHOH\rightarrow HOCCl_2OH+HOCClHOH$$

IOP Conf. Series: Earth and Environmental Science **82** (2017) 012065 doi:10.1088/1755-1315/82/1/012065

$$OH\cdot + CCl_2OH\text{-}CClHOH \rightarrow HOCClOH\text{-}CClHOH + Cl\cdot$$

$$HOCCl_2OH \rightarrow O{=}CCl_2 + H_2O$$

$$HOCClHOH \rightarrow O{=}CHCl + H_2O$$

$$O{=}CCl_2 + O_2{-}\cdot \rightarrow O{=}CCl\text{-}O\text{-}O\text{-} + Cl\text{-}$$

$$O{=}CHCl + O_2{-}\cdot \rightarrow O{=}CHO\text{-}O\text{-} + Cl^-$$

$$O{=}CCl\text{-}O\text{-}O^- + O^-\cdot \rightarrow CO_2 + Cl^- + O_2$$

$$O{=}CHO\text{-}O^- + O^-\cdot \rightarrow O{=}CHO + O_2$$

$$O{=}CHO + HO\cdot \rightarrow O{=}C{=}O(CO_2) + H_2O$$

Soil and aquifer themselves contains plenty of natural iron minerals so that they can be used in Fenton-like reaction directly. Therefore, compared with traditional "Fenton oxidation", the advantage of "Fenton-like oxidation" is that it can effectively repair the areas polluted by TCE without additional catalyst. What's more, "Fenton-like oxidation" can be used in a wide range of pH [53]. So in recent years, "Fenton-like oxidation" has been gradually used in soil and groundwater pollution control and repair and it has become a hot topic around the world.

In 2012, Oliveira et al first used pipe deposits from water networks as catalysts in paraquat peroxidation to degrade the pesticide paraquat in water. The samples used as catalysts in the oxidation studies were taken from real drinking water distribution systems. The deposits were removed from old cast iron pipes that needed to be substituted. The results revealed that the deposits exhibit good catalytic performance in the pesticide degradation and mineralization. Oliveira et al. also pointed that, depending on the types of solids present in a given water network, in situ treatment can be made without insertion of chemicals apart from the oxidant, in case of a contamination event [57].

In 2013, based on the similar principle, Gosselin et al. investigated drinking water and biofilm disinfection by Fenton-like reaction which was conducted with Fenton's reagent (H_2O_2) at pH 3 or 5 on autochthonous drinking water biofilms grown on corroded or non-corroded pipe material. These new results provide additional support for the use of Fenton's processes for cleaning drinking water distribution systems contaminated with biological agents or organics. Gosselin et al. also pointed that Fenton-like reagent is a promising alternative to conventional water disinfection treatment and should be extended to surface disinfection and biofilm inactivation [58].

6. Conclusions and outlook

So water distribution network is a complex system. The growth-ring in pipelines consists of elements such as Fe, O, C, S etc. The main compositions of growth-ing include α-FeOOH, γ-FeOOH, γ-Fe_2O_3, Fe_3O_4, a small amount of $Fe(OH)_3$ and FeS etc. There are also microorganisms including iron bacteria, sulfate reducing bacteria, Sphingobium, Microbacterium, Micrococcus, Bacillus, Pseudomonas etc. Growth-ring has strong adsorption capacity because of its large specific surface area and pore structure. The surface area changes with the existing forms, causing its adsorption capacity changes. In addition, the structure of growth-ring is loose especially in the early stage of growth-ring formation so that sediments will be stripped off from the tube wall when the flow direction or flow rate suddenly changes, releasing certain elements and changing the composition of water. The main components of growth-ring and indigenous microorganisms in growth-ring can react with TCE and PCE directly or indirectly. Therefore, in the complex system, comprehensive reactions between TCE/PCE and growth-ring/biofilm will occur during water transmission including absorption, desorption, flocculation, migration, transformation, reduction etc. As a result, TCE/PCE levels changes. What's more, incomplete degradation product of TCE/PCE such as DCE and VC could be produced which is toxic carcinogenic substance and has a great threat to human health.

Researches on TCE/PCE are concentrated in: rapid and effective removal methods of chlorinated hydrocarbons in soil and groundwater; removal efficiency and influence factors of the methods;

development of the system for removing TCE/PCE and preparation of raw materials; degradation mechanism, degradation kinetics and product analysis of TCE/PCE; migration and transformation of TCE/PCE in soil and water. However, little is known about TCE/PCE migration and transformation mechanisms and removal methods in water distribution network.

After purification, the water polluted by TCE/PCE can reach the standards of drinking water (GB5749-2006). However, water with trace amounts of TCE/PCE is still harmful to human health (US. EPA.). Long-term drinking or contacting with the water may cause cancer [59-62]. Therefore, it is vital to know migration and transformation mechanisms of TCE/PCE in water distribution network. In addition, based on the results of Oliveira et al and Gosselin et al the main components of growth-ring can act as catalysts in Fenton-like reagent (H_2O_2). Then chlorinated hydrocarbons could be degraded by Fenton-like reagent, and the potential safety hazards of TCE/PCE in the water distribution network could be solved.

Acknowledgments

This study was financially supported by Open Project of State Key Laboratory of Urban Water Resource and Environment, Harbin Institute of Technology (No. QA201611-01), and National Natural Science Foundation of China (51578180).

References

[1] Hu H Z and Mao X M 2010 Progress in research on trichloroethylene in high concentration degraded by anaerobe in groundwater *J. Sci. Technol. Rev.* **28** 112-7

[2] Deng S P, Pei Z P and Han B P 2007 Soil contamination characteristics by carbon tetrachloride in suburb source of a south city *Environ. Pollut. Control* **29** 78-80, 89

[3] Jeffrey P J, Ruth F W and Aarme V P 1998 *Environment Pollution & Control 4th* (Boston: Butterworth Heinemann)

[4] Liu Y P 2011 The research on remediation of soil contaminated with tetrachloroethylene by mixed surfactants (Beijing: North China Electric Power University)

[5] Jia N 2011 The optimum preparation of nano Fe/Ni bimetallic particles and the basic research on the removal of CAHs in water by them (East China University of Science and Technology)

[6] Liu N 2013 The feasibility study on the removal of CAHs by activated carbon with nano zero-valent iron (East China University of Science and Technology)

[7] SQUILLACE P J 1999 Volatile organic compounds in untreated ambient groundwater of the United States *J. Environ. Sci. Technol.* **33** 3012-20

[8] Zhang F G, Wang S G, Ma H, Lv C and Wang Y B 2012 The pollution of trichloroethylene and tetrachloroethylene in soil and groundwater and remediation technology *J. Sci. Technol. Rev.* **30** 65-72

[9] Xiang X P 2002 Progress in research on processing methods of trichloroethylene and tetrachloroethylene *J. Yuzhou Univ. Nat. Sci. Ed.* **19** 77-80

[10] Shen D Z 2002 *Bioremediation of Polluted Environment* (Beijing: Chemical Industry Press)

[11] Cui Y J 2002 Experimental study on organic pollutants degraded by Fenton technology in nonaqueous phase (Qingdao: Ocean University of China)

[12] Lien H L 2000 Nanoscale bimetallic particles for dehalogenation of halogenated aliphatic compounds (Lehigh, USA: Lehigh University)

[13] Bian B X, Xie Q and Zhao Y C 2005 *Resource Technology of Coal Series Solid Waste* (Beijing: Chemical Industry Press)

[14] Gu BW, Wang P M and Xiong S B 2005 The effect of mechanical grinding on characteristic of particle group of coal gangue-cement system and its mechanical property *J. Southeast Univ. Nat. Sci. Edit.***35** 167-71

[15] Vasilios B K, Brebt E S and David M B 2000 Column studies of biodegradation of mixtures of tetrachloroethene and carbon tetrachloride *J. Water Res.* **34** 4161-8

[16] Xu H 2006 The adsorption behavior of trichloroethylene in natural soil and its influencing factors (Beijing: Beijing University of Chemical Technology)

[17] Young C C, Masahiro H, Kweon J, *et al* 2000 Isolation and characterization of a tetrachloroethylene dechlorination bacterium, clostridium bifermentans DPH-1 *J. Biosci. Bioeng.* **89** 489-91

[18] Liu F, Huang Y Y and Zhang G C 2006 Study on the influencing factors of removal of chlorinated hydrocarbons by nano Ni/Fe *J. Front. Earth Sci.* **13** 150-4

[19] He J T, Li Y, Liu S and Chen H H 2005 Natural biodegradation of trichloroethylene in shallow groundwater *J. Environ. Sci.* **26** 121-5

[20] Sun J F, Zhu K and Zhang W 2008 Anaerobic dechlorination reduction degradation of chloroethylene in groundwater and soil *J. Environ. Prot. Sci.* **34** 15-8

[21] Chen C B, Yang Q and Shen Z L 2003 Progress in research on bioremediation of TCE in groundwater *J. East Chin. Geol. Inst.* **26** 10-4, 37

[22] Ma C W 2007 Research on tetrachloroethylene's migration and repair technology (Shanghai: Shanghai Jiao Tong University)

[23] Tao Y, Mi S Q and Wang S G 2008 Research on characters of Activated carbon adsorption-supercritical CO_2 desorption TCE *J. Environ. Pollut. Control* **30** 1-5

[24] Wang M and Wu Y F 1996 Trace organic pollution of groundwater *J. Front. Earth Sci.* **3** 169-75

[25] Yuan Y X, Zhao H B and Zhao M 1998 Research on growth ring in water supply network *J. Harbin Univ. Civil Eng. Archit.* **31** 72-6

[26] Li X, Wang Y P and Zhao H B 2002 Prevention and treatment of growth ring in water supply pipe *J. Harbin Univ. Civil Eng. Archit.* **35** 30-2

[27] Li S and Zhang X J 2003 Growth and development of wall biofilm in water supply pipe and its influence factors *J. Chin. Water & Wastewater* **19** 49-52

[28] Wu H W, Liu W J and He B P 1998 Forming characteristics of pipe scale in water supply pipe and its control measures *J. Chin. Water & Wastewater* **14** 37-40

[29] Zhu W P, Yang Z H and Jiang Z P 1994 Balance of carbonate in urban water supply system and scale inhibition technology *J. Technol. Water Treat.* **20** 156-62

[30] Zhao H B, Li X and Zhao M 2008 *Water Supply Pipeline Hygiene* (Beijing: China Building Industry Press) pp 5-6

[31] Wu J J 2012 Research on water quality stability of water supply pipeline in a mountainous city (Chongqing University)

[32] Fang W 2007 Study on water quality chemical stability in urban water supply system and its control method (Hunan University)

[33] Sun H F, Shi B Y and Jin D S 2011 Characteristics of inner wall biofilm in water supply pipe network and its influence on water quality *J. Chin. Water & Wastewater* **27** 40-5

[34] Xu W W 2005 Study on the cause of growth ring in water supply pipe network and its prevention measures (Harbin Institute of Technology)

[35] Jiang W C, Jiang H, Wu J J and Tian S H 2013 Experimental study on changes of water quality in corrosive cast-iron pipe *J. Huazhong Univ. Sci. Technol. (Nat. Sci. Edit.)* **41** 117-21

[36] Zhang X Y 2009 Research on biological stability of urban water supply network and its influencing factors (Harbin Institute of Technology)

[37] Devet W W J M, Dinkla I J T, Muyzer G, *et al* 2009 Molecular characterization of microbial populations in ground water sources and sand filters for drinking water production *J. Water Res.* **43** 182-94

[38] Li Y Q 2013 Research on effects of residual chlorine in water supply pipeline and tubing on biofilm formation (Harbin Institute of Technology)

[39] Yu X J 2005 Study on the growth of bacteria in the biofilm of water supply pipe wall (Harbin Institute of Technology)

[40] Bai X H, Zhou B H, Zhu B and Wang H L 2007 Analysis of microbial characteristics of the

wall biofilm in water supply pipe network in Shanghai City *J. Chin. Water & Wastewater* **23** 105-8
[41] Liu X L, Liu W J, Jin L Y and Gu J N 2008 Microbial community on inner shell of water supply pipe network of Beijing *J. Tsinghua Univ. (Sci. Technol.)* **48** 1458-61
[42] Zhang X Y, Liu W J, Pan H, Xu H F, You Z L and Zhang J S 2005 A study on growth characteristics of microorganisms in water supply network pipe wall in a certain city *J. Water & Wastewater Eng.* **31** 1-6
[43] Li J, Zhao X H, Wu Q and Huang J J 2007 Application of API in the research on bacterial diversity in water supply network *Chin. J. Health Labor. Technol.* **17** 791-2
[44] Sheng M J and Wang X Y 2006 Research and analysis on the composition of the growth ring in urban pipe network *J. Ind. Water & Wastewater* **37** 60-2
[45] Gao J L 2013 Study on influencing factors of iron release in water supply pipe (Zhejiang University)
[46] Li Y F, Jiang B Y, Zhao H B and Tao C C 2010 Corrosion test on the inner wall of gray cast-iron water supply pipes *J. Shenyang Jianzhu Univ. Nat. Sci.* **26** 321-5
[47] Chen Y and Quan X 2000 The mechanism and application of zero-value iron treating wastewater *J. Res. Environ. Sci.* **13** 24-6, 37
[48] Guo X and Wang B E 2009 Progress in research on the application of zero-value iron in environmental pollution control *J. Zhongkai Agrotechnical College* **22** 61-6
[49] Doong R-A and Wu S-C 1992 Reductive dechlorination of chlorinated hydrocarbons in aqueous solutions containing ferrous and sulfide irons *J. Chemosphere* **24** 1063-75
[50] Leah J Mathson and Paul G Tratngek 1994 Reductive dehalogenation of chlorinated methods by iron metal *J. Environ. Sci. Technol.* **14** 2045-53
[51] Kim E-J, Kim J-H, Azad A-M and Chang Y-S 2011 Facile synthesis and characterization of Fe/FeS nano particles for environmental applications *ACS Appl. Mater. Interfaces* **3** 1457-62
[52] Cui Y J, Yang S Yand Wang P 2008 Organically polluted soil and groundwater remediation by in situ Fenton oxidation *J. Progress Chem.* **20** 1196-201
[53] Zhao L Q 2009 The method study on remediation of TCE DNAPLs by Fenton-like oxidation (China University of Geosciences)
[54] Huang W Y 2011 Using iron minerals catalyzed hydrogen peroxide-sodium persulfate to remove trichloroethylene in groundwater (China University of Geosciences)
[55] Li T T 2012 Effect of natural organic matter on removal of trichloroethylene using Fenton-like reagent (China University of Geosciences)
[56] Cheng L R, Liu Y H, Ding A Z, Cui S C and Chen H Y 2012 Influencing factors of in-situ bioremediation of TCE-contaminated groundwater *J. Safe. Environ.* **12** 88-95
[57] Cátia O, Mónica S F Santos, Gabriela S, Arminda A and Luis M M 2012 Use of pipe deposits from water networks as novel catalysts in paraquat peroxidation *J. Chem. Eng.* **210** 339-49
[58] Gosselin F, Madeira L M, Juhna T and Block J C 2013 Drinking water and biofilm disinfection by Fenton-like reaction *J. Water Res.* **47** 5631-8
[59] US. EPA. Guidelines for Exposure Assessment (EPA/600/Z-92/001)
[60] US. EPA. Risk assessment guidance for superfund-vol. I, human health evaluation manual (part E, supplemental guidance for dermal risk assessment) (EPA/540/ R/99/005 http://www.epa.gov/oswer/riskassessment/rages/index.htm, 2004-7/2012-7-28)
[61] Lu B B, Huang F Y, Shan X M, Fan X Y, Zhang L X, Zhang B and Shen D H 2013 Health risk assessment of volatile chlorinated hydrocarbons in drinking water in Hefei *J. China Prevent. Medic.* **14** 260-4
[62] Han B, He J T, Chen H H and Shi J H 2006 Primary study of health-based risk assessment of organic pollution in groundwater (China University of Geosciences) **13** 224-9

IOP Conf. Series: Earth and Environmental Science **82** (2017) 012066 doi:10.1088/1755-1315/82/1/012066

Treatment of landfill leachate using an up-flow anaerobic sludge semi-fixed filter

J Hua[1,4], S Y Bai[2], Z Y Li[2] and H C Zhou[3]

[1]Wuxi Research Center for Environmental Sciences and Engineering, Wuxi, China
[2]Wuxi City College of Vocational Technology, Wuxi, China
[3]School of Environmental Sciences and Engineering, Nanjing University of information science and technology, Nanjing, China

E-mail: huaj252003@163.com

Abstract. In the present work, an up-flow anaerobic sludge semi-fixed filter (UASSF) was developed for landfill leachate treatment by packing the soft polyurethane belt as the supporting carrier. The performance of the hybrid reactor was evaluated in terms of COD removal and carbon flux distribution, also, the biomass effectiveness was investigated by restarting the reactor without the supporting carrier. The COD removal increased with the stepwise increment of the organic loading rate (OLR), while the sulfate removal decreased slowly. When the reactor was operated at design load (9 kgCOD/m^3•d), COD and sulfate removal remained around 81% and 90%, respectively. The results indicate that this kind of semi-fixed carrier is capable to form the active biofilm in the anaerobic process, and the UASSF system can perform well in the leachate treatment.

1. Introduction

Landfill leachate, a kind of high-strength wastewater, has a very hazardous impact upon human health and the environment. The concentration and composition of leachate are often determined by the age and type of landfill [1]. Impacts of leachate on environment will last for scores of years, and many treatment processes, such as physical, chemical and biological methods, have been proposed [2, 3]. Biological treatment method, especially anaerobic technology, is mainly the first stage in high-strength wastewater treatment [2-4].

As for suspended-growth biomass processes, the up-flow anaerobic sludge blanket (UASB) reactor is regarded as a successful and extensively applied device due to its advantages, including high loading rate, long solid retention time and flexibility of process operation [5, 6]. Still, UASB might be faced with the problems of slow start-up [7, 8] and sludge washout if granular sludge is not obtained [9]. Among attached-growth biomass processes, anaerobic filter (AF) is more favorable because of its higher biomass, smaller volume and shock loadings tolerance [10-12]. Types of carriers have been employed in AF to treat wastewater: natural materials (bamboo rings, coconut fibers, zeolite and perlite) [13-17], waste materials (glass, brick granules and automobile tires) [10, 18-20] and synthetic organic chemicals (polyurethane, polyethylene and nylon fiber) [21-24]. However, the major drawback of the AF is clogging of the filter, which may result in short circuiting of wastewater yielding unacceptable effluent quality [25-27]. Apart from the high suspended solids concentration in wastewater [28], another reason for filter clogging is the packed carrier itself. For most filter reactors, the carrier packed in them is of the fixed type, which is immobilized in the moving fluid. With the increase in biofilm thickness and continuous operation, the filter is clogged.

In the study, an up-flow anaerobic sludge semi-fixed filter (UASSF) was developed for landfill leachate treatment by integrating UASB and AF. The semi-fixed carrier in UASSF could move with the up-flow liquid, which contributed to avoiding biomass washout and clogging. The study focused on the performance of UASSF by COD removal efficiency (RE_{COD}) and influent carbon distribution. Additionally, the effectiveness and contribution of biomass were also evaluated by restarting the reactor without the supporting carrier.

2. Materials and methods

2.1. Leachate analysis

The leachate used in the research was collected from currently operating landfill located in Chengdu, China. The main physicochemical characteristics of leachate are listed in table 1.

Table 1. Characteristics of influent leachate.

Parameter	Mean	Range
pH	5.0	4.8-5.2
COD (mg/L)	68500	40000-75000
BOD_5 (mg/L)	28000	20000-35000
BOD/COD	0.41	0.34-0.47
SS (mg/L)	3000	1600-4000
NH_3-N (mg/L)	950	800-1100
NO_3-N (mg/L)	120	85-170
TN (mg/L)	8200	5000-9500
TP (mg/L)	148	120-175
SO_4^{2-} (mg/L)	2120	1790-2450

2.2. Reactor setup

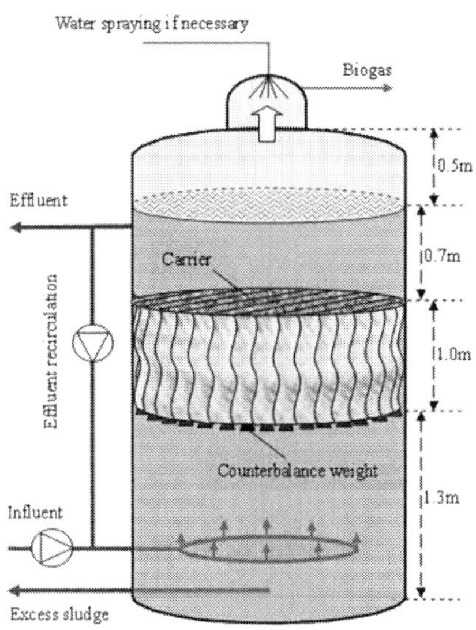

Figure 1. UASSF Schematic diagram.

Figure 2. The belt carrier used in the UASSF.

The schematic diagram of the study is shown in figure 1. The pilot scale UASSF, which is 1.95 m in diameter and 3.5 m in height, is cylindrical and made of carbon steel. The active volumes of the reactor are 9.0 m^3 and the total volumes are 10.6 m^3. Two pumps are used separately for leachate feeding and effluent recirculation, providing an up-flow velocity of 0.6 m/h for liquid in the reactor. The influent flow and gas production are measured by an electromagnetic meter and a wet-gas meter, respectively.

As carrier material, soft polyurethane belts (0.05 cm × 3 cm × 100 cm) are hung vertically in the reactor (see figure 2). The upper end of the belt is attached to a metal net which is parallel with the lateral section of the reactor, and the end lap is linked to a counterbalance weight, avoiding the belts twining with each other. Besides providing surface microbial attachment, the belts acted like a three-phase (gas–liquid–solid) separator.

2.3. Inoculation and start up

The inoculums in the reactor were from digested sludge which was obtained from a plant treating sewage. The total solids and volatile suspended solids concentration were 61200 mg/L and 10800 mg/L, respectively. The reactor structure was laid in a room whose temperature was kept at 35°C by a thermostat.

During the initial period of the operation, the suspended solid (SS) with poor settling properties was washed out by the up-flow of the liquid, while microorganisms and SS with good settleability aggregated and settled at the bottom of the reactor. Organic macromolecules began to be absorbed on the surface of the carrier, and the planktonic cells moved and adhered to the belt at the same time. It was the adherent microorganisms that constituted the first layers of the biofilm.

2.4. Analytical methods

COD, sulfate and volatile fatty acids (VFA) were analyzed by colorimetric tests on a HACH-DR 2800 photometer. Bicarbonate measurement was conducted with an analyser (GDYS-103SD, Nuoji instrument Co., Ltd, China). Biogas was checked via gas chromatography (BFRL-SP-2100, Beijing Analytical Instrument Factory, China), and total organic carbon (TOC) was determined through a TOC-500 analyser (Shimadzu Corp. Japan). Alkalinity, pH and SS were measured according to the standard methods [29].

2.5. Calculation methods

During the treatment process, part of COD would be eliminated by sulfate-reducing bacteria (SRB) [30]. 64g of COD was needed when one mol of sulfate was reduced, so the percentage of COD removed by SRB (COD_{SRB}) could be calculated based on the amount of sulfate reduced and the total COD removed [31]. The influent organic carbon is distributed in catabolic reactions (i.e. formation of biogas, bicarbonate and VFA) and anabolic reactions (i.e. biomass production) [6]. TOC and other carbon species were measured directly, so a part of carbon transformed to biomass could be deduced from the carbon mass balance.

3. Results and discussion

3.1. The performance of UASSF

The UASSF reactor was operated continuously for 112 days. Figure 3 gives the process performance of the hybrid reactor. According to the RE_{COD}, the operation could be divided into three phases roughly.

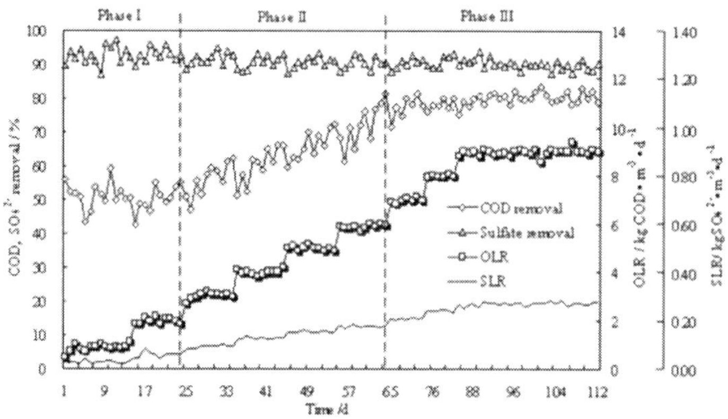

Figure 3. The performance of UASSF.

3.1.1. Phase I (day 1-24). When landfill leachate containing sulfate was introduced into the UASSF, the organic matter was removed via sulfate reduction and methanogenesis. As the organic loading rate (OLR) increased from 0.5 to 2 kgCOD/m^3•d, RE_{COD} varied between 42% and 59%. The biogas production was suppressed during this period, and methane content was lower than 38%. The biomass in the reactor seemed unable to handle the increased OLR, which was probably because some active biomass was washed out, or was due to physiological modifications resulting from physical constraints [17]. However, at sulfate loading rate (SLR) of 0.03-0.06 kgSO_4^{2-}/m^3•d in this phase, the sulfate removal was relatively high (87% to 97%), and a certain amount of the COD elimination (5.5-5.9%) was accomplished through sulfate reduction (see figure 4). It could be concluded that SRB inhibited the growth and activity of methane-producing bacteria (MPB) to some extent, though the inhibition was not strong due to the higher COD/SO_4^{2-} ratio [32].

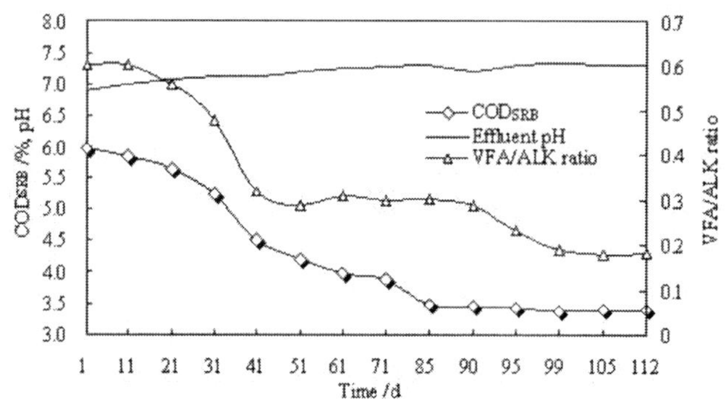

Figure 4. Variation of COD_{SRB}, pH and VFA/ALK ratio.

The organic carbon distribution in metabolism can also reflect the performance of UASSF. Figure 5 illustrates the distribution of metabolized organic carbon in the treatment process, while the unmetabolized is not included. At the initial time, there was a slow accumulation of biomass, and the corresponding carbon distribution (anabolic carbon) was only 4% at day 14. The catabolic carbon flux to VFA (45%) was much higher than that to biogas (28%) or bicarbonate (23%), which might result from the incomplete organic substrate degradation and methanogenesis inhibition. Although the influent pH was low (between 4.8 and 5.2), the corresponding value in the reactor still maintained at 6.9, and the alkalinity concentration in influent was always lower than that in effluent, indicating that the anaerobic metabolism itself produced the alkalinity [33]. Since the effluent recirculation could keep a near neutral pH in the reactor [34, 35], there was no alkalinity addition throughout this experiment. Effluent recirculation could also produce an internal dilution to reduce the impact of high OLR [7, 34] and hence improve the performance of the anaerobic reactor [36, 37]. As a result, MPB became more active and accommodated to the new circumstance in the late period of phase I. Although so, the VFA/ALK ratio was still higher than 0.4 in this period, which meant that the adequate buffering capacity had not been reached for anaerobes [38].

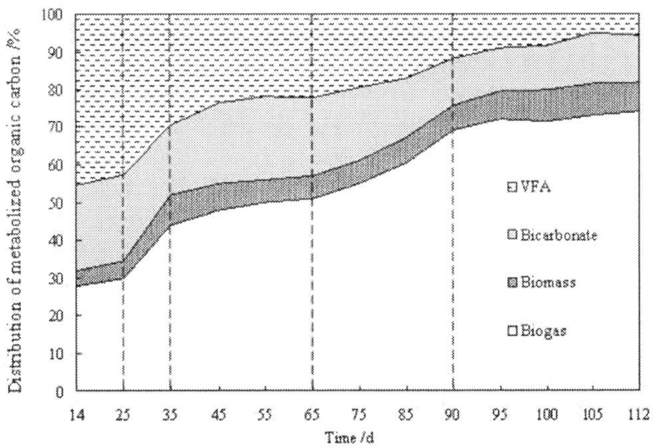

Figure 5. Distribution of metabolized organic carbon.

3.1.2. Phase II (day 25-64). With the OLR increasing from 3 to 6 kgCOD/m^3•d in this phase, the RE_{COD} grew more quickly than in phase I, indicating improvement in microbial activity of the MPB. This was supported by the continuous decrease of COD_{SRB} from 5.4 to 3.9%, though the SLR had increased from 0.08 to 0.18 kgSO_4^{2-}/m^3•d. In addition, in each stepwise increase of loading rate, the RE_{COD} increased accordingly. As for the carbon distribution from day 25 to day 35, the portion of residual VFA further decreased, while those of biomass and biogas increased from 4.5 to 8% and 30 to 44%, respectively. This aggressive increment and the rapid increase in RE_{COD} (figure 3) could be considered the result of the adjustment of microorganisms to their new growth mode [39]. In the following period (day 36-64), however, the carbon distributions did not change drastically due to the interaction between more activated MPB and increased OLR. At the end of this period, the VFA/ALK ratio and effluent pH were 0.29 and 7.2, respectively, which also suggested that the system was more stable than in phase I. This stability could be owed to the carbonate/bicarbonate cushion resulting from CO_2 generation in the anaerobic digestion process [40-42]. Another factor was the ammonia appearing when the nitrogenous compounds were degraded anaerobically. The ammonia (1350 mg/L) could buffer volatile acids by preserving a high concentration of bicarbonate [43].

3.1.3. Phase III (day 65-112). During the last phase, the removal rates of COD and sulfate did not change much, though the loading rates were still increasing in the period of day 65-84. When the OLR and SLR increased separately to 9kgCOD/m^3•d (design value) and 0.27 kgSO_4^{2-}/m^3•d from day 85, the RE_{COD} and sulfate removal remained steadily at about 81% and 90%, respectively, which portended that the competition process between SRB and MPB seemed to reach a state of equilibrium. This result was supported by the nearly constant COD_{SRB} (3.4%) from day 90. During day 65-90, the portion of carbon flux to biogas increased rapidly from 51 to 69%, indicating that the growing conditions of MPB were much better than previous phases, which was also in conformity with the high RE_{COD} and continuous decrease of carbon flux to VFA. During day 91-112, most carbon mass (72%) was transformed to biogas, but that to biomass, VFA and bicarbonate did not change much. By the end of this period, the portions of the three had reached 7.8, 8.0 and 12.2%, respectively. Therefore, the carbon distribution between anabolic reactions and catabolic reactions seemed to be in a balance. In the late period of this phase, the VFA/ALK ratio and the effluent pH were 0.18 and 7.3, respectively, this also meant that enough buffering capacity was presented in the reactor and a stable operation of the system had been performed [38, 44]. In this reactor, total alkalinity as $CaCO_3$ was 3100 mg/L, which was enough to stop the pH dropping.

With increasing loading rates, the production of methane and total gas increased throughout the experiment, though there were small fluctuations sometimes. The increase also showed the biomass in the UASSF was becoming more active, and it could handle more output of the leachate.

3.2. Effectiveness and contribution of biomass

To compare the effectiveness and contribution of biomass, the semi-fixed carrier was taken out of the reactor soon after day 112, and a new reactor without the supporting carrier was restarted immediately.

3.2.1. Determination of attached biomass. Based on visual observation, the biofilm on the belt carrier was black, dense and slimy. Thicker film was found mostly at the lower end of the carrier while thinner film was at the top of belt. Film thickness varied dependently upon the belt height where the biomass attached, which was estimated to be about 2 to 3 mm on the lower end of the belt and 1 to 2 mm on the upper end, respectively.

The belts were air-dried and cut into 20 cm-pieces. The pieces were divided into five groups based on their height on the belts, and the biofilm was then scraped from the carrier surface for biomass determination. As a result, the average concentration of attached biomass was calculated to be 1.28 g/L. The distributions of attached biomass on the carrier are presented in figure 6.

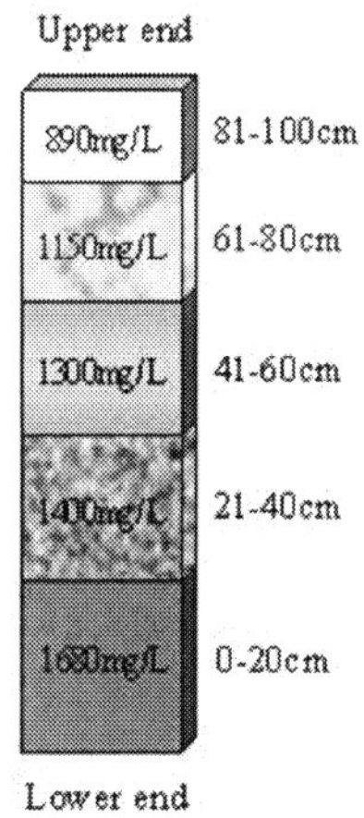

Figure 6. The distribution of attached biomass on the belt carrier.

3.2.2. Performance of suspended biomass. Owing to the acclimated anaerobic sludge, the restart of the reactor was easily accomplished within fifty days (figure 7). The reactor was operated with a stepwise increase of OLR from 2 to 8 kgCOD/m^3•d, and the RE_{COD} increased, though the reactor was instable during the first two weeks due to the high initial OLR and the new environment in the reactor. From day 15 to day 43, the RE_{COD} gradually increased from about 40 to 60%, which corresponded to the OLR of 4 to 7 kgCOD/m^3•d. However, the RE_{COD}, which nearly kept constant from day 36 to day 43, decreased sharply below about 40% when the OLR was further increased to 8 kgCOD/m^3•d from day 44. It seemed that the reactor was not able to deal with the increased OLR, i.e. to avoid the collapse of the new system without the supporting carrier, the influent OLR of 7 kgCOD/m^3•d should not be exceeded.

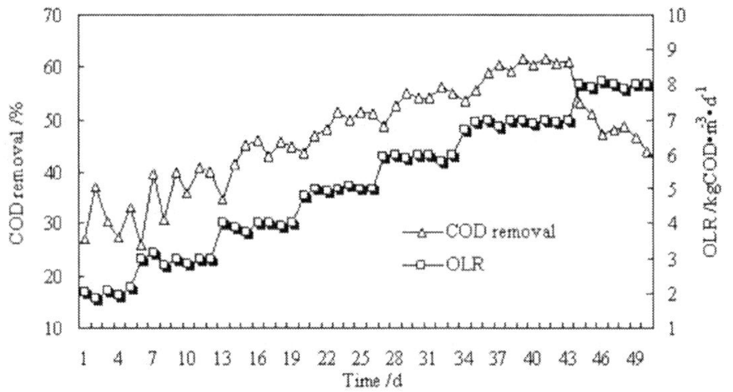

Figure 7. The performance of the new reactor without carrier.

3.2.3. Biomass comparison. After the restart test, the suspended biomass concentration (BCT) was measured to be 4.32 g/L. For the convenience of comparison, the BCT is supposed to be constant throughout the restart experiment, and the effective OLR ($OLR_{effective}$), the biomass contribution (BCB) and biomass effectiveness (BE) are defined as follows:

$$OLR_{effective} = OLR \times RE_{COD}$$

$$BCB = (OLR_{effective} / OLR_{effective} \text{ of UASSF}) \times 100\%$$

$$BE = OLR_{effective} / BCT$$

Therefore, the biomass comparison can be performed by assuming that there is a virtual reactor with only attached biomass (table 2).

Table 2. The results of biomass comparison.

Reactor	UASSF	New reactor	Virtual reactor
Carrier	Belt carrier	No carrier	Belt carrier
Biomass	Overall	Suspended	Attached
OLR	9 kg $COD/m^3 \cdot d$	7 kg $COD/m^3 \cdot d$	...
RE_{COD}	80%	60%	...
$OLR_{effective}$	7.2 kg $COD/m^3 \cdot d$	4.2 kg $COD/m^3 \cdot d$	3.0 kg $COD/m^3 \cdot d$
BCT	...	4.32 g/L	1.28 g/L
BCB	...	58%	42%
BE	...	0.97/d	2.34/d

As presented in table 2, the contribution of attached biomass accounted for 42% of the whole COD inversion. More importantly, the effectiveness of the attached biomass was 2.34/d, more than twice that of suspended biomass, which indicated that the activity of biofilm was higher than that of sludge. Though the semi-fixed carriers could act as three-phase separator to accumulate more suspended sludge, the principal role of the belt was to function as supports to immobilize bacteria. Therefore, in terms of this new carrier, the most attractive strength was its high build-up of active biomass. Other strengths, such as no clogging and sludge washout, strong resistance to environmental changes and small land area requirements were also appreciable.

4. Conclusions

In this research, a pilot scale UASSF reactor was used for landfill leachate treatment. According to the experiment, the conclusions were as follows:

- The operation of UASSF could be divided into three phases roughly. The RE_{COD} increased with the stepwise increment of OLR, while the COD_{SRB} decreased from 5.9% to 3.4% during the operation. The UASSF performed well in the treatment process, and the RE_{COD} remained around 81% at the design loading rate of 9kgCOD/m³•d.

- The metabolized carbon distributed into four parts. The portion of carbon flux to biogas increased with the operation time, while that to VFA decreased gradually. In the later stage of the operation, the portions of biogas, VFA, bicarbonate and biomass were 72, 8.0, 12.2 and 7.8% respectively, which indicated that the treatment system was operated at a steady state.
- Though the influent pH value was usually low (4.8-5.2), the effluent pH and VFA/ALK ratio were 7.3 and 0.18, respectively, in the late period of the operation, which meant that the system had a sufficient buffering capacity.
- Based on the results of reactor restart, the contribution of attached biomass accounted for 42% of the total COD conversion, while the effectiveness of it was more than twice higher than that of suspended biomass. This suggested that the semi-fixed carrier in the UASSF system was capable to form the active biofilm during the leachate treatment.
- The effluent from UASSF system needs further treatment before discharging into surface waters, which can be fulfilled using conventional aerobic treatment systems, membrane separation technology or the hybrid treatment process.

Acknowledgments
This work was supported by China Postdoctoral Science Foundation (2014M561688) and Jiangsu Planned Projects for Postdoctoral Research Fund (1301066C).

References

[1] Bilgili M S, Demir A and Özkaya B 2007 Influence of leachate recirculation on aerobic and anaerobic decomposition of solid wastes *J. Hazard. Mater.* **143** 177–83
[2] Wiszniowski J, Robert D, Surmacz-Gorska J, Miksch K and Weber J V 2006 Landfill leachate treatment methods: A review *Environ. Chem. Lett.* **4** 51–61
[3] Renou S, Givaudan J G, Poulain S, Dirassouyan F and Moulin P 2008 Landfill leachate treatment: Review and opportunity *J. Hazard. Mater.* **150** 468–93
[4] Ağdağ O N and Sponza D T 2005 Anaerobic/aerobic treatment of municipal landfill leachate in sequential two-stage up-flow anaerobic sludge blanket reactor (UASB)/completely stirred tank reactor (CSTR) systems *Process Biochem.* **40** 895–902
[5] Lin C Y, Chang F Y and Chang C H 2000 Co-digestion of leachate with septage using a UASB reactor *Bioresour. Technol.* **73** 175–8
[6] Cresson R, Carrère H, Delgenès J P and Bernet N 2006 Biofilm formation during the start-up period of an anaerobic biofilm reactor—Impact of nutrient complementation *Biochem. Eng. J.* **30** 55–62
[7] Poh P E and Chong M F 2009 Development of anaerobic digestion methods for palm oil mill effluent (POME) treatment *Bioresour. Technol.* **100** 1–9
[8] Liu Y, Xu H L, Yang S F and Tay J H 2003 Mechanisms and models for anaerobic granulation in up-flow anaerobic sludge blanket reactor *Water Res.* **37** 661–73
[9] Saravanan V and Sreekrishnan T R 2006 Modelling anaerobic biofilm reactors—A review *J. Environ. Manage.* **81** 1–18
[10] Reyes O, Sánchez E, Rovirosa N, Borja R, Cruz M, Colmenarejo M F, Escobedo R, Ruiz M, Rodríguez X and Correa O 1999 Low-strength wastewater treatment by a multistage anaerobic filter packed with waste tyre rubber *Bioresour. Technol.* **70** 55–60
[11] Wang Z and Banks C J 2007 Treatment of a high-strength sulphate-rich alkaline leachate using an anaerobic filter *Waste Manage.* **27** 359–66
[12] Van Der Merwe M and Britz T J 1993 Anaerobic digestion of baker's yeast factory effluent using an anaerobic filter and hybrid digester *Bioresour. Technol.* **43** 169–74
[13] Feng H J, Hu L F, Mahmood Q, Qiu C D, Fang C G and Shen D S 2008 Anaerobic domestic

wastewater treatment with bamboo carrier anaerobic baffled reactor *Int. Biodeter. Biodegr.* **62** 232–8
[14] Lee S, Lee H, Lee S, Chitapornpan S, Chiemchaisri C, Polprasert C and Ahn K 2007 Media configuration and recirculation of up-flow anaerobic floating filter for piggery wastewater treatment *Korean J. Chem. Eng.* **24** 980–8
[15] Fernández I, Vázquez-Padín J R, Mosquera-Corral A, Campos J L and Méndez R 2008 Biofilm and granular systems to improve Anammox biomass retention *Biochem. Eng. J.* **42** 308–13
[16] Garcia-Calderon D, Buffiere P, Moletta R and Elmaleh S 1998 Anaerobic digestion of wine distillery wastewater in down-flow fluidized bed *Water Res.* **32** 3593–600
[17] Sowmeyan R and Swaminathan G 2008 Performance of inverse anaerobic fluidized bed reactor for treating high strength organic wastewater during start-up phase *Bioresour. Technol.* **99** 6280–4
[18] Alves M M, Pereira M A, Bellouti M, Álvares Pereira R M, Mota Vieira J A, Novais J and Mota M 1998 A new method to study interactions between biomass and packing material in anaerobic filters *Biotechnol. Tech.* **12** 277–83
[19] Seth R, Goyal S K and Handa B K 1995 Fixed film biomethanation of distillery spentwash using low cost porous media *Resour. Conserv. Recycl.* **14** 79–89
[20] Zaher U, Frear C, Pandey P and Chen S 2008 Evaluation of a new fixed-bed digester design utilizing large media for flush dairy manure treatment *Bioresour. Technol.* **99** 8619–25
[21] Lai P, Zhao H Z, Ye Z F and Ni J R 2008 Assessing the effectiveness of treating coking effluents using anaerobic and aerobic biofilms *Process Biochem.* **43** 229–37
[22] Zhao H Z, Cheng P, Zhao B and Ni J R 2008 Yellow ginger processing wastewater treatment by a hybrid biological process *Process Biochem.* **43** 1427–31
[23] Chen S, Sun D Z and Chung J S 2007 Anaerobic treatment of highly concentrated aniline wastewater using packed-bed biofilm reactor *Process Biochem.* **42** 1666–70
[24] Chaiprasert P, Suvajittanont W, Suraraksa B, Tanticharoen M and Bhumiratana S 2003 Nylon fibers as supporting media in anaerobic hybrid reactors: It's effects on system's performance and microbial distribution *Water Res.* **37** 4605–12
[25] Bodkhe S 2008 Development of an improved anaerobic filter for municipal wastewater treatment *Bioresour. Technol.* **99** 222–6
[26] Jawed M and Tare V 2000 Post-mortem examination and analysis of anaerobic filters *Bioresour. Technol.* **72** 75–84
[27] Parawira W, Murto M, Zvauya R and Mattiasson B 2006 Comparative performance of a UASB reactor and an anaerobic packed-bed reactor when treating potato waste leachate *Renew Energ.* **31** 893–903
[28] Rajeshwari K V, Balakrishnan M, Kansal A, Lata K and Kishore V V N 2000 State-of-the-art of anaerobic digestion technology for industrial wastewater treatment *Renew. Sust. Energ. Rev.* **4** 135–56
[29] APHA 1998 *Standards Methods for the Examination of Water and Wastewater* 20th ed (Washington, DC, USA: American Public Health Association)
[30] Jiménez-Rodríguez A M, Durán-Barrantes M M, Borja R, Sánchez E, Colmenarejo M F and Raposo F 2009 Heavy metals removal from acid mine drainage water using biogenic hydrogen sulfide and effluent from anaerobic treatment: Effect of pH *J. Hazard. Mater.* **165** 759–65
[31] Omil F, Méndez R and Lema J M 1995 Anaerobic treatment of saline wastewaters under high sulfide and ammonia content *Bioresour. Technol.* **54** 269–78
[32] Rinzema A and Lettinga G 1988 The effect of sulphide on the anaerobic degradation of propionate *Env. Technol. Lett.* **9** 83–8
[33] Buzzini A P and Pires E C 2007 Evaluation of a up-flow anaerobic sludge blanket reactor with partial recirculation of effluent used to treat wastewaters from pulp and paper plants *Bioresour. Technol.* **98** 1838–48

[34] Sam-Soon P, Loewenthal R E, Wentzel M C, Moosbrugger R E and Marais G V R 1991 Effects of a recycle in up-flow anaerobic sludge bed (UASB) systems *Water SA.* **17** 37–46
[35] Ferguson J F, Eis B J and Benjamin M N 1984 Neutralization in anaerobic treatment of an acidic waste *Water Res.* **18** 573–80
[36] Thirumurthi D 1988 Effects of mixing velocity on anaerobic fixed film reactors *Water Res.* **22** 517–23
[37] Moosbrugger R E, Wentzel M C, Ekama G A and Marais G V R 1993 Treatment of wine distillery waste in UASB systems-feasibility, alkalinity requirements and pH control *Water Sci. Technol.* **28** 45–54
[38] Parkin G F and Owen W F 1986 The fundamentals of anaerobic digestion of waste water sludges *J. Environ. Eng.* **112** 867–920
[39] Michaud S, Bernet N, Buffière P, Roustan M and Moletta R 2002 Methane yield as a monitoring parameter for the start-up of anaerobic fixed film reactors *Water Res.* **36** 1385–91
[40] Raposo F, Banks C J, Siegert I, Heaven S and Borja R 2006 Influence of inoculum to substrate ratio on the biochemical methane potential of maize in batch tests *Process Biochem.* **41** 1444–50
[41] Borja R, Rincón B, Raposo F, Domínguez J R, Millán F and Martín A 2004 Mesophilic anaerobic digestion in a fluidised-bed reactor of wastewater from the production of protein isolates from chickpea flour *Process Biochem.* **39** 1913–21
[42] Borja R, González E, Raposo F, Millán F and Martín A 2001 Performance evaluation of a mesophilic anaerobic fluidized-bed reactor treating wastewater derived from the production of proteins from extracted sunflower flour *Bioresour. Technol.* **76** 45–52
[43] Georgacakis D, Sievers D M and Iannotti E L 1982 Buffer stability in manure digesters *Agric. Wastes.* **4** 427–41
[44] Behling E, Diaz A, Colina G, Herrera M, Gutierrez E, Chacin E, Fernandez N and Forster C F 1997 Domestic wastewater treatment using a UASB reactor *Bioresour. Technol.* **61** 239–45

Domestic sewage and secondary effluent treatment using vertical submerged biological filter

X J Yang[1], Y M Tong[1], Y Song[1], W K Qi[1,2,5], Y Y Li[3,4,5] and Y L Guo[3]

[1]School of Chemical Engineering, Beijing University of Chemical Technology, Beijing 100029, China
[2]Graduate School of Environmental Studies, Tohoku University, 6-6-06 Aza-Aoba, Aramaki, Aoba-ku, Sendai 980-8579, Japan
[3]Institute of Environment and Municipal Engineering, Qingdao Technological University, Qingdao 266033, PR China
[4]Department of Civil and Environmental Engineering, Graduate School of Engineering, Tohoku University, 6-6-06 Aza-Aoba, Aramaki, Aoba-ku, Sendai 980-8579, Japan

E-mail: qiweikang10@163.com; yyli@epl1.civil.tohoku.ac.jp

Abstract. Biological filter (bio-filter) exhibits a high potential for sustainability because construction and maintenance cost are low. When bio-filters are used as pre-treatment or post-treatment technology, other treat steps in the combined treat process could prevent clogging and get better process stability and performance efficiency. This study considered bio-filters for reducing the amounts of various wastewaters, i.e., high-strength sewage, municipal wastewater and secondary effluent water, using five kinds of media. The results obtained indicate significant improvement in organics removal when influent COD are 1500 mg/L and 2500 mg/L for anaerobic and aerobic biofilters, respectively. Sufficient alkalinity would obviously improve the substrate removal in an anaerobic bio-filter, when treating high-strength wastewater. The performances of bio-filters are adequate for further removal of the residual organic material and nutrient elements for tertiary domestic wastewater treatment.

1. Introduction

Biological filters (bio-filters) are invariably cited as a low-cost and low-maintenance technology [1, 2] , especially compared with conventional treatment technologies, such as activated sludge [3]. For example, it is practically impossible for activated sludge to treat low-strength wastewater with BOD lower than 20 mg l^{-1} as the difficulty of maintaining a normal value for MLSS and conventional efficiency. Substrate-transforming microorganisms in the bio-filter are often grown as biofilm on surfaces that can be either synthetic media [4, 5], or natural media (i.e., gravel, sand, tuff, zeolite) [6, 7]. Another important advantage of microbial film process is that interference in the bio-filter is also more easily adjusted. Because bio-film reactors contain a negative feedback control system to stabilize treatment efficiency [8]. For example, If the increment of the influent loading arises, the substrate concentration on a biofilm surface will increase correspondingly and the depth of the effective layer will also increase, resulting in the suppression of the increment in substrate concentration, and vice versa. Therefore, wastewater of a wide range of strengths can be treated effectively with microbial film processes. In addition, the bio-filter exhibits a high potential for sustainability when properly designed and maintained [9].

Bio-filters have been tested with numerous types of wastewater and it can be a valid treatment option for significant cost savings, e.g. effluent from anaerobic digesters [10, 11], nitrobenzene wastewater treated by an anaerobic baffled reactor [12], and pretreatment process for petroleum refinery wastewater [13].

It will be a dangerous risk to use secondary treated or even tertiary effluent for irrigation without disinfected for removing potential human pathogenic microorganisms. Filtration, which is regarded as one of attractive and effective processes, is usually used to remove suspended solids from wastewater effluents of biological processes to improve water quality. Therefore, Bio-filter is needed to filter effluent before discharged into the receiving waters, reused in agriculture, or utilized in body contact recreational reservoirs [14]. Bio-filters could also be applied to perform a post-treatment step for low-strength wastewater purification, e.g., secondary treatment of chromium-bearing tannery wastewaters [3], and secondary or tertiary treatment of municipal wastewater [2, 15]. The stability of bio-filter plays an important role in restraining the change of effluent water quality [16]. In order to obtain better process stability and performance efficiency, several studies have focused on combinations of anaerobic and aerobic processes for the treatment of wastewater [4, 17, 18].

The application of bio-filters should be considered to be an efficient option for reducing the amount of various wastewaters discharged into aquatic ecosystems [16]. An understanding of processes involved in pollutant removal is therefore necessary, both for improving the design and for predicting their long-term sustainability under a range of environmental conditions [19]. The objectives of this study are to evaluate the behavior of varieties of pollutant contaminants in anaerobic and aerobic bio-filters for treating wastewater at varying concentrations, i.e., high-strength domestic sewage wastewater, sewage wastewater and secondary effluent water under kinds of operation parameters, e.g. organic loading rate (OLR), hydraulic loading rate (HLR) and temperature. These results could be applied for improving the design of wastewater bio-filters and predicting their long-term sustainability for pollutant removal/reduction.

2. Material & methods

2.1. Bio-filter structure and media materials

Figure 1 shows the submerged bio-filter wastewater treatment system. The reactor consisted of a circular column with 0.48 m internal diameter and 1.2 m height. A perforated board supported the media layer from the bottom. A packing layer with a height of 0.65 m was used, and the efficient volume was 120 L. Above the packing layer there was a clay layer with a 0.20 m height for insulation and for preventing the escape of odors. The distribution and collection networks were made of perforated PVC pipe. The wastewater flowed into the reactor under the perforated board (up-flow bio-filter) by inlet pipe and was distributed by distribution pipe over the horizontal section. When the wastewater passed vertically through the packing, the biofilm could grow on a stable fixed-bed surface. After that it flowed into the collection pipe and outlet pipe. As a common filtration bed, the reactor needs backwash to maintain efficient hydraulic performance [16, 20, 21]. During these experiments, all of the bio-filters didn't clog. Therefore, they weren't backwashed.

The filter media include gravel, cinder, ceramsite, concrete block and brick. Each media has little particle size and huge specific surface area, which is good for adsorbing bacterium and high-flux for mass transfer. The layers are introduced as follows (figure 1): (1) cinder, average density of 1500 kg m^{-3}, lower strength, larger porosity; (2) brick, light weight, porous; (3) broken concrete block, average density of 2890 kg m^{-3}, irregular blocky of 2-4 cm, (4) biological ceramsite, average particle size of 5 mm, average density of 1500 kg m^{-3}, specific surface area of 4.99 cm^2 g^{-1}; (5) gravel, particle size of 2-4 cm.

IOP Conf. Series: Earth and Environmental Science **82** (2017) 012067 doi:10.1088/1755-1315/82/1/012067

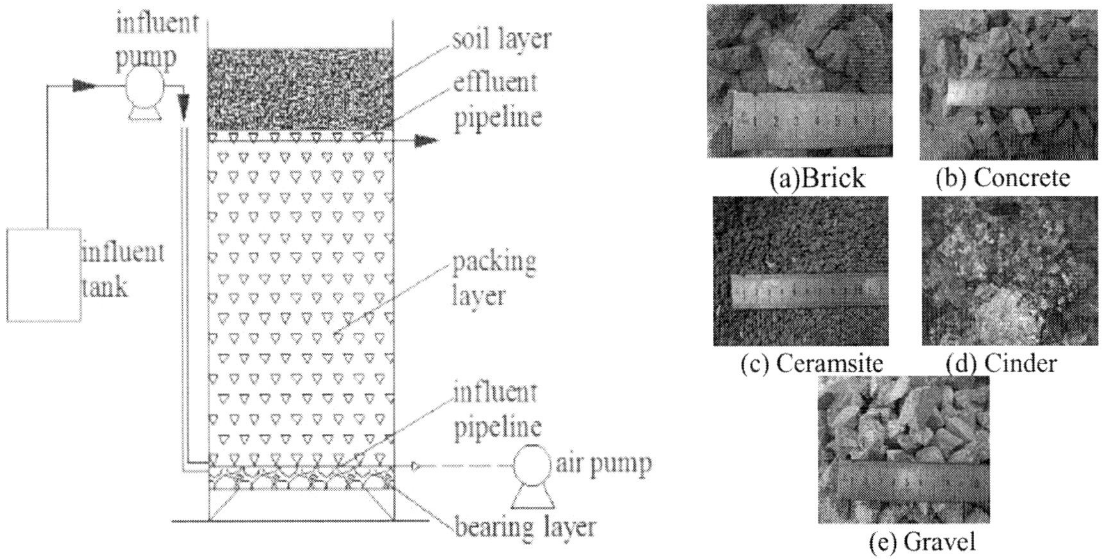

Figure 1. Diagram of experimental CW (up-flow reactor) and comparison between types of packing medium.

When the reactor is set as down-flow CW, the influent pipeline and effluent would be changed to effluent pipeline and influent, respectively. And the air pump is used for aeration which is needed in the aerobic CW. While for the anaerobic CW, the air pump is leaved unused.

2.2. Synthetic wastewater and seed sludge
The high-strength sewage and sewage was synthetic wastewater, which was mainly composed of glucose, carbamide, $(NH_4)_2CO_3$, KH_2PO_4, K_2HPO_4, $CaCl_2$, $CoCl_2$, $NiCl_2$, $ZnCl_2$, $MgSO_4$, $FeSO_4$, and $NaHCO_3$. The secondary effluent wastewater and all seed sludge that was used in the experiment came from Qingdao Haibohe WWTP.

2.3. Chemical analysis
Samples were taken from the influent tank and effluent pipe each day. The concentration of NH_4^+-N, TN, TP, COD_{Cr} and BOD were measured by using the standard method issued by the Environmental Protection Agency (EPA) of China. The pH was detected by glass electrodes connected to a METTLER TOLEDO F25 instrument. The dissolved oxygen was quantified by a HACH HQ30d apparatus.

3. Results and discussion

3.1. High-strength sewage treatment using an anaerobic bio-filter
Two anaerobic up-flow bio-filters, in which media were ceramsite and grave respectively, were used for treating high-strength wastewater and was seeded with activated sludge. The whole experiment was divided into startup stage and formal operation stage. The startup stage wastewater mainly contained approximately 220 mg l^{-1} of glucose, 40 mg l^{-1} of carbamide, 29 mg l^{-1} of $(NH_4)_2CO_3$, and 21 mg l^{-1} of KH_2PO_4. After the startup stage, which lasted about 55 days, the constituents in the synthetic wastewater were as follows: 1700 mg l^{-1} of glucose, 300 mg l^{-1} of carbamide, 217 mg l^{-1} of $(NH_4)_2CO_3$, 217 mg l^{-1} of K_2HPO_4, 175.5 mg l^{-1} of KH_2PO_4, 23.4 mg l^{-1} of $CaCl_2$, 90 mg l^{-1} of $MgSO_4$, 2.2 mg l^{-1} of $FeSO_4$, 0.1 mg l^{-1} of $CoCl_2$, 0.1 mg l^{-1} of $NiCl_2$, 0.1 mg l^{-1} of $ZnCl_2$, 0.12 mg l^{-1} of $MnSO_4$. In the startup period, influent concentration was approximately 220 mg l^{-1} of COD. In formal operation stage, influent components increased to 1500 mg l^{-1} of COD, 900 mg l^{-1} of BOD and 110 mg l^{-1} of NH_4^+-N.

At the beginning of startup stage the flow rate was 20 L d^{-1}. The COD removal rates of two bio-filters began to rise sharply after day 11 and tended to be stable on day 15 (figure 2). The average COD removal rates in ceramsite and gravel bio-filters were 74.2% and 78.4%, respectively. Subsequently, flow rate was increased to 30 L d^{-1} on day 20 and to 40 L d^{-1} (0.075 kg-COD $(m^3{\cdot}d)^{-1}$ of OLR) on day 35. COD removal rates were maintained at 79.3% for the ceramsite bio-filter and 77.4% for gravel bio-filter.

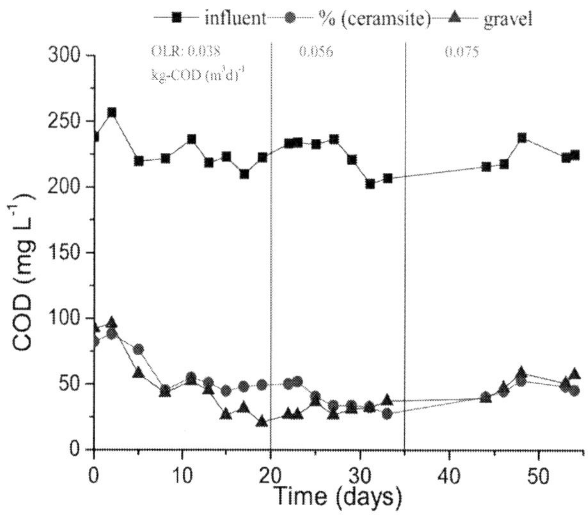

Figure 2. Startup period of anaerobic system.
The temperature is 14.2~26.5℃.

Table 1. Influence of OLR and medium on the removal rate (%) of pollution in an anaerobic reactor. The temperature is 19-23℃.

OLR (kg-COD $(m^3{\cdot}d)^{-1}$)		Startup			No alkalinity		Adding alkalinity		
		0.038	0.056	0.075	0.256	0.508	0.127	0.254	0.501
COD	ceramsite	74.2	83.5	79.3	26	28.5	83.0	78.5	73.7
	gravel	78.4	85.8	77.4	31	35.0	82.1	77.6	71.6
BOD	ceramsite	-	-	-	29.6	28.9	82.3	79.3	73.8
	gravel	-	-	-	35.2	34.9	81.5	78.6	72.1

After 55 days startup stage, the influent flow rate of wastewater was also increased as the organic material. The OLR at first was set 0.501 kg-COD $(m^3{\cdot}d)^{-1}$, six times higher than before. COD removal rates of ceramsite and gravel bio-filters were 28.5% and 35.0%, respectively (table 1). Even though OLR was decreased to 0.127 kg-COD $(m^3{\cdot}d)^{-1}$, removal rates of organic were far lower than the COD removal rates in the startup stage. Another phenomenon was that there was a net reduction in pH levels (figure 3). The influent pH was kept at approximately 7.5, but the effluent pH value decreased to only 6.2. Na_2CO_3 was added as alkalinity [22] to the influent on day 92. The Na_2CO_3 concentration varied in the experiment from 50 to 200 mg l^{-1}. The effluent pH of the two bio-filters was above 7. And the removal rates of organic recovered to above 80% in the two bio-filters after one month. As the OLR increases and HRT decreases, COD removal rate decreased from 83.0% to 73.7% for ceramsite bio-filter and from 82.1% to 71.6% for gravel bio-filter. These because that contact time between the sewage and the microbes on the biological film is shortened [23]. The trend of BOD removal rate was same with COD. When Na_2CO_3 is not added, the removal of contaminants by the gravel bio-filter is better than the removal of contaminants by the ceramsite bio-filter (figure 4). As the alkalinity increases, the difference in the removal rate of contaminants between the ceramsite bio-filter and the gravel bio-filter gradually decreases.

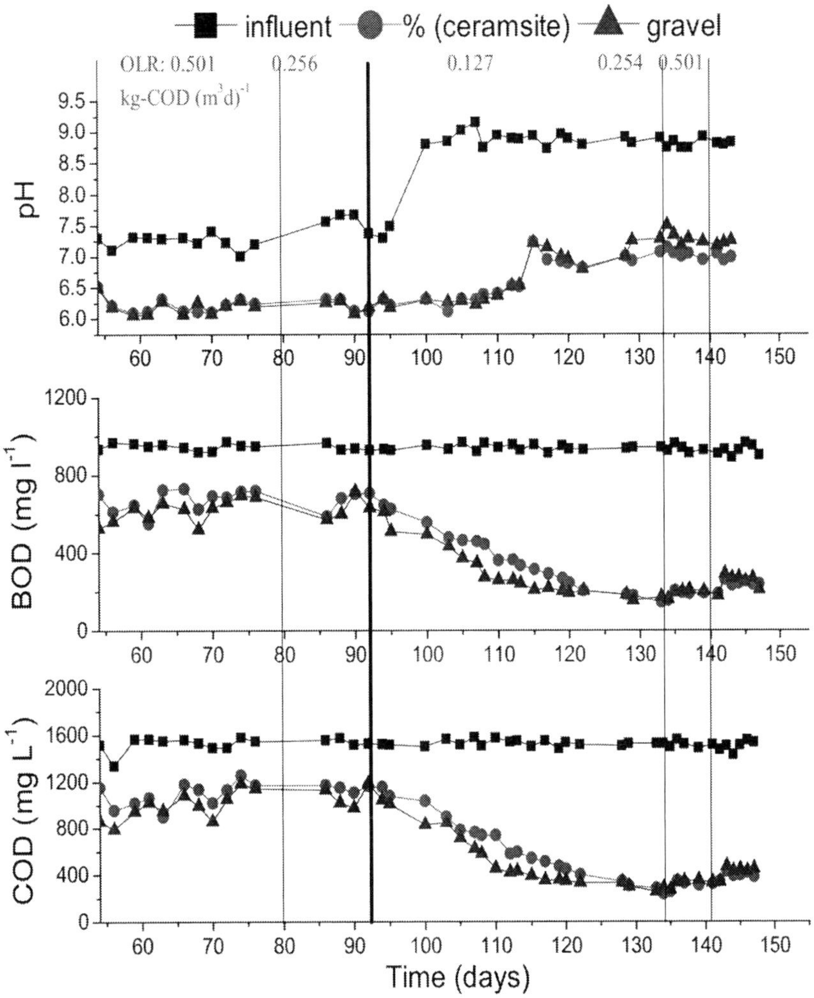

Figure 3. Overall performance of the anaerobic CW in the continuous experiment.
No Na_2CO_3 is added into the influent water before day 92. The addition of Na_2CO_3 is 200 mg l^{-1} after the day 92.
influent ceramsite gravel

In this experiment, the mechanism of using the bio-filter to remove substrate from the wastewater includes precipitation adsorption [16, 23], and the degradation of microbes in the biological film, where the function of the biological film is predominant.

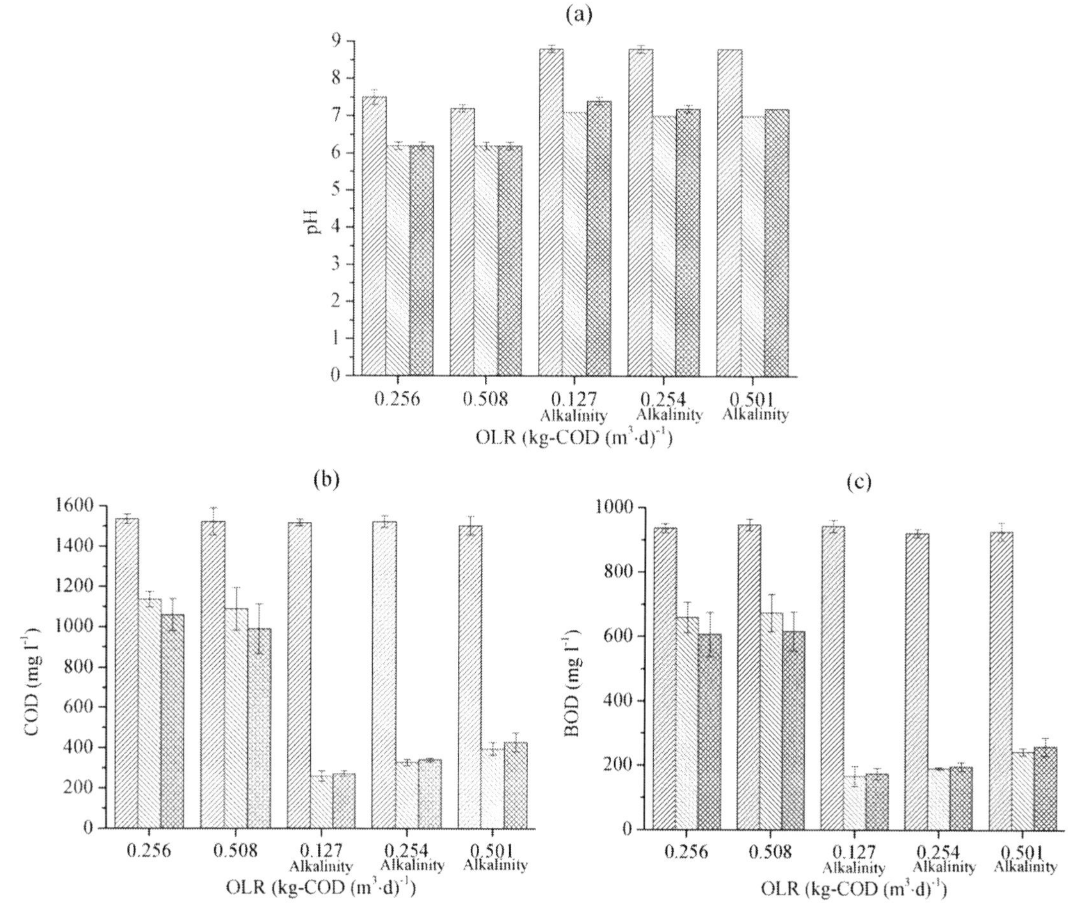

Figure 4. Influence of OLR and medium on the concentration of pollution in an anaerobic reactor. The temperature is 19-23℃.

3.2. Sewage wastewater treatment using an aerobic bio-filter

As one of the most popular domestic sewage treatment methods for small communities in rural regions, tow aerobic bio-filters with gravel as the media were used for treating municipal wastewater. The reactors were fed continuously in up-flow and down-flow modes. The air was pushed into the reactor by a micro-bubble air diffuser and the air flow rate was controlled by an air flow-meter that could provide sufficient oxygen (>2 mg/L) to ensure that the nitrifying bacteria was present. The activated sludge from the Qingdao Haibohe WWTP was further cultured during the startup stage. For continuous culture, the concentrations of influent COD and NH_4^+-N in feed water were 200-500 mg l^{-1} and 30-50 mg l^{-1}, and the flow rate was 20 L d^{-1}. The constituents in the synthetic wastewater were glucose, carbamide, $(NH_4)_2CO_3$, KH_2PO_4, K_2HPO_4, $CaCl_2$, $MgSO_4$, $FeSO_4$, and $NaHCO_3$. Through a pilot study of aerobic bio-filters in up-flow and down-flow bio-filter lands, we reviewed the influence of OLR and flow direction on the removal rate of COD and NH_4^+-N.

During the startup stage, the OLR of up-flow and down-flow bio-filters was 0.073~0.083 kg-COD $(m^3{\cdot}d)^{-1}$. After 10 days operation, the effluent COD and NH_4^+-N of two bio-filters were stable, approximately 50 mg l^{-1} and 24 mg l^{-1}, respectively (figures 5 and 6). The removal rate of substances in the aerobic up-flow and down-flow bio-filters was shown in table 2. As the OLR increased, the COD and NH_4^+-N removal rates in the aerobic bio-filters were improved. When the OLR was 0.089 kg-COD $(m^3{\cdot}d)^{-1}$, the COD removal rates were at the maximum, 96.4% and 97.5% for up-flow and down-flow bio-filters, respectively, and effluent NH_4^+-N was lowest, the removal rates are 52.4% and

80.7%. A negative feedback control system is contained to stabilize treatment efficiency. If the increment of the influent loading arises, the substrate concentration on a biofilm surface will increase correspondingly, so the depth from the film surface to which the substrate reaches, i.e., the depth of the effective layer, will also increase, resulting in the suppression of the increment in substrate concentration, and vice versa [8, 24]. A moderating effluence on the treatment efficiency can also be expected even though the fluctuations of loading. However, when the OLR increased further, the substrates removal rates began to drop. When the OLR was increased to 0.226 kg-COD $(m^3 \cdot d)^{-1}$, the COD removal rates were reduced to about 84% and NH_4^+-N removal rates were 13.1% for up-flow bio-filter and 24.5% for down-flow bio-filter. The COD removal rate of down-flow bio-filter was almost same as that of up-flow bio-filter, and the NH_4^+-N removal of down-flow bio-filter was more efficient than that of up-flow bio-filter.

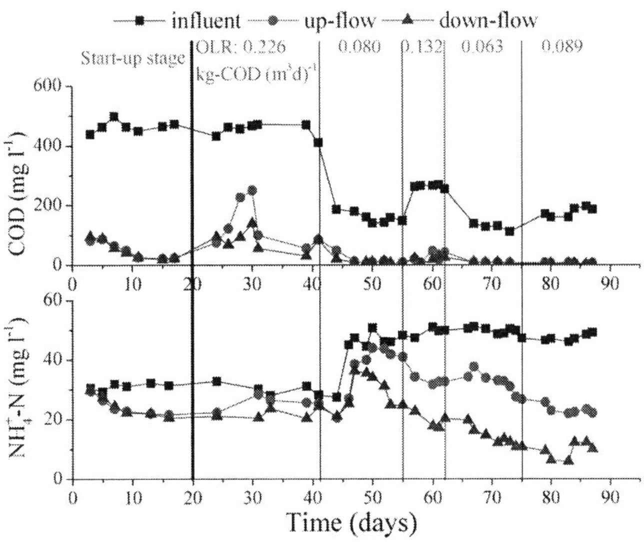

Figure 5. Overall performance of the aerobic CW in the continuous experiment.
When the OLR is 0.226 kg-COD $(m^3 \cdot d)^{-3}$, because the treatment effect of the aerobic up-flow and down-flow CWs is reduced sharply, and the system is backwashed. During operation in the remaining four stages, the system is not backwashed.

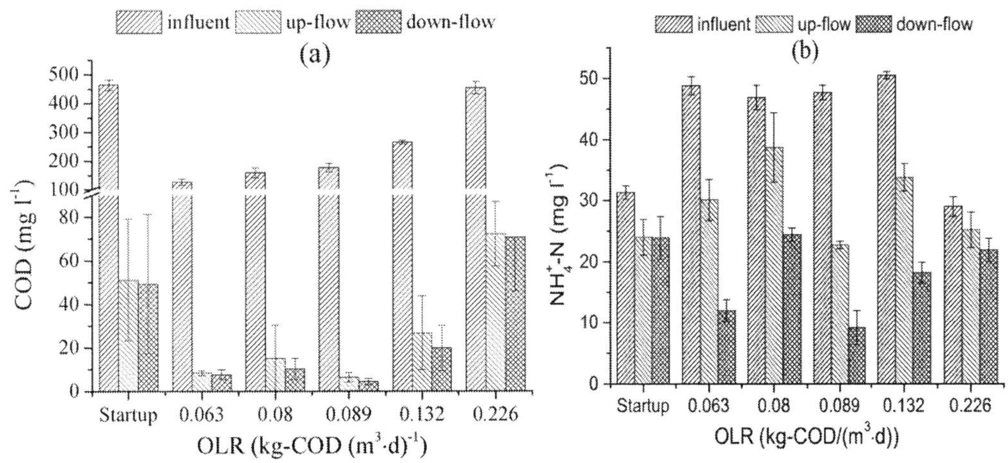

Figure 6. Effect of OLR on concentration of pollution in aerobic reactor.
The temperature is 19-26.8℃; pH is between 7.1-8.0; DO is above 2.0 mg l^{-1}.

Table 2. Effect of OLR on pollution removal rate (%) in aerobic reactor.

OLR (kg-COD $(m^3 \cdot d)^{-1}$)		Startup	0.063	0.080	0.089	0.132	0.226
COD	up-flow	89	93.4	90.5	96.4	89.9	84.1
	down-flow	89.4	93.9	93.5	97.5	92.5	84.4
NH_4^+-N	up-flow	23.3	38.3	17.5	52.4	33.1	13.1
	down-flow	23.6	75.4	48.0	80.7	64.0	24.5

The temperature is 19-26.8℃; pH is between 7.1-8.0; DO is above 2.0 mg l^{-1}.

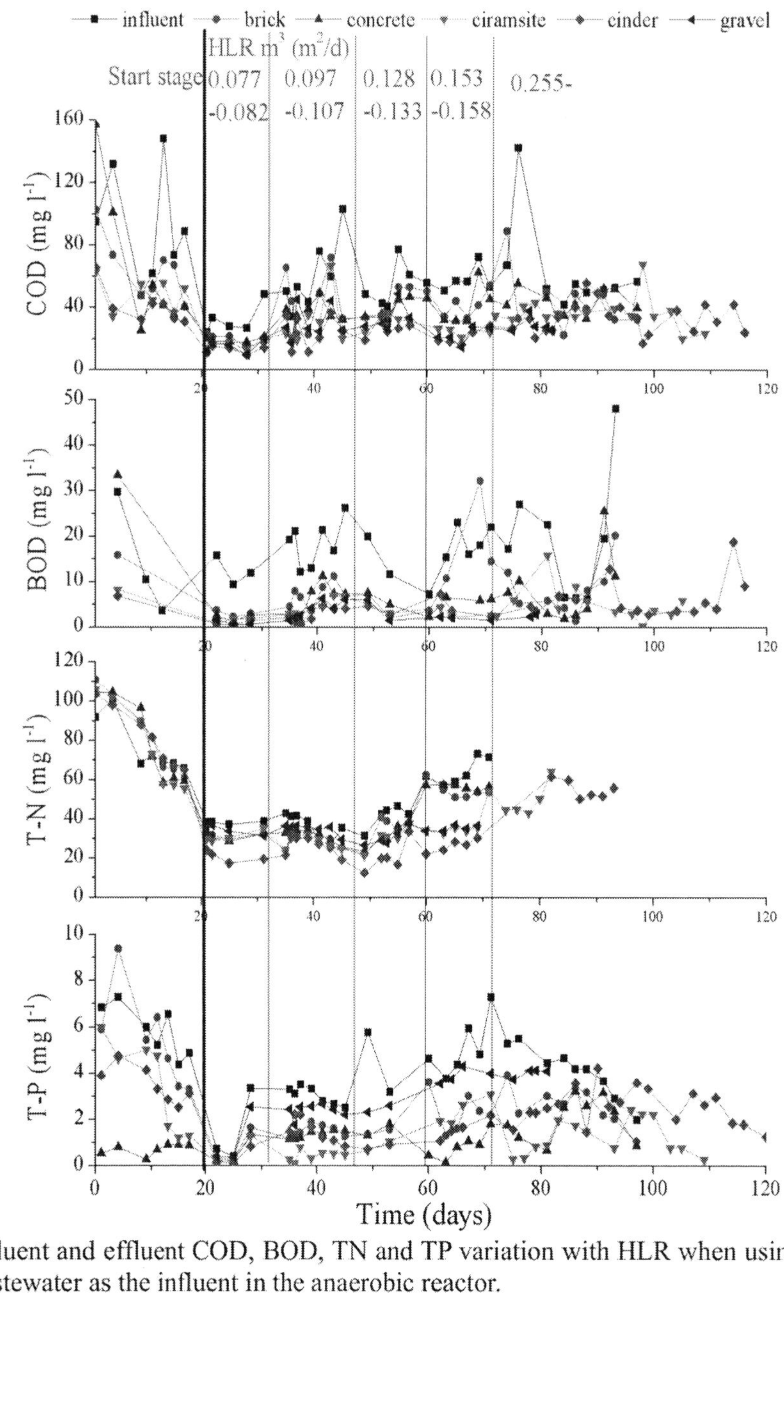

Figure 7. Influent and effluent COD, BOD, TN and TP variation with HLR when using the secondary treatment wastewater as the influent in the anaerobic reactor.

3.3. Secondary effluent wastewater post-treatment using bio-filter

Five anaerobic up-flow bio-filters were applied for treating secondary effluent wastewater with filter media of gravel, cinder, ceramsite, concrete block and brick. The influent wastewater came from settling tank effluent of Qingdao Haibohe WWTP, which was purified in this experiment. The average BOD_5/COD ration was below 0.3, meaning low biodegradability [25]. After the film formation was successful in the startup stage, the formal operation stage began.

In the startup stage, pH values were 6.6 ~ 8.2. The flow rate was 25 L d^{-1}, and the hydraulic retention time (HRT) was 115.2 h. Influent pollution concentrations varied strongly, 86.2±37.9 mg l^{-1} of COD, 14.6±13.6 mg l^{-1} of BOD, 76.7±13.8 mg l^{-1} of TN and 5.9±1.1 mg l^{-1} of TP. The startup stage went on day 20.

After startup stage, the system entered the formal operation stage (figure 7). The amounts of removed substrates in different kinds of HRT were shown in figure 8 and table 3. When the hydraulic load was 0.077~0.270 m^3 $(m^2{\cdot}d)^{-1}$, except for individual effluent water from the brick, substrate concentrations of effluent were less than 40 mg l^{-1} COD and 8 mg l^{-1} BOD. As the hydraulic load was increased, the ability of all bio-filters in removing the contaminants was reduced. When the HLR was 0.077~0.082 m^3 $(m^2{\cdot}d)^{-1}$, the effluent COD and BOD of all bio-filters were at the maximum and decreased as the hydraulic load increased. The COD average removal rates of five bio-filters are less than 60% in 120 days. The COD removal rate of gravel bio-filter was the highest, up to 52.8%. The average removal rates of cinder and ceramsite bio-filters were 51.0% and 44.8%, respectively. The concrete and brick bio-filters' efficiencies were only 33.2% and 28.6%, respectively. For five bio-filters, the BOD average removal rate of gravel bio-filter was the highest, reaching 83.9%. The next were 76.3%, 74.4% and 65.3% for ceramsite, cinder and concrete bio-filters. The BOD removal rate of brick bio-filter was the least, only 53.5%.

As a whole, TN removal rates of all bio-filters were very low in 120 days. The cinder had the best TN removal rate, and the average removal rate was 48.1%. The ceramsite was the next, and the average removal rate was 33.1%. The average TN removal rate for gravel was 27.4%. For brick and concrete, the average removal rate was less 20%. Influent water contains only nitrate nitrogen and the trend of TN removal rate was in accordance with trend of nitrate removal [26]. Complementary organic matter and suspended solids removal can be achieved in the bio-filters as the same time, but denitrification is not always obtained (even very low loading rates are applied). The process leading to TN removal is mostly bacterial denitrification transformations, under the anaerobic conditions [27]. The denitrification efficiency of treating high-nitrate waters with low organic carbon has been shown that it depends on C:N ration, and the peak efficiency occurs at C:N ration>5:1 [28-30]. So the total

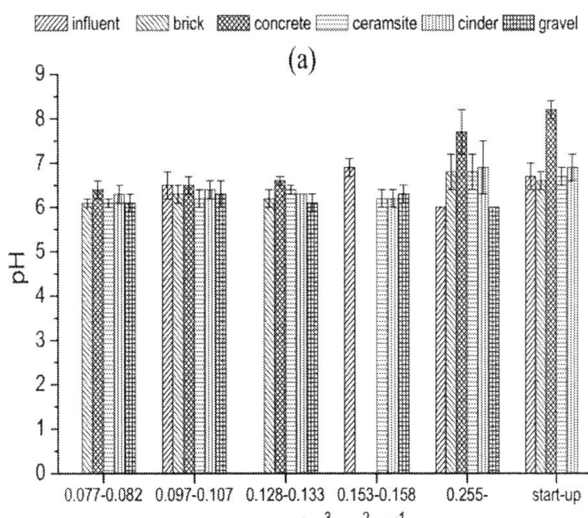

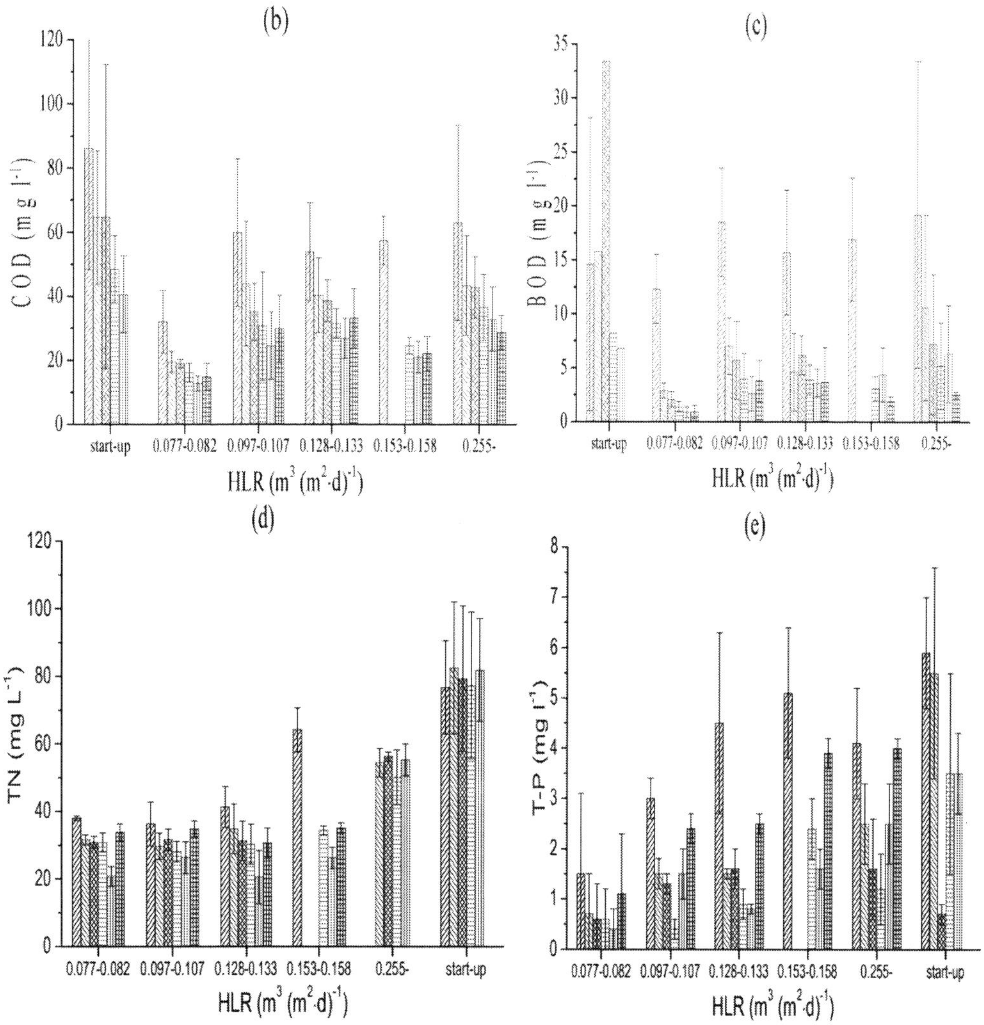

Figure 8. Concentration of pollution with HLR when using the secondary treatment wastewater as the influent in the anaerobic reactor. The temperature is between 22-30℃.

Table 3. Removal rate (%) of pollution with HLR when using the secondary treatment wastewater as the influent in the anaerobic reactor.

HLR (m^3 (m^2·d)$^{-1}$)		Startup	0.077-0.082	0.097-0.107	0.128-0.133	0.153-0.158	0.255-
COD	brick	25.1	38.8	26.6	25.4	-	31.1
	concrete	24.8	40.9	41.3	28.4	-	32
	ceramsite	43.7	49.7	48.7	41.7	57.2	42
	cinder	52.9	60.1	59	50.2	63.5	47.8
	gravel		53.7	50.2	38.7	61.6	54.5
BOD	brick	-8.2	76.3	62.1	70.7	-	44.6
	concrete	-111	82.9	69.5	60.4	-	62.4
	ceramsite	75.4	88.6	78.2	75.1	81.6	73.1
	cinder	17.1	93.1	85.9	76.9	74.1	67.1

	gravel	-	92.5	79.4	76.5	88.5	86.8
TN	brick	-7.6	16.9	18.2	15.9	-	-
	concrete	-3.4	18.7	13	24.2	-	-
	ceramsite	-0.8	18.7	22.6	26.7	46.4	-
	cinder	-6.8	45.7	27.7	50.5	59	-
	gravel	-	11.1	4.2	25.9	45.2	-
TP	brick	6.4	50.1	50.5	67.3	-	38.4
	concrete	87.8	60.4	55.6	65.3	-	61.7
	ceramsite	40.1	61.5	85.3	81	53.8	69.8
	cinder	40.1	75	49.3	81.6	68.7	37.7
	gravel	-	25.8	21.7	45.2	24.1	1.8

The temperature is between 22-30℃.

nitrogen reduction depended on the condition of the secondary effluent. However, the organic content in the influent is sufficiently low and not easily biodegradable. The bulk DO profiles decreases along height of column and the highest DO concentration was achieved at the bottom. Therefore, the removal of BOD occurs mainly in the influent part instead of denitrification. Denitrification would be inhibited due to insufficient electron acceptors even other conditions are suitable. These are also reasons that no significant negative or positive correlation between HLR and purification efficiency were found.

The phosphorus removal rates of ceramsite and concrete bio-filters were the best, and the average values were 70.7% and 63.0%, possibly due to the higher specific area of the ceramsite and its capacity to adsorb TP [7]. The cinder and brick bio-filters' efficiency were 48.7% and 47.7%, respectively. The phosphorus removal rate for the gravel was the worst, and the average removal rate was 24.4%. The low flow rate leads to higher rate of phosphorus sorption due to long contact time. However, as table 3 calculated, the effect of HLR on the phosphorus removal wasn't significant and systematic differences detected, which partly was caused by the fluctuation of influent [31].

One reason for TP removal is that the necessary anaerobic and aerobic condition for phosphate accumulating organisms was not formed in this system, which is not the main function for TP removal in this study because influent COD is relatively low and enough biodegradable substrates aren't available [20]. The major function of phosphorus removal is strongly influenced by forced precipitation and adsorbing [32, 33]. The different conditions in the bio-filter with kinds of media caused the variance removal of TP. For chemical precipitation, it needs suitable metal ions, appropriate DO and pH conditions to form insoluble substances, such as hydroxyapatite, a calcium phosphate compound [34]. Otherwise, other studies have shown that some materials are not like static adsorbent, and they may change reactive surface to influence their ability of phosphorus removal [6]. In Healy's opinion, through a filter's lifetime, the adsorption capacity decreases as the phosphate sorption sites become saturated, which is a challenge for bio-filter maintain and long-term operation [35].

4. Conclusions

This study shows that biological filters are an alternative technology for removing nutrient material from both high and low-strength sewage and secondary effluent. The substrate removal rate increases significantly after Na_2CO_3 was added in the influent of anaerobic bio-filter. In the aerobic bio-filters for treating sewage, when the OLR was increased to 0.089 kg-COD $(m^3{\cdot}d)^{-1}$, the COD removal rates were at a maximum of 96.4% and 97.5% for up- and down-flow reactors, respectively. The removal rate of NH_4^+-N was reduced as the OLR is increased. The removal rate of substrate in the aerobic down-flow bio-filter was almost same as in the up-flow bio-filter.

When bio-filters with gravel, ceramsite or cinder as media were used for treating secondary effluent, more than 70% of BOD could be remove. But TN removal was poor. Along the entire bio-filter profiles, gravel and cinder bio-filters had a higher ability to remove organic materials; the ceramsite bio-filter exhibits better performance in phosphorus removal. The application of bio-filter should be considered to be an efficient alternative for reducing the amount of various wastewaters discharged into aquatic ecosystems.

Acknowledgments
This project was supported by BUCT Fund for Disciplines Construction and Development (Project No. XK1506). The authors gratefully acknowledge the trust and support of all the members of the research team.

References

[1] Alcalde L, Folch M, Tapias J C, Martínez F, Enguídanos S and Bernácer I 2008 Secondary effluent reclamation: Combination of pre-treatment and disinfection technologies *Water Sci. Technol.* **57** 1963-8
[2] Toscano A, Langergraber G, Consoli S and Cirelli G L 2009 Modelling pollutant removal in a pilot-scale two-stage subsurface flow constructed wetlands *Ecol. Eng.* **35** 281-9
[3] Dotro G, Castro S, Tujchneider O, Piovano N, Paris M, Faggi A, Palazolo P, Larsen D and Fitch M 2012 Performance of pilot-scale constructed wetlands for secondary treatment of chromium-bearing tannery wastewaters *J. Hazard. Mater.* **239–240** 142-51
[4] Rebah F B, Kantardjieff A, Yezza A and Jones J P 2010 Performance of two combined anaerobic–aerobic biofilters packed with clay or plastic media for the treatment of highly concentrated effluent *Desalination* **253** 141-6
[5] Mollaei J, Mortazavi S B and Jafari A J 2015 Applying moving bed biofilm reactor for removing linear alkylbenzene sulfonate using synthetic media *I. J. H. E.*
[6] Pratt C and Shilton A 2009 Suitability of adsorption isotherms for predicting the retention capacity of active slag filters removing phosphorus from wastewater *Water Sci. Technol.* **59** 1673-8
[7] Sabbah I, Baransi K, Massalha N, Dawas A, Saadi I and Nejidat A 2013 Efficient ammonia removal from wastewater by a microbial biofilm in tuff-based intermittent biofilters *Ecol. Eng.* **53** 354-60
[8] Iwai S and Kitao T 1994 *Wastewater Treatment with Microbial Films* (Lancaster, USA: Technomic Publishing Company)
[9] de la Varga D, Díaz M A, Ruiz I and Soto M 2013 Avoiding clogging in constructed wetlands by using anaerobic digesters as pre-treatment *Ecol. Eng.* **52** 262-9
[10] Álvarez J A, Ruíz I and Soto M 2008 Anaerobic digesters as a pretreatment for constructed wetlands *Ecol. Eng.* **33** 54-67
[11] Comino E, Riggio V A and Rosso M 2013 Constructed wetland treatment of agricultural effluent from an anaerobic digester *Ecol. Eng.* **54** 165-72
[12] Lin Y, Yin J, Wang J and Tian W 2012 Performance and microbial community in hybrid anaerobic baffled reactor-constructed wetland for nitrobenzene wastewater *Bioresource Techn.* **118** 128-35
[13] Dai X, Chen C, Yan G, Chen Y and Guo S 2016 A comprehensive evaluation of re-circulated bio-filter as a pretreatment process for petroleum refinery wastewater *J. Environ. Sci. China* **50** 49-55
[14] Hamoda M F, Al-Ghusain I and Al-Mutairi N Z 2004 Sand filtration of wastewater for tertiary treatment and water reuse *Desalination* **164** 203-11
[15] Greenway M 2005 The role of constructed wetlands in secondary effluent treatment and water reuse in subtropical and arid Australia *Ecol. Eng.* **25** 501-9
[16] Hamoda M F, Al-Ghusain I and Al-Jasem D M 2004 Application of granular media filtration in wastewater reclamation and reuse *J. Environ. Sci. Heal.* **39** 385-95

[17] Qi W-K, Guo Y-L, Su L-M, Norton M, Qin Y and Li Y-Y 2014 An anoxic/oxic submerged constructed wetlands process for wastewater treatment: Modeling, simulation and evaluation *Ecol. Eng.* **67** 206-15
[18] Bustillo-Lecompte C F and Mehrvar M 2017 Treatment of actual slaughterhouse wastewater by combined anaerobic–aerobic processes for biogas generation and removal of organics and nutrients: An optimization study towards a cleaner production in the meat processing industry *J. Clean. Prod.* **141** 278-89
[19] Sakadevan K and Bavor H 1999 Nutrient removal mechanisms in constructed wetlands and sustainable water management *Water Sci. Technol.* **40** 121-8
[20] Yang K, He J, Dougherty M, Yang X and Li L 2009 Municipal wastewater treatment through an aerobic biofilm SBR integrated with a submerged filtration bed *Water Sci. Technol.* **59** 917-26
[21] Liu A and Liu S 2015 Study on performance of three backwashing modes of filtration media for oilfield wastewater filter *Desalin. Water Treat.* **57** 1-8
[22] Wan T, Zhang G, Du F, He J and Wu P 2014 Combined biologic aerated filter and sulfur/ceramisite autotrophic denitrification for advanced wastewater nitrogen removal at low temperatures *Front. Env. Sci. Eng.* **8** 967-72
[23] Riahi K, Mammou A B and Thayer B B 2009 Date-palm fibers media filters as a potential technology for tertiary domestic wastewater treatment *J. Hazard. Mater.* **161** 608-13
[24] Rittmann B E 1982 Comparative performance of biofilm reactor types *Biotechnol. Bioeng.* **24** 1341-70
[25] Li H S, Zhou S Q, Sun Y B, Feng P and Li J D 2009 Advanced treatment of landfill leachate by a new combination process in a full-scale plant *J. Hazard. Mater.* **172** 408-15
[26] Xiong J, Guo G, Mahmood Q and Yue M 2011 Nitrogen removal from secondary effluent by using integrated constructed wetland system *Ecol. Eng.* **37** 659-62
[27] Verhoeven J T A and Meuleman A F M 1999 Wetlands for wastewater treatment: Opportunities and limitations *Ecol. Eng.* **12** 5-12
[28] Vieira P C, Von Sperling M, Nogueira L C M and Assis B F S 2013 Performance evaluation of a novel open trickling filter for the post-treatment of anaerobic effluents from small communities *Water Sci. Technol.* **67** 2746-52
[29] Bilgin M, Şimşek İ and Tulun Ş 2014 Treatment of domestic wastewater using a lab-scale activated sludge/vertical flow subsurface constructed wetlands by using Cyperus alternifolius *Ecol. Eng.* **70** 362-5
[30] Mohan T V K, Nancharaiah Y V, Venugopalan V P and Sai P M S 2016 Effect of C/N ratio on denitrification of high-strength nitrate wastewater in anoxic granular sludge sequencing batch reactors *Ecol. Eng.* **91** 441-8
[31] Langenbach K, Kuschk P, Horn H and Kästner M 2009 Slow sand filtration of secondary clarifier effluent for wastewater reuse *Environ. Sci. Technol.* **43** 5896-901
[32] Gardner E, Morton D, Sands J, Matthews P, Cook F and Jayawardane N 2001 The filter system for tertiary treatment of sewage effluent by land application: its performance in a subtropical environment *Water Sci. Technol.* **42** 335-42
[33] Xu D, Xu J, Wu J and Muhammad A 2006 Studies on the phosphorus sorption capacity of substrates used in constructed wetland systems *Chemosphere* **63** 344-52
[34] Ke F, Li W C, Li H Y, Xiong F and Zhao A N 2012 Advanced phosphorus removal for secondary effluent using a natural treatment system *Water Sci. Technol.* **65** 1412-9
[35] Healy M G, Burke P and Rodgers M 2010 The use of laboratory sand, soil and crushed-glass filter columns for polishing domestic-strength synthetic wastewater that has undergone secondary treatment *J. Environ. Sci. Heal.* **45** 1635-41

IOP Conf. Series: Earth and Environmental Science 82 (2017) 012068 doi:10.1088/1755-1315/82/1/012068

Estimation of non-point source pollution loads by improvising export coefficient model in watershed with a modified planting pattern

F Dong[1,2], X B Liu[1,2,4], W Q Peng[1,2] and L Wang[3]

[1]State Key Laboratory of Simulation and Regulation of Water Cycle in River Basin, Beijing, China
[2]Department of Water Environment., China Institute of Water Resources and Hydropower Research, Beijing, China
[3]Bureau of Water Resources of Binzhou, Binzhou, China

E-mail: xbliu@iwhr.com

Abstract. Export coefficient model was improved to calculate and compare non-point source pollution loads in an agricultural watershed before and after implanting new cropping pattern. The modification was done by introducing the reduction coefficient in consumption amount and loss load as well as the proportion of bioactive ingredients of fertilizer and pesticide to the export coefficient model developed by Johnes in 1996. The modified export coefficient model was then applied to estimate non-point source pollution load in Gaoxi community, Yunnan Province, China where a water-saving and emission reduction technology was implemented by changing cropping pattern. Study results showed that the improved export coefficient model had a favorable flexibility in calculating the non-point source pollution loads and well applicable to the watersheds where various input data is in short. Moreover, the findings will provide scientific basis to understand the variability of non-point source pollutants in agricultural watersheds and their load estimation in order to optimize the efficiency of pollutants reduction plan implemented through agricultural adjustment.

1. Introduction

Pollutant discharged from agricultural non-point source (NPS) is one of the main causes of surface water pollution [1]. In developed countries like USA, agriculture contributes about 68-83% of NPS pollution load in surface waterbodies. Currently pollution load particularly from agricultural source become one of the main factors responsible for water pollution in some parts of China [1].

NPS pollution being diffuse in nature, it is difficult to control or manage unlike point-source pollution such as industry and sewage water. Therefore quantitative estimation of NPS pollutants load from agricultural watershed is an important tool in pollution management plan. Various simulation methods were applied in calculating NPS pollutant loads during the period of 1960-1970 in USA including physically based model and empirical model. Traditionally, physically based model has been applied to predict and evaluate the impact of land use change on water quality. However, a limited cognitive ability has made it difficult to overcome the complexity and randomness of NPS pollution during model construction. The physically based model has strict demand for input data, which leads the model construction costly and difficulty in calibration in large watershed [2]. At present, the main

IOP Conf. Series: Earth and Environmental Science 82 (2017) 012068 doi:10.1088/1755-1315/82/1/012068

methods for the calculation of NPS pollutant loads include throughput analysis, export coefficient model (ECM) and non-point source mechanism, and so on [2-4]. ECM is widely accepted model to estimate nutrient such total nitrogen (TN) and total phosphorus (TP) loads in North America, England and Wales [5]. The model is mainly dependent on field data, including coefficient of source intensity (pollutant production coefficient and pollutant emission coefficient), which can be calibrated according to rainfall, topography, transport loss and other factors. The migration and conversion property of pollutants in the aquatic environment of the watershed is also considered.

The model construction fee is less, and easy to operate [6-10]. This model has now become an effective model with the highest accuracy in estimating agricultural NPS load, especially for those watersheds area in short of long-term non-point surface pollution census data [4, 11].

The following study is focused on a watershed of Gaoxi community within Fuxian Lake basin in Yunnan province of China where a pilot project to reduce NPS load contribution through changing cropping pattern. Because, there is a lack of long-term numerical data in the study area, the ECM model is preferred to estimate the NPS pollutant load. ECM model is further modified according to the influence of NPS pollution load during pre- and post-project implementation. The objective of the study is to provide scientific basis to the understanding of the trend of agricultural NPS pollutants load in a watershed with the adjustment of agricultural pattern.

2. Model construction

2.1. Study area

The Gaoxi community situated at central Chengjiang county in Yunnan province and the north shore of Fuxian lake (102°52'24"~102°52'50" E, 24°40'40"~24°41'03" N), with an area about 12.53 km^2 and an altitude of 1755 m. The annual mean air temperature is 16.5℃ and the annual rainfall is about 928.3 mm. The rainfall distribution is extremely uneven showing clear distinction between dry and wet season. The wet season is extended from May to October that accounts for nearly 87% of the total annual rainfall, while the dry season (November to April) contributes only 13 of the total annual rainfall. The land is mainly covered by red soil. The land use in the watershed is dominated by agriculture. For example, during 2013 (before project implementation), the uncultivated farmland in Gaoxi community was 230.1 hm^2, including 216.7 hm^2 of paddy field and 13.3 hm^2 of dry field.

Gaoxi community watershed drains agricultural runoff to Fuxian lake. Fuxian lake located in Chengjiang county is the second deepest fresh water lake in the plateau of southwestern China and the recharge water mainly comes from the inflow rivers [12]. Although Fuxian lake water is categorized as Class I as per China National Water Quality Standard (CNWQS) but with the substantial change in land use and land cover in the watershed poses a threat of significant deterioration of water quality [13]. Study results showed that most of the inflow rivers are polluted and categorized as worse than CNWQS Class V [12]. Gaoxi community initiated a pilot project to reduce the pollution load on Fuxian lake by water-saving and pollution emission measures. The project site has 167.3 x 10^2 m^2 of farmland inhabited by 721 families and 2380 people. Currently, vegetable is the main crop in the project area. After project implementation, blue berry will become the main crop (118.7 x 10^2 m^2), followed by pea, green onion, Chinese chives and so on (48.7 x 10^2 m^2). This change will reduce agricultural NPS pollution loads through two pathways. Firstly, the high-efficient water-saving technology will effectively reduce the consumption of fertilizer and pesticides which will in-turn reduce the loss of agricultural NPS pollutants. Secondly, high-efficient water-saving and integrated water-fertilizer technology will increase the efficiency of fertilizer utilization and improve the migration and conversion process of agricultural NPS pollutants such as nitrogen (N) and phosphorus (P).

2.2. Modification of export coefficient model

The ECM model developed by Johnes in 1996 considered different export coefficients for each component of the land use such as type of crops, the number and distribution of livestock, f number of human population (responsible for emission and disposal of domestic wastewater), plant-fixed and atmospheric deposition of N to improve model sensitivity [14-16]. The mathematical expression of the ECM model is given in equation (1).

$$L = \sum_{i=1}^{n} E_i[A_i(I_i)] + p \tag{1}$$

In this equation L represents the total output load of a pollutant (kg/yr); E_i is the output coefficient (kg/km^2·yr) of pollutant source i; A_i is the area (km^2) occupied by land use type, or number of livestock type, or of people of source i; I_i is input of pollutants (kg) to source i; p is the input of pollutant (kg) from precipitation.

After project implementation, the cropping structure, irrigation pattern, application pattern and amount of fertilizer and pesticide were changed. Other impacting factor (e.g. number of livestock and people) were not changed significantly. The changes in cropping and irrigating pattern will lead to changes in the rate of fertilizer and pesticides, thereby affecting their loss rate. Therefore, we improvise the above ECM model (equation (1)) by considering the amount of fertilizer and pesticide and the proportion of bioactive ingredient in fertilizer and pesticide. The modified ECM model is as follows:

$$L = \sum_{i=1}^{n} E_i\left(1-\alpha_{i1}\right)\beta_i\gamma_i\left(1-\alpha_{i2}\right)A_i \tag{2}$$

where L = total output load of a pollutant (kg/yr); E_i = current consumption rate of fertilizer and pesticide in land i (kg/ km^2·yr); A_i = area (km^2) of land i; α_{i1} = consumption coefficient of fertilizer and pesticide; β_i = proportion of bioactive ingredient of fertilizer and pesticide applied in land i; γ_i = loss load coefficient of fertilizer and pesticide in land i; α_{i2} = loss coefficient of fertilizer and pesticide in land i.

2.3. Confirmation of export coefficient

The output coefficient was determined by literature search and experiment [15]. In the present study, the key components of the export coefficient model has been determined from field investigation, literature search and the annex 9 (analysis methods and partial reference parameters for surface source investigation) of the supplementary technical rules and regulations for water quality assessment of surface water and estimate of pollutant emission of water resources in China. Data concerning the consumption amount of fertilizer and pesticide were collected through on-site investigation.

According to our analysis, the main land use type in this study area was the conventional cropping pattern (crop and forest) irrespective of the high-efficient irrigation strategy. Main crops grown in the study area is vegetable, seedlings and fruits (blueberry and strawberry). The main pollutants include TN, TP and pesticides (organic phosphate and organic chloride).

2.3.1. Consumption coefficient t of fertilizer and pesticide (α_{i1}). For pre- project implementation period, the consumption amount of fertilizer and pesticide per unit area, Ei, was determined according to field investigation and the current reduction coefficient is considered as zero. For post- project implementation period, the consumption amount of fertilizer and pesticide per unit cultivation area was modified by the reduction coefficient in the consumption amount and was determined by literature search. Research results in different area showed that the consumption amount of fertilizer was reduced by 30-58% under the integrated water-fertilizer application technology compared to the conventional fertilization technology in the context of similar cropping pattern [16-18]. It was also suggested that the application of nitrogenous fertilizer in pearl, watermelon, soybean and corn was reduced by 23-30%, 66%, 27% and 22-31% respectively. Based on the above study results, we adopted a conservative strategy in studying the consumption amount of fertilizer in area under vegetable and seedlings production. The lowest value is considered as the reduction coefficient such as

20%.

When the integrated water-fertilizer application technology was used for vegetable cultivation, the amount of irrigation water becomes less. Due to the seasonal impact, evaporation and relative air humidity is reduced, the susceptibility of crops to disease is reduced, and ultimately there are fewer requirements for pesticide. It was reported that the consumption amount of pesticide per unit cultivation area was reduced by 15-30% when integrated water-fertilizer use strategy was applied [17]. According to the estimated amount of pesticide application in vegetable field and seedlings, we adopted a conservative strategy like fertilizer, and the lowest value of 15% was considered as the reduction coefficient.

2.3.2. Proportion of bio-active ingredient in fertilizer and pesticide. In the present study, we determined the proportion of bioactive ingredients in fertilizer and pesticide based on literature search and the annex 9 (analysis methods and partial reference parameters for surface source investigation) of the supplementary technical rules and regulations for water quality assessment of surface water and estimate of pollutant emission of water resources in China.

The common fertilizer applied in the study area is urea and compound fertilizer. Nitrogen (N) content in urea is 42%, while the compound fertilizer contains N, phosphorus (P) and potassium (K) in a proportion of 15:15:15. After conversion, 40% of the total amount of fertilizer applied was considered as the pure amount of fertilizer. This parameter is not affected by the water-saving irrigating technology, therefore, the pure amount of N and P before and after the implementation of the high-efficient water-saving technology can be calculated as follows:

α_i for N = (40% +42%) + (40% + 33.3%) = 30% and α_i for P = (40% + 33.3%) = 10%.

The type of pesticide used in the study area includes organic phosphate (60%), organic chloride (20%) and 20% of other kind of pesticide. The bioactive ingredient in organic phosphate was 14%, and that was 11% in the organic chloride. Similarly this coefficient is not changed by the water-saving irrigation strategy, the pure amount of N and P in pesticide before and after the implementation of the high-efficient water-saving technology can be calculated as follows,

β_i for organic phosphate or N (60%×14%) =8.4% and β_i for organic chloride or P (20%×11%) =2.2%,

2.3.3. Loss reduction coefficient (LRC) and loss load coefficient (LLC) of pollutants. According to literature search, the LRC of fertilizer and pesticide per unit area in the study site changed after the implementation of high-efficient water-saving irrigation technology. The data for loss load of fertilizer before project implementation is not available therefore the LLC of pollutant, expressed as γ_i , is determined as follows:

- LLCfor nitrogenous fertilizer: the absorption and utilization rate of nitrogen fertilizer in general crop plants is about 35%. About 65% of the nitrogen fertilizer is lost due to volatilization, leaching and leakage. The loss coefficient was calculated according to the lowest loss rate such as 20%.
- LLC for phosphorus fertilizer: the current absorption and utilization rate of phosphorus fertilizer in general plants is about 20%, and about 15% of phosphorus fertilizer is lost with water. In the present paper, the loss coefficient of phosphorus fertilizer is considered as 15%.
- LLC for pesticide: pesticide is lost mainly due to drift, volatilization, and evaporation in soil, plant and water. In particular, 25% of the consumption amount of pesticide is lost due to drift and volatilization, 3-5% due to soil evaporation and 10% due to plant evaporation. The LLC of pesticide is therefore 40%.

After project implementation, the loss load of pollutant was modified by the reduction coefficient of loss load, α_{i2}, and α_{i2} was determined by literature search. It was suggested that the integrated water-fertilizer technology can effectively reduce the consumption amount of fertilizer and pesticide and the loss load of pollutant [19]. The loss load of TN per unit yield for pear, soybean and corn in trickle fertilization was reduced by 45.2%～56.4%, 14.5%～49.7% and 26.3%～51.8% compared to the conventional irrigation, respectively. Research results showed that precise trickle fertilization can reduce the loss load of TN by 22.8~58.0% [20]. Compared to conventional irrigation, maintenance of surface soil and vegetation, mulching film coverage and precise trickle irrigation can reduce the loss load of TP in pear orchard and vegetable field by 20.6% and 12.6%, respectively [21]. It was also shown that precise trickle fertilization reduced the loss load of TP by 45.0%~75.0% [20]. The results mentioned above were summarized in table 1.

Table 1. Reduction in nutrient loss due to high-efficient water-saving irrigation strategy.

pollutants	reduction rate (%)	
TN	Minimum	14.5
	Maximum	58.0
TP	Minimum	12.6
	Maximum	75.0

When evaluating the loss rate of fertilizer in vegetable field and seedling bed, we adopted slightly conservative strategy, i.e. the lowest value documented in literature was considered as the loss rate of fertilizer, $TN\alpha_{i2}$ = 14.5% and $TP\alpha_{i2}$ =12.6% (table 1). Because the application pattern of pesticide was not changed significantly after project implementation, we didn't consider the effects of high-efficient water-saving measures on the loss load of pesticide.

2.3.4. Current state of fertilizer and pesticide consumption. The consumption amount of fertilizer and pesticide in Gaoxi community, Chengjiang county is determined through on-site investigation. The pure amount of fertilizer application in the project area is 2644.5 kg/ hm^2, which is much higher than the average consumption amount of fertilizer in China (434.3 kg/hm^2). The farmers of the project area were dependent on inorganic fertilizer which resulted in an increase in soil acidity. The mean amount of pesticide consumption in the project area is 44.7 kg/hm^2, while the mean consumption amount of pesticide in China is 13.4 kg/hm^2. The consumption amount of fertilizer and pesticides in cultivating blueberry and greening seedlings is relatively less than that of in vegetables production.

2.3.5. Adjustment in cultivation area. Before project implementation, crops grown in the 167.3 hm^2 of study area were mainly vegetables. After project implementation, 118.7 hm^2 of blueberry, 32 hm^2 of vegetable and 16.7 hm^2 of greening seedlings were grown in this 167.3 hm^2 of study area.

3. Results and discussions

The calculated results were shown in table 2.

Seen from table 2, it was clear that:

Before project implementation, the annual consumption amount of fertilizer and pesticide was 1106.06 t and 7.48 t, respectively, whereas, after project implementation, the total consumption amount of fertilizer and pesticide was 277.37 t/a and 3.17 t/a, respectively. The consumption amount of fertilizer and pesticide was significantly reduced by almost 75% and 60%, respectively.

Before project implementation, the total loss of TN, TP, Organo phosphate and organic chloride was 66.36 t/a, 16.59 t/a, 0.25 t/a and 0.07 t/a, respectively. After project implementation, the total loss of TN, TP, organic phosphate and organic chloride was 14.23 t/a, 3.64 t/a, 0.11 t/a and 0.03 t/a. The loss rate of TN, TP, organic phosphate and organic chloride was reduced by 78.6%, 78.1%, 57.6% and 57.6%, respectively.

The efficiency of reduction was shown in figure 1.

It was clear that after project implementation, the non-point source pollution loads resultant from the loss of fertilizer and pesticide tended to decrease. And the main driving force was the great adjustment in cropping structure and application of high-efficient water-saving irrigating pattern. After trial implementation, 32.0 hm^2 of the 167.3 hm^2 of vegetable field was reserved for vegetable

Table 2. Pollution load before and after project implementation.

Pollutants	Plant	Rate of fertilizer or pesticide (kg/hm^2·a)	Bioactive component proportion	Export coefficient	Loss quantity (kg/hm^2·a)	Area (hm^2)	Total Consumption of fertilizer or pesticide (t/a)	Total loss quantity (t/a)
TN	vegetables	6609.9	0.3	0.2	396.6	167.3	1106.06	66.36
TP			0.1	0.15	99.15			16.59
organic phosphorus		44.7	0.084	0.4	1.5		7.48	0.25
organic chloride			0.022	0.4	0.45			0.07
TN	vegetables	5284.5	0.3	0.17	271.05	32.0	169.1	8.68
	blueberry	900	0.3	0.17	46.2	118.7	106.8	5.48
	seedlings	88.2	0.3	0.17	4.5	16.7	1.47	0.08
sub-total					321.75	167.3	277.37	14.23
TP	vegetables	5284.5	0.1	0.13	69.3	32.0	169.1	2.22
	blueberry	900	0.1	0.13	11.85	118.7	106.8	1.4
	seedlings	88.2	0.1	0.13	1.2	16.7	1.47	0.02
sub-total					82.2	167.3	277.37	3.64
organic phosphorus	vegetables	35.7	0.084	0.4	1.2	32.0	1.14	0.04
	blueberry	15	0.084	0.4	0.45	118.7	1.78	0.06
	seedlings	15	0.084	0.4	0.45	16.7	0.25	0.01
sub-total					2.25	167.3	3.17	0.11
organic chloride	vegetables	35.7	0.022	0.4	0.3	32.0	1.14	0.01
	blueberry	15	0.022	0.4	0.15	118.7	1.78	0.02
	seedlings	15	0.022	0.4	0.15	16.7	0.25	0
total					0.6	167.3	3.17	0.03

cultivation, and the remaining 135.3 hm^2 of land was reserved for blueberry and seedlings that are characterized by low unit pollution load. Furthermore, the high-efficient water-saving irrigating technology was adopted in the experimental area. The utilization efficiency of water and fertilizer was effectively increased by integrated water and fertilizer management. Influenced by the integrated function of these two main reasons, especially the great adjustment in cropping structure, the consumption amount of fertilizer and pesticide as well as their loss rate was greatly reduced. This has significantly reduced the agricultural non-point source pollution loads in the study area.

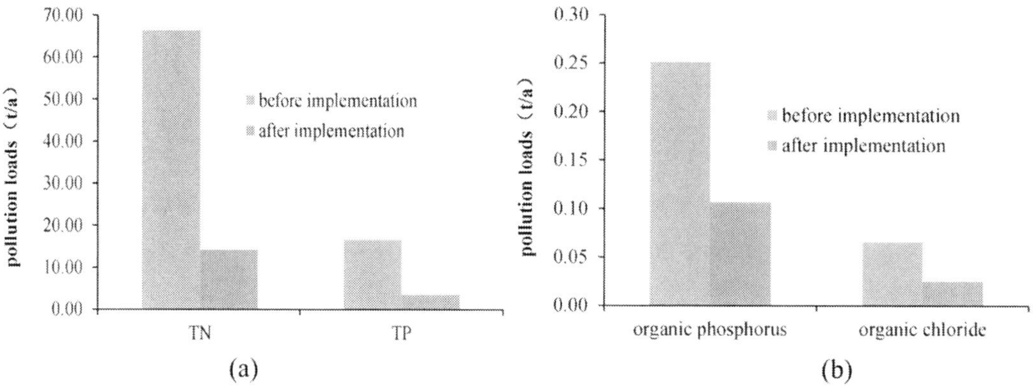

Figure 1. A comparison of pollution loads before and after project implementation.

4. Conclusions

The ECM model can't be used to calculate the non-point source load in the study area. We therefore constructed an output coefficient model to calculate the agricultural non-point source pollution loads before and after cropping pattern changes by analyzing the main factors that affect the non-point source load output before and after cropping pattern changes and introducing reduction coefficient of the consumption amount of fertilizer (pesticide), bioactive ingredients of fertilizer (pesticide), loss road coefficient of fertilizer (pesticide) in soil and reduction coefficient of the loss load in fertilizer (pesticide).

We constructed an output coefficient model based on our study and calculated the reduction amount in pollution load before and after project implementation. The results showed that the changes in cropping structure, irrigating patter, and the implication pattern as well as implication amount of pesticide had significant effects on non-point source load. TN, TP, organic phosphorate and organic chloride were reduced by 78.6^, 78.1%, 57.6% and 57.6%, respectively.

The modified output coefficient model is well suitable to the comparison and calculation of the agricultural non-point source pollution loads before and after cropping pattern change. It also showed that the modified output coefficient model has good flexibility in calculating the non-point source pollution loads. In the modified model, the main impacting factors or coefficient can be introduced freely.

Acknowledgments

This work was funded by the National Natural Science Foundation of China (No. 51479219) and the Major Science and Technology Program for Water Pollution Control and Treatment of China (No. 2013ZX07501-004 & 2017ZX07101-004).

References

[1] Ma D M, Shi H H, Feng A P, *et al* 2014 Estimation of agricultural non-point source pollution based on watershed unit: A case study of Laizhou Bay *Acta Ecologica Sinica* **34** 173-81 (in Chinese)

[2] Li H E and Li J K 2013 *Methods and Application of Quantify for Watershed Non-point Source Pollution Loads* (Beijing: Science Press) (in Chinese)

[3] Ongley E D, Xiaolan Z and Tao Y 2010 Current status of agricultural and rural non-point source Pollution assessment in China *Environ. Pollut.* **158** 1159-68

[4] Liu Z, Chao J Y, Zhang L, *et al* 2015 Current status and problems of non-point source pollution loads calculation in China *Adv. Water Sci.* **26** 432-42 (in Chinese)

[5] Johnes P J 1996 Evaluation and management of the impact of land use change on the nitrogen and phosphorus load delivered to surface waters: the export coefficient modelling approach *J.*

Hydrol. **183** 323-49

[6] Li H E, Zhuang Y T, *et al* 2003 The export coefficient modeling approach for load prediction of nutrient from nopoint source and its application *J. Xi'an Univ. Technol.* **19** 307-12 (in Chinese)

[7] Xue L H and Yang L Z 2009 Research advances of export coefficient model for non-point source pollution *Chinese J. Ecol.* **28** 755-61 (in Chinese with English abstract)

[8] Xu L H, Chen C G, Hu B W, *et al* 2015 Improvement of export coefficient model for N and P based on rainfall intensity and its application *Transactions of the CSAE* **31** 159-66 (in Chinese)

[9] Shen Z, Liao Q, Hong Q, *et al* 2012 An overview of research on agricultural non-point source pollution modelling in China *Separation and Purification Technology* **84** 104-11

[10] Shrestha S, Kazama F, Newham L T H, *et al* 2008 Catchment scale modelling of point source and non-point source pollution loadss using pollutant export coefficients determined from long-term in-stream monitoring data *J. Hydro.-Environ. Res.* **2** 134-47

[11] Zhang H B, Li J, Li X D, *et al* 2013 Study on rural non-point source pollution assessment method of regions with sparse data *J. Sichuan Univ. (Eng. Sci. Edi.)* **45** 58-66 (in Chinese)

[12] Wang Q, Wu X, Zhao B, Qin J and Peng T 2015 Combined multivariate statistical techniques, water pollution index (WPI) and Daniel Trend Test methods to evaluate temporal and spatial variations and trends of water quality at Shancheng river in the north-west basin of Lake Fuxian, China *PLoS One* **10** 1-17

[13] Yan D X, Qing Z Y, Chun M W and Guo Z L 2017 Influence of spatial variation in land-use patterns and topography on water quality of the rivers inflowing to Fuxian Lake, a large deep lake in the plateau of south-western China *Ecol. Eng.* **99** 417-28

[14] Cai M, Li H E, Zhuang Y T, *et al* 2004 Application of modified export coefficient method in polluting load estimation of non-point source pollution *J. Hydraul. Eng.* **7** 40-5 (in Chinese)

[15] Ma G W, Wang Y Y, Xiang B, *et al* 2011 Diversity characteristic and pollution load of non-point source total nitrogen and total phosphorus in Songhua River Basin *Transactions of the CSAE* **27(S2)** 163-9 (in Chinese)

[16] Locascio S J 2005 Management of irrigation for vegetables: Past, present, and future *Horttechnology* **15** 482-5

[17] Liu H C 2012 Effects of irrigation and fertilization on the growth, water and fertilizer utilization in ginger (Shandong Agricultural University)

[18] Liu J Y, Zhao H G, Zhang L Q, *et al* 2005 High-effective culture technique of un open cucumber with trickle irrigation of moisture and fertilizer integrated *Acta Agriculturae Boreali-Sinica* **20(S1)** 206-8 (in Chinese)

[19] Huang L H, Shen G X, Qian X Y, *et al* 2008 Impacts of drip fertilizer irrigation on nitrogen use efficiency and total nitrogen loss load *Transactions of the CSAE* **24** 49-53 (in Chinese)

[20] Guo C X, Shen G X, Huang L H, *et al* 2009 Control of soil salinization and reduction of N & P loss with drip fertigation in greenhouse *J. Agro-Environ. Sci.* **28** 287-91 (in Chinese)

[21] Qian X Y, Shen G X, Huang L H, *et al* 2010 Loss of soil phosphorus from rain-fed cropland and its affecting factors in dongtan of chongming *J Ecol. Rural Environ.* **26** 334-8 (in Chinese)

Flow resistance over submerged vegetation in ecological channel

Y Han[1,3], S Q Yang[2], L C Qiu[1] and N Dharmasiri[2]

[1]Collage of Water Resources & Civil Engineering, China Agricultural University, Beijing 100083, China
[2]School of Civil, Mining & Environmental Engineering, University of Wollongong, NSW 2522, Australia

E-mail: yhan@cau.edu.cn

Abstract. A discussion of flow resistance in ecological channel is a fundamental problem required by hydraulic engineers, but the well-established flow resistance mechanism is still not well discovered. This research studies on flow over submerged vegetation in ecological channel to explain mechanism of flow resistance on rough surfaces. With a detailed discussion, this research separates total resistance and unlike previous research, this study pays attention to the variation in vortices created by roughness, and new definitions of hydraulic radius and Reynolds Number are given. This study validated the proposed method can be used to simplify the friction factor calculations for hydraulic engineers.

1. Introduction

Artificial open channel are the major conveyance systems to deliver water into irrigation area. To improve the biological environment of traditional channels, some longitudinal vegetation is arranged into channel to create a proper ecological space, maintaining the biological diversity and enhance the ecological function of the irrigation channel [1, 2]. However, vegetation affect river hydraulic and its flow resistance plays an important role in an ecological channel. Flow over vegetation increases the flow resistance and further decrease the flow deliver capacity. Thus, it is important to determine the flow resistance in a channel partially covered with vegetation.

Flow over a channel partially covered with vegetation can be generalised as flow over roughness surfaces, and determination of flow resistance over roughness surfaces is a quite old topic [3]. Nikuradse [4] used two parameters to correlate the friction factor, f, i.e., Reynolds number (Re = UR / ν) and relative roughness, k_s / R, in which U = mean velocity, R = hydraulic radius, ν = kinematic viscosity, and equivalent roughness height k_s = d. Schlichting's [5] found that equivalent roughness height k_s = 0.627d when roughness packed tightly. However, Nikuradse's experimental result of k_s = d cannot be replicated in many cases, e.g., for sediment transport, k_s = 2.5d is widely used [6]. In other words, Wooding et al [7] provided detailed experimental data to describe k_s varies with roughness concentration λ, and the concept of k_s, D (pipe diameter) and δ (boundary layer thickness) was introduced by Perry et al [8], who suggested that k_s = 0.02δ. For the dependence of k_s on λ, many useful investigations have been done [9, 10].

Yang et al [11] proposed the flow resistance partitioning can be developed from the head loss (as shown in figure 1), that is

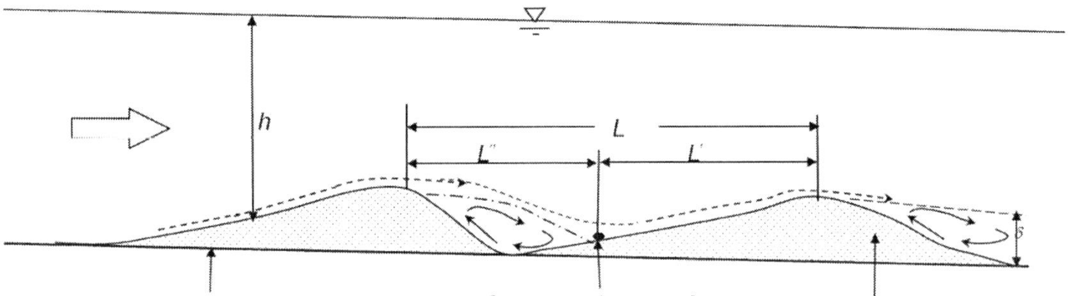

Figure 1. Combination of skin friction and form drag over a trough between roughness elements.

$$h_L = h_L^{'} + h_L^{''} \tag{1}$$

where h_L = energy loss over the length of L, h_L' = energy loss by skin friction over the length of L', and h_L" = energy loss by the large eddy over the length of L" (see figure 1). The energy slope can be determined accordingly as

$$S = \frac{h_L}{L} = S'\frac{L'}{L} + S''\frac{L''}{L} \tag{2}$$

where S = energy slope, S' = energy slope associated with skin friction, and S" = energy slope related with form drag. Alternatively, equation (2) can be rewritten as:

$$\tau_0 = \tau'\frac{L'}{L} + \tau''\frac{L''}{L} \tag{3}$$

In an effort to continue this work, this paper will investigate flow resistance over 3-D roughness distributions by extending equation (3) from the 1-D roughness depicted in figure 1. The research objectives of the paper are tried to interpret and redefine hydraulic radius followed by a new definition, so that the influence of roughness on the friction factor in laminar flows can be described.

2. New definition of hydraulic radius and friction factor in laminar flow

The hydraulic radius is one of the most important parameters in hydraulics and fluid mechanics. The conventional hydraulic radius definition has no physical implications as it only measures the area and the length. The hydraulic radius can be redefined as follows:

$$R = \frac{V}{A} \tag{4}$$

where V = fluid volume and A = the wetted boundary area. The ratio of V/A represents the potential energy contained in V dissipated as heat at boundary A. This can be demonstrated in a pipe flow, where the fluid volume in the shaded region is V, and the momentum balance equation can be written as follows:

$$\rho g V S = \tau_0 A \tag{5}$$

where ρ = fluid density and g = gravitational acceleration. Equation (5) shows that $\tau_0 = \rho gRS$, or R = V/A. It is obvious that for a pipe flow, equation (4) generates the same hydraulic radius as the conventional definition, but equation (4) implies that R is a measure of skin friction that depends on the area of solid-liquid interface area A.

IOP Conf. Series: Earth and Environmental Science 82 (2017) 012069 doi:10.1088/1755-1315/82/1/012069

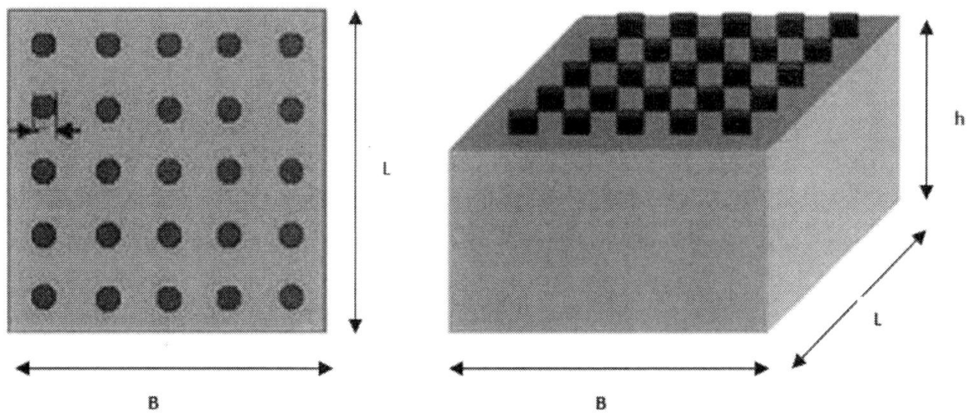

Figure 2. Sketch of flow over rigid vegetation.

Flow over rigid vegetation without sidewall effects presented in figure 2, modelled as submerged cylinders. The control volume can be calculated as $B \times L \times h$, and distance between submerged cylinders is Δx. If the number of vegetation in a flow control volume can be considered as N, the vegetation density (λ) per unit area could be;

$$\lambda = \left(\frac{\frac{\pi D^2}{4} N}{BL} \right) \tag{6}$$

where D denote the cylinder diameter of rigid vegetation. λ can be defined in a similar manner for other shapes of roughness elements.

For strips and cubes:

$$\lambda = \left(\frac{L_1^{\,2} N}{BL} \right) \tag{7}$$

where L_1 is the cross sectional length/width of the strip/square.

For 2D dunes:

$$\lambda = \left(\frac{N L_2 b}{BL} \right) \tag{8}$$

where L_2 = streamwise length of the dune, b= width of the dune.

3. Determination roughness concentration reynolds number in laminar flow

This new definition of the hydraulic radius is the same as the conventional hydraulic radius of open-channel flows without vegetation. Cheng [12] and Cheng et al [13] calculate flow resistance considered vegetation density and calculated the ratio of the cross-section area to the perimeter. Therefore, considering the aquatic plant as a cylinder, when interpreting the bed area for the form drag force, it is not consider the full contact area. Hence then the roughness concentration-related hydraulic radius for a cylindrical stem reasonably can be defined as:

$$r' = \frac{\rho g V_{water} S}{A_{water}} = \frac{(1-\lambda)BhL}{(1-\lambda)BL + NhD} = \frac{(1-\lambda)h}{\left(1+\lambda\left(\frac{4\lambda}{\pi D}h_V - 1\right)\right)} \tag{9}$$

where V water is potential energy volume in channel and A water is potential energy dissipated on the area, h_v is the height of vegetation. It should consider the " r' " is the concentration length.

With the Roughness Concentration related hydraulic radius, the roughness concentration Reynolds number can be defined as follows.

$$\mathrm{Re}_*^{'} = \frac{U_*^{'} r'}{\nu} \tag{10}$$

where we name $U_*^{'} = gr'S$ as the "roughness concentration related friction velocity.

In comparison with the previous studies, the use of r' in Reynolds number is novel. As shown subsequently, r' performs much better than other length scales, such as h, D and L for collapsing total friction factor data from disparate sources.

4. Co-existence of form drag and skin friction for a flow with submerged vegetation simulated as cylinders

Figure 3 shows a two dimensional bed form, in which L = length between two adjacent vegetation zone; h_v= vegetation height; h = flow depth; L"= length of flow separation zone caused by vegetation; and L'= skin friction length. The length L' is the resistance produced by the boundary surface to overcome the no slip condition which occurs as a result of the solid-water interface. Thus, the total energy loss over vegetation:

$$F = F'\frac{L'}{L} + F''\frac{L''}{L} \tag{11}$$

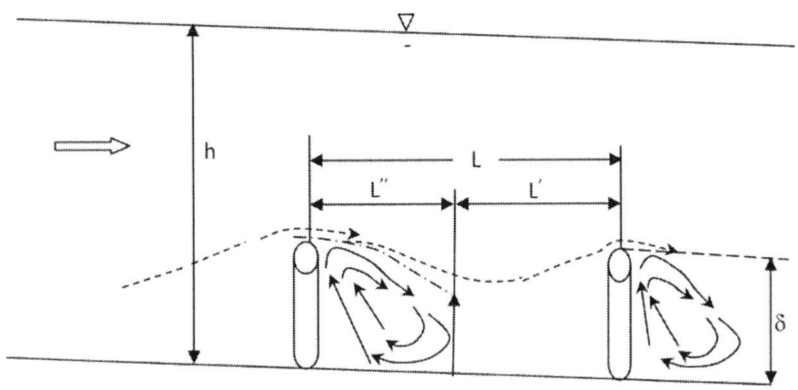

Figure 3. Flow separation behind submerged vegetation, simulated as cylinders.

Equation (11) is very similar to the fundamental equation which was developed for the mathematical model for pipe friction equation (1). So that now we have confirmed the basis for the open channel flow with discrete roughness, it is quite admissible that, if used with the correct variables, equation (11) should work fine in open channel flows. After making a qualitative connection between the pipe friction and open channel friction, it was determined in the previous section that, the pipe friction analytical model developed by Yang et al [14] can be extended for channel friction with discrete roughness.

Channels with bluff bodies usually undergo a very high turbulent condition. Therefore it is assumed that, the following equation developed in pipe friction in the fully developed turbulent region will be the stepping stone for further study regarding channel friction. This has been derived based on equation (11):

$$f = f' + \frac{L''}{L}\left(f'' - f'\right) \tag{12}$$

where $f = 8\left(\frac{U_*^{'}}{V}\right)^2$ = total friction factor, in which $U_*^{'}$ = concentration relation friction velocity and V = average velocity of the channel. f'' = friction factor in the L'', f' = friction factor in the L'. To make this an independent equation from the previous chapter, we have replaced the dimensionless eddy length (L''/L) with the dimensionless eddy volume (V_{eddy}/V_{void}) by simply multiplying by the flow area A. This generates the following equation

$$f = f' + \alpha\left(f'' - f'\right) \tag{13}$$

where $\alpha = \frac{V_{eddy}}{V_{void}}$, V_{eddy} denote the eddy volume, and V_{void} denote the fluid volume between adjacent roughness.

5. Discussion

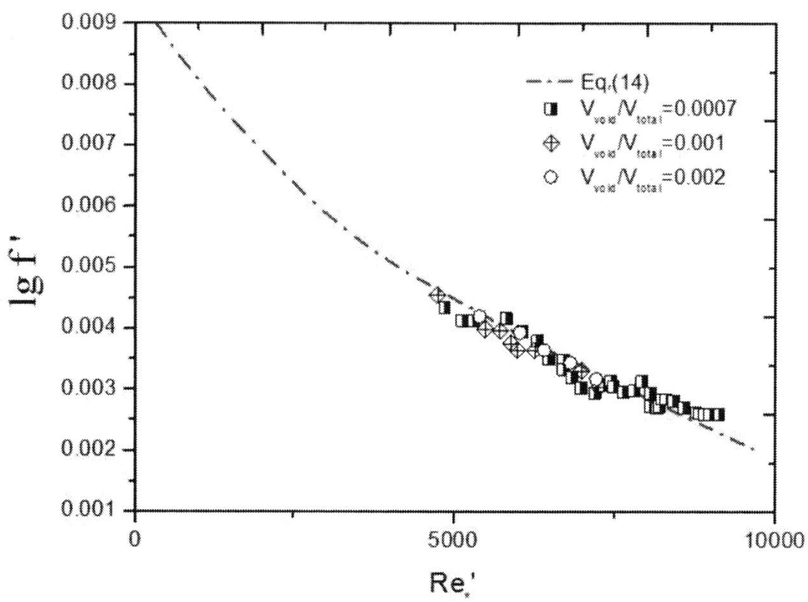

Figure 4. Variation of friction factor in turbulent smooth flow (with Nikuradse's database).

Determination of the friction factor f in turbulent smooth flow is the first step, and it is quite important to find out the correct formulation of f' and f''. In order to account for the correct formulation of f', we have adopted the method developed by Yang and Dou [15] in pipe flows with the intermittent nature of turbulence. We have slightly modified their equation on turbulent smooth pipe flows with our concentration related Reynolds number ($Re_*^{'}$) to account for the vegetated flows, as follows. Figure 4 depicts the perfect agreement of this equation with the data available in literature.

$$f' = 9.4/(2.5\ln Re_*^{'} - \frac{87.8}{Re_*^{'0.65}} + 1.5)^2 \qquad (14)$$

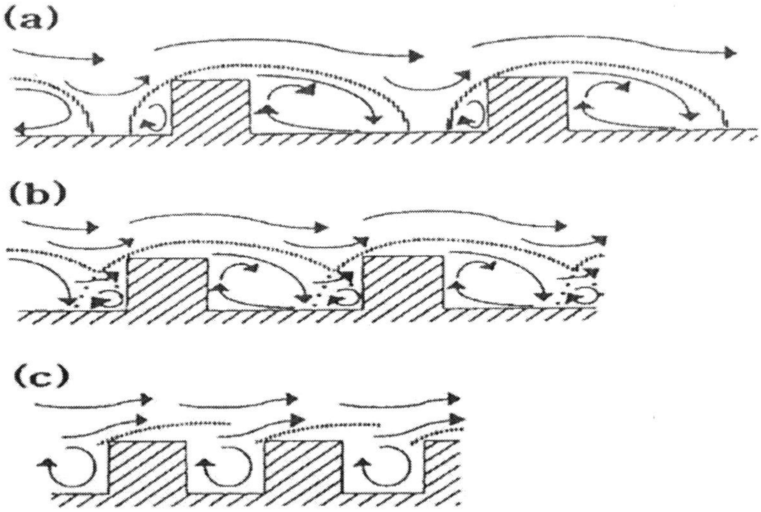

Figure 5. Classification of the change of flow patterns with the change of roughness density.

When determining the friction factor in fully rough flow (the form drag friction factor), it is quite important to consider all the factors which would contribute to this friction force. As the relative void volume of the roughness elements, V_{void}/V_{total} is varied, so the nature of the flow at the boundary varies as well. Figure 5 shows different flow patterns with artificial strip roughness.

In smooth turbulent flow, the flow is free from any roughness elements, and the total friction force is skin friction force. When the roughness elements are introduced to the flow domain with a relatively large spacing between each, then the turbulent eddies start to form at the lee side of the bluff bodies gradually. Even though the skin friction force is dominant in the flow domain initially, the form drag friction force starts to take it over gradually. If both of these forces co-exist, flow would be in the transition region between turbulent smooth to turbulent fully rough flow. According to the above classification we name this as the "semi smooth/rough turbulent flow", where is fully functioning in this region. In this region, the total spacing between the roughness elements is not fully occupied by the turbulent eddies, in fact $\frac{V_{eddy}}{V_{void}}$ is <1. However when the total friction force is dominated by the form drag friction force, then we enter the region of "hyper-turbulent flow". This is the entire spacing between the roughness elements which is occupied by the turbulent eddies forming as a result of the form drag, where $\frac{V_{eddy}}{V_{void}} = 1$. If the spacing between the elements is further reduced ($\frac{V_{void}}{V_{total}}$ is reduced), then the flow domain attains a situation of "quasi smooth flow". This illustrates how a trapped vortex shelters in the lee of the roughness elements, which we termed as "dead water". If the entire spacing between adjacent elements is not occupied this trapped vortex, it is called the "semi quasi smooth flow". The stable circulatory motion of the trapped vortex reduces the turbulent intensity of the eddy volume occurred as a result of the form drag. Higher the volume of dead water, lower the turbulent intensity, resulting in a decrease in the friction factor. The above phenomena will be formulated through the "eddy volume concept".

The formation of eddies behind bluff bodies is mainly influenced by the hydrodynamic parameters prevailing in the flow. Thus, we propose the following relationship:

$$L''=C_D\frac{V^2}{2g} \tag{15}$$

Thus the new model parameter $\frac{V_{eddy}}{V_{total}}$ can be evaluated as follows.

$$\frac{V_{eddy}}{V_{total}}=\frac{\left(A_rC_D\frac{V^2}{2g}\right)}{B\times L\times h} \tag{16}$$

It is assumed that, in vegetated flows the earliest flow region which can be achieved, even with very small vegetation density is the transitional region. As the stems possess a reasonable height, most of the time, eddies start to form behind the stems. Therefore even with a small vegetation density, it is highly unlikely a smooth flow is achieved. In order to outline the behaviour of $\frac{V_{eddy}}{V_{total}}$ with $\frac{V_{void}}{V_{total}}$, equation (17) is presented with the experimental data by Knight [16]. Figure 6 can be demonstrated in two parts, first range is $0.003<\frac{V_{void}}{V_{total}}<0.014$, the behaviour of $\frac{V_{eddy}}{V_{total}}$ depended only on $\frac{V_{void}}{V_{total}}$, thus we call it as the "fully developed region". It was quite clear that despite of the increase in Reynolds Number, $\frac{V_{eddy}}{V_{total}}$ starts to fall down after the maximum of 0.014 is achieved. In this region, Note that the maximum $\frac{V_{eddy}}{V_{total}}=0.014$ is achieved, when $\frac{V_{void}}{V_{total}}$ is also equal to 0.014, and $\frac{V_{eddy}}{V_{void}}=1$. The second part is $0.014<\frac{V_{void}}{V_{total}}<0.065$, in this region $\frac{V_{eddy}}{V_{total}}$ is influenced both by the increase in Reynolds Number and increase in vegetation density according to experimental data, and $\frac{V_{eddy}}{V_{void}}<1$. After confirming the behavior $\frac{V_{eddy}}{V_{total}}$ Vs $\frac{V_{void}}{V_{total}}$ with experimental data, we now propose the following equation to determine f'':

$$f''=\left(2\log\left[2.9\frac{V_{total}}{V_{eddy}}\right]\right)^{-2} \tag{17}$$

By substituting equations (14) and (17), to equation (13), the final equation for the submerged vegetation flow resistance model is derived as follows:

$$f=\left[9.4/(2.5\ln Re_*^{'+}-\frac{87.8}{Re_*^{'+0.65}}+1.5)^2\right]+\frac{V_{eddy}}{V_{void}}\left\{\left(2\log\left[2.9\frac{V_{total}}{V_{eddy}}\right]\right)^{-2}-\left(9.4/(2.5\ln Re_*^{'+}-\frac{87.8}{Re_*^{'+0.65}}+1.5)\right)^{-2}\right\} \tag{18}$$

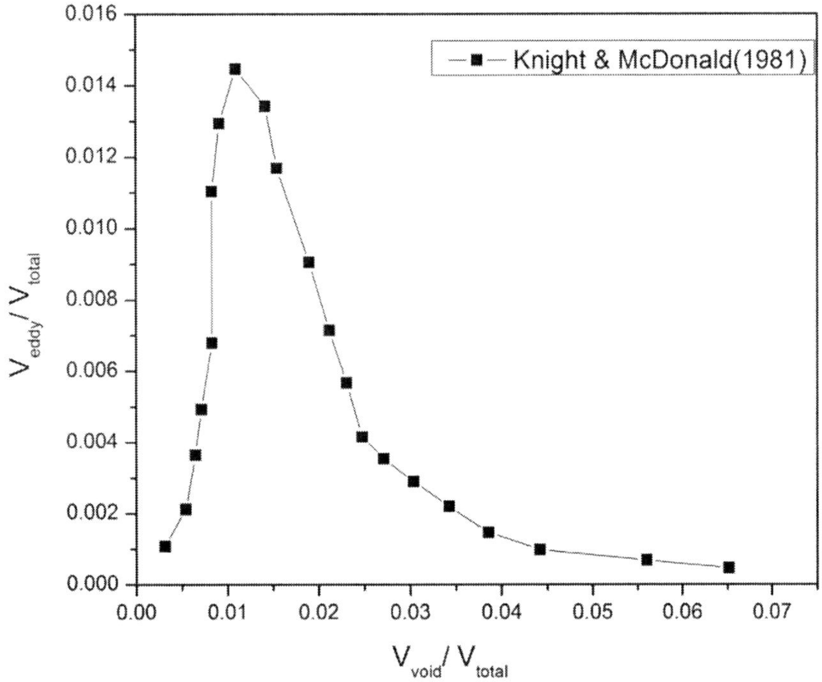

Figure 6. Variation of V_{eddy}/V_{total} with V_{void}/V_{total}.

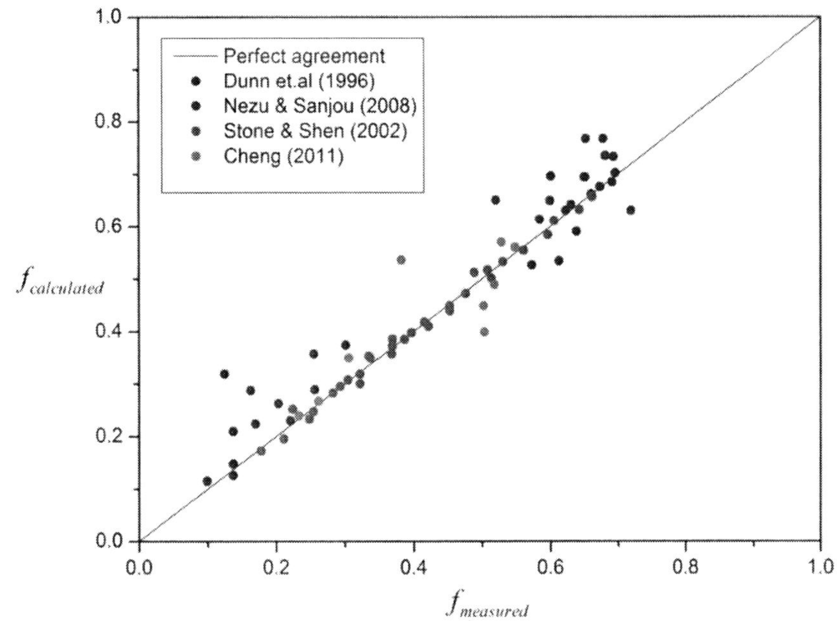

Figure 7. Comparison of $f_{measured}$ with $f_{calculated}$ ($f_{calculated}$ calculated by equation (18)).

We checked the accuracy of the developed friction factor model equation (18) in comparison to available models in literature in figure 7, with regards to vegetated flows, and observation of research results developed by Dunn et al [17], Nezu and Sanjou [18], Cheng [19] appear to be less accurate, and its accuracy is close to that of the formula by Stone and Shen [20].

This study demonstrates that the concept of hydraulic radius is useful in unifying experimental data on flow resistance in vegetated open channel flows for various roughness element configurations. By taking the responsibility to eliminate the ambiguity prevailing in submerged vegetated flows in the determination of a proper length scale, it demonstrates the derivation of a new length parameter called "roughness concentration related hydraulic radius", followed by a new definition for the Reynolds number, as roughness concentration related Reynolds number. These two parameters form the fundamental basis of a theoretical model of flow resistance in submerged vegetation flows. Further, it derives a new formula to determine the friction factor in vegetated flows, using the concept of eddy volume concentration. This model made a useful prediction on flow resistance in an ecological channel.

6. Conclusions

It investigated flow resistance over submerged fixed roughness elements in vegetated flows, and the following conclusions can be summarised:

- The hydraulic radius in vegetated flows denoted the ratio of the water volume to wetted contacted area over roughness. This study found that Nikuradse's conclusion remains valid if the hydraulic radius is defined in terms of fluid volume to solid-liquid contact area. The roughness surface area should be included in the calculation.
- If there are no roughness elements in the boundary, no flow separation occurs, thus the total resistance is equal to the skin friction. When the roughness elements are gradually introduced into the boundary, then partial separation occurs and the length of the separation region gradually increases with the increase in Reynolds Number as well as increase in relative roughness height. Therefore at this stage the total friction factor depends on both the skin friction and form drag.
- This study has linked these phenomena with flow separation and provided a general theoretical equation to calculate friction factor in flow over vegetation. This new model made a perfect prediction on flow resistance in an ecological channel.

References

[1] Cheng N S 2003 Application of Ergun equation to computation of critical shear velocity subject to seepage *J. Irrig. Drain. Eng.* **129** 278-83
[2] Cheng N S, Hao Z and Tan S K 2008 Comparison of quadratic and power law for nonlinear flow through porous media *Exp. Therm. Fluid Sci.* **32** 1538-47
[3] Cheng N S 2011 Representative roughness height of submerged vegetation *Water Resour. Res.* **47** W08517
[4] Dunn C, Lopez F and Garcia M 1996 Mean flow and turbulence in a laboratory channel with simulated vegetation (Hydrosystem Lab. Univ. of Ill. Urbana, Ill) 193-7
[5] Han Y, Yang S Q, Dharmasiri N and Sivakumar M 2014a Effects of sample size and concentration of seeding in LDA measurements on the velocity bias in open channel flow *Flow Meas. Instrum.* **38** 92-7
[6] Han Y, Yang S Q, Dharmasiri N and Sivakumar M 2014b Experimental study of smooth channel flow division based on velocity distribution *J. Hydraul. Eng.* 141 1-6
[7] Jimenez J 2004 Turbulent flows over rough walls *Annu. Rev. Fluid Mech.* **36** 173-96
[8] Knight D W and McDonald J A 1981 Open channel flow with varying bed roughness-closure *J. Hydraul. Div.* **107** 642-4
[9] Nikuradse J 1933 Stromungsgesetze in Rauhen Rohren VDI-Forsch 361
[10] Nezu I and Sanjou M 2008 Turbulence structure and coherent motion in vegetated canopy open channel flows *J. Hydro-Environ. Res.* **2** 62-90
[11] Perry A E, Schofield W H and Joubert P N 1969 Rough wall turbulent boundary layers *J. Fluid Mech.* **37** 383–413
[12] Raupach M R, Antonia R A and Rajagopalan S 1991 Rough-wall turbulent boundary layers

Appl. Mech. Rev. **44** 1–25
[13] Schlichting H 1936 Experimental investigation of the problem of surface roughness *Ingenieur-Archiv.* **7** 1–34
[14] Stoesser T, McSherry R and Fraga B 2015 Secondary currents and turbulence over a non-uniformly roughened open channel bed *Water* **7** 4896–913
[15] Stone B M and Shen H T 2002 Hydraulic resistance of flow in channels with cylindrical roughness *J. Hydraul. Eng.* **128** 500–6
[16] Wooding R A, Bradley E F and Marshall J K 1973 Drag due to regular arrays of roughness elements of varying geometry *Boundary-Layer Meteorol.* **5** 285–308
[17] Yang S Q, Han Y and Dharmasiri N 2012 Momentum balance method (MBM) and estimation of boundary shear stress distribution *J. Hydraul. Eng.* **138** 657–60
[18] Yang S Q, Nadeesha D and Han Y 2011 Flow resistance over fixed roughness elements *J. Hydraul. Res.* **49** 257–62
[19] Yang S Q, Tan S K, Lim S Y and Zhang S F 2006 Velocity distribution in combined wave–current flows *Adv. Water Resour.* **29** 1196–208
[20] Yang S Q and Dou G 2005 Drag reduction in a flat plate boundary layer flow by polymer additives *Phys. Fluids* **17** 065104

IOP Conf. Series: Earth and Environmental Science 82 (2017) 012070 doi:10.1088/1755-1315/82/1/012070

Water quality and emergy evaluation of two freshwater aquacultural systems for eutrophic water in the Controlling by Biological Chains

L M Xi[1], C Q Liu[2,3], D F Liu[1], W L Huang[1] and Y Sun[1]

[1]College of Environmental Science and Engineering, Nankai University, Tianjin, China
[2]College of Life Science, Hebei University, Hebei, China

E-mail: liucunqi@sina.com

Abstract. According to the ecological restoration theory, this experiment establishes aquaculture systems controlled by biological chains in both Xiaoxidian area and Dujiadian area of Baiyangdian Lake separately in order to improve the environment and bring economic benefits. The appearance of emergy theory provides a new method for the quantitative analysis of ecological economic system. Based on the analysis of emergy theory, this thesis compares the eco-economic systems under different polyculture models between Xiaoxidian area and Dujiadian area. The result demonstrates that Xiaoxidian ecological system is of high emergy transformity with higher emergy output and economic income per unit area compared with Dujiadian area. While Dujiadian area has higher emergy yield Rate and lower Environment Load Rate. So Dujiadian area is more sustainable due to the overload non-renewable energy of Xiaoxidian area devoted by human. Therefore, it will be better if we adjust and optimize the management of aquaculture system in Xiaoxidian area in order to find a stable equilibrium point between environmental sustainability and economic benefits.

1. Introduction

For a long period of time, people took energy as a common scale to analyze various systems. Energy analysis, however, has been effective only in the analysis of the same kind of energy. There are fundamental differences on the quality and value of different energy. That cannot be simply modified and compared [1]. So it is impossible to measure and express the essential relationship between natural environment resources and economy through general energy unit. In 1980s, the international ecological community put forward a new concept theory of emergy and a new analysis method (by multiplying a series of emergy transformity with emergy value of every emergy level to get emergy based on "Solar Joule" as unit, in order to analyze energy characteristics and sustainable development condition of system) to develop a new way for the quantitative analysis of ecological system and eco-economic system [2].

The water bloom in lakes caused by eutrophication has attracted more and more attention. Ecological restoration has become an important research direction in the area of water ecology under this kind of situation and the biomanipulation method is a very important method of ecological restoration. The theory of Biomanipulation is traditionally divided into classical [3-5] and non-classical theory [6]. Non-classical Biomanipulation Theory is the direct control of cyanobacteria

[7] by adjusting the filter feeding fishes. In addition, ecological restoration project of submerged plant mainly removes nutrients from water by the absorption and storage of aquatic organisms together with substrate sludge solidification and harvest [8]. With the further research of biomanipulation methods, the method of biological chain control springs up [9]. It controls eutrophication through the combination of submerged macrophytes, benthic animal and filter feeding fish and improves the environment without secondary pollution. The method of biological chain control solves the problem from the root by creating good economic benefits and improving the living standard of local people which lay solid mass foundation for the promotion of this method. According to the theory of biological chain control, our experiment establishes the ecological compound aquaculture system in Xiaoxidian area and Dujiadian area of Baiyangdian Lake by changing aquaculture methods and planting submerged plants in order to harvest on both environmental and economic benefits. The polyculture method of both silver carps and bighead carps can control algae [10] and bring good economic benefits. At the same time, considering economic development and industrialization of local area, we try to make full use of ecological niche to achieve maximum economic benefits. So we manage to improve the utilization efficiency of regional submerged macrophytes and benthic organisms by adding crab (Eriocheir sinensis) and planting water chestnut (Trapa bispinosa Roxb) in crab breeding area. Simultaneously, Harvest water chestnut and Myriophyllum spicatum to eliminate nitrogen and phosphorus from the aquatic ecosystem. The study compares two kinds of aquatic ecosystems through the method of emergy analysis from the perspective of both system conversion efficiency under different models and their effects on the sustainable development of Baiyangdian Lake to analyze the characteristics of two ecological systems, thus providing references for the managers and users of Baiyangdian Lake to make decisions for management and use.

2. Materials and methods

2.1. Research site overview

Baiyangdian Lake is the largest freshwater shallow lake in Northern China with severe eutrophication and frequent water bloom dominated by Chlorophyta and Cyanophyta. The evaluation of Baiyangdian Lake is based on trophic state index of modified (TSIM) and dominant species evaluation. According to the results of the evaluation, the level of nutrition is eutrophication which means that the system remains to be repaired. Aquaculture is an important factor which affects the water environment of Baiyangdian Lake. There are 9,615.8 acres of net enclosure culture, 4,602.1 acres of pond culture, 727 acres of cage culture and 20,514 acres of reclamation dam in Baiyangdian Lake (latest statistics from Marine and Fisheries Research Institute of Hebei Province and Aquatic Products Bureau of Anxin County). Most of the aquaculture modes get high output by feeding which leaves serious impact on water quality. Therefore, it is an urgent task to optimize the cultivation mode and protect the water quality influenced by cultivation.

2.2. Research sample area

Mode 1: The place (12 hectares) locates in Xiaoxidian area of Baiyangdian which is divided into 2 regions with enclosure. Zone A is a growing area of 4 hectares for 1250 kg water chestnut (*Trapa bispinosa*). Zone B is a growing area of 8 hectares for Myriophyllum spicatum with the density of 20 strains per square meter; 16000 silver carps (*Hypophthalmichthys molitrix*) (100 g/tail), 4000 bighead carps (*Aristichthys nobilis*) (100 g/tail), 1500 kg *Cipangopaludina cahayensis*, 96000 *Eriocheir sinensis* and 180000 *Macrobrachium nipponenses*, together with native mollusk specie freshwater snail(Radix auricularia).

Mode 2: The place (70 hectares) locates in Dujiadian area of Baiyangdian Lake with little hydrophyte. It mainly breeds silver carps, bighead carps, *Eriocheir sinensis* and *Cyprinus carpio* which are all fed by artificial diet with little benthic organism.

The environmental characteristics of two ecological systems are listed in table 1.

Table 1. The characteristics of two aquacultural ecosystems.

Parameters	Mode 1 (Experimental Area)	Mode 2 (Control Area)
Location	Xiaoxidian (38°51′ N,115°58′ E)	Dujiadian (38°51′ N,115°58′ E)
Mobility of Water	closed aquaculture pond	closed aquaculture pond
Area (hm^2)	12	70
Depth (m)	2～4	2～4
Aquatic Plants	*Myriophyllum spicatum* and *Trapa bispinosa* with the coverage rate of 40% in experimental area	no
Community Composition	*Eriocheir sinensis*, *Hypophthalmichthys molitrix*, *Aristichthys nobilis*, *Macrobrachium nipponense*, *Cipangopaludina cahayensis* and *Radix auricularia*	*Eriocheir sinensis*, silver carps, bighead carps, *Macrobrachium nipponense*, *Cyprinus carpio* and *Cipangopaludina chinensis*
Delivery Volume	20000 fishes that ratio of silver carps and bighead carps is 4:1, 1500 kg *Cipangopaludina chinensis*, 96000 *Eriocheir sinensis*, 1.8×10^5 *Macrobrachium nipponense*	60000 fish fry of silver carps and bighead carps (3:1), 1.125×10^5 *Eriocheir sinensis*, 2×10^5 kg *Cipangopaludina chinensis*
Delivery Volume of Baits	2000 kg concentrated fish feed, 1500 kg spiral shell (fish bait),7500 kg maize, 2000 kg potato (crab bait)	2×10^5 kg spiral shell (fish bait & production), 5000 kg potato (crab bait), 3000 kg maize (crab bait)

2.3. Water quality and trophic state index of modified (TSIM)
The measurement of water quality was done between April 8th, 2013 and October 8th, 2013.

Water transparency was measured with a 20-cm diameter black and white Secchi disk [11].

Measure Total Phosphorus by the ammonium molybdate spectrophotometric method (GB/T11893-1989 of China) [12] Measure Chlorophyll-a according to ISO10260:1992(E) Water quality-Measurement of biochemical parameters-Spectrometric determination of the chlorophyll-a concentration. TSIM is used to describe the nutritional status of water body based on the parameters such as Total Phosphorus concentration, chlorophyll-a content and transparency of water [13, 14].

$$TSIM(SD)=10\times[2.46+(3.69-1.53\times\ln SD)/\ln 2.5]$$

$$TSIM(TP)=10\times[2.46+(6.71+1.15\times\ln TP)/\ln 2.5]$$

$$TSIM(Chla)=10\times(2.46+\ln Chla/\ln 2.5)$$

$$TSIM=[TSIM(Chla)+ TSIM(TP)+ TSIM(SD)]/3$$

TSIM<30 represents poor nutrition state; 30<TSIM<50 represents middle nutrition state; TSIM>50 represents eutrophication state. Among them, eutrophication state can be divided into three levels, 50<TSIM≤60 for mild eutrophication; 60<TSIM≤70 for moderate eutrophication; TSIM>70 for severe eutrophication. The higher the index is, the higher the nutritional status of water body will be.

2.4. Emergy analysis method
Both the demonstration area and control area selected by this research are reclamation dam with clear boundary. It is easy to count resource input, resource output and emergy flow and suitable for the use of emergy analysis method to reveal the differences of emergy transformation between two systems. At the same time, we could also analyze the resource utilization efficiency and environmental effects of two aquaculture systems from the perspective of emergy in the aspect of investment, production and marketing. Then the result provides theoretical basis for reasonable application and proper promotion

IOP Conf. Series: Earth and Environmental Science **82** (2017) 012070 doi:10.1088/1755-1315/82/1/012070

of aquaculture mode.

Firstly, we need to define the boundaries of two systems and main energy sources before the emergy analysis of ecological systems. Then we could determine the main components of the systems by defining the emergy flow within each part of the system. Finally, two types of emergy flow diagram (system with aquatic plants and system without aquatic plants) can be drawn according to the communication between the ecological system and social economic system. According to different indicators, system characteristics and sustainable development condition of different systems can be analyzed [15]. Emergy flow diagram of two aquaculture models is shown in figure 1.

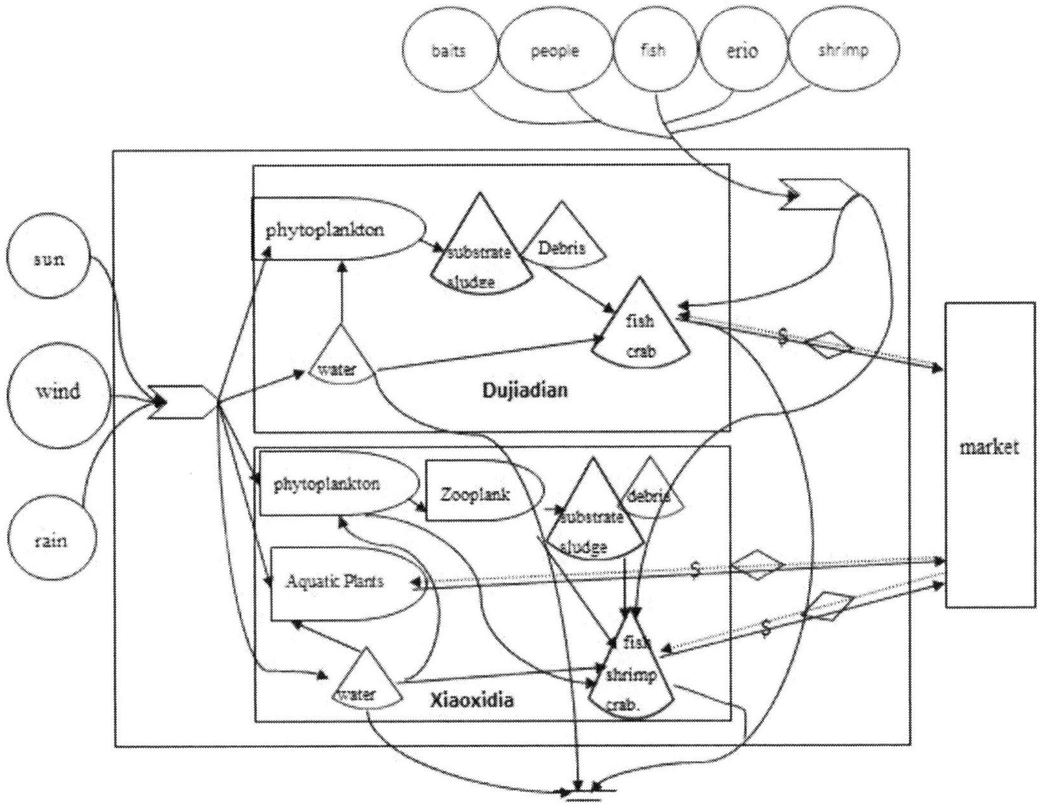

Figure 1. Emergy flows of two aquacultural ecosystems.

2.5. Emergy index

Emergy indexes is shown in table 2.

Table 2. Indices of emergy analysis.

Parameters (and abbreviations)	Calculating formula	Meaning
Solar Emergy Transformity	applied solar emergy (sej)/energy or quality (J or kg)	The amount of solar energy contained in each unit of energy or quality; an indice to measure energy qualitative level.
Emergy Yield Rate (EYR)	Yield emergy(Y)/ input emergy value(U)	degree of industrialization for ecological system; the higher the value of EYR, the higher the

		degree of industrialization.
System Emergy Consumption (Em)	Renewable emergy input (EmR)+Non-renewable emergy input (EmN)+economic feedback (EmF)	direct/indirect consumption of solar energy for the formation of emergy logistics
Emergy Per Area(EPA)	System Emergy Consumption (Em)/area	intensity and level of economic development for eco-economic system
Emergy/currency ratio (Em/$)	(EmR+EmN)/ Gross National Product (GNP)	the relationship between human labor and environmental "work"
Emergy-monetary value (Em$)	System Emergy Consumption (Em)/(Em/$)	the value of resource emergy to the economy
Emergy Investment Ratio (EIR)	economic feedback (purchase) EmF/(EmR+EmN)	the indice to measure economic development scale and the degree of environmental load
Positive Emergy Yield Rate (EYR)+	beneficial emergy yield (EmY+)/ System Emergy Consumption (Em)	measure the contribution of positive system output to the development of the system
Negative Emergy Yield Rate (EYR)-	negative emergy yield EmY-/System Emergy Consumption (Em)	measure the contribution of negative system output to the development of the system
Emergy Exchange Rate (EER)	commodity storage emergy/equivalent energy used for purchasing commodity	the relationship between natural value of commodity and people's willingness to pay
Environmental Load Rate (ELR)	(EmF+EmN)/EmR	measure the pressure of natural environment.
Emergy Sustainability Index (ESI)	EYR/ELR	evaluate the system ability for sustainable development
Emergy Index for Sustainable Development (EISD)	EYR*EER/ELR	evaluate the system ability for sustainable development

3. Results and analysis

3.1. Water quality

The comparison of water quality are shown in table 3.

Table 3. Comparison of water quality between two areas.

		Apr. 8th	Jun. 8th	Jul. 8th	Aug. 8th	Sept. 8th	Oct. 8th
TP	Modal 2	0.08	0.27	0.12	0.13	0.06	0.05
(mg/L)	Modal 1	0.07	0.13	0.12	0.05	0.07	0.04
SD	Modal 2	0.31	0.22	0.29	0.11	0.09	0.12
(m)	Modal 1	0.60	0.50	0.46	0.31	0.33	0.39
Chla	Modal 2	15.94	23.56	26.51	62.05	54.58	85.51
(mg/L)	Modal 1	16.40	13.36	17.52	23.16	25.05	24.27
TSIM	Modal 2	68.36	76.94	72.59	81.00	78.59	77.91
	Modal 1	64.43	67.18	68.24	68.20	69.13	65.71

The water transparency of Xiaoxidian area is much higher than that of Dujiadian area, which means the water of Xiaoxidian area is clearer. The temperature of summer is higher and this is suitable for the growth of plankton which results in lower transparency in summer.

The Total Phosphorus of Xiaoxidian area is slightly lower than that of Dujiadian area. Especially in June and August, the Total Phosphorus of Xiaoxidian area only reaches half of that in Dujiadian area. The reasons for the rising of Total Phosphorus in June may be: The water level drops with decreased water yield; The water temperature increases with higher intensity of aquatic organism activity and the increase of disturbance from sediment makes a large amount of phosphorus in both sediment and interstitial water diffuse into overlying water. Then the reason for the following decline includes the effect of biomanipulation. Besides, the slight increase in November may be due to the decrease of biological activities resulting from lower temperature. So the control of Total Phosphorus in Baiyangdian area is of great significance to improve the nutritional status of water body because of the phosphorus limitation in this area.

The Chlorophyll-a of Xiaoxidian area is significantly lower than that of control area in Dujiadian. Especially in August, September and October, the Chlorophyll-a of Dujiadian area reaches two to three times than that of Xiaoxidian area. This demonstrates the overall amount of phytoplankton in Xiaoxidian area is far lower than that of control area. From April to July, the low Chlorophyll-a may be due to the low temperature and other external environmental factors which could not embody the advantages of Xiaoxidian area. The biological chain control of Xiaoxidian has an extremely significant effect on reducing the total amount of phytoplankton in water body.

According to TSIM based on water transparency (figure 2), TSIM (SD) in all the areas are larger than 70 and they all belong to severe eutrophication. While it is apparent that Xiaoxidian area is obviously superior than the control area-Dujiadian.

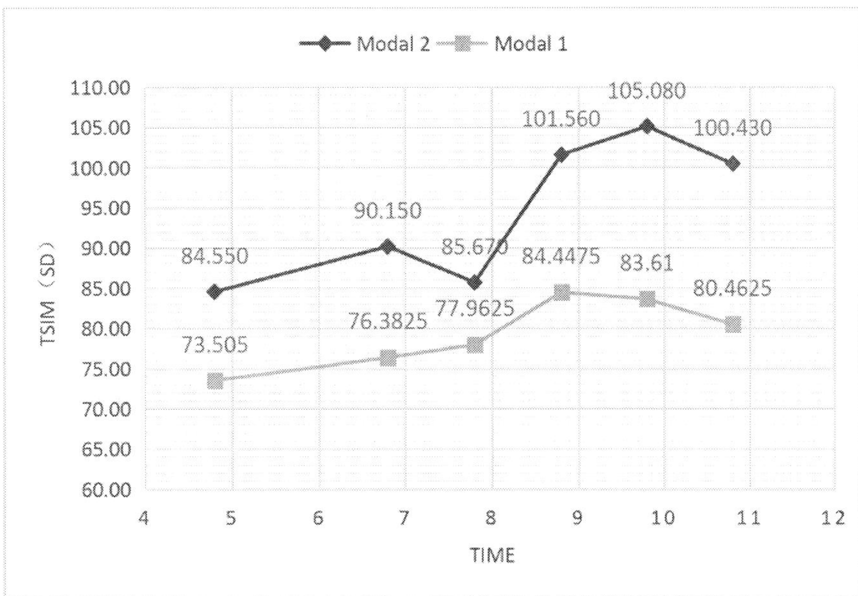

Figure 2. Comparison of TSIM Based on transparency.

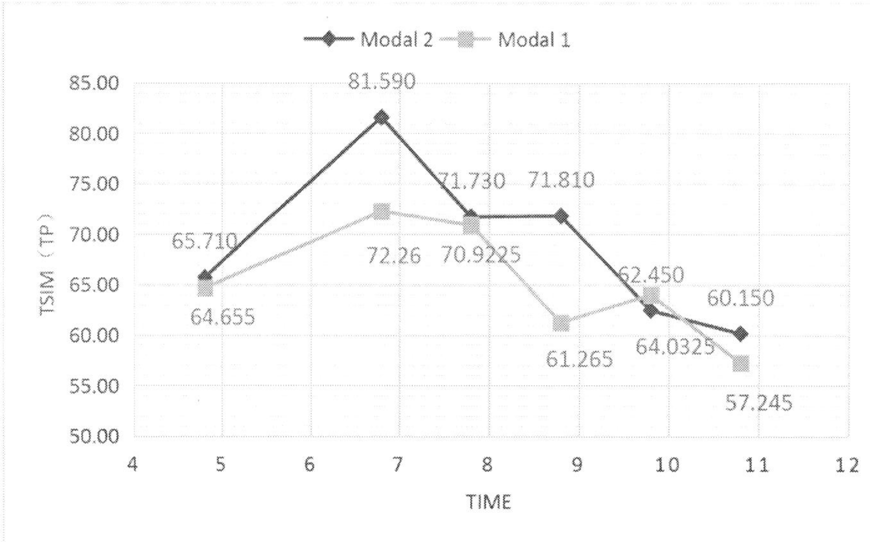

Figure 3. Comparison of TSIM based on total phosphorus.

TSIM based on Total Phosphorus is shown in figure 3. According to TSIM based on Total Phosphorus, the demonstration area is basically controlled within moderate eutrophication level and it even reaches the level of mild eutrophication in October. While the control area of Dujiadian is basically between moderate and severe eutrophication. So demonstration area of Xiaoxidian area has got fruitful results on the control of nutritional status.

TSIM based on chlorophyll-a (figure 4) shows that the demonstration area has always maintained in moderate or mild eutrophic level, while the control area changes into severe eutrophication level with increasing temperature in summer.

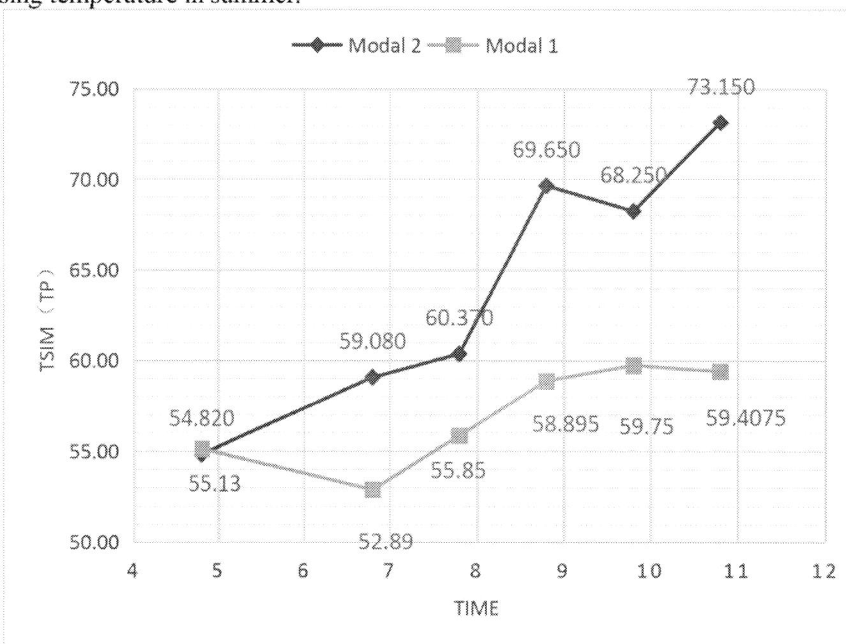

Figure 4. Comparison of TSIM Based on Chlorophyll-a.

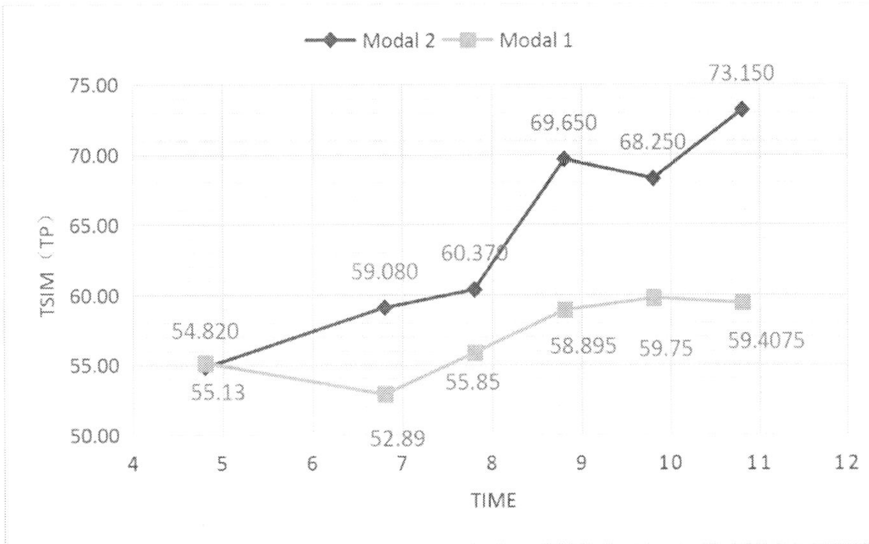

Figure 5. Integrated TSIM Comparison.

The demonstration area is controlled in moderate eutrophication level. In contrast, the control area is severe eutrophication level.

The results in figure 5 show that compared with the ordinary aquaculture model within Dujiadian, the water transparency of Xiaoxidian aquaculture system is significantly higher than that of Dujiadian. At the same time, Xioaxidian area has lower Total Nitrogen, Total Phosphorus, chlorophyll-a content and TSIM than Dujiadian area which demonstrates the aquaculture system with control of biological chain has an obvious effect on improving water quality. Judged from the physical and chemical indices, the water quality of demonstration area significantly improves because most of the factors are reduced to moderate eutrophication level from severe eutrophication level before the structural adjustment. In some individual months, the water quality even meets the standard of III（GB3838-2002 of China）. Demonstration area not only makes full use of ecological niche, but also improves water quality significantly.

3.2. The emergy input and output

The Emergy input and output are shown in table 4.

Table 4. Emergy evaluation of two aquacultural ecosystems.

No.	Item	Unit	Raw data (Dujiadian)	Raw data (Xiaoxidian)	Emergy transformity / Emergy currency ratio (Em/$) (Dujiadian)	Emergy transformity/ Emergy currency ratio (Em/$) (Xiaoxidian)	Solar emergy (Dujiadian)	Solar emergy (Xiaoxidian)	Emergy attribute
1.1	Solar radiation energy	J	3.77458E+18	6.47071E+17	1	1	3.7746E+18	6.4707E+17	R
1.2	Wind energy	J	1389835843	238257573.1	1470	1470	2.0431E+12	3.5024E+11	R
1.3	Potential energy of rain	J	2.57771E+13	4.41894E+12	10300	10300	2.6550E+17	4.5515E+16	R

	water								
1.4	Chemical energy of rain water	J	1.62422E+12	2.78438E+11	18100	18100	2.9398E+16	5.0397E+15	R
1	Total renewable natural resources						3.7746E+18	6.4707E+17	R
2.1	Bighead carp fry	$	1857.93	619.31	4.94E+12	4.94E+12	9.1782E+15	3.0594E+15	F
2.2	Silver carp fry	$	7431.72	1486.34	4.94E+12	4.94E+12	3.6713E+16	7.3425E+15	F
2.3	Shrimp seed	$	0	348.36	4.94E+12	4.94E+12	0	1.7209E+15	F
2.4	Eriocheir sinensis	$	4180.34	3963.58	4.94E+12	4.94E+12	2.0651E+16	1.9580E+16	F
2.5	Water chestnut seeds	$	0	3870.69	4.94E+12	4.94E+12	0	1.9121E+16	F
2.6	Feed	$	38397.23	4025.56	4.94E+12	4.94E+12	1.8968E+17	1.9886E+16	F
2.7	Labour cost	$	9103.86	6069.24	4.94E+12	4.94E+12	4.4973E+16	2.9982E+16	F
2	Feedback resources	$	60971.08	20383.04	4.94E+12	4.94E+12	3.0120E+17	1.0069E+17	F
	Total inpot	sej					4.0758E+18	7.4776E+17	U
3.1	Water chestnut	kg	0	13028.8		5.7393E+13			
3.2	Myriophyllum spicatum	kg	0	188696		3.9628E+12			
3.3	Silver carp	kg	40000	13500	1.0189E+14	5.5390E+13			
3.4	Bighead carp	kg	10000	4500	4.0758E+14	1.6617E+14			
3.4	River crab	kg	3500	3050	1.1645E+15	2.4517E+14			
3.6	Shrimp	kg	1000	1800	4.0758E+15	4.1542E+14			
3.7	Cyprinoid	kg	10000	0	4.0758E+14				
3	Product output	kg	64500	224574.8	6.3190E+13	3.3297E+12			Y
4.1	TN	g/a	0	3799607.501		3.73E+09		1.4173E+16	
4.2	TP	g/a	0	457881.7712		3.83E+09		1.7537E+15	
4	Pollutants of retention and purification	t/a						1.5926E+16	
	Total output	sej					4.0758E+18	7.6369E+17	Y

Emergy output per hectare	5.8225E+16	6.3641E+16

Notes: 1. Raw data is field survey data. 2. Solar energy =area (m^2)×annual average radiation ($J \cdot m^{-2} \cdot year^{-1}$); 3.Chemical energy of rain water= rainfall amount ($m \cdot year^{-1}$)×area(m^2)×Gibbs free energy ($4.94\ J \cdot g^{-1}$)×($1\times10^6\ g \cdot m^{-3}$); 4. Potential energy of rain water = rainfall amount ($m \cdot year^{-1}$)×area (m^2)×mean altitude (m)×($1000\ kg \cdot m^{-3}$)×($9.8\ m \cdot s^{-1}$); 5. Wind energy= average wind speed ($m \cdot s^{-1}$)×air storey (1000 m)×air density ($1.23\ kg \cdot m^{-3}$)×specific heat of air ($10.048\ kJ \cdot K^{-1}$)×horizontal temperature gradient ($3\times10^{-9}\ k \cdot m^{-1}$)×($3.15\times10^7\ s \cdot year^{-1}$) [16]; 6. Project emergy= Project raw data × Project conversion rate; 7. Macro-economic value of the project =project emergy/ (Emergy / currency ratio (Em/$)); 7. Emergy Transformity takes the research of Lu Hongfang et al as reference [17]; 8. The exchange rate of RMB against U.S. dollar is the average rate of experimental period.

3.3. Emergy evaluation indices

So far, there has been no unified method in high stylization to analyze emergy. Researchers need to pay attention to characteristics of the eco-economic system and the emphasis of the research before determining the specific analysis methods. The two aquaculture ecosystem are both closed ecosystem with all the input emergy converted to a co-product [18]. Nutrient elements such as nitrogen and phosphorus of water chestnut (Trapa bispinosa) and Myriophyllum spicatum are both from water and substrate sludge so they can be treated as emergy output from the retention pollutants after being purified in the system. Because the conversion efficiency of the system needs to be evaluated with no significant change in the energy of the substrate sludge, the system is expected to change all the emergy input into joint product without taking dissipation into consideration, thus treating the emergy output used to purify retention pollutants as the emergy output used for water purification which reduces the indigenous emergy of the two systems.

According to table 4, we could find that Solar emergy transformity of Dujiadian area is always higher than that of Xiaoxidian area for different products (seen as co-production) of the two systems which means that the conversion efficiency of Xiaoxidian is better. So the production efficiency of Xiaoxidian area is higher. We eliminate nitrogen and phosphorus of the system by harvesting aquatic plants, such as Trapa bispinosa and Myriophyllum spicatum in the experimental area, thus picking out the emergy used for purifying water and treating this emergy as the emergy deduction of indigenous system storage. That is because the salvage of mature Trapa bispinosa and Myriophyllum spicatum could be regarded as the deduction of indigenous nitrogen and phosphorus nutrients to a specific level. Besides, Selling Trapa bispinosa and Myriophyllum spicatum could also bring economic benefits.

Further calculations are drawn to get a series of emergy index for system evaluation (table 5).

Table 5. Comparison of emergy indices for two cultural ecosystems in Baiyangdian Lake.

Indice	Dujiadian	Xiaoxidian
EIR	0.08	0.16
ESR	0.93	0.87
EYR	13.5	7.58
EER	7.95	2.34
ELR	0.08	0.16
ESI	178.23	48.71
Output per hectare	5.82254E+16	6.36408E+16
Economic output per hectare	1481.92	5393.16
Net income per hectare (yuan)	871.02	1698.59

3.3.1. Emergy Investment Ratio (EIR): EIR=EmF/(EmR+EmN). The environmental providing emergy divided by economic feedback (purchase) reflects the degree of system utilization for natural environment. The figure for Xiaoxidian demonstration area is 0.16 and that is twice the figure of Dujiadian area (0.08). While both of them are far less than 1 which means that their dependence on natural resources for the two systems is in high level and the economic investment accounts for low proportion. Simultaneously, the artificial intervention intensity of Xiaoxidian area is slightly larger than that of Dujiadian area. That is to say, economic feedback emergy of Xiaoxidian area is larger together with smaller proportion of local resources utilization.

3.3.2. Emergy Self-sufficiency Rate (ESR). Emergy self-sufficiency rate has two concepts. In scale of country or a large area, it means the ratio between emergy input provided by the local region and the emergy input from foreign countries. In the analysis of small scale, it equals to the ratio between natural resources emergy provided by local region and the total emergy input of the system. The figure for Xiaoxidian demonstration area is 0.87 and slightly smaller than that of Dujiadian area (0.93). The result demonstrates that both of the two systems gain considerable supports from natural resources and the influence of market on Dujiadian area is smaller with higher degree of independent development.

3.3.3. Emergy Yield Rate (EYR). EYR equals to the devoted emergy (purchase) divided by emergy output which reflects the return rate of emergy investment and whether the product is competitive in price. It also symbolizes the degree of industrialization for the ecological system-the higher the value is, the higher the degree of industrialization will be. Compared with the national average level of 1.42 for agriculture, return rate of emergy investment in aquaculture industry is very good. In the two areas, we release fish fry of 60 tail / kg in Dujiadian area and 10 tail/kg in Xiaoxidian area. The actual emergy contained in unit weight of both areas is different and price will also influence the parameters to some extent. So the emergy may change in the future if we use other fish fry in the same size. In addition, due to the limitation of methods, people still have different opinions about choosing which method to purify nitrogen and phosphorus [19] and the specific conversion rate is not accurately equivalent to the plant purification function. The parameters will be more accurate if there is further research on the basis of specific emergy contribution to ecological service function.

3.3.4. Emergy-monetary value (Em$). Em$ reflects consumer's willingness to purchase the emergy contained in the product. The monetary value of emergy output for unit area of Xiaoxidian area equals to 5393.16, which is almost four times that of Dujiadian area (1481.92). As for the economic benefits, the value of Dujiadian area is 1698.59 and that is a double value of Dujiadian area (871.02). This proves that economic evaluation and emergy evaluation may gain different results within the same system. But both economic benefits and emergy benefits in Xiaoxidian area exceeds those of Dujiadian area.

3.3.5. Emergy Exchange Rate (EER). EER equals to currency emergy for purchasing commodity divided by stored emergy within commodity which indicates the relationship between the natural value of commodity and the willingness of people to pay for commodity. Dujiadian area has higher emergy exchange rate which means the products from Dujiadian area contain more emergy than Xiaoxidian demonstration area. In contrast, the product of Xiaoxidian area reflects its better natural value.

3.3.6. Environmental Load Rate (ELR): No non-renewable natural resources are provided by local area. So the Environmental Load Rate equals to emergy investment rate and both of them prove the same problem.

3.3.7. Emergy Sustainability Index (ESI): ESI=EYR/ELR: ESI evaluates the sustainable development ability of system. Dujiadian area is higher in ESI and sure to gain more advantages in sustainable development.

4. Discussion and conclusions

As we have mentioned before, aquaculture modes with different fish fry specifications and feed species may all result in a series of different parameters. Under the same aquaculture condition, the stocking of large-size fingerling is an important measure to yield large amount of fish. So managers in Xiaoxidian area purchase large-size fingerling in order to get stable short-term economic benefits. Other studies have also proved that emergy transformation rate (ETR) and Environmental Load Rate (ELR) will drop by at least 26% if the emergy input decreases 50%. At the same time, emergy sustainability index (ESI) and emergy index for sustainable development (EISD) will also increase at least 39% and 88% respectively [20]. The two indices mean that the system has more powerful sustainability. Then the next step is to fix all the aquaculture factors gradually in order to determine which management factors cause the differences in the system characteristics. It is important to take economic benefits into comprehensive consideration. Though Xiaoxidian area needs larger economic investment for large-size fingerling, the survival rate of large-size fingerling is high with lower risk of breeding. This is a kind of performance for farmers to reduce their own risk and increase environmental load.

The conversion efficiency of Xiaoxidian demonstration area is higher from the analysis of current situation with superior emergy output and economic income for unit area compared with Dujiadian area. While Dujiadian area has higher emergy output rate and lower Environmental Load Rate which reflects better sustainability. In the final analysis, farmers in Dujiadian area purchase less emergy in the process of aquaculture management. Less economic investment means the emergy purchased within feed and fry is low. So Dujiadian area shows better sustainability with emergy analysis. Besides, the area of Dujiadian is larger with more stable water condition in recent years. So there is a considerable amount of wild carps available for fishing and this also increases the emergy output rate. In contrast, Xiaoxidian area is a new reclamation dam with short formation time of water system and no reproduction of wild fish inside the area. This kind of phenomenon will be improved over time because planting grass provides a good ecological niche for snails and natural baits for fish fry. Then this increases the conversion rate of the whole system. In addition, it should be stressed that compared with aquaculture systems in other regions such as Pearl River Estuary [20], Mai Po [21], T marae Mai [22] and S Aurata [23], the environmental load rate (ELR) and emergy sustainability index (ESI) of Xiaoxidian area and Dujiadian area (as the representatives of Baiyangdian Lake aquaculture model) are much higher. So the sustainability of the aquaculture model in Baiyangdian Lake is still relatively good.

Farmers inevitably incline to choose Xiaoxidian auqaculture model with higher income for economic benefits. While from the angle of comprehensive benefits for long-term, the sustainable degree of Dujiadan area is better to maintain the stability of eco-economic system. So Dujiadan area has its unique advantages for the use and development of long-term. From the comparison of economic income, production amount for unit area of Xiaoxidian area is 3 times as much as Dujiadian area, while net profit only reaches twice of that in Dujiadian arae. This shows that we invest too much in unit area of Xiaoxidian area without getting the return in the same amount. If farmers in Xiaoxidian area agree to reduce a specific amount of economic investments, the net profit may not significantly reduce with improved emergy self-sufficient rate and output rate, thus increasing the sustainability of the system. As Vassallo has found, the results show that the environmental load rate (ELR) and emergy sustainability index (ESI) of aquaculture system depend largely on the utilization of renewable resources[18]. Based on emergy analysis, it will be better if we make full use of some management measures to improve aquaculture system such as better use of baits, reasonable fish / feed ratio and appropriate feed frequency. These measures are used to improve the utilization efficiency of baits and reduce the residual organic compounds in wastewater. In addition, the water in the aquaculture system should be frequently changed in order to provide adequate dissolved oxygen for fish. Therefore, the use of electric oxygen pump can reduce the frequency of changing water to some degree. The uncertainty, regional feature, social feature (different groups of people have different perceptions and attitudes towards environment), comprehensive feature and non-comparable feature of environmental issues decide that one of the most remarkable features for environmental policy formulation is to build consensus on stakeholders. Because the value of environment is always difficult to make analogy with economic value, only when we get supports from the whole society, can we really take effective measures to solve complex problems in the area of ecological environment. Therefore, we need to further determine the capacity of environmental load and ultimately find the balance point of economic and environmental benefits to achieve a win-win situation.

Reference

[1] Odum H T 1983 Self-organization, transformity and information *Science* **242** 1132-9
[2] Odum H T 1996 *Environmental Accounting: Emergy and Environmental Decision Making* (New York: John Wiley & Sons) pp 20-50
[3] Reynolds C S 1994 The ecological basis for the successful biomanipulation of aquatic communities *Arch. Hydrobiol.* **130** 1-33
[4] Shapiro J, Lamarra V and Lynch M 1975 Biomanipulation: An ecosystem approach to lake restoration *Symp. Water* **21** 85-96
[5] Shapiro J and Wright G I 1984 Lake restoration by biomanipulation: Round lake, Minnesota, the first two years *Freshwater Biol.* **14** 371-383
[6] Xie P and Liu J K 2001 Practical success of biomanipulation using filter-feeding fish to control cyanobacteria blooms: A synthesis of decades of research and application in a subtropical hypereutrophic lake *Sci. World* **8** 337-356
[7] Liu J K and Xie P 2002 Enclosure experiments on and lacustrine practice for eliminating microcystis bloom *Chin. J. Ocea. Limn.* **20** 113-117
[8] Gu G and Lu G F 2004 On the integrated control of water environment of Wuli lake, Lake Taihu *J. Lake Sci.* **16** 56-60
[9] Wang Y, Liu L S, Shu J M, Zhu Y Z, Li L and Zhou J 2013 Purification effect of different biological chains on eutrophic water in Baiyangdian lake *Wetl. Sci.* **11** 499-504
[10] Zhou Q, Xie P, Xu J, Ke Z X and Guo L G 2009 Seasonal variations in stable isotope ratios of two biomanipulation fishes and seston in a large pen culture in hypereutrophic Meiliang Bay, Lake Taihu *Ecol. Eng.* **35** 1603-1609
[11] Ke Z X, Xie P and Guo L G 2009 Impacts of two biomanipulation fishes stocked in a large pen on the plankton abundance and water quality during a period of phytoplankton seasonal succession *Ecol. Eng.* **35** 1610-1618
[12] Lu H, Yuan Y, Campbell D E, Qin P and Cui L 2014 Integrated water quality, emergy and economic evaluation of three bioremediation treatment systems for eutrophic water *Ecol. Eng.* **69** 244-254
[13] Aiazki M 1981 Application of modified Carson's trophic state index to Japanese lakes and its relationships to other parameters related to trophic state *Res Rep Environ Stud* **23** 13-31
[14] Goda T 1981 Comprehensive studies on the eutrophication of freshwater areas XI: Summary of researches *Environ. Stud.* **27** 59-71
[15] Xi Y G and Qin P 2009 Emergy evaluation of organic rice-duck mutualism system *Ecol. Eng.* **35** 1677-1683
[16] . Zhu H G, Qin P, Wan S W and Xie M 2001 Emergy analysis of two models of wetland utilization on seashore in Jiangsu Province *Chin. J. Ecol.* **20** 38-44
[17] Lu H F, Campbell D, Chen J, Qin P and Ren H 2007 Conservation and economic viability of nature reserves: An emergy evaluation of the Yancheng Biosphere Reserve *Biol. Conserv.* **139** 415-438
[18] Bastianoni S and Marchettini N 2000 The problem of co-production in environmental accounting by emergy analysis *Ecol. Model.* **129** 187-193
[19] Xi H Z and Kang W X 2008 Evaluation and analysis of emergy-emdollar value for wetland resources of Dongting Lake *J. Econ. Water Res.* **26** 37-44
[20] Li L J, Lu H F, Ren H, Kang W L and Chen F P 2011 Emergy evaluations of three aquaculture

systems on wetlands surrounding the Pearl River Estuary, China *Ecol. Indic.* **11** 526-534

[21] Qin P, Wong Y S and Tam N F Y 2000 Emergy evaluation of Mai Po mangrove marshes *Ecol. Eng.* **16** 271–280

[22] Brown M T, Green P, Gonzalez A and Venegas J 1992 *Emergy Analysis Perspectives, Public Policy Options, and Development Guidelines for the Coastal Zone of Nayarit, Mexico* vol 1-2 (Gainesville, FL: Center for Wetlands, University of Florida) Vassallo P, Bastianoni S, Beiso I, Ridolfi R and Fabiano M 2007 Emergy analysis for the environmental sustainability of an inshore fish farming system *Ecol. Indic.* **7** 290-298

IOP Conf. Series: Earth and Environmental Science **82** (2017) 012071 doi:10.1088/1755-1315/82/1/012071

Microscopic surface wettability electrochemical characterization of tight sandstone with infrared spectra testing

L Song[1,3], Z F Ning[1], N Li[2], B Zhang[2] and G Y Ding[1]

[1]School of Petroleum Engineering, China University of Petroleum, Beijing 102249, China

[2]Research Institute of Petroleum Production Engineering, Huabei Oilfield Company of Petro China, 062552, China

E-mail: Songleizsy@163.com

Abstract. The distribution of charge density on the surface of microscopic tight oil is studied by using Stern double electric layer theory, and the mathematical flow model of polar fluid with micro powers in tight oil reservoir is established. The Fourier transform infrared (FTIR) were used to investigate the interaction of rock surface functional groups with fluids. The results show that: (1) When the external fluid of the polar group passes through the dense micro-nano pore, it will form an electric double layer on the surface of the rock, there will be a certain thickness of the liquid membrane, the fluid migration has a certain Of the electrical viscosity effect, will have a certain flow resistance. (2) The Fourier transform infrared spectroscopy of the Chang 7 tight reservoir rock samples exists and distributes different kinds of peaks. The left peak trend determines the presence of hydroxyl groups. The four fronts and types of the right side can be used to obtain that calcium carbonate CO_3^{2-} exists. (3) There are CO_3^{2-} and hydroxyl functional minerals in the Chang 7 tight sandstone samples. It is consistent with the basic mineral analysis measured by X-ray diffraction. When the external fluid affects the rock surface, the surface will occur in the physical van der Waals force and chemical bond interaction, so it will affect the flow of water on the surface.

1. Introduction

Tight oil is an important component of unconventional oil. Because of the prospect of mining and the prospect of large reserves, it has become a hot spot in the world at home and abroad. Many scholars have carried a lot of research on the pore structure, oil and water occurrence status and permeability of tight sandstone reservoir [1].Compared with conventional oil and gas, tight oil reservoirs have low oil saturation [2], low porosity and low permeability [3]. And the pore and throat size of the tight sandstone reservoir is smaller, the pore network model is more complicated, the fluid is more complicated in the micro-nano pore throat, and the wettability of the rock surface has great influence on the seepage of the fluid.

Wettability is a major factor influencing the occurrence and flow of fluid in the pores of tight reservoirs [4], while directly restricting the relative permeability of oil and water [5], capillary force [6] as well as oil recovery [7]. In the micro-nano pores of the dense oil reservoir, the complex fluid containing the polar functional groups will have a certain effect on the surface of the rock, which affects the flow effect and the wettability is the macroscopic nature under the influence of many microscopic factors. This paper reveals the effect from the microscopic scale through mathematical models and physical experiments.

2. Materials

All the experimental samples were taken from Chang 6 to Chang 8 of Yanchang oil layer in the Ordos Basin. The basic parameters of the samples are shown in table 1. The porosity of these samples are between 6.98~13.16%, an average of 10.38%; Water based permeability is between 0.05~0.23×10-3 μm^2, an average of 0.14×10-3 μm^2, which is a typical low porosity and permeability in tight reservoirs. The X-ray diffraction results show that the mineral composition of the samples is mostly quartz, feldspar and clay. The content of quartz is 30.20~54.1%, an average of 42.31%, the content of Feldspar is 16.10~42.10%, an average of 29.54% [8] which is consistent with the parameters of the rock basin. The content of Clay mineral is high, an average of 18.21%, and chlorite is the main component of clay minerals, an average of 9.48%; the content of Mixed-layer illite and illite are the second most. In addition, there is also a small amount of kaolinite.

Table 1. Physical property of tight oil samples and mineral composition.

Sample serial number	Porosity	Permea-bility	quartz	Felds par	Cal-cite	Dolo-mite	Illite and montmo-rillonite mixed layer	Illite	Chlo rite
	/%	/10^{-3} μm^2	/%	/%	/%	/%	/%	/%	/%
YC-7	13.16	0.23	33.40	42.10	0.70	4.40	4.46	1.75	13.19

Quartz, feldspar and other minerals can support a rigid overburden pressure, which helps the preservation of the original grain holes; meanwhile, feldspar, calcite and other minerals can easily generate corrosion hole, thereby improving the quality of the reservoir [8]. Chlorite usually grows along the surface of the mineral skeleton. On the one hand the formation of chlorite envelope can prevent the growth of secondary quartz, which is an effective protection for the reservoirs; On the other hand, with the growth of a large number of chlorite, a network structure forms, and it will destroy the original pore structure. Illite mainly distributes in the pores as silky or silk thread-like bridging formula, which greatly reduces the porosity of percolation. Kaolinite and feldspar solution openings are generally associated, and Kaolinite is a book-like or lamellar shape, it has a great effect on remodeling corrosion holes transformative [9, 10].

3. Two- layer model of rock interface

The main mineralization minerals in the tight sandstone reservoir include quartz, potassium feldspar, plagioclase, calcite, dolomite and clay mineral content. In the water, the mineral surface is susceptible to water and inorganic ions, and charged particles (ions, peeling ions, etc.) in the solid-liquid interface range of power transfer, resulting in mineral and water interface and potential difference, so that the surface of mineral particles Charging, electric double layer formed in the interphase interface(As shown in figure 1). There are two main ways to change the number of mineral particles on the surface of the belt: (1) Adsorption or cleavage of H^+ [11]; (2) Ion exchange [12].

In the vicinity of the mineral surface will attract the opposite sign ions in the liquid medium, and the two-phase interface state distributes the diffusion of the opposite sign ions of the double layer. The surface charge of the surface of the electric double layer is mainly concentrated in the surface layer thickness of 1 ~ 2 atoms, forming the inner layer of the electric double layer. In the area near 1 to 2 water molecules, the number of liquid medium ions is strongly attracted by the solid surface (electrostatic interaction and van der Waals attraction) to form a strong adsorption layer, called fixed adsorption layer or Stern layer. The Stern layer is the diffusion layer of the counter ion. In the diffusion

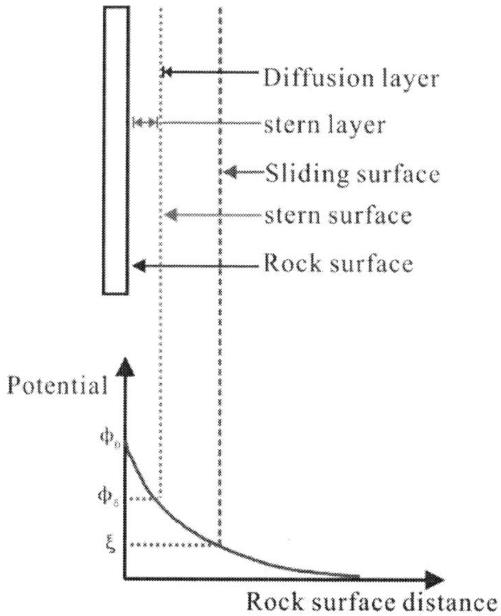

Figure 1. Double-layer model of tight reservoir surface.

layer, the counter ion concentration gradually decreases from the Stern layer to the inside of the liquid. φ_0 represents the potential difference between the solid surface and the solution, also known as the surface potential. Φ_δ denotes the potential difference between the electric center plane and the solution of the counter ion in the Stern layer, also called the Stern potential. ζ represents the potential difference between the solid surface and the solution when moving relative to the sliding surface, also known as the potential or zeta potential. In general, since the distance between the Stern surface and the sliding surface is small, the difference between the Zeta potential and the Stern potential can be ignored. For the silicate minerals with H^+ or OH^-, the pH of the solution is called zero point (PZC) at the surface potential $\varphi 0 = 0$, and the pH of the solution at zeta potential $\zeta = 0$ is called the isoelectric point (IEP). When there is no special adsorption, the chemical potential of any ion in the salt solution is [13]

$$\mu = ze\varphi x + kT\lg nx \tag{1}$$

Where φx is the potential at the distance from the solid surface x; nx is the ion number density (number of ions/L) at the solid surface x; z is the valence of the ion; e is the unit charge; k is the Boltzmann constant; T is the reaction temperature.

So the charge density ρ from the solid surface x can be expressed as

$$\rho_x = -\frac{e^2 \varphi_x}{kT} \sum_i n_{i\infty} z^2{}_i \tag{2}$$

The Poisson equation [14] is the excess charge density from the solid surface x

$$\rho_x = -\varepsilon_0 \varepsilon \left(\frac{d^2 \varphi_x}{dx^2} \right) \tag{3}$$

Where: ε_r is the relative permittivity of the solution (relative to water, 78 at room temperature); ε_0 is the vacuum dielectric constant; combined with equations (2) and (3) is [15-19]:

$$\frac{d^2 \varphi_x}{dx^2} = \varphi_x \frac{\sum_i n_{i\infty} e^2 z^2{}_i}{\varepsilon_0 \varepsilon kT} = \varphi_x \lambda^2 \tag{4}$$

Where λ is the reciprocal of the thickness of the electric double layer [38]:

$$h = \lambda^{-1} = \sqrt{\frac{\varepsilon_0 \varepsilon kT}{\sum_i n_{i\infty} e^2 z^2_i}} \tag{5}$$

The boundary condition of equation (4):

$$\varphi_\infty = 0, \varphi_{x=0} \rightarrow \varphi_0 \tag{6}$$

The relationship between the potential and the surface potential at a distance x from the surface can be described by Boltzmann's law:

$$\varphi_x = \varphi_0 e^{-\lambda x} \tag{7}$$

According to the above-mentioned double-layer theory, the potential difference between the solid surface and the solution interface due to the uneven distribution of charge. Therefore, in the electrolyte solution, the solid surface charges density and the surface potential of the corresponding relationship. For example, in the solution of NaCl and $CaCl_2$ electrolyte, a simplified model of diffusion double electric layer proposed by Gouy-Chapman can be used to calculate the surface charge density of solid as follows:

$$\rho = \sqrt{8\varepsilon_0 \varepsilon kT} \sinh\left(\frac{e\varphi_0}{2kT}\right)\left\{[Na^+]_\infty + [Ca^{2+}]_\infty\left(2 + e^{-e\varphi_0/kT}\right)\right\}^{1/2} \tag{8}$$

4. Fourier transforms infrared spectroscopy

4.1. Infrared spectroscopy features

The analysis of FTIR has the following characteristics: speed, easy operation, good repeatability, high sensitivity, and high signal to noise ratio advantages. It could rapidly and accurately qualitative analysis of solids, liquids, samples containing more than 30% of the gas content. The Vertex70 Fourier transform infrared spectrum is a third generation infrared spectrum with a spectral range of 15000 ~ 370 cm^{-1}. It has the characteristics of high accuracy, high resolution and higher accuracy than 0.01 cm^{-1} (usually In the analysis of minerals 4 cm^{-1}), high flux, low noise, fast measurement, easy operation, good repeatability, high sensitivity, low sample consumption, simple instrument structure, etc., which can effectively detect different crystals structure [20-28].

The infrared spectrum of the mineral in the 3800-3400cm^{-1} high frequency range within the original absorption curve changes, generally caused by the material with a large molecular structure of vibration, such as the molecular structure of water, organic molecules C-O, C-H key and so on. For example, the absorption peak of chlorite is 3580 cm^{-1}- 3440 cm^{-1}, the absorption peak of montmorillonite is 3620 cm^{-1}, and the absorption peak of illite is 3650 cm^{-1}.

4.2. FTIR of tight sandstone in Yanchang formation

The core of the Chang7 Formation of the extension of the Ordos Basin was tested by using the Vertex70 Fourier Transform Infrared Spectrometer, Hyperion Microscope, manufactured by Brooke Company, Germany.

As the 3600 ~ 3000 cm^{-1} band is mainly the vibration absorption band of hydroxyl, 3000 ~ 2800 cm^{-1} band is mainly methyl and methylene symmetrical and asymmetric vibrating absorption band in alkane. According to the analysis on the left side of figure 2, and there is a peak at 3650 ~ 3200 cm^{-1}. There is a strong peak at 800 cm^{-1}, which indicates that the core contains almost free hydroxyl groups and can be related to the H^+ in the external fluid, OH^- and other hydrogen bond force and chemical bond.

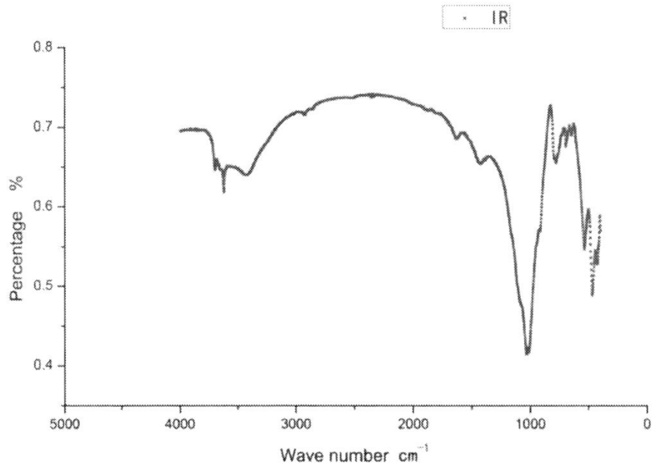

Figure 2. Chang 7 tight sandstone infrared spectra.

As can be seen from the right side of figure 2, it is observed that there are generally 4 to 5 absorption bands in the range of 1000 to 300 cm^{-1} in the infrared spectrum, including:

- including calcium carbonate CO_3^{2-} anti-symmetrical stretching vibration (v3) 800 cm^{-1}±;
- Outside bending vibration (v2) 657 cm^{-1}±,
- In-plane bending vibration (v4) 524 cm^{-1}±,
- Symmetrical stretching vibration (v1) 1050 cm^{-1}± infrared absorption peak is very weak. Among them, the anti-symmetrical stretching vibration (v3) absorption peak is strongest, and the peak width is larger;

From the above analysis, it can be seen that there are more CO_3^{2-} and hydroxyl functional minerals in the long 7-tight sandstone, which are consistent with the basic mineral analysis measured by X-ray diffraction. In the case where the external fluid affects the surface of the rock, Van der Waals force and chemical bond interaction, thus affecting the seepage.

5. Flow model with ion potential micro-nano pore

5.1. Wall potential distribution equation

As a result of the above analysis, we can see that there are electrostatic ions on the surface of the micro-rock of the dense reservoir. When the external fluid containing the dot ions passes through, the mechanism of the electric double layer will affect the distribution of charge on the surface of the rock and produce the surface potential. Since the surface charge of the rock attracts the opposite charge closer to the wall, the distribution of the charge in the fluid follows the Boltzmann distribution. The pore channel of the model is assumed to be a tube, as shown in figure 3, and because of the symmetry, Area to establish a mathematical model. The charge density of the microscopic rock surface potential φ and the solution volume per unit volume ρ is described by the following two-dimensional Poission-Boltzmann equation

$$\frac{\partial^2 \varphi}{\partial x^2} + \frac{\partial^2 \varphi}{\partial y^2} = -\frac{\rho}{\varepsilon_r \varepsilon_0} \tag{9}$$

The net charge density is

$$\rho = \left(n^+ - n^-\right) z_0 e = -2 n_0 z_0 e \sinh\left(\frac{z_0 e \phi}{k_B T}\right) \tag{10}$$

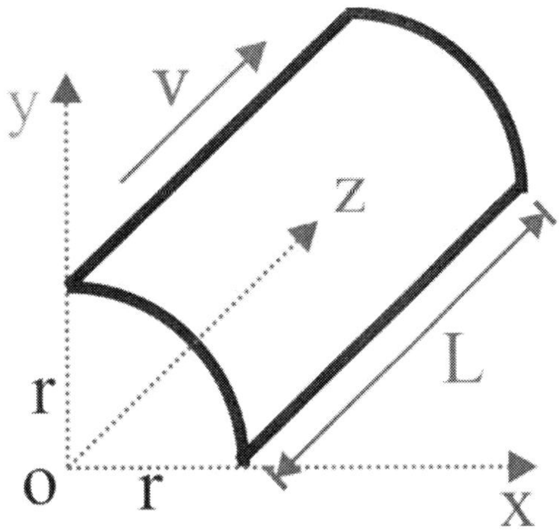

Figure 3. Model coordinate system.

where: $\mathbf{n_0}$ is a Boltzmann constant; $\mathbf{z}_0$ is the valence; $\mathbf{n}^+$, $\mathbf{n}^-$ are positive and negative ions; **e** is the basic charge (1.602×10^{-19}C); $\mathbf{k_B}$ is Boltzmann constant(1.38×10^{-23}J·K^{-1});

5.2. Electrostatic force distribution equation

The downstream current caused by the downstream current produces a conduction current in the opposite direction, which is defined as

$$\mathbf{I_s} = \int_A \mathbf{w(x,y)\rho(x,y)dA} \quad (11)$$

A is the cross-sectional area of the pore channel. Can be introduced by the above formula, the resulting electrostatic field E according to the charge balance, the conductivity σ and seepage morphology as follows

$$\mathbf{E} = -\frac{\mathbf{1}}{\sigma\mathbf{A}}\int_0^r\int_0^r \mathbf{w(x,y)\rho(x,y)dxdy} \quad (12)$$

In the absence of an applied electric field, the electrostatic force F is distributed by the potential distribution near the dense reservoir surface ϕ and the electrostatic charge density $\boldsymbol{\rho}$ is as follows

$$\mathbf{F} = \mathbf{E_z\rho(x,y)} \quad (13)$$

5.3. Momentum equation

According to the above analysis, we can see that for the incompressible NS equation, in addition to the water supply and so on to provide energy, there are dense reservoir rock surface double electric layer area by the net charge generated by the electrostatic force F, for steady flow, The control equation for v is

$$\frac{\partial^2 \mathbf{v}}{\partial \mathbf{x}^2} + \frac{\partial^2 \mathbf{v}}{\partial \mathbf{y}^2} = \frac{\mathbf{1}}{\mu}\frac{\mathbf{dp}}{\mathbf{dz}} - \frac{\mathbf{1}}{\mu}\mathbf{F} \quad (14)$$

Substituting equation (13) into the above equation

$$\frac{\partial^2 \mathbf{v}}{\partial \mathbf{x}^2} + \frac{\partial^2 \mathbf{v}}{\partial \mathbf{y}^2} = \frac{\mathbf{1}}{\mu}\frac{\mathbf{dp}}{\mathbf{dz}} - \frac{\mathbf{1}}{\mu}\mathbf{E_z\rho(x,y)} \quad (15)$$

5.4. The boundary condition of the control equation

For the potential equation (1), the boundary condition is as follows

$$x = 0, \varphi(0, y) = \varphi_s; x = r, \frac{d\varphi}{dx} = 0 \tag{16}$$

$$y = 0, \varphi(x, 0) = \varphi_s; y = r, \frac{d\varphi}{dy} = 0 \tag{17}$$

φ_s is the potential of the inner wall of the pore channel.

For the momentum equation (7), the boundary condition is

$$x = 0, v = 0; y = r, \frac{\partial v}{\partial x} = 0 \tag{18}$$

$$y = 0, v = 0; y = r, \frac{\partial v}{\partial y} = 0 \tag{19}$$

6. Conclusion

When there are polar ions of the external fluid through the dense reservoir micro-nano-pore, the rock surface will form an electric double layer, there will be a certain thickness of the liquid membrane, the fluid migration to produce a certain electrical viscosity effect.

The long 7 tight reservoir rocks measured Fourier infrared spectroscopy and the distribution of different types of peak, the left peak trend is willing to determine the hydroxyl, the right side of the four fronts and types can be obtained calcium carbonate CO_3^{2-} widely exist.

The long 7 sandstone contains more CO_3^{2-} and hydroxyl-containing minerals, and the above X-ray diffraction measured by the basic mineral analysis of the results consistent with the external fluid to the surface of the rock, the surface will occur in the physical van der Waals force And the interaction of chemical bonds, thus affecting the seepage.

The flow model of the electromechanical model is established.

References

[1] Wang Z L 2013 Research progress, existing problems and development trend of dense rock oil *Petrol. Exp. Geol.* **35** 587-95, 650
[2] Zhou S X, Xu Y B, Li S L, *et al* 2004 Physical properties prediction method and application of dense shaly sands reservoir *Nat. Gas Ind.* **24** 40-3
[3] Zou C N, Tao S Z, Hou L H, *et al* 2013 *Unconventional Oil and Gas Geology* 2nd ed (Beijing: Geological Publishing House) pp 22-32
[4] Yang S L and Wei J Z 2004 *Oil Layer Physics* (Beijing: Petroleum Industry Press)
[5] Wang F 1988 Effect of wettability alteration on water /oil relative permeability dispersion and flowable saturation in porous media *SPEJ* **3** 617-28
[6] Owens W W and Archer D 1971 The effect of rock wettability on oil-water relative permeability relationships *JPT* **23** 873-8
[7] Heiba A A, Davis H T and Scriven L E 1983 Effect of wettability on two-phase relative permeabilities and capillary pressures *SPE* 12172
[8] Yang H, Li S X and Liu X Y 2013 Characteristics and resource potential of dense oil and shale oil in Ordos Basin *Acta Petrolei Sinica* **1** 1-11
[9] Zhong D K 2012 The influence of reservoir petrology on diagenesis and pore development - A case study of the Triassic Yanchang Formation in Longdong area, Ordos Basin *PNGG* **6** 890-9
[10] Zhu H H, Zhong D K, Yao J L, Niu X B, Liang X W and Zhao Y 2014 Study on microcosmic

IOP Conf. Series: Earth and Environmental Science **82** (2017) 012071 doi:10.1088/1755-1315/82/1/012071

characteristics and mechanism of dense oil reservoirs in Chang 7 section of southwest Ordos *JCUMT* **5** 1

[11] Demir C, Bentli I, Gulgonul I, *et al* 2003 Effects of bivalent salts on the flotation separation of Na-feldspar from K-feldspar *Miner. Eng.* **16** 551-4

[12] Yukselen Y and Kaya A 2003 Zeta potential of kaolinite in the presence of alkali, alkaline earth and hydrolysable metal ions *Water Air Soil Pollut.* **145** 155-68

[13] Hiemenz P C and Rajagopalan R 1997 *Principles of Colloid and Surface Chemistry, Revised and Expanded* (Boca Raton: CRC Press) pp 468-530

[14] Israelachvili J N 2011 *Intermolecular and Surface Forces: Revised Third Edition* (San Diego: Academic Press)

[15] Zhong D K 2012 The influence of reservoir petrology on diagenesis and pore development - A case study of the Triassic Yanchang Formation in Longdong area, Ordos Basin *Petrol. Nat. Gas Geol.* **6** 890-9

[16] Zhu Z H 2011 Sandstone reservoir rock surface hydrophilicity and CO_2 (Beijing: College of Chemical Engineering, China University of Petroleum)

[17] Hiorth A, Cathles L M and Madland M V 2010 The impact of pore water chemistry on carbonate surface charge and oil wettability *Transport Porous Med.* **85** 1-21

[18] Greenwood R, Lapcikova B, Surynek M, *et al* 2007 The zeta potential of kaolin suspensions measured by electrophoresis and electroacoustics *Chem. Pap.* **61** 83-92

[19] Miller J D, Nalaskowski J, Abdul B, *et al* 2007 Surface characteristics of kaolinite and other selected two layer silicate minerals *Can. J. Chem. Eng.* **85** 617-24

[20] Brantley S L and Stillings L 1996 Feldspar dissolution at 25℃ and low Ph *AJC* **296** 101-27

[21] Qu X Y 2007 Experimental study of CO_2 fluid - sandstone interaction and its application in CO_2 gas reservoir (Changchun: College of Earth Sciences, Jilin University)

[22] Demir C, Abramov A A and Celik M S 2001 Flotation separation of Na-feldspar from K-feldspar by monovalent salts *Miner. Eng.* **14** 733-40

[23] Austad T, Rezaeidoust A and Puntervold T 2010 Chemical mechanism of low salinity water flooding in sandstone reservoirs (SPE 129767)

[24] Suijkerbuijk B, Hofman J, Ligthelm D, *et al* 2012 Fundamental investigations into wettability and low salinity flooding by parameter isolation (SPE 154204)

[25] Lager A, Webb K J, Black C J J, *et al* 2008 Low salinity oil recovery-an experimental investigation *Petrophysics* **49** 1-12

[26] Arnarson T S and Keil R G 2000 Mechanisms of pore water organic matter adsorption to montmorillonite *Mar. Chem.* **71** 309-20

[27] Sposito G 2008 *The Chemistry of Soils* (Oxford: Oxford University Press) 127-54

[28] Zhao Y X and Sun X Y 2003 *Organic Molecular Structure Spectroscopy* (Beijing: Science Press) pp 2-30

IOP Conf. Series: Earth and Environmental Science **82** (2017) 012072 doi:10.1088/1755-1315/82/1/012072

Study on a new water purification equipment with spiral lamellas

X R Feng[1,2]

[1]College of Civil Engineering and Architecture, Hainan University, Haikou 570228, China

E-mail: fxrgreat@163.com

Abstract. A new water purification equipment was introduced, especially the section of spiral lamellas. Utilization of spiral lamellas made the sedimentation space reach to 100%, not only improving sedimentation efficiency and reducing the cover space, but also saving investment. Production test results showed that the new water purification equipment with spiral lamellas had characteristics of excellent treatment efficiency and high shock resistant capacity. As the treatment water volume was 240 m^3/d, when the turbidity, COD_{Mn} and UV_{254} were 203 NTU, 1.90 mg/L and 0.030 cm^{-1} in raw water, they were 0.32 NTU, 0.72mg/L and 0.011 cm^{-1} respectively in effluent water, which could fully meet the drinking water hygiene requirement.

1. Introduction

The water purification equipments take surface water and shallow groundwater as water sources and closely combine coagulation, sedimentation and filtration processes together [1], which is equal to a small water supply treatment plant. Whether on operation and management, or on investment and scale, the water purification equipments are fit for small towns and rural areas that are shortage of farmland. As the conventional water treatment plants cover large areas and are managed complexly, their large scale construction is difficult in small towns and rural areas [2]. The development of efficient integrated water purification equipments has great significance to solve the safety problems of drinking water in rural areas.

Currently, water purification equipments on the market generally have disadvantages. Firstly, traditional oblique tube sedimentation device inevitably exist in a certain area of stagnant water [3], which obviously decreases the sedimentation efficiency. Secondly, by the volume restriction, flocculation time is short and floc particles are small and loose which are easy to be taken out of the tank [4], resulting in the deterioration of water quality. Finally, because of difficult installation of oblique tubes in circular cylinder walls, shapes of water purification equipment are mostly in rectangular, which need thicker sheets, increasing the cost of construction [5].

In order to overcome these shortcomings, this paper introduced a new integrated water purification equipment with spiral lamellas which was also sent into production trials in western mountain rural area of China.

2. Materials and methods

2.1. A new water purification equipment with spiral lamellas

This device used a new self-developed spiral lamella sedimentation technology which integrated

IOP Conf. Series: Earth and Environmental Science **82** (2017) 012072 doi:10.1088/1755-1315/82/1/012072

intensive grid coagulation and cyclone grit technology (figure 1). Raw water containing flocculant entered the bottom of the first flocculation chamber by tangential direction from inlet pipe, flowedrotating movement along the cylinder wall and then into the first flocculation chamber, of which large grain of sands were removed by cyclone grit sedimentation. There are six layers of grids in the first flocculation chamber, which are gradually decreased in perforation size from top to bottom. There are four layers of grids in the second flocculation chamber, which are gradually increased in perforation size from top to bottom. The speed of water flows was gradually slower through the grid perforation which made the floc particles obtain good flocculation conditions, while at the same time avoiding destruction of the formed floc particles.

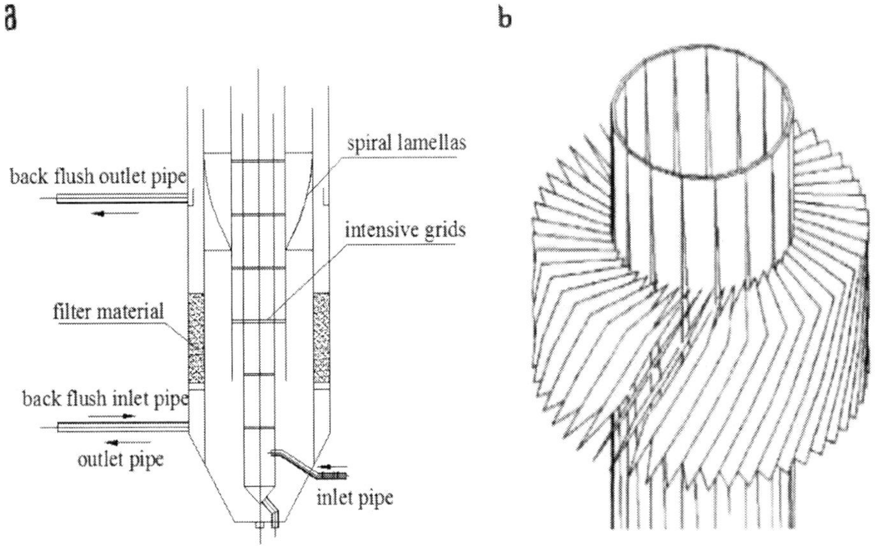

Figure 1. Integrated water purification equipment with spiral lamellas (a), and installation of spiral lamellas (b).

Lamellas were uniform spirally distributed between outer cylinder and internal cylinder walls in elliptical ring shape (figure 1 (a)). The long axis and short axis of the ellipse were determined by inner and outer cylindrical diameter of sedimentation area and spiral inclination angle. Spiral lamellas were fixed on the barrel walls of inside and outside sedimentation area. Water spirally flowed forward along the cylindrical columns (figure 1 (b)). Spiral lamellas were arranged in a ring sedimentation area, close to cylinder wall, so it would not appear short flow phenomenon and not had stagnant water region. The space utilization of sedimentation area reached to 100%. At the same time, it had advantages of high sedimentation efficiency, conducive to mud separation and easy installation of spiral lamellas.

This new water purification equipment adopted quickly downward flow filtration. Short shank filter heads were used into water distribution, which had less head loss and uniform water distribution. Filter materials adopted refining quartz sands.

2.2. Experimental section

This water purification equipment with spiral lamellas was used into production test for three months in a town of southwest China. Treatment water volume was 240 m^3/d. Seasonal water treatment tests preceded in conventional water and high turbidity water respectively. The quality of raw water was shown in table 1.

Table 1. Quality of raw water.

Indexes	Average value
Water temperature (℃)	25
pH	7.8
Turbidity (NTU)	95.9-1927
COD_{Mn} ($mg \cdot L^{-1}$)	1.90
UV_{254} (cm^{-1})	0.030

3. Results and discussion

3.1. Turbidity removal efficiency

When dosage of polyaluminum chloride (PAC) was 10 mg/L, turbidity removal efficiency was shown in figure 2. Raw water turbidity was between 40.7 NTU and 203 NTU in August for average value of 95.9 NTU. The effluent turbidity of sedimentation tank was between 0.32 NTU and 1.46 NTU for average value of 0.83 NTU, at the same time, the average removal efficiency was 99.02%. The turbidity of effluent water after filtration was between 0.09 NTU and 0.4 NTU for average value of 0.18 NTU. The turbidity removal efficiency was more than 98% in flocculation sedimentation, especially in high turbidity season which remained below 1 NTU. It can be seen that this device had the superiority of good sedimentation effect and strong shock resistant capacity.

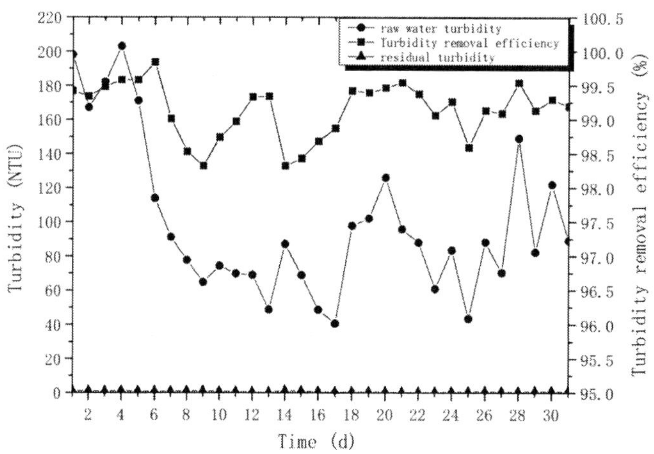

Figure 2. Turbidity removal efficiency under continuous work condition.

Due to heavy rains, raw water turbidity increased rapidly in July. Continuous tests results showed that the treatment efficiency was high by the PAC dosage of 40 mg/L. Take data on July 20 for example, when average turbidity of raw water was 1927 NTU and effluent water turbidity of sedimentation tank is 4.55 NTU, turbidity removal efficiency reached above 99.38%. Results indicated that effluent water turbidity of sedimentation chamber was stable and this equipment had excellent turbidity removal efficiency.

3.2. UV_{254} removal efficiency

UV_{254} value reflects the content of large organic matters such as lignin, humic substance, aromatic compounds and so on, which is an important control parameter of organic matters in water, at the same time, it can also be used as an alternative parameter of total organic carbon (TOC), dissolved organic carbon (DOC) and trihalomethane formation precursor (THMFP). Early as 1978, UV_{254} had been adopted into water quality monitoring conventional index in Japan.

When dosage of PAC was 10 mg/L, UV_{254} removal efficiency was shown in figure 3. During the experimental period, UV_{254} value of raw water was between 0.018 cm^{-1} and 0.064 cm^{-1} for average value of 0.030 cm^{-1}, while in sedimentated effluent water, UV_{254} value was between 0.011 cm^{-1} and 0.039 cm^{-1} for average value of 0.018 cm^{-1}. As a result, the average UV_{254} removal efficiency was 37.15%.

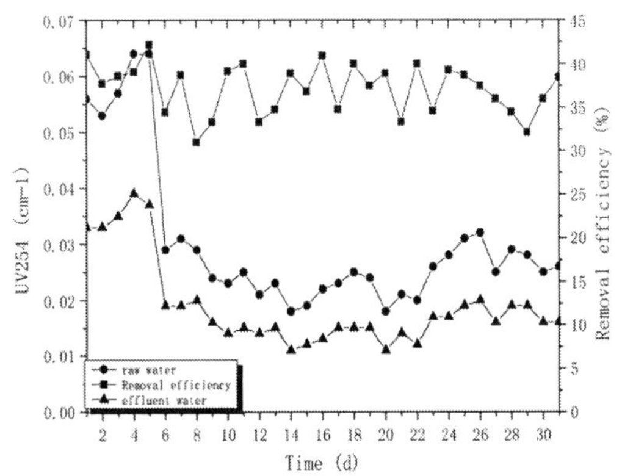

Figure 3. UV_{254} removal efficiency under continuous work condition.

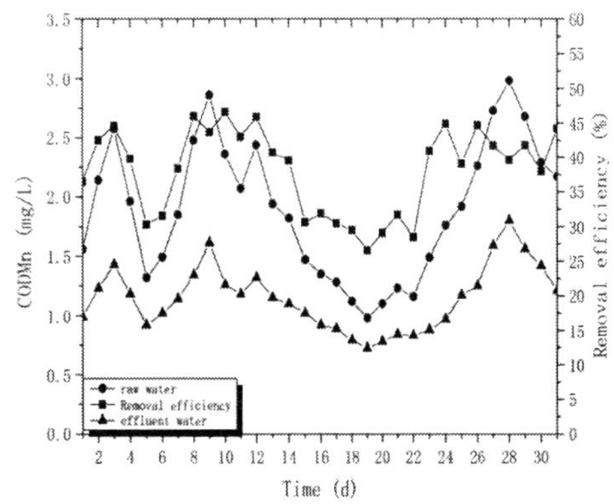

Figure 4. COD_{Mn} removal efficiency under continuous work condition.

3.3. COD_{Mn} removal efficiency

COD_{Mn} is commonly used as a composite index of pollution level caused by organic and reducing inorganic substances in surface water. It has good correlation with UV_{254}. When dosage of PAC was 10 mg/L, COD_{Mn} removal efficiency was shown in figure 4. COD_{Mn} value of raw water changed between 0.98 mg/L and 2.98 mg/L for average of 1.90 mg/L. The average value of effluent water was 1.11 mg/L. The COD_{Mn} removal efficiency reached to 39.68%.

4. Conclusions

Spiral lamellas in the new water purification equipment had superiorities of easy installation, uniform distribution, no short-flow and stagnant water areas, at the same time, with high sedimentation efficiency and 100% of space utilization. When the treatment water volume was 240 m^3/d and the dosage of PAC is 10 mg/L, turbidity removal efficiency reached to 99.6%, with 42.19% of UV_{254}, 50.7% of COD_{Mn}. In rainy season, despite that raw water turbidity was more than 3000 NTU, turbidity removal efficiency still reached to 99% by increasing the PAC dosage to 40 mg/L. The effluent water turbidity of sedimentation basin was below 5 NTU and of filtration basin was under 1 NTU. This water purification equipment had characteristics of less investment, easy manipulation and economic space occupation, fitting for small urban areas with weak economic, poor geography and low management level.

Acknowledgments

This work was financially supported by the Hainan Natural Science Foundation (No. 517061), Scientific Research Foundation of Hainan University (No. kyqd1505) and Teaching Reform Project of Hainan University (No. hdjy1758).

References

[1] Shinoda T 2014 *Desalination* **256** 203
[2] Jadhav M V and Mahajan Y S 2013 *Desalin. Water Treat.* **52** 6403-11
[3] Huang H, Yang X and Lu X 2014 *Phys. Fluids* **26** 161
[4] Chow C W K, Leeuwen J A V, Drikas M, Fabris R and Spark K M 1999 *Water Sci. Technol.* **40** 97-104
[5] Zhang K, Achari G, Sandiq R, Langford C H and Dore M H I 2012 *Water Res.* **46** 1673-83

IOP Conf. Series: Earth and Environmental Science 82 (2017) 012073 doi:10.1088/1755-1315/82/1/012073

Vertical distribution of motile phytoplankton in density currents

F Liu[1], L Zeng[1], Y H Wu[1,2], B Baoligao[1] and X Chen[1]

[1]State Key Laboratory of Simulation and Regulation of Water Cycle in River Basin, China Institute of Water Resources and Hydropower Research, A-1 Fuxing Road, Haidian District, Beijing, China

E-mail: wuyih@iwhr.com

Abstract. Frequently occurred algal blooms in the tributary areas of Three Gorges Reservoir have been demonstrated to be closely related to the density currents in earlier studies. This paper aims at studying the vertical distribution of motile phytoplankton in density currents. To achieve this objective, particle image velocimetry and planar laser induced fluorescence techniques were used to measure the velocity filed and the algal distribution in lock-exchange density currents. The vertical distribution of *Chlamydomonas reinhardtii* cells was gained, which was characterized by a lower concentration region above the interface and a higher concentration region below the interface. Moreover, the mechanism of this vertical concentration profile was discussed. The results are essential to understand the mechanisms of the appearance and disappearance of algal blooms in the Three Gorges Reservoir area.

1. Introduction

The Three Gorges Reservoir (TGR) is one of the largest reservoirs in the world, with a storage capacity of 39.9 billion m^3, a surface area of 1.080 km^2, and a control watershed area of more than 1 million km^2 [1]. Since June 2003, the backwater areas of the tributaries have suffered from the harmful and periodical algal blooms, particularly in the Xiangxi, Xiaojiang and Daning rivers [2]. The most dominant species are the motile with vesicles or flagella, such as Peridiniopsis, Chlamydomonas, Aphanizonmenon, and Microcystis [3].

At present, the mechanisms underlying the appearance and disappearance of the algal blooms in the tributaries are still not very clear. Previous studies indicate that the algal blooms in the tributary areas of TGR are sensitive to the variations in thermal stratification and hydrodynamics [4]. The bidirectional density current is the typical flow regime in the tributaries of TGR [1]. The critical depth model, proposed by Sverdrup [5], was used to explain the appearance and disappearance of algal blooms in the Xiangxi Bay [6], the largest tributary in the lower reaches of TGR.

As is known, the appropriate hydraulic condition will cause the algae to float up to the water surface and then the algae blooms occur [7]. The vertical movement and distribution of phytoplankton are governed by the hydrodynamic condition and the phytoplankton's behavior [8]. Kessler [9] presented that the swimming direction of algal cells was determined by the balance of gravitational and viscous torques. Pedley and Kessler [10] presented a continuum model to describe the behavior and distribution of motile phytoplankton. Durham et al. [11] pointed out that only when the viscous shear exceeded a critical value would the motile phytoplankton accumulate in a thin layer.

In this study, the particle image velocimetry (PIV) and planar laser induced fluorescence (PLIF)

techniques were used to measure the flow field and the distribution of *C. reinhardtii* in the lock-exchange density currents. Moreover, the relationship between the vertical distribution of phytoplankton and the depth-varying shear were analyzed.

2. Materials and methods

2.1. Experimental technique

The PIV and PLIF techniques were used for measurement. The PIV was used to measure the velocity filed. The PLIF was used to measure the distribution of algae. Figure 1 shows the schematic illustration of the experimental equipment and the PIV-PLIF system. All the windows in the laboratory were covered with black and opaque curtains to create a dark environment. An optical laser lens system (MDL-F-447) was employed to provide the 447-nm laser sheet, with approximately 0.35 cm thick. The particles (0.3 μm) of titanium oxide were used for the PIV measurements. The *C. reinhardtii* (FACHB-265) from Freshwater Algae Culture Collection at the Institute of Hydrobiology was used for the PLIF measurements. Two CCD cameras (Baumer HXC40) were used with a spatial resolution of 2048×2048 px^2. The PLIF camera was fitted with an optical high pass filter (550 nm) to collect the red fluorescence emitted by algae and to block out the 447 nm blue light.

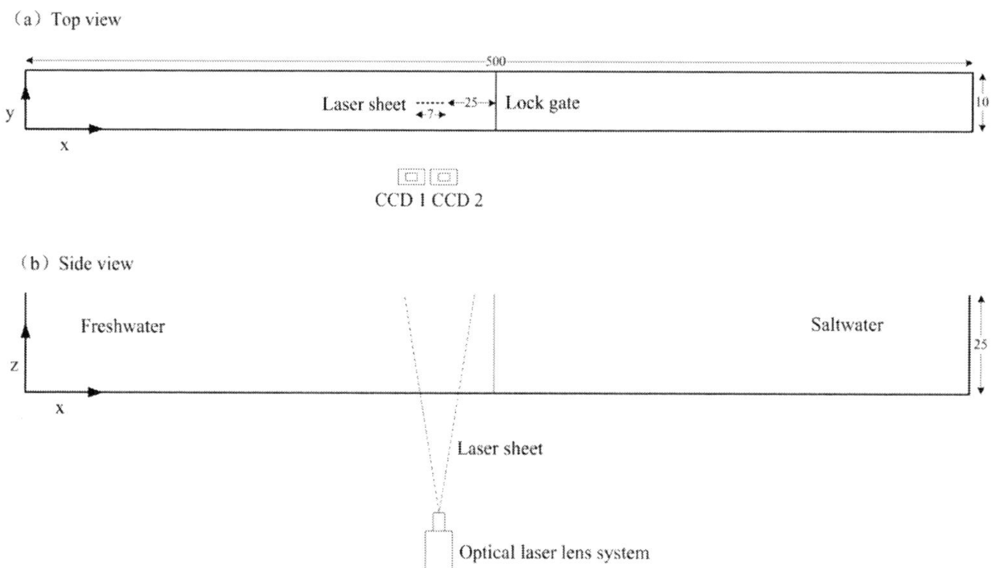

Figure 1. Schematic illustration of the experimental system (unit: cm).

2.2. Experimental setup

Density currents are very common in lakes, reservoirs and other water bodies. The lock-exchange is a widely-used laboratory-scale set-up that can create two gravity currents propagating horizontally [12]. Here the gravity currents were created in a perspex tank (the length × width × depth is 500 cm × 10 cm × 25 cm) based on the lock-exchange system shown in figure 1.

The solutions were prepared in two big storage tanks, each with a volume of 144 liters. One was filled with freshwater, and the other was filled with saltwater. The same amount of particles and living Chlamydomonas cells were added in both of the storage tanks and stirred adequately. The algae cell density was about 4×10^5 per liter in each of the prepared solution. A vertical lock gate was inserted in the middle of the perspex tank to divide the tank into two parts. The left part was filled with the prepared freshwater to a depth of 6.5 cm and the right part with the prepared saltwater to the same depth. The fluids were allowed to settle for several minutes before starting the experiment by opening the lock gate. The PLIF camera was set to single frame mode and the PIV camera to double frame mode. The two cameras recorded every 3 sec during the experiment for a total of 30 min. A reference coordinate system was obtained by recording images on the plane defined by the light sheet after every experiment. Three experiments were carried out and the salinities of saltwater were 0.3‰, 0.4‰ and 0.5‰, respectively.

2.3. Data analysis

The laser sheet through the water is an inverted trapezoidal with a base of 70 mm and a height of 65 mm. Since the intensity distribution of the laser beam is an M shape after passing through the Powell lenses used in the optical laser lens system, only the 30 mm in the middle had a relatively uniform intensity (higher than 80%). Thus, the area of approximately 30×65 mm^2 in the middle of the laser sheet was used to analyze the algae distribution. The instantaneous two-dimensional velocity fields and gray value fields were analyzed by the PIV and PLIF processing system, which is developed by Beijing Sinfotek Technology Co., Ltd. The attenuation of laser intensity along the optical path was negligible in the experiments, so the gray value fields can represent the distribution of algae cells. The analyzed area of about 30×65 mm^2 was divided into 27×57 grids (each grid was 1.107×1.107 mm^2) for both the PIV and PILE analyses. The average of 120 images represented the time averaged velocity field (or the algal distribution) for every 6 min, and 20 images for every 1 min.

3. Results and discussion

3.1. Velocity field

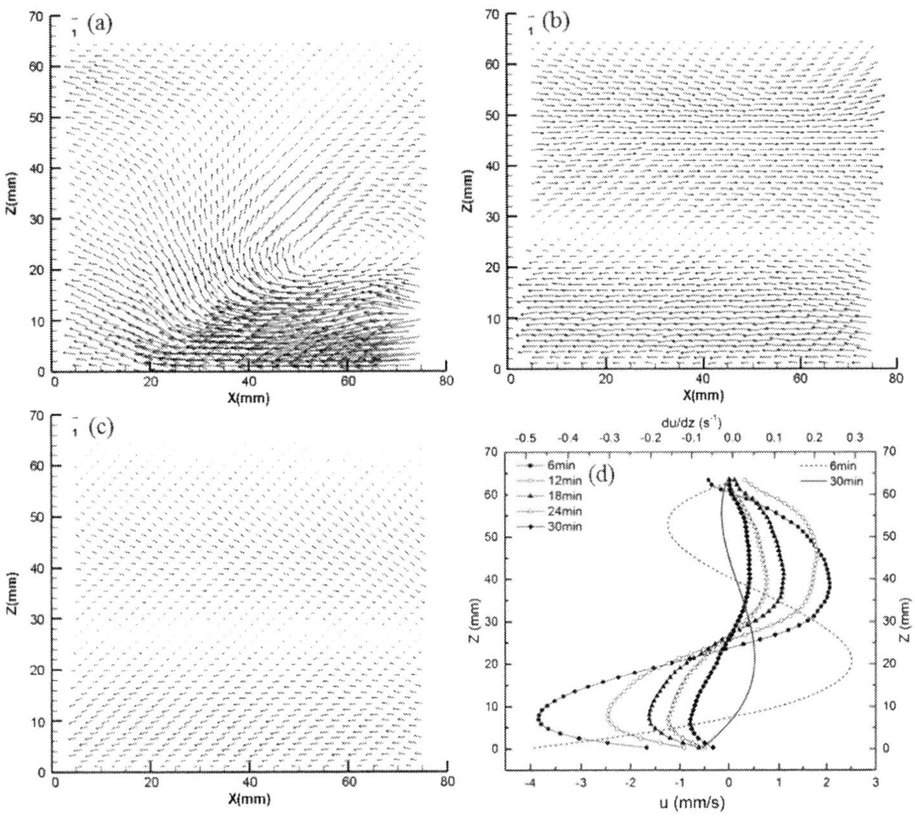

Figure 2. Velocity distribution (a: the instantaneous velocity field at t=30 sec; b: the instantaneous velocity field at t=6 min 39 sec; c: the instantaneous velocity field at t=20 min 27 sec; d: the time-dependent evolution of sectional velocity and the shear at the position X=40 mm).

The flow fields and the developing process were similar in all of the three experimental cases, and the velocity fields at different times for case 2 (the salinity of saltwater was 0.4‰) were shown in figure 2. When the lock gate was removed, the difference in the hydrostatic pressure induced the saltwater to flow to the left along the bottom of the tank, while the freshwater flow to the right along the water surface. As shown in figure 2 (a), the freshwater around the saltwater head was pushed to turn up during the traveling of the saltwater head [13]. After the head, it was the body part of the lock-exchange gravity current, and the velocity field was shown in figure 2 (b). The body part showed a typical stratified and bidirectional flow field. The horizontal velocity (u) kept decreasing over time, as shown in figure 2 (c).

Seen from the sectional velocity profiles at the position of X=40 mm (figure 2 (c)), the fluid was stratified into two layers. Above the height of Z=26mm was the freshwater layer with positive horizontal velocities while below the height was the saltwater layer with negative horizontal velocities. Between the two layers was the interface with horizontal velocities changing from negative to positive. The whole sectional velocity profile seemed as a heterochiral S. In term of the time series of the sectional velocity profiles (figure 2 (d)), the height of the saltwater layer was increasing slowly over time, and could reach 26 mm at the end of this experiment. Compared to the freshwater, the saltwater had a relatively less thickness and a higher velocity in the whole experiment. The distribution of the shear was shown in figure 2 (d). The distribution of the shear can be described as a cubic polynomial curve in all the three experiments. The shears decreased as the velocity decreased over time. The range of the time averaged shears was from -0.48 s^{-1} to 0.29 s^{-1} for t=0~6 min, and was from -0.07 s^{-1} to 0.05 s^{-1} for t=24~30 min.

3.2. Algal vertical distribution

Because the hydrodynamic condition cannot obviously change the algal vertical distribution in a short time, an average of 120 images was used to represent a time-averaged distribution for every 6 min. The algal vertical distribution and its developing process were similar in all the three experimental cases, and the algal distributions at different time for case 1 (the salinity of saltwater was 0.3‰) were shown in figure 3. Before the experiments started, algae cells were basically in a homogeneous distribution in the fluids. Thus, the distribution of gray value for the first 6 minutes was relatively uniform, as shown in figure 3 (a). For the case of stable gravity current with head passing the illuminated area, there was a region with relatively higher gray value around the height of Z=16 mm in the saltwater layer (region H) while another region with relatively lower gray value around the height of Z=28 mm in the freshwater layer (region L). The gray value of region H firstly increased to the peak value and then decreased to the gray level of other regions in saltwater as the experiments develops. Thus, the gray level of the whole saltwater was relatively uniform and the region H disappeared finally, as shown in figure 3 (e). On the contrary, the region L existed through the entire experiments. The upper and bottom boundaries of this region was expanding over time and the height of region L almost increased to 20 mm at the end.

There was also an area with relatively lower gray level around Z=50 mm near the water surface, where the horizontal velocity was relatively low and the fluid field was obviously disturbed by removing the gate. Thus, the algal distribution in the region above Z=50 mm near the water surface was not discussed in this research. The light intensity of the laser sheet was especially high at the bottom of the fluid, and the fluorescence induced by the laser was also very intense. Thus, the algal distribution in the region below Z=5 mm at the bottom was also not discussed here. In conclusion, the vertical distribution of algae was uniform at first and gradually changed into a sharp profile in density currents. The vertical distribution profile was characterized as the relatively low algal concentration region above the interface and the high algal concentration region below the interface. With the shear decreased, the low algal concentration region expanded while the high algal concentration region gradually disappeared.

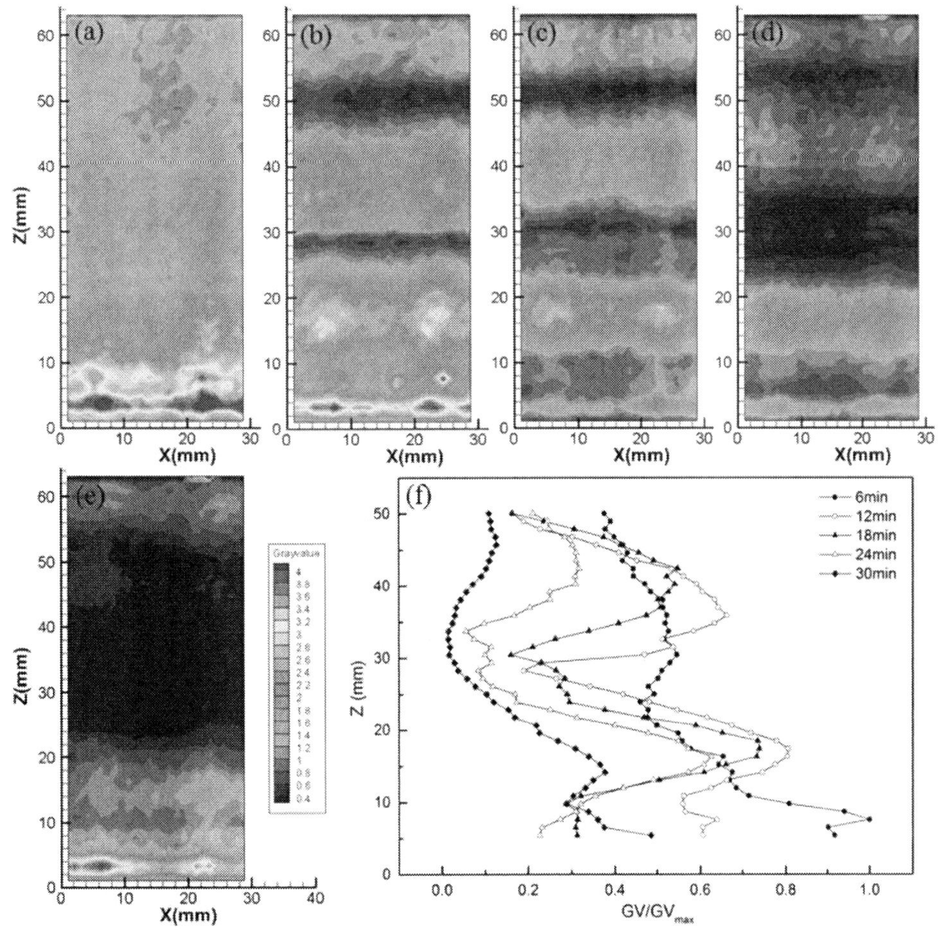

Figure 3. The time series of algal distribution in case 1 (a~e: algal vertical two-dimensional distribution for every 6 min, f: the time-dependent evolution of algal vertical distribution profile).

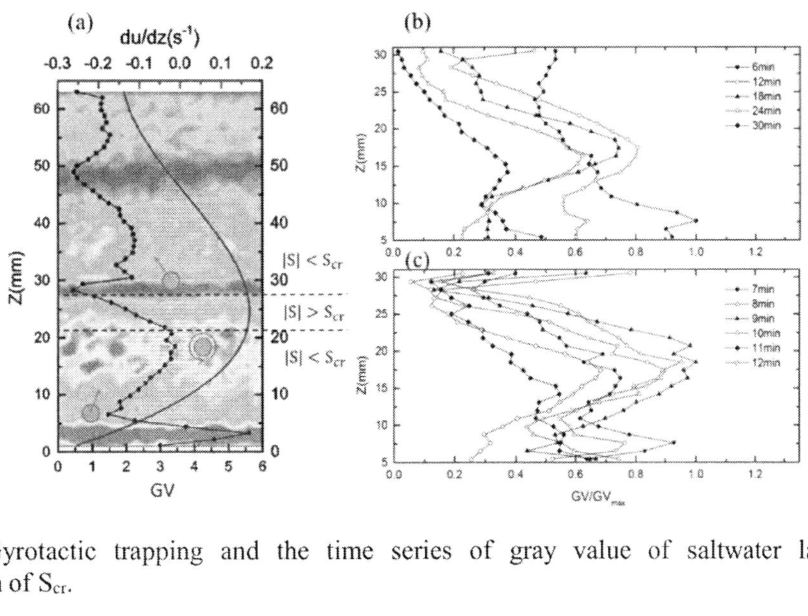

Figure 4. Gyrotactic trapping and the time series of gray value of saltwater layer and the determination of S_{cr}.

C. reinhardtii used in the experiments is a typical gyrotacitic species [14]. In density currents, the velocity gradients give the cell a viscous torque which makes cells spin. Once the cell deviates from the vertical direction, there will be a gravitational torque trying to rotate the cell back to the vertical direction. When the gravitational torque is larger than the viscous torque, the cell can swim upward generally and the direction is decided by the balance of the two torques. But when the gravitational torque is smaller than the viscous torque, cells will spin and accumulate in the layer [11], as shown in figure 4(a). Thus, there should be a critical viscous shear (Scr), above which the *C. reinhardtii* cells will spin around.

In our experiments, the hydrodynamic shear was increasing from the bottom of the saltwater layer to the interface. In the lower part of the saltwater layer, the shear was less than the critical shear and the algae cells kept swimming upward and left this region a relatively lower concentration. Then the cells kept on going upward to the region with the viscous torque equaled to the critical shear. In this region, the algae cells were trapped and spinning, thus the phytoplankton concentration in this layer was increasing over time with more and more cells from the lower part were captured here. The layer with viscous shear higher than the critical value acted as a barrier preventing algae cells from moving upward. Thus, most cells could not reach this area and forming a low concentration region here. Upon this layer with shear higher than S_{cr}, the shears were smaller than S_{cr} and cells migrated away and forming the upper boundary of the low concentration region, which moved upward with cells swimming upward. As experiments developed, the position of S_{cr} in saltwater was moving down as the whole shear decreased. Thus, the cells near the bottom boundary could swim up again and the bottom boundary seemed to be moving down.

3.3. Critical shear

The critical shear is an important parameter. Only when the shear of the interface is higher than the critical shear can this algal vertical profile be formed. For the S_{cr} appeared in the saltwater layer, the gray level and the shear were analyzed only for the region below Z=30 mm. As can be seen from figure 4 (b), the highest gray level appeared in the second 6 min. An average of 20 images was used to represent a time-averaged distribution for 1 min during the second 6 min (figure 4 (c)). The highest gray level was determined in the ninth min and the vertical distribution of gray level and shear was shown in figure 4 (a). The critical shear and the highest gray value appeared at the same position. For the three cases in this experiment, the critical shears were 0.15 s^{-1}, 0.15 s^{-1} and 0.13 s^{-1}, respectively.

4. Conclusion

The PIV-PLIF system was used to measure the velocity filed and the algal distribution profile in the lock-exchange density currents. In the whole experiments, the fluid field was stratified and bidirectional. The vertical distribution of horizontal velocity can be described with a heterochiral S shape. The distributions of shear can be approximately estimated with a cubic polynomial curve. The shear near the interface was the largest. The vertical distribution of algae can be characterized as the relatively low concentration region above the interface and the high concentration region below the interface. As the horizontal velocity and the shear decreased, the region with high algal concentration gradually disappeared while the region with low concentration gradually appeared and increased. In these experiments, the critical shear ranged from 0.13 s^{-1} to 0.15 s^{-1}.

Acknowledgments

This work was financially supported by the National Key Research and Development Program of China (2016YFC0502202), and partly by Independent Research Project of State Key Laboratory of Simulation and Regulation of Water Cycle in River Basin (2015ZY04).

References

[1] Ma J, Liu D F, Wells S A, Tang H W, Ji D B and Yang Z J 2015 Modeling density currents in a typical tributary of the Three Gorges Reservoir China *Ecol. Model.* **296** 113-25

[2] Wang L, Cai Q H, Zhang M, Tan L, Xu Y Y amd Kong L H 2011 Longitudinal patterns of phytoplankton distribution in a tributary bay under reservoir operation *Quatern. Int.* **244** 280-8
[3] China National Environmental Monitoring Centre 2015 Ecology and environmental bulletin of the Three Gorges Project (in Chinese)
[4] Yang Z J, Liu D F, Ji D B and Xiao S B 2010 Influence of the impounding process of the Three Gorges Reservoir up to water level 172.5 m on water eutrophication in the Xiangxi Bay *Technol. Sci.* **53** 1114-25
[5] Sverdrup H U 1953 On conditions for the vernal blooming of phytoplankton *ICES J. Mar. Sci.* **18** 287-95
[6] Liu L, Liu D F, Johnson D M, Yi Z Q and Huang Y L 2012 Effects of vertical mixing on phytoplankton blooms in Xiangxi Bay of Three Gorges Reservoir: Implications for management *Water Res.* **46** 2121-30
[7] Kong F X and Gao G 2005 Hypothesis on cyanobacteria bloom-forming mechanism in large shallow eutrophic lakes *Acta Ecol. Sin.* **25** 589-95
[8] Pedley T J and Kessler J O 1992 Hydrodynamic phenomena in suspensions of swimming microorganisms *Annu. Rev. Fluid Mech.* **24** 313-58
[9] Kessler J O 1985 Hydrodynamic focusing of motile algal cells *Nature* **313** 218-20
[10] Pedley T J and Kessler J O 1990 A new continuum model for suspensions of gyrotactic micro-organisms *J. Fluid Mech.* **212** 155-82
[11] Durham W M, Kessler J O and Stocker R 2009 Distribution of vertical motility by shear triggers formation of thin phytoplankton layers *Science* **232** 1067-70
[12] Hiester H R, Piggott M D and Allison P A 2011 The impact of mesh adaptivity on the gravity current front speed in a two-dimensional lock-exchange *Ocean Model.* **38** 1-21
[13] Chen G Q, Li X W and Li Z 2002 Lock-exchange turbulent density currents *Sci. China Ser. E* **32** 754-64 (in Chinese)
[14] Yoshimura K, Matsuo Y and Kamiya R 2003 Gravitaxis in Chlamydomonas reinhardtii studied with novel mutants *Plant Cell Physiol.* **44** 1112-8

IOP Conf. Series: Earth and Environmental Science 82 (2017) 012074 doi:10.1088/1755-1315/82/1/012074

The synthesis of novel thiol/amino bifunctionalized SBA-15 and application on the Cr(VI) absorption

Y Yang[1], D Wang[1,2] and J X Yang[1]

[1]The College of Materials and Chemical Engineering, Hainan University, HaiKou, 570228, China

E-mail: wangdun2004@126.com

Abstract. Novel amino/thiol organic groups modified SBA-15 materials were successfully prepared by a simple co-condensation approach. The synthesize mesoporous materials were characterized by FT-IR, SAXRD, N_2 adsorption and elemental analysis (EA). The absorption behavior of Cr (VI) in the samples has been investigated. The experiments revealed that the adsorption amount of Cr (VI) was decided by thiol groups, but the adsorption equilibrium time was mainly effected by amino groups. The 15% NH_2+20%SH sample had the best adsorption performance. Its maximum adsorption capacity can be up to 49.29 mg/g at the optimum conditions. The research of adsorption mechanism including adsorption kinetics and adsorption thermodynamics was also presented.

1. Introduction

Adsorption, as a kind of common method, was usually applied to deal with the heavy metal ions which were very harmful to the human health and environment even at very low concentration, such as Cr(VI), Pb(II). Comparison to other separation methods, it is obvious that adsorption was cheaper, more convenient and efficient. The adsorbent is the heart of the adsorption, so it was a key to find a high efficiency and inexpensive adsorbent. Recently, a type of mesoporous material, SBA-15, has attracted considerable attentions as its higher surface area, larger pore and a lot of silicon hydroxyl. Although the conventional SBA-15 has good adsorption ability, it is not a good absorbent for the heavy toxic metal ions. Functionality of SBA-15, grafting some functional organic groups into the SBA-15, as a strategy, was adopted to improve their adsorption of heavy toxic metal ions. Liu A M, et al [1] prepared two functionalized SBA-15 samples and discussed their adsorption characters for heavy toxic metal ions, the SBA-15 with thiol group exhibited a higher adsorption capacity to Hg(II), and SBA-15 with amino group can remove Zn^{2+}, Cr^{3+} and Ni^{2+} efficiently. Shahbazia A et al [2] synthesized the SBA-15 functionalized with amine ($-NH_2$) group and melamine-based dendrimer amines (MDA) and the results suggested MDA functionalized SBA-15 was the best adsorbent to remove the Pb(II), Cu(II) and Cd(II) ions. Zhang L J et al [3] grafted 1-(3-carboxyphenyl)-2-thiourea (CPTU) into SBA-15 and systematically researched the adsorption of Hg(II). This material displayed a higher adsorption capacity (64.5 mg/g), and the materials can still remove over 95% of Hg(II) after seven cycles. In our previous research [4], we found that the bifunctionalized SBA-15 which contained 30% (30% is molar ratio of (APTES+MPTMS)/TEOS) amino and thiol groups exhibited the excellence ability of removal of Cr(VI). In order to find out which organic groups play key role in the adsorption process, the SBA-15 contained different content of organic groups were synthesized to study. In this article, the bifunctionalized SBA-15 which contained fixed contents of amino group

(15% standed for molar contents of AAPTS/TEOS) and changeable thiol group contents(from 10% to 30%), as well as fixed contents of thiol group(15%) and variable contents of amino groups(from 10% to 30%) were produced via a simple co-condensation approach, and the interaction of amino group and thiol group was studied to find to synthesize the best adsorbent and selected it to investigate the adsorption kinetic and thermodynamic mechanism.

2. Chemicals and experiment

2.1. Chemicals

Poly (ethylene glycol)-block-poly (propylene glycol)-block-poly (ethylene glycol) (P123, Mn=5800), (3-Aminopropyl) triethoxysilane (APTES, 98%) and (3-Mercaptopropyl) trimethoxysilane (MPTMS, 95%) were purchased from Aldrich, Aladdin and Aladdin, respectively. Ethyl silicate (TEOS) was employed as a silica source. Commercial grade reagents and solvents were used without further purification. Distilled water was used in the aqueous solutions.

2.2. Experiments

2.2.1. Preparation of NH_2/SH bifunctionalized SBA-15 (denoted as SBA-NH_2-SH). The bifunctionalized SBA-15 was prepared as follows: 2 g P123 dissolved in 62.5 g 2M HCl solution, the resulting mixture was stirred at 40°C until the mixture become clear. Then the amino precursor (APTES) and thiol precursor (MPTMS) were slowly added after 4.25 g TEOS was added into the solution. The molar composition of the mixture solution (the fixed amino groups) was 0.02TEOS :0.003 APTES : x MPTMS : 3.05HCl : 0.0085P123 : 82.5H_2O, where x varied from 0.002 to 0.006, or MPTMS/TEOS = 10%-30%, and the molar composition of the mixture solution (fixed thiol group) was 0.02TEOS : y APTES : 0.003 MPTMS : 3.05HCl : 0.0085P123 : 82.5H_2O, where y varied from 0.002 to 0.006, or MPTMS/TEOS = 10%-30%. The reaction mixture was stirred at 40°C for 20 h, and then the mixture was aged for 24 h at 100°C under static condition. The precipitation product was recovered by filtration and dried under the air. The template was extracted by flexing for 24 h with excess acid ethanol solution. Finally, the sample was filtered, washed with ethanol, water and dried at 60°C under the air.

2.2.2. Characterization. Small angle X ray diffraction (SAXRD) patterns of the materials was measured on Bruker (German) D8 advance X-ray diffraction spectrometer using Cu Kα radiation, the scan speed kept 0.1°/s and the range of 2θ angle between 0.600° and 9.993°. The elementar Analysed system GmbH - vario EL III Element Analyser was used to examine the contents of N and S. The FT-IR spectra was recorded on a Bruker Optics Fourier Transform Infrared Spectrometer. N_2 adsorption-desorption isotherms were estimated on 3H-2000PS2 analyser at 77K.The concentration of Cr(VI) was determined by the UV-visible spectrophotometer.

2.2.3. Adsorption experiments. The Cr(VI) adsorption behaviors of bifunctionalized SBA-15 at various time and different content of -NH2 and -SH groups were determined to evaluate their Cr (VI) adsorption capacity and time under the optimization condition [4]. Briefly, 0.1 g absorbent was put in 100 mL 50 mg/L Cr(VI) solution and allowed to absorb the Cr(VI) at 25°C under pH = 2.0. The aqueous phase was collected and concentration of residue Cr (VI) was carried out on the UV-visible spectrophotometer.

2.2.4. Adsorption kinetics. The bifunctionalized SBA-15 samples (0.1 g) were respectively added to 100 mL 50 mg/L, 100 mg/L and 150 mg/L Cr(VI) solution at room temperature. 1 mL of Cr (VI) solution was taken at different time intervals to measure the residual Cr (VI) concentration.

2.2.5. Adsorption isotherms. The adsorption isotherms of bifunctionalized SBA-15 were measured by using 0.1 g absorbent in 100 mL 70 mg/L Cr (VI) solution and incubated the mixture at 313 K, 323 K and 313 K respectively for 6 h. The residual Cr (VI) concentrations and the Cr (VI) uptake per unit mass the bifunctionalized SBA-15 were determined as described above.

3. Result and discuss

3.1. Characterization of bifunctionalized SBA-15 materials

3.1.1. FT-IR. The FT-IR spectroscopy was used to determine the amino group and thiol group grafted into the SBA-15. The FT-IR of pure SBA-15 and bifunctionalized SB-15 were exhibited in figure 1. Three characteristics peaks of SBA-15 at 1085 cm^{-1}, 806 cm^{-1}, 462 cm^{-1} were easy to observe. Compared with the three curve, (2) and (3) appeared some new absorption peaks, the peak at 574 cm^{-1} was contributed to the bending vibration of C-S bond, that perfectly indicated that the thiol group have been immobilized into the SBA-15. Generally, the stretching bending of N-H bond was normally appeared at about 3380 cm^{-1} and 3310 cm^{-1}, however it is hard to observe in bifunctionalized SBA-NH_2-SH [5]. Between 3500 cm^{-1} and 3000 cm^{-1}, there was a broad band of surface hydroxyl group that was overlayed the peaks of N-H bond, but the peak around 1460 cm^{-1} could be seen a weak adsorption peaks that was the symmetric vibration bending of N-H. In summary, the FT-IR has confirmed the amino-and thiol- groups were successfully introduced into the SBA-15.

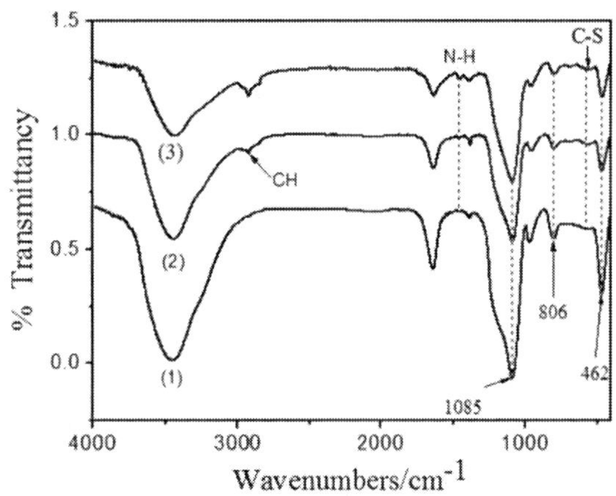

Figure 1. FT-IR spectra of three samples: (1) SBA-15, (2) bifunctionalized SBA-NH_2-SH (10%NH_2+15%SH), (3) bifunctionalized SBA-NH_2-SH (10%SH+15%NH_2).

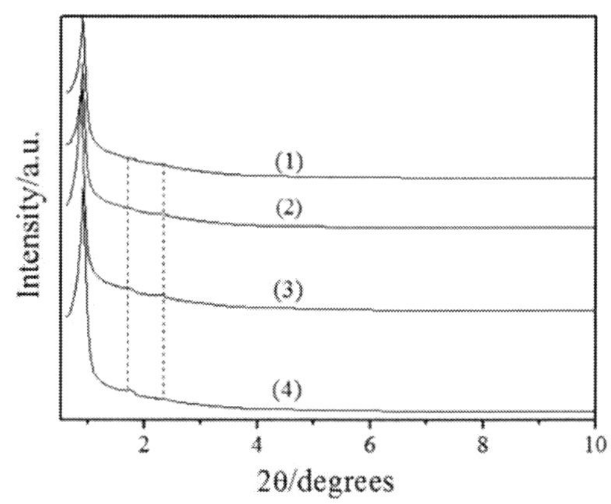

Figure 2. The SAXRD pattern of the four samples. (1) 15%NH_2+20%SH, (2) 15%SH+15%NH_2, (3) 15%SH+10%NH_2, (4) 15%NH_2+10%SH.

3.1.2. SAXRD. The SAXRD of samples was shown in figure 2. As was shown in figure 2, curve (3) and (4) appeared three characteristic diffractions peaks like pure SBA-15 that could be observed at (100), (110), and (200) [6, 7]. The other curves only could see a clearly peak at d100 and at d200, while the peak at d110 was disappeared. As was displayed in figure 2, when the total contents of amino and thiol groups were below 30%, there are three conspicuous peaks. As the total groups percentage of the concentrations were over 30%, the other two peaks turned more weak and even disappeared, that could be contribute to the distortion of the order structure of the functionalized SBA-15. The results were consistent with some previous reports like MCM-41 [8], SBA-NH2 [9] and SBA-SO_3H [10], a probable explanation was that the self-assembly of surfactant and the silica precursors aggregates were disrupted by the amino and thiol groups [11, 12], as the mixture concentrations of the synthesized samples was over a critical level, the integrity and ordering of the resultant materials were destroyed.

IOP Conf. Series: Earth and Environmental Science **82** (2017) 012074 doi:10.1088/1755-1315/82/1/012074

3.1.3. N_2 adsorption-desorption isotherms and elemental analysis. The N_2 adsorption and the pore size distributions were exhibited in figures 3 and 4. Based on the adsorption isotherm classification of the IUPAC [13], the N_2 adsorption of all samples was a typical Langmuir IV isotherm. According to the ten curves, when the total contents of the groups were below 25%, the hysteresis loop of those materials belongs to H1-type, the typical features of the mesoporous structure. While the contents were over 25%, the hysteresis loop of absorption and desorption branch were no longer parallel and gradually deviated from the H1-type. These phenomena indicated that the pore size uniformity were declined even partly appeared channel blocking after a larger number of amino and thiol groups were grafted into the SBA-15 [14, 15].

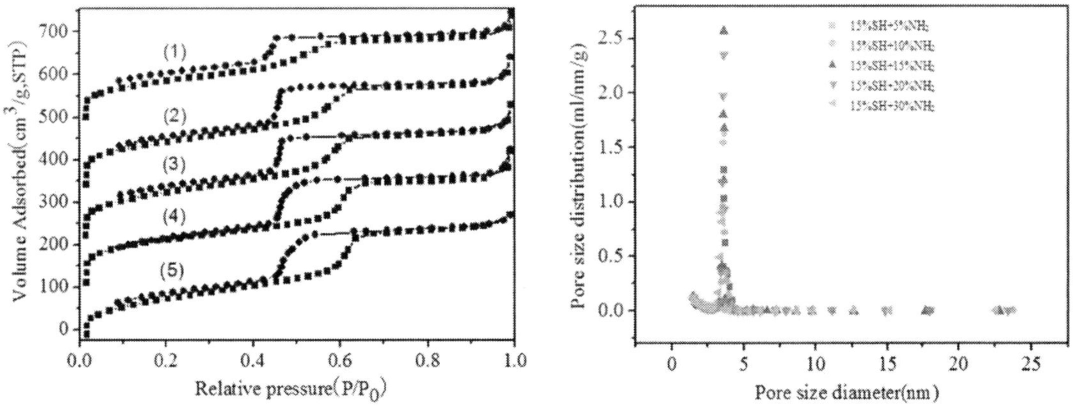

Figure 3. (a) The N_2 adsorption of five samples. (1) 15%NH_2+30%SH, (2) 15%NH_2+20%SH, (3) 15%NH_2+15%SH, (4) 15%NH_2+10%SH, (5) 15%NH_2+5%SH. (b) The pore distribution curves of the different contents of amino groups (5%-30%) and the stationary molar percentage thiol groups (15%).

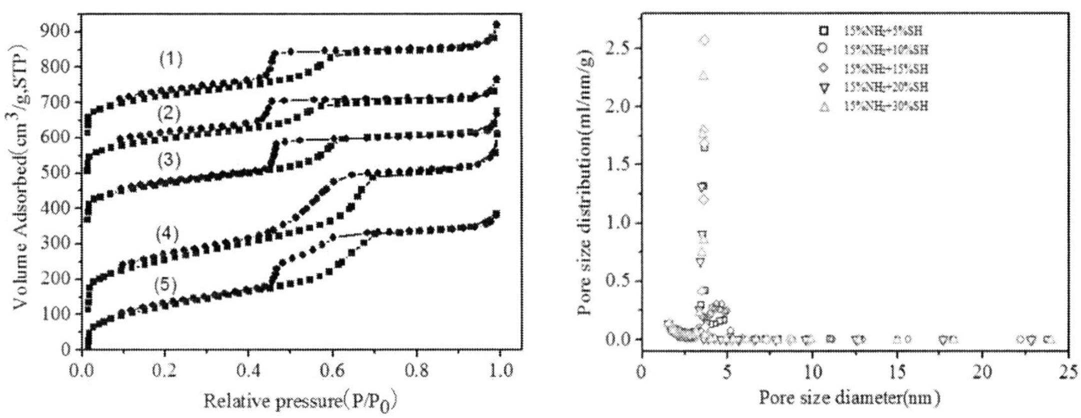

Figure 4. (a) The N_2 adsorption of five samples. (1) 15%SH+30%NH_2, (2) 15%SH+20%NH_2, (3) 15%SH+15%NH_2, (4) 15%SH+10%NH_2, (5) 15%SH+5%NH_2. (b) The pore distribution curves of the certain molar contents of amino groups (15%) and the adjustable contents of thiol groups (5%-30%).

In the table 1, the BET surface area was decreased along with the increase of the contents of N and S atoms. Through the comparison of the 15% amino group (with different contents of thiol groups) and 15% thiol group (with different contents of amino groups), there was an interesting found the BET surface area and the contents of N and S of the 15% amino group (with different contents of thiol groups) were larger than 15% thiol group (with different contents of amino groups). Based on the EA, 15% amino group (with different contents of thiol groups), the contents of N was substantially unchanged, the contents of S were increased. When the total contents were 45%, the contents of S were reached to 4.68%. 15% thiol group (with different contents of amino groups), the contents of S were firstly increased (with the percentage of N was below 15%), next they were decreased (the contents of N was over 10%), while the contents of N were always ascended. In summary, appearing this situation maybe due to the $-NH_2$ group was more hydrophilic than -SH group.

Table 1. The Summary of N_2 adsorption and elemental analysis.

Sample	BET surface area (m^2/g)	Pore volume (ml/g)	The average pore size (n m)	N content (%)	S content (%)
Pure SBA-15	662	0.83	5.98	-	-
15%SH+5%NH_2	402	0.63	3.71	0.35	3.72
15%SH+10%NH_2	399	0.8	4.57	0.44	3.87
15%SH+15%NH_2	376	0.47	3.66	0.60	3.68
15%SH+20%NH_2	369	0.39	3.51	0.79	3.39
15%SH+30%NH_2	337	0.47	3.58	0.98	3.15
15%NH_2+5%SH	594	0.48	3.73	0.60	1.51
15%NH_2+10%SH	556	0.52	3.70	0.58	2.57
15%NH_2+20%SH	399	0.47	3.63	0.56	4.34
15%NH_2+30%SH	368	0.36	3.48	0.56	4.68

3.2. The adsorption experiments of different adsorbents

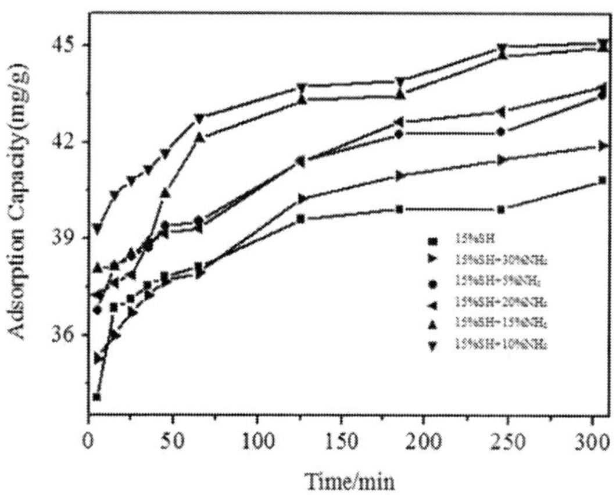

Figure 5. The relationships of time and adsorption capacity (15% SH and different contents of NH_2).

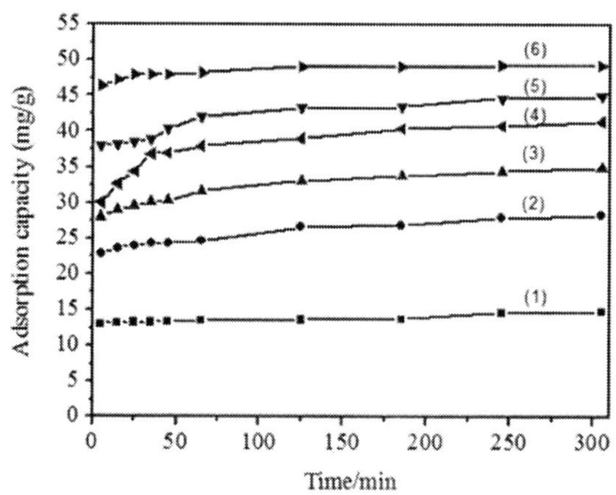

Figure 6. The time and adsorption capacity diagram, (1) 15% NH_2, (2)15% NH_2+5%SH, (3) 15% NH_2+10%SH, (4) 15% NH_2+30%SH, (5) 15% NH_2+15%SH, (6) 15% NH_2+20%SH.

As was shown in figure 5, (serie I), when the sample only contained 15% thiol groups was contrasted with other materials with different contents of amino groups, it was easy to find the both of maximum adsorption capacity of Cr(VI) and the adsorption rate were increased with the increase of the amino groups. As the number of the amino groups were added, the time of adsorption equilibrium were went down, especially 15% SH+30% NH_2 and 15% SH and 20% NH_2 .The figure 6 (serie II) displayed that for SBA-15 of 15% amino groups, a little number of thiol groups could greatly increase adsorption capacity, and the adsorption equilibrium time was kept (about 65 min). With the content of thiol groups raised, the adsorption capacity of Vr(VI) was appeared large fluctuations before adsorption equilibrium. That maybe because that more thiol groups entered into the pore channel of SBA-15, it would influence the hydrophilic of samples (-SH was a hydrophobic groups). According the result, the samples in series II could effectively and quickly remove the Cr(VI) in water. So the serie II could be took into account to widely use in water treatment of industry.

According to the serie I and serie II, the maximum adsorption capacity of the absorbent can be increased via introducing the $-NH_2$ and -SH functional organic groups into the SBA-15. The reasons maybe contribute to the thiol groups had a strong reducibility under the condition of acidity, the Cr (VI) can be turned to Cr (III) and the thiol group was oxidized to sulfonic acid group through oxidation-reduction reaction [16], and the $-NH_2$ can combine with H+ in the acidic solution to generate a number of $-NH^{3+}$ ions, which could adsorb a part of $HCrO^{4-}$ under the function of electrostatic attraction [17].

3.3. The study of adsorption mechanisms

3.3.1. Adsorption kinetic. The adsorption kinetics of Cr (VI) were investigate with the pseudo-first-order (PFO) and pseudo-second-order (PSO) models. The sample (15%NH_2+20%SH) was choose as the absorbent. The linear form of these models are described in the following equation (1) [18] and (2) [19].

$$\text{Pseudo First Order} \quad \log(q_e - q_t) = \log q_e - \frac{k_1 t}{2.303} \tag{1}$$

$$\text{Pseudo Second Order} \quad \frac{t}{q_t} = \frac{1}{k_2 q_e^2} + \frac{t}{q_e} \tag{2}$$

where K_1 (1/min) and K_2 (g/mmol/min) are the PFO and PSO rate, and q_e (g/mmol) and q_t (g/mmol) represent the adsorption capacities at equilibrium and at time t, respectively.

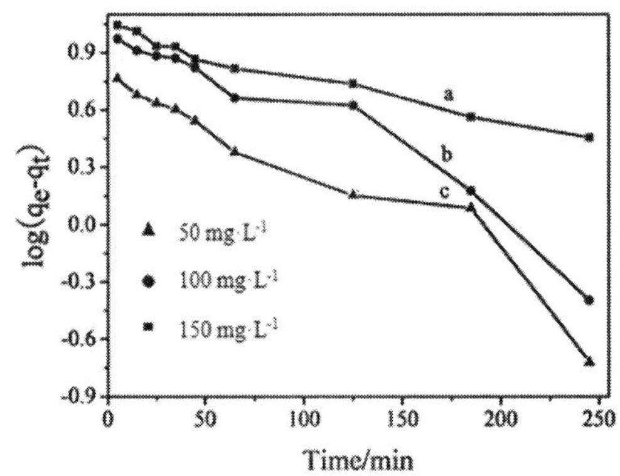

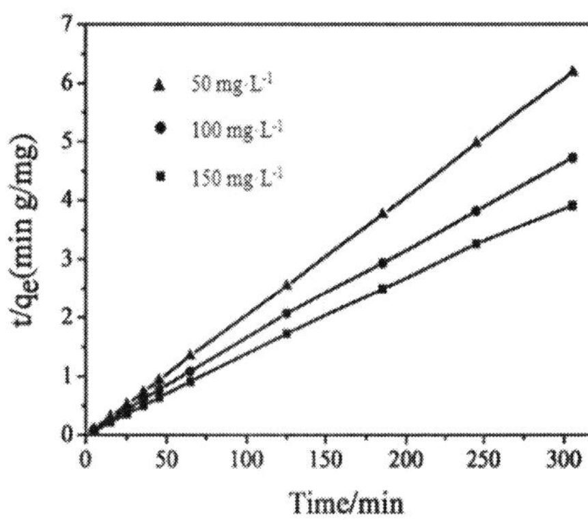

Figure 7. (a) Adsorption kinetics of Cr (VI) onto the bifunctionalized SBA-15; (b) Adsorption kinetics of Cr (VI) onto the bifunctionalized SBA-15.

On the basis of the figures 7(a) and 7(b), and the calculative value of K1, qe, R_2 in the table 2, the experimental values well fitted the PSO rate law. The R_2 (R was the correlation coefficient) value was more closed to 1. By contrast with the equilibrium adsorption capacity of the PFO and PSO model, the q_e was close to the value of adsorption capacity based on the PSO model. The k_2 was decreased along with the increase of the concentration of the Cr (VI) that indicated the adsorption rates of Cr(VI) was slower when the concentration of the Cr(VI) was up. The results indicated the PSO model was more suitable to describe the adsorption kinetics of Cr (VI) than the PFO model.

Table 2. Kinetic parameters and correlation coefficients at different concentration of Cr (VI).

Cr (VI) (mg/L)	Pseudo-first-order			Pseudo-second-order		
	k_1 (h^{-1})	q_e ($mg{\cdot}g^{-1}$)	R^2	k_2 ($(mg{\cdot}(g{\cdot}h)^{-1})$	q_e ($mg{\cdot}g^{-1}$)	R^2
50	0.0122	6.0996	0.9252	0.02267	49.5050	1.0000
100	0.0120	11.0510	0.9400	0.003615	64.9351	0.9995
150	0.0055	10.3134	0.9759	0.003238	77.5194	0.9990

3.3.2. Adsorption isotherms. The metal uptake by per unit of the adsorbent at equilibrium is defined by the adsorption isotherms. Langmuir and Freundlich's isotherms have been widely used to characterize the adsorption of solutes from aqueous solutions. The Langmuir isotherm and Freundlich's isotherm can be written in linear form as following equation (3) and (4) [19]:

$$\text{Langmuir Isotherm} \quad \frac{C_e}{q_e} = \frac{C_e}{q_m} + \frac{1}{K_L q_m} \tag{3}$$

$$\text{Freundlich's Isotherm} \quad \log q_e = \frac{1}{n}\log C_e + \log K_F \tag{4}$$

where q_e and q_m are the equilibrium adsorption capacity (mg/g) and the maximum adsorption capacity, respectively. C_e (mg/L) was the equilibrium concentration of the metal; KL and KF were the equilibrium adsorption coefficient of Langmuir and Freundlich constants, separately. And n was an empirical constant, if the adsorbent had no adsorption characteristics of mental ions, the date of n was below 1; while the n was between 1 and 10, it demonstrated the adsorbent was favourable [20] to adsorb the metal ions. The results were shown in table 3, meanwhile the figures were shown in figure 8.

As the dates of table 3, the experimental date correlated well with the Langmuir equation model (R_2>0.97). This confirmed the surface homogeneity and the monolayer surface coverage of the adsorbent and the fixed available bonding sites for metals. By the analysed the constants of the Freundlich equation model, firstly, the values of K_f was high at high temperatures, which shown that the adsorption ability of adsorbent for Cr (VI) was increased as the rise of the temperature. Secondly, the values of n were all over 1 for different temperatures, which indicated that the adsorption of Cr(VI) on adsorbent was favourable.

Table 3. Isotherm parameters obtained from the Langmuir and Freundlich's models.

T(K)	Langmuir Model			Freundlich Model		
	q_m (mg·g-1)	KL (L·mg-1)	R_2	n	KF (mg·g-1)	R_2
313	68.03	0.17	0.9962	5.15	29.04	0.9011
323	72.46	0.21	0.9789	9.87	43.76	0.6763
333	74.63	0.43	0.9888	12.80	54.26	0.8055

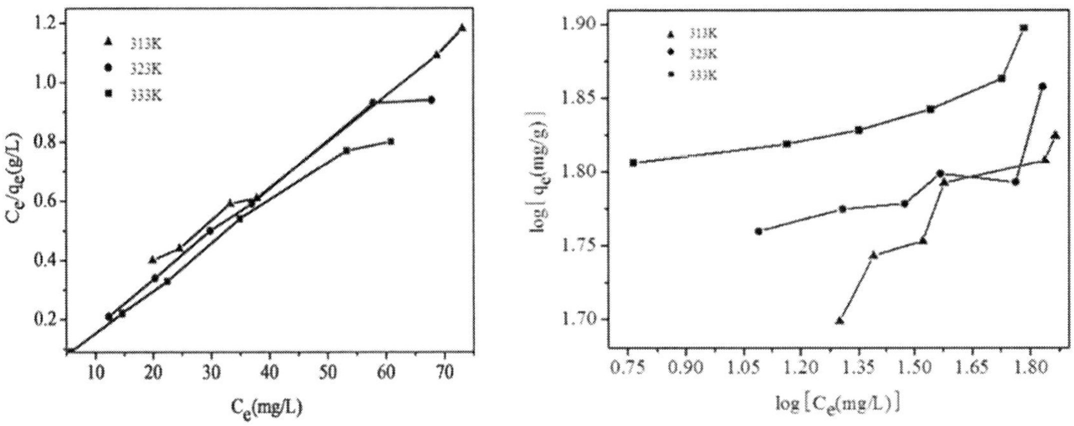

Figure 8. (a) Langmuir isotherm of Cr(VI) onto bifuncationalized SBA-15; (b) Freundlich isotherms for Cr(VI) onto bifunctionalized SBA-15.

3.3.3. Adsorption thermodynamics

The thermodynamic parameters were evaluated to investigate the influence of temperature on the adsorption of corrosion products onto the bifunctionalized SBA-15. The values for the adsorption thermodynamic parameters were calculated with the van't Hoff equation [19,20-22] as follows:

$$K_D = \frac{q_e}{C_e} \tag{5}$$

$$\Delta G^\circ = \Delta H^\circ - T\Delta S^\circ \tag{6}$$

$$\ln K_D = \frac{\Delta S^\circ}{R} - \frac{\Delta H^\circ}{RT} \tag{7}$$

where $\Delta G°$(kJmol^{-1}) is the Gibbs free energy changed in the given process; R (8.314 Jmol^{-1}K^{-1}) was the ideal gas constant, T(K) is the temperature; and the KD is the distribution constant; $\Delta S°$(Jmol^{-1}K^{-1}) is the standard entropy change, and$\Delta H°$(kJmol^{-1}) is the standard enthalpy change of the reaction. The values for $\Delta S°$ and $\Delta H°$ were indirectly calculated from the slope and intercept of the van't Hoff plot of ln Kd vs 1/T.

The plot of ln KD against 1/T (figure 9) at different temperatures showed a good linear relationship. $\Delta H°$ was was 61.46 kJ/mol, and the intercept of the plot [20] could obtained the $\Delta S°$ was 204.00 Jmol^{-1}K^{-1}, respectively. The other results were summarized in table 4. The positive value of $\Delta H°$ indicates that adsorption process is exothermic and the value of $\Delta H°$ (> 40 kJmol^{-1}) suggesting the Cr (VI) removal onto the adsorbent were both physisorption and chemisorption [23, 24]. The positive value of $\Delta S°$suggested the adsorption process of Cr(VI) on the SBA-NH_2-SH was a enthalpy promotion process. The negative value of $\Delta G°$ at all temperatures indicates the spontaneous nature of the adsorption reaction. And the $\Delta G°$ was decreased with temperature ascend, testifying the high temperature was more spontaneous than low temperature [24, 25].

Table 4. Thermodynamic parameters for adsorption of Cr(VI) onto bifunctionalized SBA-15.

Temperature (K)	ΔG° (kJ·mol-1)	ΔH° (kJ·mol-1)	ΔS° (J·mol-1·K-1)
313	-2.39	61.46	204.00
323	-4.43		
333	-6.47		

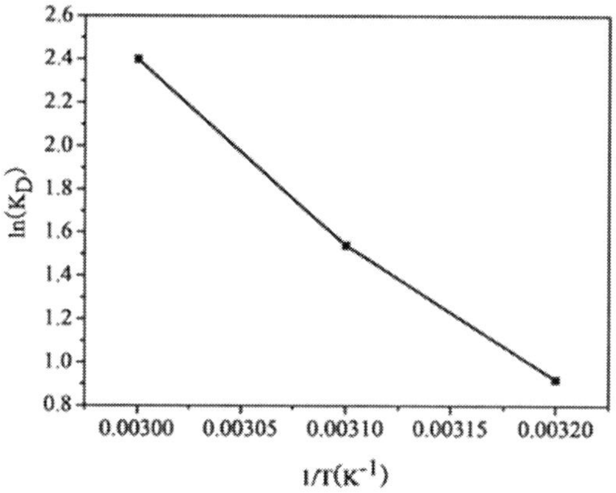

Figure 9. Van't Hoff's plots of lnKD vs. 1/T.

4. Conclusion

This study indicated that bifunctionalized SBA-15 with amino group and thiol group could remove the heavy metals of Cr(VI) efficiently from the water. The results showed that the thiol groups played a major role in the adsorption of Cr(VI), which can enhance the adsorption capacity of Cr(VI), while the amino groups could also assist to adsorb the metals but mainly accelerate the adsorption equilibrium (from 250 min down to 65 min), that means we can synthesize an ideal adsorbent by adjusting the hydrophilic character of the absorbent through the contents of $-NH_2$ and –SH groups. The study of the adsorption kinetic and isotherms demonstrated the adsorption was suitable fitted by the pseudo-second-order model, and the isotherm data was matched the Langmuir models, the isotherm parameters suggested that the adsorption process of Cr(VI) onto the bifunctionalized SBA-15 was spontaneous, endothermic, and the process was both physical and chemical adsorption.

Acknowledgments

Financial support of this work by the National Natural Science Foundation of China (No. 21202029), Hainan Natural Science Foundation (No. 513142) is gratefully acknowledged.

References

[1] Liu A M, Hidajat K, Kawi S and Zhao D Y 2000 A new class of hybrid mesoporous materials with functionalized organic monolayers for selective adsorption of heavy metal ions *Chem. Commun.* **230** 1145-6

[2] Shahbazia A, Younesia H and Badieib A 2011 Functionalized SBA-15 mesoporoussilica by melamine-based dendrimer amines for adsorptive characteristics of Pb(II), Cu(II) and Cd(II) heavy metal ions in batch and fixed bed column *Chem. Eng. J.* **168** 505-18

[3] Zhang L J, Goh S, Hu X, Crawford R and Yu A M 2012 Removal of aqueous toxic Hg(II) by functionalized mesoporous silica materials *Chem. Technol. Biotechnol.* **87** 1473-9

[4] The experiment of adsorption of Cr (VI) with bifunctioanlized SBA-15 contained *different* content of NH_2 and SH groups showed that the adsorption capacity of Cr (VI) can up to 49. 94 mg/g under following condition: 0.1 g absorbent, 100 mL 50 mg/L Cr(VI) solution, 25°C and pH = 2.0 under static condition for 6h.

[5] Song S-W, Hidajat K and Kawi S 2005 Functionalized SBA-15 Materials as Carriers for *Controlled* Drug Delivery: Influence of Surface Properties on Matrix-Drug Interactions *Langmuir* **21** 9568-75

[6] Prasetyanto E A, Sujandi, Lee S C and Park S E 2007 Highly dispersed CuO nanoparticles on SBA-16 type mesoporous silicawith cyclam SBA-16 as a precursor *Bull. Korean Chem. Soc.* **28** 2359-62

[7] Zhao D, Feng J, Huo Q, Melosh N, Fredrickson G H, Chmelka B F and Stucky G D 1998

Triblock copolymer syntheses of mesoporous silica with periodic 50 to 300 angstrom pores *Science* **279** 548-52

[8] Shang F P, *et al* 2012 One-pot cascade reactions catalyzed by acid–base mesoporous MCM-41 materials *Mater. Res. Bull.* **47** 801-6

[9] Song S-W, Hidajat K and Kawi S 2005 Functionalized SBA-15 materials as carriers for controlled drug delivery: Influence of surface properties on Matrix-Drug interactions *Langmuir* **21** 9568-75

[10] Siril P F, Shiju N R, Brown D R and Wilson K 2009 Optimising catalytic properties of supported sulfonic acid catalysts *Applied Catalysis A: General* **364** 95-100

[11] Margolese D, Melero J A, Christiansen S C, Chmelka B F and Stucky G D 2000 Direct syntheses of ordered SBA-15 mesoporous silica containing sulfonic acid groups *Chem. Mater.* **12** 2448-59

[12] Yokoi T, Yoshitake H and Tatsumi T J 2004 Synthesis of amino-functionalized MCM-41 via direct co-condensation and post-synthesis grafting methods using mono-, di- and tri-amino-organoalkoxysilanes *J. Mater. Chem.* **14** 951-7

[13] Calleja G, Sanz R, Arencibia A and Sanz-Perez E S 2011 Influence of drying conditions on amine-functionalized SBA-15 as adsorbent of CO_2 *Top Catal.* **54** 135-45

[14] Feuston B P and Higgins J B 1994 Model structures for MCM-41 materials: A molecular dynamics simulation *J. Phys. Chem.* **98** 4459-62

[15] Tian B S, Liu S H, Liu Y Y, Zhan X H and Yao X J 2011 Preparation and characterization of amino- functionalized mesoporous silica SBA-15 *Chinese J. Synthetic Chem.* **9** 460-4

[16] Shevchenko N, Zaitsev V and Walcarius A 2008 Bifunctionalized mesoporous silicas for Cr(VI) reduction and concomitant Cr(III) immobilization *Environ. Sci. Technol.* **42** 6922-8

[17] Zhu W J, Li M M, Ma W H, Zhou Y, Wu J J, Yu J and Xie K Q 2014 Synthesis of MCM- 41 mesoporous sieves and their adsorption performance of Cu^{2+} *Chinese J. Environ. Eng.* **8** 513-9

[18] Vinay K Singh and Prem N Tiwari 1997 Removal and recovery of chromium(VI) from industrial waste water *Chem. Tech. Biotechnol.* **69** 376-82

[19] Rodrigues L A, Maschio L J, Silva R E D and Silva M L C P D 2010 Adsorption of Cr(VI) from aqueous solution by hydrous zirconium oxide *J. Hazard. Mater.***173** 630-6

[20] Hameed B H, Ahmad A A and Aziz N 2007 Isotherms, kinetics and thermodynamics of acid dye adsorption on activated palm ash *Chem. Eng. J.* **133** 195-203

[21] Kulamani P, Krushna Gopal M and Suresh Kumar D 2012 Adsorption of Copper(II) on NH_2-MCM-41 and Its Application for Epoxidation of Styrene *Ind. Eng. Chem. Res.* **51** 2235-46

[22] Ren Y, Wei X and Zhang M 2008 Adsorption character for removal Cu (II) by magnetic Cu (II) ion imprinted composite adsorbent *J. Hazard. Mater.* **158** 14-22

[23] Du X, Yuan Q and Li Y 2008 Equilibrium, thermodynamics and breakthrough studies for adsorption of solanesol onto macroporous resins *Chem. Eng. Process.* **47** 1420-7

[24] Greene B and Darnall D W 1988 Temperature Dependence of Metal Ion Sorption by Spirulina *Biorecovery* **1** 27-41

[25] Romero-Gonzalez J, Peralta-Videa J R, Rodrı'guez E, Ramirez S L and Gardea-Torresdey J L 2005 Determination of thermodynamic parameters of Cr(VI) adsorption from aqueous solution onto Agave lechuguilla biomass *J. Chem. Thermodyn.* **37** 343-7

IOP Conf. Series: Earth and Environmental Science 82 (2017) 012075 doi:10.1088/1755-1315/82/1/012075

Risk assessment of urban flood disaster in Jingdezhen City based on analytic hierarchy process and geographic information system

D C Sun[1,2], J Huang[1,2,3], H M Wang[1,2], Z Q Wang[1,2] and W Q Wang[1,2]

[1]State Key Laboratory of Hydrology-Water Resources and Hydraulic Engineering, Hohai University, Nanjing, Jiangsu, China
[2]Institute of Management Science, Hohai University, Nanjing, Jiangsu, China

E-mail: j_huang@hhu.edu.cn

Abstract. The research of urban flood risk assessment and management are of great academic and practical importance, which has become a widespread concern throughout the world. It's significant to understand the spatial-temporal distribution of the flood risk before making the risk response measures. In this study, the urban region of Jingdezhen City is selected as the study area. The assessment indicators are selected from four aspects: disaster-causing factors, disaster-pregnant environment, disaster-bearing body and the prevention and mitigation ability, by consideration of the formation process of urban flood risk. And then, a small-scale flood disaster risk assessment model is developed based on Analytic Hierarchy Process(AHP) and Geographic Information System(GIS), and the spatial-temporal distribution of flood risk in Jingdezhen City is analysed. The results show that the risk decreases gradually from the centre line of Changjiang River to the surrounding, and the areas of high flood disaster risk is decreasing from 2010 to 2013 while the risk areas are more concentred. The flood risk of the areas along the Changjiang River is the largest, followed by the low-lying areas in Changjiang District. And the risk is also large in Zhushan District where the population, the industries and commerce are concentrated. The flood risk in the western part of Changjiang District and the north-eastern part of the study area is relatively low. The results can provide scientific support for flood control construction and land development planning in Jingdezhen City.

1. Introduction

The urbanization in China has been carried forward by leaps and bound, from 18.96% in 1979 to 53.7% in 2013. However, the development also brings a series of problems, for example, the rapid increase of impermeable surfaces changes the urban water cycle process [1], which leads to the increase of extreme precipitation events, the ground runoff, and eventually the urban flood risk [2]. The urban flood disaster is becoming more and more severe, which is recognized as a choke point that restricts the sustained and healthy development of economic society [3]. As a member of the Yangtze River Economic Belt and the city-cluster in the middle of Yangtze River, Jingdezhen City is in a pivotal position in the development of Jiangxi Province. However, Jingdezhen has long been suffering from floods due to its special geographical location and historical reasons. With the advance of city construction, it's necessary to evaluate flood risk to provide basic support for city development in Jingdezhen City.

The risk assessment of urban flood disaster is a process of using the assessment method to quantify the impact or loss caused by flood disaster. Many studies have been done in flood risk assessment, using historical disaster data, scenario analysis, Remote Sensing(RS) and Geographic Information System(GIS), assessment index system [4]. The research of historical disaster data is carried out by means of mathematical statistics, by analyzing the disaster results rather than analyzing the influencing factors of floods. This method is simple. However, it's necessary to add new disaster data to ensure the reliability [5]. The risk assessment based on the scenario analysis is carried out from the perspective of future scenarios, which overcomes the imprecise of the assessment from the aspect of disaster influence factors. However, it requires high modelling ability for disaster simulation [6-8]. The risk assessment approach combined RS and GIS usually analysis the spatial risk based on RS data, basic data and related socio-economic data [9]. However, the current spatial scale is large [10, 11]. The method based on assessment index system is commonly used in urban flood risk assessment. Generally, the index system is established from four aspects: disaster-causing factors, disaster-pregnant environment, disaster-bearing body, disaster prevention and mitigation ability [12-14]. Although lots of important findings have been obtained in this field, most of the current research have been done in the city regions with administrative boundaries. There are few studies on the risk assessment of small-scale floods in urban space [15-17].

Based on the theory of disaster system, considering the danger of disaster-causing factors, the sensitivity of disaster-pregnant environment, the vulnerability of disaster-bearing body, the ability of disaster prevention and mitigation from the formation process of flood disaster risk in Jingdezhen City, this paper puts forward the risk assessment index system of urban flood disaster in the small scale, constructs the risk assessment model based on the analytic hierarchy process method and analyses the spatial-temporal variation characteristics of flood disaster risk in urban region of Jingdezhen City with GIS. It provides a scientific support for making the urban flood emergency plan and strengthening the comprehensive risk management.

2. Study area

Jingdezhen City is located in the northeast of Jiangxi Province, China (see figure 1). The main river in Jingdezhen City is Changjiang River, the main tributaries are the South River and the West River. The distribution of precipitation in Jingdezhen City is extremely uneven during the year. The average precipitation during April-June is accounted for about 46% of annual precipitation, during July-September is accounted for about 22.1% of the year and during October-March is accounted for 31.8% of the precipitation. Further, Jingdezhen City is often plagued by flood disaster. According to the flood disaster in June 9, 2016, there are 528.7 thousand people affected and the direct economic losses was about $0.62 billion. There are still some problems in flood risk management in Jingdezhen City, such as lack of risk awareness and weak urban drainage facilities.

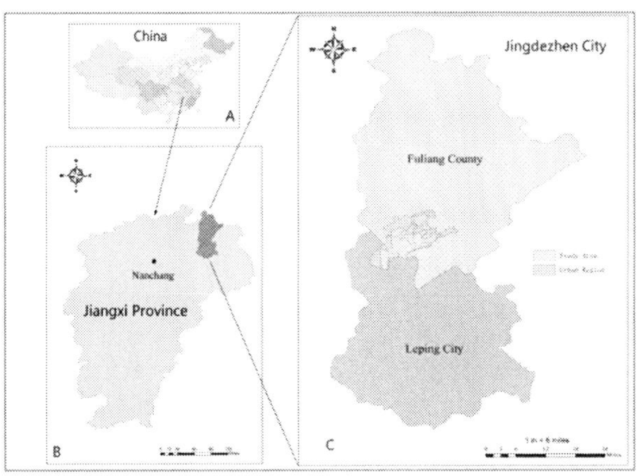

Figure 1. Study area-Jingdezhen central city.

3. Materials and methods

3.1. Formation process of urban flood risk

The urban flood risk is influenced by four aspects: disaster-causing factors, disaster-pregnant environment, disaster-bearing body, and the prevention and mitigation ability (see figure 2). The rainfall intensity in Jingdezhen is large, so there is a high probability of floods. The density of river network is large, which leads to the increasing of the sensitivity of the disaster environment. Besides, the density of urban population is large, most of the economic output concentrates in the urban region. So the vulnerability of the disaster-bearing bodies is high. Lastly, the operation of the drainage pumping station is not guaranteed and the design of the rainwater drainage is not reasonable, which result in the reduction of the disaster prevention and mitigation ability.

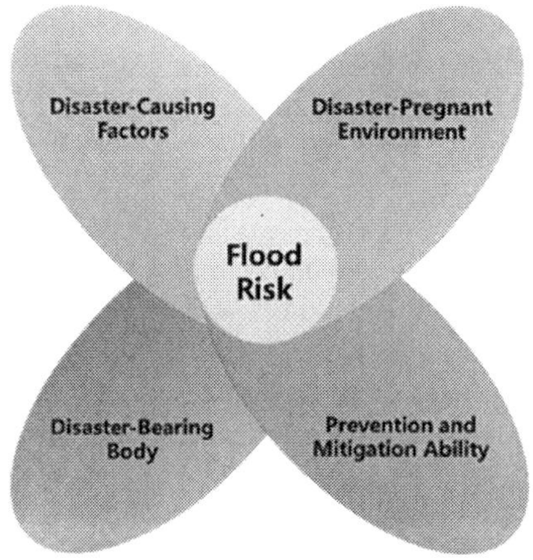

Figure 2. The formation process of urban flood disaster risk.

3.2. Risk assessment indicators

Based on the formation process of flood disaster in Jingdezhen City, the risk assessment index system of urban flood disaster is constructed from four above aspects, and 12 indicators are selected.

Precipitation is the most important factor of urban flood disaster [18], especially the short-term heavy rainfall has a greatest threat to the flood. Therefore, the average monthly precipitation, the maximum daily precipitation, the maximum continuous precipitation and the maximum continuous precipitation days were selected. Secondly, the terrain elevation and river network were selected to assess the environmental sensitivity of disaster. Through the actual investigation, it was found that terrain and topography will affect the runoff of rainfall in the ground, and the environmental sensitivity was stronger in areas closer to the river. GDP per 100 square meters, population density and land use types were selected to assess the vulnerability of the disaster-bearing body. Furthermore, in view of the medical level, drainage ability and other factors can reflect the impact of disaster prevention and mitigation on flood risk. So the radiation range of hospital, drainage pumping station and drainage pipeline are selected.

3.3. Data preparation and pre-processing

The meteorological data, population density data and gross domestic product (GDP) were collected from Jingdezhen Statistical Yearbook 2009-2013. The data were purchased from the Jingdezhen Statistical Bureau. The river network data and drainage pipeline location map were collected from Jingdezhen's Water Conservancy Bureau. The topographic (SRTM 90 m resolution) data were collected from NASA website (http://srtm.csi.cgiar.org /SELECTION/inputCoord.asp)". The land use type data is the Planning of land use map in Jingdezhen City. The map comes from the official website of Jingdezhen Planning Bureau. The hospital and the drainage pumping station location data come from the field investigation.

Precipitation indicators, population density indicators and GDP per 100 square meters' indicators use different methods for standardization as follows:

Positive indicators:

$$a = 0.1 + \frac{I - I_{\min}}{I_{\max} - I_{\min}} \times (0.9 - 0.1)$$

Negative indicators:

$$a = 0.1 + \frac{I_{\max} - I}{I_{\max} - I_{\min}} \times (0.9 - 0.1)$$

Where a is standardized data between 0.1 and 0.9. The I is the raw data. I_{max} and I_{min} refer to the maximum and minimum of the original data respectively. The data of the land use type is standardized by the analytic hierarchy process. More specifically, calculate the weight value of the four types of land use, and then judge from its weight value to determine the impact of the four types of land on the vulnerability of the disaster-bearing body.

DEM elevation data were reclassified by geometric interval classification method to determine the influence of different categories in this paper. The results are showed in table 1.

Table 1. The influence degree of terrain elevation.

Terrain elevation						
Standard deviation	<20	20~40	40~60	60~80	80~100	>100
Value	0.4	0.5	0.6	0.7	0.8	0.9

The sensitivity of the disaster-pregnant environment was determined by the distance between the river and area. Therefore, the results for influence degree of river network are shown in table 2.

Table 2. The influence degree of river network.

River network					
Distance(m)	<100	100~200	200~500	500~1000	>1000
Value	0.4	0.5	0.6	0.7	0.9

Table 3. The influence degree of disaster prevention and mitigation ability.

Medical Institutions						
Distance(m)	<50	50~200	200~500	500~1000	1000~2000	>2000
Value	0.9	0.8	0.7	0.6	0.5	0.4
Drainage Pumping Station						
Distance(m)	<50	50~100	100~200	200~500	500~1000	>1000
Value	0.9	0.8	0.7	0.6	0.5	0.4
Drainage Pipe Line						
Distance(m)	<1	1~2	2~5	5~10	10~20	>20
Value	0.9	0.8	0.7	0.6	0.5	0.4

Distance is an important indicator to access the disaster prevention and mitigation ability. The closer to the hospital, drainage pumping station, and the pipe line, the stronger the disaster prevention and mitigation ability [19-21]. The results of influence degree are shown in table 3.

3.4. Risk assessment model of urban flood

In this paper, 50 representatives from water authority department, planning department and land management department conducted a questionnaire survey. The AHP (analytic hierarchy process) method is used to get the importance of each factor, then the weights of the various indicators is calculated. The results are shown in table 4.

Table 4. The weight of each index.

Object hierarchy	Rule hierarchy	Weight	Index hierarchy	Weight
Flood Disaster Risk Assessment Index System	Disaster-causing factors	0.332	Average monthly precipitation	0.184
			Maximum daily precipitation	0.268
			Maximum continuous precipitation	0.350
			Maximum continuous precipitation days	0.198
	Disaster-pregnant environment	0.241	Terrain elevation	0.493
			River network	0.507
	Disaster-bearing body	0.287	GDP per 100 square metres	0.122
			Population density	0.320
			Type of land use	0.558
	Disaster prevention and mitigation ability	0.140	Medical institutions	0.190
			Drainage pumping station	0.412
			Drainage pipeline	0.398

In this paper, the comprehensive risk value (R) is defined as a function of disaster-causing factors (D), disaster-pregnant environment (S), disaster-bearing body (V) and disaster prevention and mitigation ability (C). The assessment model is as follows:

$$R=\sum_{i=1}^{n} W_i Q_i$$

$$Q_i=\sum_{j=1}^{n} w_j q_j$$

Among them, W_i is the weight of item i in rule hierarchy. Q_i is the index value of each rule hierarchy. W_j is the index weight of item j in each rule hierarchy. Q_j is the index value of item j in each rule. N is the number of influencing factors.

4. Result and analysis

4.1. Analysis the danger of disaster-causing factors

The changes of annual rainfall from 2009-2013 are showed in figure 3. It shows that the precipitation increases from January to June. The precipitation was concentrated in June, July and August, however the value of the extreme point is fluctuating in different year. The precipitation is particularly concentrated in June and August when the flood disasters is most likely to happen.

As seen in figure 4, the total amount of monthly precipitation is constantly fluctuating from 2009 to 2013, and the extreme points of the remaining three sub-indicators appear in different years. Besides, the extreme points of the longest continuous precipitation days and the maximum precipitation in one day are in 2012, while the extreme point of the longest continuous precipitation is in 2011. It's also found that the smaller the factor value, the greater the danger and the largest danger of disaster-causing factors are in 2011.

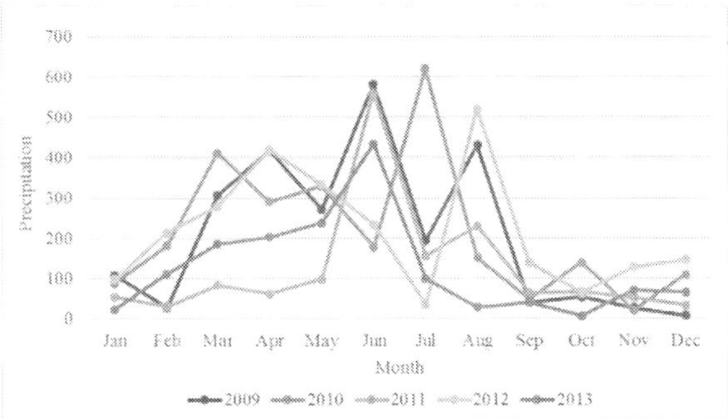

Figure 3. Monthly precipitation in Jingdezhen City (During 2009-2013).

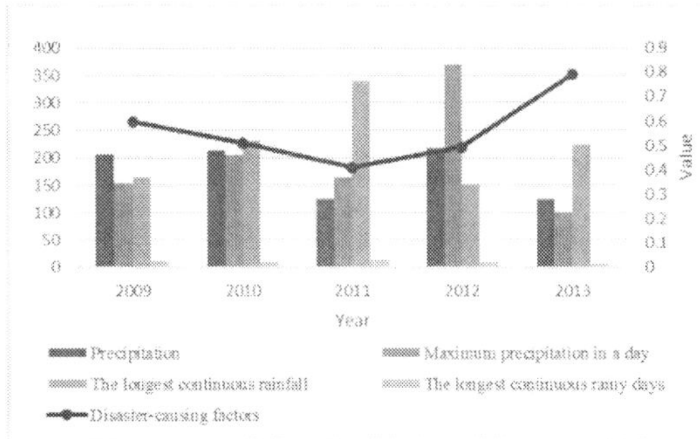

Figure 4. The changes of the precipitation in Jingdezhen City (During 2009-2013).

4.2. Analysis the sensitivity of disaster-pregnant environment

It is seen that the environmental sensitivity of Jingdezhen city is with high spatial variation (see in figure 5). The sensitivity is lower in the north and higher in the south. On the other hand, the sensitivity of the river network is the highest, and is gradually weakening from the central line of the river network to the areas along the river.

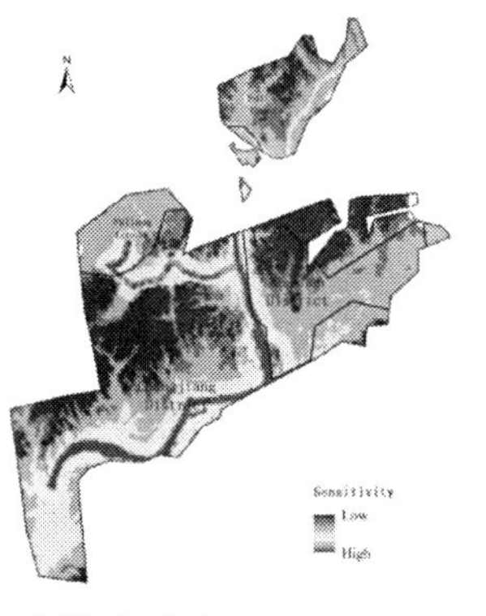

Figure 5. Sensitivity of the disaster-pregnant environment.

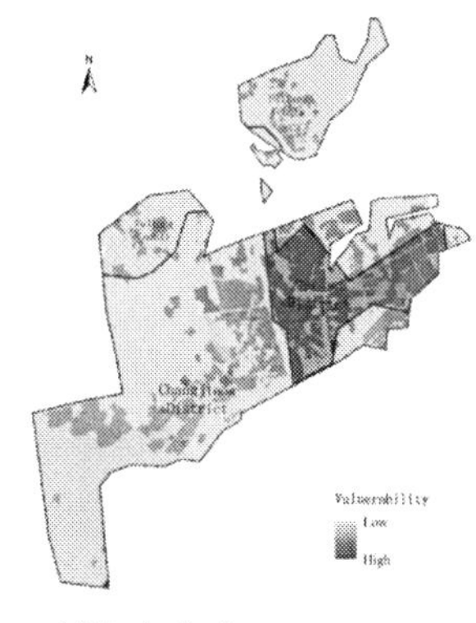

Figure 6. Vulnerability of disaster-bearing body.

4.3. Analysis the vulnerability of disaster-bearing body

As seen in figure 6, the vulnerability of the eastern city is stronger than that of the western city, and the vulnerability of the areas along Changjiang River is strong. More specifically, the vulnerability of the Zhushan District is stronger than the Changjiang District. The vulnerability in Fuliang County is small because of the small population density and small gross production value, the vulnerability is small.

4.4. Analysis the disaster prevention and mitigation ability

As seen in figure 7, the disaster prevention and mitigation ability in the northeast of the city is stronger, and it decreases gradually from the northeast to the southwest. There are obvious differences between the east and the west of Jingdezhen City. The flood disaster risk is small because the medical institutions and drainage pumping stations are gathered in the eastern part of the city.

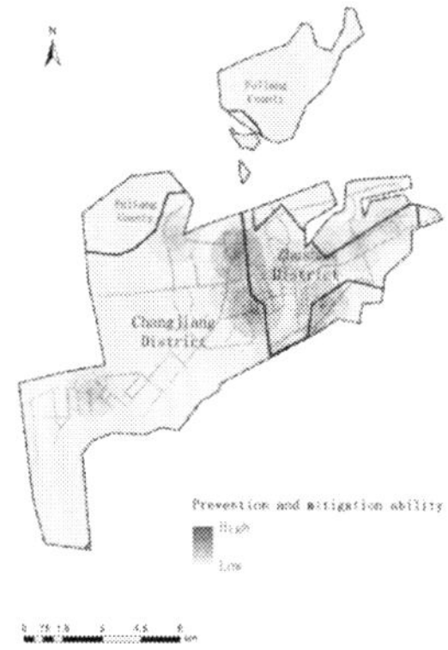

Figure 7. Disaster prevention and mitigation ability.

4.5. *The spatial-temporal variation analysis of urban flood risk*

The spatial-temporal variation of flood disaster risk in Jingdezhen City has significant differences (see in figure 8). From the view of space, the characteristics included differences in flood risk characteristics, direction characteristics and volatility characteristics. From the view of time, the area of high flood disaster risk decreased year by year, while flood disaster risk value became larger in some local areas.

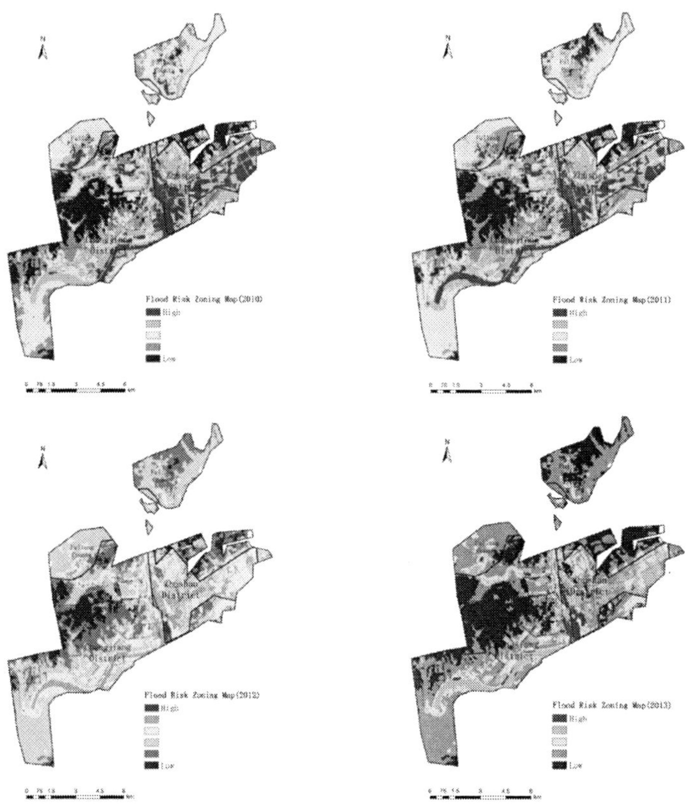

Figure 8. Risk Zoning Map in Jingdezhen City (2010-2013).

4.5.1. Spatial variation analysis. The flood risk of areas along Changjiang River as well as central and north-eastern of urban is larger. From the view of the local urban region, it has the greater risk along the east and west coast of Changjiang River. The maximum flood risk areas are the narrow river bank between Changjiang River and Zhushan District. Furthermore, the risk of flood disaster has a changing tendency from east to west divided by Changjiang River. The flood risk in Zhushan District is the largest, followed by Changjiang District.

Taking the centreline of Changjiang as the dividing line, the flood risk presents a spatial distribution characteristic which extends from the centre line of Changjiang River to the surrounding, meanwhile, the risk value of flood disaster is gradually decreasing.

The fluctuation of flood risk value in Zhushan District is the largest. And the west and southwest of Jingdezhen urban areas have higher terrain, smaller population density and GDP, the flood risk value fluctuates less.

4.5.2. Temporal variation analysis. Flood risk shows a downward trend in time, increased from 0.572~0.757 in 2010 to 0.637~0.812 in 2013. The minimum of the flood risk values are 0.572, 0.582, 0.603, 0.637 in 2010, 2011, 2012 and 2013 respectively. Further speaking, it can be concluded that the risk of high flooding area was decreasing year by year, at the same time, the peak value of flood risk value was decreasing too.

Flood risk in the eastern part of the urban region was becoming smaller, while the risk was increasing in the northwest and north of the city, and the change of flood risk in the east and west coast of Changjiang was smaller. From 2010 to 2011, the main change happened in the northeast of Jingdezhen City, called Fuliang County. Its amount of area of low flood risk was becoming larger, and the southern flood disaster risk became smaller. From 2011 to 2012, the value of flood risk reduced in the area between Changjiang and Zhushan District. Meanwhile, the risk value of western urban region increased slightly. From 2012 to 2013, although there was little change in the areas with high risk of flood disaster, the risk areas were more concentrated, and the risk of flood disaster decreased in other areas of the city.

5. Conclusion and discussion

The risk assessment of urban flood disaster is studied in this paper. Based on the understanding of the formation process of urban flood disaster risk, the indicators of danger, sensitivity, vulnerability, prevention and mitigation are constructed in this paper. The weight of those indicators was calculated by using the analytic hierarchy process. And then a small-scale flood disaster risk assessment model was developed based on AHP and GIS. Finally, the spatial and temporal distribution of flood risk in Jingdezhen was analyzed.

The results showed that the flood risk of Jingdezhen was decreasing year by year from 2010 to 2013, and the risk of flood disaster was diverse on the spatial distribution. Changjiang District and Zhushan District were had high environmental sensitivity because of the low terrain and intensive rivers. There were lots of commercial facilities and residential areas in both banks of Changjiang River which indicates high vulnerability. Inevitably, the flood disaster risk along the Changjiang River was the maximum. The flood risk of the city was decreasing when it diffused from the Changjiang River to the surrounding. The flood disaster risk in west and south of Changjiang District was small, so did the northeast of Fuliang County.

Although sufficient analysis has been done based on the urban flood risk formatting in this paper, there are also part of the important influence factors are out of consideration. It would be better to assess the risk of urban flood in dynamic concept and method. The spatial-temporal variation analysis of the flood risk in whole Jingdezhen City would be studied in the near future.

References

[1] Cheng J and Yang K 2010 Study on the impact of land use change on regional rainfall runoff in Shanghai central city *J. Nat. Resour.* **2010** 914-25

[2] Hallegatte S, Green C, Nicholls R J and Corfee-Morlot J 2013 Future flood losses in major coastal cities *Nat. Clim. Change* **3** 802-6
[3] Yuan Y and Shi P J 2003 The impact of land use change on urban flood disaster *J. Nat. Disast.* **12** 6-13
[4] Zhang Z T and Gao C 2014 Risk assessment of storm flood disaster in Huaihe River Basin under different reproduction *Geogr. Res.* **33** 1361-72
[5] Yang P G and Jin J 2016 Quantitative study on urban vulnerability based on historical storm flood disaster data: A case study of Beijing *Geogr. Sci.* **36** 733-41
[6] Cheng X F and Dai M Q 2015 Risk assessment of regional flood disaster based on scenario analysis - Taking Chaohu Lake Basin as an example *Resour. Environ. Yangtze River Basin* **24** 1418-24
[7] Liu Y L and Chen C L 2010 Risk analysis of typhoon storm - flood disaster in Wenzhou City based on China disaster prevention association *the Fourth Annual Meeting of the Professional Committee on Risk Analysis* (Changchun, Jilin, China) p 6
[8] Zhao S J, Huang C F and Guo S J 2012 Situation - driven regional natural disaster risk analysis *J. Nat. Disast.* **2012** 9-17
[9] Wei C J and Wang S X 1998 The role of remote sensing technology in the monitoring and evaluation of flood disaster in China *Bull. Chin. Acad. Sci.* **13** 443-7
[10] Cao L D and Li J L 2015 Study on comprehensive risk assessment of flood disaster in Zhejiang Province based on remote sensing *J. Nat. Disast.* **2015** 111-9
[11] Duan G Y, Zhao W J and Gong H L 2012 An improved model of regional flood risk assessment based on remote sensing data *J. Nat. Disast.* **2012** 57-61
[12] Hu B 2014 Risk division of rainstorm and flood disaster in Ningbo based on fuzzy comprehensive evaluation rainstorm *Torr. Rain Disas* **33** 380-5
[13] Chen P and Zhang L F 2015 Risk assessment index system and conceptual model of urban rainstorm and waterlogging disaster business economy *Bus. Econ.* **2015** 27-30
[14] Zhang T R 2007 Risk division of flood disaster in Chongqing based on GIS and fuzzy evaluation method *J. China West Normal Univ. (Nat. Sci.)* **28** 165-71
[15] Dai J, Pan Y N, Liu Q and Tang H O 2014 The improved AHP application in county scale rainstorm flood risk assessment *Meteorol. Sci.* **34** 428-34
[16] Wu J and Zhai G F 2014 Shantou city flood disaster risk assessment based on GIS *Risk Analysis of the China Association for Disaster Prevention Professional Committee of the Sixth Annual Meeting* (Huhehaote, Neimenggu Province, China) p 4
[17] Liang Y L and Huang D P 2013 Based on GIS of Nanning storm flood disaster risk assessment and regionalization *Meteorol. Sci. Technol.* **41** 934-9
[18] Cheng X F and Hao D D 2015 Based on OWA-GIS of Chaohu lake basin flood disaster risk assessment *Geogr. Sci.* **35** 1312-7
[19] Zhao F Y, Wang C P and Song R R 2012 Study on rationality of hospital distribution in Jinan City based on GIS Buffer Analysis *Technol. Horizon* **2012** 51-3
[20] Tang L L 2012 Risk assessment of operation and management of 11 drainage stations in Jingdezhen City (Nanchang University) p 55
[21] Tang L L 2013 Study on flood disaster assessment and waterlogging in Jingdezhen city from 2008 to 2012 *China High Tech. Enterp.* **2013** 163-4

Assistance of Ectoine on *Acinetobacter* sp. A06 of simultaneous heterotrophic nitrification and aerobic denitrification denitrifying at stress condition

P Xue[1], X W Li[1], F X Tan[1], A Qu[1], X N Yuan[1] and L H Zhang[1,2]

[1]Collaborative Innovation Center for Vessel Pollution Monitoring and Control, Dalian Maritime University, Dalian 116026, China

E-mail: dlzlh2008@163.com

Abstract. The simultaneous heterotrophic nitrification and aerobic denitrification (SND) performed by *Acinetobacter* sp. A06 could be inhibited by high salt, high ammonia nitrogen concentrations and extreme pH. To improve the SND nitrogen removal efficiency of *Acinetobacter* sp. A06 in adverse environment, we investigated the effects on nitrogen removal by *Acinetobacter* sp. A06 of five compatible solutes. The results show that Ectoine was the best effect in five osmotic compensation solutes when the salt concentration was 15 g/L, 30 g/L, 45 g/L, and 60 g/L, adding Ectoine, nitrogen removal were increased by 21.10%, 26.94%, 14.67% and 11.21%, respectively. When the NH_4^+ -N concentration was 1.5 g/L, 2 g/L and 2.5 g/L, adding Ectoine, the nitrogen removal rate increased by 15.93%, 10.07% and 7.11%, respectively,. When the extreme pH of pH was 5 or 9, adding Ectoine, the nitrogen removal rate was increased by 20.70% and 10.44%, respectively.

1. Introduction

With the large discharge of ammonia nitrogen wastewater, ammonia pollution also became serious in water. At present, the technical methods of ammonia nitrogen waste water purification are mainly physical, chemical and biological methods [1]. Simultaneous heterotrophic nitrification and aerobic denitrification (SND) was one of the biological nitrogen removal methods [2, 3]. SND has the advantages of rapid nitrification process, high nitrogen removal efficiency and pH stability of nitrogen removal system [4, 5], so it has broad application prospects. However, when we use SND to deal with high salt, high ammonia nitrogen wastewater, the nitrogen removal rate of the strain often is affected. Therefore, how to improve the efficiency of nitrogen removal in high salt, high ammonia nitrogen concentration wastewater had important significance.

In 1985, Ectoine (1, 4, 5, 6-tetrahydro-2-methyl-4-pyrimidin ecarboxylic acid, $C_6H_{10}N_2O_2$) was first identified by Galinski et al. in *Halorhodospira halochloris* (the original chlorophilus vulgaris *Ectothiorhodospira halochloris*) [6]. Ectoine can maintain intracellular osmotic pressure within the bacteria, and can resist the adverse environment. So in high temperature, high pressure, extreme pH, freezing and drying conditions, enzymes, nucleic acids, cells have a protective effect [7-9]. SND bacteria combined with Ectoine suitable for the purification of sewage wastewater with high salt and high concentration of ammonia nitrogen, which has important theoretical and practical significance.

In this paper, we investigated the effects of different osmotic pressure solutes and Ectoine assistance to *Acinetobacter* sp. A06 nitrogen removal. The effects of different osmotic pressure solutes

IOP Conf. Series: Earth and Environmental Science 82 (2017) 012076 doi:10.1088/1755-1315/82/1/012076

and different concentrations of Ectoine on the nitrogen removal of *Acinetobacter* sp. A06 are investigated, and the effects of different NaCl concentration, different NH_4^+ -N concentration, different pH Ectoine on *Acinetobacter* sp. A06 nitrogen removal assistance are investigated.

2. Materials and methods

2.1. Experimental materials

2.1.1. Bacterium. Acinetobacter sp. A06 derived from the biological laboratory.

2.1.2. Medium. LB medium (g/L): peptone 10, yeast powder 5, NaCl 5. The medium was autoclaved at 121°C for 20 min.

Nitrogen removal medium (g/L): $(NH_4)_2SO_4$ 2.83, $Na_3C_6H_5O_7{\cdot}2H_2O$ 18.38, $MgSO_4{\cdot}7H_2O$ 0.05, K_2HPO_4 2, KH_2PO_4 6, $MnSO_4{\cdot}4H_2O$ 0.01, $FeSO_4$ 0.01, pH 8. The medium was autoclaved at 121°C for 20 min.

Ectoine standard: Ectoine standard was purchased from Biomol GmbH, Germany.

2.2. Experimental methods

2.2.1. Nitrogen removal experiment. The strains were cultivated in 5 mL LB medium at 30°C and 120 rpm in a rotary shaker for 24 h. Then 1% of the cultures were inoculated in shake flasks (30/300 mL) containing nitrogen removal medium or solution at 30°C and 120 rpm in a rotary shaker for 48 h.

2.2.2. Determination of NH_4^+ -N, NO_2^- –N, NO_3^- -N concentration and calculation of nitrogen removal rate in inorganic nitrogen. NH_4^+ -N was determined by Nessler's reagent method. A reddish brown complex was formed by the reaction of NH_4^+ -N present in the form of free ammonia or ammoniumion and Nessler's reagent. According to the standard curve of NH_4^+ -N standard sample and the absorbance of the measured sample, calculate the NH_4^+ -N concentration of the tested sample [10].

NO_2^- -N was determined by diazotization-coupling reaction method. A reddish purple azo dye was formed by coupling diazotized sulfanilamide with N-(1-naphthyl)-ethylenediamine dihydrochloride under acidic conditions. According to the standard curve of NO_2^- -N standard sample and the absorbance of the measured sample, calculate the NO_2^- -N concentration of the tested sample [10].

Phenol-disulfonic acid spectrophotometric determination of NO_3^- -N concentration, nitrate in the absence of water reaction with phenol disulfonic acid to produce nitro - disulfonic acid phenol, in alkaline solution to generate yellow compounds, According to the standard curve of NO_3^- -N standard sample and the absorbance of the measured sample, calculate the NO_3^- -N concentration of the tested sample [11].

Calculation of nitrogen removal rate in inorganic nitrogen: In this paper, the total inorganic nitrogen is defined as the sum of the N concentrations of NH_4^+ -N, NO_2^- -N and NO_3^- -N. The nitrogen removal rate is defined as the percentage of the initial total inorganic nitrogen concentration in the reaction system and the initial total inorganic nitrogen concentration.

3. Results and analysis

3.1. Effect of different osmotic compensation solutes on SND by Acinetobacter sp. A06

The effect of different osmotic compensation solutes on SND by *Acinetobacter* sp. A06 was investigated. The nitrogen removal medium added Ectoine, proline, trehalose, glycerol, mannitol with a concentration of 200 mg/L and 30 g/L NaCl. After 48 h, we determined NH_4^+ -N, NO_2^- -N, NO_3^- -N concentration. NO_3^- -N, NO_2^- -N were not detected Results of SND under these conditions are shown in figure 1. Nitrogen removal for 48 h, the growth and nitrogen removal rates of *Acinetobacter* sp. A06 were significantly increased after adding different osmotic compensation solutes in the nitrogen removal medium. Among them, Added Ectoine in the nitrogen removal medium, that had best nitrogen removal efficiency, and the nitrogen removal rate increased from 3.73% to 46.35%.

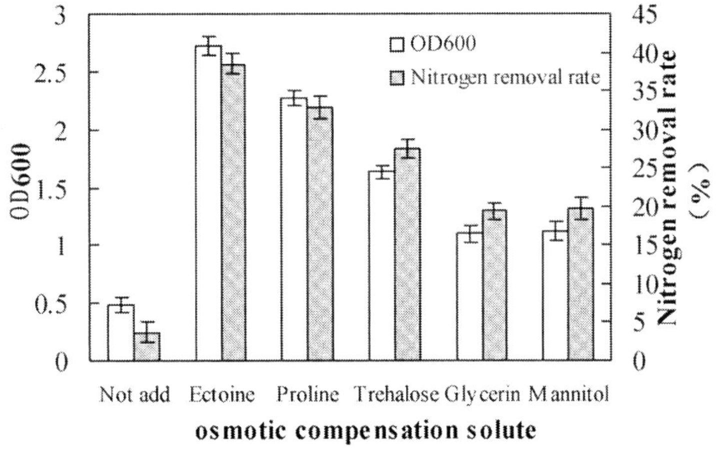

Figure 1. Effect of different osmotic pressure compensating solutes on growth and nitrogen removal.

3.2. Effects of different concentrations of Ectoine on SND by Acinetobacter sp. A06

The effect of different concentrations of Ectoine on SND by *Acinetobacter* sp. A06 was investigated. The nitrogen removal medium added Ectoine and 30 g/L NaCl, and Ectoine concentration was set to 0, 150 200 250, 300 mg/L. After 48 h, we determined NH_4^+-N, NO_2^--N, NO_3^--N concentration. NO_3^--N, NO_2^--N were not detected. Results of SND under these conditions are shown in figure 2. Nitrogen removal for 48 h, the growth and nitrogen removal rate of *Acinetobacter* sp. A06 increased with the increase of Ectoine concentration and have significantly improved. When the concentration of Ectoine was more than 250 mg/L, the increase of strain growth and nitrogen removal was stable.

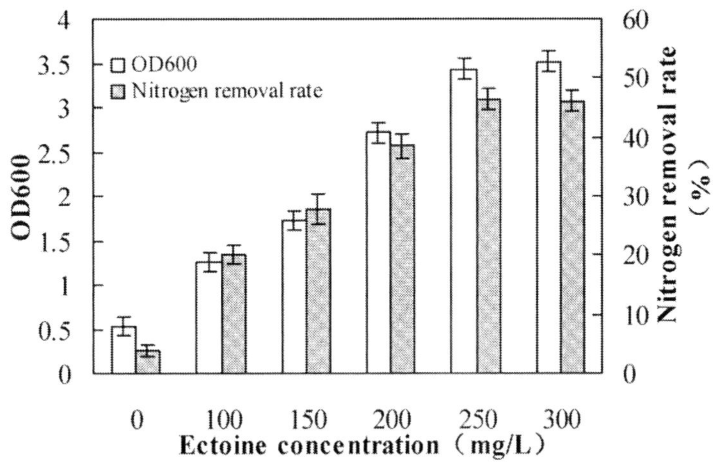

Figure 2. Effects of different concentrations of Ectoine on growth and nitrogen removal.

3.3. Effect of Ectoine on SND by Acinetobacter sp. A06 at different salt concentrations

The effect of Ectoine on SND by *Acinetobacter* sp. A06 under different salt concentrations was investigated. The nitrogen removal medium added Ectoine 250 mg/L and NaCl, and the NaCl concentration was set to 0, 15, 30, 45 and 60 g/L. After 48 h, we determined NH_4^+-N, NO_2^--N, NO_3^--N concentration. NO_3^--N, NO_2^--N were not detected. Results of SND under these conditions are shown in figure 3. Nitrogen removal for 48 h, with the increase of salt concentration *Acinetobacter* sp. A06 increment and nitrogen removal rate is lower and lower so that medium had not Ectoine, and add Ectoine medium *Acinetobacter* sp. A06 nitrogen removal rate and growth had increased significantly. When the concentration of NaCl was 30 g/L, the nitrogen removal rate increased the most, and the nitrogen removal rate increased from 19.41% to 46.35%.

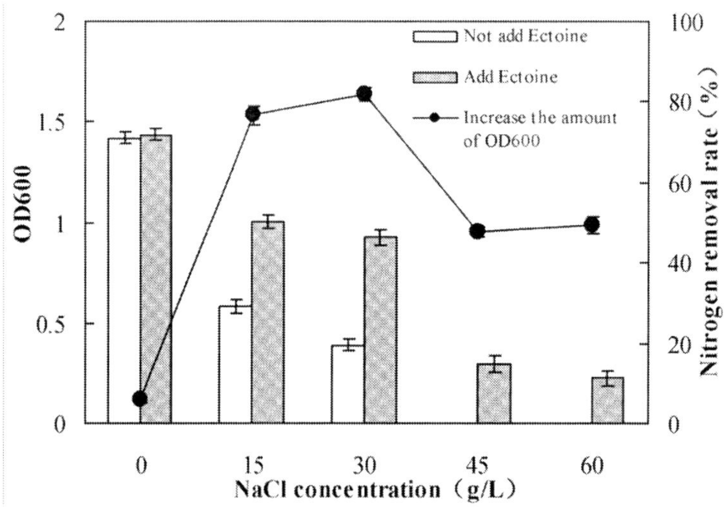

Figure 3. Effect of Ectoine on the strain growth and nitrogen removal rate under different salt concentrations.

3.4. Effect of Ectoine on SND by Acinetobacter sp. A06 at different NH_4^+-N *concentrations*

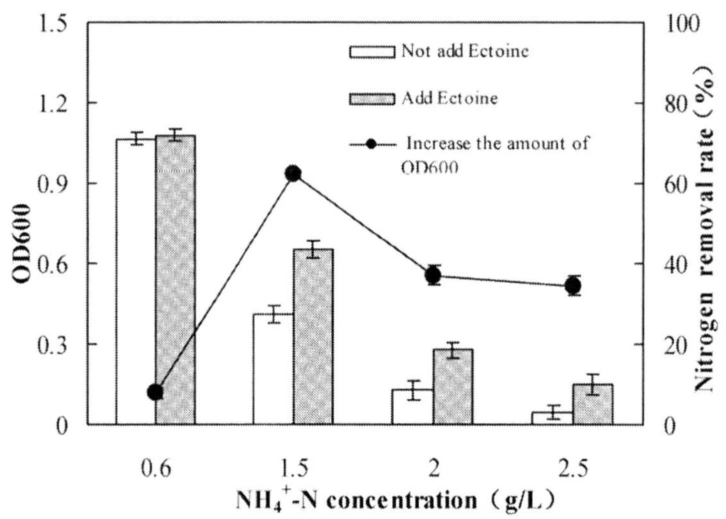

Figure 4. Effect of Ectoine on the strain nitrogen removal strains under different nitrogen ammonia concentrations.

The effect of Ectoine on SND by *Acinetobacter* sp. A06 under different NH_4^+-N concentrations was investigated. The nitrogen removal medium added Ectoine 250 mg/L and NH_4^+-N, and the NH_4^+-N concentration was set to 0.6, 1, 1.5, 2 and 2.5 g/L. After 48 h, we determined NH_4^+-N, NO_2^--N, NO_3^--N concentration. NO_3^--N, NO_2^--N were not detected. Results of SND under these conditions are shown in figure 4. Nitrogen removal for 48 h, with the increase of NH_4^+-N concentrations *Acinetobacter* sp. A06 increment and nitrogen removal rate is more and more low that medium had not Ectoine medium, and add Ectoine medium *Acinetobacter* sp. A06 nitrogen removal rate and growth had increased significantly. When the concentration of NH_4^+-N was 1.5 g/L, the nitrogen removal rate increased the most, and the nitrogen removal rate increased from 27.38% to 46.35%.

3.5. Effect of Ectoine on SND by Acinetobacter sp. A06 at different pH

The effect of Ectoine on SND by *Acinetobacter* sp. A06 under different pH was investigated. The nitrogen removal medium added Ectoine 250 mg/L, and the pH was set to 5, 6, 7, 8 and 9. After 48 h, we determined NH_4^+-N, NO_2^--N, NO_3^--N concentration. NO_3^--N, NO_2^--N were not detected. Results of SND under these conditions are shown in figure 5. Nitrogen removal for 48 h, the growth and nitrogen removal rate of the strain increased gradually, with the increase of pH. And the growth and nitrogen removal began to decrease when pH was more than 8, when Ectoine was added, *Acinetobacter* sp. A06 nitrogen removal rate and growth have improved significantly. When the concentration of pH was 6, the nitrogen removal rate increased the most, and the nitrogen removal rate increased from 22.57% to 49.78%.

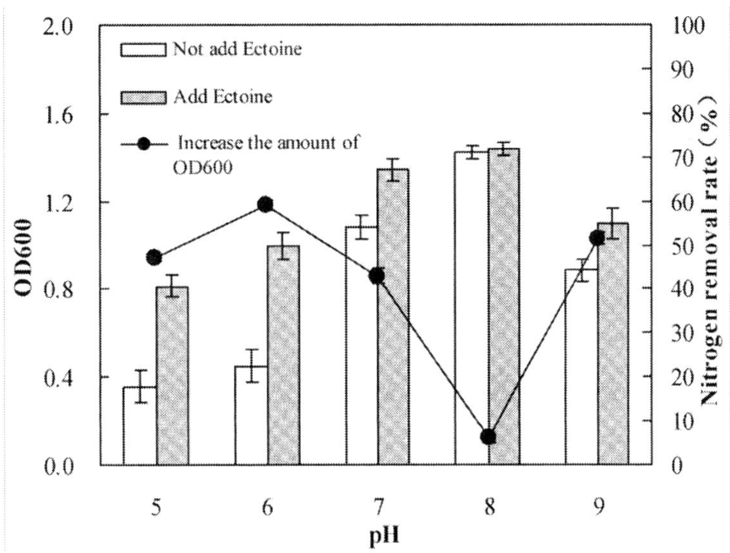

Figure 5. Effect of Ectoine on the strain growth and nitrogen removal rate under different pH.

4. Conclusion

- Different osmotic compensation solutes have a synergistic effect on the removal of *Acinetobacter* sp. A06, among of them, Ectoine contributes best to nitrogen removal.
- Increasing the concentration of Ectoine could increase the nitrogen removal rate of *Acinetobacter* sp. A06, but the rate of nitrogen removal could not be increased when Ectoine concentration was higher than or equal to 250 mg/L.
- The addition of 250 mg/L Ectoine could increase the nitrogen removal rate of *Acinetobacter* sp. A06 at high NaCl concentration, high ammonia concentration and extreme pH. When the concentration of NaCl 30 g/L, concentration of NH_4^+-N 1500 mg/L, pH 6, nitrogen removal rate increased by 26.94%, 18.97%, 27.21%, respectively. And the protective mechanism of Ectoine on *Acinetobacter* sp. A06 in the reverse environment needs further study.

Acknowledgments

This research was supported by Higher Education Development Fund (for Collaborative Innovation Center) of Liaoning Province, China (20110216001) and the Collaborative Innovation Center for Vessel Pollution Monitoring and Control @ Dalian Maritime University.

References

[1] Tian S S and Liu Q L 2013 Reatment of raw water ammonia nitrogen pollution in water supply plant *Resour. Conserv. Environ. Prot.* **7** 172

[2] Pochana K and Keller J 1999 Study of factors affecting simultaneous nitrification and denitrification (SND) *Water Sci. Technol.* **39** 61-8

[3] Chen H H, Liu S T, *et al* 2009 The development of simultaneous partial nitrification, anammox and denitrification (SNAD) process in a single reactor for denitrification *Bioresour. Technol.* **100** 1548-54

[4] Huang T L, Bai S Y, *et al* 2015 Experimental identification and denitrification of a non-nutrient heterotrophic nitrification-aerobic denitrifying bacteria *Chin. J. Environ. Eng.* **12** 5666

[5] Zhao B, He Y L, *et al* 2010 Heterotrophic ni-trogen removal by a newly isolated Acinetobacter calcoaceticus HNR *Bioresour. Technol.* **101** 5194-200

[6] Zhao B Q, Yang L F, *et al* 2007 Application of moderate halophilic bacteria in biotechnology *Acta Microbiologica Sinica* **34** 359-62

[7] Jan B, Anne U K, *et al* 2008 Synthesis and up take of the compatible solutes Ectoine and 5-Hydroxyectoine by Streptomyces coelicolor A3(2) in response to salt and heat stresses, *Appl. Environ. Microbiol.* **74** 7286-96

[8] Zhang L H, Wang Y, *et al* 2006 Supplementation effect of ectoine on thermostability of phytase *Biosci. Bioeng.* **102** 560-3

[9] Lentzen G and Schwarz T 2006 Extremolytes: Natural compounds from extremophiles for versatile applications *Appl. Microbiol. Biotechnol.* **72** 623-34

[10] 1999 *Standard Methods for the Examination of Water and Wastewater* 20th ed (Washington, D.C.: APHA)

[11] Sun H F, Wang H W and Yuan C Y 2013 Optimization of zinc-cadmium reduction method for determination of nitrate in seawater *Adv. Mater. Res.* **864-867** 1004-7

Surface grafting of styrene on polypropylene fibers by argon plasma and its adsorption-regeneration of BTX

J J Xu[1], M L Guo[2], Q G Chen[1], Z Y Lian[1,4], W J Wei[1,4], Z W Luo[3], G Xie[1], H N Chen[1] and K Dong[1]

[1]School of Environment Science and Engineering, Nanjing Tech University, Nanjing 211800, China
[2]School of Chemical Engineering, Nanjing Tech University, Nanjing 211800, China
[3]School of Safety Science and Engineering, Nanjing Tech University, Nanjing 211800, China

E-mail: lianzy@njtech.edu.cn; wjwei@njtech.edu.cn

Abstract. Active macromolecular free radicals were generated on polypropylene (PP) fibers surfaces by argon (Ar) plasma irradiation, then, PP surface modified fibers (PP-g-St fibers) were prepared by in-situ grafting reaction of styrene monomers (St). Effects of reaction parameters on grafting percentage were studied and adsorption capacities of PP-g-St fibers for benzene, toluene and xylene (BTX) were evaluated. Afterwards, regeneration adsorption efficiencies after maximum adsorption were explored. The results indicated that the optimum input power, irradiation time and grafting reaction time are 90 W, 3 min and 3 h respectively and the grafting percentage of St reached 5.7 %. The adsorption capacities of PP-g-St fibers towards toluene and xylene emulsions and solutions in water increased by 336.89 % and 344.57 % respectively, compared to pristine PP fibers. In addition, regeneration adsorption efficiencies of modified fibers remained > 90 % after six cycles of regeneration-adsorption experiments, which showed excellent regeneration ability.

1. Introduction

Benzene, toluene and xylene (BTX) are typical pollutants in waste waters, discharged either from the industries, or during transportation and storage [1-3]. They generally form emulsion with water. Due to their low degradability and highly poisonous nature, they pose a great threat to the environment and to human health if discharged without treatment [4, 5]. The main methods, till date, for treatment of BTX in wastewaters include biodegradation [6], thermal oxidation, catalytic oxidation, condensation and separation, membrane separation and adsorption [7]. Due to several advantages such as simple operation, no secondary pollution, fast processing and renewability of BTX, adsorption is considered to be one of the most efficient methods for treatment of organic pollutants in wastewaters [8-11]. The adsorption materials can be either inorganic (such as diatomite, attapulgite, volcanics), natural organic fibers (such as cotton, linen and wool) or synthetic organic nature (such as macroporous resin, polymeric fibers and polyurethane foams) [12].

Due to their characteristic large specific surface areas, good hydrophobic and lipophilic properties, good rigidity, and stable physicochemical properties, polypropylene (PP) fibers, as a kind of synthetic organic adsorption materials are widely used for the treatment of oily wastewaters [13, 14]. However, the affinity between PP fibers and BTX is low [15], due to difference in their solubility parameters which limits the adsorption of BTX, especially in case of BTX emulsion and solution in water. Hence, it is necessary to modify the PP fibers surface to improve their affinity. Melt grafting, solid grafting, liquid-phase grafting and suspension grafting are main modification methods that are practically useful [16-19], but these methods have some drawbacks such as severe degradation of PP, high cost, non-fibrous nature of the modified product, and are also environmentally harmful. Plasma-induced grafting, an excellent technics of surface modification, has been developed recently. It has many outstanding features like mild process conditions, environmentally-friendly nature, and it also causes no damage to the internal structure of the material [20-22]. The principle of this technique is to produce active sites (free radicals) by bombarding the material surface with energetic ionizing particles, where *in-situ* grafting reaction of monomers takes place [23].

Polystyrene (PS) has a large number of styrene (St) units in its molecular chain, which is similar to the molecular structure of BTX, and their solubility parameters are also close [24]. So, PS has excellent adsorption and occlusion for BTX. However, the swelling and dissolution of PS in BTX causes deformation and loss of mechanical properties of PS fiber. Moreover, it is difficult to separate the BTX occluded inside the PS fibers, and its regeneration is difficult [25].

In this paper, by combining the advantages of PP and PS for adsorption of BTX, active macromolecular free radicals were generated on PP fiber matrix using argon (Ar) plasma irradiation. The in-situ grafting reaction of St monomers occurred and PP-g-St fibers were formed. The effects of reaction parameters including input power, irradiation time and grafting reaction time on grafting percentage were studied. Adsorption capacities of PP-g-St fibers for BTX emulsion and solution in water were evaluated, and their maximum adsorption capacities for pure BTX were measured as well. Finally, regeneration methods and regeneration adsorption efficiencies after maximum adsorptions were explored.

2. Experimental

2.1. Materials

PP melt-blown fibers (laboratory prepared, fibers diameter: 2-10 μm, specific area: 1.50 m^2/g). St (CP grade), was distilled under reduced pressure prior to use. Benzene, toluene, xylene, acetone and anhydrous sodium sulfate were all analytical grade. Sodium dodecyl benzene sulfonate (SDBS) was of CP grade, carbon tetrachloride was specially produced for environmental protection, petroleum ether was of industrial grade. 500 W radio frequency power source, MSY-I and 2000 W radio frequency matcher, SP-II (plasma regenerator) were all provided by microelectronics institute of Chinese Academy of Sciences. Oil concentration analyzer ET1200 was purchased from Euro Tech Co., Ltd., Shanghai.

2.2. Preparation of PP-g-St fibers

The set-up for the preparation of PP-g-St fibers by plasma-induced grafting is shown in figure 1. The process can be described in following steps:

- PP fibers were immersed in acetone for 2 hours and dried to a constant weight at 50°C. then some PP fibers ⑩ were taken as the sample and weighed. They were then transferred to reactor ⑧, the reactor was placed between electrodes ⑨ of the plasma regenerator ② and connected to the system.
- Valve B ⑥ was closed, Solution of the St monomers was taken in the separating funnel ⑤ and Ar gas was bubbled into the solution to eliminate air.
- Valves A ③ and C ⑦ were then opened and air was removed until the pressure was 5 Pa. The air in reactor ⑧ was replaced with Ar gas for 3 times. Ar gas was again pumped in and evacuated to 5 Pa. Valves A ③ and C ⑦ were closed, and the plasma generator ② was switched on to irradiate the fibers under set plasma conditions.
- The plasma generator ② was turned off and the valves B ⑥ and C ⑦ were opened to

guide the solution of St monomers into the reactor ⑧. Valve B ⑥ should be closed when fibers sample ⑩ is totally immersed in St monomers solution. This was the beginning of grafting reaction. The grafting reaction ended when valve B ⑥ was opened again to introduce air. After that, fiber sample ⑩ was taken out and extracted with acetone for 5 hours to get rid of the remaining unreacted St and the self-polymer (PS). Finally, fiber sample ⑩ was dried at 50°C till it attained constant weight.

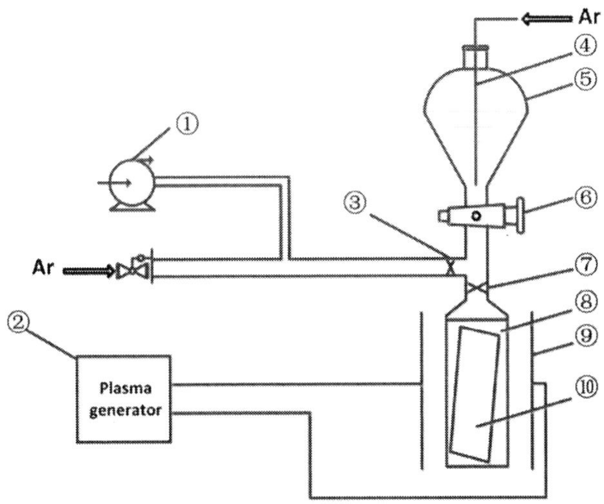

Figure 1. The set-up for the preparation of PP-g-St fibers by plasma-induced grafting. ①: vacuum pump, ②: plasma generator, ③: valve A, ④: mental air inlet pipe, ⑤: liquid separating funnel for St, ⑥: valve B, ⑦: valve C, ⑧: reactor, ⑨: electrodes, ⑩: fibers sample.

2.3. Determination of grafting percentage

The grafting percentage can be calculated using equation (1).

$$\eta = (M_2 - M_1)/M_1 \times 100\% \quad (1)$$

Where, η (%) is the grafting percentage, M_1 (g) is the weight of PP fibers, M_2 (g) is the weight of PP-g-St fibers.

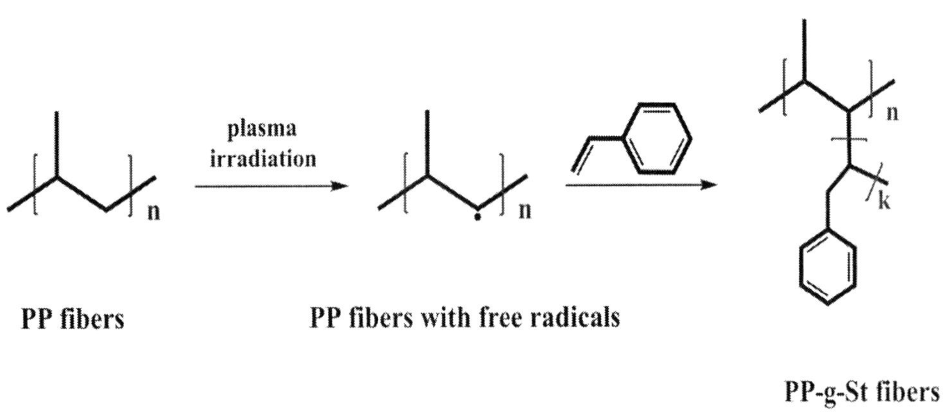

Scheme 1. Synthesis of PP-g-St fibers.

2.4. Adsorption for BTX

Adsorption capacities of PP-g-St fibers for BTX emulsion and solution in water were measured by following procedure. First, approximately 1 g of PP-g-St fibers sample was weighed and added into 2 L beaker containing 1.5 L deionized water, 20 ml BTX and 0.2 g SDBS, after stirring for 1 h and left to stand for 2h (25°C) to achieve well-distributed and stable, Excess amount of BTX floating on the water surface was removed and the initial concentration of adsorption solution was testified. Then, the PP-g-St fibers were fastened between the annular metal grids and immersed in the mixture for 1 h with low speed agitation using magnetic rotor to fully adsorb BTX emulsion and solution in water. After that, the PP-g-St fibers were taken out and transferred to a separating funnel containing a certain amount of CCl_4 to extract the BTX adsorbed, the solution was adjusted pH≤2 and oscillated for 3 min. The water present in the extract of PP-g-St fibers was removed by adding 1.0 g of anhydrous sodium sulfate (extra addition needed if they all crystallize), then filtered the solid and the volume for extraction was adjusted to 200 mL and maintained for 3 h. Finally, the concentration of BTX in the extract was determined using the oil concentration analyzer. Adsorption capacity of PP-g-St fibers can be calculated using the equation (2) (the test was repeated for 5 times to obtain mean value and the adsorption capacity of original PP fibers was measure for contrast at the same time).

$$Q' \text{ (mg/g)} = (0.2 \times C)/W \times 100\% \quad (2)$$

Where, Q' (mg/g) is the adsorption capacity of PP-g-St fibers for BTX emulsion and solution in water, W (g) is the original weight of PP-g-St fibers before adsorption, C (mg/L) is the concentration of BTX in the water.

Adsorption capacity of PP fibers for BTX emulsion and solution in the water was also measured in the same manner stated above for comparison purpose.

2.5. Regeneration

In order to research the regeneration ability of PP-g-St fibers, it is necessary to investigate the regeneration adsorption efficiency of fibers reached maximum adsorption. Due to uniform dispersion and limitation in the concentrations of BTX emulsion and solution in water, it was difficult for fibers to attain saturation. So, by using pure BTX as adsorbate, adsorption could reach maximum in a short duration of time, and the maximum adsorption capacities of the fibers were determined by following way. About 1 g of PP-g-St fibers was weighed and then immersed into pure BTX (20 ml, 25°C) for 10 min to achieve maximum adsorption. After holding the fibers corner in the air until the liquid on the surface stopped dripping, the fibers were weighed again. Maximum adsorption capacity of PP-g-St fibers was measured by equation (3) (the test was repeated 5 times to obtain mean value).

$$Q_S \text{(g/g)} = (W_1 - W_0)/W_0 \times 100\% \quad (3)$$

Where, Q_s (g/g) is the maximum adsorption capacities for pure BTX, W_0 (g) is the original weight of PP-g-St fibers, W_1 (g) is the weight of PP-g-St fibers after maximum adsorption.

Maximum adsorption capacity of PP fibers in pure BTX was also determined in a manner similar to the above procedure for comparison purpose.

Centrifugation, gravity extrusion, and solvent extraction were chosen as the regeneration methods for maximum adsorption of PP-g-St fibers, and each cycle of regeneration – adsorption was repeated for 6 times. The conditions for each method were: centrifugation – 4000 rpm for 10 min, gravity extrusion - 18.9 KPa for 10 min, solvent extraction - soaking in petroleum ether for 10 min and then drying at 65°C for 2 h. Regeneration adsorption efficiency can be calculated by equation (4).

$$K = Q_r/Q_S \times 100\% \quad (4)$$

Where, K (%) is the Regeneration adsorption efficiency, Q_r (g) is the maximum adsorption capacities after regeneration, Q_s (g) is the original maximum adsorption capacities.

3. Results and discussion

3.1. Surface profile and infrared spectroscopic analysis

The micro surface morphologies of PP fibers and PP-g-St fibers were observed by SEM (JSM-5900, produced by Japan Electronics Co., Ltd). The results are shown in figure 2.

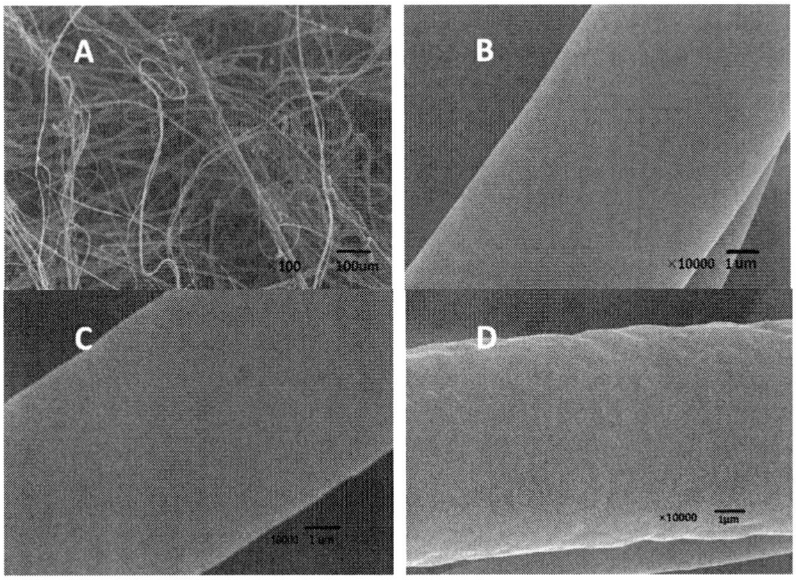

Figure 2. SEM images of PP fibers before and after plasma-induced grafting modification (A and B: PP fibers, C: PP fibers after plasma irradiation, D: PP-g-St fibers).

Figure 2 showed the images of PP fibers before and after plasma-induced grafting modification, whereas figure 2B revealed a smooth and flat surface of the PP fiber. After plasma irradiation(figure 2C), the surface of PP fibers almost remained unchanged which meant that besides creating active free radicals, plasma irradiation did not cause visible damage to the structure of PP backbone and had little effect on the mechanical properties.

The surface of PP-g-St fiber showed in figure 2D was obviously covered with a corrugated layer. Due to the unchanged smooth and flat surface of the PP fibers after irradiation, it can be inferred that St chains were introduced on the surface of PP fibers after grafting reaction. Moreover, the surface of PP-g-St fiber was rough and uneven, which could be attributed to the inhomogeneous grafting reaction. These depressions were uniformly distributed and had moderate depths, which contributed to the increased specific surface area of the fiber matrix. This was beneficial for the adsorption process, and had little effect on the mechanical properties.

The successful grafting of St onto the surfaces of PP fibers was further confirmed by FT-IR analysis (Nexus 670, Nicolet) of fiber samples. The results are shown in figure 3.

Compared to PP, the spectrum of PP-g-St showed almost the same absorption peaks expect new peaks at 702 cm^{-1} and 758 cm^{-1}, whereas the peak at 1642 cm^{-1} became stronger. The peaks at 702 cm^{-1} and 758 cm^{-1} could be attributed to the out-of-plane C-H bending vibrations of mono-substituted benzene ring [26]. The peak at 1642 cm^{-1} was also ascribed to the vibrations of the benzene ring. These results clearly indicated the grafting of St, Moreover, PP-g-St fibers were extracted in acetone to remove the unreacted St monomers and its homopolymer. Hence it became clear that a layer of St was grafted onto PP fibers to form PP-g-St fibers.

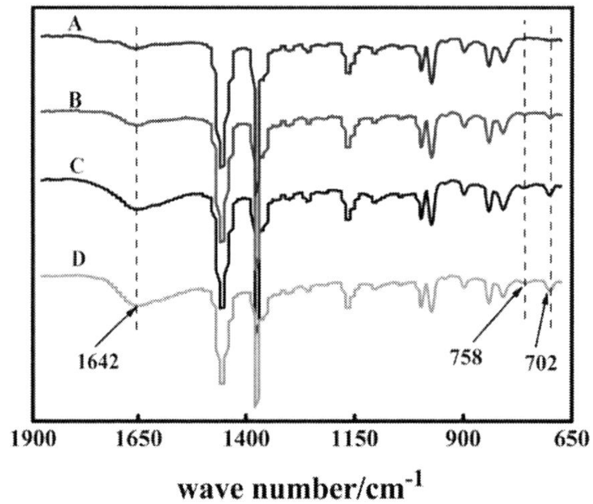

Figure 3. FT-IR spectra of PP fibers before and after modification (A: PP fibers, B, C, D are PP fibers irradiated by plasma at 50 W, 3 min and 2 h of grafting reaction, 70 W, 3 min and 2 h of grafting reaction, 90 W, 3 min and 3 h of grafting reaction respectively.).

3.2. Effects of reaction parameters on grafting percentage

Above discussion proved that St was successfully grafted onto PP fibers. To study the extent of grafting of St on PP fibers, the influences of plasma reaction parameters on grafting percentage were investigated, as shown in figure 4.

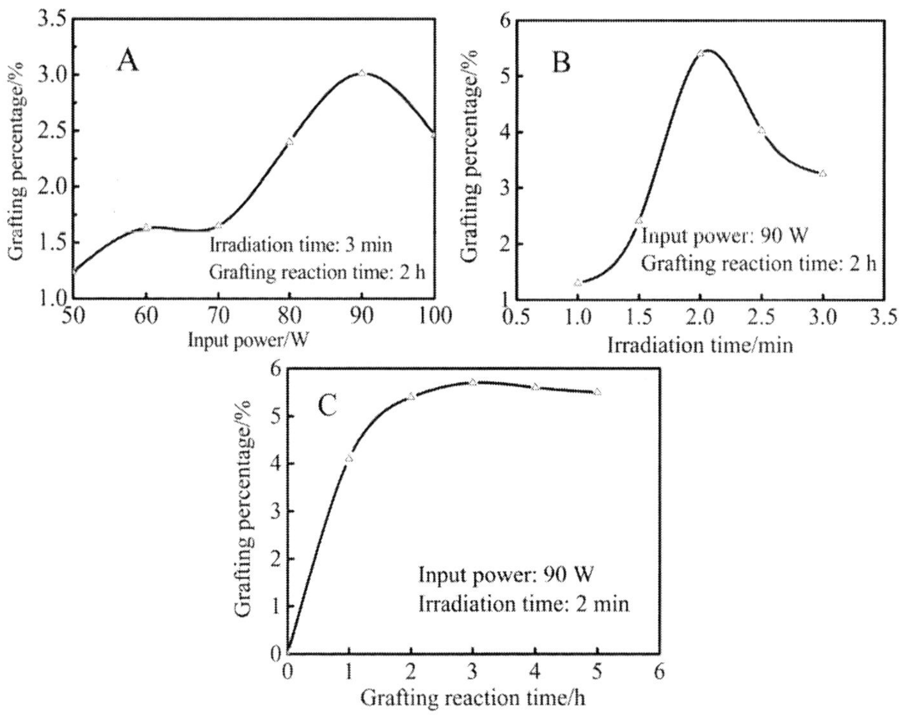

Figure 4. Effect of reaction parameters on grafting percentage of St (A: Effect of input power, B: Effect of irradiation time, C: Effect of grafting reaction time.).

First, the effect of input power of plasma generator on grafting percentage was investigated, as shown in figure 4A. It could be seen that by increasing the input power, grafting percentage increased initially and then decreased. Grafting percentage reached a maximum value of 3.01%, when input power was 90 W. This was due to the degree of dissociation of plasma atmosphere and increase in the velocities of active particles in the reactor with the increase of input power. In other words, active particles had higher density and kinetic energy, which increased the number of free radicals on the PP fiber surface and accelerated the in-situ grafting reaction. However, when input power exceeded 90 W, the grafting percentage decreased. This was due to the fact that when the density and kinetic energy of active particles were too high, the surface of PP fiber was bombarded and etched violently. This caused the fiber samples to lose weight, and so the grafting percentage decreased. Too much of etching renders the fiber locally incomplete or broken, which affects the mechanical properties of the material. Therefore, an optimal input power improves the grafting percentage, and in this paper the appropriate input power was found to be 90 W.

Secondly, the effect of irradiation time is shown in figure 4B. With increase in irradiation time, grafting percentage increased initially and then decreased. A maximum grafting percentage of 5.50 % was reached when the irradiation time was 2 min. To some extent, the effects of extended irradiation time and increased input power were equivalent. Both these factors increased the number of free radicals on the surface of PP fibers and accelerated the grafting reaction. On increasing the irradiation time, the number of free radicals increased, and grafting percentage also increased. When irradiation time was long enough, the concentration of free radicals on the surface of PP fibers reached a saturation value, and the grafting percentage was maximal. However, when irradiation time was too long, the free radical concentration became too large, and the grafting percentage decreased. This was due to the fact that the activation energies of adjacent free radicals were very low, and the coupling reaction between the free radicals by collision became more favorable, which decreased the grafting percentage.

Lastly, the influence of grafting reaction time on grafting percentage was studied and the results are shown in figure 4C. It can be clearly seen that by extending the grafting time, the grafting percentage increased initially and then remained steady. The rate of grafting percentage was high in the first 1 h, then reduced, and almost reached zero after 3 h. The grafting percentage reached a maximum value of 5.7 % after 3 h. In the first 1 h of reaction, the St monomers were fully in contact with the free radicals on the surface of PP fibers, which resulted in rapid chain growth reaction of St. On extending the grafting time, the active sites on PP fiber surface decreased, and the chain growth reaction gradually slowed down and then was terminated.

3.3. Adsorption of PP-g-St fibers for BTX

The adsorption capacity of PP fibers and PP-g-St fibers for BTX emulsion and solution in water were studied, and the results are presented in table 1.

Table 1. Adsorption capacities and increasing rate of PP fibers and PP-g-St fibers for BTX emulsion and solution in water.

BTX emulsify and dissolve in water	adsorption capacities of PP fibers (mg/g)	adsorption capacities of PP-g-St fibers (mg/g)	Increasing rate of adsorption capacities (%)
benzene	—	—	—
toluene [a]	142.30	612.70	336.89
xylene [b]	115.45	513.26	344.57

[a] initial concentration of toluene: 483 mg/L.
[b] initial concentration of xylene: 350 mg/L.

It can be seen from table 1 that compared with PP fibers, the adsorption capacities of PP-g-St fibers for toluene and xylene emulsion and solution in water were increased, which were 336.89 % and 344.57%, respectively. Solubility parameters of PP-g-St fiber and BTX were closer, which increased their affinities [18]. BTX was tightly adsorbed on surfaces having St groups by Van der Waals forces, and was occluded by the swollen St chains on the fiber surface through solvation. Moreover, the surface of PP-g-St fibers was rough and had a larger specific surface area, which also increased the adsorption capacity of PP-g-St fibers for BTX emulsion and solution in water.

The adsorption capacity of fibers for benzene in water could not be tested accurately due to the volatile nature of benzene in water. Hence, the amount of benzene volatilized could be much higher than adsorbed over a limited time period.

3.4. Regeneration of PP-g-St fibers

Both PP fibers and PP-g-St fibers are all full of meshes—web structure, which makes a quick adsorption and great adsorption capacity for pure BTX. So, research their maximum adsorption capacities for pure BTX to compare with the adsorption capacity for it in water seems to be necessary. Meanwhile, regeneration adsorption efficiencies can be obtained too basing on this foundation. Figure 5 shows the maximum adsorption capacities of PP fibers and PP-g-St fibers for pure BTX.

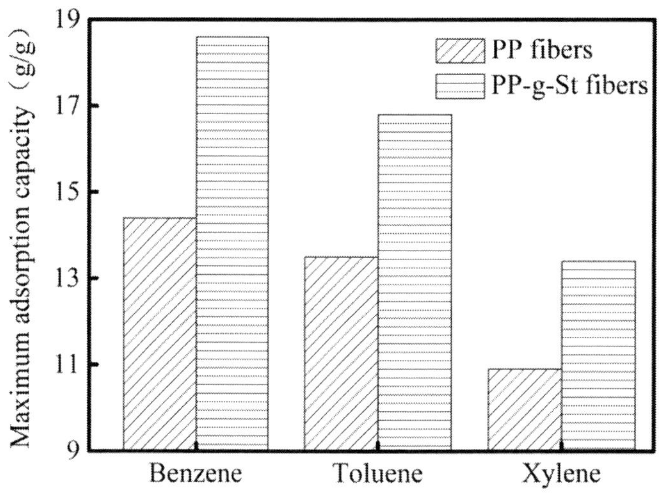

Figure 5. Maximum adsorption capacities of PP fibers and PP-g-St fibers for pure BTX.

From figure 5, it was clear that compared with PP fibers, maximum adsorption capacities of PP-g-St fibers for pure benzene, toluene, and xylene showed considerable improvement, from 14.4 g/g, 13.5 g/g, and 10.9 g/g to 18.6 g/g, 16.8 g/g, and 13.4 g/g, respectively. As mentioned previously, this was due to the fact that St was grafted onto the fiber surface and the specific surface area of fibers was increased.

The differences in increasing rates of adsorption capacities of PP fibers and PP-g-St fibers for pure BTX, BTX emulsion, and solution in water have been discussed. Firstly, maximum adsorption capacities of PP fibers and PP-g-St fibers for pure BTX were large, and they could quickly adsorb BTX, more than 10 times their own weights. Meanwhile, their adsorption capacities and adsorption rates for BTX emulsion and solution in water were small. This was due to the web-like structures of PP fibers and PP-g-St fibers that allowed them to absorb large amounts of pure BTX quickly through their gaps by capillary effect. However, for BTX emulsion and solution in water, due to the dispersion of BTX and the flow of water, BTX could not be properly absorbed inside the PP and PP-g-St fibers through the gaps.

Secondly, compared with pure BTX, the increasing rate of adsorption capacities of PP-g-St fibers for BTX emulsion and solution in water were larger than those of PP fibers. This could be explained by surface adsorption and occlusion of PP-g-St fibers, which dramatically enhanced its adsorption capacity for BTX emulsion and solution in water. Besides, the maximum adsorption capacities of PP fibers and PP-g-St fibers for benzene, toluene, and xylene showed a decreasing trend. This phenomenon could have occurred due to reduction in surface tension and surface energy, since an adsorbate with lower surface tension and surface energy is more likely to be adsorbed onto styrene in case of PP-g-St fibers [27].

Generally, synthetic fibers have better renewability than natural adsorbents. Based on the above discussion, PP-g-St fibers had a remarkable adsorption capacity for BTX by adsorption and occlusion, due to St groups present on the fiber surface. Also both, The desorption of adsorbed BTX and the regenerative efficiency of maximum adsorption of PP-g-St fibers, are extremely importance for practical applications. It could be seen from figure 5 that the adsorption capacity of PP-g-St fiber for toluene was much higher. So, using pure toluene as adsorbate (benzene was too volatile to calculate the regenerative adsorption efficiency), regenerative adsorption efficiencies of PP-g-St fibers by regeneration methods like centrifugation, gravity extrusion, and solvent extraction were tested, as shown in figure 6.

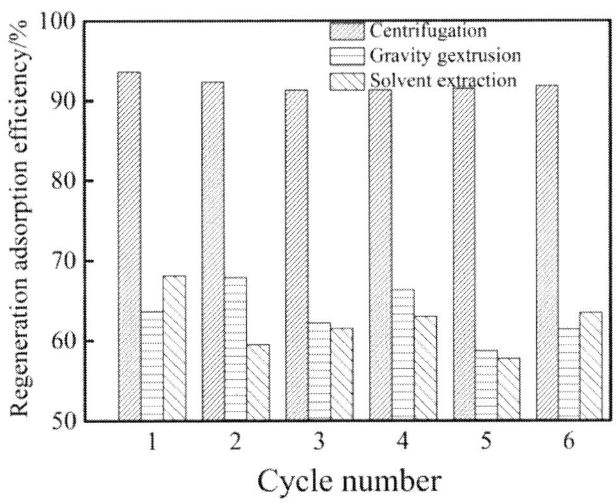

Figure 6. Regeneration adsorption efficiencies of PP-g-St fibers in three different ways.

It is evident from figure 6 that amongst the three regeneration methods, centrifugation was found to be the most efficient one. The high efficiency of this method could be attributed to the strength of centrifugal force. BTX absorbed inside the web structure could be thrown out. Moreover, BTX absorbed and occluded in St chains by Van der Waals forces could also get desorbed. But with gravity extrusion and solvent extraction methods, desorption was difficult.

After six cycles (adsorption - regeneration - adsorption) using centrifugation process, the adsorption capacity remained stable, and the regeneration adsorption efficiency was still more than 90 %. Most of the BTX absorbed on the surface of fibers was thrown out, but only a little amount of BTX, which was occluded by the St chains could not get desorbed. A moderate grafting percentage was very important. High grafting percentage could increase the adsorption capacity of PP-g-St fibers for BTX, but it would also cause problems for desorption or regeneration process.

4. Conclusion

- The optimum parameters for plasma treatment of PP fibers for grafting of St were: input power of 90 W, irradiation time of 3 min, reaction time for grafting of 3 h, to achieve a grafting percentage of 5.7%.

- Maximum adsorption capacities of PP-g-St fibers for pure benzene, toluene, and xylene showed a considerable improvement, from 14.4 g/g, 13.5 g/g, and 10.9 g/g to 18.6 g/g, 16.8 g/g, and 13.4 g/g, respectively. The increasing rate of adsorption capacity of PP-g-St fibers for toluene and xylene emulsions and solutions in water were 336.89% and 344.57%, respectively.
- The adsorption capacity of PP-g-St fibers was stable, and the regeneration adsorption efficiency was still more than 90% after six repeated cycles, using centrifugation regeneration method.

References

[1] Asadpour R, Sapari N B, Isa M H, Kakooei S and Orji K U 2015 *Fiber. Polym.* **16** 1830
[2] Wu D X, Fang L L, Qin Y M, Wu W J, Mao C M and Zhu H T 2014 *Mar. Pollut. Bull.* **84** 263
[3] Dong T, Xua G B and Wang F M 2015 *J. Hazard. Mater.* **296** 101
[4] Carvalho M N, Motta M da, Benachour M, Sales D S C and Abreu C A M 2012 *J. Hazard. Mater.* **239** 95
[5] Lin S H and Huang C Y 1999 *J. Hazard. Mater.* **70** 21
[6] Chen Y, Knappe D R and Barlaz M A 2013 *Chemosphere.* **90** 251
[7] Szala B, Bajda T, Matusik J, Zięba K and Kijak Beata 2015 *Microporous. Mesoporous Mate.* **202** 115
[8] Ranck J M, Bowman R S, Weeber J L, Katz L E and Sullivan E J 2005 *J. Environ. Eng.* **131** 434
[9] Altare C R, Bowman R S, Katz L E, Kinney K A and Sullivan E J 2007 *Microporous. Mesoporous Mater.* **105** 305
[10] Simpson J A and Bowman R S 2009 *J. Contam. Hydrol.* **108** 1
[11] Guo M L, Luo Z W and Fan C H 2016 *Asia-Pacific J. Chem. Eng.* **11** 778
[12] Korhonen J T, Kettunen M and Robin H A 2011 *Mater. Interfaces.* **3** 1813
[13] Li H Y, Wu W F, Bubakir M M, Chen H B, Zhong X F, Liu Z X, Ding Y M and Yang W M 2014 *J. Appl. Polym. Sci.* **10** 40080
[14] Rengasamy R S, Dipayan D and Praba K C 2011 *J. Hazard. Mater.* **186** 526
[15] Jin R G and Hua Y Q 2002 *Polymer Physics* (Beijing: Chemical Industry Press) vol 2 chapter 3 pp 58-9
[16] Shi D A, Yang J H, Yao Z H, Wang Y, Huang H L, Wu J, Yin J H and Costa G 2001 *Polymer.* **42** 5549
[17] Sari A, Biçer A and Alkan C 2017 *s* **161** 219
[18] Wang X Y, Yue X H, Wang X and Gou Q J 2011 *Journal of Macromolecular Science.* **50** 1882
[19] Liu W T, Wang L S, Mai J Z and Zhang A Q 2014 *Polym. Bull.* **71** 3111
[20] Li Y, Manolache S, Qiu Y P and Sarmadi M 2016 *Appl. Surf. Sci.* **364** 294
[21] Khlyustova A, Galmiz O, Zahoran M, Brablec A and Cernak M 2015 *J. Mater. Sci.* **50** 3504
[22] Samieyan E, Rahimi H and Ershad Langroudi A 2013 *Plast. Rub. Compos.* **42** 256
[23] Wavhal D S and Fisher E L 2003 *Langmuir.* **19** 79
[24] Li S N and Wei J F 2012 *J. Polym. Res.* **19** 9841
[25] Wu J Y, Alicia K A, Guo J X, Lee E J, Farid M U and Jeong S 2016 *Chem. Eng. J.* **12** 010
[26] Li Y and Chu L Y 2004 *Ind. Eng. Chem. Res.* **43** 2643
[27] Guo M L, Liang Z Y, Luo Z W, Chen Q G and Wei W J 2016 *Fiber. Polym.* **17** 257

IOP Conf. Series: Earth and Environmental Science **82** (2017) 012078 doi:10.1088/1755-1315/82/1/012078

The distribution of antibiotics in water of a river basin in South China

T Meng[1,4], W Cheng[1], M Wang[1], T Wan[1], M Cheng[1], C C Zhang[2] and Z Y Jia[3]

[1]State Key Laboratory Base of Eco-Hydraulic Engineering in Arid Area, Xi'an University of Technology, 710048, Xi'an, China
[2]Xi'an Jiaotong University, 710049, Xi'an, China
[3]Xi'an Polytechnic University, 710048, Xi'an, China

E-mail: 1570954140@qq.com

Abstract. In water environment field, one of the most attractive research topics is the determination of contamination characteristics of antibiotics in water. In order to investigate the distribution of antibiotics in surface water and drinking water of a certain river basin in southern China, we determined the types and concentrations of antibiotics that contaminated the river by performing HPLC-ESI- MS/MS method. Thus, we detected 17 antibiotics in four surface water samples (B1, B2, B3, and B4). In sampling points B3 and B4, we detected 16 antibiotics separately. The detection rates of norfloxacin, ofloxacin, and erythromycin-H_2O were 100%, and the antibiotic erythromycin-H_2O had the maximum concentration. In six drinking water samples (A1, A2, A3, A4, A5, and A6), we detected 13 antibiotics. In A5 water samples, we detected all the 13 antibiotics. The detection rate of ofloxacin and erythromycin-H_2O was 100%, and erythromycin-H_2O was the antibiotic with the highest concentration. We also found that from the upstream to the downstream of the river basin, the types of antibiotics in river increased gradually. In the upstream water samples (B1), we detected three antibiotics. Erythromycin-H_2O was the antibiotic with the highest concentration of 6.61 ng/L, and sulfapyridine had the lowest concentration of 2.82 ng/L. In the downstream water samples (B4), we detected 16 antibiotics. Erythromycin-H_2O was the antibiotic with the highest concentration of 277.58 ng/L, and the Sulfamonomethoxine was the antibiotic with the second-highest concentration of 242.1 ng/L. In addition, different membrane treatment processes could remove different amounts of antibiotics from the water samples. The study is an important reference for providing environmental protection to river water basin.

1. Introduction

An antibiotic is a class of drugs that can either inhibit the growth of bacteria or kill the bacteria. In clinical practice, antibiotics have been extensively used to treat both humans and livestock animals as they have prophylactic, therapeutic, and growth-promoting properties [1, 2]. China is an important country of interest in this study given the massive production and use of antibiotics in the country. In 2013, the total consumption of antibiotics was estimated to be 162 thousand tons in China, wherein 48% of total antibiotics were consumed by humans, while the remaining antibiotics were administered to veterinary patients [3]. In generic terms, the consumed antibiotics are continuously discharged into the natural ecosystems *via* excretion (urine and feces) following their short time of residence in the bodies of humans and animals. Then, these drugs are included in the aquatic environment when the effluents from the wastewater treatment plant are discharged into either surface water or into groundwater. The various avenues for the discharge of effluents are as follows: the leachate in landfill, leaching and runoff from farmlands fertilized with manure, sewage disposal in agricultural areas, and other industrial discharges [4, 5]. In China, environmental scientists have detected antibiotics in all kinds of aquatic environments across the country. Antibiotics in aquatic environments may pose as a potential ecological risk to aquatic ecosystems. Therefore, most current research studies focused on investigating the river basin water environment also explore the impact and concentration of antibiotics in the aquatic environment [6].

The use of antibiotics is greater in southern China than in the North because the economy is more developed and the population density is also greater in southern China [7]. Zhang Q *et al.* found that antibiotics were detected in the Huangpu River, the Yangtze River Estuary etc., with the detectable amount of sulfonamides being the greatest [8]. Ye J P *et al* found that antibiotics were also detected in the Pearl River Delta and other places, and the concentration of these antibiotics in these water bodies was in the range 10–100 ng/L [9]. Therefore, we performed a high performance liquid chromatography – tandem mass spectrometry (HPLC-MS/MS) to detect antibiotics in the water (surface water and drinking water) of a river basin in South China. Thus, we studied the kinds and the distribution of antibiotics in the river basin. The results of the analysis served as reference standards for implementing policies pertaining to water environment protection.

2. Materials and methods

2.1. Sampling

In total, we identified 10 sampling points from the region extending from the upper reaches of the river basin to the lower reaches of the river, including four rivers and six residents' drinking water source. Thus, we collected 5 L water samples from each sampling point. The river water was the surface water, while the drinking water was the residents' tap water at home. After sampling, we added 2 ml methanol and the pH was adjusted to 3 by adding 3 mol/L H_2SO_4. Then, we placed the sample in the brown bottle. The sample was carried back to the lab that day and stored at 4°C before performing detection. Figure 1 illustrates a layout of sampling points in the watershed.

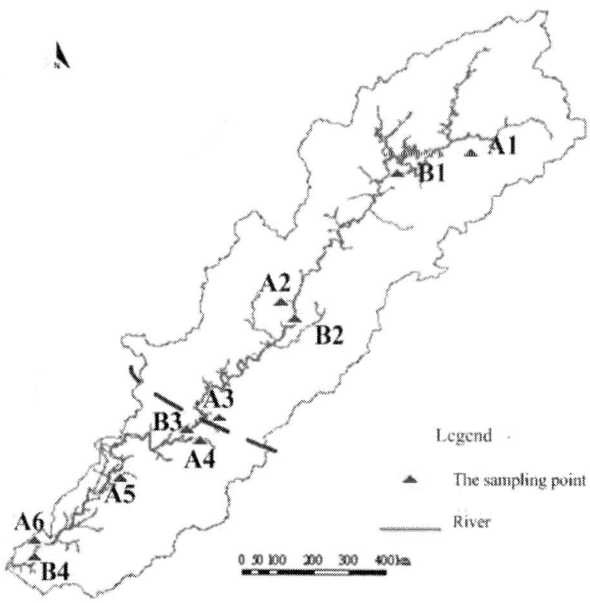

Figure 1. Map of the sampling sites in the river.

2.2. Sample preparation and analysis

The water samples (Sample volumes were 2 L.) were filtered through 0.45 μm filter membranes. We added 0.4 g disodium ethylenediamine tetraacetate (Na_2EDTA) and 200 ng of $^{13}C_3$-caffeine into filtered water; this combination of chemicals served as a surrogate monitoring the recovery of the sample from the extract. Then, the filtered water samples were extracted by performing solid-phase extraction (SPE) using Oasis HLB cartridge (500 mg; 6 mL) were provided by waters (Milford, MA, USA). The cartridge was eluted with 6 mL methanol. After extraction, we dried the HLB column under nitrogen gas at room temperature. The analytes were collected in a 10 mL brown glass vial, and then we dissolved it in 40% aqueous methanol to produce a final volume of 1.0 mL [10, 11].

An ultra-performance liquid chromatography (UPLC)–tandem mass spectrometry detection system was operated in the positive ion mode of electrospray ionization (ESI) source to detect and analyze pretreatment water samples. To perform qualitative and quantitative analysis of antibiotics, we used the multiple reflection detection method.

3. Results and discussion

3.1. Analysis of types of antibiotics in water

As shown in table 1, seventeen antibiotics were detected in total and they were classified into six types, were detected. Among them, the detection rate of ofloxacin and dehydrated erythromycin was 100%.

Table 1. Concentrations of antibiotics detected in water samples collected from drinking water and river water (ng/L).

Category	Compound	Drinking water						River water			
		A1	A2	A3	A4	A5	A6	B1	B2	B3	B4
Fluoroquinolones	Ciprofloxacin	6.27	-	4.63	-	5.02	-	-	5.49	4.82	-
	Norfloxacin	3.47	-	4.79	-	4.4	-	2.82	6.74	3.06	5.4
	Ofloxacin	12.64	4.81	11.79	7.26	8.7	11.27	6.61	15.67	6.39	9.43
Macrolides	Dehydrated erythromycin	1.84	1.39	15.73	1.91	82.52	2.62	6.47	11.6	66.48	122.7
	Kitasamycine	-	-	-	-	2.57	-	-	-	1.46	2.84
	Tylosin	-	-	-	-	1.89	-	-	-	0.26	13.08
Tetracyclines	Doxycycline	-	-	-	-	-	-	-	-	4.83	4.25
	Oxytetracycline	-	-	-	-	17.38	-	-	-	72.59	138.8
Lincosamides	Lincomycin	-	-	-	-	-	-	-	-	15.63	128.6
Polyethers	Salinomycin	-	-	-	-	9.42	-	-	-	11.26	4.46
Sulfonamides	Sulfadiazine	-	-	-	-	-	-	-	-	4.38	5.15
	Sulfamethazine	-	-	-	-	32.78	-	-	-	39.91	37.78
	Sulfamethoxazole	-	-	-	-	22.47	-	-	1.03	13.15	94.01
	Sulfamonomethoxine	-	-	-	-	60.11	-	-	5.63	27.01	242.1
	Sulfapyridine	-	-	-	-	1.65	-	-	-	1.91	2.41
	Suifoquinoxaline	-	-	-	-	-	-	-	-		12.56
	Trimethoprim	-	-	-	-	8.77	-	-	-	2.28	22.32

-: Not detected.

The results were as follows: 1) antibiotics were detected in all the sampling sites, and the types of antibiotics detected in river water were significantly more than that in drinking water; 2) in drinking water sampling points, we detected tetracyclines, polyethers, and sulfonamides in A5, but the same compounds could not be detected in other drinking water sampling points. Moreover, highest amount of dehydrated erythromycin of macrolides was detected only in A5 and not in other drinking water sampling points; 3) in river water sampling points, minimum species of antibiotics were detected in B1, and fluoroquinolones and macrolides were the antibiotics that were mainly detected in the river water sampling points.

Different kinds of water purification processes are used presently, and the extent of removal of antibiotics in water bodies depends on the method selected for water purification. Due to this reason, the types and contents of antibiotics in drinking water are significantly lower than those in rivers. In different regions of the world, the amount and variety of antibiotics in drinking water will be affected by water purification process and water quality.

3.2. Distribution of antibiotics in the river

After sampling at the upper stream, midstream, and downstream sections of the river, we determined the contents of antibiotics in every sampling site. Figure 2 illustrates the distribution of antibiotics in the river basin. As shown in figure 2, antibiotics existed in the entire reach of the river; fluoroquinolones and macrolides were the antibiotics that were mainly detected in the river. In the entire river, a 100% detection rate was observed for the following antibiotics: dehydrated erythromycin, ofloxacin, and norfloxacin. The concentrations of erythromycin, ofloxacin, and norfloxacin were in the range of 6.47–122.7 ng/L, 2.82–6.74 ng/L, and 6.39–18.85 ng/L, respectively.

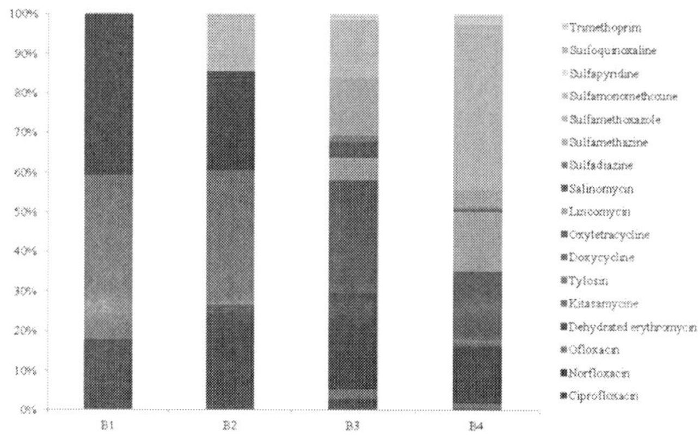

Figure 2. The proportion of antibiotics in the river sampling points.

In the upstream of river water, we detected three kinds of antibiotics. Out of the total amount of antibiotics detected in the river water, the proportion of norfloxacin was 17.7%. The concentration of norfloxacin was 2.82 ng/L. Dehydrated erythromycin accounted for 40.7% of the total amount of antibiotics detected in river water. The concentration of dehydration erythromycin was 6.47 ng/L. The proportion of ofloxacin was 41.6% in the total amount of antibiotics detected in river water; the concentration of ofloxacin was 6.61 ng/L. We detected 16 kinds of antibiotics in the middle and lower reaches of the river. The detection rate of sulfonamides was the highest, accounting for 49.21% of the total amount of antibiotics detected in the sample. The concentration of sulfamonomethoxine was 242.1 ng/L in the sample, accounting for 58.2% of the total sulfonamides. In the downstream of river water, ciprofloxacin was not detected. Sulfapyridine was the antibiotic that was present in least amounts in the sample; the concentrated of the detected sulfapyridine was 2.41 ng/L.

Based on the detection results, we conclude there is a certain relationship between the concentration of antibiotics and human activities. In the upper reaches of the watershed, villages were more dispersed and the impact of human activities on the environment was small; therefore, the types and contents of antibiotics being detected were less. In the middle and lower reaches, the impact of human activities on the environment was relatively large because of the dense population, developed industry and agriculture, and the vast distribution of many fisheries in this region. As a result, the types and contents of antibiotics being detected in this region were also more.

Along the river, there was steady increase in the kinds of antibiotics detected in water. The contents of the antibiotics being detected also increased steadily along the river, and this increase in the content and types of antibiotics is closely related to human activities; however, the influence of river water quality, hydrology, and other factors is also significant in this case. Meanwhile, the lower reaches of water also contained the afflux of a tributary. If antibiotics existed in the tributary, then it would cause a certain impact on the downstream water.

3.3. Distribution of antibiotics in drinking water

We collected drinking water samples in the rural areas along the river, and the distribution of antibiotics in the drinking water is shown in figure 3. Antibiotics were detected in the drinking water collected from different areas, and fluoroquinolones and macrolides were the two antibiotics that were primarily detected in various drinking water samples. The detection rate of dehydrated erythromycin and ofloxacin was 100% in drinking water, and the concentration of dehydrated erythromycin and ofloxacin was in the range of 1.39–82.52 ng/L, 4.81–12.64 ng/L, respectively. During the analysis of drinking water samples, we detected 2-4 kinds of antibiotics in the rest sampling sites. In addition, we also detected 13 kinds of antibiotics in A5 sampling sites.

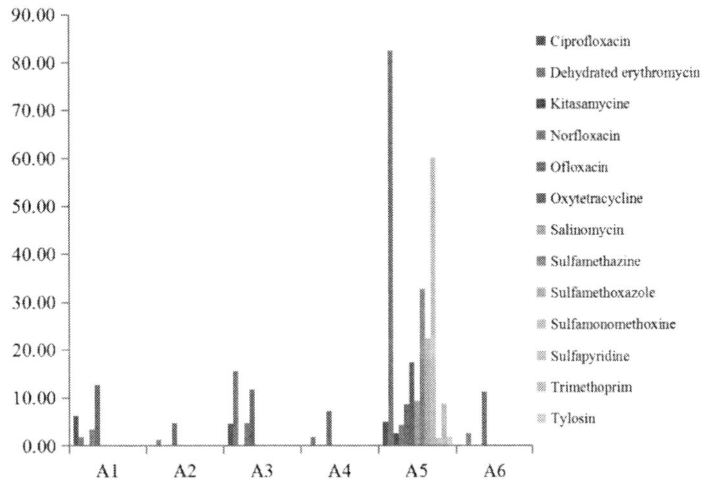

Figure 3. The distribution of antibiotics in the drinking water.

Through a survey, we found that drinking water sources of A2-A5 were all river water, and all the sources except A5 were subjected to conventional processes. An integrated water purification process was followed for the source A5. The effects were not the same as different plant processes were adopted for antibiotics treatment. Therefore, as the A5 water sample had the highest amount of antibiotics, it was more likely to be affected by the water treatment process. In addition, the water quality of water supply would have an impact on drinking water. Because A5 water plant was located in the lower reaches of others, the types of antibiotics in this water sample were significantly more than that detected in the upper reaches of the river.

The contents and types of antibiotics in drinking water were closely associated with the water quality of drinking water sources and water treatment process. Furthermore, the pollution from the municipal water supply network also affected the quality of drinking water.

4. Conclusions

- We detected 13 kinds of antibiotics in drinking water. The distribution of ofloxacin and dehydrated erythromycin was wide in drinking water, and the detection rate of these antibiotics was 100%. Moreover, water purification process certainly governed the extent of removal of antibiotics in raw water. Thus, the treatment effect of different wastewater treatment plants was not the same, the removal efficiency of conventional process was better.

- In the river water, we detected 17 kinds of antibiotics; the distribution of dehydrated erythromycin, ofloxacin, and norfloxacin was wide, and the detection rate of these antibiotics was 100%. From upstream to downstream, there was a steady increase in the content and species of antibiotics in water. This is because human activities also steadily increased as we moved from upstream to downstream of the river.

References

[1] Martinez J L 2009 Environmental pollution by antibiotics and by antibiotic resistance determinants *J. Environ. Pollut.* **157** 2893-902
[2] Wang L 2006 The current situation of antibiotics pollution and the effect on environmental microcosm *J. Pharm. Biotechnol.* **13** 144-8
[3] Guo J, Zhang Y L, Zhou X F, *et al* 2016 Occurrence and removal of fluoroquinolones in municipal sewage: A review *J. Environ. Poll. Control* **38** 75-80
[4] Shao Y R, Xi B D, Cao J L, *et al* 2013 Occurrence of antibiotics and their removal mechanism in municipal sewage treatment plants *J. Environ. Sci. Technol.* **36** 85-92
[5] Carvalho Isabel T and Lúcia S 2016 Antibiotics in the aquatic environments: A review of the European scenario *J. Environ. Int.* **94** 736-57
[6] Su H C, Ying G G, Tao R, *et al* 2012 Class 1 and 2 integrons, sul resistance genes and antibiotic resistance in Escherichia coli isolated from Dongjiang River, South China *J. Environ. Pollut.* **169** 42-9
[7] GE L K, Ren H L, Lu J J, *et al* 2015 Occurrence of antibiotics and corresponding resistance genes in the environment of China *J. Environ. Chem.* **34** 875-83
[8] Zhang Q, Xin Q, Zhu J M, *et al* 2014 The antibiotic contaminations in the main water bodies in China and the associated environmental and human health impacts *J. Environ. Chem.* **33** 1075-83
[9] Ye J P, Zou S C, Zhang G, *et al* 2007 Characteristics of selected antibiotics in the aquatic environment of the Pearl River Delta, south China *J. Ecol. Environ.* **16** 384-8
[10] Yang J F, Ying G G, Zhao J L, *et al* 2011 Spatialand seasonal distribution of selected antibiotics in surface waters of the Pearl Rivers, China *J. Envirom. Sci. Health Part. B* **46** 272-80
[11] Zhou L J, Ying G G, Liu S, *et al* 2012 Simultaneous determination of human and veterinary antibiotics in various environmental matrices by rapid resolution liquid chromatography-electrospray ionization tandem mass spectrometry *J. Chromatogr. A.* **1244** 123-38

IOP Conf. Series: Earth and Environmental Science 82 (2017) 012079 doi:10.1088/1755-1315/82/1/012079

Selection of an evaluation index for water ecological civilizations of water-shortage cities based on the grey rough set

X Y Zhang[1], J W Zhu[1,2], J C Xie[1], J L Liu[1] and R G Jiang[1]

[1]State Key Laboratory of Ecological Water Conservancy in the Northwest Arid Area, Xi'an University of Technology, Xi'an 710048, China

E-mail: xautzhu@163.com

Abstract. According to the characteristics and existing problems of water ecological civilization of water-shortage cities, the evaluation index system of water ecological civilization was established using a grey rough set. From six aspects of water resources, water security, water environment, water ecology, water culture and water management, this study established the prime frame of the evaluation system, including 28 items, and used rough set theory to undertake optimal selection of the index system. Grey correlation theory then was used for weightings in order that the integrated evaluation index system for water ecology civilization of water-shortage cities could be constituted. Xi'an City was taken as an example, for which the results showed that 20 evaluation indexes could be obtained after optimal selection of the preliminary framework of evaluation index. The most influential indices were the water-resource category index and water environment category index. The leakage rate of the public water supply pipe network, as well as the disposal, treatment and usage rate of polluted water, urban water surface area ratio, the water quality of the main rivers, and so on also are important. It was demonstrated that the evaluation index could provide an objectively reflection of regional features and key points for the development of water ecology civilization for cities with scarce water resources. It is considered that the application example has universal applicability.

1. Introduction

The aims of a water ecological civilization city are to construct a virtuous cycle that sustainably uses water resources and promotes water security, water ecosystem health and a harmonious and beautiful water environment, as well as having a profound water cultural background [1-3]. In accordance with the principles of ecology balance, a water ecological civilization city is one that is based on the characteristics of exploring the economic development level of the city, the resource conditions and the water ecological system.

Domestic scholars had proposed an evaluation system for water ecological civilizations, including basins, provinces, and regions. Liu F [4] had established a system of water ecological civilization indicators including water safety, water ecology, water environment , water utilize and so on; Tang K W [5] had established the evaluation index system of water ecological civilization from two aspects of water ecosystem and socio-economic system, and had divided the evaluation standard grade; Liu H J [6] was established the evaluation index system of ecological civilization from four part of the development and utilization of water resources, water ecological environment protection, landscape construction, water management guarantee. Due to the complex and changeable regional conditions, these indicators cannot reflect the characteristics of water resources and the social economy in cities with scarce water resources, and the evaluation criteria and classification results is difficult to be generally speaking. The problem of establishing an effective scientific index system to evaluate the water ecological civilization is worthy of greater study.

The goal of the present study was to pick up representative indices from the complex database and establish an evaluation index system for water-shortage cities that reflected the primary problems of the water environment and the primary contents in constructing an ecological civilization. It was anticipated that this might help to guide water ecological civilization in water-shortage cities and provide an early warning on key indices. Rough Set Theory was applied to analyze several evaluation indices and Xi'an City was cited as an example on which to base index weightings. This provided an index of water ecological civilization in the Xi'an is obtained, which has a certain reference value for the establishment of a water ecological civilization index system in water-shortage areas.

2. Data and methods

2.1. Selection of study area

Xi'an, the capital city of Shaanxi Province, is located in the central region of the Guanzhong Plain. With the Qinling Mountains to the south and the Weihe River to the north, Xi'an is the political, economic and cultural center of Shaanxi Province. The city's total area is 10,108 km^2, with a resident population of 8.63 million at the end of 2014 and a per capita GDP (GDP = Gross Domestic Product) of 63.8 thousand yuan. Xi'an is a typical city where the water resources are limited in gross volume and are unevenly distributed. The total quantity of the annual mean water resources equals 2.35 billion m^3, and the per capita water resource is only 275 m^3. Furthermore, the inter-annual variation in precipitation is large and its distribution is uneven during the year. The maximum recorded rainfall was 840.6 mm (in 1958), and the minimum recorded precipitation was 346.2 mm (in 1997), giving an extremes ratio of 2.43.

2.2. Construction of the evaluation index system

Based on the related literature [4-13], standard and expert opinions, the preliminary framework of water ecological civilization evaluation index was obtained. The index system was in accordance with the prominent contradiction of the water-shortage. The index system regarded water resources as the first subsystem, followed by water security, water environment, water ecology, water culture, and water management. The index system consisted of six subsystems, including the following 28 indices (see table 1).

Table 1. Evaluation index system of water ecological civilization in water-shortage cities.

Subsystem layer	Index layer			
	Code	Unit	Preliminary framework index	Final evaluation index[a]
water resource	C1	--	Standard-reaching rate of water consumption central control--	√
	C2	m^3/10^4 yuan	Water withdrawal of industrial added value per ten thousand Yuan	√
	C3	%	Leakage rate of public water supply network	√
	C4	%	Penetration rate of domestic water-saving appliances	√
	C5	%	Water saving irrigation rate of farmland	√
	C6	%	Treatment and usage rate of polluted water	√

IOP Conf. Series: Earth and Environmental Science **82** (2017) 012079 doi:10.1088/1755-1315/82/1/012079

water safety	C7	%	Standard rate of city flood control and waterlogged elimination	√
	C8	%	Standard rate of centralized drinking water source safety	√
	C9	%	Penetration rate of tap water	
	C10	%	Groundwater overdraft area ratio	
	C11	%	Standard rate of water quality	√
water environment	C12	%	Water quality compliance rate of Water function area	√
	C13	%	Urban water surface area ratio	√
	C14	%	Wastewater discharge compliance rate	√
	C15	%	Water quality status of major rivers	√
water ecology	C16	%	Satisfaction degree of river ecological basic flow	√
	C17	%	Aquatic organisms diversity	
	C18	%	Ecological revetment proportion of rivers and lakes	√
	C19	%	River longitudinal connectivity index	
	C20	%	Control rate of water loss and soil erosion	√
water culture	C21	number	Quantity of water culture carrier	√
	C22	%	Public awareness and participation in water ecological civilization	√
	C23	%	Residents' satisfaction with water ecological environment	
	C24	%	Appreciation of water area and surrounding view spot	
water management	C25	%	Monitoring rate of water quality in water function area	√
	C26	%	water metering rate	
	C27	%	Monitoring rate of pollution discharge outlets of rivers	√
	C28	%	Attention to Water ecological civilization	√

[a] "√" represents the index items that were retained after the selection.

2.3. Index grade division

According to the relevant research results and national standards, the discrete grading standards were divided into five grades, of which "5" was the best value and "1" was the worst (see table 2).

Table 2. The grading standards of indicators.

Grade / Index	5	4	3	2	1
C1	100	--	--	--	0
C2	0~35	35~60	60~100	100~150	150
C3	0~8	8~12	12~18	18~25	25~100
C4	100~90	90~75	75~60	60~40	40~0
C5	100~85	85~70	70~60	60~50	50~0
C6	100~50	50~30	30~15	15~5	5~0
C7	100~90	90~75	75~60	60~40	40~0

C8	100	100~90	90~80	80~70	70~0
C9	100~90	90~75	75~60	60~40	40~0
C10	0	1~4	4~10	10~20	20
C11	100~95	95~90	90~85	85~80	80~0
C12	100~80	80~60	60~50	50~40	40~0
C13	4~3.75	3.75~3.5	3.5~3	3~2	2~0
C14	100~95A	100~95	95~90	90~85	85~0
C15	100~90	90~75	75~60	60~40	40~0
C16	100~98	98~90	90~80	80~60	60~0
C17	100~85	85~70	70~50	50~30	30~0
C18	100~85	85~65	65~45	45~30	30~0
C19	0~0.3	0.3~0.5	0.5~0.8	0.8~1.2	1.2
C20	100~90	90~75	75~60	60~50	50~0
C21	8	7~6	5~3	2~1	0
C22	20	20~15	15~10	10~5	5~0
C23	100~85	85~70	70~60	60~50	50~0
C24	100~90	90~80	80~60	60~50	50~0
C25	100~90	90~75	75~60	60~40	40~0
C26	100~90	90~75	75~60	60~40	40~0
C27	100~90	90~75	75~60	60~40	40~0
C28	2	1~2	0.5~1	0.3~0.5	0.3~0

2.4. Evaluation index selection based on rough sets

Rough set theory is a data analysis theory proposed by the Polish mathematician Pawlak in 1982. It is a method of dealing with vagueness and uncertainty, the key point of which is connecting classification with knowledge, in the belief that knowledge comes from the classification ability of humans and other species, and representing classifications in forms of equivalence relationships. Compared with fuzzy mathematics, probability statistics and other processing methods, rough set theory does not need to use the apriority knowledge, such as membership functions and probability distribution density, when dealing with uncertain information. It is also not bound by more premise conditions, and thus is widely applicable in the fields of theory and management [14, 15].

Step 1 is the establishment of the information system and decision table.

Information system:

$$S=\{U,A,V,f\} \tag{1}$$

In the formula, U is a nonempty finite set of the research object; A is a nonempty finite set of attribute; V is the range of attribute set A; f is the set of information function of U and A.

Decision table:

$$S=\{U,C,D,V,f\} \tag{2}$$

In the formula, C is the index set, D is the comprehensive evaluation result set, which is commonly used to describe the grade or the corresponding score as well as to describe the result of the comprehension evaluation. The comprehensive evaluation decision table is actually the data table which contains the evaluation index and the evaluation result.

Step 2 is the discretization of index data

As a mathematical tool for dealing with incomplete and uncertain problems, rough set theory can deal effectively with a discrete index but it cannot deal directly with a continuous index. Therefore, discretization of the continuity index must come first before the application of the rough set methods of index screening. In this study, the equal interval method was chosen, combined with the actual data of Xi'an City.

Step 3 is the parsimonious model of the rough set

Definition: in the information system, where P belongs to C, the relationship between P and IND (P) is defined as:

$$\mathrm{IND}(P)=\{(x,y)\in U^2,\forall a\in P, x\neq y, f(x,a)=f(y,a)\} \tag{3}$$

IND (P) divides the object U into k equivalent classes, denoted as: $U/_P = \{x_1,x_2,\cdots x_k\}$.

The information system, S, is set up, where $S = \{U,A,V,f\}, R \in A, R$, R is a set of equivalence relationships, $r \in R$, if $\mathrm{IND}(R) = \mathrm{IND}(R - \{a\})$, then r is called an unnecessary part of R, otherwise r is a necessary part of R; if $r \in R$ and every r is necessary, then R is said to be independent, otherwise it will be known as dependent.

Assuming that $Q \in R$, if Q is independent and $\mathrm{IND}(Q) = \mathrm{IND}(P)$, then Q is called reduction of U in the domain of attribute set P.

The specific steps are as follows:

a: On the index system of $C = \{c_i\}(i = 1,2,3,\cdots,m)$, work out $\mathrm{IND}(C)$ and $\mathrm{POS}_C(D)$;

b: On $i = 1,2,3,\cdots,m$, work out $\mathrm{IND}(C - \{c_i\})$and $\mathrm{POS}_{C-\{c_i\}}(D)$ successively;

c: If $\mathrm{POS}_{C-\{c_i\}}(D) = \mathrm{POS}_C(D)$, then c_i is the removable redundant indicator of target system C; otherwise, c_i is the necessary indicator that cannot be eliminated from the index system.

2.5. Index weighting based on grey correlation

After selecting the index system by the above method, the grey correlation degree method was used to calculate the weight of the index. The method is a quantitative comparison between the state and the trend of a developing system. Compared with regression analysis, there is no special requirement on the number and type of sample, with low computation requirements and reliable results [16].

Step 1 (for the weighting procedure):

First determine the reference sequence of X_0, in which the bigger the better index is called $X_{0k} = \max(x_{1k},x_{2k},\cdots,x_{mk})$, $k \in n$;

The smaller the better one is called $X_{0k} = \mathrm{mim}(x_{1k},x_{2k},\cdots,x_{mk}),k \in n$. So then the judgment matrix M can be constructe:

$$M=\begin{bmatrix} x_{01} & \cdots & x_{0n} \\ \vdots & \ddots & \vdots \\ x_{m1} & \cdots & x_{mn} \end{bmatrix} \tag{4}$$

Taking into account the difference of dimensions and orders of magnitude of different indicators, there is a need for M dimensionless processing in the calculation of the weighting; in this study, normalization is used to process the data.

Step 2: the matrix of correlation coefficient is calculated: calculate the correlation coefficient $\varepsilon_i(k)$ of the K index between the calculation sequence X_i and the reference sequence X_0:

$$\varepsilon_i(k)=\frac{\min\limits_i \min\limits_k |x_0(k)-x_i(k)|+\rho \max\limits_i \max\limits_k |x_0(k)-x_i(k)|}{|x_0(k)-x_i(k)|+\rho \max\limits_i \max\limits_n |x_0(k)-x_i(k)|} \tag{5}$$

When $\rho \in [0,1]$ (generally taken 0.5), the correlation coefficient matrix E can be obtained from the formula:

$$E = \varepsilon_i(k) = \begin{bmatrix} \varepsilon_1(1) & \cdots & \varepsilon_1(k) \\ \vdots & \ddots & \vdots \\ \varepsilon_i(1) & \cdots & \varepsilon_i(k) \end{bmatrix} \tag{6}$$

Step 3: the grey relational grade is calculated:

Equation (6) can be used to calculate the grey correlation degree of different flood disaster sequences r_i:

$$r_i = \varepsilon_i(k) \times \omega_i = (x_{i1}, x_{i2}, \cdots x_{in}) \times [\varepsilon_i(1), \varepsilon_i(2), \cdots \varepsilon_i(n)]^T \tag{7}$$

Step 4: the grey correlation weight value is obtained.

After normalization of the correlation degree of each index factor, the grey correlation weight value β of the evaluation factor was obtained, so a complete evaluation index system is available.

3. Results and discussion

3.1. Index data discretization

The index data came from the "Statistical Yearbook of Shaanxi Province", "Shaanxi Water Resources Bulletin", "Xi'an Statistical Yearbook", "Xi'an Water Resources Report" and so on. The results of the discretization could be obtained by substituting the index data of Xi'an City into the discrete classification standard (see table 3).

Table 3. The discretization results of Xi'an City.

Year / Index	2006	2007	2008	2009	2010	2011	2012	2013	2014	2015
C1	0	0	5	5	5	5	5	5	5	5
C2	3	3	4	4	4	4	4	5	5	5
C3	2	2	3	3	3	3	3	3	3	3
C4	3	3	3	4	4	4	4	5	5	5
C5	4	3	4	4	5	4	5	4	4	4
C6	2	2	2	2	2	2	2	2	3	3
C7	1	1	1	1	1	1	1	2	2	2
C8	5	5	5	5	5	5	5	5	5	5
C9	3	4	4	4	4	4	4	5	5	5
C10	2	2	2	2	3	3	3	4	4	4
C11	5	5	5	5	5	5	5	5	5	5
C12	2	1	2	2	2	3	2	3	3	3
C13	2	2	3	3	4	4	4	5	5	5
C14	2	2	3	2	2	4	2	1	3	4
C15	2	2	3	3	3	3	3	3	3	3
C16	4	3	3	4	3	3	3	4	4	4
C17	4	5	4	4	5	4	5	5	5	5
C18	1	2	2	2	2	2	2	3	3	3
C19	5	5	5	5	5	5	5	5	5	5
C20	1	1	1	1	1	1	1	2	2	2
C21	4	4	4	4	5	5	5	5	5	5
C22	2	3	3	3	3	4	4	5	5	5
C23	3	3	4	3	4	4	4	5	4	5
C24	2	2	3	3	3	4	4	4	4	5

C25	4	5	5	5	5	5	5	5	5	5
C26	2	2	2	2	3	3	3	3	3	3
C27	1	2	2	2	2	2	2	2	3	3
C28	1	1	1	1	2	2	2	3	3	3

3.2. Selection of evaluation index system

According to the order of water resources, water security, water environment, water ecology, water culture, water management, these six kinds of index system were screened as follows:

- Screening of water resources index: According to the results of the discrete index and the model of the rough set index system, the index of water resources was screened. In the knowledge representation system composed of those indicators (see table 4), the domain of discourse U was a collection of all of the research objects; the condition attribute was $C = \{c_1,c_2,c_3,c_4,c_5,c_6\}$, the decision attribute was replaced by the urban water surface area ratio, that $D = c_{13} = \{d\}$.

Divided by d, U:

$$U/d = \{\{1,\ 2\},\ \{3,\ 4\},\ \{5,\ 6,\ 7\},\ \{8,\ 9,\ 10\}\}$$

Divided by U, C:

$$U/C = \{\{4,\ 6\},\{5,\ 7\},\{9,\ 10\},\{1\},\{2\},\{3\},\{8\}\}$$

According to the rough set theory, the positive region:

$$POS_C(d) = \{1,2,3,5,7,8,9,10\}$$

Divided by C2, C3, C4, C5, C6, C7, U:

$U/\{C2,\ C3,\ C4,\ C5,\ C6,\ C7\} = U/C-\{C1\} = \{\{5,7\},\{9,10\},\{1\},\{2\},\{3\},\{8\},\{4\},\{6\}\}$, there were $POS_{C-\{C1\}}(d) = \{1,2,3,4,5,6,7,8,9,10\} \neq POS_C(d)$.

Therefore, C1 was a non-redundant index and couldn't be deleted. Similarly, C2, C3, C4, C5, C6 were non-redundant indices, and couldn't be deleted.

Table 4. Knowledge representation system of water resources index

Index / Year	C1	C2	C3	C4	C5	C6	d
2006	5	3	2	3	4	2	2
2007	5	3	2	3	3	2	2
2008	5	4	3	3	4	2	3
2009	5	4	3	4	4	2	3
2010	5	4	3	4	5	2	4
2011	5	4	3	4	4	2	4
2012	5	4	3	4	5	2	4
2013	5	5	3	5	4	2	5
2014	5	5	3	5	4	3	5
2015	5	5	3	5	4	3	5

- Screening of water safety index: In the knowledge expression system of the water safety index (see table 5), the domain of discourse U was a collection of all research objects, the condition attribute $C = \{c_7,c_8,c_9,c_{10},c_{11}\}$,the decision attribute $D = c_{14} = \{d\}$.

Divided by d, U:

$$U/d = \{\{1,\ 2\},\ \{3,\ 4\},\ \{5,\ 6,\ 7\},\ \{8,\ 9,\ 10\}\}$$

Divided by U, C:

$$U/C = \{\{1\},\{2\},\{3,4,5,6,7\},\{8,9,10\}\}$$

And according to the rough set theory, the positive region:

$$POS_C(d) = \{1,2,8,9,10\}$$

Divided by C7, C8, C9, C10, C11, U:

$U/\{C8,\ C9,\ C10,\ C11\} = U/C-\{C7\} = \{1\},\{2\},\{3,4\},\{5,6,7\},\{8,9,10\}$, there were $POS_{C-\{C1\}}(d) = \{1,2,3,4,5,6,7,8,9,10\} \neq POS_C(d)$. Therefore, C7 was a non redundant index and couldn't be deleted. Similarly, C8, C11 were non-redundant indices, couldn't be deleted, and C9, C10, which were redundant indices, could be deleted.

Table 5. Knowledge representation system of the water safety index

Year \ Index	C7	C8	C9	C10	C11	d
2006	1	5	3	2	5	2
2007	1	5	4	3	5	2
2008	1	5	4	3	5	3
2009	1	5	4	3	5	3
2010	1	5	4	3	5	4
2011	1	5	4	3	5	4
2012	1	5	4	3	5	4
2013	2	5	5	4	5	5
2014	2	5	5	4	5	5
2015	2	5	5	4	5	5

- Screening of the water environment, water ecology, water culture, and water management index:

Similarly, in the index of water environment, C14, C15 were non-redundant indices, and couldn't be deleted, and C12 was a redundant index, which could be deleted. In the index of water ecology, C16, C18, C20 were non-redundant indices; In the index of water culture, C21, C22 were non-redundant indices; In the index of water management, C25, C27, C28 were non-redundant indices.

In summary, the total number of indicators of the water ecological civilization evaluation index system for Xi'an City had been reduced from 28 to 20 (see table 1).

3.3 Index weighting and results analysis

- Weighted by grey relational theory

The selected index system was put into the grey relational weight model, and the weight of the grey relational analysis was calculated:

$$\beta_i = \begin{pmatrix} 0.0349,\ 0.0416,\ 0.0704,\ 0.0522,\ 0.0503,\ 0.0722,\ 0.0405,\ 0.0311,\ 0.0357,\ 0.0682, \\ 0.0486,\ 0.0553,\ 0.0357,\ 0.0577,\ 0.0503,\ 0.0424,\ 0.0695,\ 0.0447,\ 0.0515,\ 0.0472 \end{pmatrix}$$

The selected index system was put into the grey relational weight model and the weight of the grey relational analysis was calculated.

The weighting diagram was drawn, based on β_i (see figure 1). It could be seen that the most influential index of Xi'an City's water ecological civilization are C3 (β_3 = 0.0704), C6 (β_6 = 0.0722), and these were followed by C13 (β_{13} = 0.0682), C18 (β_{18} = 0.0695), C15 (β_{15} = 0.0553), C22 (β_{22} = 0.0577). The weight of remainder indices were less than 0.055. In general, the greatest influence on the water ecological civilization index system was attributable to the indices of water resources and water environment.

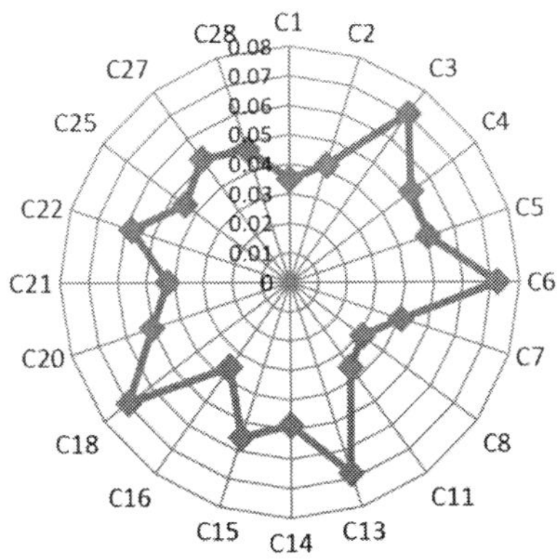

Figure 1. Radar diagram of indices weighting in Xi'an City.

- Calculation of evaluation results

According to the indices classification and the adoption of the method of five value logic partition to construct the membership function, " I " was the worst grade and "V" was the optimal grade. The value of the membership function was classified according to the index grade (see table 2). Taking 2012 as an example, the membership function was calculated from equation (8):

$$r_{ij}(x)=\begin{cases}1-\dfrac{\max\{a_{i1}-x,x-a_{i2}\}}{\max\{a_{i1}-\min\limits_j x,\max\limits_j x-a_{i2}\}} & x\notin[a_{i1},a_{i2}]\\ 1 & x\in[a_{i1},a_{i2}]\end{cases}\quad i=1,2,\cdots,n;j=1,2,\cdots,m \tag{8}$$

where a_{i1}, a_{i2} represent the upper limit and lower limit of the index value, and x indicates the index value. Then, $B=\beta\cdot F$ was calculated, and the comprehensive evaluation result for 2012 was obtained:

$$B=\beta\cdot F=\begin{bmatrix}0.0349\\ \vdots\\ 0.0472\end{bmatrix}\cdot\begin{bmatrix}r_{11} & r_{12} & \cdots & r_{1m}\\ r_{21} & r_{22} & \cdots & r_{2m}\\ \cdots & \cdots & \cdots & \cdots\\ r_{n1} & r_{n2} & \cdots & r_{nm}\end{bmatrix}=(0.0745,0.1305,0.3881,0.2573,0.1496)$$

The maximal value of the vectors, 0.3881, corresponds to section III. Therefore, the evaluation of water ecological civilization in 2012 was close to average. Similarly, comprehensive evaluation results were obtained in other years (see table 6).

From this result, it was evident that the general situation of water ecological civilization in Xi'an City was approximately average from 2006 to 2012, and achieved a better grade from 2013 to 2015. The results were consistent with the actual situation, which illustrates that the evaluation model was feasible.

Table 6. Evaluation interval and results of water ecological civilization in Xi'an City.

Year	I	II	III	IV	V	Results
2006	0.2099	0.3862	0.1983	0.1615	0.0441	II
2007	0.1955	0.3341	0.3022	0.1057	0.0625	II
2008	0.1027	0.2205	0.3305	0.3004	0.0459	III
2009	0.0952	0.2901	0.2803	0.2321	0.1023	II
2010	0.0851	0.2216	0.3562	0.2452	0.0919	III
2011	0.1161	0.0711	0.3517	0.2749	0.1862	III
2012	0.0745	0.1305	0.3881	0.2573	0.1496	III
2013	0.0351	0.1475	0.2335	0.4012	0.1827	IV
2014	0.0504	0.0705	0.2598	0.4841	0.1352	IV
2015	0.0732	0.0376	0.1809	0.5249	0.1834	IV

- Evaluation results analysis

As is typical of cities that are short of water, although Xi'an City has made certain achievements in terms of water ecological civilization in recent years, the primary problem faced by the city that needs to be solved urgently is still the negative impact of water-shortage and future deterioration the water environment for some time to come. The results shown that, during the study period, two of the most important indices that symptomized the water ecological civilization in Xi'an City were addressed by the water resources index. The city should strengthen the protection of water sources and water management, increase water recycling, accelerate the comprehensive management of soil and water loss, strengthen flood control and drainage capacity in both urban and rural areas, strengthen water pollution control etc., in order that total water resources can meet future demand.

4. Conclusions

Based on analysis of the evaluation index of water ecology civilization for Xi'an City, which was chosen as an example for the purpose of demonstrating the effectiveness of the analysis procedure, the preliminary framework of an evaluation index system was established that focused on the status, features and problems of water ecology of cities in China that are subject to water shortages. The 'grey rough set' theory was used to obtain optimization results of the evaluation index for such cities. From the specific results for Xi'an City, the conclusion was obtained that the key points of water was to guarantee the total available water resource, to improve water supply efficiency, to strengthen water pollution prevention & disposal and to renovate the water environment. Planning should be based on available water resources; synchronization of protection and utilization; broadening of income and reduction of expenditure; improvement of the overall water supply and drainage network; improvement of the water landscape and water culture; and implement a comprehensive water ecology civilization for concerned cities. In order to further verify the applicability of the analysis method, more empirical studies need yet to be carried out to refine the selection index of the evaluation indicators of water ecological civilization in other water-shortage cities.

Acknowledgments

This study has been partly supported by the National Natural Science Foundation of China (No. 51479160, 51509201, 41471451), National key R & D projects (2016YFC0401409).

References

[1] Qiu B X 2013 Situation of and measures for water security on Chinese cities *Urban Dev. Studies* **20** 1-11
[2] Wang H 2016 Theoretical basis and key problems of water ecological civilization construction

China Water Resour. **19** 5-7

[3] Zhang H, Cheng J H, Feng Y,Chen D, Ni L and Sun H 2015 An evaluation index system for ecological civilization construction in megacities and its research applications: The case of Wuhan City *Acta Ecol. Sin.* **32** 547-56

[4] Liu F and Miao W 2016 Study on the system model construction of the elements of water ecological civilization construction *China Popu. Resour. Environ.* **26** 117-22

[5] Tang K W 2013 Discussion on concept and assessment system of aquatic ecological civilization *Water Resour. Prot.* **29** 1-4

[6] Liu H J, Huang J W, Shi Y Z and Chen H W 2013 Water ecological civilization evaluation on the typical urban of the lower Yellow River *Yellow River* **35** 64-7

[7] Wang J H and Hu P 2013 Studies on evaluation system of water ecological civilization *China Water Resour.* **15** 39-42

[8] Huang Z 2013 Discussions on index system of water ecological civilization *China Water Resour.* **6** 17-9+9

[9] Yang F S and Xu C J 2014 Study on evaluation system for water eco-civilization city - A case study of Xianning City *J. Anhui Agri. Sci.* **42** 12270-3+12277

[10] Gao H, Cao X Y and Cai B G 2013 Study on evaluation system for water eco-civilization city in Shandong Province *China Water Resour.* **10** 8-10

[11] Shuang C, Jin W and Zhan Z Y 2004 Research on the integration of ecological landscape and urban morphology *Geogr. Sci. Prog.* **5** 67-77

[12] Guo X 2005 Urban ecological landscape and ecological landscape design *Ecol. Econo.* **10** 256-9

[13] Bai J and Zhang H 2011 The ecological management review and thinking of Xi'an Chanba Ecological District - focusing on Xi'an world horticultural exposition in 2011 *Ecol. Econo.* **10** 168-70+174

[14] Wang Y B, Liu G X, Feng Q, Si J H,Chang Z Q and Xi H Y 2012 Evaluation of oasis stability in Ejina based on rough set and grey theory *Resour. Sci.* **34** 1750-60

[15] Yang C J, Ge H and Wang S Z 2012 Overview of attribute reduction based on rough set *Appl. Res. Comp.* **29** 16-20

[16] Du J H Research on approach of evaluation index screening based on grey rough set *J. North Univ. China (Nat. Sci. Edit.)* **33** 559-62

IOP Conf. Series: Earth and Environmental Science **82** (2017) 012080 doi:10.1088/1755-1315/82/1/012080

A design of spectrophotometric microfluidic chip sensor for analyzing silicate in seawater

X Cao[1,2], S W Zhang[1,2], D Z Chu[1,2], N Wu[1,2], H K Ma[1,2] and Y Liu[1,2,3]

[1]Institute of Oceanographic Instrument, Shandong Academy of Science, Qingdao Shandong 266001, China
[2]Shandong Ocean Environment Monitoring Technology Key Laboratory, Qingdao Shandong 266001, China

E-mail: cx@sdioi.com

Abstract. High quality and continuous in situ silicate data are required to investigate the mechanism of the biogeochemical cycles and the formation of red tide. There is an urgently growing need for autonomous in situ silicate instruments that perform determination on various platforms. However, due to the high reagents and power consumption, as well as high system complexity leading to low reliability and robustness, the performance of the commercially available silicate sensors is not satisfactory. With these problems, here we present a new generation of microfluidic continuous flow analysis silicate sensor with sufficient analytical performance and robustness, for in situ determination of soluble silicate in seawater. The reaction mechanism of this sensor is based on the reaction of silicate with ammonium molybdate to form a yellow silicomolybdate complex and further reduction to silicomolybdenum blue by ascorbic acid. The minimum limit of detection was 45.1 nmol L^{-1}, and the linear determination range of the sensor is 0-400 μmol L^{-1}. The recovery rate of the actual water is between 98.1%-104.0%, and the analyzing cycle of the sensor is about 5 minutes. This sensor has the advantages of high accuracy, high integration, low water consumption, and strong anti-interference ability. It has been successfully applied to measuring the silicate in seawater in Jiaozhou Bay.

1. Introduction

Silicon is one of the essential nutrients that maintain biological growth in the marine environment. Silicate, a key component in diatoms, radiolarians and foraminifera and other marine organisms, is one of the greatest important sources of silicon in marine ecosystems [1]. Insufficiency of silicate may limit primary production in marine ecosystems; however, excessive silicate concentration may lead to excessive phytoplankton growth, leading to the cause of the occurrence of red tide [2]. So, an accurate and sustainable measurement of marine silicate is of great significance for marine biological resources development and ecological environment protection.

In China, in accordance with the "Specification for Marine Monitoring" (17378.4-2007), there are two methods adopted for arbitration, including a) silicomolybdenum yellow method and b) silicomolybdenum blue method. The former method is rapid but insensitive and susceptible to salinity, it is then not suitable for low silicate concentration determination; in the contrary, the later one, adopted by most of marine monitoring unit, has the advantage of high sensitivity and specificity, even though this method is of time-consuming, tedious, and difficult to achieve the on-line analysis. With the development of flow analysis technology, various of methods have been applied to measure nutrients in seawater, including flow injection [3-5], sequential injections [6-8], continuous flow [9], and stop flow analysis [8, 10, 11], etc. Among those methods, methods of flow injection, sequential injections and continuous flow are of the most common ones with a quick reaction and relatively simple structure, however with the relatively low reaction efficiency, and those consumptions of reagents or samples are usually large. Another method, stop flow method, has the advantage of excellent reaction efficiency, however the low reaction rate and adverse dispersion effect restricted its application. All these methods, adopting the precise time control and non-equilibrium state analysis which greatly shorten the analysis time, can be used to on-line analyze silicate. It is a break through against traditional conception of which the reaction must be achieved with a physical and chemical equilibrium.

Recently, with the development of Micro and Nano-technology, a "Lab on Chip" technology, an integrated fluid flow analysis system with the advantage of high integration level and speed of mass transfer and heat conduction, has become a new trend of analysis [12, 13]. This new technology contributes to the occurrence of a new generation of in situ marine nutrient sensor, such as nitrate and nitrite microfluidic sensor [14], phosphate sensor [12] and Fe and Mn sensor [15], etc., although these sensors have not yet matured enough to form reliable products. Most of prototypes are not expected to perform multi-step reactions due to deficiency of the reaction efficiency and the increment of the system complexity. In this study, A new microfluidic continuous flow analysis sensor was established and designed for determination of soluble silicate in seawater in situ based on "blue" arbitration method (multi-step reactions). Furthermore, this sensor deployed on research vessel was successfully applied in routine survey in Jiaozhou Bay in China.

2. Experimental

2.1. Reagents and chemicals

All the solutions were prepared by Milli-Q ultra-pure deionized (DI) water (Millipore, USA), followed by 30 minutes ultrasonic degassing (Shengyan, SCQ-H300A, China). All the chemical reagents were purchased from Fluka (Sigma–Aldrich, Germany). 100mg L^{-1} Silicate standards (NRCCRM, China) were used for calibration. Acidified molybdate reagent of which 11.9 g ammonium molybdate was dissolved in 250 mL water, mixing with 15 mL concentrated sulfuric acid in 150 mL water was used for color developing agent. 26.3 g citric acid dissolved in 250 mL water, followed by adding 0.2 g antimony potassium tartrate used as masking agent to avoid phosphate interference. 10.0 g ascorbic acid dissolved in 200 mL water used as reduction reagent to form silicomolybdenum blue. All above solutions were stored in 4 ℃ and kept in dark, followed by 1ml chloroform added to prevent biodegradation.

2.2. The design of microfluidic chip

The microfluidic chip (figure 1) was produced using 3.0 mm thick tinted polymethyl methacrylate (PMMA) sheets (7C83, Plexiglas, Germany). The PMMA sheets were patterned by the method of micro-milling using Circuit board plotters (LPKF Protomat S61 Micromill). Before micro-milling, sheets were thoroughly cleaned to remove dusts and grease [16]. The purpose using tinted PMMA was for reducing light, which was from ambient sources, as well as those from the LEDs not passing through the analytes, reaching the detector, as reported by Cedric F A Floquet [17]. A chloroform solvent vapor assisted bonding method was used to achieve chip sealing. Previous studies have shown that this procedure could smooth the inner surface of the microchannel and enhance the precision of optical analysis on microchip [14]. Three absorbance cells were designed in microchip, including one reference cell (25 mm length), and two measurement cells with 25 mm as the length for low concentration silicate determination and 5 mm as the length for high concentration silicate determination respectively. Also, three pairs of 810 nm LEDs (Avago Technologies, USA) and photodiodes (TAOS, USA) that directly fixed onto the chip using Norland 68 optical adhesive

(Norland, USA), were used to perform spectrophotometry analysis. For the 5 mm cell, photodiode and LED were spaced apart by 15 mm using a 7mm long "light tube" [14] to transmit light between the LED and the optical absorbance cell, and this design could reject the additional background light when the intense light from LED passing through the short absorbance cell. The light tube was produced by directly milling microchannel on the tinted PMMA sheets and filled with transparent Norland Optical Adhesive 68 with a high refractive index. Two serpentine mixers were deployed with length of 936.5 mm and 684.1 mm respectively to fulfill the reaction of coloration and reduction. All the microchannel were of 250 μm × 250 μm, except for the three optical absorbance cells which were of 400 μm × 400 μm. The total fluidic dead volume was designed as about 123.7 μL.

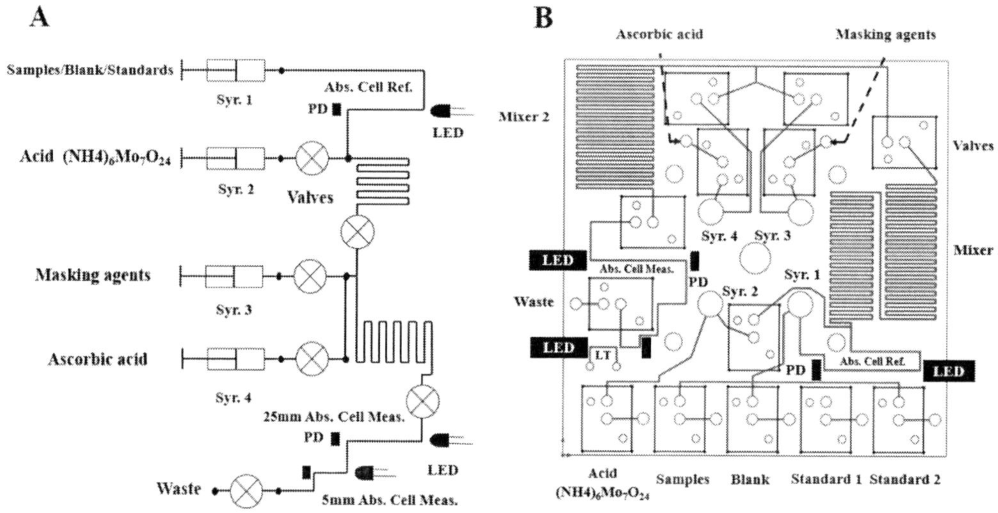

Figure 1. Microfluidic design and analysis procedure.
PD: Photodiode, Syr: Syringe, LED: Light-emitting Diode, Abs. Cell Ref: Adsorption Cell Reference, Abs. Cell Meas: Adsorption Cell Measurement, LT: Light Tube.

2.3. Description of the sensor

Silicate sensor (figures 2A and 2B) in this study contained four functional modules, including fluid driving system, microchip, control circuit and sensor peripheral. Fluid driving system consisted of syringe pump and 13 microinert solenoid valves (100R10, SMC, Japan) which were directly mounted on the microfluidic chip sealed with inert O ring. The syringe pump was custom-designed with a stepper motor (Dingzhi, China) driving four 2 mL PMMA syringes controlled by two infrared limit switches. The distance between these two switches on the active plate of the pump were pre-calibrated, and this structure formed a syringe pump. The total fluid dead volume (include valves and microchip) of the sensor was about 383.7 μL. Control circuit consisted of MCU, power module, stepper-motor-based linear actuator, data processing and memory module, etc. MCU used in this sensor was STM32 chip (TI, USA). Power module was considered to perform in two operation mode, which were self-contained and externally powered monitoring. Stepper-motor-based linear actuator adopted DS-5045-003 commercial actuator (Dingzhi, China) to enable and control the stepper motor. A 24 bit A/D converter (ADS1675, Texas Instruments, USA) was used to acquire, convert and amplify data form photodiodes. A 2 GB flash memory card was used to store the signals from the photodiodes and the silicate concentration data. A fault tolerant system was established to guarantee data was not missing or corrupted. The whole micro-analyzer system was enclosed in a watertight PVC tube (figure 2B) with a diameter of 15 cm and 45 cm high.

An external filter unit which consisted of peristaltic pump and 0.45 μm capsule filter and several reagent bags was equipped as sensor peripherals.

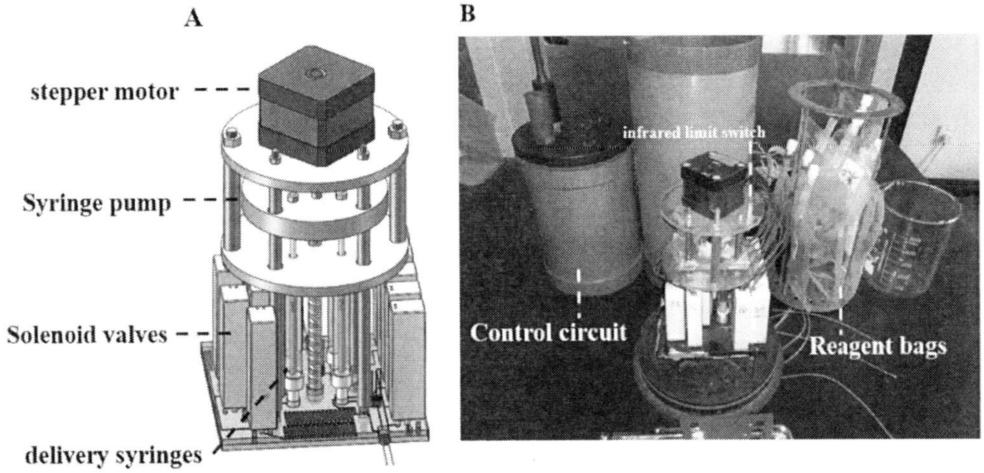

Figure 2. Construction model and real product of silicate sensor.

2.4. Analytical process
The analytical procedures involved continuous five steps: (1) sampling of the seawater sample using syringe form environment; (2) injection of the sample and relative reagents in micro-system; (3) the first hybrid reaction to form yellow silicomolybdate; (4) the second hybrid reaction to form blue silicomolybdate, by mixing with masking agent and ascorbic acid; and (5) determination of the absorbance. Before addition of the reagents, the absorbance of the sample, blank, and/or standards were measured in the reference absorbance cell. This step was essential for deduction any background. Both the sample and the reagent flowed through the micro-system at a rate of 80 mL min-1, with a ratio of 1:1. The solenoid valve after the mixer were closed when the first mixer was full of the mixture of sample and reagent, and then jerked the syringe for several times to enhance the reaction efficiency. The flow was stopped for 150 s so as to allow color formation in consideration of its low reaction rate for forming yellow silicomolybdate. Blank (artificial seawater) were determined prior to sample in every determination cycle to inspect the baseline, and meanwhile to wash the microchannel to prevent the residue from last determination. The measurement period of the whole determination was about 300 s (including the blank before sampling). Besides, a high (200 μmol L^{-1}) and low (5 μmol L^{-1}) concentration standards were carried as inner standard when this sensor was deployed for field test. For ensuring the accuracy and avoid baseline drift, these two standards were periodically determined once every 15 samples.

3. Results and discussion

3.1. Calibration, precision, and limit of detection
The average concentration of the soluble silicate in seawater is about 36 μmol L^{-1}, with a dramatic variation [1]; in flourish season of diatoms phytoplankton, silicate concentration significantly is reduced to 0.13 μmol L^{-1} [18] due to the uptake of the large amounts of phytoplankton; but in the deep ocean, the concentration of silicate is about 100-200 μmol L^{-1}. Such a large concentration range greatly increases the difficulty to perform analysis of silicate in situ.

In order to pay attention to high and low concentration of the silicate determination, two different optical lengths (25 mm and 5 mm) of absorbance cells were designed on the microfluidic chip. As shown in figures 3A and 3B, the calibration curves for each of the two absorbance cells. For 25 mm absorbance cell, an obvious absorbance change against concentration gradient change is obtained from 0 to 90 μmol L^{-1}, with the best linear measurement range was 0-40 μmol L^{-1} having a regression equation $A=0.203C_{Si}+0.0112$ (n=7, R2=0.9984); for 5 mm absorbance cell, the concentration gradient change is obtained from 0 to 800 μmol L^{-1}, with the best linear measurement range 0-400 μmol L^{-1}

having a regression equation $A=0.0023C_{Si}+0.0171$ (n=7, R2=0.9985). For the quality of the data, the sensor procedure set for measurements below 30 μmol L-1, data from 25 mm absorbance cell is adopted; while above 30 μmol L^{-1}, 5 mm absorbance cell is adopted. The total linear dynamic range is 0-400 μmol L^{-1}.

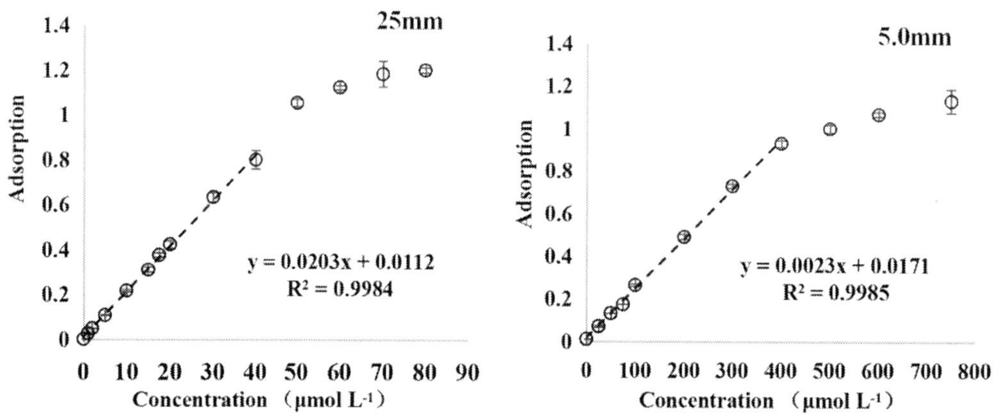

Figure 3. Calibration of sensor.
A for Calibration of 25mm optical path length, B for Calibration of 5mm optical path length.

In this study, continuous determination of 7.7 μmol L^{-1} and 132.1 μmol L^{-1} samples was made for 11 times, with concentration calibrated using the method of blue silicomolybdate from "Specification for Marine Monitoring" (17378.4-2007). Accordingly, as calculated, the relative standard deviation (RSD%, n=11) is 2.1% and 1.4% respectively, indicating that the repeatability and precision are good enough for deployment in seawater; besides, the limit of detection of this sensor, defined as three times the standard deviation of blank sample (artificial seawater without silicate [19]) divided by the number of calibration slope, is 45.1 nmol L^{-1} and 1.6 μmol L^{-1} responding to the optical length 25 mm and 5 mm respectively; the limit of quantitation, defined as fifteen times the standard deviation of blank sample (artificial seawater without silicate) divided by the number of calibration slope, is 225.5 nmol L^{-1} and 8.0 μmol L^{-1}.

3.2. Spike recovery

Recovery experiments were conducted using samples form Qingdao Zhongyuan Wharf (120°18′ E, 36°03′ N) and Zhoushan Banshengdong Wharf (122°19′ E, 29°56′ N). After 0.45 μm filtration, 50% and 100% background concentration of samples were spiked using silicate standard into the samples. As shown in table 1, 98.1%-104.0% recoveries have been obtained by using high and low background concentration samples, indicating the considerate validity and accuracy of this method.

Table 1. Determination of seawater sample and recovery tests (n=3).

Samples	Salinity	Spikes Conc. (μmol L^{-1})	Recovery Conc. (μmol L^{-1})	Recovery%
Qingdao Zhongyuan Wharf	30.1	0	1.7±0.2	-
		1.8	3.6±0.1	103.1%
		3.6	5.2±0.2	98.1%
Zhoushan Banshengdong Wharf	25.4	0	62.3±0.2	-
		32.0	95.3±0.5	101.1%
		64.0	131.3±0.4	104.0%

IOP Conf. Series: Earth and Environmental Science 82 (2017) 012080 doi:10.1088/1755-1315/82/1/012080

3.3. Application for routine monitoring

The microfluidic sensor was also deployed from November 29th, 2015 on an oceanographic cruise (about 50 sites) in the southern of Jiaozhou bay Qingdao, and the specific sites and route of this survey showed in figure 4A. The sensor was fixed on a frame of CTD profiler with water bottles, and measured each sample three times to obtain the average and deviation, meanwhile water samples were collected and frozen. Afterwards the samples were taken away to the lab for analyzing as comparisons. Figure 5 shows the comparisons of the data form sensor and lab analysis. As shown by the results that the trend of the data from sensor are consistent with data from lab analysis. Furthermore, significance test (t test) was also conducted to verify the data consistency. In t test, $\alpha=0.05$, $n=3$, the critical value of $t(a, f)$ was obtained as 2.776, indicating that data from sensor is not significantly different from lab analysis, due to all of the t values obtained from this experiment are below the critical value.

In Sea trial, the concentration of silicon conversed from the concentration silicate in seawater were 32.2-69.1 μg L^{-1} in the southern of Jiaozhou bay in late November, with 47.1 μg L^{-1} as the average concentration; the maximum value and minimum value occurred in *ZY06* site in Qianwan port and HS06 site in Zhucha island respectively. As shown in figure 4B, the silicate contour map indicates that silicate concentration in nearshore is higher than in open sea.

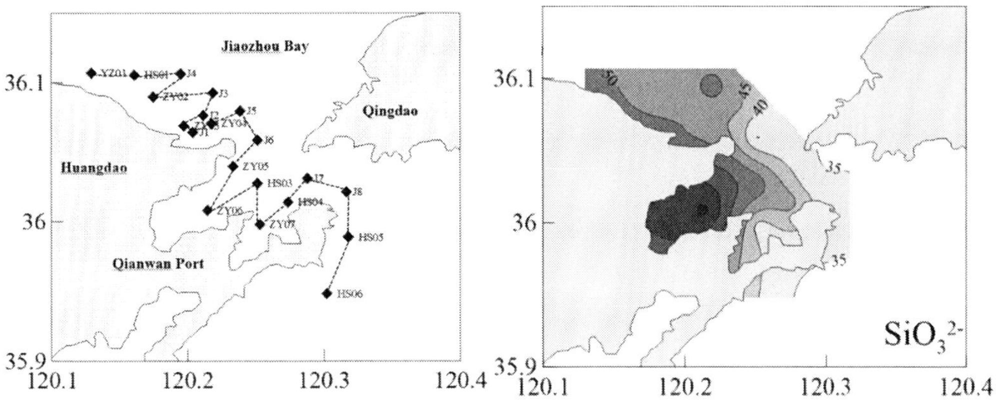

Figure 4. Sea trial route and silicate distribution around sea trial area (μg L^{-1}).

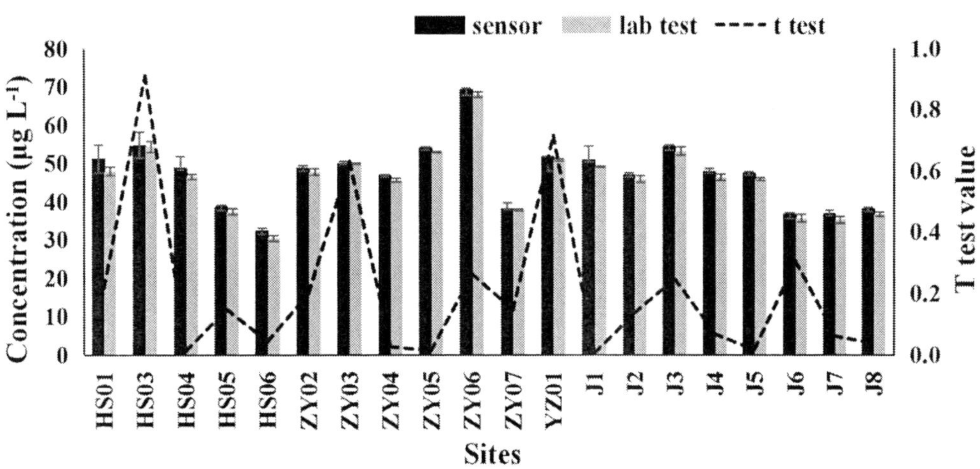

Figure 5. Silicon data from sea trial and comparison with national standard method (μg L^{-1} Si, n=3).

4. Conclusion

Based on the reaction of silicate with ammonium molybdate to form a yellow silicomolybdate complex and further reduction to silicomolybdenum blue by ascorbic acid, the new microfluidic continuous flow analysis sensor is designed for determination of soluble silicate in seawater. The measurement period of the whole determination may be about 300s, much less than the time cost (3 h) using lab method described in "Specification for Marine Monitoring" (17378.4-2007). With double light path in situ analysis adopted, this microfluidic sensor has an expanded linear range of 0-400 μmol L^{-1} with 45.1 nmol L^{-1} as the minimum limit of detection. Offshore experiment with this sensor shows that (1) data from sensor are consistent with lab analysis; (2) this microfluidic may be deployed in situ for continuous analysis, and (3) it has the advantages of high precision and robustness.

Acknowledgments
The author acknowledges funding from National Natural Science Foundation of China (No: 41206076 and 41506110), National key research and development program of China (2016YFC1400803 and 2016YFC1600801), Natural Science Foundation of Shandong Province, China (ZR2014DM012 and ZR2014YL006), and Key research and development program of Shandong Province, China (2015GGX103031).

References
[1] Guo J *Chem. Oceanogr.* 1997 (Xiamen, China: Xiamen University Press)
[2] Yang D, Yu Z, Zhang K, Li M and Jiang D 2008 *Marin. Environ. Sci.* **27** 547-53
[3] Wang S, Lin K, Chen N, Yuan D and Ma J 2016 *Talanta* **146** 744-8
[4] Kortazar L, Alberdi S, Tynan E and Angel Fernandez L 2016 *Microchem. J.* **124** 416-21
[5] Kozak J, Latocha K, Kochana J, Wieczorek M and Koscielniak P 2015 *Talanta* **133** 150-4
[6] Lapa R, Lima J and Pinto I 2000 *Anal. Sci.* **16** 1157-60
[7] Ribeiro Mesquita R B, Oliveira Barbosa Ferreira M T S, Alves Segundo R L, Carvalho Pinheiro Teixeira C F, Bordalo A A and Santos Silva Rangel A O 2009 *Anal. Methods-UK* **1** 195-202
[8] Somnam S, Motomizu S, Grudpan K and Jakmunee J 2014 *Chiang Mai J. Sci.* **41** 606-17
[9] Lin K, Ma J, Pai S, Huang Y, Feng S and Yuan D 2017 *Anal. Lett.* **50** 510-29
[10] Mas-Torres F, Munoz A, Estela J M and Cerda V 2000 *Int. J. Environ. An. Ch.* **77** 185-202
[11] Somnam S, Motomizu S, Grudpan K and Jakmunee J 2014 *Chiang Mai J. Sci.* **41** 606-17
[12] Legiret F, Sieben V J, Woodward E M S, Bey S K A K, Mowlem M C, Connelly D P and Achterberg E P 2013 *Talanta* **116** 382-7
[13] Pessoa-Neto O D, Silva T A, Dos Santos V B and Fatibello-Filho O 2015 *Aust. J. Chem.* **68** 1108-14
[14] Beaton A D, Cardwell C L, Thomas R S, Sieben V J, Legiret F, Waugh E M, Statham P J, Mowlem M C and Morgan H 2012 *Environ. Sci. Technol.* **46** 9548-56
[15] Milani A, Statham P J, Mowlem M C and Connelly D P 2015 *Talanta* **136** 15-22
[16] Ogilvie I R G, Sieben V J, Floquet C F A, Zmijan R, Mowlem M C and Morgan H 2010 *J. Micromech. Microeng.* **20** 0650166
[17] Floquet C F A, Sieben V J, Milani A, Joly E P, Ogilvie I R G, Morgan H and Mowlem M C 2011 *Talanta* **84** 235-9
[18] Mizerkowski B D, Hesse K, Ladwig N, Machado E D C, Rosa R, Araujo T and Koch D 2012 *Ocean Dynam.* **62** 1409-24
[19] Ma J, Yuan D, Zhang M and Liang Y 2009 *Talanta* **78** 315-20

A new type of polymeric heavy metal complexing precipitant used as fishery disinfectant and antiparasitic drug

Y C Liu[1,2], A N Zhang[2], X B Wang[1,2], J Xu[2], X H Zeng[1,2,3] and H M Wang[1,2,3]

[1]Laboratory of Chinese Herbal Pharmacology, Renmin Hospital, Hubei University of Medicine, Shiyan 442000, China
[2]Hubei Key Laboratory of Wudang Local Chinese Medicine Research, School of pharmacy, Hubei University of Medicine, Shiyan, 442000, China

E-mail: zengken@126.com; meirwang@126.com

Abstract. This paper presents a technique to produce a new kind of fishery drug that is water emulsion suspending agent containing polymeric calcium-iron-dithiocarbamate with heavy metal complexing precipitate ability, good disinfection and auxiliary insecticidal efficacy. The product has good dispersion, high efficiency and low toxicity, as well as no pollution and no harmful residues. It not only can be used in the pond waters and ornamental waters, but also can meet the high requirements of the aquaculture waters. There is non-pollutant emission in the production, which is a green environment-friendly technique without three waste discharges. This technology belongs to the ecological and environmental protection.

1. Introduction

With the development of society and economy, heavy metal pollution of the environment is more and more serious. The world's most famous minamata disease and itai-itai disease are caused by heavy metal pollution. In the past twenty years, owing to the heavy metal in excess of the standards of aquatic products caused by the existing low standards of aquaculture water environment, Chinese export of aquatic products has been repeatedly frustrated.

Table 1. The effects of polymeric N- methyl calcium-iron-dithiocarbamate.

No.	Water area	Symptom	Concentration of this drug	Effect
1	1m^3	Pd^{2+} 0.1 mg/L, Cu^{2+} 0.1 mg/L	0.15 mg/L	Pd^{2+}≤0.05 mg/L, Cu^{2+}≤0.01 mg/L
2	12 acres of land	dark green and odor, Cu^{2+} 0.1 mg/L	0.12 mg/L	Color returned to normal and odor elimination, Cu^{2+}≤0.01 mg/L
3	8 acres of land	dark green and odor	0.09 mg/L	Color returned to normal and odor elimination
4	10 acres of land	dark green and odor	0.08 mg/L	Color returned to normal and odor elimination

The technique is to produce a water emulsion suspending agent containing polymeric N- methyl calcium-iron-dithiocarbamate that is stereo mesh structure. From the original agents to production of a uniform dispersion of water emulsion suspending agent, there is no pollutant emission with the method of one-pot reaction, to reach 100% atomic utilization ratio. This product has good dispersion, good disinfection and auxiliary insecticidal efficacy, with high heavy metal complexing precipitation capacity and low toxicity, as well as no pollution and no harmful residues. The effects of polymeric N-methyl calcium-iron-dithiocarbamate are revealed (table 1). It not only can be used in the pond waters and ornamental waters, but also can meet the high requirements of the aquaculture waters.

Meanwhile, heavy metal as a kind of widespread pollution is a common chemical poison to fish acute poisoning death. Unfilled D orbit in the outermost electron shell mostly of heavy metal ion as a good electron acceptor has a strong affinity with many active groups of enzymes. Heavy metal ion can combine with the thiol group of the enzyme to form an insoluble thiolate, inhibit the activity of the enzyme and interfere with the metabolism after being absorbed into fish.

The Ministry of Agriculture in August 2003 issued No. 31 Decree of Regulation on Quality and Safety Management of Aquaculture, "aquaculture quality and safety management requirements". It regulates that the pollution-free aquaculture has become the minimum standards in the provisions, and aquatic drugs are under strict supervision. At the same time, a promoting plan was issued. The main content is that China's aquaculture industry has to implement standardized management and improve the quality testing system of aquatic products in a short period of 3-4 years, through the development of aquaculture standards. And in the main production areas monitoring system should be strengthened, with quality certification system of aquatic products, to comprehensively improve the international competitiveness of China's aquatic products. Therefore, research and development of pollution-free fishery drug has become the focus of the current aquaculture industry [1, 2].

In order to deal with the heavy metal pollution especially copper pollution caused by excessive use of copper sulfate and improve aquaculture water environment to meet the high requirements of Chinese aquaculture industry in aquaculture waters. Copper is an essential trace element for organism, and it is an important component of biological serum protein and cytochrome oxidase. But too much copper can have a toxic effect on organisms and cause extensive damage to the branchial gland of fish, such as increasing mucus, hypertrophy, hyperplasia, being disintegrated and covering each other, finally cause death of fish by suffocation. The Symptoms are as follows. The mouth vomits, the gill is pale green and the fish body is pale. Its autopsy revealed that a large number of cell debris and mucus in the lumen, and the columnar epithelium necrosis and ulceration were found in the digestive tract of the his body. The copper ion is above 1.0 mg/L, which will make water turbidity and odor also.

Dithiocarbamate is widely used in industry and also known as insecticidal and bactericidal drugs in agriculture [3-9]. For example, pesticides containing N, N-dimethyl dithiocarbamate, or ethylene-dithiocarbamate are also available. However, so far, these pesticides are mainly used for crops, to prevent all kinds of pests and diseases.

The existing dithiocarbamate can not meet this requirement. There have been no reports on the use of dithiocarbamate as water disinfectants and antiparasitic fishery drugs, except two papers reported by our team [1, 2]. Herein we wish to report a new type of polymer heavy metal complexing precipitant that can be used as fishery disinfectant and antiparasitic medicine. This technology belongs to the ecological and environmental protection.

2. Materials and methods

The aqueous emulsion suspending agent containing polymeric N-methyl calcium-iron-dithiocarbamate is a kind of high heavy metal complexing precipitant with good disinfection and auxiliary insecticidal efficacy. It contains 10-34% N-methyl calcium-iron-dithiocarbamate, 8-16% carrier dispersant, 5-12% multifunctional emulsifier, 0.2-5% colloidal dispersant, 1-3% antioxidant and 0.2-3% antiseptic agent and 2-3% antifreeze.

First, mix N-methylamine (870 g, 30 mol) and $Ca(OH)_2$ (2442 g, 33 mol) solution, drop CS_2 (1320 g, 30 mol) slowly, after stirring 2-3 hours at room temperature, N-methyl calcium-dithiocarbamate is generated. Then add green vitriol (4170 g, 15 mol) to the calcium salt, after metathesis reaction and polymerization catalyzed by hydroxylamine hydrochloride (208.5 g, 3 mol) an aqueous medium product of polymeric N-methyl calcium-iron-dithiocarbamate is produced. Third， the raw aqueous product polymeric N-methyl calcium-iron-dithiocarbamate is ground to particles of less than 300 microns with a ball mill, add ultra-fine kaolin (600 g), Tween-80 (600 g), xanthan gum (250 g), propyl gallate (500 g), sodium benzoate (150 g), glycerol (250 g), and water to 20000 g, along with stirring intensively. Last, with a high-speed shearing emulsifying machine, shear and emulsify it to obtain 10% of uniform dispersed water emulsion suspending agent of polymeric N-methyl calcium-iron-dithiocarbamate, with the yield of about 91%.

From the original agents to production of a uniform dispersion of water emulsion suspending agent, there is no pollutant emission with the method of one-pot reaction, to reach 100% atomic utilization ratio (figure 1). This is a green environment-friendly production technique without three waste discharges. This technology belongs to the ecological and environmental protection.

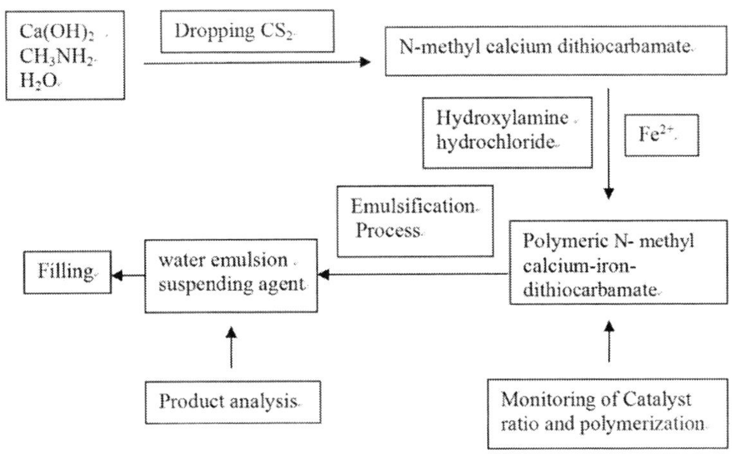

Figure 1. The production technique of water emulsion suspending agent containing polymeric N-methyl calcium-iron-dithiocarbamate.

3. Results and discussion

Not only to consider the effect of aquaculture water modifying agent, but also to avoid poisoning the human body and cultured animals, which is different from the using of industrial wastewater treatment chemicals. Therefore, water modifying agent must be efficient, no residue and good dispersion. Two low molecular dithiocarbamates (sodium and potassium salts) have strong coordination ability to heavy metal ions, can be used for complexing metal ions in wastewater treatment, but its instability and toxicity limits its application range.

We have reported that water emulsion suspending agent containing iron dithiocarbamates or copper dithiocarbamates can be used as fishery disinfectant and antiparasitic drug, but the metal complexing capability is weak.

In order to develop a kind of fishery drug that is polymeric heavy metal complexing precipitant. The structure modification to the lead compounds is changing the N-group and combining with macroelement and trace elements of the human body, so as to improve the precipitate ability but not to reduce complexing ability.

The technique is to produce a water emulsion suspending agent containing polymeric N- methyl calcium-iron-dithiocarbamate that is stereo mesh structure. From the original agents to production of a uniform dispersion of water emulsion suspending agent, there is no pollutant emission with the method of one-pot reaction, to reach 100% atomic utilization ratio. This product has good dispersion, good disinfection and auxiliary insecticidal efficacy, with high heavy metal complexing precipitation capacity and low toxicity, as well as no pollution and no harmful residues. The effects of polymeric N-methyl calcium-iron-dithiocarbamate are revealed (table 1). It not only can be used in the pond waters and ornamental waters, but also can meet the high requirements of the aquaculture waters.

Acknowledgments
We gratefully acknowledge financial support of this work by the Open Project of Hubei Key Laboratory of Wudang Local Chinese Medicine Research (Hubei University of Medicine (Grant No. WDCM009), the Hubei Province Health and Family Planning Scientific Research Project (Grant No. WJ2017M215, No. WJ2015Z113), Shiyan Municipal Science and Technology Bureau Science and Technology Project (Grant No. 15Y12), and the scientific research innovation team of Hubei University of Medicine (Grant Nos. 2014QDJZR10, 2014CXZ01). Y C Liu and A N Zhang are first coauthors and contributed equally to this work, X H Zeng and H M Wang are Corresponding authors.

References

[1] Zeng X, Gao H, Wang H and Wang X 2016 Iron -dithiocarbamate used as fishery drug *Lect. Notes Earth Sci.* **4** 70-1
[2] Wang H, Li H, Zeng X and Wang X 2016 A new type of copper salt used as disinfectant and fishery antiparasitic medicine *Lect. Notes Earth Sci.* **4** 72-3
[3] Goodhue L D and Tissol C E 1952 Thiosulfenyl dithiocarbamates as grain insect repellents *US US 2621143 A*
[4] Eng G, Song X, Duong Q, Strickman D, Glass J and May L 2003 Synthesis, structure characterization and insecticidal activity of some triorganotin dithiocarbamates *Appl. Organomet. Chem.* **17(4)** 218-25
[5] Tiekink E R T 2008 Tin dithiocarbamates: Applications and structures *Appl. Organomet. Chem.* **22(22)** 533-50
[6] Nobel C I, Kimland M, Lind B, Orrenius S and Slater A F 1995 Dithiocarbamates induce apoptosis in thymocytes by raising the intracellular level of redox-active copper *J. Biol. Chem.* **270(44)** 26202-8
[7] Clarke D, Baum H, Stanley E and Hester W 2002 Determination of dithiocarbamates *Anal. Chem.* **23(12)** 1842-6
[8] Szolar O H 2007 Environmental and pharmaceutical analysis of dithiocarbamates *Anal. Chimica Acta* **582(2)** 191-200
[9] Wang X, Gaowei L I, Meng T, Liu L and Zhao W 2015 Research progress in the bioactivities of the dithiocarbamates *Chinese J. Appl. Chem.* **32(2)** 123-33

IOP Conf. Series: Earth and Environmental Science **82** (2017) 012082 doi:10.1088/1755-1315/82/1/012082

Acoustic assessment of fish resources in subtropical run-off-river hydropower plant in the Pearl River

Z Wu[1], S L Zhu[1], J Li[1], J P Yang[1], Y F Li[1] and X H Li[1,2]

[1]Pearl River Fisheries Research Institute, Chinese Academy of Fishery Science, No. 1 Xingyu Road, Liwan District, Guangzhou, Guangdong Province, China

E-mail: lxhui01@aliyun.com

Abstract. As the utmost downstream dam in the main stream of Xijiang River, Changzhou Hydro-Junction has changed the original hydrological characteristics and ecological environment of Xijiang River. In order to grasp the fish distribution and standing crops after reservoir impoundment, a split beam echo-sounder EY60 (120 kHz) was used in the river channel from Guiping to Wuzhou in April 2014, 2015, and 2016. The average fish density in the survey area was 0.1215±0.4198 ind.m^{-3} in 2014, 0.0190±0.1232 ind.m^{-3} in 2015 and 0.0107±0.0186 ind.m^{-3} in 2016. According to the horizontal distribution data, fish density showed significant difference in different area and the density in the upper area was significantly higher than that in other areas. Moreover, fish density was negatively correlation with water depth. As a whole, the Target Strength (TS) decreased in the past three year, TS was significantly different among the three areas and showed a decreasing tendency from the dam area to the upper area. Fish resources decreased gradually, as indicated by the decreased fish density and miniaturized fish sizes in the past three years.

1. Introduction

Changzhou Hydro-Junction is located in Xijiang River, the main branch of the Pearl River. It was constructed in 2003 and water impoundment started in 2007. It stretches across the Xijiang River is 3.47 km long and 34.6 m high. As a run-off river hydropower plant, Changzhou Hydro-Junction affects the hydrological conditions and water ecological system, leads to the river fragmentation [1-4], blocks the migrant fish, and threatens fish populations who producing drifting eggs [5, 6]. A large-scale fishery resource investigation was conducted in the 1980s in Xijiang River and 136 fish species were collected [7]. In recent years, the increased river utilization projects, such as water conservancy and hydropower projects, have destructed fish habitats, decreased the diversity and threatened fish species [7, 8].

The Changzhou Dam led to the decrease in fishery resources, reduced complementary groups and postponed annual spawning time [9-11]. In order to reduce its adverse impact on fish, a fish passage was constructed and showed a good fish pass efficiency [12-14] and fish could swim towards the upstream of the dam when the spillways were open [15]. Previous studies on the Xijiang River mainly discussed the downstream area of the Changzhou Dam [15-17]. The dam has been operated for about 10 years and the fish distribution or standing crop in the reservoir was unknown. Fish is the climax community in the water ecosystem and may respond to any change of the water ecosystem, so it is necessary to explore fish resources and their changing trends.

IOP Conf. Series: Earth and Environmental Science **82** (2017) 012082 doi:10.1088/1755-1315/82/1/012082

2. Materials and methods

2.1. Survey area

Hydroacoustic surveys were carried out between Guiping (N 23.4455°, E 111.2422°) and Wuzhou (N 23.4298°, E 110.5321°) in late March 2015 and 2016 and were performed in the dam area in 2014. The survey area is 156 km long with the mean depth of 12.29 m and the maximal depth of 65 m. Five sampling sites were set along the survey area (figure 1). Two spillways (Neijiang and Zhongjiang) are across two branches of the river and the other branch Waijiang is regarded as a ship lock.

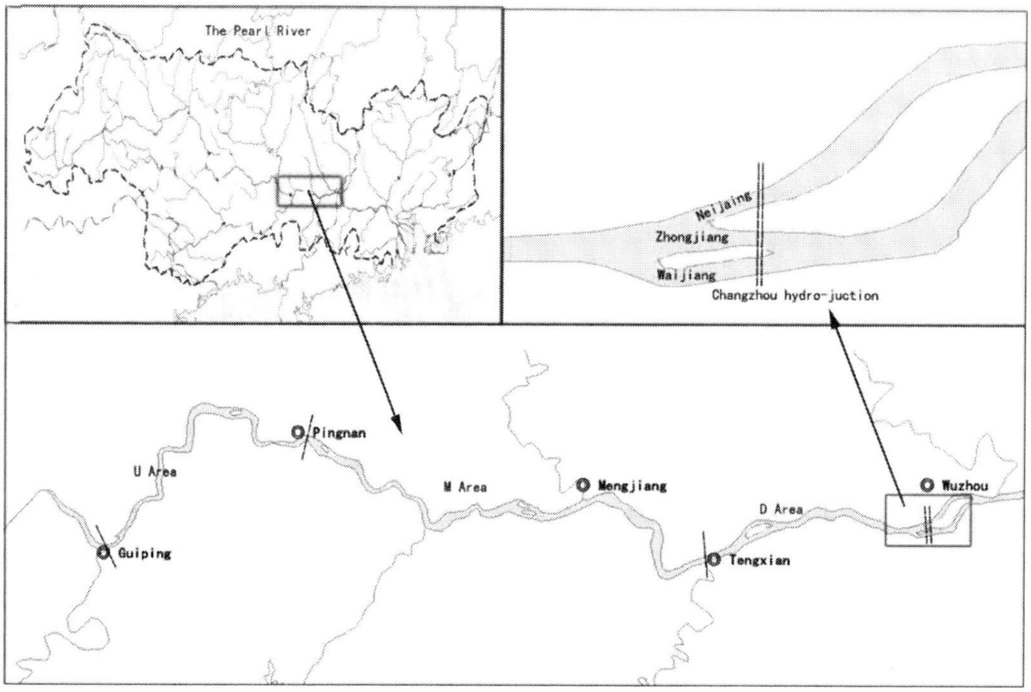

Figure 1. Locations of the reservoirs studied. Five cycles represent catch sampling site, and the survey area divided into three areas by the dashed line.

2.2. Acoustic sampling

The survey area was canyon-shaped and acoustic sampling was performed according to a zigzag trajectory in the area with the depth exceeding 3 m and the allowable width. Considering the influences of sailing, submerged reef and cages in watercourse, cruise line was adjusted real time during the survey. All the surveys were divided into transects with different lengths, so that for every survey, 450–500 transects with independent estimates of biomass, density and size distribution were available [18].

The transducer was attached to a pole and placed vertically in the water depth of 0.5 m. The boat velocity along water flow direction was between 8 and 10 km.h^{-1}.

A split-beam SIMRAD EY60 echo-sounder was used. Its main characteristics are provided as follows: a frequency of 120 kHz, a sampling interval of 16 ms and 40logR TVG (time varied gain) amplification. The acoustic system was routinely calibrated with a tungsten carbide sphere (-40.4 dB; Diameter: 23 mm) before each measurement series [19].

2.3. Fish sampling

Fish assemblages were collected randomly in early April 2015 and 2016 from the upper area to the dam area. At each site, we sampled fish in three methods: gill nets, fly-fishing, and crayfish traps, respectively. Fish samples were measured according to the standard length (SL) to the nearest 0.1 cm and total weight (TW) to the nearest 0.1 g. All scientific names, authors and years were checked against FISHBASE (www.fishbase.org).

2.4. Data processing
Acoustic data were processed in Echoview Version4.9 [20]. When the fish density was low, the single target echoes did not overlap. Therefore, echo counting method was adopted [21]. A target threshold of −60 dB was set and all other parameters were set according to previous results [22]. Fish length was calculated from TS according to the empirical formula [23]:

$$TS = 19.1\lg(F) - 62.0 \tag{1}$$

We used the ordinary Kriging interpolation method to simulate the spatial distribution of fish density. In comparison with other interpolation techniques, ordinary Kriging allows an optimal unbiased simulation and the simulation procedure can be obtained under the normal distribution conditions [24, 25]. Fish density was subjected to natural logarithmic transformation.

Four diversity indices were calculated based on the composition data obtained from typical trawl survey: Species richness *S* (the number of species), Pielou's Evenness index J' [26], which measures how evenly abundance is distributed across species, the Shannon–Wiener diversity index H' [27], and the Simpson index D. The indices are calculated as:

$$H' = -\sum_{i=1}^{S} P_i \ln P_i \tag{2}$$

$$J' = \frac{H'}{\ln(S)} \tag{3}$$

$$D = 1 - \sum_{i=1}^{S} P_i^2 \tag{4}$$

where *S* is the total number of species; *n* is the individual number; P_i is the proportion of the abundance of species *i* in the total abundance of all fish species.

3. Results

3.1. Feature of fish horizontal distribution

Table 1. Fish density distribution in different times and areas.

Year	Survey area	Average density (ind.m^{-3})	Standard error	Maximum	Minimum
2014	Dam Area	0.1215	0.5945	3.9940	0.000227
2015	Dam Area	0.0372	0.1952	2.3718	0.000025
	Middle Area	0.0069	0.0273	0.2718	0.000040
	Upper Area	0.0089	0.0277	0.3502	0.000207
2016	Dam Area	0.0037	0.0125	0.0838	0.000048
	Middle Area	0.0036	0.0048	0.0317	0.000048
	Upper Area	0.0155	0.0215	0.1592	0.000183

The average fish density was 0.1215±0.4198 ind.m^{-3} in 2014, 0.0190±0.1232 ind.m^{-3} in 2015, and 0.0107±0.0186 ind.m^{-3} in 2016, respectively. The lowest density was recorded in the dam area and the highest density was recorded in the area close to the branch river in the upper area (table 1). Test of Homogeneity of Variance showed fish density did not conform to a normal distribution (P <0.05), so nonparametric test was used to analyze the results. Fish density in 2014 was significantly larger than that in 2015 and 2016 in dam area (P<0.01 for both) and fish density showed no significant difference in 2015 and 2016. In the survey area, fish showed the scattered distribution (figure 2(a)) except that some fish aggregation areas were detected in Zhongjiang and Neijiang (figure 2(b)). Fish density in Waijiang close to the dam area was very low.

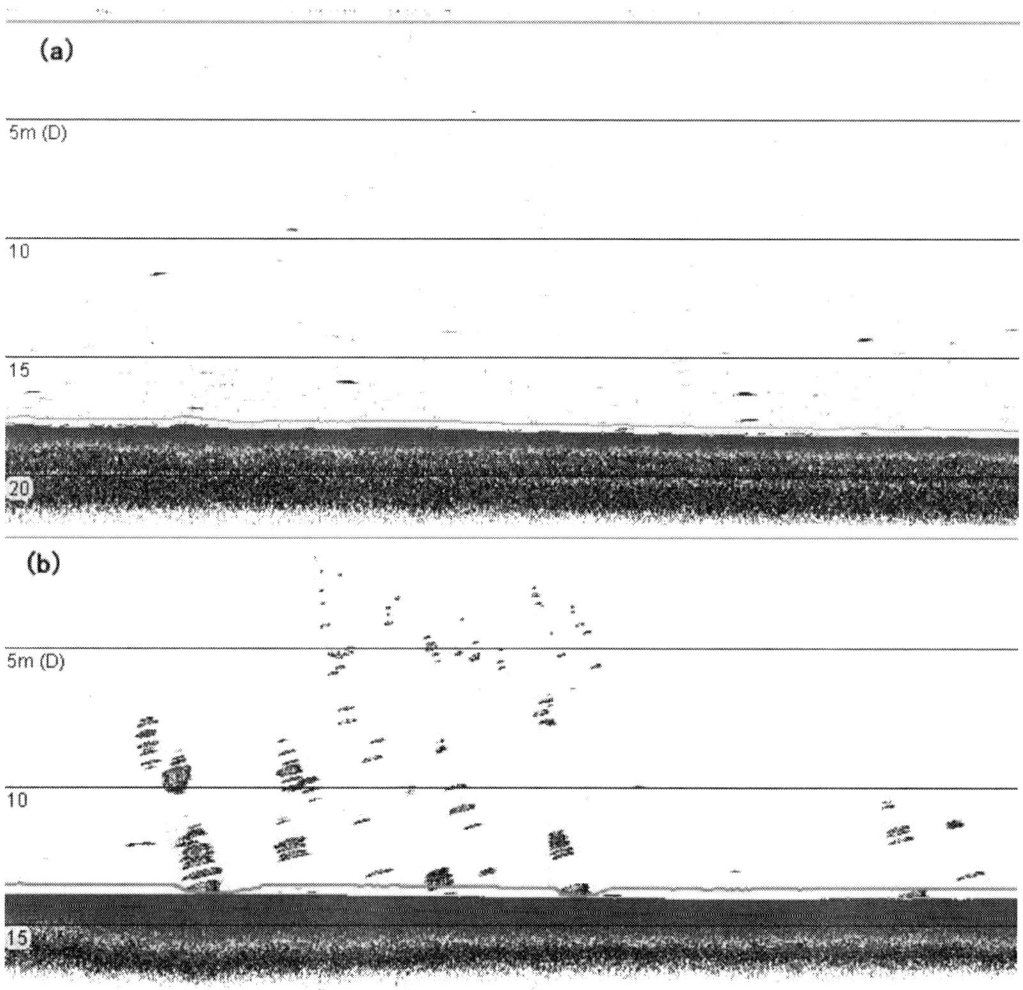

Figure 2. The features of fish echo distribution, (a) fish dispersed, (b) fish assembled.

The spatio-temporal distribution pattern of fish was largely similar in 2015 and 2016, displaying patchy spatial distribution (figure 3). In 2015 and 2016, fish density in the upper area was significantly higher than that in the dam area and middle area and the other patches with the higher fish density were distributed in the river sections which were close to towns.

3.2. *Fish vertical distribution and water depths*

Water depth ranged from 3.1 m to 59.7 m and average water depth was 12.29 m. Target depths differed significantly among different areas in the two years (2016: F=131.99, P<0.01; 2015: F=277.25, P<0.01). Fish were mainly distributed in various water depth ranges: 5-15 m and 25-35 m in the dam area, 8-28 m in the middle area, and 10-20 m in the upper area. It revealed that the vertical distribution of fish varied with water depth. Spearman correlation analysis results revealed that fish density was negatively correlated with water depth and target depth (P<0.05, r>-0.2 for both) and that target depth was significantly positively correlated with water depth (*P*<0.01, *r*>0.75). Figure 4 shows a negative linear relationship between fish density and two depth parameters (average water depth and target depth).

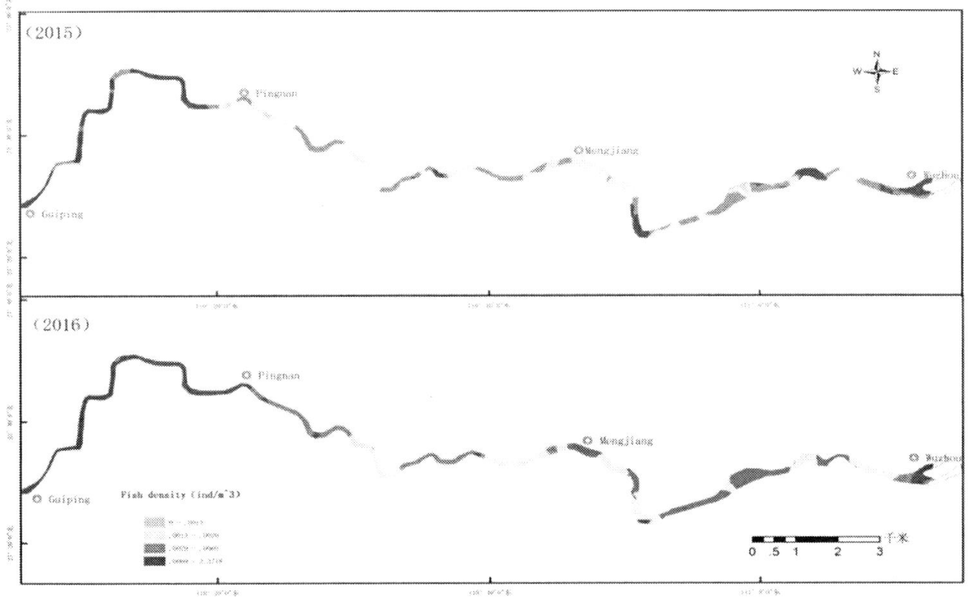

Figure 3. Fish density horizonal distribution in Xijiang river, two pictures use same legend.

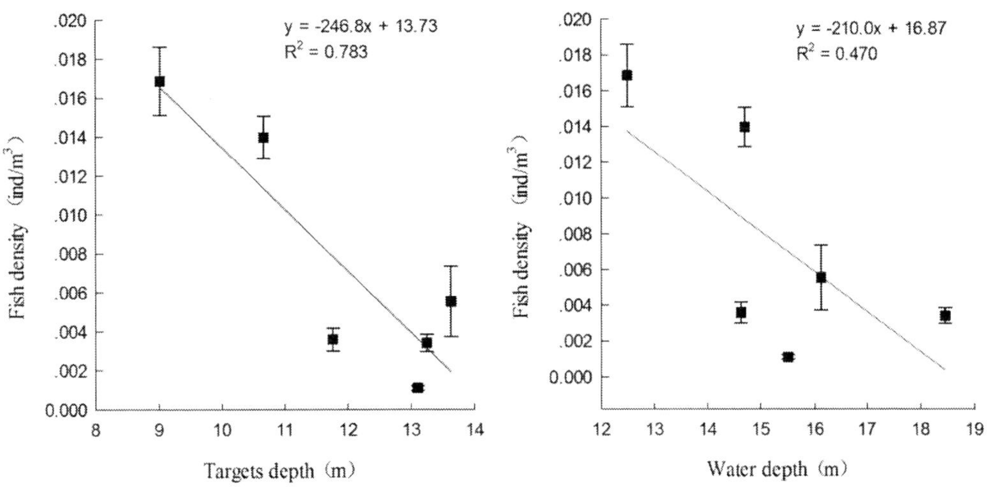

Figure 4. Correlation diagram between average fish density and target depths and water depths in 2016. Fish density represent with average and standard error, the red line is trendline.

3.3. Fish target strength

Single Target Detection and Fish Tracks were used to obtain the fish TS in Echoview. The average TS had an annually decreasing tendency and an increased distribution tendency from the dam area to the upper area (figure 5). One-way ANOVA analysis revealed that TS in various years in dam area was significantly different (*F*=200.27, *P*<0.01). The average TS was -51.45±3.46 dB with the aggregation of 85.32% in -57~-53 dB (transformed into fish length of 2.3 cm~3.7 cm) in 2014; Average TS was respectively -52.06±4.77 dB in 2015 and -55.18±4.83 dB in 2016 and showed significant differences among the three areas (2015: F=79.24, P<0.01; 2016: F=536.46, P<0.01).

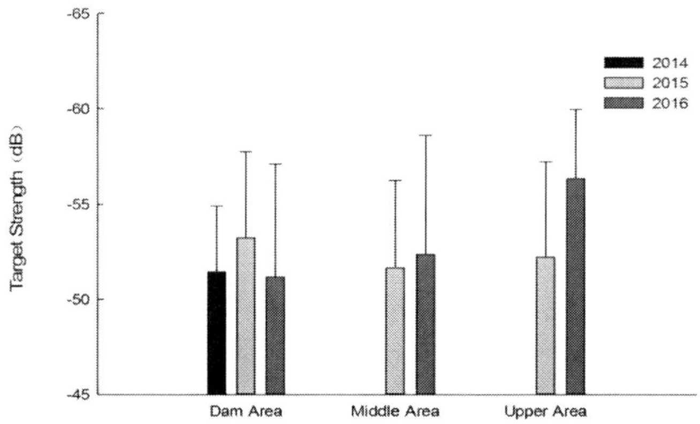

Figure 5. Target strength distribution.

3.4. Fish sampling information

A total of 1564 fish was caught, and 63 fish species were the identified. The 63 fish species belonged to 3 orders, 12 families, and 40 genera. The species mainly belonged to Cypriniformes (41), Siluriformes (15), and Perciformes (7). The top ten species were *Leiocassis crassilabris*, *Pelteobaggrus vachelli*, *Pseudolaubuca sinensis*, *Toxabramis houdemeri*, *Squalidus argentatus*, *Ptychidio jordani*, *Cyprinus carpio*, *Carassius auratus*, *Squaliobarbus curriculus*, and *Hemiculter leucisculus*. Species richness S, Pielou's evenness index J', Shannon–Wiener index H', and Simpson index D were 63, 0.797, 1.433, and 0.948, respectively.

4. Discussion

The study revealed a clear fish distribution pattern. Estimated fish density was the highest in the upper area (tributary area) and gradually decreased (not monotonically) towards the dam area. The same distribution pattern had been reported by other researchers [18, 28], so this area was considered an important part of the reservoir. Dongta Spawning Ground was the largest natural propagation site in the Pearl River [10], and it had entered the fish breeding season in April [11]. The bigger fish were detected in the area near the dam, whereas the smaller fish were detected in the area close to the tributary (i.e. the upstream area). Small fish may benefit from the better trophic conditions. The tributary area may be the preferred option because some species in the reservoir may spawn in the upstream area.

Another fish distribution feature in the Changzhou reservoir is the high density and large size found in the dam area of Zhongjiang and a very low density in the dam area of Waijiang. The difference may be interpreted as follows. When the spillways were open during flooding discharge, many fish flushed into the reservoir from the downstream area [15] or passed through the Changzhou Fishway [12, 13], Fish are prone to avoid ships, thus leading to the underestimate results [29, 30]. In the ship lock of Waijiang, the avoidance reactions of fish lead to the low density. Moreover, some species can pass through the ship lock, but he ship lock cannot serve as the fish migration channel [31-33]. Compared to downstream area and Zhongjiang with the high fish density, Changzhou dam ship lock cannot be regarded as a channel for fish migration.

Hydropower projects changed the intrinsic ecotope and influenced aquatic organisms especially fish. Fish density declined year by year and fish length (TS) showed a decreasing trend. Some policies such as seasonal fishing bans from April 1st to June 1st and the release and enhancement programs of rare and economically valuable fishes contributed to the increase in fish biomass and purified water through the biomanipulation action. The compensation mechanism and the relationship between fish resources and environmental and hydrological factors remain largely unclear. Further studies should be carried out to further understand the relationship.

Biodiversity is changing at an unprecedented rate on a global scale as a complex response to several anthropogenic changes [27]. Dam construction alters the intrinsic features of rivers, thus causing significant changes in fish communities during and after impoundment, such as habitat fragmentation and blocking, flow regime modification, and hypolimnetic discharges. Compared to previous results [7, 8], the fish ecotype was significantly changed. The most species in the dam area were settling fishes, whereas only less migrating fishes reside in the tributary area, such as Chinese carps (*Mylopharyngodon piceus*, *Ctenopharyngodon idellus*, *Hypophthalmichthys molitrix*, and *Aristichthys nobilis*). Species richness S, Pielou's evenness index J', and Shannon–Wiener index H' were decreased compared to the previous survey results [34].

Acknowledgments

We thank the staffs of the fishery administration and fishing harbor superintendence agency of Wuzhou City and Guiping City and local fishermen for their help in field investigations. This research is funded by the Central Public-interest Scientific Institution Basal Research Fund, CAFS (No. 2017HY-ZC0705).

References

[1] Baumann P and Stevanella G 2012 Fish passage principles to be considered for medium and large dams: The case study of a fish passage concept for a hydroelectric power project on the Mekong mainstem in Laos *Ecol. Eng.* **48** 79-85
[2] Lytle D A and Poff N L R 2004 Adaptation to natural flow regimes *Trends Ecol Evol* **19** 94-100
[3] Cheng F, Wei C L, Murphy B R and Xie S 2015 Potential effects of dam cascade on fish: Lessons from the Yangtze River *Rev. Fish Biol. Fish.* **25** 569-85
[4] Djemali I, Laouar H and Toujani R 2010 Distribution patterns of fish biomass by acoustic survey in three Tunisian man-made lakes *J. Appl. Ichthyol.* **26** 390-6
[5] Holcomb J M, Nichols R B and Gangloff M M 2016 Effects of small dam condition and drainage on stream fish community structure *Ecol. Freshwater Fish* **25** 553-64
[6] Marchetti M P and Moyle P B 2000 Spatial and temporal ecology of native and introduced fish larvae in lower Putah Creek, California *Environ. Biol. Fishes* **58** 75-87
[7] Li J, Li X, Jia X, Li Y, He M, Tan X, Wang C and Jiang W 2010 Evolvement and diversity of fish community in Xijiang Rive *J. Fish. Sci. China* **17** 298-311
[8] Li J, Li X, Tan X, Li Y, He M, Huo J, Lin J and Su S 2009 Species diversity of fish community of Provincial Xijiang River Rare Fishes Natural Reserve in Zhaoqing City, Guangdong Province *J. Lake Sci.* **21** 556-62
[9] Li Y, Li X, Tan X, Yang J, Shuai F and Li J 2013 Occurrence and interannual variability of larval Xenocyprininae in the mid and lower reaches of the Pearl River *J. Fish. Sci. China* **4** 816-23
[10] Shuai F, Li X, Li Y, Yang J and Li J 2016 Analysis of the echohydrological demands of bighead carp breeding in the Dongta spawning grounds of the Pearl River, via cross-wavelet analysis *Acta Ecol. Sin.* **19** 1-8
[11] Tan X, Li X, Lek S, Li Y, Wang C, Li J and Luo J 2010 Annual dynamics of the abundance of fish larvae and its relationship with hydrological variation in the Pearl River *Environ. Biol. Fishes* **88** 217-25
[12] Zhi W, Tan X C, Tao J P,Yong T, Li X H and Shuli Z 2013 Hydroacoustic monitoring of fish migration in the Changzhou Dam fish passage *J. Appl. Ichthyol.* **29** 1445-6
[13] Tao J, Tan X, Yang Z, Wang X, Cai Y, Qiao Y and Chang J 2015 Fish migration through a fish passage associated with water velocities at the Changzhou fishway (Pearl River, China) *J. Appl. Ichthyol.* **31** 72-6
[14] Tao J, Wang X, Tan X, Cai P, Qiao Y and Chang J 2015 Diel pattern of fish presence at the

Changzhou fishway (Pearl River, China) during the flood season *J. Appl. Ichthyol.* **31** 451-8

[15] Tan X, Kang M, Tao J, Li X and Huang D 2011 Hydroacoustic survey of fish density, spatial distribution, and behavior upstream and downstream of the Changzhou Dam on the Pearl River, China *Fish. Sci.* **77** 891-901

[16] Tan X, Li X, Chang J and Tao J 2009 Acoustic observation of the spawning aggregation of megalobrama hoffmanni in the Pearl River *J. Freshwater Ecol.* **24** 293-9

[17] Zhou L, Zeng L, Fu D, Xu P, Zeng S, Tang Q and Li G 2016 Fish density increases from the upper to lower parts of the Pearl River Delta, China, and is influenced by tide, chlorophyll-a, water transparency, and water depth *Aquat. Ecol.* **50** 59-74

[18] Draštík V, Kubečka J, Tušer M, Čech M, Frouzová J, Jarolím O and Prchalová M 2008 The effect of hydropower on fish stocks: Comparison between cascade and non-cascade reservoirs *Hydrobiologia* **609** 25-36

[19] Foote K G, Knudsen H P, Vestnes G, Maclennan D N and Simmonds E J 1987 Calibration of acoustic instruments for fish density estimation *ICES Coop. Res. Rep.* **144** 1-69

[20] Higginbottom I, Woon S and Schneider P 2008 Hydroacoustic data processing for standard stock assessment using Echoview *Technical Manual* (Hobart: Myriax Software Pty Ltd publication) pp 40-59

[21] Simmonds J and MacLennan D N 2008 *Fisheries Acoustics: Theory and Practice* (New York: John Wiley & Sons)

[22] Rudstam L G, Parker S, Sandra L, Sullivan P J and Warner D M 2009 Towards a standard operating procedure for fishery acoustic surveys in the Laurentian Great Lakes, North America *ICES J. Mar. Sci.* **66** 1391-7

[23] Love R H 1971 Dorsal-aspect target strength of an individual fish *J. Acoust. Soc. Am.* **49** 816-23

[24] Adams C F, Harris B P and Stokesbury K D E 2008 Geostatistical comparison of two independent video surveys of sea scallop abundance in the Elephant Trunk Closed Area, USA *ICES J. Mar. Sci.* **65** 995-1003

[25] Agterberg F P 1992 An introduction to applied geostatistics *Earth-Sci. Rev.* **32** 211-3

[26] Pielou E C 1966 Species-diversity and pattern-diversity in the study of ecological succession *J. Theor. Biol.* **10** 370-83

[27] Yemane D, Mafwila S K and Kathena J 2015 Spatio-temporal trends in diversity of demersal fish species in the Benguela current large marine ecosystem region *Fish. Oceanogr.* **24** 102-21

[28] Bowersox B J, Scarnecchia D L and Miller S E 2014 Distribution, abundance, and vertical migrations of Leptodora kindtii in a mainstem Missouri River reservoir, Montana *J. Freshwater Ecol.* **29** 171-86

[29] Xie Y, Michielsens C G J, Gray A P, Martens F J and Boffey J L 2008 Observations of avoidance reactions of migrating salmon to a mobile survey vessel in a riverine environment *Can J. Fish Aquat. Sci.* **65** 2178-90

[30] Robertis D A and Handegard N O 2012 Fish avoidance of research vessels and the efficacy of noise-reduced vessels: A review *ICES J Mar Sci* **70** 34-45

[31] Tao J, Gao Y, Qiao Y, Zheng H, Wang X, Wan L and Chang J 2010 Hydroacoustic observation of fish spatial patterns and behavior in the ship lock and adjacent areas of Gezhouba Dam, Yangtze River *Acta Ecol. Sin.* **30** 233-9

[32] Klinge M 1994 Fish migration via the shipping lock at the Hagestein barrage: Results of an indicative study *Water Sci. Technol.* **29** 357-61

[33] Yang D, Kynard B, Wei Q, Chen X, Zheng W and Du H 2006 Distribution and movement of Chinese sturgeon, acipenser sinensis, on the spawning ground located below the Gezhouba Dam during spawning seasons *J. Appl. Ichthyol.* **22** 145-51

[34] Su W, Xue Y, Zhang C and Ren Y 2015 Spatio-seasonal patterns of fish diversity, Haizhou Bay, China *Chin. J. Oceanol. Limnol.* **33** 121-34

The start-up of denitrifying phosphorus removal system by using nitrite as electron acceptor

W Li[1,4], J Liu[1], H Z Sun[1], J X Fu[1], Y Gao[2] and J Sun[3]

[1]Municipal and Environmental Engineering Institute, Shenyang Jianzhu University, Shenyang 110168, China
[2]Shenyang Xuqiang Business Management Co., Ltd 110168, China
[3]Guangdong Zhongtian Municipal Engineering Design Co., Ltd, 550081, China

E-mail: liweilengjinyue@163.com

Abstract. The inoculation of short-cut denitrifying polyphosphate-accumulating organisms (DPAOs) mainly included two-phase inoculation and three-phase inoculation. The short-cut denitrifying phosphorus removal bacteria were quickly inoculated by sequencing batch reactor (SBR) to treatment domestic wastewater. The results showed that the average effluent concentration of TP was 0.85 mg/L after 132 cycles by 44 days in two-phase inoculation. The removal rates of NO_2^--N, TP and COD were 94.73%, 95.47% and 89.96% after 126 cycles by 42 days in three-phase inoculation, and the effluent concentrations were separately 1.31 mg/L, 0.45 mg/L and 17.07 mg/L, which reached the first A class requirement of Urban sewage treatment plant pollutant discharge standard. It was indicated that the efficiency of three-phase inoculation was higher. Anoxic phosphorus uptake was influenced seriously by anaerobic residual carbon, and it was the difference of the two inoculations.

1. Introduction

In an anaerobic / anoxic alternating operating environment, the sludge in the system can use the nitrite as electron acceptor for simultaneous nitrogen and phosphorus removal. People named this type of bacteria as denitrifying polyphosphate accumulating organisms (DPAOs). Compared with O_2 and NO_3-N as electron acceptor, DPAOs with nitrite nitrogen as electron acceptor has the advantages of short operating cycle, high phosphorus uptake rate, saving more oxygen consumption and energy, and shortening reaction time [1-5]. This has led some scholars' extensive research on short-cut denitrifying phosphorus removal technology. Some scholars studied on domestication of short-cut denitrifying phosphorus removal sludge, and explored the process operating conditions, performance and practical application [6-8]. Some scholars discussed the metabolic mechanism, analyzed the influence of substrate concentration, the quality concentration of the nitrite nitrogen, SRT, pH, temperature and other environmental factors on short-cut denitrifying phosphorus removal [9-11].

Up to now short-cut denitrifying phosphorus removal technology has been reached in the laboratory, and the application and enrichment of DPAOs are the keys to short-cut denitrifying phosphorus removal technology research. This study was conducted by two acclimation methods (two-phase inoculation and three-phase inoculation) to find a method for rapid application of short-cut denitrifying. It aimed to provide support to promote the practical application of short-cut denitrification and phosphorus removal technology.

2. Materials and methods

2.1. Experimental setup

The DPAOs was quickly domesticated by SBR as shown in figure 1. It was made of double plexiglass. The outer layer was connected to the electric constant temperature water bath, so that the operation of system was in a constant temperature state. The reactor had a inside diameter of 14cm, a height of 85 cm, and the effective volume of 12 L. Sewage and sludge were mixed completely by electric mixer in the reaction process. The lower part of the reactor was equipped with microporous aeration diffusers that connected the gas flow meter. To ensure the appropriate pH range, the system was equipped with a pH on-line monitor. When exceeding the set range, the system will send an alarm signal and add acid or lye through the peristaltic pump to adjust the pH in the allowable range. The bottom of the reactor was arranged with a mud discharge port, and a plurality of outlet ports were arranged on the side. Steps of water distribution, mixing, aeration and drainage were controlled by time-control switch and solenoid valve to achieve the automatic switching.

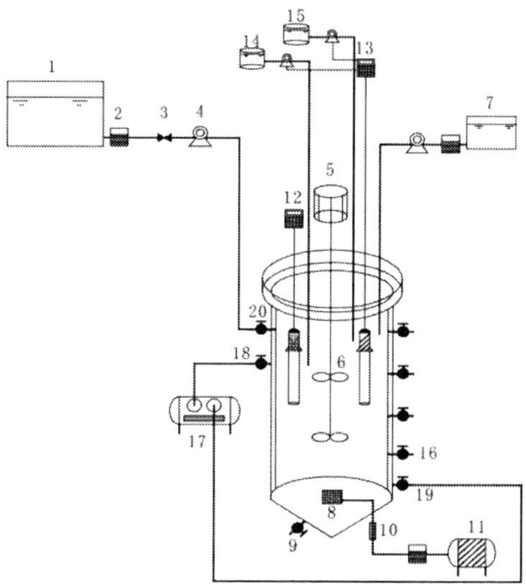

Figure 1. Experimental setup of denitrifying dephosphatation.
1-inlet tank; 2-time-control switch; 3-solenoid valve; 4-water pump; 5-electric mixer; 6-Stirring paddle; 7-nitrite box; 8-microporous aeration diffuser; 9-mud discharge port; 10-gas flow meter; 11-air pump; 12-DO on-line monitor; 13-pH on-line monitor; 14-acidic buffer; 15-alkaline buffer 16- outlet ports; 17-electric constant temperature water bath; 18- Water bath inlet; 19-Water bath outlet; 20-water inlet.

2.2. Test water quality

The experiment used simulated domestic sewage; the main ingredients were KH_2PO_4, NH_4Cl, $CaCl_2$, $MgSO_4$, CH_3COONa and trace element. The system used $NaHCO_3$ to adjust the pH in the range of 7.5 ~7.8.

Table 1. Composition of simulative domestic sewage.

COD (mg/L)	N (mg/L)	P (mg/L)	$MgSO_4$ (mg/L)	$CaCl_2$ (mg/L)	Trace element (ml/L)
170.000~220.000	15~20	9.000~12.000	7~9	20~25	1

Trace element composition: $FeCl_3$ 1.50 g/L, $H_3BO_3$0.15 g/L, $CoCl_2$ 0.15 g/L, $CuSO_4$ 0.03 g/L, $MnCl_2$ 0.06 g/L, $Na_2MoO_4$0.06 g/L, $ZnSO_4$ 0.12 g/L, EDTA 10.00 g/L.

Water specific composition was shown in table 1.

2.3. Analyzed items and methods

Routine testing items and analysis methods were shown in table 2. The analysis and detection methods were from the "water and wastewater monitoring and analysis methods" (fourth edition) which was promulgated by the State Environmental Protection Administration.

Table 2. The essential analyzed items and methods.

Analyzed Iitems	Methods
pH	PHS-25 pH on-line monitor
NO_2^--N	Spectrometer
TP	Spectrometer
COD	Spectrometer
MLSS	Filter paper weighing method
Temperature	Electric constant temperature water bath
DO	HQ40d DO on-line monitor

2.4. Running processes

The inoculation of DPAOs mainly included two-phase inoculation (anaerobic/aerobic, anaerobic/anoxic) and three-phase inoculation (anaerobic/aerobic, anaerobic/precipitation drainage/anoxic, anaerobic/anoxic). For two-phase inoculation, the traditional polyphosphate accumulating organisms (PAOs) were enriched in anaerobic/ aerobic environment; then PAOs were inoculated DPAOs in anaerobic/anoxic environment. In the anoxic phase, the nitrite solution was added to the system continuously to make nitrite become electron acceptor of phosphorus uptake. When the phosphate removal rate in sewage was high and continuous stability, it can be considered that the DPAOs were domesticated successfully. For three-phase acclimation, the operation mode was anaerobic/precipitation drainage/anoxic, sewage with residual carbon source needed to be drained at the end of the anaerobic reaction and then sewage with no external carbon source re-entered the system. It aimed to avoid the unconsumed carbon source at the end of anaerobic impact on anoxic denitrifying dephosphatation, and limit the growth and reproduction of conventional denitrifying bacteria.

Table 3. Running mode of two acclimation means.

Means	Phase	Running mode
two-phase acclimation	Phase I	inlet—anaerobic 2 h—aerobic 2 h—sedimentation 0.5 h—outlet 0.5 h
	Phase П	inlet—anaerobic 2 h—anoxic 2 h—sedimentation 0.5 h—outlet 0.5 h
three-phase acclimation	Phase I	inlet—anaerobic 2 h—aerobic 2 h—sedimentation 0.5 h—outlet 0.5 h
	Phase П	inlet—anaerobic 2 h—sedimentation 0.5 h—outlet 0.5 h—second-inlet—anoxic 2 h—sedimentation 0.5 h—outlet 0.5 h
	Phase Ш	inlet—anaerobic 2 h—anoxic 2 h—sedimentation 0.5 h—outlet 0.5 h

The sludge used in the experiment was taken from the aeration tank of Sambaotun Sewage Treatment Plant in Fushun City, Liaoning Province. The sludge was added to two reactors named A and B in equal amounts after 48 hours of exposure. The two reactors, A and B, were domesticated in two stages of domestication and three stages of domestication. During the operation period, the water quality of the two reactors was the same. The volume ratio of sludge to water was 1:4, and MLSS is controlled at around 3500 mg/L, pH was controlled from 7.5 to 7.8, SV=30% ,SRT=24 d. The system ran three cycles per day and ran 5 to 6 hours per cycle. The specific operating mode was shown in table 3.

3. Results and discussion

3.1. Accumulation of traditional PAOs

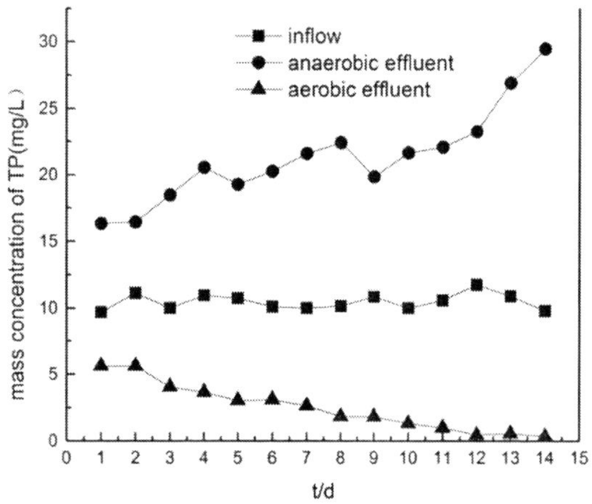

Figure 2. Concentration variation of TP of A reactor in first stage.

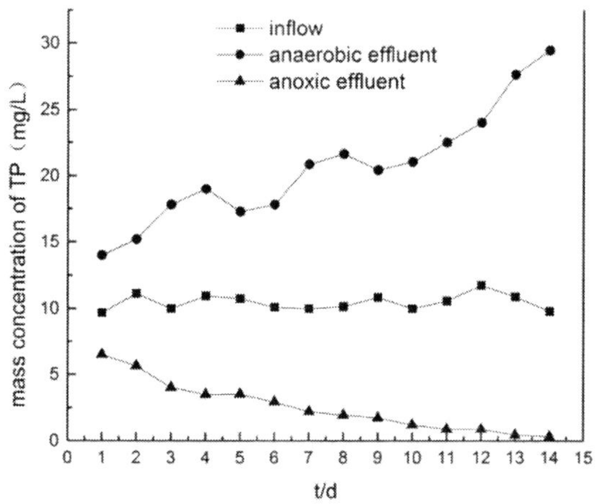

Figure 3. Concentration variation of TP of B reactor in first stage.

The system was running for 14 days at this stage and the average influent concentration of TP was 10 mg/L. The concentrations variation of TP in reactor A and B were shown in figures 2 and 3, concentration variation of COD was shown in figure 4. The water quality and running mode of the two reactors were the same, so the variation tendency was similar. PAOs hydrolyzed the polyphosphate (poly-P) in the cells to produce energy by using the nutrients in the sewage, so that they could convert the external carbon source into internal carbon source poly-β-hydroxybutyrate (PHB) and store in cells. This process removed most of the organic in the sewage in the form of phosphorus-release; and in the aerobic stage PAOs used PHB to absorb phosphorous excessively to achieve phosphorus removal. As shown in figures 2 and 3, the initial phosphorus removal of the system was not good. Figure 4 showed that the external carbon source was mainly consumed at the anaerobic phase. After the end of the first day, the mass concentrations of TP in the two reactors were 5.61 mg/L and 6.50 mg/L, respectively. With the system running, TP removal efficiency gradually increased. Figure 3 showed that the anaerobic phosphorus release decreased in the 10th day of a reactor, which is due to the trouble with solenoid valve. The problem caused water outflow from solenoid valve, which affected anaerobic phosphorus release owing to system's lack of nutrients. The system ran smoothly at the next 4 days, and the phosphorus release and uptake were steadily rising. On the 14th day, the maximum anaerobic phosphorus release in the two systems reached about 19.00 mg/L, and TP mass concentration in the effluent was less than 0.50 mg/L, which reached the "pollutant discharge standard for urban sewage treatment plants" (GB18918-2002) level 1 A standard. The results showed that the traditional PAOs became the dominant species in the two systems, that is, the enrichment of the PAOs with O_2 as the electron acceptor provided favorable conditions for the domestication and enrichment of DPAOs.

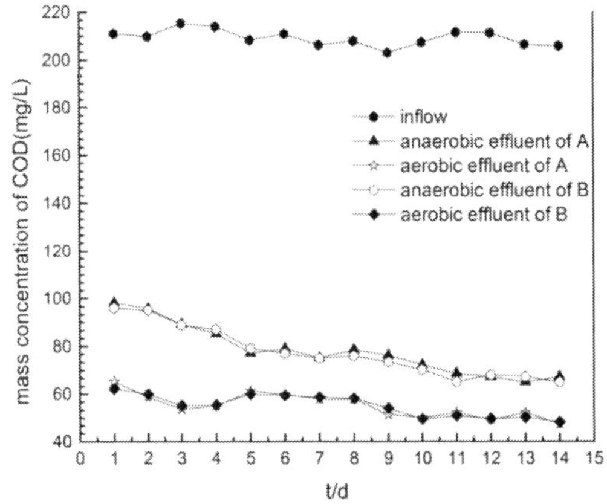

Figure 4. Concentration variation of COD of A and B reactors in first stage.

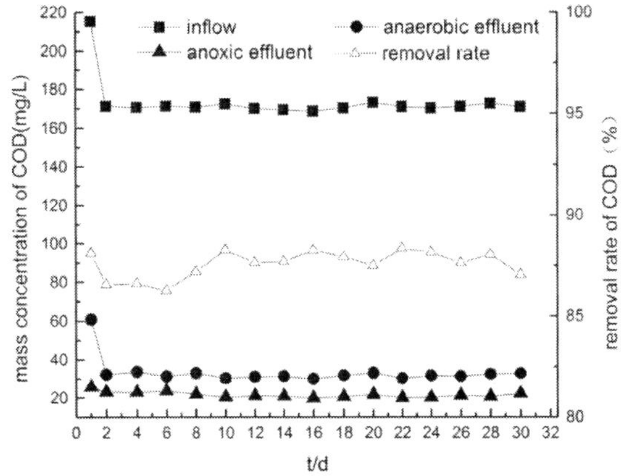

Figure 5. Concentration variation and removal rates of COD of A reactor in second stage.

3.2. Domestication and enrichment of DPAOs

There were already a large number of traditional PAOs in the two systems after the domestication of the first stage. Then change the system operating conditions, namely to stop the aerobic aeration and use the nitrite as electron acceptor of anoxic dephosphorus, to achieve DPAOs enrichment. High concentration of nitrite solution had certain toxicity to microbial growth, which lead to the deterioration of biological dephosphorization system [12-14]. If the concentration of nitrite was too low, the electron acceptor of the denitrifying phosphorus removal was deficient, which made the system could not achieve the optimum state of phosphorus removal [15]. Therefore, nitrite nitrogen solution was dropped to the reactors. In the anoxic phase, the concentration of nitrite nitrogen increased from 10.00 mg/L to 25.00 mg/L, which gave the microbial a buffer adaptation process. The results were discussed in terms of two-phase acclimation and three-phase acclimation.

3.2.1. Two-phase Acclimation. After completing the operation of the first stage, reactor A worked with the next phases of anaerobic/anoxic directly, which lasted 30 days. In the anaerobic, DPAOs used the energy generated by phosphorus release to absorb the external carbon source and stored it in the form of PHB. Meanwhile the resedual carbon source entered the anoxic stage. At the anoxic stage, DPAOs used PHB to absorb phosphorous excessively without carbon source, therefore, excessive carbon source would be used by denitrifying bacteria for denitrification, which inhibited the growth, reproduction and other life activities of DPAOs, and prevented DPAOs to become a systemic dominant bacteria. The concentration and removal rate of COD in this stage were shown in figure 5. In the first day, the COD mass concentration of the influent was 215.37 mg/L, and the COD concentration in the system remained at 60.83 mg/L at the end of anaerobic stage and 25.63 mg/L in the end of anoxic stage. It was concluded that the supply of carbon was much larger than that of the absorption in the anaerobic influent, so that a large amount of the external carbon source was not used, which led to the high COD concentration of effluent after the anaerobic stage. The nitrite nitrogen solution added in the anoxic stage was denitrified by denitrifying bacteria using the remaining external carbon source, so that the COD content was further reduced. The concentration of COD in the influent was reduced to about 170.00 mg/L, which made the concentration of COD was basically stable at about 32.00 mg/L at the end of anaerobic reaction and 20.00 mg/L after the anoxic stage. Analysis suggested that part of the carbon source met the needs of microbial growth and metabolism, and the other part was used by the regular heterotrophic denitrifying bacteria in the system. The average removal rate of COD was 87.57% at this stage.

Figure 6 showed the variation on concentration and removal rate of TP and nitrite nitrogen in the second stage of A reactor. The initial mass concentration of nitrite nitrogen was 10.00 mg/L. As shown in figure 6, nitrite nitrogen removal rate was 89.61% in first day, whereas the removal rate of nitrite nitrogen in third day dropped to 32.33%, and after the fourth day, the removal rate increased gradually. This was because in first day, there was an excess carbon source at the end of anaerobic, and most of the nitrite nitrogen was removed by denitrification. Figure 5 showed that the amount of carbon source dosage began to be reduced after two days. In the anoxic stage, the mass concentration of COD was only 32.17 mg/L. Therefore the nitrite nitrogen could only be removed by a small amount of DPAOs at this point, and the removal rate dropped sharply. As the reaction progressed, the DPAOs gradually became a dominant strain in the system, and nitrite nitrogen was used as an electron acceptor for phosphorus removal, and its removal rate increased. The mass concentration of nitrite nitrogen increased to 25.00 mg/L, the concentration of nitrite nitrogen in effluent was 1.71 mg/L, and the removal rate was steady above 92.00%. The amount of phosphorus release decreased in the initial four days, and began to rise gradually five days later. This was because DPAOs were non-dominant species in the system, and the traditional PAOs could not play a good role in removing phosphorus due to the lack of O_2 electron acceptor. As the domestication progressed, the number of DPAOs in the system gradually increased. DPAOs became the dominant bacteria of the system, and the removal rate of TP increased above 92.00% from the 28th day. After 30 days' domesticated cultivation, The mass concentration of COD, nitrite nitrogen and TP of effluent were 19.86 mg/L, 1.99 mg/L, 0.79 mg/L respectively, and the removal rate of them reached 88.32%, 92.03% and 92.15% respectively, which reached the "pollutant discharge standard for urban sewage treatment plants" (GB18918-2002) level 1 B standard.

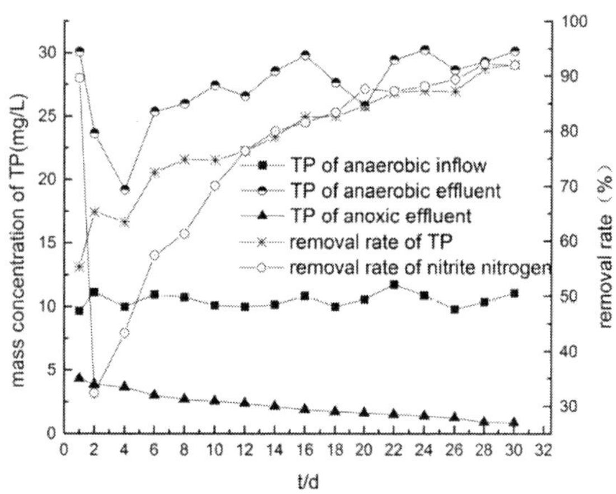

Figure 6. Concentration variation, removal rate of TP and removal rate of NO_2^--N of A reactor in second stage.

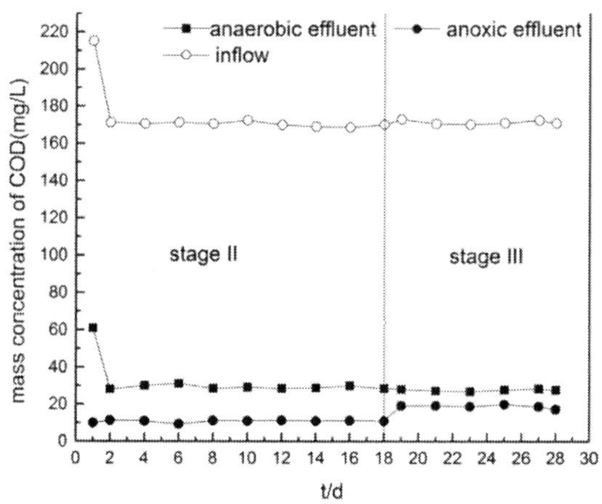

Figure 7. Concentration variation of COD of B reactor in third stage.

3.2.2. Three-phase Acclimation The three-phase acclimation method lasted for 28 days, of which the second phase lasted 18 days and the third stage lasted for 10 days. Figure 7 showed the variation on concentration of COD in the reactor B for the second phase and the third phase. It could be seen that in the first day, the COD concentration was 215.37 mg/L in influent and was still 60.84 mg/L after anaerobic stage. Analysis suggested the content of residual carbon was too much, so the mass concentration of COD in influent was reduced to as low as 170.00 mg/L. No extra carbon source was added to the anoxic stage. The concentration of COD in the system was lower than 12.00 mg/L. Figure

8 showed the variation on concentration and removal rate of TP, and removal rate of nitrite nitrogen in the reactor B for the second phase and the third phase. The initial removal rate of TP and nitrite were 50.34% and 34.70% respectively. The reason for this analysis was that the number of DPAOs in the system with nitrite nitrogen as electron acceptor was very small, so the denitrifying dephosphatation was poor. The conventional anti-nitrification could not take place because of the deficiency of the external carbon source and only a small part of nitrite was used by DPAOs. So the removal rates of TP and nitrite nitrogen were not high. According to the test results, DPAOs with nitrite nitrogen as the electron acceptor were present in the system after the domestication of first phase. As the reaction proceeded, the removal rate of TP and nitrite nitrogen increased gradually. The graph showed that the removal rates of nitrite nitrogen and TP were highly relevant, it was indicated that there were already a significant amount of DPAOs in the system. The removal rates of TP and nitrite nitrogen were more than 94.00% as the reaction was progressing into its fifth day. DPAOs accumulated a large number of PHB as an electron donor for phosphorus-uptake after phosphorus-release in anaerobic stage, at the same time, using nitrite nitrogen as electron acceptor to remove phosphate and nitrite in the system successfully.

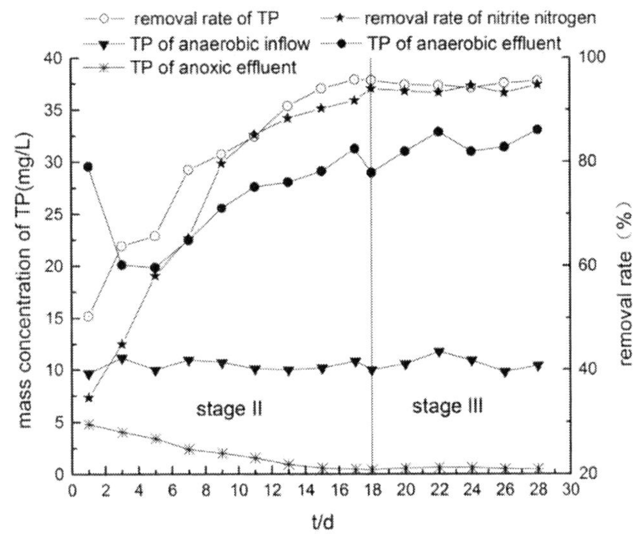

Figure 8. Concentration variation,removal rate of TP and removal rate of NO_2^--N of B reactor in second and third stages

Through the domestication of the second phase, DPAOs had become the dominant population in the system. In the third phase, the secondary influent part of the anoxic was canceled, and the concentration ratio of COD and TP in the anaerobic influent was controlled to 17, Under these conditions, the system steadily ran for 10 days, and the removal rates of TP, COD and nitrite nitrogen reached 95.47%, 94.73% and 94.73% respectively, and the mass concentration of them were 0.45 mg/L, 1.31 mg/L and 17.07 mg/L, which reached the "pollutant discharge standard for urban sewage treatment plants" (GB18918-2002) level 1 A standard. The stable running of the system illustrated the success of DPAOs domestication.

3.3. Analysis of result

The results of the two acclimatization methods showed that DPAOs can be successfully acclimated by methods of anaerobic/anoxic, and the three-phase acclimation method was shorter and more efficient, and the better sewage treatment effect can be achieved. Analysis suggested that the remaining external carbon source after the anaerobic stage was discharged in the second stage of the three-phase acclimation and add no external carbon source water, which made the process of phosphorus uptake was not affected by the external carbon source. It could be considered that the concentration of exogenous carbon source had a great influence on the effect of denitrifying phosphorus. If the amount of residual external carbon source was too much at the end of anaerobic stage, it would be used by the oxygen-depleting organic presented in the later of anaerobic stage. In that case, it was beneficial to the denitrification reaction of the traditional denitrifying bacteria in the anoxic stage, which formed the competition with the DPAOs on the electronic receptor, and resulted in inhibition of phosphorus absorption. Ultimately, the growth and reproduction of DPAOs got damaged. Under the condition of low COD concentration, DPAOs would use internal carbon sources for denitrification while removing phosphorus, so it could be conclude that low concentration of COD environment would benefit DPAOs. This might be the main reason for the better results of the three-phase acclimation method.

4. Conclusion

- In the process of periodic culture, the domestication of sludge was carried out. First cultivated the traditional PAOs and made them be the dominant species in the system; then domesticated and enriched DPAOs by continuous dropping nitrite nitrogen solution. The short-cut denitrifying phosphorus removal system could be start up successfully.
- The results of the two reactors showed that the three-stage acclimation method was more suitable for the domestication of DPAOs. The difference between the two methods was that the influent quality in the second stage of the three-stage acclimation method, and it was the critical stage. The concentration of COD in the system after the end of the anaerobic stage was the main factor affecting the enrichment of DPAOs.

Acknowledgments

This work was supported in part by the Doctoral Scientific Research Foundation of Liaoning Province, China (No. 201501069), and Foundation of Liaoning Educational Committee (No. LJZ2016014). In addition, the study was sponsored by National Nature Science Foundation of China (No. 51508342). National water pollution control and control scientific and technological special project (2014ZX07202-011). The authors declare that they have no conflict of interest.

References

[1] Zheng X L, Sun P D, Han J Y, *et al* 2014 Inhibitory factors affecting the process of enhanced biological phosphorus removal (EBPR) *Mini-review Process Biochem.* **49** 2207-13
[2] Jahari P, Munz G and Oleszkiewicz J A 2014 Selection of denitrifying phosphorous accumulating organisms in IFAS systems: Comparison of nitrite with nitrate as an electron acceptor *Chemosphere* **109** 20-7
[3] Li W, Wang Y J and Zhu L 2015 Effect of SRT on the denitrifying dephosphatation capacity of A/A process *Ind. Water Treat.* **35** 55-9
[4] Pourria J, Giulio M and Jan A O 2014 Selection of denitrifying phosphorous accumulating organisms in IFAS systems: Comparison of nitrate with nitrate as an electron acceptor *Chemosphere* **109** 20-7
[5] Wang Z, Meng Y, Fan T, *et al* 2015 Phosphorus removal and N_2O production in anaerobic/anoxic denitrifying phosphorus removal process: Long-ternimpact of influent phosphorus concentration *Bioresour. Technol .* **179** 585-94
[6] Hu X M, Li W and Liu J L 2013 Influence of pH on denitrifying phosphorus removal using nitrite as electron acceptor *J. Cent. S. Univ. (Sci. Technol.)* **44** 144-2149
[7] Wang Y Y, Zhou S, Wang H, *et al* 2015 Comparison of endogenous metabolism during long-term anaerobic starvation of nitrite/nitrate cultivated denitrifying phosphorus removal sludges *Water Res.* **68** 374-86
[8] Liu S L and Li J Z 2015 Accumulation and isolation of simultaneous denitrifying

polyphosphate-accumulating organisms in an improved sequencing batch reactor system at low temperature *Int. Biodeter. Biodegr.* **100** 140-8

[9] Wang Y Y and Zhou S 2014 Nitrite survival and nitrous oxide production of denitrifying phosphorus removal sludges in long-term nitrite/nitrate-fed sequencing batch reactors *Water Res.* **1354** 60-1

[10] Li Y 2013 Study on model and technology of shortcut denitrifying phosphorus removal (Beijing: China University of Mining and Technology)

[11] Li X F, Yan B G and Li M 2014 Research and development on the influencing factors in the process of denitrifying dephosphorization *Ind. Water Treat.* **34** 10-3

[12] Zhai Y, Zhang Y Q and Li Y 2014 Anaerobic/anoxic domestication short - range denitrifying phosphorus accumulating organisms *Jiangsu Agric. Sci.* **42** 332-4

[13] Sun L, Zhao X X and Zhang H F 2015 Biological characteristics of a denitrifying phosphorus-accumulating bacterium *Ecol. Eng.* **81** 82-8

[14] Han C, Wang Z and Yang W J 2016 Effects of pH on phosphorus removal capacities of basic oxygen furnace slag *Ecol. Eng.* **89** 1-6

[15] Liu G Q and Wang J M 2015 Modeling effects of DO and SRT on activated sludge decay and production *Water Res.* **80** 169-78

Dynamic characteristics of nitrogen and phosphorus in the representative input tributaries of Miyun Reservoir

L F Du[1], J Liu[1], S J Li[1,2], X R Li[1], J F Yang[1], T K Zhao[1] and M T Ma[1]

[1]Institute of Plant Nutrition and Resources, Beijing Academy of Agriculture and Forestry Sciences, Beijing 100091, China

E-mail: shunjiangli@163.com

Abstract. The study analyzed dynamic characteristic of nitrogen and phosphorus and influencing variables in three representative input tributaries of Miyun Reservoir from 2012 to 2016. The results showed that Chaohe River in the study period had relatively high TN, NO_3^--N and NH_4^+-N concentrations. While high phosphorus content was found in Qinghe River. Basing on environmental quality standard for surface water, NO_3^--N, NH_4^+-N and TP met the drinking water criterion in most samples. However, TN content met the drinking water criterion only in less than 15% samples. In Qingshuihe River, pH and TSS were significantly variable due to nitrogen variations and TOC was significantly variable for phosphorus variations. However, water parameters (BOD, COD, TSS, TOC, EC and pH) had no significant effects on nitrogen and phosphorus dynamics in Chaohe River and Baihe River.

1. Introduction

Miyun reservoir is the only surface drinking water source for Beijing, which dedicates to control the floods and provides water for hydraulic electrogenerating and supply industrial and agricultural water and offer recreation. Its main input tributaries are Chaohe River, Baihe River and Qingshui River. Before 1990s, the water quality of Miyun Reservoir belonged to the middle nutritional status [1-4]. After 1990s, the total nitrogen (TN) and COD index were within eutrophication standards. In 2002, cyanobacteria formed blooms [2, 5]. The nutrient inflow load in Miyun Reservoir is mainly from the flood season, and the inorganic nitrogen and phosphorus pollutants are mainly from Chaohe River, Baihe River and Qingshui River.

Water quality monitoring in Miyun Reservoir began in 1975 [6]. The former researches mainly focus on the water quality of Miyun Reservoir. And most of the water quality parameters reported in the literature, such as TN and total phosphorus (TP), are the average values measured in the two periods or periods of the year. However, there are few reports about the continuous variation of water quality parameters in Miyun Reservoir Basin.

In this study, almost 4 years' field observation was conducted in three input tributaries in Miyun Reservoir basin. The objectives of this study were to (a) investigate dynamic characteristics of nitrogen and phosphorus, (b) study their influence factors and discuss the relationship among them in these three tributaries. This study should be helpful for understanding the biogeochemical cycling and environmental effects of nitrogen and phosphorus in Miyun Reservoir Basin.

2. Materials and methods

2.1. Study sites and sampling
Chaohe River, Baihe River and Qingshui River are the main input tributaries of Miyun Reservoir. According to on-the-spot investigation and literatures, twelve representative sampling sites were selected for this study, including five sites in Chaohe River (Bakeshiying, Gubeikou Bridge, Beidianzi Bridge, Xiahui hydrometrical station and Xinzhuang Bridage), three in Baihe River (Sihetang Bridge, Zhangjiafen hydrometrical station and Old Daguan Bridge) and four in Qingshuihe River (Huangyankou, Dongzhuangkou, Dongzhuang and Taishidun Bridge). The field survey and sample collection were conducted in every month from October 2012 to June 2016. Water samples about 1000 ml were collected from 0.3 meters below the water surface. All samples were transported in an ice-cooled cabin to the laboratory and stored in a cooler before analysis.

2.2. Analytical methods

2.2.1. Determination of nitrogen and phosphorus characteristics and other variables. All parameters analyzed in study are based on national standards or industry standards. TN was determined by alkaline potassium persulfate digestion with UV Spectrophotometry (GB11894-198) [7], ammonium nitrogen (NH_4^+-N) was determined by colorimetric method and nitrate (NO_3^--N) analyzed by UV Spectrophotometry. TP content was determined by ammonium molybdate spectrophotometric method (GB11893-1989) [8].

Additionally, some water variables, such as BOD, COD, TSS, TOC, EC and pH were analyzed using the national standard analysis. BOD was determined with water quality automatic analyzer (HJ/T 99-2003) [9], COD with potassium dichromate method (GB11914-89) [10], TSS with gravimetric determination (GB11901-89) [11], TOC by total organic carbon analyzer (HJ 501-2009) [12], EC with water quality automatic analyzer of conductivity (HJ/T 97-2003) [13], pH with glass electrode method (GB/T 6920-86) [14].

2.2.2. Statistical analyses. All statistical analyses in study were all conducted with software R (version 3.0.2). Among these, analysis for the least significant difference (LSD) has been done with "agricolae" packages in R [15], and constrained ordination analysis with "GPArotation" [16] and "Vegan" [17] packages in R.

Before constrained ordination analysis, we carried out initial de-trended correspondence analyses for the most appropriate ordination method with the gradient length [18]. Results of de-trended correspondence analyses showed that gradient in nitrogen and phosphorus data among sites was short (<3) [19], which means we should use redundancy analysis (RDA) to determine the relationship between nitrogen and phosphorus characteristics and explainable water variables. Additionally, 999 permutation test was used to test statistical significance of each analysis. And forward selection by residuals permutation at the 5% significance level was conducted for determining significant variables. Variation partitioning was performed with these significant variables and results are presented in an ordination diagram. In ordination diagram, all of the variables are represented with arrows. Additionally, smaller angle between arrows representing high correlation between variables, and the direction and the length of the arrows meaning positive or negative correlations and their importance in explaining the variation of nitrogen and phosphorus.

3. Results and analysis

3.1. Nitrogen and phosphorus characteristics in three tributaries
In this study, there are 1260 samples for 105 sampling times from October 2012 to June 2016. TN, NO_3^--N, NH_4^+-N and TP concentration in three input tributaries of Miyun Reservoir (Baihe River, Chaohe River and Qingshuihe River) are shown in figure 1.

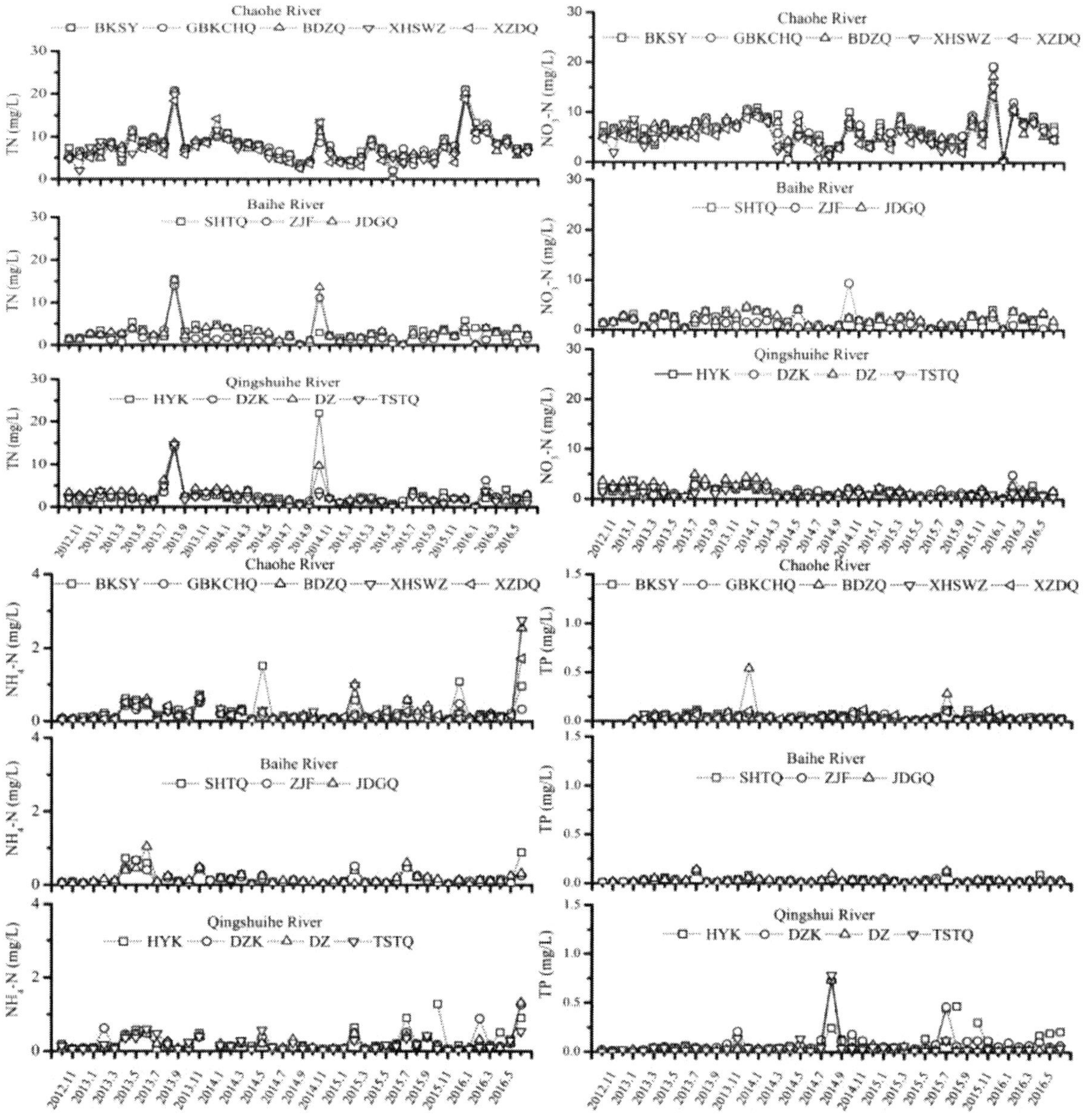

Figure 1. Comparison of TN, NO_3-N, NH_4-N and TP concentration in three rivers from 2012 to 2016.

3.1.1. TN characteristics. TN concentration in Chaohe River was significantly higher than in Baihe and Qingshuihe River with ANOVA analysis ($p<0.05$), while there are no difference between Baihe and Qingshuihe Rivers. Mean concentration of TN in Chaohe River is 7.66 mg/L with the highest content in 230 samplers 21.2 mg/L and the lowest 0.081 mg/L (table 1). According to environmental quality standard for surface water (GB 3838-2002) [20], TN contents in about 99.57% samples is above the Vcriterion (>2 mg/L) and only about 0.43% samples meet the drinking water criterion. Additionally, among five sampling sites in Chaohe River, there was no significant difference.

Mean TN content in Baihe River and Qingshuihe River was 2.74 and 2.56 mg/L, respectively. In Baihe River, TN contents in about 60.99% samples are above the V criterion and only about 12.06% samples meet the drinking water criterion. In Qingshuihe River, about 52.15% samples above the V criterion and 14.52% samples meet the drinking water (figure 2).

Comparing the variation of TN from 2012 to 2016, results showed TN content in 2013 was significantly higher than in 2015 in Qingshuihe River, TN content in 2012 and 2014 significantly higher than those in other years, while there was no significant difference in Baihe River.

Table 1. Overall monitoring results of three of Miyun reservoir.

River	Parameter	Num. of sample	Mean (mg/L)	Max (mg/L)	Min (mg/L)	CV (%)
Baihe River	TN	139	2.74	15.6	0.01	91.05
	NO_3-N	139	1.87	9.45	0.05	71.47
	NH_4-N	139	0.17	1.04	0.002	108.22
	TP	139	0.03	0.141	0.001	87.94
Chaohe River	TN	230	7.66	21.2	0.081	46.87
	NO_3-N	230	6.21	19.2	0.252	45.91
	NH_4-N	224	0.23	2.76	0.01	141.77
	TP	220	0.04	0.543	0.0008	112.57
Qingshuihe River	TN	185	2.56	22	0.09	104.14
	NO_3-N	185	1.56	6.996	0.041	73.05
	NH_4-N	185	0.21	1.32	0.01	109.73
	TP	185	0.06	0.787	0.0008	177.76

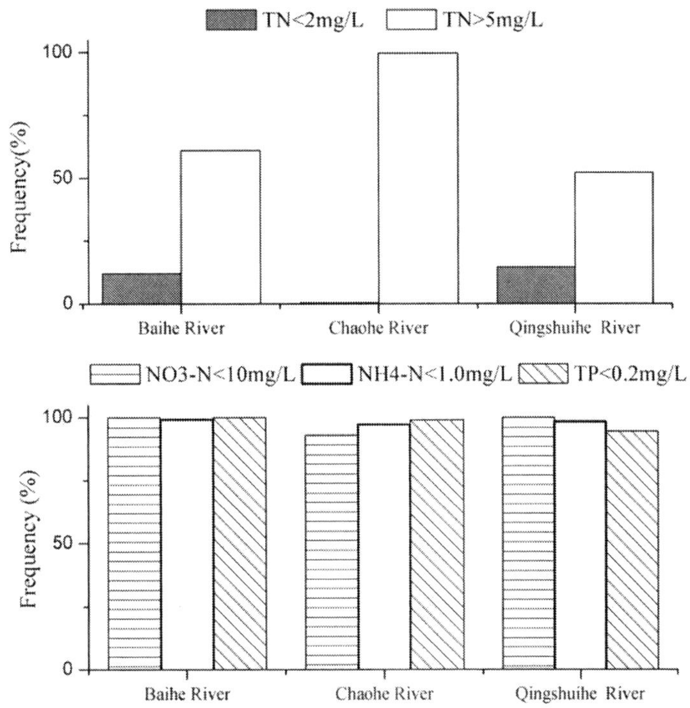

Figure 2. Frequency distribution for TN, NO_3^--N, NH_4^+-N and TP concentration in three rivers from 2012 to 2016.

3.1.2. NO_3^--N characteristics. NO_3^--N concentration in Chaohe River was significantly higher than in Baihe and Qingshuihe River with ANOVA analysis ($p<0.05$), while there are no difference between Baihe and Qingshuihe River.

Mean concentration of NO_3^--N in Chaohe River is 6.21 mg/L with the highest content in 230 samplers 19.2 mg/L and the lowest 0.252 mg/L (table 1). According to environmental quality standard for the centralized drinking water sources (GB 3838-2002), NO3-N contents in Chaohe River in about

6.96% samples are beyond the drinking water criterion (< 10 mg/L) , while all samples in Baihe and Qingshuihe River meet the drinking water criterion (figure 2). Additionally, there was no significant difference among five sampling sites in Chaohe River, while NO_3^--N in Zhangjiafen hydrometrical station was higher than other two sites in Baihe River.

Comparing the variation of NO3-N from 2012 to 2016, results showed NO_3^--N content in 2012 and 2013 was significantly higher than in other year in Qingshuihe River, while there was no significant difference in Baihe River and Chaohe River.

3.1.3. NH4-N characteristics. For NH4-N contents, there were no significant difference among Chaohe River, Baihe and Qingshuihe River with ANOVA analysis ($p>0.05$). Mean concentration of NH4-N in Chaohe River is 0.23 mg/L with the highest content in 224 samplers 2.76 mg/L and the lowest content 0.01mg/L (table 1). According to environmental quality standard for surface water (GB 3838-2002), NH_4-N contents in about 97.31% samples meet the drinking water criterion (<1 mg/L) and only about 0.90% samples are above the V criterion (>2mg/L) (figure 2).

Mean NH_4-N content in Baihe River and Qingshuihe River was 0.17 and 0.21 mg/L, respectively. NH_4-N contents in about 99.29% samples meet the drinking water criterion (<1 mg/L) in Baihe River, while 98.35% samples in Qingshuihe River (figure 2).

Comparing the variation of NH_4^+-N from 2012 to 2016, results showed NH_4^+-N content in 2016 was significantly higher than in 2012 and 2014 in Qingshuihe River, NH_4^+-N content in 2013 significantly higher than those in 2012 and 2014, while NH4+-N content in 2016 significantly higher than those in 2012 and 2014.

3.1.4. TP characteristics. TP concentration in Qingshuihe River was significantly higher than in Baihe and Chaohe River with ANOVA analysis ($p<0.05$), while there are no difference between Baihe and Chaohe River. Mean concentration of TP in Qingshuihe River is 0.06 mg/L with the highest content in 185 samplers 0.787 mg/L and the lowest 0.0008 mg/L (table 1). According to environmental quality standard for surface water (GB 3838-2002), TP contents in about 3.28% samples are above the V criterion (>0.4 mg/L) and about 94.54% samples meet the drinking water criterion (<0.2 mg/L) (figure 2). Additionally, among four sampling sites in Qingshuihe River, there was no significant difference.

Mean TP content in Baihe River and Chaohe River was 0.03 and 0.04 mg/L, respectively. About 99.09% samples in Chaohe River and 100% samples in Baihe River meet the TP criteria for drinking water (<0.2 mg/L) (figure 2). Comparing the variation of TP from 2012 to 2016, results showed TP content in 2013 was significantly higher than in 2012 in Baihe River and Chaohe River, while there are no significant differences in Qingshuihe River from 2012 to 2016.

3.2. Multivariate analysis for effecting variables

Redundancy analysis showed the relationship of nitrogen and phosphorus concentration with water variables in Chaohe River, Baihe River and Qingshuihe River, respectively. However, results showed there were significant correlations between nitrogen and phosphorus concentrations and water variables only in Qingshuihe River ($p=0.017$). Eigenvalues of the first two RDA axes accounted for 9.83% of the variance of nitrogen and phosphorus contents (figure 3). The permutation tests showed that three variables, such as pH ($p=0.001$), TOC ($p=0.001$) and TSS ($p=0.004$), had statistically significant effects on variation of nitrogen and phosphorus contents. Along RDA axis 1, NO_3^--N and TN stand out with a positive RDA axis 1 and are thus negatively related with TSS and pH. Along RDA axis 2, TP stands out with a negative RDA axis 2 and is associated with TOC, while NH_4^+-N stands out with a positive RDA axis 2 and shows a positive score correlating with TSS and pH.

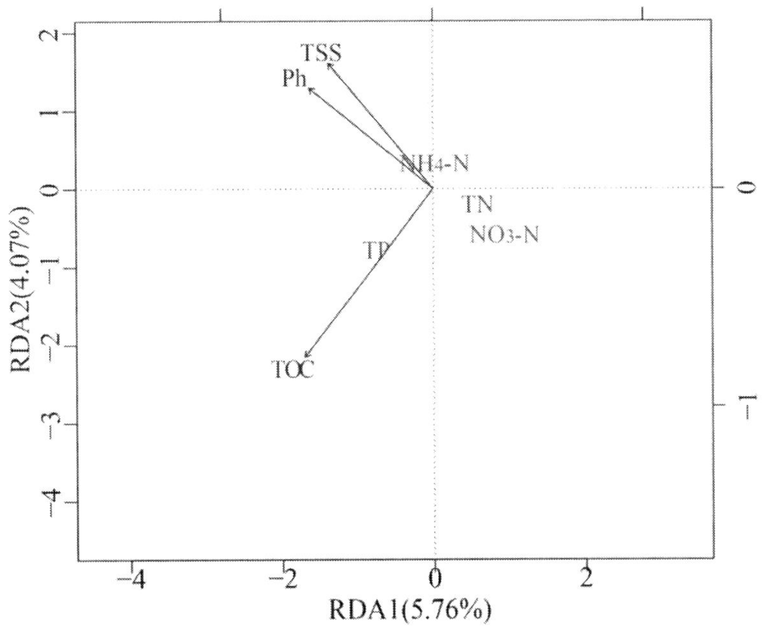

Figure 3. Triplot from RDA showing the relationship of nitrogen and phosphorus with water variables.

4. Discussion and conclusion

The results of this study were similar to those of previous studies, showing that nitrogen in Chaohe River was higher than in Baihe and Qingshuihe [21, 22]. The reason for higher nitrogen is probably related to intensive agriculture around Chaohe River basin, such as intensive croplands and aquaculture. The higher degree of intensive agriculture and the greater is the influence on water quality. Some reports showed that nitrate mainly come from runoff from fertilizer and manure applied to agricultural land [23, 24], and the contribution rate of nitrate nitrogen to TN is about 93%, which is the reason that TN in Chao River is significantly higher than ones in Baihe River and Qingshuihe River. Higher TP was found in Qingshuihe River. The reason for this is due to the increase of population density, which increases phosphorus load with domestic sewage. Baihe River has relative lower nitrogen and phosphorus load, which is related to its Mountainous Areas sources. Nevertheless, new tourist sites around the Baihe River may be nitrogen and phosphorus pollution sources. Additionally, it is worth noting that there are no significant differences among sampling site in every river, except for AJFWS sampling points in the Baihe River, which is significantly lower than the other 2 sampling points.

There are many variables effecting nitrogen and phosphorus contents in many study, including physical and chemical parameters in water body and ecological factors around river basins. In study, pH and TSS were significant variables for nitrogen variation and TOC was a significant variable for phosphorus variation in Qingshuihe River. However, water variables, such as BOD, COD, TSS, TOC, EC and pH, have no significant effects on variation of nitrogen and phosphorus in Chaohe and Baihe Rivers, which showed variables outside water body, such as landscape, agricultural planting, aquaculture, domestic sewage discharge and so they probably have effects on nitrogen and phosphorus variation.

Eutrophication has become the primary water quality issue for most of the freshwater in the world and Miyun reservoir is no exception. The present study analyzed dynamic characteristic of nitrogen and phosphorus and influencing variables in three representative input tributaries of Miyun Reservoir from 2012 to 2016. The results showed that Chaohe Rivers in the study period showed being with relatively high TN, NO_3^--N and NH_4^+-N concentrations. An even higher phosphorus content was found in Qinghe River. According to environmental quality standard for surface water (GB 3838-

2002), NO_3^--N, NH_4^+-N and TP meet the drinking water criterion in most of the samples. However, TN content in less than 15% samples meets the drinking water criterion. In Qingshuihe River, pH and TSS were significant variables for nitrogen variation and TOC was significant variables for phosphorus variation. However, water variables (BOD, COD, TSS, TOC, EC and pH) have no significant effects on nitrogen and phosphorus in Chaohe River and Baihe River.

Acknowledgments
This work was supported by the National Key Research and Development Program of China in the 13th Five-Year Program (No. 2016YFD0801001) and the National Key Research and Development Program of China in the 13th Five-Year Program (No. 2017YFD0801401) and Postdoctoral foundation of Beijing Academy of agriculture and Forestry Sciences (2016-2017).

References

[1] Du G S 1991 Study on water quality of Miyun Reservoir *Journal of Beijing Teachers College (Natural Sciences Edition)* **12** 82-5
[2] Liu J, Du G S, Liu X D, Song F, Liu J and Hu Z L 2004 Study on plankton community in Miyun reservoir *Acta Bot Boreal -Occident. Sin.* **24** 1485-8
[3] Song F, Zhang J Q, Wang X H, *et al* 1986 Discussion on water quality in Miyun Reservoir *China Environ. Sci.* **6** 52-5
[4] Zhang W H, Wang J T, Wu D W and Du G S 1991 Nutrient status and development trend in Miyun Reservoir *Water Resour. Prot.* **2** 40-2
[5] Wang L, Yang M, Guo Z H, Zhang Y, Jiang Y and Fan K P 2006 Study on water quality transformation in Miyun Reservoir *China Water Wastewater* **22** 45-8
[6] Chang C Z 1990 Evaluation of water quality of Miyun Reservoir *Water Resour. Prot.* **2** 49-55
[7] The Ministry of Environmental Protection in China 1989 Water quality-determinatlon of total nitrogen-alkaline potassium persulfate digestion-UV spectrophotometric method (GB11894-1989)
[8] The Ministry of Environmental Protection in China 1989 Water quality-determination of total phosphorus-ammonium molybdate spectrophotometric method (GB 11893-89)
[9] The Ministry of Environmental Protection in China 2003 The technical requirement for water quality automatic analyzer of dissolved oxygen(DO)(HJ/T 99-2003)
[10] The Ministry of Environmental Protection in China 1989 Water quality-determination of the chemical oxygen demand-dichromate method(GB11914-89)
[11] The Ministry of Environmental Protection in China 1991 Water quality-determination of suspended substance -gravimetric method (GB11901-89)
[12] The Ministry of Environmental Protection in China 2009 Water quality-determination of total organic carbon-combustion oxidation nondispersive infrared absorption method (HJ 501-2009)
[13] The Ministry of Environmental Protection in China 2003 The technial requirement for water quality automatic analyzer of electroconductivity (HJ/T 97-2003)
[14] The Ministry of Environmental Protection in China 1986 Water Quality-oetermination of pH value—glass electrode method (GB/T 6920-86)
[15] Mendiburu F 2013 Agricolae: Statistical procedures for agricultural research (R package version 1) pp 1-4
[16] Bernaards C A and Jennrich R J 2005 Gradient projection algorithm and software for arbitrary rotation criteria in factor analysis *Educ. Psychol. Meas.* **65** 676-96
[17] Oksanen J, Blanchet F G, Kindt R and Legendre P 2013 Vegan: Community ecology package (R package version 2.0-7)
[18] Ter Braak C J F and Prentice I C 1988 A theory of gradient analysis *Adv Ecol Res* **18** 271-317
[19] Lepš J and Šmilauer P 2003 *Multivariate Analysis of Ecological Data Using CANOCO* (Cambridge, UK: Cambridge University Press)

[20] The Ministry of Environmental Protection in China 2002 Environmental quality standards for surface water (GB 3838-2002)
[21] Yu Y L and Wang Q S 2008 Analysis of seasonal changes of water quality in Miyun Reservoir and reaches of its main influents *Chinese J. Agrometeorol* .**29** 432-5
[22] Wang Q S, Sun D B, Hao W P, Gu Y, Li Y Z, Mei X R and Zhang Y Q 2011 Nitrate concentration distribution in groundwater of the Miyun Reservoir Watershed *Acta Pedol. Sin.* **48** 141-50
[23] Wang Q S, Mei X R, Zhang Y Q, Sun D B and Yu Y L 2009 Review of water quality of Miyun Reservoir *J. Agri. Sci. Technol.* **11** 45-50
[24] Zamparas M and Zacharias I 2014 Restoration of eutrophic freshwater by managing internal nutrient loads: A review *Sci. Total Environ.* **496** 551-62

IOP Conf. Series: Earth and Environmental Science 82 (2017) 012085 doi:10.1088/1755-1315/82/1/012085

Feasibility analysis of marine ecological on-line integrated monitoring system

D Z Chu[1,3], X Cao[1], S W Zhang[1,2], N Wu[1,2], R Ma[1], L Zhang[1,2] and L Cao[1]

[1]Institute of Oceanographic Instrumentation, Shandong Academy of Sciences, No. 37 Miaoling Road, Laoshan District, Qingdao City, Shandong Province, 266061, China
[2]Shandong Technological Center of Oceanographic Instrumentation, No. 37 Miaoling Road, Laoshan District, Qingdao City, Shandong Province, 266061, China

E-mail: cdz0303@163.com

Abstract. The in-situ water quality sensors were susceptible to biological attachment. Moreover, sea water corrosion and wave impact damage, and many sensors scattered distribution would cause maintenance inconvenience. The paper proposed a highly integrated marine ecological on-line integrated monitoring system, which can be used inside monitoring station. All sensors were reasonably classified, the similar in series, the overall in parallel. The system composition and workflow were described. In addition, the paper proposed attention issues of the system design and corresponding solutions. Water quality multi-parameters and 5 nutrient salts as the verification index, in-situ and systematic data comparison experiment were carried out. The results showed that the data consistency of nutrient salt, PH and salinity was better. Temperature and dissolved oxygen data trend was consistent, but the data had deviation. Turbidity fluctuated greatly; the chlorophyll trend was similar with it. Aiming at the above phenomena, three points system optimization direction were proposed.

1. Preface

The impact of human offshore activities on the marine environment should not be overlooked, "2013 China Marine Environmental Quality Bulletin" pointed out: coastal seawater pollution was still serious, and it was China's marine ecological environment damage "hardest hit". Marine environmental pollution was increasing, marine ecological structure had been destroyed. Red tide and oil spills and other ecological disasters occurred frequently. Economic losses of aquaculture often occurred. In addition, sustainable development of marine economy was facing severe challenges. "National Science and Technology to revitalize the marine planning outline (2008 - 2015)" proposed that it was necessary to carry out stability test and popularization of the marine ecological environment monitoring equipment, and construct long-term ecological environment observation platform. China established ocean station in the coastal since the early 20th century. Then the country continued to invest special funds. Now we possessed more than 100 marine observatories (points) and more than 120 in-situ buoy over 6 meters. But the observation parameters were mainly based on the hydrology and meteorology, seawater ecological monitoring was almost blank. This caused the waste of platform resources. With the support of the National 863 Program, water quality sensors developed rapidly, and appeared a number of sensors which were based on optical method, spectrometer, wet chemical system, electrochemical system and so on [1]. This provided a strong hardware support for the marine ecological monitoring.

2. Current research

The development trend of marine environmental monitoring was from single parameter to multi-parameter integration. Finally, a regional Marine monitoring network was formed. The development of the regional monitoring system in abroad was earlier, such as: the America Maine Bay Ocean Observing System (GoMOOS), the Chesapeake Bay Monitoring System (CBOS), the Mexico Coast Marine Observing System (GCOOS), Norway EAWATCH EUROPE marine integrated observing system, the German MERMAID system and the small multi-parameter integrated monitoring system (Ferrybox, Integrated monitoring system construction in domestic behind foreign about 10 years. Since the 21st century, we had built "Shanghai marine three-dimensional monitoring system (demonstration area)", "Fujian sea real-time three-dimensional monitoring system" and "Pearl River estuary marine environment comprehensive monitoring system" [2, 3].

Water quality monitoring system was an important part of the regional marine monitoring network. A large number of researches had been done in domestic. The research work mainly focused on the multi-sensor data acquisition circuit and the host computer software development [4-7], establishment of evaluation model and system [8-11], integrated monitoring and management [12-14], and so on. Zhou Bing [15] proposed a water quality automatic monitoring laboratory integrated management system, his ideas were more open. Except the management of online water quality sensor, the laboratory management work was included to ensure the normal operation of the system. In recent years, the study of biological toxicity for seawater environmental assessment had rapid development [16], this development direction can't be ignored.

On the whole, in the domestic, marine water quality on-line integrated ecological monitoring system was mainly based on in-situ sensor, single parameter or multi-parameters combination. The drawbacks of this program were that the sensors were relatively discrete, not easy to maintain. In addition, the in-situ sensors which were long-term laid underwater, would be susceptible to biological attachment, corrosion, wave impact. The effect was particularly serious for optical sensors. The paper will propose a highly integrated marine ecological on-line integrated monitoring system to solve the above problems. The paper mainly introduced the system design ideas, as well as the system construction plan and the considerations. By comparing the data obtained by the in-situ sensor and the integrated system, the paper analyzed the feasibility of marine ecological on-line integrated monitoring system which was established inside monitoring station. The establishment of the integrated monitoring system would accelerate the application of independently developed water quality sensor, and form a rich marine ecological observation network system to obtain comprehensive monitoring data, and provide reliable basis for ecological disaster warning and marine environmental protection and management. This had far-reaching significance.

3. System construction

Marine ecological on-line comprehensive monitoring system could simultaneously monitor various parameters to meet the necessary of regional marine monitoring, and obtain continuous and complete data for assessment of regional water quality or disaster warning. It is a highly integrated system, the main design idea was: "put up" as much as possible in-situ sensor from underwater, then reasonably classified, and placed in cabinet within the monitoring station or buoy, in accordance with the similar in series, the overall in parallel way, weaken the maintenance difficulty. Ultimately, the filtered water was analyzed. At the same time, the system could set measure frequency according to the change law of seawater quality parameters, and achieve sampling, analysis, pipeline cleaning, data processing and transmission automation.

Firstly, hydrodynamic unit was necessary. It overcame the tidal effects, then provided seawater samples for the system. The seawater samples were pumped into integrated monitoring system. Secondly, the system formed detect branch through reasonable pipeline design. The water quality sensor was classified according to the principle of testing and the requirement of water samples, then laid in the detection flow path. The host computer obtained the water quality analysis data through the serial port. Finally, the data was displayed in real time on the system display, and synchronously processed and remotely transmitted to the monitoring center station or other supervisor units.

According to the above ideas, the system should contain at least the water distribution unit, various water quality sensors, data acquisition unit, control unit, data storage and transmission unit. About the water sample collection and distribution unit, the water can't be filtered for some sensors (such as turbidity sensors). Otherwise, it would cause data distortion, couldn't truly reflect the water quality of the tested sea area. However, for some sensors that use micro-channel detection (such as nutrient sensors), filtering is necessary. Otherwise, the instrument couldn't be used normally. Therefore, the allocation unit should contain at least two paths. The control unit used Advantech's small industrial computer and configuration display. The industrial computer performance was stable, with a wealth of serial resources. In addition, the use of configuration software greatly simplified the system control process, could effectively shorten the development cycle. The control unit could set the sensor running mode (continuous operation or regular operation) according to the scheduled program. When the component failed, the control unit could respond to fault diagnosis, and the fault alarm information was sent to the management center, in order to timely repair. According to the agreed protocol, the data acquisition unit obtained the data of the sensor through the serial port after receiving the completed instruction, then stored it in the storage unit. When all sensor data were acquired, the DTU module (model H7110), which was preset IP address and port number, was connected with the monitoring center station or other supervisor, then remotely transmit data using the TCP protocol.

Figure 1 showed the system workflow, with two detection branches, using parallel design. One of the branches was a loop, composed of a single parameter sensor (7, 8, 9) in series, without filtering. The other was a nutrient analyzer branch, with two-stage filter, the former using gauze, the latter using a stainless steel filter. The system had a buffer tank, PC controlled solenoid valve switching and pump start or stop during operation to achieve sensor measurement on the corresponding branch. In addition, in order to prevent the pipeline biological attachment, the system had chemical reagent cleaning and water rinse. Cleaning work could maintain the best filtration capacity, extend the service life.

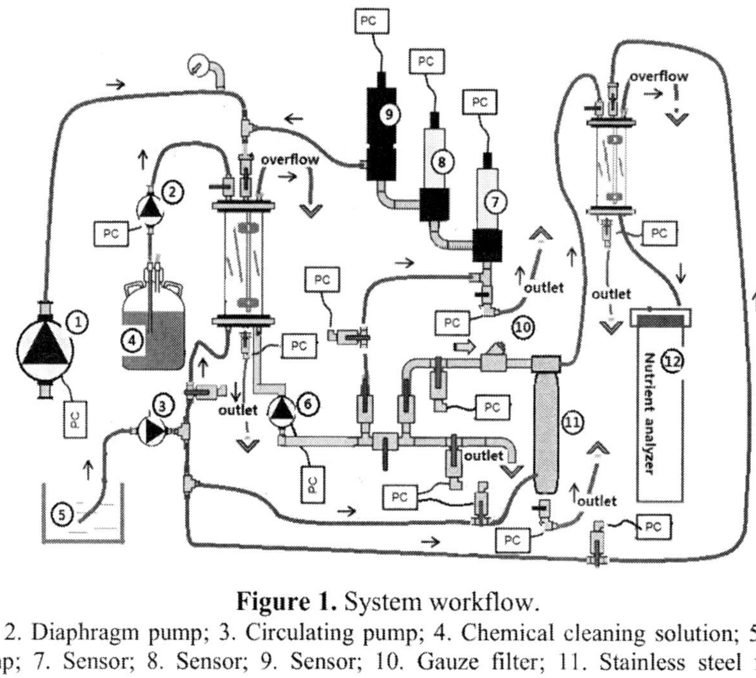

Figure 1. System workflow.

1. Water pump; 2. Diaphragm pump; 3. Circulating pump; 4. Chemical cleaning solution; 5. Water bucket; 6. Circulating pump; 7. Sensor; 8. Sensor; 9. Sensor; 10. Gauze filter; 11. Stainless steel filter; 12. Nutrient analyzer.

4. Technical solutions

To ensure the system normal operation, the collected data was true and effective, except the measurement of the sensor itself was stable and reliable, the system design also need to pay attention to the following questions: 1. Sampling unit must be able to fully reflect the water quality characteristics of the sampling area, a reasonable sampling point was very important. 2. Need to avoid the residual water samples causing cross-pollution, flooding the real data. 3. The sensor detection chamber on the flow path must be reasonable designed to avoid producing optical pollution, that may cause measurement distortion for optical sensor. 4. The filter system should be reasonable and reliable. The article will take Zhoushan monitoring station as an example, and present the solution of the above problems.

About problem 1, the Zhoushan monitoring station was located near the pier, often had boat in nearby, the normal tidal range within 5 meters. The sampling site should be fully considered, the water exchange was smooth, the water was even. There are two ideas for the design of the pump: submersible pump or vacuum pump. Vacuum pump could use dual-pump structure, alternating use or a failure to start another. This solution could increase the reliability of the water system, this is the advantage. But the suction of vacuum pump was limited, it couldn't be used in some large tidal stations. For Zhoushan monitoring station, the system maximum sampling height was 7-8 meters, and about 2 meters after into the room. Finally, large lift submersible pump program was used. It was made of 316 stainless steel, 15 meters lift. To overcome the tidal effects, and collect a fixed depth of water samples, a small special floating structure was designed. The submersible pump was fixed at the bottom. At the same time, sampling pipeline was reserved sufficient length to meet the needs of the tide, using U-type layout program. The floating body was easy to float with the sea wave, or suffer disturbance of the passing vessel. So it was difficult to maintain the same sampling point. Therefore, the system design drew on the design ideas of the temperature and salt wells, using PE / UPVC pipe to construct an artificial sampling well. The well could fix and protect floating components. Without affecting the strength of the sampling well, the sidewalls should be as much as possible to drill water exchange holes. This could ensure that water quality uniform inside and outside of the sampling well. Sampling well and floating body design shown in figures 2 and 3.

Figure 2. Sampling wells.

Figure 3. Floating body.

The solution to the problem 2 was: Firstly, optimized the flow path design, all pipelines outlet slightly lower than inlet, reduced unnecessary turning, as far as possible using vertical pipe layout, shorten the length of the horizontal pipeline. The bottom of the system designed water collection cavity. The water samples centralized discharged from it. In order to avoid the solenoid valve failure, the overflow pipeline design was necessary. Secondly, open all the solenoid valves before each measurement, using the fresh seawater to wash the last detection residual samples, and ensure that the data were valid in real time.

For the problem 3, the bottom of the detection chamber to the optical sensor head (such as water quality multi-parameter sensor) should reserve sufficient distance, at least 12 cm. In order to reduce air bubbles, water samples should inject from below and discharge from above.

For the problem 4, the filtration system used multistage filtration. Submersible pump set primary filtration, filtered out algae, weeds and so on. The secondary filter was set according to the requirements of the sensor. For example, the pre-filtration of the nutrient sensor used 0.45 micron filter diameter. Prior to the use of powder sintered stainless steel filter. The strength was high, the pressure at the dead end was strong, the higher recoil pressure ensured that washing effect was better.

5. Comparison experiment

After the system was built, we carried out mass experiment to study the feasibility of the system installed in Zhoushan, respectively using the ALEC sensor and the nutrient analyzer (independent research and development). The two kind instruments were laid in different flow paths, the former didn't need filter, the latter need filter twice. The test protocol was that each instrument prepared two sets, one set placed in the sampling well, the other set placed in the system. Figures 4-9 showed the data for the ALEC multi-parameter, obtained at 2016.5.20 evening and the morning of 5.21. Figures 10-14 showed the data obtained by the nutrient analyzer.

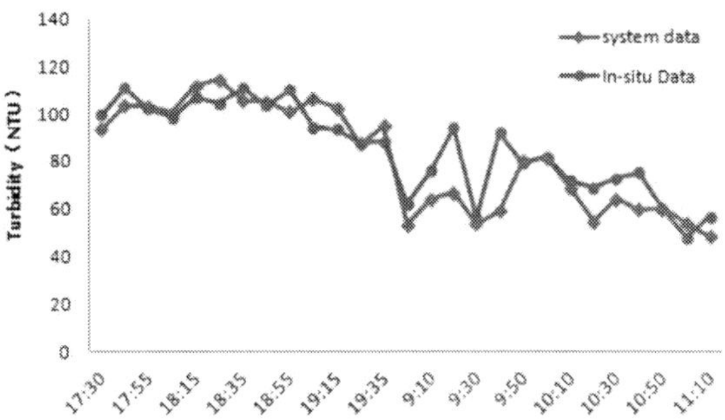

Figure 4. Turbidity.

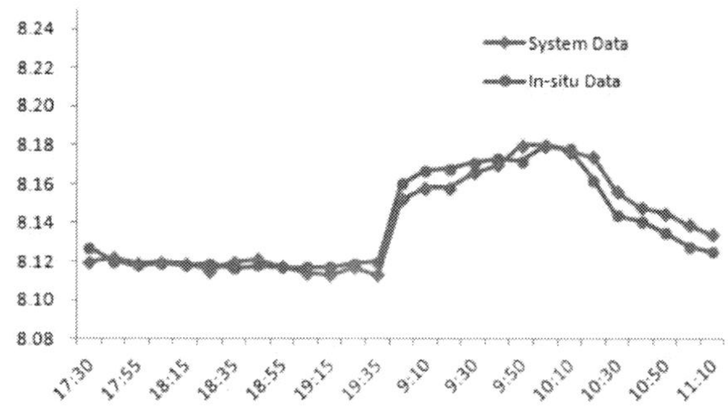

Figure 5. PH.

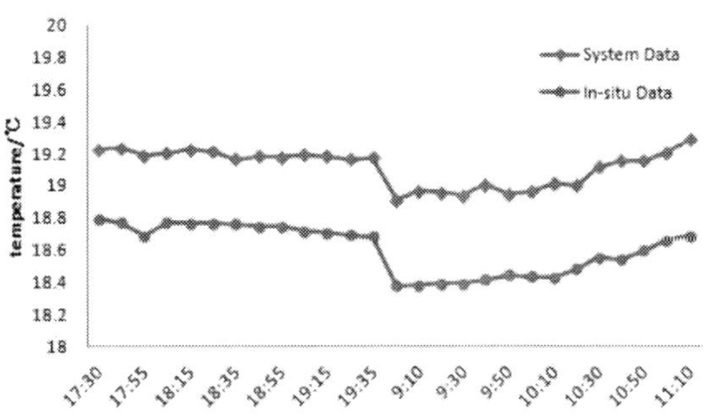

Figure 6. Temperature.

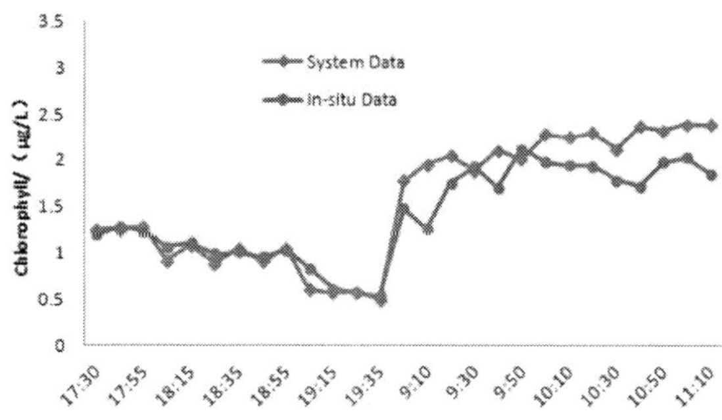

Figure 7. Chlorophyll.

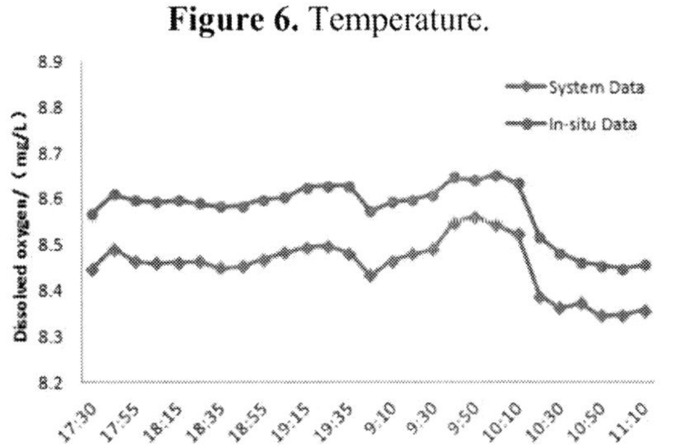

Figure 8. Dissolved oxygen.

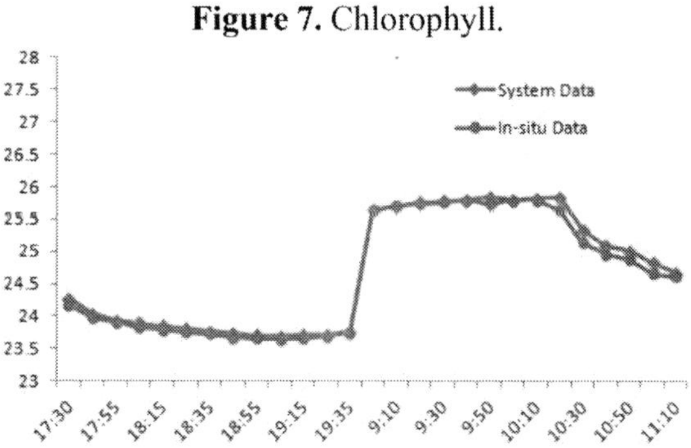

Figure 9. Salinity.

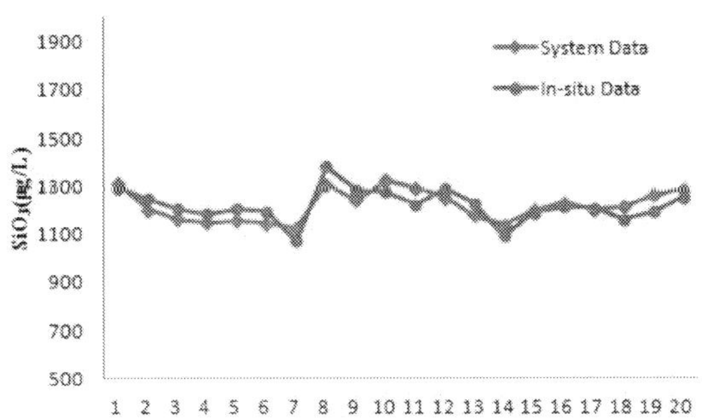

Figure 10. SiO_3.

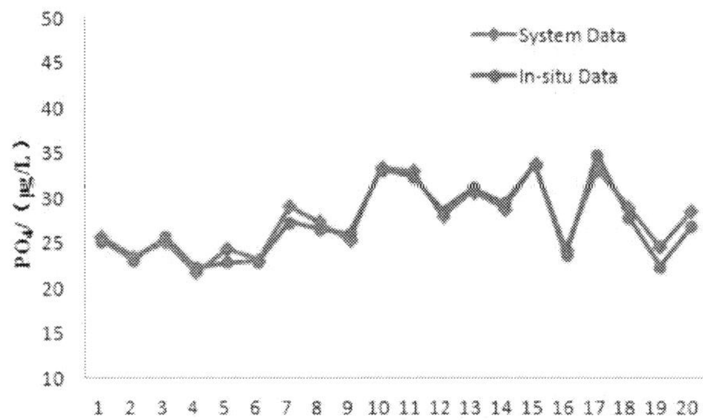

Figure 11. PO_4.

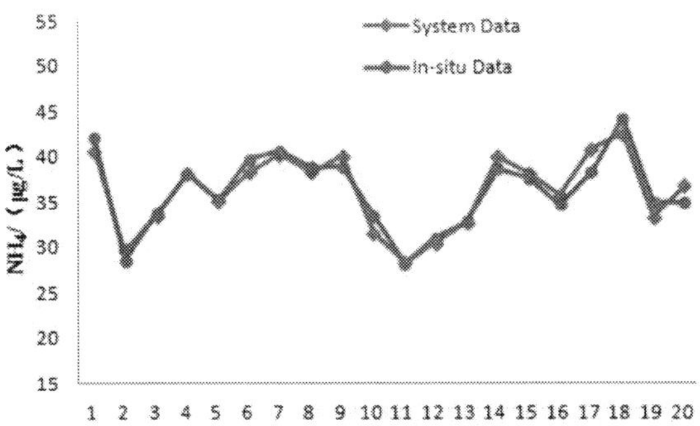

Figure 12. NH_4.

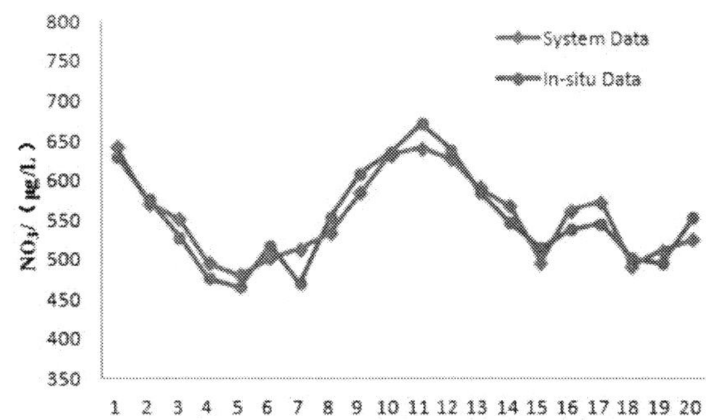

Figure 13. NO_3.

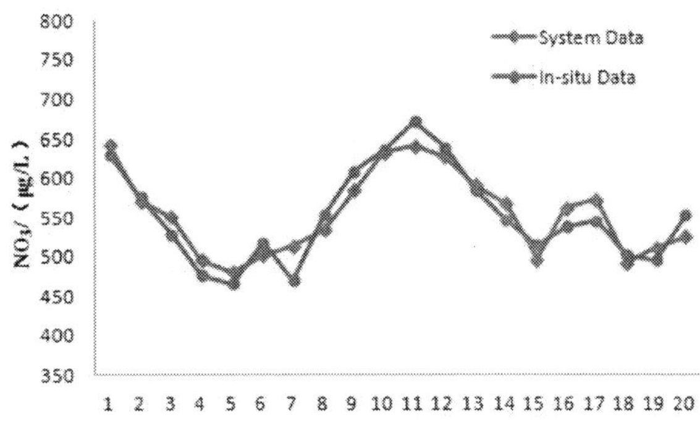

Figure 14. NO_2.

Through the data comparison, we found PH and salinity data consistency was very good. Although the temperature and dissolved oxygen data trend was consistent, but the system temperature data higher than situ data 0.4-0.5℃, the system dissolved oxygen data lower than situ data 0.1-0.15 mg/L, the turbidity and chlorophyll data on 5.20 was consistent, but more volatile on 5.21.

The temperature data phenomenon may have two reasons: 1. The sampling well was constructed by the black PE pipe, which was easy to absorb heat causing temperature rise. The flow hole of the sampling well may be poorly designed. 2. Under the action of the submersible pump, the seawater was subject to shock and friction. Perhaps was the result of combination of two factors. The reason for dissolved oxygen data may be the occurrence of escape during the sample and measurement process. Turbidity and chlorophyll data on 5.20 were obtained in the evening, no fishing boat passes, the water was relatively stable, while the data on 5.21 were affected by fishing boats and passenger ships. A mass of muddy water was around the monitoring station, the water quality was uneven. Obviously, the chlorophyll data had similar fluctuations with turbidity.

SiO_3 salt instrument performance indicators was ± 5% error, while figure 10 showed 15% of data exceed the error, the maximum deviation was 6.03%. PO_4 salt instrument performance indicators is ± 3% error, while figure 11 showed 20% of data exceed the error, the maximum deviation was 7.82%. NH_4 salt instrument performance indicators was ± 5% error, while figure 12 showed 15% of data exceed the error, the maximum deviation was 6.53%. NO_3 salt instrument performance indicators was ± 5% error, while figure 13 showed 10% of data exceed the error, the maximum deviation was 9.04%. NO_2 salt instrument performance indicators was ± 3% error, while figure 14 showed 10% of data exceed the error, the maximum deviation was 3.64%.

6. Conclusion

Feasibility analysis showed that the same kind of nutrient data was consistent, limited by the number of samples and the stability of the nutrient analyzer itself, need long-term verification. The consistency of pH and salinity data in the multi-parameter was good, the temperature data obtained by the system was 0.4-0.5 ℃ higher than in-situ. The dissolved oxygen data obtained by the system was 0.1-0.15 mg/L lower than in-situ. Turbidity and chlorophyll data was not ideal, turbidity volatility was large, causing chlorophyll fluctuations. In general, the marine ecological on-line integrated monitoring system was feasible. The system need to be optimized, optimization direction: 1.Reasonably set the submersible pump and circulating pump flow rate, reduce air bubbles. 2. Improv the sampling well program, open more exchange holes without affecting the strength of well. In addition, may consider removing the sampling wells, using open sampling program. 3. Adjust the measurement process, increase the emptying time to ensure that there is no residual water sample or recoil tap water.

Acknowledgments

The authors acknowledge funding from National Key R & D Plan (No. 2016YFC1400801, 2016YFC1400802, 2016YFC1400803, 2016YFC1400804), Natural Science Foundation of Shandong Province (No. ZR2016DM17, ZR2014YL006), and Special Funds for Scientific Research of Marine Public Welfare Industry (No. 201505007-2).

References

[1] Cao X, Liu Y and Cao L 2015 *J. Oce. Technol.* **34** 48-53
[2] Du L B, Wang J C and Sun J C 2009 *Shandong Sci.* **22** 1-6
[3] Hui S T 2000 *J. Oce. Technol.* **19** 1-17
[4] Wang F, Huang L, Wu S, Li K, Li J, Gao Y and Cao Y 2015 *Chin. Soc. Agri. Eng.* **31** 148-53
[5] Hasegawa M, Higashijima A, Nakamura K, Hanada K, Sato K N, Sakamoto M, Idei H, Kawasaki S and Nakashima H 2008 *Fusion Engineering & Design* **83** 588-93
[6] Ding Y, Han H and Liu F 2010 *Knowl-based Syst.* **23** 61-9
[7] Hua H, Ding Y S and Liu F M 2009 *J. Donghua Univ.(English Edition)* **26** 499-502
[8] Yu G, Chen J, Zhang X and Li Z 2013 *China Environ. Monitor. Assess.* **185** 6793-807
[9] Nan H L, Jing H W, Ding N and Zhang Y 2006 *For. Invent. P* **31** 35-9
[10] Foley M M, Armsby M H, Prahler E E, Caldwell M R, Erickson A L, Kittinger J N, Crowder L B and Levin P S 2013 *Bioscience* **63** 619-31
[11] Mahan C G, Young J A, Miller B J and Saunders M C 2015 *Environ. Manage.* **55** 508
[12] Fletcher P J, Kelble C R, Nuttle W K and Kiker G A 2014 *Ecol. Indic.* **44** 11-25
[13] Ballance L T and Whitty T 2010 *Restor. Ecol.* **18** 780-1
[14] Piet G J, Jongbloed R H, Knights A M, Tamis J E, Paijmans A J, Sluis M T V D, Vries P D and Robinson L A 2015 *Biol. Conserv.* **186** 158-66
[15] Stelzenmüller V, Breen P, Stamford T, Thomsen F, Badalamenti F, Borja Á, Buhl-Mortensen L, Carlstöm J, D Anna G and Dankers N 2013 *Mar. Policy* **37** 149-64
[16] Ansah Y B, Frimpong E A and Amisah S 2012 *Environ. Manage.* **50** 166-80

Research on signal processing method for total organic carbon of water quality online monitor

R Ma[1,2], Z X Xie[1,3], D Z Chu[2], S W Zhang[2], X Cao[2,3] and N Wu[2]

[1]Engineering College, Ocean University of China, No. 238 Songling Road, Laoshan District, Qingdao City, Shandong Province, 266100, China
[2]Institute of Oceanographic Instrument, Shandong Academy of Science, No. 37 Miaoling Road, Laoshan District, Qingdao City, Shandong Province, 266061, China

E-mail: mr47@sohu.com

Abstract. At present, there is no rapid, stable and effective approach of total organic carbon (TOC) measurement in the Marine environmental online monitoring field. Therefore, this paper proposes an online TOC monitor of chemiluminescence signal processing method. The weak optical signal detected by photomultiplier tube can be enhanced and converted by a series of signal processing module: phase-locked amplifier module, fourth-order band pass filter module and AD conversion module. After a long time of comparison test & measurement, compared with the traditional method, on the premise of sufficient accuracy, this chemiluminescence signal processing method can offer greatly improved measuring speed and high practicability for online monitoring.

1. Introduction

With the development of modern industry, a large number of poisonous and harmful materials are directly discharged into rivers and sea, the water environment and ecological system caused great impact [1]. At present, water quality of more than half of the water function area is not up to standard in China. In the place of moderate and severe eutrophication, some river regions are called 'all rivers have been polluted' [2]. Therefore, water pollution and ecological security problems are increasingly outstanding, sudden environmental pollution accidents occur frequently [3].

Online monitoring of water quality is always lack of long-term effective environmental monitoring method in China [4]. The Solubility of the pollutants, particularly the persistent organic pollutants, is very low in water, but these pollutants are carcinogenic, that can cause deformity. After a series of low-dose & long-term intake, it will become a potential threat for human health. Therefore the pollution of water quality monitoring is imminent [5, 6].

Total organic carbon (TOC) is a kind of quickly comprehensive indicators for water quality, which could be defined as the containing carbon quantity of the total amount organic matter in the water [7]. TOC is able to fully reflect kinds of pollutants in water. The national standard method of TOC analysis is combustion oxidation - nondispersive infrared absorption method. The advantage is a single, simple process, good reproducibility, high sensitivity. But the disadvantage is that: the probe need frequent calibration, the instrument need large volume and long preheating time, the reaction process must use acid, catalyst and carrier gas. Other common methods include: UV oxidation method, the conductance measurement method, etc. At present, TOC analysis has been widely used in surface water monitoring, such as drinking water, industrial water, pharmaceutical water and so on, which has become the key means of the water quality control.

Nowadays the TOC online analyzers made by European or American manufacturers are already very advanced. The commercial TOC online analyzers including: a) Sievers-500 series, made by GE Analytical Instruments Co (United States), adopted the membrane conductivity measurement method, high accuracy, no reagent, but the measuring range is limited, which is mainly used in pharmacy, electric power and pure water quality occasions. b) Multi N/C ®UV HS, made by Analytik Jena AG(Germany), adopted the UV oxidation - non-dispersive infrared test, need regular replacement reagents, mainly used in biological, pharmaceutical, research and other industries. But there are still no rapid, stable and effective domestic methods of TOC online automatic measurement in the aspect of water pollution monitoring at present. So that is an urgent need to a series of independent research and development of TOC online monitoring instrument.

The emerging chemiluminescence fast measurement method can realize the TOC online, real-time and continuous measurement [8]. This chemiluminescence is produced by the oxidation reaction of ozone and organic matter in the sea water, and the luminous signal could be collected and recorded as digital signal by photoelectric conversion technology in the process of reaction. The advantage of this method: no reagent, no secondary pollution, fast response speed. So that, on the one hand, this method can avoid the big particles impact on the accuracy of the water, on the other hand, it can long-term work at the scene of the harsh environment. The method could be suitable for ship, monitoring stations and other occasions to use, which can realize real-time and continuous measurement on the scene [9]. But the disadvantage of this method is that it can't achieve the laboratory instrument measurement accuracy [10]. Therefore, the emphasis and difficulty in existing technology needs to solve is how to improve the measurement accuracy of the signal.

For this purpose, this paper proposes and implements a Photomultiplier tube (PMT) test at the scene of the chemiluminescence reaction of TOC analysis method of high speed, no pollution, and to ensure the enough accuracy of measurement. This paper mainly introduces the principle of PMT and its application in the analyzer, analyzes in detail the design of the signal processing circuit and the design of filter algorithm, finally provides the scene environment test result. Through a large amount of experimental data show that the proposed method compared with other traditional methods is convenient maintenance and high reliability, rapid measurement, no pollution and other characteristics, very suitable for online measurement [11].

2. The research methods

The key point of the signal processing method research of the proposed TOC online monitoring technology is the weak chemiluminescence detection method, which could avoid the complex process of laboratory analysis, and can truly reflect the complex and changeable situation at the scene [12]. In order to achieve the purpose of the TOC measurement technology research, based on the maturely existing technology: technology of ozone oxidation kinetics of chemiluminescence curve, we adopt a novel and mature thread of technology for environmental pollution monitoring technology development. The research emphasis is as follows:

2.1. In order to ensure the validity of the data, the scheme adopt multistage amplifier technology & bandpass filtering technology

Because the signal of chemiluminescence reaction is very weak, the PMT voltage signal is very weak (only 1 3 mV). It needs to be amplified and converted to digital signal output through phase-locked amplifier, programmable gain amplifier, bandpass filter and AD converter. Through chemiluminescence signal superposition response of the different organic matter in water samples, we can get the total luminous intensity curve of the dynamics with the change of time. This method uses the chemiluminescence time-resolved technology to capture the signal during the reaction process, and achieve high efficiency light intensity time series integral. At last we can get total chemiluminescence intensity.

2.2. In order to ensure the reliability of data, the scheme improved the compound digital filter algorithm
A kind of disturbance signal of large particle of microorganisms in the water often influences the reliability of the measurement. Therefore, technical scheme improves the compound digital filter algorithm and data correction techniques, after the original analog signal be amplified & filtered. And a series of piecewise analysis and calculation for this detection curve is designed, which can effectively remove such disturbance signal.

3. The technical scheme of chemiluminescence for TOC online measurement
This paper puts forward a kind of special used for online monitoring of TOC content, it can solve the existing methods technology of complex operation, not on-line monitoring, and poor reliability problems. In order to solve these problems, we present designs of the technical scheme as follows:

- Stage of sampling control: Firstly, the process will set the three-way valve in fully open station and open peristaltic pump A (water pump), so that the water samples can flow into the circulation pool through a filter membrane of 50 μm diameter(filter out most of the grains of sand and other inorganic pollutants in water samples).
- Stage of reaction: Secondly, the process will open the peristaltic pump B (Solution pump) to inject ozone solution into the circulation pool by certain velocity and blend with the water samples, so that an oxidizing luminous reaction can be happened with the TOC & the ozone solution. Weak chemiluminescence intensity signal produced by this oxidizing luminous reaction could be transmitted to the photoelectric detection module through an optical fiber.
- Stage of analysis: At the same time, the three -way valve should be closed, ensuring that the new water samples will not enter the circulation pool, so that ozone solution will deplete TOC of the water samples in the circulation pool. The light intensity of curvilinear integral calculation, through the correlation algorithm can be concluded that the content of TOC in sea water samples;
- Stage of back flush: At last, the solenoid valve B should be closed, and the three-way valve should be switched to the normally open state. The ozone solution can back flush the filtration system on time, so that the flow system could keep free from contamination & attachment.

The technical scheme of TOC online measurement is shown in figure 1:

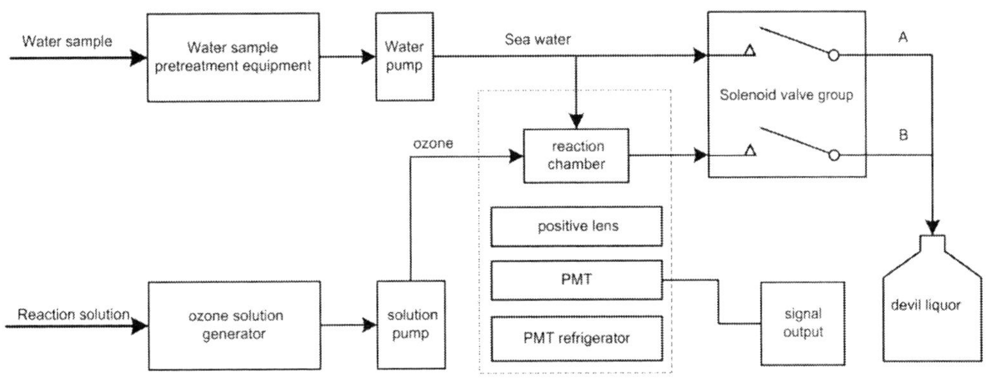

Figure 1. The technical scheme of TOC online measurement.

4. Design of signal processing method for TOC online measurement

4.1. Rectifier filter circuit design scheme
Because the signal of chemiluminescence reaction is very weak, so the voltage output from photomultiplier tube is very low [13]. Therefore, the voltage is needed to become a digital signal output through phase-locked amplifier, bandpass filter and analog-digital conversion, and many details

needed to pay attention to in the design. The circuit diagram of the signal acquisition processing part is shown in figure 2:

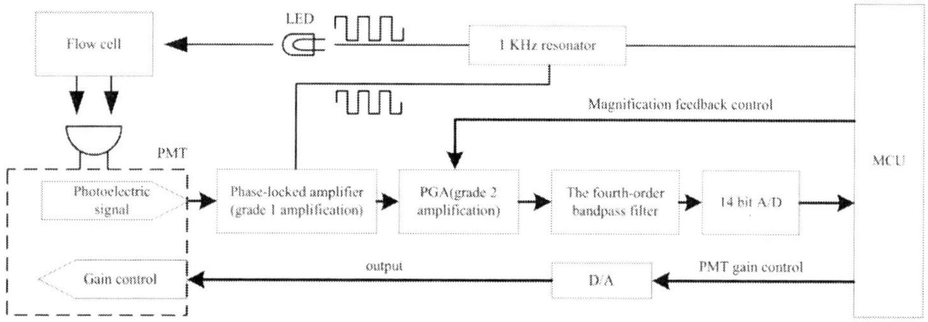

Figure 2. The circuit diagram of the signal acquisition processing part.

The signal processing circuit of the core research includes the following three modules:

4.1.1. The precision rectifier module. This design chooses AD630 phase-locked amplifier output signal of photomultiplier tubes for the first level of front phase-locked amplifier. The AD630 is a high precision balanced modulator/demodulator that combines a flexible commutating architecture with the accuracy and temperature stability afforded by laser wafer trimmed thin film resistors. A network of on-board applications resistors provides precision closed-loop gains of ±1 and ±2 with 0.05% accuracy. These resistors may also be used to accurately configure multiplexer gains of 1, 2, 3, or 4. External feedback enables high gain or complex switched feedback topologies. AD630 within the two op-amp constitutes the direction and reverse amplifier gain of 2, and using the analog switch to switch. When the input signal is on time, analog switched to synthetic amplifier side, but when the input signal is negative, the analog signal is switched to reverse amplifier side. So that the output impedance of signal source has certain requirements, if the input signal of the output impedance is higher, need to increase the level of the buffer level [14].

4.1.2. The signal filtering module. After two-stage amplification of signal, only amplify the signal and noise at the same time, did not improve the signal-to-noise ratio of the input signal. Therefore a bandpass filter is required before the signal getting into A/D. If we set a suitable center frequency of bandpass filter to the signal frequency, a large amount of high frequency and low frequency noise and interference would be filtered. Therefore, AD795 operational amplifier is selected as a bandpass filter n this design. The AD795 is a low noise, precision, FET input operational amplifier. It offers both the low voltage noise and low offset drift of a bipolar input op amp and the very low bias current of a FET input device. The 1014 Ω common-mode impedance insures that input bias current is essentially independent of common mode voltage and supply voltage variations. It has both excellent dc performance and a guaranteed and tested maximum input voltage noise. It features 2 pA maximum input bias current and 500 μV maximum offset voltage, along with low supply current of 1.5 mA maximum. Furthermore, the AD795 features a guaranteed low input noise of 3.3 μV p-p (0.1 Hz to 10 Hz) and a 11 $nV/\sqrt{Hz}$. maximum noise level at 10 kHz. AD795 application can be used in many high input impedance and low noise.

4.1.3. The signal conversion module. In order to improve the resolution of the measurement, we need a 14-bit high precision analog-digital conversion (ADC) chip. The AD7940 is a 14-bit, fast, low power, successive approximation ADC. This chip operates from a single 2.5 V to 5.5 V power supply and features throughput rates up to 100 kSPS. It contains a low noise, wide bandwidth track-and-hold amplifier that can handle input frequencies in excess of 7 MHz, and it uses advanced design techniques to achieve very low power dissipation at fast throughput rates. The reference for the part is taken internally from VDD, which allows the widest dynamic input range to the ADC. Thus, the analog input range for this part is 0 V to VDD. The conversion rate is determined by the serial clock (SCLK) frequency. The conversion rate is determined by the SCLK frequency. The conversion process and data acquisition process are both controlled by chip select (CS) signal and SCLK signal, interfaced with microprocessors.

Through the above, a series of signal processing, the output of the photomultiplier weak voltage signal into digital signal input MCU, the signal needs in a program for later analysis.

4.2. Research of digital filtering processing algorithms

As is known to all, all of the filtering algorithm can't both sensitivity and perfectly smooth degree. Therefore, we have to find a balance between an acceptable sensitivity and possible smooth degree. When the data changed fast, filtering results can follow up sensitivity in time(prefer sensitivity), but when the data tends to be stable, oscillated at a fixed point up and down, the filtering results can be smoothly (prefer smooth degree).

A first-order filter algorithm formula is shown as follow:

$$Y(n)=\frac{A\times X(n)}{256}+\frac{(256-A)}{256}\times Y(n-1)$$

Description: A - filter coefficients (integer data); $X(n)$ - the sampling value; $Y(n-1)$ - the last filter output value; $Y(n)$ - the filter output values.

This algorithm could get an effective filter value through weighting the sampling value and the last filtering output value, and then the output signal could supply a feedback to the input signal. The formula needs to use loop addition & subtraction to implement multiplication & division for the microprogrammed control unit (MCU) without multiplication & division instructions. But it will increase the computational complexity of system and affect the calculation efficiency by too many times of multiplication & division.

In order to reduce the computational complexity of MCU, the formula must be simplified. Based on this principle, a kind of automatic identification algorithm is put forward, the calculation efficiency could be doubled. The improved algorithm is as follows:

If X(n)< Y(n-1),

$$Y(n)=Y(n-1)-\frac{\left(Y(n-1)-X(n)\right)\times A}{256};$$

Else if X(n)>Y(n-1),

$$Y(n)=Y(n-1)+\frac{\left(X(n)-Y(n-1)\right)\times A}{256};$$

Description: A - filter coefficient (value range: 0 ~ 255). The value of the coefficient is the determinant of the weights about $X(n)$ in the filtering results, and could realize automatic calculation in the process by the following dynamic adjustment algorithm shown in figure 3.

The principle of this dynamic adjustment algorithm is as follows:

- When the data is changing toward inconsistent direction, it means that the data is not stable, it is necessary to reset the filter coefficient and ignore the new sampling value;
- When the data is changing toward one direction, it is necessary to gradually improve the filter coefficient and improve the weight coefficient of the new samples values;
- When the data is changing rapidly, it is necessary to accelerate improve the filter coefficients.

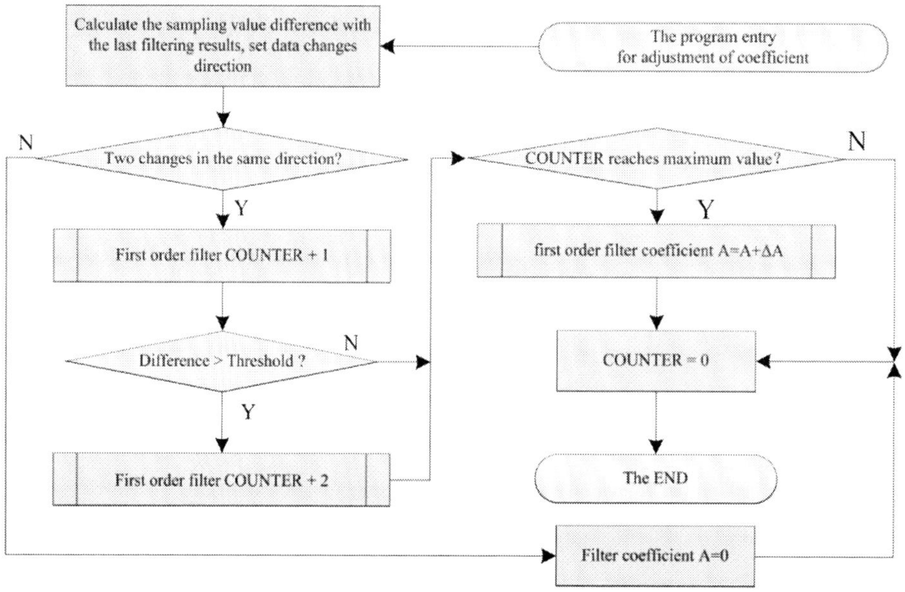

Figure 3. The algorithm of the filter coefficients dynamic adjustment process.

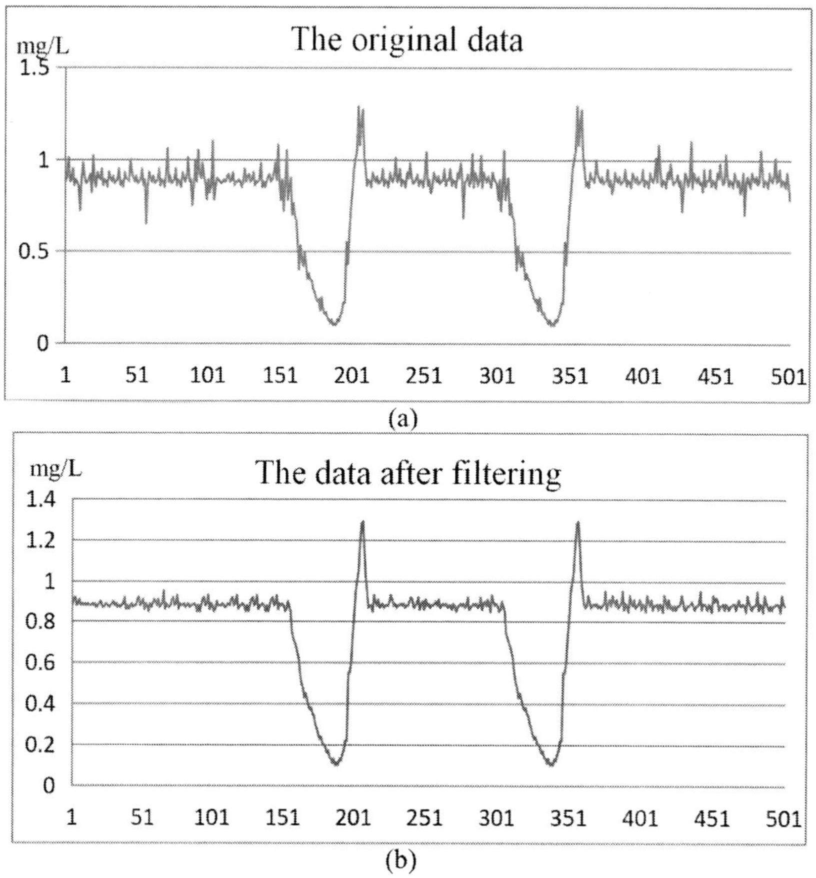

Figure 4. The signal filter processing result contrast.

4.3. Simulation of the filter signal
Most of the signal noise could be removed by the noise reduction filter processing. To test and verify the feasibility of algorithm, the same sample water was chosen for testing experiment, the measured signal curve as shown in figure 4.

Curve A is original measurement signal without filtering algorithm, curve B is measurement signal with filtering algorithm, Obviously, compared with the curve A, the smoothness of curve B has been greatly promoted, and could provide the reliable data for the next step oxidation aeration by integral calculation of photoelectric signal.

5. Sea trials experimental data analysis and about the effects of salinity

5.1. Sea trials experimental data analysis
In order to test and verify the data parallelism and the feasibility of the proposed signal processing method, we installed the instrument on the experimental vessel and joined multiple voyages. Comprehensive comparison of measurement experiments have been carried out in the appointed position coordinate. The comparison measuring method is as follows: In each voyage and gathering the same samples of laboratory methods of TOC data measured at the same time.

The real-time data of TOC online analysis is collected and recorded in each coordinate of the voyage. At the same time, seawater sample should be collected and sealed up at the position. It would be measured by laboratory equipment in accordance with the standard of GB 2007-17378.4 (The specification for marine monitoring-Part 4: Seawater analysis). At last, the data achieved from the lab would be compared with the data of TOC online analysis instrument at the same time and the same position.

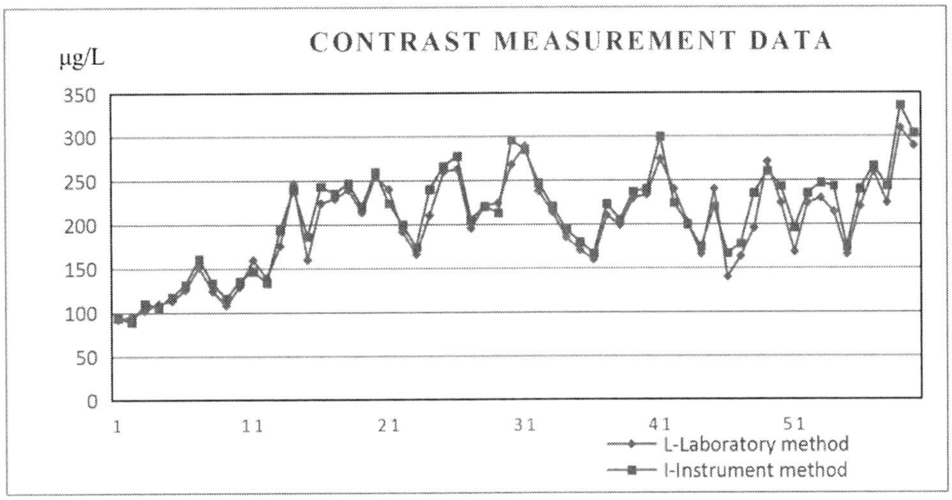

Figure 5. TOC comparison measurement data.

After the comparison measurement, the trend chart is as shown in figure 5: the diagram ordinate unit is 'μg/L', abscissa unit is 'Times', the I-TOC line is the instrument method real-time data, L-TOC line is laboratory method measuring data. A conclusion can be clearly drawn by observing the graph: the two kinds of data have a great match with each other, and have a same trend. Data parallelism is higher than 85% through a discrete degree calculation, which can meet the requirement of online analysis instrument index in China.

5.2. Monitoring results analysis of the effects of salinity
To test the effect of salinity of different sea area on the TOC concentration, we went ahead with two

voyages: the Jiaozhou Bay voyage in February 2017, and the Luzon Strait voyage in August 2016. At last, we completed a gradient analysis of seawater samples with different salinity. Several measurement results are as follows: In February 2017, the measured value is (2.4 ± 0.22) mg/L in the sea area of 15.3 salinity, and the measured value is (1.8 ± 0.12) mg/L in the sea area of 22.2 salinity. In August 2016, the measured value is (0.72 ± 0. 15) mg/L in the sea area of 33.9 salinity. The experimental data of suffer diagram is as shown in figure 6. The experiment results show that: the measurement method of the proposed TOC online analyzer can be applied whether in high salinity areas or low salinity area.

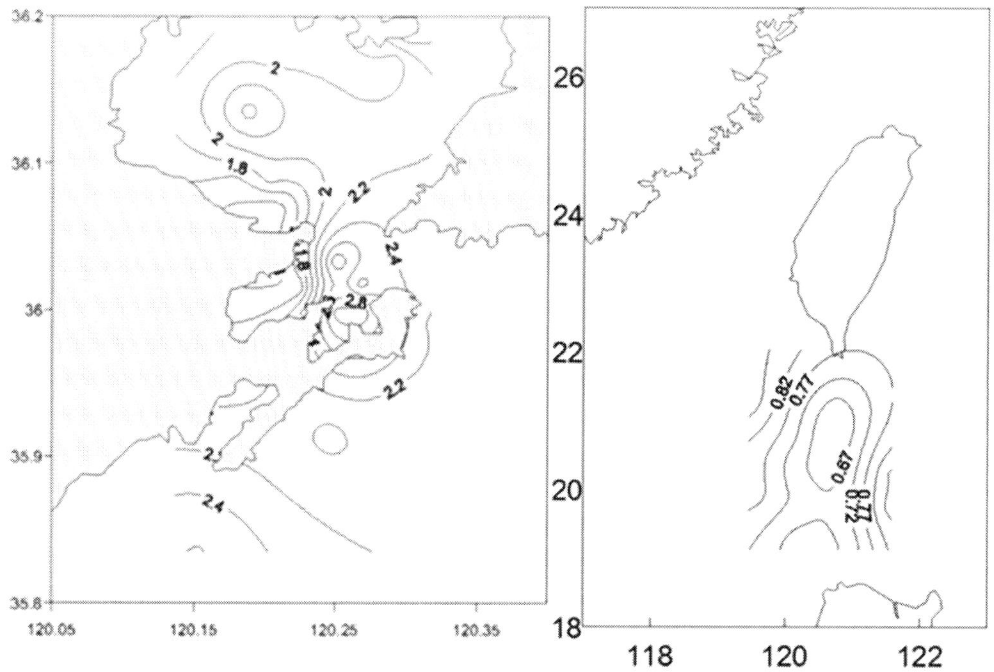

Jiaozhou Bay voyage in February 2017 **Luzon Strait voyage in August 2016**

Figure 6. The influence of salinity on the gradient of TOC measuring data.

6. Conclusions

This paper proposes a kind of photoelectric signal processing method for total organic carbon (TOC) online analysis: using chemiluminescence fast measurement method combined with dynamic curve model, through the photomultiplier tubes testing weak light signal, and a series of hardware, software, signal processing, finally realizes the TOC concentration measurement. Through long-term test, the method of compare the measurement data and laboratory data, the data parallel performance enough to achieve the online analytical instrument standard.Compared with the existing various TOC analysis method, this method is more suitable for ship, buoys, ocean stations, used in situations such as offshore oil platform, greatly improve the field monitoring ability of Marine ecological environment, has good popularization and application value, in the ocean of eutrophic ecological disaster monitoring has very important significance, and also provide technical support to protect the health of the Marine ecological environment.

Acknowledgments

The authors acknowledge funding from National Key R & D Plan (No. 2016YFC1400801, 2016YFC1400802, 2016YFC1400803, 2016YFC1400804), Natural Science Foundation of Shandong Province (No. ZR2016DM17, ZR2014YL006), and Special Funds for Scientific Research of Marine

Public Welfare Industry (No. 201505007-2).

References

[1] Algeo T J and Lyons T W 2006 *Paleoceanography* **21** 1016-23
[2] Boxaspen K 2006 *ICES J. Mar. Sci.* **63** 1304-16
[3] Clifton L, Clifton D A, Pimentel M A F, Watkinson P J and Tarassenko L 2014 *IEEE J. Biomed. Health* **18** 722-30
[4] Cooley S R 2012 *Curr. Opin. Env. Sust.* **4** 258-63
[5] Dezwart D, Kramer K and Jenner H A 1995 *Environ. Toxicol. Water Qual.* **10** 237-47
[6] Foy R H, Lennox S D and Smith R V 2001 *Water Res.* **35** 3004-12
[7] Fu W, Wu Y, Sun L and Zhang W 2007 *Biotechnol. Bioeng.* **97** 1387-97
[8] Kennedy P, Kennedy H and Papadimitriou S 2005 *Rapid Commun. Mass. Sp.* **19** 1063-8
[9] McKinna L I W 2015 *Prog. Oceanogr.* **131** 177-99
[10] Panetta R J, Ibrahim M and Gelinas Y 2008 *Anal. Chem.* **80** 5232-9
[11] Reeburgh W S 2007 *Chem. Rev.* **107** 486-513
[12] Shoham-Frider E, Azran S and Kress N 2012 *Arch. Environ. Con. Tox.* **63** 495-502
[13] Silva N and Ines Astorga M 2010 *Lat. Am. J. Aquat. Res.* **38** 1-13
[14] Tyson R V 2004 *J. Geol. Soc. London.* **161** 667-73

IOP Conf. Series: Earth and Environmental Science 82 (2017) 012087 doi:10.1088/1755-1315/82/1/012087

Software for marine ecological environment comprehensive monitoring system based on MCGS

X H Wang[1,2,3], R Ma[1,2], X Cao[1,2], L Cao[1,2], D Z Chu[1,2], L Zhang[1,2] and T P Zhang[1,2]

[1]Shandong Provincial Key Laboratory of Ocean Environment Monitoring Technology, No. 37 Miaoling Road, Laoshan District, Qingdao City, Shandong Province, 266061, China

[2]Institute of Oceanographic Instrumentation, Shandong Academy of Sciences, No. 37 Miaoling Road, Laoshan District, Qingdao City, Shandong Province, 266061, China

E-mail: wxh16@163.com

Abstract. The automatic integrated monitoring software for marine ecological environment based on MCGS configuration software is designed and developed to realize real-time automatic monitoring of many marine ecological parameters. The DTU data transmission terminal performs network communication and transmits the data to the user data center in a timely manner. The software adopts the modular design and has the advantages of stable and flexible data structure, strong portability and scalability, clear interface, simple user operation and convenient maintenance. Continuous site comparison test of 6 months showed that, the relative error of the parameters monitored by the system such as temperature, salinity, turbidity, pH, dissolved oxygen was controlled within 5% with the standard method and the relative error of the nutrient parameters was within 15%. Meanwhile, the system had few maintenance times, low failure rate, stable and efficient continuous monitoring capabilities. The field application shows that the software is stable and the data communication is reliable, and it has a good application prospect in the field of marine ecological environment comprehensive monitoring.

1. Introduction

It is the prerequisite and basis for the research and protection of marine ecological environment to obtain continuous and long-term comprehensive observation data and to objectively analyze the present situation and dynamic trend of marine ecological environment. At present, more than 100 marine environmental monitoring stations built in China are mostly for hydrometeorological observation, and there is no comprehensive observation of marine ecological environment. Moreover, the integrated monitoring equipments used in monitoring stations are heavily dependent on import and they can't be fully applied to China's complex marine environmental monitoring Condition [1-3]. It is of great significance to the marine environmental protection and marine ecological science research in China to make full use of the existing marine station infrastructure, observation equipment and security conditions, to carry out the long-term comprehensive observation system integration technology research of the marine station ecological environment, and to develop stable, reliable, automatic, easy to operate and highly integrated long-term comprehensive observation system, Which can make up for the shortcomings of the existing marine environmental observation system in the long-term continuous online observation.

To this end, the project team developed a set of integrated monitoring system for China's marine environment. This paper introduces the design, development and real-time application of MCGS-based PC software applied to the system. The software enables real-time online data monitoring, display, storage and network transmission of a wide range of marine ecological parameters. It is simple to operate and flexible in configuration and it can ensure ocean monitoring information collection, processing and control operation more systematically and efficiently.

2. System overall design

2.1. System composition

The marine ecological environment monitoring system designed and developed by this project is mainly composed of seawater acquisition and distribution subsystem, monitoring sensors subsystem, data acquisition hardware subsystem and host computer software subsystem. The system integrates common parameters of seawater quality, such as temperature, salinity, turbidity, dissolved oxygen, chlorophyll, pH, total organic carbon, silicate, ammonia nitrogen, nitrite, nitrate, phosphate and so on, and it has the characteristics of high reliability, simple operation and low maintenance frequency. The overall structure of the system is shown in figure 1.

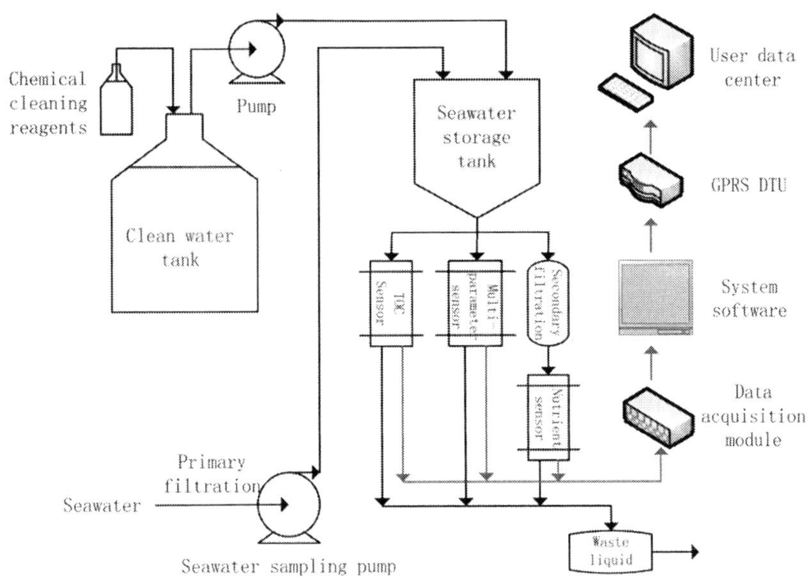

Figure 1. The overall structure of the system.

When a certain analysis process begins, the seawater sampling pump collects surface seawater into the storage tank of the detection system. After seawater acquisition is completed, the software assigns the seawater in the storage tank to every sensor inspection cabin according to each sensor's demand for specific parameter detection. After the end of an analysis process, the water samples in the module are emptied and the test data is uploaded to the system software via the data acquisition module. Then, the software will process, display and save the data. Depending on different situations, users can set different data transmission frequencies. The data can be transmitted to the user data center immediately after the end of each detection process, or it can be packed and transmitted after the whole day detection completed to the user data center, via GPRS Data Transfer Unit (DTU).

In order to prevent the impact of biological attachment on data detection, users can regularly clean the system according to seawater quality. When the system chemical cleaning process begins, system will fill storage tank and sensor detection cabins with cleaning fluid which is mixture of chemical reagents and clean water. Then, empty the cleaning fluid after a few minutes and repeat above process several times to wash off the biological attachment. At last, use seawater to flush the system storage tank and sensor detection cabins several times.

2.2. Seawater acquisition and distribution subsystem

The seawater acquisition and distribution subsystem is mainly composed of water collection module, water distribution module and filter module. The water collection and distribution module are responsible for collecting the surface seawater and distributing the water samples to the monitoring sensor subsystem. The filter module mainly includes a primary filter on the pump and a secondary nutrient filter with a diameter of 1 um.

2.3. Monitoring sensors subsystem

The monitoring sensors subsystem is mainly composed of ecological sensors, used to monitor the marine ecological parameters. At present, the system mainly integrates the self-developed nutrient sensor, multi-parameter sensor and TOC sensor, and reserves interfaces for other sensors. The sensors communicate with the host computer software according to the system's communication protocol, receive software instructions to monitor the parameters, and upload monitoring data to the software, to achieve the real-time monitoring of marine ecological environment parameters.

2.4. Data acquisition hardware subsystem

Data acquisition hardware subsystem uses modular architecture and bus design and its core control system uses high-performance embedded integrated industrial computer TPC1561Hi. The subsystem is mainly responsible for power supply of the system and sensors and state control of a variety of pumps, valves and other devices. The subsystem achieves interaction with monitoring sensors through the serial port and complete data transmission and communication connection with the host computer software by Mod-bus protocol, to ensure data stability and reliability.

2.5. Host computer software subsystem

The host computer software is mainly composed of data acquisition and processing module, graphical display interface, parameter configuration module, flow control module, data storage module and network transmission module. It is responsible for data transmission and communication connection with data acquisition hardware subsystem. It collects, processes, displays, saves and calls the eco-environment parameters, and transmits the monitoring parameters to the user data center over the wireless network.

3. System software design

3.1. Software design overview

The host computer software is the intelligent core of the system and it uses Monitor and Control Generated System (MCGS) based on windows system as the development platform. MCGS can solve the practical engineering problems in a variety of ways such as animation display, process control, alarm processing, report output, etc., and has been widely used in the field of automation control [4, 5]. The software adopts multi-thread technology, multi-layer architecture and functional modular design, which can realize the operation control of subsystems, interactive communication between subsystems, data acquisition, analysis and processing, data graph display and database storage, historical data calls, external data communication and other functions.

The system software mainly includes four modules, the system parameters setting module, the analysis process control module, the data collection and processing module, and the data transmission module. The system parameters setting module includes system initialization, parameter settings of manually run and parameter settings of automatically run. It is mainly responsible for setting process trigger time and system valve status. The analysis process control module is the core of the software

IOP Conf. Series: Earth and Environmental Science **82** (2017) 012087 doi:10.1088/1755-1315/82/1/012087

which can control the operation of each analysis processes of the system. It is divided into automatically run process control and manually run process control, both of which include TOC, multi-parameter and nutrient analysis process, system chemical cleaning process, filtration system recoil process and force interrupt of process. The data collection and processing module is responsible for achieving the detection of marine ecological parameters, data processing and display, data storage and invocation. The data transmission module can achieve local data exchange and network data communication. GPRS DTU equipment is used to achieve the network communication with user data center to realize real-time remote data transmission. The system software structure design is shown in figure 2.

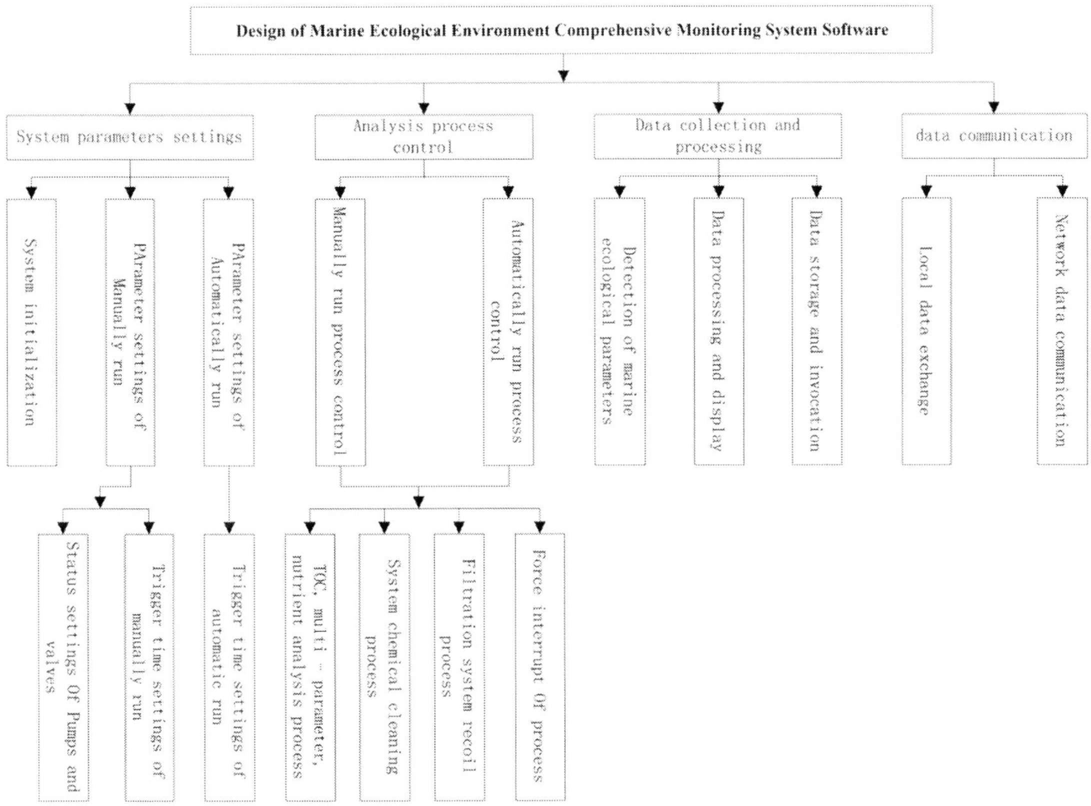

Figure 2. Structure of the software for marine ecological environment monitoring system.

3.2. Software function design

3.2.1. Software process design. The software is modularized design and is divided into four independent processes: TOC analysis process, multi-parameter analysis process, nutrient analysis process and chemical cleaning process. Among them, the nutrient analysis process is divided into nutrient pretreatment process, separate analysis process of the five nutrient parameters and nutrient post-processing process. Every process can be both automatic and manual controlled. The automatic operation process is triggered by time scanning. Users can set the trigger time of the analysis flow every day in the time setting interface. When the system time meets the time that users set, the corresponding analysis process will be triggered. The manual operation flow is triggered by the trigger button corresponding to each process. The trigger block diagram of each analysis process is shown in figure 3. With the automated operation of nutrient analysis process for example, the operation flow chart is shown in figure 4.

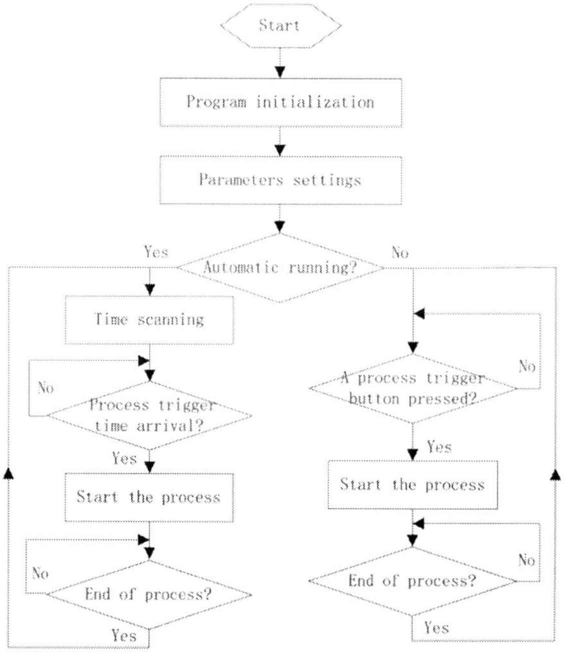

Figure 3. Block diagram of analysis process triggers.

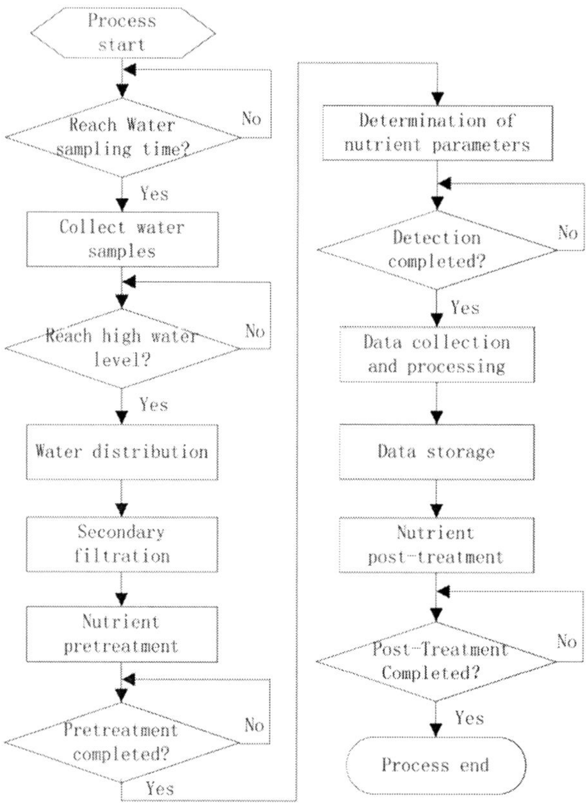

Figure 4. Automatic detection and analysis flowchart of the nutrient.

3.2.2. Software communication design. The software achieves remote network communications with user data center through the Data Transfer Unit (DTU). DTU is a wireless terminal device that is used to convert serial data into IP data or to convert IP data to serial data by a GPRS wireless communication network. It doesn't need background computer support and utilizes GPRS network. It has the characteristics of stability, speed, always online and etc.. It is widely used in meteorological, geological, hydrological and hydrological related industrial monitoring, environmental monitoring and other fields [6-8]. DTU can convert the raw data on the serial port into TCP/IP packets for transmission without having to change the original data content. Therefore, the DTU can be connected with various user devices that use serial communication, and there is no need to make changes to the user equipment.

The software communicates with the DTU via RS232 serial port, and it sends the detected marine ecological environment parameters to the DTU according to the format requested by the user. DTU encapsulates the data into TCP/IP format data packets through the GPRS module. The data packet is sent from the GPRS network to the user data center which has fixed public IP address or fixed domain name. User data center can directly read and analyze the data to achieve remote monitoring and analysis of marine ecological environment parameters. The data communication principle of DTU is shown in figure 5.

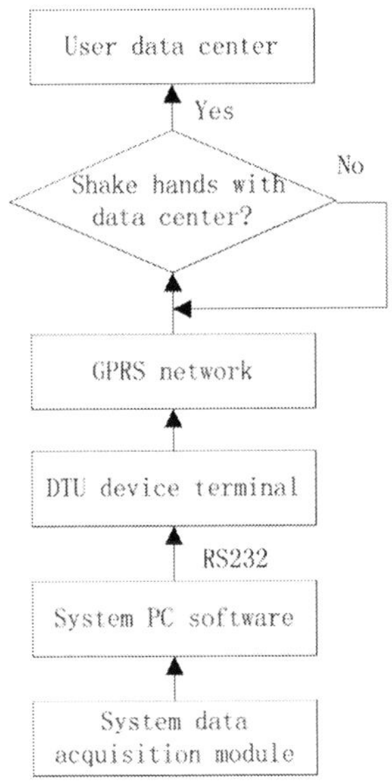

Figure 5. DTU data communication schematic.

4. Experimental validation

The system was installed at Zhoushan Tidal Station of the East China Environmental Monitoring Center. S.O.A.. From March to August of 2016, a 6-month follow-up monitoring test was conducted at Zhoushan Tidal Station. During the six-month trial, the system was run continuously for 24 hours a day. Test data was uploaded regularly to user data center according to the requirements of the East China Environmental Monitoring Center. S.O.A.. A site comparison test lasting six to nine days was performed every month, to compare the results of the system monitoring with the standard manual method. As shown in figures 6-9, this paper selected the site comparison test results of salinity, pH, dissolved oxygen and $SiO4^{2-}$ from May 23 to May 31, 2016 as a representative. The results showed that the relative error of the parameters such as temperature, salinity, turbidity, pH, dissolved oxygen with the standard method was controlled within 5% and the relative error of the nutrient parameters was within 15%. Therefore, the test results showed that the system had good stability and accuracy. The continuous operation of 6 months showed that the system had few maintenance times, low failure rate, stable and efficient continuous monitoring capabilities. Meanwhile, the system was easy to operate and easy to maintain. Therefore, the system can be well applied in the field of marine ecological environment monitoring.

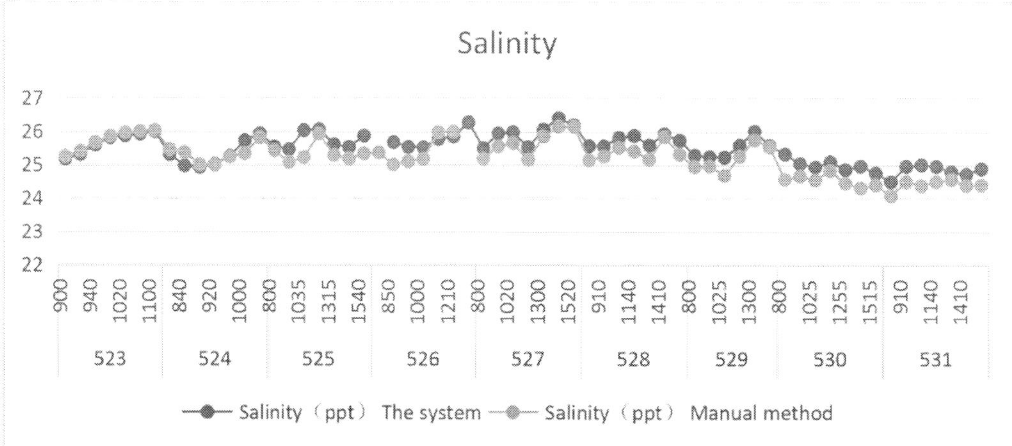

Figure 6. Site comparison test results of salinity.

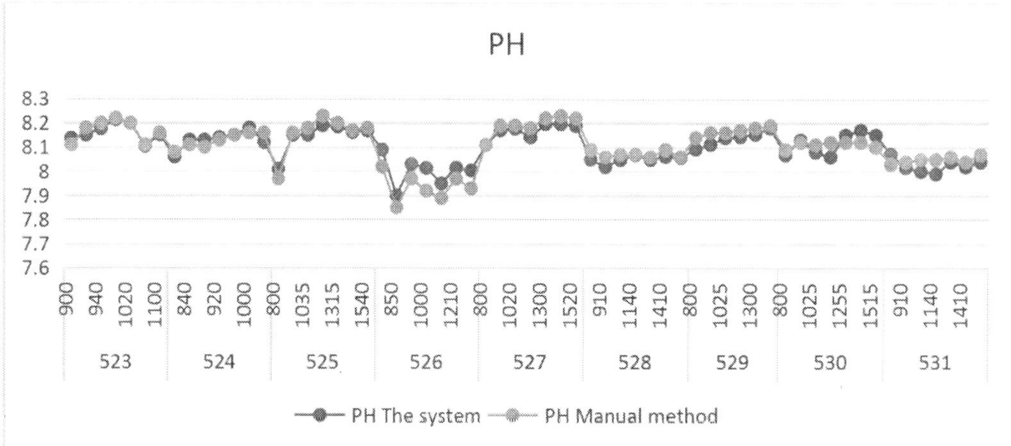

Figure 7. Site comparison test results of pH.

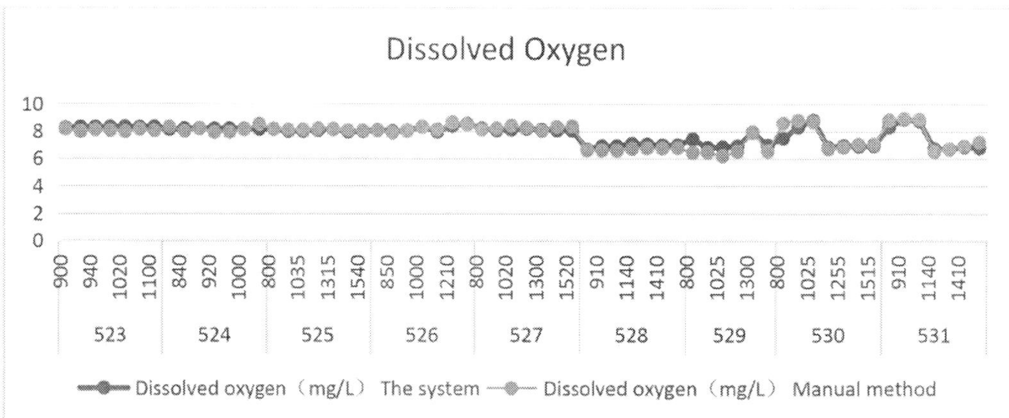

Figure 8. Site comparison test results of dissolved oxygen.

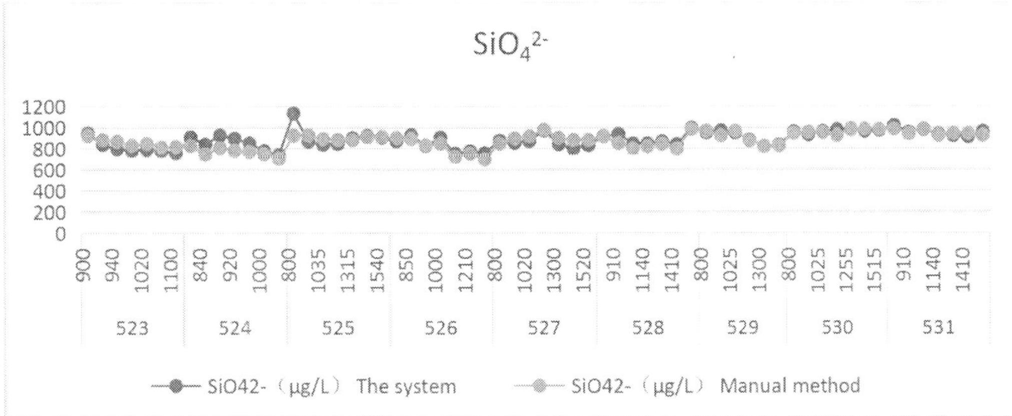

Figure 9. Site comparison test results of SiO_4^{2-}.

5. Conclusions

The software realizes the functions of automatic data acquisition and processing, real-time display and preservation of marine ecological parameters stably and efficiently. It provides a simple and quick historical data display and query interface, and completes the multi-network data transmission.

The software has the following main innovations:

- MCGS configuration software has the advantages of stable data structure, flexibility, versatility, portability and scalability, clear and easy user interface, simple user operation and so on.
- The software has flexible data acquisition and control methods, high data storage and transmission efficiency. It can be set according to the actual needs of different users to achieve real-time data efficient storage and daily data network transmission.

After the actual engineering application, the software can accurately display, store and regularly transmit the marine ecological environment parameters. In the absence of external interference, it can stably and continuously work. Therefore, it can be applied to marine ecological environment comprehensive monitoring field.

Acknowledgments

This study was supported by National Key R & D Plan (No. 2016YFC1400804, 2016YFC1400803, 2016YFC1400801), Natural Science Foundation of Shandong Province (No. ZR2014YL006, ZR2016DM17), National Natural Science Foundation of China (No. 41206076), Project supported by Shandong Provincial Natural Science Foundation of China (No. ZR2014DM012), Shandong Province Key Research and Development Program (No. 2015GGX103031).

References

[1] Xu L and Wang X 2003 *Mar. Environ. Sci.* **22** 63-8
[2] Mills G and Fones G 2012 *Sensor. Rev.* **32** 17-28
[3] Chang N, Imen S and Vannah B 2015 *Crit. Rev. Env. Sci. Tec.* **45** 101-66
[4] Chernykh E V, Gorbachev K V, Mikhaylov V M, Nesterov E V and Stroganov V A 2015 *IEEE T. Plasma. Sci.* **43** 2064-9
[5] Zhang C W, Zhou X G, Zhang P and Zhang LP 2014 *Appl. Mech. Mater.***528** 334-8
[6] Guo S, Wang C, Si Q, Gui X, Kong D and Han L 2016 *J. Am. Coll. Cardiol* **68S** C187-8
[7] Horcas S G, Debrabandere F, Tartinville B, Hirsch C and Coussement G 2017 *Wind Energy* **20** 619-36
[8] Meng X, Wang Y and Sun J 2005 *Modern Elec. Tec.* **19** 31-33

IOP Conf. Series: Earth and Environmental Science 82 (2017) 012088 doi:10.1088/1755-1315/82/1/012088

Marine radioactive field monitoring sensor based on NaI (Tl)

G X Ren[1,2,3], Z Q Wei[1], D Y Liu[2] and Y Y Zhang[2]

[1]College of Information Science and Engineering, Ocean University of China, No. 238 Songling Road, Laoshan District, Qingdao City, Shandong Province, 266100, China
[2]Institute of Oceanographic Instrument, Shandong Academy of Science, No. 37 Miaoling Road, Laoshan District, Qingdao City, Shandong Province, 266061, China

E-mail: rgx422@hotmail.com

Abstract. There are many deficiencies in traditional laboratory means, which make it difficult to meet the real-time monitoring requirements of nuclear radiation on marine field. In this paper, a radioactive monitoring sensor for marine field was proposed, which is based on NaI (Tl) scintillation crystal, while energy calibration and resolution calibration are conducted by employing a standard radioactive source, and curve fitting is conducted by employing MATLAB. Through the test under seawater in Qingdao wharf, the results are in good agreement with the laboratory test results.

1. Introduction

The radioactive material leakage of Fukushima Nuclear Power Plant in Japan occurred on March 2011 which caused large-scale nuclear pollution, and the marine nuclear radiation monitoring began to draw attention. The traditional monitoring method is to bring a large number of samples back to the laboratory for processing and analysis, which includes burdensome process and waste time and energy. The test results can often not be obtained until 3-4 months later [1, 2], this non-real-time mode of work needs the skill of chemical pretreatment of the samples and is restricted by half-time of nuclide, therefore, the research and application of on-site, real-time and continuous marine radioactive monitoring technology have been highly valued [3]. In this paper, γ spectrometer detection technique based on NaI (Tl) scintillation crystal is used to realize a monitoring sensor on marine nuclear radiation field based on NaI (Tl) detector, which can rely on buoy system for long-term stable operation in harsh marine environment, and can achieve real-time monitoring of marine radioactivity.

2. System implementation

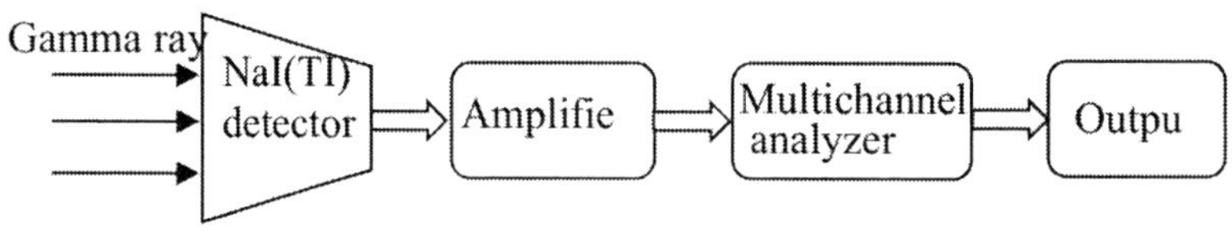

Figure 1. Radioactive detection model.

The basic principle of using NaI (Tl) scintillation crystal is the interaction between γ-ray and NaI (Tl) detector, so as to convert different energy of γ photons into electric pulse signals in different amplitudes, collect electrical pulse signals and process data to get the energy spectrum diagram [4]. As shown in figure 1, the schematic diagram of the system implementation was given.

The system consists of NaI (Tl) detector (including NaI (Tl) crystal, photomultiplier, preamplifier), main amplifier, digital multi-channel pulse amplitude analyzer, power supply, monitoring system, data acquisition and storage system, energy spectrum analysis and application display software and other components. As shown in figure 2, the relative luminous efficiency of NaI (Tl) crystal scintillation detector is higher and is better matched with the photomultiplier, and γ photons with different energy are converted into electric pulse signals in different amplitudes for preamplifier processing; the main amplifier further amplifies and shapes voltage signals from tens mV to hundreds mV; the multi-channel analyzer is used for pulse amplitude analysis, and the count on each channel constitutes γ energy spectrum diagram; the energy spectrum analysis uses algorithm of Gaussian Fitting, which obtains the type of radionuclides and calculates the nuclide concentration automatically.

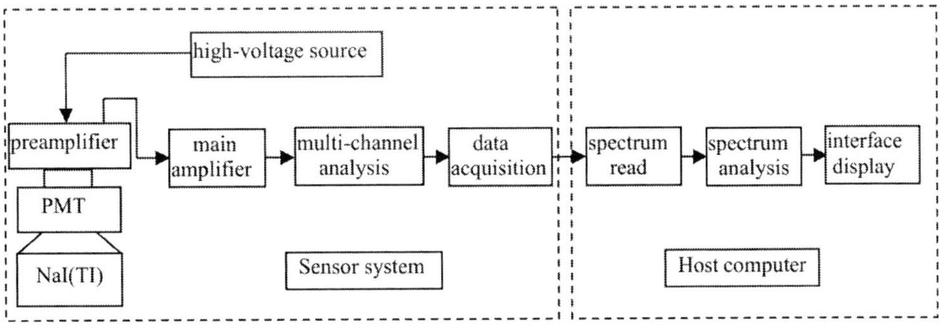

Figure 2. Block diagram of marine nuclear radiation monitoring sensor system.

3. Energy calibration

The corresponding energy peak position is detected by using the known energy γ radiation source to to obtain the relation curve between energy and peak position (channel) [5]. The correspondence between the peak position and the energy is the energy calibration, and the curve determined by the relationship between the peak and the energy is the energy calibration curve. Through the energy calibration curve, the energy of the characteristic peak can be determined by the peak position (channel), and then the nuclide matched with it can be found in the nuclide library to identify the nuclide type. Therefore, the energy calibration is crucial in the qualitative identification of nuclides. The energy resolution characterizes the resolution ability of the detector on different nuclides, whose value is the ratio of the full width at half maximum of the characteristic peak to the peak position [6-8]. The functional relationship curve between the energy resolution and energy is the energy resolution curve.

The sensor uses the radiation standard source: ^{241}Am, ^{133}Ba, ^{60}Co, ^{137}Cs, ^{152}Eu for energy calibration and resolution calibration [9], the detector uses NaI crystal with the size of 3 * 3 inches, the point source is 25 cm from the detector's front end, and the multi-channel energy spectrum analysis meter is 1024 channel, with the energy range of 0-2000 keV.

The energy calibration curve with confidence ratio of 95% is calculated by matlab fitting:

$$y = 1.891x - 10.4$$

In the formula, x is the peak position, y is the energy of the nuclide characteristic peak [10], in keV. The standard deviation of fitting is 0.9999. The energy calibration curve is shown in figure 3, and it can be seen that the curve is well linear.

The parabola fitting is made to data on the rectangular coordinate paper with characteristic γ ray energy and the corresponding full-energy peak position track address, and the energy resolution curve calculated with matlab fitting when the fiducial probability is 95%:

$$Y = 0.008453 + 0.06973 * \sqrt{x + 0.009146x^2}$$

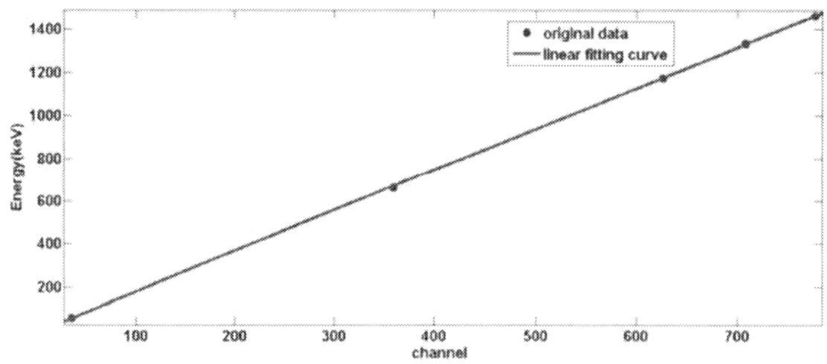

Figure 3. Energy calibration curve.

In the formula, x is Energy, in MeV, Y is half width FWHM, in MeV. The fitting standard deviation is 0.9941. The energy resolution curve is shown in figure 4. It could be seen from the experimental data in table 1 that when the voltage is as high as 800v, FWHM of ^{137}Cs is 45.3 keV when the energy is 661.6 keV, and the energy resolution is 6.8%.

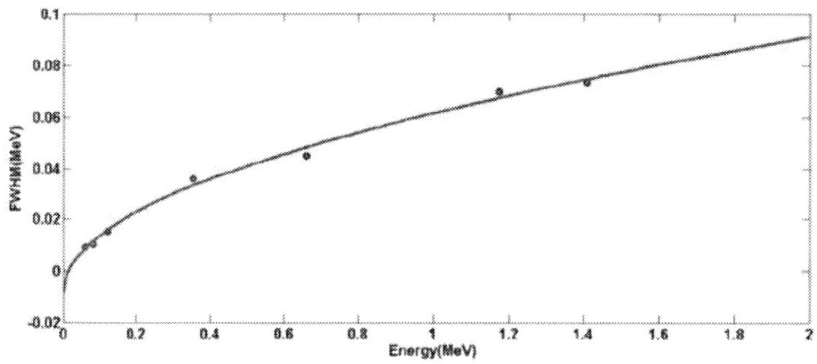

Figure 4. Energy resolution curve.

Table 1. Source characteristics used in energy calibration.

No.	Nuclide	Major γ ray energy (keV)
1	241Am	59.54
2	133Ba	81.0,302.85,356.02
3	137Cs	661.662
4	60Co	1173.24,1332.5
5	152Eu	121.782,344.29,778.2,964.11,1085.89,112.08,1408.0

4. Experimental results and analysis

Figure 5 shows that the sensor is connected to the upper computer in the laboratory to test the copying machine. The sensor is experimented in the Eight Great Passes in Qingdao, the water at the wharf is about 8 m deep, in order to eliminate the interference of cosmic rays and submarine material, the sensor is suspended 3 meters below the water surface by crane, and the cumulative time is 24 hours, as shown in figure 6. The spectrum at 19:00 on April 19, 2015 is shown in figure 7, from which the 40 K characteristic peak can be clearly identified.

The data of 24 hours from it for analysis were detected. The device of the upper computer of the sensor configures software to automatically find the peak, use Gaussian fitting method to fit the 40 k spectral line curve, as shown in figure 8:

Figure 5. The sensor is connected to the upper computer in the laboratory.

Figure 6. Field measurement of detector ready to be immersed into seawater.

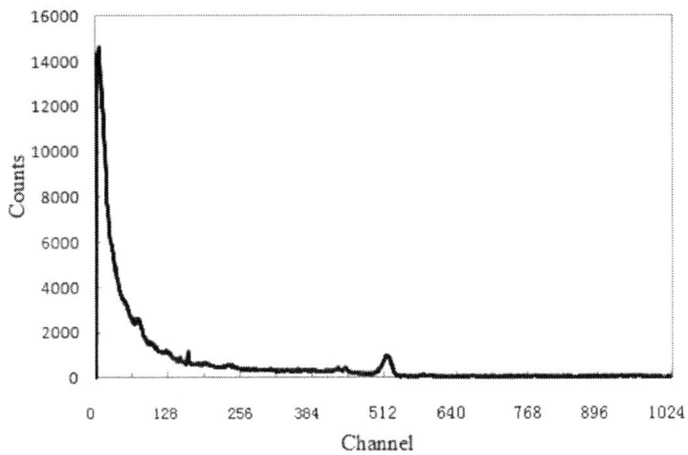

Figure 7. Spectrum below 3 meters of water on April 19, 2015.

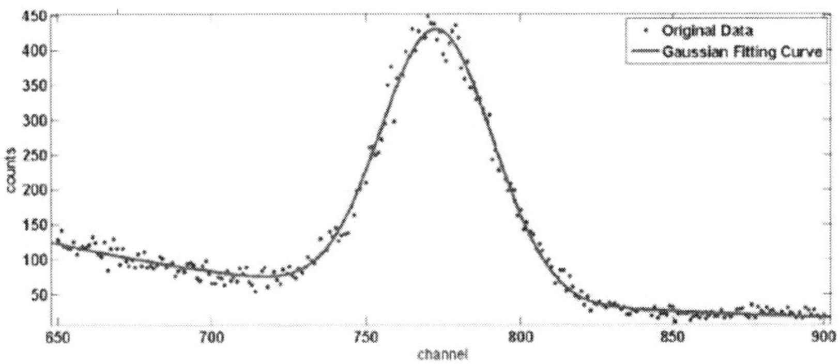

Figure 8. The instrumental spectrum and the original spectrum after seawater 40 k characteristic peak Gaussian fitting.

The fitting curve is: $f(x) = a1 * \exp(-((x-b1)/c1)^2 + a2 * \exp(-((x-b2)/c2)^2)$

The coefficient values are as follows (confidence probability: 95%):
A1 = 383.3, b1 = 772.9, c1 = 25.67, a2 = 2.811e + 015, b2 = -7191, c2 = 1414,

According to the fitting curve, the characteristic peak of 40K seawater is 18650 (24 hours), according to the formula of activity [11]

$$A = \frac{S}{T * \varepsilon * I}$$

The activity of 40k is 12.42 Bq/L. In the formula, S is the net count of nuclide characteristic peak, T is the measurement time, ε is the detection efficiency of the instrument, and I is the fractional ratio.

In order to verify the accuracy of the 40 K measurement of the sensor, the seawater is sampled in the same position at the wharf at the same time, and then the sample is taken back to the laborator, where the salinity of the seawater is measured with SYC2-2 salinity meter produced by China Ocean University to the, and then calculates the concentration of 40 K according to the proportional relationship [12], which is 12.33 Bq/L, and the comparison result shows that the measurement result of the sensor is consistent with the laboratory measurement result.

Through the experiments on the waters of the Eight Great Passes in Qingdao, it can be seen that there is a significant 40 K peak near the 526 channel, which is in line with the natural properties of seawater. The two spikes near 170-channel is caused by the noise of the circuit. The subject will study the elimination of these noises in the next work.

5. Conclusions

The sensor described in this paper will mainly be used for long-term, continuous real-time monitoring of the marine environment on marine fields, such as Marine Data Buoy and ocean station [13, 14], and can identify a variety of target radionuclides and calculate concentrations. In the next step, the energy efficiency calibration algorithm and low power consumption will be improved, to improve the sensor optimization and realize the on-line monitoring of marine radioactivity.

Acknowledgments

Authors wish to acknowledge assistance or encouragement from colleagues, special work by technical staff or financial support from organizations. Moreover, the authors acknowledge funding from International Science &Technology Cooperation Program of China (No. 2013DFR90220) and Science and Technology Plan of Shandong Province (No. 2015GSF115001).

References

[1] Bagatelas C, Tsabaris C, Kokkoris M, Papadopoulos C T and Vlastou R 2010 *Environ. Monit. Assess.* **165** 159-68
[2] Caffrey J A, Higley K A, Farsoni A T, Smith S and Menn S 2012 *J. Environ. Radioactiv.* **111** 120-5
[3] Clifton L, Clifton D A, Pimentel M A F, Watkinson P J and Tarassenko L 2014 *IEEE J. Biomed. Health* **18** 722-30
[4] Ocone R, Kostezh A, Kurinenko V, Tyshchenko A, Derkach G and Leone P 2004 *Appl. Radiat. Isotopes* **61** 129-32
[5] Tsabaris C and Ballas D 2005 *Appl. Radiat. Isotopes* **62** 83-9
[6] McKinna L I W 2015 *Prog. Oceanogr.* **131** 177-99
[7] Tsabaris C, Bagatelas C, Dakladas T, Papadopoulos C T, Vlastou R and Chronis G T 2008 *Appl. Radiat. Isotopes* **66** 1419-26
[8] Vlastou R, Ntziou I T, Kokkoris M, Papadopoulos C T and Tsabaris C 2006 *Appl. Radiat. Isotopes* **64** 116-23
[9] Casanovas R, Morant J J and Salvado M 2012 *Radiat. Meas.* **47** 588-95
[10] Xu P, Di Y M and Qiu X L 2007 Research on digital measurement and analysis technology of γ radiation *Nucl. Electron. Detect. Technol.* **27** 234-8

[11] Dezwart D, Kramer K and Jenner H A 1995 *Environ. Toxicol. Water Qual.* **10** 237-47
[12] Tsabaris C and Prospathopoulos A 2011 *Appl. Radiat. Isotopes* **69** 1546-53
[13] Liu Y, Zhu P and Tan L J 2010 Study of luminescence technology to measure total organic carbon (TOC) in sea water *Environ. Sci. Technol.* **33** 123-6
[14] Wang B, Li M and Liu S Xuan 2014 Current status and trend of ocean data buoy observation technology applications *Chinese J. Sci. Instrum.* **11** 2401-14

IOP Conf. Series: Earth and Environmental Science 82 (2017) 012089 doi:10.1088/1755-1315/82/1/012089

Study on the water retention effect of compound soil of arsenic sandstone and sand under the condition of typical crop planting

S Y Liu[1], N Wang[1,2], J C Xie[1], R G Jiang[1] and M L Zhao[1]

[1]State key Laboratory Base of Eco-hydraulic Engineering in Arid Area, Xi'an University of Technology, Xi'an 710048, China

E-mail: wangni@xaut.edu.cn

Abstract. Arsenic sandstone is the main reason of soil erosion in the Mu Us Sandy Land, simultaneously was proved to be a kind of good water retaining agent. In order to provide references for the utilization of water and soil resources and the prevention and control of desertification and soil erosion of the southern margin of Mu Us Sandy Land, on the basis of earlier studies the farmland experiments of compound soil with three ratios of 1:1, 1:2 and 1:5 between arsenic sandstone and sand under maize planting patterns were designed, whose experimental process was divided into six stages according to the crop growth status. The results showed that the soil moisture content was highest in the layer of 0~40cm where the compound soil mainly concentrated in, which was related to the potent water retention of arsenic sandstone and strong water permeability of undisturbed sandy soil. The variation coefficients in the soil of 1:1 and 1:2 were more stable and evenly distributed. The compound soil can effectively improve the soil water retention capacity, and prolong the storage time of soil water. Among them, water loss rate in soil of 1:1 and 1:2 were lower. The coefficient of variation also confirms that the water distributions of the two types of soil were more uniform and stable. Besides illustrating the effects of the soil amelioration measures on spatial and temporal variation of soil moisture content and the improvement of soil water regime, the study provides some references for the development and utilization of agriculture in Mu Us Sandy Land.

1. Introduction

Ecological environment has been widely considered as the basic condition of human survival and development. There are three main factors, including abundant loose gravel source, bare land surface and sand driving wind with high frequencies, which were regarded as the potential factors leading to desertification in arid and semi-arid area. As one of the four Chinese deserts, the Mu Us Sandy Land (MUSL) has huge mining potential; while sandy soil erosion and land desertification seriously restricted the sustainable development of the region. In order to make desertification in sandy land controlled and achieve economic benefits simultaneously, various integrative desertification control methods or models with farming and animal husbandry in consideration have been developed in recent years.

The early study of sandy land was mainly confined to the control of desertification and the protection of the ecological environment. In this period, the goal of sand control was often the pursuit of ecological effect, and the economic benefit was not taken into account. Therefore, the enthusiasm of public to participate was not high, and the difficulty for large-scale ecological management and sustainable development were the two main problems existed [1-4]. Since the 80~90's in twentieth century, the research on sandy land gradually began to move towards integration and diversification, in order to seek a new sand control mode to be able to obtain the social, ecological and economic benefits. Desertification due to human activity and limited natural resource is the most serious problem on agricultural development in sandy land, which was highly concerned on account of its threats to human survival and social development [5, 6]. However, the sustainable development of any model requires the support of water in the sandy land. In various desertification control and development methods, water was known as the pivotal point. Hellden et al [7] pointed out that water resources were very important limiting factors in the growth of sandy land, and it was also an important cause of desertification.

In Mu Us Sandy Land, arsenic sandstone and sand are two kinds of main components, which are the major causes of soil erosion and land desertification relatively. Arsenic sandstone and sand on the physical properties are complementary. Han et al [8, 9] proposed that the arsenic sandstone and sand could be mixed into the soil by using permeability of sand and water holding characteristics of arsenic sandstone, in order to improve the sandy soil physical and chemical properties and the production capacity of soil. Wang et al [10, 11] studied the physicochemical properties of compounds of arsenic sandstone and sand at the ratios of 1:0,5:1,2:1,1:1,1:2,1:5 and 0:1 respectively, put forward that the compounds of 1:1, 1:2,and 1:5 has better water binding capacity and soil quality.

Based on the related research results of combination schemes of arsenic sandstone and sand, three kinds of compound soil (the ratio of 1:1, 1:2 and 1:5 of arsenic sandstone and sand) were taken as culture medium in the field experiment, with maize, which was the most important food crops for local people planted in it, to explore the water retention effect in different proportions of compound soil, and to provide reference for the development and utilization of agriculture in MUSL.

Table 1. The physical and chemical properties of soils.

Index		Arsenic sandstone	Sand	1:1	1:2	1:5
Texture		Sandy loam	Sand	Loamy soil	Sandy loam	Sandy loam
Volume weight of soil/(g·cm^{-3})		1.42~1.67	1.61	1.56	4.52	1.37
Capillary porosity/%		44.94	4.28	3.86	3.94	4.11
Compactness		Hard	Loose	Hard	Hard	Hard
Particle size proportion /%	sand particle (≥ 0.05~<0.2mm)	19.57	91.39	46.84	64.67	74.79
	powder particle (≥ 0.002~<0.05mm)	72.94	5.51	44.92	30.04	20.08
	clay particle (<0.002mm)	7.49	3.10	8.24	5.29	5.13
Nutrient content/(g·kg^{-1})	Organic matter	3~5	1~3	2.26	2.61	2.97
	Total nitrogen	0.13	0.75	0.44	0.54	0.65
	Total phosphorus	0.38	0.63	0.50	0.55	0.59
	Total potassium	19.20	26.51	22.26	23.67	25.09
Water parameters	Field capacity/%	17.32	4.46	11.72	9.92	7.65
	wilting coefficient/%	8.96	1.29	6.53	5.06	3.30
	Saturated hydraulic conductivity (mm.min^{-1})	0.07	7.10	0.26	0.49	1.61
	Water stability aggregate content/% (>0.25 mm)	32.30	18.13	29.58	26.32	22.54

IOP Conf. Series: Earth and Environmental Science **82** (2017) 012089 doi:10.1088/1755-1315/82/1/012089

2. Materials and methods

2.1. Study area

Yuyang District, which lies in the Middle North region of Yulin City in Shaanxi Province, is located in the southern margin of Mu Us Sandy Land. The experimental area belongs to Dajihan Village of Xiaojihan Country in Yuyang District (E 109°29′50″-109°31′44, N 38°26′23″-″38°27′02″), where is drought and lack of rain. The multi-year mean temperature is 6.0~8.5℃ and the average temperature in January is within 9.5~12℃ while 22~24℃ in July. The annual rainfall is 250 mm~440 mm, in which the amount of July, August and September accounts for 60~75% of the whole year. The intra-annual precipitation changes a lot, cause that the rainfall of rainy year is 2~4 times of dry year. Maximum daily rainfall amount is up to 100~200mm. Precipitation often appears in the form of a shower, which lasted a short time. As one of the most serious soil erosion areas in the upper and middle reaches of the Yellow River, Yuyang District has a large sediment concentration in surface runoff. Surrounded by lots of arsenic sandstone and sand [12], this region is a typical wind sand area. The physical and chemical properties of arsenic sandstone and sand are showed in table 1.

2.2. The experimental design

The main type of experimental area is aeolian sandy soil. The arsenic sandstone and sand was compounded in the ratio of 1:1, 1:2 and 1:5 (arsenic sandstone: sand), and the three types of soil were abbreviated to TC1, TC2, TC3 respectively. In addition, the original soil (total sand) was abbreviated to TS. Maize were chosen as the experimental crops for the reason that it was the most important grain crop for local people and were suitable to the local climate.

Twelve experimental plots with dry compounds of arsenic sandstone and sand at the ratios of 1:1, 1:2 and 1:5 respectively were divided into two groups (figure 1). Each plot was set up as $12m \times 5m$. Compound soils of different mixing ratios were covered to the surface of each plot to the depth of 40 cm, under that was the local original sand. Before planting, the field is fertilized with diammonium phosphate (DAP) of 335 kg/ha and compound fertilizer (N-P_2O_5-K_2O=12-19-16) of 375 kg/ha, then with a rotary tiller the compound soil and fertilizers were mixed uniformly within surface layer of 15 cm. The same fertilization measures were used to deal with 6 residential areas.

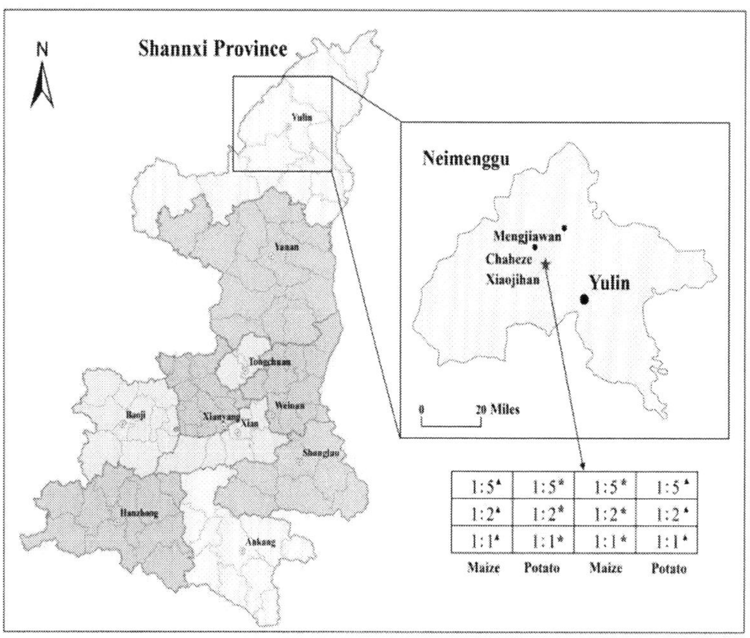

Figure 1. The position of the study area and the distribution of monitoring points.

Neutron moisture meter (CNC100B) was used as experimental measuring instrument in one of the groups of experimental plots, and the other group was processed by oven-drying method as a comparison. The data measured by the neutron moisture meter will be verified and corrected by oven-drying method. In the following study, the data measured by neutron moisture meter was analyzed and calculated in the results. Six moisture content measuring tubes (marked by *) were installed in the experimental area in different proportions of compound soil. During the experiment, neutron moisture meter was used to detect the soil moisture content of 9 soil layers (0, 10, 20, 40, 60, 80, 100, 120, 130 cm).The arrangement of the measuring tubes was shown in figure 1:

Before starting the experiment, the whole growth period of the crops were divided into 6 stages according to the growth status of crop. The sowing date was May 9th, 2016, and the harvest date was September 17th and 18th. The entire crop growth period was taken as research time. Refer to the existing research results, combined with the local crop cropping system, the growth stages of maize were divided as table 2.

Table 2. Division of corn growth stages.

Crop	Period (d)	Germination stage	Seeding stage	Jointing stage	Heading stage	Filling stage	Mature stage
Maize	130	10	30	35	20	20	15
Growth status		Sowing-Emergence	Emergence-Jointing	Jointing-Heading	Heading-Filling	Filling-Ripening	Ripening-Harvest
Date	5.09-9.17	5.09-5.18	5.19-6.18	6.19-7.24	7.25-8.14	8.15-9.04	9.04-9.18

2.3. The irrigation scheduling

Sprinkler irrigation has significant water saving effect, whose water utilization rate could reach 80%.In order to save water, sprinkler irrigation was used for irrigation in this study. According to the "Quota of agriculture water use (Shaanxi provincial standard DB 61/T 943-2014)" in the Northern Shaanxi area along the Great Wall wind beach area (including Jingbian, Hengshan, Dingbian, Yulin, Shenmu, Fugu), the irrigation schedule for maize was shown in table 3. In this study, all the crops were irrigated with a uniform irrigation schedule.

Table 3. Irrigation schedule for Maize under sprinkler irrigation.

Crop	Growth stages	Irrigation quota (m^3/mu)	Quota (m^3/mu)	times
Maize	Germination stage	290	28	1
	Seeding stage		28	1
	Jointing stage		28	2
	Heading stage		30	3
	Filling stage		24	2
	Mature stage		20	2

3. Results and discussion

3.1. Dynamic changes of soil moisture in the vertical soil profile

In the early stage of crop growth, the water content was less than that of late stages, as shown in figure 2 which illustrates the dynamic changes of soil moisture in the vertical soil profile. As the upper soil was easily affected by the weather conditions, the average water content of each soil layer decreased gradually with the increase of soil depth. Soil moisture in the 1:2 compounds was the highest and the most stable one, showing a better integrity and continuity of water, as well as the distribution at different growth stages was relatively uniform, followed by 1:5. The moisture within the layer of 0-40 cm of all the three proportions have always been in a higher state, which was associated with that compound soil of arsenic sandstone and sand were mainly stored within the depth of 40 cm. This showed that the arsenic sandstone effectively increased the water holding characteristics of soil water retention and the dynamic distribution of soil moisture was influenced by soil texture. Influenced by distribution of soils of different textures, all the three kinds of compound soil showed the same characteristics: soil moisture was highest in the 0~40 cm layer, at lower value in the layer of 40~80 cm, and increased gradually in the depth of 80~130 cm under the influence of infiltration effect and groundwater. Inflection point appeared at the depth of 100 cm in the soil of compound ratio of 1:1 and 1:5, while it appeared in the 1:2 compounds at the depth of 80 cm inflection point with minor changes.

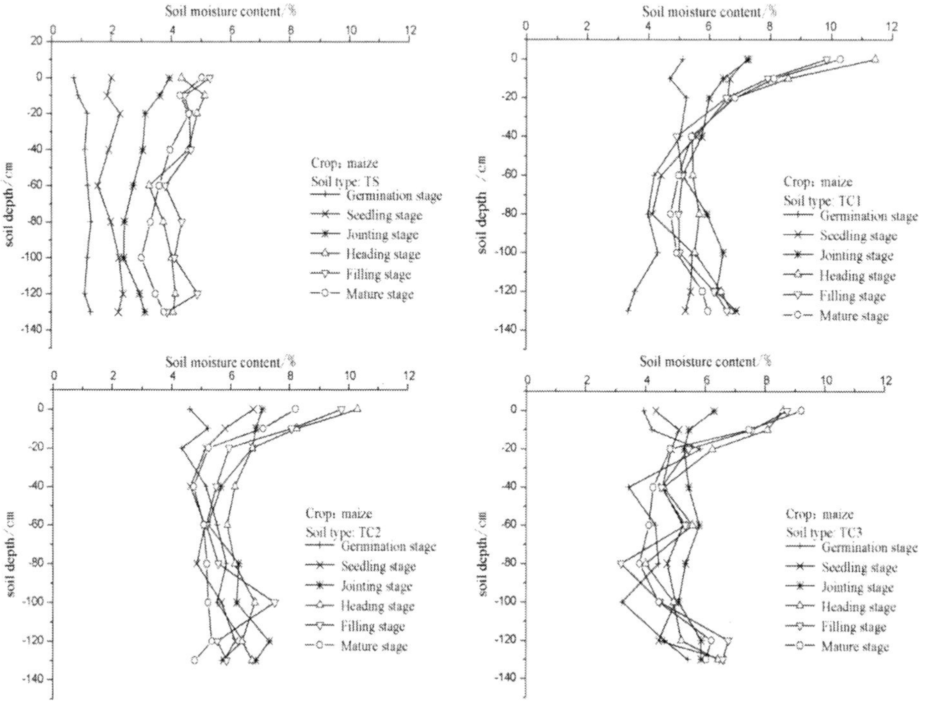

Figure 2. Dynamic changes of soil moisture in the vertical soil profile.

3.2. *Variation of soil moisture content over time*

Heading stage of maize is a critical period to determine the growth status and the yield of maize. Taking the Heading stage as a typical period to study the variations of soil moisture content over time, different types of soil within the soil layers of 40 cm were stratified sampled and measured every 24 hrs, and the change of soil moisture content with time in 300 h was observed. No rainfall was occurred in the period, and soil has not been supplied by other water supplies, namely all losses are evaporation losses and crop root absorption. The results showed that the soil moisture content of different types of soil has a high consistency with time, showing a trend of rapid decline over time, especially in the former 75 h. Soil moisture content of TS was less than 3% after 75 h, showing a poor water retention performance, which could not provide and maintain the water needed for crop growth for long-term. The loss of water within 72 h accounted for 78.8% (TS), 49% (TC1), 44% (TC2), and 49.7%% (TC3) (figure 3). The amount of water loss decreased gradually over time, as well as the loss rate.

Soil moisture loss rate was the moisture dropped in per 24 hour. The moisture content that dropped every 24 hours within 300 hours were shown by Box-plot (figure 4), which illustrated that the upper and lower range of TC1 and TC2 were significantly less than TS and TC3, showing that the decline process of moisture content was more stable. The median line of TC1, TC2, TC3 were adjacent (TC3 was larger slightly), while that of TS was significantly higher than the other three, indicating that the rate of TS was fastest significantly. Compared with TS, it showed that the compound soil can effectively improve the soil water retention capacity, and prolong the storage time of soil water, in which the ratio of 1:1 and 1:2 had the best performance. "×"stands for extreme values, and in view of this point, the data of TC2 were much more centralized and better than TC1.

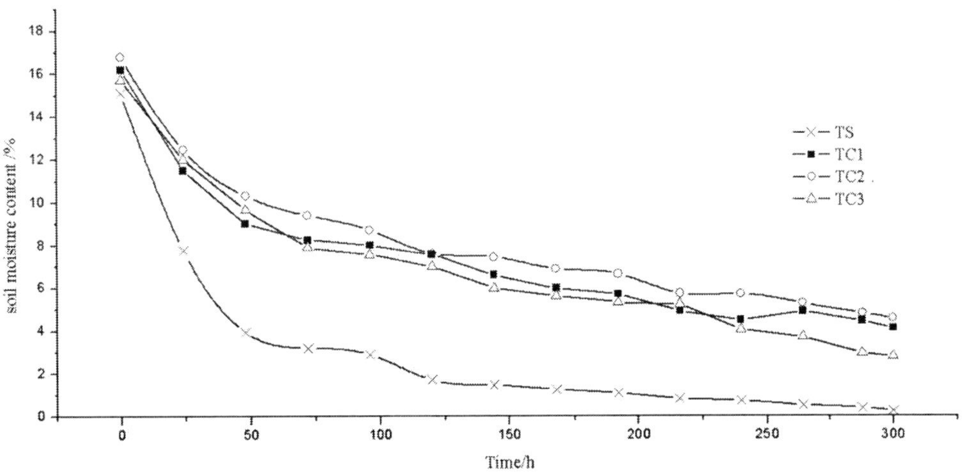

Figure 3. Changes of soil moisture of four types of soil over time.

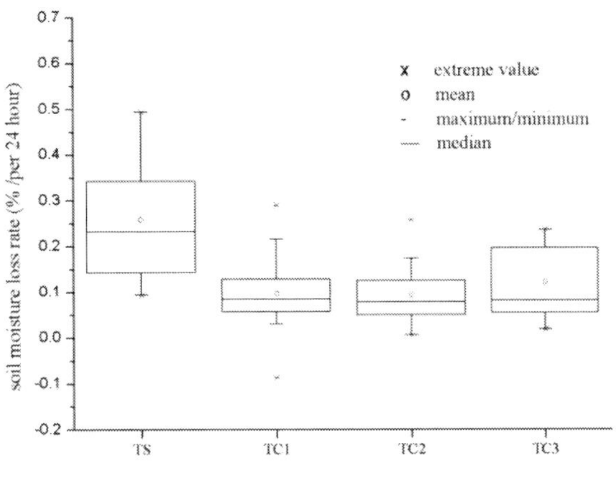

Figure 4. Soil moisture loss rate of four types of soil.

3.3. Variation characteristics of soil moisture in different layers

The coefficient of variation characterizes the stability of water in the soil. Previous analysis has shown that the compound soil has better water retention than total sand; therefore the TS were no longer discussed here. In order to explore variation characteristics of soil moisture content in different soil depths, the compound soil of same depth and growth stages were analyzed by the degree of dispersion, using the variation coefficient method to measure the variation of the observed values. The data of three types of soil were calculated in the same way. The calculation formula of coefficient of variation [13] (C.V) is:

$$C.V = \frac{S}{x} \tag{1}$$

Among them, S stands for the standard deviation of a certain layer of soil during certain growing period, $\bar{x}$ stands for the average value.

The value of the coefficient of variation represents the degree of dispersion of the measured value, and the calculated results are shown in figure 5. The variation coefficient $C.V \geq 0.15$ was used as the criterion for judging the degree of dispersion. As can be seen from figure 5, the coefficient of variation which were higher than 0.15 mostly concentrated in the deep layer of 0~40 cm and 100~130 cm. In the three ratios of compound soil, the highest value was 0.244 of the layer of 0~10 cm in the mature stage, 0.251 of the layer of 120 cm in the germination stage and 0.271 of the layer of 20 cm in the mature stage respectively. The main reason was that since the beginning of middle and late growing stages, the leaf area of maize increased, which caused the decrease of sunlight on surface of soil, and then evaporation reduced. Owing to the groundwater in Yuyang area is shallow (only 1.2 m~2.9 m in average) [14, 15], the layer of 100~130 cm was affected by the groundwater more strongly. And the growth of maize root is more vigorous, especially in Yulin which belongs to the arid area, thus roots of maize could grow to about 1m underground. The maximum mean value of the coefficient of variation all concentrated in the layer of 20~40 cm in the three kinds of soil (0.0298, 0.0238 and 0.0424 respectively), which represents that the soil moisture in the layer of 40cm was the most stable one with less fluctuations.

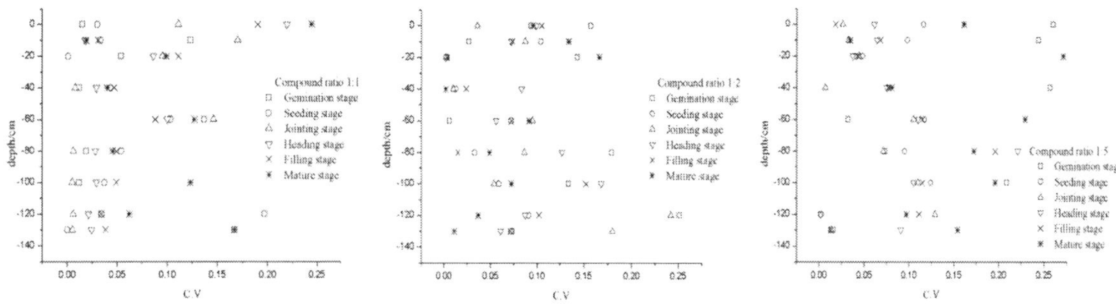

Figure 5. Variation coefficient of soil moisture in different soil layers.

3.4. Crop yield of maize in compound soil

From the overall point of view, crop yield planted in the TC1 and TC2 were higher, while it in the TC3 was low (table 4). Among them, the output of maize in 1:2compounds soil was the highest. This may be related to that the soil moisture under rate of 1:1 and 1:2 was the higher and more stable, and the soil moisture distribution were relatively uniform. Grain output is closely related to grain security. From the point of view of production, maize is more suitable for planting in TC2.

Table 4. Crop yield of maize in four types of soil.

Soil Type	Crop yield (t/ha)
TS	5.74
TC1	8.25
TC2	8.96
TC3	7.5

4. Conclusions

Through the measurement and analysis of soil moisture content in different depths, the dynamic variation characteristics were studied under maize planting patterns in compounds of arsenic sandstone and sand with ratios of 1:1, 1:2 and 1:5 in the southern margin of MUSL. Affected by the soil texture, the surface layer of 0~40 cm depth was a higher water content area, where the compounds primarily concentrated in, layer of 40~80 cm belonged to the lower value area, and the soil moisture content in the layer of 80-130 cm increased with the increase of depth, from which it can be seen that the vertical distribution curve showed a "C" type. Relatively, the soil moisture content of corn planting patterns under the ratio of 1:2 compounds of arsenic sandstone and sand was higher and more stable, and evenly distributed.

Under the same tillage and irrigation system, the water content of the compound soil was obviously higher than that of the whole sand. Soil moisture content will decrease with time, while the rate of water loss of compounds were relatively flat, indicating that compound soil can effectively improve the soil water retention capacity, and prolong the storage time of soil water. Among them, water loss rate of TC1 and TC2 were lower. The coefficient of variation also confirms that the water distributions of TC1 and TC2 soil were more uniform and stable.

It can be concluded that incorporation of arsenic sandstone and sand significantly enhances the soil water retention effect and water holing capacity, and the compound soil has a positive effect on crops growth. Besides illustrating the effects of the soil amelioration measures on spatial and temporal variation of soil moisture content and the improvement of soil water regime, the study provides some references for the development and utilization of agriculture in MUSL. This new mode of sand land agriculture development could greatly improve the ecological environment in MUSL, and bring more income for local farmers, which was a sustainable comprehensive development mode of sandy land.

Acknowledgments
This work was funded by the Ministry of Land and Resources of China Public Science and Technology Research Funds Project (201411008).

References

[1] Bai B T R, Guo K Z, Yang Y S 2005 Comprehensive management of Mu Us sand fixing technology and application *Inner Mongolia Water Conserv.* **3** 36-9
[2] Zhu Z D, Liu S and Yang Y L 1984 Discussion on the possibility and reality of desertification in the cross area of the farming and animal husbandry in northern China *Geogr. Sci.* **8** 198-206
[3] Ma Y X 2002 Causes of desertification and the prevention and treatment measures in agro-pastoral areas in Northern China *Chinese J. Agrometeor.* **2** 6-8
[4] Niu L L, Zhang T Y and Ding G D 2006 Actuality and countermeasure of ecological rehabilitation in Maowusu Sandy Land *Res. Soil Water Conserv.* **6** 239-42
[5] Huang Y Q 2008 Study on agricultural development in north agro-pastoral transitional zone based on circular economy *J. Shanxi Agric. Sci.* **12** 7-9
[6] Li Q S and Qiu S W 1997 The Songnen Plain sandy land resources development and agricultural development research *Geogr. Sci.* **3** 259-64
[7] Ulf H 2008 A coupled human-environment model for desertification simulaton and impact studies *Global Planet. Change* **64** 158-68
[8] Han J C, Xie J C and Zhang Y 2012 Potential role of feldspathic sand-stone as a natural water retaining agent in Mu Us Sandy Land Northwest China *Chinese Geogr. Sci.* **5** 550-5
[9] Han J C, Liu Y S and Luo L T 2012 Research on the core technology of remixing soil by coft rock and sand in the Mu Us Sandy Land Region *China Land Sci.* **8** 87-94
[10] Wang N, Xie J, Han J and Luo L 2014 A comprehensive framework on land-water resources development in Mu Us Sandy Land *Land Use Policy* **40** 69-73
[11] Ni W, Jiancang X and Jichang H 2013 A sand control and development model in sandy land based on mixed experiments of arsenic sandstone and sand: A case study in Mu Us Sandy Land in China *Chinese Geogr. Sci.* **6** 700-7
[12] Chai M M, Han J C, Luo L T, *et al* 2013 Effects of soft rock and sand mixing ratios and crop planting frequency on soil and crop yield *J. Northwest A & F Univ. (Nat. Sci. Ed)* **10** 179-84, 192
[13] He Q H, He Y H and Bao W K 2003 Research on dynamics of soil moisture in arid and

IOP Conf. Series: Earth and Environmental Science **82** (2017) 012089 doi:10.1088/1755-1315/82/1/012089

semiarid mountainous areas *J. Mountain Sci.* **2** 149-56

[14] Liao Y C and Fu Z G 2002 Causes and technical systems for prevention of land desertificationin the interlock zone of farming and pasturing of North China *Agric. Res. Arid Areas* **2** 95-8.

[15] Liu J Q 2010 The causes and countermeasures of desertification in Mu Us Sand Land *Inner Mongolia Forestry Incestigation and Design* **6** 18-21

IOP Conf. Series: Earth and Environmental Science 82 (2017) 012090 doi:10.1088/1755-1315/82/1/012090

Study on denitrifying dephosphatation process, influence factors and mechanism

W Li[1,3], H Z Sun[1], W Wei[1], J Liu[1], J X Fu[1] and J Wang[2]

[1]Municipal and Environmental Engineering Institute, Shenyang Jianzhu University, Shenyang, China
[2]China Northeast Municipal Engineering Design and Research Institute Co., Ltd., Shenyang, China

E-mail: liweilengjinyue@163.com

Abstract. Introduced the typical denitrifying dephosphatation process and compared them. Influence factors and reaction mechanism were discussed in this paper. The results could provide reliable theoretical basis for denitrifying phosphorus removal process in actual application.

1. Introduction

Denitrifying polyphosphate accumulating bacteria (DPB) under anaerobic/anoxic alternate environment used oxygen, nitrite and nitrate as terminal electron acceptor to replace the traditional phosphorus accumulation organisms, through their metabolism to complete phosphorus uptake and denitrification, it was considered as a sustainable wastewater treatment process because it can solve the problem of carbon source competition, reduce sludge production and the reaction time.

This paper comprehensive analyzed the typical process of denitrifying phosphorus removal, the characteristic of carbon source, nitrate, nitrite, pH, SRT and the reaction mechanism of denitrifying phosphorus removal. All above is to solve the problems of the typical process and also have important theoretical and practical significance to innovate and develop more reasonable technological process.

2. The typical progress of denitrifying dephosphatation

Typical denitrifying phosphorus removal processes is divided into single and dual sludge system. In single sludge system, DPB, nitrifying bacteria and other heterotrophic bacteria are stay in same reactors. In dual sludge system, DPB and nitrifying bacteria exist alone in immobilized membrane bioreactor or aerobic nitrification sequencing batch reactor [1, 2].

2.1. Single sludge system

2.1.1. UCT. In 1980s, Cape Town University in South Africa developed UCT process could improve the biological phosphorus removal, to avoid the effect of nitrate in the return sludge on anaerobic phosphorus uptake, it was based on the traditional biological nitrogen and phosphorus removal process by changing the reflux method [3, 4], but the increase of reflux system made the operation became complicated, and the operating costed also increased. Process flow of UCT was shown in figure 1. The disadvantage of UCT was the competition between DPB, nitrifying bacteria and other heterotrophic bacteria when they stayed in the same environment [4].

Anaerobic pool: DPB used small molecular weight VFA from degrading macromolecular organic compounds to synthesize large numbers of their own cytoplasm.

Anoxic pool: DPB used nitrate in recirculation solution from aerobic pool and the organic matrix in sewage to realize denitrifying dephosphataion.

Aerobic pool: DPB used the residual carbon in sewage to decompose the PHB and sustain its growth, excessive intake dissolved phosphorus in the environment at the same time. Nitrifying bacteria converted the ammonia nitrogen of the wastewater into nitrate by nitrification.

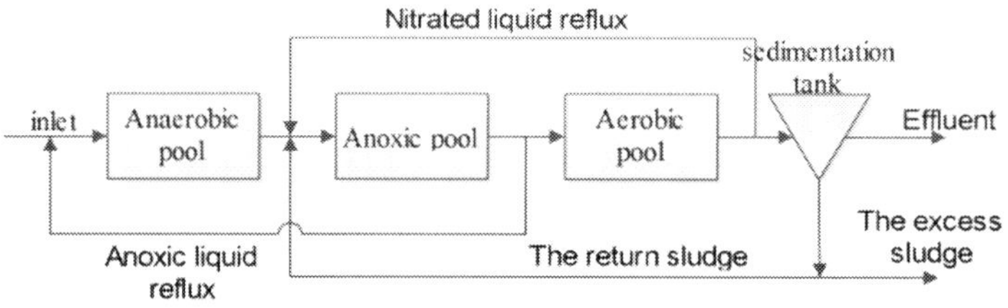

Figure 1. Process flow of UCT.

2.1.2. BCFS. In 1996, professor Mark of Delft university in Holland developed a new process named BCFs, which was based on the Pasveer oxidation ditch and used the principle of UCT process, it could improve the biological phosphorus removal and add a chemical phosphorus removal unit in anaerobic pool [3, 5]. The BCFs process was consisted of five reactors with independent function(Anaerobic pool, Contact pool, Anoxic pool, Anoxic/aerobic pool, Aerobic pool) and three circulation systems. Circulating A makes sludge release phosphorus. Circulating B adds nitrate nitrogen into the anoxic pool. Circulating C increases the nitrification or the simultaneous nitrification and denitrification. Process flow of BCFs was shown in figure 2.

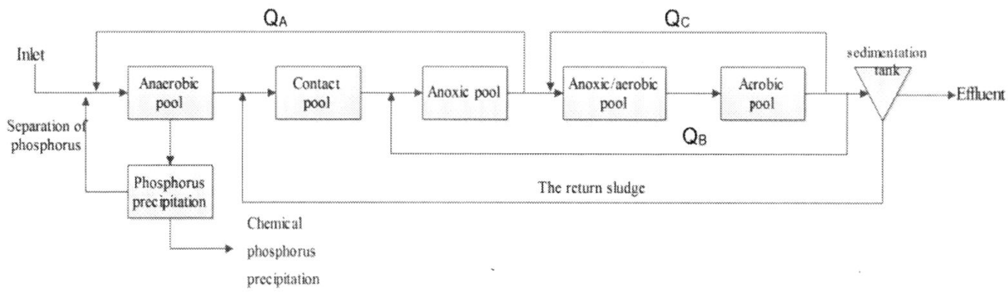

Figure 2. Process flow of BCFs.

BCFS increased a contact pool and an anoxic/aerobic pool, it is the main difference between BCFs and UCT. Returning sludge and the mixture of anaerobic pool mixed in contact pool, it could consume the residual COD in anaerobic effluent. Meanwhile, the nitrate nitrogen in sludge was also reduced to nitrogen [3]. The anoxic/aerobic pool was used to remove nitrogen, compared with the usual oxidation ditch process it was in low dissolved oxygen environment. However, the disadvantage of the BCFS was the large area of the anoxic/aerobic pool which was 1/3 of system [4].

2.2. Dual sludge system

2.2.1. Dephanox. In 1992, Wanner found that Dephanox was a typical dual sludge system for nitrogen and phosphorus removal by organic substrate, the process flow of Dephanox was shown in figure 3.

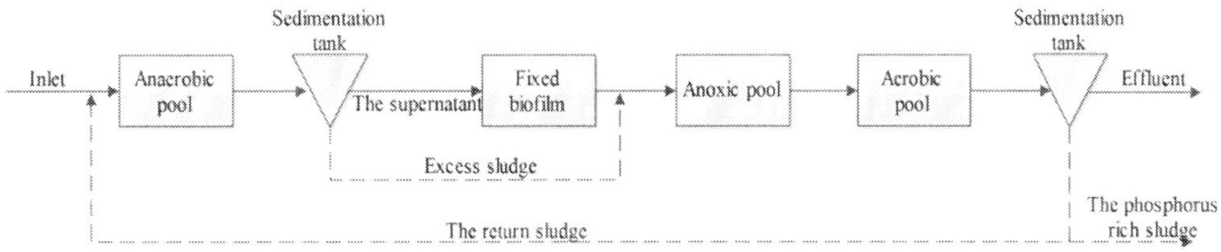

Figure 3. Process flow of Dephanox.

The characteristics of Dephanox were following:

- In the end of anaerobic pool, the muddy water mixture was separated; the nitration reaction provided sufficient electron acceptor for subsequent reactions.
- After the anaerobic phosphorus released, the sludge in sedimentation tank crossed into anoxic pool used nitrate as electron acceptor to begin the denitrifying dephosphataion.
- The residual phosphorus was further removed in second oxidation tank [4, 5].

2.2.2. A^2N. A^2N process was proposed by Kuba, et al [6], it placed DPB and nitrifying bacteria into different sludge systems. The process flow was shown in figure 4.

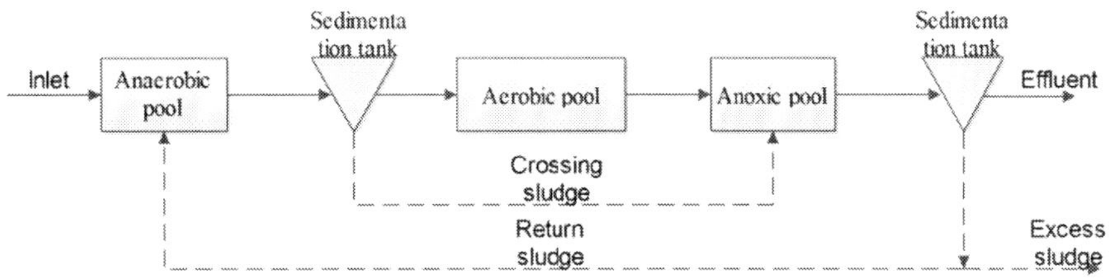

Figure 4. Process flow of A^2N.

Return sludge and sewage were mixed completely and flew to the anaerobic pool, DPB used small molecular weight VFA to synthesize PHB and stored in the cell, at the same time to release the phosphorus. The mixture was separated in middle sedimentation tank, and then the supernatant of rich ammonia entered into the aerobic pool for nitrification. The sludge in middle sedimentation tank directly crossed into the anoxic pool and used NO_3^- as electron acceptors to denitrifying dephosphataion [2].

A^2N was suitable for low C/N sewage. The phosphorus removal was not completely for lack of electron acceptor to the high C/N sewage, so it needed to add an aerobic pool after the anoxic pool to consume the residual PHB and removed phosphorus furtherly using O_2 as electron acceptor. Finally, the sludge and sewage were separated in the secondary sedimentation tank and the sludge was returned to anaerobic pool for reaction.

The disadvantages of the process were:

- The ammonia nitrogen concentration was high. Because the sludge without nitrification directly into anoxic pool would carry ammonia nitrogen and the ammonia nitrogen treatment effect was very low at later stage, so the ammonia nitrogen concentration was too high.
- The stability of system was difficult to control. When the anoxic phase lack nitrate would lead phosphorus uptake effect was poor. When nitrate was excessive would make excess nitrate added into sludge, further affected on anaerobic phosphorus released and PHB synthesized.
- The process was rather lengthy [4, 6].

2.3. Comparison of two processes

Through many years of development, the progress of denitrifying dephosphataion experienced from single sludge to dual sludge system, the single sludge system was most improved from the traditional phosphorus and nitrogen removal process, it mostly used for theoretical and experimental research. There were several defects:

(1) In low temperature (generally below 15 degrees) condition. In order to satisfy the completely nitration reaction it needed long SBT, so the structure would be larger. (2) There were competitions among the phosphorus accumulating bacteria, DPB and nitrifying bacteria for the substrates. (3) The amount of sludge and returned flow was large, also needed higher requirements for operation [4, 5].

The dual sludge system compared with the single sludge system had following advantages: (1) With low energy requirements and oxygen demanded. (2) Solved the competitions among the phosphorus accumulating bacteria, DPB and nitrifying bacteria for the substrates, made the DPB had the best growth environment. (3) Reduced the number of sludge recirculation systems and the amount of excess sludge [2, 3, 5].

3. The influence factors of denitrifying dephosphataion process

3.1. *Carbon source*

3.1.1. Types of the carbon sources. Carbon source is one of the most important nutrient elements in microbial growth. In the process of nitrogen and phosphorus removal, carbon sources are most used for phosphorus release, denitrification reaction and heterotrophic bacteria metabolism. Many research results had indicated that different types of carbon sources had significant differences in phosphorus release rate, because carbon source is related to phosphorus release and intracellular storage. Generally, DPB only could use VFA for anaerobic phosphorus release, different carbon sources hydrolytic acidified into different degrees of VFA, therefore the anaerobic phosphorus release effect was also different [7]. The experimental results of Lv X M et al [8] showed that the effect of acetic acid on phosphorus release was the best using glucose, acetic acid and propionic acid as carbon source. Fu J X et al [9] found that the sewage was easily to acidification and accompany by a large number of foam floating, and the pH also decreased when glucose was used as carbon source. When acetic acid was used as carbon source, it only had bubbles generation but no acidification problem. Therefore, acetic acid could be used for phosphorus release in the anaerobic section and provided favorable conditions for the subsequent anoxic phosphorus absorption.

3.1.2. Concentration of the carbon sources. In denitrifying dephosphataion process, the carbon source concentration has effect on the maximum amount of phosphorus release directly. Li Y J et al [10] used intermittent test to study the effect of organic carbon source on nitrogen and phosphorus removal in denitrifying dephosphataion process, experimental results showed that the higher COD concentration in anaerobic zone (150 ~ 250 mg/L) the more phosphorus released, the more denitrifying phosphorus uptake, but when COD concentration exceeds 200 mg/L, the residual organic matter remained in the anoxic stage and inhibited phosphorus uptake. In anoxic zone the influence of the carbon source concentration on denitrifying phosphorus uptake mainly reflected in: (1) PHB degradation rate by DPB was lower than carbon source utilization efficiency by common nitrifying bacteria, made the denitrification preferentially, thus inhibited excessive phosphorus uptake and make the effect of denitrifying phosphorus and nitrogen removal was reduced. (2) Carbon source could reduce NO_2^-N accumulation, and inhibitory effect of NO_2^-N on anoxic phosphorus uptake [11].

3.2. *pH*

The permeability of the cell membrane and the surface charge of the microorganism are affected by pH, the growth and reproduction of microorganism are also inseparable with the pH, and therefore, it was very important to find a suitable pH for the functional bacteria. The pH has affected anaerobic phosphorus release and anoxic phosphorus uptake. Under anaerobic conditions, pH increased would lead the amount of phosphorus released raised, because pH related to cell membrane charge could affect activity of enzymes in metabolic process. At the same time, the phosphorus accumulation organisms and the glycogen accumulation organisms compete with each other, inappropriate pH would make phosphorus accumulation organisms lose their advantage in flora. Under anoxic conditions, low pH would lead the amount of anaerobic phosphorus released reduced, and also inhibited the anoxic phosphorus uptake. When pH was increased, the amount of anoxic phosphorus uptake also raised, but when pH was too high ineffective phosphorus release would happen, it had adverse effect on PHB synthesized, the amount of PHB as electron donor was reduced, and thus it had a negative effect on anoxic phosphorus absorption [12]. The test results of Li Y F et al [13] showed that phosphate was probably precipitated on the surface of micelle with the raise of pH, it inhibited the absorption of carbon source and the anaerobic phosphorus release. But too low pH would have inhibitory effect on DPB. In conclusion pH too high or too low both had an adverse effect on denitrifying phosphorus and nitrogen removal process.

The domestic and foreign scholars' research results on the optimal range of pH can been seen in table 1.

Table 1. The results of the optimal range of pH.

Author	The test method	the optimal range of pH
Li Yafeng [11]	The static test	7.0~8.0
Hu Xiaomin [10]	The method of SBR	6~8
Ren Nanqi [24]	The method of SBR	7.0~8.0
Xu Wei [25]	The method of SBR	7~8.5
Adeline S M Chua [17]	The method of SBR	7.0~8.0

3.3. The NO_3^- and NO_2^- mass concentration

3.3.1. NO_3^-. The results showed that NO_3^- and organic carbon source coexisted in anaerobic stage would make the denitrification preferentially, inhibited the release of phosphorus and the synthesis of PHB and then further affected the phosphorus uptake [11, 14].

On the other hand, in the case of sufficient carbon source, the NO_3^- concentration would determine whether the phosphorus uptake completely. Because the denitrification rate of using intracellular storage by DPB was much lower than the denitrification rate of degrading COD by conventional denitrifying bacteria, therefore, only when the NO_3^- concentration was higher than the demand of ordinary denitrifying bacteria, the DPB could make full use of NO_3^- to grow, otherwise would be suppressed [15, 16].

Zhu W T et al [16] found that when NO_3^- concentration was too low (5 mg/L), the DPB would lack electron acceptor to uptake phosphorus; When the NO_3^- concentration was too high (20 mg/L), it would lead the reduction of nitrogen removal rate. The experiment results [17] showed that in anaerobic section the nitrate nitrogen concentration should be below 10 mg/L and in anoxic section the nitrate nitrogen concentration should be controlled at about 40 mg/L, the effect of denitrifying phosphorus removal was best.

3.3.2. NO_2^-. The results showed it had no inhibition on anoxic phosphorus uptake and it could be used as electron acceptor, when NO_2^- concentration was low, but if NO_2^- concentration reached a certain degree would inhibit the reaction [6, 11]. Until now there has been no consistent conclusion about nitrite inhibition concentration, but most scholars believed that the upper limit value was related to the characteristics of phosphorus accumulating sludge [1, 18]. Yang Y Y, et al [18] suggested there were three kinds of theories about the inhibition of nitrite: (1) Free nitrous acid (FNA) had the function of uncoupling. The existing research believed that nitrite was an uncoupler, it could increase the permeability of the membrane to destroy the proton driving force. If the proton driving force as motivity for ATP synthesis was inhibited, then the energy produced by oxidation cannot be used for ADP phosphorylation, finally the coupling process of oxidative phosphorylation was separated. (2)

FNA inhibited the phosphorus absorption and the microorganism growth, reducing the energy requirements of microorganism. (3) FNA inhibited nitrite reductase.

Li W, et al [19] used anoxic sludge in A^2O process to study the effect of NO_2^- as electron acceptor on anoxic phosphorus uptake found that when nitrite concentration was too low, it couldn't provide enough electron acceptor for denitrifying phosphorus accumulating reaction; But when nitrite concentration was too high, most of denitrifying bacteria in the system became dominant strain and had poor effect in the whole process. The experimental results of Xu R [20] and Li Y F, et al [13] showed that under anaerobic-anoxic-aerobic running mode when NO_2^- concentration was 30 mg/L, the effect was the best. The study of Kang T T, et al [21] showed that under anaerobic-anoxic-aerobic running mode, gradually increased or reduced NO_2^- concentration could induce the DPB to tolerate high NO_2^- concentration, and considered the long carbon source adding time was beneficial to carbon source consumption in anaerobic section. The NO_2^- sudden input whether caused inhibition, it depended on the adaptability of DPB to nitrite, the continuous NO_2^- injection method was beneficial to relieve the inhibition of the reaction.

3.4. Sludge retention time

Sludge retention time (SRT) reflects the basic characteristics of microbial growth, growth conditions, generation cycle in the activated sludge system. It is an important technical parameter in the denitrifying nitrogen and phosphorus removal process [3]. Denitrifying phosphorus and nitrogen removal process can be divided into single sludge system and dual sludge system, because they have different nitrification section set mode and different requirements on SRT. In the single sludge system, DPB, nitrifying bacteria and other heterotrophic bacteria exist in the same environment, there are competitions between different strains. The SRT can be controlled in narrow range which can meet the needs of nitrogen and phosphorus removal. The longer the SRT, the more obvious nitrification; on the other hand, the shorter the SRT, the better effect of phosphorus removal [11]. In dual sludge system, the denitrifying bacteria and the DPB exist independently, the SRT can be selected according to their own respective conditions and they don't affect each other.

Li Y J, et al [10] used SBR experimental studied the effects of three different SRT on denitrifying phosphorus and nitrogen removal found that when the SRT was 15 d, the reaction effect was best, and SRT had a significant effect on the COD, nitrogen and phosphorus removal in SBR. The study of Fu J X, et al [9] showed when the SRT was too long would make the sludge renew slowly and the sludge mass concentration decrease gradually, which would lead serious influence on the sludge activity; When the SRT was too short, the small amount of denitrifying sludge would eliminate the phosphorus accumulating bacteria in the reactor, and excessive sludge discharged would lead to a serious reduction of sludge mass concentration.

3.5. Glycogen accumulation organisms

The PAOs and GAOs metabolic patterns are similar, the main difference is that the energy of PAOs is derived from Poly-p hydrolyze, while the energy of GAOs is derived from glycolysis [22]. In anaerobic stage, there is a competition between PAOs and GAOs for organic substrates, resulting in unstable operating system, and GAOs can't absorb phosphate in matrix like PAOs, the effect of phosphorus removal reduced. Therefore, it is necessary to control the competition between GAOs and PAOs on the carbon sources, maintain the efficient and stable phosphorus removal capacity [23].

Ren L, et al [23] believed that the influent C/P was a significant factor to competition between PAOs and GAOs. PAOs metabolic process needed more carbon sources to provide the energy, so the influent with low carbon sources could affect the phosphorus removal capacity. The high carbon source concentration was not conducive to PAOs growth, because the dephosphorus particles of PAOs were reduced and inhibited its growth in condition of high C/P, but the GAOs needed less phosphorus to grow. Wang Y Y, et al [1] studies showed that the temperature was also an important factor affecting the competition between PAOs and GAOs. At the temperature of 20 degrees, PAOs was the dominant strain in the phosphorus removal system. But when the temperature rose to 30 degrees, GAOs replaced

PAOs as the dominant strain in the system.

4. Denitrifying phosphorus removal mechanisms

Under anaerobic/anoxic alternate operating conditions, DPB used NO_3^--N or NO_2^--N as electron acceptor for phosphorus and nitrogen removal, it can be seen from figure 5. Compared with the traditional biological nitrogen and phosphorus removal process, it had advantages like low running cost, low energy consumption and less oxygen demand.

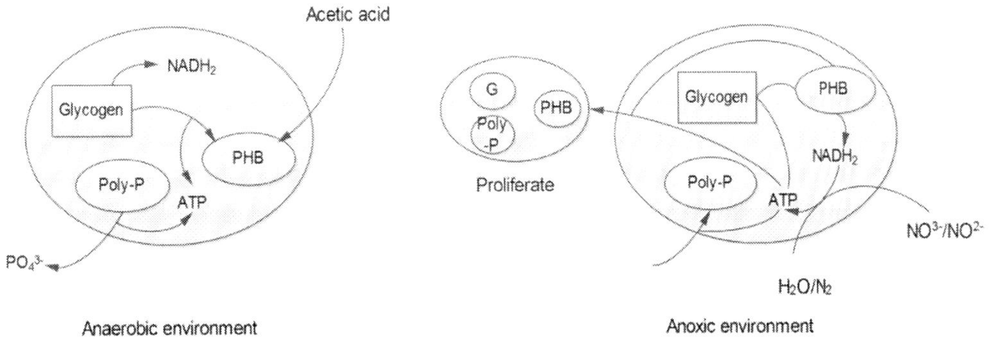

Figure 5. Metabolic model of denitrifying phosphorus removal.

Under anaerobic conditions, acetic acid entered DPB by active transport to make it multiply. This process consumed the proton motive force, so it was necessary to reconstruct or restore the proton motive force by decomposing the intracellular polyphosphate and releasing in the form of inorganic phosphate [14, 22, 24]. Acetic acid entered the cell and stored in the form of PHB. PHB was a kind of high molecular polymer which was a straight chain lipid compound consisted of beta hydroxybutyrate monomer. PHB required reducing power and energy to synthesize, in which the degradation of polyphosphate could provide the ATP, the glycogen degradation and TCA cycle could produce reduction force. Polyphosphate is a linear chain or cyclic polymer which was orthophosphate monomer formed by ester bonds. In anaerobic conditions, the polyphosphate in DPB was hydrolyzed, part of the energy made ADP translate to ATP which was provided to PHB synthesized. And another made ADP translate to ATP, the orthophosphate formed by hydrolyzing was excreted. Some scholars [25] believed that reducing power was produced by intracellular glycogen degraded. Under anaerobic conditions, the glycolytic pathway degraded glycogen to pyruvate and at the same time produced the NADH. A part of pyruvate directly entered into the TCA cycle and eventually converted to carbon dioxide, another was converted to acetyl coenzyme A for the synthesis of PHB. Some scholars [26] believed that the reducing power was generated under anaerobic conditions by the TCA cycle. However, some researchers have suggested that under anaerobic conditions DPAOs had only a part of TCA cycle been used. The TCA cycle generated NADH through a series of oxidation reaction and produced acetyl coenzyme A through part of the reduction reaction [26].

Under anoxic conditions, DPAOs used NO_3^- or NO_2^- as electron acceptor to oxidize PHB which was stored in cell to produce acetyl coenzyme A as carbon source for microbial growth, and produced the ATP, a part of it was used for its own synthesized and sustain life activities, another part was used for phosphorus uptake and polyphosphate synthesized. In order to stabilize the proton flow driven produced by the consumption of PHB, DPAOs excessive absorbed inorganic phosphate to synthesize Poly-p and store in cells, at the same time NO_3^- or NO_2^- was reduced to N_2 and discharged from system, finally achieve the effect of phosphorus and nitrogen removal [14, 20, 24]. Although NO_2^- can be used as electron acceptor but it was also an inhibitor for microbial growth and metabolism, so it was very important to control the NO_2^- concentration in denitrifying phosphorus removal process.

5. Conclusions

- Denitrifying phosphorus removal dual sludge system had low energy needs and oxygen requirement. Solve the competition on matrix and nutrient between species, and reduced the number of sludge return system, compared with single sludge system.
- Under anaerobic conditions, add acetic acid as substrate can provide favorable conditions for the subsequent anoxic phosphorus uptake. When pH was controlled between 7.0 and 8.0, it was suitable for the growth of DPB. Under anaerobic conditions, the NO_3^- mass concentration should be less 10 mg/L. Under anoxia conditions, the NO_3^- mass concentration was about 40 mg/L. When NO_2^- was used as electron acceptor, the NO_2^- mass concentration was about 30 mg/L, the effect of denitrifying phosphorus removal was best. SRT of 15 days was favorable for denitrifying phosphorus removal.
- In the anaerobic phase, DPB absorbed and degraded carbon source to synthetize PHB and store it in cell. In the anoxic phase, the excessive phosphorus absorption was accomplished by PHB oxidation, and the NO_3^- or NO_2^- was restored, which made the two opposing processes phosphorus absorption and denitrification occur together.

Acknowledgments
This work was supported in part by the Doctoral Scientific Research Foundation of Liaoning Province, China (No. 201501069), and Foundation of Liaoning Educational Committee (No. LJZ2016014). In addition, the study was sponsored by National Nature Science Foundation of China (No. 51508342). National Water Pollution Control and Governance Scientific and Technological Special Project (2014ZX07202-011). The authors declare that they have no conflict of interest.

References

[1] Wang Y Y, Wang S Y and Peng Y Z 2005 The influence of MLSS, pH and NO_2^--N on denitrifying dephosphatation *China Water Wastewater* **7** 47-51
[2] Qiu L P, Sun C J, Wang J B, *et al* 2015 Research advances of denitrifying phosphorous removal technology *J. Univ. Jinan* **3** 161-6
[3] Zhang J 2016 The development and regulation factors of denitrifying phosphorus removal process *Environ. Eng.* **1** 266-9
[4] Wang J Y and Li W G 2015 Technology and research progress of denitrifying dephosphataion *J Shandong Jianzhu Univ.* **3** 271-6
[5] Sun L and Qian Y H 2015 Advance of research on denitrifying phosphorus accumulating organisms *Water Sav. Irrig.* **2** 40-4
[6] Kuba T, van Loosdrecht M C M and Heijnen J J 1997 Biological dephosphatation by activated sludge under denitrifying conditions: pH influence and occurrence of denitrifying dephosphatation in a full-scale waste water treatment plant *Water Sci. Tech.* **36** 75-82
[7] Tang Q F 2012 Study on denitrifying phosphorus removal process and its influencing factors *Mod. Busi. Tra. Ind.* **18** 197-8
[8] Lv X M 2014 Microbial community structure and process control strategy of denitrifying phosphorus removal (Harbin: Harbin Institute of Technology Environmental Science and Engineering)
[9] Fu J X, Zhao L, Chi F Q, *et al* 2009 Research on the affecting factors of denitrifying phosphorus removal with nitrite *J. Shenyang Jianzhu Univ.* **3** 531-4
[10] Li Y J 2013 Compare of enrichment and metabolism characteristics between denitrifying phosphorus removal bacteria and traditional phosphorus accumulating organisms (Tianjin, China: Tianjin University)
[11] Li X F, Yan B, Li M, *et al* 2014 Research and development on the influencing factors in the process of denitrifying dephosphorization *Ind. Water. Treat.* **4** 10-3
[12] Hu X M, Li W, Liu J L, *et al* 2013 Influence of pH on denitrifying phosphorus removal using nitrite as electron acceptor *J. Cent. S. Univ.* **5** 2144-9

IOP Conf. Series: Earth and Environmental Science **82** (2017) 012090 doi:10.1088/1755-1315/82/1/012090

[13] Li Y F, Wang X, Gao Y, *et al* 2013 Study on effects of different organic matter, nitrite and pH for denitrification nitrogen and phosphorus removal *J Shenyang Jianzhu Univ* **3** 531-7
[14] Bai S Y, Zhang H, Chen S T, *et al* 2013 Factors influencing denitrifying phosphorus removal characteristics of activated sludge *China Water Wastewater* **15** 46-9
[15] Han Y, Xu S, Dong T, *et al* 2015 Effects of carbon sources, temperature and electron acceptors on biological phosphorus removal *Environ. Sci.* **5** 590-6
[16] Zhang J L, Wen Y B, Li L B, *et al* 2007Discussion on the affecting factors of denitrifying phosphorous removal *Safe* and *Environ. Eng.* **4** 67-70
[17] Xu W, Xie S B, Tian F X, *et al* 2007 Study on the influence factors of short-cut denitrification *Environ. Sci. Technol* .**1** 5-8
[18] Yang Y Y, Zeng W, Liu J R, *et al* 2010 Effect of nitrite on biological phosphorus removal in wastewater *Microbiol. China* **37** 586-93
[19] Li W, Liu J, Meng H T, *et al* 2016 Shortcut denitrifuing phosphorus removal In A2N-SBR process *Environ. Eng.* **8** 62-7
[20] Xu R, Wang J F, Chen C J, *et al* 2015 Ingluence fsctors Of denitrifuing phosphorous removal using nitrite as an electron acceptor *Technol. Water Treat.* **3** 11-4
[21] Kang T T, Wang L, He Y Y, *et al* 2016 The performance of DPAO using nitrite as electron acceptor and its practical application *China Environ. Sci.* **36** 1705-14
[22] Ye L H, Li D, Zhang J, *et al* 2016 Advance of research on the technology of nitrite-denitrifying phosphorus removal *J. Beijing Univ. Technol.* **4** 585-93
[23] Ren L, Xing L Z, Sun F F, *et al* 2014 Research progress of affecting factors of competition between phosphate accumulating organism and glycogen accumulating organism in biological phosphorus removal *J. Shandong Jianzhu Univ.* **29** 170-5
[24] Kang T T and Wang L 2016 The performance of DPAO using nitrite as electron acceptor and its practical application *China. Environ. Sci.* **6** 1705-14
[25] Mino T, van Loosdrecht M C M and Heijnen J J 1998 Microbiology and biochemistry of the enhanced biological phosphate removal process *Water Res.* **32** 3193-207
[26] Comeau Y, Hall K, Hancock R, *et al* 1986 Biochemical model for enhanced biological phosphorus removal *Water Res.* **20** 1511-21

IOP Conf. Series: Earth and Environmental Science **82** (2017) 012091 doi:10.1088/1755-1315/82/1/012091

Spatial-temporal variation of ecosystem water use efficiency in Beijing's suburban region

F Mi[1], Q Zhang[1], X C Zhang[1], S B Yuan[2], N Lu[3] and N Na Yan[4]

[1]School of Economics and Management, Beijing Forestry University, 35 Qinghua East Rd, Haidian District, Beijing, 100083, China

[2]Afforestation Office, Beijing Gardening and Greening Bureau, 1 Xiaohuangzhuang North St, Dongcheng District, Beijing, 100013, China

[3]State Key laboratory of Urban and Regional Ecology, Research Center for Eco-Environmental Science, Chinese Academy of Sciences, 18 Shuangqing Rd, Haidian District, Beijing, 100085, China

[4]Institute of Remote Sensing and Digital Earth, Chinese Academy of Sciences, West Beichen Road, Chaoyang District, Beijing, 100101, China

E-mail: mifengsun@163.com

Abstract. Suburban ecosystem has multiple functions such as soil conservation and water regulation, which are critical for the welfare of human beings in the city. Water use efficiency (WUE) is an important indicator of ecosystem function that represents the amount of productivity per unit mass of evapotranspiration (ET). Improving WUE of suburban ecosystem is significant to climate regulation by carbon sequestration and water consumption, especially for cities with severe water shortage like Beijing, the capital of China. Based on remote sensing data, this paper examined the spatial and temporal variations in WUE in Beijing's suburban region from 2002 to 2010. The results showed that the average annual WUE was 0.868 g C mm^{-1} m^{-2}. It has large spatial variation with the minimum of 0.500 g C mm^{-1} m^{-2} in the Miyun District. During the study periods, the area with significant increasing trend of WUE was 63.2% of the total suburban region. In terms of ecosystem type, the value of WUE was following the sequence, deciduous coniferous forest (0.921g C mm^{-1} m^{-2}) > mixed forest (0.887g C mm^{-1} m^{-2}) > deciduous broadleaf forest (0.884 g C mm^{-1} m^{-2}) > shrubland (0.860 g C mm^{-1} m^{-2}) > evergreen coniferous forest (0.836 g C mm^{-1} m^{-2}) > grassland (0.830 g C mm^{-1} m^{-2}). As ET was similar among the ecosystems, the difference in WUE was mainly due to the discrepancy of NPP. We found that NPP significantly correlated with the diversity of ecosystem type (represented by Shannon-Wiener index). Our results suggest that ecological engineering construction, scientific ecosystem type selection, ecosystem diversity improvement and drought-resistant species cultivation are conductive to improve ecosystem WUE in Beijing's suburban region.

1. Introduction

Socioeconomic development and population growth accelerated the urbanization process, while the neglect of ecological environment protection caused the deteriorating eco-environment [1]. Frequent urban environment problems, such as air pollution, water pollution, and etc., give serious threat to the sustainable development of cities [2]. Urban green space is an important part of urban ecosystem, which plays an instructive role in maintaining urban ecological balance and improving eco-environment quality [3]. The development of green space in central city is restricted by the shortage of urban land, therefore, it is obviously important to construct suburban green space to protect urban ecological environment. In fact, green space needs to consume large quantities of water during the exertion of ecological function [4]. The fast development of suburban green space would aggravate contradiction between water deficient and green space water use [5]. Therefore, improving ecosystem water use efficiency in suburban region is contributed to relief urban water shortage.

Water use efficiency (WUE) is the amount of CO_2 fixed (or the amount of organic matter produced) by per unit mass of water consumption, which is an important indictor to evaluate the level of carbon and water coupling in urban ecosystem, and a quantitative indictor to assess the efficiency of water input and output in green space [6]. Researchers adopted different approaches to analyze the ecosystem WUE from multiple scales. The early research scale was primarily the leaf scales and vegetation scales, and researchers measured the transpiration rate and photosynthetic rate of leaf or vegetation to calculate the WUE by the experimental methods or measuring methods. The eddy covariance development advanced the WUE analysis in community scales and local scales. Using the eddy covariance technology, the data of CO_2 fluxes and evapotranspiration (ET) nearly the site can be measured, which can be intended for the WUE calculation. Meanwhile, the WUE analysis of region scales has been widely used with the development of remote-sensing technology. The investigators use the high precision remote-sensing date to assess the spatiotemporal dynamic of regional WUE [7]. The previous researches indicated that using the remote-sensing method can assess the regional WUE accurately, but study on the ecosystem WUE of suburban region is still in its relative infancy.

Deterioration of urban ecological environment has aroused the widespread attention of the community in Beijing. To solve the increasingly serious environmental problems, Beijing municipal government identifies the western and northern suburban region as the forest reserve [8]. The strengthening of forest protection has prompted the increased area of suburban ecosystems. The suburban ecosystems were 744.956 thousand hectares at the end of 2015, which has increased 125.713 thousand hectares comparing to 2005. The growth of suburban ecosystem contributed to the improvement of eco-environment quality, yet the water consumption of ecosystem has also increased significantly. Beijing is one of regions with severe water shortage in the world, and water safety problem has already restricted sustainability of the city. The improvement of water consumption of suburban ecosystem may have great influence on the water inflow of mountainous reservoirs [9]. The mountainous reservoirs are the most important water resource for Beijing, whose water storage may impact the amount of water available. Therefore, studying on spatial-temporal variation of ecosystem WUE in suburban region may help to understand the rule of ecosystem water consumption and then improve its water use efficiency.

In this study, we aimed to improve the ecosystem WUE in Beijing's suburban region by reasonable ecosystem collocation. Specifically, we attempted to explore following questions: 1) the temporal variation of ecosystem WUE from 2002 to 2010 in Beijing's suburban region; 2) the influence of ecosystem type (deciduous coniferous forest, evergreen coniferous forest, deciduous broadleaf forest, mixed forest, shrubland and grassland) and ecosystem diversity on WUE.

2. Materials and methods

2.1. Study area

The suburban region of Beijing (39°26-41°03'N, 115°25'-117°30'E) distributes in the western and northern of Beijing (figure 1), including 7 districts (Pinggu, Miyun, Huairou, Yanqing, Changping, Mentougou and Fangshan). The area of suburban region is 1041.750 thousand hectares, which accounts for 62% of Beijing's total area [10]. The climate of Beijing's suburban region is warm-temperate continental monsoon climate. The average temperature is 10℃-12℃ and average annual precipitation amount is 585 mm, and the precipitation is mainly concerned in July and September [6]. The typical ecosystem type in Beijing's suburban region mainly includes deciduous coniferous forest, mixed forest, deciduous broadleaf forest, evergreen coniferous forest, shrub and grass (figures 1(b) and 1(c)). The area of suburban ecosystem had increased nearly 5423.490 hectares from 2000 to 2010. The area of deciduous broadleaf forest (change area: 15459.390 hectares) and evergreen coniferous forest (change area: 1178.370 hectares) increased rapidly, while the area of shrubland (change area: 95059.220 hectares) and grassland (change area: 2214.360 hectares) decreased rapidly.

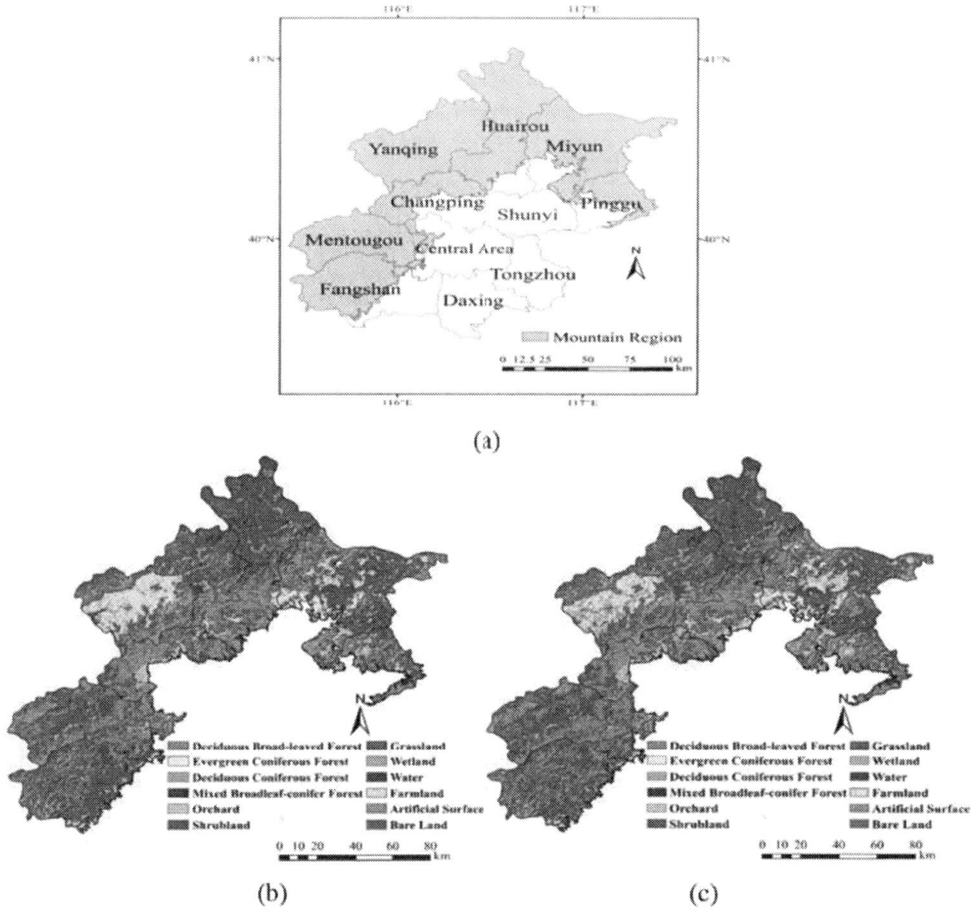

Figure 1. The location (a) and land use/cover distributions in 2000 (b) and 2010 (c) of Beijing's suburban region.

2.2. Methods

The paper used the water use efficiency equation to calculate the ecosystem WUE [11,12].

$$WUE_i = NPP_i \,/\, ET_i \tag{1}$$

Where, WUE_i is the mean annual ecosystem water use efficiency (g C mm^{-1} m^{-2}); NPP_i is the mean annual net primary productivity (g C m^2); ET_i is the mean annual evapotranspiration (mm); i is the study year, and 1 for 2002, 2 for 2003, …, and 9 for 2010.

Later, we used the ordinary least square (OLS) estimation to analyze the changing trend of ecosystem ET, NPP and WUE in each pixel from 2002 to 2010 [13]. In addition, a F-test method was used to test the significance level of changing trend.

$$slope = \left(n\sum_{i=1}^{n} i \times WUE_i - \sum_{i=1}^{n} i \sum_{i=1}^{n} WUE_i \right) \left[n\sum_{i=1}^{n} i^2 - \left(\sum_{i=1}^{n} i \right)^2 \right]^{-1} \quad (2)$$

In addition, a F-test method was used to test the significance level of changing trend. The *slope* > 0 and P-value < 0.01 represents its extremely significant increased; the slope > 0 and P-value < 0.05 represents its significant increased; the *slope* > 0 and P-value > 0.05 represents its non-significant increased; and vice versa.

Finally, the paper used the Shannon-Wiener equation (3) to calculate ecosystem diversity index in each pixel. Meanwhile, mean NPP of pixels would be calculated when the Shannon-Wiener index was same in multiple pixels, which used to indicate the NPP in this this Shannon-Wiener index. Then, we used the Pearson correlation coefficients and image analysis methods to study the relationship between NPP and ecosystem diversity index [14].

$$H' = -\sum_{j=1}^{s} \left(N_j N^{-1} \right) \log_2 \left(N_j N^{-1} \right) \quad (3)$$

Where, H' is ecosystem diversity index; N_j is the jth ecosystem type area; N is the total ecosystem area; j is the ecosystem type.

2.3. Data and information sources

The land use data (resolution: 30 m) in 2000 and 2010 were originated from the study results of Chinese Ten Years Eco-Environment Remote Sensing Monitoring project, and classification system is IPCC and LCCS. Annual actual ET data (resolution: 1km) for Beijing from 2002 to 2010 were came from ETWatch products. The ETWatch data is calculated by using SEBS model and SEBAL model, the precision of which was verified in Hai Basin using by the eddy correlation instrument and ground-based lysimeter flux observations [15]. The NPP data (resolution: 250 m) for Beijing from 2002 to 2010 were originated from MODIS – NPP products (https://modis.gsfc.nasa.gov/). The DEM data (resolution: 30 m) was originated from USGS (https://www.usgs.gov/).

3. Results and discussions

3.1. The spatial variation of WUE

The spatial distribution of ecosystem WUE from 2002 to 2010 was shown in the figure 2(a). The average annual WUE was 0.868 g C mm^{-1} m^{-2} (ranging from 0.217 to 1.907 g C mm^{-1} m^{-2}), which matched well with the results reported by Yu, et al (0.853 g C mm^{-1} m^{-2}) and Xia, et al (0.873 g C mm^{-1} m^{-2}) [16,17]. The average annual WUE had obvious difference in spatial distribution. The ecosystem WUE in southwest region was significantly higher than the northeast region. The ecosystem WUE in Mentougou, Fangshan and central region of Yanqing were more than 1.000 g C mm^{-1} m^{-2}, while the ecosystem WUE in Miyun was less than 0.500 g C mm^{-1} m^{-2}. The ecosystem WUE in other regions mainly ranged from 0.500 g C mm^{-1} m^{-2} to 1.000 g C mm^{-1} m^{-2}, and the ecosystem WUE in sporadic regions was more than 1.000 g C mm^{-1} m^{-2}.

Overall, the spatial distribution of ET had no significant difference, which mainly remained in the range of 550 – 700 mm (figure 2(b)) while the NPP had obviously different in spatial distribution, which basically coincided with the spatial distribution of WUE (figure 2(c)). It showed that the spatial distribution difference of WUE was mainly caused by different NPP. The ecosystem NPP was more than 550 g m^2 in Mentougou, Fangshan and Yanqing, which caused WUE in these regions were higher than 1.000 g C mm^{-1} m^{-2}. The ecosystem WUE in Miyun was lower than 0.500 g C mm^{-1} m^{-2}, which due to the regional NPP was lower than 450 g m^2. Meanwhile, the spatial distribution difference of NPP was closely related to the spatial distribution of ecosystem type. By comparing the spatial distribution of land use (figure 1(b)) and NPP (figure 2(a)), the regions with the higher NPP (NPP ≥

550 g C m^2) mainly covered by deciduous broadleaf forest and mixed forest, while the shrubland mainly distributed in the regions with the lower NPP (NPP ≤ 450 g C m^2).

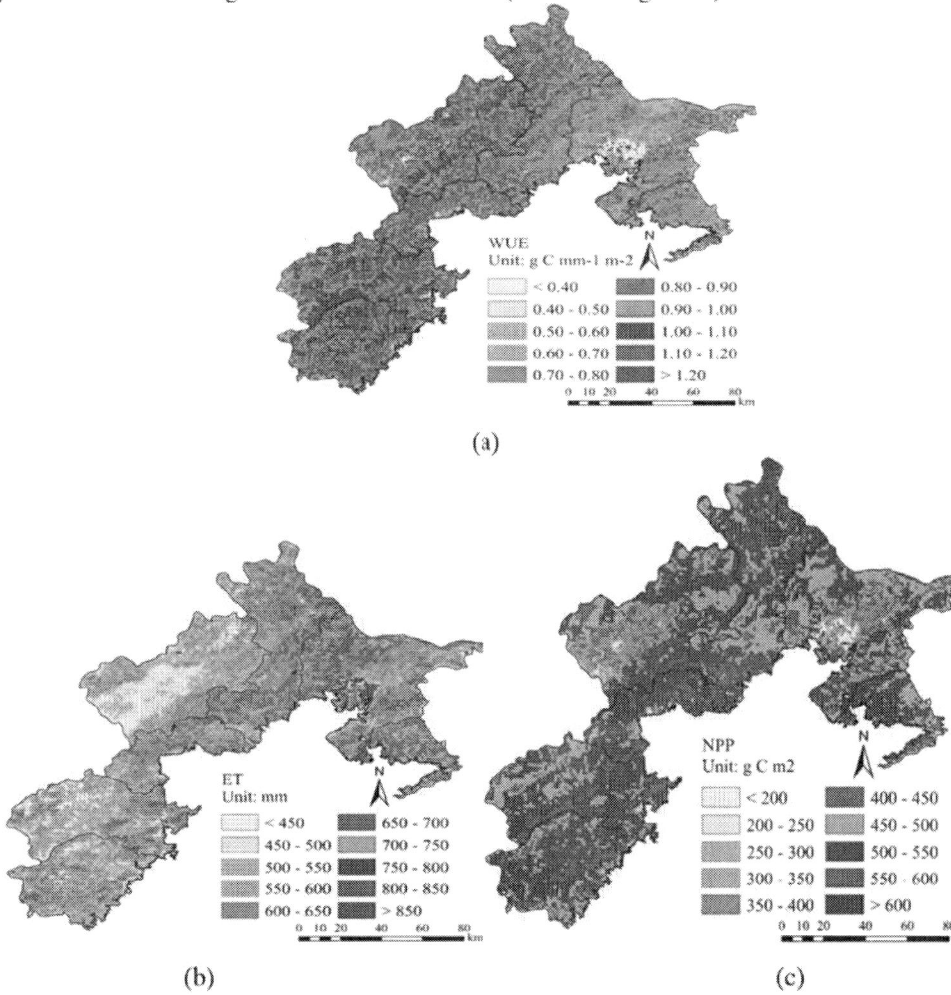

Figure 2. The spatial distribution of average annual WUE (a), average annual ET (b), and average annual NPP (c) from 2002 to 2010 in Beijing's suburban region.

3.2. The trend of change in WUE

The temporal variation of ecosystem WUE from 2002 to 2010 was showed in figure 3(a). There were nearly 63.19% of ecosystem WUE increased during the study period. Through the significance test of variation coefficient, 6.21% of ecosystem experienced a significant WUE increase ($P < 0.05$). The ecosystem that experienced an extremely significant increase of WUE ($P < 0.01$) accounted for 1.22% of the total ecosystem cover. The ecosystem WUE in Miyun and the western region of Pinggu showed a significant increasing trend, and the regions of Huairou and Mentougou's western region experienced a non-significance increase of WUE ($P > 0.05$). In addition, the regions experiencing the decrease of WUE were mainly distributed in the northern region of Yanqing, the western region of Changping and Fangshan, yet decreased trend was not significant ($P > 0.05$).

The temporal variation of ecosystem ET and NPP led to the WUE variation. During the study, the ecosystem with the decreasing ET accounted for nearly 73.03% of the total ecosystem (figure 3(c)), and the ecosystem with the increasing NPP was nearly 28.19% (figure 3(d)). The decreasing ET and increasing NPP caused the WUE experiencing an increased trend in most regions. The variation of ET and NPP may be influenced by the increase of CO_2 concentration and the human protection activities [18,19]. The CO_2 emissions was more than 150 million ton, which rose by about 15% compared with 2002 (nearly 121 million ton) in Beijing. The rapidly increasing CO_2 emissions caused the improvement of CO_2 concentration, which directly affected the ecosystem NPP increase. The projects of Conversion of Cropland to Forest Program (CCFP) and Sand Control Program for Areas in the Vicinity of Beijing & Tianjin (SCP) were carried out in Beijing suburban regions since 2002, which may advance the restoration of natural hydrological balance and then contributed to the decline of ET. In addition, the regions with significant increase WUE may closely related to NPP increased. Comparing with the figures 3(b) and 3(d), we can find that the regions with the significant increase WUE mainly have the significant increase NPP ($NPP_{slope} > 5.0$).

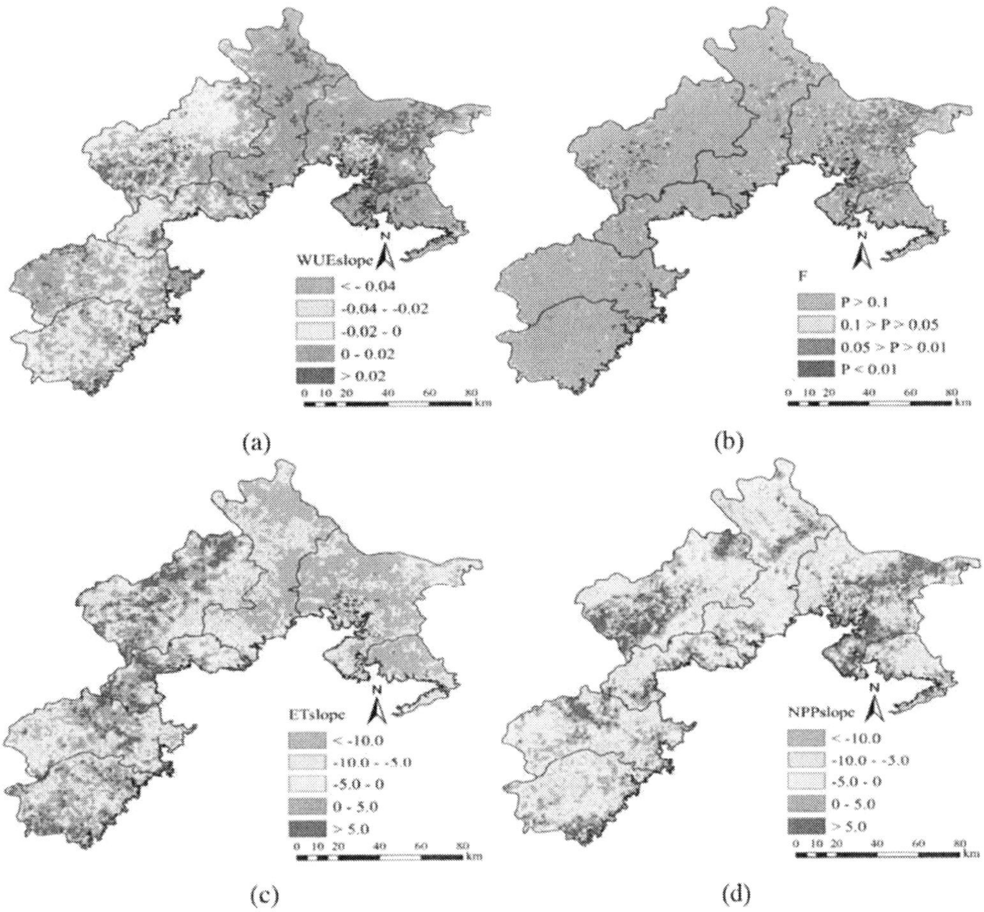

Figure 3. The temporal variation (a) and significant test (b) of WUE, the temporal variation of ET (c) and NPP (d) from 2002 to 2010 in Beijing's suburban region.

3.3. The difference of WUE among ecosystem types

Based on the analysis of spatial-temporal of ecosystem WUE, the paper calculated different ecosystem WUE. The average annual WUE of each ecosystem type was different (figure 4). The maximum average annual WUE occurred in the deciduous coniferous forest (0.921 g C mm^{-1} m^{-2}), yet the minimum average annual WUE was found in the grassland (0.830 g C mm^{-1} m^{-2}). The average annual WUE of other ecosystem types were respectively 0.887 g C mm^{-1} m^{-2} (mixed forest), 0.884 g C mm^{-1} m^{-2} (deciduous broadleaved forest), 0.860 g C mm^{-1} m^{-2} (shrubland) and 0.836 g C mm^{-1} m^{-2} (evergreen coniferous forest). Result of sorting each ecosystem WUE was consistent with the other regions in the same latitude [20,21].

The difference among each ecosystem WUE was caused by the ET and NPP. Each ecosystem NPP was in the range of 500 – 520 g C m^2 except grassland. Ecosystem ET made the WUE different in each ecosystem types. Deciduous coniferous forest had the physiological characteristic of defoliation and smaller LAI, which made the ET (567.7 mm) lower than other ecosystem types. Due to high LAI, the ET of mixed forest and deciduous broadleaf forest were higher than deciduous coniferous forest, which caused the mixed forest WUE and deciduous broadleaf forest WUE were lower than deciduous coniferous forest. Although evergreen coniferous forest LAI was same as deciduous coniferous forest, the physiological characteristic of evergreen made it had higher ET in spring, which directly resulted in the lower WUE. Grassland WUE was lower than other ecosystems WUE due to the lower NPP. Grassland ET (582.66 mm) was same as mixed forest (579.12 mm), yet the NPP (480.054 g C m^2) was significantly lower than other five ecosystem types. In addition, the WUE was smaller than 1.000 g C mm^{-1} m^{-2} because the average annual NPP of each ecosystem types was significantly lower than the ET. Thus, the improvement of ecosystem NPP was of great importance for regional WUE in Beijing's suburban region.

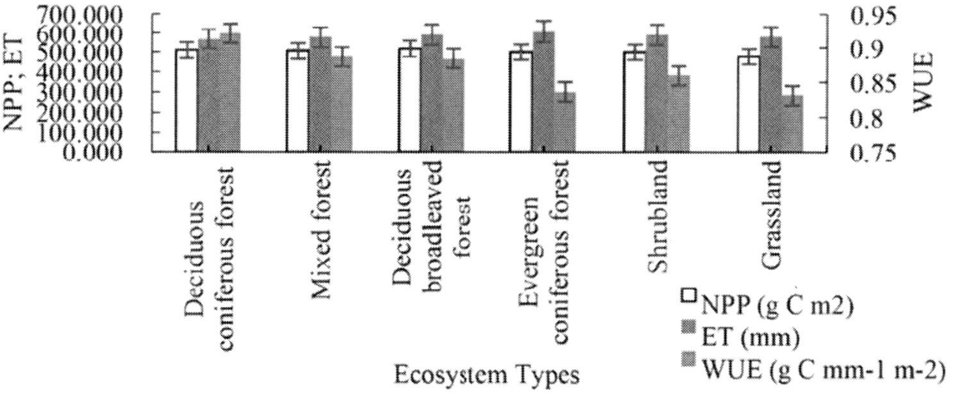

Figure 4. The difference of WUE among ecosystem types in Beijing's suburban region.

3.4. The correlation between ecosystem diversity and NPP

The Shannon-Wiener index indicated the degree of ecosystem diversity in study objectives. The range of Shannon-Wiener index in each pixel was 0 – 2.264, which showed that the degree of ecosystem diversity had certain difference in Beijing's suburban region. However, the Shannn-Wiener index in nearly 37.786% of pixel was 0, which indicated that the distribution of ecosystem types was relatively simple in most regions. The previous researches had demonstrated the ecosystem diversity affects the level of ecosystem utilization resources and the ability to adapt to environment changes. Therefore, increasing the degree of ecosystem diversity was of great significance to the construction of regional ecological environment.

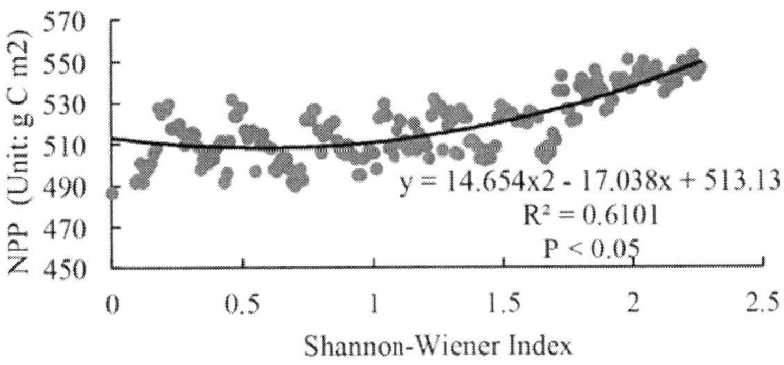

Figure 5. The relation between Shannon-Wiener Index and ecosystem NPP in Beijing's suburban region.

The ecosystem NPP was the key influencing factor of WUE and the foundation of ecosystem service function, which was closely related to ecosystem diversity [22]. Through the analysis of correlation coefficients and fluctuation trend, we found that the Shannon-Winner index and ecosystem NPP had strongly positive correlation. The correlation coefficient of Shannon-Winner index and ecosystem NPP was 0.704 ($P < 0.01$), which showed that the ecosystem NPP might be impacted with the ecosystem diversity level. Meanwhile, figure 5 showed that ecosystem NPP and Shannon-Winner index were fitted by the second order polynomial, and the relationship expression was $y = 14.654x^2 - 17.038x + 513.13$. The results showed that the improvement of ecosystem diversity could help increase ecosystem NPP, and then increase ecosystem WUE.

3.5. The measures of improving suburban ecosystem WUE

Although WUE of Beijing's suburban ecosystem region remains consistent with mean WUE of eastern China, improving suburban ecosystem WUE is imperative with water resource crisis serious in Beijing. Therefore, the study puts forward some measures about improving the ecosystem WUE based on the above analysis. Firstly, the management departments should continue to promote ecological engineering construction and strengthen ecological protection in Beijing's suburban. The ecological engineering construction will contribute to improve the suburban ecosystem cover rate. Meanwhile, the ecological protection will reduce the behaviors of excessive deforestation and blind land reclamation. Secondly, forestry departments should take the ecosystem WUE in consideration in the process of selection and cultivation of ecosystem types. The ecosystem types with high WUE should be developed gradually, such as deciduous coniferous forest and mixed forest. However, grassland areas should be dropped because grassland WUE is obviously lower than other ecosystem types. Thirdly, management departments should improve the structure of ecosystem and realize reasonable ecosystem collocation. The results showed the ecosystem diversity had positive impact on ecosystem WUE, so the management departments should optimize the allocation proportion of arbor-shrub-grass and construct the arbor-shrub-grass compound land. Lastly, scientific institutions should accelerate the process of drought-resistant species cultivation, which may effectively reduce the water consumption of ecosystem and then improve suburban ecosystem WUE.

4. Conclusion

This paper studies ecosystem WUE in Beijing suburban region from 2002 to 2010. WUE has obvious difference in spatial distribution, with the northeast region was lower than the southwest region, which was consistent with the spatial distribution of NPP. The mean WUE gradually increased during the study period, with 63.19% of area had significant increasing trend. The result of sorting the WUE values according to ecosystem type was that deciduous coniferous forest > mixed forest > deciduous broadleaf forest > shrubland > evergreen coniferous forest > grassland. As ET was similar among the ecosystem types, the differences in WUE was mainly caused by that of NPP. The further analysis between NPP and ecosystem diversity index showed that diverse ecosystem types were facilitated to have higher NPP. Therefore, the study proposed some measures of improving WUE based on the above results, including the construction of ecological engineering forest, scientific selection of ecosystem types, improvement of ecosystem diversity and cultivation of drought-resistant species. This will be helpful for improving water conservation function of suburban ecosystem and relieving urban water shortage in Beijing.

Acknowledgments

Financial support for this research was provided by the Project of Social Science Youth Foundation of Beijing Municipal (Grant No. 15JGC148), the Project of the General Program of National Social Science Foundation of China (Grant No. 15BGL130), and the Project of Humanity and Social Science Youth Foundation of Education Ministry of China (Grant No. 13YJCZH131). Feng Mi, Qin Zhang and Xuanchang Zhang conceptualized the article, conducted background research, and edited the content. Shibao Yuan, Nan Lu and Nana Yan provided the ETWatch data and edited the content.

References

[1] Wolch J R, Byrne J and Newell J P 2014 *Landscape Urban Plan.* **125** 234-44
[2] Colding J and Barthel S 2013 *Ecol. Econ.* **86** 156-66
[3] Pataki D E, Carreiro M M, Cherrier J, Grulke N E, Jennings V, Pincetl S, Pouyat R V, Whitlow T H and Zipperer W C 2011 *Front. Ecol. Environ.* **9** 27-36
[4] Barthel S and Isendahl C 2013 *Ecol. Econ.* **86** 224-34
[5] Hornbeck J W, Adams M B, Corbett E S, Verry E S, Lynch J A, Hornbeck J W, Adams M B, Corbett E S, Verry E S and Lynch J A 1993 *J. Hydrol.* **150** 323-44
[6] Mu S, Zhang Y and Dhruba B G C 2016 *J. Resour. Ecol.* **7** 68-76
[7] Hu Z M, Yu G R, Wang Q F and Zhao F H 2009 *Acta. Ecologica. Sinica.* **29** 1498-507
[8] Warren-Thomas E, Zou Y, Dong L, Yao X, Yang M, Zhang X, Qin Y, Liu Y and Sang W 2014 *Forest Ecol. Manag.* **334** 369-76
[9] Zheng J K, Sun G, Li W H, Yu X X, Zhang C, Gong Y B and Tu L H 2015 *Hydrol. Earth Syst. Sci.* **12** 7785-819
[10] Gao C, Cheng J and Chen X 2016 *J. Nature Dis.* **25** 69-77
[11] Ito A and Inatomi M 2012 *J. Hydrometeorol.* **13** 681-94
[12] Tian H, Lu C, Chen G, Xu X, Liu M, Ren W, Tao B, Sun G, Pan S and Liu J 2011 *Ecohydrology.* **4** 322-40
[13] Gang C, Wang Z, Zhou W, Chen Y, Li J, Chen J, Qi J, Odeh I and Groisman P Y 2015 *J. Agron. Crop. Sci.* **202** 343-54
[14] Lemieux J and Cusson M 2014 *Plos ONE.* **9** 109261
[15] Wu B F, Xiong J, Yan N N, Yang L D and Du X 2008 *Adv. Water Resour.* **19** 671-8
[16] Xia L, Wang F, Mu X, Jin K, Sun W, Gao P and Zhao G 2015 *J. Earth. Syst. Sci.* **124** 921-31
[17] Yu G, Song X, Wang Q, Liu Y, Guan D, Yan J, Sun X, Zhang L and Wen X 2008 *New Phytol.* **177** 927
[18] Keenan T F, Hollinger D Y, Bohrer G, Dragoni D, Munger J W, Schmid H P and Richardson A D 2013 *Nature.* **499** 324-7
[19] Lake J A, Quick W P, Beerling D J and Woodward F I 2001 *Nature.* **411** 154
[20] Law B E, Falge E, Gu L, Baldocchi D D, Bakwin P, Berbigier P, Davis K, Dolman A J and Falk M 2002 *Agric. For. Meteorol.* **113** 97-120
[21] Zhu Q, Jiang H, Peng C, Liu J, Wei X, Fang X, Liu S, Zhou G and Yu S 2011 *Ecol. Model.* **222** 2414-29
[22] Kessler M, Salazar L, Homelier J and Kluge J 2015 *J. Ecol.* **102** 1623-33

Linkages between precipitation cycle and slope stability: A case study in the Caijiapo landslide among Weihe River Basin

Y L Zhu[1,2,3] and G P Qi[1]

[1]College of water conservancy and Hydropower Engineering, Gansu Agricultural University, Lanzhou, Gansu 730070, China
[2]School of Water Resources and Hydropower, Xi'an University of Technology, Xi'an, Shaanxi 710048, China

E-mail: yuelu_zhu@163.com

Abstract. It is of great importance to forecast landslide in future certain time interval. Hence, this study proposed a method combination precipitation period with variety of slope stability factor based on macro scale. This study designed 5 grades of rainfall strength simulation and corresponding 5 grades of slope state using Markoff prediction technique analysis change of slope factor K in the short time. Results indicated that: (1) the rainfall in Baoji area has three periods of 5a, 13a and 25a, in which 5a is the primary period; (2) the slope safety factor K has a similar fluctuation trend compared to precipitation state, thus division slope factor K of 5 grade as safety state, basically safety state, stable state, partial risk state and risk state, respectively; (3) in the previous year if slope in the safety, basically safe and stable state, in the next year slope factor K above 1.15 has 60%, 75% and 72% probability, respectively; (4) when the previous year at partial risk or risk situation, there is high probability that value of the K is always more than 1.15, it is means to a great extent the landslide have already occurred, thus defined a concept which between "the safety state after destruction" and "safety state have no destruction".

1. Introduction

Slope instability is a geologic hazard that usually occurs after earthquake, and the water content of soil is one of the major factors for this, thereby continuous precipitation leads to engineering landslide is very common. Landslide is one of the most costly non repetition natural hazards, causing greatly disadvantageous economic, life and property and social influences. So research of linkages between precipitations with slope stability is very meaningful.

A large number of experiments and data show that [1-3], main effect of landslide is rainfall, because infiltration leads to soil water content of slope increasing and farther results decline in shear strength. Rainfall from the slope surface to inside, and additional penetration is generated by percolation which increases landslide sliding force. In the rainy season, temporary additional water and groundwater movement are instability factors for strength of unsaturated soil. However, in fact, precipitation will reduce the shear strength, so this is another important factor to influence slope stability. But the previous researches focus on seepage in rock/soil, adverse mechanical property cause by seepage, slope failure mechanism and so on [4-6]. The traditional research program is an effective method to evaluate stability of slope but the calculations and analysis on slope coefficient are ignoring rainfall accumulation of time and space.

IOP Conf. Series: Earth and Environmental Science **82** (2017) 012092 doi:10.1088/1755-1315/82/1/012092

Based on above understanding, this paper combines the rainfall cycle and probability of precipitation in the study area with assessment risks of project, and predicts the probability of landslide. It is viewing from mocro long time sequence to explore the slope stability evaluation. The primary objectives of this study are to (1) obtain the rainfall cycle in the study area; (2) reveal linkages between precipitation state with safety factor K of slope from mocro long time sequence.

2. The study area and data

The landslide is located in North village area in Shaanxi province. Geographic coordinates of landslide center is about N34°36′, E107°28′. Landslide is mainly composed of Quaternary sedimentary rock, one county highway through the slope leading edge in its bottom. Affected by long-term rainfall of historical the landslide boundary produced deformation and trailing edge appears as tensile cracks. At present, the slope has not produced large sliding, slope is in stage of creep deformation, if it is not treated timely, it will be direct threat to the safety of residents in north village and more than 100000 square meters region will be in danger. The maps of study areas are shown in figure 1.

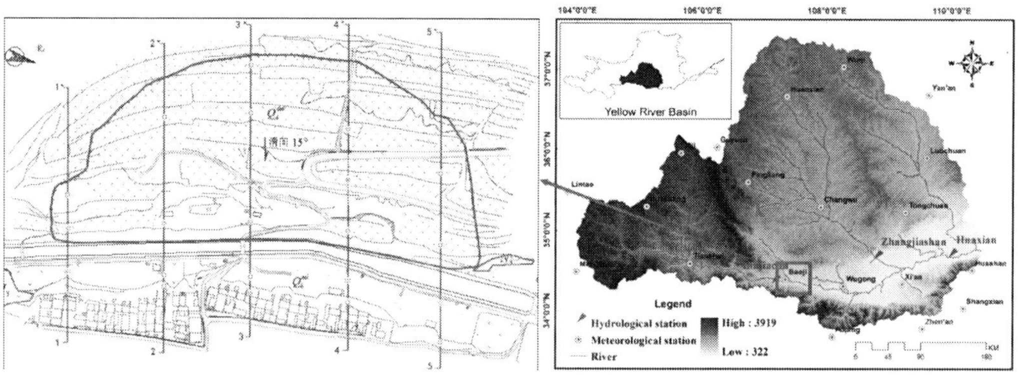

Figure 1. The Weihe River Basin and the location of the slope.

3. Methods

3.1. Analysis of regional precipitation patterns

Based on data of Boji meteorological station, this study is from annual and inter-annual scale to analyze precipitation time series in 1960-2006, in order to evaluation possibility of landslide in engineering. The precipitation results of annual and inter-annual scale are shown in figure 2.

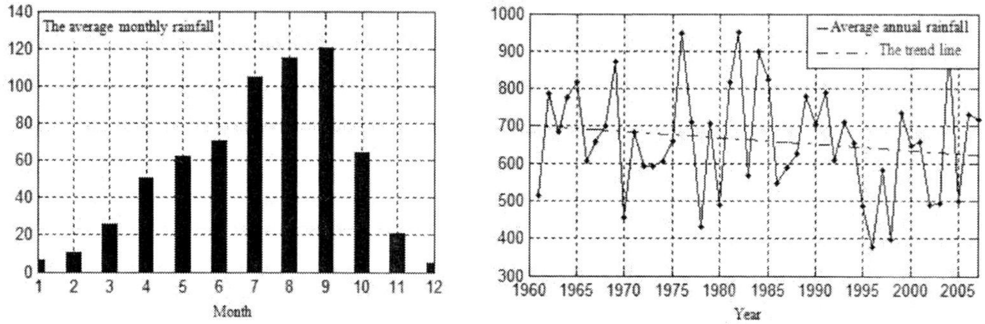

Figure 2. Precipitation time measured curve.

Figure 2 is traditional precipitation time series, and its description rainfall from time domain. Figure 2 can only analyze the relationship between time and rainfall intensity but also can analyze the frequency and cycle of precipitation, thus, this paper uses wavelet transform to study precipitation time series [7], because it can convert time domain description of series to frequency domain description that can effectively find the period. Wavelet transform is continuation and development of Fu Liye transform, the basic theory is through translation and scaling basic wavelet to approximation function or signal.

If basic wavelet function is $\psi(x)$, then wavelet function family $\psi_{ab}(x)$ is defined as follow:

$$\psi_{a,b}(x) = \frac{1}{\sqrt{a}}\psi\left(\frac{x-b}{a}\right) \quad a,b \in R, a \neq 0 \tag{1}$$

Where a is telescopic factor, b is translation factor, x is variable.

Hence, one signal or function such as $f(x) \in L^2(R)$ which wavelet transform is defined as follow:

$$w_f(a,b) = \int_{-\infty}^{+\infty} f(x)\psi_{a,b}dx = \frac{1}{\sqrt{a}}\int_{-\infty}^{+\infty} f(x)\psi\left(\frac{x-b}{a}\right)dx \tag{2}$$

Where $w_f(a,b)$ is wavelet transform coefficients.

Mexican hat wavelet is one commonly basic wavelet, which has two order derivative of Gauss function. Mexican hat wavelet is often written as MHF and expressed as follows:

$$\psi(x) = \frac{1}{\sqrt{2\pi}}(1-x^2)e^{\frac{-x^2}{2}} \tag{3}$$

Combine (2) and (3), and to change parameter a and b can get wavelet transform coefficients $w_f(a,b)$. Take b as abscissa and a as ordinate, we can get two-dimensional isoline which is called coefficient diagram of wavelet transform. By wavelet transform coefficient diagram can get evolution characteristics and mutation characteristics of time series: the positive wavelet coefficients correspond to partial multi periods and the negative wavelet coefficients correspond to partial less periods and zero point correspond to mutation-point. The larger absolute value of wavelet coefficients the more significant change in corresponding time scale, hence, wavelet coefficients can display time series with multi-resolution.

To integral wavelet coefficients get wavelet variance and expressed as follows:

$$\text{var}(a) = \int_{-\infty}^{+\infty}\left|w_f(a,b)\right|^2 db \tag{4}$$

Wavelet variance can show scale of wave energy distribution and determine the scales of main time series, which is main cycle.

This paper studies 1960-2006 precipitation series in Baoji meteorological station by MEF wavelet time scale analysis. To choice MEF wavelet is based on precipitation evolution process which has characteristics of multi time scale and continuously changing. Thus, it should use continuous wavelet transform (such as MEF) to analysis this case.

Because observed precipitation data are limited time series, both initial boundary and trailing end of this series may have a boundary effect [8], in order to eliminate or reduce this effect, this paper extended both boundary of the data and to abandon the minor coefficients after wavelet transform, in this case the scale factor is 1/2 of the data length.

3.2. Analysis of regional precipitation cycle

The result of MEF wavelet analysis in regional precipitation is shown in figure 3, wavelet coefficients has upper and lower isoline distribution, the upper contour corresponding long time periodic oscillation (low frequency oscillation) and lower contour corresponding short time periodic oscillation (high frequency oscillation), it shows from time distribution that oscillation is not uniform, hence, precipitation has different cycle, in this case, different precipitation cycle lead to landslide has similar cycle as precipitation. In figure 3, there is no strong vibration in the center and cycle can divide into 5a, 13a, 25a. In the smaller time scale (less than 5a) the change is complex that study is not considered.

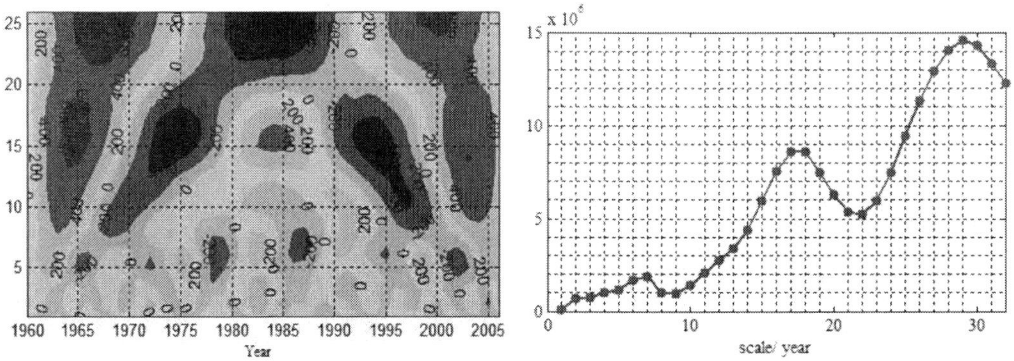

Figure 3. Contour map of wavelet transform.

In trends of one series, the zero point of wavelet coefficient transform is the series variation points. Thus in figure 3 the zero point is the inflection point of the precipitation, so it can judge mutation point of precipitation under different time scales and get the law of variation. Further from figure 3, the precipitation cycle of the area shows a law as alternate of two less and one more, where the 60-74 is less period and 74-84 is more period and after 1984 is less period. Quantitative research of rainfall plays a very important role to safety evaluation of landslide hidden danger engineering.

Wavelet variance result shows rainfall sequence in 1960-2010 has maximum fluctuations as 5a, which has strongest shock, it's shows in this interval that variation characteristics of precipitation is significantly which 5a is mainly period. In addition, although the rest of years have vibration and less energy, they can be divided into auxiliary periods. The precipitation cycle judgment is basic index for evaluation landslide hazard in Engineering.

3.3. Division of the precipitation state

Not all rainfall affects the stability of the slope, when the rainfall intensity is lightly because of too little water infiltrate that rainfall can not influence slope stability. Southwest Jiao Tong University Professor Huang Tao [9] based on similar slope rainfall designed one simulation experiment that result shows the observing horizontal displacement Y of slope leading edge and water infiltration index X has relationship of exponential function, the expressed as following:

$$Y = ae^{bx} \tag{5}$$

Huang's research showed that infiltration of water reaches a certain state then slope will be instable. Based on this, it puts forward the concept of dangerous slope infiltrate critical volume: in the actual engineering that whether landslide has occurred is irrespective form of infiltration and time of infiltration. As long as the infiltration amount reached threshold value, the slope may cause damage. This theory provides a favorable method for forecast slope collapse by continuous rainfall.

Hence, amount of rainfall determines dangerous level of slope, based on this, use mean standard deviation method to study 5a precipitation cycle, this paper divided the 5 state respectively: $[0,\bar{x}-1.0s]$, $(\bar{x}-1.0s,\bar{x}-0.5s]$, $(\bar{x}-0.5s,\bar{x}+0.5s]$, $(\bar{x}+0.5s,\bar{x}+1.0s]$, $[\bar{x}+1.0s,+\infty)$; The $\bar{x}$ is mean of precipitation sequence and s is variance of precipitation sequence.

3.4. Establishing and solving of the model calculation

Field sampling survey shows that slope layers approximately composed of 5 kinds of soil: the upper slope overburden is Quaternary Pleistocene alluvium Q_3^{eol} , Middle Pleistocene soil and slide accumulation layer Q_3^{del} the lower slope overburden Pleistocene alluvium soil Q_2^{al} and contains part of artificial fill Q^{ml} , the bottom of slope is third Ji mudstone bedrock *N*. In this study, calculation selection 3-3 profiles have decisive effect to landslide body stability state. 3-3 profiles are shown in the second section, and parameters of slope soil are shown in table 1.

Table 1. The finite element calculation parameter table.

Soil/ parameters	E	Poisson	C	φ	C'	φ'	Bulk density	Buoyant density	Saturation density	K
	MPa	—	kPa	°	kPa	°	kN/m^3	kN/m^3	kN/m^3	k
Q^{mal}	15	0.3	22.5	10.5	15	9.5	16.73	7.75	17.56	0.0035
Q_3^{eol}	8	0.3	24.5	14	13.5	11	17.14	8.97	18.78	0.0040
Q_3^{del}	24	0.3	16.5	12.4	14.3	11.7	17.99	9.18	18.99	0.002
Q_2^{al}	24	0.3	25	15	15	14.8	17.28	8.93	18.74	0.001
N	8000	0.2	30	15	27	13	18.44	9.63	19.44	0.000

According to the mean standard deviation method, this study design artificial rainfall which has simulation 5 different states of rainfall intensity All different rainfall intensities are shown in table 2. It is known by stochastic hydrology [10] that precipitation is positively related to rainfall intensity. According to the meteorological department divide 5 kinds of state on rainfall intensity, this research hypothesis 5 grades rainfall status (measured daily rainfall data). For limit equilibrium principle that simulation applying finite element software to calculation, which result shows no effect on the slope stability factor in the lightly rainfall state, therefore this study only selected 4 states to calculate, the rainfall duration is 48 h and the results are shown in figure 4. The purpose of this experiment is numerical simulation variety of safety factor *K* under different rainfall state.

Table 2. Finite element calculation of precipitation strength design value.

year	less	M-less	Middle	M-more	more
LV	Light rain	Less rain	Middle rain	Heavy rain	Rainstorm
Rainfall intensity	4.9 mm/d	16.9 mm/d	24.9 mm/d	49.9 mm/d	74.9 mm/d

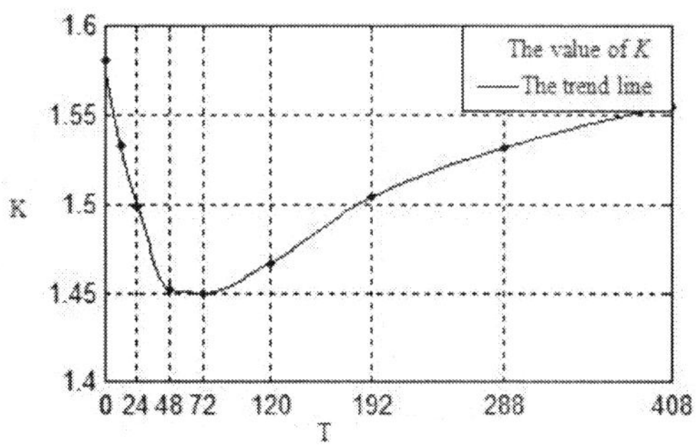

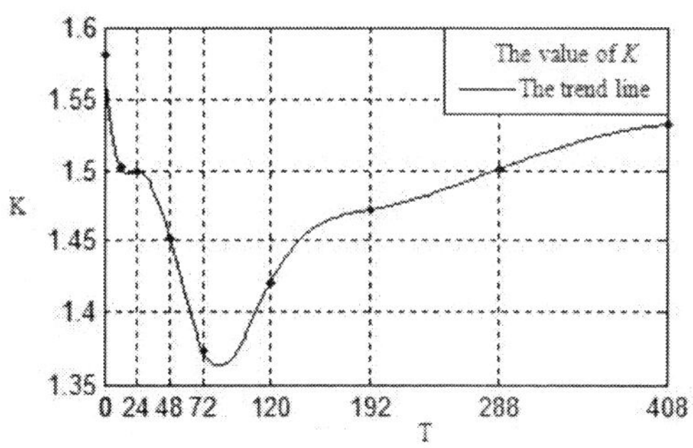

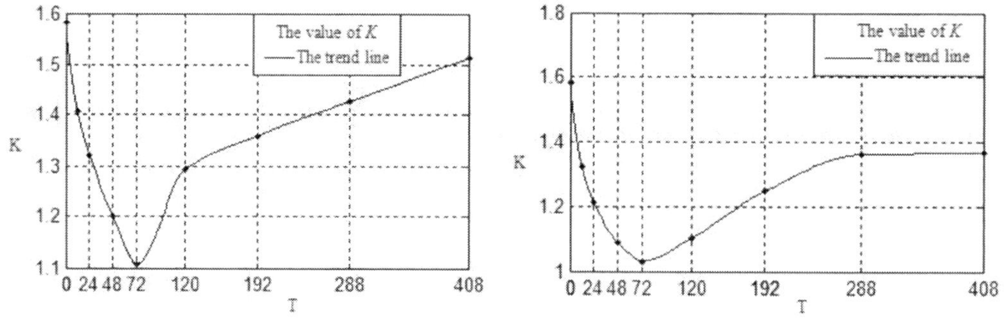

Figure 4. Change of slope coefficient K with different rainfall strength.

3.5. Based on the precipitation state to evaluation of landslide

The specification [11, 12] shows slope is on safe when safety factor $K \geq 1.15$, slope is on critical state when $K \in [1,1.15]$, and slope is in dangerous state when $K \leq 1$. Based on the specification, combination stability factor and precipitation state, calculate result divided stability factor K to 5 grades such as safe, less safe, stable, dangerous and too dangerous. Then choose the minimum K in every state (72 hours after the rainfall) as calculated value. Calculate results are shown in table 3.

Table 3. The results of slope stability evaluation.

Precipitation state	less	M-less	middle	M-more	more
Stability coefficient	1.477	1.373	1.168	1.107	1.032
Evaluation grade	Safety	Partial safety	Stable	Partial risk	DANGER

4. Result

4.1. Analysis method

Each kind of precipitation state corresponds to a certain year that accords rainfall to calculate safety factor, in this way, slope factor K evaluation grade and precipitation state are linked. This paper uses the Malco J prediction technique [13] to predict slope stability factor K in the different precipitation condition, moreover, this method can judge the stability of the slope in different years. Because landslide prediction has complexity characteristic, this study is only for short-term forecast.

Markoff prediction technology is the method to study the time series of which basic principle is variation of Markoff chain. Markoff chain linked change of future and present, thus it can accord present state or past state to estimating future trends. If the time series is expressed a finite function such as N, then from the state of i by one step state transition to the state of j, in the process of transfer that has different state may be occur, if using P_{ij} shows this possibility, which P_{ij} is called one step state probability. These probabilities are arranged in order and formed a matrix that called one step state transition probability matrix. Denoted P as follow:

$$P = \begin{bmatrix} P_{11} & P_{12} \cdots P_{1N} \\ P_{21} & P_{22} \cdots P_{2N} \\ \vdots & \vdots \quad \vdots \\ P_{N1} & P_{N2} \cdots P_{N3} \end{bmatrix} \tag{6}$$

The matrix has two characteristics: one is the matrix elements sum of each row equals to 1; the second is non-negative element in the matrix. Therefore, if the sequence at time T_0 is in state i, after n step transition to time T_N that state change to j, the possibility of this transfer quantity index called n

step transition probability, denoted n step transition probability as follow:

$$P(t_n = j / t_0 = i) = P_{ij}(n) \tag{7}$$

The n step transition probability matrix is denoted:

$$P = \begin{bmatrix} P_{11}(n) & P_{12}(n) \cdots P_{1N}(n) \\ P_{21}(n) & P_{22}(n) \cdots P_{2N}(n) \\ \vdots & \vdots & \vdots \\ P_{N1}(n) & P_{N2}(n) \cdots P_{N3}(n) \end{bmatrix} \tag{8}$$

Base on this theory, if the current state of the slope were known, one step state transition probability matrix can be used for prediction slope stability in the coming year and n step state transition probability matrix can be used to prediction slope state of next n year. This study is only about statistics element value on one step state transition probability matrix, to predict change of slope stability in the next year (short-term). This method belongs to probability application in prediction of slope stability.

Use mean standard deviation method, precipitation data corresponding slope stability factor of K respectively as follow: lightly rainfall→safety, less rainfall→safe table, moderate rainfall→critical state, more rainfall→dangerous, heavy rainfall→too dangerous, in this paper five levels respectively labeled as 1, 2, 3, 4 and 5, each state appears year as shown in table 4.

Table 4. Study area slop state statistics.

year	1960	1961	1962	1963	1964	1965	1966	1967
state	1	4	3	4	4	3	3	4
year	1968	1969	1970	1971	1972	1973	1974	1975
state	4	1	3	3	3	3	3	5
year	1976	1977	1978	1979	1980	1981	1982	1983
state	3	1	3	1	4	5	2	5
year	1984	1985	1986	1987	1988	1989	1990	1991
state	4	2	2	3	4	4	4	3
year	1992	1993	1994	1995	1996	1997	1998	1999
state	3	3	1	1	2	1	4	3
year	2000	2001	2002	2003	2004	2005	2006	
state	3	1	1	5	1	3	3	

One step state transition as follows:

$$F = (f_{ij})_{mxn} = \begin{bmatrix} 1\to1 & 1\to2 & 1\to3 & 1\to4 & 1\to5 \\ 2\to1 & 2\to2 & 2\to3 & 2\to4 & 2\to5 \\ 3\to1 & 3\to2 & 3\to3 & 3\to4 & 3\to5 \\ 4\to1 & 4\to2 & 4\to3 & 4\to4 & 4\to5 \\ 5\to1 & 5\to2 & 5\to3 & 5\to4 & 5\to5 \end{bmatrix} \tag{9}$$

The matrix is the slope stability coefficient K, which is calculated in accordance with the precipitation status in study area, by one-step transition matrix risk is divided into 5 grades respectively as: safe state, stable state, basically stable state, partial risk state and risk state. The transition frequency matrix is calculated from formula (9) and is shown as:

$$F = (f_{ij})_{m\times n} = \begin{bmatrix} 2 & 1 & 3 & 3 & 1 \\ 1 & 1 & 1 & 0 & 1 \\ 4 & 0 & 9 & 4 & 1 \\ 1 & 1 & 5 & 2 & 1 \\ 1 & 1 & 1 & 1 & 0 \end{bmatrix} \tag{10}$$

In formula (10), f_{ij} is defined as frequency which is i state transition to j state.

The transfer probability is as follows:

$$p_{ij} = \frac{f_{ij}}{\sum_{j=1}^{m} f_{ij}} \tag{11}$$

According to the formula (11), the result of one step transfer matrix shows as:

$$P^{(1)} = \begin{bmatrix} 0.2 & 0.1 & 0.3 & 0.3 & 0.1 \\ 0.25 & 0.25 & 0.25 & 0 & 0.25 \\ 0.22 & 0 & 0.5 & 0.22 & 0.05 \\ 0.1 & 0.1 & 0.5 & 0.2 & 0.1 \\ 0.25 & 0.25 & 0.25 & 0.25 & 0 \end{bmatrix} \tag{12}$$

Base on this, according to the slope stability state in a given year, we can predict slop stability state of coming year and the calculation formula as follows:

$$P_{t+1} = P_t P^{(1)} \tag{13}$$

4.2. Results and discussion

- If the slope evaluation standards in previous year is safety state, then the unconditional probability distribution as follows:

$$P_{last} = [1,0,0,0,0]$$

In the next year the probability of slope state may appear as follows:

$$P_{next} = [1,0,0,0,0] \times P^{(1)} = [0.2,0.1,0.3,0.3,0.1]$$

This shows that, when a slope for safety state in this year as the next year probability for safety state is 20%, the probability for stable state is 10%, the probability for basically stable state is 30%, the probability of a partial risk state is 30%, and the probability of risk state is 10%. According to the classification standard of this study, the value of safety factor K has 60% probability above to 1.15.

- If the previous year of slope evaluation standards is stable state then the unconditional probability distribution as follow:

$$P_{last} = [0,1,0,0,0]$$

In the next year the probability of slope state may appear as follow:

$$P_{next} = [0,1,0,0,0] \times P^{(1)} = [0.25,0.25,0.25,0,0.25]$$

This means that when a slope for stable state in this year, in the next year probability for slope safety state is 25%, the probability for stable stare is 25%, porbability for basically stable state is 25%, the probability for partial risk state is 0%, and the probability for risk state is 25%. Calculation result shows that the value of safety factor K has 75% probability above to 1.15.

- If the previous year of slope evaluation standards is basically stable state then the unconditional probability distribution as follows:

$$P_{last} = [0,0,1,0,0]$$

In the next year the probability of slope state may appear as follows:

$$P_{next} = [0,0,1,0,0] \times P^{(1)} = [0.22,0,0.5,0.22,0.05]$$

This means that when a slope as basically stable state , slope has safety state in the next year is 22% probability , the probability for stable state is 0%, the probability for basically stable is 50%, the probability for partial risk is 22%, the probability for risk is 5%. The value of safety factor K has 72% probability above to 1.15.

- If the previous year of slope evaluation standards is partial risk state then the unconditional probability distribution as follows:

$$P_{last} = [0,0,0,1,0]$$

In the next year the probability of slope state may appear as follows:

$$P_{next} = [0,0,0,1,0] \times P^{(1)} = [0.1,0.1,0.5,0.2,0.1]$$

This means that when the slope in a partial risk state, in the next year probability of safety state is 10%, stable state probability is 10%, basically stable state probability is 50%, the probability of a partial risk state is 20%, and the probability of risk state is 10%. But in this case it has different from above the situation (1)~(3), although the safety factor K above to 1.15 has 70% probability, this is likely to be a "safety after landslide", it means if landslide occurred, adverse factors of slope accumulation will be released, thus the slop will be reach a new state of safety, so in the next year, in a certain degree safety state and stable state due to whether landslide has occurred, partial risk state and risk state due to whether the slop keep steady, but in seemingly steady state will further accumulate adverse factor, thus in the study, we must pay attention to this phenomenon.

- If the previous year of slope evaluation standards is risk state then the unconditional probability distribution as follows:

$$P_{last} = [0,0,0,0,1]$$

In the next year the probability of slope state may appear as follows:

$$P_{next} = [0,0,0,0,1] \times P^{(1)} = [0.25,0.25,0.25,0.25.0]$$

To be similar as situation (4), when slope in a risk state, in the next year the safety state, stable state, basically stable state and partial risk state have the same probability as 25%, sum of those probability is 100% meanwhile the risk state probability is 0%. It shows that the landslide may be take place already, so the adverse factors are to be consumed, hence the landslide can not occur in the next year.

5. Conclusion

To analyze long series precipitation, this paper examined rainfall period in the study area and pointed out the change in precipitation may cause geological body get more adverse factor for landslide. The main conclusions are shown as follow:

- The rainfall in the Baoji area has three periods of 5a, 13a and 25a, where 5a is the main period of precipitation, and it characteristic shows alternate of "two less and one more" as fluctuation.

- According to different rainfall intensities states, this paper designed different simulated precipitation, and the stability of slope is simulated by the finite element software. Simulation results show the slope safety factor *K* has similar fluctuation trend with precipitation state. Therefore, this study based on the specification to considering the change of precipitation has similar regularity with slope factor, dividing slope factor *K* of 5 grades, that is, safety, basically safety, stable, partial risk and risk, respectively.
- Based on periodic precipitation, this paper used Malco prediction technique, considering the stability state of the slope in the influence of precipitation factors. Results indicated that: in the previous year if the slope is in the safety state, in the next year *K* has 60% probability that higher than 1.15; If the slope in previous year is in basically safe state, in the next year *K* has 75% probability that higher than 1.15; If the slope in previous year is in stable state, in the next year *K* has 72% probability that higher than 1.15.
- When the previous year slope is in partial risk state or risk state, although slope factor *K* has larger probability than 1.15, the mechanism is different from the first three. There is high probability that stability factor *K* is always higher than 1.15, which means to a great extent that the slope have already occurred landslide. In that case, the adverse factor accumulation has been released, thus the slope being in safety state in the next year.

The landslide system has many uncertain factors. In many cases, the prediction of safety factor *K* is safe, but the safety state has difference means between "The safety state that after destruction" and "safety state that have no destruction". In this tow situation they are completely different in mechanism. Landslide has characteristic of complexity, openness, gradual change, mutation, and its entropy is expected to increase. Hence, to study and absolute prediction such as a chaotic system is impossible. Therefore, only at the macro scale, combine gradient with mutation and use continuous monitoring, reinforce in time, along with taking avoidance tactics when necessary, as much as possible to reduce the losses caused by landslides.

Acknowledgments
This research was supported by National key R & D plan (2016YFC0400306), and National Natural Science Foundation of China (51409048). Sincere gratitude is extended to the editor and anonymous reviewers for their professional comments and corrections, which greatly improved the paper.

References

[1] Liao H J and Ji J 2008 Stability analysis of soil slopes considering saturated and unsaturated seepage effect *J. Rock Soil Mech.* **29** 3230-4 (in Chinese)
[2] Zhou L X and Tong H W 2010 Deformation monitoring analysis on diaphragm wall *J. Guangzhou Univ. (Nat. Sci. Ed.)* **9** 68-73 (in Chinese)
[3] Song K Q, Yuan J G, *et al* 1994 The Study on the characteristic and prediction of landslides in Xian Area *J. Soil Water Conserv.* **8** 52-7 (in Chinese)
[4] Fu H Y, Zeng L, *et al* 2012 Stability analysis of soft rock slope under rainfall infiltration *J. Rock Soil Mech.* **33** 2360-5 (in Chinese)
[5] Qian J H 1988 *Soil Mechanics* 2nd ed (Jiangsu, Nanjing Province: Hohai University Press)
[6] Liao H J and Zhao S D 2008 *Advanced Engineering Geology* (Beijing: Higher Education Press)
[7] Wei F Y 2007 *Modern Climatic Statistical Diagnosis and Prediction Technology* (Beijing: Meteorological Press)
[8] Xiong S B and Huang C Y 2010 *Fundamentals of Mechanical Engineering Testing Technology* (Beijing: Machinery Industry Press)
[9] Huang T and Luo X Y 2004 Model testing study on slope stability under environment of surface water permeation *Chin. J. Rock Mech. Eng.* **23** 2671-5 (in Chinese)
[10] Zhan D J 2011 *Engineering Hydrology* (Beijing: Chinese Water Conservancy and Hydropower

IOP Conf. Series: Earth and Environmental Science **82** (2017) 012092 doi:10.1088/1755-1315/82/1/012092

Press)
[11] Xie D Y, Yao Y P and Dang F N 2008 *Advanced Soil Mechanics* (Beijing: Higher Education Press)
[12] Xie D Y, Li B H and Shao S J 2008 *Geotechnical Engineering Advanced Soil Mechanics* (Beijing: Higher Education Press)
[13] Freedman D 1983 *Markov Chains* (New York: Spring-Verlag)

IOP Conf. Series: Earth and Environmental Science **82** (2017) 012093 doi:10.1088/1755-1315/82/1/012093

Research on the response of the water sources to the climatic change in Shiyang River Basin

Y Z Jin[1], J J Zeng[2,4], X Q Hu[1], D Y Sun[1], Z F Song[1], Y L Zhang [1], S C Lu[1] and Y Q Cui[3]

[1]Gansu Research Institute for Water Conservancy, Lanzhou 730000, China
[2]Key Laboratory of Western China's Environmental Systems (Ministry of Education), College of Earth and Environmental Sciences, Lanzhou University, Lanzhou, Gansu 730000, China
[3]College of Teacher's Education, Honghe University, Honghe 654400, China

E-mail: zengjj15@lzu.edu.cn

Abstract. The influence of the future climate change to the water resource will directly pose some impact on the watershed management planning and administrative strategies of Shiyang River Basin. With the purpose of exploring the influence of climate change to the runoff, this paper set Shiyang River as the study area and then established a SWAT basin hydrological model based on the data such as DEM, land use, soil, climate hydrology and so on. Besides, algorithm of SUFI2 embedded in SWAT-CUP software is adopted. The conclusion shows that SWAT Model can simulate the runoff process of Nanying River well. During the period of model verification and simulation, the runoff Nash-Sutcliffe efficient coefficient of the verification and simulation is 0.76 and 0.72 separately. The relative error between the simulation and actual measurement and the model efficient coefficient are both within the scope of acceptance, which means that the SWAT hydrological model can be properly applied into the runoff simulation of Shiyang River Basin. Meantime, analysis on the response of the water resources to the climate change in Shiyang River Basin indicates that the impact of climate change on runoff is remarkable under different climate change situations and the annual runoff will be greatly decreased as the precipitation falls and the temperature rises. Influence of precipitation to annual runoff is greater than that of temperature. Annual runoff differs obviously under different climate change situations. All in all, this paper tries to provide some technical assistance for the water sources development and utilization、 assessment and optimal configuration.

1. Introduction

The influence and the adaptability of the climate change to the forming process of water circulation and spatial and temporal distribution of water resources is one of the hot issues in the field of hydrological research, whose research and technical methods mainly include statistic analysis, cause analysis, scenario hypothesis, downscaling, model simulation, coupled model and so on [1]. The ways that the climatic change influence water cycle processes are multiple, which have the characteristics of complex, dynamic and no-linear [2, 3]. The application of hydrological model in the study of climatic change influence to the water resources is a good solution and is also a trend of current research. Distributed watershed hydrological model SWAT (soil and water assessment tool) is developed by the

Agricultural Research Center of USDA, and has become one of the widely used hydrological models [4-6]. As a kind of distributed watershed hydrological model which is based on the procedure, SWAT integrates various digital techniques such as RS (Remote Sensing), GIS (Geographic Information System) and DEM (Digital Elevation Model) and can simulate and predict different climate, runoff under the underlying surface, silt and hydrological process of non-point source pollution. In the source region of the Yellow River, Hanjiang River Basin, Taihu Lake Basin, watershed of Miyun Reservoir, Huai River Basin, and Haihe Basin, SWAT Model has gained great effect in some research fields such as water and sediment simulation, assessment and management of the water resources, impact of the climatic change and soil utilization to water resources, and non-point source pollution simulation. However, in arid and semi-arid region of China, little research has ever been done about the application of SWAT Model. Besides, exploration on the algorithm of model parameter calibration and uncertainty analysis is also seldom involved [7, 8].

As an inland river, Shiyang River is one of the basins that have the most dense population, the highest degree development and utilization of water resources, the greatest contradiction of water use, the most serious problem of ecological environment. The current development and utilization of water resources in this basin has far outweigh its bearing capacity, which further deteriorate the ecological environment of this region. Currently, research based on SWAT Model in Shiyang River Basin is mainly doing some investigation in simulating the variation characteristics of the runoff. By using SWAT Model, Wang Junde did some simulation study on the runoff of Zamu River within Shiyang River Basin, and then discussed the hydrological characters of a model of 7 different kinds of vegetation combination within basin and micro terrain scale. Based on SWAT Model and set upper and middle reach basin of Shiyang River as his study object, Zhang Y F did some investigation on the variation characteristics of the mountain-pass runoff under various climate change scenarios from the year 1988 to 2005, and then discussed the impact of climatic change on the hydrological regime and water resources. Zhang L Y did a quantitative analysis of Gulang River Basin about the influence of climatic changes and land-utilization or land-cover changes to water cycle elements such as water evaporation, surface runoff and so on. However, few researchers have ever used this high-precision and comprehensive model to simulate the surface runoff and to study the future changes and trend of hydrological factors by supposing different climate change scenarios.

With the development of economy and society, shortage and uncertainties of the water resources caused by future climatic changes will bring new challenges to sustainable utilization and management of water resources and the socioeconomic system. The model is used to simulate monthly runoff of Nanyang River Basin in order to study the applicability of SWAT Model in Shiyang River, which will provide fundamentals for a quantitative analysis of runoff simulation in future climate change scenarios. By doing a scenario-based design of the climatic changes in Shiyang River Basin, this paper aims to make a response analysis of water resources and provide some technical assistance for the water sources development and utilization, assessment and optimal configuration.

2. Introduction of the study area

Shiyang River Basin is located on the ancient Silk Road—east of Hexi Corridor of Gansu Province, west of Wushao Ling, north of Qilian Mountains, and the latitude is between 36°29′～39°27′ N and 101°41′～104°16′ E. Connecting with Baiyin and Lanzhou City in southeast, adjoining Zhangye City in northeast, neighboring Qinghai Province in southwest, and bordering on the Nei Monggol Autonomous Region, this basin totally involves 4 cities and 9 counties/districts, and covers an area of 41,600 square kilometers. Lying in hinterland of China, this region generally belongs to temperate continental arid climate, which can be further divided into three types from south to north: semi-arid and semi-humid alpine area in the south Qilian Mountains, arid and cool area in the middle Hexi Plain region, and warm and arid area in the north. With an annual precipitation of only 150-300 mm, the annual evaporation of the plain is as high as 1300-2600 mm, which has far outweigh its amount of precipitation, therefore the drought index has reached as high as over 52. Besides many brooks, Shiyang River Basin is mainly composed by 8 main rivers named Dajing River, Gulang River,

Huangyang River, Zamu River, Jinta River, Xiying River, Dongda River, and Xida River. These rivers and streams are mainly supplied by atmospheric precipitation and snowmelt water from the mountain area, whose runoff area and annual runoff are 1.11×104 km^2 and 1.56×109 km^2 separately. Types of soil include sierozem, grey desert soil, grey-brown desert soil, and so on. Types of vegetation include kobresia pygmaea, picea crassifolia in medium high mountains, sabina przewalskii, artemisia frigida, reaumuria soongorica, stipa breviflora and so on.

3. Data and methods

3.1.Basic data

The input data in SWAT Model include spatial data, attribute data, meteorological data, and hydrological data (table 1).

Table 1. Data and format needed in SWAT model of Shiyang River.

Data	Item	accuracy	format	source
spatial data	DEM	1:100,000	GRID	National Natural Science Foundation
	LUCC	1:100,000	GRID/shape	“Environmental and Ecological Data
	soil type	1:1,000,000		Center of Western China”
attributes data	soil attributes data		DBF	Soil-type Records of Gansu Province, Soil Records of Gansu Province and soil records of every county around the basin.
meteorological data	Precipitation, maximum and minimum temperature, radiation, humidity, wind speed	date	DBF	China Meteorological Administration, Meteorological Bureau of Gansu Province
hydrological document	quantity of flow	date, month, year	DBF	Hydrological Bureau of Gansu Province

- Meteorological data

The measured data are collected from 5 meteorological stations (Yongchang, Minqin, Wuwei, Menyuan, Wushao Ling) and mountain-exit hydrological stations of 4 Rivers from 1956 to 2010. They include solar radiation, wind speed, humidity, maximum temperature and humidity, and daily precipitation from precipitation stations.

- DEM data

DEM data is obtained firstly by projection transform with the method of Albers projection and then undergo processes of ArcGIS splicing, depression detention and basin demarcation. After a series of digital processing, a DEM Model is established (figure 1).

- Soil data

Soil-type data derive from soil shp data of Chinese Mainland. A soil-type map for this study is achieved by projection transform and cut with the help of GIS software and then stored it into a format of grid or shape. The soils are divided into 8 orders, 11groups, and 12types. Types and area of soil can be seen in table 2. As to soil attributes data, depth value of vegetation root systems, depth of the whole soil layer, quantity of layers, depth of soil surface to every layer, contents of clay particles, efutriation, sand particles, gravel, organic matters of every layer, they all come from Soil-type Records of Gansu

Province, Soil Records of Gansu Province and such records of every county within the study area. The study adopted documents that have gained achievements in soil texture conversion with different particle diameters, and conducted transformation by applying the method of linear interpolation. Table 3 shows the soil texture transformation of the main types of soil in Shiyang River Basin. *SOL_BD* (wet density of soil (mg/m^{3}or g/cm^3)), *SOL_AWC* (soil available water content (mm)), and *SOL_K* (saturated hydraulic conductivity (mm/h)) can be calculated by SPAW software [9, 10].

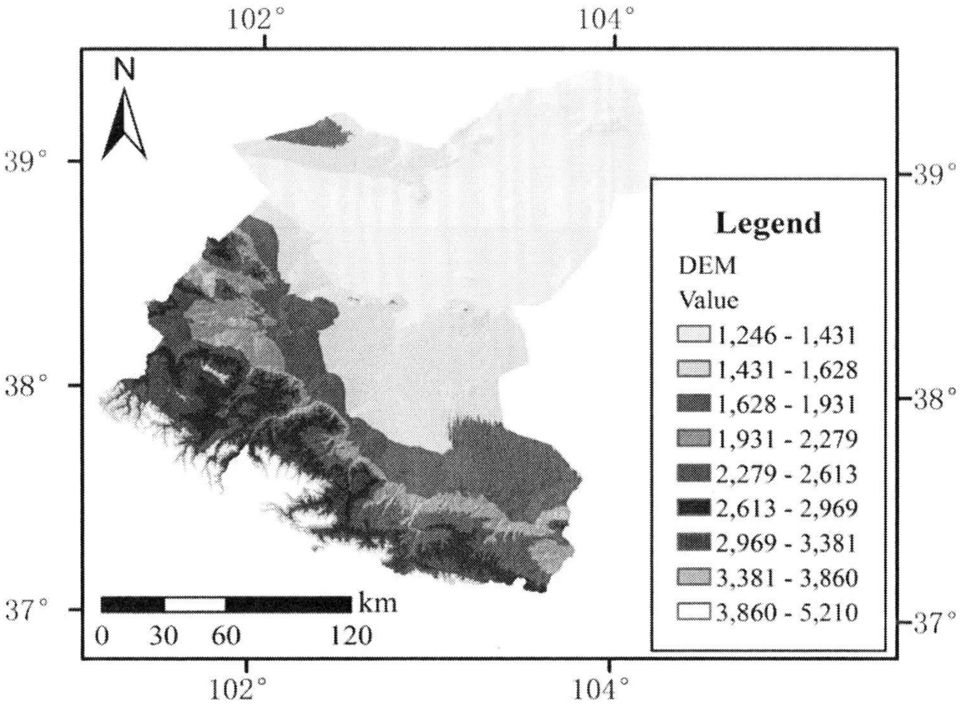

Figure 1. Digital elevation model of Shiyang River Basin.

Table 2. Types and area of soil in Shiyang River Basin.

Code	Sub Type	Soil Type	Soil Order	Area (km^2)
23118100	inland salty soil	salty soil	saline-alkali soil	4076.512
23115141	semifixed aeolian sandy soil	aeolian sandy soil	primitive soil	6610.549
23120151	frigid desert soil	frigid desert soil	alpine soil	566.518
23120102	frigid calcic soil	frigid calcic soil	alpine soil	683.547
23112111	castanozem	castanozem	pedocal	2467.572
23111121	grey cinnamon soil	grey cinnamon soil	semi-Luvisols	1838.687
23114104	salty gray-brown desert soil	gray desert soil	desert soil	10418.459
23112103	(shajiang black soil)	chernozem	pedocal	474.531
23114101	oasis spodosol	gray desert soil	desert soil	4443.778
23120102	felty soil	felty soil	alpine soil	2291.078
23113113	meadow sierozem	sierozem	aridisol	3023.46
23112101	chernozem	chernozem	pedocal	3662.313

Table 3. Soil texture conversion of the main types of soil in Shiyang River Basin.

Types	Depth (cm)	Particle Diameter (mm) >2	2-0.2	0.2-0.02	0.02-0.002	<0.002	Texture
chernozem	0-18			48.50	33.40	18.10	clay loam
(062)	18-28			42.20	40.30	17.50	clay loam
23112101	28-110			50.00	33.00	17.00	clay loam
	110-132			44.60	35.20	20.20	clay loam
meadow sierozem	2-22	0	46.85		37.41	15.74	clay loam
(109)	22-57	0	53.04		33.45	13.51	loam
23113113	57-74	0	55.12		44.38	0.50	sandy loam
	74-107	0	48.76		50.24	1.00	silt loam
	107-163	0	46.04		50.69	3.27	silt loam
gray desert soil	0-26		5.10	42.14	36.67	16.09	clay loam
(123)	26-50		3.38	39.34	44.43	12.85	loam
23114101	50-78		3.38	35.79	44.50	16.33	clay loam
	78-105		6.43	33.60	42.90	17.07	clay loam
	105-130		9.59	40.10	37.21	13.01	loam
	0-23		0.21	42.65	37.00	20.14	clay loam
calcic chernozem	23-57		0.53	37.37	39.45	22.56	clay loam
(064) 23112103	57-90		0.46	19.64	51.26	28.64	silty clay loam
	90-120		6.21	30.83	37.32	25.64	loamy clay loamy clay
salty gray-brown desert soil (129) 23114104	8-12	37.03	65.54	9.04	12.58	12.84	gravelly sandy loam loamloam gravelly sandy loam
	12-26	13.41	85.76	4.82	4.91	4.51	gravelly sandy loam loamloam gravelly sandy loam
	26-36	65.92	96.77	0.50	1.54	1.19	gravelly sandy loam loamloam gravelly sandy loam
	36-54	4.82	97.89	0	1.54	0.57	gravelly sandy loam loamloam gravelly sandy loam
	54-105	77.01	98.37	0	1.37	0.36	gravelly sandy loam loamloam gravelly sandy loam

grey cinnamon soil (042) 23111121	10-37	14.50		46.18	36.21	17.61	gravelly sandy loam loamloam gravelly sandy loam
	37-75	13.60		49.05	33.77	17.18	gravelly clay loam loamloam gravelly sandy loam
	75-135	39.60		56.24	28.20	15.50	gravelly sandy clay loam
Castanozem (072) 23112111	0-18		17.55	15.58	29.86	37.01	clay loam
	18-40		14.17	10.63	32.76	42.44	clay loam
	40-60		69.88	2.66	10.89	16.77	silty clay loam
semifixed aeolian sandy soil (164) 23115141	0-0.5		81.2	5.4	3.8	9.6	loamy sand
	0.5-23		89.3	0.5	1.9	8.3	loamy sand
	23-56						sandy soil
	56-86						sandy soil
inland salty soil (225) 23118100	0-14		0.1	34.9	54.0	11.0	silt loam
	14-34		0	37.0	50.0	13.0	silt loam
	34-85		1.5	66.0	23.5	9.0	sandy loam
	85-140		0	51.0	37.0	12.0	loam

- Land use data

Land use that can be recognized by SWAT Model is based on the classification of that of America, which is encoded by 4 English letters. Through a previous splicing processing by applying GIS software, this study worked out a map of types of land use (figure 2). Types of land use are classified into 6 genera and 31 subgenera, which is based on the classification method of land sources of China. However, land use data should be reclassified and converted into the English letters as the codes that are regulated by SWAT Model. Finally, by using these codes, a database of the types of land use in study area and vegetation attached to the model is generated, which will be connected with agricultural management database [11].

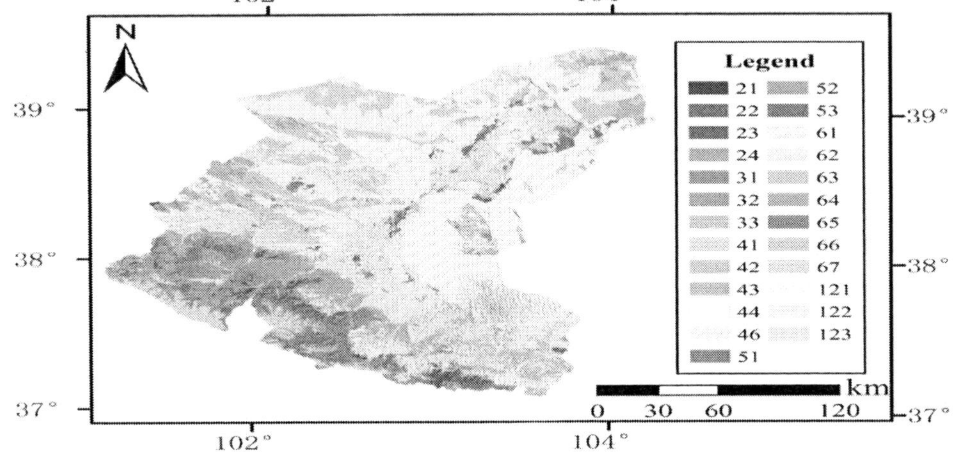

Figure 2. Land use of Shiyang River Basin.

- Hydrological Data

All the relative data are measured data that come from hydrological stations from the year 1956 to 2010. Besides setting surface area of the reservoir, storage capacity, upper limit of discharged flow, and evaporation coefficient, the reservoir parameters also include the setting of water outflow and supply according to the amounts of actual water demand.

3.2. Research methods

- SWAT Model

SWAT Model is adopted in this hydrological simulation. Based on DEM data, the whole basin is divided into several sub-basins according to the location of outlets. Considering the spatial and temporal variability of underlying surface, these sub-basins are further divided into several hydrological response units (HRU) according to soil types, land use types and slopes. Runoff is calculated on every sub-basin and HRU by applying traditional conceptual model, and flow quantity of the whole basin will be worked out by calculation of flow concentration. According to the data collected, the runoff simulation in this paper adopts the method of SCS, the river flow routing adopts method of Muskingum, and for potential evapotranspiration, it takes formula of Penman-Monteith [12].

- Methods of model calibration and verification

Model calibration and verification are critical during the process of modeling, which will directly determine the accuracy of the result. SWAT Model is complicated in structure and multivalued in parameters, in which there are more than 20 parameters only in the aspect of runoff. In order to improve the efficiency of model calibration and decrease the uncertainty of the model, sensitivity analysis (SA) is always carried out in the practical operation. According to the results of the analysis, values (sensitivity parameters) that have relatively greater impact on the model result should be adjusted in order to recognize and validate distributed hydrological model. Sensitivity analysis of the model often uses the method of LH-OAT (Latin-Hypercube-One-Factor-At-a-Time) which is fixed in SWAT software, because in the aspect of parameter choice, this method well integrates the advantages of the wholeness in LH (Latin Hypercube Sampling) and certainty in OAT (One-Variable-At-a-Time). However, the efficiency is a little bit low while doing analyzing. Therefore, SWAT-CUP, developed by Swiss Federal Institute of Water Biological Science and Technology, is a program specially designed for SWAT Model to do SA, calibration, verification and uncertainty analysis. It is embedded with algorithms of SUFI2 (Sequential Uncertainty Fitting Version 2), PSO (Partial Swarm Optimization), MCMC (Markov Chain Monte Carlo), ParaSOL (Parameter Solution), GLUE (Generalized Likelihood Uncertainty), which make it has the advantages of high-efficiency and visualization. This paper applies SUFI2 to make sensitivity analysis and do model calibration and verification [13].

4. Results and analysis

After being loaded and processed by GIS, a DEM map is used to calculate flow direction and accumulation of the basin, and the whole basin is divided into 127 sub-basins by integrating the actual situation of river network and taking the calculating efficiency into account. A simulation will be implemented after loading the reclassified data of land-use types, soil types and the slope is classified into 2 types.

4.1. Sensitivity analysis of model parameters

Calibration of parameters is to find a most consistent parameter between the simulation value and observed value. According to the result of sensitivity analysis, this paper finally chooses the following parameters to calibrate the model: *CN2, SOL_AWC, ECSO, EPCO, Canmx, SOL_K, SOL_Z, GW_DELAY, ALPHA_BF*.

Among these parameters, runoff curve number (*CN2*) is a very important parameter in SCS runoff model, which can describe the relationship between precipitation and runoff and comprehensively reflect runoff capacity of underlying surface (land-use/land cover, soil types, hydrological conditions, moisture situation in earlier stage). Theoretically, value range of curve number should be within 0～100, which has a positive correlation with runoff volume; soil available water content (*SOL_AWC*) reflects the water storage capacity, which has an inverse relationship with runoff generation. The bigger the coefficient is, the greater the soil storage capacity will be, and the less the runoff volume will become. *ESCO* is a coefficient used to adjust water compensation among different soil layers. The evaporation in the deep soil will be reduced as this coefficient rises, and the runoff volume will thus increase; *SOL-K* relates soil water-flow rate with hydraulic gradient and can measure the difficulty level of water movement in the soil, which has a positive correlation with runoff generation; *GW_DELAY* and *ALPHA_BF* have great impact on hydrograph. *ALPHA_BF* is achieved by inputting data of daily runoff from the year 1993 to 2004 and applying digital filtering method to segment baseflow, which means to make a segment calculation to the ten-year daily runoff data by running bflow.exe module under DOS. Detailed description of these 9 parameters can be seen in table 4.

Table 4. Sensitivity analysis and data of SWAT model in Shiyang River Basin.

Parameters	Definition	Sensitivity Level	documents	Data Range
CN2	Curve Number	1	Management (mgt)	(59.85, 78.8)
SOL_AW	Soil Available Water Content	2	Soil (.sol)	0.010
ESCO	Soil evaporation correction coefficient	3	HRU (.hru)	0.957
EPCO	Plant transpiration correction factor	4	HRU (.hru)	0.766
CANMX	Maximum Canopy Interception	5	HRU (.hru)	0.568
SOL_K	Saturated Hydraulic Conductivity	6	Soil (.sol)	(52.09, 86.84)
SOL-Z	depth of soil layer	7	Soil (.sol)	(138.38, 222.69)
GW_DELAY	Groundwater delay time	8	Groundwater (.gw)	(23.25, 31)
ALPHA_BF	baseflow coefficient	9	Groundwater (.gw)	(0.048, 0.176)

4.2. Calibration and verification of parameters

Model calibration is to modify the parameters and compare the simulated value with the measured value until the predetermined objective function is achieved. This study chooses correlation coefficient (R^2), relative error (*RE*) and efficient coefficient of Nash-Sutcliffe Model as the index of assessing simulating result. Correlation coefficient R^2 can assess the coincide degree between the simulated value and the measured value. When R^2=1, it means the two values coincide with each other quit well; when $R^2 < 1$, the smaller the value is, the worse the coincide degree will be.

Calculation formulas of relative error RE and model efficient coefficient equations are (1) and (2) separately:

$$RE = \frac{Q_P - Q_0}{Q_0} \tag{1}$$

$$E_{ns} = 1 - \frac{\sum_{i=1}^{n}\left(Q_0 - Q_p\right)^2}{\sum_{i=1}^{n}\left(Q_0 - Q_{avg}\right)^2} \tag{2}$$

In these formulas, Q_O is the measured value, Qp is the simulated value, Q_{avg} is the measured average value, and n is the number of the value. $RE > 0$ means the simulated value is larger than needed; $RE < 0$ means the value is smaller than needed; RE=0 means the two values coincide with each other well. The more close to 1 the *Ens* value is, the better the simulated result will be; *Ens*=1 means the simulated result is coincide with the measured result; *Ens* < 0 means the application of simulated values is much less credible than the direct usage of measured average value.

According to the wholeness of data acquisition, this paper selected Nanying Hydrologic Station of 1993-2004 to do the runoff simulation. Years 1993 to 1999 are taken as the period of data calibration, and years 1999 to 2004 are taken as the period of data verification. Simulated results of yearly and monthly runoff are showed in table 5.

Table 5. Simulated result of monthly runoff in Nanying River.

Period	Efficient Coefficient (*Ens*)	Relative Error (*RE*)	Correlation (R^2)	Coefficient
Verification (1993-1999)	0.76	2.8	0.89	
Simulation (1999-2004)	0.72	5.6	0.86	

Calibration (1993～1999) and verification (1999～2004)of Nanying River, see figure 3.

If the expected value of simulated result can reach about 0.6, it indicates that SWTA Model can well generalize the parameters of the study area and accurately describe the hydrological process. Therefore, in geographical environment of Shiyang River, it is feasible to apply SWAT Model to operate runoff simulation. Besides, as is presented in parameter calibration, all of the 9 parameters are correlated with underlying surface of the basin. Although good results can be achieved after the model calibration, some of the parameters such as *CN2, SOL_AW, ESCO, EPCO, CANMX, SOL_K, SOL-Z, GW_DELAY, ALPHA_BF* are close to or have reached the critical values that are set in the model [13].

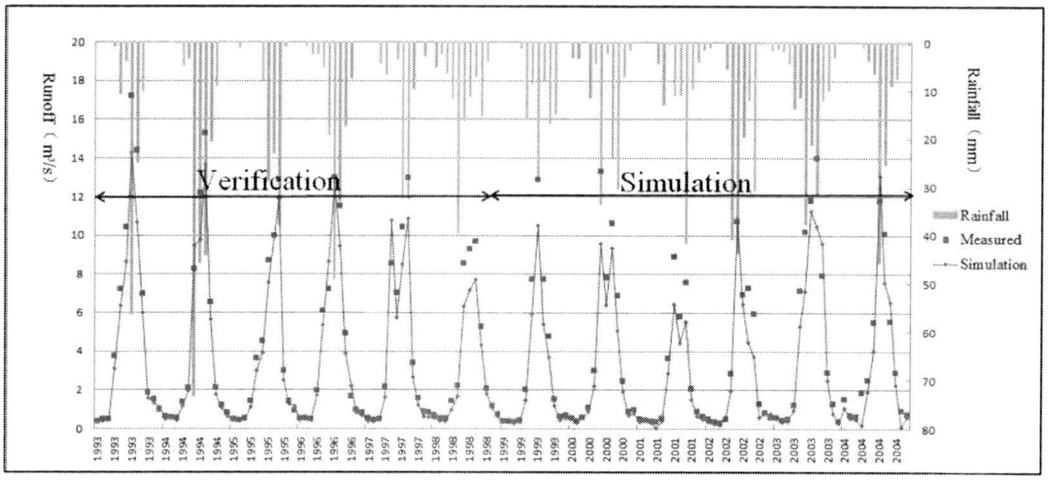

Figure 3. Simulation and verification of runoff process in Nanying hydrological station.

5. Response analysis on the water sources to the climatic change in Shiyang River Basin

5.1. Runoff effect under climatic change

Based on the measured data of Nanying Hydrological Station from the year 1956 to 2010, the study made a precipitation variation analysis in Shiyang River Basin an worked out the following two pictures:

Precipitation of Nanying Reservoir is continuously rising with an obvious increase in 1960s, and no point of abrupt change appeared. Precipitation and runoff of the reservoir are also in a completely opposite tendency, which can be seen in figure 4.

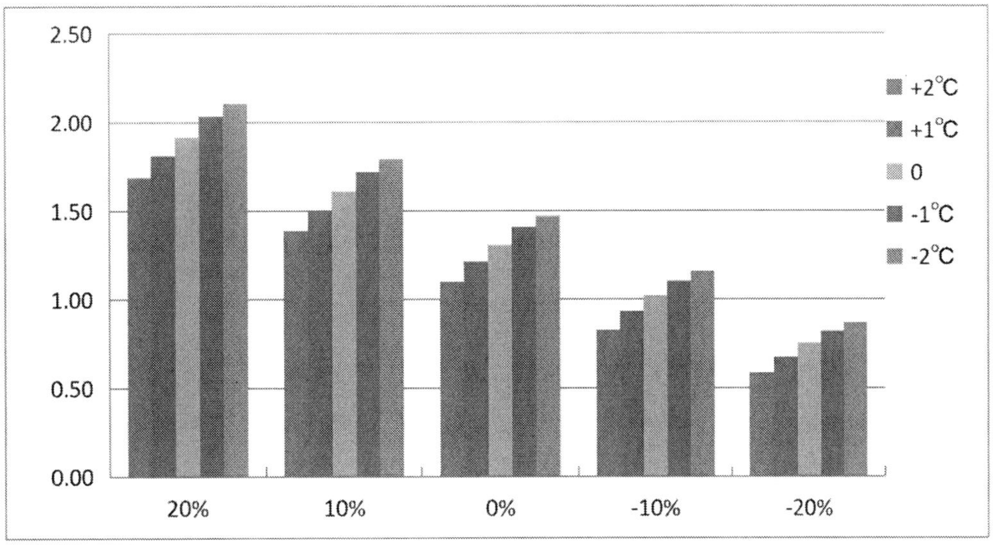

Figure 4. Results comparison of runoff simulation under different climate change situations in Nanying River.

5.2. Situation design of climate change

In order to analyze the great influence of climate change to hydrological water resources, a quantitative research is conducted by integrating the method of climate situation design with hydrological model simulation. Based on meteorological data of 1976-2010 in Nanying River, the future climate change situation are designed as follows: precipitation increases by +20%, +10%, -20%, -10% or remains unchanged; climate change increases by 1 ℃, 2 ℃, - 1 ℃, - 2 ℃ or remains unchanged [14]. Data of land use adopts the land cover data of 2012. There are 24 kinds of different situation combination according to the situation design of climate change, which can be seen in the following table 6.

Table 6. Design of climate change situations.

Precipitation Change						
Temperature Change		20%	10%	0	-10%	-20%
	+2℃	situation1	situation2	situation3	situation4	situation5
	+1℃	situation6	situation7	situation8	situatio9	situation10
	0	situation11	situation12	situation	situation13	situation14
	-1℃	situation15	situation16	situation17	situation18	situation19
	-2℃	situation20	situation21	situation22	situation23	situation24

5.3. Simulated results and analysis

Taking the result of future climate change prediction under a global model as a reference, this paper simulates the monthly runoff of Nanying River under the above-mentioned 24 kinds of climate change situations separately within the range of possible climate changes in the study area [15]. After a runoff simulation based on SWAT Model, the annual runoff, runoff variation and change rate will be achieved.

Table 7. Simulated results of annual runoff under different precipitation and temperature situations.

	Precipitation Change						
	annual runoff (m^3)		20%	10%	0	-10%	-20%
temperature change		+2℃	1.5419	1.3766	1.0273	0.8535	0.5312
		+1℃	1.6613	1.4180	1.0833	0.9548	0.6416
		0	1.7651	1.5564	1.2201	1.0407	0.7347
		-1℃	1.8972	1.6690	1.2905	1.0753	0.8735
		-2℃	2.0177	1.7549	1.4647	1.0923	0.8816
	runoff change (m^3)		20%	10%	0	-10%	-20%
		+2℃	0.2777	0.1124	-0.2368	-0.4107	-0.7329
		+1℃	0.3971	0.1539	-0.1808	-0.3094	-0.6226
		0	0.5009	0.2922	0.0000	-0.2235	-0.5295
		-1℃	0.6331	0.4048	0.0264	-0.1889	-0.3907
		-2℃	0.7535	0.4907	0.2005	-0.1718	-0.3826
	change rate (%)		20%	10%	0	-10%	-20%
		+2℃	21.9696	8.8925	-18.7334	-32.4871	-57.9788
		+1℃	31.4130	12.1705	-14.3046	-24.4734	-49.2468
		0	39.6220	23.1134	0.0000	-17.6803	-41.8817
		-1℃	50.0767	32.0198	2.0854	-14.9393	-30.9039
		-2℃	59.6038	38.8199	15.8600	-13.5932	-30.2622

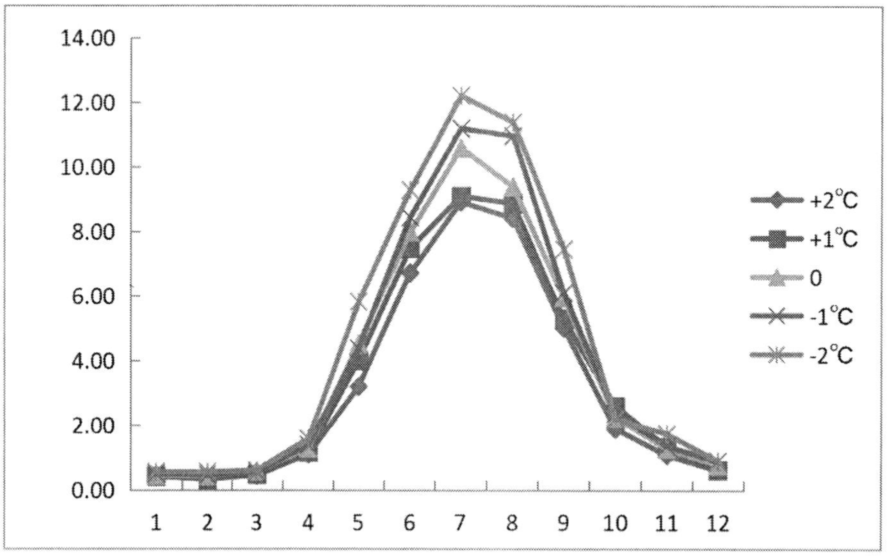

Figure 5. Simulated results of monthly runoff under the same precipitation and different temperatures situations.

From table 7 and figures 4 and 5, this study can draw conclusions as follows:

- Impact of climate change on runoff is remarkable.

Annual runoff will be greatly decreased as the precipitation falls and the temperature rises. That's because precipitation is the foundation of runoff formation, and is also the direct source of water resources. When the precipitation decreases, the water supply amount of the whole basin will thus be decreased, which will in turn reduce the runoff volume. Temperature rising will increase the evaporation of the basin, which will cause decline of the runoff under the premise of invariable water supply.

- Influence of precipitation to annual runoff is greater than that of temperature.

With the temperature being unchanged, 10% of increase or decrease of the precipitation will cause a change of 23.1% or－17.7% to annual runoff; while with the precipitation being unchanged, 1℃ of rise or fall of the temperature will cause a change of－14.3% or 2.09% to annual runoff. Consequently, change rate of the annual runoff that caused by 10% precipitation change is almost 2.49 times that of temperature change; With temperature being unchanged, 20% of increase or decrease of the precipitation will cause a change of 39.6% or－41.9% to annual runoff; while with the precipitation being unchanged, 2℃ of rise or fall of the temperature will cause a change of－14.3% or 15.9% to annual runoff. Therefore, change rate of the annual runoff that caused by 20% precipitation change is almost2.7 times that of temperature change.

- Annual runoff differs obviously under different climate change situations.

Different climate conditions will generate different annual runoff. When the precipitation increases by 20% and the temperature falls 2℃, the annual runoff will increase the most by 59.6%; on contrary, when the precipitation decreases by 20% and the temperature rises 2℃, the annual runoff will reduce the most by 57.9%. Therefore, to predict accurately the climate change is of great importance for analyzing hydrological water resources in the basin.

6. Conclusion

The introduction of SWAT Model provides a very useful tool for the scientific research and management of water resources in Shiyang River Basin. Its distinct distributed structure and operating mode also widen the horizon for developing distributed hydrological model in China. By applying SWAT Model and integrating GIS, RS, a distributed hydrological model of Shiyang River Basin is established based on spatial data, attribute data, meteorological data, and hydrological data. Besides, the model also goes through sensitivity experiment, calibration and verification. The results indicate that SWAT Model can properly simulate runoff process of Nanying River Basin, which is a sub-basin of Shiyang River Basin. Coefficient of linear regression and efficiency of the simulated runoff and the measured runoff can meet the requirements of regulation. During the period of model verification and simulation, the runoff Nash-Sutcliffe efficient coefficient of the verification and simulation is 0.76 and 0.72 separately. Comparing the simulation and actual measurement, the relative error is within the range of 10%. In Nanying River, both the relative error and model efficiency coefficient are within the range of acceptance, which means that SWAT Model can well be applied into runoff simulation of Shiyang River Basin. Analysis on the response of the water resources to the future climate change in Shiyang River Basin implies that the impact of climate change on runoff is remarkable under different climate change situations and the annual runoff will be greatly decreased as the precipitation falls and the temperature rises. Influence of precipitation to annual runoff is greater than that of temperature. Annual runoff differs obviously under different climate change situations.

7. Discussion

SWAT Model is constructed under databases abroad. Therefore, there is a necessity to modify the database when applying the model, especially need to establish the user's own soil attribute database and the code of land use also needs to be transformed. Data that are needed in constructing SWAT Model in China is hard to achieve, and the accuracy is also not high enough. Lack of some of the parameters is also very common. Consequently, there is an urgent need to strengthen the data supervision and integration and establish a scientific, comprehensive, and unified basic database.

In the runoff simulation process of Shiyang River Basin, data are more reliable and accurate in this paper as it establishes the spatial database and attribute database such as soil attribute data, land use attribute data and so on. Results of SWAT Model simulation well reflect the runoff hydrological process of Shiyang River Basin, which can provide technical assistance for the water sources development and utilization, assessment and optimal configuration. It is suggested that attention should be paid to the soil concrete condition of the study area while establishing soil database and certainty of every parameter in the meteorological database while establishing meteorological database, such as dew-point calculation and so on. As to situation design of climate change, the overall changing trend of the global climate change should be fully simulated. In the future work, a more scientific, comprehensive, and unified basic database should be established in order to provide research support in the aspects of non-point source pollution, water and soil erosion, land use, agricultural management, and so on, which will lay a solid foundation for the comprehensive management of water resources by further application of SWAT Model.

Acknowledgments
This research was supported by the National Natural Science Foundation of china (51369003, 51369004), Project Supported by the special foundation for Gansu Province international scientific cooperation (1604WKCA002), and the Fundamental Research Funds for the Central Universities (lzujbky-2016- 862516, lzujbky-2017-it90).

References

[1] Lamba J, Thompson A M, Karthikeyan K G, Panuska J C and Good L W 2016 Effect of best management practice implementation on sediment and phosphorus load reductions at subwatershed and watershed scale using SWAT model *Int. J. Sediment Res.* **4** 386-94

[2] Liu M, Li C L, Hu Y M, Sun F Y, Xu Y Y and Chen T 2014 Combining CLUE-S and SWAT models to forecast land use change and non-point source pollution impact at a watershed scale in Liaoning Province *Chin. Geog. Sci.* **5** 540-50

[3] Zhang Z, Lu W X, Chu H B, Cheng W G and Zhao Y 2014 Uncertainty analysis of hydrological model parameters based on the bootstrap method: A case study of the SWAT model applied to the Dongliao River Watershed, Jilin Province, Northeastern China *Sci. Chin. (Technol. Sci.)* **1** 219-29

[4] Khan K, Lu Y L, Khan H, Zakir S, Khan I S, Khan A A, Wei L and Wang T Y 2013 Health risks associated with heavy metals in the drinking water of Swat, northern Pakistan *J. Environ. Sci.* **10** 2003-13

[5] Sophocleous M 2011 Two-way coupling of unsaturated-saturated flow by integrating the SWAT and MODFLOW models with application in an irrigation district in arid region of West China *J. Arid Land* **3** 164-173

[6] Chin D A, Fan X H and Li Y C 2011 Validation of growth and nutrient uptake models for tomato on a gravelly south Florida soil under greenhouse conditions *Pedosphere* **21** 46-55

[7] Lee M S, Park G A, Park M J, Park J Y, Lee J W and Kim S J 2010 Evaluation of non-point source pollution reduction by applying best management practices using a SWAT model and QuickBird high resolution satellite imagery *J. Environ .Sci.* **6** 826-33

[8] Mingxing L I 2010 Regional soil moisture simulation for Shaanxi Province using SWAT model validation and trend analysis *Sci. Chin. (Earth Sci.)* **4** 575-90

[9] Lai G, Yu G and Gui F 2006 Preliminary study on assessment of nutrient transport in the Taihu Basin based on SWAT modelling *Sci. Chin. (Series D: Earth Sci.)* **S1** 135-45

[10] Neitsch S L, Arnold J G, Kinlry J R, *et al* 2002 Soil and water assessment tool theoretical documentation: Version 2000.M. (Texas: Texas Water Resources Institute, College Station) pp 3-5

[11] White K L and Chaubey I 2005 Sensitivity analysis, calibration, and validations for a multisite and multivariable SWAT model *J. Am. Water Resour. Asso.* **41** 1077-89

[12] Moriasi D N, Arnold J G, van Liew M W, *et al* 2007 Model evaluation guidelines for systematic quantification of accuracy in watershed simulations *Trans. ASABE* **50** 885-900

[13] Ficklin D L, Zhou L Y, Elike L, *et al* 2009 Climate change sensitivity assessment of a highly agricultural watershed using SWAT *J. Hydrol.* **374** 16-29

[14] Yasir S A, Li A, Crosato A, Mohamed Y A, Abdalla S H and Wright N G 2014 Sediment balances in the Blue Nile River Basin *Int. J. Sedi. Res.* **3** 316-28

[15] Gui F and Yu G 2008 Numerical simulations of nutrient transport changes in Honghu Lake Basin, Jianghan Plain *Chin. Sci. Bull.* **15** 2353-63